3RD EDITION

The Blue PLANET

An Introduction to Earth System Science

BRIAN J. SKINNER
Yale University

BARBARA MURCK
University of Toronto

JOHN WILEY & SONS, INC.

VICE PRESIDENT AND EXECUTIVE PUBLISHER Jay O'Callaghan
EXECUTIVE EDITOR Ryan Flahive
ASSOCIATE EDITOR Veronica Armour
PRODUCTION SERVICES MANAGER Dorothy Sinclair
SENIOR PRODUCTION EDITOR Janet Foxman
MARKETING MANAGER Margaret Barrett
CREATIVE DIRECTOR Harry Nolan
SENIOR DESIGNER Maureen Eide
SENIOR PHOTO EDITOR Mary Ann Price
PHOTO RESEARCHER Elle Wagner
ILLUSTRATION EDITOR Anna Melhorn
EDITORIAL ASSISTANT Darnell Sessoms
SENIOR MEDIA EDITOR Lynn Pearlman
PRODUCTION SERVICES Furino Production
COVER IMAGE NASA/Science Source/Photo Researchers, Inc.

This book was set in Sabon by MPS Limited, a Macmillan Company, Chennai, India and printed and bound by Courier Companies. The cover was printed by Courier Companies.

This book is printed on acid-free paper. ♾

Library of Congress Cataloging-in-Publication Data

Skinner, Brian J., 1928–
The blue planet: an introduction to earth system science/—3rd ed. Brian J. Skinner, Barbara Murck.
p. cm.
Includes bibliographical references.
ISBN 978-0-471-23643-6 (hardback)
I. Murck, Barbara Winifred, 1954– II. Title.
QB631.S57 2011
550—dc22

2010032269

Main Book ISBN 978-0-471-23643-6

Binder-Ready Version ISBN 978-0-470-55648-1

Printed in the United States of America

10 9 8 7 6

Preface

The first people to leave Earth's orbit and see the far side of the Moon were the astronauts of Apollo 8. On Christmas Eve 1968, when Apollo moved out from behind the Moon, crew members Borman, Lovell, and Anders saw a wonderful sight: it was Earth, rising above the barren lunar landscape. Photographs of the scene have become iconic reminders of our planet's splendid isolation in space. Many years earlier, in 1948, a famous English scientist, Sir Fred Hoyle, predicted that the first images of Earth from space would change forever the way we think about our planet. How prescient that prediction was. It was the first giant step in the development of a holistic view of Earth. From space the atmosphere seems like a surface smear, the oceans are great blue blotches, and the brown continents are crossed by bands of green vegetation. The oneness of all the parts of Earth is immediately apparent.

This book is an introduction to the science of the holistic view of Earth. It is about the interactions between the different parts of Earth—how matter and energy move between the atmosphere, hydrosphere, biosphere, and geosphere (by which we mean the solid Earth). The assemblage of parts and interactions has come to be called the *Earth system*. This book is about the *science of the Earth system*. It is also a book about the way we humans, now 7 billion of us, are influencing the way the Earth system works.

PURPOSE OF THE BOOK

Earth system science is rapidly changing the way people study and think about Earth. People have always been concerned with local climate and weather; now they are also concerned with the global climate, and whether humans are causing it to change. People have always been concerned with local water availability and quality; now they are also concerned with the status of water resources and aquatic ecosystems—both freshwater and marine—around the world. From space it can be seen that winds blow dust from the Sahara desert across the Atlantic Ocean; knowing where to look, scientists can detect dust in ocean sediments and discover that it affects life in the Caribbean islands. This holistic way of thinking about interconnections and interrelationships is changing the way scientists study Earth. We have written this book to introduce students to the developing science of the Earth system.

Courses about Earth system science are being taught with increasing frequency in different academic departments. Such courses may have titles such as Global Change, Earth Science, Biospherics, The Global Environment, Planet Earth, or even Environmental Science, but the approach for all of them is increasingly that of studying Earth as an assemblage of interacting parts and processes.

THE SYSTEMS APPROACH

The key to understanding Earth as a system of many parts is to appreciate the interactions between the parts and to understand how energy and matter move around the system. Earth is, to a very close approximation, a closed system; by this we mean that it neither gains nor loses matter, but energy can both enter and leave the system. For the sake of study and measurement we divide Earth into a large number of subsystems, each of which is an open system, meaning that both matter and energy can move back and forth between them. Earth system science, then, is the study of Earth as an assemblage of open systems, and the goal of the science is to eventually understand the interactions among all parts of the assemblage. In this way, the effects throughout the system caused by a perturbation in one part of the assemblage—say, a volcanic eruption, or a rise in the carbon dioxide content of the atmosphere—can be estimated and forecast.

The traditional way to study Earth was to consider the various parts in isolation from each other. One group of scientists studied the atmosphere, another group the oceans, still another the geosphere, and yet another, the assemblage of life forms. Communication and interaction between the different groups were once rare. Earth system science is removing barriers, and interdisciplinary interactions today are common among those who study the Earth system.

THE BOOK'S ORGANIZATION

Reflecting this emphasis on a systems approach, the book is organized into six parts, each containing three chapters, except for Part One, The Earth System: Our Place in Space, which has four chapters. Parts Two through Six address the principal subsystems of the Earth system: Geosphere,

Hydrosphere, Atmosphere, Biosphere, and Anthroposphere, in that order.

The chapters in Part One start with a The Earth System, a discussion of systems, cycles and feedbacks. The second chapter, Energy, starts by introducing the Laws of Thermodynamics, then moves to the sources of Earth's energy and how energy cycles through the Earth system. Chapter 3 discusses Matter, and new in this edition is a section on organic matter. The final chapter in Part One, Space and Time, concerns Earth's place in the solar system, including a discussion of the structure and dynamics of the Sun, the energy from the Sun that reaches Earth, and time scales of Earth history.

The three chapters of Part Two are concerned with The Geosphere: Earth Beneath Our Feet. Chapter 5, The Tectonic Cycle, discusses the outflow of Earth's internal heat energy and the resulting motions of the mantle and lithosphere. The nature, locations and dynamics of volcanic eruptions are the focus of Chapter 6 (Earthquakes and Volcanoes), and Chapter 7 (The Rock Cycle) examines the collective interactions at Earth's surface between the atmosphere, hydrosphere, biosphere and the geosphere.

The chapters of Part Three examine The Hydrosphere: Earth's Blanket of Water and Ice, and its essential role in the Earth system. Chapter 8 deals with The Hydrologic Cycle, Chapter 9 with The Cryosphere, and Chapter 10 discusses The World Ocean. The three chapters of Part Three pay special attention to the role of the hydrosphere in the climate system, and in meeting the needs of both natural systems and human society.

Part Four, The Atmosphere: Earth's Gaseous Envelope, comprises three chapters devoted to the nature and role of the atmosphere. Chapter 11, The Atmosphere, discusses the structure and dynamics of the atmosphere. Chapter 12, Wind and Weather Systems, discusses both global and local circulation patterns. Chapter 13, The Climate System, examines in detail what we know about past climates and the causes of climatic changes.

The three chapters of Part Five, The Biosphere: Life on Earth, discuss the characteristics of the biosphere and the role of life in the Earth system. Chapter 14, Life, Death, and Evolution, discusses the basic processes of life, and how life has adapted to and altered the Earth system over the course of the planet's history. Chapter 15, Ecosystems, Biomes, and Cycles of Life, discusses the importance of material recycling in ecosystems, the minimum characteristics of a life-supporting system, and how biogeochemical cycles can be influenced by human activities. Chapter 16, on Populations, Communities, and Change, considers carrying capacity, and factors that affect the health and limit the growth of populations; special attention is given to biodiversity and to current threats to diversity.

The final section, Part Six, concerns The Anthroposphere: Humans and the Earth System. Chapter 17 addresses The Resource Cycle, with a particular focus on the different roles of renewable and nonrenewable resources in the growth and health of the human population. Chapter 18 discusses Mineral and Energy Resources, and how their use affects various parts of the Earth system. The final chapter in the book pulls together the many lines of evidence discussed in earlier chapters in assessing The Changing Earth System as a result of human activities.

Although we have given careful consideration to the organization of the book, we realize that not all instructors may favor the one we have adopted. Therefore, the parts and chapters have been written so that, so far as possible, they stand alone, and that some reorganization of topics is possible without serious loss of continuity.

THE ILLUSTRATIONS

As with previous editions of this text, special attention has been devoted to the artwork and photographs that illuminate discussions in the text. Because no country or continent holds a monopoly on relevant and interesting examples, we have selected photographs, maps, and illustrations from around the world in order to provide a global perspective of Earth system science. The art program has benefited from talented artists who have worked closely with the authors to make their illustrations both attractive and scientifically accurate. Many of the illustrations and photographs in this edition are new to the text, and we think both instructors and students will find them engaging and educational, as well as beautiful.

FEATURES

- *Part Opener.* Each of the six parts of the book opens with a brief statement that outlines the part of the Earth system discussed in the part, and the connections with other parts of the system.
- *Chapter Overview.* Each chapter opens with a bulleted list of topics discussed in the chapter followed by a brief statement of the purpose of the chapter.
- *"A Closer Look" Boxes.* Within chapters, specialized and detailed topics are boxed under the heading "A Closer Look". The boxed material can be included or deleted at the discretion of the instructor.
- *"The Basics" Boxes.* Topics that need special explanation, such as "Electromagnetic Radiation" in Chapter 2, are boxed under the heading "The Basics".
- *"Make the Connection"*. In each chapter one or more questions are inserted in the text, asking students to make the connection between some item in the chapter and the larger Earth system. For example, in Chapter 16, following a discussion of populations, the student is asked to think of a population of insects, animals or plants, then to list the number of things that might limit the growth of the population, and to identify whether the limitations come from the hydrosphere, atmosphere, geosphere, or the anthroposphere. In some cases there is no "one"

correct answer for the question; the goal is to get students to think about connections and relationships.

- *Summary and Review.* Each chapter closes with a summary of in-chapter material, a list of key terms, and a series of questions. The questions are of two kinds: (1) review questions that relate strictly to material in the chapter, or, under separate headings, to material in *A Closer Look* or *The Basics*; and discussion questions, which are intended for class or section discussion, sometimes calling for a bit of additional research. In most cases the discussion questions raise broader issues than those in the specific chapter to which they are attached.
- *Appendices and Glossary.* Three useful Appendices provide students with reference materials on units and conversions, naturally occurring elements and isotopes, and the properties of common minerals. The Glossary has been expanded and improved in this edition, and we think students will find it to be a very useful study tool.

NEW TO THIS EDITION

The most important change in this third edition of *The Blue Planet* is the addition of a new author, Barbara Murck of the University of Toronto. Professor Murck brings broad experience of fieldwork and research in the Earth and environmental sciences to the author team, and she is an award-winning teacher.

The third edition has been extensively reorganized based on constructive input from users of the previous two editions. For example, the two chapters on the solar system in the second edition have been combined into one chapter, and earthquakes and volcanoes are covered in a single chapter instead of two. In addition, the four chapters on the biosphere in the second edition have been extensively reorganized, tightened up, and improved, and are now three chapters. Even though a new chapter (Energy) has been added to the book, the condensing and rearrangement has produced a volume of 19 chapters instead of the 20 in the second edition.

Instructors' Companion Site (www.wiley.com/college/skinner). This comprehensive website includes numerous resources to help you enhance your course. These resources include:

- **Image Gallery**. We provide online electronic files for the line illustrations in the text, which the instructor can customize for presenting in class (for example, in handouts, overhead transparencies, or PowerPoints).
- A complete collection of **PowerPoint presentations** available in beautifully rendered, 4-color format, and have been resized and edited for maximum effectiveness in large lecture halls.
- A comprehensive **Test Bank** with multiple-choice, fill-in, and essay questions. The test bank is available in two formats: Word document and Respondus.
- **Pre-Lecture Clicker/PRS Questions** based on the "A Closer Look" and "The Basics" boxed features allows the instructor to connect the readings to the classroom lectures.
- **GeoDiscoveries Media Library**. This easy-to-use website offers lecture launchers that helps reinforce and illustrate key concepts from the text through the use of animations, videos, and interactive exercises. Students can use the resources for tutorials as well as self-quizzing to complement the textbook and enhance understanding of Earth System Science. Easy integration of this content into course management systems and homework assignments gives instructors the opportunity to integrate multimedia with their syllabi and with more traditional reading and writing assignments. Resources include:
  - Animations: Key diagrams and drawing from our rich signature art program have been animated to provide a virtual experience of difficult concepts. These animations have proven influential to the understanding of this content for visual learners.
  - Videos: Brief video clips provide real-world examples of geographic features, and put these examples into context with the concepts covered in the text.
  - Simulations: Computer-based models of geographic processes allow students to manipulate data and variables to explore and interact with virtual environments.
  - Interactive Exercises: Learning activities and games built off our presentation material. They give students an opportunity to test their understanding of key concepts and explore additional visual resources.
- **Google Earth™ Tours.** Virtual field trips allow students to discover and view geospheric landscapes around the world. Tours are available as .kmz files for use in Google Earth™ or other virtual atlas programs.
- Online **Case Studies** provide students with cases from around the world in which to see and explore the interaction of people and their environment. It was revised by Robert Ford.
- An online database of photographs, **www.ConceptCaching.com**, allows professors and students explore the atmosphere, hydrosphere, lithosphere, and biosphere. Photographs and GPS coordinates are "cached" and categorized along core concepts of geography and geology. Professors can access the images or submit their own by visiting www.ConceptCaching.com.

Student Companion Website (www.wiley.com/college/skinner). This easy to use and student-focused website helps reinforce and illustrate key concepts from the text. It also provides interactive media content that helps students prepare for tests and improve their grades. This website provides additional resources that compliment the textbook and enhance your students' understanding of Physical Geography:

- **Chapter Review Quizzes** provide immediate feedback to true/false, multiple-choice, and short answer questions based on the end-of-chapter review questions.
- **Online Case Studies** provide cases from around the world in which to see and explore the interaction of people and their environment.
- **Google Earth™ Tours.** Virtual field trips allow students to discover and view geospheric landscapes around the world. Tours are available as .kmz files for use in Google Earth™ or other virtual atlas programs.

ACKNOWLEDGMENTS

Our many colleagues who prepared essays did so with grace and professional acumen. We are very grateful to them. We are extremely grateful for the guidance and judgment provided by colleagues who discussed this project in encounter groups and who reviewed all or part of the manuscript. They are:

David J. Anastasio, *Lehigh University*
Lisa Barlow, *University of Colorado*
John M. Bird, *Cornell University*
Stephen Boss, *University of Arkansas*
James W. Castel, *Clemson University*
Steven Dickman, *Binghampton University*
Bruce Fegley, *Washington University—St. Louis*
Robert Ford, *Westminster College*
Karen H. Fryer, *Ohio Wesleyan University*
Bart Geerts, *University of Wyoming*
H.G. Goodell, *University of Virginia*
Dr. Ezat Heydari, *Jackson State University*
Gregory S. Holden, *Colorado School of Mines*
Julia Allen Jones, *Oregon State University*
Dr. Adrienne Larocque, *University of Manitoba*
Keenan Lee, *Colorado School of Mines*
Thomas Lee McGehee, *Texas A&M University*
Harold L. Levin, *Washington University—St. Louis*
Dan L. McNally, *Bryant College*
Chris Migliaccio, *Miami-Dade Community College*
Robert L. Nusbaum, *College of Charleston*
Roy E. Plotnick, *University of Illinois*
Gene Rankey, *University of Kansas*
Doug Reynolds, *Central Washington University*
Kathryn A. Schubel, *Lafayette College*
Kevin Selkregg, *Jamestown Community College*
Lynn Shelby, *Murray State University*
Leslie Sherman, *Providence College*
Christian Shorey, *Colorado School of Mines*
John T. Snow, *University of Oklahoma*
K. Sian Davies-Vollum, *Amherst College*
Cameron Wake, *University of New Hampshire*
Nick Zentner, *Central Washington University*

We would like to express our thanks to Steve Porter and Dan Botkin, who contributed so much effort and expertise to the production of the first two editions of *The Blue Planet*.

The third edition was a long time coming, and many people have anticipated its arrival—thank you so much, all of you, for your patience and loyalty. That the third edition is now here is due in no small part to the perseverance of Ryan Flahive, Executive Editor and overseer of all things Geoscience-related for John Wiley and Sons. Many thanks also to Jay O'Callaghan, VP and Publisher, for his continued support of our work.

For this project it was great to be reunited with many old Wiley friends from earlier projects. Cliff Mills (Developmental Editor), it was wonderful to see your face again (virtually, at least). Jeanine Furino (Furino Production), you're a task-master of the highest order (that's a compliment), and thanks to you the book is complete, on time (more or less), and of extremely high quality. Anna Melhorn (Illustration Coordinator), MaryAnn Price (Photo Editor), and Lynn Pearlman (Senior Media Editor), your work is crucial in producing a book that is both visually stunning and educationally sound. Veronica Armour (Associate Editor, Geosciences) helped keep us on track, and Darnell Sessoms (Editorial Assistant) was always there to lend a hand. Margaret Barrett (Marketing Manager), thanks for all of your work to bring this new edition to our adopters and readers.

Finally, it shouldn't go without saying that we are grateful to our families, our colleagues, and our students at Yale University and the University of Toronto—and, of course, to each other. We love working together and we are extremely proud of the 3rd edition of *The Blue Planet*.

About *the* AUTHORS

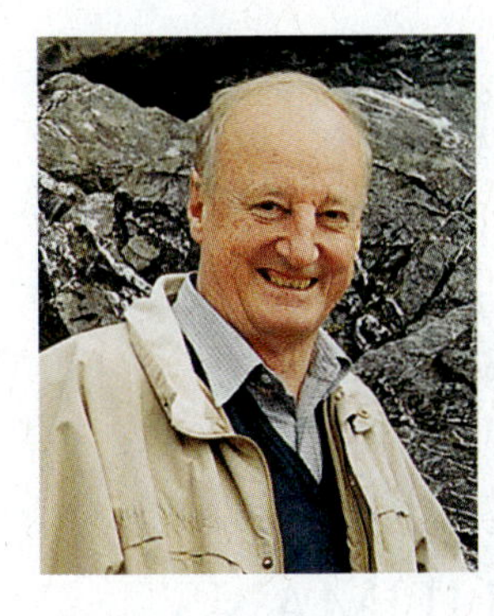

Brian J. SKINNER

Brian Skinner was born and raised in Australia, studied at the University of Adelaide in South Australia, worked in the mining industry in Tasmania, and in 1951 entered the Graduate School of Arts and Sciences, Harvard University, from which he obtained his Ph.D. in 1954. Following a period as a research scientist in the United States Geological Survey in Washington D.C., he joined the faculty at Yale in 1966, where he continues his teaching and research as the Eugene Higgins Professor of Geology and Geophysics. Brian Skinner has been president of the Geochemical Society, the Geological Society of America, and the Society of Economic Geologists, He holds an honorary Doctor of Science from University of Toronto, and an honorary Doctor of Engineering from the Colorado School of Mines.

Barbara MURCK

Barbara Murck is a geologist and senior lecturer in environmental science at the University of Toronto, Mississauga. She completed her undergraduate degree in Geo-logical and Geophysical Sciences at Princeton University and then spent two years in the Peace Corps in West Africa, before returning to Ph.D. studies at the University of Toronto. Her subsequent teaching and research has involved an interesting combination of geology, natural hazards, environmental science, and environmental issues in the developing world, primarily in Africa and Asia. She also carries out practical research on pedagogy and was recently awarded the President's Teaching Award—the highest honor for teaching given by the University of Toronto. She has co-authored numerous books, including several with Brian Skinner.

Brief CONTENTS

CONTENTS

The Blue PLANET

An Introduction to Earth System Science

PART ONE

Our home in space
The Sun, an average-size, middle-aged star, emerges over planet Earth in this digitally-generated image.

The Earth System: Our Place *in* SPACE

Earth: *from the Old English* eorðe, *meaning the material world (as opposed to the heavens and the underworld); related to Old Saxon* ertha, *Dutch* aerde, *and German* erda.

System: *from the Greek and then the Latin word* systema, *meaning organized whole or arrangement compounded of several parts.*

Earth system science is rapidly changing the way we study and think about Earth and about life on this planet. The approach to the study of our home planet that we now call Earth system science has grown out of the impressive scientific and technological advances of the past few decades that have made it possible to measure and monitor the smallest changes in Earth systems, on the grandest scales. This approach permits scientists to achieve an unprecedented degree of connectivity. It allows us to adopt a holistic view of the planet, focusing not just on the individual parts but on the system as a whole.

Key to advancing our understanding of this planet is the need to develop an appreciation of the interactions between the different parts of the **Earth system**—the geosphere, hydrosphere, atmosphere, biosphere, and anthroposphere. These provide the structural template for the rest of this book—Parts II through VI. But here, in Part I, we look at Earth as a whole and consider our place within the solar system. In Chapter 1 we introduce the concept of systems and cycles, feedbacks, reservoirs, and fluxes—the basic types of processes that connect Earth systems to each other. We also discuss the scientific method and how science works. Then in Chapter 2 we look at the fundamental nature of energy—the driver of all processes on Earth and elsewhere in the solar system. In Chapter 3 we consider the nature of the materials of which Earth (and everything else) is composed. Finally, in Chapter 4 we examine Earth as one among many in the system of planetary objects that clusters around our Sun. We look at how the characteristics of this planet were inherited from the processes whereby the solar system originated. We also consider the age of Earth and the measurement of time and change through Earth history.

The chapters of Part I are as follows.

- Chapter 1. The Earth System
- Chapter 2. Energy
- Chapter 3. Matter
- Chapter 4. Space and Time

CHAPTER 1

The Earth SYSTEM

OVERVIEW

In this chapter we:

- Introduce the Earth system and Earth system science
- Learn what systems are and why they are important
- Identify the nature of the major reservoirs and fluxes of the Earth system
- Describe the cycles of materials and energy through the Earth system
- Learn how science works and how models are used in Earth system science

Earth and Moon
The Moon rises over Earth in this photo, part of NASA's famous Blue Marble series. The original Blue Marble photographs were taken in 1972. This particular version was taken by a Geostationary Operational Environmental Satellite (GOES) in 1997 and it is one of the most detailed images ever made of Earth. The prominent storm visible off the west coast of North America is Hurricane Linda.

The global interconnectedness of air, water, rocks, and life has become a focus of modern scientific investigation. As a result, a new approach to the study of Earth has taken hold. The traditional way to study Earth has been to focus on separate units—a population of animals, the atmosphere, a lake, a single mountain range, soil in some region—in isolation from other units. In the new, holistic approach, Earth is studied as a whole and is viewed as a system of many separate but interacting parts. Nothing on Earth is isolated; research reveals numerous interactions among all of the parts.

Those interactions, the materials and processes that characterize them, and our scientific understanding of them are the subject of this book.

WHAT IS EARTH SYSTEM SCIENCE?

Earth system science is the science that studies the whole planet as a system of innumerable interacting parts and focuses on the changes within and among those parts (FIG. 1.1). Examples of these parts are the ocean, the atmosphere, continents, lakes and rivers, soils, plants, and animals; each can be studied separately, but each is dependent on and interconnected with the others. Earth system science is a new approach to the study of Earth—a new science—and a new science requires new tools.

A New Science and a New Tool

Indeed, a new science may arise *because* new tools allow new kinds of observation and measurement, and these in turn lead to new ways of thinking about some phenomena. Earth system science requires observations of Earth at various scales and the handling of large amounts of data from many different locations; new scientific tools are required, both to generate and to manage the data.

Several decades ago, scientists from NASA, the National Aeronautics and Space Administration, realized that they were uniquely positioned to turn the scientific instrumentation originally developed for space exploration to a new use—that of studying our home planet and the changes being made by humans from a more comprehensive perspective. Spurred on by a 1988 report of the National Research Council recommending a space program focused on a study of Earth, NASA began a formal study program in 1991 called Mission to Planet Earth. Now known as Earth Science Enterprise (ESE), the mission is a comprehensive program for studying environmental changes from space and provides a mechanism for advancing the new discipline of Earth system science.

FIGURE 1.1 Earth's interacting parts
Earth system science is the study of the whole planet as a system of many interacting parts, with a particular focus on the changes within and among those parts, including the impacts of human activities.

NASA's Science Plan for 2007–2016 summarized the current status of the Earth-observing endeavor in this way:

> *From space we can view the Earth as a planet, seeing the interconnectedness of the oceans, atmosphere, continents, ice sheets, and life itself. At NASA we study planet Earth as a dynamic system of diverse components interacting in complex ways—a challenge on a par with any in science. We observe and track global-scale changes, and we study regional changes in their global context. We observe the role that human civilization increasingly plays as a force of change. We trace effect to cause, connect variability and forcing with response, and vastly improve national capabilities to predict climate, weather, and natural hazards.*

Scientists all over the world have had the opportunity to use data from a variety of Earth-observing instruments to gain a greater understanding of Earth's natural processes on a global scale. This has contributed to our basic understanding of the Earth system and its cycles, such as clouds, water and energy cycles, oceans, atmospheric chemistry, land surface, water and ecosystems, glaciers and polar ice, and solid Earth. It also has improved the effectiveness of natural hazard prediction, natural resource management, and monitoring of human impacts on the environment. Viewing and recording the Earth system from space provides a grand-scale template, a context within which smaller-scale, land-based observations can be understood. Thus space-based observation not only provides a new view of Earth, it complements and enhances traditional land-based approaches.

Earth Observation

Observations of all types, on all scales, contribute to our understanding of the Earth system, but the quintessential tool of Earth system science is satellite-based remote sensing. **Remote sensing** is the continuous or repetitive collection of information about a target—Earth, in this case—from a distance (**FIG. 1.2**). Remote sensing, more than any other new technology, has made possible observations on a grand scale, and many kinds of measurement and monitoring that could not otherwise have been accomplished. For example, the "ozone hole" over Antarctica—the decline in the concentration of ozone high in the atmosphere—is measured by remote sensing, using detectors carried on satellites. Other types of remote sensing technologies allow scientists to closely monitor changes in deserts, forests, and farmlands, as well as growth in human settlements, roads, and other parts of the built environment.

When measurements are made remotely from satellites, scientists use the data in many areas of specialization. Satellite observations, above all other ways of gathering evidence, continually remind us that each part of Earth interacts with, and is dependent on, all other parts. Modern Earth system science was born from the realization of that interdependence and the availability of satellites to make measurements. The health of waterways

FIGURE 1.2 Studying Earth from space
The exploration of space had an unexpected side-benefit: the opportunity to turn space-based instruments around and take a closer look at our own planet. Landsat, shown here in an artist's rendition, was one of the first satellites used by NASA in the 1970s to begin collecting data about Earth by remote sensing.

and coastal zones, the impacts of pollutants, and the onset of natural disasters—all of these are now much easier to study and monitor, thanks to remote sensing technologies. (See *A Closer Look: Monitoring Earth from Space.*)

Whereas satellite-based remote sensing has given us a new perspective on our home planet, new ways to explore other, previously inaccessible areas of Earth, have also added greatly to our knowledge. For example, starting in the 1960s small deep-sea remote-controlled and robotic submarines have allowed scientists to travel to the depths of the ocean. These submarines led to the discovery of life near deep-sea vents, revealing entirely new species, food chains, and ecosystems in formerly unknown and unimagined environments.

Make the CONNECTION

How can life manage to survive in the deepest parts of the ocean, where there is little or no light available for photosynthesis?

As important as new methods of measurement are new ways to store and analyze the vast amounts of data that scientists continue to accumulate about the Earth system. **Geographic Information Systems (GIS),** which are computer-based software programs, allow massive amounts of spatially referenced data points to be stored,

A Closer LOOK

MONITORING EARTH FROM SPACE

Scientists use data from satellites to study the Earth system, its many parts, and its interactions. Remote sensing encompasses the collection of information—of any kind and by any means—about an object from a distance. It can even refer to the use of seismic surveying to determine the locations of sites of archaeologic or geologic interest under the ground (this is remote sensing because the observing instrumentation is on the surface, while the object of interest is deep underground). However, the most common use of the term *remote sensing* refers to the collection of information using instrumentation carried by satellites in orbit. The applications of remote sensing to understanding and monitoring the Earth system and the environment are virtually limitless. These include monitoring of forest health and deforestation, crop yields, soil moisture, natural hazards (including volcanic eruptions, floods, and storms), air pollution, the composition and other characteristics of the atmosphere, temperatures of the surfaces of land and sea, changes in the urban environment, and many others (**FIG. C1.1**).

There are two basic types of satellite orbits, and these influence the kind of information that can be gathered by detectors carried on the satellites. *Geostationary* satellites are fixed in their orbits above one point on Earth's surface. They orbit at high altitude

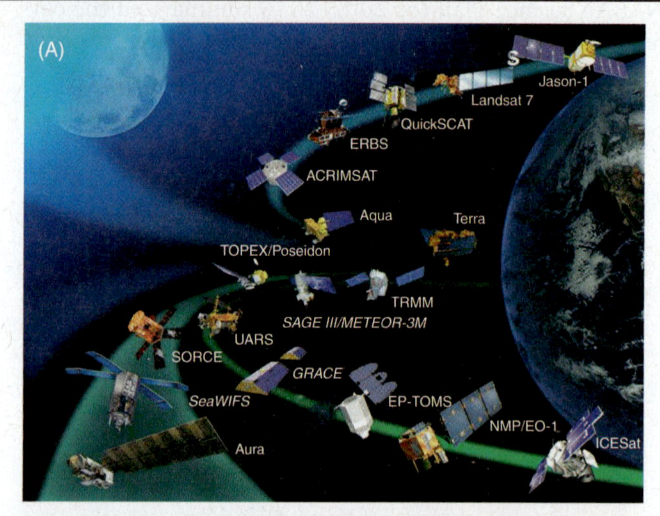

FIGURE C1.1 Earth from space
(A) Many different types of satellites are currently employed in the monitoring of Earth from space, including these from the American fleet. (B) This is a composite of two satellite images, showing dense smoke billowing from forest fires (red spots) in the Kalimantan region of Borneo, Indonesia, in September of 2009. The fires were set for the purpose of clearing land. (C) This Advanced Spaceborne Thermal Emission and Reflection Radiometer (ASTER) image of Mt. Vesuvius, Italy was acquired September 26, 2000. The image covers an area of 36 by 45 km. Warmer ground temperatures are shown in reds, and cooler temperatures in blue. In 79 CE, Vesuvius erupted cataclysmically, burying all of the surrounding cites with up to 30 m of ash. Vesuvius is intensively monitored for potential signs of unrest that could signal the beginning of another eruption.

along with their characteristics. From these, maps can be produced and sets of information of different kinds can be compared. For example, satellite images of a forest based on several types of remotely sensed data might show different types of vegetation, moisture content of the soil, surface temperature, road systems and buildings, and even the locations of human settlements. Several images can be layered on top of each other, and new data and images can be derived from the quantitative comparison of the various layers (**FIG. 1.3**, on page 10).

Not all of systems research is remote in the sense of remote in space. A lot is also remote in the sense of time. Earth has a long history, and that history has involved many changes of the land surface, of the locations of continents and oceans, and of the global climate. One of the most striking records of past climates is contained in the great ice sheets in Greenland and Antarctica, which are sampled by cores drilled into the ice. The cores record a story of warmings and coolings of the climate over the last million years.

and have a very broad perspective on the surface. For example, the Geostationary Operational Environmental Satellites (GOES) of the National Oceanic and Atmospheric Administration (NOAA) circle the Earth in geostationary or *geosynchronous* orbits, 35,800 km above the equator. Their instrumentation observes Earth from the same place all the time and is therefore continuously monitoring a single position on the surface. Because they are so high above the surface, these satellites take in a large field of view—GOES West observes almost the entire western hemisphere at once, and GOES East does the same in the eastern hemisphere. This makes them particularly useful for applications such as the monitoring of regional storm systems (**FIG. C1.2**).

The other basic kind of satellite orbit is *sun-synchronous* or *polar-orbiting*. Satellites in sun-synchronous orbit (sometimes called POES for Polar Orbiting Environmental Satellites) are designed to circle Earth from pole to pole at much lower altitudes. Therefore they are able to return data and images of much more detailed resolution than the high-altitude GOES satellites. These satellites can only observe a thin strip of the ground surface at one time, but because they are constantly moving with respect to the surface (like the Sun, hence sun-synchronous), they eventually cover the entire planet with overlapping strips, called *swaths*. Examples of polar-orbiting satellites include Landsat and SPOT, which produce satellite images for commercial uses, and the satellites of the DMSP (Defense Monitoring Satellite Program of the United States). The latter orbit at just 830 km above the surface (compare this to the altitude of the GOES satellites) and return images in which objects less than 1 m across can be distinguished.

Satellites are *platforms*; they act as carriers for instruments that do the real work of remote sensing: a wide range of *detectors*, which collect information, primarily different kinds of electromagnetic radiation (light of various wavelengths), about an object, or *targets*. We will consider the properties of electromagnetic radiation in greater detail in Chapter 2. In the meantime, let's just say that one of the real strengths of satellite-based remote sensing is the ability to carry instruments that can detect and collect a much greater variety of electromagnetic radiation than the visible light that can be detected by the human eye.

FIGURE C1.2 Hurricane Katrina
This satellite image of Hurricane Katrina in August, 2005 was taken by a high-altitude, geostationary satellite.

NASA has been involved in satellite deployment since the very beginnings of space exploration and has been the central player in turning satellite-based technologies to the study of our home planet. NASA describes its Earth Observing System (EOS) as "the world's most advanced and comprehensive system of instrumentation and technologies dedicated to the measurement of global change." The EOS is a coordinated series of satellites designed for long-term global observations of the land surface, biosphere, solid Earth, atmosphere, and oceans. The central goal of the EOS is to enable an improved understanding of Earth as an integrated system.

Systems

We have used the word "system" to talk about Earth as an integrated whole. The system concept allows scientists to break down a large, complex problem into smaller, more easily studied pieces. A **system** is any portion of the universe that can be isolated from the rest of the universe for the purpose of observing and measuring changes. By saying that *a system is any portion of the universe*, we mean that the system can be whatever the observer defines it to be. That is why a system is only a *concept*; you choose its limits for the convenience of your study. It can be large or small, simple or complex. You could choose to observe the contents of a beaker in a laboratory experiment. Or you might study a flock of birds, a lake, a small sample of rock, an ocean, a volcano, a mountain range, a continent, or an entire planet. A leaf is a system; it is part of a larger system (a tree), which in turn is part of an even larger system (a forest). The mountain–river–lake landscape shown in **FIGURE 1.4** is a system; some of the smaller subsystems

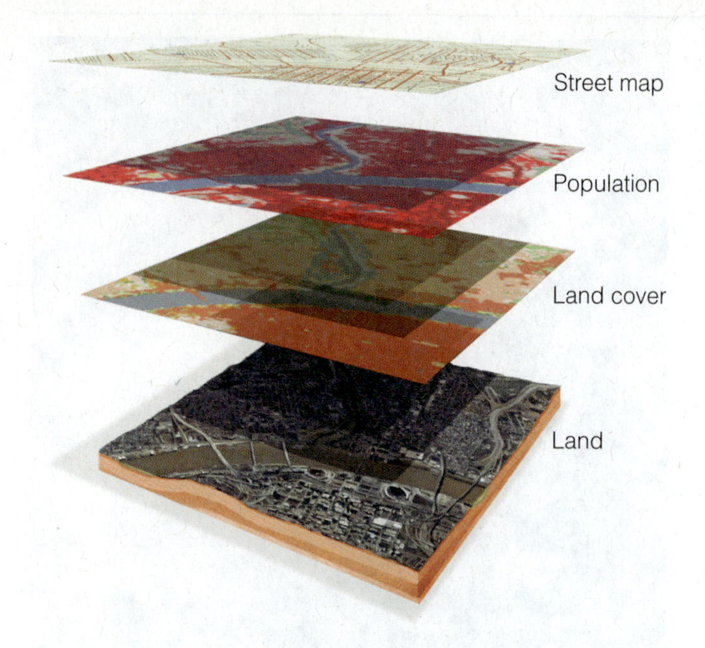

FIGURE 1.3 Geographic information systems
Geographic information systems allow for the storage of large volumes of spatially referenced data points, along with their characteristics. The data can be used to produce maps showing the distribution of specific characteristics. The map layers can be combined and compared quantitatively, to yield derivative data and images.

the BASICS

Types of Systems

The most important defining characteristic of a system is the nature of its boundaries. System boundaries differ in terms of what they will and won't allow to pass through them, that is, to move into or out of the system. On the basis of boundary differences, we define three basic types of systems, shown in **FIGURE B1.1.**

The simplest kind of system to understand is an **isolated system** in which the boundary prevents the system from exchanging either matter or energy with its surroundings. The concept of an isolated system is easy to understand—both the matter and energy within the system are fixed and finite because none can enter and none can leave the system. Although seemingly simple, such a system is only imaginary. In the real world it is possible to have boundaries that prevent the passage of matter, but it is impossible for any real boundary to be so perfectly insulating that energy can neither enter nor escape. Nevertheless, isolated systems have proven to be useful to scientists in the conceptual study of some kinds of natural changes.

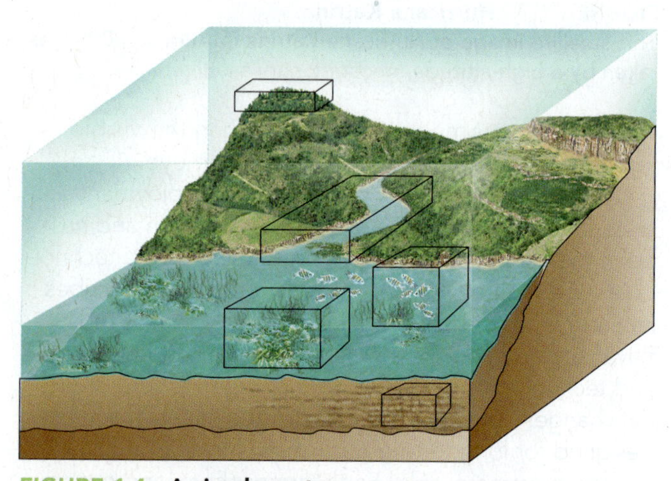

FIGURE 1.4 A simple system
The mountain–river–lake landscape shown here is an example of a system. Some of its component subsystems are outlined by boxes.

that compose it—a volume of water, a volume of bottom sediment, and a hilltop, among others—are outlined with boxes.

The fact that a system is *isolated from the rest of the universe* means that it must have a boundary that sets it apart from its surroundings. The nature of the boundary is one of the most important defining features of a system, allowing us to establish three basic kinds of systems—isolated, closed, and open—with different types of boundaries.

By *observing and measuring changes*, we mean that we use the systems concept to study complex problems. This might mean observing what happens in a natural system under changing conditions, such as what happens in a wetland during a drought, or what happens to a dead organism as it decays on a forest floor, or what happens when magma rises in a volcano until it erupts. Or it might mean imposing changes on an artificial system in a laboratory, such as heating up a rock in a special crucible so that we can observe what happens as it melts.

Models

When we do the latter type of investigation—imposing changes on an artificial system in a laboratory setting—we are studying the natural system indirectly, by building and examining a model. A **model** is a representation of something. Most models are simplifications of complex originals, and they are typically created at a more manageable scale than the original. If you build a tiny ship or car, or sculpt a dinosaur or a human figure, or build the solar system in miniature, you are creating a model—a representation of the original item, at a more manageable scale. A model can be quite detailed; it might even be a working model, but it will always be a simplified representation of the original. (Models are not necessarily *smaller* than the original object, though; if you build a model of an atom or a cell, they will be much *larger* than the actual objects, but they are still models. The important characteristics of models are the simplification and the manageable scale.)

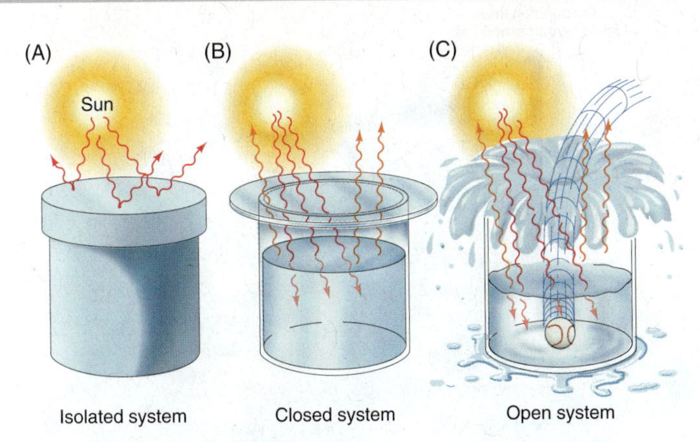

FIGURE B1.1 Systems
The three basic types of systems are: (A) An isolated system. (B) A closed system. (C) An open system.

The nearest thing to an isolated system in the real world is a **closed system** in which the boundary permits the exchange of energy, but not matter, with the surroundings. It's difficult to think of systems in the real world that have *perfectly* sealed boundaries, allowing *no* matter—not even one atom—to enter or escape, but we can imagine some examples that are *nearly* closed systems. An example would be a tightly sealed solar oven, a closed metal box that does not allow the contents to leak out but does allow for the contents to be heated by sunlight beating down on the exterior. Another example of a closed system would be a box with a moveable lid that could be pushed in or pulled out like a piston, compressing or expanding the space inside the box; the changing pressure on the contents of the box represents *work*, a form of physical energy (Chapter 2).

The third kind of system, an **open system**, is one that can exchange both energy and matter across its boundary. An island is a simple example of an open system: Water is matter that comes into the system as rain. Water also freely leaves the island via streams flowing to the ocean, absorption by plants, or evaporation back to the atmosphere. Energy also comes into the system in the form of sunlight and leaves the system as heat radiated by plants, rocks, and soils. Open systems are more difficult to study mathematically than closed or isolated systems because they have more potential for uncontrolled variation. However, most natural systems are, in fact, open systems, allowing both energy and matter to move freely in and out.

We can build models of objects; we also can build models of *processes*. If you put some gravel, plants, and a small puddle of water together in a fish tank, with a heat lamp to simulate sunlight, you are building a model of the water cycle (**FIG. 1.5A**). Some models are physical—the model car and the water-cycle-in-a-fish-tank, for example. Others are pictorial or graphical, illustrating what we believe to be the important functional parts of the original in a picture format (**FIG. 1.5B**). Computer games are models, too—they are graphical representations based on numerical representations (computer programs) that simulate the real world. In games such as *The Sims* and its predecessors *SimEarth*, *SimCity*, and *SimAnt*, these models can be quite realistic—if you introduce a predator or limit the food supply, your colony or your city (or even your planet) will die, just as it would in the real world. Scientists develop numerical models of this type, making them as realistic as possible, and then use them to test what might happen in the real world if certain types of changes were imposed.

The storage and movement of materials and energy in a group of interacting systems are commonly depicted in the form of box models. A **box model** is a simple, convenient graphical representation of a system (**FIG. 1.5C**). A box model can be used to show the following essential features of a system:

1. The processes by which matter (or energy) enters and leaves the system, and the rates at which they do so.
2. The processes by which matter (or energy) moves among the various parts of the system internally, and the rates at which this happens.
3. The amount of matter (or energy) in the system at a given time and its distribution within the system.

Fluxes, Reservoirs, and Residence Times

One of the keys to understanding the Earth system is to measure how volumes and exchanges of materials and energy between Earth's reservoirs change over time. The next challenge is to figure out *why* the changes happen and how quickly. In Figure 1.5B and C, the processes by which matter is transferred from one part of the system to another are depicted by arrows. The arrows represent processes—in this case, processes such as evaporation and precipitation. The amount of matter (or energy) that is transferred along any one of those arrows, and the rate at which it is transferred, is called a **flux**.

If we want our box model to be *quantitative*—that is, numerically specific—then we can denote the fluxes numerically, using units of *mass or volume per time*. So, for example, we might make observations and measurements showing that 2 million cubic meters (units of *volume*) are moving from the atmosphere to the ocean each year (units of *time*), by falling as rain. To display this result quantitatively using our box model, we would simply put that number—2×10^6 m^3/yr—next to the arrow that points the way from the atmosphere to the ocean, labeled "Rain"

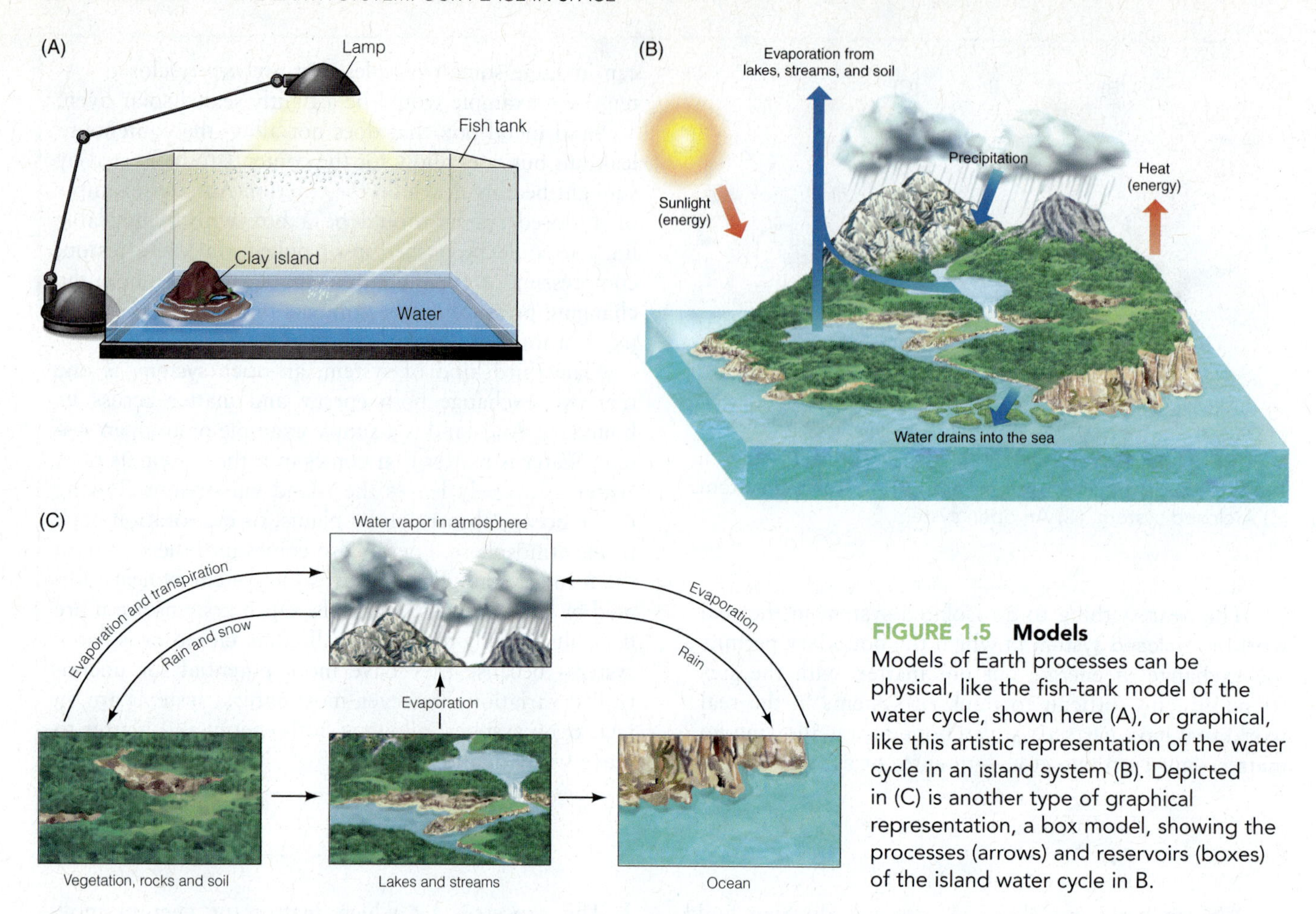

FIGURE 1.5 Models
Models of Earth processes can be physical, like the fish-tank model of the water cycle, shown here (A), or graphical, like this artistic representation of the water cycle in an island system (B). Depicted in (C) is another type of graphical representation, a box model, showing the processes (arrows) and reservoirs (boxes) of the island water cycle in B.

on Figure 1.5C. The capacity to be quantitative—that is, to ascribe numbers to specific processes and fluxes—is one of the strengths of box models.

The "boxes" in a box model represent the places where water (or energy, or whatever might be the material of interest) is stored for a period of time within the system. These storage places are called **reservoirs.** In our box model of the island water cycle (Fig. 1.5C), the reservoirs shown are the atmosphere (where water resides as clouds and as water vapor); the biosphere (where water is an important constituent of living organisms); rocks and soil (where water resides for varying lengths of time as groundwater and as soil moisture); lakes and streams (where water collects in pools and channels on the surface); and the ocean. So when we consider rain, for example, we are specifically referring to the transfer of water from the atmospheric reservoir to the oceanic reservoir (or to a land-based reservoir, such as a lake or soil). The movements, or fluxes, represented by the arrows in Figure 1.5C may be fast or slow, and so an essential part of Earth system science is the measurement of rates of movement.

When the flux of matter into a reservoir matches the flux out of that reservoir, we say that the reservoir is at **steady state.** Flows between reservoirs, and even within one reservoir, never cease, but the rates of flow may change. When this happens, the volumes of the reservoirs that are supplying or receiving matter must change, too. If the flux of some substance into a reservoir is greater than the flux of that substance out of the reservoir, then we refer to the reservoir as a **sink.** On the other hand, if more of a substance is coming *from* a reservoir than is flowing *into* it, the reservoir is called a **source.** The characteristics of sources and sinks are extremely important in controlling the cycling of matter (and energy) through the Earth system.

Returning to our box model of the water cycle, Figure 1.5C, the average length of time water spends in any of these reservoirs is called its **residence time.** The residence time of any material in any particular reservoir is determined by the interaction of many factors, including the physical, chemical, and biologic properties of the material itself, the properties of the reservoir, and any external forces or processes acting on either the material or the reservoir. Water typically has very short residence times in plants and animals (how long does it take a glass of water to move through your body?), but somewhat longer residence times in the atmosphere (days), rivers (months), lakes (tens of years), groundwater (tens to hundreds of years), and glaciers and the deep ocean (tens to many thousands of years).

The issue of residence time—how long a material can be expected to remain in a reservoir, and the processes whereby it may be caused to leave the reservoir—is one of the fundamental concerns of Earth system science and of modern environmental science. To explore the concept of residence time a bit further consider the example of smoke, which consists of gas with billions of tiny solid particles emitted as a result of combustion (fire), either natural or artificial. The gaseous component and the lighter-weight solid particles in smoke can remain aloft in the atmosphere for a long time, sometimes years, and can be transported great distances, sometimes globally, by atmospheric circulation. But some of the solid particles in smoke are heavier than others; lead and lead-bearing compounds, which are common products of combustion and many industrial processes, provide an example. Because particles of lead are heavy (*dense*) relative to the air that surrounds them, they tend to settle to the surface rather quickly. Therefore, the residence time for particles of lead in the atmosphere is much shorter than for some of the other components of smoke—typically only 10 days or less.

Note, though, that the physical characteristics of the lead particles themselves (such as, in this case, the density of the lead-bearing particles) are not the only factor that determines their residence time in the atmospheric reservoir. The properties and processes that characterize the reservoir are also important. For example, lead particles might remain for a longer time in the atmosphere if winds are very strong, or if the air is very warm (thus providing more buoyant updrafts), as it would be in the summer or in the tropics. Precipitation can play a role, too; solid particles can wash out of the atmosphere much more quickly when they become attached to water droplets.

Some materials have such long residence times in certain reservoirs that they are isolated from the rest of the Earth system for very long periods; examples include water frozen for thousands of years inside long-lived glaciers, and fossil fuels (the organic remains of plants and animals, converted into coal, oil, or natural gas) preserved for hundreds of millions of years in rocks deep underground. To describe this situation we use the term **sequestration**, the same term that is applied to the burial of carbon dioxide captured from the smokestack of a power plant; it means that the material is isolated from any contact with the rest of the world. Similarly, materials that reside naturally in long-term reservoirs are referred to as *sequestered* because they are isolated from contact with the rest of the Earth system.

Living in a Closed System

The Earth system comprises four vast reservoirs, with constant flows of energy and matter among them (FIG. 1.6). The four great reservoirs are the atmosphere, the hydrosphere, the biosphere, and the geosphere. Each of these complex reservoirs functions as a subsystem on its own.

As a whole, Earth is a closed system—or at least very close to being a closed system (FIG. 1.7). Energy reaches

FIGURE 1.6 Earth's interacting parts
This is a diagrammatic representation—essentially a simple box model—of Earth as a system of interacting parts. Each character represents one of the four major reservoirs (or subsystems), and each arrow represents a flow of materials or energy.

FIGURE 1.7 Earth as a closed system
Earth essentially operates as a closed system. Energy reaches Earth from an external source and eventually returns to space as long-wavelength radiation, but the matter within the system is basically fixed. The subsystems within Earth are open systems, freely exchanging matter and energy.

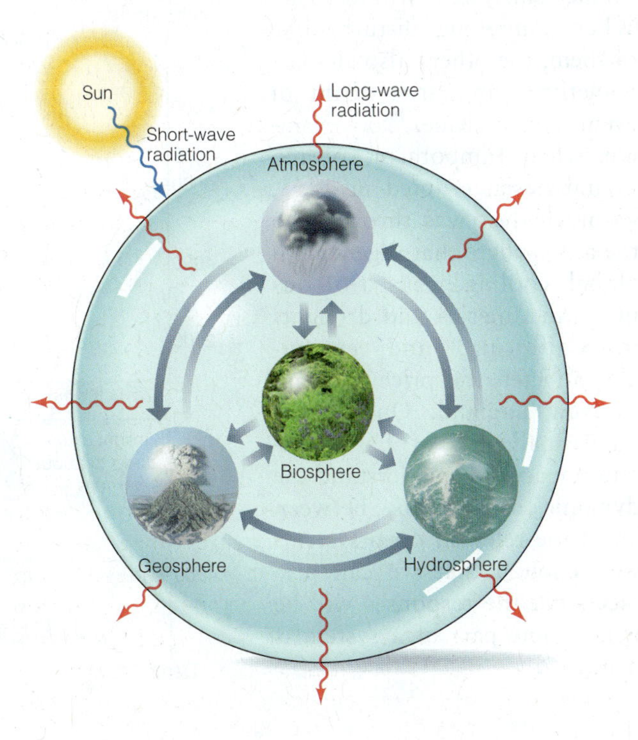

Earth in abundance in the form of solar radiation. Energy also leaves the system in the form of longer-wavelength infrared radiation. Matter, on the other hand, is largely confined within the system. It is not quite correct to say that *no* matter crosses the boundaries of the Earth system; we lose a small but steady stream of hydrogen atoms from the upper part of the atmosphere, and we gain some incoming material in the form of meteorites. However, the amount of matter that enters or leaves the Earth system is so minuscule compared with the mass of the system as a whole that for all practical purposes Earth is a closed system.

The fact that Earth is a closed system has two important implications for those of us who occupy its surface. First, because *the amount of matter in a closed system is fixed and finite*, the mineral resources on this planet are all we have and—for the foreseeable future—all we will ever have. Someday it may be possible to visit an asteroid for the purpose of mining nickel and iron; there may even be a mining space station on the Moon or Mars at some time in the future. For now, however, it is realistic to think of Earth's resources as being finite and therefore limited. A further consequence of a fixed and finite closed system is that waste materials remain within the confines of the Earth system. Or, as environmentalists say, "There is no *away* to throw things to."

Second, *if changes are made in one part of a closed system, the results of those changes eventually will affect other parts of the system.* Earth is a closed system, but all of its innumerable smaller parts are interconnected; they are open systems, and both matter and energy can be transferred between them. The atmosphere, hydrosphere, biosphere, and geosphere are all open systems, and so is every smaller subsystem within them. When something disturbs one of them, the others also change. Sometimes an entire chain of events may ensue. For example, when Tambora, a volcano in Indonesia, erupted in 1815, so much dust was thrown into the atmosphere that it generated global cooling, caused floods in South America and droughts in California, and eventually affected the price of grain in New England. One of the main challenges of Earth system science is to understand the dynamic interactions between all of the relevant open systems well enough that we can predict what the responses will be when some part of a system is disturbed.

EARTH SYSTEM RESERVOIRS

As mentioned earlier, a convenient way to think about Earth as a system of interdependent parts is to consider it as four vast reservoirs of material with flows of matter and energy between them. Each of these major systems can be further subdivided into smaller, more manageable study units. For example, we can divide the hydrosphere into the ocean, glacier ice, streams, and groundwater.

The place where Earth's four reservoirs interact most intensively is a narrow zone that we might call the *life zone* because its most important characteristic—from a human perspective, certainly—is that it supports life and allows life to exist on this planet. The life zone is a region no greater than about 10 km above Earth's surface and 10 km below the surface (**FIG. 1.8**). In this narrow zone all known forms of life exist because it is only here that conditions favorable for life are created by interactions between the lithosphere, hydrosphere, and atmosphere, and modified by the biosphere.

Earth is habitable and is able to offer a life-supporting zone, by virtue of its particular relationship with the Sun. Earth is the only planet we know of where water exists at the surface in solid, liquid, and vapor forms. This happens because Earth is just the right distance from the Sun—not too near (where it would be too hot), and not too far (where it would be too cold). Other planets have water, but none other that we know of (yet) has exactly the right combination of temperatures and materials to support liquid water—and thus to provide a life zone—near the surface, although there is growing evidence that Mars might have had suitable conditions for life a few billion years ago.

FIGURE 1.8 The life zone
All life on Earth lives within a zone no wider than 20 km. It is the zone where interactions between the geosphere, hydrosphere, and atmosphere create a habitable environment.

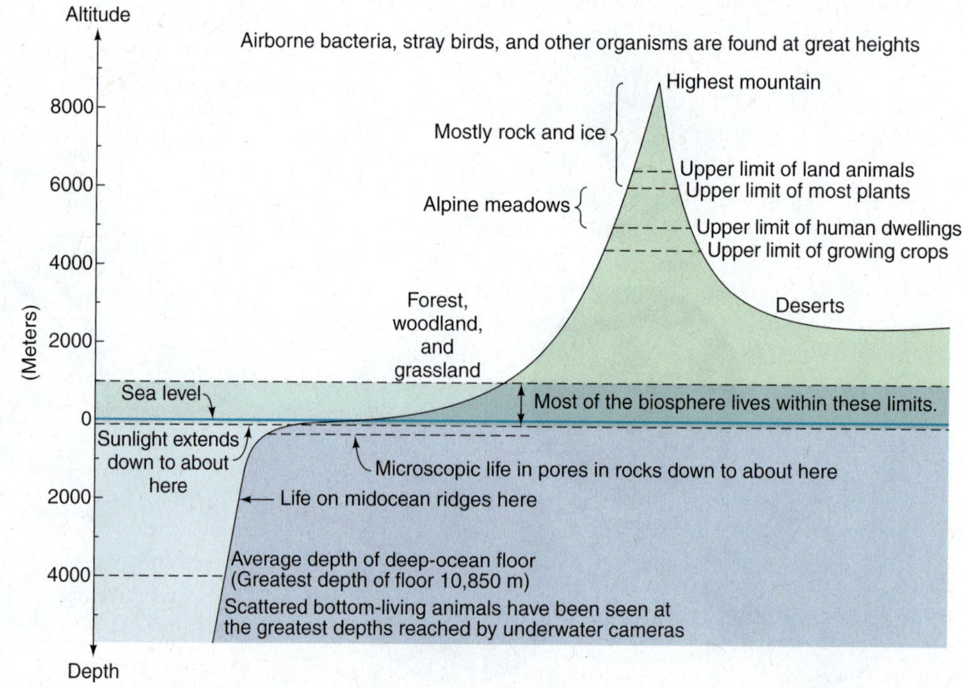

The study of Earth's four major reservoirs and the interactions among them forms the foundation of Earth system science and informs the content of this book. To our study of these four natural systems, we will add a fifth focus: the anthroposphere, which gives us the opportunity to consider resource use in the built and technological environments, as well as human influences on other parts of the Earth system.

The Geosphere

The **geosphere** (FIG. 1.9A) is the solid Earth, composed principally of *rock* (by which we mean any naturally formed, nonliving, firm coherent aggregate mass of solid matter that constitutes part of a planet) and *regolith* (the irregular blanket of loose, uncemented rock particles that covers the solid Earth). The surface of the geosphere is a particularly interesting and dynamic place, where energy that comes into the Earth system from outside sources meets energy that comes from within the planet. These forces combine and compete to build up and wear down the materials at Earth's surface, creating the enormous diversity of landscapes around us. The dynamic nature of the geosphere can also be hazardous for human interests, by way of processes such as landslides, earthquakes, and volcanic eruptions. The geosphere will provide the central focus for Part II of this book.

The Hydrosphere

The **hydrosphere** (FIG. 1.9B) is the totality of Earth's water, including oceans, lakes, streams, underground water, and all the snow and ice. The perennially frozen parts of the hydrosphere are collectively referred to as the **cryosphere**. Much water also resides in the atmosphere, but we generally consider atmospheric water to be separate from the water of the hydrosphere. Like all of Earth's subsystems, however, these two great water reservoirs are connected; through the process of rain, for example, water moves from the atmospheric reservoir into the hydrosphere. The hydrosphere and atmosphere provide critically important services for the environment because they store, purify, and continually redistribute water. There is also considerable water contained inside the planet, most of which has never been in contact with the atmosphere and has never been part of the processes of the hydrosphere; we call this *juvenile* or *primordial water*, and it, too, is considered to be separate from the water of the hydrosphere. We will consider the hydrosphere and its various processes and characteristics in Part III.

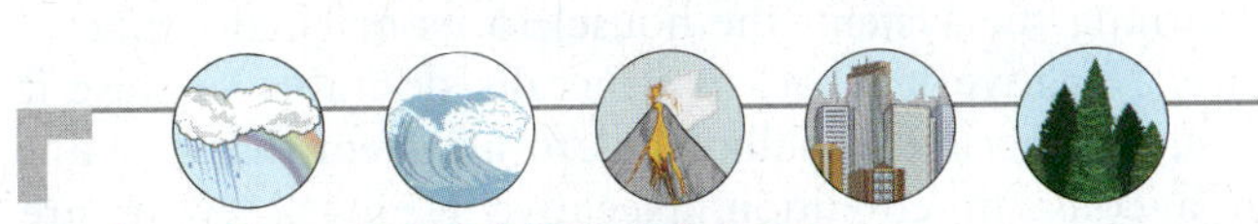

Make the CONNECTION

By what processes do you think juvenile or primordial water might find its way out of Earth's interior, and where would it end up?

The Atmosphere

The **atmosphere** (FIG. 1.9C) is the mixture of gases—predominantly nitrogen, oxygen, argon, carbon dioxide, and water vapor—that surrounds Earth. The atmosphere seems very thick to us, but in the context of the whole planet it is a very, very thin layer, indeed. The atmosphere provides many crucial services, such as protecting life from damaging solar radiation, and being the reservoir for oxygen and carbon dioxide, two gasses that are essential for the biosphere. The outermost layer of the atmosphere is, in effect, the boundary of the Earth system, separating us from our surroundings in space. In Part IV of the book we will look more closely at the atmosphere and its processes.

The Biosphere

The **biosphere** (FIG. 1.9D) includes all of Earth's organisms, as well as any organic matter not yet decomposed. The existence of the biosphere obviously makes Earth a unique planet—so far, we have yet to discover another planet that hosts life, either in this solar system or outside of it, although there are some promising candidates among the moons of Jupiter and Saturn. In Part V we will consider the characteristics and processes of the biosphere in detail.

One of the new scientific recognitions—new in the past 30 years—is the great extent to which life affects the other major parts of the Earth system. The chemical composition of Earth's atmosphere is very different from what would be found on a lifeless planet. For example, the atmospheres of Venus and Mars, Earth's nearest planetary neighbors, are more than 95 percent carbon dioxide and less than 4 percent nitrogen. Earth's atmosphere, in contrast, is 79 percent nitrogen, 21 percent oxygen, as well as a small amount of carbon dioxide. The difference is principally the result of life's processes over billions of years of Earth history. Photosynthesis by green plants, algae, and photosynthetic bacteria has removed carbon dioxide from the atmosphere and added oxygen. Oxygen is a highly reactive gas that rapidly combines with many other chemical elements and so does not remain in its free form for a long time. To counteract the removal of oxygen by chemical reactions, life acts as an oxygen pump, continuously returning oxygen to the atmosphere. This means that *free oxygen in Earth's atmosphere is the result of at least 3 billion years of photosynthesis and is therefore a product of life.* This is just one of the profound ways that life has changed Earth. In the course of this book, you will discover many other ways that life has affected other major components of the Earth system.

The Anthroposphere

The **anthroposphere** (FIG. 1.9E) is the "human sphere" (from the Greek root *anthro-*, human). It comprises people and their interests, as well as human impacts on the

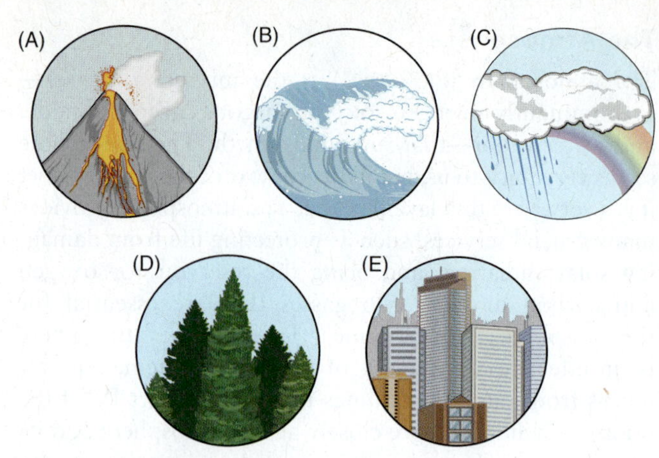

FIGURE 1.9 Earth's subsystems
The major subsystems of the Earth system are: (A) geosphere, (B) hydrosphere, (C) atmosphere, (D) biosphere, and (E) anthroposphere. Throughout the book we will be emphasizing the connections among these subsystems.

natural Earth system. The anthroposphere is the part of the natural system that has been modified by humans, for human purposes or as a result of human activities. Another term that applies to the anthroposphere is *technosphere*, which refers specifically to technology, machines, and the built environment.

Humans have always changed their local environments, but when the human population was small, these changes happened so slowly that they did not alter the Earth system. Now the population is large and growing ever larger. At the time these words are being written, in 2010, the world's population is nearing 6.9 billion and increasing by about 81 million each year. There are now so many of us that we are changing Earth just by being alive and going about our business. In doing so we are taxing the resources of the Earth system and the capacity of the system to manage impacts.

The impacts of human activities on Earth systems will be a focus throughout this book, but we also will give specific consideration to the anthroposphere in Part VI.

DYNAMIC INTERACTIONS AMONG RESERVOIRS

The causes and effects of disturbances in a complex closed system are very difficult to predict. Consider the regional weather pattern called El Niño, which occurs every few years off the west coast of South America but can influence the weather all around the world. El Niños (discussed in greater detail in Chapter 12) are characterized by weakening of the trade winds, anomalously warm sea-surface temperatures, and the suppression of upwelling cold ocean currents. The result is worldwide abnormalities in weather and climatic patterns, and widespread incursions of biologic communities into areas where they do not normally occur. These features of El Niños are reasonably well known; what is not known is the triggering event. The interactions among processes in the atmosphere, hydrosphere, geosphere, and biosphere are so complex, and these subsystems are so closely interrelated, that scientists cannot pinpoint exactly what it is that begins the whole El Niño process. One new hypothesis suggests that El Niño may be a result of ocean–atmosphere interactions caused by differences in their properties as liquid and gas. Another hypothesis suggests that a small change in the geosphere—specifically, submarine volcanic activity that causes localized heating of seawater—may create enough of a thermal imbalance to trigger an El Niño.

From an environmental point of view, the significance of interconnectedness is obvious: When human activities produce changes in one part of the Earth system, their effects—often unanticipated—will eventually be felt elsewhere. When sulfur dioxide is generated by a coal-fired power plant in Ohio or England, it can combine with moisture in the atmosphere and fall as acid rain in Ontario or Scandinavia. When pesticides are used in the cotton fields of India, the chemicals can find their way to the waters of the Ganges River and thence to the sea, where some may be ingested by fish and stored in their body tissues. The fish, in turn, may be caught and eaten. In this way, pesticides can end up in the breast milk of mothers halfway around the world from the place where they were applied. Such processes can take a long time to happen, and that is why they have been all too easy to overlook in the past.

Feedbacks

Because energy flows freely into and out of systems, all closed and open systems respond to inputs and, as a result, have outputs. A special kind of system response, called **feedback**, occurs when the output of the system also serves as an input and leads to changes in the state of the system. A classic example of feedback is a household central heating system (FIG. 1.10A). When room temperature cools, a metal strip in the thermostat cools and contacts an electric circuit, turning on the furnace. When the temperature rises, this strip warms and bends away from the electric contact, turning off the furnace. The metal strip senses the temperature change and sends a signal to the furnace; hence, feedback occurs.

The household central heating system is an example of **negative feedback**, in which the system's response is in the opposite direction from the initial input. In this case, cooling is the initial change; the response to this disturbance on the part of the furnace is to initiate warming, returning the system (the house) to its original temperature. Negative feedback is generally desirable because it is stabilizing and usually leads to a system that remains in a constant condition. Negative feedback cycles are often described as being self-limiting or self-regulating. A system that is self-regulating is said to have the property of *homeostasis*, which implies a state of **equilibrium**, or balance. However, it is a *dynamic equilibrium*; that is, the system isn't just static or unchanging. Instead, it is constantly responding to small changes and disturbances,

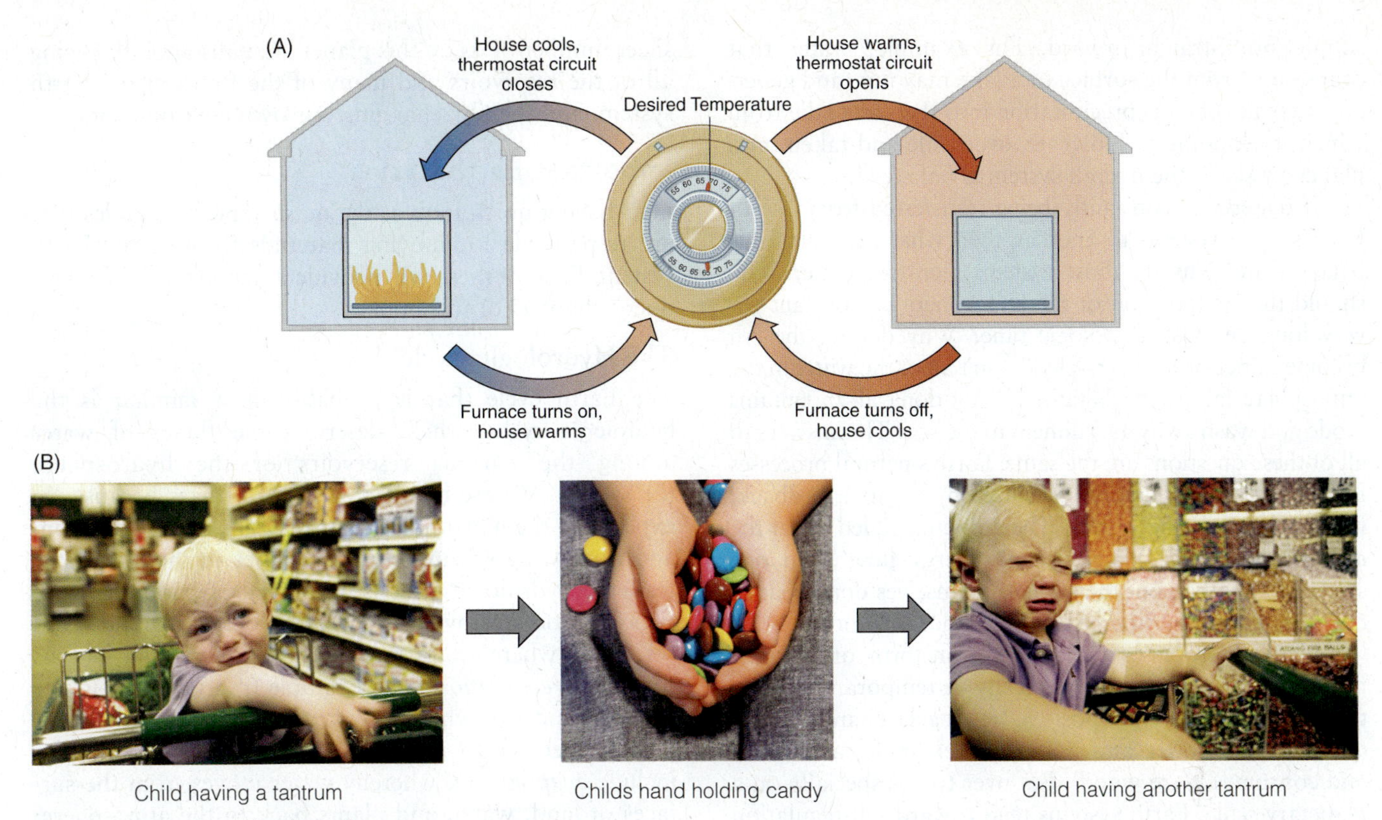

FIGURE 1.10 Feedback cycles
(A) Here is a familiar example of negative feedback. A change in temperature in one direction leads the thermostat to send a signal that makes the heating/cooling system change in the opposite direction. Hence, the feedback is negative. (B) Here is a familiar example of a positive feedback. A child who wants candy throws a temper tantrum in the store. In response, the parent gives the child some candy. This leads the child to have another temper tantrum the next time he wants candy. The response leads to a reinforcement or repetition of the initial condition; hence the feedback is positive. In a true positive feedback cycle, the child's tantrums would get worse and worse each time.

in such a manner that it returns to a state of equilibrium. The human body, for example, displays homeostasis by producing sweat to help cool down the body if it becomes overheated. The body overheats; sweat is produced as a result; the body cools: negative feedback.

With **positive feedback**, on the other hand, an increase in output leads to a further increase in the output. A familiar example (FIG. 1.10B) would be a child who throws a temper tantrum. If the parent gives the child some candy in response to the tantrum, the child will likely have more intense and more frequent tantrums each time he wants candy, leading the parent to be even quicker to hand over the candy. Positive feedback, sometimes called a "vicious circle," is destabilizing; instead of returning the system to equilibrium, a positive feedback amplifies the original disturbance.

A fire starting in a forest provides an environmental example of positive feedback. The wood may be slightly damp at the beginning and not burn well, but once a small fire starts, wood near the flame dries out and begins to burn. This, in turn, dries out more wood, leading to a larger fire. Serious problems can occur when our interactions with the natural environment lead to positive feedbacks. We will discuss positive and negative feedback cycles in greater detail in subsequent chapters.

Cycles

The state of dynamic equilibrium arising from feedbacks is typical of subsystems of the Earth system. The subsystems are finely balanced, complex systems; when a disturbance occurs or a change is induced, the system will react to that change, generally inducing further changes. Reactions to a disturbance typically involve the movement of material from one part of a system or reservoir to another. For example, sunlight shining on a lake causes evaporation, a process in which water moves from the lake to the atmosphere. This changes the state of moisture in the local atmosphere. Eventually that water will be transferred back out of the atmosphere, via the process of precipitation. Lots may happen to that water in the interim; it may travel quite a distance in the atmosphere, or it may even turn to snow or ice before it falls to the surface. But eventually the atmosphere will readjust to its original state of equilibrium by releasing the water back to the surface as precipitation.

The constant movement of material from one reservoir to another—such as water moving from the lake to the atmosphere and back again, over and over—is called a **cycle**. Because of the complexity of the interacting parts of the Earth system, natural cycles are generally neither simple nor straightforward. For example, water that evaporates from the surface of a lake may not (and generally does not) fall as precipitation into the same lake from which it evaporated. However, over time and taken on a planetary scale, the overall system is balanced.

If material is constantly being transferred from one of Earth's open systems to another, then what maintains this balance, and why do those systems seem so stable? Why should the composition of the atmosphere be constant for very long periods of geologic time? Why doesn't the sea become saltier or fresher? Why doesn't all of the water in the atmosphere fall as precipitation? Why don't all mountains erode and wash away as sediment to the sea? The answers to all of these questions are the same: Earth's natural processes follow cyclic paths that are stabilized by negative feedback; it is a self-regulating system. The amounts added equal the amounts removed. Materials and energy flow from one system to another, but the systems themselves don't change overall because the different parts balance each other.

Natural processes (such as a rainstorm or a forest fire) may rage out of control locally or temporarily, sometimes even displaying positive feedback characteristics (the fire gets hotter and hotter, for example, as it dries and consumes more wood). But over time, especially on a planetary scale, Earth systems tend toward self-regulation and a state of equilibrium. This is lucky for us; if Earth were not self-regulating, the entire planet would have been engulfed by fire or rain or any of the countless other local disturbances that are constantly happening. The cycling and recycling of materials and the dynamic interactions among subsystems have been going on since Earth first formed, and continue today. One of the challenges of Earth system science is to understand the relationships between temporary local disturbances and the long-term dynamic equilibrium of the system as a whole.

If natural systems exist in a state of dynamic equilibrium, what happens when human activities cause changes or disturbances in those systems? Such changes often affect the rates at which materials move from one reservoir to another. For example, carbon moves through the Earth system in a complex natural cycle in a state of dynamic equilibrium. One of the reservoirs where carbon is sequestered on a long-term basis is in deposits of fossil fuels underground. When we tap into this long-term reservoir by unearthing and burning the coal, oil, or natural gas, we release that carbon back to the atmosphere at a rate that is *much* faster than the rate at which it would have been released by natural processes. The result of changing the flux of carbon from land-based reservoirs to the atmosphere has been a dramatic change in the composition of the atmosphere, which, in turn, is affecting the global climate system. By virtue of the sheer number of us on this planet, humans are influencing all of the reservoirs and many of the flows in the Earth system, and thereby changing our own environment.

Important Earth Cycles

Let's have a quick look at the most important cycles that are responsible for moving materials through the Earth system. Each of these will provide a major focus in subsequent chapters of this book.

The Hydrologic Cycle

The Earth cycle that is probably most familiar is the **hydrologic cycle**, which describes the fluxes of water among the various reservoirs of the hydrosphere (FIG. 1.11). We are familiar with these fluxes because we experience them as rain and other forms of precipitation, and as flowing streams. Like all cycles in the Earth system, the hydrologic cycle is composed of pathways—the various processes by which water is cycled around—and reservoirs where water may be held for varying lengths of time. *Precipitation* is an example of a pathway in the hydrologic cycle, whereby water moves from the atmospheric reservoir to the land or ocean. Other examples include *evaporation*, whereby water moves from the surfaces of land, water, and plants back to the atmosphere; and *surface runoff*, whereby water coalesces into channels and runs off the land surface toward the oceans. The reservoirs in which water is stored in the hydrologic cycle are the reservoirs that comprise the hydrosphere, including surface water bodies, clouds, the ocean, glacier, groundwater, and living organisms. We will discuss these and the other pathways, processes, and reservoirs that make up the hydrologic cycle in Chapter 8.

FIGURE 1.11 The global hydrologic cycle
The hydrologic cycle is probably the most familiar of Earth's important cycles. It traces the movement of water from one reservoir to another throughout the Earth system. Here the global hydrologic cycle is portrayed as a simple box model. Compare this to Fig. 1.5B, C.

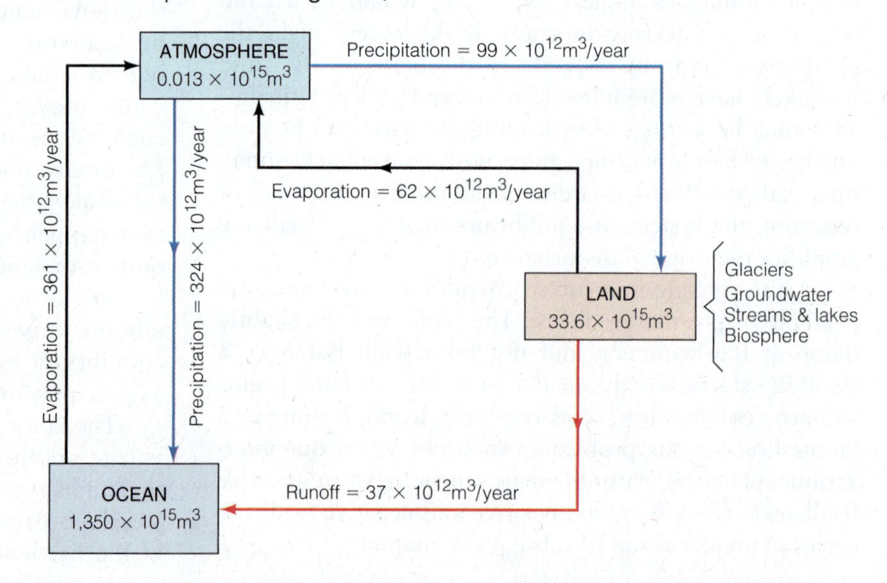

The hydrologic cycle illustrates, once again, the idea of balance or dynamic equilibrium in Earth cycles. Because Earth is a closed system, the total amount of water in the hydrologic system is fixed. However, there can be quite large fluctuations in the quantity of water held in a local reservoir at any given time. These fluctuations can cause floods in one area and droughts in another, but on a global scale they do not change the total volume of water on Earth. The system is balanced overall, but it is a dynamic balance because water can move so readily from one part of the system to another.

The hydrologic cycle also provides a clear example of the interconnectedness of Earth's subsystems. For example, water moves as groundwater in the subsurface, which immediately connects the hydrologic cycle with the rock cycle through soil, gravel, and other Earth materials. We also know that biologic organisms are reservoirs for water—without it, they could not live—and this immediately connects the hydrologic cycle with the biosphere and with the cycles of materials and processes that support life on this planet.

Make the CONNECTION

By what processes do you think water might move from liquid form in the hydrosphere into a frozen form in the cryosphere, and back again?

The Energy Cycle

The **energy cycle** encompasses the great "engines"—the external and internal energy sources of the planet—that drive the cycles of the Earth system. The energy cycle is a bit different from the other Earth cycles because it describes the movement of *energy* through the system, rather than the movement of *materials*.

We can think of Earth's energy cycle as a "budget." Energy may be added to or subtracted from the budget and may be transferred from one storage reservoir to another, but overall the additions and subtractions and transfers must balance. When a balance does not exist, Earth's near-surface environment either heats up or cools down until a balance is reached. This has happened in the past, as exemplified by changes in Earth's average surface temperature during ice ages. Today we know that as a result of the buildup of carbon dioxide, too much heat energy is being retained in the lower part of the atmosphere near Earth's surface, leading to an increase in surface temperature.

Energy in the Earth system differs from matter in one important aspect—matter can be cycled from one reservoir to another, back and forth, endlessly, but energy cannot be endlessly recycled. This is because the flow of energy involves degradation and increasing disorganization as the energy becomes dispersed as heat. This means that there must be a source, or sources, of energy coming into the system that replenishes the energy budget on a continuous basis. This is indeed the case; energy from external sources (primarily the Sun) and internal sources (geothermal heat) is constantly flowing to Earth's surface. The energy drives a wide variety of Earth processes, including mountain-building, wind, waves, and photosynthesis (**FIG. 1.12**). In the process, the energy is degraded and is eventually lost through radiation to outer space, only to be replenished at the surface by incoming energy from the Sun and from Earth's own internal energy sources.

Like the other cycles, the energy cycle has storage reservoirs—places where energy resides for various lengths of time in the Earth system. For example, living organisms are reservoirs for energy, as you can readily feel from the warmth if you hold a small animal such as a dog or a cat in your arms. The solid ground is also a reservoir for energy; try lying on a flat rock that has been in the Sun all day. Although less obvious, the ocean and even glaciers are also energy reservoirs. They feel cold to our touch because their temperatures are lower than

FIGURE 1.12 The energy cycle
Energy from both internal and external sources cycles through the reservoirs of the Earth system, driving processes from wind and waves to photosynthesis.

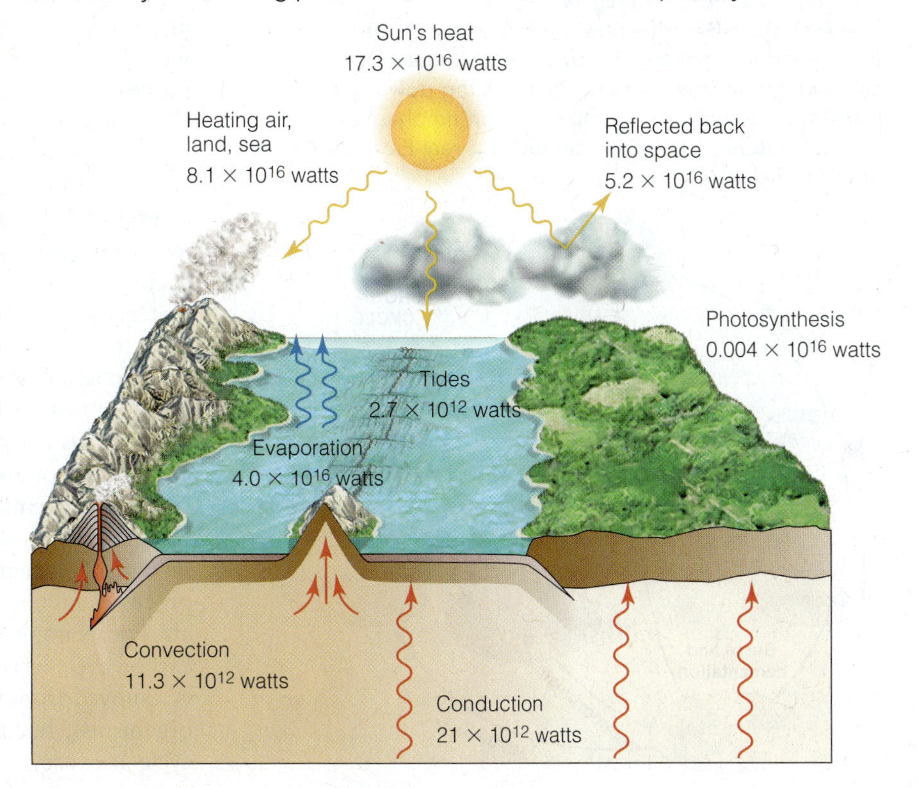

our body temperature, but they still contain heat and are important reservoirs for energy in the Earth system. We will consider some of the basic properties of energy, the laws governing the flow of energy, and more of the specifics of how energy cycles through the Earth system in Chapter 2.

The Rock Cycle

The **rock cycle** describes the results of competing internal and external forces that meet at Earth's surface, continually building up, breaking down, transporting, and transforming rocks. Some of the processes involved in the rock cycle, which will be the specific focus of Chapter 7, include weathering, erosion, transport, deposition, metamorphism, melting, crystallization, volcanism, and uplift of mountains (FIG. 1.13).

Earth scientists know, from many years of careful observation, that most erosional processes are exceedingly slow. An enormously long time is needed to erode a mountain range, for instance, or for huge quantities of sand and mud to be transported by streams, deposited in the ocean, then cemented into new rocks, and the new rocks deformed and uplifted to form a new mountain. Slow though it is, this cycle has been repeated many times during Earth's long history. This has led to an important discovery: Earth is extremely old. We will consider the materials of which Earth is composed in Chapter 3 and subsequent chapters, and we will look more closely at time, the origin and age of Earth, and how the ages of objects are determined in Chapter 4.

FIGURE 1.13 The rock cycle
The rock cycle describes the processes by which competing internal and external forces meet at Earth's surface, continually building up, breaking down, and transforming rocks. This simple version emphasizes the "cyclic" nature of these processes, but real Earth cycles are not this simple.

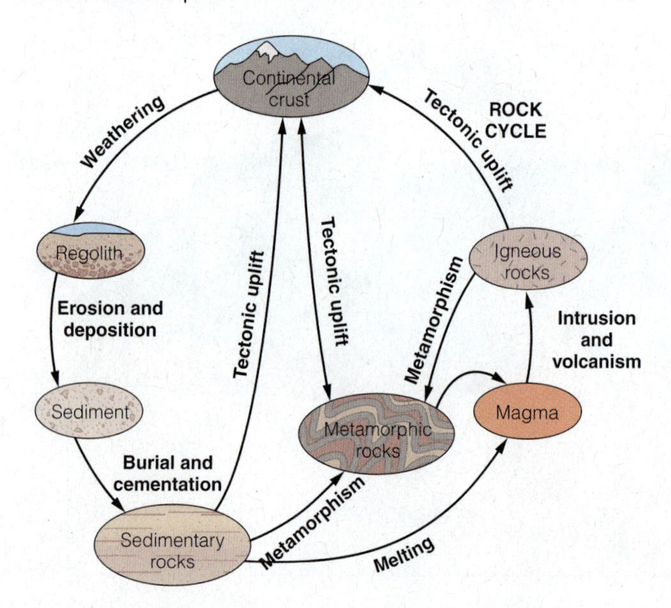

The Tectonic Cycle

The **tectonic cycle** describes the processes whereby Earth's major geologic features are formed, including mountain ranges, continents, deep-ocean trenches, and ocean basins. The tectonic cycle provides a unifying context for understanding these processes; it links Earth's surface with the interior of the planet. The tectonic cycle also explains the geographic distribution of geologic hazards, such as earthquakes and volcanic eruptions (FIG. 1.14). In Chapter 2 you will learn how the flow of energy from inside the planet to the surface drives these processes. We will explore the tectonic cycle and its consequences for life at Earth's surface in Chapters 5 and 6.

Biogeochemical Cycles

A **biogeochemical cycle** describes the movement of any chemical element or chemical compound that cycles through the biosphere and plays a role in its stability, as well as cycling through other Earth reservoirs. The biogeochemical cycles nitrogen, sulfur, oxygen, carbon, and phosphorus are particularly important because each of these elements is critical for the maintenance of life. Of interest as well are many additional elements and compounds that participate in and are influenced by some biologic processes and that have impacts (sometimes negative, sometimes positive) on living organisms. A good example is mercury, which builds up in the tissues of organisms, is influenced to change its chemical form by various biologic processes, and can be toxic to humans if it becomes too concentrated in the body. Biogeochemical cycles are of particular interest for humans because they trace the pathways whereby elements and compounds move through the physical environment and into our bodies.

Biogeochemical cycles involve biologic processes such as respiration, photosynthesis, and decomposition, as well as a variety of enzymatic and bacterially mediated processes. They also involve and provide links to nonbiologic processes such as weathering, soil formation, and sedimentation. In a biogeochemical cycle, living organisms are important storage reservoirs. The other major reservoirs—the atmosphere, oceans, surface water bodies, soils, and rocks—are common to all biogeochemical cycles, but their relative importance varies with different materials. For example, the atmosphere is an extremely important reservoir in the nitrogen cycle (recall that the atmosphere is approximately 79% nitrogen), but it is not at all important in the phosphorus cycle. This is because phosphorus does not commonly occur in a gaseous form and so cannot enter the atmosphere very easily.

Biogeochemical cycles are complex as well as complicated, and it is difficult to generalize about the processes involved in them. The reservoirs, chemical forms, and processes involved tend to be specific to particular materials. For example, photosynthesis plays a crucial role in the carbon cycle (it is the principal mechanism whereby carbon is removed from the atmosphere and used in the tissues of organisms), but it is not at all important in the mercury cycle because mercury is not involved in the process of

FIGURE 1.14 Tectonic processes
The geologic processes of the tectonic cycle link Earth's surface with the interior of the planet. The tectonic cycle provides a unified context for processes like earthquakes (A) in Chile, 2010 and volcanic eruptions (B) Eyjafjallajökull erupting in Iceland, 2010, and explains their geographic distribution.

photosynthesis. Elements often change their chemical form as they move through the cycle, as well, changing from organic to inorganic forms, from elements to a wide variety of compounds, and from liquid to solid to gas.

Human Impacts on Earth Cycles

We can extend the concept of cycles to include human-controlled cycles that involve or affect natural processes. Significant changes are now taking place in many of the fluxes of materials between Earth's reservoirs, and as a result the reservoirs are changing in some unexpected ways. Some of the changes have become daily news—the ozone hole, the increase of carbon dioxide in the atmosphere, the dispersal of pesticides throughout the ocean, the rate at which we are consuming nonrenewable resources such as oil, and the extinction of plant and animal species, to name several examples. Human activities most commonly influence biogeochemical cycles through atmospheric emissions of pollutants, which can dramatically change the fluxes of materials from one reservoir to another. For example, it is estimated that the flux of sulfur throughout the Earth system that results from human activities (mainly the burning of coal) is now greater than the flux of sulfur that results from all natural processes combined.

We, the human population, are the cause of these and other recent changes. Many kinds of large animals have, at various times, lived on Earth. Throughout all of Earth's long history, however, no large animal species has ever been as numerous as humans are today (FIG. 1.15A). Our collective activities have become so pervasive that there is no place on Earth we haven't changed. We go almost everywhere to seek the resources we need. In the process, we have made rainfall more and more acidic, we have caused fertile topsoil to erode, and we have changed the composition of the soil that remains. We have caused deserts to expand (FIG. 1.15B), and we have changed the composition of the atmosphere, oceans, streams, and lakes. Even the environment of the remote polar regions shows the influence of our activities (FIG. 1.15C).

Scientists have coined a special term to describe the changes produced in the Earth system as a result of human activities: **global change.** Measuring, monitoring, and understanding global change is a topic of intense study. A crucial part of Earth system science is to investigate how the collective actions of the human population are changing reservoirs and fluxes, and to determine what the consequences of these changes will be. We will address issues related to human impacts on the Earth system throughout this book.

HOW SCIENCE WORKS: HYPOTHESIS AND THEORY

Earth system science, like all other forms of science, is a method of learning and understanding natural phenomena. It advances by application of the **scientific method**, a logical research strategy that has developed through trial and error over many years (FIG. 1.16). The scientific method is based on observations and the systematic collection of evidence that can be seen and tested by anyone with resources who cares to do so. It

FIGURE 1.15 Human activity and global change
(A) The lights of human settlements—visible from outer space—give an idea of the extent of human impact on the Earth system. (B) Deserts expand and retreat as a result of natural processes, but human influences have greatly accelerated the rate of advance of deserts in some parts of the world. (C) Even the most remote parts of Earth have been affected by human activity. Polar bears accumulate pesticides in their fat, even though the nearest pesticide use is thousands of kilometers away, and the sea ice on which they depend may be melting as a result of global climate change.

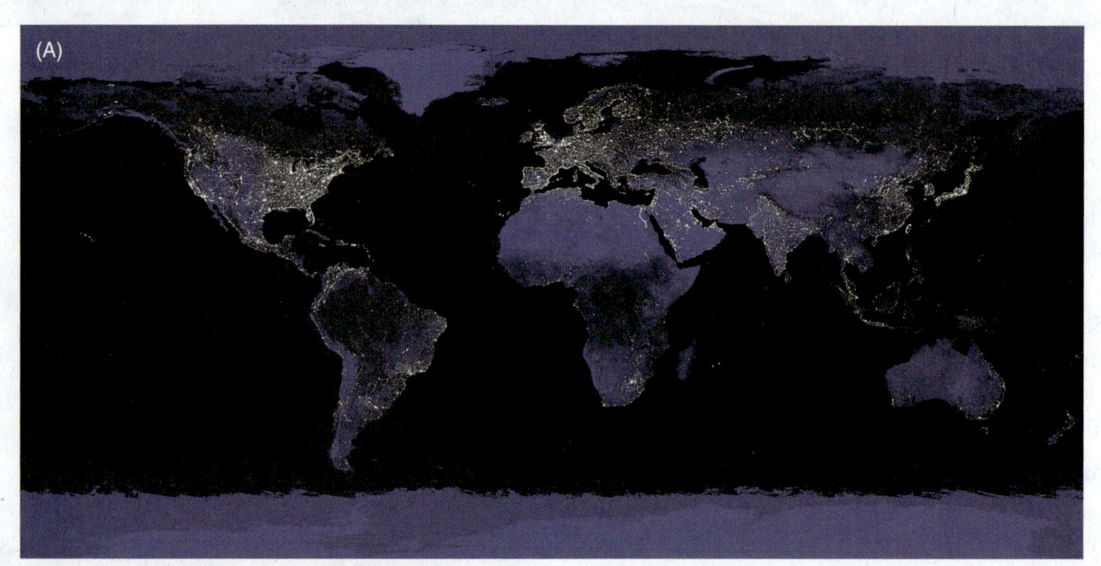

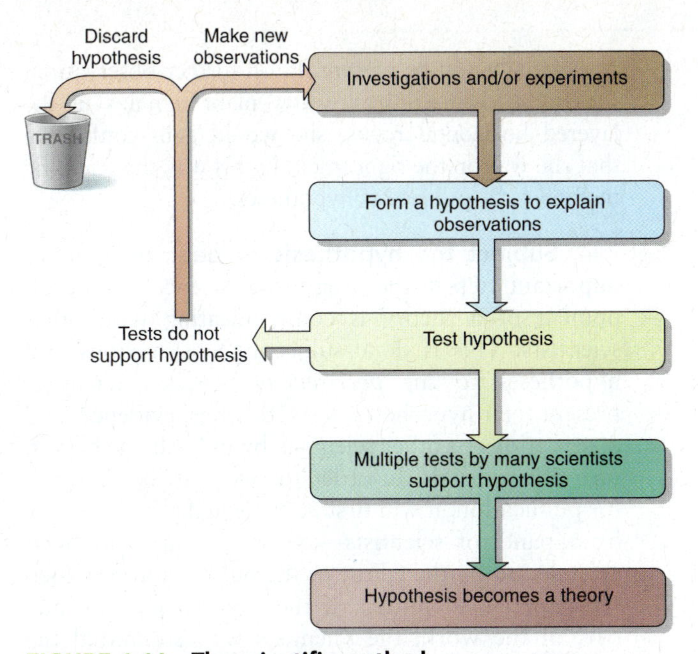

FIGURE 1.16 The scientific method
Science advances by way of the scientific method, the basic steps of which are shown here.

is a *collective process* in which scientists examine, comment on, test, and verify or disprove each other's ideas. It is also an *iterative process*, involving examination, reexamination, and yet more examination, testing, and refinement of ideas. Although it is not always practiced in all scientific settings in exactly the same way, the scientific method can be viewed as consisting of several basic steps.

Formulating and Testing a Hypothesis

Scientists start with the observation of a phenomenon and seek to acquire trustworthy evidence about it through measurement and **experimentation**. Scientists try to explain their observations by developing a **hypothesis**—a plausible but as yet unproven explanation for the way things happen. Hypotheses are based on our prior scientific understanding of the natural world and how it works.

Scientists use their hypotheses to make predictions, which are then used to test the hypothesis. If a hypothesis makes a prediction that turns out to be right, it tends to strengthen the hypothesis. An incorrect prediction, though, will greatly weaken or even disprove the hypothesis, which would then need to be changed, refined, or discarded altogether. Tests may involve controlled experiments in a laboratory, further observations and measurements, and possibly the development of a mathematical model. Whenever it is possible, scientists like to test their hypotheses against observations in the real world.

Developing and Refining a Theory

When a hypothesis has been examined and found to make successful predictions and withstand numerous tests, scientists become more confident in its validity. It may then become a **theory**, which is a generalization about nature. (In everyday speech, people often misuse the term *theory* to mean "hypothesis" by saying, "That's just a theory." What they really mean is, "That's just a hypothesis." In science, by the time a statement attains the stature of a theory, it is very substantial and must be taken seriously.)

Theories don't result from the work of an individual scientist, no matter how brilliant or hardworking. Because a theory is a generalization about nature, it has to be applicable in a wide range of cases. Often it takes a brilliant scientist to "connect the dots," to recognize the commonalities in many different observations about the natural world, and to state them in the form of a hypothesis that eventually becomes a theory; Darwin's statement of the theory of evolution and Einstein's statement of the theory of general relativity come to mind as obvious examples. However, the fact that these brilliant scientists made these statements did not make their hypotheses into theories; that happened as a result of many years of testing and refinement by many different scientists.

A theory, by definition, has been validated through experimentation and empirical observation. It is important to understand, however, that even theories are open to further testing and refinement. In fact, the key to the scientific method is *disprovability*. Sometimes theories are disproved by careful observation, measurement, and experimentation. Any theory that cannot, at least possibly, be disproved, is not truly a scientific theory.

The Laws of Science

Eventually, a theory or a group of theories whose applicability has been decisively demonstrated may become a **law** or a **principle**, which are the fundamental rules of science. A scientific law is a statement that some aspect of nature is always observed to happen in the same way and that no deviations from the rule have ever been seen. An example of a law is the statement that heat always flows from a hotter body to a cooler one. No exceptions have ever been found. (Actually, the flow of heat from a hot body to a cold body is a consequence of an even more fundamental law, the second law of thermodynamics, discussed in Chapter 2.)

Scientific laws and principles become part of our fundamental understanding of how the natural world functions, and they inform future scientific investigations. The assumption that underlies all of science is that everything in the material world is governed by scientific laws. However, even theories and laws are open to question when new evidence is found. This means that scientists, by virtue of the fact that they are constantly reexamining results obtained through observation and experimentation in the scientific method, are accustomed to dealing with uncertainty on an everyday basis. In science, everything is questionable; our ideas can never be conclusively *proved*—they can only be conclusively *disproved*.

the BASICS

The Scientific Method in Practice

Let's follow the steps of the scientific method, using the practical example of a geologist who is attempting to discern the mode of formation of a particular set of rocks.

STEP 1. Observe and gather data. In FIGURE B1.2, our scientist observes and measures a sequence of layered rocks. She sees that the layers are *horizontal* and *parallel,* important clues. Further, each layer consists of innumerable *small* grains, and the size of the grains *varies* from layer to layer but is approximately the same within each layer. It may require taking many detailed measurements before the scientist's observations are complete.

STEP 2. Formulate a hypothesis. A hypothesis is a plausible but as yet unproven explanation for how something happens. The scientist hypothesizes that these rocks were formed from sediment that was transported by some natural geologic process and deposited in the location where she has found it. But how was it transported? Hypothesis 1 is that a *glacier* transported the sediment. Hypothesis 2 is that *wind* did the transporting. Hypothesis 3 is that *water* transported the sediment.

STEP 3. Test the hypothesis. Our scientist uses her hypothesis—or in this case, multiple hypotheses—to make predictions and develop tests.

- To test Hypothesis 1, the scientist travels to a modern glacier and studies the sediment it deposits. She notes that the grains are different sizes, all mixed up, and not in neatly defined layers, and concludes that transportation and deposition by a glacier would not successfully replicate the sediments in the rock she is studying. So, Hypothesis 1 fails.
- Then she goes to a desert region where she sees wind-transported material deposited in dunes. She observes that particle sizes are approximately the same, but they aren't in parallel and horizontal layers—the layers are at odd angles. Again, transportation and deposition by wind would not yield the observed results, so Hypothesis 2 fails.
- Finally our scientist visits a lake and observes sediment deposited in water. Now she sees horizontal layers that are parallel, and the particles in each layer are approximately the same size. Hypothesis 3 has potential. However, more testing is needed; a visit to another lake might be in order, or an examination of observations made in similar environments by other scientists. Our scientist also notes that plants are growing in the lake. A responsible scientist, she develops another type of test: If the sediment that formed the rocks really was deposited in a lake, then the remains of aquatic plants might still be present. If, on further observation, she finds fossilized fresh-water plant remains in the layered horizontal rocks, she would gain confidence that she was on the right track. In this way the scientific method tests and retests hypotheses.

STEP 4. Subject the hypothesis to peer review. An important step in the progress of science and development of a theory is communicating with other scientists. This is done subjecting observations and hypotheses to the *peer-review* process. Scientists present their hypotheses, tests, theories, evidence, and observations to other scientists by publishing them in scientific journals. In order for work to be accepted for publication, it will first be subjected to peer review by a panel of scientists—experts in the field, who are selected by the editor of the publication to which the paper is submitted. If the experts question any part of the work, the scientists who submitted the work must respond to the queries satisfactorily before the paper is considered publishable. Once published, all of the material in the paper is open to examination and testing by anyone interested in the topic. Reviewing, testing, revising, and critiquing are essential steps in the progress of science.

STEP 5. Formulate a theory. Sometimes the observations and hypotheses of many scientists come together to form the basis for a coherent, well-supported understanding of a natural process or phenomenon. This doesn't result from the work of an individual scientist. Our geologist, for example, is focusing on explaining the mode of formation of a particular set of rocks. One of her hypotheses (number 3) has proven to be plausible and has withstood several different tests; it may well become the accepted explanation for this type of scientific phenomenon. To become a theory, though, her work would need to be tested and reconfirmed many times through the peer-review process. It would have to be valid as a generalization about the natural world, broadly applicable to similar sets of rocks, and verified by the work of many other scientists.

STEP 6. Formulate a law or principle. Laws and principles are statements that some natural phenomenon invariably is observed to happen in the same way. For example, in geology the *Law of Original Horizontality* states that sediment deposited in quiet water is always in horizontal layers (or nearly so, because a lake bottom might have slight irregularities). Once the Law of Original Horizontality had been stated and confirmed, it became possible for geologists to apply it to relevant situations and build upon it. Our geologist won't need to reinvent the law each time she studies a new rock sequence—if the appropriate characteristics are present in the rocks, she will be able to use this basic principle to support her conclusion that the sediment was deposited in quiet water.

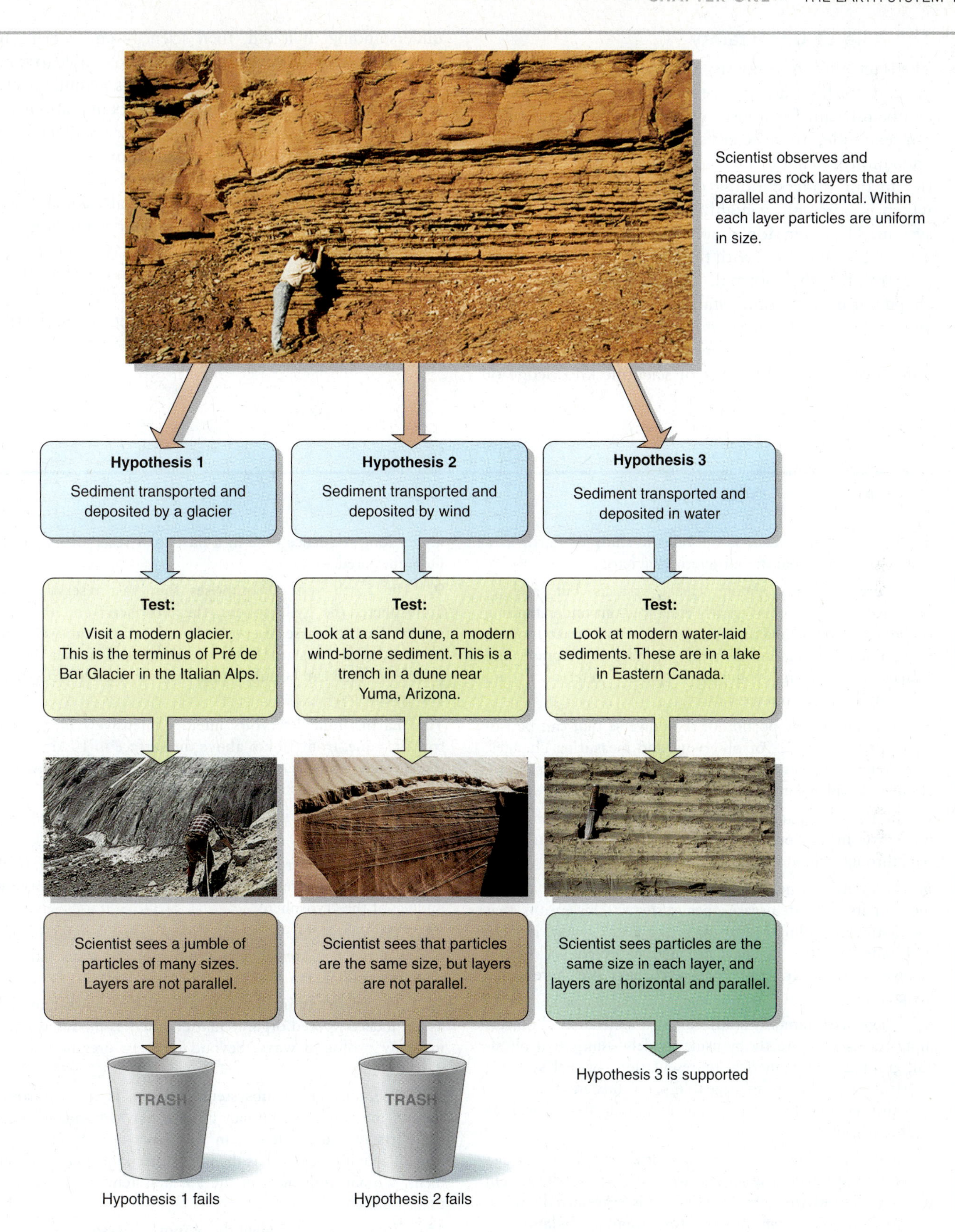

Scientist observes and measures rock layers that are parallel and horizontal. Within each layer particles are uniform in size.
Hypothesis 1
Sediment transported and deposited by a glacier
Hypothesis 2
Sediment transported and deposited by wind
Hypothesis 3
Sediment transported and deposited in water
Test:
Visit a modern glacier. This is the terminus of Pré de Bar Glacier in the Italian Alps.
Test:
Look at a sand dune, a modern wind-borne sediment. This is a trench in a dune near Yuma, Arizona.
Test:
Look at modern water-laid sediments. These are in a lake in Eastern Canada.
Scientist sees a jumble of particles of many sizes. Layers are not parallel.
Scientist sees that particles are the same size, but layers are not parallel.
Scientist sees particles are the same size in each layer, and layers are horizontal and parallel.
TRASH
TRASH
Hypothesis 1 fails
Hypothesis 2 fails
Hypothesis 3 is supported

FIGURE B1.2 The scientific method
The scientific method is used to test the various hypotheses that have been proposed to explain a phenomenon, as shown here.

The Role of Uncertainty

The fact that nothing is absolutely certain in the natural world is not problematic for scientists, but it can be difficult for nonscientists to comprehend fully. For example, to a scientist it is a straightforward and unproblematic fact that there are uncertainties in our understanding of Earth's climate system, which is, after all, a highly complex and changeable system. However, for a policymaker or government official trying to deal with the issue of climate change, uncertainty—the normal, ongoing questions that shape our quest to understand more about the climate system—can be a source of immense frustration.

It is important to understand, however, that uncertainty does not imply a lack of scientific knowledge or understanding. If it did, then science—and society, by extension—would become paralyzed and unable to function. For example, the law of gravity is a fundamental scientific principle. It is subject to reexamination and refinement, just like any other aspect of our understanding of the natural world. But we still know, for sure, that the apple will fall from the tree to the ground, or that the ball we throw up in the air will eventually come down. In the same way, we rely on the understanding and application of fundamental scientific laws and principles, through the scientific method, to inform the social, economic, and political actions that we take on behalf of the natural environment. Learning about those scientific principles helps us make better decisions, and take more responsible and more effective action.

SUMMARY

1. Earth system science is the holistic study of Earth as a system of many separate but interacting parts.

2. There are many remote sensing systems, but satellite-based remote sensing has greatly enhanced our understanding of the Earth system and our ability to monitor human impacts Geographic Information Systems facilitate the storage and management of large volumes of spatially referenced data points and their characteristics.

3. A system is any portion of the universe that can be isolated for the purpose of observing and measuring changes. The system concept allows scientists to break down large, complex problems into smaller, more easily studied pieces.

4. In a closed system, energy but not matter can cross the boundary. In an open system, both energy and matter can pass through the boundary.

5. Earth as a whole approximates a closed system, but most of its subsystems are open systems. This means that the matter (including resources) in the Earth system is fixed and finite, and its subsystems are interconnected, such that a change in one subsystem will cause changes elsewhere in the system.

6. Scientists commonly study complex systems (and objects and processes) indirectly by using models—simplified physical, graphical, or numerical representations of real systems, constructed at a more manageable scale. Box models represent the essential features of systems graphically, as well as quantitatively.

7. A flux is the amount of matter (or energy) that is transferred into or out of a system, or from one part of the system to another, and the rate at which it is transferred. When the fluxes into a system (or part of a system) are balanced by fluxes out of the system, it is said to be at steady state.

8. Places within a system where material (or energy) is stored for a period of time are called reservoirs. The residence times of different materials in different reservoirs are controlled by the properties of both the material and the reservoir, and by the processes and forces acting upon them. A material with a very long residence time in a particular reservoir is said to be sequestered.

9. The Earth system comprises four vast reservoirs: the atmosphere, the hydrosphere, the biosphere, and the geosphere. Each of these reservoirs functions as a subsystem on its own, with constant flows of energy and matter among them, and each can be subdivided into yet smaller subsystems and reservoirs.

10. Earth's four reservoirs interact intensively in the narrow life zone, from 10 km above the surface to 10 km below the surface, which has unique characteristics that allow this planet to support life.

11. The geosphere is the solid Earth, composed principally of rock and regolith.

12. The hydrosphere is the totality of Earth's water, including oceans, lakes, streams, underground water, and the snow and ice of the cryosphere.

13. The atmosphere is the mixture of gases—predominantly nitrogen, oxygen, argon, carbon dioxide, and water vapor—that surrounds Earth.

14. The biosphere includes all of Earth's organisms, as well as all undecomposed organic matter. Life has modified Earth in many profound ways, beyond its mere presence on the planet.

15. The fifth great subsystem of Earth is the anthroposphere, the sphere of human influence, which encompasses the parts of the Earth system that have been modified by human activities, as well as the built or technological environment. Human influence on the Earth system is greater now than it has ever been.

16. Feedback occurs when the output of a system also serves as an input and leads to changes in the state of the system. In negative feedback, the system's response is in the opposite direction from the initial input; negative feedback is self-regulating and stabilizing. In positive feedback, an increase in output leads to a further increase in the output, destabilizing the system by amplifying the original disturbance.

17. The constant movement of material from one reservoir to another is called a cycle. Earth cycles are driven by changes and adjustments as the system seeks to maintain a state of dynamic equilibrium. Because of the complexity of the interacting parts of the Earth system, natural cycles are generally not simple.

18. Important Earth cycles include the hydrologic cycle, which describes the movement of water through the Earth system; the energy cycle, which examines the pathways followed by energy from both external and internal sources; the rock cycle, which describes the building up and breaking down of rock as a result of competing internal and external forces; the tectonic cycle, which describes the processes by which Earth's major geologic features are formed; and biogeochemical cycles, which trace the movement of chemical elements and compounds among interrelated biologic and geologic systems.

19. A crucial part of Earth system science is to investigate how the collective actions of the human population are changing the reservoirs and flows of the Earth system, and to determine what the consequences of these global changes will be.

20. All of science—including Earth system science—advances by application of the scientific method. The steps of the scientific method are to systematically observe phenomena and gather data; formulate a hypothesis; test the hypothesis, numerous times and in various ways; formulate a theory; and finally, formulate a law or principle.

IMPORTANT TERMS TO REMEMBER

anthroposphere *15*
atmosphere *15*
biogeochemical cycle *20*
biosphere *15*
box model *11*
closed system *11*
cryosphere *15*
cycle *17*
Earth system science *6*
energy cycle *19*
equilibrium *16*
experimentation *23*
feedback *16*
flux *11*
Geographic Information System (GIS) *7*
geosphere *15*
global change *22*
hydrologic cycle *18*
hydrosphere *15*
hypothesis *23*
isolated system *10*
law (of science) *23*
model *10*
negative feedback *16*
open system *11*
positive feedback *17*
principle (of science) *23*
remote sensing *7*
reservoir *12*
residence time *12*
rock cycle *20*
scientific method *22*
sequestration *13*
sink *12*
source *12*
steady state *12*
system *9*
tectonic cycle *20*
theory *23*

QUESTIONS FOR REVIEW

1. How does Earth system science differ from physics, biology, or any other specialized area of science?

2. What is remote sensing? What has been the role of remote sensing in the emergence of Earth system science as a discipline?

3. What consequences arise from the fact that Earth is a closed system?

4. What is a model? How are models used in the study of the natural world? Give examples of physical, graphical, and numerical or computer-based models.

5. What does it mean when we say that a system or a reservoir is at "steady state"? Relate the concept of steady state to the definition of "flux."

6. What is residence time, and what factors control the residence time of a material in a particular reservoir?

7. What is the difference between a state of "equilibrium" and a state of "dynamic equilibrium," and which one better describes the state of most Earth systems?

8. Define geosphere, hydrosphere, atmosphere, biosphere, and anthroposphere.

9. What and where is Earth's "life zone"?

10. Why is a positive feedback sometimes called a "vicious cycle"?

11. In what ways does the energy cycle differ from the other important Earth cycles?

12. What is a biogeochemical cycle, and why are they of particular interest in Earth system science?

13. Suggest three human activities that affect Earth's external activities in a noticeable manner.

14. What is meant by the term *global change*? How do your own activities contribute to global change?

15. What are the five basic steps of the scientific method?

QUESTIONS FOR RESEARCH AND DISCUSSION

1. Identify three human activities, in the area where you live, that are causing big changes in the environment. Can you recognize some of the ways in which these activities are influencing the movement of materials from one reservoir to another in the Earth system? For example, driving a car moves materials—in the form of polluting emissions—into the atmosphere. Can you think of some other examples?

2. Draw a box model representing the movement of water into, out of, and within a building, such as your own house. Remember to consider all possible reservoirs (living and

nonliving), and all forms of water—solid, liquid, and vapor. When considering water, would your house be an open or a closed system? Now reconsider Figure 1.5C, the box model of the water cycle in an island system. Is this an open system or a closed system in nature? Is it portrayed as an open system in Figure 1.5C? What about in Figure 1.5A? Have a closer look at the boxes showing the reservoirs in Figure 1.5C. Do these boxes show all of the possible reservoirs in the system? Are they the best choices, or are there other ways that you can think of to represent the reservoirs in this system? What about the arrows, representing the various processes in the water cycle; are they adequate, or are there some processes that are missing from this representation? Consider why these particular choices were made in deciding how to represent the system, and whether you would choose to represent it differently. Try drawing your own box model to represent the processes and reservoirs in this island water cycle.

3. How do you think politicians and policymakers should deal with scientific uncertainty when they are making decisions that may affect our management of the natural environment? (We will return to this question in the final chapter of the book.)

QUESTIONS FOR *THE BASICS*

1. Why are isolated systems conceptual, not real?

2. What are the differences between closed and open systems?

3. Explain why the two final steps of the scientific method—formulating a theory, and formulating a law or principle—are never based on the work of a single scientist, even though one person may be responsible for actually *stating* the theory or law.

QUESTIONS FOR *A CLOSER LOOK*

1. What is the difference between a geostationary satellite and a polar-orbiting satellite?

2. The uses and applications of high-altitude satellites such as GOES differ substantially from the applications of lower-altitude satellites like Landsat and SPOT. In what ways do they differ? Give some examples.

3. Visit NASA's Earth Observing System (EOS) website and find satellite images to illustrate the application of remote sensing to environmental monitoring and the study of Earth systems.

CHAPTER 2

Energy

OVERVIEW

In this chapter we:

- Consider the nature of energy and define some different types of energy
- Introduce the fundamental laws that govern the flows of energy in the Earth system
- Look at the sources of energy that power the Earth system
- Examine how energy moves through the Earth system
- Consider how humans have tapped into various energy sources to power our technologies.

Energy drives the Earth system
The Strokkur Geyser in Geysir, Iceland (the location from which geysers got their name) gushes skyward, propelled by heat energy from deep underground.

FIGURE 2.1 Energy in Earth's subsystems
Energy is present in every part of the Earth system and is required for the functioning of every Earth process, whether natural or human.

Energy is fundamentally important. Everything that happens in and on Earth is driven by energy. Without energy there would be no Earth system (**FIG. 2.1**), and certainly there would be no possibility of life. In this chapter we will consider the nature of energy itself, the laws that govern the flow of energy, and the types of processes that are driven by energy as it moves through the reservoirs of the Earth system. Everything you learn in this chapter can be applied to every topic throughout the entire remainder of this book.

WHAT IS ENERGY?

Energy is the capacity to do work, to move matter, to make things happen. Energy exists in several forms, such as the energy of a moving body and heat energy. The sum of the different kinds of energy in a system is called the *internal energy* of the system. To change the internal energy of a system, energy must be added to or taken away from the system, which is called **work**. Work done *on* a system involves the transfer of energy into the system; work done *by* the system involves the transfer of energy out of the system.

Energy can take lots of forms, but all of them (including heat) can be described as either *potential energy* or *kinetic energy*, or a combination of these. **Potential energy** is energy that is stored in a system. It can take the form of:

- *chemical energy*, the energy that holds the molecules and compounds of our world together; to change a molecule or compound, energy must either be added to do so or be released as a consequence; the burning of methane (CH_4) to form H_2O and CO_2 is an example, so natural gas is an example of stored chemical energy.
- *nuclear energy*, the energy that holds atomic nuclei together and is released when a heavy atom is split, or when two light atomic nuclei are fused to make a heavier nucleus.
- *stored mechanical energy*, such as the energy that resides in a compressed spring or a tree sapling bent over in the wind.
- *gravitational energy*, or energy of position; the energy that resides, for example, in a ball poised at the top of a steep hill, or in water at the top of a waterfall; the energy that arises from the gravitational force between two objects, such as the Moon and Earth.

Kinetic energy is energy that is expressed in the movement of electrons, atoms, molecules, materials, and objects. It includes:

- *radiant energy*, or electromagnetic radiation (discussed below), which travels in waveform photons of light, of which visible light is one example.
- *electrical energy*, the movement of electrons or other charged particles, of which lightning and the aurora borealis ("northern lights") are two natural examples.
- *thermal energy*, or **heat**, the vibrational movement of atoms and molecules.

- *sound*, which results when an object is caused to vibrate; specifically, the alternating compression and expansion of a material as energy passes through it in the form of sound; and finally.
- *motion* of objects, as in moving water or wind or a rolling stone.

These categories are not strict; rather, they are descriptive, and serve to help us understand the behavior of different types of energy. For example, thermal energy typically consists partly of potential thermal energy (called *latent heat*) and partly of kinetic thermal energy (called *sensible heat*).

Fundamental Laws of Thermodynamics

The transfer of energy from one form to another and one body to another is subject to a set of natural laws known as the laws of **thermodynamics** (from the Greek *therme*, heat, and *dynamis*, power). The laws of thermodynamics govern the behavior of energy as it moves through the Earth system, into and out of the Earth system, and among its reservoirs. Scientists first began to articulate the laws of thermodynamics in the early 1800s, as part of their efforts to solve problems related to the design and efficiency of steam engines during the Industrial Revolution.

First Law: Conservation and Transformation

The *first law of thermodynamics* involves the conservation of energy and is stated as follows: *In a system of constant mass, the energy involved in any physical or chemical change is neither created nor destroyed, but merely changed from one form to another.* When the first law was originally discovered, it was not known that matter can be changed to energy according to Einstein's famous equation $E = mc^2$ (where E is energy, m is mass, and c is the speed of light in a vacuum). This is the reason that we now add the words "in a system of constant mass" when we are interested in studying systems in which matter is not being transformed into energy.

The first law says that one form of energy can be transformed into another form of energy. This is a very common occurrence in Earth system processes. For example, a growing wheat plant utilizes radiated energy (sunlight) to make carbohydrates, a form of stored chemical energy, via the process of photosynthesis. When we eat the wheat, digestion transforms the chemical energy from the plant tissues into a source of energy that is usable by our bodies, allowing us to walk, run, breathe—in short, the energy is transformed once again, into energy of motion (FIG. 2.2). As we perform these actions, some of the energy is transformed yet again, into heat, which is lost to the atmosphere through our skin. And so on.

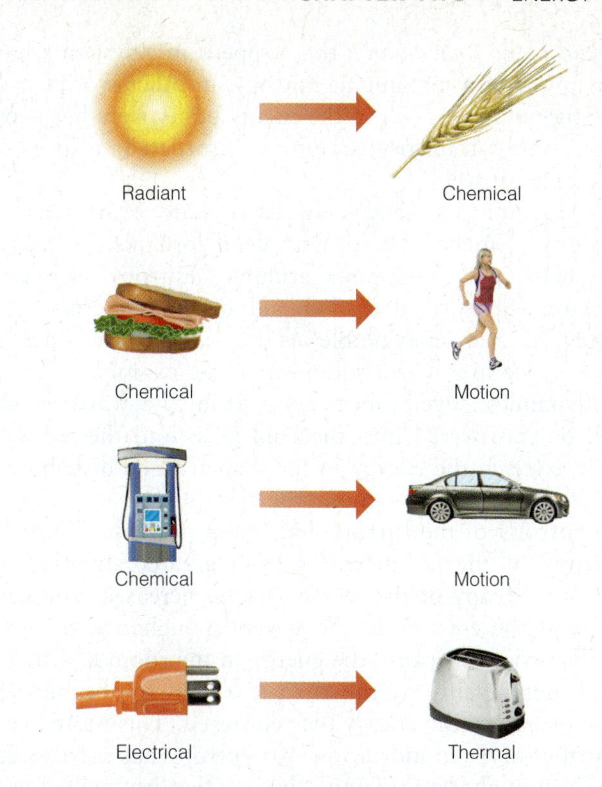

FIGURE 2.2 Energy transformations
Energy can (and commonly does) change from one form to another as it moves through the Earth system. Some common examples of energy transformations are shown here. (*Source:* Adapted from U.S. Energy Information Administration).

Second Law: Efficiency and Entropy

The *second law of thermodynamics* is especially important because it also clarifies the relationship, mentioned above, between work and heat. According to the second law: *Energy always changes from a more useful, more concentrated form to a less useful, less concentrated form.* An important consequence of the second law is that energy cannot be completely recycled to its original state of usefulness. Complete recycling is impossible because whenever work is done, some energy is inevitably converted to heat. The energy needed to collect and recycle all of the energy dispersed as heat requires more energy than can be recovered. In other words, no real process can ever be 100 percent efficient—some heat energy is always lost. (It was important for scientists and technologists in the early 1800s to understand the second law, as they were trying to design the most efficient steam engine they possibly could.)

The second law of thermodynamics has other important consequences. One consequence is that the flow of energy involves **degradation**, the transformation of energy into a form that is less useful, or less available for work. Degradation leads to increasing disorganization in the system, as energy becomes dispersed as heat. The measure of this disorganization is called **entropy**. The more ordered energy is, the more *available* it is to do useful work; this happens in a system when entropy is at a minimum. When energy is completely disordered, on the other hand, no more useful work can be extracted, and the energy is said to be *unavailable*; this happens in a system when entropy is at a maximum. Entropy, in other words, is a measure of how much of the energy of a system has been irretrievably transformed into a disordered, dispersed, unusable form.

The universe tends toward a state of increasing entropy. Clocks run down, dead organisms decay, ice melts, and mountains crumble. Entropy increases because energy is degraded and becomes increasingly dispersed and unavailable as it is used. Consider a glass of ice in a warm room—a classic example in thermodynamics. Over time, energy from the warm room will be transferred into the cold glass and the ice will melt, causing the energy in the system to be distributed more smoothly throughout the system. In the process, the entropy of the surrounding room decreases, but the entropy of the ice increases by a greater amount, so the net entropy of the whole system increases. You can tell that the entropy in the system is higher at the end of the process because the energy in the room is distributed more evenly overall than it was at the beginning, when some of the energy was clustered. This more even distribution is an indication that entropy has increased.

You might be thinking, "but ice that has melted can be refrozen." This is certainly true, but to do that you would need to utilize energy from outside of the system to perform the work of refreezing the ice. In so doing, the entropy of the outside system, and the unavailability of its energy, would necessarily increase, again conforming to the second law.

A consequence of the second law is that there is a universal tendency to smooth out clumps, clusters, bumps, and concentrations; in other words, there is a tendency toward increasing smoothness, randomness, and disorder. Where there is unevenness or nonrandomness in a system there must be boundaries or **gradients**— that is, variations in properties from one location to another within the system. A gradient is like a slope, but not necessarily a topographical slope; it could be a temperature pressure, or chemical slope. For example, there is a very steep temperature gradient between a frozen chunk of ice and the surrounding warm air of the room. The steepness of gradients is crucial and has an impact on the amount of work and the vigor of the work that can be extracted from a system.

All natural processes move "downslope" or "downgradient" because of the tendency toward increasing entropy and smoothing out of gradients. Therefore, heat flows from the warm air of the room "down" the temperature gradient into the cold ice, evening out the temperature and smoothing out the gradient. If the opposite were to occur—heat moving from the cold ice into the warm air—it would make the gradient even steeper, causing the ice to become even colder, and decreasing the entropy of the system. We know from everyday experience that heat does not spontaneously behave in this way.

Third Law: Absolute Zero

The *third law of thermodynamics* postulates the existence of the state of *absolute zero* temperature. **Temperature** is a measure of heat, the vibrational motion of particles. In a state of absolute zero temperature, all of the microscopic molecular motion on which we based our definition of heat (above) would cease. Another way to look at it is that the entropy of a material at absolute zero would disappear because the material is in a perfectly ordered, perfectly still state. (Quantum physics tells us that even at absolute zero there would still be a tiny amount of residual motion in the system, but that discussion is beyond our present purpose.)

Absolute zero also provides the foundation point (that is, the zero point) for the **Kelvin temperature scale.** The Kelvin scale is an *absolute* temperature scale because it is defined on the basis of a fundamental thermodynamic constant, the point at which entropy is equal to zero (FIG. 2.3A). All other temperature scales, including the Celsius (or centigrade) and Fahrenheit scales, are *relative* scales because they use an arbitrarily chosen reference point to define their zero. For example, in the case of the **Celsius temperature scale,** the zero reference point is the freezing temperature of water at surface atmospheric pressure, and the boiling point of water is assigned a value of 100° C. The freezing temperature of water, 0° C, is equivalent to 273.16 degrees on the Kelvin scale.

It is generally thought to be physically impossible to actually cool a material to absolute zero, although scientists on the forefront of *cryogenics*—the science of cold—have come very close. These scientists have discovered that materials display interesting physical behaviors when they are close to absolute zero, such as *superconductivity*, the ability to conduct an electric current extremely efficiently and without any resistance (that is, without heat loss) (FIG. 2.3B).

The laws of thermodynamics have fundamental implications for all Earth processes, including life itself and life-supporting processes (which are discussed in more detail in Chapter 15), as well as for a wide range of technologic applications and innovations. As you read through the rest of the book, think about energy, work, heat, entropy, and thermodynamics, and try to discern how the laws of thermodynamics work through Earth processes in the world around us.

Make the CONNECTION

Look around you right now. Are there sources of energy being utilized to power your surroundings? What is the source of that energy? Can you think of some other sources for energy that can power our modern technologies?

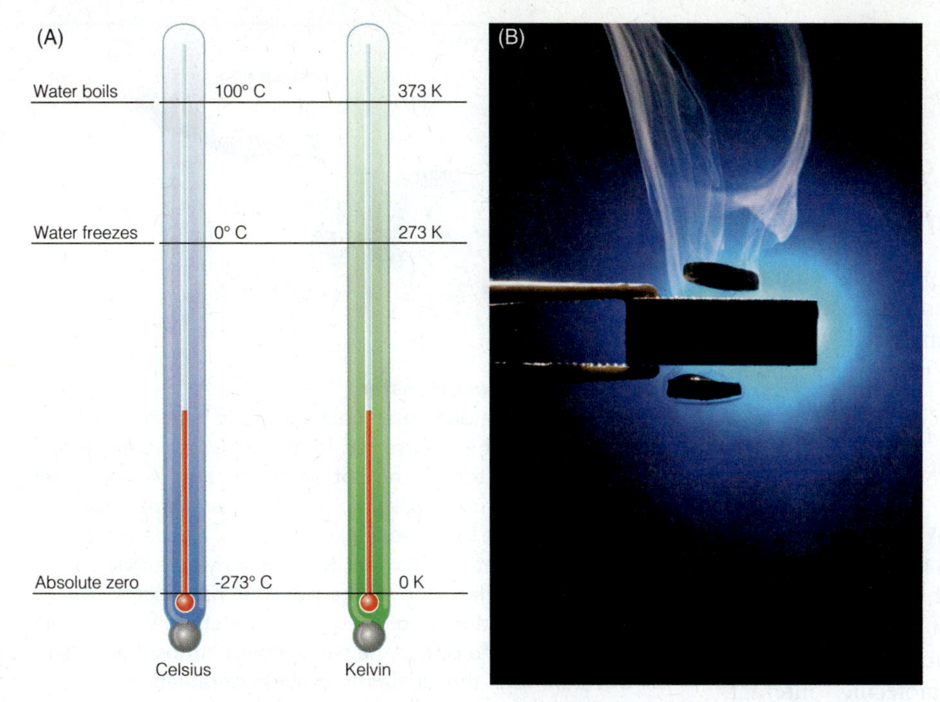

FIGURE 2.3 Temperature
(A) The Kelvin and Celsius temperature scales are based on the same incremental scale, but their zero reference points are different. The Kelvin scale uses absolute zero, or zero entropy, as its zero reference point, whereas the Celsius scale uses the freezing point of water at surface atmospheric pressure as the zero reference point. (B) At temperatures close to absolute zero, some materials become "superconductors." Superconductivity involves a lowering of electrical resistivity (in other words, the material conducts an electric current with extreme efficiency, without loss of heat), as well as the active exclusion of magnetic fields. The latter, called the *Meissner effect*, is responsible for the levitation of the small magnets shown floating near the superconductor in this photograph.

EXTERNAL ENERGY SOURCES

We know that all processes in the Earth system are driven by energy. Where does the energy come from? Some of it comes into the Earth system from external sources, and some originates from sources in the planet's interior. We will look at each of these sources, and then consider what happens in the course of the cyclic movement of energy through the various reservoirs of the system.

The Sun

The Sun is by far the main source of energy coming into the Earth system from an external source. In Chapter 4 we will look more closely at the role of the Sun as the center of our solar system and how it has influenced Earth's evolution as a planet. For now, let's consider where the Sun's energy comes from, what form that energy takes, and how it reaches Earth.

Source of the Sun's Energy

The Sun is a **star**—an enormous ball of ionized gas, or *plasma* (Chapter 3), that radiates heat as a result of thermonuclear reactions in its core. The term *nuclear* indicates the involvement of *atomic nuclei*, the dense, positively charged cores of atoms. Atoms, elements, and the basic types of materials of which Earth and all other objects in our solar system are composed will be discussed in greater detail in Chapter 3. The nuclear reactions that occur inside stars, including our Sun, involve the **fusion**, or merging, of lightweight chemical elements, particularly hydrogen, to form heavier elements such as helium and carbon. Under very special circumstances, these elements can undergo subsequent fusion processes that can lead to the formation of still heavier elements. The process of element formation via fusion, which happens primarily in the cores of stars, is referred to as *nucleosynthesis* (essentially, the synthesis, or making, of nuclei).

Fusion, which happens only at exceedingly high temperatures, converts matter to energy, according to Einstein's equation $E = mc^2$. There are many possible fusion reactions, but the Sun and most other stars produce their energy by two of them: the proton-proton (PP) chain and the carbon-nitrogen-oxygen (CNO) chain. In the Sun, the PP chain accounts for about 88 percent of the energy produced and the CNO chain the remaining 12 percent. In the PP chain hydrogen protons ($^1_1$H) fuse directly to form helium ($^4_2$H), whereas the CNO chain has intermediate steps that involve carbon, nitrogen, and oxygen. In both processes, however, the net result is the same—four hydrogen protons fuse while absorbing two electrons. The two electrons combine with two of the protons to form two neutrons; the result is a single helium nucleus containing two protons and two neutrons, with the release of some energy. During the fusion process some mass is lost; it is this lost mass that is converted to energy according to $E = mc^2$. The amount of energy released by the fusion of four $^1_1$H to produce one $^4_2$H is exceedingly small (about 4.2×10^{-12}J), but 4.5×10^6 metric tons of hydrogen are converted to helium every second in the Sun. This explains why the Sun's **luminosity**—the amount of energy (or light) radiated into space per unit of time—is so enormous.

the BASICS

Heat Transfer: Conduction, Convection, and Radiation

Energy transfer in the form of heat is particularly important in driving a wide variety of Earth processes. There are three ways that heat can be transferred from one object to another, or from one place to another among the reservoirs of the Earth system: *conduction*, *convection*, and *radiation*. These transfer mechanisms often operate together in a given circumstance, though one will typically be dominant.

Conduction is the process by which heat moves directly from one object to an adjacent object with which it is in contact. Recall that heat or thermal energy is the vibrational motion of atoms and molecules. The transfer of heat by conduction occurs when hot, rapidly vibrating atoms and molecules interact with adjacent atoms and molecules that are cooler and, therefore, vibrating less rapidly. The hot particles transfer some of their kinetic energy to the cooler, more slowly moving particles, which therefore heat up. In FIGURE B2.1, conduction is the process by which heat moves from the electric stove element directly into the bottom of the metal pot. Note that the atoms and molecules involved in conduction are merely passing along their vibrational kinetic energy—the hot, rapidly moving particles are jostling the neighboring cooler, slower-moving particles, passing along some heat energy with no actual movement of material from one place to another. Conduction is an important mechanism for heat transfer in the Earth system, notably for the transfer of heat from the interior of Earth, through the lithosphere,

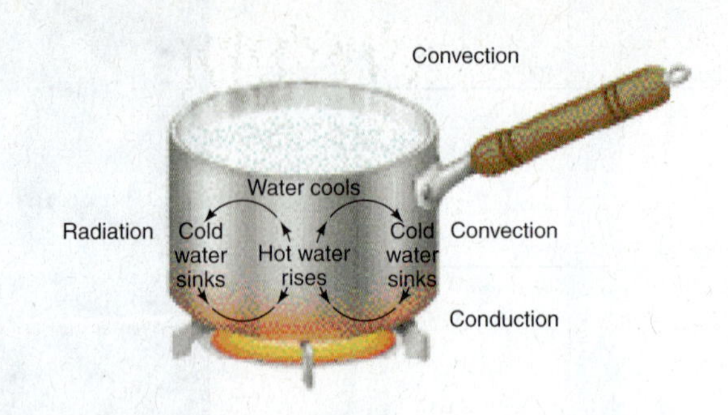

FIGURE B2.1 Heat transfer
There are three fundamental mechanisms of heat transfer. In this example, heat is transferred from the hot stove element directly to the bottom of the pot via conduction. As the water inside the pot warms, it rises; heat is thus transferred from the bottom of the pot to the top by convection. At the top surface, the warm water loses heat to the overlying air via conduction. The warmed air over the pot then rises, transferring heat upward into the atmosphere via convection. Warmth lost from all parts of the pot to the surrounding atmosphere also will travel through the air via the process of radiation.

Because such a huge amount of its hydrogen is continuously being converted to helium, the Sun will eventually run out of hydrogen fuel. Fortunately, its supplies of hydrogen are enormous, and scientists calculate that there is enough hydrogen in the Sun's interior to keep the nuclear fusion reactor operating for another 4 to 5 billion years.

From Sun to Earth

The energy released by fusion reactions in the Sun is in the form of *gamma rays* (γ *rays*), which are extremely short, high-energy electromagnetic waves (see *The Basics: Electromagnetic Radiation*), and *neutrinos*, which are electrically neutral, nearly massless particles that move at the speed of light. Of the total energy, 2 percent is in the form of neutrinos and 98 percent is γ rays. Neutrinos escape from the Sun's core so easily that they are gone about 2 seconds after they are formed. They can pass, unchanged, through the planet and do not play any part in bringing solar energy to Earth. Gamma rays, however, cannot easily get free from the Sun, and it is the γ rays, as we shall see later in this chapter, that are responsible for the energy that reaches Earth.

The Sun radiates energy equally in all directions, but only a tiny fraction of the total energy emitted reaches Earth. From orbiting satellites that get their energy via solar panels, we know that the flux of solar energy that reaches Earth is 1370 watts per square meter (W/m^2) of surface. Note that a *watt* is a unit of *power* rather than a unit of *energy*; it is the power required to do work at the rate of one joule per second. A *joule*, on the other hand, is a basic unit of thermal energy. It can be defined in terms of heat (*calories*), or it can be defined in the physical terms of work, which makes sense based on what we know about the interchangeability of heat and work. (Refer to Appendix A for definitions of these and other fundamental units of measurement.)

The energy flux to Earth's surface can be used to calculate the Sun's luminosity in the following manner: Picture an imaginary sphere with a radius equal to Earth–Sun distance (1.5×10^{11} m) and centered on the Sun (FIG. 2.4). Such a sphere has a surface area of 2.8×10^{23} m^2. The energy flux through every square meter on the surface of the sphere is 1370 watts. The total energy output of the Sun must therefore be the number of square meters multiplied by the flux:

$$2.8 \times 10^{23} \text{m}^2 \times 1370 \text{ W/m}^2$$

or

$$3.8 \times 10^{26} \text{ watts}$$

to the surface. Conduction works most efficiently in materials that have high *thermal conductivity* (the ability to conduct or transfer heat energy), which is a common property of metals.

Convection is a mechanism of heat transfer in which heat content is physically carried from one location to another. This occurs as a result of the heating and subsequent movement of hot material. Most materials (not all, but by far the majority) become less dense when heated. This reduction in density (typically due to expansion) causes hot material to become buoyant relative to the material that surrounds it. The hot material will then rise, to be replaced by colder (and thus denser), downward-flowing materials. The end result is that a volume of material has been transported upward, carrying its heat content along with it. In Figure B2.1, the water in the bottom of the pot heats and expands. It moves toward the surface, carrying heat upward as it goes, and is replaced at the bottom by cooler water flowing in from the sides and top of the pot. Note that, unlike conduction, convection involves the physical transfer of material from one location to another—the moving material is the vehicle for the transfer of the heat, which simply comes along for the ride. Convection is a very efficient way to transfer heat from one place to another because it doesn't depend on good thermal conductivity in the intervening material. One notable example of convection in the Earth system is the transfer of heat from Earth's core, through the mantle, to the base of the lithosphere, which is the driving force for plate tectonics. Another example is the upward circulation of warm air, which is the basis for most of Earth's weather-forming processes.

Radiation is the process in which energy in the electromagnetic spectrum is transferred from a radiating or emitting body through a surrounding gas, liquid, or vacuum. Radiation is how thermal energy reaches Earth from the Sun. When radiant energy from the Sun hits an object—let's say a broad, flat rock on the surface of Earth, for example—the short-wavelength solar energy is absorbed by the object, which heats up as a result. In Figure B2.1, heat is radiating outward from all sides of the pot. The effectiveness of radiation in transferring heat depends on the medium through which the transfer is happening. Radiation passes rapidly through a gas, like air, but much less effectively through a liquid, like water. Thermal energy is not transmitted through solids very effectively (if at all) by radiation. Instead, the radiated energy that falls on the surface of the solid will be either reflected, absorbed, or transmitted, as discussed in the text. A good deal of the energy that is absorbed by the solid will be transformed into heat, which may travel through the solid object by conduction, and may eventually be reradiated by the object. For example, if you put a rock in sunlight, it will absorb some of the sunlight, which will be transformed into heat. Later, the rock will reradiate that heat energy back to the surrounding atmosphere.

FIGURE 2.4 The Sun's luminosity
As discussed in the text, we can calculate the Sun's luminosity by imagining a sphere centered on the Sun. The energy that falls on Earth's surface, which we know to be 1370 W/m^2, is a portion of the whole sphere and can be used to calculate the total energy reaching the sphere.

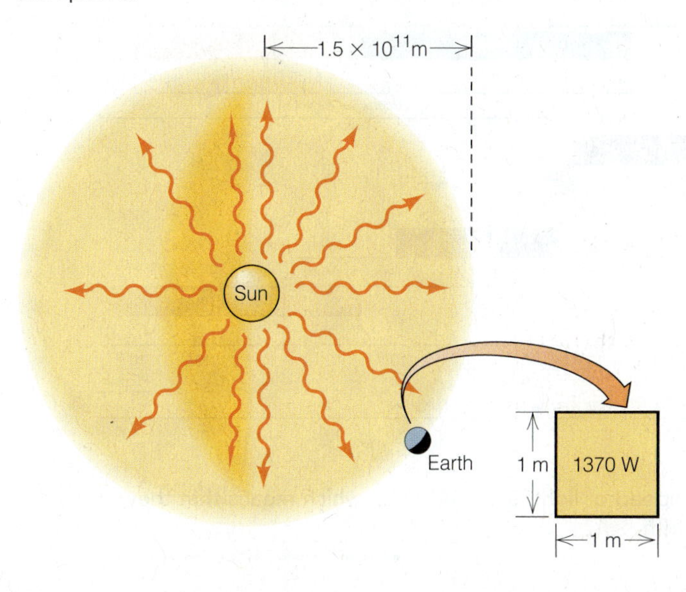

which is the Sun's luminosity. (If this doesn't strike you as a *vastly* large number, consider the luminosity of an average light bulb, at 40 W; then add 25 zeros to that number, and remember that each one of those zeros involves multiplying by ten.)

Earth receives only a tiny fraction of the Sun's luminosity. Viewed from the Sun, Earth is a disc with a radius of 6.4×10^6 m (FIG. 2.5). The surface area of this disc is 13×10^{14} m^2, so the luminosity that hits the whole Earth is $1.3 \times 10^{14}\text{m}^2 \times 1370\ \text{W/m}^2 = 1.8 \times 10^{17}$ watts. Thus, Earth receives only one 2 billionth of the total solar output of energy! Although this fraction is tiny, it is sufficient to supply all the energy needed to drive Earth's external processes and to keep the biosphere growing.

Astronomers monitor the Sun's luminosity very carefully. A 1 percent decrease in the flux of electromagnetic radiation, from 1370 W/m^2 to 1356 W/m^2, is estimated to reduce Earth's average temperature by 1 K. Similarly, a 1 percent increase in luminosity to 1384 W/m^2 would probably increase the average temperature by 1 K. A change of 1 K may seem small, but even a mighty volcanic eruption such as the 1816 eruption of Tambora volcano, which put so much debris into the atmosphere that it caused the "year without a summer," probably did not cause Earth's

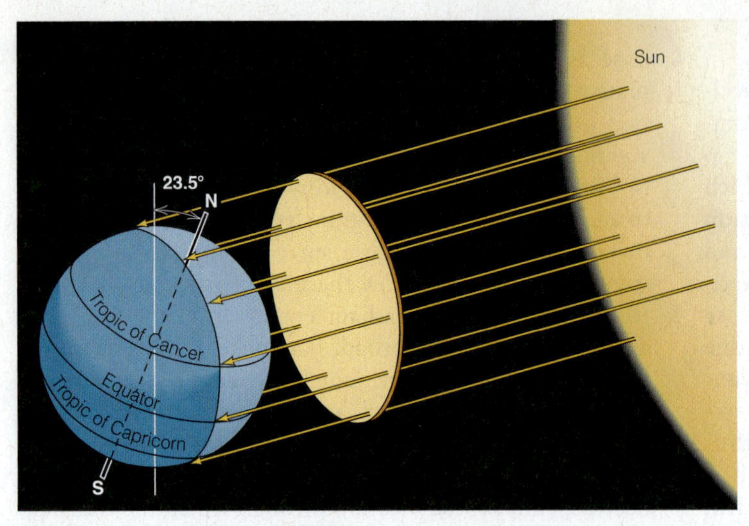

FIGURE 2.5 Solar energy flux
Energy from the Sun passes through an imaginary disc that has a diameter equal to Earth's diameter. The flux of energy through the disc is 1370 watts per square meter. The amount of energy that hits a square meter on Earth's surface is a maximum at the point where the incoming radiation is perpendicular to Earth's surface (that is, where the Sun is directly overhead at midday). This point changes daily because Earth's axis is tilted at 23.5° to the ecliptic, the plane of the solar system.

average temperature to drop by any more than 1 K. Clearly, changes in the Sun's luminosity have the potential to cause significant changes in the Earth system. Since about 1980 the Sun's luminosity appears to have decreased by about 0.3 percent, or 4 W/m^2. Where climate changes are concerned, therefore, changes in luminosity, as well as natural and anthropogenic changes to the atmosphere, must be considered (see Chapter 13).

Structure of the Sun

FIGURE 2.6A shows that the Sun consists of six concentric layers—four inner regions that make up the sphere we see and two gaseous outer layers that we cannot see. The *core*, the site of all the nuclear fusion reactions, is about 170,000 km in radius. The temperature of the core ranges from 8×10^6 K (the minimum temperature required for PP fusion) at its outer margin to more than 15×10^6 K (the temperature required for CNO fusion) at its center. The composition of the core is estimated to be about 62 percent helium and 38 percent hydrogen by mass.

the BASICS

Electromagnetic Radiation

Whenever an electrically charged particle is accelerated, it radiates energy in the form of **electromagnetic radiation.** Visible light is the most familiar form of electromagnetic radiation, but X rays, γ rays, infrared rays, and radio waves are also electromagnetic radiation, which differ from one another in wavelength. A group of electromagnetic rays arranged in order of increasing or decreasing wavelength is called a **spectrum.** The most familiar example is the *visible spectrum*, which is the range of wavelengths to which our eyes are sensitive.

Waves have three essential properties: the *wavelength* λ, which is the distance between two successive crests of the wave; the *speed* v, which is the distance traveled by a crest in one second; and the *frequency* f, which is the number of crests that pass a given point each second.

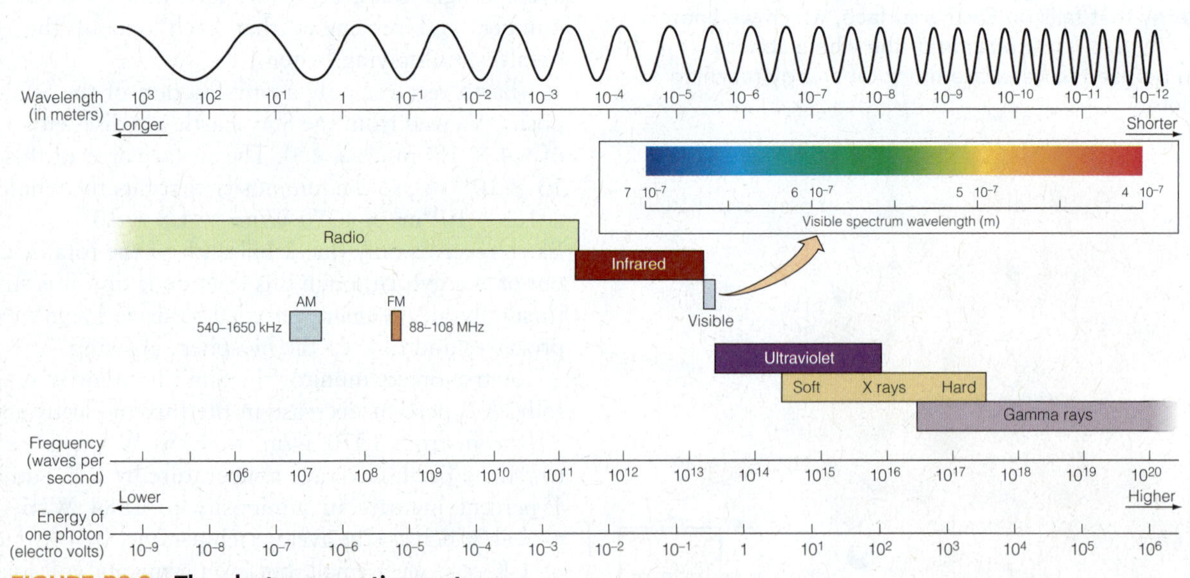

FIGURE B2.2 The electromagnetic spectrum
All electromagnetic waves travel at the same speed (the speed of light, 3.0×10^8 m/s), which means that they can be discussed in terms of either frequency or wavelength.

Surrounding the core is a region that is also very hot, but not hot enough for fusion to occur, called the *radiative layer*. The energy released from the core moves across this layer by radiation. Electrons in the radiative layer absorb the γ-radiation, and this is what makes the escape of energy from the Sun such a slow process. Above the radiative layer is the *convective layer*, across which energy moves by convection. In a sense the radiative and convective layers are kept gassy and are prevented from collapsing into the core by the intense pressure created by γ-radiation attempting to move outward.

Above the convective layer is the visible portion of the Sun, called the *photosphere*, an intensely turbulent zone that emits the light that reaches Earth. The photosphere has an average temperature of about 5800 K. The photosphere passes into the *chromosphere*, a low-density layer of very hot gas. The chromosphere is transparent to light passing through and therefore is very difficult to see. The chromosphere merges into the outermost layer of the Sun, the *corona*, a zone of even lower density gas than the chromosphere. Because both the chromosphere and the corona are transparent, we are able to observe and measure them only during a *solar eclipse*, when light from the body of the Sun is obscured by the Moon (FIG. 2.6B).

The Solar Spectrum

The radiation energy released in the Sun's core has a very short wavelength and is extremely energetic. As γ rays from the core move out through the radiative layer, they are repeatedly absorbed and reemitted by electrons; in the process, they are converted to longer-wavelength, lower-energy radiation. No energy is lost in the process; it is just parceled out into a greater number of less energetic rays. By the time the radiation reaches

The relation between the three wave properties is $f\lambda = v$. All wavelengths of electromagnetic radiation travel with the speed of light, denoted c, which in a vacuum is 299,793 km/s. For electromagnetic radiation, therefore, $f\lambda = c$. Because all electromagnetic radiation has exactly the same speed, c, it is possible to refer to electromagnetic radiation in terms of either wavelength or frequency.

Wavelengths of electromagnetic waves are usually measured in meters (FIG. B2.2). The unit of frequency is the *hertz* (Hz). A frequency of 1 Hz is one wave crest passing a given point each second. Scientists prefer to use the term *cycle* rather than wave crest when referring to frequency. One cycle per second is just another way of saying one wave crest per second. Radiation in the long-wavelength end of the electromagnetic spectrum, such as radio waves, is much less energetic than radiation in the short-wavelength end, such as gamma radiation and X rays.

Electromagnetic radiation can be described equally well in terms of waves or in terms of packets of radiant energy called *quanta* or *photons*. Sometimes it is more convenient to deal with the wave properties of the radiation; at other times it is better to deal with the packets of energy properties. The relationship between these two "forms" of electromagnetic radiation is $E = hf$, where E is the energy of a photon, f is the frequency of the corresponding electromagnetic wave, and h is a constant known as Planck's constant. The higher the frequency (and therefore the shorter the wavelength of the corresponding electromagnetic wave), the greater the amount of energy carried by a photon.

FIGURE 2.6 The Sun's interior
The Sun consists of six concentric zones, as shown in (A). The chromosphere and the corona can only be clearly seen during a solar eclipse, when the Moon blocks the light coming from the photosphere. This photo (B) taken during an eclipse, shows faintly glowing gas streaming hundreds of thousands of kilometers out from the corona.

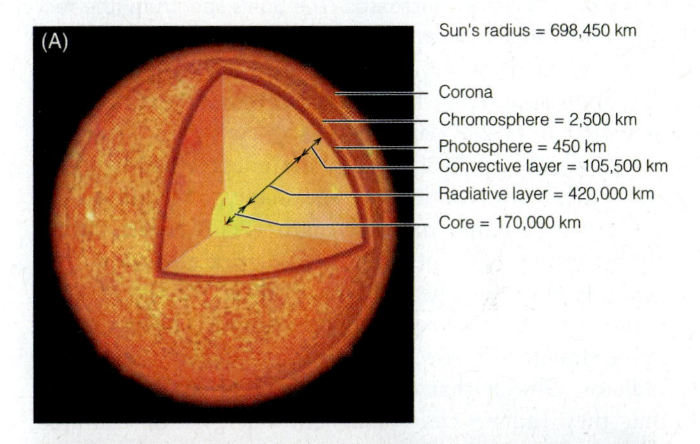

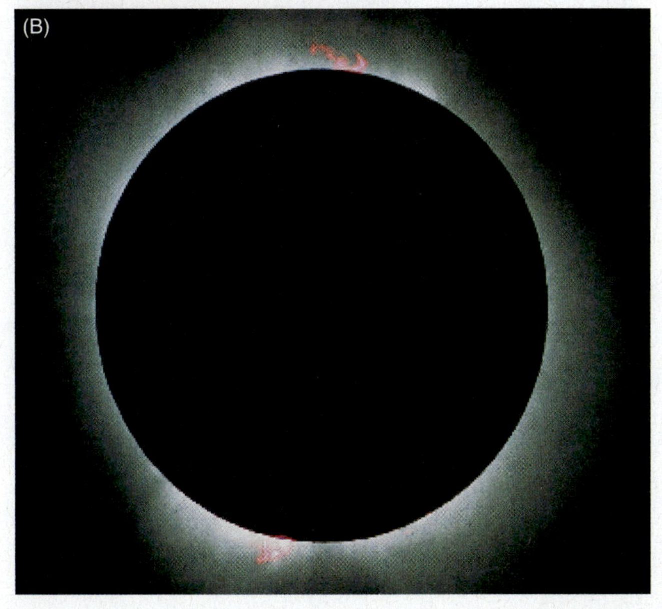

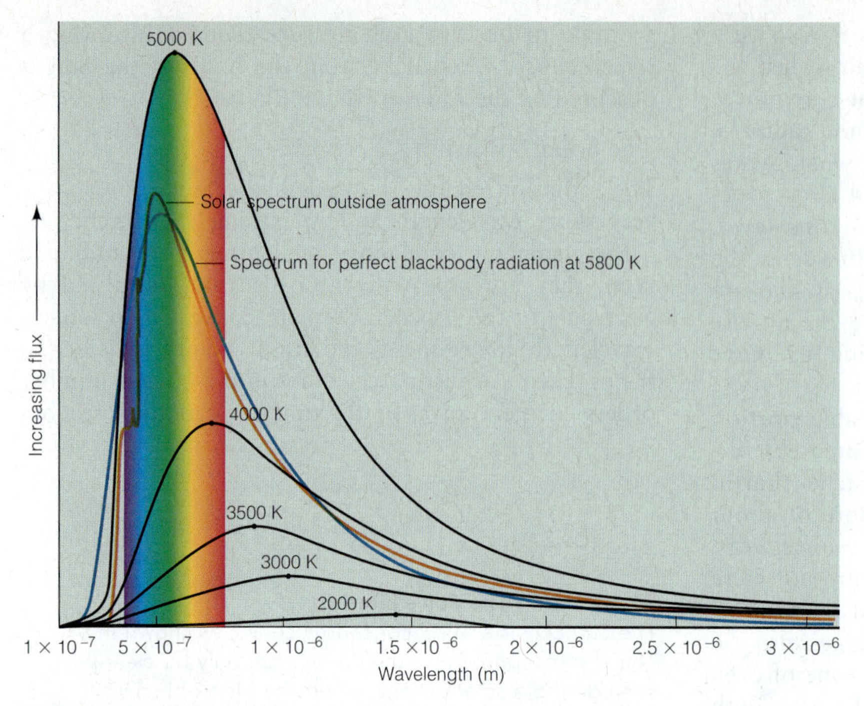

FIGURE 2.7 The Sun's spectrum
The black curves show the energy flux from blackbody radiators at different temperatures. Note how the radiation peak moves to shorter wavelengths and the total energy flux (the area under the curve) increases as the temperature of the radiating body increases. The Sun's spectrum (the red curve) is nearly identical to that of a perfect blackbody radiator at 5800 K.

the photosphere, it has wavelengths in the range of 3×10^{-6} m to 2×10^{-7} m.

As shown in **FIGURE 2.7**, the energy flux from the Sun varies with wavelength. The shape of the Sun's *spectral curve* (a graph of wavelength versus flux) matches almost exactly the spectrum of radiation given off by a perfectly blackbody heated up to 5800 K, the average temperature of the photosphere. Any body that has a spectral curve similar to those in Figure 2.7 is called a **blackbody radiator.** Objects that are hot become *radiators*, meaning that they radiate electromagnetic energy. For example, when a piece of metal is heated in an intensely hot flame, the metal first starts to glow a dull red. As it gets hotter and hotter, it becomes more brightly red, then orange, yellow, white, and finally bluish white. The color of the metal at any given time while it is being heated is a measure of the temperature. If we measure the electromagnetic radiation emitted by the hot metal, we will find that the spectral curve is similar to the smooth curves in Figure 2.7.

The term *blackbody* seems confusing when applied to something that is glowing brightly and emitting electromagnetic radiation. In fact, the term refers to the *radiation-absorbing* properties of a body, and a perfect blackbody is one that absorbs all light that strikes it, and reflects none. If you directed a very powerful beam of light at the Sun, almost none would be reflected back, making the Sun a nearly perfect blackbody. (The two terms *blackbody* and *blackbody radiator* mean the same thing and are interchangeable. They refer to "ideal" radiators, from which "real" objects vary, more or less.)

The hotter a radiating body, the shorter the wavelength of its radiation peak (Fig. 2.7). The peak of radiation for the Sun is at a wavelength of about 5×10^{-7} m, corresponding to a temperature of 5800 K; this wavelength is in the visible part of the electromagnetic spectrum. As well, the hotter an object is the more energy it radiates. We know this is true from such simple observations as the amount of energy given off by a hot stove versus a cold stove. The same relationship is apparent in Figure 2.7, where we see that the higher the temperature of the blackbody, the greater the energy flux (the space underneath the curve) at any given wavelength.

The spectrum of electromagnetic radiation emitted by the Sun is not exactly the same as the spectrum of solar radiation that actually reaches Earth's surface (**FIG. 2.8**). This is because gases in Earth's atmosphere selectively absorb some wavelengths from the solar radiation as it passes through on its way to the surface. Notably, ozone (O_3) in the **stratospheric ozone layer** absorbs radiation in the very short-wavelength, ultraviolet portions of the spectrum. This atmospheric filter has the effect of removing some of the most energetic (and biologically harmful) radiation from the incoming solar spectrum, making it possible for life to exist on the surface of Earth. The absorbed energy is transformed into heat, warming the stratosphere from within. In Chapter 11 we will look more closely at how sunlight interacts with materials in Earth's atmosphere.

Gravity and Tides

The Sun is by far Earth's largest external source of energy, but energy also comes into the system as a result of **gravity,** the mutual physical attraction between the masses of the Moon and Earth, and to a lesser extent the Sun and Earth (lesser because the distance is much greater). The source of the energy is not gravity itself, but the influence of gravity on Earth's rotation. Here's how it works: The gravitational pull that the Moon exerts on Earth is balanced by an equal and opposite *inertial force* created by Earth's movement. At Earth's center, the gravitational and inertial forces are balanced (equal but opposite), but they are not balanced from place to place on the surface. The magnitude of the Moon's gravitational pull varies over Earth's surface, but the inertial force that tends to keep objects moving straight ahead, in a direction away from the center of mass of the Earth–Moon system, has the same strength at any point on the surface. (This is because it is a *centrifugal force* created by the planet's rotation, similar to the outward pull you feel if you sit on the edge of a fast-moving merry-go-round.)

Consequently, on the side of Earth nearest the Moon the magnitude of the gravitational attraction exceeds the inertial force, and the resulting pull is in the direction of the Moon. On the opposite side of Earth, the magnitude of the inertial force exceeds the Moon's gravitational pull, and the resulting pull is directed away from the Moon (FIG. 2.9A). This difference causes *strain* in both objects, producing a periodic or cyclic distortion called a **tide**. In a sphere (such as Earth or the Moon), the distortion takes the form of flattening with no change in volume—the sphere becomes distorted into an ellipsoid with two bulges, one pointing toward the other body and the other pointing away from it (FIG. 2.9B). Luckily for us, Earth is *elastic*; that is, it has the capacity to deform reversibly. If Earth were inelastic, the strain and distortion caused by gravitational interaction with the Moon could cause the planet to break up. Instead, the internal resistance, or *friction*, caused by elastic distortion within the planet is translated into heat, and this is one of the main sources of Earth's internal heat energy. The distortions of the solid body of Earth are referred to as *Earth tides* or *body tides*. The heat generated by Earth's body tides is small compared to other sources of Earth's internal heat, and by itself is not sufficient to cause melting and create volcanoes. But on Jupiter's moon Io, body tides are huge and so much heat energy is generated that Io is the most volcanically active body in the solar system.

The gravitational and inertial forces that cause body tides act on *all* particles within the Earth system, with notable effects on large bodies of water. Each water particle on the side of the planet facing the Moon is pulled toward the Moon as a result of gravitational attraction. This creates a distortion or bulge of ocean water in the direction of the Moon. On the opposite side

FIGURE 2.8 Outer-space and sea-level spectra
The outer-space spectral curve for the Sun is different from the sea-level spectral curve, because gases in Earth's atmosphere selectively absorb some of the wavelengths of solar radiation. Notably, ozone (O_3) in the stratospheric ozone layer absorbs radiation in the very short-wavelength (ultraviolet) portion of the Sun's spectrum, thus preventing some of the highest energy and most biologically harmful solar energy from reaching the surface.

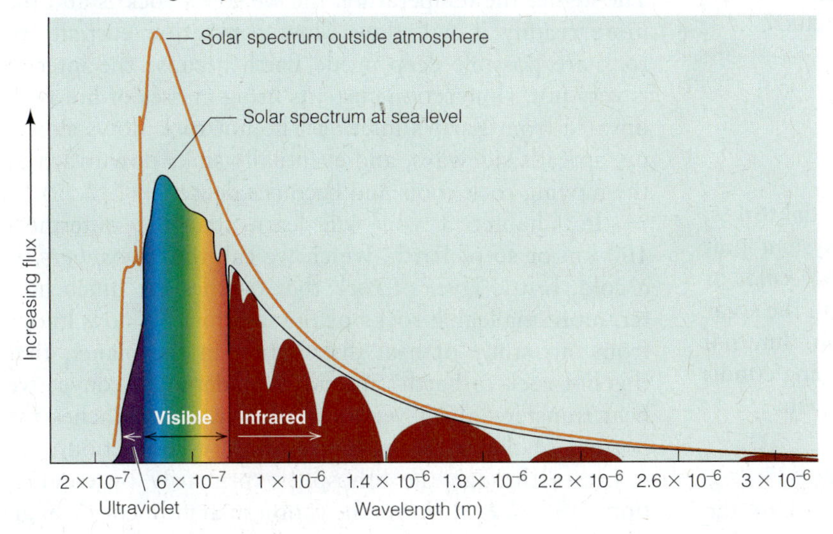

FIGURE 2.9 Earth tides
Earth's shape is distorted into an ellipsoid as a result of gravitational interaction with the Sun and (mainly) the Moon. In this example, the Moon is positioned over 30° N or 30° S. Red indicates material that has moved up relative to the reference sphere, and blue indicates material that has moved down relative to the reference sphere. One of the bulges points toward the Moon, the other points away. Much of the strain energy involved in tidal distortion of the planet's interior is translated into heat energy; this is a significant source of Earth's internal heat.

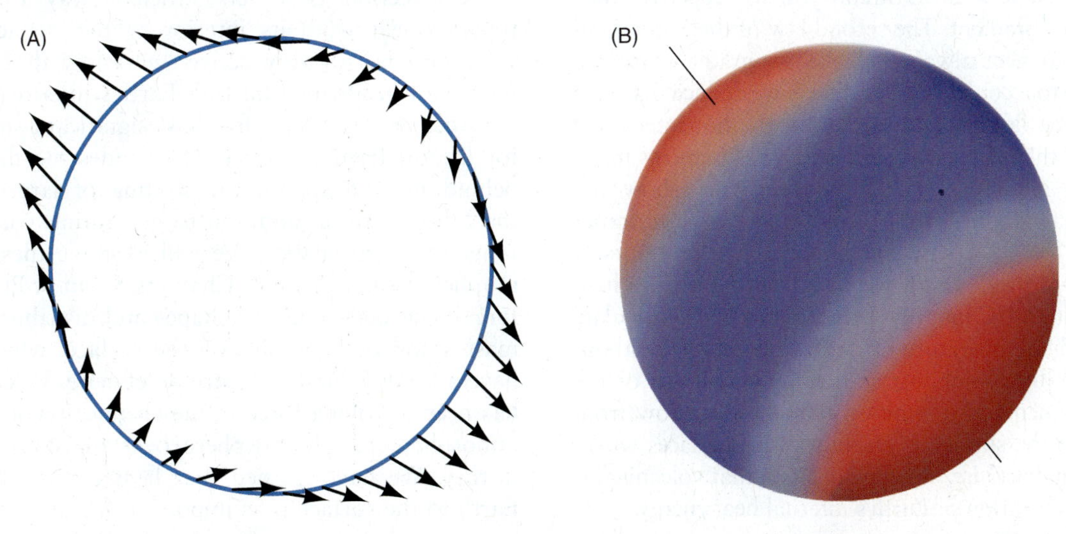

of Earth, farther away from the Moon, a water particle on the surface of the ocean experiences a much weaker gravitational pull, but the same inertial force. This creates a bulge of ocean water pointing *away* from the Moon. The daily rotation of Earth and its continents past these two oceanic bulges—which remain pointing toward and away from the Moon—generates the twice-daily *ocean tides*. We will discuss ocean tides in greater detail in Chapter 10.

Make the CONNECTION

What would Earth be like if there was no Moon? In what ways are Earth and Earth's Moon connected?

INTERNAL ENERGY SOURCES

In addition to energy from the Sun and from tidal forces, there are sources of energy into the Earth system that come from within the planet itself. These *geothermal* or **terrestrial energy** sources are much smaller than the solar input, but larger than the tidal contribution. Internal energy sources have much to do with determining conditions at the surface of the planet, as you will learn.

The Distribution of Terrestrial Energy

Volcanic eruptions, unlike winds, are unrelated to the Sun's energy output. No matter how hot it gets on a summer's day, the Sun's heat is insufficient to melt rocks, and even frigid Antarctica has active volcanoes. We therefore reason that the heat energy needed to form the molten lava that spews from a volcano must come from somewhere inside the planet. This reasoning is not difficult to test. If you went down into a mine and measured rock temperatures, you would find that the deeper you went, the higher the temperature would become.

The increase in temperature as you go deeper is called the **geothermal gradient.** The second law of thermodynamics tells us that heat always flows down-gradient, from a warmer place to a cooler one; therefore, we deduce that heat energy must be flowing outward from the hot interior of Earth toward the cool surface. Careful measurements made in deep mines and in drill holes around the world show that the geothermal gradient varies from place to place, ranging from 15° to 75°C/km, and becomes less pronounced with depth. Far inside Earth we calculate that the gradient is only about 0.5°C/km. By extrapolation, we calculate further that the temperature of Earth's core must be about 5000°C, or 5300 K—almost as hot as the surface of the Sun. Measurements also establish that the heat flow from the interior to the surface is greatest in those places where there is volcanic activity. We can conclude that volcanism is indeed a manifestation of Earth's internal heat energy.

Heat energy that flows out through the solid rocks of Earth's outermost layer does so by *conduction*, the transfer of heat directly from one object to an adjacent object (refer to *The Basics: Heat Transfer: Conduction, Convection, and Radiation*). However, volcanoes obviously involve the movement of hot material from inside the planet to the surface; therefore, we conclude that at least some heat energy reaches Earth's surface by *convection*, in which the heat is physically carried from one location to another by the movement or flow of the material involved. The hypothesis that convection occurs in a seemingly rigid solid body like Earth may seem odd. But rocks, if sufficiently hot, lose strength before they begin to melt. In that state they can flow like sticky liquids, although the rates of flow are exceedingly slow. The higher the temperature, the weaker a rock is and the more readily it will flow. Slow convection currents of rock are possible deep inside Earth because the interior is very hot. Convection currents bring masses of hot rock upward from Earth's interior. The hot rock flows slowly up, spreads sideways, and eventually sinks downward as the moving rock cools and becomes denser (**FIG. 2.10**).

In Chapter 3 you will learn that the outermost 100 km or so of Earth, which we call the *lithosphere*, is a cold, brittle layer of rock that overlies the much hotter, more malleable rocks of the interior. We now know, from our study of heat distribution in the planet, that this hot rock in Earth's interior is undergoing convective heat transfer—albeit very, very slowly. Heat reaches the bottom of the lithosphere primarily by convection, and it passes through the lithosphere primarily by conduction. **FIGURE 2.11** shows the geothermal gradient through the lithosphere of the continents and ocean basins; they differ because the lithosphere is thinner under the ocean basins than under the continents. At the bottom of the lithosphere, at a depth of about 100 km, the temperature is about 1300° C. At this temperature, rock strength has declined so greatly that convection is possible. The change in the geothermal gradient at about 100 km depth marks the place where heat moves by convection (below 100 km) and starts moving by conduction (above 100 km).

Convection is a very efficient way for Earth to transfer heat from its interior to the surface. Because volcanism is still very active, we know that this transfer is still occurring and that Earth's interior is still hot. Convective heat transfer has significant implications for life on Earth's surface. It provides the driving force behind *plate tectonics*, the shifting of large chunks of the lithosphere, in addition to generating volcanic eruptions and earthquakes. We will examine these processes in much greater detail in Chapters 5 and 6. Plate tectonic forces also determine the shapes and distribution of land masses and ocean basins on the surface, which, in turn, have a fundamental influence on climate. Volcanism itself has been a driving force in the chemical evolution of the atmosphere and hydrosphere over the course of Earth history. We have learned that heat coming from inside Earth to the surface is as important for life on this planet

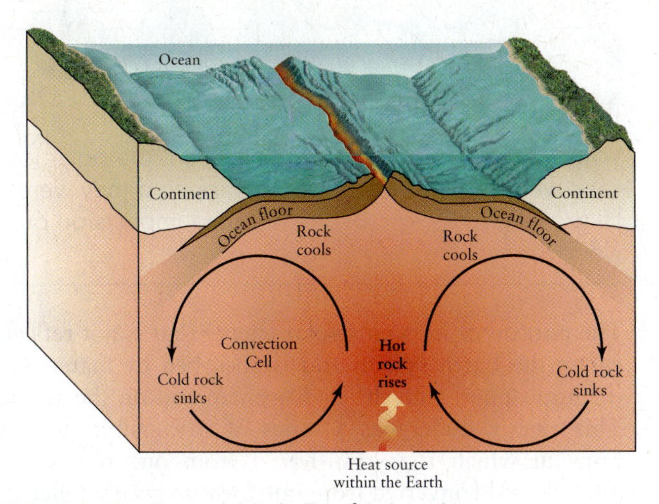

FIGURE 2.10 Convection in Earth's interior
Hot rock rises slowly from deep inside Earth, cools, flows sideways, and sinks. The rising hot rocks and sideways flow are the source of plate tectonic motion, and have an enormous influence on the shapes and distribution of land masses and ocean basins on Earth's surface.

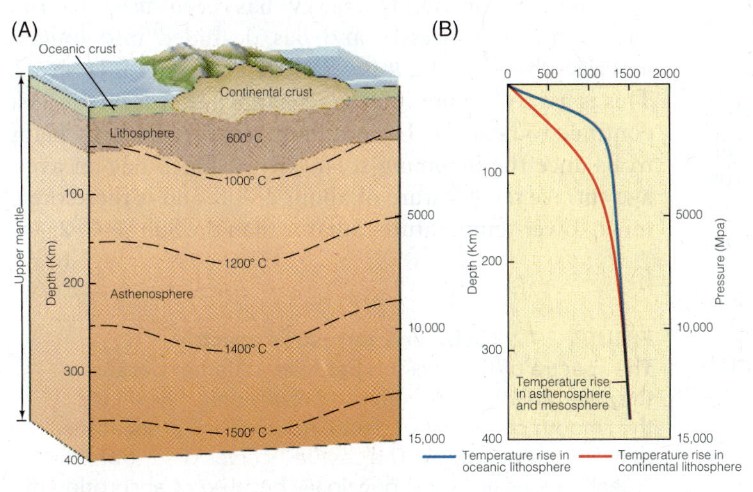

FIGURE 2.11 Geothermal gradient
Temperature increases with depth. (A). The dashed lines are *isotherms,* lines of equal temperature. Temperature increases more slowly under the continents than under the oceans. The lithosphere is thicker under the continents, so heat flows more slowly to the surface in those areas. (B). This is the same information as shown in (A) but in graph form. Earth's surface is at the top, so depth (and corresponding pressure) increases downward. Temperature increases from left to right.

as the energy that comes from external sources. Without either of these energy sources, life on Earth as we know it would not be possible.

Sources of Terrestrial Energy

What, then, is the source of this internal terrestrial energy? We know that Earth is nowhere near massive enough or dense enough in its core to undergo nuclear reactions like those in the core of the Sun. It turns out that a number of processes and sources contribute to Earth's internal energy. These include:

- *Radiogenic heat.* This is the main source of terrestrial energy, accounting for about two-thirds of heat flow from the interior of the planet. **Radiogenic heat** is produced by the spontaneous breakdown, or decay, of radioactive elements such as uranium and thorium that are naturally occurring inside Earth. Like the fusion processes in the core of the Sun, radioactive decay is a nuclear process. There are many variations, but the common theme is that some mass is lost in the conversion from one form of the element to another. This "lost" mass is converted to energy, which is released as heat. Some naturally occurring elements take billions of years to decay; thus, they are still providing a steady supply of internal heat.
- *Accretionary heat.* This is internal heat left over from the origins of Earth and other planets of our solar system, when particles of many sizes collided with one another, eventually sticking together or *accreting* to form the larger planetary bodies (Chapter 4). The kinetic energy of those countless collisions was converted into heat. Because it is a relatively large planet, Earth still retains some of the heat from the accretionary process early in the history of the solar system.
- *Tidal heating.* Gravitational interactions of Earth with the Moon and Sun cause body tides. Internal friction from the constant distortion of the planet is translated into heat energy.
- *Core formation.* Earth's *core* (Chapter 3) formed as a result of the segregation of dense material (mainly iron) during a period when the planet was partially molten. The gravitational potential energy of the dense material that formed the core was converted to heat as it sank to the center of the planet. Further heat was released as the innermost material—the part that now constitutes Earth's solid *inner core*—solidified. The latter came from the latent heat (mentioned earlier in our discussion of kinetic and potential energy) released in the course of *crystallization*, the change from liquid to solid iron. This change is still going on, as the liquid *outer core* slowly crystallizes, so latent heat is still being released today as a result of the solidification of the core.

EARTH'S ENERGY CYCLE

Of most interest to us, in the context of Earth system science, is what all of this energy from external and internal sources does when it reaches the life zone, where all life exists and where Earth's major reservoirs interact. The **energy cycle** (Chapter 1) encompasses the inputs and outputs, pathways, and reservoirs for the energy that drives all of the other cycles of the Earth system. Like other cycles, Earth's energy cycle functions like a "budget." Energy may be added to or subtracted from the budget and may be transferred from one storage reservoir to another, but overall the additions and subtractions and transfers must balance. If the balance is not maintained, Earth's life zone must either heat up or cool down.

Energy In

The rate at which energy flows into Earth's energy budget is greater than 174,000 terawatts (or 174,000 $\times$ 10^{12} watts). This quantity completely dwarfs the rate at which humans use energy, which is 10 terawatts. Incoming short-wavelength solar radiation overwhelmingly dominates the flow of energy into Earth's energy budget, accounting for about 99.985 percent of the total. Part of this vast influx powers the winds, rainfall, ocean currents, waves, and other processes in the hydrologic cycle. Another part of solar radiation is used for photosynthesis, which converts the solar energy into chemical energy and stores it temporarily in the biosphere as organic matter. When plants, algae, and bacteria die and are buried, a small portion of this energy is stored as coal, oil, and natural gas. When we burn fossil fuels, we release this stored solar energy.

The second most abundant source of energy, at 23 terawatts or 0.013 percent of the total, is Earth's internal heat energy. As discussed previously, terrestrial energy plays an important role in the tectonic cycle; it is the source of the energy that causes the lithospheric plates to shift, uplifts mountains, causes earthquakes and volcanic eruptions, and generally shapes the face of Earth.

The smallest source of energy for Earth is the energy produced by gravity, tides, and Earth's rotation. Tidal energy accounts for approximately 3 terawatts, or 0.002 percent, of the total energy flow into the budget.

Energy Out

That accounts for the inputs. What about the outputs? Earth loses energy from the energy cycle in two main ways: by **reflection**, and by *degradation and reradiation*.

About 40 percent of the 174,000 terawatts of incoming solar radiation is simply reflected unchanged, back into space, by the top of the atmosphere, the clouds, ocean surfaces, continents, and ice and snow. For any planetary body, the percentage of incoming radiation that is reflected unchanged is called the **albedo**. A high albedo means a highly reflective surface. Each material has a characteristic reflectivity. For example, ice is more reflective than rocks or pavement; water is more highly reflective than vegetation; and forested land reflects light differently than agricultural land. Thus, if large expanses of land are converted from forest to plowed land or from forest to city, the actual reflectivity of Earth's surface, its albedo, will be altered. Any major increase in albedo will of course have an effect on Earth's energy budget, since it would cause a greater percentage of the incoming solar radiation to be reflected.

Make the CONNECTION

What would happen to Earth's albedo if the polar ice caps were to melt? What would happen if they were to grow? What might be some of the impacts on our world?

The portion of incoming solar energy that is not reflected enters the Earth system and is absorbed by materials in the atmosphere and the hydrosphere and on the surface. This energy undergoes a series of irreversible degradations in which it is transferred from one reservoir to another and converted from one form to another. Because of the second law of thermodynamics, the energy that is absorbed, utilized, transferred, and degraded eventually ends up as heat, in which form it is reradiated back into space. Weather is a manifestation of energy transfer and degradation; so are winds, waves, the growth of plants, and many other processes of the Earth system.

After the incoming energy has been used to run various Earth processes and has degraded into longer-wavelength radiation, it is reradiated back to outer space. This is a good thing; if it didn't happen, Earth would just continue to heat up. The outgoing terrestrial energy helps to balance the incoming solar energy. Earth has an average surface temperature of about 288 K and is therefore a much lower-temperature radiator than the Sun (FIG. 2.12).

FIGURE 2.12 Solar and terrestrial spectra
The spectra of the Sun and Earth are different because they are radiating at different temperatures. Shown are the smooth curves for perfect blackbody radiators at the relevant temperatures. The "actual" curves differ from the "ideal" curves (in black) principally because of absorption of radiation in certain wavelengths by Earth's atmosphere.

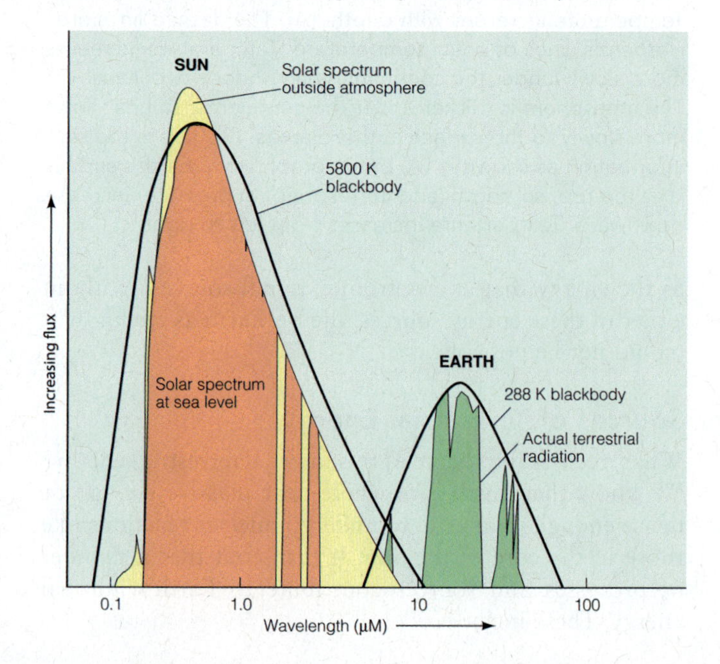

As a result, the radiation Earth emits has a peak at a wavelength of 1×10^{-5} m, much longer than the Sun's peak wavelength of 5×10^{-7} m. Earth's peak wavelength for emitted radiation is thus in the infrared region of the electromagnetic spectrum and cannot be seen by the human eye.

Figure 2.12 also shows that Earth's outgoing radiation, like incoming solar radiation, is selectively absorbed by gases in the atmosphere. This is of crucial importance for life on this planet. Just as ozone absorbs incoming radiation in the short-wavelength part of the solar spectrum, other gases, notably water vapor (H_2O) and carbon dioxide (CO_2), selectively absorb radiation in the longer-wavelength, infrared portions of the spectrum of outgoing terrestrial energy. This has the effect of retaining some of the outgoing radiation from the thermal portion of the spectrum near the surface, forming a "blanketing" layer of warmed air (FIG. 2.13). This is called the **greenhouse effect**, and it makes life possible on the surface of this planet.

It is worth noting that a "greenhouse" is not a particularly accurate analogy for this warming process in the lower atmosphere. The glass of a greenhouse causes warming inside the building by allowing sunlight in to warm the air, and then preventing the warmed air from rising and leaving the building. In other words, the warmed air is trapped inside, preventing *convective* heat loss. In contrast, the gases in the lower part of the atmosphere cause warming by absorbing outgoing radiation, thus preventing the *radiative* loss of heat. Today there are concerns that the large quantities of infrared-absorbing **radiatively active gases** that have been injected into the atmosphere by human industrial activities—primarily carbon dioxide from the burning of fossil fuels—is causing an enhancement of the natural greenhouse effect, which scientists fear is leading to a global increase in surface temperatures. In other words, the lower part of the atmosphere may be getting *too* efficient at absorbing outgoing radiation. In Chapter 13, we will look much more closely at the natural and enhanced greenhouse effects when we consider the forces that control climate and climatic change.

Pathways and Reservoirs

Like other cycles, the energy cycle involves pathways, processes, and storage reservoirs—places where energy resides for various lengths of time in the Earth system. For example, living organisms are reservoirs for energy, as you can readily feel from the warmth if you hold a baby or a furry animal in your arms. The solid ground is also a reservoir for energy; try lying on a flat rock that has been in

FIGURE 2.13 Greenhouse effect
Some of the short-wavelength solar radiation reaching Earth is absorbed by land, oceans, clouds, and atmospheric dust and gases, and some is reflected back into space by reflective surfaces that include snow, ice, clouds, and dust. Some of the shortest-wavelength (ultraviolet) radiation is absorbed by ozone (O_3) in the ozone layer. Earth radiates longer-wavelength radiation back into space. Radiatively active greenhouse gases, including water vapor and carbon dioxide (CO_2), absorb some of the outgoing long-wave (infrared) radiation, retaining it near the surface and causing the air temperature of the lower atmosphere to rise. This natural greenhouse effect makes it possible for life to exist, as we know it, on Earth's surface.

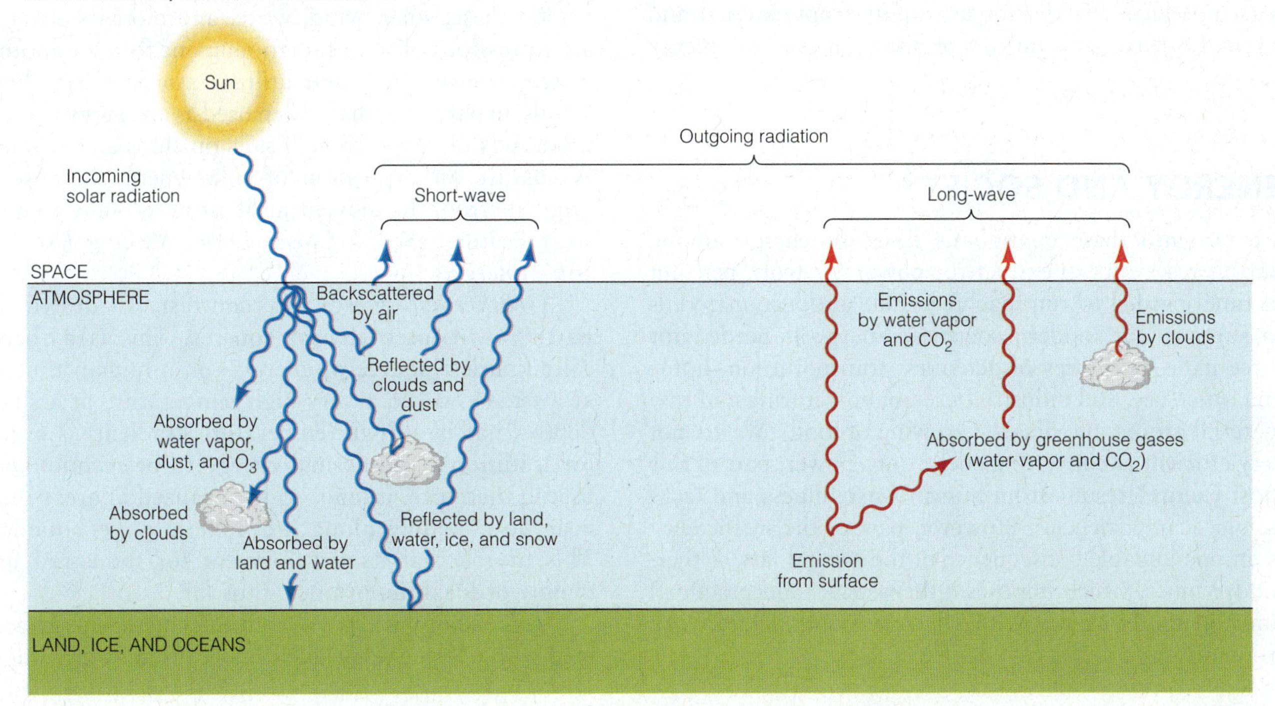

the Sun all day. The atmosphere also holds and transports a significant amount of solar energy, via the global winds and climate system.

Though less obvious, the ocean and even glaciers are also energy reservoirs, and as such they are of huge significance in Earth's energy cycle. The water of the ocean and the ice of a glacier feel cold to our touch because their temperatures are lower than our body temperature, but they still contain heat and are important reservoirs for energy. They are low-concentration, or *diffuse*, reservoirs. Think of salty water as an analogy. You can imagine a glass of water that is extremely salty—*much* saltier than ocean water. The *concentration* of salt (that is, the amount of salt per unit of water) in the glass is much higher than the concentration of salt in the ocean. Yet the ocean, overall, is a *much* larger reservoir of salt than the glass of salty water. Even though the concentration of salt is low in the ocean, because it is spread out through such a large volume of water, the *total amount* of salt is much, much higher than in the glass of salty water. In the same way, ocean water and even glaciers are diffuse but very large reservoirs for heat. Even though heat is present in low concentrations in these reservoirs, the volumes of the reservoirs are so enormous that the overall quantity of heat contained in them is very large. This is why oceans play such a fundamentally important role in determining weather patterns. For example, hurricanes form and grow by extracting heat energy from ocean water.

The energy cycle differs from the various material cycles in the Earth system because matter can be cycled from one reservoir to another, back and forth, endlessly, but energy cannot be endlessly recycled. We know this, again, from the second law of thermodynamics because the flow of energy always involves degradation and increasing disorganization as the energy becomes dispersed as heat. Luckily for us, energy from external and internal sources constantly replenishes the surface energy budget.

ENERGY AND SOCIETY

We tap into these continuous flows of energy among Earth's reservoirs to extract the **power** (or work per unit of time) needed to run machinery and produce materials to support our modern societies. Energy is needed for three main categories of activities: transportation, home and office use, and industrial use (manufacturing and raw material processing, plus the growing of food). We are not very efficient at converting energy into power; part of this "lost energy" results from human wastefulness and technological inefficiencies. However, part of the inefficiency is an unavoidable consequence of the second law of thermodynamics, which imposes a theoretical (and practical) limit on the efficiency with which we can extract work from any system.

How much energy do all the people of the world use? The total is enormous. The energy drawn annually from the major fuels—coal, oil, and natural gas—as well as that from nuclear power plants, is 2.6×10^{20} J. Nobody keeps accurate accounts of all the wood and animal dung burned in the cooking fires of Africa and Asia, but the amount has been estimated to be so large that when it is added to the 2.6×10^{20} J figure, the world's total energy consumption rises to about 3.0×10^{20} J annually. This is equivalent to the burning of 2 metric tons of coal or 10 barrels of oil for every living man, woman, and child, each year!

Energy Sources and the Energy Cycle

It is interesting to think about present-day energy sources in terms of Earth's energy cycle (FIG. 2.14). *Fossil fuels*, for example, come from the remains of plants and animals that have been trapped in sedimentary rocks and chemically altered to form fuels such as coal, oil, and natural gas. Since all fossil fuels are the remnants of organisms that once lived, they are ultimately an expression of stored solar energy, since their organic matter originally formed by the process of photosynthesis, driven by solar energy.

Three sources of energy other than fossil fuels have already been extensively developed: they are Earth's plant life (so-called *biomass energy*); hydroelectric energy; and nuclear energy. Together with fossil fuels, these energy sources have accounted for the vast majority of the modern world's energy needs, although the search is now on for alternative sources. Five others—the Sun's heat (*solar energy*), winds, waves, tides, and Earth's internal heat—have been tested and developed on a limited basis, but have not yet been developed on a large scale.

Of these, solar, wind, wave, and biomass energy are all expressions of solar radiation in one form or another—stored, transformed, and utilized more or less directly. Winds, in particular, have been used as an energy source for thousands of years through sails on ships and windmills. Winds are an expression of solar energy because they originate from the movement of air in response to uneven solar heating. (See *A Closer Look: Making Use of the Sun's Energy*.)

Geothermal energy, in contrast, is drawn from Earth's own internal heat sources. This is a theoretically limitless source of energy—on a human timescale, at least. However, a very high temperature at a shallow depth (that is, a high-temperature gradient) is required for traditional geothermal energy to be technologically useful; therefore, it tends to be localized in areas of high heat flow, mainly along lithospheric plate boundaries. This means that its development for industrial use is mainly of local importance, thus far.

Tidal energy is drawn from the kinetic energy associated with oceanic tides, ultimately caused by gravitational

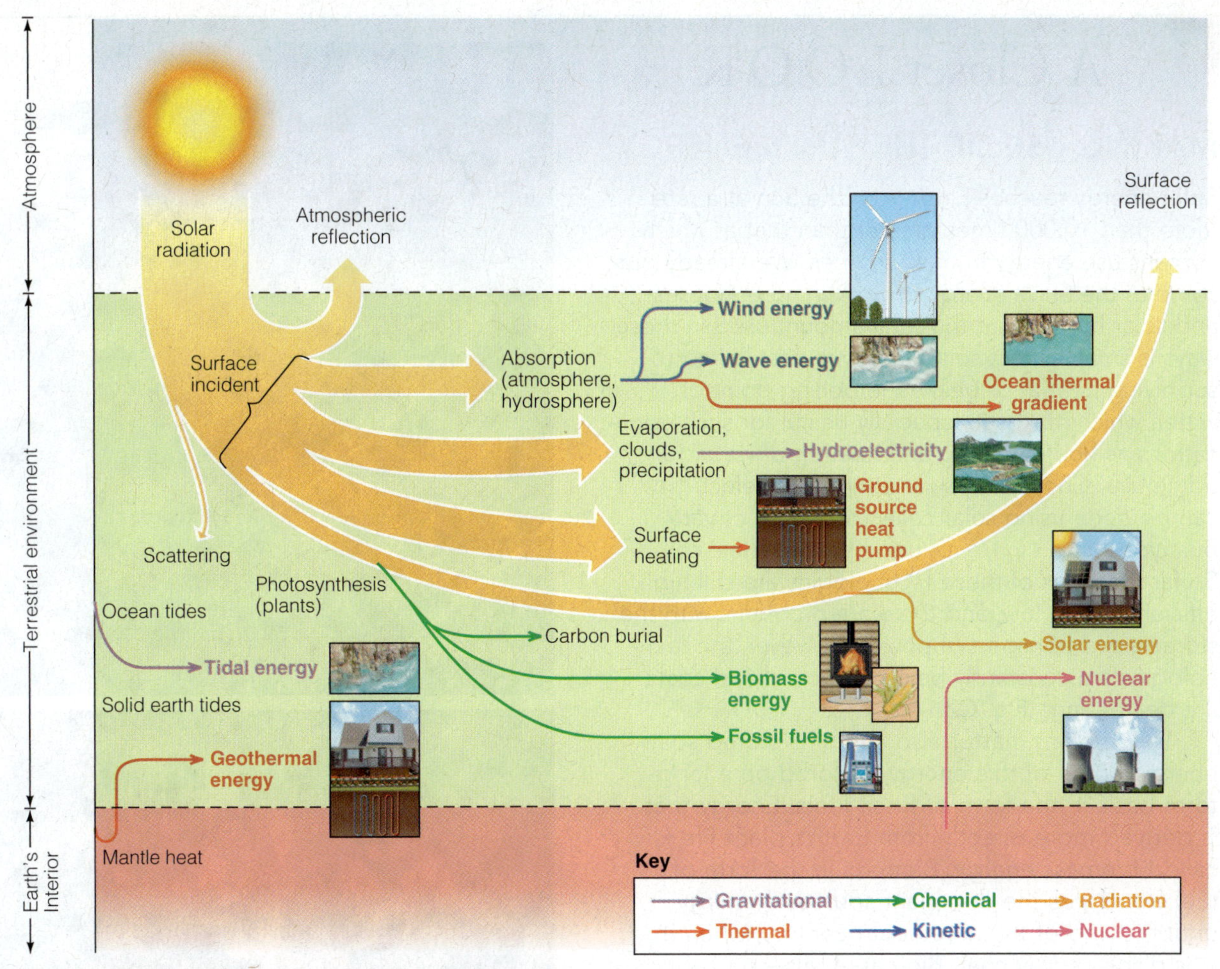

FIGURE 2.14 Sources of energy for human use
This is a simple diagram of the energy sources of modern society, from the perspective of Earth's energy cycle. There are many different types and sources of energy, and many times more energy delivered to the surface than is required for human use. The challenge is to find energy sources that are sustainable and environmentally benign.

interactions between Earth, Sun, and Moon. If a dam is put across the mouth of a bay so that water is trapped at high tide, the outward flowing water at low tide can drive a turbine. Unfortunately, few places around the world have tides high enough to make tidal energy feasible. *Hydroelectric power* is also recovered from the potential energy of water—in this case, stream water as it flows to the sea. If all the possible hydropower in the world were developed, we could satisfy only about one-third of the present world energy needs.

Nuclear energy is the heat energy produced during controlled splitting, or **fission**, of radioactive isotopes. Three of the radioactive atoms that keep Earth's interior hot by spontaneous radioactive decay—^{238}U, ^{235}U, and ^{232}Th—can be mined and used in this way. When ^{235}U fissions, it not only releases heat and forms new elements but also ejects some neutrons from its nucleus, which induce more ^{235}U atoms to fission. A continuous chain reaction occurs, and a tremendous amount of energy can be obtained in the process.

Theoretically, it should be possible to extract energy from nuclear fusion, as well—the same energy-generating process that occurs in the Sun—using hydrogen from readily available sources as the primary fuel. However, fusion in the Sun happens at very high temperatures and pressures that are technologically unattainable here on Earth. So far, scientists have been unable to overcome this limitation.

It is clear that there are numerous sources of energy available in the Earth system, and far more energy exists than we can use or would ever need. Unfortunately, all of these energy sources have drawbacks and limitations, and all of them, too, have potential negative impacts on the quality of the environment. We will look more closely at each of these energy sources, their technological limitations, and the environmental impacts of their use in Chapter 18. What is not yet clear is when, or even whether, we will be clever enough to learn how to tap the different energy sources in ways that will be economically feasible and socially equitable, and will not dramatically disrupt the environment.

A Closer LOOK

MAKING USE OF THE SUN'S ENERGY

Solar energy reaches Earth from the Sun at a rate more than 10,000 times greater than that at which humans use energy from all sources. We already put some of the Sun's energy to work in greenhouses and solar homes, but the total amount used in these ways is small. Direct solar energy is best suited to supplying heat at or below the boiling point of water, which makes it especially useful for such applications as home heating and heating of water for home use. Converting solar energy into electricity can be done using solar collector panels of various types (**FIG. C2.1A**), or using photovoltaic cells. So far the costs of these technologies are still high, efficiencies are low, and there are problems with the storage and transport of power. However, the technologies are constantly being improved and costs are decreasing. (Fig. C2.1A).

Living plant matter also contains stored solar energy. Some of this energy is stored on a long-term basis in the form of fossil fuels. Energy that is derived more directly from Earth's plant life is called *biomass energy*. Biomass in the form of fuel wood was the dominant source of energy until the end of the nineteenth century, when it was displaced by coal. Biomass fuels—primarily fuel wood, peat, animal dung, and agricultural wastes—are still widely used throughout the world (**FIG. C2.1B**), mainly in less economically developed countries. Another form of biomass energy involves the extraction of alcohol fuels (such as ethanol) from plant matter. Corn and sugarcane have been used for this purpose. Although these are "sustainable" energy sources in the sense that they are renewable, some concerns have been raised about the potential impacts of diverting food crops for use as energy sources.

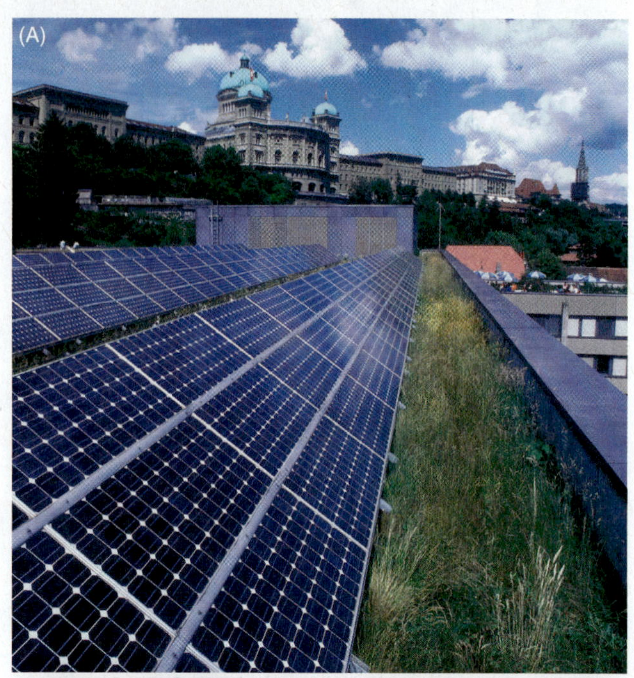

FIGURE C2.1 Energy from the Sun
(A) Solar panels like these collect and concentrate energy from the Sun, making it available to power human technologies. (B) Biomass fuels, such as the wood being burned here, are indirect forms of solar energy. They make use of the matter generated by plants by photosynthesis, using energy from the Sun.

SUMMARY

1. All processes and flows of material that happen in and on Earth are driven by energy in one form or another. Energy—the capacity to do work—is typically added to or taken away from a system either by work done to or by the system, or by heat flow.

2. Heat, or thermal energy, is the random vibrational motion of microscopic particles, and temperature is the measure of that motion. Other forms are chemical energy, nuclear energy, mechanical energy, gravitational energy, radiant or electromagnetic energy, electrical energy, sound, and motion of objects.

3. Potential energy is energy that is stored in a system. Kinetic energy is energy that is expressed in the movement of electrons, atoms, molecules, materials, and objects. It is common for energy to be partly potential and partly kinetic.

4. The laws of thermodynamics govern the behavior of energy. The first law states that in a system of constant mass, the energy involved in any physical or chemical change is neither created nor destroyed, but merely changed from one form to another.

5. The second law of thermodynamics governs all energy flows. It states that energy always changes from a more useful, more concentrated form to a less useful, less concentrated form. One consequence of the second law is that the flow of energy involves degradation, the transformation of energy into a form that is less available for work.

6. The third law of thermodynamics postulates the existence of the state of absolute zero, in which all molecular motion would cease and entropy would disappear. Absolute zero provides the baseline for the Kelvin temperature scale.

7. The three fundamental mechanisms of heat transfer are conduction, convection, and radiation. In conduction, heat moves directly from one object into an adjacent object by the transfer of kinetic energy. In convection, heat is moved from one location to another as a result of the heating and subsequent movement of materials. In radiation, electromagnetic energy is transferred from a radiating body through a surrounding gas, liquid, or vacuum.

8. The Sun is a star, in which nucleosynthesis occurs by fusion. The Sun produces most of its energy by thermonuclear reactions in which hydrogen protons fuse to form helium. During the fusion process, some mass is converted to energy. Each individual fusion reaction produces only a tiny amount of energy, but such an enormous amount of hydrogen fuel is continuously converted to helium that the Sun's overall luminosity is huge. Only a tiny fraction of the total energy emitted by the Sun reaches Earth; even so, a 1 percent increase or decrease would have a significant impact on climate.

9. Visible light, ultraviolet light, X rays, γ rays, infrared rays, and radio waves are forms of electromagnetic radiation that differ from one another in wavelength. Radiation in the long-wavelength end of the electromagnetic spectrum is much less energetic than radiation in the short-wavelength end. Electromagnetic radiation can also be described in terms of quanta or photons. The higher the frequency, the greater the amount of energy carried by the photon.

10. The Sun has six concentric layers: the core, in which the nuclear fusion reactions occur; the radiative layer, across which energy from the core moves by radiation; the convective layer, across which energy moves by convection; the photosphere, an intensely turbulent zone which emits the light that reaches Earth; the chromosphere, a transparent, low-density layer of very hot gas; and the corona, a zone of even lower density gas than the chromosphere.

11. The Sun's spectral curve matches that of a blackbody radiator emitting at 5800 K, the average temperature of the photosphere. The peak is at a wavelength of 5×10^{-7} m, in the visible part of the electromagnetic spectrum. Gases in Earth's atmosphere selectively absorb some wavelengths from the solar radiation as it passes through on its way to the surface. Notably, ozone absorbs radiation in the very short-wavelength, ultraviolet portions of the spectrum.

12. The gravitational attraction between the Moon and Earth, and to a lesser extent the Sun and Earth, causes distortions—tides—in the solid body of Earth, as well as in Earth's oceans.

13. Earth has its own internal sources of terrestrial or geothermal energy, as evidenced by the geothermal gradient, the increase in temperature with depth. The geothermal gradient varies from place to place; it is less pronounced under the continents than under the ocean basins, where the crust is thinner. The core has a temperature of about 5000°C.

14. Heat flow from the interior to the surface is greatest in locations with volcanic activity. Heat reaches the bottom of the lithosphere primarily by convection, and passes through the lithosphere primarily by conduction. Slow convection currents of solid rock deep in Earth's interior provide the driving force behind plate tectonics, in addition to generating volcanic eruptions.

15. Earth's internal energy comes from several sources, including radiogenic heat from the spontaneous decay of naturally occurring radioactive elements; accretionary heat from the collision and accretion of particles during the formation of Earth; tidal heat generated by the distortions of body tides; and gravitational and latent heat of crystallization associated with core formation.

16. The total amount of energy flowing into Earth's energy budget is more than 174,000 terawatts (or $174{,}000 \times 10^{12}$ watts). Solar radiation dominates the flow of energy into Earth's energy budget, at 99.985 percent of the total. The second most powerful source, at 23 terawatts or 0.013 percent of the total, is Earth's internal heat energy. Tidal energy accounts for approximately 3 terawatts, or 0.002 percent of the total.

17. About 40 percent of incoming solar radiation is reflected back to space as a result of the planet's albedo. The remainder, along with tidal and geothermal energy, is absorbed by materials at the surface. It undergoes a series of irreversible processes in which it is transferred from one reservoir to another and converted from one form to another, degraded, and eventually reradiated back to outer space.

18. Earth, a lower-temperature radiator than the Sun, has a peak at a wavelength of 1×10^{-5} m, in the infrared part of the electromagnetic spectrum. Earth's outgoing radiation is selectively absorbed by radiatively active gases, notably water vapor and carbon dioxide. This causes the greenhouse effect, in which some outgoing radiation from the thermal portion of the spectrum is retained near the surface, forming a "blanketing" layer of warmed air.

19. Even diffuse reservoirs like the ocean and glacial ice are important as reservoirs for heat in Earth's energy cycle. Heat is present in very low concentrations in these reservoirs, but their volumes are so enormous that the overall quantity of heat contained in them is very large.

20. Abundant energy is available in the Earth system in a variety of forms, but all sources for human use have drawbacks, limitations, and potential environmental impacts. Fossil fuels, by far the main source of energy for modern industrial society, come from the ancient remains of plants and animals; they are an expression of stored solar energy. Wind energy, wave energy, and biomass energy are also secondary expressions of solar radiation. Geothermal energy comes from Earth's internal heat sources. Tidal and hydroelectric energy come from the kinetic energy of water. Nuclear energy is produced by controlled fission of radioactive isotopes.

IMPORTANT TERMS TO REMEMBER

albedo 44
blackbody radiator 40
Celsius temperature scale 34
conduction 36
convection 37
degradation 33
electromagnetic radiation 38
energy 32
energy cycle 43
entropy 33
fission 47
fusion 35
geothermal gradient 42
gradients 34
gravity 40
greenhouse effect 45
heat 32
Kelvin temperature scale 34
kinetic energy 32
luminosity 35
potential energy 32
power 46
radiation 37
radiatively active gas 45
radiogenic heat 43
reflection 44
spectrum 38
star 35
stratospheric ozone layer 40
temperature 34
terrestrial energy 42
thermodynamics 33
tide 41
work 32

QUESTIONS FOR REVIEW

1. What is the difference between "heat" and "temperature"?

2. What are the three laws of thermodynamics? Summarize them, and explain the basic importance of each in the context of Earth system science.

3. Describe the transfer of heat by conduction, convection, and radiation.

4. Explain the source of the Sun's energy. What is luminosity, and how do scientists measure it?

5. Sketch a cross section through the Sun and label the various layers.

6. What is a blackbody radiator? What is the difference in the radiation spectrum of a blackbody radiator at 6000 K and one at 2000 K? How does the spectral curve of the Sun compare to that of a blackbody radiator?

7. Why and how does the Sun's spectrum of electromagnetic radiation in space differ from that measured at the surface of Earth?

8. How does gravitational interaction between Earth and the Moon cause body tides and ocean tides on Earth? Include one sketch to illustrate your answer.

9. What are the four main sources of Earth's internal terrestrial heat energy? Which one is the most important?

10. What is the geothermal gradient? Why is it different under the continents than under the oceans?

11. How does heat move to the surface from inside the planet?

12. How is it possible that solid rock deep inside Earth can undergo convection, and what are the results of that convection?

13. What are radiatively active gases, and what impact do they have on Earth's outgoing radiation? Give two examples of important radiatively active gases.

14. Explain why ocean water and glacial ice are important reservoirs for heat, even though they feel very cold to the touch.

15. For each of the following energy sources, discuss its "ultimate" source of energy in the context of Earth's energy cycle: natural gas and oil; geothermal energy; hydrothermal energy; wood fuels; nuclear power.

QUESTIONS FOR RESEARCH AND DISCUSSION

1. In this chapter we have presented the Sun as if it were a steady, reliable, smoothly operating body. This is the case most of the time, but the Sun can undergo some dramatic changes on both short and long timescales. Do some research on short-term active phenomena, such as sunspots, solar prominences, and storms, and long-term changes in solar luminosity. Report on the impacts of these variations on Earth and human society.

2. On the basis of the discussion on world energy use in this chapter, do you think society is in imminent danger of running out of energy sources? Why, or why not? Support your argument with specific examples, and be sure to discuss any of the drawbacks of your position.

3. Do some research to find examples of processes in the Earth system—either natural or anthropogenic—with which you can complete the following table. Some of these processes will be very familiar to you; others will be unfamiliar, and you will learn more about them as the book progresses:

Kinetic energy forms and processes		Potential energy forms and processes	
radiant energy	1. sunlight 2. 3.	chemical energy	1. coal 2. 3.
electrical	1. lightning 2. 3.	stored mechanical energy	1. crouching tiger 2. 3.
thermal energy	1. volcanism 2. 3.	nuclear energy	1. power plant 2. 3.
sound	1. thunder 2. 3.	gravitational energy	1. ocean tides 2. 3.
motion	1. rolling stone 2. 3.		

QUESTIONS FOR *THE BASICS*

1. Consider a lake in the sunlight. Which heat transfer process—conduction, convection, or radiation—would govern the temperature of the water just below the surface of the lake? Is it just one process, or do you think it would be a combination of processes?

2. What is electromagnetic radiation? Is the speed of short-wavelength radiation different from the speed of long-wavelength radiation?

3. What is the relationship between the frequency and the wavelength of electromagnetic radiation?

QUESTIONS FOR *A CLOSER LOOK*

1. One of the big problems with many alternative energy sources is that they tend to be heavily reliant on conditions in a specific locality. Choose several of the energy sources discussed here, and talk about their pros and cons with respect to their local or more regional applicability.

2. What are some of the potential drawbacks of using crops such as corn and sugarcane to produce alcohol fuels?

3. *All* energy sources have potential negative environmental impacts. What are some of the environmental impacts that may be associated with solar or wind energy?

CHAPTER 3

Matter

OVERVIEW

In this chapter we:

- Introduce the nature of matter and some of its fundamental characteristics
- Examine materials, both organic and inorganic, that are important in Earth system processes
- Summarize Earth's overall composition and internal structure
- Define the basic properties of minerals and rocks
- Describe regolith and its vital component, soil

Earth from the inside out
Lava—molten rock—from Hawaii's Kilauea Volcano cascades over a cliff on its way to the ocean.

Matter is substance. Unlike energy, matter has mass and occupies space. Matter and energy are related in some very interesting ways that are of particular interest to physicists. In this chapter, we are mainly concerned with identifying and investigating the types and properties of matter in general, and the materials of which the Earth system is made.

EARTH'S MATERIALS

We begin our study of Earth's materials by looking at some of the fundamental characteristics of matter. Familiarity with the basic properties and characteristics of matter can help us better understand Earth processes and give us an understanding of how different materials behave under different circumstances. We will also look at the overall composition of Earth and how the various materials of Earth are distributed throughout the planet.

Common States of Everyday Matter

We encounter different forms, or physical **states**, of matter frequently. The three common states of matter with which we are most familiar in our everyday lives are *solid*, *liquid*, and *gas*. The chemical compound we know as water (H_2O) is somewhat unusual in that it commonly occurs in all three of these states in the near-surface environment of Earth. In surface water bodies and in the form of precipitation, it occurs in the liquid state (water), but H_2O in the gaseous state (steam or water vapor) is also very common, as are various solid forms (ice and snow).

Materials that occur in the same state can still differ substantially from one another in properties. Consider, for example, two common liquids: lava and rainwater. Both have the fundamental characteristics of liquids, flowing freely under the influence of gravity, but as liquids they could hardly be more different (FIG. 3.1). Lava (molten rock) occurs in liquid form only at very high temperatures, typically over 1000°C, and freezes very quickly to form solid rock below 1000°C. Rainwater, on the other hand, completely vaporizes at temperatures even one-tenth this hot, remaining liquid only between 0° and 100°C, and requiring temperatures below 0°C to be transformed into its solid form (ice).

Even when they are both in the liquid state, lava and water differ substantially. They differ in both composition (for rainwater, hydrogen and oxygen with minor impurities; for lava, a complex mixture of oxygen, silicon, and many other components) and in physical properties. Under the pressure conditions of Earth's surface, lava is a relatively *viscous* liquid, meaning that it resists flow and is rather thick and syrupy. Rainwater, on the other hand, is a relatively low-viscosity, runny liquid at surface and near-surface conditions. Both of these types of liquids—**aqueous** (water-based) liquids, like rainwater, and non-aqueous liquids, like lava—are very important in a wide variety of Earth processes.

Change of State

When water turns to steam, or ice melts to water, or water freezes to ice, or steam condenses to form water droplets, we say that the material has undergone a *change of state*

the BASICS

Solids, Liquids, and Gases

Everyone is familiar with the common states in which matter occurs at the surface of Earth (FIG. B3.1). Let's review their basic properties and what distinguishes them from one another:

- In the **solid** state, matter is firm or compact in substance with a definite volume and density, and tends to retain its shape even if it is not confined, because its constituent atoms are fixed in position relative to each other.
- In the **liquid** state, matter has a definite volume but its constituent atoms are able to flow freely past one another; the material does not retain its own shape but conforms to the shape of its container, taking on a free surface under the influence of gravity.
- In the **vapor** (or **gas**) state, matter takes on the shape of the container in which it is contained, filling the container completely (or escaping into space if it is not confined), while its constituent atoms can move freely and acquire a uniform distribution within the container.

FIGURE B3.1 The common states of matter
(A) Solids retain their shape and volume, and their constituent atoms have little mobility relative to one another.
(B) Liquids take on the shape of their container, developing a free surface under the influence of gravity, and their constituent atoms move freely relative to one another.
(C) Gases mix thoroughly and fill their containers completely (or escape into space if unconfined).

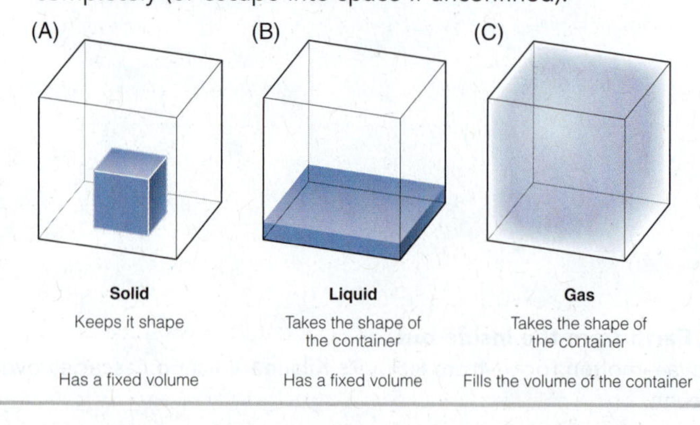

(A) 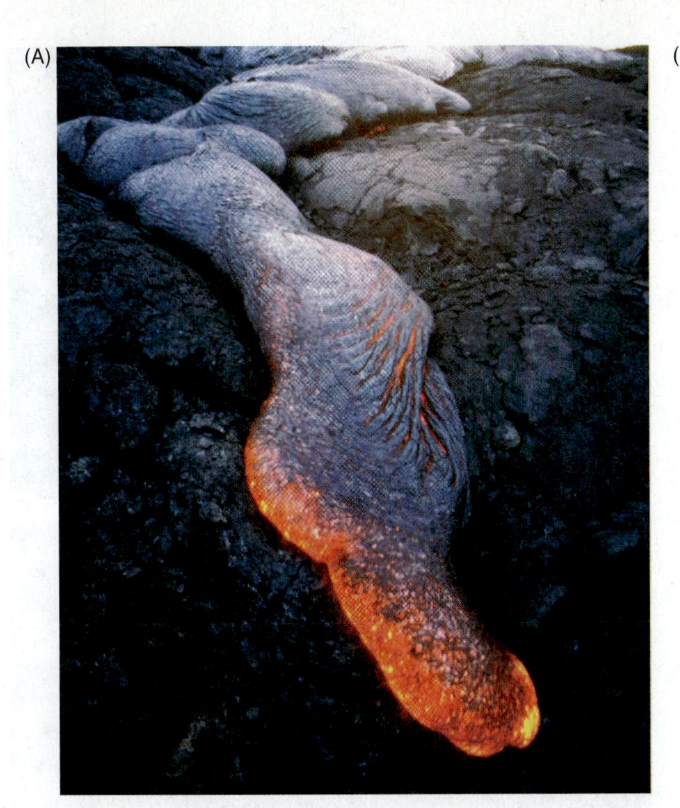

(B) 

FIGURE 3.1 Two liquids
Although lava (A) and rainwater (B) both occur in the liquid state at Earth's surface, they are very different in chemical composition and physical properties.

or *change of phase*. What is the difference between a *state* and a *phase* of matter? Materials in different states can exist together at the same time, in the same container (FIG. 3.2A). For example, you could have a glass of water (liquid) with ice cubes (solid) floating in it. Even though they have the same chemical composition, the liquid water and the solid ice in your glass are in different states; they are also different **phases**—homogeneous masses of material that can be separated from one another by a definable boundary. Coexisting materials that are in different states (such as water and ice) are separated by defined boundaries, so they are also distinct phases, by definition.

The reverse is not necessarily true, however; it is possible for some materials to coexist in two different phases but in the same state (FIG. 3.2B). For example, oil and water can coexist in the same container; they are both liquids (same state), but they have distinct properties and are separable from one another by a definable boundary (different phases). The only liquids that can coexist as two separate phases must be, like oil and water, *immiscible* ("unmixable"), meaning that they do not mix spontaneously and thoroughly with one another. If you were to pour two *miscible* ("mixable") liquids together into the same glass—water and lemon juice, for example—they would mix to form just one liquid phase (lemonade).

It is also possible (indeed, common) for multiple solid phases to coexist, as long as the solids have distinct properties and can be separated from one another along their boundaries (Fig. 3.2B). A cookie with nuts in it would be

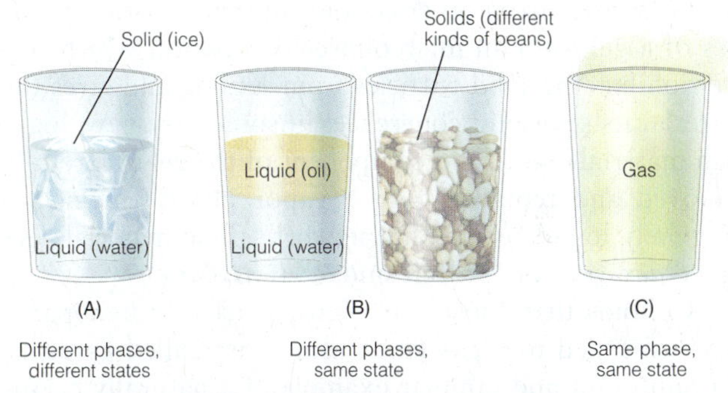

FIGURE 3.2 States and phases
Matter can coexist in various states and phases. In (A), matter of uniform composition (H_2O) coexists in two different states (liquid and solid) and two different phases (water and ice), separated by physical boundaries. In (B), different phases coexist in the same state (oil and water, both liquids; different types of beans, all solids). In (C), there is only one phase and one state, as is always the case with gases.

a familiar example of a mixture of different coexisting solid phases. The two materials (cookie, nuts) are both solids (same state), but they have distinct properties and can be separated from one another (different phases).

Unlike solids and liquids, gases are always completely miscible (by definition). This means that all of the gaseous material in a given system will necessarily occur as just one gaseous phase (FIG. 3.2C). This phase may be a mixture of several different gases, but because gases are miscible, they always mix thoroughly to form a single gaseous phase.

Other States of Matter

Although we are most familiar with solids, liquids, and gases, in fact the most common state of matter in the universe is none of these; it is **plasma**, an ionized (electrically charged) gas with unique properties and characteristics. Plasma is a common state of matter in the Sun and other stars, as well as in interstellar and intergalactic materials. We do see plasmas on Earth, in phenomena such as lightning and the polar auroras (Aurora Borealis and Aurora Australis, the northern and southern lights), and in some manufactured items such as neon and incandescent lamps, arc welders, and plasma televisions.

Other familiar materials—both natural and manufactured—occur in forms that seem to have the properties of more than one state of matter. A good example of this is materials that occur in *gel* form. Gels seem solid, holding their shape to a certain extent, but they are also jellylike; they tend to flow and are easily deformed, thus displaying some of the properties of liquids. In detail, gels are **colloids**, consisting of extremely fine particles dispersed in a continuous medium, usually a liquid. The particles in colloids are so fine—typically less than 100 μm (μm = micrometer or micron, one-millionth of a meter), but sometimes even measured in nanometers (nm = one-billionth of a meter)—that they do not settle out but remain suspended in the surrounding medium.

Gels differ from other colloids in that the solid particles of a gel form an interconnected network, which further inhibits the solid particles from settling and gives the material its semisolid coherence. In some very new, high-tech materials the liquid component of the gel is carefully removed and replaced with a vapor phase, leaving an extremely low-density solid network; these materials are called *aerogels*, or "frozen smoke" (**FIG. 3.3A**).

Colloids that consist of microparticles or nanoparticles suspended in a gas (such as air) are called **aerosols**. An important and familiar example of a naturally occurring aerosol is fog; if you have asthma, or know someone who does, then you are probably aware that medicines are often delivered in the form of aerosols, with extremely fine liquid droplets of medicine suspended in and transported by a pressurized stream of gas. Many other gels, colloids, and aerosols are important in Earth processes (**FIG. 3.3B**), including blood, milk, smoke, smog, slush, humus, mud, opal, some volcanic emissions, and sea foam. Some common manufactured substances that occur in gel, colloid, and aerosol forms include gelatins (such as Jell-O), toothpaste, asphalt, butter, cheese, glue, styrofoam, ink, paint, hair spray, and room deodorizer.

FIGURE 3.3 Aerosols and colloids
Many important materials, both natural and manufactured, occur as aerosols and colloids. (A) Aerogel or "frozen smoke" is an ultra-low-density manufactured colloidal gel that consists of solid nanoparticles in a vapor medium (instead of a liquid medium, as in most colloidal gels). (B) Mud is an example of a naturally occurring colloidal mixture of extremely fine clay particles in water.

Atoms, Elements, Ions, and Isotopes

Chemical elements are the most fundamental substances into which matter can be separated by ordinary chemical means. Each element is identified by a symbol, such as H for hydrogen and Si for silicon. Some symbols, such as that for hydrogen, come from the element's English name. Other symbols come from other languages. For example, iron is Fe from the Latin *ferrum*, and sodium is Na from the Latin *natrium*. All of the naturally occurring elements and their symbols, together with elements that have been synthesized in laboratories, are listed in Appendix B.

Even a tiny piece of a pure element consists of a vast number of identical particles of that element, called atoms. An **atom** is the smallest individual particle that retains the distinctive properties of a given chemical element. Atoms are so tiny they can be seen only by using the most powerful microscopes ever invented, and even then the image is imperfect because individual atoms are only about 10^{-10} m in diameter. In other words, 100 million atoms laid side by side would form a line just 1 cm long. (But note that atoms don't sit still, so you could not actually put so many atoms in a straight line.) Atoms are built of *protons*, which have positive electrical charges; *neutrons*, which, as their name suggests, are electrically neutral; and *electrons*, which have negative electrical charges that balance the positive charges of protons.

The protons and neutrons join together to form the core, or *nucleus*, of an atom (FIG. 3.4). Protons give the nucleus a positive charge, and the number of protons in the nucleus of an atom is called the *atomic number* of the atom. The number of protons in the nucleus (the atomic number) is what gives the atom its special characteristics and what makes it a specific element. Thus, any and all atoms containing one proton in the nucleus are atoms of hydrogen; atoms containing two protons in the nucleus are helium; and so on. All atoms having the same atomic number are atoms of the same element. The atomic numbers of all elements are listed in Appendix B.

FIGURE 3.4 Carbon atom
This is a schematic diagram of an atom of the element carbon. The nucleus contains six protons and six neutrons. Electrons orbiting the nucleus are confined to specific orbits called energy-level shells. (A) This three-dimensional representation shows the first energy-level shells. The first shell can contain two electrons, the second eight. (B) This two-dimensional representation of the carbon atom shows the number of protons and neutrons in the nucleus and the number of electrons in the energy-level shells.

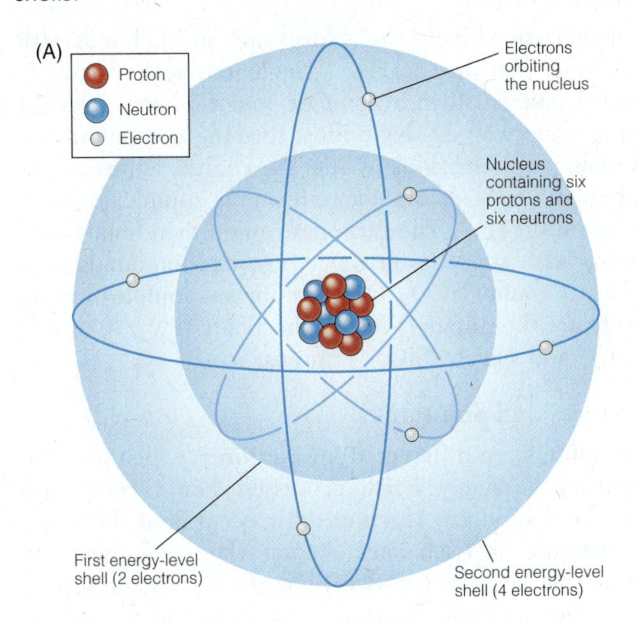

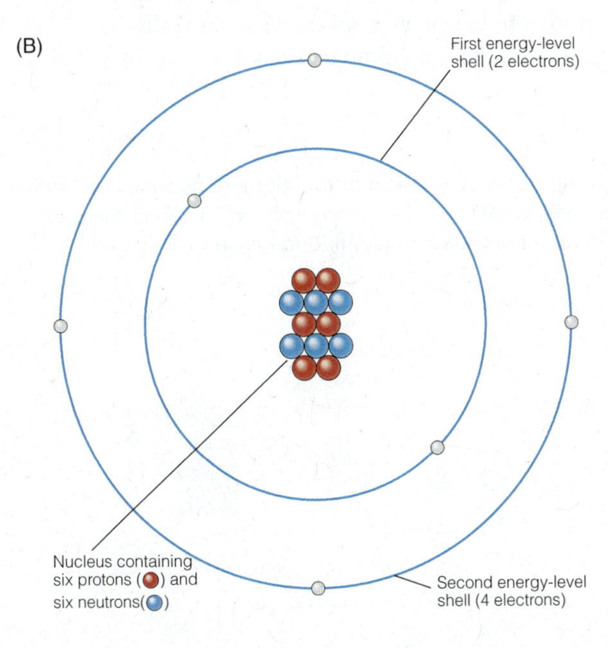

Electrons are much smaller than protons or neutrons; in fact, they are considered to be almost "mass-less." In atoms the electrons move in a distant and diffuse cloud around the nucleus. Within these clouds the electrons are constrained to specific energy levels, which we refer to as *energy-level shells*. Different atoms have different numbers of electrons and different numbers of energy-level shells. The maximum number of electrons that can occupy a given energy-level shell is fixed. As shown in Figure 3.4, energy level 1 can only accommodate 2 electrons; level 2 can accommodate up to 8 electrons; level 3, up to 18 electrons; and level 4, up to 32 electrons.

Because neutrons are electrically neutral, they cannot change the atomic number of an element. Neutrons can change the mass of an atom, however, and the sum of the neutrons plus protons in the nucleus of an atom is called the *mass number*. The number of protons in the nucleus (the atomic number) is designated by a subscript before the chemical symbol, whereas the sum of the protons plus neutrons (the mass number) is indicated by a superscript. Thus helium, with atomic number 2 and mass number 4, is written $^{4}_{2}He$. Hydrogen, occupying the very first position in the periodic table of the elements with its one proton and one electron, is written $^{1}_{1}H$.

Isotopes

Most elements have several **isotopes**, atoms with the same atomic number and hence essentially the same chemical properties, but different mass numbers. Carbon, for example, has three naturally occurring isotopes, $^{12}_{6}C$, $^{13}_{6}C$, and $^{14}_{6}C$. Some isotopes are **radioactive**, which means they transform spontaneously to another isotope of the same element or to an isotope of a different element. (In Chapter 2 we discussed the radioactive decay of isotopes as a source of some of the heat energy inside Earth.) Among the carbon isotopes only $^{14}_{6}C$ is radioactive; it transforms to an isotope of nitrogen, $^{14}_{7}N$ by the spontaneous transformation of a neutron to a proton. There are 25 naturally occurring radioactive isotopes. Some, like $^{14}_{6}C$ and $^{40}_{19}K$, are in common substances that we eat, drink, or breathe every day. Others, like $^{238}_{92}U$ and $^{232}_{90}Th$, are much less likely to make it into our food chain or common environment unless introduced by human activities.

Ions

An energy-level shell filled with electrons is very stable and is like an evenly loaded boat. To fill the outermost energy-level shell and so reach a stable configuration, atoms may either share or transfer electrons among themselves. In its natural state, an atom is electrically neutral because the positive electrical charge of its protons is exactly balanced by the negative electrical charge of its orbiting electrons. When an electron is transferred as part of the stabilizing of energy-level shells, this balance of electrical forces is upset. An atom that loses an electron has lost a negative electrical charge and is left with a net positive charge. An atom that gains an electron has a net negative charge. An atom that has excess positive or negative charges caused by electron transfer is called an **ion**. When the excess charge is positive (meaning that the atom gives up electrons), the ion is called a *cation*; when negative (meaning an atom adds electrons), the ion is an *anion*.

The most convenient way to indicate ionic charges is with superscripts. For example, Ca^{2+} is a cation (calcium) that has given up two electrons, and F^- is an anion (fluorine) that has accepted an electron. (When the ionic charge is 1, we omit the number; the symbol F^- really means F^{1-} and the symbol Li^+ really means Li^{1+}.) Because the formation of ions involves only the energy-level shell electrons and not the protons and neutrons in the nucleus, it is common practice to omit the atomic and mass number symbols for reactions involving ions.

Compounds and Mixtures

Compounds form when one or more kinds of anion combine chemically with one or more kinds of cation in a specific ratio. For example, hydrogen cations (H^+) combine with oxygen anion (O^{2-}) in the ratio 2:1, making the compound H_2O (water). In a compound, the sum of the positive and negative charges must be zero. We write the formula of a compound by putting the cations first and the anions second. The numbers of cations or anions are indicated by subscripts, and for convenience the charges of the ions are usually omitted. Thus, we write H_2O rather than $[H_2]^{2+}O^{2-}$.

FIGURE 3.5 presents an example of how electron transfer leads to formation of a compound for the elements lithium and fluorine. A lithium atom has energy-level 1 occupied by two electrons but has only one electron in level 2, even though level 2 can accommodate up to eight electrons. The lone electron in level 2 can easily be transferred to an element such as fluorine, which already has seven electrons in level 2 and needs only one more to be completely filled (and thus stable). In this fashion, both a lithium cation and a fluorine anion end up with filled shells, and the resulting positive charge on the lithium and negative charge on the fluorine draw, or **bond**, the two ions together. This particular type of bond, in which electrons are exchanged, is called an *ionic bond*.

Lithium and fluorine form the compound lithium fluoride, which is written LiF to indicate that for every Li ion there is one F ion. The combination of one Li ion and one F ion is called a molecule of lithium fluoride. A **molecule** is the smallest unit that retains all the properties of a compound. Properties of molecules are quite different from the properties of their constituent elements. The elements sodium (Na) and chlorine (Cl) are highly toxic, for example, but the compound sodium chloride (NaCl, table salt) is essential for human health.

Complex Ions

Sometimes two kinds of ions form such strong bonds with each other that they act like a single ion. Such a strongly bonded pair is called a *complex ion*. Complex ions act in the same way as single ions, forming compounds by bonding with other ions of opposite charge. For example, carbon and oxygen combine to form the complex carbonate anion $(CO_3)^{2-}$. The carbonate anion then bonds with cations such as Na^+ and Ca^{2+} to form compounds such as Na_2CO_3 and $CaCO_3$. Other important complex ions in nature are the sulfate $(SO_4)^{2-}$, phosphate $(PO_4)^{3-}$, nitrate $(NO_3)^-$ and silicate $(SiO_4)^{4-}$ anions.

Mixtures and Solutions

Compounds are different from *mixtures*, which are also important in many Earth processes. For a compound to form, a chemical reaction must occur, and there will be changes or exchanges among the constituents at the atomic level. When two or more liquids, or several gases, blend thoroughly without undergoing a chemical reaction, the result is a mixture; technically, it is called a *solution* if it is a homogeneous mixture in which the

FIGURE 3.5 A compound
To form the compound lithium fluoride, an atom of the element lithium combines with an atom of the element fluorine. The lithium atom transfers its lone outer-shell electron to fill the fluorine atom's outer shell, creating an Li^+ cation and a F^- anion in the process. The electrostatic force that keeps the lithium and fluorine ions together in the compound lithium fluoride is an ionic bond.

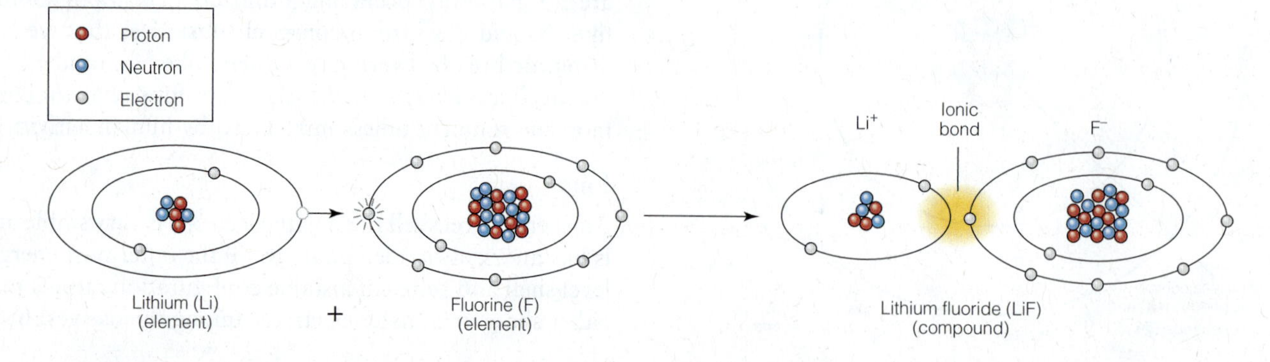

different constituents are not easily distinguishable from one another. Mixtures and solutions are not compounds because the individual constituents retain their identities, with no changes on an atomic level—in other words, no chemical reactions are involved. If you mix lemon juice with water the liquids will mix thoroughly, but it is not a compound because there are no chemical reactions between the two liquids. If you dissolve sugar in water you will obtain a homogenous mixture—a solution—but the sugar and water molecules still retain their identities. The gels and colloids discussed previously also are mixtures, not compounds.

Mixtures and solutions don't have to be liquid or gaseous. For example, a bowl of chocolate chips mixed with peanuts is a mixture (consisting of two phases, both in the solid state). If you threw the chocolate and peanuts into a blender and mashed them into a very smooth, homogeneous mixture, you could say that you had made a solution—but it would still be a mixture, not a compound, because the peanuts and chocolate would still be distinct at the molecular level.

You can often tell a mixture or solution from a compound because it is usually possible to separate the components of a mixture by some physical or mechanical means, whereas the components of compounds can only be separated chemically. For example, you could separate the sugar from the water by evaporating the water. Even the lemon juice and water mixture is separable; if left to sit, the lemon juice will begin to settle to the bottom (which is why fruit juice mixtures often must be stirred or shaken before drinking). A mixture of flour with fine magnetic iron powder could be separated using a magnet to attract the iron particles. Separation by physical or mechanical means—filtering, evaporation, settling, or magnetic attraction—demonstrates that these combinations are mixtures or solutions, not compounds.

ORGANIC MATTER

So far, we have looked at a variety of states, forms, and combinations in which matter occurs on Earth. Materials are also distinguished from one another on the basis of differences in composition. Perhaps the most fundamental compositional distinction—but one that is not always clear-cut, as you will learn—is the distinction between *organic* and *inorganic* materials.

Organic and Inorganic Compounds

The distinction between material that is "organic" and material that is "inorganic" seems straightforward, but in practice it can be quite ambiguous. One thing to keep in mind is that the word "organic," when used in a scientific context, definitely does not mean "green" or "environmentally friendly"; in fact, some organic compounds are highly toxic to humans.

Originally, organic matter was defined with strict reference to its derivation from living organisms. That definition still holds, to a certain extent, but it has been broadened quite a bit. Some scientists would say that any compound that contains carbon is organic; however, this definition is so broad that it fails to catch some important distinctions. Today it is common to apply the term **organic** specifically to compounds consisting of carbon atoms that are joined to other carbon atoms by a strong type of bond called a *covalent bond.* Covalent bonds involve the sharing of electrons among neighboring atoms, rather than exchange of electrons as in ionic bonding. Organic carbon compounds also commonly contain hydrogen (H), thus forming carbon-hydrogen compounds, or **hydrocarbons.** Carbon-hydrogen (C-H) bonds are sometimes explicitly included as part of the definition of organic compounds.

The term *organic* usually—but not invariably—also implies that the compound is **biotic**, that is, once or presently alive, or of biologic origin. **Inorganic** compounds, in contrast, are **abiotic**, that is, nonliving and of nonbiologic origin. Inorganic compounds may contain carbon as a constituent, but they lack the carbon-carbon covalent bonds that are characteristic of organic compounds.

Based on these definitions, the distinction between organic and inorganic materials may seem straightforward, but it isn't. For example, consider diamond. Diamond is a mineral that consists of covalently bonded carbon (chemical formula C). It contains the carbon-carbon bonds required for classification as an organic compound, but it comes from deep within Earth's mantle. It is a mineral, it is of nonbiologic origin, and it would not be classified as an organic compound.

On the other hand, consider carbon dioxide (chemical formula CO_2). Carbon dioxide is a gas (at surface temperatures and pressures) that is a minor constituent of air. It is crucial in the process of photosynthesis, and it is the source of carbon for the organic compounds that form plant matter. Carbon dioxide comes from biologic sources and is used in biologic processes, yet it lacks carbon-carbon bonds. It doesn't even contain hydrogen, the other common characteristic of organic compounds. To further muddy the waters, carbon dioxide can also come from sources that are clearly nonbiologic and inorganic, such as volcanic eruptions. According to our definitions, it doesn't seem straightforward to classify carbon dioxide as an organic compound, and yet it is fundamentally important in many biologic processes.

Important Organic Molecules

One common characteristic of organic compounds is their tendency to occur in long, chainlike structures called **polymers.** In *polymerization,* small molecules (called *monomers*) link together to form long chains or three-dimensional networks. Polymerization is important in both organic and inorganic molecules, as you will learn when we discuss the crystal structures of minerals. Polymers that consist of organic compounds or are of biologic origin are called **biopolymers** (**FIG. 3.6**). Several categories of organic compounds play key roles in biologic processes in the Earth system, and three of them are

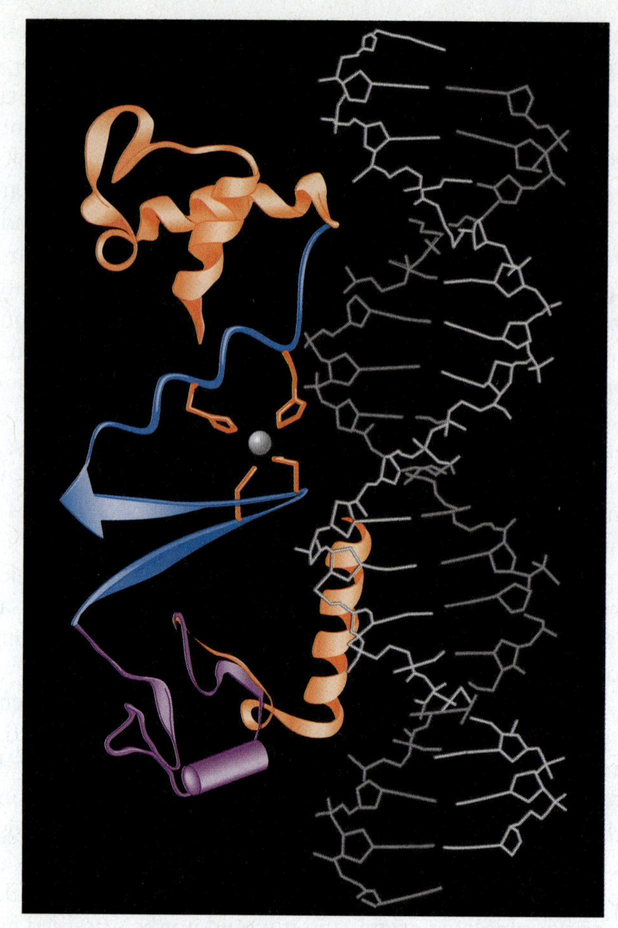

FIGURE 3.6 Biopolymers
Biopolymers such as proteins and nucleic acids are large organic molecules formed by polymerization—the stringing together of small molecules to make large chainlike or sheetlike molecules.

biopolymers: *proteins*, *nucleic acids*, and *carbohydrates*. A fourth important type of organic molecule that is not a polymer, but still fundamental to life, is *lipids*. Let's look briefly at each of these.

Proteins

The building blocks of proteins are organic molecules called *amino acids*. In an amino acid a central carbon atom shares electrons with, and is therefore bonded to, a hydrogen atom, an acidic carboxyl group (-COOH), a basic amine group ($-NH_2$), and a side group of atoms unique to each kind of amino acid. **Proteins**, then, are long, chainlike polymers made of amino acids bonded together. Up to 20 different amino acids may be combined into a protein polymer. The identity of a specific protein is determined by the sequence of the amino acids and by the shape the protein molecule assumes as the chain folds. Within a folded protein molecule, certain parts of the polymeric chain will be exposed while other parts will be hidden inside. As a result, the folding pattern determines the function of the protein because the position of each amino acid determines how it interacts with cell surfaces and with other molecules.

In life, proteins have many roles. They provide structural support for organisms and help in the growth of tissues. Animals, for example, generate skin, hair, tendons, and muscles with the help of proteins. Proteins also help in the storage of energy and in the body's defense against foreign organisms and pathogens. Still other proteins act as *hormones*, molecules that serve as chemical messengers, and as *enzymes*, molecules that speed up chemical reactions such as the digestion of food.

Nucleic acids

How is it possible for living organisms to produce proteins? As you will discover in Chapter 14, the production process is directed by **nucleic acids.** There are two nucleic acids—*deoxyribonucleic acid (DNA)* and *ribonucleic acid (RNA)*—and they are responsible for storing and passing along an organism's hereditary information.

Nucleic acids are giant organic polymers built from molecules called *nucleotides*, each of which contains a sugar group, a phosphate group, and a nitrogenous base. The sugar group in DNA is called *deoxyribose*, and in RNA it is *ribose*. DNA is a double-stranded polymer. The two strands can be pictured as sides of a ladder, with the nucleotides holding the sides together, like the ladder's steps. The DNA ladder is twisted into a spiral, giving the entire molecule a shape called a *double helix*. RNA is similar but single-stranded.

Hereditary information is encoded, like a bar code, by the order in which the nucleotides are stacked in a DNA molecule. The information is passed along by being rewritten to a molecule of RNA. The RNA then directs the order in which amino acids assemble to build proteins, which go on to influence the structure, growth, and maintenance of the organism. Genetic information from DNA is passed from one generation to another as the strands replicate during cell division and egg or sperm formation. **Genes** are regions of DNA coded for specific proteins that perform particular functions. In most organisms, the *genome*—the set of all of an organism's genes—is divided among specialized areas called *chromosomes*, which act as carriers of genetic information. Different types of organisms have different numbers of genes and chromosomes. Most bacteria have a single circular chromosome, for instance, whereas humans have 46 linear ones.

Carbohydrates

A third type of biopolymer, **carbohydrates**, is a group of chemical compounds that are the basis of much of the food we eat. Carbohydrates, which include sugars, starches, and cellulose, are found in all organisms. They are molecules composed of carbon, hydrogen, and oxygen, with the general formula $C_x(H_2O)_y$. Simple carbohydrates, called *sugars* (or *monosaccharides*), have formulas that are small multiples of CH_2O. For example, *glucose* ($C_6H_{12}O_6$) is one of the most common and important sugars. Glucose is the sugar in our blood, and it provides the energy that fuels plant and animal cells.

Glucose, in turn, is the building block for the more complex carbohydrate molecules, or *polysaccharides* $(C_6H_{10}O_5)_n$, found in starches and celluloses. *Starch* is a polysaccharide used by plants to store energy; it is the main carbohydrate in the human diet because it is present in foods such as wheat, rice, corn, and potatoes. *Cellulose* contains hundreds to thousands of glucose molecules bonded together. Cellulose is the most abundant organic compound on Earth; it is what gives strength to the roots, stems, bark, and leaves of plants. It is also the material from which insects and crustaceans build hard shells of *chitin*. Cellulose is tough, fibrous, insoluble, and indigestible, except to cattle and other ruminants, which have stomachs in which specialized bacteria break down the large polymers into smaller molecules that can be digested. Cellulose is used in the manufacture of many products that incorporate fibers, such as papers and textiles.

Lipids

A fourth important family of organic molecules includes a group of compounds called **lipids**. Lipids, which are not polymers, are chemically diverse, but they are commonly grouped together because they do not dissolve in water. They include:

- *Fats* and *oils*. The hydrocarbon structures of fats and oils somewhat resemble gasoline, a similarity echoed in their function: to effectively store energy and release it when burned. This is a convenient form in which to store energy for later use, especially for mobile animals.
- *Phospholipids*. These have molecular structures that consist of one water-repellent (or *hydrophobic*) side and one water-attracting (or *hydrophilic*) side. Phospholipids arranged in a double layer make up the primary component of animal cell membranes.
- *Waxes*. Waxes are long-chain hydrocarbon molecules with a wide variety of constituents that determine their physical and chemical properties, such as their hardness and melting temperature. They can play structural roles (think of beeswax in bees' hives), and are often produced as secretions by organisms. For example, *lanolin* is a waxy secretion from glands in a sheep's skin, which helps the sheep to shed water from its wool, and it is used commercially for many purposes related to this water-repellent property.
- *Steroids*. These important chemicals are present in animal cell membranes and are used in the production of hormones, including some that are crucial in sexual maturation and reproduction.

All of these organic molecules, and many other types not detailed here, are crucial in many biologic processes, but they play other important roles in the Earth system too. For example, consider some of the differences between coal (a solid) and oil (a liquid). Coal forms from the remains of terrestrial plants that collect in a swampy environment and are buried and subsequently altered. Because it originates from plant matter, the basic building blocks of coal tend to be carbohydrates, specifically cellulose. Coal is an exceedingly complex material; it contains many other types of chemical compounds that are inherited from its plant precursors (such as resins, oils, and waxes) and generated during chemical alteration process by which it is formed. However, the abundance of plant cellulose in its source material has a significant influence on its final properties. Oil, on the other hand, forms from the remains of marine animals, so the nature and proportions of its starting materials are very different from those of coal. Both oil and natural gas are composed mainly of lighter, longer-chain hydrocarbon biopolymers (notably *naphthenes*). This difference in starting materials makes the final properties of oil and natural gas very different from those of coal.

COMPOSITION AND INTERNAL STRUCTURE OF EARTH

The rules that govern the properties of the various states of matter; the characteristics of fundamental particles such as electrons, protons, and neutrons; the formation of compounds and molecules, both organic and inorganic, by combinations of atoms; and the transformation of atoms into isotopes or ions—all are fundamental to all matter in the universe (although there are additional forms and behaviors of matter that we have not considered here). Of particular interest to us is the matter of which Earth, specifically, is composed, and how the overall composition of Earth has determined the material properties of this planet and its constituent parts.

Overall Composition and Internal Structure

Earth inherited its overall composition from its location in the solar nebula, from which the solar system formed. Earth is one of the *terrestrial planets* (Chapter 4), which means that its overall composition is predominantly rocky and metallic. Like the other terrestrial planets, Earth has internal layering that originated early in solar system history as a consequence of the chemical differentiation of the partially molten planet. The internal layers of Earth can be distinguished from one another on the basis of differences in composition, differences in rock strength, and even differences in the state of matter in the layers.

Layers of Different Composition

At the center of Earth is the densest of the three major compositional layers (FIG. 3.7): the **core** is a spherical mass composed largely of metallic iron, with smaller amounts of nickel and other elements. The thick shell of dense, rocky matter that surrounds the core is called the **mantle**. The mantle is less dense than the core but denser than the outermost layer. Above the mantle lies the thinnest and outermost layer, the **crust**, which consists of rocky matter that is less dense than mantle rock.

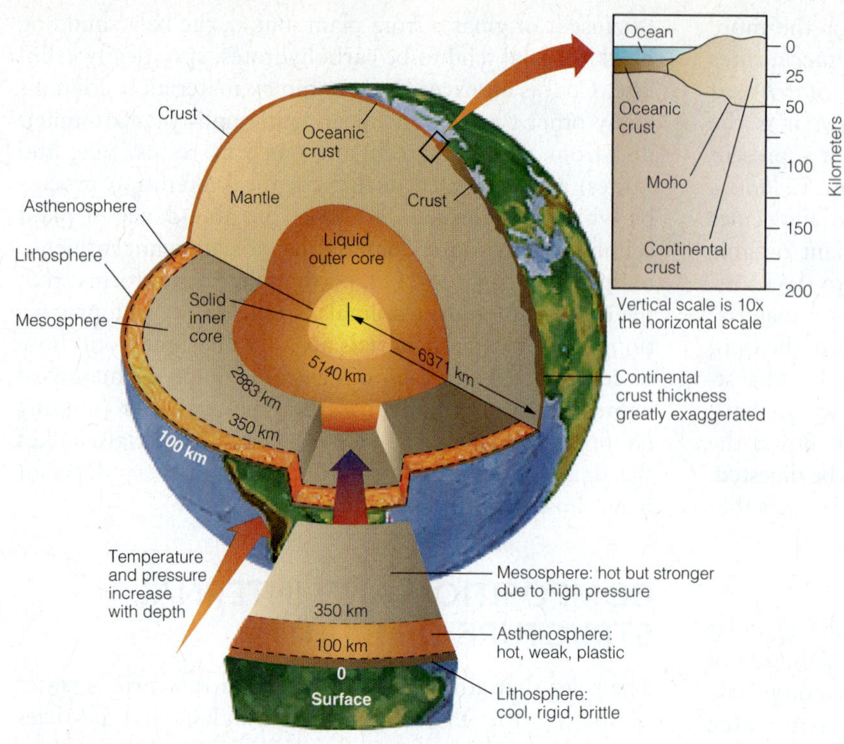

FIGURE 3.7 Earth's interior
A sliced view of Earth reveals layers of different composition and zones of different rock strength. The compositional layers, starting from the inside, are the core, the mantle, and the crust. Note that the crust is thicker under the continents than under the oceans. Note, too, that boundaries between zones of different physical properties—lithosphere, asthenosphere, mesosphere—do not coincide with the compositional boundaries.

The core and the mantle of Earth have nearly constant thicknesses, but the crust is far from uniform and differs in thickness from place to place by a factor of nine. The crust that underlies the ocean basins, the **oceanic crust**, has an average thickness of about 8 km, whereas the **continental crust** averages 45 km in thickness and ranges from 30 to 70 km. The two types of crust differ fundamentally in composition. Oceanic crust is typically composed of basalt, a type of rock that is relatively dense. Continental crust tends to be much more variable in composition, but typically consists of rocks that are much lower in density. The two different kinds of crust are the result of plate tectonic processes that continually shape and rearrange Earth's surface. The differences in their characteristics—particularly density—fundamentally affect how oceanic crust and continental crust behave in the tectonic cycle; we will discuss this in greater detail in Chapter 5.

We have direct access to Earth's surface and, therefore, can study significant parts of the crust directly. However, we cannot see and directly sample either the core or the mantle. Therefore, it is valid to ask how we know anything about their composition. The answer is that we must use a combination of scientific reasoning, indirect sampling, and indirect measurement—essentially, remote sensing. In Chapter 1 we discussed how instruments carried on satellites allow scientists to collect information about Earth's surface from a distance; similarly, instruments located on the surface can allow scientists to collect useful information about the planet's interior. For example, one way to determine composition indirectly is to measure how the densities of materials change with depth below the surface. We can do this by using a *seismometer* (Chapter 6) to detect earthquake waves. The speed at which an earthquake wave passes through the planet's interior is influenced by the density and (therefore) composition of the materials. An abrupt change in the speed of an earthquake wave indicates a change in density. By studying such changes, scientists can infer that the mantle consists of distinct layers with different densities, and can estimate the compositions of the different layers.

Sometimes samples of Earth's interior are made available to scientists at the surface indirectly—indeed, accidentally. For example, lava that originates in the mantle may rip off fragments of mantle rock and carry them to the surface during a forceful volcanic eruption. Such rocks, called *xenoliths* (literally, from the Greek, "foreign rocks") provide rare and scientifically valuable samples, and a chance to directly study an otherwise inaccessible part of the planet.

Make the CONNECTION

Volcanic eruptions bring magma and solid rocks to the surface, but they also release abundant volcanic gases from the deep interior. Volcanism has been going on since the very beginnings of Earth history. How has this affected the composition of Earth's hydrosphere and atmosphere?

The composition of the core presents the greatest difficulty. The temperatures and pressures in the core are so high that materials there probably have unusual properties. Some of the best evidence concerning core composition comes from comparisons with iron meteorites. Such meteorites are believed to be fragments from the core of a large asteroid that was chemically differentiated, like Earth, then shattered by a gigantic impact early in the history of the solar system. Evidence shows that this now-shattered object must have had compositional layers similar to those of Earth, including a metallic iron core of similar composition.

Unlike the mantle and the core, we can see and directly sample the crust. Extensive sampling has shown that the crust is quite varied in composition, its overall composition and density are very different from those of the mantle, and the boundary between them is distinct.

Layers of Different Rock Strength

In addition to compositional layering, Earth contains layers with differences in the strength of the rock making up each layer (see Fig. 3.7). The strength of a solid is controlled by both temperature and pressure, and we know that both temperature and pressure increase with depth inside the planet (Chapter 2). When a solid is heated, it loses strength; when it is compressed, it gains strength. Differences in temperature and pressure divide the mantle and crust into three distinct strength regions. In the lower part of the mantle, the rock is so highly compressed that it has considerable strength even though the temperature is very high. Thus, a region of high temperature but also relatively high rock strength exists within the mantle from the core–mantle boundary (at 2883 km depth) to a depth of about 350 km; this zone is called the **mesosphere** ("middle sphere").

Within the upper mantle, from 350 to about 100 km below the surface, is a region called the **asthenosphere** ("weak sphere"), where the balance between temperature and pressure is such that rocks have little strength. Rocks in the asthenosphere are weak and easily deformed, like butter or warm tar. As far as geologists can tell, the compositions of the mesosphere and the asthenosphere are essentially the same. The difference between them is one of physical properties; in this case, the property that differs is strength.

Above the asthenosphere, and corresponding approximately to the outermost 100 km of Earth, is a region where the rocks are much cooler, stronger, and more rigid than those in the plastic asthenosphere. This hard outer region, which includes the uppermost mantle and all of the crust, is called the **lithosphere** ("rocky sphere"). It is important to remember that it is rock strength—not rock composition—that differentiates the lithosphere from the asthenosphere.

The boundary between the lithosphere and the asthenosphere is caused by differences in the balance between temperature and pressure. Rocks in the lithosphere are strong and can be deformed or broken only with difficulty; rocks in the asthenosphere below can be easily deformed. One analogy is a sheet of ice floating on a lake. The ice is like the lithosphere, and the lake water is like the asthenosphere. As we shall see in subsequent chapters, the difference in strength between the lithosphere and the asthenosphere plays an important role in determining Earth's topography.

Layers of Different Physical State

Most of the material in Earth's interior is solid, but the metallic iron of the core exists in two physical states. The solid center is the **inner core.** Pressures in this region are so great that iron is solid despite its high temperature. Surrounding the inner core is a zone where temperature and pressure are balanced such that the iron is molten and exists as a liquid. This is the **outer core.** The difference between the inner and outer cores is not one of composition. (The compositions are believed to be the same.) Instead, the difference lies in the physical states of the two: one is a solid, the other a liquid.

Abundant Elements

Of the 92 naturally occurring chemical elements, only 12 occur in Earth's outermost layer, the crust, in amounts equal to or greater than 0.1 percent. In fact, the crust is dominated by just two chemical elements: oxygen and silicon make up more than 70 percent of continental crust, by weight. These 12 elements—oxygen, silicon, aluminum, iron, calcium, magnesium, sodium, potassium, titanium, hydrogen, manganese, and phosphorus—are called the *abundant elements*, and together they comprise 99.23 percent of the continental crust (TABLE 3.1). All of the remaining elements are referred to as *scarce elements*.

The 12 abundant elements are largely responsible for determining the composition of the common materials of Earth. For example, more than 4000 different minerals are known, but Earth's crust is overwhelmingly dominated by just 40 of them, most of which contain one or more of the 12 abundant elements. Notice, too, that with two exceptions—carbon and nitrogen—the elements essential for life are also abundant elements. These elements are oxygen, phosphorus, hydrogen, and potassium. Both carbon and nitrogen are almost abundant elements, which is fortunate, because it means that life on Earth is not limited by a lack of key life-supporting elements. In the next section we examine how the abundant and scarce elements combine to determine the compositions of common Earth materials.

Make the CONNECTION

Luckily for us, the major life-supporting elements such as carbon, nitrogen, oxygen, hydrogen, phosphorus, and potassium are relatively abundant. How might life on this planet differ if one or more of these elements was present in limited supply?

TABLE 3.1 The Most Abundant Chemical Elements in the Continental Crust

Element	Ion	Percent by Weight
Oxygen (O)	O^{2-}	45.20
Silicon (Si)	Si^{4+}	27.20
Aluminum (Al)	Al^{3+}	8.00
Iron (Fe)	Fe^{2+} and Fe^{3+}	5.80
Calcium (Ca)	Ca^{2+}	5.06
Magnesium (Mg)	Mg^{2+}	2.77
Sodium (Na)	Na^{+}	2.32
Potassium (K)	K^{+}	1.68
Titanium (Ti)	Ti^{4+}	0.86
Hydrogen (H)	H^{+}	0.14
Manganese (Mn)	Mn^{2+} and Mn^{4+}	0.10
Phosphorus (P)	P^{3+}	0.10
All other elements		0.77
	TOTAL	100.00

MINERALS

Minerals are the building blocks of which the geosphere is made. To be called a **mineral**, a substance must meet five requirements:

1. It must be *naturally formed.* This excludes the vast numbers of substances that can be produced in the laboratory but are not found in nature.
2. It must be *inorganic.* This excludes the vast number of organic compounds produced by living matter.
3. It must be a *solid.* This excludes all liquids and gases.
4. It must have a *specific chemical composition.* This excludes solidssuch as with compositions that cannot be expressed by an exact chemical formula; an example is **glass**, a non-crystalline solid with a very wide range of possible compositions. This requirement also means that minerals are either chemical compounds or chemical elements.
5. It must have a *characteristic crystal structure*, that is, an orderly internal arrangement of atoms. This excludes all liquids and gases because their atoms flow freely and are not arranged in an orderly, fixed manner. It also excludessuch as glass (again) because it has a disordered internal arrangement of atoms resembling that of a liquid.

Let's once again take diamond as an example. Diamond is a mineral. It is naturally formed, it is a solid, it is made of the chemical element carbon, and all the carbon atoms are packed together in a regular geometric array that is characteristic of diamonds. Coal resembles diamond in that it is largely made of carbon, but that's where the resemblance ends. Coal is not a mineral; it is a rock, made of bits and pieces of various solid materials stuck together. In addition to carbon, coal contains many chemical compounds, many of which are organic (biologic in origin), and its composition varies from sample to sample so that it has neither a specific composition nor a characteristic crystal structure.

Coal is not a mineral, but it is a rock. Rocks are collections or aggregates of minerals and other natural materials such as volcanic glass and organic matter. They are nature's books and in them is recorded the story of the way Earth works: how continents move, how mountains form and then erode, how ocean basins form and disappear. Minerals are the words used in nature's books. Minerals and rocks go together naturally, but before we can study rocks in greater detail, we need to know more about minerals.

Mineral Compositions and Structures

The most convenient way to study and classify minerals is on the basis of their two most important characteristics:

1. *Crystal structure*, which is the way the atoms of the elements are packed together in the mineral; and,
2. *Composition*, which is the major chemical elements that are present and their proportions in the mineral.

The ions in most solids are organized in the regular, geometric patterns of a crystal structure, like eggs in a carton, as shown in FIGURE 3.8. Solids that have such a crystal structure are said to be *crystalline*. Solids that lack a crystal structure are *amorphous* (Greek for "without form"). Glass and amber are examples of amorphous solids. All minerals are crystalline, and the crystal structure of a mineral is a unique property of that mineral. All specimens of a given mineral have identical crystal structure.

As mentioned earlier, scientists have identified more than 4000 minerals, and the number is rising slowly because a few new ones are found every year. Most occur in the rocks of the continental crust, but a few have been identified only in meteorites, and two new ones were

FIGURE 3.8 Atomic structure of a mineral
These figures show the arrangement of ions in the most common lead mineral, galena (PbS). Lead, Pb, is a cation with a charge of 2+, and sulfur, S, is an anion with a charge of 2−. To maintain a charge balance between the ions, there must be an equal number of Pb and S ions in the structure. Ions are so small that a cube of galena 1 cm on its edge contains 10^{22} ions each of lead and sulfur. (A) Ions at the surface of a galena crystal are revealed with a scanning-tunneling microscope. Sulfur ions are the larger lumps, and the smaller ones are lead. (B) The packing arrangement of ions is repeated continuously through a crystal. The ions are shown pulled apart along the black lines to demonstrate how they fit together.

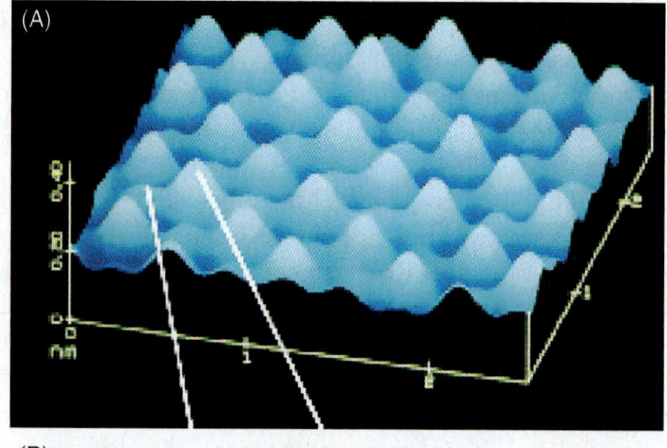

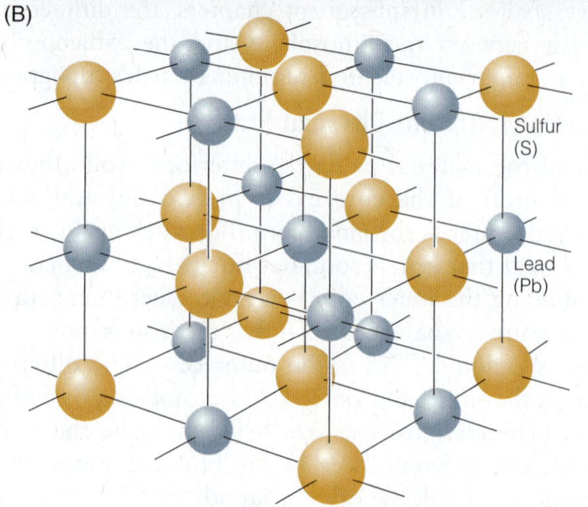

discovered in the Moon rocks brought back by the astronauts. The total number of minerals may seem large, but it is tiny compared with the astronomically large number of ways a chemist could combine the 92 naturally occurring elements to form compounds. The reason for the disparity is that the compositions of the 40 common rock-forming minerals are mostly dominated by the 12 abundant elements of the crust. Minerals containing scarce elements do occur, but only in small amounts, and they form only under special and restricted circumstances. Some scarce elements, such as hafnium, are so rare that they are not known to form minerals under any circumstances; they occur only as trace impurities in common minerals.

The Common Minerals

Because oxygen and silicon make up more than 70 percent of the continental crust, it stands to reason that these two elements will form the basis for the most common and abundant rock-forming minerals. Oxygen forms a simple anion, O^{2-}, and compounds that contain the O^{2-} anion are called *oxides*. Silicon forms a simple cation, Si^{4+}, and oxygen and silicon together form a strong complex ion, the silicate anion $(SiO_4)^{4-}$. Minerals that contain the silicate anion are complex oxides, and to distinguish them from simple oxides they are called *silicates*. The compound MgO is an oxide, but Mg_2SiO_4 is a silicate.

Silicates are the most abundant of all minerals, and simple oxides are the second most abundant group. Other mineral groups, all of them important but all less common than silicates and oxides, are *sulfides*, which contain the simple anion S^{2-}, *carbonates* $(CO_3)^{-2}$, *sulfates* $(SO_4)^{-2}$, and *phosphates* $(PO_4)^{-3}$.

Silicates

The silicate anion $(SiO_4)^{4-}$ has the shape of a tetrahedron. The four relatively large oxygen anions surround and bond to the much smaller silicon cation as shown in FIGURE 3.9. All silicates contain the silicate anion as an integral part of the crystal structure. In many silicates, however, the anions actually join together by sharing their oxygen atoms, and so they form chains, sheets, and three-dimensional networks of tetrahedra. This is polymerization (shown in FIGS. 3.10 and 3.11), discussed previously in the context of carbon atoms linking together to form

FIGURE 3.9 Silicate tetrahedron
This is the tetrahedron-shaped silicate anion, $SiO_4{}^{4-}$. (A) The anion has four oxygen atoms touching each other. A silicon atom (dashed circle) occupies the central space. (B) This exploded view shows the relatively large oxygen anions at the four corners of the tetrahedron, equidistant from the relatively small silicon cation.

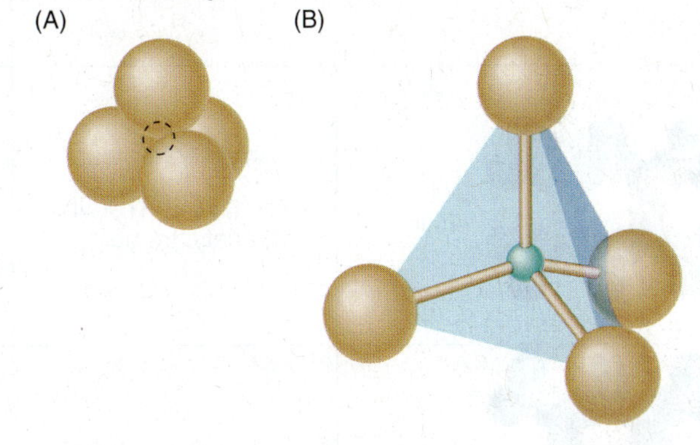

FIGURE 3.10 Silicate polymerization
Complex silicate anions form by polymerization. (A) This is a polymer chain in which each silicate anion shares two of its oxygen atoms with adjacent anions. A geometric representation of the chain is on the right. The formula of each basic unit in the chain is $(SiO_3)^{2-}$. (B) This is a double polymer chain for which the formula of the basic unit is $(Si_4O_{11})^{6-}$. Again, a simplified geometric representation is shown on the right.

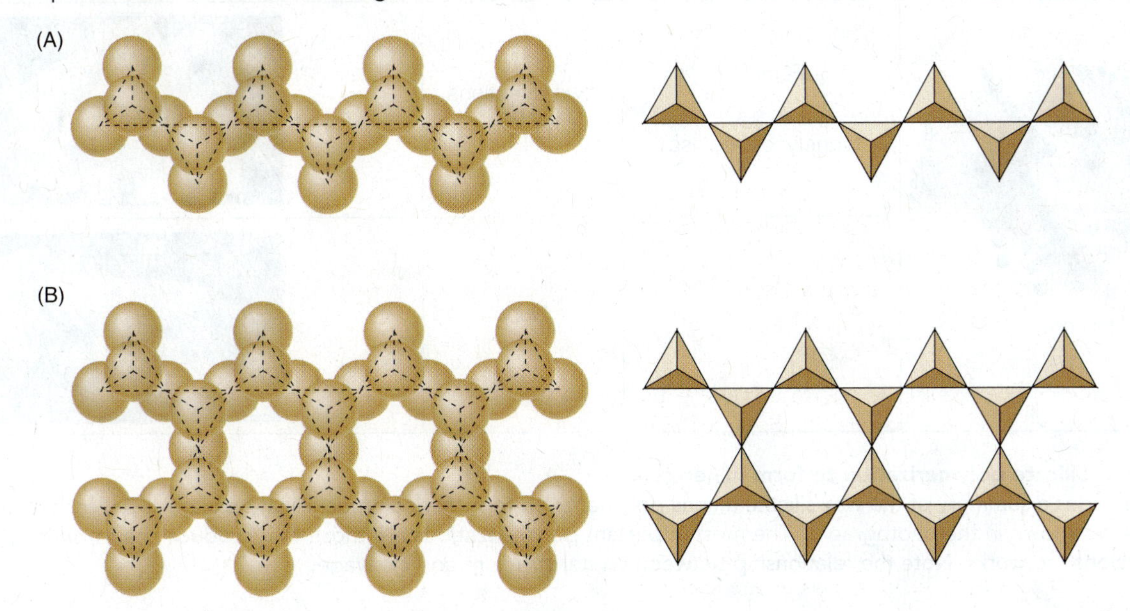

Silicate structure	Mineral/Formula	Cleavage	Example of a specimen
 Single tetrahedron	Olivine Mg_2SiO_4	None	
 Hexagonal ring	Beryl (Gem form is emerald) $Be_3Al_2Si_6O_{18}$	One direction	
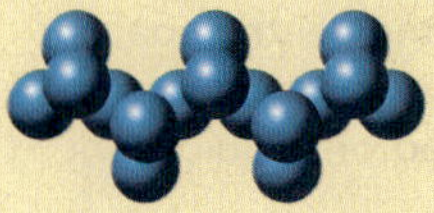 Single chain	Pyroxene group $CaMg(SiO_3)_2$ (variety: diopside)	Two directions at 90°	
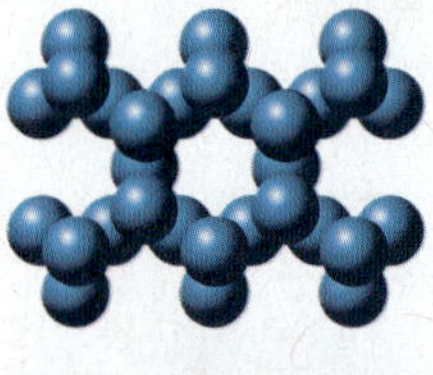 Double chain	Amphibole group $Ca_2Mg_5(Si_4O_{11})_2(OH)_2$ (variety: tremolite)	Two directions at 120°	
 Sheet	Mica $KAl_2(AlSi_3O_{10})(OH)_2$ (variety: muscovite) $K(Mg,Fe)_3(AlSi_3O_{10})(OH)_2$ (variety: biotite)	One direction	
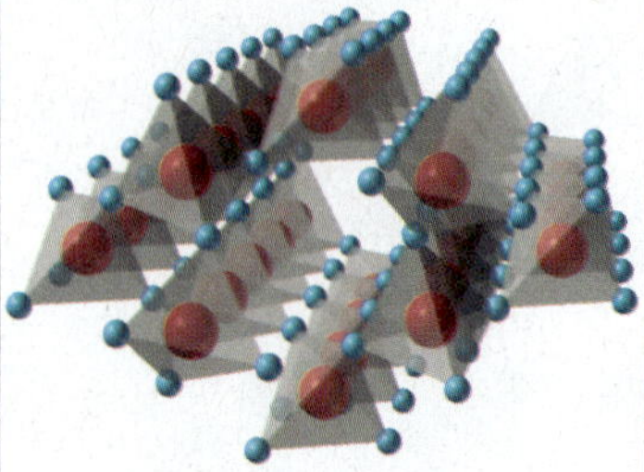 Network	Feldspar $KAlSi_3O_8$ (variety: orthoclase)	Two directions at 90°	
	Quartz SiO_2	None	

FIGURE 3.11 Silicate polymerization to form minerals
This table provides a summary of the way silicate anions polymerize to form the common silicate minerals. Typical examples of each type are shown in the photographs. The most important polymerizations are those that produce chains, sheets, and three-dimensional networks. Note the relationship between crystal structure and cleavage.

long-chain organic molecules. Polymerization plays a major role in determining the properties of silicates.

The silicates are by far the most abundant minerals in the continental crust, and among them the *feldspars* are the predominant variety—approximately 60 percent of all minerals in Earth's crust are feldspars (FIG. 3.12). Indeed, the very name reflects how common feldspars are. The name is derived from two Swedish words, *feld* (field) and *spar* (mineral). Early Swedish miners were familiar with feldspar in their mines. The miners were also farmers, and they found the same minerals in the rocks they had to clear from their fields before they could plant crops. Struck by the abundance of feldspar, the miners chose a name to indicate that their fields seemed to be growing an endless crop of the minerals.

The second most abundant mineral in the crust is the silicate called *quartz*. Together, feldspar and quartz account for 75 percent of the continental crust. All the silicates added together make up 95 percent or more of both the continental crust and the oceanic crust, and an even larger percentage of the mantle.

Nonsilicates

The nonsilicate minerals are numerous and widespread, and may at first sight seem more abundant than they actually are. Three oxides of iron—hematite (Fe_2O_3), magnetite (Fe_3O_4), and goethite ($FeO{\cdot}OH$)—are estimated to be the most abundant nonsilicates. Other important nonsilicate mineral groups are the carbonate minerals calcite ($CaCO_3$) and dolomite ($CaMg(CO_3)_2$), the sulfate mineral gypsum ($CaSO_4{\cdot}2H_2O$), and the sulfide minerals pyrite (FeS_2), sphalerite (ZnS), galena (PbS), and chalcopyrite ($CuFeS_2$). Many of the less common nonsilicates are the minerals sought by miners for the production of metals such as gold, silver, iron, copper, and zinc.

FIGURE 3.12 The most common minerals
Crystals of feldspar (green) and quartz (gray) from Pikes Peak, Colorado represent the two most common minerals in Earth's crust. This specimen is 20 cm across.

Identifying Minerals

Mineral properties are determined by composition and crystal structure. To discover its identity, it is not necessary to analyze a mineral for its chemical composition or determine its crystal structure. Once we know which properties are characteristic of which minerals, we can use those properties to identify the minerals. The properties most often used to identify minerals are crystal form, growth habit, cleavage, luster, color, hardness, and specific gravity. Appendix C lists the properties of common minerals.

Crystal Form and Growth Habit

Ice fascinated the ancient Greeks. When they saw glistening needles of ice covering the ground on a frosty morning, they were intrigued by the fact that the needles were six-sided and had smooth, planar surfaces. The ancient Greeks called ice *krystallos,* and the Romans latinized the name to *crystallum.* Eventually, the word **crystal** came to be applied to any solid body that grows with planar surfaces. The planar surfaces that bound a crystal are called *crystal faces*, and the geometric arrangement of crystal faces, called the *crystal form*, became the subject of intense study during the seventeenth century.

Seventeenth-century scientists discovered that crystal form could be used to identify minerals, but some aspects of crystal form were difficult for them to explain, such as why the size of crystal faces differs from sample to sample. Under some circumstances, a mineral may grow as a thin crystal; in other cases, the same mineral may grow as a fat one, as FIGURE 3.13 shows. Superficially, the two crystals of quartz in FIGURE 3.13A look very different, and this photograph illustrates that neither crystal size nor crystal face size is a unique property of a mineral.

The person who solved the mystery was a Danish physician, Nicolaus Steno. In 1669 Steno demonstrated that the unique property of crystals of a given mineral is not the relative face sizes, but rather the angles between the faces. These angles give each mineral a distinctive crystal form. The angle between any designated pair of crystal faces is constant, he wrote, and is the same for all specimens of a mineral, regardless of overall shape or size. Steno's discovery that interfacial angles are constant is made clear by the numbering in Figure 3.13A. The same faces occur on both crystals. All the sets of faces are parallel: face 1 on the left is parallel to face 1 on the right, face 2 is parallel to face 2, and so forth. Therefore, the angle between any two equivalent faces must be the same on both crystals.

Steno speculated that constant interfacial angles must be a result of internal order, but the ordered particles—ions—were too small for him to see. Proof of internal order was only achieved in 1912 when Max von Lauë, a German scientist, demonstrated, by use of X-rays, that

FIGURE 3.13 Crystal form and habit
Minerals have characteristic crystal structures, but these are not always manifested in the same way in the crystal form and habit of the mineral. (A) These two crystals are both quartz and have the same crystal form. Although the sizes of the individual faces differ markedly between the two crystals, each numbered face on one crystal is parallel to an equivalent face on the other crystal. (B) These quartz grains (colorless) grew in an environment where other grains prevented the development of well-formed crystal faces. The amber-colored grains are iron carbonate ($FeCO_3$). (C) Pyrite (FeS_2) characteristically forms crystals with faces at right angles and with pronounced striations. The largest crystals in this photograph are 3 cm across. The specimen is from Bingham Canyon, Utah. (D) Some minerals have distinctive growth habits, even though they do not develop well-formed crystal faces. The mineral chrysotile sometimes grows as fine, cottonlike threads that can be separated and woven into fireproof fabric. When used for this purpose, it is referred to by its industrial name, asbestos.

crystals are made up of ions packed in fixed geometric arrays, as shown in Figure 3.8.

Crystals form only when a mineral can grow freely in an open space. They are uncommon in nature because most minerals do not form in open, unobstructed spaces. Compare Figures 3.13A and 3.13B. The crystals in Figure 3.13A grew freely into an open space, and so, well-developed crystal faces were able to form. The quartz in Figure 3.13B, however, grew irregularly, without developing crystal faces, because it grew in an environment restricted by the presence of other minerals. We call such irregularly shaped mineral particles *grains*. Using X-ray techniques, it is easy to show that in both a crystal of quartz and an irregularly shaped grain of quartz, all the atoms present are packed in the same strict crystal structures. That is, both the quartz crystals and the irregular quartz grains are crystalline.

Every mineral has a characteristic crystal form. Some minerals have such distinctive forms that we can use the property as an identification tool without having to measure angles between faces. For example, the mineral pyrite (FeS_2) is commonly (but not always) found as intergrown cubes (FIG. 3.13C) with marked *striations* (lines) on the faces. Cube-shaped crystals with striated faces are a reliable way to identify pyrite.

A few minerals develop distinctive growth habits when they grow in restricted environments, and these growth habits can be used for identification. For example, FIGURE 3.13D shows asbestos, a variety of the mineral serpentine that characteristically grows as fine, elongate threads.

Cleavage

A mineral's tendency to break in preferred directions along bright, reflective planar surfaces is called *cleavage*. If you break a mineral with a hammer or drop a specimen on the floor so that it shatters, you will probably see that the broken fragments are bounded by cleavage surfaces that are smooth and planar, just like crystal faces. In exceptional cases, such as sodium chloride, which is the mineral halite (NaCl), as shown in FIGURE 3.14A, all of the breakage surfaces are smooth planar surfaces. (Don't confuse crystal faces and cleavage surfaces, however, even though the two often look alike. A cleavage surface is a breakage surface, whereas a crystal face is a growth surface.)

Many common minerals have distinctive cleavage planes. One of the most distinctive is found in mica (FIG. 3.14B and Fig. 3.11). Clay also has a distinctive cleavage; that is why it feels smooth and slippery when rubbed between the fingers.

Luster, Color, and Streak

The quality and intensity of the light reflected from a mineral produce an effect known as *luster*. Two minerals with almost identical color can have quite different lusters. The most important lusters are described as *metallic*, like that on a polished metal surface, and *nonmetallic*.

FIGURE 3.14 Cleavage
(A) The relationship between crystal structure and cleavage is shown by this mineral, halite, NaCl, which has well-defined cleavage planes and always breaks into fragments bounded by perpendicular faces. (B) The perfect cleavage of mica (variety muscovite) is illustrated by the planar flakes into which this specimen is being split.

Nonmetallic lusters are divided into *vitreous*, like that of glass; *resinous*, like that of resin; *pearly*, like that of pearl; and *greasy*, as if the surface were covered by a film of oil.

The color of a mineral, though often striking, is not a reliable means of identification. Color is determined by several factors, one of which is chemical composition, and even trace amounts of chemical impurities can produce distinctive colors.

Color in opaque minerals having a metallic luster can be particularly confusing because the color is partly a property of grain size. One way to reduce errors of judgment with regard to color is to prepare a *streak*, which is a thin layer of powdered mineral made by rubbing a specimen on a nonglazed porcelain plate, called a streak plate. The powder gives a reliable color effect because all the grains in a powder streak are very small, and so the grain-size effect is reduced. Red streak characterizes hematite (Fe_2O_3), even though the specimen looks black and metallic (FIG. 3.15).

Hardness

The term *hardness* refers to the relative resistance of a mineral to being scratched. It is a distinctive property of minerals. Hardness, like crystal form and cleavage, is governed by crystal structure and by the strength of the bonding forces that hold the atoms of the crystal together. The stronger the forces, the harder the mineral.

Relative hardness values can be assigned by determining the ease or difficulty with which one mineral will scratch another. Talc, the basic ingredient of most baby ("talcum") powder, is the softest mineral known, and diamond is the hardest. A scale called the *Mohs relative hardness scale* is divided into 10 steps, each marked by a common mineral (see Table C3.1). These steps do not represent equal intervals of hardness; rather, any mineral on the scale will scratch all other minerals on the scale that have a lower number. Minerals on the same step of the scale can only scratch each other.

FIGURE 3.15 Color and streak
Hematite and its streak show different colors. Massive hematite is opaque, has a metallic luster, and appears black. On a porcelain plate, however, this mineral gives a red streak.

A Closer LOOK

STEPS TO FOLLOW IN IDENTIFYING MINERALS

The following steps, used in conjunction with TABLE C3.1, TABLE C3.2, and Appendix C, will help you identify common minerals.

1. Decide whether the mineral has a metallic or non-metallic luster. If the mineral has a metallic luster, use the streak, hardness, and cleavage to decide which mineral it is.
2. If it has a nonmetallic luster, determine whether it is harder or softer than the blade of a pocket knife. (If harder, the mineral will scratch the blade; if softer, the blade will scratch the mineral.)
3. Once you determine hardness relative to the knife, decide whether the sample is dark or light in color. Go to the appropriate section of the table and use the cleavage data to determine which mineral you have.

TABLE C3.1 Mohs Scale of Relative Hardness[a]

		Mineral	Common Objects
Hardest	10	Diamond	
	9	Corundum	
	8	Topaz	
	7	Quartz	
	6	Potassium feldspar	
			Pocketknife; glass
	5	Apatite	
	4	Fluorite	
			Copper penny
	3	Calcite	
			Fingernail
	2	Gypsum	
Softest	1	Talc	

[a]Named for Friedrich Mohs, an Austrian mineralogist, who chose the 10 minerals of the scale.

TABLE C3.2 Reference Chart for the Identification of Common Minerals and a Guide to the Rock Types in which the minerals might be found

Metallic Luster[a]

Mineral	Streak	Rock Type[b]
Chalcopyrite	Greenish yellow	O, I, M
Galena	Lead gray	O
Hematite	Reddish brown	O, M, S
Limonite	Yellow to brown	S, W
Magnetite	Black	I, M, S
Pyrite	Brass yellow	O, M, I, S
Sphalerite	Yellow to brown	O

Nonmetallic Luster[a]

Mineral	Cleavage	Rock Type
A. Harder Than a Knife Blade		
Dark Colored		
Amphibole	Perfect, two planes at 120°	I, M
Garnet	None	M, I
Olivine	None	I
Pyroxene	Perfect, two planes at 90°	I, M
Quartz	None	I, M, O, S
Light Colored		
Feldspar	Perfect, two planes at 90°	I, M
Quartz	None	I, M, S, O
B. Softer Than a Knife Blade		
Dark Colored		
Chlorite	Perfect, one plane	M, S
Hematite	None	O, S, M
Limonite	None	W, S
Mica (var. biotite)	Perfect, one plane	I, M, S
Light Colored		
Apatite	Poor, one plane	I, M, S
Calcite	Perfect, three planes	S, M, O, I
Clay (var. kaolinite)	Perfect, one plane	W, S
Dolomite	Perfect, three planes	S, M, O
Fluorite	Perfect, four planes	O, S
Gypsum	Perfect, one plane	S, W
Halite	Perfect, three planes at 90°	S
Mica (var. muscovite)	Perfect, one plane	I, M, S, O
Talc	Perfect, one plane	M, S

[a]See Appendix C for additional properties.
[b]I = igneous, M = metamorphic, O = ore, S = sedimentary, W = weathering product.

Density and Specific Gravity

We know from everyday experience that two identical baskets have different weights when one is filled with feathers and the other with rocks. The property that causes this difference is **density**, or the average mass per unit volume. The common units of density are grams per cubic centimeter (g/cm^3).

Because density is difficult to measure accurately, we usually measure a property called specific gravity instead. *Specific gravity* is the ratio of the weight of a substance to the weight of an equal volume of pure water. Specific gravity is a ratio of two weights, and so it does not have any units. Because the density of pure water is 1 g/cm^3, the specific gravity of a mineral is numerically equal to its density.

The densities of some minerals are distinctive and can help in their identification. For example, gold has a density (or specific gravity) of 19.3 g/cm^3, which makes a handheld sample feel considerably heavier than the typical crustal rock, which has an average density of only about 2.7 g/cm^3. Galena, a lead sulfide mineral (PbS), is another example of a mineral with a distinctive density; even at just 7.6 g/cm^3, galena is noticeably heavier than most rock and mineral samples.

ROCKS

A **rock** is any naturally formed, nonliving, firm, coherent aggregate mass of solid matter that constitutes part of a planet (or asteroid, moon, or other related planetary object). Minerals are the most common building blocks of rocks; however, the word "mineral" does not appear in the definition. This is because rocks can be made of materials that are not minerals, such as natural glass (in the rock called obsidian, for example), or bits of organic matter (in the rock called coal, for example). Nevertheless, most rocks are made entirely or predominantly of minerals.

The Three Families of Rocks

There are three large families of rock, each defined by the processes by which the rocks are formed:

1. **Igneous rock** (named from the Latin *igneus,* meaning fire) is formed by the cooling and consolidation of magma. **Magma** is molten rock; it is one of the important liquid Earth materials. Magma forms deep underground. When it reaches the surface and pours out through a vent or crack, we refer to the molten rock as **lava**, and the vent through which it emerges is called a *volcano* (Chapter 6).
2. **Sedimentary rock** (from the Latin *sedimentum,* meaning a settling) is formed either by chemical precipitation of material carried in solution in sea, lake, or river water, or by the deposition of mineral particles transported in suspension by water, wind, or ice (Chapter 7).
3. **Metamorphic rock** (from the Greek *meta*, meaning change, and *morphe*, meaning form; hence, change of form) is either igneous or sedimentary rock that has been changed as a result of high temperatures, high pressures, or both (Chapter 7).

Earth's crust is mainly igneous rock or metamorphic derived from igneous. However, as FIGURE 3.16 shows, most of the rock that we actually see at Earth's surface is sedimentary. The difference arises because sediments are products of all the changes brought about by reactions between rainfall, ice, and the atmosphere, changes we

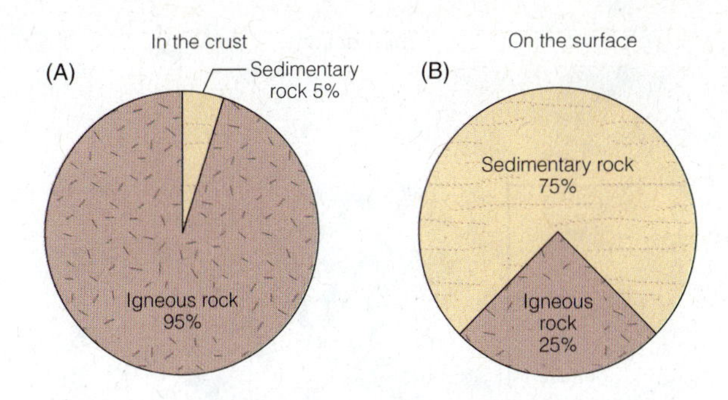

FIGURE 3.16 Rock of the crust and surface
These graphs show the relative amounts of sedimentary and igneous rock in Earth's crust and at the surface. (A) Most of the crust consists of igneous rock (95%), with sedimentary rock (5%) forming a thin veneer at the surface. (B) Because the sedimentary rock veneer covers so much of Earth's surface, it is mainly what we see. Thus, 75 percent of the surface is sedimentary rock. Igneous formations pushing through the sedimentary veneer account for the other 25 percent.

refer to as *weathering*. Sediments and sedimentary rocks are draped as a relatively thin veneer over the largely igneous and metamorphic crust below.

Features of Rocks

At first glance, rocks seem confusingly varied. Some are distinctly layered and have pronounced, flat surfaces; others are coarse and evenly grained and lack layering; yet, they may contain the same kinds of minerals present in the layered rock. In some rocks, the individual mineral grains are large and easily identifiable; in others, it is difficult or impossible to distinguish the individual grains. Studying a large number of rock specimens soon makes it clear that no matter what kind of rock is being examined—sedimentary, metamorphic, or igneous—the differences between samples can be described in terms of two main features: *texture* and *mineral assemblage.*

Texture is the overall appearance of a rock, which results from the size, shape, and arrangement of its constituent mineral grains. Commonly, examination of a microscopic texture requires the preparation of a *thin section* of rock that must be viewed through a microscope. A thin section is prepared by grinding and polishing a smooth, flat surface on a small fragment of rock. The polished surface is glued to a glass slide, and then the rock is ground away until the glued fragment is so thin that light passes through it easily. A polished surface and a thin section are shown in FIGURE 3.17.

The second feature used in describing rocks is its **mineral assemblage**, that is, the kinds and relative amounts of minerals present. A few kinds of rock contain only one mineral, but most rocks contain two or more minerals. Both the mineral assemblage and the texture of a rock reflect the conditions under which it was formed and/or later modified, so they are important in interpreting the origin and geologic history of the rock. The minerals found most commonly in the three rock families are listed in TABLE 3.2.

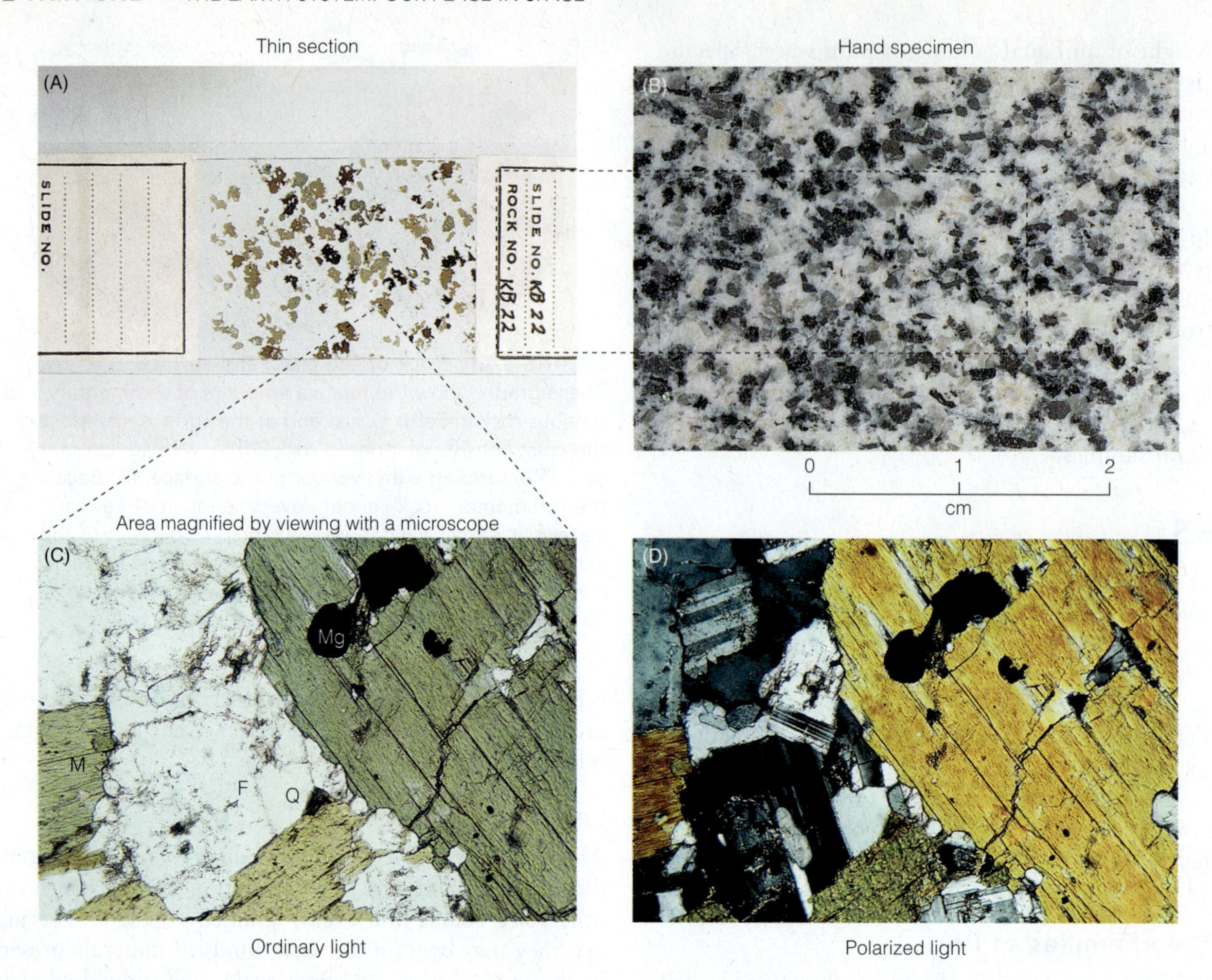

FIGURE 3.17 Thin sections of rock
Polished surfaces and thin slices reveal rock textures and mineral assemblages to great advantage. The specimen here is an igneous rock containing quartz (Q), feldspar (F), amphibole (A), mica (M), and magnetite (Mg). (A) A thin slice of rock is mounted on glass. The slice is 0.03 mm thick, and light can pass through the minerals. (B) This is the polished surface of the hand sample of rock. The dashed rectangle indicates the area used to make the thin slice shown in part (A) (C) An area of the thin slice is viewed under a microscope at a magnification of 25x. (D) The same view as in part (C) is here seen through polarizing lenses in order to emphasize the shapes and orientations of individual grains.

Basic Rock Identification

The two basic features that we use to describe rocks—texture and mineral assemblage—are the features on which scientists base most of the common approaches to classifying the vast array of rock types on this planet. To identify or classify a rock, we must first place it into one of the three families of rocks: igneous, metamorphic, or sedimentary. (We will look more closely at the details of rock classification within each of the rock families in Chapters 6 and 7.)

TABLE 3.2 Minerals Most Commonly Found in the Three Rock Families

Rock Family	Common Minerals
Igneous	Feldspar, quartz, olivine, amphibole, pyroxene, mica, magnetite
Sedimentary	Clay, chlorite, quartz, calcite, dolomite, gypsum, goethite, hematite
Metamorphic	Feldspar, quartz, mica, chlorite, garnet, amphibole, pyroxene, magnetite

Identifying Igneous Rocks

Igneous rocks can generally be identified as such on the basis of a specific textural feature—the presence of interlocking crystals—that is the result of their formation through the crystallization of magma. We differentiate between two basic kinds of igneous rocks by the sizes of their constituent mineral grains, which are controlled by the rates at which the magmas cooled and crystallized. Those that crystallize from magma under the ground are referred to as **plutonic rocks** (FIG. 3.18A); a common and familiar plutonic rock type is *granite*, which roughly approximates the composition of continental crust. Rocks that crystallize from magma that has emerged onto the surface through a volcano or volcanic vent (thereby becoming lava) are called **volcanic rocks** (FIG. 3.18B). The most common volcanic rock type on

FIGURE 3.18 The major rock families
(A) Plutonic rocks, like the granodiorite shown here, are igneous rocks that form from the solidification of magma underground. They cool slowly and therefore have time to grow individual mineral grains that are coarse enough to be visible by eye. (B) Volcanic rocks are igneous rocks that tend to be fine-grained because they cool quickly and therefore do not have time to form large crystals. Shown here is volcanic glass that solidified so quickly that it didn't grow any crystals at all, and has an amorphous (noncrystalline) structure. (C) Metamorphic rocks often show evidence of the mineralogical and textural features that result from exposure to high temperatures and pressures, as seen in the highly contorted banding of this sample of gneiss. (D) Sedimentary rocks are often identifiable by the individual accumulated grains of sediment, as shown in this sample of conglomerate. (E) Some rocks, like this limestone, contain fossils, which provide good evidence of sedimentary origin.

Earth is *basalt*, the rock of which the oceanic crust is largely composed.

The interlocking crystals in volcanic rocks tend to be very small—sometimes even indistinguishable to the eye—because they come from lava that cools extremely rapidly when exposed to the cool temperatures of the surface, and thus do not have time to grow large crystals. In some cases the cooling lava has no time to grow crystals, and so it solidifies into a disordered, amorphous solid: This is naturally occurring *volcanic glass*, or *obsidian*. (In the photograph of lava in Fig. 3.1A, for example, you can see that the edges of the liquid lava are already starting to solidify into a dark, glassy-textured volcanic rock.) In contrast, magmas that solidify underground cool slowly and have lots of time—often millions of years—in which to grow large, interlocking crystals. For this reason, it is usually possible to distinguish and identify the relatively coarse mineral grains in a plutonic igneous rock.

Identifying Metamorphic Rocks

Metamorphic rocks are those that have been chemically and/or physically altered as a result of exposure to very high temperatures and pressures inside Earth. Metamorphism happens in the solid state and does not involve melting and the formation of magma. Metamorphism is analogous to the process that occurs when a potter fires a clay pot in an oven. The tiny mineral grains in the clay undergo a series of chemical reactions as a result of the increased temperature. New compounds form, and the formerly soft clay molded by the potter becomes hard and rigid. We can often recognize metamorphic rocks, and identify them as such because they show signs of this alteration (**FIG. 3.18C**). For example, we may see evidence that the minerals originally present in the rock have been chemically altered or *recrystallized* to form new minerals, as in the example of the potter's clay. Or we may find that all of the elongate mineral grains in a particular rock have been rotated and lined up so that they are all oriented in the same direction. In other cases, flat minerals such as micas may all line up to create a platy texture, like a deck of cards. Textures of this type are the result of directional stresses that operated on the rock during the metamorphic process. They allow us to distinguish metamorphic from igneous and sedimentary rocks, both of which tend to display more even, or *equigranular* textures.

Identifying Sedimentary Rocks

Sedimentary rocks are formed near Earth's surface, usually by deposition or precipitation of mineral and rock particles called *sediment*. Typically, we can identify sedimentary rocks on the basis of textural evidence showing that they are composed of accumulations of individual particles that have become stuck together. For example, loose sediments can be transformed into sedimentary rocks by the deposition of a cementing material. Water circulating slowly through the open spaces between the loose particles deposits new materials, such as calcite, quartz, or goethite, which cement the sediment grains together (**FIG. 3.18D**). When we find individual mineral grains held together by cementing minerals, we can generally conclude that we are looking at a sedimentary rock. The presence of fossils, or broken bits of organisms (such as shells), also can reveal the sedimentary origin of a rock (**FIG. 3.18E**); these materials typically cannot survive the intense temperatures and pressures that are characteristic of the environments of formation of metamorphic and igneous rocks.

REGOLITH

When rock is exposed at Earth's surface, it is susceptible to alteration by the action of water, wind, and other agents that may act to alter the composition of the minerals present or to break the rock apart. Rocks of all kinds are physically broken up and chemically altered throughout the zone where the geosphere, hydrosphere, biosphere, and atmosphere mix. This zone extends from the ground surface downward to whatever depth air and water can penetrate—a few meters to many hundreds of meters. Rock in this zone is full of fractures, cracks, and other openings, some of which are very small, but all of which make the rock vulnerable to chemical attack. Given sufficient time, the result is conspicuous decomposition and disintegration of the rock by processes known collectively as *weathering*, which will be discussed in greater detail in Chapter 7. The product of rock weathering is a layer of broken-up, disintegrated rock matter called **regolith**.

The Surface Blanket

The broken-up material that constitutes the regolith—literally, "blanket rock" from the Greek words—forms a blanketing layer draped over most of Earth's surface. In some areas, such as in high mountains or where bedrock is exposed as a result of the recent passage of glacial ice, the regolith can be very thin or absent. In other places, such as warm, wet regions where rocks are subject to chemical alteration and disintegration, the regolith layer can be tens or even hundreds of meters deep. Regolith contains many different types of materials, including broken-up bits of plants and animals, as well as fragments of minerals and rocks, but we can describe most of the various materials of the regolith by using three categories. *Saprolite* (**FIG. 3.19A**) is rock that has been weathered and broken up, but is still *in situ*, that is, still in the place where it originally formed. The other two categories of regolith materials are sediment and soil.

Sediment

The unconsolidated (loose) rock and mineral particles that constitute the regolith are collectively referred to as **sediment** (**FIGURE 3.19B**). When sediment is picked up by wind, water, or ice, transported to another location, and deposited, it is called *alluvium*. There are two important families of sediment; the principal difference between them is the way the sediment was transported. **Clastic sediment** (from the Greek word *klastos*, meaning broken) is simply bits of broken rock and minerals that are moved as solid particles. Any individual particle of clastic

FIGURE 3.19 Regolith
Regolith consists of a variety of unconsolidated materials that blanket Earth's surface. (A) Saprolite is rock that has been weathered and broken up but is still in place. The saprolite shown here resulted from deep weathering of rock in a tropical environment. (B) Sediment is loose, unconsolidated particulate matter, regardless of its location or origin. When sediment has been picked up by wind, water, or ice, transported to another location, and deposited, it is called alluvium. This satellite image shows the Betsiboka River in Madagascar carrying a substantial quantity of sediment—the reddish-brown material—to the ocean. (C) Soil is a special type of sediment that has been altered chemically and biologically, such that it can support rooted plant life.

sediment is a *clast*, and clasts tend to be the rock-forming minerals, such as quartz and feldspar, that are most durable during weathering. **Chemical sediment** is dissolved material that is transported in the form of a solution, and is deposited when the dissolved minerals are precipitated.

Soil

Soil is a special type of regolith, which contains organic matter mixed with the mineral component. This organic content is an essential part of the definition of **soil**: the part of the regolith that can support rooted plants (FIG. 3.19C). The physical and chemical weathering of solid rock to form regolith is the initial step in soil formation. The organic matter in soil is derived from the decay of dead plants and animals. Living plants are nourished by the nutrients released from decaying organisms, as well as by the nutrients released during weathering. Plants draw these nutrients upward, in water solution, through their roots. Therefore, throughout their life cycle, plants are directly involved in the manufacture of the fertilizer that will nourish future generations of plants.

The activities of soil formation and plant growth are an integral part of a continuous cycling of nutrients through the regolith and biosphere. With its partly mineral, partly organic composition, soil forms an important bridge between the geosphere and the biosphere. We will look more closely at soils, their formation, their properties, and their classification in Chapter 15.

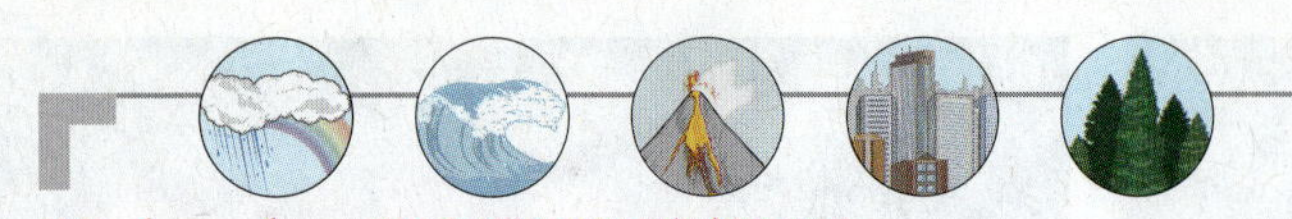

Make the CONNECTION

Soil is often described as a "bridge" between the geosphere, biosphere, hydrosphere, and atmosphere. What do you think this means? What are some of the other linkages between the geosphere and plants that we grow for food?

HOW MATTER MOVES THROUGH THE EARTH SYSTEM

In Chapter 2 you learned about the internal and external sources of energy in the Earth system. Energy drives the continuous cyclical movement of materials among Earth's four reservoirs—the geosphere, hydrosphere, atmosphere, and biosphere. The cycling and recycling of materials and the dynamic interactions among the subsystems has been going on since Earth first formed, and it continues today, even though the Earth system has changed dramatically over geologic time. Thermodynamics (Chapter 2) tells us that matter can be neither created nor destroyed (although it can change in form). As a result, the materials within the closed Earth system can cycle indefinitely among the four major reservoirs. The same is not true of energy, which also cannot be created or destroyed but *can* be changed, degraded, and eventually lost from the system. Luckily for us, energy from both outside and inside Earth is abundant, and its supply to the surface is constantly renewed. If it were not, the cycling of materials through the Earth system would cease, and so would life on this planet.

The movement of any chemical element or compound through the Earth system that involves biological activity can be described in terms of *biogeochemical cycles*, introduced in Chapter 1. Biogeochemical cycles involve biologic processes such as respiration, photosynthesis, and decomposition, as well as nonbiologic processes such as weathering, soil formation, and sedimentation in the cycling of chemical elements or compounds. In a biogeochemical cycle, living organisms act as important storage reservoirs for some elements. The cycles of nitrogen, sulfur, oxygen, carbon, and phosphorus are particularly important because each of these elements is critical for the maintenance of life. Water is also crucially important in biogeochemical cycles because of its ability to dissolve and transport a wide range of materials, and because of its central role in many biologic processes and materials.

In its most general form, a biogeochemical cycle is the complete pathway that a chemical element follows through the Earth system—from the biosphere to the atmosphere, to oceans, to sediments, soils, and rocks, and from rocks back to the atmosphere, ocean, sediments, soils, and biosphere. We will look more closely at biogeochemical cycles in Chapter 15.

Many kinds of questions can only be answered through understanding the movement of chemical elements through the biogeochemical cycles. This chapter on Earth's materials does not answer these questions, but it sets the stage for your exploration to find the answers to the following, and to other related questions in the coming chapters.

GEOLOGIC QUESTIONS

- How do geologic processes contribute to the cycling of materials through the Earth system?
- How do the properties of Earth materials influence these movements?
- What processes create mineral resources? Using this understanding, how can we better locate such resources?

HYDROLOGIC QUESTIONS

- What are some of the fundamental roles water plays in supporting life on Earth?
- How does water facilitate the movement of chemical elements through Earth's biogeochemical cycles?
- When a body of water becomes contaminated, can we adjust the biogeochemical cycle that involves the contaminant in order to reduce its level and its effects?

ATMOSPHERIC QUESTIONS

- What factors determine the concentrations of elements and compounds in the atmosphere, and by what processes are these materials exchanged with the other major subsystems?
- What is the role of the atmosphere in the movement of elements through the Earth system? Is the atmosphere more important for some biogeochemical cycles than for others?
- Where the atmosphere is polluted as the result of human activities, how does this affect the natural biogeochemical cycles?

BIOLOGIC QUESTIONS

- What factors, including the availability or absence of particular chemical elements, might promote or limit the abundance and growth of organisms?
- What are the sources of the chemical elements required for life?
- What problems occur in biologic systems when an element is too abundant or too scarce?
- How have chemical and geologic processes influenced the evolution of life on Earth through geologic time, and vice versa?

We will explore the science behind these and other related questions in subsequent chapters.

SUMMARY

1. The common states of matter on Earth are solid, liquid, and gas. Other forms and states of matter, including plasmas, gels, colloids, and aerosols, are also important in Earth processes.

2. Chemical elements are the most fundamental substances into which matter can be separated by ordinary chemical means. An atom is the smallest individual particle that retains the properties of a given element. Atoms are made of protons, neutrons, and electrons, which together determine the properties of the element. An atom that loses or gains an electron is called an ion.

3. Compounds form when one or more kinds of ion combine chemically in a specific ratio, typically through the sharing or exchange of electrons. The type and configuration of this sharing or exchange, called bonding, helps determine the properties of the compound. The smallest unit that retains the properties of a given compound is a molecule.

4. Mixtures and solutions of materials are also important in Earth processes. They are not the same as compounds because the materials in the mixture do not undergo chemical reactions and they can usually be separated from one another by physical means.

5. Organic compounds are generally of biologic origin and typically contain carbon-carbon and carbon-hydrogen bonds. Organic compounds, especially certain types of biopolymers, are extremely important in Earth processes. Four major categories of biologically important compounds are proteins, nucleic acids, carbohydrates, and lipids.

6. The overall composition of Earth is inherited from its location in the solar nebula. As a terrestrial planet, the composition of Earth is predominantly rocky and metallic.

7. Earth has internal layers that originated as a consequence of the chemical differentiation of the partially molten early planet. The layers can be distinguished from one another on the basis of differences in composition—core, mantle, crust; rock strength—mesosphere, asthenosphere, lithosphere; and state of matter—inner core (solid), outer core (liquid).

8. Just 12 elements—oxygen, silicon, aluminum, iron, calcium, magnesium, sodium, potassium, titanium, hydrogen, manganese, and phosphorus—comprise 99.23 percent of the continental crust. These are the abundant elements, which control the compositions of many common Earth materials.

9. A mineral is a naturally formed solid compound or element with a specific chemical composition and a characteristic crystal structure. Minerals are conveniently identified and classified on the basis of their physical properties, which, in turn, depend on their chemical compositions and crystal structures.

10. Oxygen and silicon form the basis for silicates, the most common and abundant rock-forming minerals. Simple oxides are the second most abundant group; other important but much less common mineral groups are sulfides, carbonates, sulfates, and phosphates.

11. A rock is any naturally formed, nonliving, firm, and coherent aggregate mass of solid matter that constitutes part of a planet. Most—but not all—rocks are made entirely or predominantly of minerals.

12. There are three families of rock: igneous rocks, which form by the cooling and consolidation of molten rock; sedimentary rocks, which form by the chemical precipitation of material carried in solution by lake, river, or ocean water, or by the deposition of particles transported in suspension by water, wind, or ice; and metamorphic rocks, which have been changed as a result of high temperatures, high pressures, or both.

13. A rock can be identified and classified on the basis of its texture (the size, shape, and arrangement of constituent mineral grains), and mineral assemblage, which reflect the conditions under which the rock was formed and/or later modified.

14. Volcanic rocks, which form as a result of the rapid cooling and solidification of lava at the surface, can be recognized by their very fine grain sizes or glassy texture. Plutonic rocks form as a result of the slow cooling and crystallization of magma underground, and therefore tend to be much coarser-grained, with interlocking crystals.

15. Metamorphic rocks often are recognizable because they show evidence of chemical and/or physical alteration due to exposure to very high temperatures and pressures, such as recrystallization or layering.

16. Sedimentary rocks form near Earth's surface by the deposition or precipitation of sediment. They can usually be identified on the basis of textural evidence of individual accumulated particles that have become stuck or cemented together. Fossils also provide strong evidence of sedimentary origins.

17. When rocks are exposed at or near Earth's surface, they become chemically altered or physically broken as a result of weathering. The layer of broken, disintegrated rock and mineral matter thus formed is called regolith. Regolith is composed of saprolite (decomposed rock *in situ*); sediment (loose, unconsolidated particulate matter); and soil (regolith that contains organic matter and can support rooted plant life).

18. There are two fundamental types of sediment: clastic, composed of bits of broken rock and mineral particles; and chemical, derived from material transported in solution and deposited when the dissolved minerals are precipitated.

19. Our modern world is totally dependent on an adequate supply of mineral, rock, and soil resources. Without them industrial society would cease to function. Experts cannot tell how long Earth's supplies of mineral resources will last because there is no way to know exactly how much of a given mineral remains.

20. Energy drives the continuous cycling of materials among Earth's reservoirs: geosphere, hydrosphere, atmosphere, and biosphere. Biogeochemical cycles are the pathways that chemical elements follow as they move through the Earth system. The key that unifies biogeochemical cycles is the involvement of the four principal components of the Earth system: rocks, air, water, and life.

IMPORTANT TERMS TO REMEMBER

abiotic 59
aerosol 56
aqueous 54
asthenosphere 63
atom 56
biopolymer 59
biotic 59
bond 58
carbohydrate 60
chemical element 56
chemical sediment 75
clastic sediment 74
colloid 56
compound 58
continental crust 62
core 61
crust 61
crystal 67
density 70
gas 54
gene 60
glass 64
hydrocarbons 59
igneous rock 71
inner core 63
inorganic 59
ion 58
isotope 57
lava 71
lipid 61
liquid 54
lithosphere 63
magma 71
mantle 61
matter 54
mesosphere 63
metamorphic rock 71
mineral 64
mineral assemblage 71
molecule 58
nucleic acid 60
oceanic crust 62
organic 59
outer core 63
phase 55
plasma 56
plutonic rock 72
polymer 59
protein 60
radioactive 57
regolith 74
rock 71
sediment 74
sedimentary rock 71
soil 75
solid 54
state 54
texture (of a rock) 71
vapor 54
volcanic rock 72

QUESTIONS FOR REVIEW

1. What is the difference between an element and an atom? Between a cation and an anion? What is a molecule, and how is a molecule different from an atom?

2. How are compounds different from mixtures and solutions?

3. Describe the compositional layering of Earth's interior. What characteristics, aside from composition, distinguish the various layers inside Earth?

4. What are four categories of important organic molecules? What are some characteristics that they have in common and some characteristics that distinguish them from each other?

5. What is a mineral? What is the difference between a rock and a mineral?

6. Approximately how many common minerals are there? Which is the most common one in Earth's crust?

7. Describe the structure of the silicate anion and how silicate anions join together to form silicate minerals.

8. Name five minerals that are not silicates, and name the anion each contains.

9. Describe the three major rock families.

10. Can a rock be uniquely defined on the basis of its mineral assemblage? If not, what additional information is needed?

11. Revisit Figure 3.22. Without looking at the text or captions, describe the features shown in each photograph that would allow you to identify that rock as belonging to one of the major rock families.

12. What is regolith? Describe three different types of regolith.

13. What are the two main types of sediment, and how do they differ from each other in their mode of formation?

14. What is soil? Why is soil such an important resource?

15. What are biogeochemical cycles? Describe some of the *bio-*, *geo-*, and *chem-* processes that the term refers to.

QUESTIONS FOR RESEARCH AND DISCUSSION

1. Within the room in which you are sitting, identify all the objects that are derived in some way from minerals. What would happen to society if all mining were stopped?

2. Name five minerals that are found in the area in which you live. Are any minerals mined in the area in which you live? What are they, and what are they mined for?

3. Would you expect other planets in the solar system to have the same kinds of minerals as on Earth? How about planets around other suns? Be sure to say why you think there may be similarities or differences.

QUESTIONS FOR *THE BASICS*

1. What is the difference between a "state" and a "phase" of matter?

2. What are the three familiar states of matter on Earth?

3. In what fundamental ways do the three common states of matter differ from one another?

QUESTIONS FOR *A CLOSER LOOK*

1. What is a crystal? How is a crystal different from a mineral grain?

2. Color is such an obvious characteristic of mineral samples; why is it not particularly reliable for the purpose of identifying minerals?

3. Explain how you would determine the hardness of a mineral in a hand sample.

CHAPTER 4

Space *and* TIME

OVERVIEW

In this chapter we:

- Examine the role of the Sun in our solar system and in supporting life on Earth
- Consider how the solar system formed and how its formation influenced Earth's characteristics as a planet
- Tour the solar system and its wide variety of objects
- Look at the evidence for planets around other suns
- Discuss how scientists determine the ages of events in the distant past

Earthlike planet?
CoRoT-7b is a rocky planet about 480 light years from Earth, shown here in an artist's rendition with its sun, the star CoRoT-7, rising in the background. Although there have now been many observations of exoplanets, or planets outside of our own solar system, CoRoT-7b is so far the most earthlike in its physical characteristics.

In Chapter 2 we considered the Sun as a provider of energy into the Earth system. We now turn our attention to the larger system of which Earth is a small part, the **solar system.** The solar system includes the group of planets, moons, and other natural objects in orbit around the Sun. The astronomical characteristics of the solar system, including the characteristics of the Sun, the positions of the planets, the shape of Earth's orbit, and its distance from the Sun, determine the properties of Earth as a planet, its materials and processes, and its ability to support life. The Earth system works the way it does because of Earth's position in the solar system, so we begin with a look at the characteristics of the Sun as a star.

THE SUN: AN ORDINARY STAR

With the exception of nearby planets and their moons, each visible point of light in the night sky—each "star"—is actually a sun or a collection of suns. A collection that consists of a billion or more suns is called a **galaxy.** On a clear, dark night, it's easy to count as many as 5000 such clusters. With an ordinary pair of binoculars, almost a million stars are visible; with the aid of the most powerful telescopes, the number of stars that can be seen rises to the billions. The number of stars and galaxies is so large that no one has ever tried to make an exact count; the best we can do is estimate. Our Sun is just one among billions and billions of stars in the Universe.

The Sun and the planets that circle it are part of a galaxy that we call the *Milky Way* (FIG. 4.1). The number of stars in the Milky Way is mind-boggling enough—estimated between 200 and 400 billion—but modern telescopes reveal an estimated 100 billion other galaxies in the universe! Most of the other galaxies are so far distant from us that they appear as single, tiny, fuzzy-looking stars through all but the largest telescopes. With those largest telescopes, astronomers can confirm that each fuzzy "star" is indeed a galaxy. Multiply the estimated number of galaxies (10^{11}) by the minimum number of stars in a typical galaxy (10^9). The result, a hundred billion billion (10^{20}), is an estimate of the minimum number of stars in the sky and therefore of the minimum number of suns in the universe.

The Sun's Vital Statistics

Like all other stars, our Sun is a massive, glowing ball of gas (or, more accurately, *plasma,* Chapter 3). In its core, as in the cores of all other glowing stars, is a huge nuclear reactor that is constantly generating vast amounts of energy, as described in Chapter 2. As stars go, the Sun is fairly ordinary, but compared to the planets and other objects in our solar system it is vastly larger. The volume of the Sun is close to 1.3 million times the volume of Earth, and its diameter is approximately 109 times that of Earth (FIG. 4.2). We say approximately because it is hard to measure the edge of a ball of gas precisely. In the Sun's case, the "edge" is generally considered to be the limit of the visible photosphere, although a transparent blanket of gas, comprising the chromosphere and the corona (Chapter 2), extends for several thousand kilometers beyond the glowing photosphere. The temperature of the photosphere is about 5800 K, but deep in the Sun's core the temperature soars to more than 15×10^6 K.

FIGURE 4.1 A galaxy like our own
This photograph by the Hubble Space Telescope shows Spiral Galaxy NGC 3949, a galaxy that is quite similar to the Milky Way in both shape and structure, located in the constellation Ursa Major about 50 million light-years from Earth. Because our solar system is embedded in the Milky Way galaxy, it is impossible to get the right perspective to photograph the large-scale features of our own galaxy.

FIGURE 4.2 Earth's Sun
The Sun is almost unimaginably large compared to Earth. This photo shows a coronal mass ejection from the Sun, which took place in October, 2003.

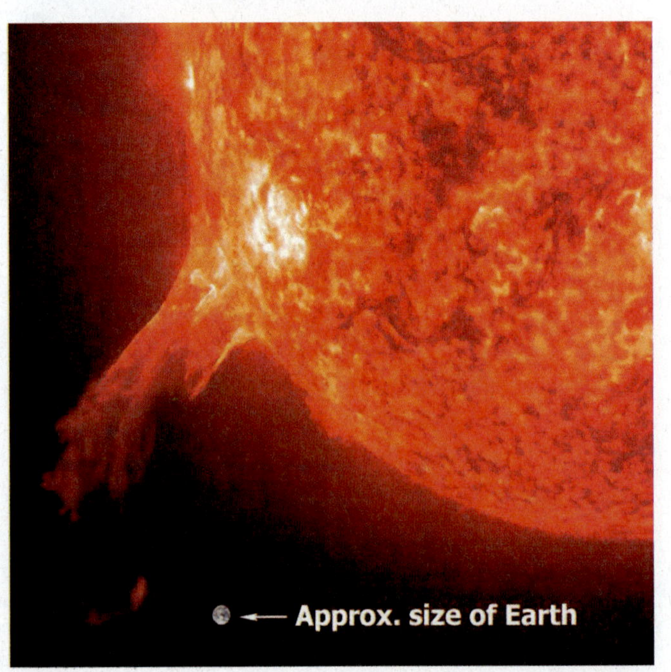

Because of its vast size, the Sun's mass, 2×10^{30} kg, is 300,000 times greater than Earth's mass. However, because the Sun is gaseous, its density is only about 1.4 g/cm^3—one-fourth that of Earth. Matter in gaseous form is still matter, though, and the Sun accounts for almost the entire mass—about 99.8 percent—of our solar system.

The Sun's composition is overwhelmingly dominated by the two lightest elements, hydrogen and helium, at more than 98 percent of its mass. This means, by extension, that the composition of the solar system also is overwhelmingly dominated by hydrogen and helium, with only traces of all of the heavier elements. Some of these heavier elements (mainly carbon, nitrogen, and oxygen) originate by nuclear fusion (nucleosynthesis, Chapter 2) in the cores of stars. Astronomers' theories about the origin of the rest of the heavy elements give us important clues about the origin of our Sun and the solar system. We will consider those clues shortly.

The average distance from Earth to the Sun is 149,600,000 km (FIG. 4.3). This distance is known as an **astronomical unit (AU)**, and it is used as a unit for measuring distances across the solar system. We talk about *average* distance because the Earth–Sun distance is constantly changing, primarily because Earth's orbit around the Sun is slightly elliptical rather than exactly circular. For comparison, consider that the Moon is only about 382,000 km from Earth. If we were to draw a sphere the size of the Sun, centered on Earth, the edge of the sphere would be about 315,000 km beyond the Moon.

FIGURE 4.3 Solar system distances
The average distance from Earth to the Sun is approximately 150 million km (defined as 1 AU). This distance varies because Earth's orbit around the Sun is not exactly circular. As shown here, the inner planets are orbiting *much* closer to the Sun than the outer planets. On the scale of this diagram, the Moon's orbit around Earth is so close that the two objects would not be distinguishable. The orbits of objects in the Asteroid Belt lie in the gap between the orbits of Mars and Jupiter.

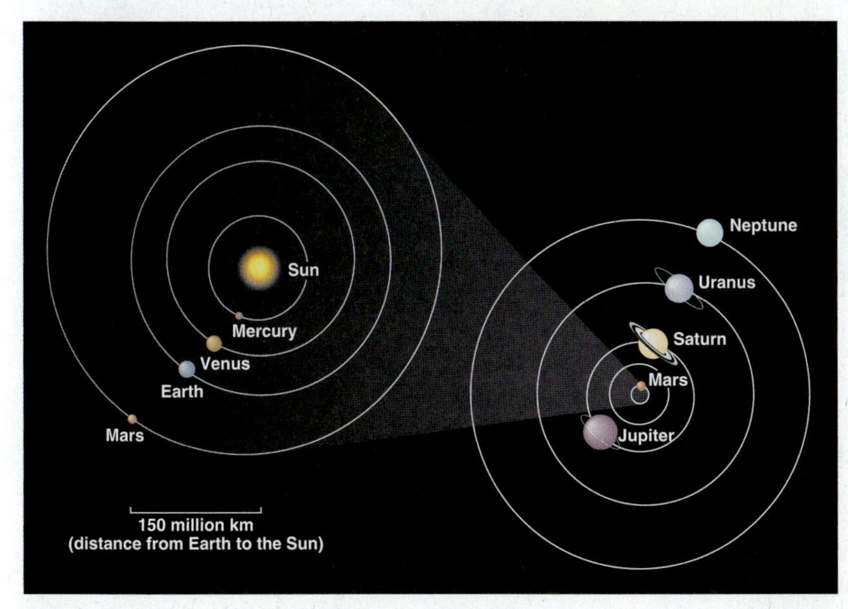

The Life-Supporting Properties of the Sun

As discussed in Chapter 2, nuclear fusion reactions in the core of the Sun convert mass into energy, accounting for the Sun's *luminosity* or energy output of 3.8×10^{26} watts. The light and heat thus generated and radiated by the Sun control Earth's climate and make Earth a habitable planet. As a result of its distance from the Sun (one AU), Earth receives just the right amount of light and heat to support life. Venus and Mercury, the two planets closest to the Sun, are too hot and too dry for life to exist. Mars is so far from the Sun that it is probably too cold and also too dry; H_2O does not exist there as water, only as ice and water vapor. Mars may once have been a warmer planet, and evidence sent back by unmanned rovers on the surface of Mars indicate that it must have had bodies of liquid water on the surface. If the evidence in favor of an earlier, more hospitable Mars turns out to be correct, forms of life may once have existed there, but it is unlikely that life exists there today.

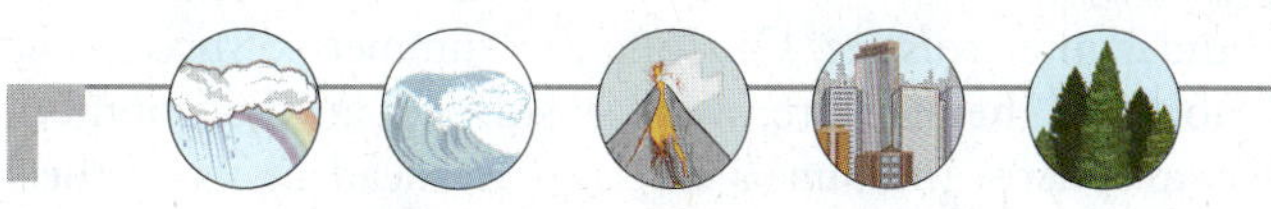

Make the CONNECTION

What would Earth be like if the amount of light and heat were slightly more or slightly less, or if the dominant energy source were something other than the Sun (such as tidal-gravitational or chemical energy)? How might this have affected the hydrosphere, atmosphere, and biosphere?

With only a few exceptions, sunlight is the source of energy that sustains life on Earth. Plants, algae, and certain bacteria utilize the Sun's energy to produce organic matter. Other organisms obtain the Sun's energy indirectly when they feed on plants, algae, or photosynthetic bacteria. No matter where we look on this planet, life and life-supporting processes are overwhelmingly dependent on energy from the Sun.

Earth's Warming Blanket

As discussed in Chapter 2, solar energy comes into the Earth system principally in the form of short-wavelength electromagnetic radiation and visible light. Earth then reradiates energy back to outer space in a longer wavelength form. Radiatively active gases in the lower part of the atmosphere (notably water vapor and carbon dioxide, CO_2) selectively absorb outgoing longer-wavelength terrestrial radiation. Thus some of the outgoing radiation in the thermal part of the electromagnetic spectrum is retained near Earth's surface, forming a "blanketing" layer of warmed air.

the BASICS

Seasons on Earth

We experience seasons because Earth's spin axis does not stand upright with respect to the plane of its orbit around the Sun, called the *ecliptic*. If Earth's spin axis were perpendicular to the plane of its orbit, the equator would always be oriented directly toward the Sun and there would be no seasons. But the axis is tilted 23.5° away from perpendicular, and the tilt remains essentially fixed as Earth moves around the Sun.

Because of the axial tilt, at one point in the orbit the northern hemisphere faces the Sun, and at another the southern hemisphere faces the Sun (**FIG. B4.1**). Our calendars have been designed around the two days each year when Earth's axis is tilted either directly toward or directly away from the Sun; they are called the summer and winter *solstice*. On June 21 (summer solstice in the northern hemisphere, winter solstice in the southern hemisphere) the Sun is directly overhead at its furthest north position, and on December 21 (summer solstice in the southern hemisphere, winter solstice in the northern hemisphere) it reaches its furthest south position. This means that the northern hemisphere gets more of the Sun's heat and illumination in June and less in December.

A second factor contributes to seasonality. Earth's orbit around the Sun is an ellipse, so that the distance between Earth and the Sun changes day by day. Earth is closest to the Sun in early January and farthest away in early July. Thus, Earth overall gets more of the Sun's rays in January, during the southern hemisphere summer, and less overall in July during the northern hemisphere summer.

Although the tilt of the axis, the rotation of Earth around its axis, and Earth's orbit around the Sun appear to be fixed, there are slight, long-term variations that lead to climate changes over timescales of tens of thousands of years. As will be discussed in Chapter 13, axial rotational and orbital changes are the major reasons that continental ice sheets advance and retreat. Earth's orbital and rotational motions are major features of the Earth system.

FIGURE B4.1 Seasons
Seasons occur on Earth because the tilt of Earth's axis keeps a constant orientation as the planet revolves around the Sun, with the result that the northern hemisphere is pointed directly toward the Sun during June and the southern hemisphere is pointed toward the Sun during December.

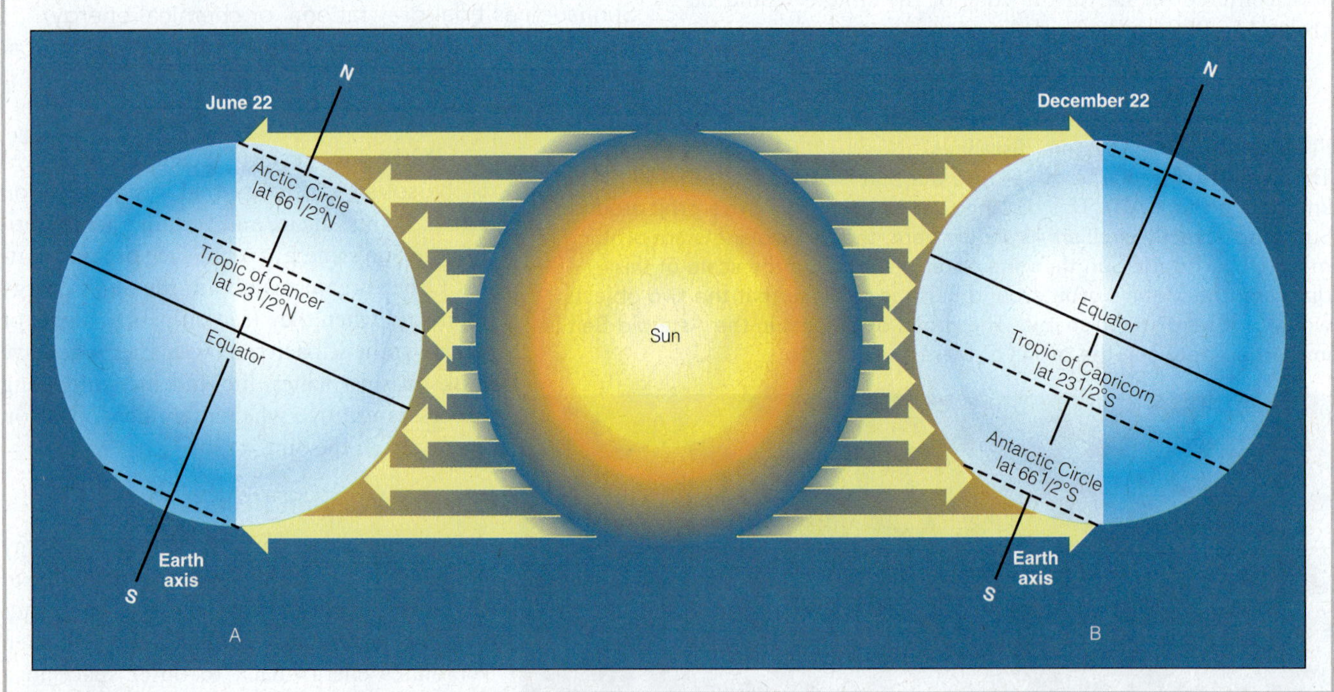

This naturally occurring "greenhouse" effect makes life possible on the surface of this planet. Without greenhouse warming of the lower portion of the atmosphere, the average temperature at Earth's surface would be a chilly −18°C—33 degrees colder than the actual average surface temperature of +15°C. Would life have managed to take hold and evolve on this planet in the absence of greenhouse warming? Perhaps; but it certainly would not have been life as we know it today. Other planets in our solar system and even some of the moons have atmospheres that provide greenhouse warming, but in each case, for one reason or another, the conditions are not ideal for the existence of life.

Interestingly, the principal gas that is responsible for the natural greenhouse effect on Earth is water vapor, ordinary H_2O. Water vapor is important in greenhouse warming both because it is highly effective at absorbing energy in the thermal part of the electromagnetic spectrum, and because it is abundant in the lower part of the atmosphere. In contrast, anthropogenic carbon dioxide (CO_2), primarily from the burning of fossil fuels, is the gas that is most often implicated in the *enhanced* greenhouse effect that may be responsible for present-day global climatic warming on Earth. We will revisit these topics in greater detail in later chapters.

THE SOLAR SYSTEM

There are eight planets in the solar system, and we live on the third planet from the Sun. Five planets—Mercury, Venus, Mars, Jupiter, and Saturn—are visible to the unaided eye, but the other two, Uranus and Neptune, are so distant they can only be seen through telescopes. Humans have never set foot on any other planet. We may do so some day, but so far the only other place we have reached with manned vehicles is the Moon. Unmanned space vehicles have visited all of the planets, landed on two—Venus and Mars—and parachuted into the dense atmosphere of another—Jupiter.

Prior to the 1960s, no one had ever actually seen Earth as a planet in its entirety; then came the first image of Earth from space. There it was, the whole strikingly blue planet in one sweeping view, the clouds, the oceans, the polar ice fields, and the continents, all at the same time and in their proper scale (FIG. 4.4). For the first time in human history it was abundantly clear just how small and isolated Earth is, and how different from the other planets in our solar system.

Tour of the Solar System

The solar system consists of the Sun, eight planets, at least five dwarf planets, a vast number of smaller rocky bodies called asteroids, millions of comets, innumerable small fragments of rock and dust called meteoroids, and at least 170 known moons (as of this writing, but the number is increasing rapidly as more discoveries are made). All of the objects in the solar system travel along trajectories that are determined by gravity. The planets, asteroids, comets, and meteoroids orbit the Sun, whereas the moons orbit the planets.

The planets of our solar system can be separated into two groups based on density and closeness to the Sun (FIG. 4.5A). The innermost planets—Mercury, Venus, Earth, and Mars—are small, rocky and metallic in composition, and dense (FIG. 4.5B). Because they are similar in composition to our Earth, they are called the **terrestrial planets** (*terra* is Latin for Earth). Compare these to Figure 4.3, which shows the relative sizes of the planets' orbits and their distances from the Sun. The orbits of the *asteroids*—rocky bodies that are too small to be called planets—fall mainly in the gap between Mars and Jupiter. Astronomers hypothesize that the asteroids are rocky fragments that failed to join together to make a larger planet.

The planets farther from the Sun than Mars are much larger than the terrestrial planets, yet much less dense because they are largely gaseous, with thick hydrogen- and helium-rich atmospheres; for this reason they are referred to as the *gas giants*. These **jovian planets**—Jupiter, Saturn, Uranus, and Neptune—take their name from *Jove*, the Roman god Jupiter. The jovian planets contain a much greater abundance of lightweight, icy, and volatile materials, compared to the terrestrial planets.

Origin of the Solar System

Before we look more closely at the characteristics of the planets and other objects of our solar system, we should step back and consider how the system came to be. For much of human history, people believed that everything in the universe revolved around Earth—for good reason. They observed the Sun rising and setting, while Earth seemingly stood still, and so they reasoned that the universe was *geocentric*, with the Sun revolving around a stationary Earth (FIG. 4.6). Today we know that Earth revolves around the Sun—a *heliocentric* model of the solar system. Any hypothesis for the origin of this system must account for as many of its features as possible.

Features to Be Explained

If you could view the solar system from afar, you would notice that almost all objects—planets, moons, and the Sun—revolve and rotate in the same direction. Furthermore, all of the planets revolve around the Sun, and all of the moons revolve around their respective planets in the same plane. From a vantage point high above the North Pole, the rotations would be counterclockwise. Moons revolve around the planets, and planets revolve around the Sun in the same direction. Furthermore, the Sun and most of the planets (Venus is an exception) rotate around their

FIGURE 4.4 Earth from space
The first photographs of Earth taken from space showed us a small, fragile, beautiful, and isolated planet full of color and life. This photo was taken in 1968 by astronauts on Apollo 8, the first human spaceflight to leave Earth's orbit.

(A)

(B)

	Mercury	Venus	Earth	Mars	Jupiter	Saturn	Uranus	Neptune
Diameter (km)	4880	12,104	12,756	6787	142,800	120,000	51,800	49,500
Mass (Earth=1)	0.055	0.815	1	0.108	317.8	95.2	14.4	17.2
Density, g/cm^3 (water=1)	5.44	5.2	5.52	3.93	1.3	0.69	1.28	1.64
Number of moons	0	0	1	2	16	18	15	8
Length of day (in Earth hours)	1416	5832	24	24.6	9.8	10.2	17.2	16.1
Period of one revolution around Sun (in Earth years)	0.24	0.62	1.00	1.88	11.86	29.5	84.0	164.9
Average distance from Sun (millions of kilometers)	58	108	150	228	778	1427	58	4497
Average distance from sun (astronomical units)	0.39	0.72	1.00	1.52	5.20	9.54	0.39	30.06

FIGURE 4.5 The solar system
(A) The eight planets of our solar system are shown here in their correct relative sizes and order outward from the Sun (but not the correct distances from the Sun). The Sun is 13 times larger in diameter than Jupiter, the largest planet. (B) This table of data about the planets shows that the terrestrial planets are much smaller as well as much denser than the outer jovian planets.

FIGURE 4.6 Celestial spheres
Aristotle pictured the Sun, the Moon, the five visible planets, and the stars as being suspended on concentric, hollow spheres that rotate about an imaginary axis extending outward from the two poles of Earth. Beyond the star sphere lay the realm of the gods.

axes in the same counterclockwise direction. Venus is an exception because it rotates in a clockwise or *retrograde* direction. (The prevalent hypothesis for this odd behavior is that Venus was knocked completely end-over-end by a collision with a large asteroid early in solar system history, which left the planet spinning slowly in the opposite direction from its original sense of rotation.)

The revolutions and rotations of the planets and moons are so regular and consistent that all societies in history have used them as a way of keeping track of the passage of time. Any hypothesis for the origin of the solar system must be able to explain the form of the solar system (a flat disk) and the remarkably consistent motions of the planets and moons, including the fact that both rotation and revolution are in the same direction, as well as in the same plane. The hypothesis must also explain the grouping of the planets into two classes—terrestrial and jovian—as well as the existence and characteristics of the great variety of other objects in our solar system.

The Solar Nebula

The origin of the Sun was probably similar to the origins of billions of other stars; thus, a lot of the ideas concerning the origin of the solar system come from observations made by astronomers studying star formation. The prevailing model for the origin of the solar system is the **nebular hypothesis**, which proposes that the Sun and its planets formed from a huge, swirling cloud of cosmic gas and dust.

Stars, including the Sun, consist largely of the two lightest chemical elements, hydrogen and helium, with only a tiny amount of heavier elements. Rocky planets like Earth, Mars, and Venus, on the other hand, consist largely of heavier elements such as carbon, oxygen, silicon, and iron, with very small amounts of hydrogen and helium. Thus the cloud from which the solar system formed had to contain not only the light elements hydrogen and helium that are found in the Sun, but also the heavier elements that make up the bulk of the rocky planets. Elements heavier than hydrogen and helium are only forged within stars; therefore, scientists believe that they must have been released into the cosmic gas cloud when an old star exploded as a supernova (FIG. 4.7).

Astronomers have discovered and photographed the scattered remains of many exploded stars, and they observe that the hydrogen, helium, and heavier elements from the star are scattered into space in a vast, slowly rotating cosmic gas cloud. Using the Hubble Space Telescope, astronomers have discovered that many young stars have rotating discs of gas around them.

We don't know whether all the atoms now in the Sun and in the planets came from the remnants of one ancient star or several, but scientists have estimated that the atoms now in the Sun and Earth were part of a cosmic cloud about 6 billion years ago. Thinly spread, the scattered atoms formed a tenuous, slowly rotating cloud of gas (FIG. 4.8A). Over a very long period of time, the gas thickened as a result of a slow regathering of the atoms. The gathering force was gravity, and as the atoms moved closer together, the gas became hotter and denser as a result of compression. Near the center of the gathering cloud of gas, the atoms eventually became so tightly compressed, and the temperature so high, that nuclear fusion started and a new star was born—our Sun. Surrounding the new sun was a flattened, rotating disc of gas and dust, and to this rotating disc the name **solar nebula** is given (FIG. 4.8B).

Condensation and Accretion

The nebular hypothesis successfully explains why the objects in the inner part of the solar system are rocky and those in the outer part contain a higher proportion of gas and ice. By the time the Sun started burning, about 4.7 billion years ago, the cooler outer portions of the solar nebula had become so compressed that solid particles and liquid droplets began to condense from the gas in the same way that ice condenses from water vapor. The condensates became the building blocks of the planets, moons, and all the other solid objects of the solar system (FIG. 4.8C).

Distance from the Sun and the condensation temperatures of materials explain why the terrestrial planets consist mainly of high-temperature rocky materials that have high densities, whereas the jovian planets and their moons consist primarily of low-temperature, low-density condensates. Planets and moons nearest the Sun, where temperatures were highest, consist mostly of substances that can only condense at high temperatures, mainly compounds containing oxygen and elements such as silicon, aluminum, calcium, iron, and magnesium. Meanwhile, a strong solar wind stripped much of the lighter gaseous material, such as helium and hydrogen, from the innermost regions. Farther away from the Sun, where the temperature was lower, sulfur-bearing compounds, water, ice, and methane ice also condensed.

Condensation of material from a cosmic gas cloud is only one piece of the planetary birth story. Condensation formed a cosmic snow of innumerable small rocky fragments. The cosmic snow revolved around the Sun in the same direction as the rotating gas cloud from which the snow particles had formed. One further step was needed in order to form the cosmic snowballs that we call planets. The step involved impacts between fragments of cosmic snow drawn together by gravitational attraction. The growth process—a gathering of more and more bits of solid matter from surrounding space—is called **planetary accretion**. Scientists estimate that condensation of the solar nebula and planetary accretion was complete by about 4.56 billion years ago, and when it was complete, the rotational motion of the ancient gas cloud had been preserved in the revolutions and rotations of the Sun, planets, and moons.

FIGURE 4.7 Supernovae and nebulae
The cloud of interstellar gas from which our solar system eventually formed probably originated in the death of a very massive star through a supernova, such as the one shown here, SN 2004dj, in a distant galaxy called NGC 2403.

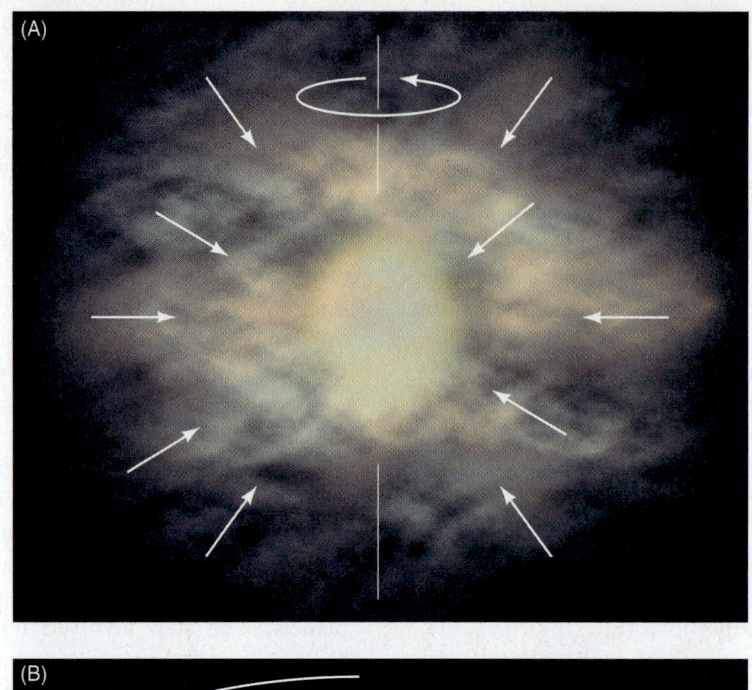

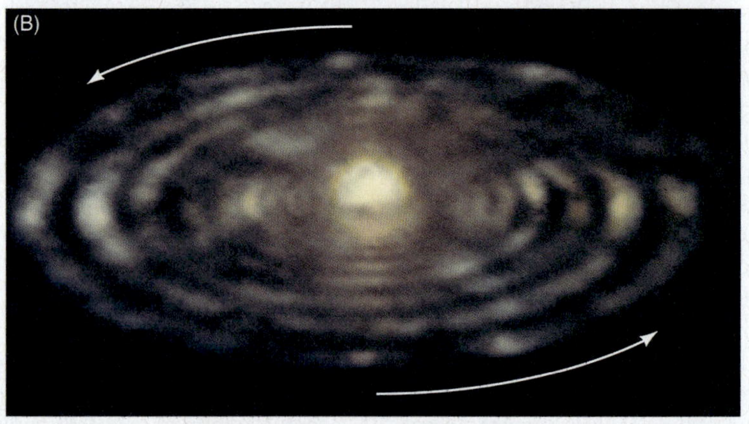

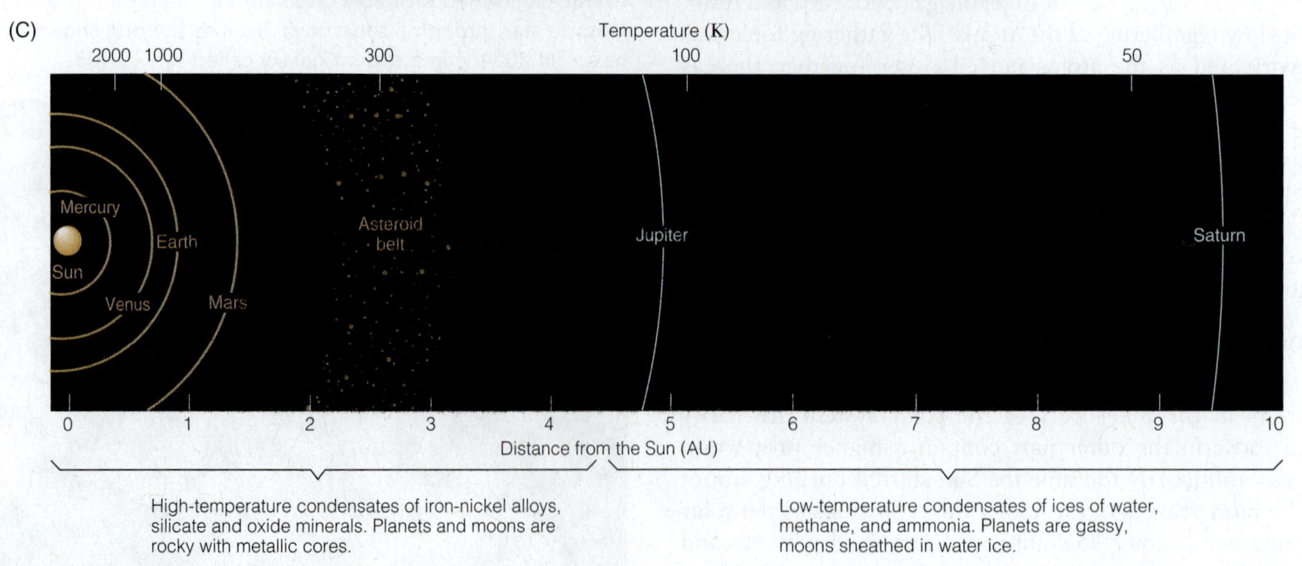

FIGURE 4.8 Nebular hypothesis
A gas cloud slowly swirling in space marks the beginning of the nebular hypothesis (A). Slowly the thinly spread gas began to collapse inward under the force of gravity; eventually, the center became so compressed and hot that nuclear fusion began and a star (our Sun) was born. Gravitational forces caused the nebula around the young star to flatten into a rotating disk (B), as materials began to condense from the cloud. The temperature gradient in the nebular disk—from very hot near the center to very cold on the outer edges—controlled the nature of the materials that condensed in various parts of the cloud, thus influencing the final compositions of the inner and outer planets (C).

Giant collisions such as the one that may have reversed the rotation of Venus were the inevitable final stage of planetary accretion, when most of the loose debris had been swept up (either attracted to the larger bodies by gravity or swept to the outer reaches of the solar system by the strong solar wind that characterizes young stars) and only larger objects remained. Signs and remnants of giant collisions from this period abound in the solar system. One such collision probably caused Uranus to tip over on its side; another may have flipped Venus completely end-over-end. Still another collision of a large object with Earth was responsible for the origin of our Moon (discussed in further detail, below). We see the scars of such impacts in the heavily pockmarked and cratered surfaces of Earth's Moon, Mercury, and other planets and natural satellites (FIG. 4.9).

Subsequent Planetary Evolution

Space missions continue to provide evidence indicating that all the objects in the solar system formed at the same time and from a single solar nebula. This evidence will be important when we attempt to become more specific about the timescales of these processes and the ages of various materials and bodies in our solar system. The nebular hypothesis for the origin of the solar system agrees well with the factors that needed to be explained, including the form and structure of the solar system, the uniform sense of rotation and revolution of solar system objects, and the distribution of material compositions among the bodies of the solar system. However, the solar system—particularly the planets—continued to evolve past the point at which we left the story of the nebular hypothesis. Five key factors played determining roles in the subsequent evolution of the terrestrial planets; let's look at each of them.

Melting, Impacts, and Differentiation. The first key factor in planetary evolution involves melting (FIG. 4.10A). When moving bodies collide, the energy of their motions (kinetic energy) is converted to heat energy. As planetary accretion approached its climax about 4.56 billion years ago, bigger and bigger collisions meant that more and more kinetic energy was converted to heat—so much so that the terrestrial planets started to melt. They probably did not melt completely, but sufficient melting occurred that the denser, iron-rich liquids sank to the center of the planets, and less dense liquids, rich in light elements essential for life, such as potassium, sodium, phosphorus, aluminum, silicon, and oxygen, floated to the surface. This process of chemical segregation is called **planetary differentiation**. During and after the phase of partial melting, the Moon and the terrestrial planets continued to be struck by remaining large fragments of cosmic debris. Such events still occur, but the time of nearly continuous, massive impacts ended more than 4 billion years ago.

Volcanism. The second key factor in the evolution of the terrestrial planets involves volcanism (FIG. 4.10B). After partial melting, the planets remained hot inside because radioactive elements were and still are present. Every time a naturally radioactive element undergoes a radioactive change, it releases some heat energy, which is an important internal source of energy. All of the terrestrial planets are cooling down, but the rates of cooling are determined by the sizes of the planets, and the cooling rates vary greatly as a result. The largest planets, Venus and Earth, are cooling very slowly and therefore are still relatively hot today. One important indication of high internal temperature is volcanism, which continues on Earth and probably on Venus. As we will discuss later in the book, volcanism has been, and continues to be, very important for life. The gases in the atmosphere, for example, escaped from inside Earth as volcanic emissions.

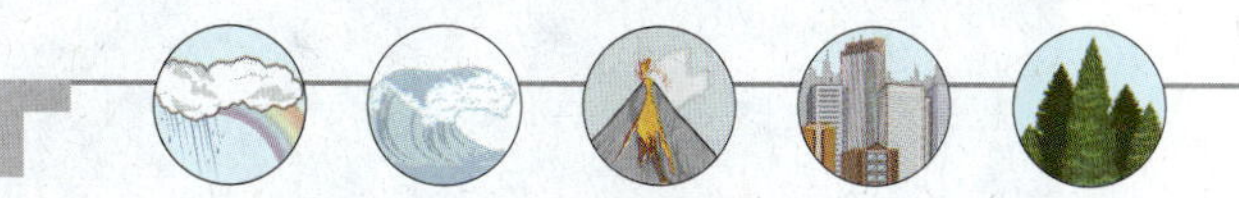

Make the CONNECTION

Summarize the ways in which Earth's hot interior changes life on the surface. For example, it causes volcanism; how has volcanism changed the Earth system over time? You will add to your understanding of these influences in subsequent chapters.

Planetary Mass. The mass (both the size and density of the materials that comprise the planet) is obviously important in determining things such as the orbit of a planet and how many moons it captures. For our purposes the mass of a planet has an even more important role: It determines whether the planet has sufficient gravitational pull to "hold onto" the gaseous envelope that surrounds it. In the case of Earth, as you will learn, the original gases

FIGURE 4.9 **Scars of a violent past**
The heavily crated surface of Mercury, photographed by the Messenger spacecraft in 2008, displays the scars of violent impacts from early in the history of the solar system.

(A)

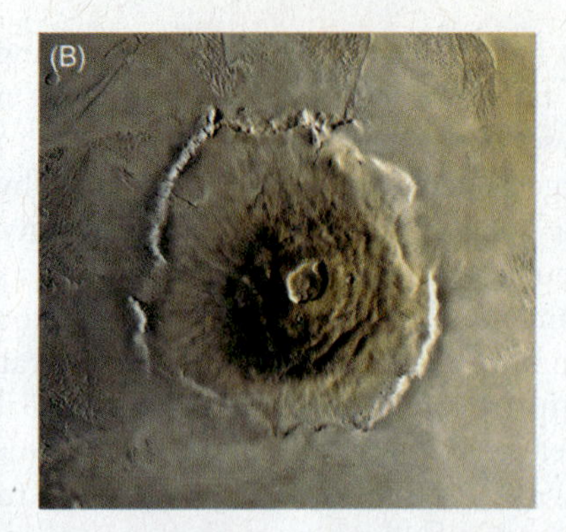

FIGURE 4.10 **Key factors in planetary evolution**
The five key factors that influenced planetary evolution subsequent to the earliest formation of the solar system were (A) impact cratering and resultant partial melting, as seen here in an artist's conception of a major impact event early in Earth history; (B) volcanism, as seen here in a photograph of the Martian volcano Olympus Mons—the largest known volcano in the solar system; (C) planetary mass, which determines the orbital characteristics of the planet as well as its ability to retain an atmosphere; Mercury (left), shown here in comparison to Earth, is too small and has insufficient gravitational pull to retain an atmosphere; (D) distance from the Sun, as seen here in an image of 55 Cancri, a star with a family of five known planets, one of which (depicted in blue) is the correct distance from the star to potentially support liquid water; and (E) the development of a biosphere and its subsequent influence on the chemical evolution of the atmosphere, hydrosphere, and regolith. Photosynthetic organisms like this filamentous alga (*Cladophora*) have greatly influenced the composition of the atmosphere over geologic time by contributing oxygen.

that surrounded the planet at its formation were stripped away. Luckily for us, Earth was just massive enough to hold onto the volcanic gases that were subsequently released from its interior, giving us an atmosphere. Mercury, in contrast, is an example of a planet that is so small that its gravitational pull is insufficient to allow it to retain an atmosphere (FIG. 4.10C).

DISTANCE FROM THE SUN. The fourth factor that controlled the way the planets evolved is their distance from or proximity to the Sun (FIG. 4.10D). The Sun–planet distance determines whether or not H_2O can exist as liquid water, and hence whether or not there can be oceans. Liquid water also is an essential condition for life. The two planets closest to the Sun—Mercury and Venus—are too hot for liquid water to exist. Venus does have water vapor in its atmosphere, but the temperature at the surface of Venus is close to 500°C. Mars, which is farther from the Sun than is Earth, is too cold to have liquid water but does have water ice.

BIOSPHERE. The fifth factor determining the evolutionary paths of the planets has been the presence or absence of a biosphere (FIG. 4.10E). The hydrosphere and the biosphere have played essential roles in the development of the biogeochemical cycles that control the composition of Earth's atmosphere. On Earth, plants and microorganisms have been the means whereby carbon dioxide and water have been combined, through photosynthesis, to make solid organic matter and oxygen gas. The burial of organic matter in sediment removes carbon from the atmosphere and at the same time adds a balancing amount of oxygen to the atmosphere. Because life did not develop on Venus, there is no buried organic matter and all of the CO_2 is still in the atmosphere.

Now that you have an understanding of the processes that led to the formation and subsequent evolution of the solar system, let's look a bit more closely at the characteristics of the planets and other important solar system objects.

The Terrestrial Planets Today

One of the reasons that the accretion and subsequent evolutionary history of the planets is so important is that it explains the different layers found in Earth and the other terrestrial planets. As you learned in Chapter 3, during partial melting early in solar system history Earth differentiated into three layers: a relatively thin, low-density, rocky *crust*; a rocky, intermediate-density *mantle*; and a metallic, high-density *core* (as shown in Fig. 3.7). Earth's crustal rocks, soils, atmosphere, and oceans, are all made from mixtures of the chemical elements that came to the surface during this early period of chemical differentiation. Similar layers are present in Mercury, Venus, Mars, and our Moon, and even in the largest asteroids, although each has different proportional sizes and compositions of layering. The most remarkable is Mercury, with a core that makes up 42 percent of its volume and an estimated 80 percent of the mass.

We not do know if any of the terrestrial planets besides Earth have molten or partially molten cores. The molten outer core and the relatively rapid rotation of Earth give rise to Earth's strong magnetic field. (According to the *dynamo theory*, Chapter 5, motion in a conducting fluid, such as the molten iron outer core, can generate a magnetic field that is self-sustaining.) The other terrestrial planets do have magnetic fields, but they are much weaker than Earth's magnetic field.

There are other important similarities among the terrestrial planets. All of them have experienced volcanic activity, which means that they either have, or have had, an internal heat source. The volcanism of terrestrial planets is dominated by the formation of the volcanic rock *basalt*—it is the principal rock of Earth's oceanic crust, and it forms most of the surface of Venus, Mars, and the dark-colored "seas" on Earth's Moon. Venus and Earth continue to be volcanically active today. Mars was volcanically active for billions of years, as shown by the existence on Mars of the largest volcano known, Olympus Mons; however, it appears that it is no longer active. The Moon and Mercury, smallest of the terrestrial planetary objects, have been volcanically dead for billions of years.

All of the terrestrial planets also have been through intense collisions and impact cratering. This bombardment was heaviest during the waning period of formation of the solar system, but collisions and the resultant cratering still occasionally occur even today. The scars of these impacts are clearly visible on the heavily cratered surfaces of the Moon and Mercury (FIG. 4.11), but they have been well hidden or erased on the more geologically active planets, particularly Venus and Earth, by the combined action of weathering, erosion, and volcanism.

One additional feature that appears to be unique to Earth is plate tectonic activity. Observations of planets and their moons reveal that rocks on the surface of each terrestrial planet fracture and deform as they do on Earth, which suggests that the terrestrial planets have *lithospheres*. What little evidence we have suggests that *asthenospheres* and lithospheres probably are present in each terrestrial planet, but that the asthenosphere of Earth is unusually near the surface and the lithosphere is unusually thin. Very likely, Earth is such a dynamic planet because its lithosphere is so thin. The other terrestrial planets seem to have much thicker lithospheres and to be much less dynamic than Earth.

Make the CONNECTION

More than any other characteristic, what makes Earth unique is the dynamic nature of its surface. See if you can list some processes in the geosphere, atmosphere, hydrosphere, and biosphere that would erase the scars of a meteorite impact on Earth's surface. You will add to your understanding of these processes in subsequent chapters.

Finally, all of the terrestrial planets have lost their **primordial** or **primary atmospheres**, the original envelopes of hydrogen and helium with which they were surrounded early in the history of the solar system. Earth, Mars, and Venus have evolved new **secondary atmospheres**—envelopes of gaseous volatile elements that leaked from their interiors via volcanoes and were trapped by the planets' gravity. Mercury and the Moon are too cold and too small to have held on to these gases by gravity, so they lack atmospheres.

The Jovian Planets Today

The outer planets are shrouded by thick, impenetrable atmospheres; therefore, scientists' understanding of their evolution is based on inference and deduction rather than on direct observation of their surfaces. However, it is reasonable to assume that the early histories of the outer

FIGURE 4.11 Impacts on Earth
Although all of the planets experienced intense meteorite bombardment early in solar system history, the scars of these impacts on Earth have been hidden or erased by the subsequent action of weathering, erosion, and plate tectonics. Here, however, is a crater caused by an impact at Meteor Crater in Flagstaff, Arizona, just 50,000 years ago. The crater is 1.2 km in diameter and 200 m deep.

planets were similar to those of the terrestrial planets and that inside each of the outer planets there is a core of rocky material similar to a terrestrial planet.

We cannot see anything that lies below the thick blankets of atmosphere that cover the jovian planets, but we can work out their internal structures based on remote-sensing measurements of the planets and their moons. For example, we can calculate that the masses of Jupiter and Saturn are so great that none of their atmospheric gases has been able to escape their gravitational pulls (FIG. 4.12A). This is true even for the two lightest gases, hydrogen and helium, which made up the bulk of the planetary nebula. This means that the two largest jovian planets must have bulk compositions that are about the same as the composition of the solar nebula from which they formed. The composition of Jupiter is estimated to be 74 percent hydrogen, 24 percent helium, and 2 percent heavy elements—very similar to the composition of the Sun.

Huge storm systems, high-speed winds, and even lightning systems are common in the deep, turbulently flowing atmospheres of the gas giants. An example of a particularly long-lived storm is the so-called Great Red Spot of Jupiter (FIG. 4.12B), an anticyclonic storm twice the size of Earth, which is thought to have persisted for more than 300 years.

All four of the gas giants probably have small rocky cores inside their vast, thick atmospheres. The rocky cores of Jupiter and Saturn may be as large as 20 or more Earth masses. Surrounding the rocky cores is possibly a layer of ice, analogous to the ice sheaths seen on many of the moons of the outer planets (FIG. 4.13). Pressures inside the jovian planets must be enormous, so that deep in the interiors hydrogen may be so tightly squeezed that it is condensed to a liquid.

Still deeper inside Jupiter and Saturn, pressures equivalent to 3 million times the pressure at the surface of Earth are reached. Under such conditions, the electrons and protons of hydrogen become less closely linked and hydrogen becomes metallic (FIG. 4.14). In Jupiter it is possible that pressures may even reach values high enough for solid metallic hydrogen to form a sheath around the ice core. The presence of metallic hydrogen within Jupiter and Saturn is further supported by the fact that both planets have magnetic fields, which suggests (as in Earth) the presence of a moving, electrically conducting fluid near the core. (Note, however, that in Earth the fluid responsible for generating the magnetic field is iron; in Jupiter and Saturn, it is metallic hydrogen.) Neptune and Uranus are thought to be similar to Jupiter and Saturn, although neither is large enough for pressures to be high enough to form metallic hydrogen.

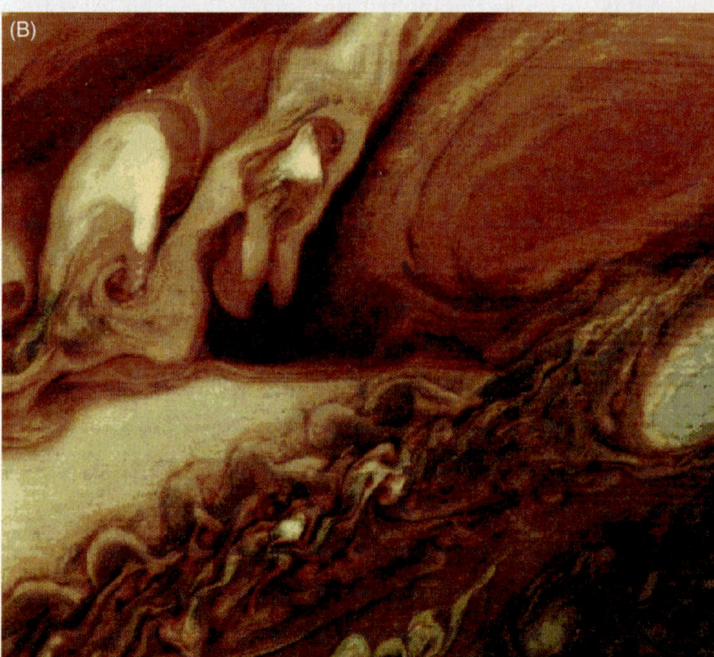

FIGURE 4.12 Jupiter, gas giant
Jupiter is the largest of the gas giants. It is seen here (A) with a shadow (small dark spot) cast by Europa, one of its more than 60 moons. Like the other gas giants, the bulk composition of Jupiter is approximately the same as the composition of the solar nebula from which it formed. Jupiter is estimated to be 74 percent hydrogen, 24 percent helium, and 2 percent heavy elements. Storms are common in the turbulent atmospheres of the gas giants (B) Visible in this photograph is the famed Great Red Spot on Jupiter, twice as wide as Earth, an anticyclonic storm that is thought to have been active for over 300 years.

Other Solar System Objects

The solar system contains an enormous number and variety of objects. One of the strengths of the nebular hypothesis is that it accounts for much of this variety. Let's briefly consider some of the more important subplanetary objects of our solar system.

Moons

The definition of "moon" is somewhat contentious; for the most part, any natural object in a regular orbit around a planet can be called a **moon** or *natural satellite*. There has been some discussion about the size required for an object to be classified as a moon, but no formal constraints have

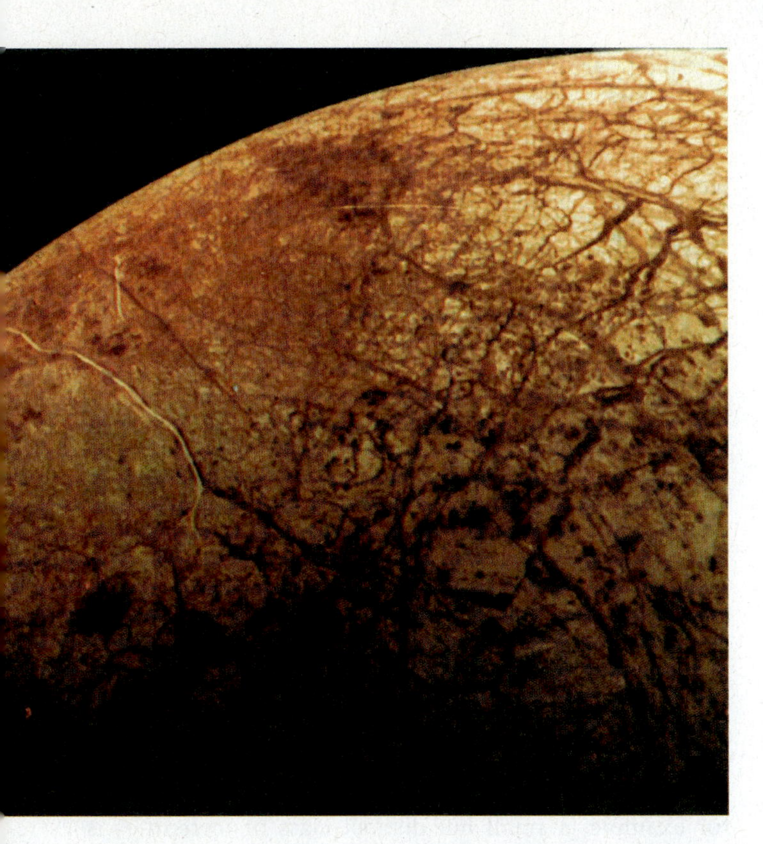

FIGURE 4.13 Europa, a moon of Jupiter
Europa, the smallest of the four large moons of Jupiter, has a low density, indicating that it contains a substantial amount of ice. The surface is mantled by ice to a depth of 100 km.

been defined. Hence, objects as small as 1 km in diameter (S/2003 J 12, in the Jupiter family) and as large as 5268 km (Ganymede, one of the four largest moons of Jupiter) have been called moons. Objects smaller than 1 km in diameter that meet the other criteria of natural satellites may be called *moonlets*. The majority of the known moons in the solar system belong to the Jupiter family (63 of them), followed closely by the 61 moons of Saturn.

Only the largest of the known moons—19 of them, to be precise—are large enough to be approximately spherical in shape; the smaller moons range from almost spherical to extremely irregular. Gravity tends to pull planetary objects into a spherical shape, a condition known as *hydrostatic equilibrium*. However, for this to happen the object must have sufficient mass for its own gravity to be strong enough to overcome the rigidity of its composite materials.

At least one moon (Titan, the largest moon of Saturn) has a substantial atmosphere. Here, too, gravity plays a role. To retain an atmospheric envelope of gases, a moon or planet must be both massive (to have a strong gravitational pull) and cold (so the gases are not moving vigorously). If these conditions are not met, the gases will escape the pull of the object and float away into space, like the atmospheres of the Moon and Mercury. Titan is of particular interest to scientists because conditions at its surface may be very similar in composition to those on Earth when primitive organisms first appeared about 3.6 billion years ago.

There are several possible modes of origin for the various moons in our solar system. Many of these natural satellites, especially those whose orbits lie very close to their parent planets, are thought to have coalesced locally from the same mass material within the solar nebula that gave rise to the parent planet. Some, such as the two small Martian moons Phobos and Deimos, may have been captured gravitationally as they strayed too close to the planet. Still others are thought to be the fragmentary remnants of great collisions that happened early in the history of the solar system.

Earth's Moon

At 3476 km in diameter, Earth's Moon (also known by its Latin name, *Luna*) is not the largest moon in the solar system. However, the Moon is approximately one-fourth the size of Earth, which makes it the largest natural satellite in comparison to the size of its parent planet (with the exception of the Charon–Pluto dwarf planet system). This suggests that the origin of the Moon may differ from the origins of other moons, many of which were likely "stray" objects captured by the gravitational pull of their much larger parent planets. Most scientists concur that the Moon's origin was much more violent.

It was only after the Moon landings in 1969–1972 that scientists began to grasp the great importance of violent collisions in the history of the solar system. Every crater on the Moon, as far as we know, was formed by an impact. (The astronauts looked for volcanic craters too but never found a single one.) Scientists recognize impact craters on

FIGURE 4.14 Jupiter's interior
Like the other gas giants, Jupiter probably has a relatively small core of rock that may be mantled by ice. Pressures in the interior of these massive planets are so intense that hydrogen may even exist in a solid state.

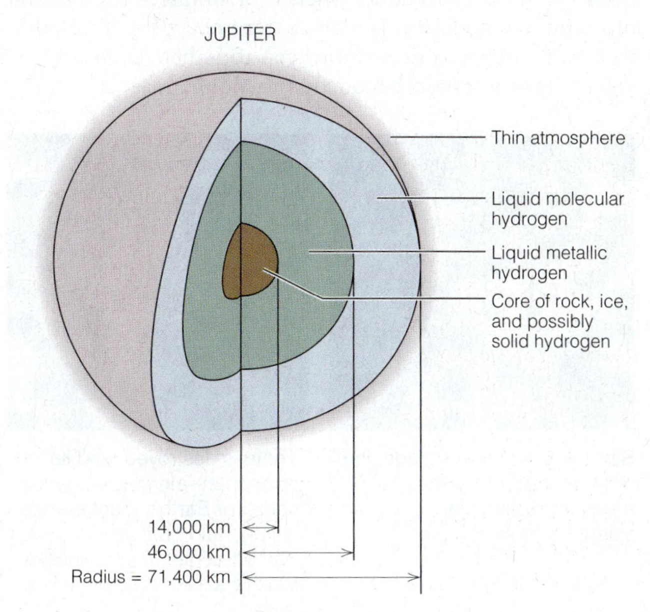

Earth, too, although they are harder to find because erosion and other processes cover them up or erase them.

Most planetary scientists now conclude that Earth collided with another planetary body, roughly the same size as Mars (FIG. 4.15), about 4.5 billion years ago, and they have given that object the name Theia. The impact with Theia was responsible for tilting Earth's axis of rotation at an angle of 23.5 degrees to the plane of its orbit around the Sun. This tilt explains why we have seasons (as discussed earlier in this chapter in *The Basics: Seasons on Earth*). The impact must also have melted most of Earth's surface due to the tremendous amount of energy released. (At the hyper-speeds typical of cosmic impacts, every ton of impactor strikes Earth with an energy equivalent to 100 tons on dynamite.) The collision completely destroyed the Theia and blasted so much debris into orbit that for a while Earth had rings much denser than Saturn's. Eventually the rings of debris accreted, forming the Moon. This hypothesis explains the existence of a magma ocean early in lunar history (shown by rocks retrieved from the Moon). It also explains our Moon's relatively large size in contrast to other moons, which are many times smaller than their parent bodies, and accounts for certain chemical discrepancies and similarities between Earth and its Moon.

Asteroids and Meteorites

Asteroids are subplanetary objects orbiting the Sun. They are commonly rocky and/or metallic, and can range from a few hundred meters to several hundred kilometers across. The orbits of the asteroids are concentrated in orbits that lie between the orbits of Mars and Jupiter, in the *Asteroid Belt* (FIG. 4.16). At 950 km diameter, Ceres was the largest of the asteroids but is now technically classified as a dwarf planet (see below). Most asteroids are objects or fragments of objects that appear to have undergone differentiation, like the terrestrial planets. The study of asteroid fragments that have fallen to Earth has been extremely helpful to scientists in unraveling the history and timescale of formation of objects in the near-Earth portion of the solar system.

According to the planetary accretion hypothesis, the planets assembled themselves from rocky, metallic, and icy debris 4.56 billion years ago, shortly after the Sun itself was formed. Today some of the debris that has never been swept up by any planet still exists in the form of *meteoroids*—small, solid objects (generally from less than 1 km down to dust-sized particles) orbiting the Sun. Sometimes pieces of this debris fall to Earth as **meteorites.** When small meteorites pass through Earth's atmosphere, they typically burn up and incandesce as a result of friction. The glowing streak that results is the well-known "falling star," technically called a *meteor*.

Many—though not all—meteoroids and meteorites are related in origin to the asteroids of the main Asteroid Belt. Others are fragments of comets that may have broken up as a result of a passage too close to the Sun. Still others are the fragmentary remnants of giant impacts; for example, a small but distinct class of meteorites is thought to have originated from an impact that broke fragments of material off of the surface of Mars. Another class of meteorites, the *carbonaceous chondrites*, consists of objects that are thought to be among the oldest and most primitive materials in our solar system, virtually unaltered by geologic processes since the time of their formation. Whatever their particular origin, meteorites are fascinating relics of the early days of the solar system.

FIGURE 4.15 Formation of Earth's Moon
(A) Impact: Some 4550 million years ago, a still-forming Earth runs into another growing planet, which scientists have dubbed Theia. (B) Impact + 8 hours: Theia is obliterated, and its remnants—along with a good chunk of Earth's mantle—are blasted into orbit around Earth. The off-center impact knocks Earth's axis of rotation askew. (C) Impact + 24 hours: The debris spreads itself into a ring and begins to clump together. (D) Impact + 1 year: The largest clump starts to attract other fragments and is well on its way toward becoming the Moon.

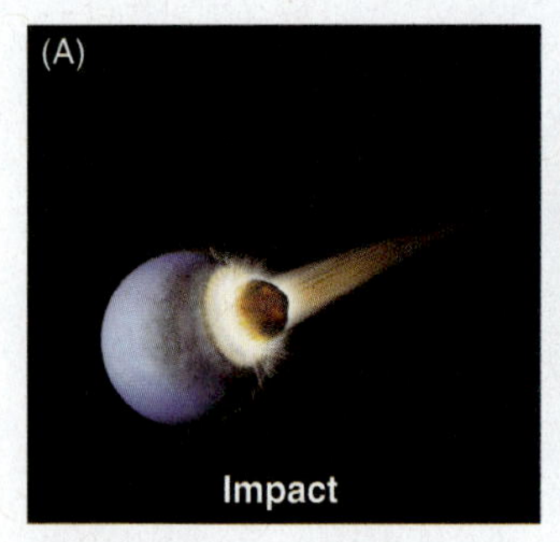

Some 4.5 billion years ago, the still-forming Earth runs into another growing planet, which scientists have dubbed Theia.

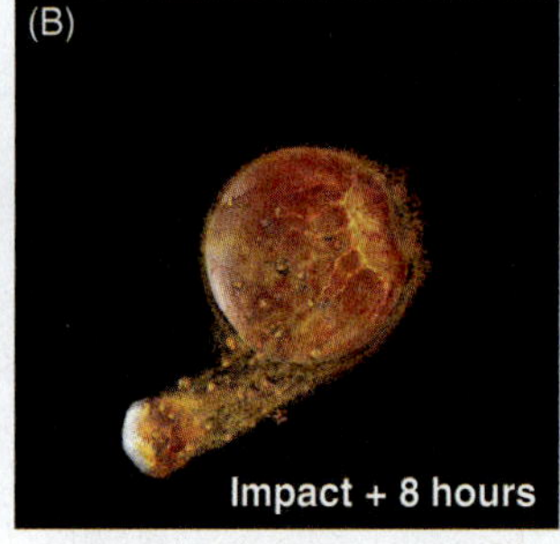

Theia is destroyed, and its remnants—along with a good chunk of Earth's mantle—are blasted into orbit around Earth. The off-center impact knocks Earth's axis of rotation askew.

The debris spreads itself into a ring and begins to clump together.

The largest clump starts to attract other fragments and is well on its way toward becoming the Moon.

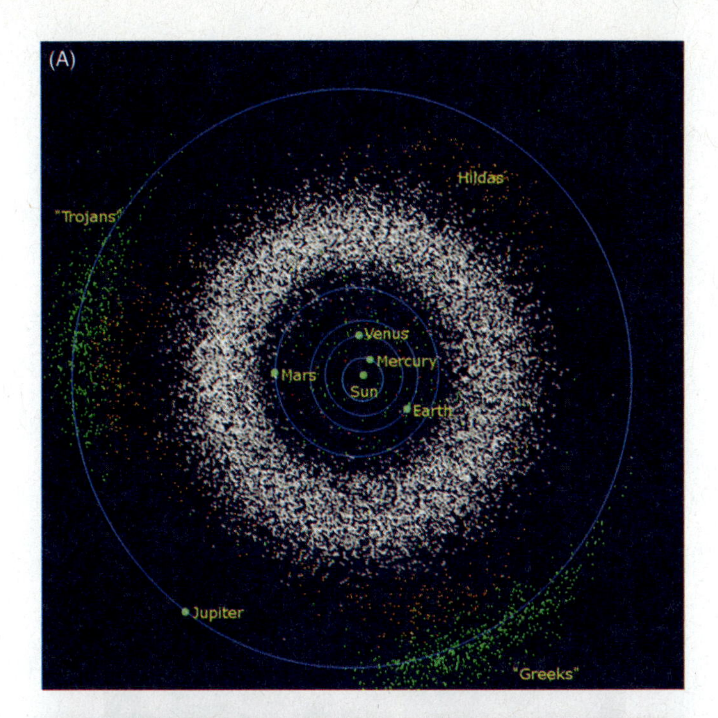

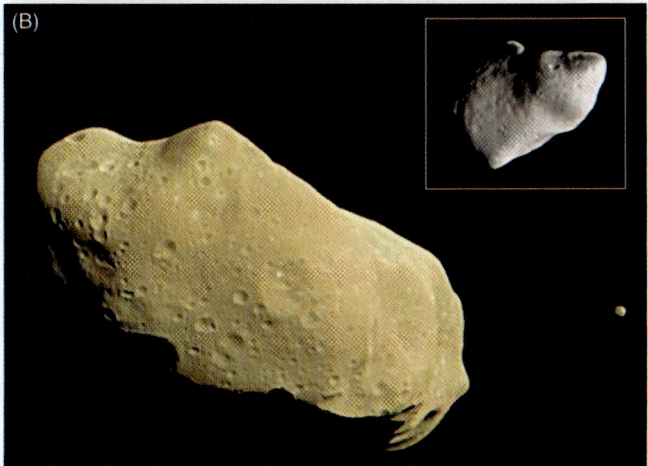

FIGURE 4.16 Asteroids
Asteroids are small, mostly rocky objects that orbit the Sun. Although there are no specific "rules" regarding the minimum or maximum size, asteroids tend to be smaller than dwarf planets, and they populate an orbital gap between Mars and Jupiter (A). The Trojan and Greek groups are asteroids whose orbits are perturbed by gravitational interaction with the massive planet Jupiter. (B) This small asteroid, named 243 Ida, is about 54 km in its longest dimension. Ida has a tiny moon named Dactyl (seen to the right of Ida and in the inset photo), only 1.4 km across, which may have originated as a fragment that broke off of Ida as a result of a collision.

Pluto and the Dwarf Planets

Pluto, which used to be considered the ninth and smallest planet of our solar system, and the farthest from the Sun, is little larger than Earth's Moon. Pluto lacks the massive atmosphere of the larger jovian planets and has a density of 2.06 g/cm^3, intermediate between the high densities of the terrestrial planets and the low densities of the other jovian planets. The most probable explanation for the density of Pluto is that the planet has a structure like those of the moons of Jupiter and Saturn—a rocky center with a thick outer layer of ice.

After its discovery in 1930, Pluto took its place as the ninth planet, but it was always the "odd planet," differing from the other planets, particularly the gas giants, in fundamental and puzzling ways. Some astronomers hypothesized that it formed in another solar system and was captured by the Sun's gravitational field sometime after the formation of our solar system. Today, however, it is recognized as more likely that Pluto is related to the icy objects of the Kuiper Belt, discussed below.

The discovery, in 2003, of an object now known as Eris, larger than Pluto (at 2550 km diameter) and orbiting even further out toward the edge of the solar system, opened up the possibility for many more small bodies to be designated as planets. Many scientists saw this development as problematic for the definition of what actually constitutes a planet. In 2006, therefore, as part of the International Astronomical Union's reconsideration of the definition of the term *planet,* Pluto was demoted from its status as a planet. It was reclassified as a *dwarf planet,* a minor planet or small body that is orbiting the Sun (and is therefore not a moon); is massive enough to be spherical (in a state of hydrostatic equilibrium); but is not massive enough to have cleared its orbital path (the official definition actually says that it has "cleared the neighborhood") of other objects by virtue of its gravitational dominance. In contrast, a **planet** would meet all three of these criteria—orbiting the Sun, spherical shape, and gravitational dominance. An object that orbits the Sun (rather than orbiting a planet, for example) but does not meet the other two criteria is now classified as a *small solar system body.*

As of 2009, four other dwarf planets in addition to Pluto have been officially recognized, although many more are certain to follow them into this classification in the near future. The four are Eris (now the largest of the dwarf planets), Haumea, Makemake, and Ceres, formerly the largest asteroid. Another possible candidate for dwarf planet status is Pluto's natural satellite, Charon. As moons go, Charon is unusual. At almost half the size of Pluto, Charon is the largest natural satellite relative to the size of its parent planet—or in this case, parent dwarf planet—leading to the suggestion that this is not a planet–moon system but rather a *double dwarf planet* system. The Pluto–Charon system itself has two known much smaller moons, called Nix and Hydra (FIG. 4.17).

Comets, the Kuiper Belt, and the Oort Cloud

Pluto, Charon, Haumea, Makemake, Eris, and the thousands of similar objects whose orbits are outside of the orbit of Neptune belong to a group of objects known as the *Kuiper Belt.* The Kuiper Belt is similar to the Asteroid Belt in that it consists of a swarm of relatively small objects orbiting the Sun, but there the similarities end. Unlike the predominantly rocky and metallic objects that occupy the Asteroid Belt, Kuiper Belt objects consist mainly of

FIGURE 4.17 **Pluto, dwarf planet**
Pluto, formerly the smallest of nine planets, was demoted to dwarf planet status in 2006. Pluto's natural satellite, Charon (just below and right of Pluto), is almost 50 percent of Pluto's size, making it more like a companion dwarf planet. Nix and Hydra, the two smaller moons of Pluto, are also seen here.

FIGURE 4.18 **Comet**
Comets are small, loosely packed, icy bodies—like dirty snowballs. They travel periodically to the inner part of the solar system, following highly elongate, elliptical orbits. When they approach the Sun, the ices in their cores volatilize and incandesce, creating the glowing head long tail that we associate with comets, as seen here in Comet Hale-Bopp, 1997. The tail always points away from the Sun.

ices. In short, they are akin to **comets**—small solar system bodies composed primarily of ice with some dust and rock particles, which typically orbit the Sun in elliptical (rather than spherical) orbits. When comets closely approach the Sun, the heat causes the ices of which they are composed to volatilize (i.e., turn to gas) and become ionized, creating a glowing atmosphere (called a *coma*) and the characteristic long, glowing *tail* that we commonly associate with comets (FIG. 4.18). In fact, the Kuiper Belt is thought to be a vast repository for *short-period comets* (those with orbital periods less than 200 years).

Much farther out—a thousand times farther away than Pluto—is another, even more vast store of cometary material, surrounding the solar system in a spherical mass called the *Oort Cloud*. The Oort Cloud is thought to be the source for *long-period comets*, those whose orbits only take them into the inner part of the solar system every few hundreds to thousands or even millions of years. Icy comets that were originally orbiting in the inner part of the solar system could have met one of several fates:

- the loss of their icy constituents due to volatilization, and the breakup of the rubbly, rocky residue that was left behind;
- being swept to the outer part of the solar system by a strong solar wind early in solar system history; or
- collision with a planet or planetesimal.

The possibility of collision is particularly important with regard to the Earth system. As an inner planet, Earth has retained more than its share of volatile constituents, including water. It is now thought that much of the water currently in the Earth system may have been delivered to the surface through impacts with icy comets.

OTHER SUNS AND PLANETARY SYSTEMS

Because all stars aside from our Sun are so far away from Earth, it is not possible to measure how big they are or to see all the detail we can see on the Sun. Almost everything we know about stars comes by way of the electromagnetic radiation they emit. The way we decode and interpret the messages carried by starlight depends to a large degree on our understanding of how the Sun works. Fortunately, a lot of information can be gathered from some straightforward measurements.

Classifying Stars

When you look at the stars on a clear night, two things are quickly apparent. The first observation is that the colors of stars are not all the same; they range from red through yellow to bluish-white. For example, in Orion, one of the most familiar and easily recognized constellations, there is a bright but distinctly red star called Betelgeuse and an equally bright star called Rigel that is a striking bluish-white (FIG. 4.19). Astronomers classify stars based on color, and hence on temperature. The peak wavelengths of stars, which are blackbody radiators, are determined by temperature. The higher the temperature,

FIGURE 4.19 Classifying stars
In the constellation Orion (the hunter), the reddish star at the upper left is Betelgeuse; the bright, bluish-white star at the lower right is Rigel. Because this photo is a time exposure, many faint stars not visible to the eye are shown.

the shorter the wavelength, and the bluer the star; the lower the temperature, the longer the wavelength, and the redder the star. Rigel, with its peak at the blue end of the spectrum, is hotter than red Betelgeuse. Because the Sun's peak is in yellow wavelengths, the temperature of the Sun falls between the temperatures of Betelgeuse and Rigel. Each color—or as it is more commonly called, *spectral class*—is further split into 10 subdivisions ranging from 9 (hottest) to 0 (coolest). The Sun is a G5 star.

The second observation we can make is that stars vary greatly in their brightness. Some, like Rigel, draw attention because they are so bright. If you look closely, however, you can find stars so faint you can hardly be sure they are there. The brightness of a star in the sky is a function of both the star's luminosity, the total amount of energy emitted per second, and its distance from Earth. To calculate luminosity, therefore, we need to know the Earth–star distance. This distance is difficult to measure, but it can be determined with some degree of accuracy for stars out to a distance of 300 light-years from Earth. Once the distance to a star is known, its luminosity can be calculated.

Once the temperature and luminosity of a star are known, they can be compared with the values for other stars through a plot of luminosity versus temperature, called a *Hertzsprung-Russell (H-R) diagram*. About 85 percent of the stars plotted on an H-R diagram fall on or close to a smooth curve (FIG. 4.20) that astronomers refer to as the **main sequence.** The rest of the stars fall into two groups—one above, the other below the main sequence. Stars that plot above the main sequence are very luminous, but also cooler than main-sequence stars of equal luminosity. In order to have such high luminosities at lower temperatures, the stars above the main sequence must be very large; they are *red giants*. About 2 to 3 percent of all stars are giants or, in the case of the very largest stars, such as Betelgeuse, *super giants*. Betelgeuse is so large that if it were the Sun, its edge would be beyond the edge of Mars! Below the main sequence is a small group of stars that are much less luminous and therefore much smaller than the main sequence stars; they have come to be called *white dwarfs*.

Stellar Evolution

The H-R diagram can be used to explain the history of a star. All stars have lives; they are born, they age, and they die. Their life cycle has much to do with their size at any given time, and the smaller a star the longer it can live. Very massive stars live for hundreds of millions of years, intermediate-mass stars like the Sun live up to 10 billion years, and the smallest stars can live for 20 billion years

FIGURE 4.20 Star luminosity and temperature
A Hertzsprung-Russell diagram is a plot of star luminosity versus surface temperature. The vertical axis is a comparative one based on the Sun having a luminosity of 1. The horizontal axis shows surface temperatures increasing to the left. The Sun is a middle-range, main-sequence star.

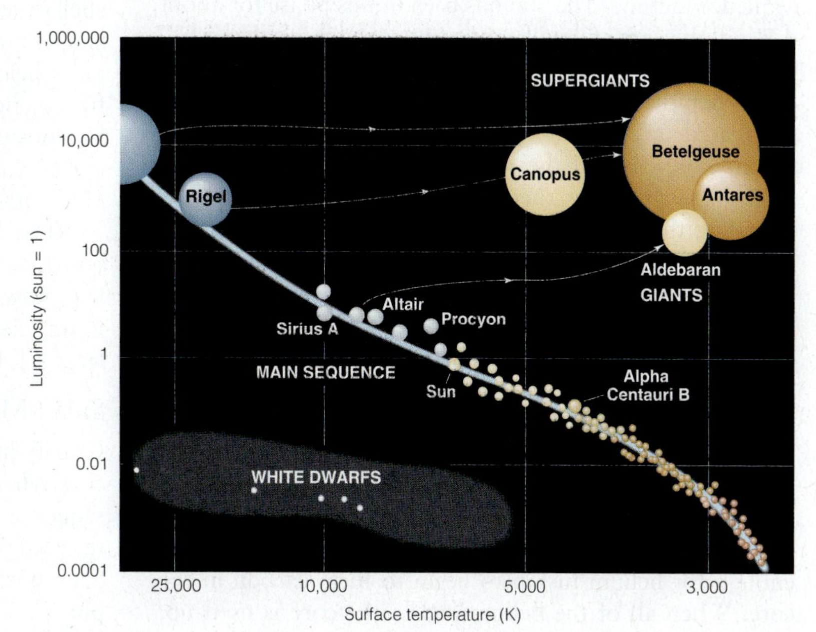

or longer. The reasons for the differences lie in the balance of forces inside a star.

The hot gas that makes up a star does not drift off into space because the force of gravity continuously pulls it *inward*. The greater the mass of gas in a star, the stronger the inward pull. In the absence of a counterbalancing *outward* force, gravity would cause a star to collapse into a small, dense mass. The principal force that counterbalances gravity is electromagnetic radiation. The outward flux of electromagnetic radiation generated by nuclear fusion creates an outward push that prevents gravitational collapse. The balance between the inward gravitational force and the outward radiation force determines the size of the star. A fusion reactor requires fuel, and just as an automobile no longer runs when its gasoline supply is used up, so a fusion reactor can no longer operate when its nuclear fuel is depleted. When its fuel supply runs low, the balance of forces in a star changes, and as a result the size, temperature, and luminosity all change too. When and how the changes occur are a function of how much fuel was present to start with and how fast it was used up.

For the lifetime of most stars, a balance is reached between the gravitational and radiation forces, and the star maintains the stable luminosity and temperature that place it on the main sequence of the H-R plot (as seen in Fig. 4.20). The evidence that stars spend most of their lives on the main sequence is straightforward—most of the stars in the sky plot on the main sequence. Stars appear to be born and to die at about the same rate, yet 85 percent of all stars plot on the main sequence. That can happen only if stars have, through most of their lives, a temperature and luminosity that plot on the main sequence.

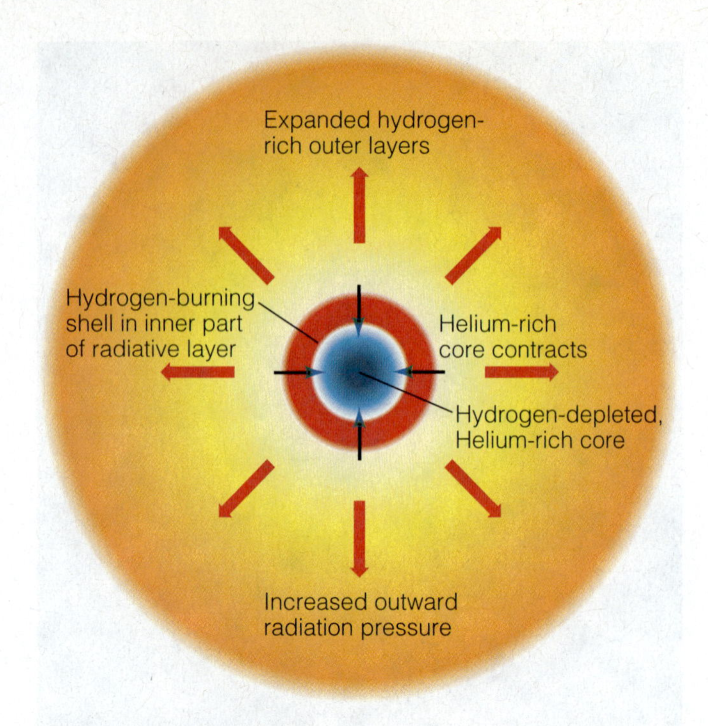

FIGURE 4.21 Shell fusion
When a star that is roughly the mass of our Sun (1 S) uses up its hydrogen fuel and moves off of the main sequence on the H-R diagram, it commences a phase called shell fusion. The now helium-rich core contracts, generating a lot of heat and outward pressure. Hydrogen in the innermost part of the radiative layer begins to undergo fusion, contributing to the outward radiative pressure. The result is that the star increases greatly in size, becoming a red giant.

1-S Stars

Let's first consider the life story of a star roughly the size of our Sun, that is, one solar mass (1 S). The hydrogen in the core of a 1-S star will fuel nuclear fusion to form helium for about 10 billion years, which is thus its expected main-sequence lifetime. The Sun has been in this phase for about 4.7 billion years and will continue for about another 5 billion years or so, making it a middle-aged star.

When the hydrogen fuel in the core of the 1-S star has been used up, nuclear fusion in the core ceases, gravity asserts control, and the now helium-rich core contracts (FIG. 4.21). However, there is still abundant hydrogen surrounding the core in the radiative layer. As the core collapses and becomes even hotter, a shell of hydrogen in the inner part of the radiative layer starts the nuclear fusion process in what is called *shell fusion*. Core collapse heats the star's interior by compression; increasing temperature speeds up the rate of nuclear fusion in the radiative-layer shell. Such an immense amount of heat is generated that the star expands, moving off the main sequence to become a red giant.

The helium-rich core of the red giant continues to contract. Eventually, the core temperature becomes hot enough for helium fusion to begin to form carbon in the core. When all of the helium fuel in the core is used up, shell fusion in the radiative layer starts again, but this time with the fusion of helium. The now carbon-rich core continues to contract gravitationally, but the mass of a star of this size is too small for contraction to raise the temperature to the point at which carbon fusion can occur. The carbon-rich core, surrounded by a slowly diminishing shell of helium, becomes a white dwarf, a small, very dense star that is slowly cooling. On the H-R plot a white dwarf plots below the main sequence (as seen in Fig. 4.20). As fusion ends and cooling proceeds, the white dwarf loses its luminosity, moves toward the lower right-hand corner of the H-R plot, and eventually becomes a dead star called a *black dwarf*. The star's life cycle is now complete.

The scenario just described is approximately that which our Sun will follow. There is nothing for humans to fear, however, because the hydrogen fuel in the Sun's core is sufficient to keep the Sun on the main sequence for at least 5 billion more years.

Small-Mass Stars

Gravity in a very-small-mass star (0.1 S) is so weak that the core where nuclear fusion commences is small; as a consequence, the small amount of electromagnetic radiation released just counteracts the gravitational contraction force. Such stars have low luminosities and low temperatures and plot on the lower-right-hand end of the main sequence (as seen in Fig. 4.20). The hydrogen fuel in a small-mass star is used up so slowly that it lasts an incredibly long time. The fuel will eventually be depleted, however, and shell fusion, core contraction, and helium fusion will follow. Beyond that stage, however, observation tells us nothing about what will happen in the future. The main-sequence lifetime of such small-mass stars may be 20 billion years or more—longer than the age of the universe itself, so we have not yet had the opportunity to observe what happens when they die. Objects with masses less than 0.1 S are too small for nuclear fusion to begin in their cores; these are "almost-stars," and the planet Jupiter is an example.

Massive Stars

The life cycle of a very massive star (100 S) is different. The initial gravitational contraction of a massive star is so intense that the temperature is very high. Massive stars burn fuel very rapidly, have very high luminosities, plot on the upper-left-hand end of the main sequence (as seen in Fig. 4.20), and quickly deplete their fuel. As a result, massive stars have short lives on the main sequence. Once off the main sequence, massive stars go though the same steps of burn-out, core contraction, shell fusion of hydrogen, core fusion of helium, burnout, and contraction of a carbon-rich core.

The next event, however, is very different. As the carbon core contracts and shell fusion of helium proceeds, a temperature is reached where carbon starts to fuse to heavier elements; this is the environment in which all the heavy elements now in Earth are believed to have formed. Such heavy-element formation is thought to happen in a flash and to release so much energy that the star blows up in a *supernova*. After a supernova, what remains of the core collapses into an immensely dense mass, a *black hole*. Scientists speculate that the matter now in the Sun and the planets was blasted into space in one or more supernovas about 10 billion years ago. In a very real sense, all the atoms in our bodies and in everything around us are stardust from an ancient supernova.

Discovery of Other Planetary Systems

We know now that our Sun is an ordinary, medium-sized, middle-aged star with properties and characteristics identical to those of billions of other ordinary stars. What then, if anything, is special or different about "our" sun? How many of the other suns in our galaxy or in other distant galaxies have planets that orbit them? Astronomers now believe that somewhere between 5 and 10 percent of the 200 to 400 billion stars in the Milky Way galaxy have characteristics similar to those of our Sun. It is likely that many of these stars have planetary systems surrounding them, and that some of the planets will be small and rocky like the terrestrial planets of our own solar system.

By the early 1990s, technology had advanced sufficiently for astronomers to pick out signs of planetary systems orbiting other stars. Such planets, known as extrasolar planets or *exoplanets*, are extremely difficult to spot. Unlike stars, they do not shine, so they are not easily visible. In most cases, astronomers cannot see exoplanets directly but can infer that they are orbiting stars because their gravitational pull causes their parent star to wobble in a characteristic way. As of June 2009, 353 exoplanets have been found, according to NASA, but that number is increasing quickly; none of these are believed to be Earthlike. Most exoplanets found thus far are giant planets because of the limits of detection capabilities, and most of them probably resemble Jupiter.

In 2007, astronomers announced the discovery of Gliese 581c, a small exoplanet orbiting the star Gliese, about 20 light-years away from Earth, near the constellation Libra. Gliese 581c is one of the best candidates for an Earthlike exoplanet. It also lies within its star's *habitable zone*, which means that its distance from the star is such that its temperature might be in the right range to potentially harbor liquid water. The size of a habitable zone and its location relative to any particular star depend on the size and lumninosity of that star. Although Gliese 581c does lie within its sun's habitable zone, shortly after its discovery it was shown that the planet may have a runaway greenhouse effect and so might not actually be habitable. Another candidate for an Earthlike exoplanet is one of a family of five planets orbiting a star called 55 Cancri (**FIG. 4.22**), discovered in November 2007. This latest discovery is a Neptune-like planet orbiting within 55 Cancri's habitable zone.

The European Space Agency's CoRoT telescope, launched in 2006, is currently searching for rocky exoplanets. NASA's Kepler Mission, launched on March 6, 2009, will tell us whether planets like Earth are common or rare in our galaxy. The hunt for other Earths continues, and whether any other planets have ever hosted life remains to be determined (see *A Closer Look: Extraterrestrial Life*).

FIGURE 4.22 Exosolar planets
This is an artist's rendition of the exosolar family of five planets (only four are shown here) around the star 55 Cancri, the brightest dot. The blue planet that looms large in the foreground, which is about the size of Neptune, is thought to be orbiting within the habitable zone of the star and may thus be a candidate for an Earthlike exoplanet. This is the same planet as the one shown in Figure 4.10C.

A Closer LOOK

EXTRATERRESTRIAL LIFE

Life is possible on Earth because of the Sun's energy and because Earth lies within the Sun's habitable zone and is thus capable of sustaining liquid water. For centuries, human beings have pondered the question of whether life might exist on other planets. Research in our own solar system has demonstrated the unique qualities of Earth; it seems unlikely that life now exists anywhere else within this solar system. Robot explorers Spirit and Opportunity, launched in 2003, are still sampling and exploring the surface of Mars as of this writing (2010—long past the date when they were due to expire). They are looking for evidence to help scientists better understand the wetter and warmer past of this small, cold planet, and to determine whether life may once have existed there (FIG. C4.1).

Given that the terrestrial planets are similar in composition and overall structure, why don't we see abundant life on Mars, or Venus? The answer probably lies in two factors: first, the size of Earth, and the fact that it has a still hot interior, leads to volcanism, plate tectonics, and a continual addition of new nutrients to the surface in the form of lava; and second, the distance of Earth from the Sun is such that H_2O can exist as water, ice, and water vapor. There is a possibility that life may have existed on Mars at some time in the distant past, but space research has not yet found any evidence of life on Mars today. Venus—Earth's twin in most respects—is just a little bit closer to the Sun.

FIGURE C4.1 Seeking extraterrestrial life
NASA's twin robotic geologists, the Mars Exploration Rovers named Spirit and Opportunity, one of which is depicted here, carried a variety of scientific instruments to search for clues about the history of water (and therefore, possibly, life) on Mars.

Perhaps this was the factor that limited the capacity of Venus to support liquid water on the surface (although there is abundant water vapor in the atmosphere). Without life, as mentioned above, there was no way to remove carbon from the atmosphere and sequester it by burying organic materials. The result on Venus was a CO_2-rich atmosphere, a runaway greenhouse effect, and an environment that is generally hostile to life.

Is there a possibility that one of the exoplanets somewhere else in our galaxy has Earthlike qualities in size, composition, and proximity to its sun, and might it therefore be capable of sustaining liquid water and hosting life? We don't know the answer to this question, but given the number of stars in the galaxy, the odds seem to be pretty good. With the discovery of planets like Gliese 581c and the family of 55 Cancri, we seem tantalizingly close to finding such a planet.

Astronomer Frank Drake made an attempt in 1960 to systematically assess the chances that there is another civilization within our galaxy with which we might one day communicate. His efforts yielded the now-famous *Drake equation*, which states that:

$$N = R^* \times f_p \times n_e \times f_l \times f_i \times f_c \times L$$

where:

N is the number of civilizations in our galaxy with which we might hope to be able to communicate

and

R^* is the average rate of star formation in the Milky Way galaxy

f_p is the fraction of those stars that have planets

n_e is the average number of planets that can potentially support life per star that has planets

f_l is the fraction of the above that actually go on to develop life at some point

f_i is the fraction of the above that actually go on to develop intelligent life

f_c is the fraction of civilizations that develop a technology that releases detectable signs of their existence into space

L is the length of time such civilizations release detectable signals into space

Obviously this equation provides a "guesstimate" rather than a true estimate, and there has been much debate—then, as now—about the exact numbers that should be plugged in for the variables. Drake originally came up with an answer of 10 civilizations within our own galaxy with which we might hope to communicate. More recent solutions, based on the best-known values for the each variable, yield answers from 2 to 5000. But the most intriguing thing about the equation is the idea that it offers—the idea of the *possibility* of an intelligent civilization other than our own.

TIME AND CHANGE

In this chapter, and later in the rest of the book, we consider Earth processes that operate on spatial scales from atomic to planetary to solar system to universal, and on timescales that range from nearly instantaneous (a single radioactive decay or the fusion of two atomic nuclei) to hours or days (a major impact, like the one that formed the debris from which our Moon coalesced), to hundreds or thousands of years (the orbital periods of typical comets), to millions and billions of years (the formation of the planets and evolution of the solar system). It is worthwhile to consider the significance of these widely varying time and spatial scales (**FIG. 4.23**) in terms of the history of Earth, the solar system, and the universe, and to think about the tools and concepts that scientists use to determine the ages of materials and events.

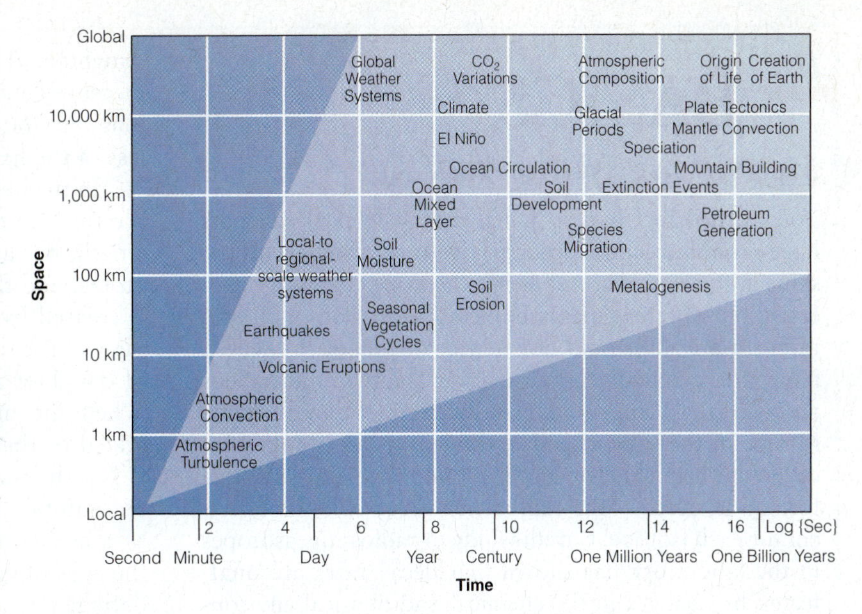

FIGURE 4.23 Time and space scales of Earth processes Earth processes happen on widely varying scales in time and space, from submicroscopic atomic and nuclear processes that happen almost instantaneously to the growth and evolution of planets and solar systems over billions of years.

Origins

Scientists estimate the age of the universe by looking at the rate at which objects in the universe are moving apart from each other. The hypothesis is that everything in the universe originated at one location—not a physical location, so much as a mathematical singularity—in an explosion called the *Big Bang*. The materials generated during the Big Bang and afterward are still moving away from that common center; the universe is expanding. Thus, the age of the universe can be defined as the time that has elapsed between the present day and the time of the Big Bang. If we can measure the rate of expansion of the universe (one of the original tasks of the Hubble Space Telescope when it was first deployed), and reverse that movement until everything in the universe theoretically comes back together in a single location, we can find not only the center of the universe but the time that has elapsed since its formation. Currently, the most widely accepted value for the age of the universe is around 13.7 billion years.

The universe is at least twice and possibly three times as old as the Sun and the solar system. The age of the Sun (about 4.7 billion years), on the other hand, is not too different from the age of the solar system (about 4.56 billion years). This makes sense, since—according to the nebular hypothesis—their origins are linked. The age of the solar system has been determined by direct measurement of the ages of the oldest and most primitive objects that we have been able to find in our solar system, the class of meteorites called carbonaceous chondrites, mentioned above. The ages of the carbonaceous chondrites all cluster closely around 4.56 billion years; furthermore, these objects show little evidence of having been altered by geologic processes subsequent to their formation. Thus, we infer that they are among the oldest materials of the solar system and that the solar system itself (as well as Earth, the other planets, and most of the other objects in our solar system) is close to this age.

The age of the Sun cannot be measured directly, but by considering the nebular theory and what is known about star-forming processes, scientists have concluded that the Sun probably existed as a young star for about 100 million years prior to the formation of the early solar system; hence we obtain a theoretical age for the Sun of about 4.7 billion years—a mere 100 million years older than the solar system. We also can deduce something about the age of the Sun by looking at its life cycle in the context of the H-R diagram. We know from observations of the births and deaths of other 1-S stars that the Sun will have a main-sequence lifetime of about 10 billion years, and it is about halfway through this process.

Relative and Numerical Age

In dealing with the ages of materials both within the Earth system and elsewhere in the universe, scientists make use of two different concepts of time and age. **Relative age** refers to the order in which a sequence of past events occurred (**FIG. 4.24**). The age of an object, material, or event is determined by comparison to an older or younger object or event. For example, we know from our hypothesis of formation of the solar system that the Sun must be at least a little bit older than the rest of the solar system, in relative terms. Closer to home, if a flood carries a massive load of silt, depositing it on top of a preexisting layer of silt, we can conclude that the relative age of the underlying layer is older than the new one. This is

the BASICS

Measuring Numerical Age

You learned in Chapter 3 that most chemical elements have several naturally occurring isotopes (atoms with the same atomic number and hence the same chemical properties, but different mass numbers). Most isotopes found in Earth are stable and not subject to change. However, a few are radioactive because of instability in the nucleus and will transform spontaneously to either a more stable isotope of the same chemical element or an isotope of a different chemical element. The rate at which this transformation occurs—the *radioactive decay rate*—is different for each isotope. Careful study of radioactive isotopes in the laboratory has shown that decay rates are unaffected by changes in the chemical and physical environment. This is important; if the rate of radioactive decay is not changed by geologic processes, then it can be used to determine the numerical age of a geologic material.

All decay rates follow the same basic law (shown in FIG. B4.2). Stated in words, this law is that the *proportion* of parent atoms that decay during each unit of time is always the same. The number of decaying parent atoms continuously decreases, while the number of daughter atoms continuously increases. The rate of radioactive decay is measured by the **half-life**, which is the time needed for the number of parent atoms to be reduced by one-half.

For example, if the half-life of a radioactive isotope is 1 hour and we started an experiment with a mineral containing 1000 radioactive atoms, only 500 parent atoms would remain at the end of an hour and 500 daughter atoms would have formed. At the end of a second hour there would be 250 parent and 750 daughter atoms, and after hour three, 125 parents and 875 daughters. The half-lives of radioactive isotopes used to measure geologic time are thousands to millions of years long, but the decay law is the same for all isotopes regardless of the half-life.

In the illustration of radioactive decay in Figure B4.2, the time units are half-lives. The time units are of equal length, but at the end of each unit the number of parent atoms, and therefore the radioactivity of the sample, has decreased by exactly one-half of the value at the beginning of the unit. Figure B4.2 also shows that the growth of daughter atoms matches the decline of parent atoms. When the number of remaining parent atoms (N_p) is added to the number of daughter atoms (N_d), the result is N_0, the number of parent atoms with which the mineral sample started. That is the key to using radioactivity as a means of measuring geologic time and determining the ages of geologic materials, which we call radiometric dating.

To consider a real example of radiometric dating, potassium (K) has three natural isotopes: ^{39}K, ^{40}K, and ^{41}K. Only one, ^{40}K, is radioactive, and its half-life is 1.3 billion years. The decay of ^{40}K is interesting because two different decay processes occur. Twelve percent of the ^{40}K atoms decay to ^{40}Ar, an isotope of the gas argon, and the remaining 88 percent of the ^{40}K atoms decay to ^{40}Ca. The fraction of ^{40}K atoms that decays to ^{40}Ar is always 12 percent; the percentage is not affected by changes in physical or chemical conditions.

When a potassium-bearing mineral crystallizes, it includes some ^{40}K in its crystal structure. As soon as the mineral is formed, ^{40}Ar and ^{40}Ca daughter atoms start accumulating in the mineral because they are trapped, like the parent ^{40}K atoms, in the crystal structure. Because the ratio of ^{40}Ar to ^{40}Ca daughter atoms is always the same, it is only necessary to measure *either* the ^{40}Ar *or* the ^{40}Ca daughter atoms in order to know how many ^{40}K atoms have decayed. It is more convenient to measure ^{40}Ar

FIGURE 4.24 Relative and numerical age
A good analogy for relative age is a stack of papers placed one on top of the other, day by day. The paper on the bottom is older, in relative age, than any of the papers that overlie it. However, we cannot know the actual age of any of the papers, or the rocks in a geologic sequence unless we look at the date printed on the newspaper or determine the numerical age of the rock by some means such as radiometric dating.

the assumption that underlies the illustration of relative age in Figure 4.24—that rock units are laid down one on top of another in sequence, with the oldest units on the bottom. This is a useful but overly simplistic view; in Chapter 7 we will look more closely at how scientists use the rock record and an understanding of geologic processes to determine the relative ages of materials and events in Earth history.

Numerical age, in contrast, is the actual time, usually expressed in years, when a specific event happened or a specific material formed or was deposited. The discovery of radioactivity in 1896 provided the first way to measure the numerical age of geologic materials quantitatively and reliably. Radioactive decay is a process that runs continuously, that is not reversible, that operates

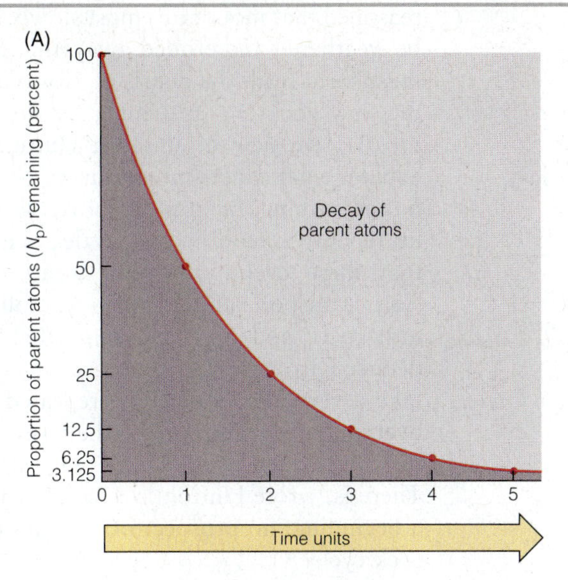

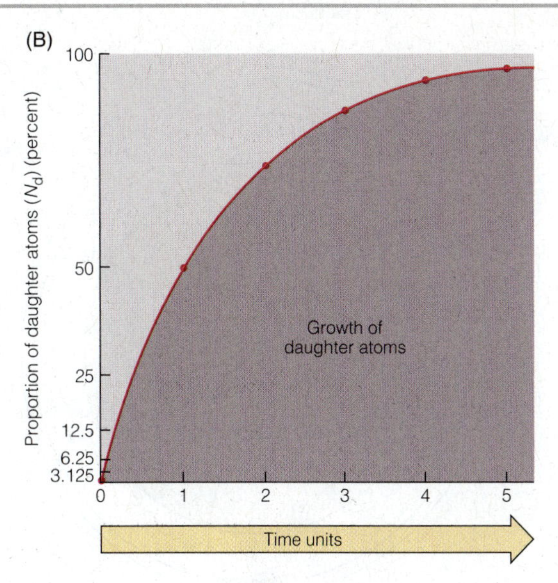

FIGURE B4.2 Radioactive decay
(A) At time zero, a sample consists of 100 percent radioactive parent atoms. During each time unit, half the atoms remaining decay to daughter atoms. (B) At time zero, no daughter atoms are present. After one time unit corresponding to a half-life of the parent atoms, 50 percent of the sample has been converted to daughter atoms. After two time units, 75 percent of the sample is daughter atoms, and 25 percent is parent atoms. After three time units, the percentages are 87.5 and 12.5, respectively. Note that at any given instant N_p, the number of parent atoms remaining, plus N_d, the number of daughter atoms, equals N_o, the number of parent atoms at time zero.

because argon is an element that can be measured very accurately.

All that has to be done to determine the time of formation of an igneous rock (that is, the numerical age of the rock) is to select a potassium-bearing mineral and measure the amount of parent ^{40}K that remains, as well as the amount of trapped ^{40}Ar. Because the half-life of ^{40}K is known, it is a straightforward matter to calculate the radiometric age—the length of time a mineral has contained its built-in radioactivity clock. What is actually measured is the time since the mineral crystallized. Because the time of mineral formation is effectively the same as the time at which the igneous rock was formed, the mineral age and the rock age are the same.

Many naturally radioactive isotopes can be used for radiometric dating, but six predominate in geologic studies. These are the two radioactive isotopes of uranium and the single radioactive isotopes of thorium, potassium, rubidium, and carbon. These isotopes occur widely in different minerals and rock types, and they have a wide range of half-lives, so that many geologic materials can be dated by radiometric means.

the same way and at the same speed everywhere, and that leaves a continuous record without any gaps. For a discussion of how radioactivity is used to measure rock and mineral ages, see *The Basics: Measuring Numerical Age*. This approach, **radiometric dating**, has been used to determine the ages of primitive objects in our solar system, like the carbonaceous chondrites, and by extension the ages of the solar system, Earth, and other solar system objects, as well as many recent materials and events in Earth history.

How are these concepts of relative and numerical time and the scientific tools for measurement of time utilized in the study of Earth systems? Through the painstaking application of the tools and techniques of age determination, and the worldwide comparison and correlation of rock units, geologists have assembled a **geologic column** that summarizes in chronological order the succession of known rock units (FIG. 4.25). This geologic timescale has been divided into the major chronological divisions of Earth history: the *Hadean*, *Archean*, *Proterozoic*, *Paleozoic*, *Mesozoic*, and *Cenozoic Eons*. The scientists who first began to delineate the geologic column in the nineteenth century were particularly challenged by the question of numerical time.Since then, through a combination of geologic fieldwork, radiometric dating, and other approaches, scientists have been able to divide the geologic column into much finer chronological divisions and to fit a detailed scale of numerical time to the geologic column. In Chapter 7 we will examine in greater detail how they have addressed these challenges.

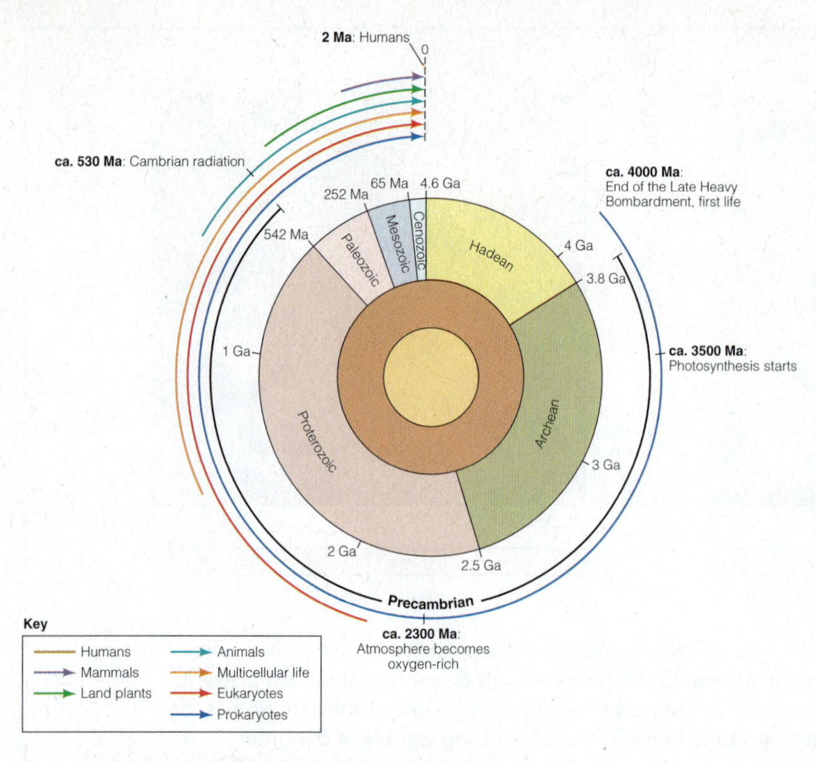

FIGURE 4.25 Geologic time in perspective
Through a combination of radiometric dating and other tools used to determine numerical ages, and geologic observations used to determine relative ages, geologists have established a geologic column that outlines the major periods and events of Earth history. Note that most of Earth history—88 percent of it—is "Precambrian" time, a period about which we still know relatively little.

Uniformitarianism and Catastrophism

One of the great debates in the history of scientific thinking about time is the relative importance of large, catastrophic events as compared to the cumulative effects of smaller, slower events. During the seventeenth and eighteenth centuries, people hypothesized that all of Earth's features—mountains, valleys, and oceans—had been produced by a few great catastrophic events. This idea came to be known as *catastrophism*, and some people also believed that these events had occurred recently and fit a chronology of catastrophic events recorded in the Bible. In the late eighteenth century, this hypothesis was compared with geologic evidence and was found wanting. The person who used the scientific method to assemble the evidence and propose a counter theory was James Hutton (1726–1797), a Scottish physician and "gentleman farmer." Hutton was intrigued by what he saw in the environment around him, especially around Edinburgh, where he lived and studied. He wrote about his observations, offered hypotheses, and then used tests and observational evidence to develop theories.

Hutton observed the slow but steady effects of erosion, by which rock particles are carried great distances by running water and ultimately deposited in the sea. He reasoned that mountains must slowly but surely be weathered and eroded away, that new rocks must form from the debris of erosion, and that the new rocks in turn must be slowly thrust up to form new mountains. Hutton couldn't explain what causes mountains to be thrust up, but everything, he argued, moves slowly along in repetitive, continuous cycles. He realized that these cycles of uplift, weathering, erosion, transport and deposition, solidification into rock, and renewed uplift that could be deduced from visible evidence in rocks must take a very long time to be repeated over and over again. He did not use the term *rock cycle* for this sequence of events, but today we do. There is, wrote Hutton in 1795, "no vestige of a beginning, no prospect of an end" to Earth's rock cycle.

Hutton's ideas evolved into what we now call the **principle of uniformitarianism**, which states that natural laws do not change and therefore the processes that we see in action today have been operating in essentially the same way throughout most of Earth's history. We can examine any rock, however old, and compare its characteristics with those of similar rocks forming today in a particular environment. We can then infer that the old rock likely formed in the same sort of environment. In short, *the present is the key to understanding the past*. For example, in many deserts today we can see gigantic sand dunes formed from sand grains transported by the wind. Because of the way they form, the dunes have a distinctive internal structure (FIG. 4.26A). Using the principle of uniformitarianism, we infer that any rock composed of cemented grains of sand and having the same distinctive internal structure as modern dunes (FIG. 4.26B) is the remains of an ancient dune.

Geologists who followed Hutton have been able to explain Earth's features in a logical manner using the principle of uniformitarianism and the concept of the rock cycle. In so doing, they have also made an outstanding discovery about time and origins: *Earth is incredibly old*. It is clear that most erosional processes are exceedingly slow. An enormously long time is needed to erode a mountain range, for instance, or for huge quantities of sand and mud to be transported by streams, deposited in the ocean, then cemented into new rocks and the new rocks deformed and uplifted to form a new mountain. Slow though it is, this cycle has been repeated many times during Earth's long history.

The concept of uniformitarianism is important to all branches of science, not just geology. For example, astronomers' theories of how stars form, pass through a long life cycle, and then die, as discussed earlier in this chapter, are based on uniformitarianism. Because the lifetime of a

FIGURE 4.26 Principle of uniformitarianism
The internal structure of sand dunes, ancient and modern, demonstrates the power of uniformitarianism. (A) A distinctive pattern of wind-deposited sand grains can be seen in a hole dug in this dune near Yuma, Arizona. (B) The same distinctive pattern appears in ancient rocks in Zion National Park, Utah. We can infer that these rocks were once sand dunes.

star is measured in billions of years, it is not possible to make all needed observations by watching a single star. Instead, astronomers study the billions of stars in the sky, observe examples at various stages of development, and find that the cycle of birth, growth, and death follows a predictable pattern. Whenever a new star is examined, uniformitarianism allows the observer to use previous observations to estimate where the new star is in its life cycle.

Uniformitarianism is a powerful principle, but should we abandon catastrophism as a totally incorrect hypothesis? The answer is no, because we now know that events we consider to be catastrophic can be readily explained by well-understood, ordinary processes, and therefore by uniformitarianism. These are not the catastrophes perceived by seventeenth-century biblical scholars, who had to call on supernatural forces to explain things. Rather, they are events that can be readily explained but are so large and damaging that they caused catastrophic change. An example of such a rare event is a major meteorite impact, such as the one that created Meteor Crater in Arizona. When we view Earth's history as a combination of continual small changes at a wide range of scales, spatial and temporal, as well as a series of repeated but rare events, we have to conclude that uniformitarianism can describe even the rare events and that there is every reason to believe that similar events will occur again in the future.

SUMMARY

1. The Sun is a star, the central body of our solar system. It is part of the Milky Way galaxy, one of about 100 billion galaxies, each of which contains between 200 and 400 billion stars. Though an ordinary star, the Sun overwhelmingly dominates our solar system, containing 99.8 percent of its mass—mostly hydrogen and helium, with small amounts of the heavier elements.

2. Earth is just the right distance from the Sun to take advantage of its life-supporting properties. Sunlight supports and sustains almost all life on Earth, through photosynthesis. Radiatively active gases in the lower part of the atmosphere absorb outgoing longer-wavelength terrestrial radiation, forming a "blanketing" layer that raises the average surface temperature, making life possible on this planet.

3. We experience seasons because Earth's rotational axis is tilted with respect to the ecliptic. This means that at certain times the northern hemisphere points toward the Sun and at other times of the year the southern hemisphere points toward the Sun, thus receiving the most direct solar energy.

4. The solar system consists of the Sun, eight planets, at least five dwarf planets, a vast number of small rocky bodies called asteroids, millions of comets, innumerable small fragments of rock and dust called meteoroids, and 170 known moons, all of which travel along trajectories determined by gravity. The planets, asteroids, comets, and meteoroids orbit the Sun, whereas the moons orbit the planets.

5. The innermost terrestrial planets—Mercury, Venus, Earth, and Mars—are small, dense, and rocky and metallic in composition. The outer, jovian planets—Jupiter, Saturn, Uranus, and Neptune—are much larger and more massive, though much less dense than the terrestrial planets since they are composed dominantly of hydrogen and helium.

6. Almost everything in the solar system—planets, moons, and the Sun—revolves and rotates in the same direction and in the same plane. (Venus is a notable exception, spinning in a retrograde direction, perhaps because of an enormous collision that knocked the planet end-over-end.)

7. The prevailing model for the origin of the solar system is the nebular hypothesis, which proposes that the Sun and its planets formed from a huge, swirling, thinly dispersed cloud of cosmic gas and dust. Elements heavier than hydrogen and helium came from the remains of an older star that exploded in a supernova.

8. Over a very long period of time, the gas in the nebular cloud thickened under the influence of gravity. As the atoms moved closer together, the gas became hotter and denser. Near the center, the atoms became so tightly compressed and the temperature so high that nuclear fusion started and a new star was born. Surrounding the new star, our Sun, was a flattened, rotating disc of gas and dust, the solar nebula.

9. Condensation in the solar nebula produced a cosmic "snow" of high-temperature, rocky condensates in the inner part and low-temperature, volatile, icy condensates in the outer part, which revolved around the Sun in the same direction as the rotating gas cloud from which they formed. Impacts occurred between particles drawn together by gravitational attraction. Planetary bodies grew by accretion, gathering bits of solid matter from surrounding space. Giant collisions occurred when most of the loose debris had been swept up and only larger objects remained.

10. The main factors that have influenced the subsequent evolution of the terrestrial planets were impact cratering and resultant partial melting; volcanism; distance from or proximity to the Sun; and absence or presence of a biosphere. Although we cannot see the surfaces of the jovian planets through their thick atmospheres, it is reasonable to conclude that these factors also influenced their evolution as planets.

11. The terrestrial planets underwent partial melting and differentiation into low-density, rocky crusts; rocky, intermediate-density mantles; and metallic, high-density cores. All terrestrial planets have experienced volcanism and intense meteorite collisions. Plate tectonics may be unique to Earth, which also has a stronger magnetic field than the other terrestrial planets. All of the terrestrial planets lost their primary atmospheres, but Earth, Mars, and Venus evolved and retained secondary atmospheres.

12. The outer planets are gas giants. Jupiter and Saturn are so massive that their atmospheric gases have not escaped—even the lightest gases, hydrogen and helium. Thus, they have compositions very similar to that of the solar nebula from which they formed. Huge storms, high-speed winds, and lightning are common in their deep atmospheres. They probably have small rocky cores mantled by ice. Deep in their interiors, pressures are so intense that hydrogen condenses, becoming a liquid or even a metal inside Jupiter and Saturn.

13. A planet orbits the Sun, is massive enough to have attained hydrostatic equilibrium, and has gravitational dominance over its neighborhood. There are eight planets in our solar system. A moon is a natural satellite in a regular orbit around a planet. Earth's Moon originated as a result of a massive collision that tilted Earth's axis of rotation. Meteoroids are small fragments of rocky debris, which sometimes fall to Earth as meteorites. Asteroids are larger rocky objects orbiting the Sun in the Asteroid Belt. Thousands of predominantly icy objects orbit in the Kuiper Belt, a vast repository for comets. Much of Earth's water may have been delivered to the surface by impacts with icy comets.

14. Stellar classes are based on luminosity and color. The balance of inward gravitational forces and outward radiation forces inside a star determines its size, which determines its main-sequence lifetime. When the fuel supply of a star runs low, its size, temperature, and luminosity all change. When a small (1 S) star ends its main-sequence lifetime, it undergoes core contraction and shell fusion of hydrogen, becoming a red giant and then a white dwarf. Very massive stars (100 S) have very high temperatures and luminosities, quickly deplete their fuel, and have short lives on the main sequence. Once off the main sequence, massive stars go though burnout, core contraction, and shell fusion of hydrogen and helium. Eventually, fusion begins to form elements heavier than carbon, releasing so much energy that the star blows up in a supernova.

15. Five to ten percent of the stars in the Milky Way have characteristics similar to those of our Sun; some have planetary systems. Many exoplanets have been found, including some within a theoretical habitable zone. It is unlikely that life exists anywhere else within our solar system, but it is possible that one of the exoplanets may have Earthlike qualities in size, composition, and proximity to its sun, and might be capable of sustaining liquid water and hosting life.

16. The widely accepted value for the age of the universe is 13.7 billion years. The age of the solar system (4.56 billion years) has been determined by direct measurement of the ages of the oldest and most primitive known objects in our solar system, carbonaceous chondrites. The Sun probably existed as a young star for about 100 million years prior to the formation of the solar system; hence its age is about 4.7 billion years.

17. Relative age refers to the order in which a sequence of past events occurred, and the age of an object or event in comparison to another older or younger object or event. Numerical age is the actual time when a specific event happened or a material formed or was deposited.

18. The rate of decay of radioactive materials is not changed by geologic processes, so it can be used to determine numerical age. Using a radioactive isotope of known half-life, scientists measure how much of the parent isotope remains in the rock or mineral and how much of the daughter isotope has been produced. It is then possible to determine how long the decay process has been operating.

19. Through worldwide correlation of rock units, on the basis of fossils and other evidence of relative age, geologists have assembled the geologic column, the succession of known rock units in chronological order. Radiometric dating has allowed scientists to attach numerical ages to many of the units and events in the geologic timescale.

20. The principle of uniformitarianism states that the present is the key to understanding the past. Natural laws do not change, and the processes that we see in action today have been operating throughout Earth's history. Scientists in many fields use uniformitarianism to explain a wide range of processes in the natural world, even rare catastrophic events such as massive floods or meteorite impacts.

IMPORTANT TERMS TO REMEMBER

asteroid *94*
astronomical unit (AU) *83*
comet *96*
galaxy *82*
geologic column *103*
half-life *102*
jovian planets *85*
main sequence *97*
meteorite *94*
moon *92*
nebular hypothesis *87*
numerical age *102*
planet *95*
planetary accretion *87*
planetary differentiation *89*
primary (primordial) atmosphere *91*
principle of uniformitarianism *104*
radiometric dating *103*
relative age *101*
secondary atmosphere *91*
solar nebula *87*
solar system *82*
terrestrial planets *85*

QUESTIONS FOR REVIEW

1. How many stars are there, altogether? Is the Sun different from most other stars, or similar, and in what respects?

2. The Sun is not only much *larger* than Earth, it is also much more *massive*. What's the difference? How do the differences in size and mass translate into the difference in *density* between Earth and the Sun?

3. Summarize and describe the various groups of objects of our solar system.

4. The eight planets can be divided into two groups on the basis of their properties and proximity to the Sun. What are the two groups called? Summarize the common characteristics of each group.

5. Describe the steps in the formation of the solar system, as per the nebular hypothesis.

6. How does the nebular hypothesis account for the observed features of our solar system, such as the directions of rotation and revolution of planets, moon, and the Sun; the common plane of revolution; and the compositional distribution of bodies in the solar system?

7. What were the main factors that influenced the evolution of the planets subsequent to the formation of the solar system?

8. Among Mercury, Venus, Earth, Mars, and the Moon, describe some features that are common to all and some features that differ.

9. Among Jupiter, Saturn, Uranus, and Neptune, describe some features that are common to all and some features that differ.

10. How do astronomers classify stars? How do they use the H-R diagram to illustrate the life cycles of stars?

11. Why and in what way do the life cycles of very massive (100 S) stars differ from the life cycles of very small (0.1 S) stars and of stars that are similar to our own Sun?

12. What is the "habitable zone" of a star? Have any exoplanets been discovered so far that are within the habitable zone of a sun?

13. What is the Drake equation, and how does it help scientists quantify the possibility that life may exist outside of our solar system?

14. What is the difference between relative age and numerical age?

15. What is the principle of uniformitarianism, and how is it applied by different kinds of scientists to explain natural phenomena?

QUESTIONS FOR RESEARCH AND DISCUSSION

1. What are carbonaceous chondrites? Find out more about their special characteristics. Why are they so useful to scientists in studying the age of formation of the solar system and planets?

2. Why is the largest volcano in the solar system (Olympus Mons) located on Mars, which apparently isn't even volcanically active any more? Are there volcanoes on Earth that are similar to Olympus Mons?

3. How does the principle of uniformitariansm account for major catastrophic events that occur suddenly, like a major meteorite impact or an enormous flood? What about catastrophic events that occur more slowly, like a major glaciation or widespread extinction of species? Does the principle of uniformitarianism account for the impacts of human actions?

QUESTIONS FOR *THE BASICS*

1. What is the ecliptic? How does the ecliptic relate to the nebular hypothesis and the idea of the solar nebula as a flat, disk-shaped cloud?

2. Why is Earth's axis not oriented perpendicular to the ecliptic, and what impact does this have on seasonality?

3. Sketch a diagram or use a physical model to show why the seasons are opposite in the northern and southern hemispheres.

QUESTIONS FOR *A CLOSER LOOK*

1. What is radioactivity? What is an isotope?

2. What is the basic law that governs radioactive decay rates?

3. Explain how K-Ar radiometric dating allows scientists to determine the numerical age of an igneous rock.

PART TWO

Volcanic contributions
Ecuadoran volcano Tungurahua erupts, sending gases from Earth's interior into the atmosphere. Throughout Earth history, volcanic eruptions have connected the

The Geosphere: Earth Beneath

Our Feet

Geosphere: *from the Greek words* geo- *or* ge, *meaning Earth, and* sphaira, *meaning globe, ball, or sphere.*

Earth's principal solid reservoir, the **geosphere**, may seem constant and unchanging, but nothing could be further from the truth. In Chapter 1 we discussed how space technologies have enabled scientists to observe Earth from space more closely than ever before, and one of the triumphs of remote sensing with satellites has been the ability to make detailed measurements of the surface. Such measurements show that the geosphere is ever changing. Mountains are slowly rising, continents are shifting, and ocean basins are continually changing their shapes and sizes.

As you learned in Chapter 1, the tectonic cycle and the rock cycle are the principal mechanisms by which such changes happen to the face of Earth. Tectonic processes, the focus of Chapter 5, are driven by forces deep within the planet. Through earthquakes and volcanic eruptions, which we examine in detail in Chapter 6, Earth's surface is slowly but continually being reshaped, renewed, rebuilt, and recycled from the inside. These processes represent the interface—they connect the geosphere and the interior of the planet to the other subsystems. Through weathering, erosion, and other processes of the rock cycle, which we explore in Chapter 7, Earth's surface is continually being rearranged, redistributed, and worn down from the outside. The constantly shifting balance between the internal and external forces is what makes Earth's surface such a dynamic place to live.

Understanding the theory of plate tectonics, its relationship to the rock cycle, and their interactions with the other great systems is the key to understanding the geosphere and its role in the Earth system. This is the main focus of Part II.

The chapters of Part II are as follows.

- Chapter 5. The Tectonic Cycle
- Chapter 6. Earthquakes and Volcanoes
- Chapter 7. The Rock Cycle

CHAPTER 5

The Tectonic CYCLE

OVERVIEW

In this chapter we:

- Look at the history and development of the theory of plate tectonics
- Consider the possible driving forces for plate motion
- Examine the way plate tectonics shapes Earth's landscapes
- Investigate the structure and geologic history of continents
- Consider the influence of plate tectonics on the Earth system as a whole

The top of the world
The Himalayan Mountains and Tibetan Plateau are actively rising as a result of the ongoing collision between the Indo-Australian and Eurasian tectonic plates. Seen emerging through the clouds is the glacier-covered peak K2 in the Kunlun Mountains of China.

Mountains such as the Alps or Appalachians that seem changeless to us are only transient wrinkles when viewed from the perspective of geologic time. Mountain ranges grow when fragments of moving lithosphere collide, heaving masses of twisted and deformed rock upward. Then the mountains are slowly worn away, leaving only the eroded roots of an old mountain range to record the ancient collision (FIG. 5.1). The continents are slowly moving, sometimes bumping into each other and creating a new mountain range, sometimes splitting apart so that a new ocean basin forms.

These processes have gone on through much of Earth's 4.56-billion-year history, and they continue today. For example, the Himalaya is a range of geologically young mountains that began to form when the Indian subcontinent collided with Asia about 45 million years ago; these mountains are still being uplifted today (FIG. 5.2A). The Red Sea (FIG. 5.2B) is a young ocean that started forming about 30 million years ago when a split developed between the Arabian Peninsula and Africa as the two landmasses began to move apart; this new ocean is still widening today.

Clearly these great forces that operate within the **geosphere**—the solid Earth—are responsible for shaping Earth's surface. What is the origin of these forces? What drives them? Have they operated throughout Earth history? In what ways do these processes influence the rest of the Earth system? Is the geosphere actually solid, or is it more complex than it appears on the surface? In this chapter we will seek answers to these and related questions.

FIGURE 5.1 Continental collision
This satellite image (about 70 km in width) shows the twisted, deeply eroded roots of the Appalachian Mountains in southwestern Virginia. The mountains were formed as a result of a collision between two masses of continental crust several hundred million years ago. The collision folded, twisted, and fractured the rocks, which have since been weathered and eroded by the forces of the rock cycle.

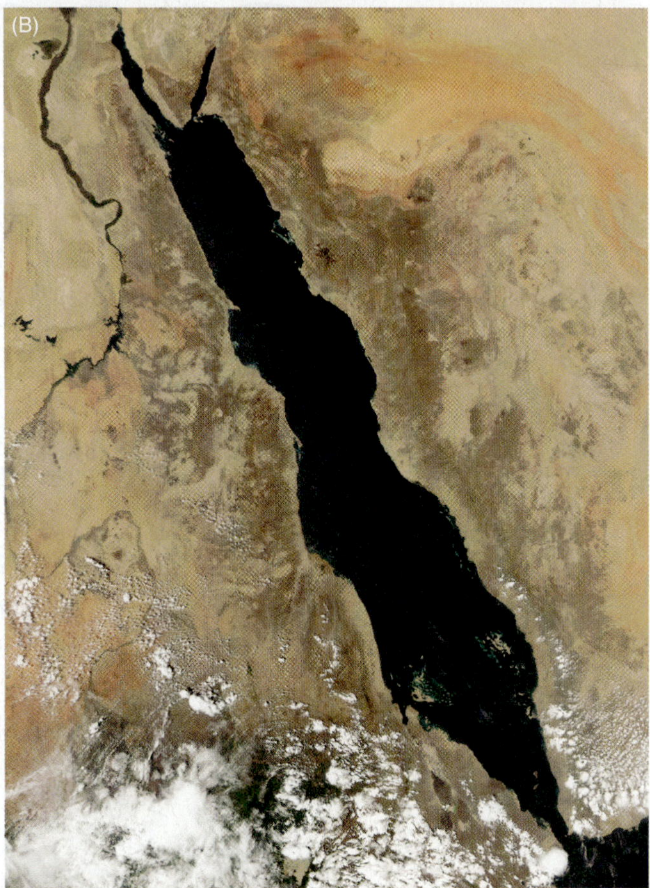

FIGURE 5.2 Tectonics today
The Himalaya (A) is a geologically young mountain range that is still being actively uplifted, as a result of a collision between the Indian and Asian continents. The Red Sea (B) is a new, linear ocean that is forming where Arabia is separating and moving away from Africa.

PLATE TECTONICS: A UNIFYING THEORY

The continents, the ocean basins, and everything else on the surface of Earth are moving along like passengers on large rafts. The rafts are huge slabs or *plates* of lithosphere that float on the underlying material of the mantle. Tectonics is the study of the movement and deformation of these lithospheric plates, and the branch of tectonics that deals with their interactions is called **plate tectonics.** All of the major topographic features on Earth's surface, whether submerged beneath the sea or exposed on land, arise as a direct result of the motion and interactions of lithospheric plates, and subsequent modifications by the external forces of the rock cycle.

Plate tectonics has emerged as a unifying theory that explains hundreds of years of independent observations of Earth's topographic features and the distribution and characteristics of rock units. The plate tectonic model integrates our scientific understanding of the processes of rock formation, mountain building, and terrain modification. It also provides a highly effective framework for our discussion of the geologic processes that affect people and shape the environments in which we live. Let's first consider how scientists amassed the evidence and developed the concepts that support the theory of plate tectonics.

The Development of an Idea

Like all scientific theories, the theory of plate tectonics did not arise overnight and was not the work of just one scientist. Many brilliant scientists have contributed both their ideas and their hard work to test the theory, to amass evidence in support of it, and to shape it into a robust explanation for many of the features and processes of Earth's geosphere.

The Back Story of Plate Tectonics

In 1508, Leonardo da Vinci discovered fossil seashells high in the mountains of Italy and realized that he must be looking at an ancient seafloor. But how and why were the shells so high up? Had the ocean once covered the mountains, or had the seafloor been raised? Because seashells are not found everywhere across the land surface (as they would be if it had been covered uniformly by water), Leonardo reasoned that part of the former seafloor must have been uplifted. Almost three centuries later, James Hutton (who proposed the principle of uniformitarianism in the late 1700s; see Chapter 4) realized that such uplifting would expose rocks to weathering and erosion, and he built these concepts into his model of the rock cycle.

Then Charles Darwin, whose name is forever entwined with the theory of evolution by natural selection (Chapter 14), made an important observation about the rock cycle. In the 1830s Darwin made a globe-encircling voyage on *HMS Beagle*. In Patagonia, southern Argentina, he saw vast sediment-covered plains that stretch from the Andes to the Atlantic Ocean. The sediments had apparently been formed by erosion of the Andes. How, Darwin wondered, could "any mountain chain have supplied such masses, and not have been utterly obliterated?" Then in 1835, on the Chilean coast, Darwin experienced a great earthquake (FIG. 5.3). Investigating the effects of the quake, he discovered "putrid mussel shells still adhering to rocks ten feet above high water mark." Darwin realized that the ground had been elevated and that the Andes were slowly being pushed upward, earthquake by earthquake. Although he could not explain the nature of the "force which has upheaved the mountains," Darwin understood that this continual slow elevation answered the question he had posed in Patagonia—tectonic forces raise mountains, while the erosive forces of the rock cycle wear them away.

By the middle of the nineteenth century, the idea that Earth's crust is subject to large vertical movements, or *uplift*, was widely accepted. Some evidence—mainly the parallelism of the coastlines on either side of the Atlantic Ocean—suggested the possibility of sideways (lateral) movements, too. However, most scientists of

FIGURE 5.3 Darwin's inspiration
These are the steep-sided spires of the Torres del Paine in the southern Andes Mountains of Chile. It was in the Andes that Charles Darwin realized that a balance exists between the tectonic forces that elevate the Andes and the weathering forces that slowly wear them down.

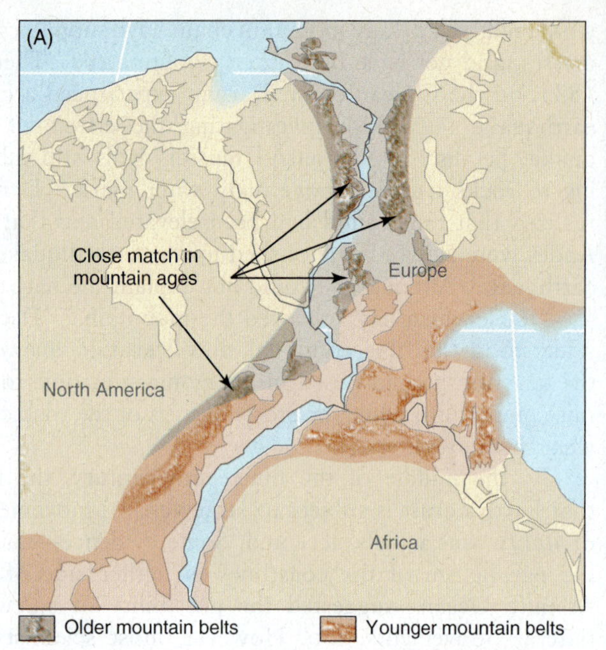

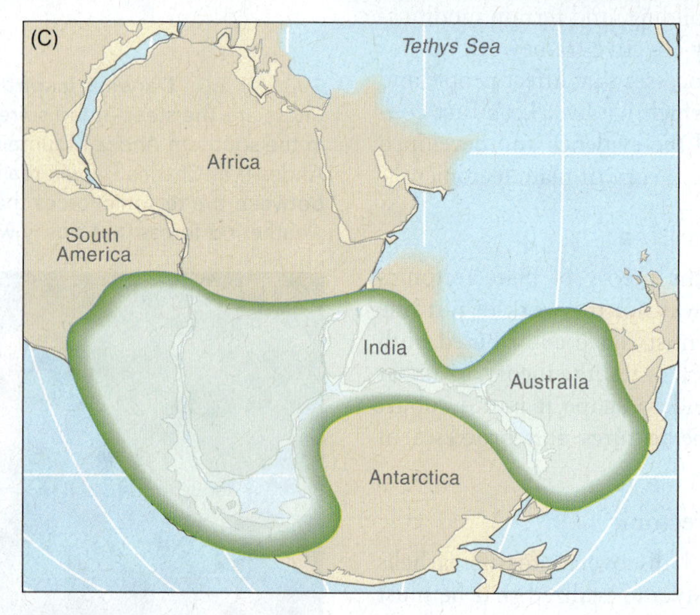

FIGURE 5.4 Evidence for continental drift
Alfred Wegener and supporters of the hypothesis of continental drift amassed significant evidence that the continents had once been joined in a supercontinent that they called Pangaea. This included evidence that mountain ranges of similar age and geology appear to match from one continent to another (A). Evidence from fossils suggested that some ancient organisms ranged freely among continents now widely separated by ocean basins. An example is the ancient tree *Glossopteris,* whose fossil leaves are shown in (B). The map (C) shows where *Glossopteris* fossils have been found; the areas match well when the continents are moved back to their probable locations 300 million years ago.

the time found the idea of laterally moving continents unacceptable.

The Continental Drift Hypothesis

In 1914, a German scientist named Alfred Wegener published a book in which he tried to explain such phenomena as the parallel shapes of the Atlantic coastlines of Africa and South America, glaciated landscapes and mountain ranges that appeared to match up from one continent to another across oceans, and the occurrence of very similar plant and animal fossils on continents separated by wide ocean basins (FIG. 5.4). At some time in the distant past, Wegener suggested, all of the world's landmasses were together in a single huge continent, so that plants and animals could spread freely. He gave this ancient "supercontinent" the name *Pangaea* (pronounced Pan-JEE-ah, from root words meaning "all lands"). According to Wegener's hypothesis, Pangaea was somehow disrupted, and its fragments (the continents of today) slowly drifted to their present positions. This process came to be known as **continental drift.**

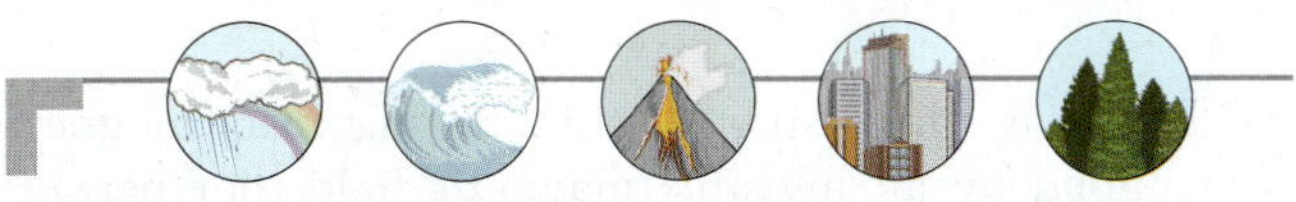

Make the CONNECTION

Imagine what Earth would be like if today's continents were reassembled into the supercontinent Pangaea. What would be the main impacts on climate, ecosystems, and oceanic processes?

Wegener and other proponents of the hypothesis of continental drift likened the process to the breaking up of a sheet of ice that floats in a pond. The broken pieces, they argued, should all fit back together again, like pieces of a jigsaw puzzle. But even though Wegener presented impressive observational evidence that the continents had once been joined together, the hypothesis was not widely accepted in his lifetime. No one could explain how a solid, rocky continent could possibly overcome friction and slide across the oceanic crust. The process, said his critics, would be like trying to slide two sheets of coarse sandpaper past each other. The mechanism for continental drift—that is, *how* the continents moved—as well as the driving force behind it, were missing.

Wegener died in 1930. Debate about continental drift slowed down because some of the supporting evidence gathered by Wegener was found, on close examination, to be open to doubt—for example, plant seeds might have floated from one continent to another, rather than spreading by land when the continents were joined. By 1939, when World War II broke out, the continental drift hypothesis had few supporters.

What revived the debate on a worldwide basis in the 1950s and led, eventually, to the hypothesis of plate tectonics were some fundamental scientific discoveries about radioactivity, radiometric dating, and Earth's magnetism (see *The Basics: Earth's Magnetism*). It turned out that Wegener was right—the continents do move, but not in the manner he hypothesized.

The Search for a Mechanism

Alfred Wegener's hypothesis of continental drift had been largely abandoned by the 1940s because the movement of an entire continent seemed to be physically impossible. However, as happens so often in science, the hypothesis was revived because of discoveries made in another field—studies of Earth's magnetism.

Apparent Polar Wandering

During the 1950s, scientists started using volcanic rocks—ancient lavas—to measure the past north–south directionality, or **polarity**, of Earth's magnetic field. As discussed in *The Basics: Earth's Magnetism*, **paleomagnetism**, the ancient magnetism locked in minerals, preserves a record of Earth's magnetic field at the time of rock formation. Two important pieces of information can be obtained from paleomagnetism. The first is the polarity of the magnetic field at the time the rock became magnetized. The second is the magnetic inclination, the angle from the horizontal assumed by a freely swinging bar magnet. Magnetic inclination varies with latitude; thus, paleomagnetic inclination is a record of the location between the pole and the equator (that is, the *magnetic latitude*) where the rock was formed. Once we know the magnetic latitude of a rock and the polarity of Earth's magnetic field at the time the rock was formed, we can determine the exact position of the magnetic poles at that time.

Geophysicists studying paleomagnetic pole positions during the 1950s found evidence suggesting that the poles had wandered all over the globe. They referred to the strange plots of ancient pole positions as *apparent polar wandering*. Geophysicists were puzzled because they knew that Earth's magnetic and geographic poles should always be close together. When it was discovered that the path of apparent polar wandering measured in North America differed from that in Europe (FIG. 5.5), geophysicists were even more puzzled. They knew that it was highly unlikely that the magnetic poles had moved—and how could they have moved differently for rocks in different locations? Somewhat reluctantly, they concluded that the continents themselves must have moved, carrying the magnetized rocks with them.

From apparent polar wandering, scientists had independent confirmation that the continents had shifted, and so the hypothesis of continental drift was revived.

FIGURE 5.5 Wandering poles?
The lines show the apparent wandering path of the north magnetic poles over the past 600 million years, as determined from paleomagnetic measurements. The numbers indicate millions of years before present. The path determined for North America (red) differs from the path determined for Europe (black). Scientists concluded that the continents, rather than the poles, had moved.

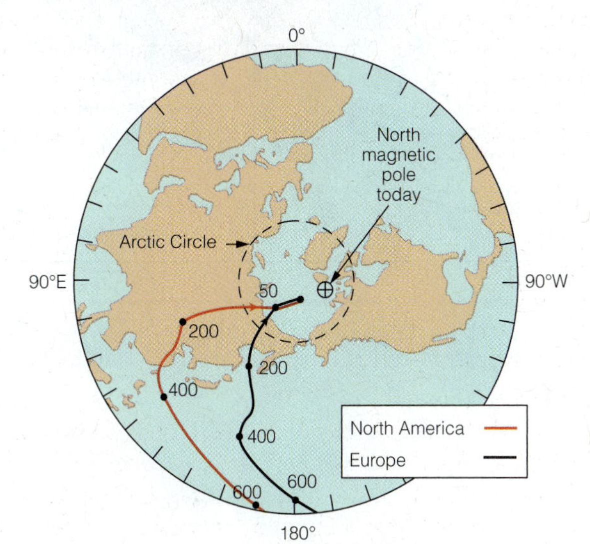

the BASICS

Earth's Magnetism

The source of Earth's magnetic field lies in its molten outer core (Chapter 4). As a result of Earth's rotation, the molten iron of the outer core is continually flowing around the solid inner core. The flowing stream of molten iron causes an electrical current to flow in the outer core, and the electrical current, in turn, generates the magnetic field. The idea that a rotating, conducting fluid can generate and sustain a magnetic field is an application of the *dynamo theory* of electromagnetism. Because the magnetism is a result of Earth's rotation, the magnetic north and south poles are close to the geographic North and South Poles, which represent Earth's axis of rotation.

Earth is thus sort of like a gigantic magnet, surrounded by an invisible **magnetic field** that permeates everything placed in the field. If a small magnet is allowed to swing freely in Earth's magnetic field, the magnet will become oriented so that its axis points to Earth's magnetic North Pole (**FIG. B5.1**). In addition to pointing toward the poles, the angle of a freely swinging magnet with respect to the planet's surface varies regularly with latitude, from zero at the magnetic equator to 90° at the magnetic pole. This angle, measured from the horizontal surface, is the *magnetic inclination*. The freely swinging magnet follows the magnetic field lines, orienting horizontally with respect to the surface

FIGURE B5.1 Earth's magnetic field
Earth is surrounded by a magnetic field generated by the magnetism of the outer core. A magnetic needle, if allowed to swing freely, will always line up parallel to the magnetic field and point to the magnetic North Pole. The magnetic poles are close to the geographic (i.e., rotational) North and South poles because Earth's magnetism results from its rotation. The freely swinging magnet follows the magnetic field lines, orienting horizontally with respect to the surface at the equator and varying to a more steeply dipping inclination with increasing latitude, until it points directly toward the surface in a vertical orientation at the poles. This angle is the magnetic inclination.

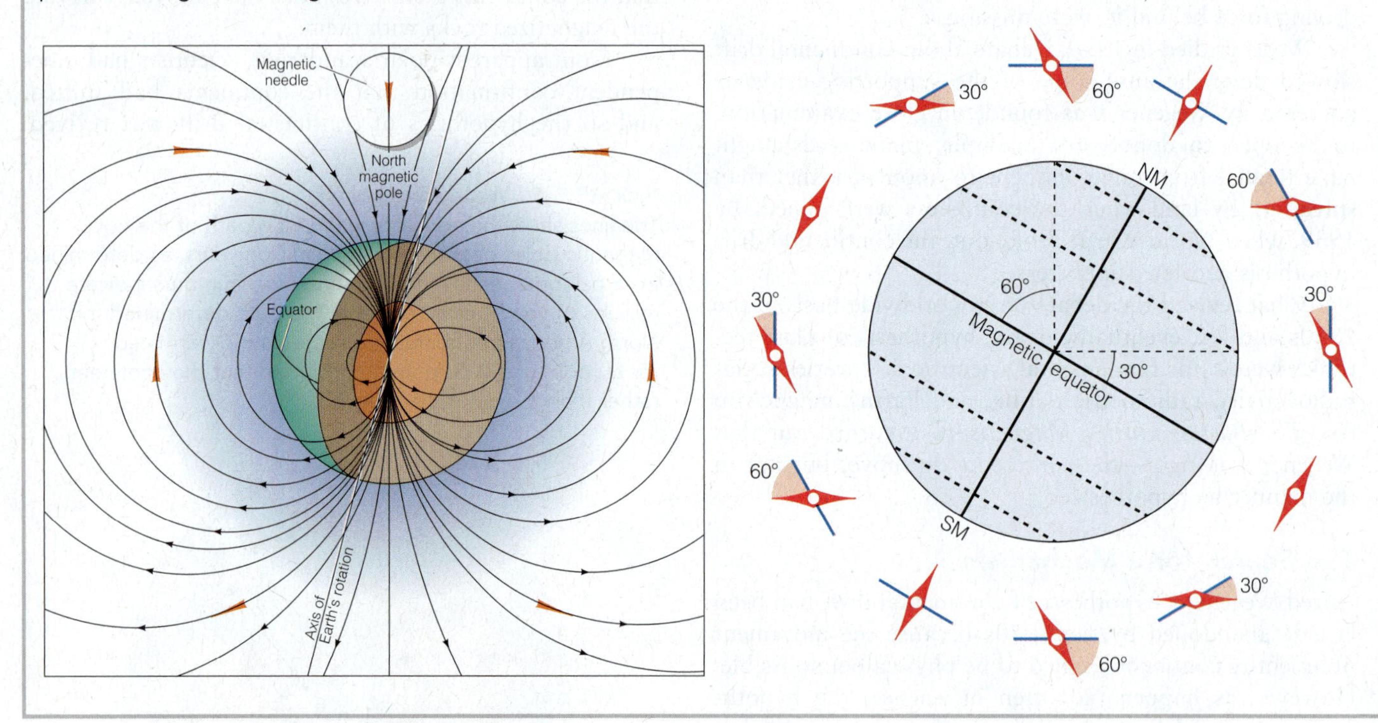

at the equator and varying to a more steeply dipping inclination with increasing latitude, until it points directly toward the surface in a vertical orientation at the poles.

ROCK MAGNETISM AND PALEOMAGNETISM

Some rocks are magnetic. This property arises because the electrons spinning around an atomic nucleus act like tiny atomic magnets; in some minerals, these atomic magnets line up in parallel arrays called *domains*, which reinforce each other. Magnetite (Fe_3O_4) and certain other iron-bearing minerals can become permanently magnetized in this way. In nonmagnetic minerals, in contrast, the tiny atomic magnets within them are oriented in random directions.

Here's how it happens. Above the *Curie point* (a characteristic temperature for each material with the capability of acquiring permanent magnetism), the thermal agitation of atoms is such that permanent magnetism is impossible. The Curie point for the mineral magnetite is 580°C; above 580°C, the atomic magnets within the magnetite are randomly oriented. If the magnetite cools to below 580°C, however, the tiny atomic magnets begin to align to whatever magnetic field is influencing them (**FIG. B5.2**). The domains that are parallel to Earth's magnetic field become larger and dominant, expanding at the expense of the nonparallel domains. Eventually they are "pinned" within the crystal structure, so they won't move even if the influence of the surrounding magnetic field is removed. A permanent magnet is the result.

Consider what happens when lava cools. All of the minerals crystallize at temperatures above 700°C—well above the Curie point of magnetite. As the crystallized lava (now a volcanic rock) continues to cool and the temperature drops below 580°C, the magnetite grains become tiny permanent magnets with the same north–south orientation, or *polarity*, as Earth's magnetic field. As long as that volcanic rock lasts (until destroyed by weathering, metamorphism, or melting), it will carry its *paleomagnetism*, or "ancient magnetism," providing a record of Earth's magnetic field long ago.

Sedimentary rocks also can acquire a weak permanent magnetism. As grains of sediment settle through ocean or lake water, or even as dust particles settle through the air, any particles of a magnetic mineral, such as magnetite, act as freely swinging magnets and orient themselves parallel to Earth's magnetic field. Once locked into a sedimentary rock, the magnetically oriented grains make the whole rock a weak permanent magnet.

Investigation of the properties of natural magnetism in rocks led to the revival of the continental drift hypothesis in the 1950s by revealing seafloor spreading as a mechanism for the movement of continents. The study of paleomagnetism also has provided a tool for determining the relative ages of rock layers and the fossils they contain; we will discuss this in greater detail in Chapter 7.

FIGURE B5.2 Magnetic magnetite
The mineral magnetite can acquire permanent magnetism. Above 580°C (the Curie point), the vibration of atoms is so great that the magnetic poles of individual atoms (small arrows) point in random directions. Below 580°C, the atoms begin to align and form tiny magnets. In the presence of an external magnetic field, most domains will be parallel to the external field, and the material becomes permanently magnetized.

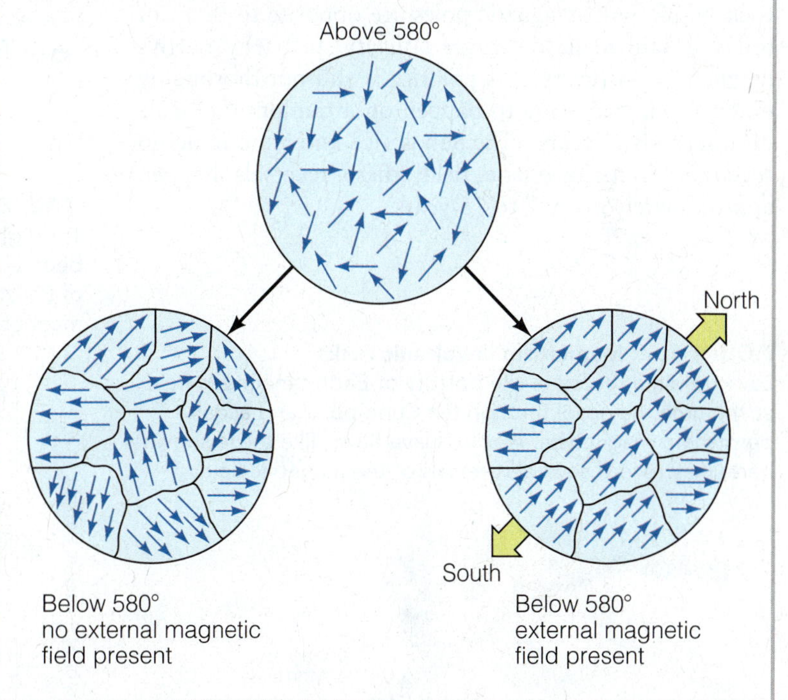

However, a mechanism to explain *how* the movement occurred was still lacking, and so was a theory to account for the driving force behind the movement.

Magnetic Reversals and Seafloor Spreading

Help came from an unexpected quarter. All the early debate about continental drift, and even the data on apparent polar wandering, had centered on evidence drawn from the continental crust. But if continental crust moves, why shouldn't oceanic crust move too?

In the 1950s, geologist Harry Hess of Princeton University (along with other scientists) began to focus on the long, linear oceanic ridges that run the lengths of many of the world's seafloors, like enormous snakelike scars. Hess hypothesized that the topography of the seafloor could be explained if the oceanic crust were moving laterally, splitting apart along the oceanic ridges. Hess could not explain what made the oceanic crust move, but he nevertheless proposed that it did. His hypothesis came to be called **seafloor spreading**, and strong evidence in favor of it was soon found.

Through studies of paleomagnetism, geophysicists had discovered an extraordinary and—even today—poorly understood phenomenon: Some rocks contain a record of **magnetic reversals**; that is, they indicate a south magnetic pole where the north magnetic pole is today, and vice versa. A rock whose paleomagnetic poles match those of Earth today is said to have *normal polarity*; a rock whose paleomagnetic poles are opposite to those of today is said to have *reverse polarity*. Just why Earth's magnetic field reverses its polarity is still not thoroughly understood; it appears to happen on a random timescale of thousands of years. (The Sun is also known to undergo reversals of its magnetic field; these reversals happen approximately every 7 to 15 years.)

The paleomagnetic record of Earth's magnetic reversals provided some information that proved crucial to the scientific debate about continental drift. Using a combination of radiometric dating (Chapter 4) and magnetic polarity measurements in thick sequences of lava flows (FIG. 5.6), scientists have determined exactly when magnetic polarity reversals have occurred in the past. FIGURE 5.7 shows a detailed chronology of magnetic reversals for the past 20 million years, and ongoing work is providing evidence of reversals much deeper in Earth history.

What does this chronology of magnetic reversals have to do with seafloor spreading? The hypothesis of seafloor spreading postulated that oceanic crust is splitting apart and moving laterally away from the oceanic ridges, and that magma rises from the mantle to fill the gap, forming new oceanic crust along the ridges. One important consequence of this splitting and spreading process is that the oceanic crust far from any ridge should be older than crust nearer to the ridge. When lava is extruded into the gap along the oceanic ridge, the new rock that is formed becomes magnetized and acquires the polarity of Earth's magnetic field. The oceanic crust then splits and moves slowly off the ridge in either direction, allowing a new strip of crust to form along the ridge. If a magnetic reversal occurs in the interim, the new strip of crust will acquire this reverse polarity. As the new oceanic crust splits and moves apart, it acts like a very slowly moving magnetic recorder and conveyor belt, carrying with it the rock record of magnetic

FIGURE 5.6 Magnetism in volcanic rocks
Lavas retain a record of the polarity of Earth's magnetic field at the time they cool through the Curie point and acquire permanent magnetism. A pile of lava flows, like those shown here, may record several reversals of the magnetic field.

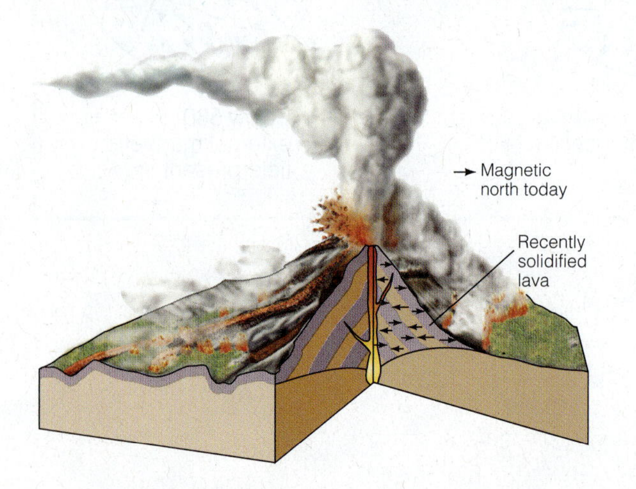

FIGURE 5.7 Magnetic reversals
Through detailed paleomagnetic studies, scientists have been able to determine a chronology of magnetic reversals of Earth's magnetic field. Shown here is 20 million years of magnetic reversal information.

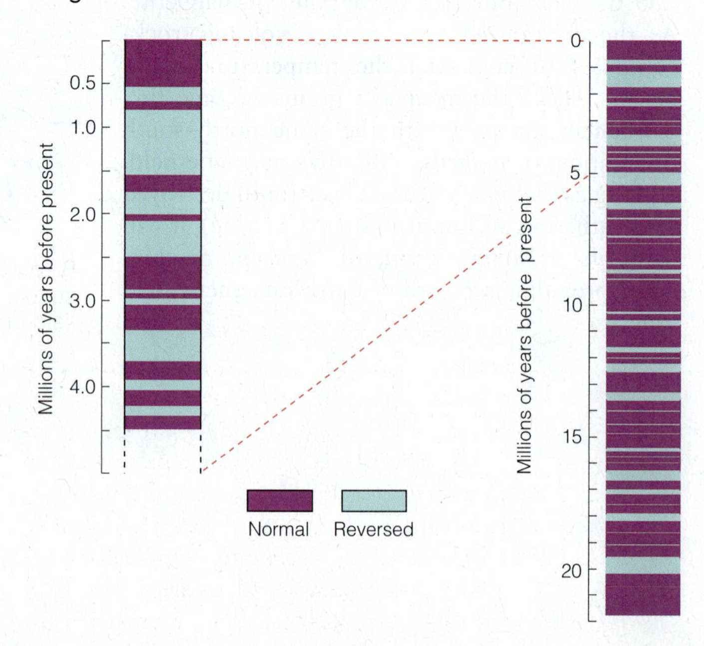

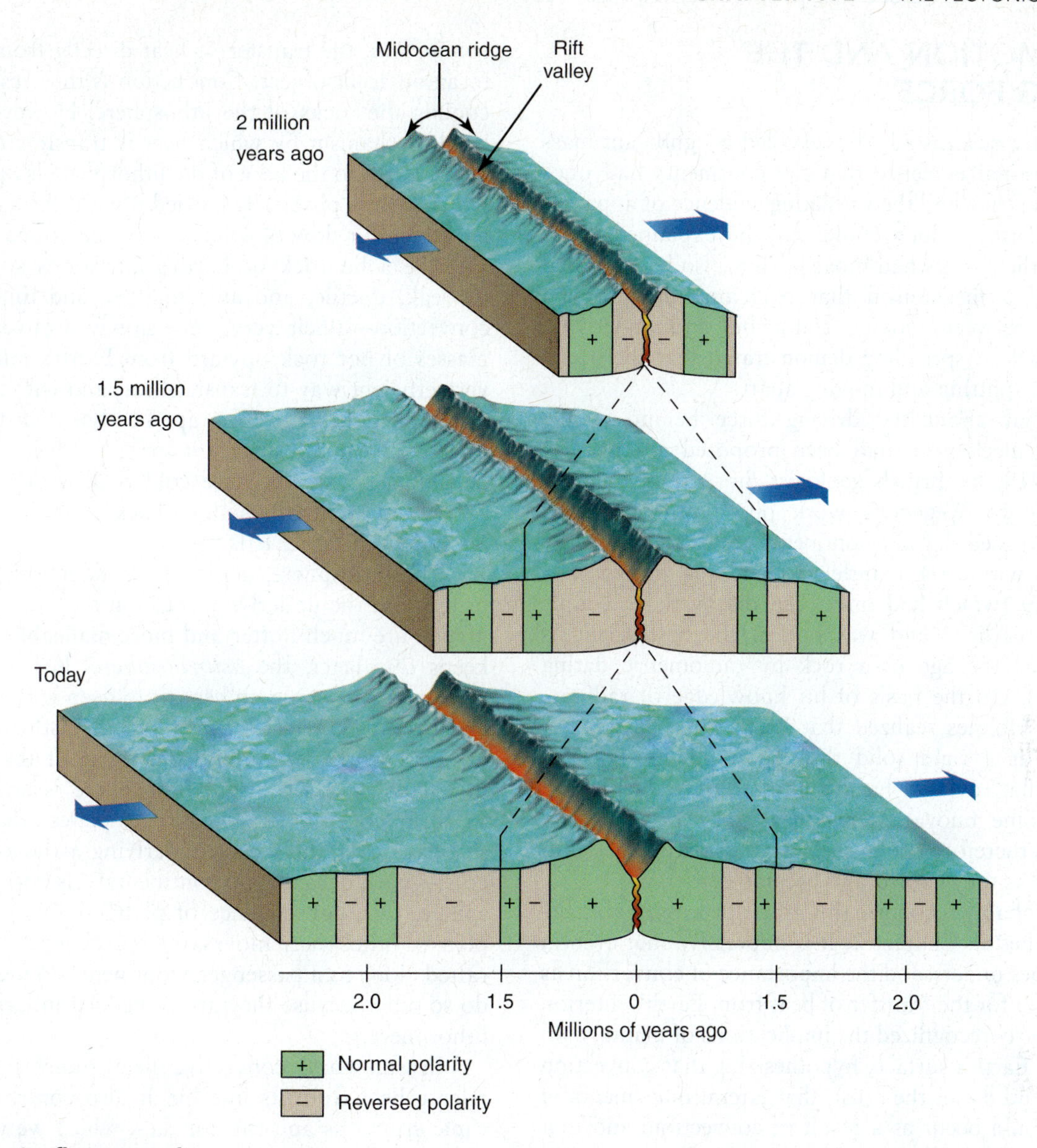

FIGURE 5.8 Seafloor spreading
Lava extruding along a midocean ridge forms new oceanic crust. As the lava cools, it becomes magnetized with the polarity of Earth's magnetic field at the time. As the plates on either side of the midocean ridge move apart from one another, successive bands of oceanic crust have alternating normal and reversed polarities. The resulting magnetic bands are symmetrical on either side of the midocean ridge, and the rocks increase in age with distance from the ridge.

polarities in both directions away from the ridge. Continued movement over time, through several magnetic reversals, yields a sequence of magnetic strips with alternating normal and reverse polarity, oriented parallel to the oceanic ridge (FIG. 5.8).

Scientists have been able to match this pattern of parallel magnetic strips with the known chronology of magnetic polarity reversals, like the one shown in Figure 5.7. This allowed the ages of seafloor rocks to be determined, and it provided scientists with a picture of the seafloor spreading processes that occur along midocean ridges. A crucial piece of information was quickly revealed: Just as the seafloor spreading hypothesis predicted, rocks closest to the oceanic ridges are young, and the ages of seafloor rocks increase with distance from the ridges (as shown in Fig. 5.8). Because the chronology of magnetic polarity reversals had been so carefully determined, the ages derived from the magnetic striping of seafloor rocks also provided a means of estimating the speed with which the seafloor had moved. In some places, such movement is remarkably fast: as high as 10 cm/yr.

Confirmation of the hypothesis of seafloor spreading resolved the problem of *how* the continents were able to move, establishing that they don't have to move across the top of the oceanic crust, a process that geophysicists simply couldn't accept. Rather, the continents are carried passively along, atop the same fragments of lithosphere that are carrying seafloor crust away from the oceanic ridges in spreading ocean basins.

PLATE MOTION AND THE DRIVING FORCE

By the 1960s, scientists had assembled a significant mass of observational evidence that the continents had once been joined. They had the validating evidence of apparent polar wandering, which could only be explained if the continents themselves had moved. They also had detailed evidence of a mechanism that overcame the problem of friction between moving continents and underlying oceans—seafloor spreading demonstrated that the ocean basins were splitting and moving apart.

But what about the driving force behind it all? Actually, a mechanism had been proposed all the way back in 1919, by British geologist Sir Arthur Holmes. In response to Wegener's work on continental drift, Holmes proposed that the continents are carried along by movements within the mantle. Holmes was an expert in radioactivity (which had only been discovered a couple of decades earlier), and was one of the first scientists to determine the age of a rock by radiometric dating (Chapter 4). On the basis of his knowledge of radioactive decay, Holmes realized that Earth must be too hot to behave as a rigid solid throughout its interior. He suggested that rocks in Earth's interior were ductile and would become buoyant when heated by radioactivity; they would therefore rise to the surface, sinking back into the interior as they cooled and became denser.

You probably recognize this as the process of *convection*, described in Chapter 2. It is impressive that even in 1919 Holmes understood the importance of convection as a mechanism for the transfer of heat from Earth's interior. He also clearly recognized the implications of mantle convection for Earth's surface, hypothesizing that convection might lift and break the crust; that lateral movements of the crust could occur as a result of convection; and that where convection cells turned downward, the crust might crumple up to form mountains.

Holmes's work early in the twentieth century helped lay the groundwork for our present-day understanding of these processes. At the time, however, his ideas were an extraordinary departure from conventional thinking about processes such as uplift, lateral movement, and mountain building. Decades passed before scientists were able to produce the hard evidence to support Holmes's fundamental ideas about the driving forces behind crustal movements and continental drift. Since then, research in a number of fields—notably the remote sensing of Earth's interior regions through the study of earthquakes, and the numerical modeling of heat transfer in Earth's interior—has allowed scientists to greatly refine their understanding of convection and how it drives plate motion.

Heat Flow in the Mantle: Review

In Chapter 2 you learned about the three basic modes of energy transfer: radiation, conduction, and convection. You also learned that Earth's interior is very hot and that heat energy flows through the lithosphere to the surface by *conduction*, the transfer of heat directly from one object to an adjacent object. Conduction is thus responsible for cooling the rocks of the lithosphere. However, the principal mechanism by which heat is transferred from deep inside Earth to the base of the lithosphere is *convection*, in which heat is physically carried toward the surface by the movement or flow of solid rock, as described by Holmes.

The solid rock of Earth's interior is so hot that it is weak, ductile, and able to flow, and thus undergoes convection—albeit very, very slowly. Convection brings masses of hot rock upward from Earth's interior. It is a very efficient way to transfer heat, and without it Earth's interior would be heating up. The hot rock flows slowly upward. Near the surface it spreads sideways and cools, becoming denser (because cold rock is denser than hot rock), and eventually sinking back down into the mantle (as shown in Fig. 2.10).

The lithosphere, cooled by conduction, is cold and brittle, but the underlying rocks in the upper part of the mantle are much hotter and more malleable. Convection keeps this layer, the *asthenosphere*, hot and weak by continually bringing up heat from deep within the mantle and the core. (You learned about the asthenosphere and how its properties differ from those of the lithosphere in Chapter 3.) The rigid lithosphere is strong enough to form coherent fragments, or plates, that can slide sideways over the weak, underlying asthenosphere. The crust—both oceanic and continental—is part of the lithosphere. One consequence of plate tectonics is that as a plate of lithosphere moves, the crust capping that plate is rafted along as a passenger. Continents do move, but they do so only because they are embedded in larger plates of lithosphere.

In summary, convective heat transfer helps Earth release heat from its interior. It also contributes to volcanic eruptions and earthquakes, which we will examine in greater detail in Chapter 6. And finally, it transfers heat to the asthenosphere and provides the driving force behind the movement of lithospheric plates. Let's look more closely at how this occurs.

Convection as a Driving Force

Just as Wegener felt sure that continents had drifted but could not explain how, today we are sure that plates move and that convection currents in the mantle play a role in that movement, but we are still unable to say precisely what role. The situation is analogous to recognizing the shape, color, size, and speed capability of an automobile and knowing that gasoline supplies the energy needed for movement, but not knowing exactly how the gasoline makes the engine work. Until the driving mechanism is explained, plate tectonics must remain only an approximate description. Meanwhile, we can hypothesize about the causes of the motion and test the hypotheses by making detailed calculations based on the laws of nature.

We know that the lithosphere is cold, hard, and brittle, and that the underlying asthenosphere is hotter and more mobile. We also know that the lithosphere and the asthenosphere are closely bound together. If the asthenosphere moves, the lithosphere must move too, just as the movement of a sticky molasses or tar will move a piece of wood floating on its surface. Conversely, movement of the lithosphere causes movement in the asthenosphere. Scientists cannot yet separate the relative importance of the two effects; however, on two points they are quite certain: (1) the lithosphere has energy of motion, and (2) the source of this energy is Earth's internal heat, which reaches the base of the lithosphere by convection in the mantle. What has not yet been figured out is the precise way the heat energy brought up by convection causes the plates to move.

Three forces might play a role in moving the brittle slabs of lithosphere. The first is that rising magma and the formation of new oceanic crust at a spreading ridge may be *pushing* the plates away from each other (FIG. 5.9A). A second way the lithosphere could be made to move is by *dragging*. Lithosphere that breaks and starts to sink through the asthenosphere is cold, and therefore denser than the hot asthenosphere through which it is sinking. The sinking fragment of lithosphere may act like a heavy weight, pulling or dragging the entire plate downward (FIG. 5.9B). A third possible mechanism is for the whole plate to be *sliding* downhill away from the spreading ridge. The lithosphere is thinnest near a spreading ridge, where hot magma is rising from the mantle, and grows cooler and thicker away from the ridge. Consequently, the boundary between the lithosphere and the underlying asthenosphere slopes away from the ridge. Even if the slope is extremely gentle, the lithosphere's own weight could cause it to slide at a rate of several centimeters per year (FIG. 5.9C).

All of these proposed mechanisms have pros and cons. At present, there is no way to choose among them. Calculations suggest that each of them operates to some extent, so that the entire process is possibly more complicated than we now imagine. The prevailing hypothesis is that old, cold lithosphere breaks and begins to sink, pulling downward on the plate. Once this movement starts, downhill slide and ridge push combine to keep the process going.

In addition to the discussion about convection as a driving force for plate motion, there is ongoing scientific debate about the exact configuration of convection cells in the mantle. Some scientists believe that convection occurs in the whole mantle, stretching all the way from the core-mantle boundary to the base of the lithosphere (FIG. 5.10). However, we know from seismic studies and

FIGURE 5.9 Mechanisms of plate motion
These are three possible mechanisms whereby mantle convection and the resulting movement in the asthenosphere may drive the motion of lithospheric plates. (A) Magma rising at a spreading center may exert enough pressure to *push* the plates apart. (B) A slab of cold, dense lithosphere sinking into the mantle may *drag* the rest of the plate along with it. (C) A plate of lithosphere may *slide* away from the spreading center under its own weight, along the sloping lithosphere–asthenosphere boundary. It is likely that some combination of these forces is responsible for lithospheric plate motion.

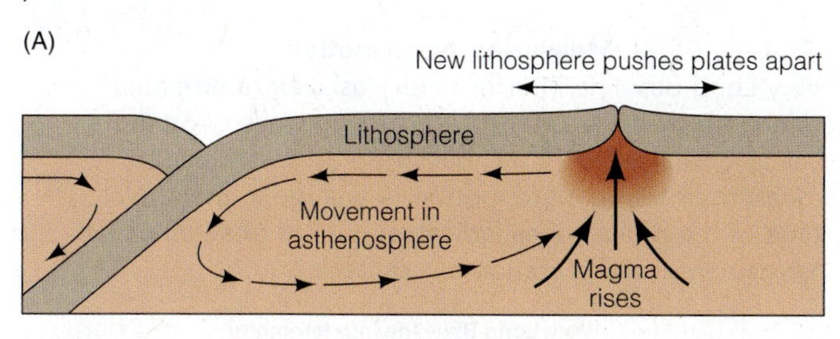

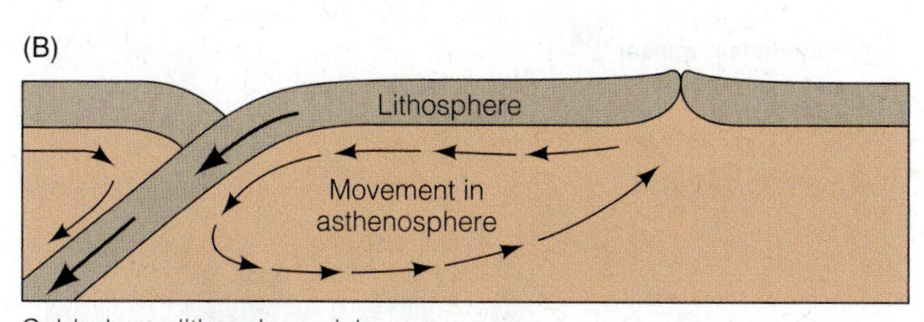

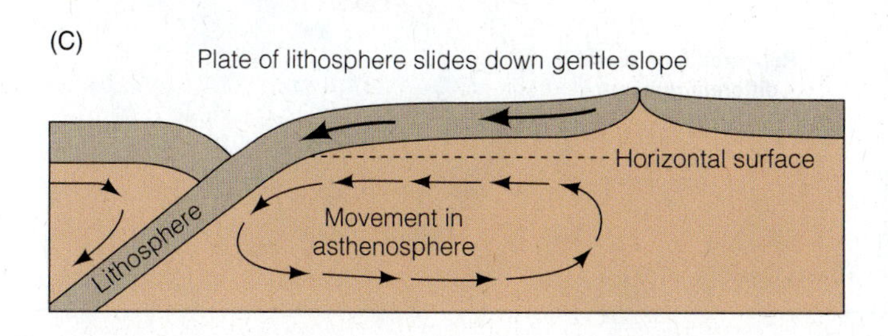

FIGURE 5.10 Modeling mantle convection
The results of computer models of the mantle, like the one shown here, have revealed much useful information about heat flow and convection in the mantle. Here, the colors show the relative temperatures of mantle rocks, with red indicating higher temperatures and blue indicating cooler temperatures. This model shows whole-mantle convection, but other models suggest that the mantle may be convecting in layers.

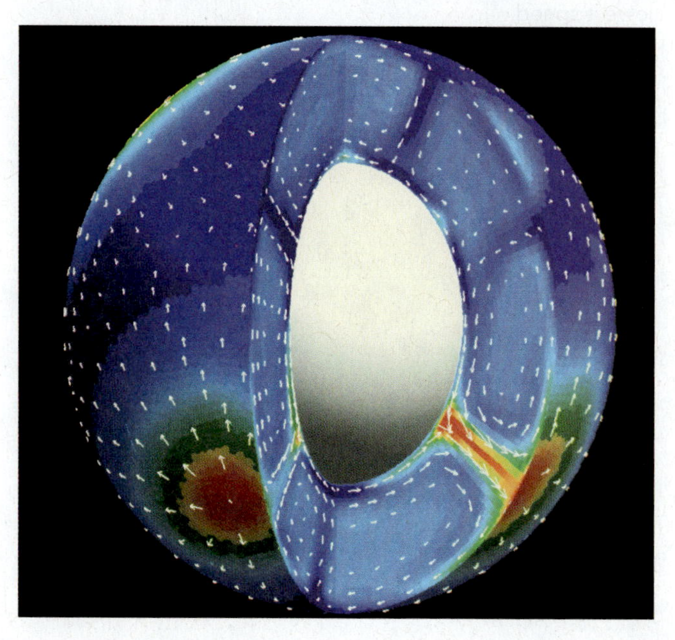

A Closer LOOK

MEASURING THE ABSOLUTE SPEED OF PLATE MOTION

Measuring the rate and direction of movement of lithospheric plates presents a couple of tricky challenges. You might think, intuitively, that all points on a plate move with the same speed, but your intuition would be incorrect. It would be correct only if the plates were flat and moved over a flat asthenosphere (like a flat piece of wood floating on water). Tectonic plates are pieces of a shell on a spherical Earth; in other words, they are curved, not flat. Any movement on the surface of a sphere is not a simple translational movement, but a rotation about an axis of the sphere. A consequence of such a rotation is that different parts of a plate move with different speeds, as shown in **FIGURE C5.1**.

The fact that the entire lithosphere is moving presents an additional challenge. It's not so difficult to measure *relative* motion—the movement of a plate in comparison to a neighboring plate. This can be done with considerable precision using laser beams to measure the distance between two points on adjacent plates. As this distance changes over time, the movement of one plate relative to the other can be calculated. However, it is much more challenging to measure *absolute* motion, for which a fixed or stationary frame of reference is needed. A familiar example of absolute versus relative motion occurs when one automobile overtakes another (**FIG. C5.2**). If the drivers

FIGURE C5.1 Movement of a curved plate Earth is a sphere, so lithospheric plates are curved. This means that each point on a plate moves at a different rate. The movement of a curved plate can be described as a rotation about the plate's own spreading axis. Point *P* has zero speed because it is the fixed point around which rotation occurs. Point *A'*, at the edge of the plate farthest from *P* has the highest speed; Point *A*, closest to *P*, has the lowest speed.

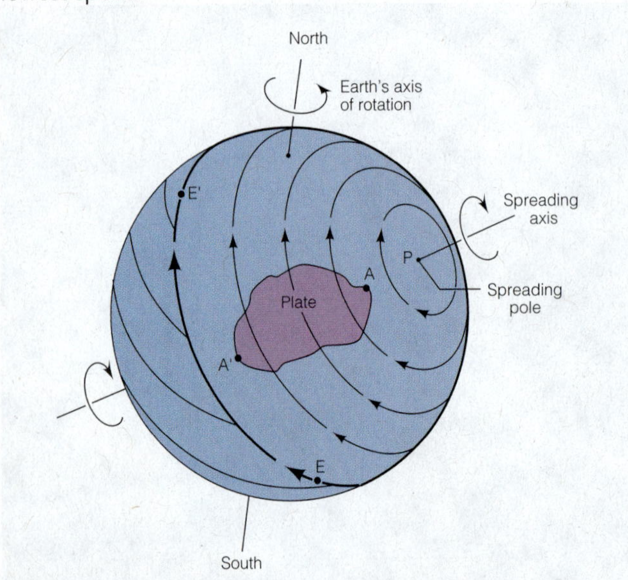

FIGURE C5.2 Absolute and relative speed If you stand on the North American Plate, it is a fairly simple matter to measure the rate of movement of the neighboring Pacific Plate, but this is a measure of relative speed, as shown here for two cars moving at different speeds. To measure the absolute speed of the cars, a stationary reference point is required, like the stoplight shown here; the same is true for plate motion.

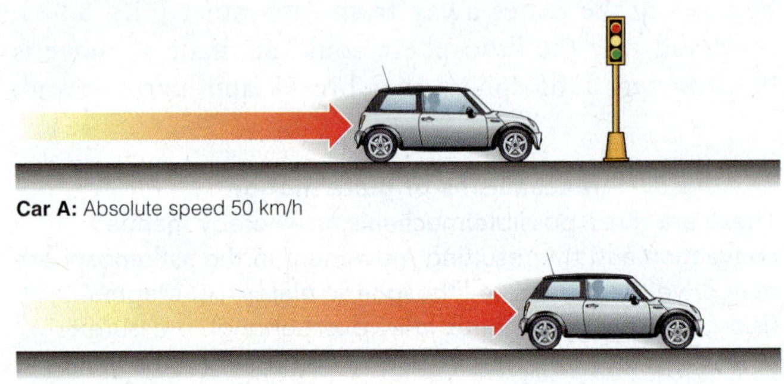

Relative speed (the difference between them): 5 km/h

Absolute speed must be measured in comparison to a stationary point of reference, such as the stoplight.

FIGURE C5.3 Measuring plate motion Very Long Baseline Interferometry uses extraterrestrial reference points—distant starlike radio wave generators called quasars—as a fixed backdrop against which to measure absolute plate motion. A change in the arrival time of the radio waves indicates that the position of the terrestrial receiving station has shifted.

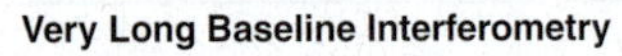

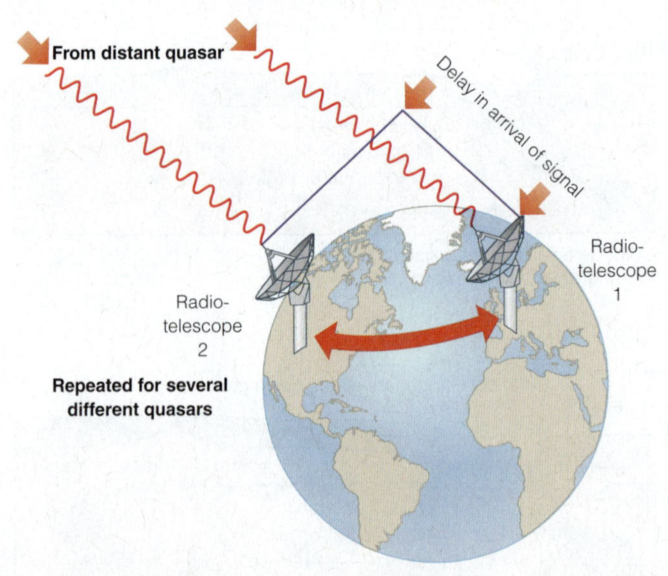

could see only each other and not any fixed objects outside their cars, they could judge only the *difference* in speed between the two cars. For example, one car could be traveling at 50 km/h and the overtaking car at 55 km/h, but the observers would only be able to determine the *relative speed* difference: 5 km/h. On the other hand, if the drivers could measure their speed with respect to a stationary reference, such as a stoplight, they could determine that their *absolute speeds* were 50 and 55 km/h, respectively.

We would be constrained to determine only relative speeds of plate motion if a stationary reference point or framework did not exist; fortunately, such reference points do exist. One set of approaches makes use of extraterrestrial objects that are very far away from Earth, such as distant stars or satellites, as a reference framework. For example, *Very Long Baseline Interferometry (VBLI)* is a tool that relies on a network of radio antennae at different stations around the globe to measure tiny differences in arrival times of radio waves from *quasars*, distant star-like objects that constantly generate radio waves. Since the source of the radio waves is constant and fixed in space, and the velocity of radio waves through space is also constant, even the smallest difference in arrival time at a station on Earth indicates that the station (and hence the plate on which it sits) has changed its position (**FIG. C5.3**). *Satellite Laser Ranging (SLR)* utilizes reflective satellites in Earth orbit as an extraterrestrial reference point. Short bursts of laser light are sent to the satellites and reflected back to Earth. Since, again, the velocity of light is constant, any small difference in travel time is an indication that the position of the terrestrial station has shifted relative to the satellite. Increasingly, *Global Positioning Satellites (GPS)*—the same satellites that provide directions to the GPS unit in your car or cell phone—are being used in a similar manner. By taking measurements in multiple locations against this fixed external framework, and comparing the measurements over time, small differences can be detected.

There is also a fixed framework *within* the Earth system that can be used as a reference against which to measure absolute plate motion. During the nineteenth century, American geologist James Dwight Dana observed that the ages of volcanoes in the Hawaiian chain, some now submerged beneath the sea, increase from southeast to northwest (Fig. C5.3). Apparently, a long-lived magma source, called a *hotspot*, is rooted somewhere in the mantle underneath Hawaii. A volcano remains in contact with the magma source for only about a million years before being carried off by the moving Pacific Plate. A volcano that is carried off of a hotspot should cease to exhibit volcanic activity and begin to erode. Continued movement of a lithospheric plate over a fixed hotspot should produce a chain of extinct volcanoes increasing in age with distance from the hotspot, and this is exactly what we see in the Hawaii volcanic chain. If long-lived hotspots do exist in the mantle, they can provide a series of fixed points against which to measure absolute plate speeds. More than a hundred hotspots have now been identified, although controversy and discussion still remain about how long and at what depth these hotspots are rooted in the mantle.

FIGURE C5.4 Volcanic hotspots
The Hawaiian islands are an expression of volcanism related to a mantle hotspot, a long-lived source of magma that is rooted in place. The island of Hawaii and the underwater volcano Loihi are the youngest volcanoes in the chain and are currently sitting atop the hotspot. The other volcanoes in the chain each spent a brief period—about a million years—on top of the hotspot and were then carried off by the moving Pacific Plate. The numbers show the ages, in millions of years, of the volcanoes in the chain; they are older and more eroded with increasing distance from the hotspot. A bend in the chain of volcanoes occurred because the Pacific Plate changed its direction of movement about 43 million years ago; prior to that time it was moving in a more northerly direction. As the Pacific Plate continues to move toward the northwest, Hawaii and Loihi will eventually be carried off, and a new volcano will form over the hotspot.

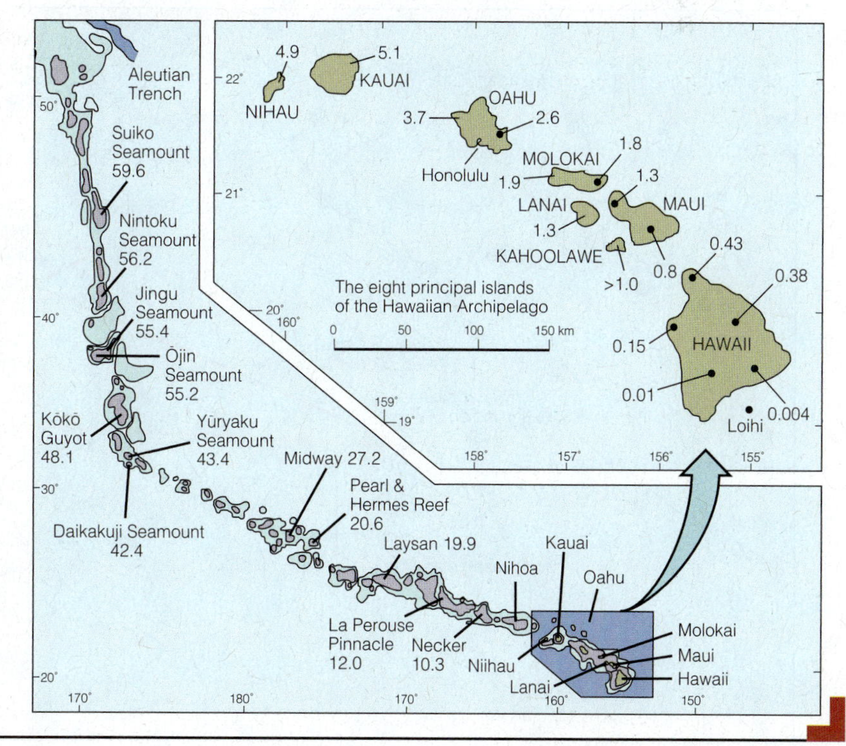

computer modeling that the asthenosphere is not the only layer present within the mantle; there are additional layers with differing properties in deeper parts of the mantle. Perhaps this layering indicates that the upper part of the mantle is convecting separately from the lower part of the mantle. Hopefully, future research will resolve these questions about mantle convection and its relationship to plate motion.

Measuring Plate Motion

We know that convection in the mantle is a pretty slow process, but how fast do the lithospheric plates move across the surface? As a plate moves, everything on it moves too. If the top of the plate is partly oceanic crust and partly continental crust, then both the ocean floor and the continent move with the same speed and in the same direction. The first clear evidence that a seafloor and continent on the same plate of the lithosphere move together came from paleomagnetism; a series of remarkable measurements now provide even more convincing evidence about plate motion.

The newest evidence of plate motion comes from satellites. Using laser beams bounced off satellites, scientists can measure the distance between two points on Earth with an accuracy of about 1 cm. Thus they can monitor any change in distance between, for example, Los Angeles on the Pacific Plate and San Francisco on the North American Plate. By making distance measurements several times a year, they can measure present-day plate velocities directly and accurately. As seen in **FIGURE 5.11**, plate speeds based on satellite measurements closely agree with speeds calculated from paleomagnetic measurements. The close agreement between different measurement techniques implies that plates tend to move steadily rather than by starts and stops. For additional discussion of plate speeds and how they are measured, see *A Closer Look: Measuring the Absolute Speed of Plate Motion.*

The combined results of the various approaches used to measure plate motion have revealed some basic information. Plates that carry lots of continental crust, such as the African, North American, and Eurasian plates, move relatively slowly. For example, the African Plate is nearly stationary (evidenced by the fact that volcanoes there seem to be very long-lived). In contrast, plates without any continental crust, such as the Pacific and Nazca plates, move relatively rapidly; in fact, these are the fastest moving plates. The point on the Pacific Plate that is farthest from its rotational axis (the fastest moving point on a curved plate, analogous to point *A′* on Fig. C5.1) is moving at a speed of about 10 cm/y, about twice as fast as your fingernails grow. The Nazca Plate is adjacent to the Pacific Plate, and moving in roughly the opposite direction; its fastest moving point also has a speed of almost 10 cm/y. This means that the distance between these two

FIGURE 5.11 Plate speeds
Lithospheric plate speeds can be determined in several different ways. The numbers along the midocean ridges are average speeds indicated from paleomagnetic measurements. A speed of 18.3, as shown in the eastern Pacific Ocean, means that the distance between a point on the Nazca Plate and a point on the Pacific Plate increases, on average, by 18.3 cm per year in the direction of the arrows. The long red lines connect stations that have been used to determine plate motion by means of satellite laser ranging (SLR). The speeds determined in this way (L) are very close to the speeds estimated from magnetic measurements (M).

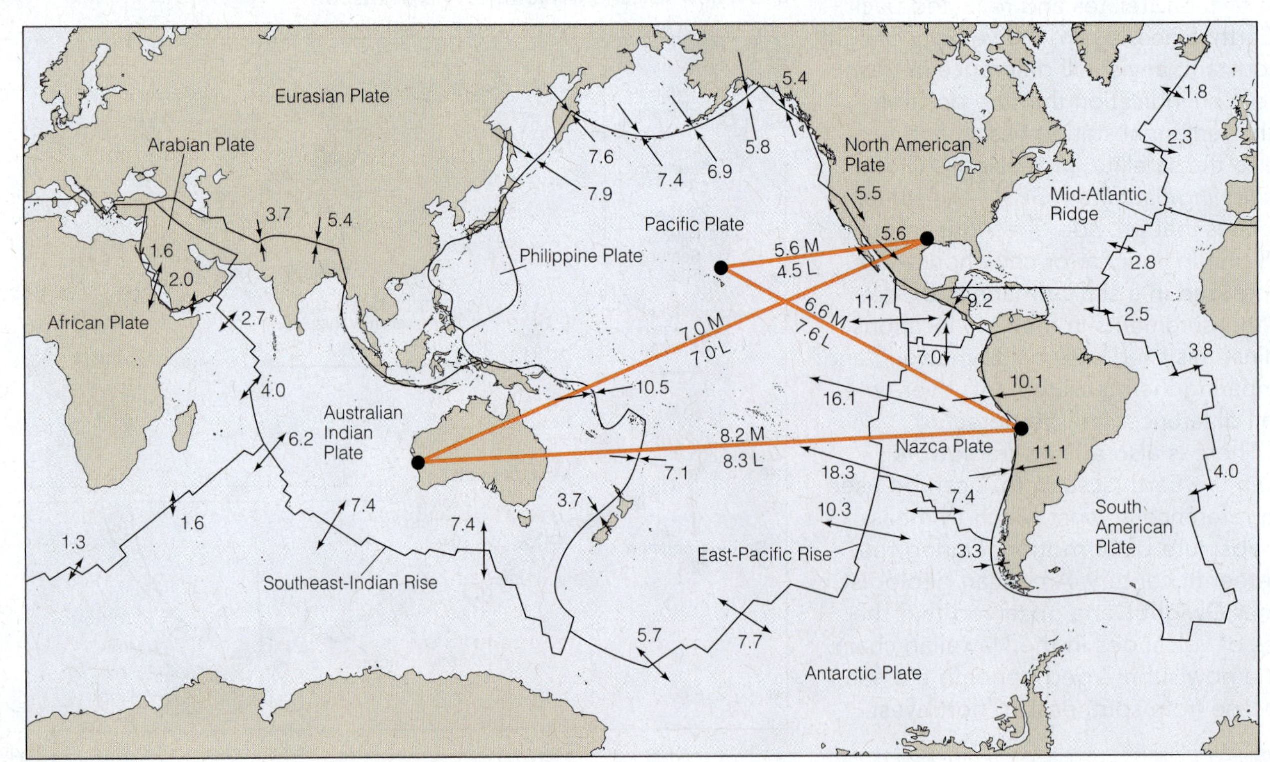

points, one on the Pacific Plate and one on the Nazca Plate, grows larger by almost 20 cm/yr.

Recycling the Lithosphere

A further complication in the measurement of plate motion is that some plates are changing in size. If, as the theory of seafloor spreading requires, new oceanic crust is created by magma rising along the midocean ridges, then either Earth's surface area and diameter must be expanding and the ocean basins getting larger, or else an equal amount of old crust must be getting destroyed somewhere else.

Most scientists reject the idea that Earth's diameter has increased significantly over geologic time. Expansion would affect Earth's rotation and other astronomical characteristics, and there is insufficient evidence to indicate that this has occurred. On the other hand, most of the great mountain chains of the world, like the Alps and Himalayas, contain abundant evidence of *compression*—more consistent with a *shrinking* Earth than an expanding Earth.

Another factor to consider is the age of oceanic crust. All oceanic crust on Earth is geologically young—the very oldest oceanic crust is no more than about 200 million years old. Contrast this with the ages of some continental cores, which are billions of years old. Some process must be destroying or consuming oceanic crust before it gets to be very old.

Scientists now recognize that *constructive* plate boundaries, where new crust is being generated, are balanced by *destructive* plate boundaries, where old crustal material is being consumed or destroyed. These are locations where slabs of lithosphere capped by old, cold oceanic crust are sinking into the asthenosphere along the downgoing limbs of mantle convection cells, in a process called **subduction.** The destruction of old oceanic crust and the creation of new oceanic crust are thus balanced—providing another example of a cyclic process.

How does this affect the sizes of lithospheric plates? A plate that has a spreading ridge but no subduction zone must grow in size; the African and North American plates are examples. To keep things in balance, plates with subduction zones must be slowly shrinking. Most modern subduction zones are located around the edge of the Pacific Plate; thus, much of the oceanic lithosphere now being destroyed is in the Pacific. The Indian Ocean, the Atlantic Ocean, and most other oceans are growing larger, while the Pacific Ocean is steadily getting smaller. It is estimated that about 200 million years in the future the Pacific Ocean will have disappeared and Asia and North America will have collided as a result.

Scientists postulated the existence of destructive plate boundaries where subduction and the recycling of lithosphere must be occurring, and some topographic evidence pointed to this process even before the plate tectonic model provided an explanation. Confirmation was eventually provided by studies of very deep-seated earthquakes under the ocean basins; we will examine this evidence in greater detail in Chapter 6.

PLATE INTERACTIONS AND EARTH'S LANDSCAPES

One of the great strengths of the plate tectonic model is that it provides a coherent framework for hundreds of years' worth of observations about Earth's surface and the features of our landscape. These features and their connection to the plate tectonic model are most easily visualized by considering the processes that occur at different kinds of plate margins—that is, the locations where lithospheric plates interact with one another by moving together, splitting apart, or sliding past one another (FIG. 5.12). We now understand that the dynamic forces of plate tectonics, particularly the processes at work in these areas of active plate interaction, have fundamentally shaped the landscapes of Earth's surface.

Plate Margins

Jostled by the underlying mobile asthenosphere, the lithosphere has broken into nine major plates and a large number of smaller plates. Because the lithosphere is rigid, the plates move as individual, coherent units. Interactions between plates occur along their edges, which are called *plate boundaries* or *plate margins* (Fig. 5.12).

Plate boundaries are the most geologically active places on Earth. This is not too surprising—if you think about a plate the size of the Pacific Ocean rubbing past a plate the size of North America, it should be pretty clear that some epic geologic activity is likely to occur in such a location. Scientists have deciphered most of what we know about plate margins and the interactions that occur along them through the study of earthquakes and volcanism. In Chapter 6 we will look in greater detail at these processes and how they are related to the tectonic cycle.

In the meantime, here is a summary of the basic types of plate boundaries:

1. **Divergent plate margins** are fractures in the lithosphere where two plates move apart. They are also called **spreading centers,** and they can occur in plates capped by either continental crust (FIG. 5.12A) or oceanic crust (FIG. 5.12B).
2. **Convergent plate margins** occur where two plates are moving toward each other. Convergent margins can occur between plates carrying either oceanic crust (FIG. 5.12C), continental crust (FIG. 5.12D), or both (FIG. 5.12E); because of their very different properties, the results of the convergence will be fundamentally different.
3. **Transform fault plate margins** are fractures in the lithosphere where two plates slide past each other, grinding and abrading their edges as they do so. Transform fault margins (FIG. 5.12F) can occur in oceanic crust or in continental crust.

These types of interactions have been occurring along plate boundaries for as long as plate tectonics has been active on this planet—that is, for much of Earth

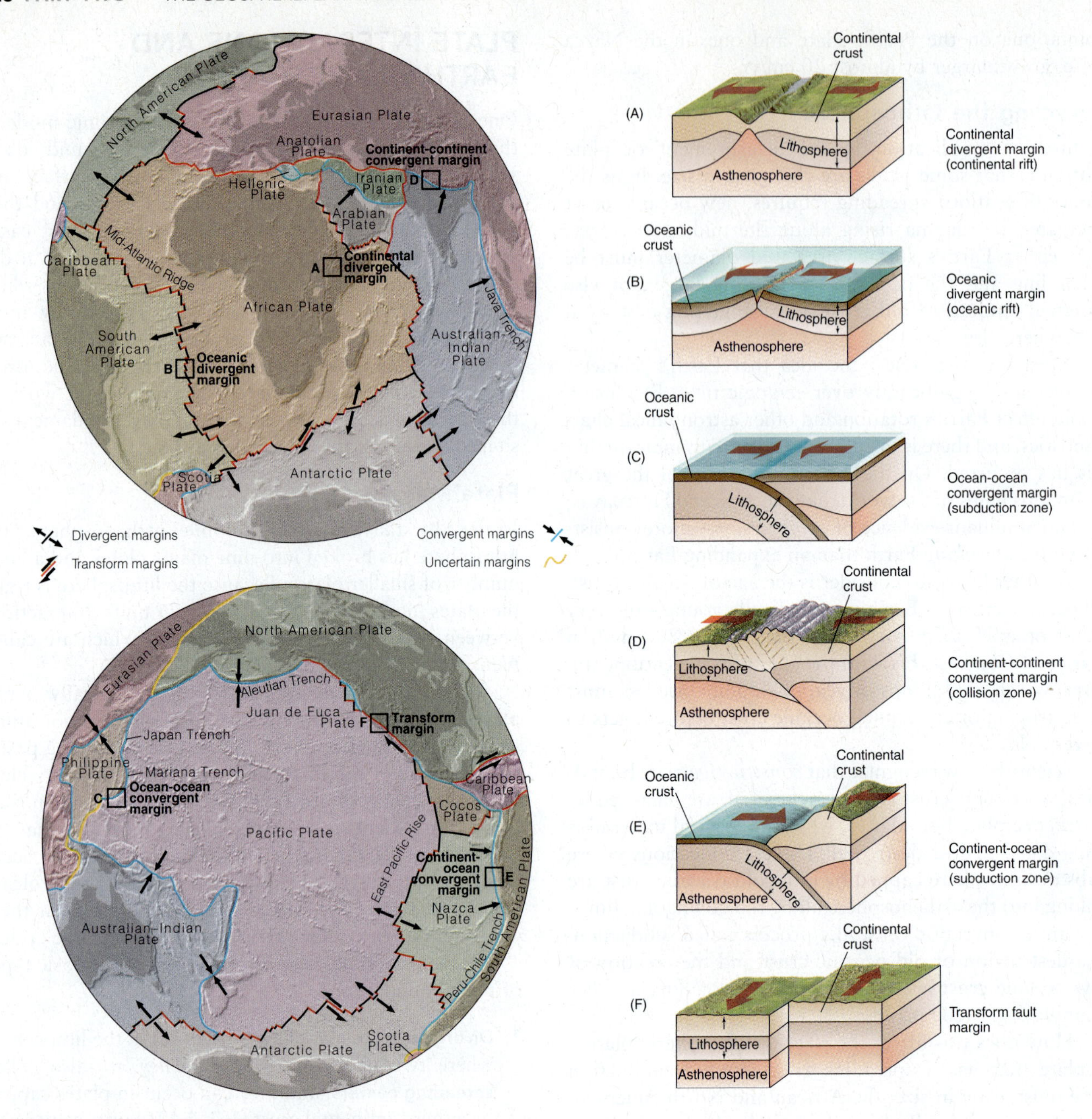

FIGURE 5.12 Plates and plate boundaries
Lithospheric plates interact along plate boundaries in three principal ways, as shown by the arrows on this diagram. They can move apart along divergent margins, they can come together along convergent margins, or they can slide past one another along transform fault margins. The black boxes on the maps and the diagrams labeled (A–F) correspond to the different types of plate boundaries discussed in the text.

history. Let's look at some of the characteristics of each type of plate boundary and the processes that occur along them.

Divergent Plate Margins

Curious as it may seem, divergent plate margins start on a continent (Fig. 5.12A) and become an ocean (Fig. 5.12B). The sequence of events is illustrated in FIGURE 5.13. Huge continental masses act like thermal blankets, slowing down the escape of heat from the interior. A plate capped by a large continent, such as the African Plate, slowly heats up from below, expands to form a broad plateau, and eventually splits to start a cycle of spreading. Today we can see evidence of this process in the form of a great fracture, or **rift**, cutting across the eastern part of the African continent—the East African Rift.

A continental rift that continues to grow will eventually become a long, linear sea, like the Red Sea (Fig. 5.13 and Fig. 5.2B). The earliest rifting of what is now the Red Sea must have looked much the way the East African Rift Valley looks today. As the rift that eventually became the Red Sea widened, a time was reached where seawater

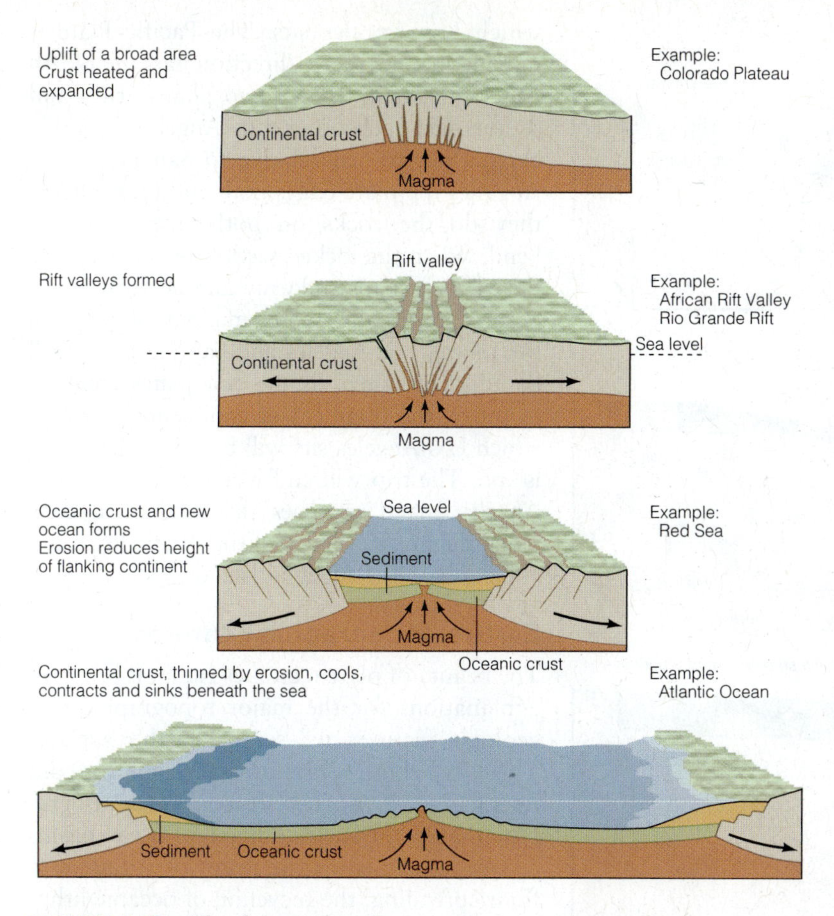

FIGURE 5.13 The formation of an ocean
The rifting of continental crust (top) begins the formation of a new ocean basin, which may continue by the process of seafloor spreading until a mature ocean basin develops (bottom).

entered. Over time, the young, linear sea will grow by the process of seafloor spreading, as its basin widens and magma wells up to form new oceanic crust along the spreading rift. The Atlantic Ocean today is an example of a mature ocean basin with a long, linear spreading center, the Mid-Atlantic Ridge, running its entire length (as shown in Fig. 5.12). The two halves of the Atlantic Ocean are still moving apart from one another on either side of the Mid-Atlantic Ridge, as the ocean basin continues to expand.

Convergent Plate Margins

Convergent margins are plate boundaries where two lithospheric plates move toward one another. Because of the very different physical characteristics of oceanic and continental crust, plates carrying different types of crust behave very differently when they converge. Let's first consider what happens when plates carrying oceanic crust come together.

Recall from Chapter 3 that oceanic crust is typically composed of a relatively dense volcanic rock called basalt, which has an average density of about 3.0 g/cm^3. When two plates capped by oceanic crust converge (an *ocean–ocean convergent margin*), one of the plates will undergo subduction and descend into the mantle underneath the other plate (Fig. 5.12C). The location where this occurs is called a **subduction zone.** Oceanic crust is dense, but mantle rocks are even denser; this supports the hypothesis that oceanic crust does not simply founder into the mantle (it couldn't sink under its own weight if it is less dense than the underlying mantle). The fact that the lithosphere is cold contributes to a higher density in the subducting plate, relative to the underlying hot asthenosphere. It also appears that convection drives the downgoing motion of the lithospheric plate, either by dragging, pushing, sliding, or some combination, as discussed earlier. Which one of two converging oceanic plates will be subducted and which one will ride over the top of the other seems to be primarily a matter of the velocity and angle at which the plates approach each other.

In contrast, continental crust is *much* less dense and more compositionally varied than both oceanic crust and the mantle. The average composition of continental crust is roughly that of the plutonic rock granodiorite. With an average density of about 2.7 g/cm^3, continental crust is simply too buoyant to be dragged or pushed down into a subduction zone. When a plate capped by continental crust meets a plate capped by oceanic crust (an *ocean–continent convergent margin*), therefore, the continental plate will ride up and over, and the oceanic plate will undergo subduction (Fig. 5.12E). This is happening today along the western coast of South America, for example.

When *both* plates of a converging pair are capped by low-density continental crust (a *continent–continent convergent margin*), neither can undergo subduction, and the eventual result is a collision (FIG. 5.14). A **continental collision zone** marks the final disappearance of an ocean basin between converging continents (Fig. 5.12D). Continental collisions result in crumpling, folding, and uplift of continental crust, and the formation of spectacular mountain ranges. The Alps, the Urals, the Himalaya, and the Appalachians all resulted from continental collisions, and therefore each is the graveyard of an ancient ocean basin. (The place where the ocean basin finally disappears, sandwiched between the colliding continents, is called a *continental suture zone.*)

Whereas divergent margins are constructive, producing new crust and the opening of an ocean basin, convergent margins are destructive, consuming old oceanic crust and ending with the disappearance of the ocean basin. This entire process—rifting and the opening of a new ocean basin; its widening into a mature ocean; convergence and the recycling of oceanic crust in a subduction zone; and the eventual disappearance of the ocean basin in a continental collision zone—is referred to as a *Wilson cycle*, in honor of J. Tuzo Wilson, one of the great contributors to the development of the theory of plate tectonics.

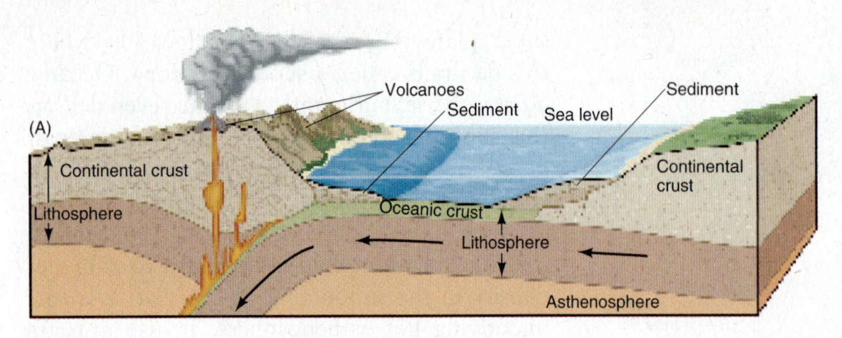

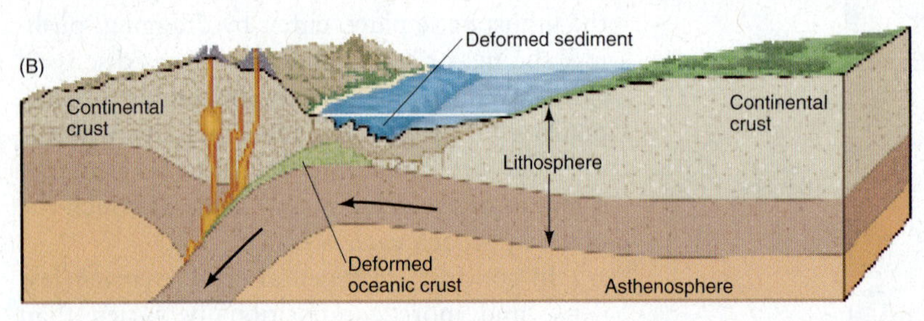

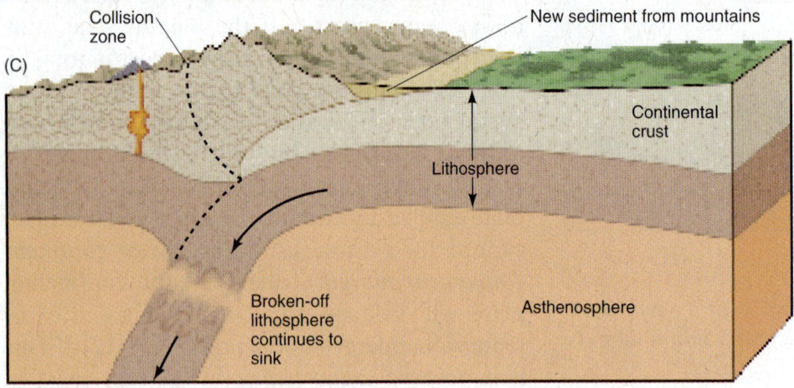

FIGURE 5.14 Continental collision
Mountains form when two plates capped by continental crust collide. (A) Subducting oceanic lithosphere leads to the closure of an ocean basin. (B) As the ocean basin closes, the rocks and sediments along the converging margin begin to be deformed. (C) When the two continental masses collide, the lithosphere becomes thickened. A mountain range is uplifted where the ocean basin once was.

Transform Fault Plate Margins

Plates come together along convergent margins and move apart along divergent margins, but they also can slide past one another (Fig. 5.12F). *Transform faults* are great vertical fractures that cut right down through the entire lithosphere. They occur along boundaries where two plates are sliding past one another laterally, in a horizontal or **strike-slip** motion. Whereas divergent margins are constructive and convergent margins are destructive, transform margins are said to be *conservative*, because crust is neither created nor consumed in this geologic environment.

One transform fault margin that has been much in the public eye because of the threat of earthquakes along it is the San Andreas Fault in California (FIG. 5.15). This fault, which runs approximately north–south, separates the North American Plate on the east, on which San Francisco sits, from the Pacific Plate on the west, on which Los Angeles sits. The Pacific Plate is moving in a northerly direction past the North American Plate. As the two plates grind and scrape past each other, Los Angeles is slowly carried northward relative to San Francisco. At times the plate edges grab and lock, and as they do the rocks on both sides flex and bend. When the locked section breaks free, the flexed rock snaps suddenly and an earthquake occurs. Eventually, many millions of years in the future, Los Angeles and San Francisco will be adjacent. Then, as the two plates continue to move, the fragment of continental crust on which Los Angeles sits will become a long thin island. The trip will end when the future "Los Angeles Island" reaches the subduction zone along the northern edge of the Pacific Plate and collides with the Aleutian Islands.

Earth's Topographic Features

The beauty of plate tectonics is that it provides explanations for the major topographic and geologic features we see at Earth's surface. This is true both for ocean basins and for continents (FIG. 5.16). The structure and *topography* (that is, the surface shape) of ocean basins can be explained as the result of rifting, seafloor spreading, the recycling of oceanic lithosphere in subduction zones, and the eventual closure of the ocean basin between two converging plates. Continents, on the other hand, do not undergo subduction; they ride along like passengers on the lithosphere through geologic time. Continental rocks record the history of this ride in the repeated crumpling, folding, and uplift of the continents into great mountain chains in collision zones. Even after the mountains have been worn down by weathering and erosion, clear evidence of these massive tectonic events can still be seen in the twisted remnants of ancient mountain ranges.

Prior to the development of the theory of plate tectonics, there were other hypotheses to explain features such as the distribution of volcanoes, the continuity of mountain chains, the shapes of coastlines, and the topography of the ocean basins. However, none of them have stood the test of time or done as good a job at unifying scientific observations of Earth's topographic and geologic features as the theory of plate tectonics has done. Let's examine some of these features, and consider how they fit into the theory of plate tectonics.

Topography of Ocean Basins

Oceans begin to form where rifting occurs along a divergent plate boundary. Heated from below, the lithosphere swells upward to form a broad plateau. The lithosphere thins, and its boundary with the asthenosphere comes close to the surface (Fig. 5.12A, B and 5.13). This thinning occurs

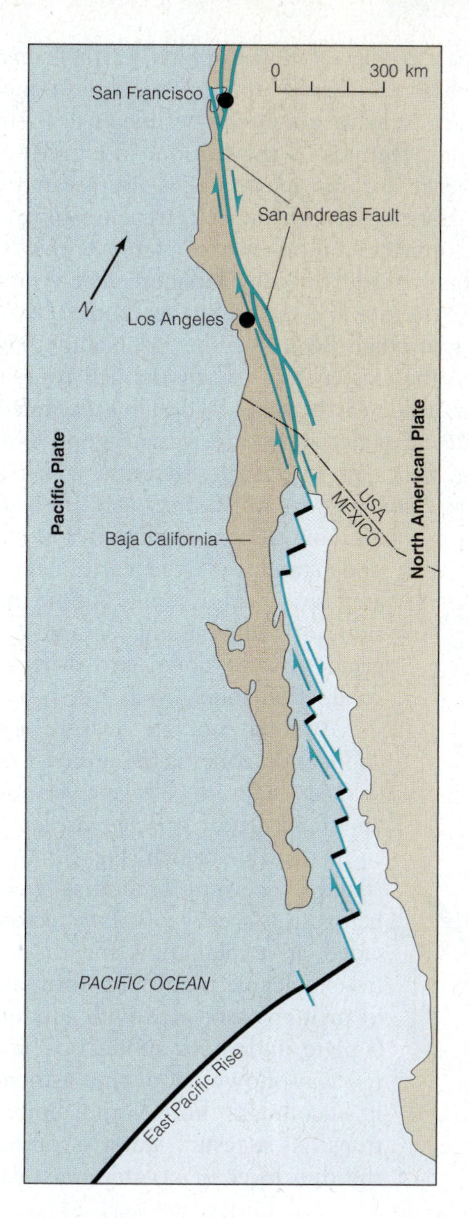

FIGURE 5.15 Transform fault margin
The San Andreas Fault is a transform fault margin that separates the Pacific Plate from the North American Plate. Directions of motion are shown by the blue arrows. Los Angeles, on the Pacific Plate, is moving north, while San Francisco is moving south, bringing the two closer together at a speed of 5.5 cm/yr. As the two plates grind past one another, earthquakes occur.

because magma rising toward the spreading center heats the lithosphere so that only a thin layer near the top retains the strength properties of the lithosphere. Tensional stress from the divergence of the plates eventually causes the lithosphere to break. The resulting rift is typically steep-sided, with pronounced topographic relief between the plateau and the floor of the rift valley. Seafloor spreading centers are typically raised, curvilinear ridges, like a long, jagged gash running the length of the ocean basin (as seen in Fig. 5.16); the Mid-Atlantic Ridge is typical.

As divergence proceeds, the two sides of the original rift move outward to form the edges of the newly expanding ocean basin. This is why the coasts of Africa and South America look similar—they are the two sides of the original rift along which the Atlantic Ocean first began to split and expand. However, modern continental shorelines usually don't coincide exactly with the original split that started the rifting process. This is because some ocean water may spill out of the ocean basin onto the edges of the continent (FIG. 5.17), covering the true edge of the continent. The flooded margin of a continent is known as the **continental shelf.** At the seaward edge of the continental shelf is a sharp drop-off called the **continental slope.** At the base of the steep continental slope is the **continental rise,** a region of gently changing slope where the seafloor flattens out and continental crust meets oceanic crust. This is the true geologic edge of the continent.

Between the continental rise and the spreading ridge lies the rarely seen world of the deep ocean floor. Large, flat areas known as **abyssal plains** are a major topographic feature of the seafloor (Fig. 5.17). These plains generally are found at depths of 3 to 6 km below sea level and range in width from about 200 to 2000 km. Abyssal plains form as a result of the mud settling through the ocean water and burying the seafloor topography beneath a blanket of fine debris. They are most common in the Atlantic and Indian oceans, which have large, mud-laden rivers entering them.

FIGURE 5.16 Topography of continents and oceans
The plate tectonic model accounts for most of the topographic features of the continents and ocean floors. If we could drain all of the water from the ocean, we would see the seafloor as vast, flat areas with long chains of underwater mountains and deep trenches. Note the difference in elevation between the ocean basins and the continents, the long, snakelike midocean ridges, and the high mountain ranges that characterize many of the continents.

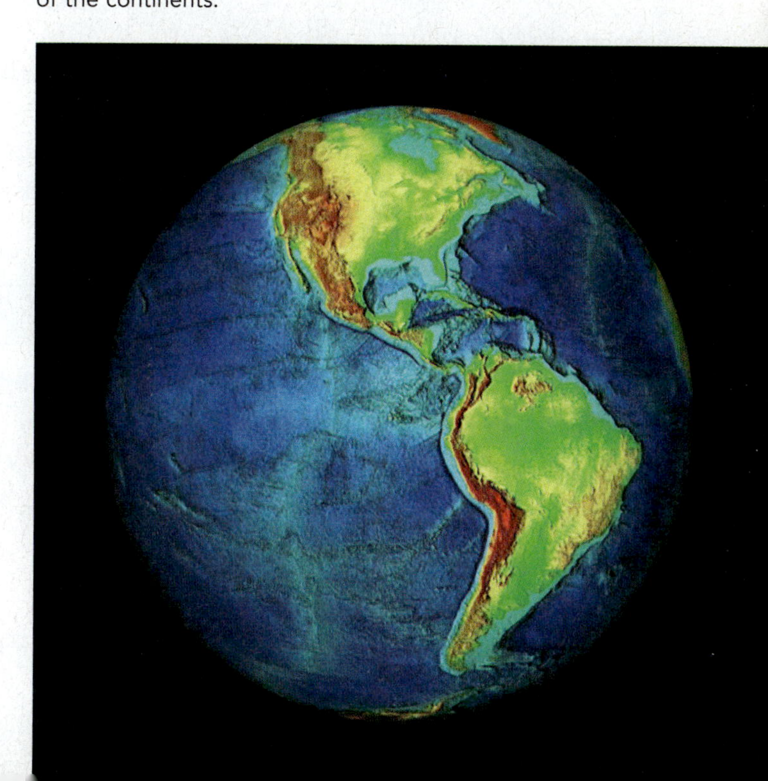

Topography of Subduction Zones

The above topographic features are typical of oceanic basins that are in the process of opening, like the Atlantic. But what about ocean basins that are closing? Oceans close along subduction zones, where old oceanic crust is consumed and recycled to the mantle. The geographic features of a typical subduction zone are shown in FIGURE 5.18. Particularly striking is the deep **oceanic trench** that marks the place where oceanic-capped lithosphere sinks (or is dragged or pushed) into the asthenosphere. Trenches like this are the deepest topographic features of the world's ocean floors. The Mariana Trench, which reaches a depth of nearly 11 km below sea level, is the deepest topographic feature on Earth—considerably deeper than the height of even the tallest mountains. These deep trenches are clearly visible as dark, low-lying topographic features on the seafloor in Figure 5.16.

Evidence for the existence of deep-seated geologic activity associated with oceanic trenches long predated the development of the theory of plate tectonics. The evidence was first identified by Japanese geophysicist Kiyoo Wadati in the late 1920s, and further studied by American geophysicist Hugo Benioff in the 1940s and 1950s. They identified the existence of clearly defined zones of deep, high-magnitude earthquakes underlying oceanic trenches (FIG. 5.19). For decades there was no good explanation for these zones of earthquake activity, which are now known as *Benioff zones* or *Wadati-Benioff zones*. In the unifying context of plate tectonics, we now understand that the earthquakes associated with Benioff zones outline the dimensions of a downgoing oceanic plate as it grinds its way down into the mantle in a subduction zone.

Another notable feature typical of subduction zones is the presence of an arc-shaped chain of volcanic islands (called an *island arc* or a *magmatic arc*) parallel to the oceanic trench (Fig. 5.18). Volcanic island arcs were of course well known before the theory of plate tectonics provided an explanation for their distribution. Volcanic island arcs form as a result of melting associated with subduction. As a plate undergoes subduction and grinds its way down into the asthenosphere, it encounters hotter and hotter conditions. As a result, water is driven out of the thin layer of oceanic crust carried on

FIGURE 5.17 The "true" edge of a continent
The true edge of a continent is where continental crust meets oceanic crust (shown by the dashed line on this diagram). This doesn't always correspond to the shoreline of the ocean basin, as defined by the edge of the water. This is because some of the water in the ocean basin may spill out onto the shallowly-sloping continental shelf.

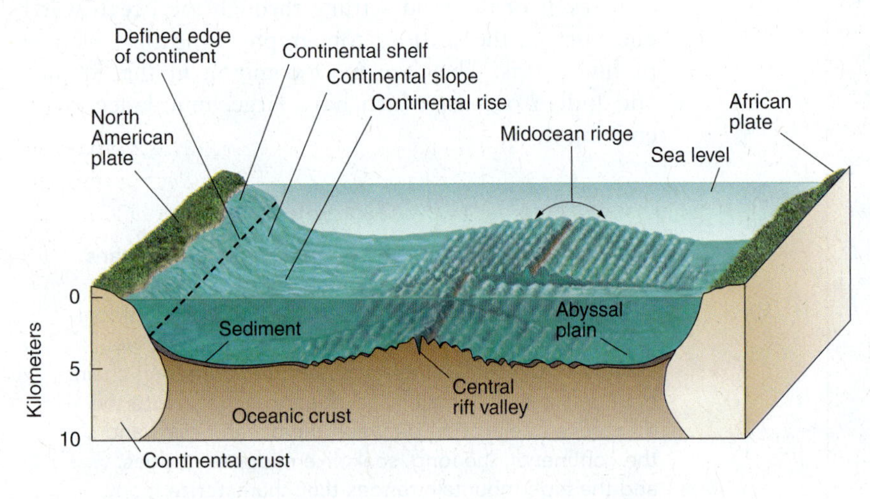

FIGURE 5.18 Subduction zone
When oceanic crust is involved in a convergent plate margin, subduction will occur. The features typical of subduction zones include deep-ocean trenches and island arc volcanism.

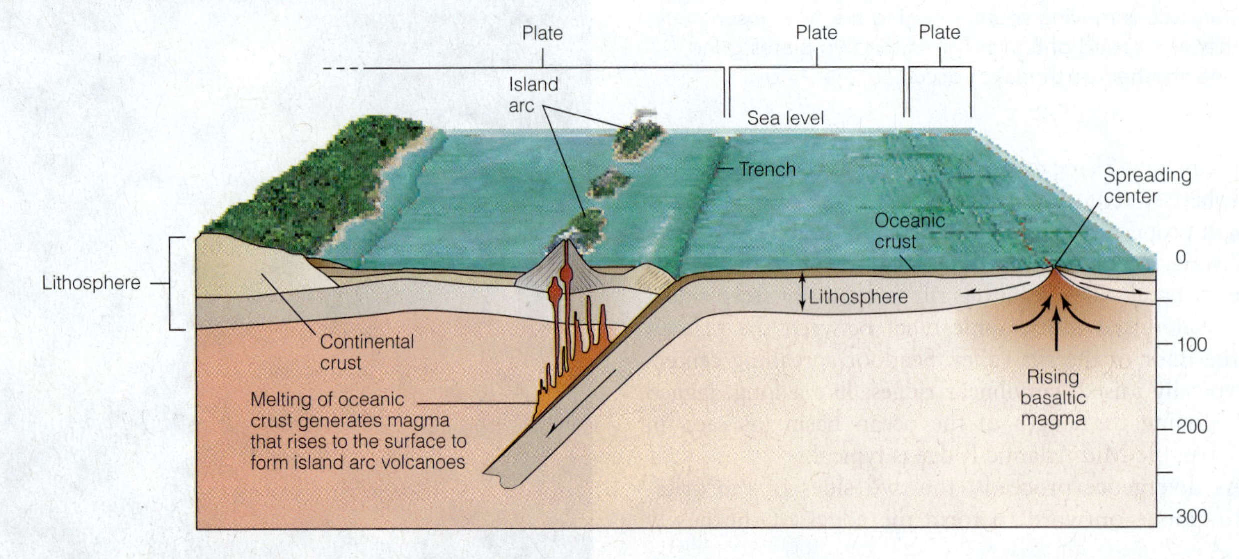

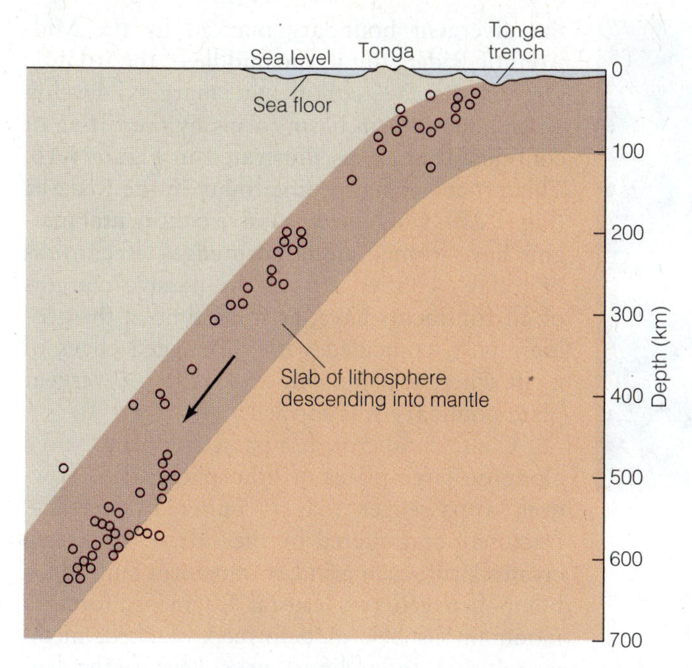

FIGURE 5.19 Benioff zone
Deep earthquakes define the dimensions of a downgoing slab of oceanic lithosphere in a Wadati-Benioff zone located in the Pacific Ocean near the island of Tonga. Each circle represents a single earthquake in a given year. The earthquakes are generated by the downward grinding movement of the comparatively cold slab of lithosphere.

the subducting plate. Water has the effect of lowering the melting temperature of rocks. At a depth of about 100 km, with the addition of water, conditions are hot enough that melting commences in the mantle rocks that overlie the subducting plate. Some of the magma generated by this melting rises to the surface, forming the volcanoes of the magmatic arc. Note that the arc of volcanic islands is located behind the trench, rather than right on top of it; this is because the downgoing plate doesn't start to melt until it reaches a depth of about 100 km, where the temperature is sufficiently high. FIGURE 5.20 shows a line of volcanic islands associated with an active subduction zone along an ocean–ocean convergent plate margin in Java, Indonesia.

Topography of Continents and Collision Zones

The plate tectonic model does a good job of accounting for the topographic features of ocean basins. Oceanic crust is quite uniform in composition, density, thickness, and other characteristics; as a result, the topographic expression of tectonic activity in oceanic crust is relatively straightforward. However, the same is not true of continental crust, which is extremely heterogeneous. The relationship between continental topography and plate tectonics is correspondingly more complex. Let's look first at the characteristics of continental margins.

Some continental margins coincide with the geologically active edges of tectonic plates; these are called *active continental margins*. The Andean coast of South America is an example of an active continental margin that results from an ocean–continent convergence. On this coast, the Nazca Plate (capped by oceanic crust) is being subducted beneath the South American Plate (capped by continental crust). The most distinctive feature of an active continental margin where subduction is occurring is the *continental volcanic arc*. This line of volcanoes (the Andes in this case, FIG. 5.21) is similar to a volcanic island arc in ocean–ocean subduction zones. The volcanoes are formed when melting occurs and the magma rises to the surface, creating a line of volcanoes on the continental side of the subduction zone.

A continental collision zone is another type of active continental margin, where the edges of two continents, each on a different plate, come into collision (as shown in Fig. 5.14). A modern example is the line of collision between the Australian–Indian Plate and the Eurasian Plate. India, on the Australian–Indian Plate, and Asia, on the Eurasian Plate, are colliding and the Alpine–Himalayan mountain chain is the result (Fig. 5.2A). Active continental margins can also occur as transform faults. The most striking example of a modern transform fault continental boundary is the San Andreas Fault, discussed previously. Major transform faults on continents often look as though a huge knife was dragged along, splitting the land (FIG. 5.22).

FIGURE 5.20 Volcanic island arc
The string of volcanoes that defines the spine of the island of Java and the islands to the east is typical of an ocean–ocean convergent margin. The line of volcanoes forms when the subducting oceanic plate reaches a depth of about 100 km and begins to melt, generating magma that returns to the surface through volcanoes. Some of the more famous volcanoes are labeled.

FIGURE 5.21 Continental volcanic arc
This line of steep-sided, snow-capped volcanoes in the Andes of Ecuador sits above the subduction zone where the Nazca Plate is sinking below the western edge of the South American Plate.

FIGURE 5.22 Transform topography
The topographic expression of a continental transform fault margin sometimes looks like a huge knife was dragged through the crust. This is the San Andreas Fault in California.

Other continents sit in the stable interiors of plates, with edges that are far from the geologically active plate margins. The edges of these continents are *passive continental margins*. The Atlantic Ocean margins of the Americas, Africa, and Europe are examples. The eastern coast of North America, for example, is in the stable interior of the North American Plate, far from any plate margins. (The nearest plate margin is the divergent boundary marked by the Mid-Atlantic Ridge, out in the middle of the Atlantic Ocean.) Passive continental margins develop when a new ocean basin forms by the rifting of continental crust, as illustrated in Figure 5.13. This process is happening today in the Red Sea (Fig. 5.2B), where new, passive continental margins have formed along both edges. Regardless of today's configurations, the passive margins of all continents have, at some time in the geologic past, coincided with the jagged edges of a rift where a plate split apart and a divergent plate boundary was formed.

Continental crust is a passenger being rafted along on large plates of lithosphere; it is, however, a passenger that is buffeted, stretched, fractured, and altered by the ride. Each stretch creates a rift; each grind, a transform fault. Each bump between two crustal fragments forms a mountainous belt of deformed, stacked, metamorphosed, and uplifted rocks. Because the density of continental crust is too low for it to sink into the mantle, much of the evidence of plate motion is recorded in the scars that remain from these ancient collisions. It is fortunate that such ancient evidence is preserved in continental rocks. Oceanic crust is geologically young because it is routinely destroyed by subduction; therefore, the only direct evidence of ancient geologic events must come from the continental crust. To get a better understanding of how continents have been shaped by plate tectonics, we need to step back and take a broader look at the characteristics of continents and how they have evolved over geologic time.

BUILDING THE CONTINENTS

One of the main areas of current research in plate tectonics concerns how and why the continents first formed, and how (or if) this process of stabilization of the continents was related to the onset of plate tectonics. Why did the crust become separated into two fundamentally different types, oceanic and continental? How long has plate tectonics been operating on this planet? Have the processes and impacts associated with plate tectonics always been the same? How have the continents changed over geologic time? The answers to these questions are still speculative. Much of the evidence of processes that operated very early in Earth history has been lost from the rock record as a result of tectonic recycling combined with the external forces of weathering and erosion. However, sufficient evidence remains that geologists are beginning to build a more coherent picture of these earliest periods of Earth history.

Before we consider how the continents have evolved over geologic time, it will be helpful to look first at some of the regional characteristics of continents.

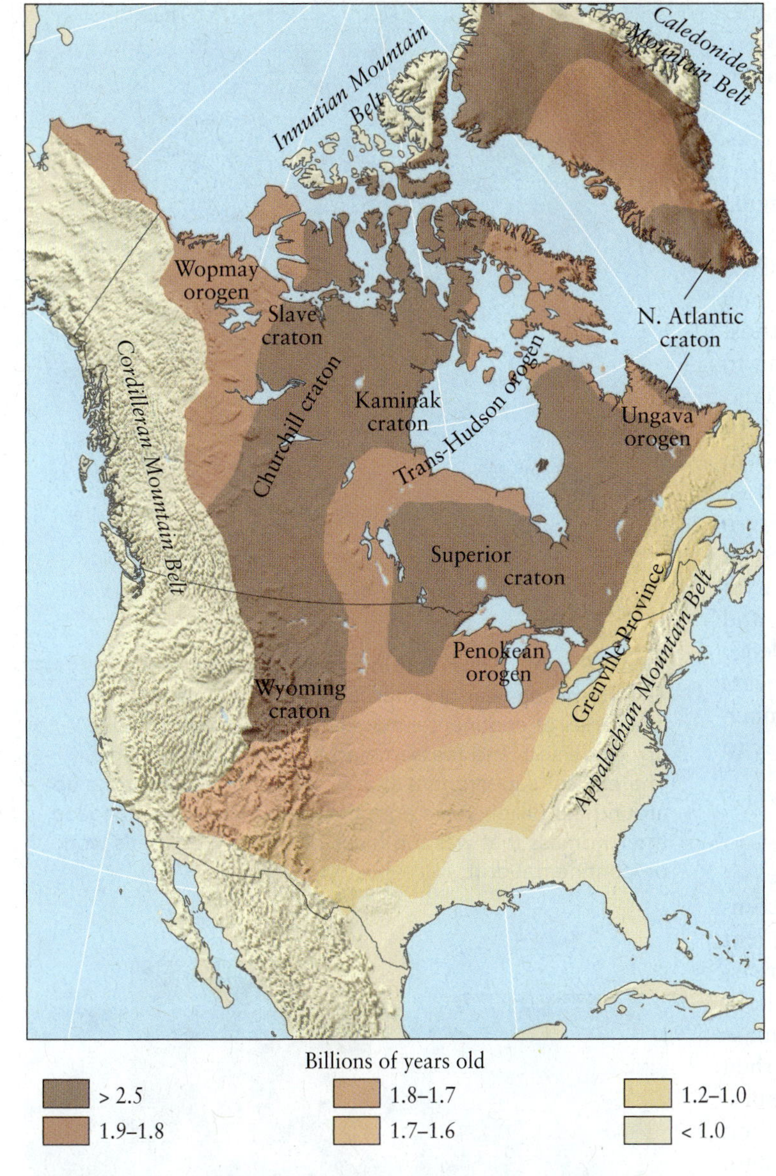

FIGURE 5.23 North American cratons
The North American continental shield is an assemblage of very old, stable continental masses called cratons, interspersed with younger orogenic belts.

Regional Structures of Continents: Cratons and Orogens

On the scale of a continent, two kinds of structural units can be distinguished in the continental crust: cratons and orogens. A **craton** is a stable core of very ancient rock (**FIG. 5.23**). It is impossible to assign a single age to any particular craton because cratons are complex, heterogeneous assemblages of rock that formed, evolved, and stabilized over very long periods. However, all cratons are old; the earliest appear to have stabilized at least 3 billion years ago, and the youngest around 0.6 to 0.5 billion years ago.

Draped around the cratons are the second kind of continental crustal unit, **orogens**, which are elongate regions of crust that have been intensely bent and fractured during continental collisions (Fig. 5.23). Orogens are the eroded roots of ancient mountain ranges that formed as a result of collisions between cratons. Orogens differ from each other in age, history, size, and details of origin; however, all were once mountainous terrains, and all are younger than the cratons they surround. Only the youngest orogens are still mountainous today. Ancient orogens, now deeply eroded, reveal their history through the kinds of rock they contain and the way the rocks are twisted and deformed.

An assemblage of cratons and orogens is called a **continental shield.** The Canadian Shield that forms the core of the North American continent is a familiar example (Fig. 5.23); it is an assemblage of several cratons with interspersed orogenic belts.

The existence of orogens—ancient collision zones—sandwiched between cratons in various locations around the world is the best evidence available to support the idea that plate tectonic interactions have operated throughout much of Earth history. The oldest known orogens that join cratonic masses are about 2 billion years old. This means that plate tectonics and the geosphere portion of the Earth system must have been functioning much as it does today, for *at least* 2 billion years. But what was happening before that time? How did the continents first stabilize, and how have they changed since then?

Stabilization of Continental Crust

Recall from Chapter 4 that the earliest time division we recognize in Earth history is the Hadean Eon, from the origin of the planet (4.56 billion years ago) to 3.8 billion years ago. By that time, the transition to the Archean Eon, there is evidence that a substantial body of continental crustal rocks existed. This evidence comes from studies of the oldest known rocks, some as old as 4.28 billion years, from the Isua Greenstone Belt in Greenland and the Acasta Gneiss Complex in Canada. Individual mineral grains even older than this (4.4 billion years) have been identified, suggesting the existence of at least some continental crust at that time, but no solid rock or mineral evidence predating this has yet been identified. Evidence suggests that the continental mass that was present by the end of the Hadean Eon continued to grow through the Archean Eon, reaching 80 to 90 percent of today's continental mass by about 2.5 billion years ago (the beginning of the Proterozoic Eon).

Exactly how the continental mass first stabilized, how it became chemically differentiated from oceanic crust, and how it grew into a stable craton are matters of continuing speculation. During the Hadean Eon and into the Archean, heat flow from Earth's interior must have been much greater than it is today (probably about three times as high), the lithosphere must have been thinner, and convection must have been more vigorous than it is today. If plate tectonics was operative early in Earth history, it is likely that the plates were small, rapidly moving, and constantly recycled back into the mantle. It is possible that continental crust first began to form as a result of magma rising directly to the surface from mantle hotspots. Some of this material would have been remelted and remixed back into the mantle, but some eventually accumulated on top of the thin lithospheric plates. Eventually, the lithosphere would have begun to cool and thicken, leading to the development of plates as we understand them today.

By at least 3.4 billion years ago, long-lived masses of continental crust had begun to form. The oldest known cratons include the Kaapvaal Craton in southern Africa (3.1 billion years), and the Pilbara Craton in wWestern Australia (3.4 billion years); even older fragments of continental crust are known from the Slave Craton in northern Canada, the Dharwar Craton in India, and elsewhere. By the advent of the Proterozoic Eon, evidence for collisions between blocks of crust is strong. It appears that by at least 2 billion years ago—and perhaps much earlier—plate tectonics was operating in a style similar to what we observe today.

Supercontinents

The subsequent history of continental evolution involves the assemblage of cratons into large continental complexes, called **supercontinents**, and their subsequent breakup, dispersal, and reassemblage. There is increasingly good evidence that the Kaapvaal and Pilbara cratons were welded together into a continental mass called *Vaalbara* by about 3.1 billion years ago, and then broke apart about 2.8 billion years ago. Another supercontinent called *Kenorland* formed around 2.7 billion years ago and lasted until about 2.5 billion years ago. After that there was a cyclic succession of assemblages, breakups, and dispersals of supercontinents (FIG. 5.24). Most recently, about 300 million years ago, the continental masses reassembled into the supercontinent we know as *Pangaea*—the landmass that was the focus of Alfred Wegener's evidence that the continents had once been joined together. The breakup of Pangaea into what would eventually become our modern-day continents started almost 200 million years ago. This breakup and dispersal continues today, as seen in the splitting of the African continent along the East African Rift and the widening of the Atlantic Ocean.

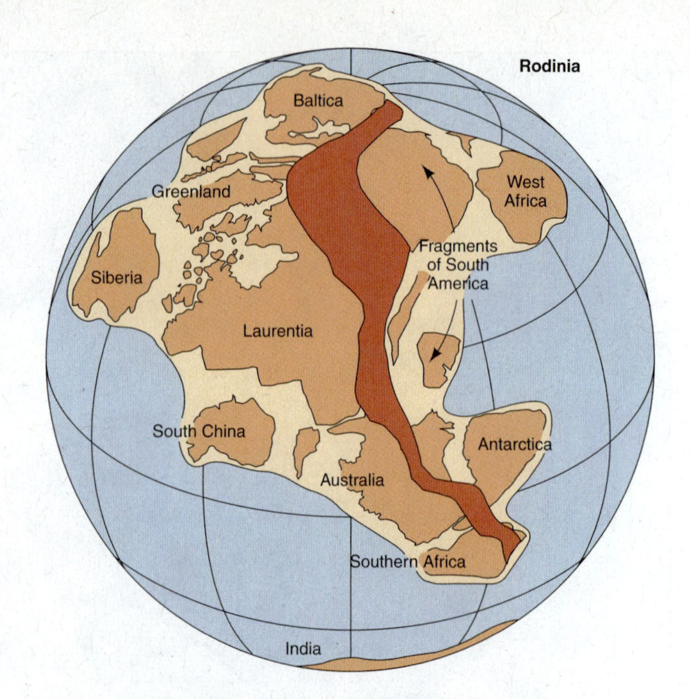

FIGURE 5.24 Supercontinents before Pangaea
The 1.1-billion-year-old supercontinent Rodinia was one in a series of continental masses that assembled, broke up, dispersed, and reassembled through geologic time. The earliest supercontinent, Vaalbara, most likely broke up around 2.8 billion years ago; the most recent was Pangaea, the landmass that was the focus of Alfred Wegener's work on continental drift.

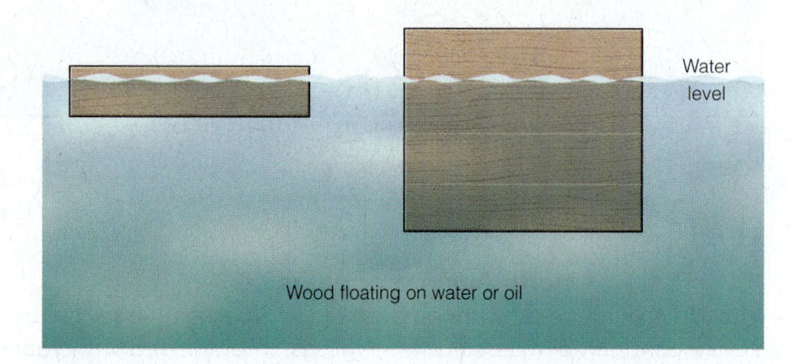

FIGURE 5.25 Isostasy
The principle of isostasy tells us that a floating object, such as an iceberg or a piece of wood floating on water, will have more mass below the level of the water than above. The flotational height of the object depends on its overall mass and density. The basic principle of isostasy applies to the lithosphere, which is "floating" on top of the malleable asthenosphere.

Isostasy, Gravity, and the Roots of Mountains

One of the defining characteristics of cratons is "stability," but what do we really mean by this? The definition of stability as applied to cratons is based on the concept of **isostasy**, the property whereby the lithosphere maintains a flotational balance as it rides along on top of the asthenosphere. Cratons, the ancient cores of continents, are said to have attained *isostatic stability*. Let's consider what this means.

Isostasy

The lithosphere behaves as if it were "floating" on the asthenosphere, much like a piece of wood or an iceberg floats on water (FIG. 5.25). (*Floating* is not exactly the correct word because the asthenosphere is mostly solid, but the lithosphere is buoyant and acts as though it were floating. The great ice sheets of the last ice age provide an impressive demonstration of isostasy FIG. 5.26A.

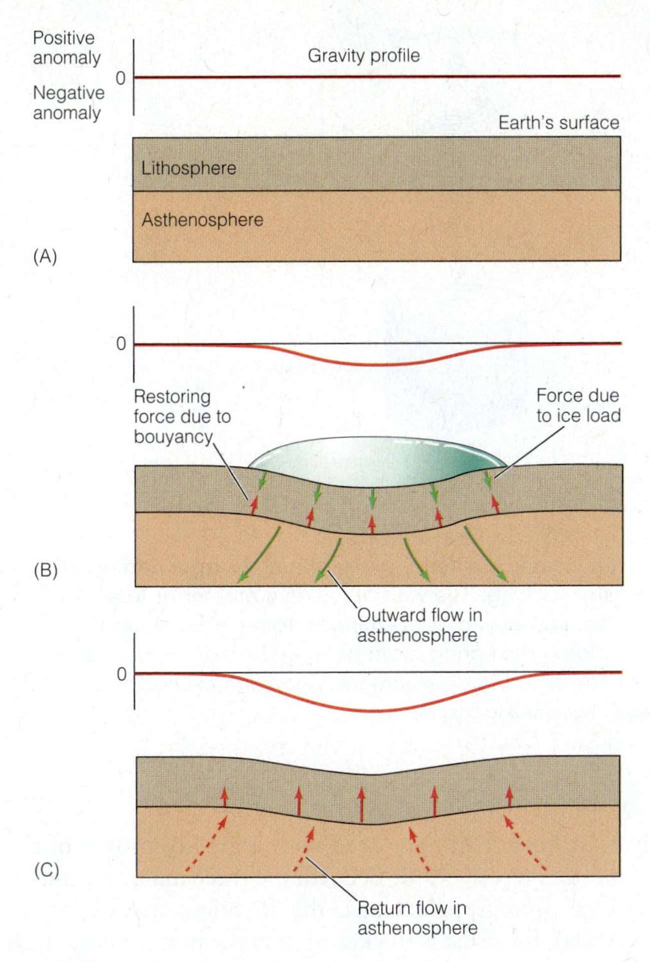

FIGURE 5.26 Isostatic rebound
Great thicknesses of glacial ice weighed down the continental lithosphere in North America and Scandinavia during the last ice age, depressing the land surface and causing the asthenosphere to flow outward (B). When the ice melted (C), the land began to rebound and the asthenophere began to flow into the underlying areas. However, the land is still rebounding, even though the ice disappeared many thousands of years ago. This suggests that flow in the asthenosphere is very slow.

The weight of a large continental ice sheet, which may be 3 to 4 km thick, will depress the lithosphere (FIG. 5.26B). When the ice melts, the land surface slowly rises again. The effect is very much like pushing a block of wood into a bucket of oil or some other thick, viscous fluid. When you stop pushing, the wood slowly rises to an equilibrium position determined by its density. The speed of its rise is controlled by the *viscosity* (that is, the thickness, or resistance to flow) of the fluid.

Glacial depression and rebound of the lithosphere mean that rock in the asthenosphere must flow laterally away when the ice depresses the lithosphere, and then must flow back again when the ice melts (FIG. 5.26C). The balance between temperature and pressure in the asthenosphere is such that the rock is very close to its melting temperature; that, apparently, is why the asthenosphere is so weak. However, we can observe that the land in parts of northeastern Canada and Scandinavia is still rising, even though most of the ice that covered these areas during the last ice age had melted away by 7000 years ago. From this observation, we infer that the flow in the underlying asthenosphere must be very slow and that the asthenosphere itself must be extremely viscous.

The density of lithospheric fragments determines their flotational balance on the asthenosphere. Hence, just as a light piece of wood floats higher than a weighted-down piece of wood, the low-density continental lithosphere floats higher than the high-density oceanic lithosphere. Continents and the mountains on them are composed of low-density rock, and they stand high because they are thick and light. Ocean basins are topographically low because the oceanic crust is composed of denser rock. Isostasy and the fact that the continental crust is less dense than the oceanic crust are the reason Earth has continents and ocean basins.

When we say that a craton has attained a state of isostatic stability or equilibrium, it means that this mass of ancient continental rock is in a stable flotational state with respect to the underlying asthenosphere and is neither rising nor sinking. Where mountain ranges are forming in active orogens, however, the lithosphere is locally thickened as a result of the continental collision. This active uplifting and thickening of the lithosphere is supported by the underlying, easily deformed asthenosphere. Even as the mountains are uplifted, they are already beginning to be worn down by the forces of weathering and erosion (FIG. 5.27). The removal of

FIGURE 5.27 Mountain roots and isostatic adjustment
Mountains have roots made of continental rock, extending deep into the malleable asthenosphere below. When the mountain begins to be worn down by erosion, the orogen will rise as a result of isostatic adjustment and returning flow in the asthenosphere.

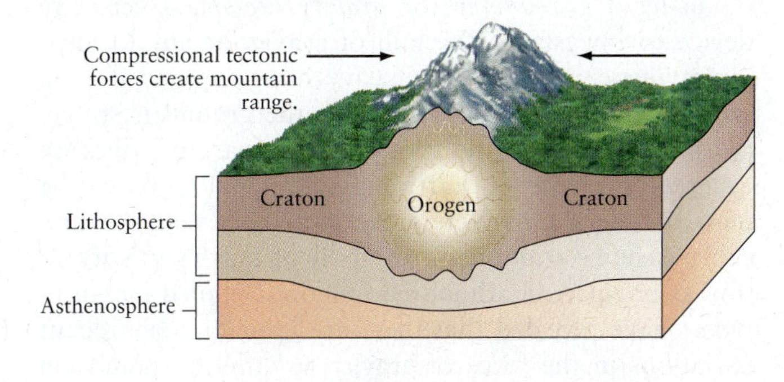

material by erosion takes weight off of the lithosphere, just like the melting of the glacial ice. Return flow in the underlying asthenosphere adds buoyant support, allowing the mountainous terrain to rebound and be uplifted once again.

Gravity Anomalies

A further interesting characteristic of mountains is that they have deep roots that balance their topographically high peaks. Just as icebergs float in isostatic balance on water, with most of their mass below the water line, mountains are thickened fragments of lithosphere with deep roots of continental rock extending down into the malleable underlying asthenosphere (Fig. 5.27). The existence of deep roots beneath mountains makes sense in the context of what we know about isostasy, but it also has been verified by studies of variations in Earth's gravitational pull.

The pull exerted by Earth's gravity on a body at the surface is slightly greater at the poles than at the equator. This is because Earth *looks* round, but in fact it's not a perfect sphere; it is an ellipsoid that is slightly flattened at the poles and bulged at the equator. The pull of gravity (which is inversely proportional to the square of the distance between the center of mass of two bodies) is stronger at the poles because the distance to the center of Earth is slightly less than at the equator. Recall, as well, that your weight is a measure of how strongly Earth's gravitational force is pulling your body toward the center of Earth. Thus, a person who weighs 90.5 kg at the North Pole would weigh only 90 kg at the equator. If this weight-conscious person made very exact measurements while traveling from the pole to the equator, it would become apparent that weight varies irregularly, rather than smoothly, indicating that the pull of gravity varies irregularly between the pole and the equator.

We can confirm this irregular variation of gravity using a *gravimeter* (or *gravity meter*), a sensitive device for measuring the pull of gravity at any locality. A gravimeter consists of a heavy mass suspended by a sensitive spring (FIG. 5.28). When the ground is stable, the pull exerted on the spring by the mass provides an accurate measure of Earth's gravitational pull on the mass. Modern gravimeters are incredibly sensitive and can measure variations in the pull of Earth's gravity as tiny as one part in a hundred million. Detailed measurements have revealed that there are large and significant variations in the force of gravity at different points on Earth; scientists call them **gravity anomalies**, and a great deal of important information about Earth's interior can be derived from them. So-called *positive gravity anomalies* result from concentrations of high-density rock under the surface, and *negative gravity anomalies* result from concentrations of low-density rock.

As an example, a profile of the crust beneath the United States (determined by seismic studies) is shown in FIGURE 5.29A, compared with a gravity profile of the same area (FIG. 5.29B). Beneath the three major mountain systems (the Appalachians, the Rockies, and the Sierra Nevada), the crust is thicker than in the non-mountainous regions. Gravity measurements almost everywhere across the high-standing continent are less than expected, considering the distance from the center of Earth, but they are least where the crust is thickest. These regions of low gravity pull are the major negative gravity anomalies caused by masses of low-density rock that form the roots of these high mountain ranges.

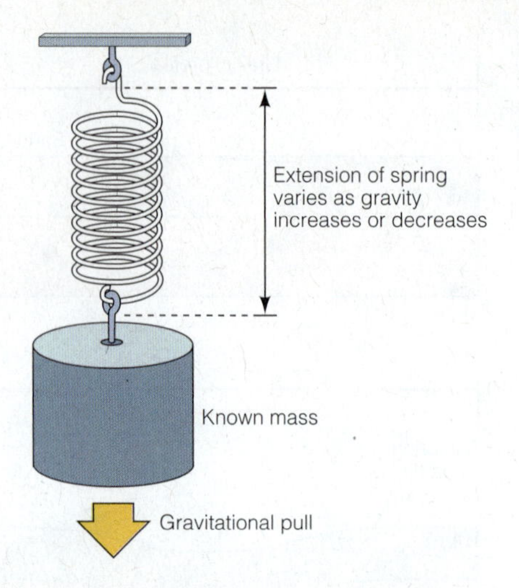

FIGURE 5.28 Gravimeter
A gravimeter is a heavy piece of metal suspended on a sensitive spring. The weight exerts a greater or lesser pull on the spring as gravity changes from place to place, extending the spring more or less. The weight and spring are contained in a vacuum together with exceedingly sensitive measuring devices.

Make the CONNECTION

Events in the geosphere, such as the opening of an ocean basin or the rising of a mountain range, happen very slowly. In contrast, events in the atmosphere, biosphere, and hydrosphere tend to happen very rapidly. What do you think causes this difference?

Plate Tectonics And The Earth System Today

Almost every part of the Earth system is influenced directly or indirectly by plate tectonics driven by Earth's internal heat source. To complete our discussion of plate tectonics, let's consider some examples of such influences.

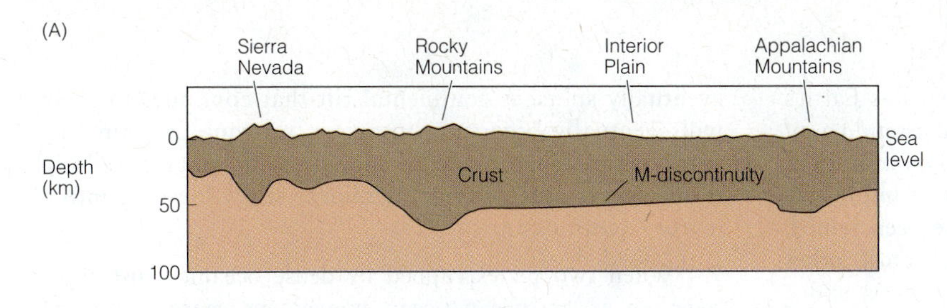

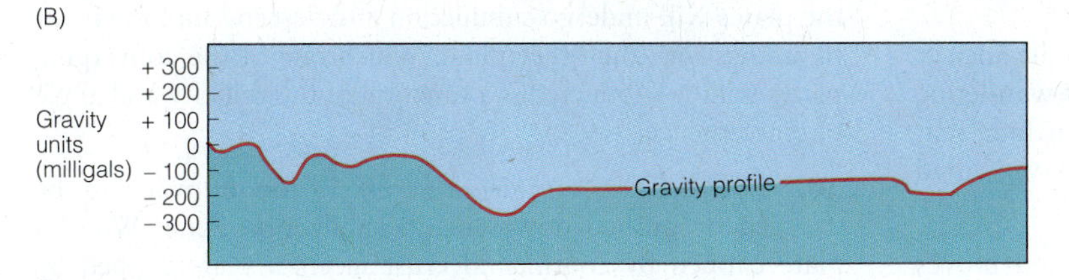

FIGURE 5.29 Gravity profile of North America
(A) This profile shows the variation in thickness of the continental crust across the North American continent from the Appalachians to the Sierra Nevada. The thickness of the crust (determined by seismic studies) is greatest underneath the three major mountain ranges. (B) A gravitational profile of the same region shows distinct negative gravity anomalies at the Sierra Nevada, Rockies, and Appalachian Mountains, where the crust is thickest. These gravity anomalies are caused by masses of low-density continental rocks that lie beneath the mountains.

Climate and Ecosystems in Modern Orogens

Today's great mountain ranges are young orogens—active continental collision zones that have formed during the last few hundred million years. Occurring in great arc-shaped systems a few hundred kilometers wide, these mountain systems commonly reach several thousand kilometers in length. Within a collision system, the rocks are compressed, fractured, folded, and crumpled, commonly in an exceedingly complex manner. Examples are widespread: the Alps, the Himalaya, and the Carpathians are all young mountain systems still being actively uplifted, whereas the Appalachians and the Urals are older systems that are slowly being eroded down.

Both young and old, mountain ranges are distinctive and impressive features of the landscape, formed by plate tectonics and modified by the rock cycle. These great mountain ranges fundamentally influence the way the world's wind and weather systems work. As a result, they also exert a significant influence on the modern distribution of climatic zones, and therefore on the distribution of plant and animal communities.

Composition of Ocean Water

The great chain of submarine volcanoes along the midocean ridges is the site of intense chemical reaction between the waters of the ocean and the hot rocks of the seafloor. Some chemical elements are extracted from seawater into the rocks, and others are released from the rocks into the seawater. In this manner the hot rocks of the oceanic crust play an important role in determining the composition of seawater.

Composition of the Atmosphere

Volcanoes, both along plate spreading edges and adjacent to subduction zones, release vast amounts of gas into the atmosphere. As you learned in Chapter 4, Earth's primordial envelope of gases was stripped away, early in the history of our solar system. Without the release of volcanic gases over geologic time, Earth would not have evolved the secondary atmosphere that supports life on the planet today.

The vigor of volcanic activity is one of the fundamental controls on the concentration of greenhouse gases in the atmosphere, and hence on the global climate. During periods of Earth history when volcanic activity has been at a minimum, atmospheric greenhouse gases are correspondingly low and the resulting global climate is cool; these are referred to as *icehouse* periods. Conversely, when volcanic activity is particularly vigorous, atmospheric greenhouse gas concentrations increase and the global climate is consequently warmer; these are known as *hothouse* periods. Volcanic gases continue to play an important role in the climate and the long-term composition of the atmosphere. Very large eruptions can also cause dramatic short-term, irregular climatic disturbances.

Sequestration of Carbon

Sediment that is formed as a result of processes in the rock cycle accumulates in the low-lying ocean basins, where it is converted to rock. Such sediments and sedimentary rocks are the most important places in which carbon-rich organic matter is sequestered. Carbon sequestration is one of the most fundamentally important processes on Earth. Without such sequestering the atmosphere would be essentially devoid of oxygen but rich in carbon dioxide, like the atmosphere of Venus. Such an Earth would not be habitable. Even the rate of opening and closing ocean basins is important in this context. The repositioning of the continents and ocean basins influences everything from sea level to the rates of weathering, erosion, and sedimentation; these, in turn, influence the rate of carbon sequestration and hence the global climate.

In all of these ways, and many others, plate tectonics has influenced virtually every part of the Earth system over the history of this planet, and it continues to do so today. Without plate tectonics, the surface of Earth would be still and lifeless, like the surface of Mercury.

SUMMARY

1. Plate tectonics is a unifying theory that explains Earth's topography and integrates our scientific understanding of rock formation, mountain building, and terrain modification. Early proponents of continental drift amassed an impressive body of evidence that the continents had once been joined; however, most scientists remained skeptical until other sources provided solid evidence for the mechanisms and a driving force behind continental movement.

2. Studies of rock paleomagnetism have revealed the ancient positions of Earth's magnetic poles. The apparent wandering paths of the ancient poles led geophysicists to conclude that the continents, rather than the poles themselves, have shifted over geologic time.

3. In seafloor spreading, oceanic crust splits and moves laterally away from spreading ridges. Magma rises from the mantle to fill the gap, forming new oceanic crust. Oceanic crust far from a spreading ridge should thus be older than crust nearer to the ridge. This has been confirmed by paleomagnetic studies, which have revealed the ages of seafloor rocks through studies of the chronology of reversals of Earth's magnetic field. Successive reversals are recorded in new oceanic crust as it forms and moves laterally away from the ridge.

4. The driving force for lithospheric plate motion is convection in the mantle. Hot rock in the mantle flows slowly upward, spreads sideways, and cools, becoming denser, and eventually sinking back down into the mantle. The cold, rigid lithosphere moves as coherent plates over the hot, weak asthenosphere, and oceanic or continental crust capping the plate is carried along.

5. Measuring the speed and direction of movement of lithospheric plates requires a fixed frame of reference, such as distant stars. By taking distance measurements in multiple locations over time, small differences can be detected. Deep-seated mantle hotspots provide a fixed framework *within* the Earth system, against which to measure plate motion. In general, plates that carry continental crust move slowly, whereas plates without continental crust move more rapidly.

6. Earth's diameter is not increasing or decreasing significantly, so the generation of new crust along constructive plate boundaries must be balanced by the loss of material along destructive plate boundaries. In subduction zones, slabs of lithosphere capped by oceanic crust are recycled back into the mantle, sinking along the downgoing limbs of convection cells.

7. Jostled by the underlying mobile asthenosphere, the lithosphere has broken into nine major plates and a large number of smaller plates. Interactions occur along plate boundaries, which are marked by earthquakes and volcanic activity. The basic types of plate boundaries are: divergent margins or spreading centers, where plates move apart; convergent margins, where plates move toward each other; and transform fault margins, where plates slide past each other.

8. Divergent plate margins start in continents and become oceans. A plate capped by a large continent slowly heats up from below, expands to form a broad plateau, and eventually splits. A continental rift that continues to grow will eventually become a linear sea, growing over time by seafloor spreading. As the basin widens and magma wells up to form new oceanic crust, the linear sea will evolve into a mature ocean basin.

9. When two plates capped by dense oceanic crust come together at an ocean–ocean convergent margin, one of the plates will undergo subduction and descend into the mantle underneath the other plate. Which one of the converging plates will be subducted is a function of the velocity and angle of convergence.

10. Low-density continental crust is too buoyant to be dragged or pushed down into a subduction zone. When a plate capped by continental crust meets a plate capped by oceanic crust (an ocean–continent convergent margin), the oceanic plate will undergo subduction.

11. When both plates of a converging pair are capped by low-density continental crust (a continent–continent convergent margin), neither can undergo subduction, and the eventual result is a continental collision. Collisions result in the formation of mountain ranges and the disappearance of an ocean basin in the suture zone between the converging continents. The cycle of rifting, opening and widening of an ocean basin, convergence in a subduction zone, and the eventual disappearance of the ocean basin in a collision zone is called a Wilson cycle.

12. Transform faults are great vertical fractures that occur along boundaries where two plates are sliding past one another laterally, in a strike-slip motion. As the plates grind past one another their edges may grab and lock; as they do, the rocks on both sides flex and bend. When the locked section breaks free, the flexed rock snaps suddenly and earthquakes typically occur.

13. Modern continental shorelines usually don't coincide exactly with the edges of the original oceanic rift because some ocean water may spill out of the ocean basin onto the flat-lying continental shelf. The true edge of a continent is typically hidden underwater, where the continental slope meets the continental rise, and continental crust meets oceanic crust.

14. Subduction zones are marked by deep trenches where ocean-capped lithosphere sinks (or is dragged or pushed) into the asthenosphere. Active trenches are associated with Benioff zones, where deep, high-magnitude earthquakes outline the dimensions of the downgoing plate. Volcanic island arcs result from melting that begins when the sinking plate reaches a depth of about 100 km. Magma generated by this melting rises to the surface, forming a line of volcanoes.

15. Continents located in the stable interiors of plates have passive margins that originated in the geologic past as the jagged edges of the rift where a plate began to split apart. In contrast, active continental margins coincide with the geologically active edges of tectonic plates. These can be continental collision zones, transform fault margins, or ocean–continent convergent boundaries where subduction is occurring. The most distinctive feature of an active ocean–

continent convergent margin is a continental volcanic arc, which forms when the subducting plate melts and magma rises to the surface on the continental side of the trench.

16. Continental rocks carry a record of the deformation they have undergone through geologic time, whereas oceanic crust is geologically young and does not preserve a long record of geologic activity. The stable, ancient cores of continents are called cratons. Draped around them are younger orogens, elongate regions of crust that have been intensely bent and fractured in continental collisions; they are the eroded roots of ancient mountain ranges. The oldest orogens are about 2 billion years old, indicating that plate tectonics must have been functioning for at least this long.

17. If plate tectonics operated early in Earth history, it is likely that the plates were small, rapidly moving, and constantly recycled back into the mantle. Eventually the lithosphere cooled and thickened, leading to the formation of plates as we understand them today. Continental evolution involves the assemblage of cratons into large complexes, called supercontinents, and their subsequent breakup, dispersal, and reassemblage. Most recently, the continental masses assembled into the supercontinent Pangaea. The breakup of Pangaea to form our modern-day continents started almost 200 million years ago, and continues today.

18. Cratons are isostatically stable. Isostasy is the property whereby the lithosphere maintains a flotational balance on the asthenosphere. The density and thickness of lithospheric fragments determine their flotational balance. The malleable asthenosphere flows outward very slowly from areas where the lithosphere is weighed down, slowly returning if the weight is removed, as in the melting of glacial ice or erosion of a mountain. Isostasy and the fact that continental crust is less dense than oceanic crust are the reason Earth has topographically high continents and low-lying ocean basins.

19. Topographically high mountains have deep roots of low-density continental rock extending into the underlying asthenosphere. This has been confirmed by studies of variations in the pull of gravity at Earth's surface. Large mountain ranges are characterized by significant negative gravity anomalies because of their deep, low-density roots.

20. Today's great mountain ranges are young orogens—active continental collision zones that have formed during the last few hundred million years. They are great systems of compressed, fractured, and folded rocks, thousands of kilometers in length. The influence of these modern orogens on the Earth system is profound; even our climatic zones and the distribution of plant and animal communities are fundamentally affected by their presence.

IMPORTANT TERMS TO REMEMBER

QUESTIONS FOR REVIEW

1. What is the difference between the lithosphere, the crust, the asthenosphere, and the mantle, and what role does each play in plate tectonics?

2. Who was Alfred Wegener, and what revolutionary idea did he promote? Why were scientists reluctant to accept Wegener's idea when he first proposed it?

3. Explain how the apparent wandering of magnetic poles throughout geologic history can be used to help demonstrate that continental drift has occurred.

4. What are the main features of the seafloor spreading model? What critical evidence demonstrated that seafloor spreading actually occurs?

5. What is a "plate" in plate tectonics?

6. Briefly describe the three basic types of plate margins.

7. Describe the sequence of events that leads to the opening of a new ocean basin flanked by two passive continental margins.

8. Describe what happens when two plates capped by oceanic crust converge. Compare this with what happens when two plates capped by continental crust converge.

9. Identify the major topographic features of ocean basins, and state how they are related to plate tectonics.

10. The Himalaya and the Alps are said to be the "graveyards" of ancient oceans. Why?

11. What is the driving force for plate motion? How does it work, and how does it move the plates?

12. Why are ocean basins places of low elevation on Earth's surface and continents are places of high elevation?

13. What causes gravity anomalies, and how can they be measured? How have studies of gravity anomalies provided evidence of the deep roots of mountains?

14. What are cratons, and how do they differ from orogens?

15. What are supercontinents, and what do they tell us about the evolution of continents over geologic time?

QUESTIONS FOR RESEARCH AND DISCUSSION

1. If Earth were to undergo a magnetic reversal, what would be the effects? How likely is this to happen in your lifetime?

2. Earth's surface topography (Fig. 5.16) reveals much about plate tectonics and mantle convection processes that operate in the interior of the planet. What key observations would you plan to make if you were sending a spaceship to another planet and wished to find out whether plate tectonics operated on that planet? Assume that the spaceship cannot land and that all observations have to be made remotely.

3. Try making some calculations of plate motion:

   A. In the vicinity of Los Angeles, the Pacific Plate is moving northward relative to the North American Plate at a speed of 5.5 cm/y. Determine how long it will be before Los Angeles and San Francisco are side by side. Draw a map of the way the west coast of North America might look when Los Angeles and San Francisco are side by side. Now draw a map showing what it will look like 10 million years after that.

   B. The place where New York City now stands was once attached to North Africa at approximately the position of Marrakech in Morocco. New York City and Marrakech are now 5700 km apart. The North Atlantic is moving away from the African Plate at a speed of 2 cm/yr. How long has it taken for the Atlantic Ocean to reach its present width?

QUESTIONS FOR *THE BASICS*

1. What is the source of Earth's magnetic field?

2. How do minerals become magnetized? How does lava carry a record of Earth's changing magnetic field?

3. Explain how paleomagnetic studies can be used to determine ancient magnetic pole positions.

QUESTIONS FOR *A CLOSER LOOK*

1. What is the difference between the relative speed and the absolute speed of a plate?

2. How are the absolute speeds of plates determined? Describe one method based on an external frame of reference and one method based on an internal frame of reference.

3. Which plates are moving the fastest, and which are moving the slowest?

CHAPTER 6

Earthquakes *and* Volcanoes

OVERVIEW

In this chapter we:

- Describe how and why earthquakes happen, and how scientists measure them
- Examine earthquake-related hazards and efforts by scientists to predict earthquakes
- Consider the processes that lead to rock melting and volcanism
- Identify different types of volcanic eruptions and the hazards associated with them
- Explain the connections among earthquakes, volcanoes, plate tectonics, and the other parts and processes of the Earth system

Can you say "Eyjafjallajökull"?
Icelandic volcano Eyjafjallajökull erupted for several months in the spring of 2010. The eruption was accompanied by earthquake activity and lava extrusions. It sent a large volume of volcanic gas and ash into the atmosphere, causing havoc for air travel throughout much of Europe. The heat of the eruption also caused melting of the overlying glacier, which led to meltwater flooding (called *jökulhlaup* in Icelandic).

In Chapter 5 you learned that many scientists have contributed to the development of the theory of plate tectonics. Countless observations of the geosphere can now be understood to be the result of plate motions. Without plate tectonics Earth as a planet and the Earth system would be very different. Much of the detailed evidence used by scientists to develop and refine the theory of plate tectonics has come from the study of earthquakes and volcanic eruptions. In this chapter we will examine these processes and the contribution they have made to our understanding of the tectonic cycle and its impacts on the other cycles and spheres of the Earth system.

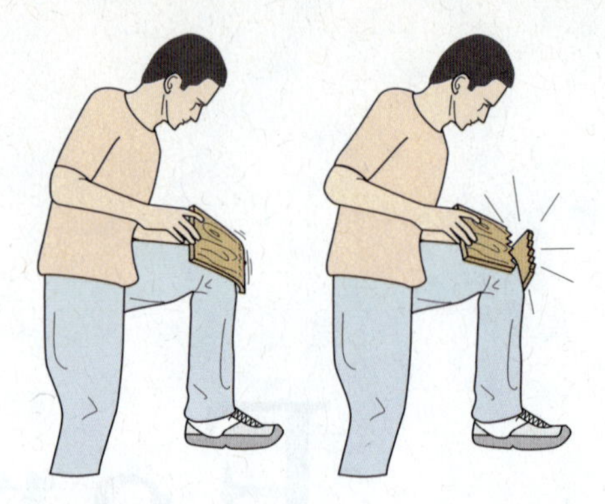

FIGURE 6.1 Elastic energy
If you bend a piece of wood over your knee, it will bend and store elastic energy. If you push the wood too far, it will break and release the stored energy all at once. This is analogous to what happens when rock masses bend and break, causing earthquakes.

EARTHQUAKES: WHEN ROCKS SHIFT

They can be devastating in terms of human impacts, but earthquakes are of enormous scientific importance. They are by far the most powerful way to study Earth's interior and the workings of the geosphere. The way Earth vibrates after a large quake is controlled by the properties of the rocks inside. Earthquake vibrations can be used to study Earth's interior, in the same way that a doctor uses X-rays or CT scans to study the inside of a human body—they are remote sensing probes that allow scientists to sense and measure the characteristics of the materials beneath our feet. Earthquake activity, or **seismicity**, also plays an essential role in outlining the boundaries of tectonic plates. The study of earthquakes, together with the information derived from volcanism, has allowed scientists to decipher the mechanisms by which plates are formed at spreading centers and consumed in subduction zones.

What Is an Earthquake?

Place a thin stick of wood across your knees. Press both ends downward, and the wood bends. Stop pressing and the wood springs back to its original shape. A change in shape or size of a body is called **deformation**, and any change of shape or size that reverses when the deforming force is removed is called *elastic deformation*. The muscle energy you used to bend the wood doesn't disappear—it is stored as **elastic energy** in the wood. When the bending force is removed, it is this elastic energy that restores the wood to its original shape. Consider what happens, however, when the pressure is so great that the elastic limit of the wood is exceeded: The stick breaks with a sudden snap, and the stored elastic energy is released all at once (FIG. 6.1). The elastic energy is converted partly to heat at the breaking point in the wood, partly to sound waves that make the snapping noise, and partly to vibrations in the wood.

Earthquake vibrations are the same kind of vibrations that you feel when the wood breaks. Of course, breaking sticks are not the cause of earthquakes, but rocks are elastic and—within limits—they can be bent. The most widely accepted hypothesis concerning the origin of earthquakes involves the storage of elastic energy in rock masses, followed by breaking of the rock mass, or slipping of two rock masses along an earlier break, and the sudden release of the stored elastic energy. Just how elastic energy is stored and built up in rock, and exactly how and why it is suddenly released are the subjects of intensive research.

Origin of Earthquakes

When slippage of rock occurs along a break or fracture in a rock, the fracture is called a **fault**. The cause of most earthquakes is thought to be sudden movement along faults, but it's not that simple. Some earthquakes are millions of times stronger than others. The same amount of energy that in one case is released by thousands of tiny slips and tiny earthquakes is, in another case, stored and released in a single immense earthquake. What causes these differences? It must have something to do with the way elastic energy is stored and built up in rock masses.

The predominant hypothesis of earthquake formation, called the **elastic rebound hypothesis**, suggests that some fault surfaces are rough, so that the rock masses on either side of the fault become locked against one another rather than slipping easily past. If two rock masses are locked but tectonic forces are still pushing on them, the rocks on either side of the fault will undergo deformation. By deforming, the rock masses store elastic energy, just as bending caused the stick of wood to store elastic energy. When the locked fault finally does slip, allowing the rock masses to lurch past one another, the deformed rocks on either side rebound to their original shapes and the stored energy is released all at once in an earthquake. The longer the lock persists, the more elastic energy is stored, and the more energy is released when the earthquake finally occurs.

The first measurements supporting the elastic rebound hypothesis came from studies of the San Andreas Fault in California. The San Andreas Fault is a complex vertical transform fault that cuts down through the lithosphere, separating the Pacific Plate to the west from the North American Plate on the east (Fig. 5.15). The two plates are sliding in opposite directions—the Pacific Plate is moving toward the north-northwest

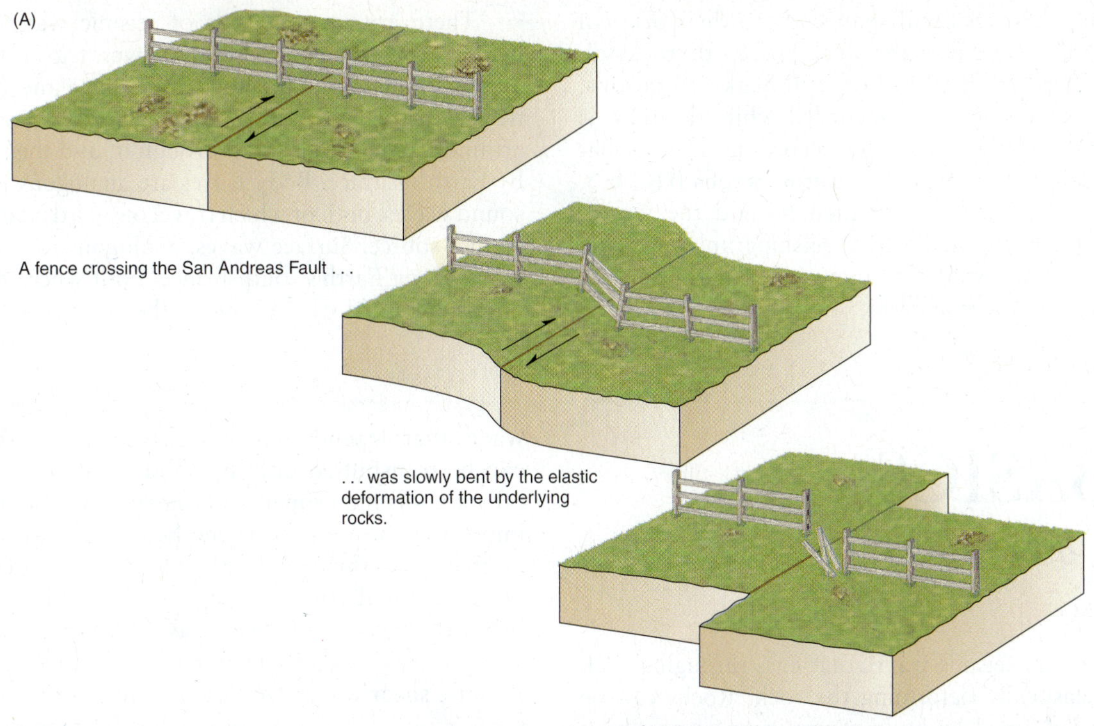

(A)
A fence crossing the San Andreas Fault . . .
. . . was slowly bent by the elastic deformation of the underlying rocks.
After the great San Francisco earthquake of 1906, the two sides of the fence had separated by 7 m.

FIGURE 6.2 Movement along a fault
An earthquake is caused by the sudden release of elastic energy stored in rocks. (A) This sketch of a fence crossing the fault is based on surveys near the San Andreas Fault in California, before and after the great earthquake of 1906. (B) Along the San Andreas Fault near Point Reyes, California, sections of a fence show the dramatic displacement that occurred in the 1906 quake. This fence was constructed in the 1970s along the track of an old fence that was wrenched apart by the quake.

relative to the North American Plate. In some places the sliding is smooth and continuous, but at other locations the fault is locked. On April 18, 1906, near San Francisco, the two sides of the locked fault shifted abruptly (FIG. 6.2). The elastically stored energy was released as the bent crust rebounded, thereby creating a violent earthquake. The part of the fault where the 1906 earthquake occurred is now locked again. Other damaging earthquakes have occurred elsewhere along the fault since 1906—notably the Loma Prieta (near San Francisco) earthquake of 1989 and the Northridge (near Los Angeles) earthquake of 1994—and eventually another earthquake will occur in this location. The evidence is clear; faults slip in recurring steps, and as a consequence earthquakes are also recurrent.

The 1906 San Francisco earthquake caused an enormous amount of damage, but it also launched a great deal of research. Building on that research, scientists have developed our present-day understanding of how earthquake damage can be minimized and how earthquakes can be used to study the internal structure and properties of the geosphere. Before proceeding with a discussion of these points, let's look more closely at how earthquake vibrations travel through rock.

Seismic Waves

When an earthquake occurs, the elastically stored energy is carried outward from the focus to other parts of Earth by vibrations called **seismic waves.** The seismic waves released in the earthquake spread out spherically in all directions, just as sound waves spread out spherically in all directions from a sound source. Seismic waves are elastic disturbances, so unless the elastic limit of the rock is exceeded there will be no permanent deformation; the rocks will snap back to their original shape after the waves pass through. Therefore, rocks do not carry a permanent record of earthquake vibrations, and seismic waves must be recorded while the rock is still vibrating. Many continuous recording devices that can detect seismic waves, called **seismographs** (**FIG. 6.3**, on page 148), have been installed around the world. The actual recording made by a seismograph, showing the various seismic waves as they arrive at the seismic station, is called a **seismogram.**

There are several kinds of seismic waves, and they belong to two families: **Body waves** travel outward in all directions from the focus and have the capacity to travel through Earth's interior. **Surface waves** travel around Earth, rather than through it, and they are guided by Earth's surface. Body waves are analogous to light and sound waves, both of which travel outward in all directions from a source. Surface waves, analogous to ocean waves, travel along Earth's solid surface, both where it meets the atmosphere and where it meets the ocean.

the BASICS

Types of Seismic Waves

BODY WAVES

Body waves are seismic waves that pass through a rock mass by elastically deforming the rock. Rocks can be elastically deformed by body waves in two ways: by a change in *volume* (like squeezing a tennis ball or blowing up a balloon) or by a change in *shape* (like bending or twisting a piece of wood).

Body waves that cause volume changes consist of alternating pulses of compression (squeezing) and expansion (stretching) acting in the direction of wave travel (**FIG. B6.1A**). Sound waves are a familiar example of such *compressional waves*. A sound wave passes through air by alternating compressions and expansions of the air. Our ears sense the pulses of compression and expansion, and our brains transform these pulse signals into sound. Compression/expansion waves can pass through gases, liquids, and solids. That is why we can hear sounds not only in the air but also through the walls of houses and under water. Compression/expansion waves pass easily through rocks, and they have the greatest velocity of all seismic waves—6 km/s is a typical value near Earth's surface. They are the first to be recorded by a seismograph after an earthquake and are therefore called **P (for primary) waves.**

Body waves that deform materials by change of shape are called *shear waves*. As a shear wave travels through a material, each particle in the material is displaced perpendicular to the direction of wave travel (**FIG. B6.1B**). Liquids and gases don't have shapes; they simply flow freely to fill any container we put them in. In other words, liquids and gases cannot transmit waves that depend on a change in shape, so shear waves can be transmitted only by solids. This is a fundamentally important property of shear waves. It is crucially important in interpreting the behavior of seismic waves as they pass through Earth, because the presence of a body of liquid (such as magma) below the surface will block the passage of shear waves. A typical velocity for shear waves in rocks near Earth's surface is 3.5 km/s. Because shear waves are slower than P waves and reach a seismograph later, they are called **S (for secondary) waves.**

SURFACE WAVES

Surface waves are a little more difficult to understand than body waves. They are caused by the same phenomenon that makes a bell ring. When a bell is rung, the surface of the bell rises and falls in complex ways and makes the air around the bell vibrate. The sounds

FIGURE B6.1 Seismic body waves
There are two types of seismic body waves. (A) P waves cause volume changes in the rock the wave is passing through, by alternate compressions and expansions. An individual point in a rock moves back and forth parallel to the direction in which the P wave is traveling. (B) S waves cause a change in rock shape because they result in a shearing motion. Any individual point in the rock moves up and down, perpendicular to the direction in which the S wave is traveling.

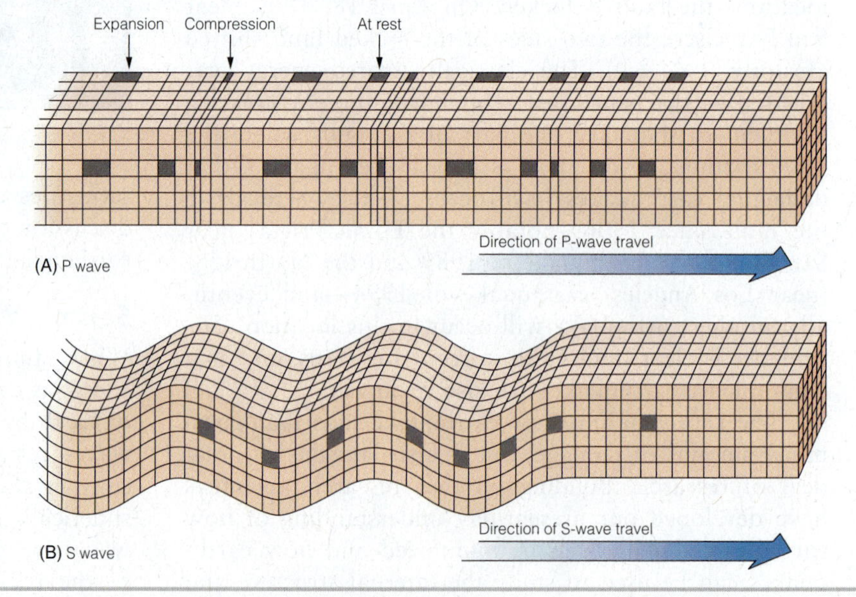

Determining Earthquake Locations

The point where energy is first released during an earthquake, and from which the seismic waves travel outward, is called the **focus** (plural *foci*). Because most earthquakes are caused by movement along a fault, the focus may not be a simple point, but rather a region that may extend for several kilometers along the fault plane (FIG. 6.4). Most earthquake foci lie at some depth below the surface, so it is more convenient to identify an earthquake site by referencing the **epicenter**, the point on Earth's surface that lies vertically above the focus (Fig. 6.4). The usual way to describe the location of an earthquake is to state its focal depth and the location of its epicenter.

The location of an earthquake's epicenter can be determined from the arrival times of seismic body waves at a seismograph. P and S waves travel through different rocks at different velocities, but P waves travel the

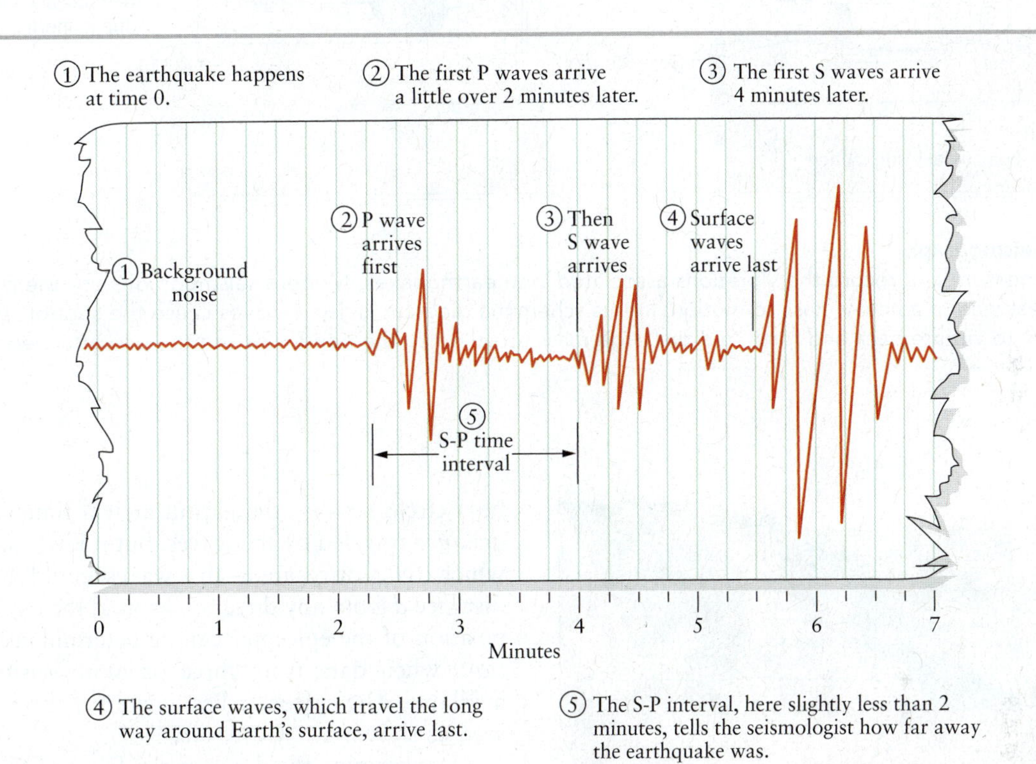

FIGURE B6.2 Seismogram of a typical earthquake
This is a typical seismogram, made by a seismograph. All three types of seismic waves leave the earthquake's focus at the same time. The P waves travel the fastest and arrive at the seismic station first, followed by the S waves. The surface waves travel even more slowly around the outside of Earth rather than through the interior, and they arrive last at the seismic station. The lag times between the arrivals of the different types of seismic waves are a measure of how far they have traveled.

we hear from a ringing bell come from the vibrating air. The same is true of earthquakes. An earthquake makes the whole Earth ring like a bell, although the vibrations don't create sounds we can hear. When Earth "rings," the surface rises and falls, owing to bell-like vibrations. To an observer on the surface of Earth, the bell-like vibrations appear to be waves, and are recorded as waves on a seismograph in the same way that body waves are recorded. Surface waves travel more slowly than P and S waves. They also pass *around* Earth, on the surface, rather than through it—in other words, they take the long way around. For both of these reasons, surface waves are the last to arrive and the last to be detected by a seismograph (FIG. B6.2).

Surface waves can have very long *wavelengths* (the distance from crest to crest of the wave)—up to hundreds of kilometers. (See Chapter 10 for a more detailed description of basic wave terminology.) The *amplitude*, or wave height, is a function of the displacement of material by the passing wave. Just as an ocean wave disturbs the water to some distance beneath the ocean surface, so do seismic waves traveling along Earth's surface cause rocks below the surface to be disturbed. The greater the amplitude, the greater the surface disruption and the deeper the wave motion reaches. Because they last a long time and have large amplitudes and wavelengths, surface waves can be extremely damaging to buildings and other structures on the surface.

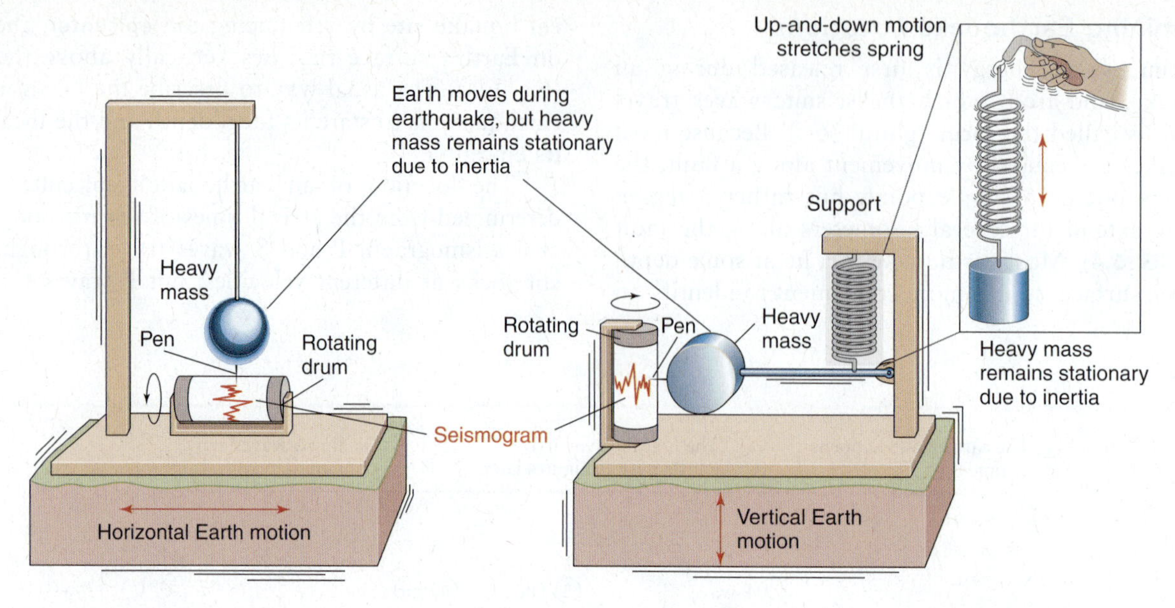

FIGURE 6.3 Seismograph
Seismographs measure and record the vibrations associated with earthquakes. Modern seismographs use the principle of inertia—the resistance of a heavy mass to motion. In this schematic diagram, seismic waves cause the support post and the roll of paper to vibrate back and forth, but the large mass attached to the pendulum and the pen attached to it barely move at all.

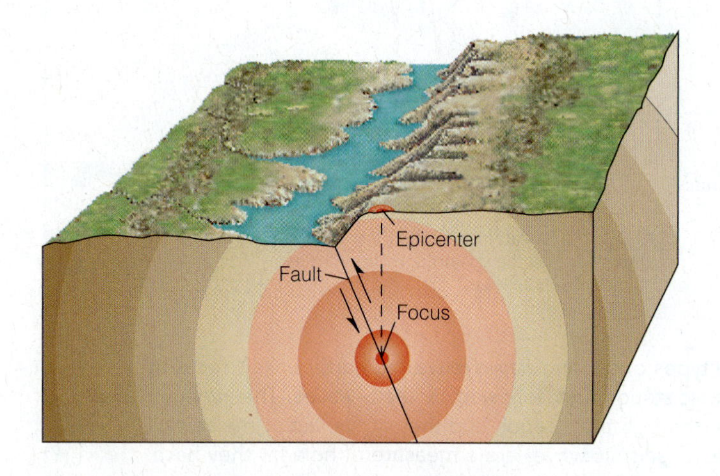

FIGURE 6.4 Earthquake focus and epicenter
The focus of an earthquake is the site of first movement on a fault, the site where energy is released, and the location from which the seismic waves travel outward in all directions. The focus is often not a simple point, but an area on the fault plane where the initial movement of rock masses takes place. The epicenter is the point on Earth's surface that lies directly above the focus.

fastest (as shown in Fig. B6.2). The farther the waves travel, the more the S waves lag behind the P waves, and the greater will be the gap between their arrival times at a seismograph. Using the lag time as an indicator of the distance traveled, the seismologist calculates the distance and draws a circle on a map around the seismic station, with radius equal to the calculated distance, *d*. It has to be a circle because the gap in arrival times indicates the *distance* traveled by the waves, but not the *direction* from which they originated—the waves could have traveled distance *d* from any direction around the circle. The exact position of the epicenter can be determined by *triangulation*, when data from three or more seismographs are available. The epicenter lies where the three circles intersect, as shown in FIGURE 6.5.

Determining the depth to the focus of the earthquake is a bit more complicated than determining the location of the epicenter. Seismic waves actually originate at the focus, not at the epicenter of the quake. If the focus is very deep and far from the seismic station, the calculated distances will overshoot the epicenter, and the three circles will overlap in a small triangle rather than at a single point. The characteristics of surface waves arriving at the seismograph can also offer clues to the focal depth; in very deep-focus earthquakes the surface waves typically have small amplitudes by the time they reach the surface. Seismologists also look for signs of body waves that have been reflected by different layers in the interior of the planet; calculating the travel times, path distances, and arrival times of these waves can reveal the depth at which the waves originated.

Measuring Earthquake Magnitudes

The location of the epicenter and focal depth are only part of the information that can be read from seismograms. Of equal importance is the calculation of the amount of energy released during an earthquake or, as it is commonly stated, the **magnitude** of the earthquake.

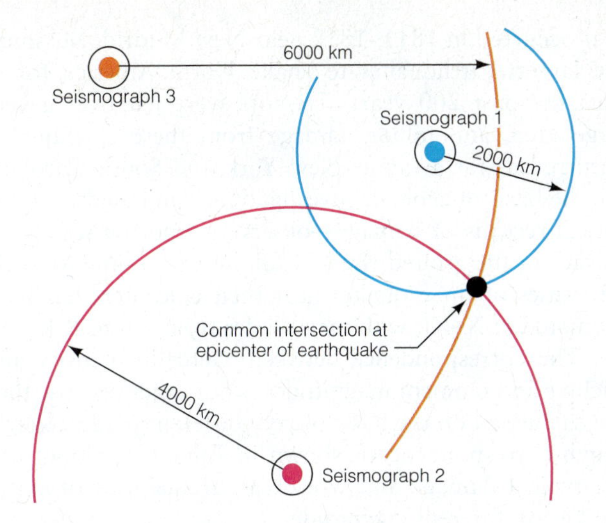

FIGURE 6.5 Locating an earthquake
The location of an earthquake's epicenter can be calculated from seismic wave measurements at three seismic stations. The time interval between the arrival of the first P wave and the first S wave depends on the distance from the seismic station to the epicenter. Let's say that seismic station 1 calculates a distance of 2000 km to the earthquake's point of origin, on the basis of the time lag between P and S wave arrivals. Seismic station 2 calculates a distance of 4000 km, and seismic station 3 calculates a distance of 6000 km. On a map, a circle of appropriate radius (equal to the distance) is drawn around each station; it has to be a circle because the arrival times indicate the distance traveled by the waves, but not the direction from which they came. The epicenter is where the three circles intersect on the map.

How Big Can Earthquakes Be?

Very large earthquakes are, fortunately, relatively infrequent (TABLE 6.1). In earthquake-prone regions, very large earthquakes occur about once a century. Small earthquakes may occur along a fault during this time as a result of local slippage, but even so, elastic energy continues to accumulate because most of the fault remains locked. When the lock is finally broken and an earthquake occurs, the stored elastic energy is released during a few terrible minutes.

Measurements of the bending of elastically deformed rocks before an earthquake, and of those same rocks after an earthquake has released the deforming force, can provide an accurate measure of the amount of the energy released. By careful measurement of elastically deformed rocks along the San Andreas Fault, seismologists have found that about 100 joules (J) of elastic energy can be accumulated in 1 m^3 of deformed rock. This is not very much—100 J is equivalent to only about 25 calories of heat energy—but when billions or trillions of cubic meters of rock are deformed, the total amount of stored energy can be enormous. The amount of elastically stored energy released during the Loma Prieta earthquake of 1989 was about 10^{15} J, and the 1906 San Francisco earthquake released at least 10^{17} J. For comparison, the two atomic bombs dropped on Japan during World War II each released on the order of 10^{13} J, about 10,000 times less than the 1906 San Francisco earthquake.

How big can earthquakes be? It does seem that there is an upper limit to the size of earthquakes. Rocks only have enough strength to store a finite amount of elastic energy before they reach the breaking point and release the energy. The largest earthquake ever measured occurred in a subduction zone off the coast of Chile in 1960; it had a magnitude of 9.5. Only two other quakes have ever been measured with magnitudes greater than 9.0—the Great Alaska earthquake of 1964 (M = 9.2), and the Sumatra-Andaman earthquake of 2004 (M = 9.1). Seismologists employ a number of scales to quantify the magnitudes of earthquakes. The most common are the Richter magnitude scale, the moment magnitude scale, and the modified Mercalli intensity scale.

TABLE 6.1 Earthquake Magnitudes, Frequencies, and Effects

Richter and Moment Magnitude*	Number per year	Modified Mercalli Intensity Scale*	Characteristic Effects in Populated Areas
<3.4	800,000	I	Recorded only by seismographs
3.5–4.2	30,000	II–III	Felt by some people who are indoors
4.3–4.8	4,800	IV	Felt by many people; windows rattle
4.9–5.4	1,400	V	Felt by everyone; dishes break, doors swing
5.5–6.1	500	VI–VII	Slight building damage; plaster cracks, bricks fall
6.2–6.9	100	VIII–IX	Much building damage; chimneys fall; houses move on foundations
7.0–7.3	15	X	Serious damage, bridges twisted, walls fractured; many masonry buildings collapse
7.4–7.9	4	XI	Great damage; most buildings collapse
>8.0	<1	XII	Total damage; waves seen on ground surface, objects thrown in the air

*The correspondence between Richter and moment magnitudes and the Mercalli intensity is not exact because they are calculated on the basis of very different parameters.

Richter Magnitude

The energy released during an earthquake can be calculated from the recorded amplitudes of seismic waves on seismograms. The familiar **Richter magnitude** is calculated from the maximum recorded amplitudes of seismic waves (that is, the heights of the waves on a seismogram), with a correction for distance from the epicenter. The correction for distance means that all seismic stations will calculate the same magnitude for a given earthquake, no matter how near or far they are from the quake's epicenter. The Richter scale is logarithmic; it is divided into steps, starting with magnitude 1, and each unit of increase corresponds to a tenfold increase in the amplitude of the seismic wave signal. Thus, a magnitude 2 earthquake has a recorded wave amplitude that is ten times larger than a magnitude 1, and a magnitude 3 is a hundred times larger than a magnitude 1. Importantly, though, each step in the Richter scale corresponds to roughly a 32-fold increase in the amount of energy released. Thus, the difference in energy released between an earthquake of magnitude 4 and one of magnitude 7 is $30 \times 30 \times 30 = 27,000$ times!

Seismic Moment Magnitude

Seismologists today determine magnitudes using the **seismic moment magnitude** scale, which is calculated using different starting assumptions from the Richter magnitude scale. For a Richter magnitude calculation it is assumed that the focus of an earthquake is a point. Therefore, the Richter scale is best suited for earthquakes in which energy is released from a relatively small area of a locked fault. In contrast, the calculation of seismic moment takes account of the fact that energy may be released over a large area. A classic example was the Sumatra-Andaman earthquake of 2004, in which a 1300-km length of the fault moved. Although the method of calculation is different, the scales are similar because they measure the same thing—the amount of energy released. On either scale, an earthquake of magnitude 9 is catastrophic, whereas an earthquake of magnitude <3 is imperceptible to humans.

Mercalli Intensity

Because damage to the land surface and to property is so important, another scale that is commonly used to quantify the magnitude of earthquakes is based on the strength of vibrations that people feel, and the extent of damage that occurs during earthquake; this is called the **modified Mercalli intensity** (or MMI). The MMI scale ranges from I (not felt, except under unusual circumstances) to XII (visible waves of the ground surface, most buildings destroyed). Unlike the Richter and moment magnitudes, there is no distance correction in the MMI scale, so Mercalli intensity varies with distance from the epicenter. An earthquake could have an intensity of X near the epicenter, whereas a hundred kilometers away the intensity would be only II.

The modified Mercalli intensity scale is particularly useful for the study of earthquakes that occurred prior to the installation of the modern global network of seismographs. An example is a series of three great earthquakes that occurred in 1811–1812 near New Madrid, Missouri, the largest earthquakes to shake North America for at least the past 200 years. Tremors were felt over a very large area, and minor damage from these earthquakes occurred as far away as New York and South Carolina. On the basis of geologic investigations combined with historical records of damage—observers reported seeing the surface of the ground rise and fall in waves—the Mercalli intensities of these quakes near their epicenters has been estimated at X–XI, with magnitudes from 7.8 to 8.1.

The correspondence between Mercalli intensity and Richter and moment magnitudes is not exact because they are calculated on the basis of very different parameters. A rough correspondence is shown in Table 6.1, along with the typical damage and worldwide frequencies of earthquakes of different magnitude.

EARTHQUAKE HAZARD AND RISK

Most people in North America think immediately of California when earthquakes are mentioned; Alaska, the Pacific Northwest, and Mexico are not far behind. This makes sense because of the relative frequency of large earthquakes in these areas. You learned in Chapter 5 that most seismic activity takes place along active tectonic plate boundaries. The California coast is an active transform fault plate boundary; in Alaska the Pacific Plate is undergoing subduction in the Aleutian Trench; in the Pacific Northwest the small Juan da Fuca Plate is subducting under the North American Plate; and the Cocos Plate is subducting under the west coast of Mexico (refer to Fig. 5.12). However, the largest earthquakes in North America in over 200 years were the New Madrid earthquakes mentioned above—right in the middle of the North American Plate.

Understanding the tectonic settings of earthquakes is of central importance when we seek to quantify earthquake hazards, that is, the potential physical effects and where and when they may strike, and the risks to human interests in terms of life, health, and property damages. **FIGURE 6.6** is a seismic hazard map for North America, based on Peak Ground Acceleration, a measure of ground motion from seismic activity. Notice that the most seismically active areas are located along the plate boundaries mentioned above, although a few areas of seismic activity are located far from plate boundaries.

Make the CONNECTION

Look at a seismic hazard map to determine whether you live in an area of seismic risk. Are there other types of natural hazards to which your home is vulnerable, such as flooding, tornadoes, drought, or landslides? Do you know the proper steps to protect yourself from these hazards?

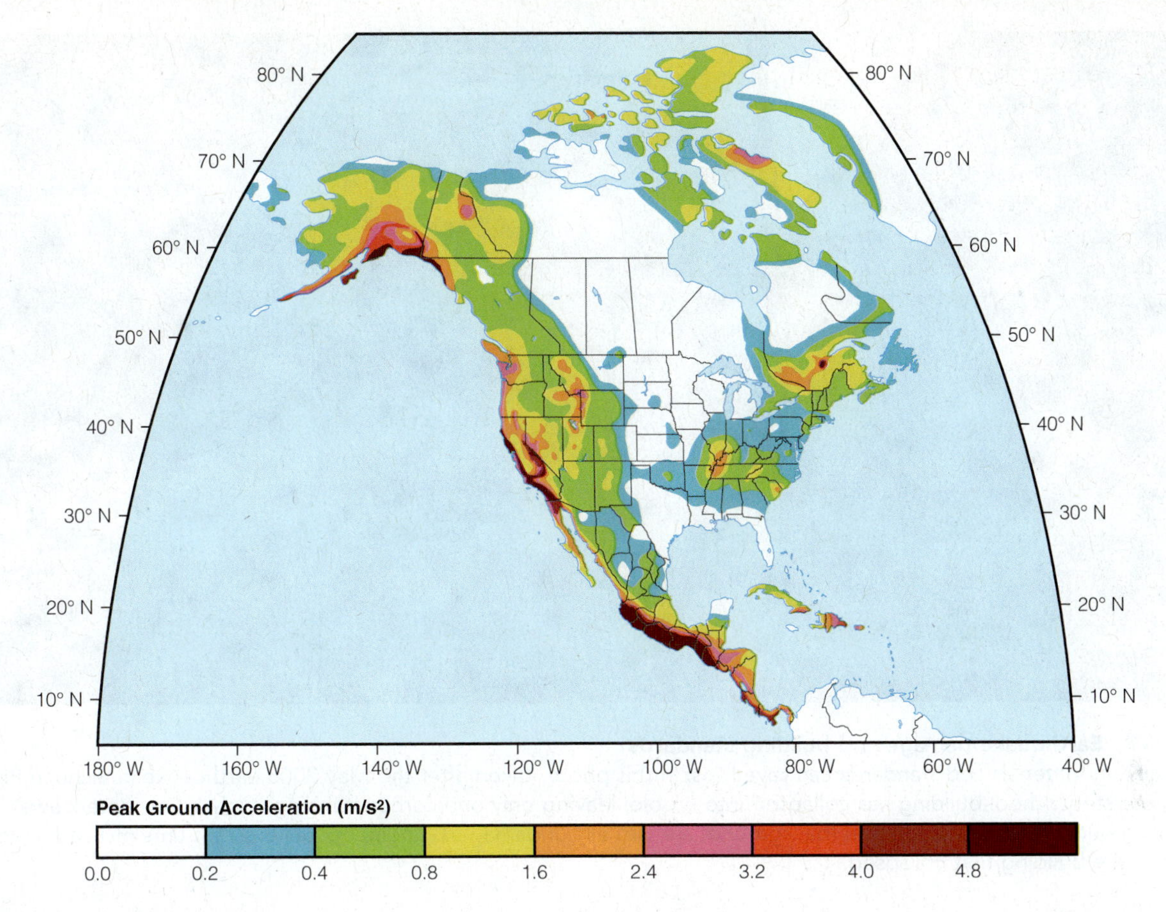

FIGURE 6.6 Seismic hazard map
This is a seismic hazard map for North America, based on Peak Ground Acceleration, a measure of ground motion due to seismic activity. The red and brown colors reflect more seismic activity, and they are concentrated along the active plate margins of Central America, California, the Pacific Northwest, and Alaska. However, there are a few other areas of high seismic risk located far from active plate margins.

Earthquake Disasters

There are hundreds of thousands of earthquakes each year. Fortunately, only one or two are large enough and close enough to major population centers to cause loss of life. Certain areas are known to be earthquake-prone, and special building codes in such places require structures to be as resistant as possible to earthquake damage. All too often, however, an unexpected earthquake will devastate an area where buildings are not adequately constructed (FIG. 6.7). TABLE 6.2 lists the earthquakes of the past 800 years that are known to have killed at least 50,000 people. The most disastrous one on record occurred in 1556 in Shaanxi Province, China, where an estimated 830,000 people died. Many of those people lived in cave dwellings excavated in soft, wind-deposited sediment called loess, which collapsed as a result of the quake. Since 1900, there have been eight earthquakes worldwide in which 50,000 or more people have died.

Earthquake Damage

The dangers of earthquakes are profound, and their effects are of two principal kinds. Ground motion and disruption of the ground surface are *primary effects* that cause direct damage to buildings, infrastructure, and the landscape. *Secondary effects* cause damage indirectly as a result of processes set in motion by the earthquake.

Primary Effects

GROUND MOTION. Ground motion results from the movement of seismic waves, especially surface waves, through surface-rock layers and soil. The motion can damage and sometimes completely destroy buildings (FIG. 6.8A, on page 153). Proper design (including such features as a steel framework and a foundation tied to bedrock) can do much to prevent such damage, but in a very strong earthquake even the best buildings may suffer some damage.

SURFACE RUPTURE. Where a fault breaks the ground surface, buildings can be split, roads disrupted, and any feature that crosses or sits on the fault broken apart. If movement along a fault disrupts the surface, it can create an open *fissure* (FIG. 6.8B, on page 153) or a *fault scarp*, a small cliff or escarpment that is the topographic expression of movement along the fault.

FIGURE 6.7 **Earthquake damage and building standards**
Proper building materials and standards can save lives. In this photo, taken after the May 2008 earthquake in Sichuan Province, China, a four-story school building has collapsed into rubble, leaving only one corner standing. Only a few meters away, a kindergarten building (at left) and hotel (background) are virtually undamaged. Engineers believe that unreinforced concrete was used in the building that collapsed.

TABLE 6.2 Earthquakes during the Past 800 Years That Have Caused 50,000 or More Deaths

Place	Year	Estimated number of deaths
Silicia, Turkey	1268	60,000
Chili, China	1290	100,000
Naples, Italy	1456	60,000
Shaanxi, China	1556	830,000
Shemaka, Russia	1667	80,000
Catania, Italy	1693	60,000
Beijing, China	1731	100,000
Calcutta, India	1737	300,000
Lisbon, Portugal	1755	60,000
Calabria, Italy	1783	50,000
Messina, Italy	1908	160,000
Gansu, China	1920	180,000
Tokyo and Yokohama, Japan	1923	143,000
Gansu, China	1932	70,000
Quetta, Pakistan	1935	60,000
T'ang Shan, China	1976	240,000
Sumatra-Andaman, Indian Ocean	2004	283,000*
Sichuan, China	2008	69,000
Port-au-Prince, Haiti	2010	50,000–100,000**

*Most of the deaths due to tsunami.

**The death toll is uncertain and may be much higher, according to the United Nations, which places the death toll at 250,000–300,000; many thousands also died in the aftermath of the earthquake.

Secondary Effects

FIRES. A secondary effect, but one that is sometimes a greater hazard than moving ground, is fire (FIG. 6.8C). Ground movement displaces stoves, breaks gas lines, and loosens electrical wires, thereby starting fires. Ground motion also breaks water mains, which usually means that there is no water available to put out fires. In the earthquake that struck San Francisco in 1906, more than 90 percent of the building damage was caused by fire.

LANDSLIDES. In regions of steep slopes, earthquake vibrations may cause soil to slip and cliffs to collapse (FIG. 6.8D). This is particularly true in Alaska, parts of Southern California, China, the Andes, and hilly places such as Iran and Turkey. Houses, roads, and other structures are destroyed by rapidly moving soil flows.

LIQUEFACTION. The sudden disturbance of water-saturated sediment and soil can turn seemingly solid ground to a liquid-like mass of quicksand. This process is called **liquefaction**, and it was one of the major causes of damage during the earthquake that destroyed much of Anchorage, Alaska, during the great earthquake of 1964, and the earthquake that caused apartment houses to sink and collapse in Niigata, Japan, that same year (FIG. 6.8E).

TSUNAMI. A **tsunami** is a seismic sea wave. Tsunamis are often erroneously called tidal waves, but they have nothing to do with tides. They are initiated by sudden movement of the seafloor caused by an earthquake, volcanic eruption, or underwater landslide, and they have been

FIGURE 6.8 Earthquake impacts
(A) Ground motion during an earthquake is a direct cause of building collapse, as seen here after the major earthquake of January, 2010, in Port-au-Prince, Haiti. (B) Disruption of the ground surface can result from movement along a fault, as seen in this open fissure that resulted from an earthquake off the coast of Chile in February, 2010. (C) Fire is a significant secondary impact of earthquakes. These fires in Mexico City resulted from an earthquake of magnitude 8.5 that occurred in 1985. (D) In hilly and mountainous regions, landslides are commonly triggered by earthquakes. This massive landslide caused by an earthquake in Peru buried two villages. (E) Shaking can cause liquefaction of the ground, here resulting in the collapse of several apartment buildings in Niigata, Japan.

A Closer LOOK

THE SUMATRA-ANDAMAN EARTHQUAKE AND TSUNAMI

On December 26, 2004, the most powerful earthquake in 40 years—and the third most powerful of the last century—shook the Indian Ocean floor, about 160 km off the island of Sumatra in a subduction zone called the Sunda Trench. Earthquakes on the ocean floor are common, but are usually noticed only by seismologists. However, this earthquake—the Sumatra-Andaman earthquake of 2004—will be long remembered because it caused the largest and deadliest tsunami in history.

The quake began when part of the Indian Plate, which is subducting under the Eurasian Plate in the Sunda Trench, suddenly slipped downward approximately 15 m. Unlike most earthquakes, which are over in seconds, the slippage continued for 10 minutes as the fault broke, section by section, for 1300 km toward the north. The five most powerful earthquakes since 1900 have occurred in similar tectonic environments, where a subducting plate is being overridden by another plate; these massive quakes are referred to as *megthrust earthquakes*.

On the surface of the ocean, waves generated by the sudden movement of the ocean floor during the megathrust quake swept toward Indonesia, Thailand, Sri Lanka, and India (**FIG. C6.1**). In the open ocean, tsunamis are barely noticeable; the wave velocity may be hundreds of kilometers per hour and the wavelength many kilometers, but the peak wave height is typically less than a meter. However, upon reaching the shore, the wave begins to "hit bottom," slows down, and builds to colossal heights. Wave heights of 20 or 30 m were reported in Sumatra during this event. The waves swept far inland, obliterating everything in their path (**FIG. C6.2**).

FIGURE C6.1 Progress of a killer tsunami
This series of computer-simulated maps shows the progress of the tsunami across the Indian Ocean from 0 to 3 hours after the Sumatra-Andaman earthquake of 2004. The satellite images show a coast in Sri Lanka before and during the tsunami.

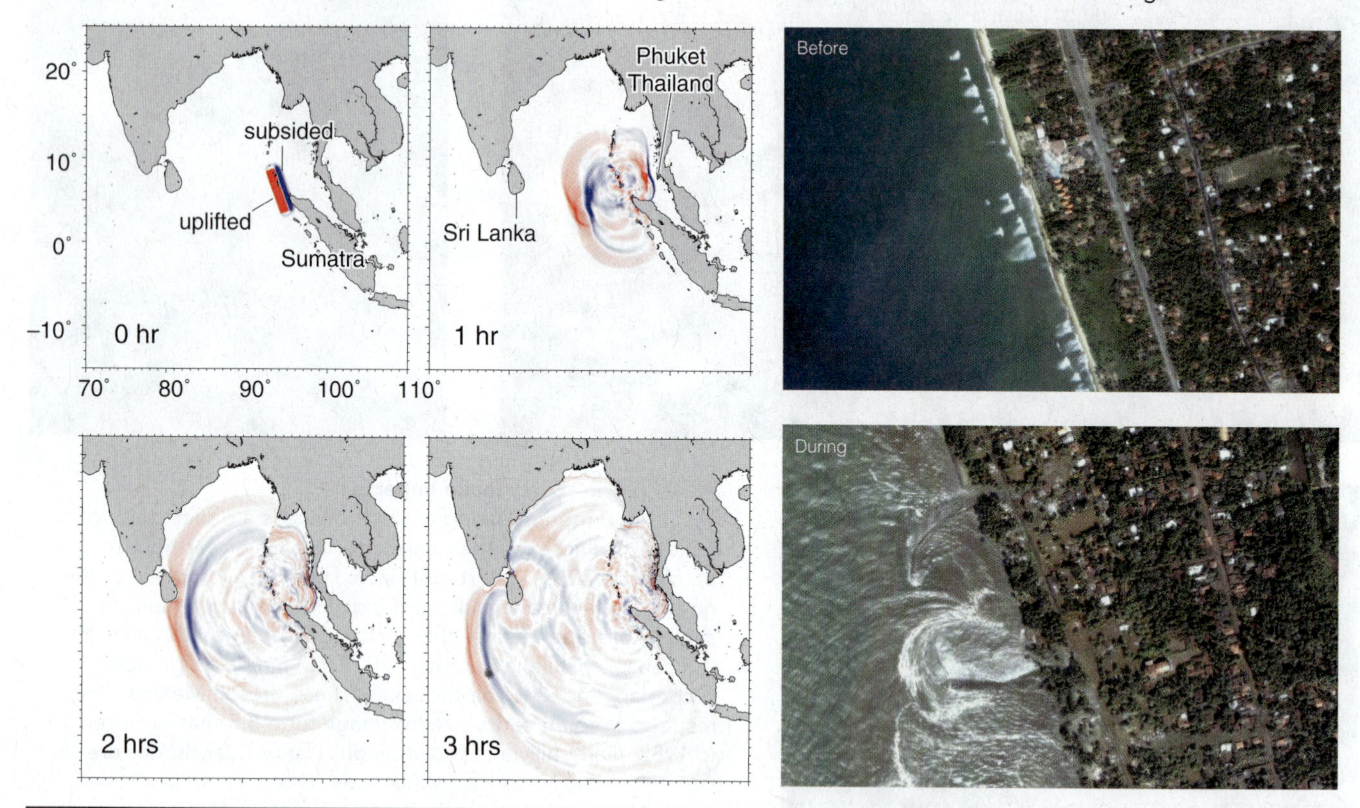

particularly destructive in the Pacific and Indian oceans. Seismic sea waves travel at speeds up to 950 km/h and can have wavelengths up to 200 km. Amplitudes of the waves are so low they can rarely be seen in the open ocean, but as a wave approaches the shore the water piles up rapidly to heights of 30 m or more. About 5 hours after a severe submarine earthquake near Unimak Island, Alaska, in 1946, for instance, a tsunami struck Hawaii. The wave had traveled at a speed of 800 km/h. Although the amplitude of the wave in the open ocean was less than 1 m, the

Although earthquakes and tsunamis are common in the Indian Ocean, there was no warning system in place when this disaster occurred (in contrast to the Pacific Ocean, which has had an extensive warning system in place for a number of years). The resulting devastation caused hundreds of billions of dollars in damages and at least 283,000 deaths. Tsunamis can be devastating, but their travel time across the open ocean also means that they can be amenable to the issuance of an early warning, as long as appropriate systems for detection, decision making, and communications have been established. Research is now being directed to the installation of a coordinated early warning system for the Indian Ocean, as well as on ways to reduce the damage cause by tsunamis.

FIGURE C6.2 The Sumatra-Andaman Tsunami of 2004
This photo shows a ground-level view of the destruction in western Sumatra caused by the tsunami.

amplitude increased dramatically as the wave approached land, so that when it hit Hawaii, the wave had a crest 18 m higher than normal high tide. This destructive wave demolished nearly 500 houses, damaged a thousand more, and killed 159 people; residents of Hawaii still remember it.

The phenomenon of tsunamis entered the world's consciousness dramatically when an earthquake on December 26, 2004 caused a devastating tsunami, described in *A Closer Look: The Sumatra-Andaman Earthquake and Tsunami.*

Earthquake Prediction

Some of the most dreadful natural disasters have been caused by earthquakes. It is hardly surprising, therefore, that a great deal of research around the world focuses on the prediction of earthquakes. In the context of earthquakes, *forecasting* means determining the long-term likelihood of an earthquake striking in a particular location; it is based principally on developing an understanding of the tectonic setting and history of seismic activity in that location. The ability of scientists to issue accurate long-term forecasts has improved greatly. However, success with issuing shorter-term predictions and accurate, specific early warnings remains elusive.

Prediction and Early Warning

China has suffered many terrible earthquakes, and Chinese scientists have tried everything they can think of to predict quakes. On one occasion they even used observations of animal behavior to successfully foretell a quake. In July 18, 1969, zookeepers at the People's Park in Tianjin observed highly unusual animal behavior. Normally quiet pandas screamed, swans refused to go near water, yaks did not eat, and snakes would not go into their holes. The keepers reported their observations to the earthquake prediction office, and at about noon on the same day a magnitude 7.4 earthquake struck.

There have been many informal reports of strange animal behavior before earthquakes, but the Tianjin quake is the only well-documented case. The scientific basis for modern prediction efforts is the observation of precursor anomalies—any odd or unusual occurrences that might signal an imminent seismic event. Unfortunately, most quakes do not seem to be preceded by anything odd, and the ones that do show precursor anomalies are highly inconsistent. This greatly complicates scientists' efforts.

Most research on earthquake prediction today is based on the study of properties that might be expected to change in rocks that are undergoing elastic deformation, including rock magnetism, electrical conductivity, and porosity. Even simple observations, such as changes in the level of water in a well, might indicate a change in the porosity of rocks, for example. Tilting of the ground or slow rises and falls in elevation may also indicate that deformation is occurring. Most significant are the small cracks and fractures that can develop in severely bent rock. These openings can cause swarms of tiny earthquakes called *foreshocks* that may be a clue that a big quake is coming.

One of the most successful cases of earthquake prediction, made by Chinese scientists in 1975, was based on a combination of observations of slow tilting of the land surface, fluctuations in magnetism, and swarms of small foreshocks that preceded a M7.3 quake that struck the town of Haicheng. Half the city was destroyed, but authorities had evacuated more than a million people before the quake. As a result, only a few hundred were killed. Unfortunately, the seemingly erratic nature of earthquake precursors, combined with the difficulties inherent in monitoring events that occur deep underground and at unexpected times and places, have led to limited progress in earthquake prediction since the success at Haicheng.

Forecasting

In contrast to the minimal progress in short-term prediction, long-term forecasting of the likelihood of earthquake occurrence in a given location has improved substantially over the past 20 or 30 years. In places where earthquakes are known to occur repeatedly, such as around the margins of the Pacific Ocean, seismologists can sometimes discern patterns in the timing of recurrence of major earthquakes. If such a pattern suggests a recurrence interval of, say, a century, it may be possible to forecast where and when a large quake is likely to occur. Furthermore, studies of recurrence patterns have identified a number of **seismic gaps** in tectonically active areas; these are places where, for one reason or another, earthquakes have not occurred for a very long time, even though tectonic plate motion is still known to be occurring and where deformation has been increasing. Seismic gaps receive a lot of research attention because they indicate the places most likely to experience large earthquakes.

EARTHQUAKES AND EARTH'S INTERIOR

Despite their impact on people, earthquakes are of enormous scientific significance. The occurrence and distribution of earthquakes and the behavior of seismic waves as they travel through Earth's interior provide the most powerful set of tools available to scientists for the study of plate tectonics and the inside of this planet.

Earth's Internal Layering

Seismic body waves behave like light waves, which is to say that, in addition to being able to pass through materials, they can also be reflected and refracted when they encounter the surfaces or boundaries of materials (FIG. 6.9). **Reflection** is the familiar phenomenon of light bouncing off a mirror or other shiny surface. Similarly, seismic body waves are reflected by numerous boundaries inside the planet. A less familiar process, **refraction,** occurs when the speed of a wave changes as it passes from one medium to another, causing the wave path to bend. The same phenomenon can be observed in a pencil half-immersed in a glass of water; the light waves are refracted as they pass from the water to the air, causing

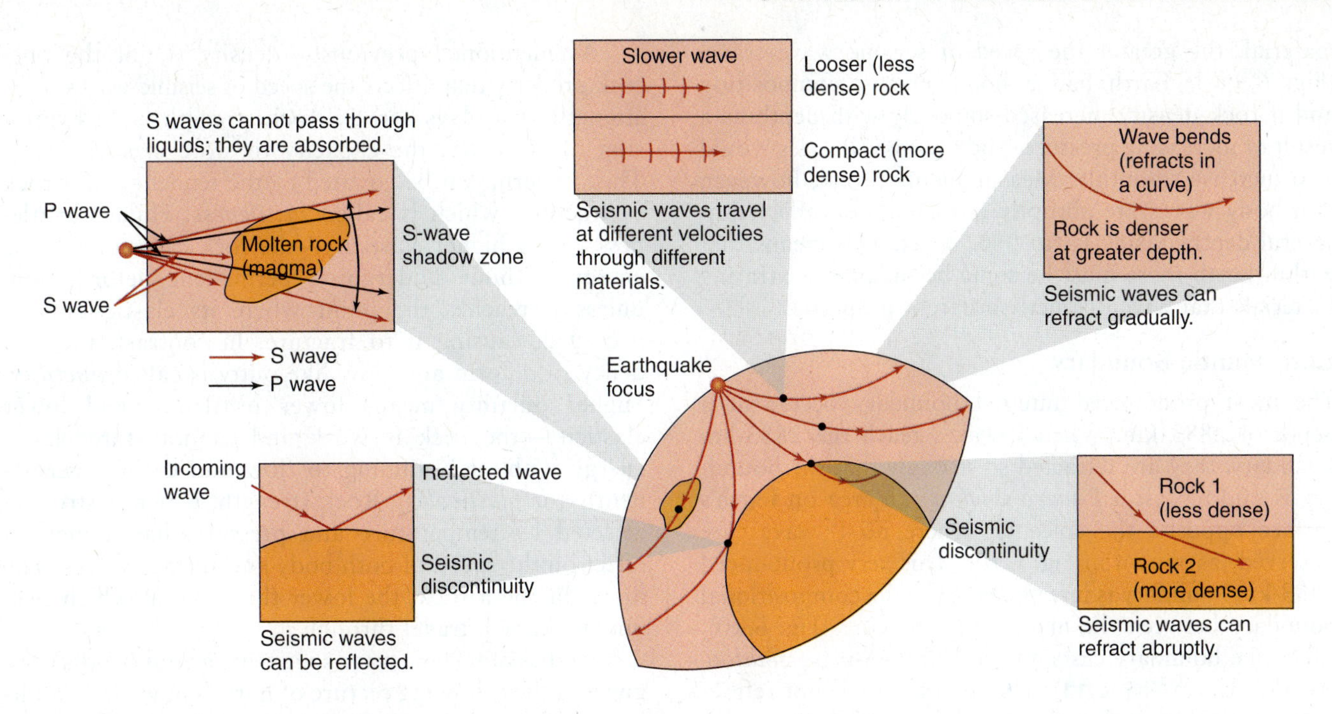

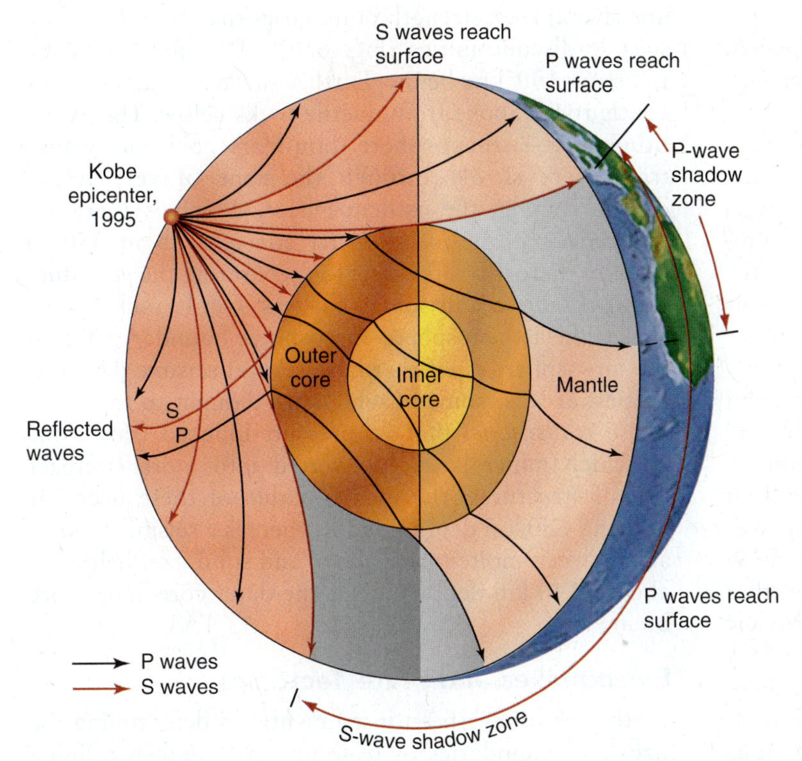

FIGURE 6.9 Seismic waves in Earth's interior Seismic body waves are reflected and refracted as they travel through the planet and encounter materials with different physical properties. The example shows the epicenter of the major earthquake of 1995 in Kobe, Japan. As the S and P waves travel outward from the earthquake's focus they are variously reflected, refracted, or even blocked (for example, S waves are blocked by the liquid outer core). All of these affect the arrival times of the waves at seismic stations, allowing scientists to make inferences about the materials through which the waves have traveled.

the pencil to appear bent. In a similar manner, when a seismic wave passes from one rock mass into another mass with different physical characteristics, its travel path will be refracted.

Refraction and reflection of seismic body waves as they pass from one material to another have been extremely useful in building up a picture of Earth's interior, as introduced in Chapter 3. This is because the speed of body waves is determined partly by the density of the rocks they are passing through, and partly by other properties such as the strength or rigidity of the rocks. In general, the higher the density of the material, the greater the speed of seismic wave travel (Fig. 6.9). If Earth had a homogeneous composition and if rock density increased smoothly with depth as a result of increasing pressure, body wave velocities would also increase smoothly. Measurements reveal, however, that body waves are abruptly refracted and reflected at several depths inside Earth (**FIG. 6.10**). This means that within Earth there must be some boundaries separating materials that have distinctly different properties.

Core-Mantle Boundary

The most pronounced internal boundary occurs at a depth of 2883 km. When P waves reach the 2883-km boundary, they are refracted so strongly that the boundary is said to cast a *P-wave shadow*, an area on Earth's surface opposite the epicenter where no P waves are observed (as shown in Fig. 6.10). This very pronounced 2883-km boundary is interpreted to be the compositional boundary between the mantle and the core (Fig. 6.10). The same boundary casts an even larger *S-wave shadow* (also shown in **FIG. 6.11**). The reason here is not refraction, but the fact that S waves cannot travel through liquids. S waves are blocked by the outer core, casting a huge S-wave "shadow" on the other side of the planet, which allows us to conclude that at least the outer portion of the core must be molten.

Mantle-Crust Boundary

Early in the twentieth century, Andrija Mohorovičić, a Croatian scientist, demonstrated the existence of a compositional boundary between the crust and the mantle. Mohorovičić noticed that, for earthquakes whose focus lay within 40 km of the surface, seismographs about 800 km from the epicenter recorded two sets of body waves that arrived at the seismograph at different times. He concluded that the set that arrived second must have traveled from the focus to the station by a direct path through the crust, whereas the set that arrived first must have been refracted into rock that was denser than crustal rock. These refracted waves, moving through the denser zone, traveled faster within that zone and so reached the surface first. Mohorovičić hypothesized that a distinct compositional boundary must separate the crust from this underlying zone of denser composition. Scientists now refer to this boundary as the *Mohorovičić discontinuity* and recognize it as the boundary that marks the base of the crust (or, put another way, the top of the mantle). The feature is commonly called the *M-discontinuity* and in conversation it is shortened to **moho**.

Seismic Discontinuities

The core-mantle boundary and the mantle-crust boundary (the moho) are **seismic discontinuities**—boundaries where the velocity of seismic wave travel changes suddenly, rather than smoothly. These are Earth's principal seismic discontinuities, but there are several others (Fig. 6.10).

As mentioned previously, density is not the only rock property that affects the speed of seismic waves; rock strength also plays a role. Rock strength is an expression of *elasticity*, the capacity to store elastic energy. This, in turn, can be equated to the tendency of a rock to fracture, which is called *brittleness*. Higher brittleness means higher elasticity—the rock can store elastic energy without undergoing permanent deformation, unless it reaches the point where its elastic limit is exceeded, causing it to fracture. In contrast, the tendency to deform and flow like putty is called *ductility*. Higher ductility means lower brittleness and lower elasticity—the rock is weak and cannot store elastic energy without beginning to flow and deform permanently (or *plastically*). Rock strength, which is strongly affected by temperature and pressure, has a marked effect on the speed of both body and surface waves. The more ductile a rock, the lower the speed at which seismic waves will travel through it.

Studies of seismic wave speeds at various depths have given us the following picture of Earth's interior. In addition to the moho and the core-mantle boundary, there are several rock strength boundaries that are marked by seismic discontinuities (Fig. 6.10). The first boundary is about 100 km below Earth's surface and separates brittle rocks above from ductile rocks below. This is the lithosphere–asthenosphere boundary. Seismic waves travel very slowly through the zone of very weak, ductile rock in the asthenosphere, which is therefore referred to as the *low-velocity zone*. At about 350 to 400 km there is a diffuse region separating the very ductile rock above and somewhat less ductile rock below. This is the asthenosphere–mesosphere boundary. Within the mesosphere itself there are several seismic discontinuities, at least some of which are thought to represent *phase transitions*, that is, pressure-induced transitions in which minerals are converted into more compact crystal structures; the most notable of these occur at 400 and 650 km depth. Finally, there is a seismic boundary between molten iron above and solid iron below, at about 5140 km depth. This is the outer core–inner core boundary.

Earthquakes and Plate Tectonics

Earthquakes have been instrumental in determining the sizes and boundaries of tectonic plates and in refining our understanding of tectonic processes such as subduction. Recall from Chapter 5 that earthquakes and volcanic eruptions occur primarily along plate margins and are the most obvious manifestation of active plate interactions. As you have learned, earthquakes occur along faults where huge blocks of rock are grinding past each other. Tectonic motions produce directional pressure, which causes rock masses to move. The locations of earthquakes—both their epicenters and their focal depths—provide important information about the dimensions of tectonic plates and the nature of their interactions.

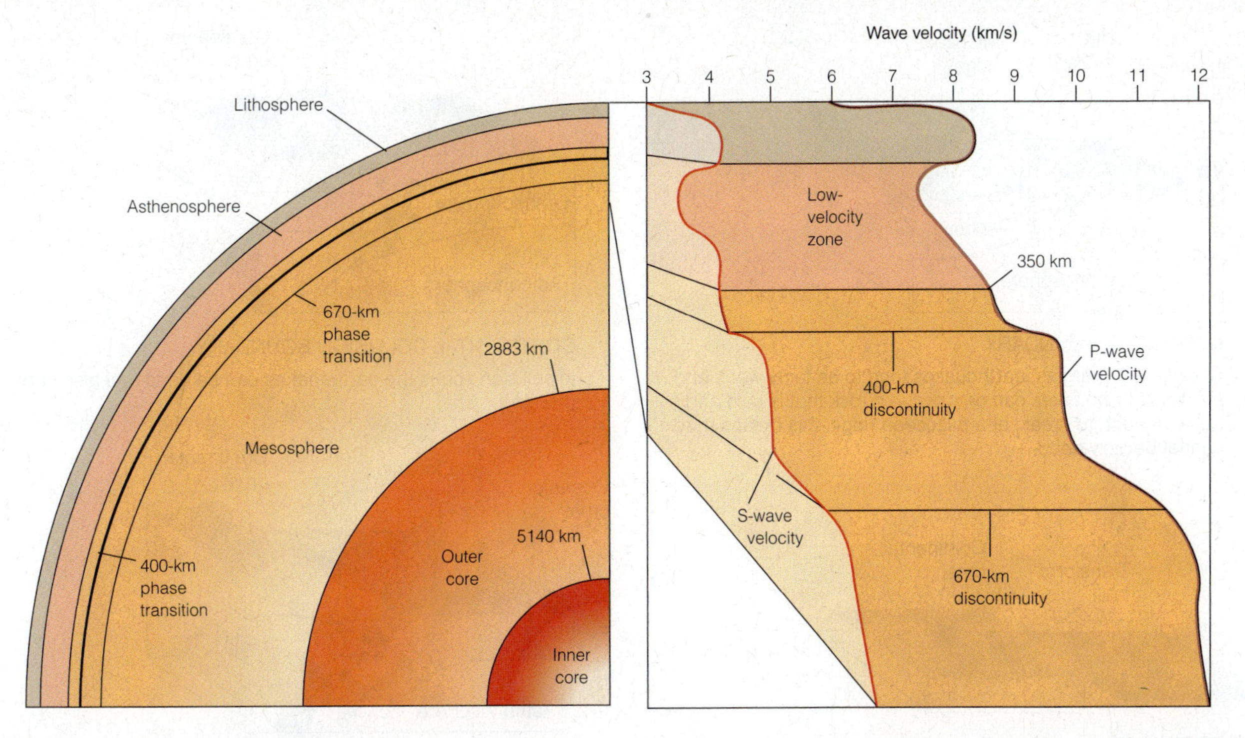

FIGURE 6.10 Layering in Earth's interior
Earth's interior is not homogeneous, but has several boundaries within it, marked by seismic discontinuities. P waves and S waves slow down or speed up abruptly and are refracted or reflected at these boundaries. However, the exact nature of some of the boundaries is still not completely understood.

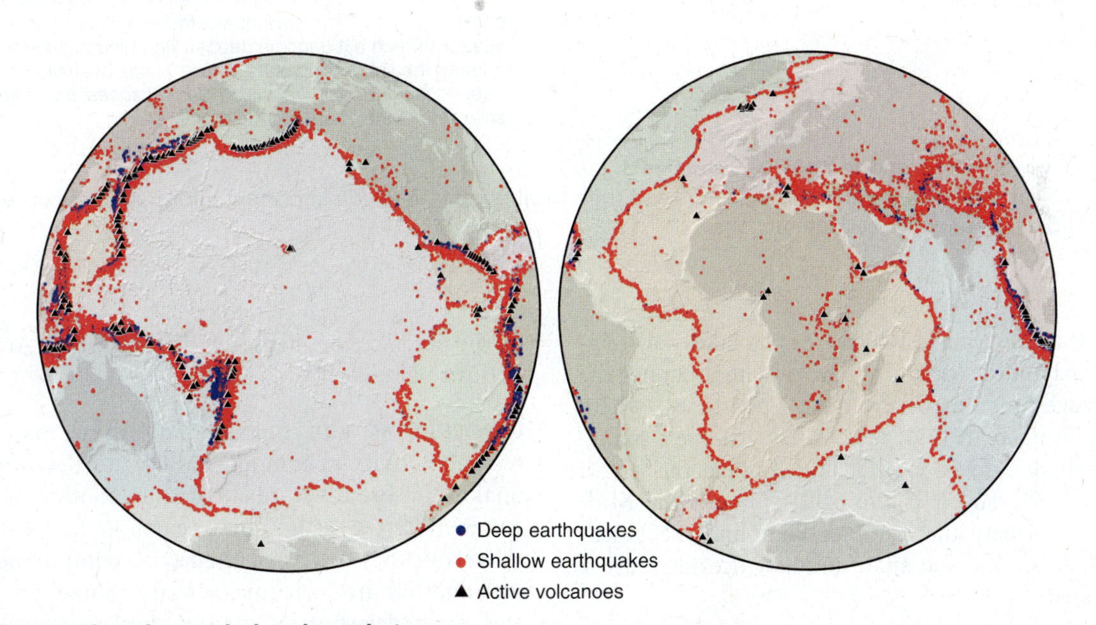

FIGURE 6.11 Earthquakes and plate boundaries
This map shows the epicenters of large earthquakes worldwide, over a period of about twenty years. The areas with the greatest amount of seismic activity and active volcanism (black triangles) outline the boundaries of tectonic plates. The deepest earthquakes occur in the deep oceanic trenches associated with subduction zones.

Seismicity and Plate Margins

As shown in Figure 6.11, the locations of large earthquakes, compiled over a period of some years, neatly outline the boundaries of the tectonic plates. Scientists have determined that there are four different types of seismic activity, characteristic of the basic types of plate boundary environment (FIG. 6.12):

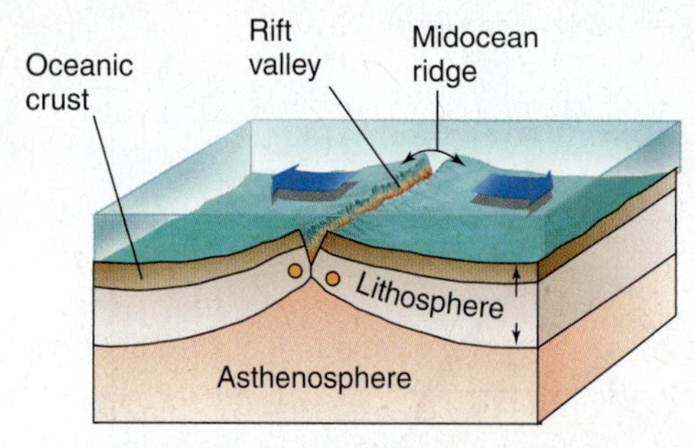

DIVERGENT BOUNDARY

At divergent margins, earthquakes tend to be fairly weak and shallow. Earthquakes can only occur in rock that is cold and brittle enough to break; at a midocean ridge, this means they cannot be very deep.

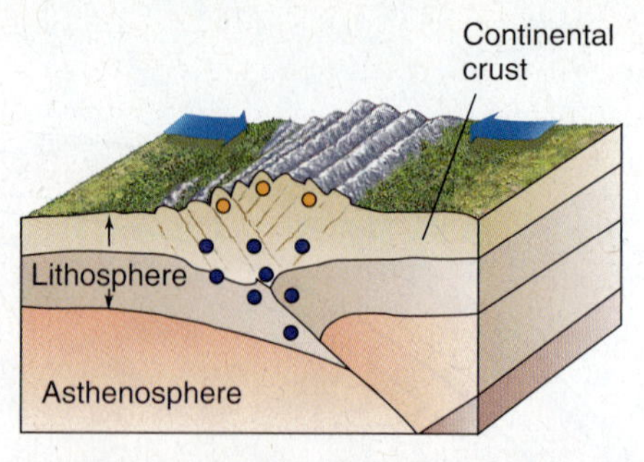

CONTINENTAL COLLISION BOUNDARY

In collision zones the earthquakes can be deep and also very powerful.

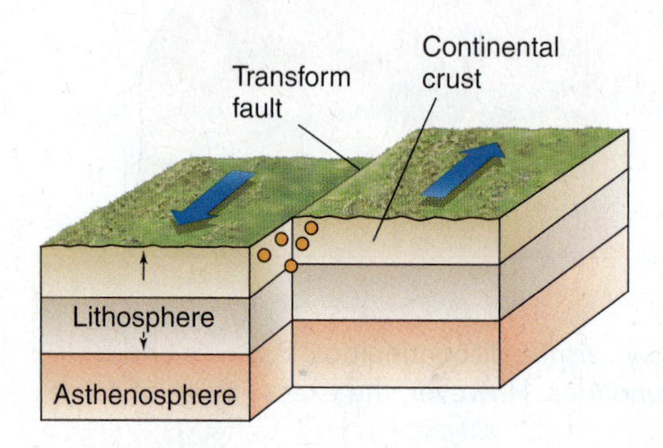

TRANSFORM FAULT BOUNDARY

Transform fault margins have shallow earthquakes, but they can be very powerful.

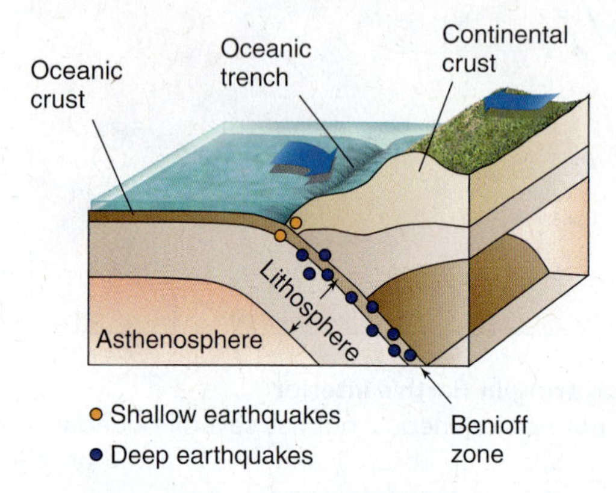

SUBDUCTION ZONE BOUNDARY

The deepest and most powerful quakes occur in subduction zones. Here, an oceanic plate moves downward relative to a continental plate. The earthquake foci are shallow near the oceanic trench but become deeper along the descending edge of the subducting plate. These zones of shallow- and deep-focus earthquakes, called *Benioff zones*, first alerted scientists to the phenomenon of *subduction*.

FIGURE 6.12 Earthquakes and tectonic environments
The locations of earthquakes—both their epicenters and their focal depths—provide important information about tectonic processes and plate boundaries and interactions.

Spreading Ridges. Shallow-focus earthquakes of relatively low magnitude occur in narrow lines along the spreading ridges that mark divergent boundaries, where plates are splitting apart (FIG. 6.12A). You can see a snake-like line of shallow-focus earthquakes running the length of the Atlantic Ocean on Figure 6.11; it marks the Mid-Atlantic Ridge, where the Eurasian Plate and the North American Plate are moving apart and new oceanic crust is being generated.

Transform Faults. Shallow-focus but sometimes very powerful earthquakes delineate transform fault plate boundaries (FIG. 6.12B). These earthquakes have no associated volcanic activity. We have previously discussed the seismic activity associated with the San Andreas Fault, the boundary between the Pacific Plate and the North American Plate. Another example of a transform fault plate boundary where major earthquakes have occurred is the Anatolian Fault in northeastern Turkey. This fault marks a boundary of the small Anatolian Plate, which lies at the intersection between the Eurasian, Arabian, and African plates.

Continental Collisions. The third major type of seismic activity occurs as shallow- to deep-focus earthquakes in broad bands marking continental collision zones (FIG. 6.12C). These earthquakes can be very powerful; they are associated with compressional deformation, but no volcanic activity. Some of the largest and most devastating earthquakes ever recorded have occurred in the broad zone of seismic activity associated with the convergence of the Indian Plate and the Eurasian Plate. This massive continental collision is still going on; it causes periodic great earthquakes and has resulted in the formation and continuing uplift of the Tibetan Plateau and the Himalayan Mountains.

Subduction Zones. The deepest and most powerful earthquakes occur in the subduction zone environment (FIG. 6.12D), where a plate carrying oceanic crust converges with a plate carrying either oceanic or continental crust. These largest of all earthquakes are sometimes called *megathrust* earthquakes. Since 1900 there have been five megathrust earthquakes with moment magnitudes greater than 9.0, including the Sumatra-Andaman earthquake of 2004. Since they are associated with subduction, these earthquakes occur in or near the ocean. This means that megathrust and other large subduction-related earthquakes have the potential to cause tsunamis.

Focal Depths

Let's examine the phenomenon of earthquake focal depths in greater detail, and consider what it reveals about the different plate tectonic environments.

Earthquakes along spreading ridges are typically very shallow. Recall from Chapter 5 that molten rock from the mantle wells up along such rifts, creating new crust. The crust-mantle boundary (the moho) is very close to the surface under a spreading ridge. This means that rocks immediately along the ridge are hot and too weak to store much elastic energy. Earthquakes can occur only in rocks that are strong enough to store elastic energy and brittle enough to break (rather than deform) when their elastic limit is exceeded. This is why earthquakes along divergent plate boundaries typically have shallow foci; only the rocks closest to the surface are cold enough and strong enough to behave in this way.

In contrast, earthquakes in collision zones can range from shallow-focus to very deep-seated. Continental collisions are associated with complex faulting, buckling and deformation of the crust, and uplift and stacking of continental rock masses to form great mountain ranges. During this process of compression and shortening, the crust itself undergoes massive thickening. The roots of mountain ranges typically extend to significant depths, isostatically balancing their great height. The focal depths of earthquakes in continental collision zones thus vary from shallow to very deep, reflecting both the thickness of the crust and the complexity of the geologic environment.

Earthquakes that occur in subduction zones also range from shallow to deep focus, but with an interesting difference. Note from the colored dots on Figure 6.11 that the subduction-zone boundaries of the Pacific Plate are marked by earthquakes that vary from shallow (on the oceanic side) through intermediate to very deep-seated (on the continental side, or on the other side of deep oceanic trenches). What is the cause of these zones of earthquakes with varying focal depths? Even before the theory of plate tectonics was developed, geophysicists were well aware of these *Benioff zones* or *Wadati-Benioff zones* (refer to Fig. 5.19). Scientists now understand that the earthquakes associated with Benioff zones outline the dimensions of the downgoing lithospheric plate. As the subducting plate grinds its way down into the mantle, great earthquakes occur along its length. As long as the plate remains cold enough and strong enough to store elastic energy, earthquakes can occur. The foci of such earthquakes can be as deep as 700 km, indicating that subducting lithospheric plates retain their strength to very great depths. Such observations have allowed scientists to refine their understanding of the process of subduction.

VOLCANOES: WHEN ROCKS MELT

It should be clear by now that a strong relationship exists between some types of earthquakes and volcanic activity, and that this relationship is best understood in the context of plate tectonics. Let's now shift our attention from earthquakes to volcanism.

A **volcano** is a vent from which a combination of melted rock, solid rock debris, and gas is erupted. Underlying every active volcano is a reservoir of molten material, called a *magma chamber*. **Magma** is sometimes described simply as molten rock, but in fact it is a much more complex material; it is the mixture of molten rock, suspended mineral grains, and dissolved gas that forms in the crust or mantle when temperatures are sufficiently high. The term *volcano* itself comes from the name of the Roman god of fire, Vulcan, and it conjures up visions of streams of **lava**—magma that reaches Earth's surface—pouring out over the landscape. Some lava does flow as hot streams (FIG. 6.13), but magma can be erupted in

FIGURE 6.13 Flowing lava
This stream of glowing-hot lava moving smoothly away from the eruptive vent shows how fluid and free-flowing lava can be. This eruption occurred in Hawaii in 1983.

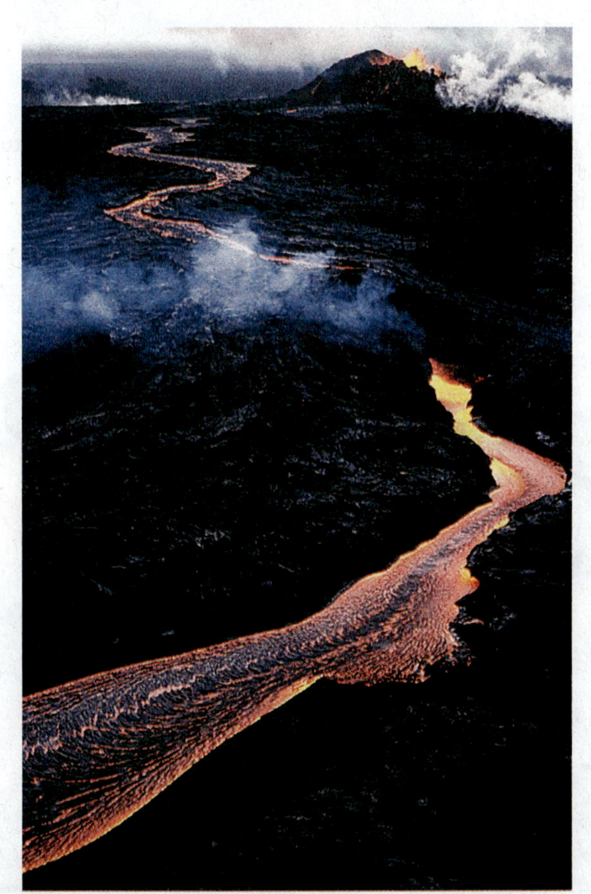

other forms, too, such as in massive clouds of tiny, hot fragments.

Volcanoes are one of the obvious ways by which the geosphere influences the atmosphere and hydrosphere and, through them, the biosphere and the anthroposphere. For example, when magma reaches Earth's surface, dissolved gases escape into the atmosphere; most volcanic gases are somewhat acidic, and acidic rain results. Over geologic time, the chemistry of the atmosphere and hydrosphere has evolved as a result of the addition of volcanic gases. Another example involves the massive clouds of tiny fragments erupted from some volcanoes; the fragments can be blasted so high into the stratosphere that they stay aloft for more than a year and block so much sunlight that the global climate is affected.

Why Do Rocks Melt?

Understanding volcanism involves an appreciation of how and why rock melts, and how melted the rock behaves. It is very difficult to study melted rock in nature (FIG. 6.14A) because rocks melt only at very high temperatures (see *The Basics: The Characteristics of Magma and Lava*). Fortunately, rock can be melted artificially as well as naturally (FIG. 6.14B). We can thus learn about the behavior of molten rock from laboratory experiments, too. Scientists have combined field- and laboratory-based sources of information to build and refine their understanding of magma, lava, and the properties of molten rock.

At Earth's surface (or in a laboratory at atmospheric pressure), rock begins to melt when heated to a temperature between about 800°C and 1000°C—the exact temperature depends on the composition. However, rock (unlike ice, for example) typically consists of many different minerals, each with its own characteristic melting temperature. Thus we cannot talk about a single melting point for a rock. Complete melting happens across a temperature range, but is commonly attained by about 1200°C. Besides rock composition, two factors strongly affect the temperatures at which rock melting occurs: pressure and the presence of water.

Pressure and Rock Melting

As discussed in Chapter 2, temperature increases with depth below the surface along the geothermal gradient; the increase occurs more quickly in and under oceanic crust than in the much thicker continental crust (FIG. 6.15). Beneath the continental crust, a temperature of 1000°C is reached at a depth of about 150 km. Underneath the ocean crust, temperature rises much more rapidly, reaching 1000°C at a comparatively shallow depth of about 80 km. As you can see in Figure 6.15, the temperature in the upper part of the mantle is higher than the temperature at which most rocks melt at Earth's surface. Yet the upper mantle is mostly solid; how is this possible?

The answer is that *pressure* also increases dramatically with increasing depth, and increasing pressure causes rock to resist melting (FIG. 6.16A). For example, *albite*, a common rock-forming mineral in the feldspar group, melts at 1104°C at the surface. At a depth of 100 km, the pressure is 35,000 times greater than it is at sea level.

FIGURE 6.14 Magma and lava are hot

(A) A geologist in a protective suit measures the temperature of lava erupting from Mauna Loa, Hawaii. Bright orange, yellow, and white lava is hotter, whereas dull red, brown, and black colors indicate cooler lava. (B) In a refinery, workers heat metal ores to the melting point.

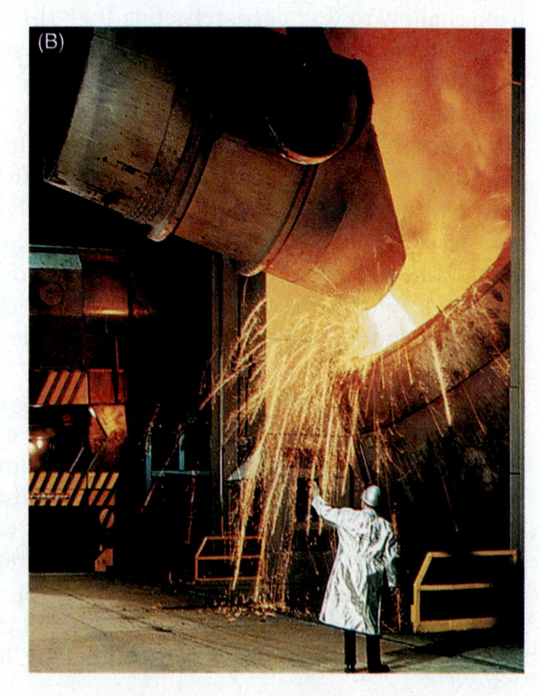

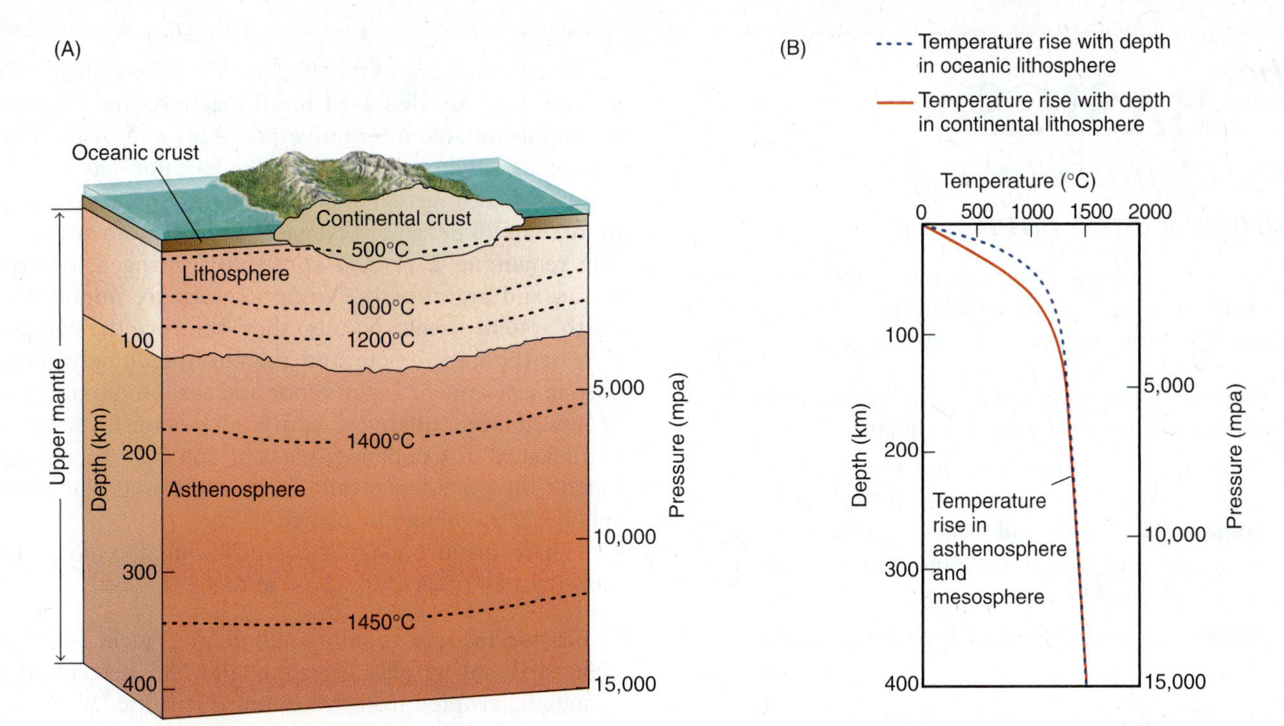

FIGURE 6.15 Geothermal gradient
(A) Temperature increases with depth, along the geothermal gradient. The dashed lines are *isotherms*, lines of equal temperature. Notice how the lines "sag" underneath the continental crust because the rate of increase of temperature with depth is slower there. (B) This graph represents the same information as shown in A. Earth's surface is at the top, so depth (and pressure) increase downward. The dashed curve shows the geothermal gradient under the oceans, and the solid curve shows the gradient under continental crust. Note that the two curves merge below about 200 km.

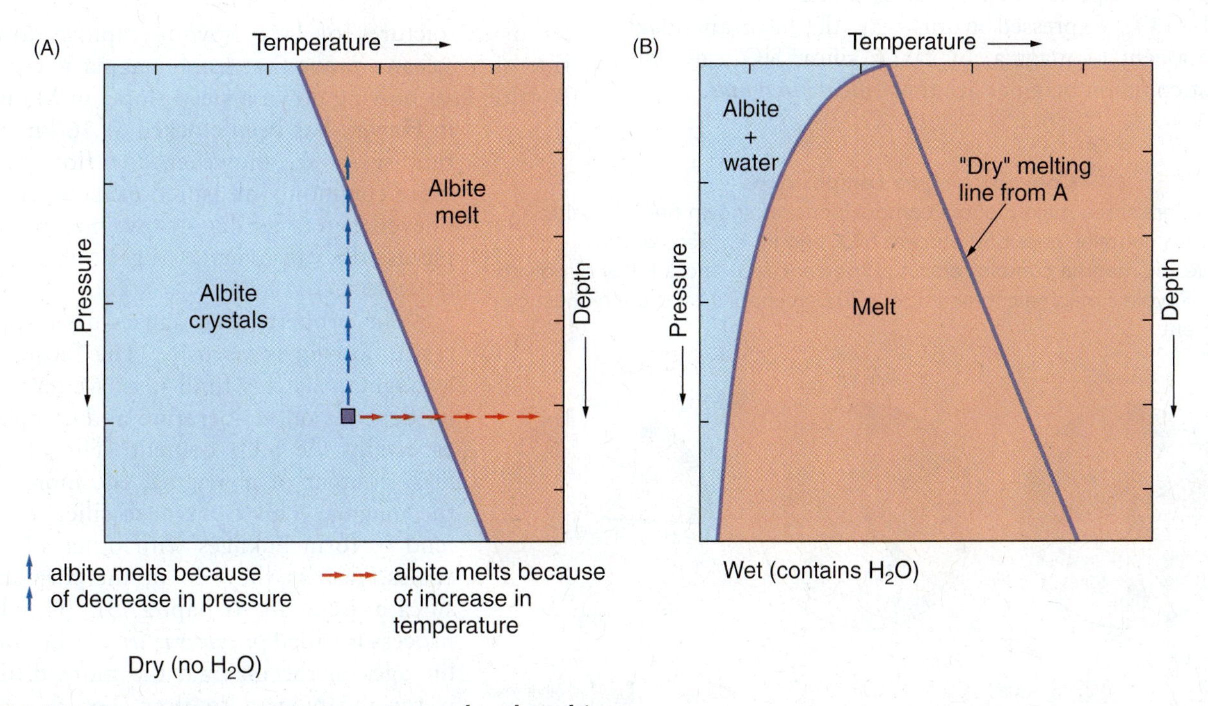

FIGURE 6.16 Temperature, pressure, water, and rock melting
(A) The melting temperature of a dry mineral (albite, in this case) increases with increasing pressure. A mineral at depth (shown by the small square) can melt in two different ways: either by an increase in temperature (red arrows) or by a decrease in pressure (blue arrows). (B) The melting temperature of a mineral in the presence of water typically decreases as pressure increases. This is exactly the opposite of what happens to dry minerals. Note that a mineral under the temperature and pressure conditions of the small square, which was solid in the dry case (A), would be melted if water were present (B).

the BASICS

The Characteristics of Magma and Lava

By studying and observing volcanic eruptions and molten rock in nature and in the laboratory, it is possible to draw three important conclusions concerning magma:

1. Magma is characterized by a *range of compositions* in which silica (SiO_2) is predominant.
2. Magma has the properties of a liquid, including the *ability to flow*. This is true even though most magma contains solid rock and mineral particles as well as dissolved gases, and in some instances is almost as stiff as window glass.
3. Magma is characterized by *high temperatures*.

RANGE OF COMPOSITIONS

The chemical compositions of magmas vary, but they are dominated by Earth's common chemical elements (Chapter 3)—silicon (Si), aluminum (Al), iron (Fe), calcium (Ca), magnesium (Mg), sodium (Na), potassium (K), hydrogen (H), and oxygen (O). Because O^{2-} is by far the most abundant anion, scientists usually express magma composition in terms of oxides, such as SiO_2 and Al_2O_3. Expressed in this way, the most abundant component of magma, by far, is silica, SiO_2, and the most common magmas are thus *silicate magmas*.

Small amounts of gas (0.2 to 5% by weight), called **volcanic gas**, are dissolved in all magmas, and they play an important role in eruptive processes and in the Earth system as a whole. The principal volcanic gas is water vapor, which, together with carbon dioxide, accounts for more than 98 percent of all gases emitted from volcanoes. The remaining 2 percent consists of nitrogen, chlorine, sulfur, and argon gases. Volcanic gases are important in Earth system science because they influence the composition of the atmosphere and thereby the climate; this is particularly true of water vapor and carbon dioxide, both of which are radiatively active (greenhouse) gases. As you learned in Chapter 4, without volcanic degassing of Earth's interior there would be no atmosphere or hydrosphere surrounding our planet.

Three distinct magma compositions are more common than all others (**FIG. B6.3** and **TABLE B6.1**):

1. **Basaltic magma** contains about 50 percent SiO_2 and little dissolved gas. Approximately 80 percent of all magma erupted from volcanoes is basaltic.
2. **Andesitic magma** contains about 60 percent SiO_2 and a lot of dissolved gas.
3. **Rhyolitic magma** contains about 70 percent SiO_2 and the highest gas content.

FIGURE B6.3 Common magma compositions
There are three main magma compositions, as shown here. Basaltic magma contains about 50 percent SiO_2 and little dissolved gas. Andesitic magma contains about 60 percent SiO_2 and a lot of dissolved gas. Rhyolitic magma contains about 70 percent SiO_2 and the highest gas content.

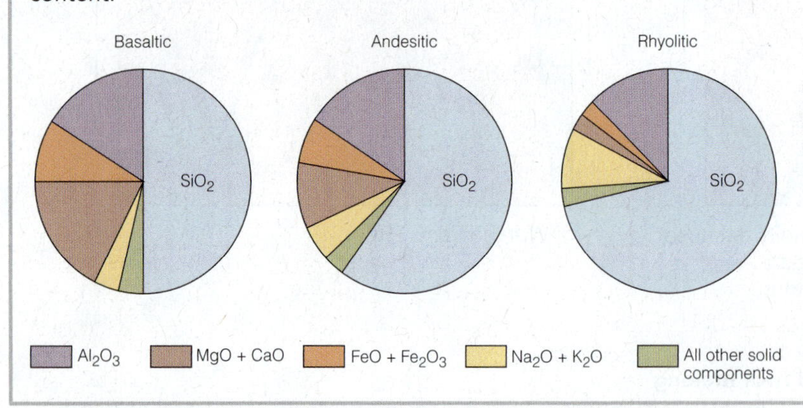

ABILITY TO FLOW

Dramatic pictures of lava flowing rapidly down the side of a volcano prove that some magma is very fluid. Basaltic lava moving down a steep slope on Mauna Loa in Hawaii has been clocked at 16 km/h. Such fluidity is rare, however, and flow rates are more commonly measured in meters per hour or even meters per day—slow enough that people usually can manage to get out of the way (**FIG. B6.4A**).

The property that causes a substance to resist flowing is **viscosity**. The more viscous a magma, the less fluid it is. Magma viscosity depends on temperature and composition, especially the SiO_2 content. The higher the SiO_2 content of a magma, the more viscous the magma. This is because silica molecules tend to form linkages with other silica molecules, just as they do in the formation of silicate minerals (Chapter 3). The linking process is called *polymerization*, and the more linkages in the magma, the more difficult it is for the magma to flow. For this reason,

FIGURE B6.4 Viscosity of lava
(A) This house in Kalapana, Hawaii, is about to succumb to the slow but unstoppable advance of a lava flow in June 1989. The grass of the lawn burns on contact with the molten rock. (B) The smooth, ropy-looking volcanic rock on which the geologist is standing formed from pahoehoe lava, a very hot, low-viscosity, freely flowing lava that erupted from Kilauea volcano in 1959. The much rougher-looking upper flow (the one being sampled by the geologist) formed from an *aa* lava that was cooler and therefore more viscous and slower-moving; it has the same chemical composition as the earlier flow.

TABLE B6.1 Magmas and Associated Volcanic and Plutonic Rocks

Magma Type	Silica Content	Melting Temperature	Viscosity	Plutonic Rock Type	Volcanic Rock Type
Basaltic	low (~50%)	high (~1200°C)	low (runny)	Gabbro	Basalt
Andesitic	intermediate (~60%)	intermediate (~1000°C)	intermediate	Diorite	Andesite
Rhyolitic	high (~70%)	low (~800°C)	high (sticky)	Granite	Rhyolite

high-silica rhyolitic magma is always more viscous than low-silica basaltic magma, and andesitic magma is intermediate between the two. Like the gas content, magma viscosity influences the violence of an eruption; a high-viscosity, high-silica magma typically will erupt with more difficulty (and hence more explosively) than a runny, low-viscosity, low-silica magma.

HIGH TEMPERATURE

All magmas are hot. Because volcanoes are dangerous places, and because scientists who study them are not eager to be roasted alive, measurements of magma temperatures during eruptions must be made from a distance using optical devices (as shown in Figure 6.15A). Magma temperatures determined in this manner range from 1000° to 1200°C. Experiments on rock melting carried out in laboratories have confirmed that rock melting is influenced by a variety of factors, but only rarely occurs at temperatures below about 800°C.

Temperature influences the properties of magmas, including the viscosity. The higher the temperature, the lower the viscosity of a magma and the more readily it flows. In **FIGURE B6.4B**, the smooth, ropy-surfaced lava on which the geologist is standing, called *pahoehoe* (a Hawaiian word pronounced pa-hó-e-hó-e), formed from a hot, fluid, runny basaltic magma. The rubbly, rough-looking lava piled up at the center and left formed from a cooler basaltic magma that had a higher viscosity. Hawaiians call this rough lava *aa* (pronounced ah'-ah). No matter how hot magma is when it exits a volcano, the lava soon cools, becomes more viscous, and eventually slows to a complete halt, solidifying to form a volcanic rock.

At that pressure, the melting temperature of albite is 1440°C, which is higher than the normal temperature at a depth of 100 km; thus albite remains solid at that depth.

Pressure plays such an important role in melting that a decrease in pressure on hot rock can lead to melting, just as an increase in temperature can do. Melting caused by a decrease in pressure is called *decompression melting* (Fig. 6.16A). It can happen, for example, if rock is being carried toward the surface by mantle convection. The decrease in pressure can induce melting, even though no additional heat is being added to the rock as it approaches the surface.

Water and Rock Melting

In contrast to the effect of pressure on melting, the presence of water (or water vapor) in a rock dramatically reduces its melting temperature (FIG. 6.16B). Anyone who lives in a cold climate knows that salt can melt the ice on an icy road because a mixture of salt and ice has a lower melting temperature than pure ice. Similarly, a mineral-and-water mixture has a lower melting temperature than the dry mineral alone. The effect of water on the melting of a rock becomes particularly important in subduction zones, where water is carried down into the mantle by wet oceanic crust, as described in chapter 5. Note from Figure 6.16B that water additionally causes the melting temperature of a mineral to increase, rather than decrease, with decreasing pressure; this is opposite to what happens in the case of the dry mineral (Fig. 6.16A). It means that a magma containing dissolved water may actually solidify as it moves toward the surface under conditions of decreasing pressure.

Make the CONNECTION

What are some of the ways that water makes its way from Earth's interior to become part of the hydrosphere or atmosphere? What are some of the ways that water might be returned to Earth's interior after having been at the surface?

Fractional Melting

Because rocks are composed of many different minerals, melting typically occurs over a temperature range of 200 degrees or more. This means that the boundary between solid and melt is not crisp as in the melting of an ice cube, but blurry, as in FIGURE 6.17. When temperature increases enough for part of the materials in a rock to melt and part to remain solid, it becomes a **partial melt** or *fractional melt*. Only if the temperature continues to increase or the pressure decreases will the rock melt completely. The process of partial or fractional melting can make it possible for a melt of one composition to become separated from residual rock of a different composition. This is a critical step in the development—from just a few common magma starting compositions—of the great diversity of magmas and resulting igneous rock types that exist on Earth.

Volcanic Eruptions

Now that you have some understanding of how and why rocks melt to form magma, we can consider how the magma makes its way to the surface and erupts. Magma, like most other liquids, is less dense than the solid matter from which it forms. Therefore, once formed, lower-density magma exerts an upward push on any enclosing higher-density rock that lies above it, and slowly forces its way up. There is, of course, pressure on the rising mass of magma as a result of the weight of the overlying dense rock. Because pressure is proportional to depth, it decreases as the magma rises, allowing additional decompression melting to occur and gases dissolved in the magma to be released. The process of ascent and the characteristics of the magma (especially its silica content, gas content, and temperature) combine to determine its behavior as it emerges from the volcano, which determines the *eruptive style* of the volcano. Some volcanoes erupt explosively; others erupt more gently. Several factors lead to these differing eruptive styles.

Nonexplosive Eruptions

People imagine that any volcanic eruption is hazardous, but geologists have learned that basaltic volcanoes, such as those in Hawaii, are comparatively safe to study because they generally erupt nonexplosively, with characteristic streams of quietly flowing lava like those shown in Figures 6.13 and B6.4A.

The differences between nonexplosive and explosive eruptions are largely a function of magma viscosity and dissolved gas content. Nonexplosive eruptions are characteristic of low-viscosity magmas and low-dissolved gas levels. Basaltic magma has a lower SiO_2 content, a higher temperature, and therefore a lower viscosity (that is, it is runnier) than andesitic or rhyolitic magmas; it also has a lower content of dissolved gas than either of the other magma types. Eruptions of basaltic magma, often called *Hawaiian-type eruptions*, are rarely explosive.

Basaltic magma eruptions can still be spectacular, however. Pressure controls the amount of gas a magma can dissolve—more gas can be dissolved at high pressure, less gas at low pressure. Gas dissolved in a rising magma acts the same way as gas dissolved in soda. When a bottle of soda is opened, the pressure inside the bottle drops, gas comes out of solution, and bubbles form. Similarly, gas dissolved in an upward-moving magma comes out of solution and forms bubbles as the pressure decreases. In a low-viscosity basaltic magma the released gas bubbles will rise very rapidly, like the bubbles in the soda. If basaltic magma moves rapidly toward the surface, the pressure exerted by the overlying rock declines very quickly. In that case, the volcanic gas can come out of solution so rapidly that the froth of bubbles causes spectacular fountaining in the erupting lava (FIG. 6.18A).

FIGURE 6.17 Rock melting and partial melting

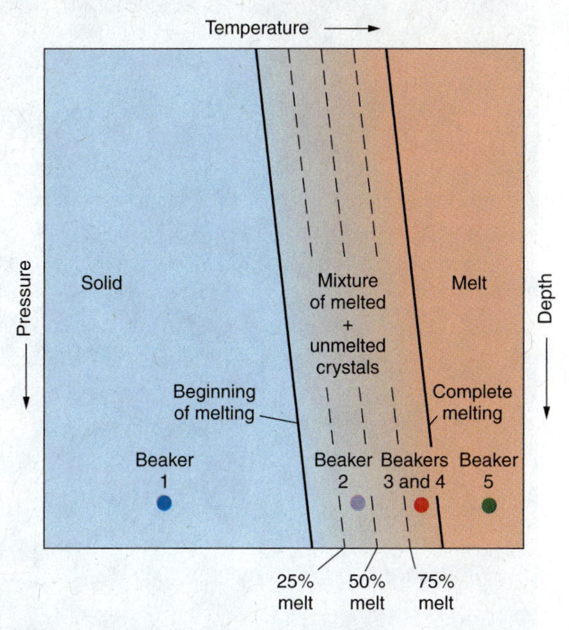

◀ Because almost all rocks contain a mixture of materials, they do not melt all at once at a single temperature; instead, there is a range of temperatures and pressures in which they contain a mixture of melted and unmelted crystals. The dots on this diagram show the stages in the melting process, which are illustrated further below.

▼ The process of fractional melting can lead to another effect called *fractionation* (illustrated for convenience in a laboratory beaker rather than buried in Earth's mantle).

Mineral 1

Mineral 2

Melt, rich in mineral 1 is squeezed out

Remaining solid is rich in mineral 2 (high-temperature component)

1. The first beaker shows a mixture of two minerals. At a low temperature, both are solid.
2. As the temperature increases, mineral 1 (the dark mineral) begins to melt and as it does so, it dissolves some of mineral 2.
3. Mineral 1 has totally melted and has dissolved a lot of mineral 2 in the process; remainder of mineral 2 remains solid.
4. At a constant temperature, we have mechanically compressed the sample, separating the solid from the melt.*
5. If temperature were to continue to increase, the material in the beaker would eventually become completely melted.

*In the lithosphere, this mechanical separation can occur as a result of the tectonic motion of plates. If the melt now cooled again, we would have two separate deposits, one of mineral 1 plus a little of mineral 2, crystallizing from the melt, and one of pure mineral 2, the solid remnant in Beaker 4.

When fountaining dies down because most of the dissolved gas has come out of solution and escaped, the hot, fluid lava emerging from the vent flows rapidly downslope. As the lava cools and continues to lose dissolved gases, its viscosity increases and the character of flow changes. The very fluid initial lava forms thin pahoehoe flows, but with increasing viscosity the rate of movement slows and the cooler, stickier lava is transformed into a slow-moving *aa* flow (as mentioned in *The Basics: Characteristics of Magma and Lava*). Thus, during a single Hawaiian-type eruption, pahoehoe and *aa* may be formed from the same batch of basaltic magma (Fig. B6.4B).

As basaltic lava cools and the viscosity increases, gas bubbles find it increasingly difficult to escape. When the lava finally solidifies to rock, the last-formed bubbles become trapped and their form is preserved (**FIG. 6.18B**). These bubble holes are called *vesicles*, and the texture they produce in an igneous rock is said to be *vesicular*.

FIGURE 6.18 Nonexplosive but gas-rich eruptions
(A) Hawaii's volcanic eruptions, such as this eruption of Kilauea that began in 1983, are generally nonexplosive, but the rapid release of gas can cause spectacular lava fountains, as shown here. When the lava from a gas-rich eruption cools and solidifies, it may trap some of the gas bubbles, leading to the formation of vesicles and vesicular texture, as shown here (B) in a sample from Hawaii. Each of these pea-sized vesicles is a gas bubble that was trapped when the lava solidified.

Explosive Eruptions

Viscous magmas—both andesitic and rhyolitic—have higher silica content and erupt at lower temperatures than basaltic magma; they also have high dissolved-gas contents, because water vapor is involved in the melting process. The combination of silica-rich composition, low temperature, high viscosity, and high dissolved-gas content is a recipe for an explosive eruption. As the gas-charged, viscous magma rises, the gas comes out of solution and bubbles form, but they cannot escape from the sticky, viscous magma. If the rate at which the magma rises (and hence the rate at which bubbles form) is rapid, the bubbles can shatter the viscous magma into a cloud of tiny, red-hot fragments that erupt violently from the volcanic vent.

A fragment of hot, shattered magma or any other fragment of rock ejected during an explosive volcanic eruption is called a **pyroclast** (from the Greek words *pyro*, meaning fire, and *klastos*, meaning broken). A deposit of unconsolidated (loose) pyroclasts is called **tephra**, and when the pyroclasts are consolidated or cemented together the result is a **pyroclastic rock**. Abundant pyroclastic material is an important characteristic of explosive volcanic eruptions. Some of the terms used to describe tephra of different size are illustrated in FIGURE 6.19.

Volcanologists quantify the explosiveness of an eruption with reference to the *Volcanic Explosivity Index*, or *VEI*, a logarithmic scale ranging from 0 (for nonexplosive eruptions) to 8, applied to the most cataclysmic eruptions known. Measures used to quantify eruptions include the volume of volcanic products ejected and the height of the eruption cloud, combined with qualitative observations of the nature of the eruption. Note that the VEI is applied to the eruption, not to the volcano itself; a given volcano can erupt many times, with many different outcomes. Three of the most devastating types of explosive volcanic eruptions, with correspondingly high VEI, involve the formation of eruption columns and tephra falls, pyroclastic flows, and lateral blasts.

Eruption Columns and Tephra Falls. As rising magma approaches Earth's surface, the rapid drop in pressure causes dissolved gas to bubble furiously, like a

FIGURE 6.19 Pyroclasts and explosive eruptions

Pyroclasts or tephra come in a variety of sizes. (A) Volcanic *bombs* are fragments of rock that are fist-sized and larger. (B) Intermediate-sized tephra are called *lapilli*; these fragments cover the Kau Desert in Hawaii. (C) Volcanic ash, the smallest tephra, blankets a farm in Oregon after the eruption of Mount St. Helens in 1980. Volcanic ash isn't the same as wood ash; it consists of microscopic pieces of volcanic glass. (D) Volcanic eruptions are rated according to the Volcanic Explosivity Index (VEI), which takes into account the volume of ejected pyroclastic material, height of the eruption column, duration of the eruption, and other observational factors. The scale, which ranges from 0 to 8, is a logarithmic scale. There are no known eruptions with VEI greater than 8.

FIGURE 6.20 Eruption column
The massive eruption of Mount St. Helens in May 1980 produced an eruptive column of ash and released destructive pyroclastic flows and debris avalanches down the steeply sloping sides of the volcano.

violently shaken bottle of soda. As a result of the bubbling, a viscous magma can break into a mass of hot, glassy pyroclasts. The resulting mixture of hot gas and pyroclasts produces a violent upward thrust that culminates in an explosive eruption. The hot, turbulent mixture rises rapidly in the cooler air above the volcano to form an **eruption column** that may reach as high as 45 km in the atmosphere (FIG. 6.20). The rising, buoyant column is driven by heat energy released from hot, newly formed pyroclasts. At a height where the density of the column equals that of the surrounding atmosphere, the column begins to spread laterally to form a mushroom-shaped eruption cloud. This type of explosive eruption is often called a *Vesuvian eruption*, after the famous Mediterranean volcanoof which are Mount Vesuvius. Vesuvian eruptions are also called *Plinian eruptions*, in reference to the Roman naturalist and philosopher Pliny the Elder, who was killed while attempting to rescue some friends from the devastating eruption of Mount Vesuvius in 79 CE.

As an eruption cloud begins to drift with the upper atmospheric winds, the pyroclasts fall out and eventually accumulate on the ground as tephra deposits. During exceptionally explosive eruptions, tephra—especially **volcanic ash**, the smallest tephra particles—can be carried as far as 1500 km or more. Some eruption columns reach such great heights that winds are able to transport the pyroclasts, gases, and extremely tiny particles and droplets called *volcanic aerosols* completely around the world. This was the case in the massive eruptions of the Indonesian volcanoes Tambora in 1815 (VEI 7) and Krakatau in 1883 (VEI 6), the Philippine volcano Mount Pinatubo in 1991 (VEI 6), and several other historic eruptions. Such atmospheric pollution can block incoming solar radiation, resulting in lower average temperatures at the land surface for a year or more, as well as causing spectacular sunsets as the Sun's rays are refracted by the airborne particles.

PYROCLASTIC FLOWS. A hot, highly mobile mass of tephra that is denser than the atmosphere will not rise upward, but instead rushes down the flank of a volcano as a **pyroclastic flow**. Such flows, often referred to by the French term *nuée ardente* (glowing cloud), are among the most devastating and lethal forms of volcanic eruptions. Observations of historic pyroclastic flows show that they can travel 100 km or more from source vents and reach velocities of more than 700 km/h. One of the most destructive of such eruptions, on the Caribbean island of Martinique in 1902 (VEI 4), produced a pyroclastic flow that rushed down the flanks of Mount Peleé volcano and overwhelmed the city of Saint Pierre, instantly killing 29,000 people.

LATERAL BLASTS. The 1980 eruption of Mount St. Helens, a volcano in the Cascades Range of northwestern North America, displayed many features of a typical large, explosive eruption, including a major eruptive column and pyroclastic flows (Fig. 6.20). Nevertheless, the magnitude of the event caught geologists by surprise. The events leading to this eruption are shown diagrammatically in FIGURE 6.21. As magma moved upward under the volcano, the northern flank of the mountain began to bulge upward and outward. Finally, the bulging flank became unstable, broke loose, and quickly slid toward the valley as a gigantic landslide of rock and glacier ice. The landslide exposed the mass of hot magma in the core of the volcano. With the lid of rock removed, dissolved gases bubbled so furiously that a mighty *lateral blast* resulted, blowing a mixture of pulverized rock, pyroclasts, and hot gases sideways as well as upward. The sideways blast, initially traveling at the speed of sound, roared across the landscape. Within the devastated area, extending as much as 30 km from the crater and covering some 600 km^2, trees were blasted to the ground and covered with hot debris. The 1980 eruptions of Mount St. Helens (VEI 5) is the best documented recent example of a lateral blast, but a similar eruption of Bezmianny volcano in Kamchatka (eastern Russia) in 1956 produced a devastating lateral blast, eruption column, and associated pyroclastic flows.

Types of Volcanoes

The name *volcano* is applied to any vent, crack, or opening from which lava, pyroclasts, or gases are erupted. Many volcanoes are *subaerial*, meaning "under the air" and referring to volcanoes and eruptions (as well as other geologic processes) that occur on the land surface. Other volcanoes are *submarine*, or underwater, and still others are *subglacial*. Today, the latter term refers principally to the active volcanoes located under the glaciers and ice sheets of Iceland and Antarctica.

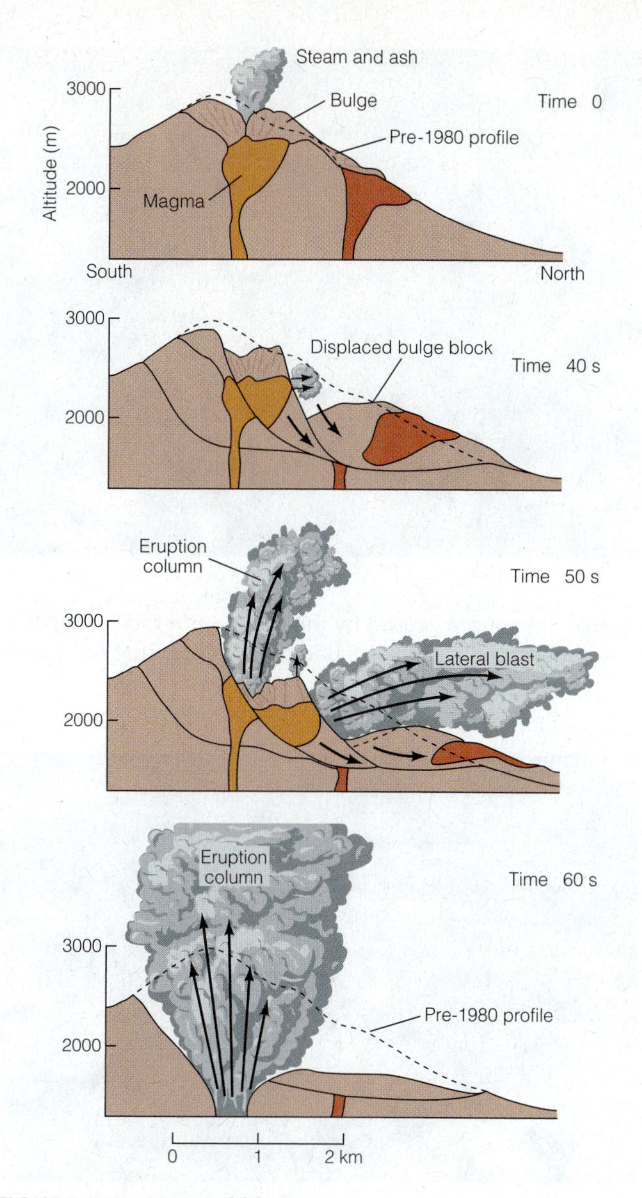

FIGURE 6.21 Lateral blast
This is the sequence of events that led to the major eruption of Mount St. Helens in 1980. Time = 0: earthquakes and puffs of steam and ash indicate that magma is rising; the north face of the mountain bulges. Time = 40 seconds: an earthquake shakes the mountain, and the bulge breaks loose and slides downward; this reduces the pressure on the magma and initiates the lateral blast eruption. Time = 50 seconds: the violence of the eruption causes a second block to slide downward, exposing more of the magma and initiating an eruption column. Time = 60 seconds: the eruption increases in intensity, carrying volcanic ash as high as 19 km into the atmosphere.

An example is Eyjafjallajökull, the Icelandic volcano that disrupted air travel with a significant pyroclastic eruption (VEI 4) in 2010.

The shape of a volcanic structure has a lot to do with the kind of magma erupted and with the relative proportions of lava and pyroclasts. Scientists apply three shape-related terms to different types of volcanoes: shield volcano, tephra cone volcano, and stratovolcano.

FIGURE 6.22 Shield volcano
Mauna Kea is a 4200-m-high shield volcano on Hawaii, here seen from Mauna Loa. Note the gentle slopes formed by highly fluid basaltic lava flows. The pahoehoe flow in the foreground is on Mauna Loa.

Shield Volcanoes

The kind of volcano that is easiest to visualize is one built up of successive flows of very fluid lava. Such lavas are capable of flowing great distances down gentle slopes and of forming thin sheets of nearly uniform thickness. Eventually, the pile of lava builds up a **shield volcano**, a broad, roughly dome-shaped mound with an average surface slope of only 5° near the summit and about 10° on the flanks (FIG. 6.22). Shield volcanoes are characteristically formed by successive eruptions of basaltic lava; the proportions of ash and other pyroclasts are small. Hawaii, Tahiti, Samoa, the Galápagos, and many other oceanic islands are the upper portions of large shield volcanoes.

Tephra Cones and Stratovolcanoes

Rhyolitic and andesitic volcanoes tend to eject a large proportion of pyroclasts and therefore to be surrounded by layers of tephra. As the debris showers down, a steep-sided volcano called a **tephra cone** builds up around the vent (FIG. 6.23). The slope of the cone is determined by the size of the pyroclasts.

Larger, long-lived volcanoes, particularly those of andesitic composition, emit a combination of lava flows and pyroclasts. **Stratovolcanoes** are steep conical mountains that consist of layers of both lava and tephra. The volume of pyroclastic material may equal or exceed the volume of the lava. The slopes of stratovolcanoes, which may be thousands of meters high, are steep like those of tephra cones, about 30° near the summit and 6° to 10° at the base. The beautiful, steep-sided cones of stratovolcanoes are among Earth's most picturesque sights. The snow-capped peak of Mount Fuji in Japan (FIG. 6.24) has inspired poets and writers for centuries. Mount Rainier and Mount Baker in Washington and Mount Hood in Oregon are majestic examples in North America.

FIGURE 6.23 Tephra cones
(A) Two small tephra cones are erupting andesitic lava in Kivu, Zaire. Arcs of light are caused by the eruption of red-hot lapilli and bombs. (B) These tephra cones in Arizona consist of lapilli-sized fragments; note the small basaltic lava flow that emanated from the base of one of the cones.

FIGURE 6.24 Stratovolcanoes
Mount Fuji in Japan is a snow-clad giant that towers over the surrounding countryside, displaying the classic steep-sided profile of a stratovolcano.

FIGURE 6.25 Caldera
Crater Lake, Oregon, occupies a caldera 8 km in diameter that crowns the summit of a stratovolcano, posthumously called Mount Mazama. Wizard Island is a small tephra cone that formed after the collapse that created the caldera.

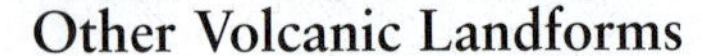

Other Volcanic Landforms

Many large volcanoes, especially shield and stratovolcanoes, are marked near their summit by a large depression. This is a **caldera**, a roughly circular, steep-walled basin several kilometers or more in diameter. Calderas form after the partial emptying of a magma chamber in an explosive volcanic eruption. Rapid eruption of magma can leave the magma chamber empty or partly empty. The unsupported roof of the chamber slowly sinks under its own weight, like a snow-laden roof on a shaky barn, dropping downward on a ring of steep vertical fractures. Subsequent eruptions commonly occur along these fractures. Crater Lake, Oregon (FIG. 6.25), occupies a circular caldera 8 km in diameter, formed after an explosive pyroclastic eruption (VEI 7) about 6600 years ago. The volcano that erupted has been posthumously named Mount Mazama.

Sometimes lava reaches Earth's surface through a vent that is an elongate fracture in the crust; this type of eruption is called a **fissure eruption**. Such eruptions, which can be very dramatic, are characteristically associated with very fluid basaltic magma, and the lavas resulting from a fissure eruption on land tend to spread widely and create flat lava plains. The Laki fissure eruption in Iceland occurred along a fracture 32 km long (FIG. 6.26). Lava flowed 64 km outward from one side of the fracture and nearly 48 km outward from the other side. The eruption lasted from 1783 to 1784, and the lava flows eventually covered an area of 588 km², making this the largest lava flow in historic times. It was

FIGURE 6.26 Fissure eruption
The eruption of Laki, a fissure volcano in Iceland, lasted from 1783 to 1784 and was the largest flow of lava in recorded history.

also one of the most deadly, destroying homes and food supplies and covering vast areas of farmlands. Famine followed, and 9336 people died. There is good evidence that even larger fissure eruptions occurred in prehistoric times. For example, the Roza flow, a great prehistoric sheet of basaltic lava in eastern Washington State, can be traced over 22,000 km^2.

VOLCANIC DISASTERS

Volcanic eruptions are not rare events, although not all eruptions are hazardous to humans. Volcanoes that have erupted within historic times are said to be **active**. Those that have not erupted in recent memory but still exhibit some signs of volcanic activity are said to be **dormant**, and those that appear to be completely dead are referred to as **extinct**. In addition to all the submarine eruptive centers along midocean ridges, there are at least 500 active volcanoes, many of them located in the circum-Pacific region, and every year about 50 volcanoes erupt somewhere on Earth. Basaltic volcanoes are far more common than the more typically explosive andesitic and rhyolitic stratovolcanoes. Though less common, explosive pyroclastic eruptions do occur and can be disastrous, since millions of people live on or close to stratovolcanoes. Since 1800 there have been 19 volcanic eruptions in which a thousand or more people died (FIG. 6.27).

Volcanic Hazards

Eruptions of stratovolcanoes present five major kinds of hazards:

1. Hot, rapidly moving pyroclastic flows (*nuées ardentes*) and lateral blasts may overwhelm people before they can run away. The tragedies of Mount Peleé in 1902 and Mount St. Helens in 1980 are examples.
2. Tephra and hot, poisonous gases may bury people or suffocate them. Such a tragedy occurred in 79 CE when Mount Vesuvius burst to life. Hot, poisonous volcanic gases killed people in the nearby Roman cities of Pompeii and Herculaneum, and then tephra buried them (FIG. 6.28, on page 175). Another major gas release occurred in 1984 at Lake Nyos, Cameroon, when a sudden release of carbon dioxide from the volcano's crater lake caused the suffocation deaths of 1700 people and countless cattle.
3. Tephra can be dangerous long after an eruption has ceased. Rain or meltwater from snow can loosen tephra piled on a steep volcanic slope and start a deadly mudflow. In 1985, following a small and otherwise nondangerous eruption of the Colombian volcano Nevado del Ruíz, massive hot mudflows called *lahars* were formed when the ash mixed with water from melting glaciers. The mudflows moved swiftly down the mountain and killed 20,000 people. Lahars continue to be a problem in the Philippines, many years after the major 1991 eruption of Mount Pinatubo.
4. Violent undersea or coastal eruptions can cause tsunamis. Like underwater earthquakes, an underwater volcanic eruption can suddenly displace ocean water, generating a tsunami. A major tsunami set off by the eruption of Krakatau in 1883 killed more than 36,000 coastal dwellers on the islands of Java and Sumatra in Indonesia.
5. A tephra eruption may wreak such havoc on agricultural land and livestock that people die from famine. Tephra can also overwhelm cities and other sites of human activities. The 1991 eruption of Mount Pinatubo in the Philippines is an example; more people died in the aftermath of the eruption and in refugee camps than during the eruption itself.

Additional volcanic hazards can include fires; flooding caused by the bursting of crater lakes or blockage of

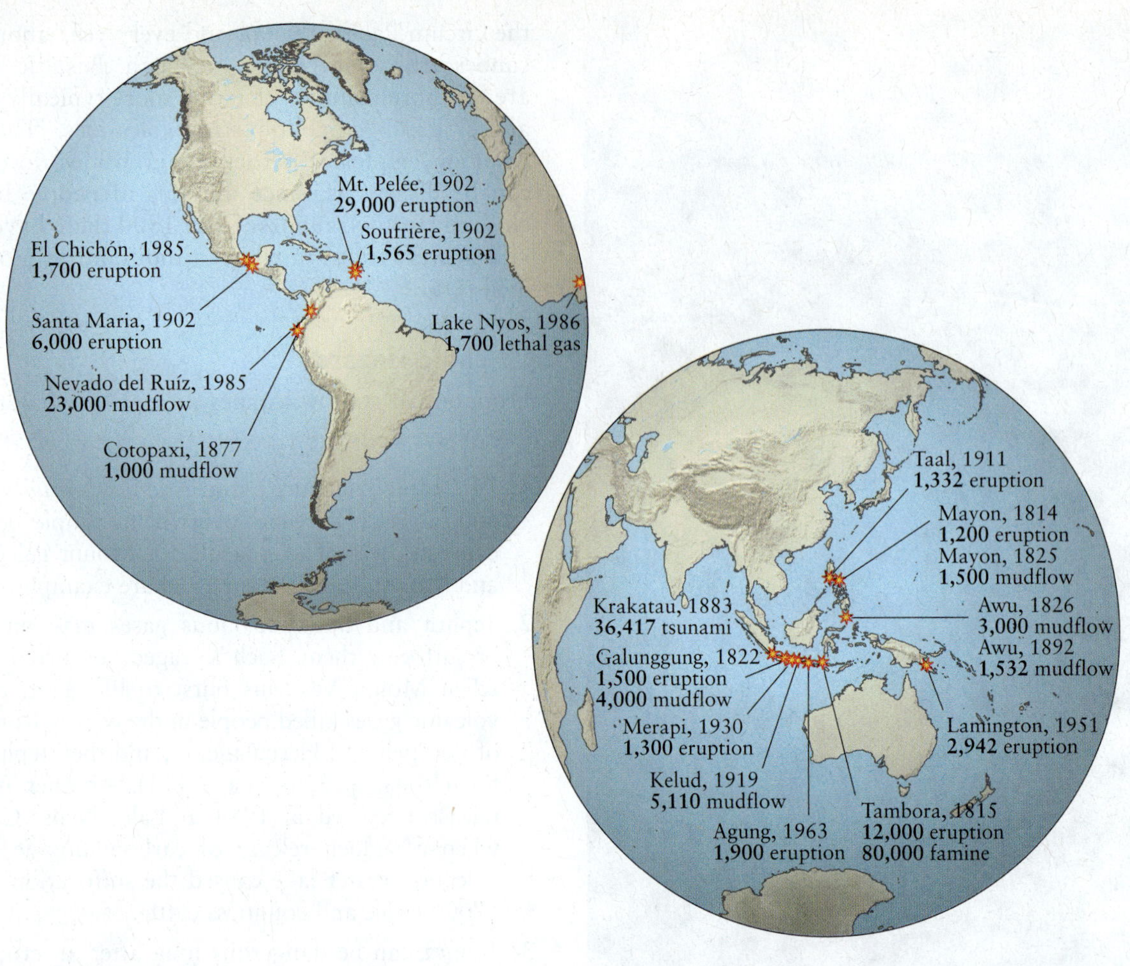

FIGURE 6.27 Deadly eruptions
Since 1800, there have been 19 volcanic eruptions in which 1000 or more people have died from eruption-related causes. One major event that doesn't appear on this map is the 1991 eruption of Mount Pinatubo in the Philippines, because accurate scientific monitoring and predictions and timely evacuations saved thousands of lives.

river channels by lava or tephra; acidic precipitation and other climatic impacts, both local and global; and *volcanic tremors*, a kind of seismicity that is caused by the movement of magma in a magma chamber.

Predicting Volcanic Eruptions

Earth is a volcanically active planet, and it is certain that violent and dangerous eruptions will occur in the future. Long-term prediction of major volcanic eruptions is based on the geologic history of eruptions, combined with an understanding of the tectonic environment in which volcanism occurs, leading to the identification of volcanoes that are likely to erupt explosively. These are mostly (though not exclusively) andesitic and rhyolitic stratovolcanoes in subduction zone environments. Likely candidates for dangerous eruptions in North America are the Cascade Range stratovolcanoes of the Pacific Northwest, such as Mount Rainier. Potentially dangerous volcanoes are also to be found in Japan, the Philippines, New Guinea, New Zealand, Indonesia, the countries of Central and South America, Central Africa, and the Caribbean islands.

To some extent, volcanic hazards can be anticipated, provided experts can gather data before, during, and after eruptions. As in attempts to predict earthquakes, the prediction of volcanic eruptions is based on careful observation and monitoring, with the goal of detecting precursor anomalies that may point to an imminent eruption. Volcanoes are somewhat more cooperative than earthquakes in this regard. For one thing, since the volcano itself exists as a landform, it is clear where to place the scientific equipment with which to monitor the volcano's precursory behavior (unlike earthquakes, which may occur anywhere and whose precursory behavior is completely hidden under the ground).

Volcanic eruptions, though highly complex events, often exhibit precursory behavior that can allow for a prediction to be made. Common signs of an imminent eruption include swarms of small earthquakes of increasing frequency or intensity; other significant changes in seismicity, such as a dramatic change in the focal depths of local earthquakes; sudden changes in the amount or composition of gases being emitted from the volcano;

FIGURE 6.28 Victims of Mount Vesuvius
These five citizens of Pompeii, Italy, were killed during the eruption of Mount Vesuvius in 79 CE. Their deaths were caused by poisonous volcanic gases, and the bodies were buried by pyroclastic material. Over the centuries the bodies decayed, but their shapes were imprinted in the tephra blanket. Excavators who discovered the imprints carefully recorded them by making these plaster casts.

changes in the depth, composition, or temperature of groundwater; bulging or heating of the slopes of the volcano; and signs of magma moving upward or beginning to emerge from the volcano. Some of these precursory events are monitored using ground-based equipment, but much monitoring can now be carried out using remote sensing (FIG. 6.29).

Scientists have only been able to issue reliable predictions for volcanic eruptions for the past few decades, but improvements are steadily being made as each new volcanic eruption is carefully studied. Once a reliable prediction of an imminent eruption has been made, experts can then advise civil authorities when to implement hazard warnings and when to move endangered populations to areas of lower risk. As for any natural hazard, issuing and implementing an effective early warning are dependent on having a preexisting evacuation plan and communications strategy in place.

After an Eruption

When a volcanic eruption occurs, a blanket of fresh new lava or tephra spreads across the land. In this manner volcanism renews the land surface. Because volcanism occurs, or has occurred in the past, on each of the terrestrial planets, surface renewal by volcanism is an important planetary process. Surface renewal is an especially important process for life on Earth because it replaces nutrients that are removed by weathering.

It is remarkable how quickly the land recovers after an eruption (FIG. 6.30). Within a year of the Mount St. Helens eruption, trees and other plants had started to sprout in the area devastated by the lateral blast of 1980; animals started to return to the area to graze as soon as the plants were big enough. Agricultural land can also be worked very quickly after an eruption. Although the eruption of Mount Pinatubo caused great distress to the local population, local farmers planted crops in the volcanic ash as soon as the eruption stopped. Plants can even grow on recently erupted lava. In Hawaii, papaya trees have been reported to grow on basaltic lava within two years after the lava was erupted.

Make the CONNECTION

What conditions need to be in place and what kinds of processes need to occur in order for plants and animals to begin to recolonize a landscape devastated by a major volcanic eruption?

FIGURE 6.29 Volcano monitoring from space
This false-color satellite image shows the area around Mount Vesuvius (center right) and the Bay of Naples, Italy, a densely populated region. Recent lava flows appear bright red in this image, which records infrared radiation (an indication of heat). Older lavas and volcanic ash appear in shades of yellow and orange. The dark blue and purple region at the head of the bay is the city of Naples. West of Naples lies a cluster of smaller volcanoes called the Flegreian Fields. By comparing successive satellite images, scientists can detect changes in ground temperature.

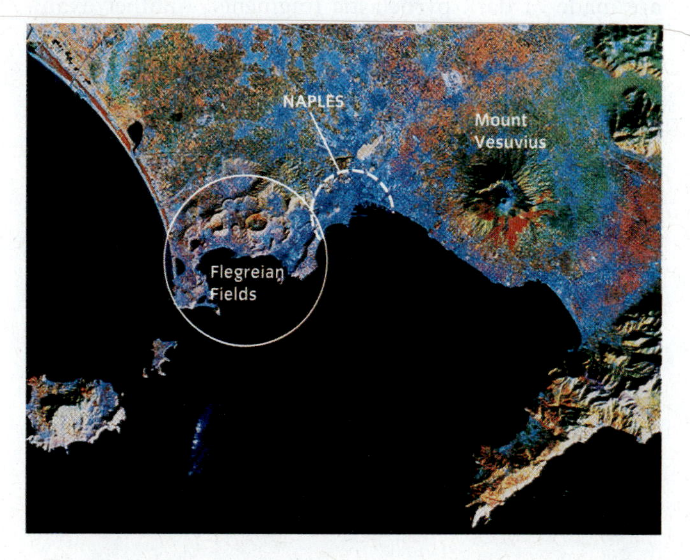

FIGURE 6.30 New volcanic soils
This photo shows the devastated, ash-covered landscape surrounding Mount St. Helens after the massive eruption of May 1980. Plants recolonized the new volcanic soil very rapidly, as early as the summer of 1980; this photograph of fireweed growth was taken in 1984.

Volcanic Benefits

From the human viewpoint, volcanism has bad features and good. The bad are the hazards and dangers associated with eruptions and the effects of volcanic dust and gases on the climate. The good is the production of volcanic rock from which rich volcanic soils develop. New rock means new supplies of the fertilizer elements needed by plants. Lava and volcanic ash, when subjected to weathering, produce very fertile soils. Some of the richest agricultural land in Italy, near Naples, has volcanic soil developed on tephra. Japan, the North Island of New Zealand, the Hawaiian Islands, the Philippines, and Indonesia are other places where rich volcanic soils produce high agricultural yields.

Volcanic eruptions in coastal areas also can produce new land, like the black sand beaches of Hawaii, which are made of dark pyroclastic fragments. Another example of new volcanic land is the island of Surtsey, which emerged from the ocean near Iceland in 1963 and is composed of lava flows and pyroclastic cones. Volcanism also provides geothermal energy and plays an important role in the formation of some kinds of mineral deposits. And finally, as you have learned, without the process of volcanic degassing occurring throughout Earth history we would have no atmosphere or hydrosphere and no way to support life on this planet.

MAGMA UNDERGROUND

Now that we have considered what happens to magma when it erupts, we turn briefly to the question of what happens to magma that never reaches the surface. Beneath every volcano lies a complex of chambers and channels through which magma moves toward the surface. We cannot study the magmatic channels of an active volcano, but we can look at ancient cones that have been laid bare by erosion, as seen in FIGURE 6.31. These ancient channels are filled by igneous rock because they are the underground sites where magma solidified (FIG. 6.32).

As introduced in Chapter 3, igneous rocks that cool and solidify at or very near the surface are called *volcanic rocks*, whereas those that cool and solidify under the ground are called *plutonic rocks*. All bodies of igneous rock that solidify underground, regardless of shape or size, are called **plutons**, after Pluto, the Greek god of the underworld. The

FIGURE 6.31 Volcanic neck
Shiprock, New Mexico is the eroded remnant of a volcanic conduit. The tephra cone that once surrounded this volcanic neck (A) has been removed by erosion. (B) This diagram shows how the original volcano may have appeared prior to erosion.

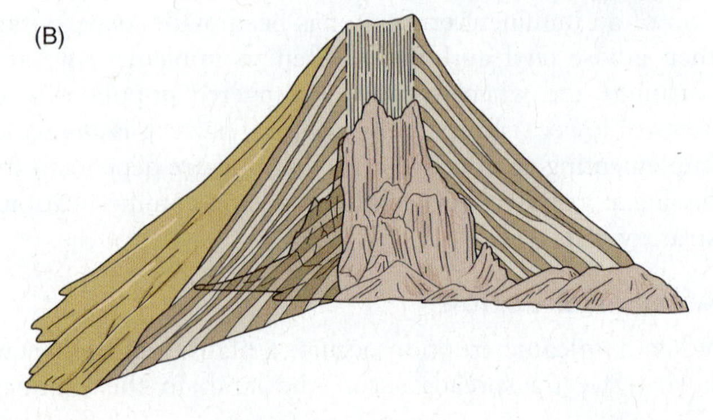

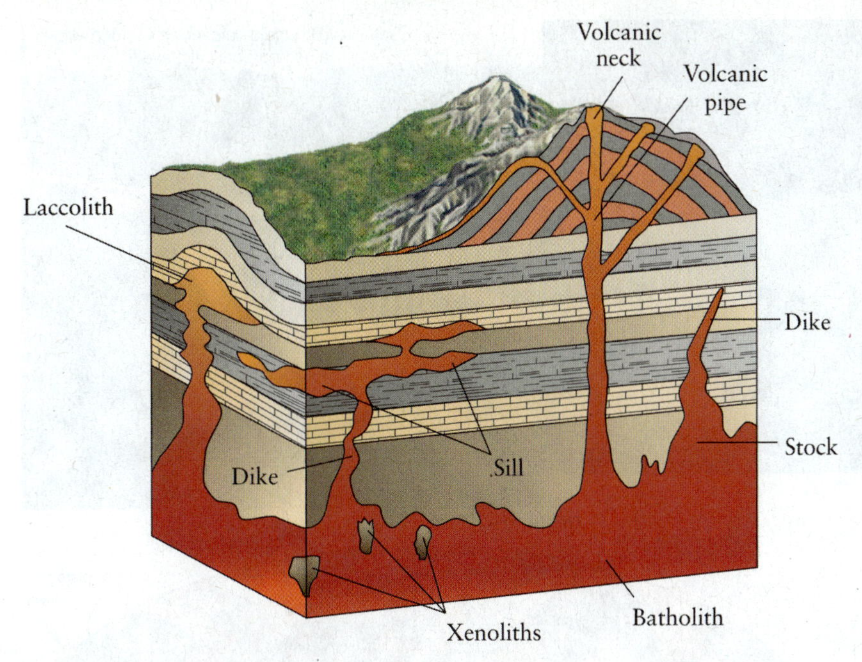

FIGURE 6.32 Plutons
This diagram shows the various forms of plutons, bodies of magma that have cooled and solidified underground.

magma that forms a pluton was squeezed upward from the place where melting originated. Magmas thus intrude into the overlying rock, sometimes cutting across rock formations and sometimes squeezing between layers. (This is the origin of the term *intrusive igneous rock*, applied to plutonic rocks, as opposed to *extrusive igneous rocks*, applied to volcanic rocks that are extruded at the surface.)

Plutons are given special names depending on their shape and size (Fig. 6.32). Dikes, sills, and laccoliths are tabular, parallel-sided sheets of igneous rock. Specifically, *dikes* cut across the layering of the intruded rock, *sills* are parallel to the layering, and *laccoliths* are sills that cause the intruded rocks to bend upward. A *volcanic pipe* is a cylindrical conduit of igneous rock below a volcanic vent; a *volcanic neck* is a pipe laid bare by erosion (an example is Shiprock, shown in Fig. 6.31). *Stocks* are intrusive igneous bodies of irregular shape that are less than 10 km in maximum dimension, and cut across the layering of the intruded rock. **Batholiths** are similar to stocks, but they are huge (up to 1000 km in length and 250 km wide); these enormous plutonic bodies are fundamental building blocks of continental crust.

All of these plutonic forms demonstrate the intimate connections between lavas that erupt at the surface and the magmas that fed the eruptions through systems of channels and conduits under the ground. Each of the three common magma types we have introduced—basaltic, andesitic, and rhyolitic—forms a characteristic plutonic rock type if it solidifies under the ground and a characteristic volcanic rock type if it solidifies at the surface (Table B6.1). The common plutonic rock associated with basaltic magma is *gabbro*; the volcanic rock is *basalt*. The common plutonic rock associated with andesitic magma is *diorite*; the volcanic rock is *andesite*. The common plutonic rock associated with rhyolitic magma is *granite*; the volcanic rock is *rhyolite*.

The different conditions under which magmas and lavas cool and solidify—very slowly in the case of intrusive magmas, very rapidly in the case of extrusive lavas—have much to do with determining the characteristics of the plutonic and volcanic rocks that form as a result, particularly their texture. As discussed in Chapter 3, plutonic rocks tend to be coarse-grained because they cool slowly and have time to grow large mineral grains. Volcanic rocks, in contrast, cool so rapidly that they cannot grow large mineral grains and therefore tend to be fine-grained or even glassy. In Chapter 7 we will look more closely at the rocks that form as a result of plutonic and volcanic processes.

THE TECTONIC CONNECTION: ORIGIN AND DISTRIBUTION OF MAGMAS AND VOLCANOES

The location of a volcano—that is, its tectonic environment—has a great deal to do with the type of magma that is formed there and, hence, with the type of volcano and volcanic eruptions that may result. As discussed in Chapter 5, volcanoes are mostly found in two tectonic settings: near plate margins and above "hotspots" in the mantle. **FIGURE 6.33** illustrates the types of magma and, therefore, the types of lava and volcanic eruptions associated with different tectonic settings.

Midocean Ridges, Hotspots, and Basaltic Magmas

At divergent plate margins, such as midocean spreading ridges, the oceanic crust is quite thin and the geothermal gradient is steep. This setting favors the eruption of hot, low-viscosity basaltic lavas. Basaltic magma generated in the underlying mantle by decompression melting rise through fissures along the midocean ridge, creating new oceanic crust. Lava that erupts underwater often forms characteristic pillowed flows, as shown in **FIGURE 6.33A**.

What can we deduce about the source region for basaltic magma? Volcanoes erupt basaltic magma on both oceanic and continental crust, with little variation in composition of the magma. The source of basaltic magma must, therefore, be within the mantle (because the crust varies greatly in composition from place to place, so magmas generated in the crust should be much more variable

FIGURE 6.33 Magmas, lavas, and plate tectonics
Different magma types are generated in different tectonic settings, and this leads to different types of volcanic eruptions. Yellow dots on the map indicate mantle hotspot volcanism, and red triangles indicate plate boundary volcanoes.

in composition). The geographic distribution of basaltic volcanoes does not, however, seem to be related to specific features of the crust or plate tectonics, such as subduction zones. This suggests that basaltic magma must be formed by melting of the mantle itself, and that the magma rises regardless of what lies above.

The most likely hypothesis for the generation of basaltic magma, therefore, is that it is formed by partial melting of the mantle as a result of deep-seated, upward-moving convection currents. The mantle is low in silica; when it melts, the result is basalt, a relatively silica-poor magma. Further, the mantle is thought to be gas-poor, although there are areas of exception to this; for this reason basaltic magma is usually gas-poor.

There is another tectonic environment in which basaltic magmas are formed and reach the surface—it is the *intraplate* (midplate) environment associated with hotpots over mantle plumes. As discussed in Chapter 5, **hotspots** are locations where lava is fed to the surface by upwelling masses of hot material called **plumes.** The locations of hotspots do not bear any particular relationship to plate boundaries; they are thought by some (but with ongoing discussion) to be rooted in place at the core-mantle boundary, and thus may serve as a fixed frame of reference against which to measure the motion of tectonic plates. The lavas found above mantle hotspots tend to be hot and basaltic, and they build giant shield volcanoes by layering one fluid lava flow on top of another (FIG. 6.33B). The most famous examples are the great shield volcanoes that form the Hawaiian Islands, but more than 100 long-lived hotspots have now been identified. The origin of basaltic magmas at hotpots, like midocean ridge basaltic magmas, is thought to be partial melting of the mantle as a result of decompression.

Continental Rifts and Rhyolitic Magmas

The lavas that are erupted at continental divergent margins tend to be high in silica content because the magma passes through thick continental crust on its way to the surface, melting and assimilating silica-rich crustal material as it goes. The magmas formed in this way are rhyolitic; the example shown in FIGURE 6.33C is from Yellowstone National Park. The vast magma chamber underlying Yellowstone last erupted cataclysmically about 640,000 years ago, creating the Yellowstone Caldera. The eruption is given a VEI of 8, and the magma chamber remains an active heat source.

Where are rhyolitic magmas formed? It has long been known that volcanoes that erupt rhyolitic magma are abundant on the continental crust but are not known on the oceanic crust. This observation suggests that the processes that form rhyolitic magma do not occur in the mantle and must be confined within the continental crust. (If the processes that form rhyolitic magma did occur in the mantle, rhyolitic magma would rise to the surface regardless of the kind of crust above and would therefore be found in volcanoes on oceanic crust.) Rhyolitic magma is thus thought to form as a result of the partial melting of relatively silica-rich continental crust. Rhyolitic magma tends to be gas-rich because rocks of the continental crust contain both water vapor and carbon dioxide, and during melting these gases become concentrated in the magma. The vast outpourings of rhyolitic lava and tephra that created the volcanic rocks of Yellowstone National Park, for example, apparently occurred because the base of the continental crust was heated from below by basaltic magma in the mantle. The heating caused partial melting and the production of rhyolitic magma from the continental crust.

Subduction Zones and Andesitic Magmas

At ocean–continent and ocean–ocean subduction zones (FIG. 6.33D and E), the subducting rocks have high water content and therefore melt at a lower temperature. As subduction proceeds, water is released into the wedge of overlying mantle, inducing wet melting. Magmas formed under wet conditions often solidify before reaching the surface, but they may stay molten for a variety of reasons. For example, partial melting may separate the minerals with a higher melting point from melt formed at a lower temperature, allowing the minerals to erupt to the surface. The andesitic magmas that result from melting in subduction zones are cooler and more viscous, they have higher silica content than basaltic lavas, and they produce a different and more varied suite of rocks. They are more likely to erupt explosively, producing pyroclastic deposits and building stratovolcanoes.

What can we conclude about the source regions for andesitic magma? Volcanoes that erupt andesitic magma are found on both the oceanic crust and the continental crust, wherever subduction occurs. This suggests that andesitic magma must form in the mantle and rise up regardless of the nature of the overlying crust. However, andesitic volcanoes have a restricted geographic distribution. For example, a ring of andesitic volcanoes surrounds the Pacific, forming the so-called *Ring of Fire.* The Ring of Fire, which geologists also call the *Andesite Line*, is exactly parallel to the Pacific Plate's subduction margins. This confirms that andesitic magma forms as a direct consequence of specific plate tectonic processes, by the wet partial melting of subducted oceanic crust and the wedge of mantle that immediately overlies the subducting plate. Andesitic magma is gas-rich, and its volcanic gas composition is dominated by water vapor. This makes sense, too, given that seawater is carried into the mantle by the downgoing plate, and becomes incorporated into the magma once melting begins.

Now On to the Rock Cycle

Once again, the theory of plate tectonics has provided us with a context within which we can understand the processes that lead to both earthquakes and volcanic eruptions. These processes influence all spheres of the Earth system, both directly and indirectly. It should also be apparent from the topics discussed in this chapter that rock melting and the subsequent intrusion of magma or extrusion of lava are essential processes in the rock cycle. In the next chapter we will look more closely at these and other processes associated with the rock cycle, and at the rocks and other Earth materials that form as a result of these processes.

SUMMARY

1. Earthquakes occur when rock masses break and slip past one another along a fault. Earthquakes allow scientists to obtain information about Earth's interior, determine the dimensions of tectonic plates, and decipher the mechanisms by which plates are formed at spreading centers and consumed in subduction zones.

2. The elastic rebound hypothesis suggests that rock masses on either side of a fault can become locked, storing elastic energy. When the locked fault slips, the rocks rebound and the stored energy is released in the form of seismic waves. Body waves travel outward in all directions, elastically deforming rock through which they pass. P (primary) waves are compressional waves, deforming rock by a temporary change in volume. S (secondary) waves are shear waves, deforming rock by a temporary change in shape; they are slower than P waves and cannot be transmitted by liquids or gases. Surface waves travel around Earth, rather than through it, and are slower than P and S waves; they can be extremely damaging.

3. The focus is the point on a fault where energy is first released during an earthquake. An earthquake's location is described in terms of its focal depth and epicenter, the point on the surface directly above the focus. The lag time between P- and S-wave arrivals at a seismic station indicates the distance traveled by the seismic waves. When distances from three or more seismographs are available, the position of the epicenter can be determined by triangulation.

4. Large-magnitude earthquakes are infrequent, but a single one can release as much energy as thousands of smaller quakes. There is an upper limit to the size of earthquakes because rocks can store only a limited amount of elastic energy before they fracture. Scientists quantify earthquake magnitudes using the moment magnitude, calculated from the recorded amplitudes of seismic waves and corrected for distance from the epicenter. The scale are logarithmic; each step corresponds to a 10-fold increase in the recorded wave amplitude and a 32-fold increase in the amount of energy released. The modified Mercalli intensity scale is based on the felt vibrations and extent of damage during an earthquake, which typically decreases with distance from the epicenter.

5. Most seismic activity takes place along active tectonic plate boundaries, although very large quakes sometimes occur far from plate boundaries. Primary effects of earthquakes include ground motion (which may result in building collapse) and surface rupture. Secondary effects include fires, landslides, liquefaction, and tsunamis. Tsunamis are generated by sudden movements of the seafloor and have been particularly destructive in the Pacific and Indian oceans. They travel at speeds up to 950 km/h with wavelengths up to 200 km, and as they approach the shore the water can pile up to heights of 30 m or more.

6. Long-term forecasting of earthquakes is based on understanding tectonic environments, determining patterns in recurrence intervals, and identifying seismic gaps where elastic energy is accumulating. Shorter-term prediction and early warning remain elusive. The scientific basis for prediction lies in the observation of precursor anomalies, but earthquakes behave erratically. Research focuses on properties that might be expected to change in rocks undergoing elastic deformation.

7. The higher the density of a material, the faster seismic body waves travel through it. If Earth were homogeneous, and if rock density increased smoothly with depth, body wave velocities would increase smoothly; however, they are refracted and reflected in several places, indicating the presence of boundaries in Earth's interior. The most pronounced of these seismic discontinuities marks the core–mantle boundary. Refraction at this boundary is so strong that it casts a large P-wave shadow; an even larger S-wave shadow results from the blocking of S waves by the liquid outer core. The moho is the seismic discontinuity that marks the compositional boundary between the crust and the mantle. The speed of seismic waves is also affected by rock strength; the more ductile (less brittle) a rock, the slower the speed at which seismic waves can travel through it. For example, seismic waves travel very slowly through the weak, ductile rock of the asthenosphere, called the low-velocity zone.

8. Shallow-focus earthquakes of relatively low magnitude occur along the spreading ridges that mark divergent boundaries, where plates are splitting apart and new oceanic crust is created by seafloor volcanism. Shallow-focus but sometimes very powerful earthquakes delineate transform fault plate boundaries. Very powerful, shallow- to deep-focus earthquakes occur in broad bands marking continental collision zones; they are associated with compressional deformation and no volcanic activity. The deepest and most powerful earthquakes occur in subduction zone environments; these are called megathrust earthquakes.

9. Earthquakes can occur only in rocks that are strong enough to store elastic energy, and brittle enough to break (rather than bend) when their elastic limit is exceeded. This is why earthquakes along divergent plate boundaries typically have shallow foci; only the rocks closest to the surface are cold enough and strong enough to behave in this way. Earthquakes in collision zones can be very deep-seated because continental collisions are associated with massive thickening of the cold, brittle lithosphere. Subduction zones are marked by earthquakes that vary from shallow to very deep-seated; these Benioff zones outline the dimensions of the downgoing lithospheric plate.

10. Magma is the mixture of molten rock, suspended mineral grains, and dissolved gas that forms in the crust or mantle when temperatures are sufficiently high. Lava is magma that reaches Earth's surface. A volcano is a vent, crack, or opening from which lava, pyroclasts, or gases are erupted. Volcanoes can be subaerial, submarine, or subglacial.

11. Magma occurs in a range of compositions in which silica (SiO_2) is typically dominant. Small amounts of volcanic gas, principally water vapor and carbon dioxide, are dissolved in all magmas; they play an important role in eruptive processes and in the Earth system. The three common magma types are basaltic (low silica); andesitic (intermediate silica); and rhyolitic (high silica). The higher the SiO_2 content of a magma, the more viscous the magma.

Temperature also influences the properties of magmas, including viscosity; the higher the temperature, the lower the viscosity.

12. Rocks typically melt over a temperature range. Partial or fractional melting occurs when part of a rock melts and part remains solid. At atmospheric pressure, rock begins to melt when heated to a temperature between about 800°C and 1000°C, depending on composition; complete melting is commonly attained by about 1200°C. The temperature in the upper part of the mantle is higher than the temperature at which most rocks melt at the surface, but the upper mantle is mostly solid; this is because pressure increases with depth, and increasing pressure causes rock to resist melting. A decrease in pressure can lead to decompression melting; this can happen as rock rises to the surface. The presence of water (or water vapor) in rock reduces its melting temperature. This is important in subduction zones, where water is carried down into the mantle by oceanic crust and released at depth.

13. Magma is less dense than solid rock and rises because it exerts an upward force against the weight of the overlying rock. Pressure from the weight of overlying rock decreases as the magma rises, allowing decompression melting to occur and gases dissolved in the magma to be released. The ascent process and the characteristics of the magma combine to determine the eruptive style of a volcano. Basaltic magma has a low SiO_2 content, high temperature, low viscosity, and low dissolved gas content; all of these factors typically lead to nonexplosive, Hawaiian-type eruptions, characterized by streams of flowing lava and relatively low VEI.

14. Viscous andesitic and rhyolitic magma have higher silica and dissolved-gas contents, and erupt at lower temperatures than basaltic magma. This combination of factors typically leads to explosive eruptions, with abundant pyroclastic material and high VEI. Three of the most devastating types of explosive eruptions involve eruption columns and tephra falls; pyroclastic flows; and lateral blasts. Some Vesuvian or Plinian eruption columns reach such great heights that winds transport the pyroclasts, gases, and volcanic aerosols around the world, blocking incoming solar radiation and affecting the global climate.

15. The shape of a volcano is determined by the kind of magma it erupts and the relative proportions of lava and pyroclasts. Shield volcanoes have gentle slopes; they form by successive eruptions of fluid basaltic lava. Rhyolitic and andesitic volcanoes tend to eject a large proportion of pyroclasts, and therefore to be surrounded by tephra deposits. Larger volcanoes, particularly those of andesitic composition, form steep-sided stratovolcanoes with interlayered lava flows and pyroclastics. A caldera can form at the summit of a volcano as a result of the partial emptying and subsequent collapse of the magma chamber. Fissure eruptions occur when fluid lava reaches Earth's surface through an elongate fracture in the crust.

16. There are about 500 active volcanoes today, many of them in the circum-Pacific region. Explosive eruptions are less common but more dangerous than nonexplosive eruptions. Hazards include pyroclastic flows; poisonous gas emissions; volcanic mudflows, or lahars; and tsunamis. Tephra eruptions may wreak havoc on agricultural land, livestock, and air travel. Long-term prediction of eruptions is based on the identification of volcanoes that are likely to erupt explosively, mostly andesitic and rhyolitic stratovolcanoes in subduction zone environments. Short-term prediction is based on the observation of precursor anomalies indicating movement of magma within the volcano.

17. Beneficial aspects of volcanism include the renewal of the surface and the production of new volcanic rock from which rich soils develop, as well as the creation of new land in coastal areas. Volcanism provides geothermal energy and plays an important role in the formation of some mineral deposits. Without volcanic degassing we would have no atmosphere or hydrosphere, and no way to support life on this planet.

18. Beneath every volcano lies a complex of chambers and channels through which magma moves toward the surface. Plutons are bodies of magma that never reached the surface. Dikes are tabular plutons that cut across the layering of the intruded rock; sills are parallel to layering, and laccoliths are sills that cause the intruded rocks to bend upward. A volcanic pipe is a cylindrical conduit of igneous rock below a volcanic vent; a volcanic neck is a pipe laid bare by erosion. Stocks and batholiths are large intrusive igneous bodies of irregular shape that cut across the layering of the intruded rock.

19. Igneous rocks that cool and solidify at or near the surface are volcanic rocks; those that cool and solidify underground are plutonic rocks. The common plutonic rock associated with basaltic magma is gabbro; the volcanic equivalent is basalt. The common plutonic rock associated with andesitic magma is diorite; the volcanic equivalent is andesite. The common plutonic rock associated with rhyolitic magma is granite; the volcanic equivalent is rhyolite. The conditions under which magmas and lavas cool and solidify determine the texture of the rocks that are formed. Plutonic rocks are coarse-grained because they cool slowly and have time to grow large mineral grains. Volcanic rocks cool rapidly and are therefore fine-grained or glassy.

20. Basaltic magmas form by partial melting of the mantle. At oceanic divergent margins, the crust is quite thin and the geothermal gradient is steep, which favors the eruption of hot, low-viscosity lavas. Hot, fluid lavas also erupt at hotspots, where basaltic lava is fed to the surface by upwelling mantle plumes. Lavas formed at continental divergent margins tend to be rhyolitic; they are high in silica content because the magma passes through thick continental crust on its way to the surface, melting and assimilating silica-rich crustal material. Rhyolitic magma is gas-rich because the continental crust contains water and carbon dioxide, which become concentrated in the magma during melting. At ocean–continent and ocean–ocean subduction zones, the subducting rocks have high water content. As subduction proceeds, water is released into the wedge of overlying mantle, inducing wet melting. These andesitic lavas are water-rich and erupt explosively, producing pyroclastics and building stratovolcanoes. Volcanoes that erupt andesitic magma are found on both oceanic and continental crust, wherever subduction occurs. This indicates that andesitic magma must form in the mantle as a direct result of the process of subduction.

IMPORTANT TERMS TO REMEMBER

active volcano *173*
andesitic magma *164*
basaltic magma *164*
batholith *177*
body wave *146*
caldera *172*
deformation *144*
dormant volcano *173*
elastic energy *144*
elastic rebound hypothesis *144*
epicenter *147*
eruption column *170*
extinct volcano *173*
fault *144*
fissure eruption *172*
focus *147*
hotspot *179*
lava *161*
liquefaction *152*
magma *161*
magnitude *148*
modified Mercalli intensity *150*
moho *158*
partial melt *166*
P (primary) wave *146*
plume *179*
pluton *176*
pyroclast *168*
pyroclastic flow *170*
pyroclastic rock *168*
reflection *156*
refraction *156*
rhyolitic magma *164*
Richter magnitude *150*
S (secondary) wave *146*
seismic discontinuity *158*
seismic gap *156*
seismic moment magnitude *150*
seismic wave *145*
seismicity *144*
seismogram *146*
seismograph *146*
shield volcano *171*
stratovolcano *171*
surface wave *146*
tephra *168*
tephra cone *171*
tsunami *152*
viscosity *164*
volcanic ash *170*
volcanic gas *164*
volcano *161*

QUESTIONS FOR REVIEW

1. Explain why and how earthquakes occur, according to the elastic rebound hypothesis. Why is there an upper limit on the intensity of earthquakes?

2. What is the relationship between an earthquake's focus and its epicenter?

3. How are seismic waves recorded and measured? What is the difference between a seismograph and a seismogram?

4. How do scientists use seismic waves to determine the location of an earthquake? Describe the process.

5. Describe the important features of the Richter magnitude scale, the seismic moment magnitude scale, and the modified Mercalli intensity scale, and comment on the differences between them.

6. What are reflection and refraction, and how do they help scientists determine what's inside Earth?

7. Earthquakes can cause damage in many ways; name and briefly describe four.

8. How have scientists used earthquakes to determine the dimensions of tectonic plates and the characteristics of tectonic processes such as subduction?

9. What is the difference between magma and lava? What is the difference between pyroclasts and tephra?

10. What are the differences between shield volcanoes and stratovolcanoes? What are the differences between stratovolcanoes and tephra cones? Why does a shield volcano like Mauna Loa in Hawaii have gentle slopes, whereas a stratovolcano like Mount Fuji in Japan has steep slopes?

11. Name some ways in which volcanoes affect life on Earth.

12. What are the main types of plutons, and how do they form? How do plutons illustrate the connection between magma beneath the surface and lava on the surface?

13. Where inside Earth does basaltic magma form? Where is it typically erupted?

14. What is the origin of rhyolitic magma? Where would you expect to find rhyolitic volcanoes?

15. What is the origin of andesitic magma? With what type of volcano and what tectonic setting are andesitic magmas associated?

QUESTIONS FOR RESEARCH AND DISCUSSION

1. Why is short-term prediction of earthquakes so much less successful than long-term prediction (forecasting)? Why do you think seismologists are extremely cautious about making predictions and issuing early warnings? Do you think it will ever be possible to predict earthquakes accurately? Research your answer.

2. Use the elastic rebound hypothesis to describe what happens to rocks at the focus just before, during, and after an earthquake. What precursor phenomena would you expect to detect as a result of these changes in the rocks? Aside from the elastic rebound hypothesis, are there any other widely accepted hypotheses about the physical mechanisms of earthquakes? How do they differ?

3. The slopes of active volcanoes tend to be populated. What reasons can you think of for living near a volcano? Do you think the advantages outweigh the disadvantages? Why or why not? What should government officials and scientists do to alert those who live near active volcanoes to the hazard?

QUESTIONS FOR *THE BASICS*

1. What are the important differences between seismic body waves and surface waves? Between P waves and S waves?

2. Describe the effects of SiO_2 content on the properties of magmas. How are temperature and viscosity related in magmas and lavas?

3. What are the most important volcanic gases, and how do they affect the properties of magmas and lavas?

QUESTIONS FOR *A CLOSER LOOK*

1. What is a megathrust earthquake?

2. In what tectonic environment did the Sumatra-Andaman earthquake of 2004 occur?

3. Tsunamis can be devastating, but under the right circumstances they can be very amenable to the issuance of effective early warnings. What would those circumstances need to be?

CHAPTER 7

The Rock CYCLE

OVERVIEW

In this chapter we:

- Explain how rock breaks down to become sediment and how sediment becomes sedimentary rock
- Examine sedimentary rock layering and explain how stratigraphy can be used to determine geologic ages
- Define the kinds and processes of metamorphism and classify the main types of metamorphic rock
- Review the processes involved in rock melting and investigate how magma solidifies into igneous rock
- Consider how the rock cycle and the tectonic cycle are interrelated and how internal and external processes interact at Earth's surface to shape our landscape

Carbon storage
These pure white cliffs overlook Lago di Braies (also known as Pragser Wildsee) in the Dolomites of South Tyrol, Italy. The Dolomites are a section of the Alps that are composed almost entirely of carbonate rock. Carbonate rock is an extremely important component of Earth's carbon cycle, because it acts as a long-term reservoir for carbon dioxide. If all of the carbon dioxide currently stored in carbonate rock were returned to the atmosphere, we would have an atmosphere dominated by carbon dioxide and a runaway greenhouse effect.

The **rock cycle**, first introduced in Chapter 1, involves the cycling of material through the geosphere as a result of the formation, breakdown, rearrangement, and reformation of rock (FIG. 7.1). Tectonic forces driven by Earth's internal heat energy have raised mountains throughout most of Earth history. However, most of those mountains no longer exist because external forces, driven by the Sun's energy, continually break down and wear away the uplifted rock. Water, wind, and ice then transport and deposit the broken-down rock debris to form sediment, which eventually becomes sedimentary rock. When continents collide, sedimentary rock caught up in the collision is subjected to elevated temperatures and pressures, and sedimentary rock becomes metamorphic rock. If temperatures become high enough, rock melting will occur, leading to the formation of igneous rock.

The history of all the tectonic bumps and grinds, indeed the long history of Earth itself, can be read from the accumulated debris of erosion, from altered and deformed rock, and from the products of rock melting. The evidence of this history is recorded in the sedimentary, metamorphic, and igneous rock formed as a result of the rock cycle.

James Hutton (Chapter 4) and the other scientists who described the rock cycle two centuries ago were unaware of plate tectonics, but now we realize that plate tectonics answers many of the questions that puzzled those early scientists. We now also realize that the rock cycle and the tectonic cycle are the cause of the continuous cycling of material through the geosphere, leading to the continual renewal of Earth's surface. Figure 7.1 illustrates the rock cycle and hints at some of the interactions between the cycles. It is clear even from this simple sketch that the rock and tectonic cycles are involved with and influence everything that goes on at Earth's surface. Both cycles play major roles in the Earth system.

FIGURE 7.1 **The rock cycle**
The rock cycle comprises the processes whereby materials within and on top of Earth's crust are weathered, transported, deposited, deformed, altered, melted, and reformed into new rock. Rock doesn't always take the long way around this circle—there are many "shortcuts" from one part of the rock cycle to another. The rock cycle is closely interrelated with the other cycles of the Earth system.

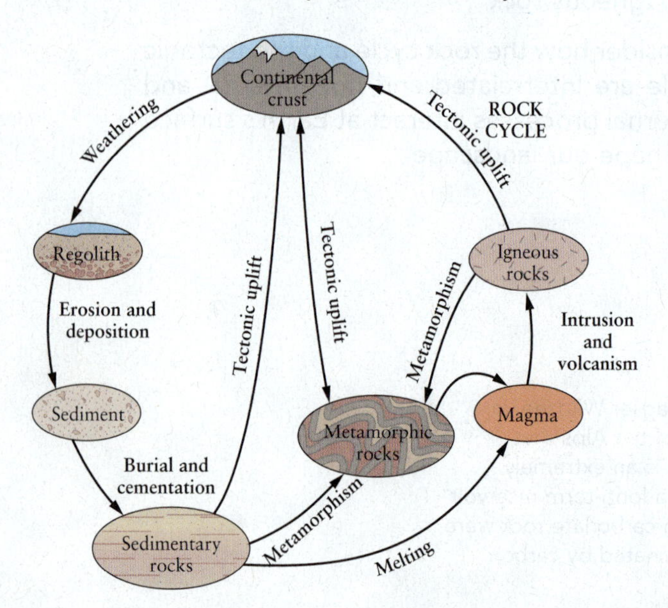

FROM ROCK TO REGOLITH

We begin our examination of the rock cycle by looking at the processes that wear away the great mountain ranges and all other exposed rock at Earth's surface. As soon as fresh rock of the geosphere is exposed, it is attacked by the hydrosphere, atmosphere, and biosphere. Rocks of all kind are physically broken apart and chemically altered throughout the zone where the geosphere, hydrosphere, biosphere, and atmosphere mix. This zone extends from the ground surface downward to whatever depth air and water penetrate—many kilometers (FIG. 7.2). Within that zone, rock is porous because it contains a network of fractures, cracks, and other openings, some of which are very small but all of which make the rock vulnerable to attack by air, water, and microbes. Given sufficient time, the result is conspicuous decomposition and disintegration of the rock by processes known collectively as **weathering**. The end result of weathering on a global scale is the formation of **regolith**, the irregular blanket of loose, uncemented rock particles that covers Earth's surface. *Soil* is that part of the regolith that can support plants, as you will learn in Chapter 15.

We have all seen weathering in action. You may have visited a cemetery and strained to read the inscription on an old marble tombstone so modified by weathering that the characters were barely legible (FIG. 7.3). Or you may have been seated around a roaring campfire and seen a rock next to the fire explode suddenly, because it was composed of minerals that expand at different rates when heated. Such examples show that weathering can involve both chemical and physical processes.

Chemical Weathering

Minerals in igneous and metamorphic rocks form at pressures and temperatures much higher than those of the surface; as a result, most such minerals are chemically unstable at the lower temperatures and pressures of the surface. When rock is uplifted and eventually exposed, therefore, their mineral components are vulnerable. Through exposure to the atmosphere, hydrosphere, and biosphere they are chemically changed into new, more stable minerals. **Chemical weathering**, then, is the decomposition of rocks and minerals as chemical reactions transform them into new chemical compounds that are stable at or near Earth's surface.

The principal agent of chemical weathering is a weak solution of carbonic acid (H_2CO_3), which is formed when falling rainwater dissolves small quantities of atmospheric carbon dioxide:

$$H_2O + CO_2 \rightarrow H_2CO_3$$

FIGURE 7.2 The zone of weathering
In this photograph taken in South Africa, sedimentary rock is undergoing the process of disintegration. Water and air penetrate through the fractures in the rock and react with the minerals there. The rock near the surface is more heavily weathered than the lower layers because it has been exposed to more water and air.

All rainwater is thus weakly acidic. As the acidic water moves downward and laterally on and beneath the ground surface, additional carbon dioxide is dissolved from decaying vegetation. Thus, chemical weathering is the result of interactions of the atmosphere, the hydrosphere, and the biosphere, which produce an acidic solution that attacks the upper part of the geosphere. The dissolution of the marble tombstone shown in Figure 7.3 is an example of chemical weathering.

The chemical reaction that decomposes the common rock-forming mineral potassium feldspar ($KAlSi_3O_8$) provides a good example of chemical weathering. A molecule of carbonic acid *dissociates* (breaks apart chemically by splitting into smaller molecules) in water to form a hydrogen ion (H^+) and a bicarbonate ion [$(HCO_3)^-$]. The H^+ ions enter the potassium feldspar and replace potassium ions (K^+), which then leave the feldspar and pass into solution (FIG. 7.4). Water combines with the remaining aluminum silicate molecule to create kaolinite ($Al_2Si_2O_5[OH]_4$), a new clay mineral not present in the original rock. Weathering of rock that contains feldspars and Fe-Mg bearing minerals produces clay minerals and goethite (FeO·OH), a weathering product of the common iron mineral magnetite. When granite weathers, clay minerals and goethite are also produced from the feldspar and mica it contains. However, the mineral quartz, which is also present in granite, is resistant to chemical weathering; it survives the chemical weathering process unaltered.

Physical Weathering

Some regolith consists of rock fragments identical to the underlying bedrock that are unweathered or only slightly weathered, indicating that little or no chemical alteration has occurred. Instead, the fragments have experienced **physical weathering** (also known as **mechanical weathering**), that is, the disintegration of rock as a result of physical breakup.

Several processes are effective in physical weathering. A rock mass buried deep beneath the land surface is subjected to immense confining pressure as a result of the weight of overlying rock (FIG. 7.5). As erosion gradually removes the overlying rock, the pressure is reduced and the buried rock mass adjusts by expanding upward. In the process, sheetlike fractures called **joints** develop (FIG. 7.6A, on page 189). Joints become important passageways for water; this can facilitate weathering through *frost wedging*. When water freezes, its volume increases by about 9 percent. If water freezes in a confined crack, such as a joint, the resulting pressure can be so great that the rock is wedged apart (FIG. 7.6B, on page 189). Material dissolved in the groundwater moving through fractured rock can precipitate out to form salts. The enormous forces exerted by salt crystals growing in cavities or along the boundaries between mineral grains can lead to breakage of the rock. Fire, too, can be a very effective agent of weathering. An intense forest or brush fire can overheat

FIGURE 7.3 Impacts of weathering
This marble tombstone, which has stood in a New England cemetery since the early nineteenth century, has gradually been dissolved by rainwater. Over the years, the once sharply chiseled inscriptions have become barely legible.

FIGURE 7.4 Ion exchange in chemical weathering
This photomicrograph (right) shows a feldspar grain that has been altered by ion exchange. The clay residue has been removed to make the pattern of alteration more visible. The diagrams show how the alteration process proceeds.

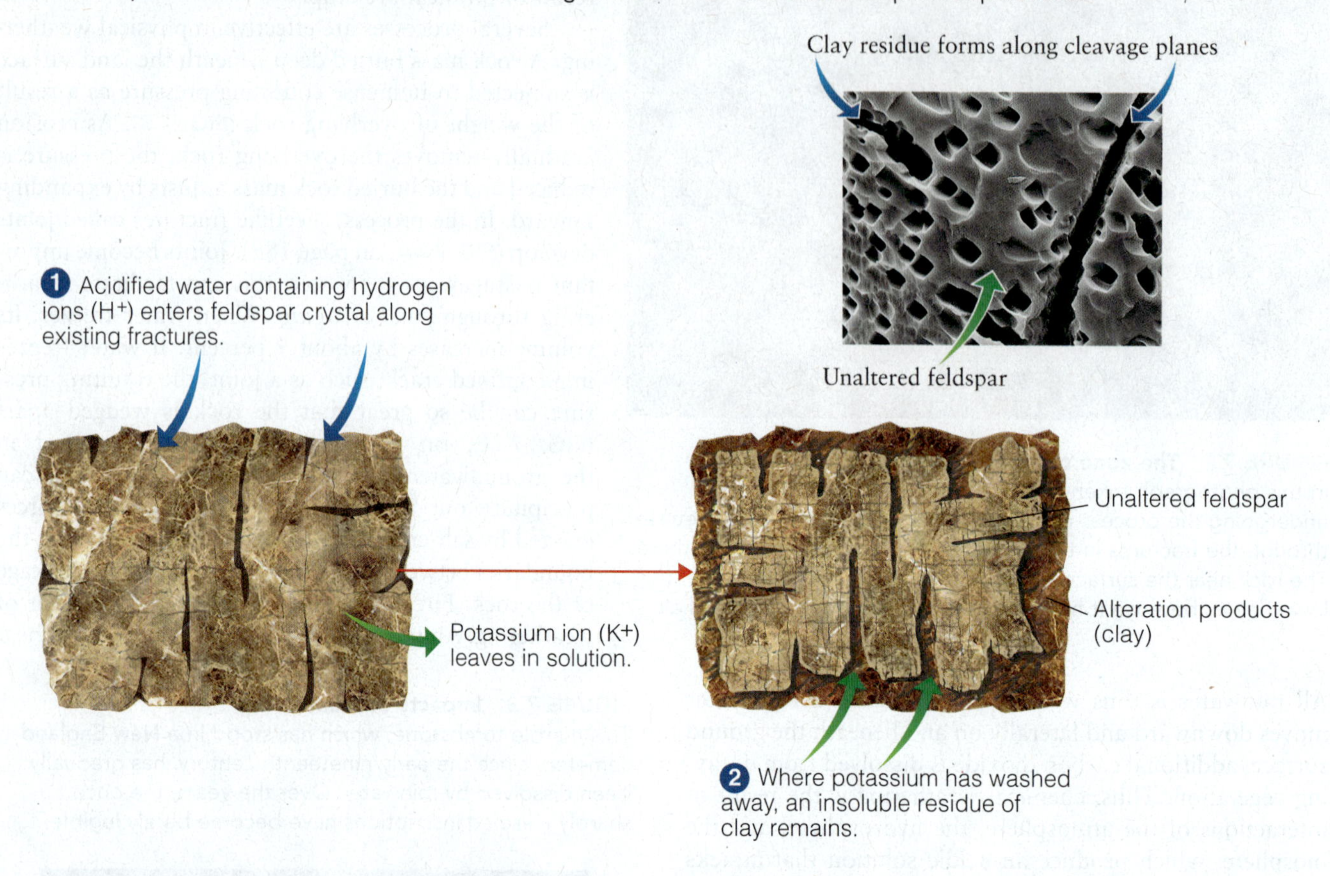
Clay residue forms along cleavage planes
Unaltered feldspar
1 Acidified water containing hydrogen ions (H^+) enters feldspar crystal along existing fractures.
Potassium ion (K^+) leaves in solution.
Unaltered feldspar
Alteration products (clay)
2 Where potassium has washed away, an insoluble residue of clay remains.

(A)

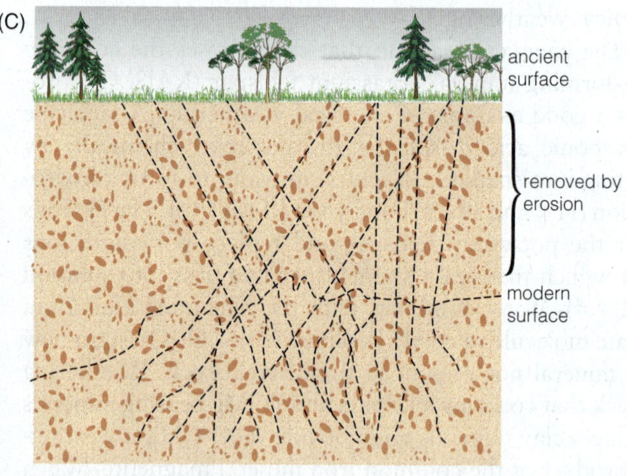
(C)
ancient surface
removed by erosion
modern surface

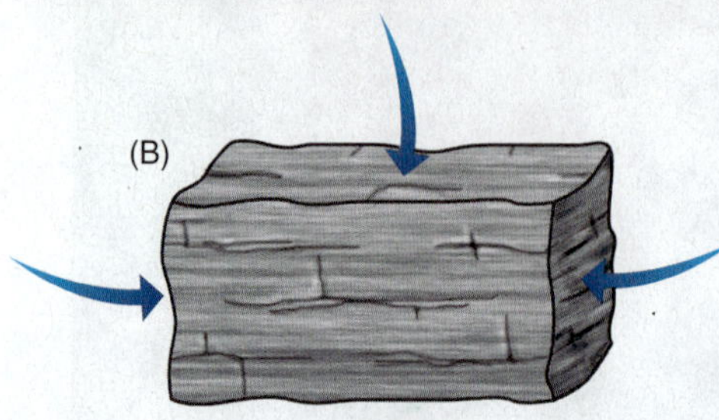
(B)

FIGURE 7.5 Joint formation
Jointing is a form of physical weathering. (A) These heavily jointed rocks are found in the Joshua Tree National Monument in California. Erosion has widened some of the joints to the point of detaching stones from the main rock body. (B) The rock originally formed underground, where it was subject to great pressure from the overlying and surrounding rock. (C) As the rock rose to the surface and the overlying rock was eroded, the pressure decreased, causing the rock to expand and crack. Later, erosion rounded off and widened the joints even further.

(A)
Smooth, rounded surface where sheets have exfoliated (peeled off) in the past.

(B)
Frost wedging has widened this joint in the rock.

(C)

FIGURE 7.6 Mechanical weathering
(A) Sheet jointing: Sheet jointing results in curved domes, like the famous Half Dome in Yosemite National Park, California. (B) Frost wedging: This granite boulder in the San Andres Mountains of New Mexico has been split apart by repeated freezing and thawing of water that penetrated along the joints. (C) Root Wedging: The tree began growing into a crack in an outcrop of rock. The growing roots caused the crack to expand. Eventually, a mass of rock broke away, exposing the tree's root system.

the outer part of a rock, causing it to expand, fracture, and break away. Repeated fires can thereby significantly reduce the size of a rock. Finally, plant roots extending along cracks can slowly wedge the rock apart (FIG. 7.6C).

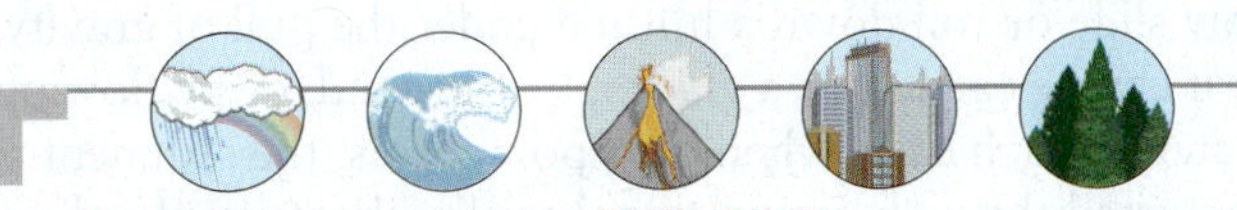

Make the CONNECTION

Weathering is an interesting intersection between processes of the atmosphere, hydrosphere, biosphere, and geosphere. Humans can also be agents of weathering. What kinds of human activities cause rock to break apart into smaller fragments?

Although physical weathering is distinct from chemical weathering, the two processes generally work hand in hand, and their effects are inseparable. The effectiveness of chemical weathering increases as the exposed surface area increases, and surface area increases greatly whenever a large unit is divided into successively smaller units (FIG. 7.7). One cubic centimeter of rock has a surface area of 6 cm^2, but when subdivided into particles the size of the smallest clay minerals, the total surface area exposed to weathering increases to nearly 40 million cm^2. Thus a physical process such as jointing can greatly increase the vulnerability of the rock to other physical processes like frost wedging, as well as increasing its exposure to chemical weathering.

Sediment and Its Transport

Weathering breaks rock into smaller, disaggregated particles of the regolith. Then, like a perpetually restless

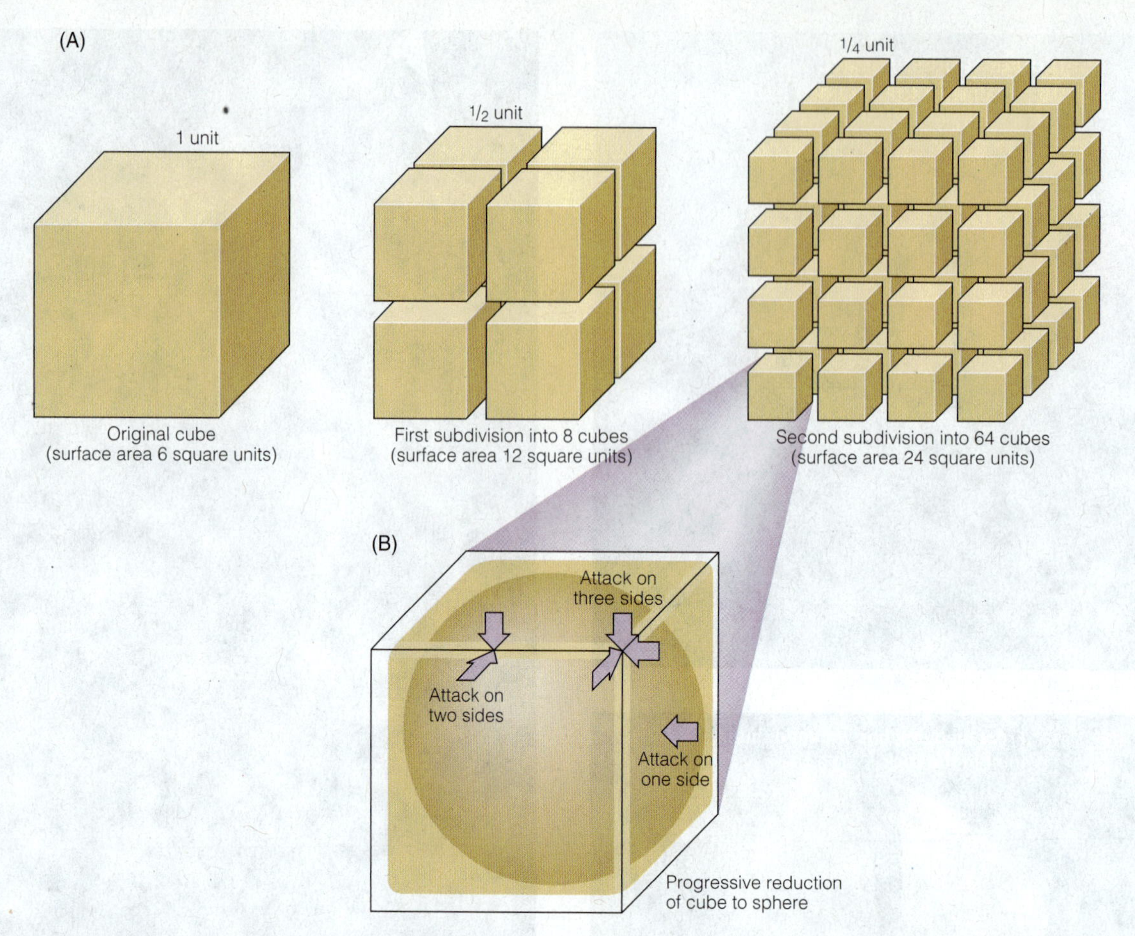

FIGURE 7.7 Rock disintegration
Weathering causes progressive breakdown of rock into smaller units. Each time a cube (A) is subdivided by slicing through the center of each edge, the surface area doubles (B). This increases the effectiveness of chemical attack. Solutions moving along the joints that separate the fragments of rock attack the corners and edges, causing the fragments to eventually become rounded.

housekeeper, nature ceaselessly sweeps the particles of regolith off the solid rock beneath, carrying the sweepings away and dropping them in river valleys, lakes, and innumerable other low-lying places. The weathered particles sometimes move under the influence of gravity, but often they are carried or pushed by a medium such as water, wind, or ice. We refer to this set of transport processes as **erosion**, and regolith that has undergone erosional transport is called **sediment.** After transport, the sediment accumulates in its new location as a result of **deposition.** We can see sediment being transported by trickles of water after a rainfall and by every wind that carries dust. The mud on a lake bottom, the sand on a beach, even the dust deposited on a windowsill is sediment. Because erosion takes place almost continuously, we find sediment nearly everywhere.

There are three principal families of sediment: clastic, chemical, and biogenic. They differ in the ways in which the sediment is formed, transported, and deposited.

Clastic Sediment

Clastic sediment (from the Greek word *klastos*, meaning broken) is simply bits of broken rock and minerals that are moved as solid particles. An individual fragment is a *clast*, and clasts tend to be the rock-forming minerals that are the most durable during chemical weathering, notably quartz. Clast size is the primary basis for classifying clastic sediment, and there are four main classes: from coarsest to finest, they are **gravel, sand, silt,** and **clay** (FIG. 7.8). Gravel may be further classified into *boulder gravel*, *cobble gravel*, and *pebble gravel* on the basis of the sizes of the clasts. As you can see from Figure 7.8, specific rock types form from each of these types of clastic sediment.

Clastic sediment can be transported in many ways. It may slide or roll down a hillside under the pull of gravity, or it can be carried by a glacier, by the wind, or by flowing water. In each case, when transport ceases, the sediment is deposited through gravitational settling because of a drop in energy. For example, sediment transported by wind or water is deposited when the moving air or flowing water slows to a speed at which clasts can no longer be carried. In a general way, clast size in sediment moved by wind or water is related to the speed of the transporting agent: the faster the speed, the larger the clasts that can be moved (FIG. 7.9). During transport, individual clasts become

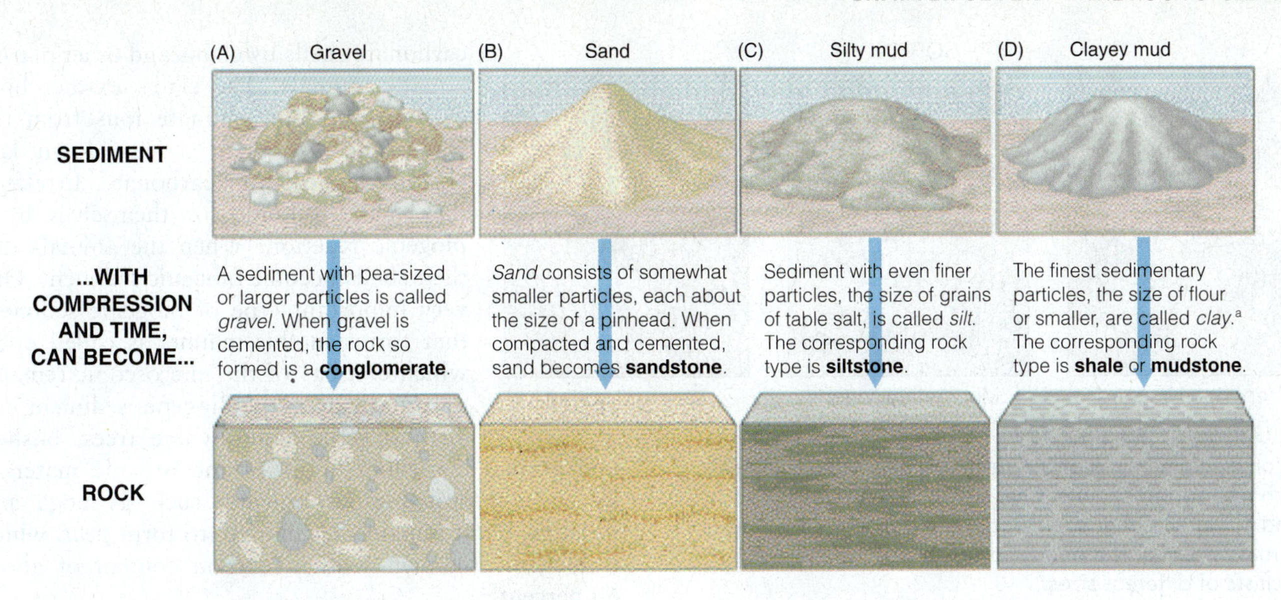

(A) Gravel
(B) Sand
(C) Silty mud
(D) Clayey mud
SEDIMENT
...WITH COMPRESSION AND TIME, CAN BECOME...
A sediment with pea-sized or larger particles is called gravel. When gravel is cemented, the rock so formed is a conglomerate.
Sand consists of somewhat smaller particles, each about the size of a pinhead. When compacted and cemented, sand becomes sandstone.
Sediment with even finer particles, the size of grains of table salt, is called silt. The corresponding rock type is siltstone.
The finest sedimentary particles, the size of flour or smaller, are called clay.a The corresponding rock type is shale or mudstone.
ROCK

FIGURE 7.8 **Clastic sediment and associated rocks**

Name of Particle	Size (mm)	Sedimentary Rock
Boulder	More than 256	Boulder conglomerate[b]
Cobble	64 to 256	Cobble conglomerate[b]
Pebble	2 to 64	Pebble conglomerate
Sand	1/16 to 2	Sandstone
Silt	1/256 to 1/16	Siltstone or Mudstone
Clay[a]	Less than 1/256	Shale

[a]Clay, used in the context of this table, refers to particle size. The term should not be confused with clay minerals, which are distinct mineral species.

[b]If the clasts are angular, the rock is called a *breccia* rather than a conglomerate.

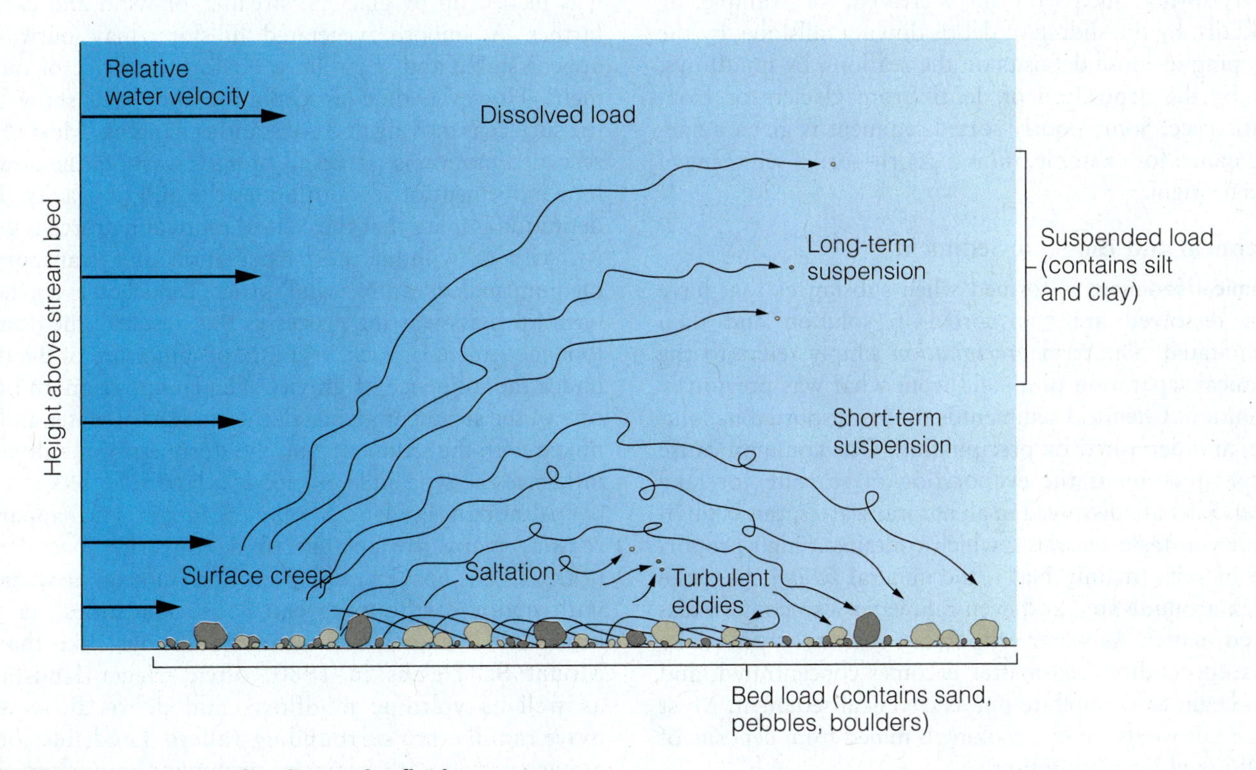

FIGURE 7.9 **Transport of sediment by fluids**
In general, the faster a fluid is moving, the heavier the particles it can transport. This diagram can be interpreted to represent either flowing wind or flowing water. In either case, the finest particles are aloft, where the velocity of fluid flow is the highest. The coarsest particles are left on the bottom, too heavy to be lifted by the fluid. (Flowing ice, although it does behave as a fluid, is different because of its very high viscosity.)

SORTING

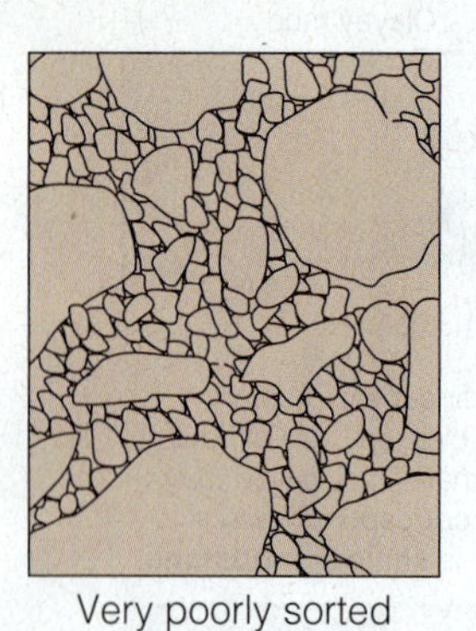

Very poorly sorted

Moderately sorted

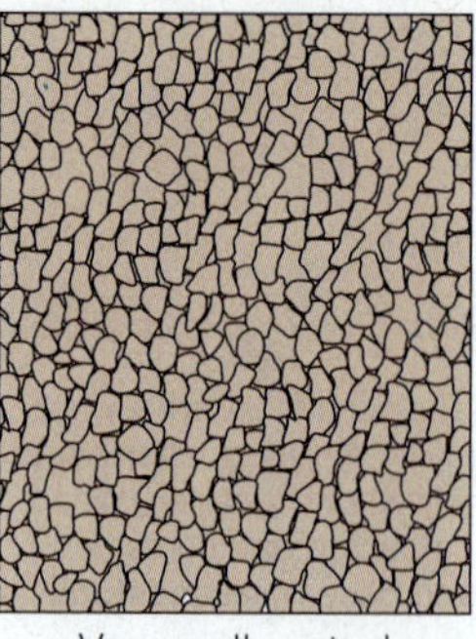

Very well sorted

FIGURE 7.10 Sorting
In some sediment, all of the clasts are nearly the same size. Such well-sorted sediment has usually been transported by wind or water. Other sediment, often transported by ice or by mass wasting, is a poorly sorted or unsorted jumble of clasts of different sizes.

broken down even further, and eventually may become smooth and rounded as a result of abrasion against other particles and against bedrock.

When the speed of the transporting medium fluctuates, clast sorting occurs (FIG. 7.10). For example, a rapidly flowing river will pick up all of the fine particles, leaving behind only the largest clasts. As the speed of the flowing water slows, smaller and smaller clasts are dropped. The resulting sediment is said to be *well sorted*. Sediment with a wide range of clast sizes is said to be *poorly sorted*. Such sediment is created, for example, by rockfalls, by the sliding of debris down a hillslope, by the slumping of loose deposits on the seafloor, by mudflows, and by the deposition of debris from glaciers or from floating ice. Some poorly sorted sediment is given a specific name; for example, *till* is a poorly sorted sediment of glacial origin.

Chemical and Biogenic Sediment

Chemical sediment is formed when substances that have been dissolved are transported in solution and then precipitated. The term *precipitation* simply refers to the chemical separation of a solid from what was previously a solution. Chemical sediment is thus transported in solution, and deposited by precipitation. One common cause of precipitation is the evaporation of seawater or lake water. Salts are dissolved in all natural waters; this is obvious if you taste seawater, which contains a high proportion of salts (mainly NaCl, the mineral *halite*), but lake water, groundwater, and even rainwater also contain dissolved matter. As water evaporates, such as when a lake dries up, its dissolved matter becomes concentrated, and salts begin to precipitate out as chemical sediment. Most of the salt we use in our cooking is mined from deposits of salt formed by evaporation.

Chemical sediment also can be formed as a result of biochemical reactions in water; the resulting sediment is called **biogenic sediment.** One common example of a biogenic reaction involves the formation of calcium carbonate shells by clams and other marine aquatic animals. The clams extract both calcium and biocarbonate ions from the water and use them to lay down layers of solid calcium carbonate. In effect, they make a home for themselves by a biogenic reaction. When the animals die, their shells become biogenic sediment. One very important type of biogenic sediment that forms in this manner is called *ooze*, which consists of the microscopic remains of aquatic animals. Biogenic sediment can also form on land, where trees, bushes, and grasses contribute most of the organic material. In water-saturated environments, such as bogs and swamps, plant remains accumulate to form *peat*, which is biogenic sediment with a carbon content of about 60 percent.

Mass Wasting

You have learned that erosion is facilitated by the flow of fluids—water, air, or ice—that lifts clasts and transports them from their source region to a deposition site. (Even glacial ice behaves as a fluid, flowing slowly downslope, as you will learn in Chapter 9). Both the flow of fluids and the downslope movement of rock fragments are controlled by gravity. Under the pull of gravity, weathered debris falls from mountainsides and tumbles downslope, where it is picked up by glaciers, streams, or wind and carried farther. A smooth, vegetated hillslope may outwardly appear stable and show little obvious evidence of movement. However, time-lapse photography will show that the surface is moving and constantly changing. Most of the recorded motion is the result of **mass wasting**, the downslope movement of regolith under the pull of gravity. This definition implies that the central motivating force is gravity, with or without the participation of a transporting medium such as water, wind, or ice. **Landslide** is a general term for mass-wasting processes that result in the downslope movement of rock, regolith, or a mixture of the two, under the influence of gravity. The composition and texture of the sediment involved, the amount of water and air mixed with the sediment, and the steepness of the slope all influence the type and velocity of a landslide (FIG. 7.11).

Although landslides may occur for no apparent reason, many are related to a triggering occurrence (FIG. 7.12). For example, ground motion associated with major earthquakes can cause landslides, as you learned in Chapter 6. Volcanic eruptions, like that at Mount St. Helens in 1980, often trigger landslides, as well as volcanic mudflows and debris flows that move rapidly into surrounding valleys. Landslides often are associated with heavy or prolonged rains that saturate the ground and make it unstable. A stream wearing away at its banks, or pounding storm surf along a cliffed seacoast, can also produce landslides when steep slopes are undercut too far. Human activities that

FIGURE 7.11 Landslides
Three kinds of landslides are illustrated here: a fall, a slide, and a slump. (A) Kaibito Canyon in Arizona has been the site of repeated rockfalls, as you can see from the debris at the base of the cliff and from the scars on the cliff face where large rock masses have detached themselves in the past. (B) In this rockslide in the Andes Mountains in Argentina, the rocks moved in a roughly straight line from the point of detachment to the valley floor. (C) A slump is a type of landslide in which the debris moves on a curved surface, as shown by the arrow. This slumping slope failure occurred in central California.

FIGURE 7.12 Triggers for landslides
Landslides can be triggered by earthquakes and volcanic eruptions, as shown here. (A) The Great Alaska Earthquake of 1964 caused major landslides in Anchorage and surrounding areas. Here, a landslide triggered by the earthquake caused part of the Chugach Mountains to collapse, covering part of the Sherman Glacier. (B) The eruption of Mount St. Helens in 1980 triggered the largest debris avalanche ever recorded, part of which is shown here.

modify natural slopes also can cause landslides; for example, slides frequently occur where road construction oversteepens a natural slope.

Tectonic Environments of Sedimentation

Under the influence of gravity, sediment accumulates in low-lying areas—troughs, trenches, and basins of various types. Locations where clastic sediment is commonly deposited are largely controlled by plate tectonics. Deposition of chemical and biogenic sediment is not as strongly influenced by plate tectonics because such sediment typically forms in open aquatic environments. However, the location of ocean basins, where such sediment often occurs, is controlled by plate tectonics.

Clastic sediment originates by weathering and erosion of continental rock and by pyroclastic volcanic eruptions. The settings where clastic sediment is most likely to form and to accumulate are strongly influenced by plate tectonics. Low-lying rift valleys are formed when continental crust splits apart because of tensional forces. Rifts such as the East African Rift hold deep wedges of clastic sediment deposited by streams. Such rifts may continue to widen, developing into linear seas like the Red Sea, and eventually into mature oceans like the Atlantic. Today, sediment continues to be washed from the continents that border the Atlantic Ocean, and is deposited on the passive continental margin. The continental shelf slowly subsides as the pile of sediment grows thicker and thicker (FIG. 7.13A).

Various low-lying regions are also located within and along the edges of high mountain ranges. Many of these regions are structural basins or troughs caused by faulting and folding associated with mountain building (FIG. 7.13B). Coarse sediment carried by mountain streams are deposited and accumulate in these areas.

The lowest-lying points on Earth's surface are the long, deep trenches that form on the ocean floor in subduction zones (FIG. 7.13C). Some subduction zones are located along the margins of continents, such as the subduction zone on the Pacific coast of South America, where the Nazca Plate is subducting under the continent. In such settings, the elevation of Earth's surface plunges many thousands of meters from the top of the nearby line of arc volcanoes on the continent to the bottom of the oceanic trench, all within the lateral distance of a few hundred kilometers. Because erosion proceeds most rapidly where slopes are the steepest, we would expect to find rapid erosion and deposition of sediment here—and we do. The adjacent volcanoes ensure that a lot of volcanic debris is present in the sediment. The great thicknesses of sediment that accumulate in subduction trench environments are called *accretionary wedges*.

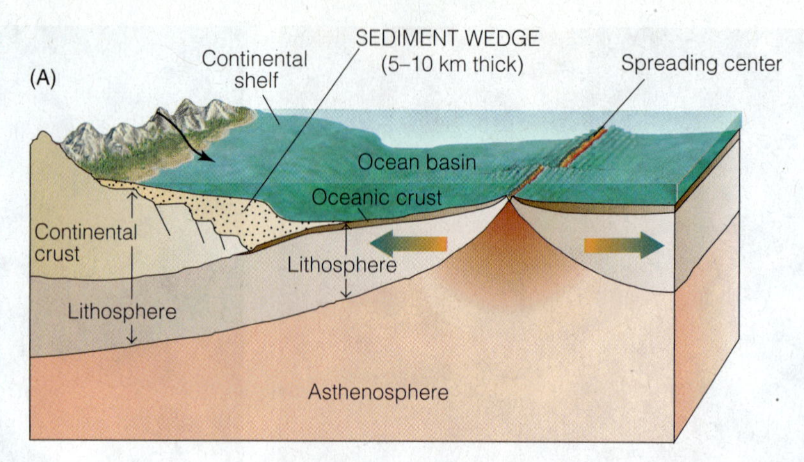

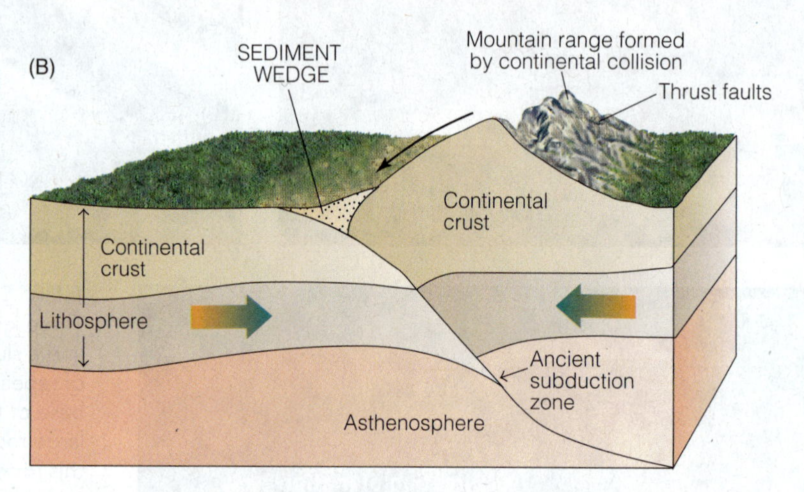

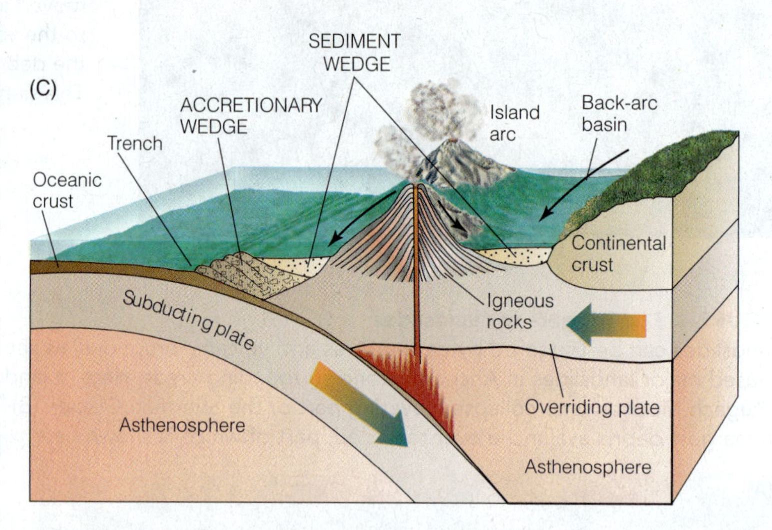

FIGURE 7.13 Tectonics and sedimentation
Clastic sediment accumulates in low-lying areas, which occur in specific locations that are strongly controlled by plate tectonics. (A) Thick sedimentary piles accumulate in rift valleys and along passive continental margins that are formed when continental crust rifts and a new ocean basin opens up. (B) Sediment accumulates in structural basins along the edge of mountain ranges thrust up by continental collisions. (C) Sediment is shed from continents into deep-sea trenches in subduction zones. The sediment forms a wedge that is compressed and crushed as the oceanic plate subducts under the continental plate. The *back-arc basin* that forms behind the subduction zone is another location where sediment accumulates.

The continued accumulation of sediment in basins, trenches, and other low-lying areas leads to the eventual deep burial of the sediment that was first deposited there. This sets the stage for conditions that can lead to the next step in the rock cycle, through which sediment can be turned back into rock.

FROM REGOLITH TO ROCK

In addition to the sediment-forming processes of weathering, erosion, and deposition, the rock cycle includes a set of processes whereby loose sediment can be turned back into solid rock. The process by which sediment or regolith becomes rock is called **lithification** (from the Greek *lithos*, meaning stone; hence stone-making), and the lithification of sediment yields **sedimentary rock.**

When you look at an outcrop of sedimentary rock, such as the one shown in FIGURE 7.14, one of the first things you will notice is bedding. **Bedding** is a banded appearance that is due to sediment being deposited in distinct layers. The presence of bedding is a clear indication that a rock is sedimentary; that is, it formed from an accumulation of sediment (although not *all* sedimentary rock shows bedding). Let's look more closely at what happens to sediment after deposition and how lithification happens. Then we will consider what we can learn through the detailed study of sedimentary rocks and sedimentary layering.

Diagenesis and Lithification

In order for loose sediment to become sedimentary rock, the individual particles must somehow be bound together into a cohesive unit. After a layer of sediment accumulates it may be buried, either by the accumulation of more sediment or by tectonic processes, or both. An example is the accumulation of great thicknesses of sediment on the continental shelf of a passive continental margin such as the Atlantic coast of North America.

FIGURE 7.14 Sedimentary bedding
Each of these colorful layers in Capital Reef National Park, Utah, is a separate sedimentary bed or stratum.

As sediment accumulates, the continental shelf subsides and the pile grows thicker. As a result of burial, the sediment is subjected to higher pressure. This results in **compaction**, which is generally the first step in lithification (FIG. 7.15).

For lithification to occur, the loose grains of sediment must adhere tightly to one another. This can occur in several ways. For example, lithification can occur through **cementation,** in which the loose particles of sediment are bonded together by a cementing material. This can happen if groundwater evaporates, leaving behind chemicals such as silica, calcium carbonate, or iron hydroxides that precipitate and cement the grains of sediment together. Another way that lithification can occur is through **recrystallization**, in which mineral grains that were once separate can grow to become interlocked. This is especially common in limestones formed from coral reefs. Calcium carbonate grains in the fossilized reef change from their original form (the mineral *aragonite*) to a more stable form, the mineral *calcite*. In the process, the grains interlock and grow together. The same process occurs when ice crystals in a snow pile recrystallize to form a compact mass of ice.

Most of the changes that initiate the process of lithification occur at pressures that are higher than those at Earth's surface, but they are still relatively low in geologic terms. The various low-temperature and low-pressure changes that happen to sediment after deposition are collectively referred to as **diagenesis.** They include lithification, as well as processes involving chemical reactions or microbial activity that may happen during lithification but are not technically part of the process.

Sedimentary Rock

The end result of lithification is sedimentary rock. By examining sedimentary rock, we can learn about the source region where the sediment originated, the rock from which the constituent mineral particles were derived, and even the weathering processes by which the particles were formed. Sedimentary rock can also reveal information about the sediment transport and deposition process, as well as the details of diagenesis and lithification. Scientists use clues from sedimentary rock to build an understanding of the environments in which these processes occurred, long ago in Earth history.

If sedimentary rock is made up of particles derived from the weathering and erosion of igneous or metamorphic rock, it may contain many of the same minerals as the source rock. How, then, can we tell that it is sedimentary, rather than igneous or metamorphic? In addition to the presence of bedding, *fossils*—the preserved remains of living organisms—are an important indicator (FIG. 7.16A). Fossils cannot survive the high temperatures and (in some cases) pressures associated with the formation of igneous and metamorphic rock, so their presence is diagnostic of sedimentary rock. The texture of the rock, the presence of cement, and the nature and shape of the constituent mineral grains provide additional clues to its sedimentary

FIGURE 7.15 Lithification

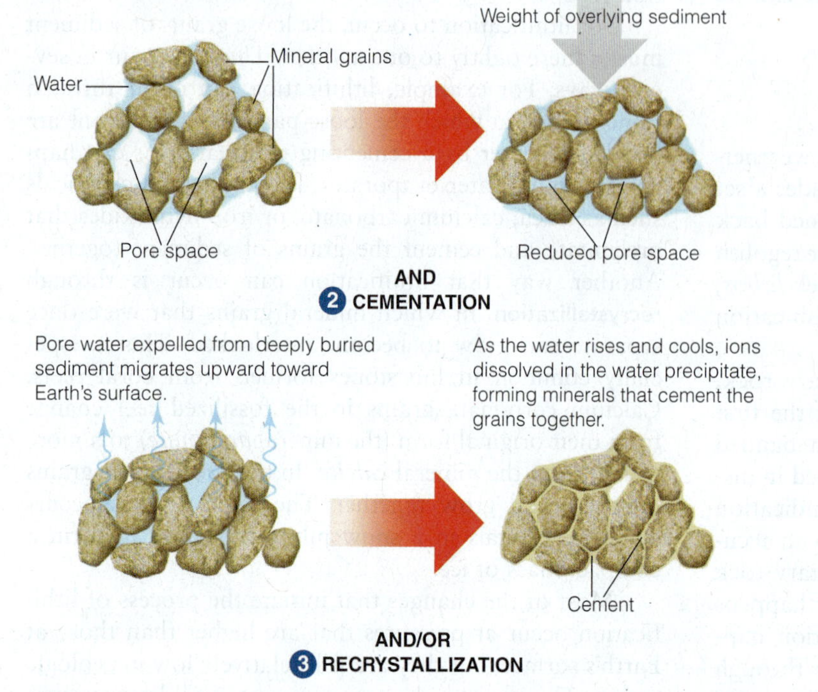
1 COMPACTION
The weight of accumulating sediment forces the grains together, thereby reducing the *pore space* and forcing water out of the sediment.
Weight of overlying sediment
Mineral grains
Water
Pore space
Reduced pore space
AND
2 CEMENTATION
Pore water expelled from deeply buried sediment migrates upward toward Earth's surface.
As the water rises and cools, ions dissolved in the water precipitate, forming minerals that cement the grains together.
Cement
AND/OR
3 RECRYSTALLIZATION
Pressure causes less stable minerals to rearrange crystals into more stable forms. Aragonite is present in the skeletal structures of living corals and other marine invertebrates.
Over time, aragonite recrystallizes and becomes calcite, which has a different crystal structure.
Shells made of aragonite
Aragonite in the shells has been transformed into calcite.

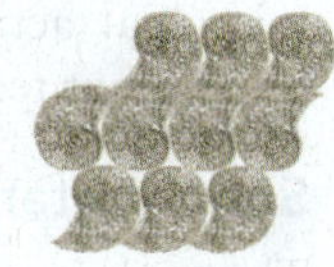

origin (FIG. 7.16B). The classification and naming of sedimentary rock, whether clastic, chemical, or biogenic, is based on these properties.

Clastic Sedimentary Rock

When clastic sediment is lithified, the result is clastic sedimentary rock whose properties reflect the sediment it came from. The four classes are conglomerate, sandstone, siltstone, and shale, derived from gravel, sand, silt, and clay, respectively, as shown in Figure 7.8.

To be classified as *conglomerate*, sedimentary rock must have clasts that are rounded and larger than 2 mm in diameter. Typically, large clasts are surrounded by much finer-grained material, called a *matrix*. If the clasts are angular, with sharp rather than smoothly rounded edges and corners, the rock is called a *breccia*. The presence of angular clasts means that the sediment was only transported a short distance and was not subjected to a long process of rounding by abrasion during transport.

Grains in *sandstone* range from 0.05 to 2 mm in size. They are usually dominated by quartz because quartz is a tough mineral that resists chemical weathering. If the sediment has not been transported very far from its source, it may still contain other minerals or rock fragments.

Siltstone consists primarily of silt- and clay-sized particles; these include tiny pieces of rock and mineral grains, as well as clay minerals. *Shale* contains an even higher proportion of clay-sized particles. Clay minerals

(A)

(B)

FIGURE 7.16 Sedimentary clues
(A) The fossils in this rock (a limestone) give away its sedimentary origin. Fossils cannot survive the high temperatures at which igneous and metamorphic rock are formed. (B) In this highly magnified view, we see some mineral grains rounded by the sedimentary transport process. The gaps between them have been filled by cement, another indication of its sedimentary origin.

are flat or *platy* in structure. When they settle quietly out of a lake or stream, they tend to lie flat on the bottom. This typically gives the resulting shale the property of being *fissile*, which means that it splits into sheet-like fragments. Shale that is not fissile is called *mudstone*.

Chemical and Biogenic Sedimentary Rock

Chemical sedimentary rock results from the lithification of chemical sediment. Such sediment is formed by the chemical precipitation of minerals from water, and the mode of precipitation forms one of the bases for classification of sedimentary rock. For example, sedimentary rock that forms as a result of evaporation is called *evaporite*. The most important evaporite minerals from seawater are halite (NaCl) and gypsum ($CaSO_4 \cdot 2H_2O$), and the corresponding rocks are salt and gyprock, respectively. Evaporation of lake water can yield more exotic minerals, such as sodium carbonate and borax. Many evaporite minerals are mined for industrial purposes.

An unusual but important kind of chemical sedimentary rock is called *banded iron formation*. Such rocks are the source of most of the iron mined today. Not only are they valuable for their ore, but they tell the story of a critical period in Earth's history (see *A Closer Look: Banded Iron Formation*).

Limestone is the most important rock that forms by lithification of biogenic sediment. Limestone is formed from lithified shells and other skeletal material from marine organisms (as seen in Fig. 7.16A). Like many other chemical and biogenic sedimentary rocks, limestone is typically almost *monomineralic*—that is, it is dominated by just one mineral. Some aquatic organisms build their shells or skeletons from the mineral *calcite*, but most construct them from *aragonite*, which, like calcite, is a form of calcium carbonate. Newly formed carbonate sediment is typically dominated, therefore, by either calcite or aragonite. However, during diagenesis, aragonite is transformed into the more stable mineral, calcite, which

A Closer LOOK

BANDED IRON FORMATIONS

Banded iron formations, such as this 2.5-billion-year-old example in the Hamersley Range in Australia (FIG. C7.1), formed when iron that was dissolved in seawater precipitated as chemical sediment. Almost all banded iron formations are between 2.5 and 1.8 billion years old; this strongly suggests that unique conditions existed on Earth during that period. Today, seawater contains only slight traces of iron because the oxygen in the atmosphere reacts with it to form insoluble iron compounds. If the ocean was once rich in dissolved iron, there must have been very little oxygen in the atmosphere at that time.

How did Earth make the transformation from an oxygen-poor atmosphere 2.5 billion years ago to the oxygen-rich atmosphere of today? In some 2.5-billion-year-old rocks there are microscopic fossils of cyanobacteria. These are thought to be the first organisms on Earth to extract energy from sunlight by photosynthesis, a chemical process that releases oxygen. Scientists hypothesize that algae might have oxygenated Earth's atmosphere very rapidly—and possibly more than once, as they proliferated and then poisoned themselves by producing too much oxygen. Each time the oxygen concentration changed, an iron mineral would precipitate out of the seawater, creating a new band of iron-rich sediment. Eventually, about 1.8 billion years ago, the oxygen level of the atmosphere reached a point where the ocean could no longer retain much iron, and banded iron formations could no longer form.

Banded iron formations are economically important, because they provide almost all of the iron that is currently mined. They are also scientifically significant; not only do they provide a glimpse into a unique period of Earth's history, but they also tell a story of dynamic interconnections between the ocean, atmosphere, biosphere, and geosphere.

FIGURE C7.1 Record of a changing atmosphere
Banded iron formation, like this one from the Hamersley Range in Australia, is chemical sedimentary rock that preserves a record of the change of Earth's atmosphere from oxygen-poor to oxygen-rich, prior to 1.8 billion years ago.

Black layers are rich in reduced iron (Fe^{2+}).
Red layers are rich in oxidized iron (Fe^{3+}).
White layers are rich in silica.
Close-up of layers in banded iron formation

then becomes the main ingredient of limestone. During diagenesis, calcite may sometimes be replaced by the mineral *dolomite* (also a carbonate mineral, which contains both calcium and magnesium); the resulting rock is called *dolostone*.

Another important monomineralic biogenic rock, which consists almost entirely of extremely tiny particles of quartz, is called *chert*. The quartz in chert does not come from beach sand, but from the siliceous skeletons of microscopic sea animals called radiolaria. As mentioned earlier, peat is another important biogenic sediment; when lithified, it becomes the biogenic sedimentary rock *coal*, which we will examine in Chapter 18.

Sedimentary Strata

Deposition of sediment forms distinct layers. Sedimentary bedding, or *stratification*, results from this layered arrangement of sedimentary deposits. Each sedimentary **stratum** (plural = **strata**) is a distinct layer (or *bed*) of sediment that accumulated at Earth's surface. Each bed in a succession of sedimentary strata can be distinguished from adjacent beds by differences in thickness or some character such as the size or shape of the clasts, or the color of the rock.

Scientists can tell a lot about ancient environments through the careful examination of sedimentary strata and the surfaces that separate adjacent beds. If you examine Figure 7.14, for example, you will see differences in the thicknesses and colors of the many layers. The differences arise from changes in the environment as the sediment accumulated. Because sedimentary rock carries important clues about past environments at Earth's surface, the sequence and age of strata provide the basis for reconstructing much of Earth's environmental history.

The Principles of Stratigraphy

The study of sedimentary strata is called **stratigraphy.** Three straightforward and simple, but nevertheless very powerful, principles underlie stratigraphy.

1. The *principle of original horizontality* states that sediment is deposited in a layer that is horizontal, or nearly so, and parallel to Earth's surface (FIG. 7.17). This is so because gravity is the principal force that influences the deposition of sediment. From this generalization we can infer that strata that are now steeply inclined, buckled, broken, or bent have been disturbed by tectonic forces since the time they were deposited and lithified.
2. The *principle of stratigraphic superposition* states that in any sequence of sedimentary strata the order in which the strata were deposited is from the bottom to the top (as shown in Fig. 4.24). This means that the youngest stratum in a sedimentary sequence will be on the top, and the oldest will be on the bottom. This principle is particularly useful for determining the relative ages of sedimentary rock units.

FIGURE 7.17 Strata and the principle of original horizontality
All of these strata were originally deposited as horizontal layers; the strata in (B) and (C) were disrupted by later tectonic activity. (A) These horizontal strata are from Badlands State Part, South Dakota. (B) These tilted strata are part of the Telfer Gold Mine, Great Sandy Desert, Western Australia. The tiny figure of a person walking along the mining road, to the left of center of the photo, gives an idea of the scale. (C) These folded strata are from Hamersley Gorge, Western Australia.

3. The *principle of lateral continuity* is based on the observation that sediment is deposited in more or less continuous layers. A layer of sediment will extend horizontally as far as it was carried by the water, air, or ice that deposited it. However, strata are commonly

not simple, perfectly flat layers of uniform thickness. Individual beds may thin or thicken, reflecting the accumulation of greater or lesser thicknesses of sediment on an uneven ground surface. Layers of sediment do not terminate abruptly; rather, they get thinner and ultimately pinch out altogether at their farthest edges. The same is true of sedimentary rock layers.

Certain processes can lead to irregularities in sedimentary bedding. For example, *cross-bedding* refers to sedimentary beds that are inclined with respect to a thicker stratum within which they occur. Cross beds (shown in both modern sediment and ancient sedimentary rock in Fig. 4.26) are the work of turbulent flow in streams, wind, or ocean waves. As they are moved along, the clasts tend to collect in ridges, mounds, or heaps in the form of ripples, waves, or dunes that migrate slowly forward in the direction of the current. Clasts accumulate on the downstream slope of the pile to produce beds with slopes as steep as 35°. The inclination of cross bedding records the direction in which the transporting current of water or air was flowing at the time of deposition. Cross bedding is an example of the evidence that scientists use to learn about the environment in which sediment was deposited.

Stratigraphic Correlation

Any distinctive stratum or group of strata that differ from the strata above and below is given a name. The strata in Figure 7.14, for example, are called the Navajo Sandstone. Early in the nineteenth century an English land surveyor, William Smith, while surveying for the construction of new canals, realized that distinctive sedimentary strata in southern England lay, as he put it, "like slices of bread and butter" in a definite, unvarying sequence. He became familiar with the characteristics of each layer, especially the fossils each contained, and with the sequence of the layers. By looking at a specimen of sedimentary rock collected from anywhere in southern England, he could name the stratum from which it came and the position of the stratum in the sequence.

William Smith did not believe that his discovery reflected any particular scientific principle; he considered it purely practical. Nevertheless, it opened the door to the correlation of sedimentary strata over increasingly wide areas. Stratigraphic **correlation** is the determination of equivalence in age of the succession of strata found in two or more different areas. Smith initially correlated strata over distances of several kilometers and later over tens of kilometers. By comparing the fossils and other characteristics of the sedimentary strata, it ultimately became possible for scientists to correlate across hundreds and then thousands of kilometers. This is possible, of course, because of the lateral continuity of strata. Correlation, along with the principle of superposition, makes stratigraphy a powerful tool for determining the relative ages of rock units (FIG. 7.18). This has allowed scientists to build and refine the geologic column and timescale.

Breaks in the Stratigraphic Record

A sequence of strata deposited layer after layer without any interruption is said to be *conformable*. Commonly, however, there are substantial breaks or gaps in a pile of strata. These represent times of nondeposition, to which the term **unconformity** is applied. Some unconformities record changes in environmental conditions that caused deposition to cease for a considerable time. Other unconformities record periods during which erosion resulted in the removal of part of the earlier depositional record.

There are three important kinds of unconformities. The first, labeled (1) in FIGURE 7.19, on page 202 is a *nonconformity*, where strata overlie igneous or metamorphic rock. This indicates that the original layer of sediment was deposited on a base of igneous or metamorphic rock. The second type is an *angular unconformity*, labeled (2) in Figure 7.19. An angular unconformity implies that the older strata were deformed, uplifted, and partially removed by erosion, after which younger sediment was deposited on top of the erosional surface. The third kind of unconformity is called a *disconformity*; it is an irregular surface of erosion between parallel strata. A disconformity, labeled (3) on Figure 7.19, implies that sedimentation ceased for a period of time, possibly accompanied by erosion, but with no deformation or tilting of the underlying strata.

FIGURE 7.18 Correlating sedimentary strata
This drawing demonstrates stratigraphic correlation and other techniques for determining the ages of rock units. The assumption is made that the rock units were laid down one on top of the other horizontally, so the oldest is on the bottom, based on the principles of original horizontality and stratigraphic superposition. Sequence B has the same rock units as sequence A, although their thickness differs. By matching patterns and characteristics, scientists can correlate one unit to another, even across great distances, because of the principle of lateral continuity. Unit 3 is older than unit 1, and unit 2 lies between them. If the absolute ages of units 1 and 3 can be determined using numerical techniques, scientists can bracket the relative age of unit 2—its age must be around 1.1 million years old.

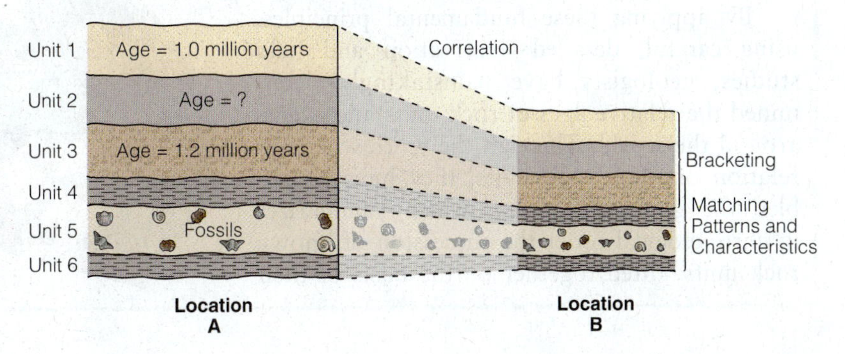

the BASICS

Using Strata to Measure Geologic Time

In Chapter 4 you learned that there are two basic ways that scientists specify the age of a geologic material or event: numerical age and relative age.

Numerical age is a determination of the age, in numbers of years, when the material was formed or the event took place. The discovery of radioactivity in 1896 provided a reliable way to measure numerical age. Radioactivity is a process that runs continuously, that is not reversible, that operates the same way and at the same speed everywhere, and that leaves a continuous record without any gaps. Although other tools can provide measurements of numerical age, *radiometric dating* (Chapter 4) is today the most common and most accurate tool for determining the numerical ages of geologic and archaeological materials.

The second basic way to specify the age of a material or an event is to establish its *relative age*, that is, its age in comparison to another material or event (refer to Fig. 4.24). The determination of relative age in a stratigraphic sequence is based on the fundamental principle that sediment is deposited in layers, one on top of another—that is, the principle of superposition. This principle allows us to conclude that the youngest rock will be found at the top of an undisturbed stratigraphic sequence and the oldest rock at the bottom.

To this we must add one more fundamental principle that allows us to determine relative ages. The *principle of cross-cutting relationships* says that a stratum (or any geologic feature) must be older than any feature that cuts or disrupts it. If a stratum is cut by a fracture, for example, the stratum itself must be older than the fracture that cuts it. If magma intrudes and fills a fracture, the result is a vein of igneous rock that cuts across a stratum (FIG. B7.1); in this case, too, the stratum must be older than the cross-cutting vein. Similarly, a foreign rock (called a *xenolith* or an *inclusion*) that is encased within another rock unit must predate the rock that encloses it.

By applying these fundamental principles, using careful, detailed observation and field studies, geologists have painstakingly determined the relative ages of rock units and events around the world. Through the worldwide correlation of these rock units, they have assembled the *geologic column*, which summarizes in chronological order the succession of known rock units, fitted together on the basis of their relative ages (FIG. B7.2). One of the great successes of nineteenth-century geologists was the demonstration that the relative ages of specific sedimentary rock sequences are equivalent on all continents.

As discussed in Chapter 4, however, the scientists who worked out the geologic column were challenged by the question of numerical time. They could figure out the relative time order in which the strata of the geologic column had formed, but they also wished to know how long it had taken for the sediment to accumulate. Answering such questions as the age of Earth, the age of the ocean, how fast mountain ranges rise, how long it takes for a sedimentary sequence to be deposited, or how long humans have inhabited Earth requires a determination of numerical age. Through radiometric dating, geologists have determined the dates of solidification of many bodies of igneous rock (which are generally more amenable than sedimentary rock to radiometric dating). Many igneous bodies have clearly identifiable positions in the geologic column; as a result, by comparison, it becomes possible to assign approximate dates to sedimentary layers in the column. Thus, through a combination of geologic fieldwork and radiometric dating, scientists have been able to fit a scale of numerical time to the geologic column of relative ages that was first established in the nineteenth century. This **geologic timescale** is being continuously refined, so the numbers given in Figure B7.2 should be considered the

FIGURE B7.1 Determining relative age
The principle of cross-cutting relationships allows us to determine that the intrusive igneous body shown here to be cutting across units 1, 2, and 3 must therefore be younger than those units. By combining the basic principles of stratigraphy with radiometric dating, scientists have built up a detailed picture of relative and numerical ages of rock units and geologic events worldwide.

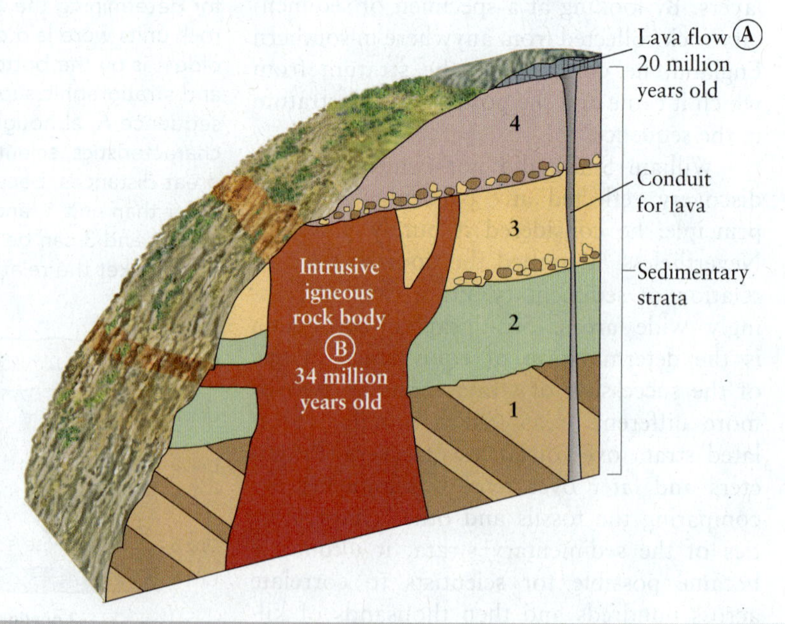

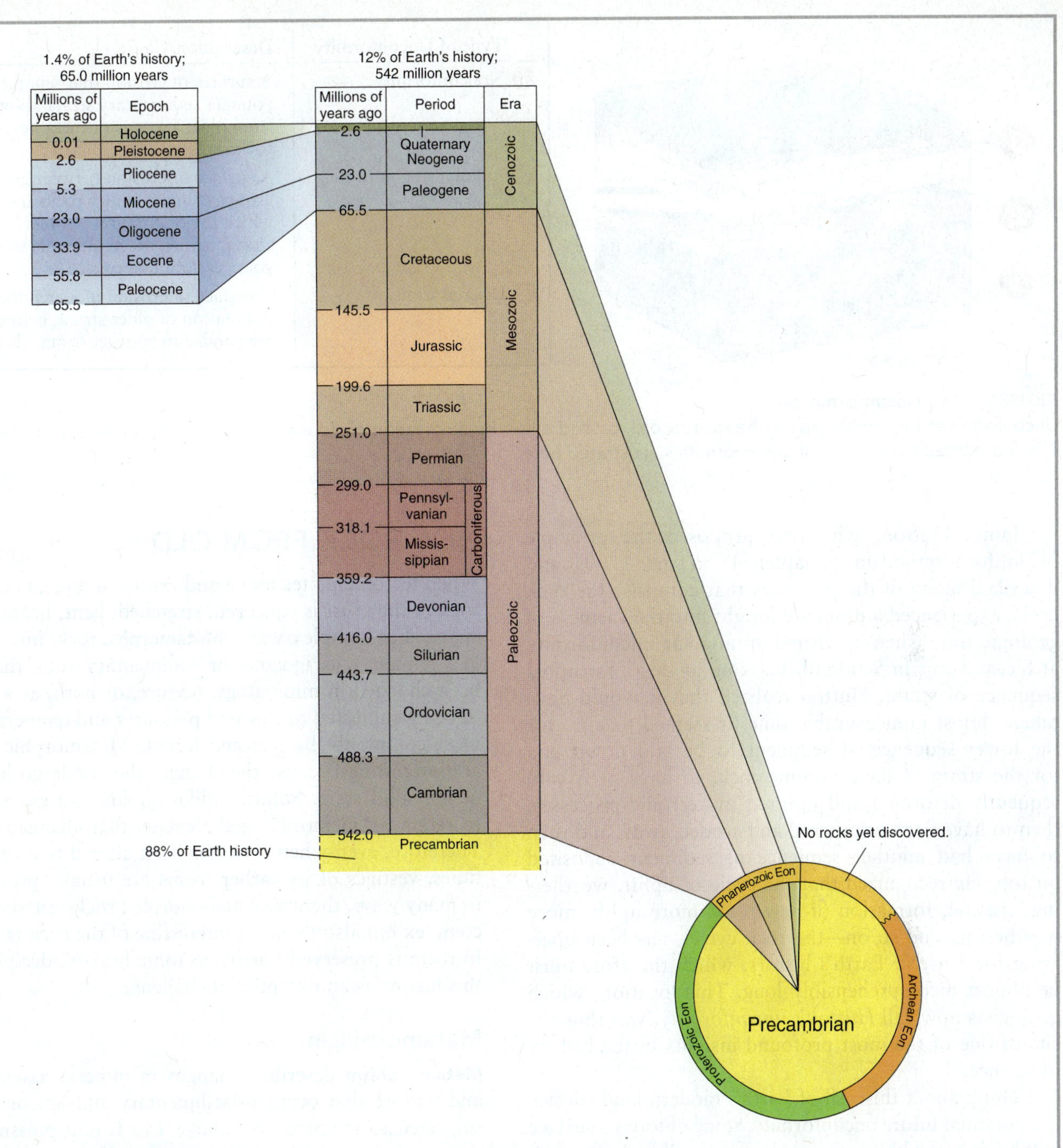

FIGURE B7.2 Geologic column
The geologic column has been constructed painstakingly by many scientists working for many years to correlate the relative ages of rock units in one part of the world to others. Radiometric dating has allowed a geologic timescale to be established and linked to the relative ages of the geologic column. These ages are always being refined by new and more accurate determinations of numerical ages.

best available now. Further work will continue to improve their accuracy.

Standard names have been assigned to the subdivisions of the geologic column and the geologic timescale. The subdivisions are *eons*, *eras*, *periods*, and *epochs*, as shown in Figure B7.2. The time units have mostly been assigned names on the basis of the rock types that typify them or the location where rock units of that age were first described. Boundaries between units generally signify major geologic events. For example, the end of the Cretaceous Period is marked by one of the greatest mass extinctions of species in Earth history—the extinction of the dinosaurs. The name *Cretaceous* is derived from the Latin word *creta*, meaning "chalk," because the period is characterized by extensive deposits of the biogenic sedimentary rock *chalk*.

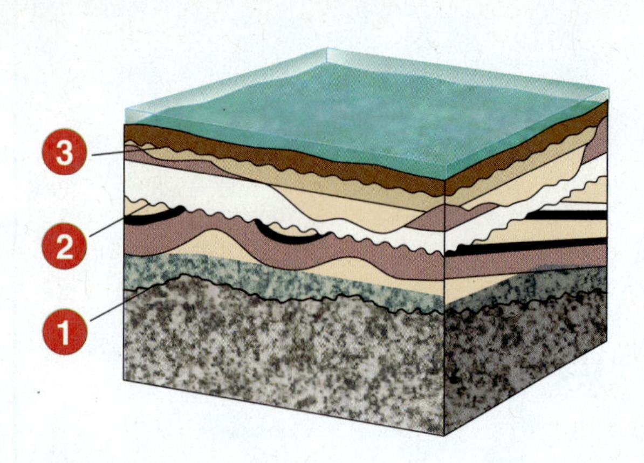

Type of Unconformity	Description/Cause
1 Nonconformity	A surface of erosion that separates younger sedimentary strata above from older igneous or metamorphic rocks below.
2 Angular unconformity	A surface of erosion between two groups of sedimentary rocks in which the orientation of older strata, below, are at an angle to younger strata, above.
3 Disconformity	A surface of erosion in which the orientation of older strata, below, are parallel to younger strata, above.

FIGURE 7.19 Unconformities
Unconformities represent gaps in the rock record, caused either by a cessation of deposition or a period of erosion, or both. There are three basic types of unconformities, illustrated here.

James Hutton, who first proposed the principle of uniformitarianism (Chapter 4) in the 1700s and described many of the processes that comprise the rock cycle, experienced a dramatic insight into the vastness of geologic time when he visited an angular unconformity at Siccar Point in Scotland. Looking at this interrupted sequence of strata, Hutton realized that it would have taken almost immeasurable time (it seemed to him) for the lower sequence of sediment to be laid down and for the strata to have become rock, to have been subsequently deformed and uplifted by tectonic processes, then to have been weathered and eroded away, and then to have had another sequence of sediment deposited on top. He recognized that the cycle of uplift, weathering, erosion, formation of new rock, more uplift, more weathering, and so on—the rock cycle—has been operating for most of Earth's history, which therefore must be almost incomprehensibly long. This location, which geologists now call *Hutton's unconformity*, was thus the site of one of the most profound insights in the history of science.

Think about this: All of Earth's modern land surface is a potential future unconformity. Some of today's surface will be destroyed by erosion, but some will be covered by sediment and preserved as a record of the present landscape. For example, the Swiss Alps, which were elevated by plate tectonic movements, are now being eroded away. The eroded material is continuously carried away by streams and deposited in the Mediterranean Sea. The Mediterranean seafloor was once dry land, but tectonic forces depressed it, just as tectonic forces elevated the Alps. A surface of unconformity separates the young, river-transported sediment and the older rock of the seafloor on which the sediment is piling up. In a sense, accumulation in one place compensates for destruction in another. As Hutton recognized, unconformities provide powerful evidence that interactions between the various spheres of the Earth system have been going on throughout Earth's long history.

NEW ROCK FROM OLD

When tectonic plates move and crustal fragments collide, rock in the crust is squeezed, stretched, bent, heated, and changed in complex ways. **Metamorphic rock**, introduced in Chapter 3, is igneous or sedimentary rock that has been changed in mineralogy, texture, or both, as a result of being subjected to elevated pressures and temperatures, most commonly by tectonic forces. Metamorphic rocks are interesting because the changes they undergo happen in the solid state. Solids, unlike liquids and gases, tend to retain a "memory" of the events that changed them. Therefore, even when rock has been altered two or more times, vestiges of its earlier forms are usually preserved. In many ways, therefore, metamorphic rocks are the most complex but also the most interesting of the rock families. In them is preserved Earth's tectonic history; deciphering this history is an exceptional challenge.

Metamorphism

Metamorphism describes changes in mineral assemblage and texture that occur in sedimentary and igneous rock subjected to temperatures above 150°C and pressures in excess of about 300 MPa (the pressure caused by a few thousand meters of overlying rock). There is, of course, an upper limit to metamorphism because at sufficiently high temperatures rock will begin to melt (starting at a temperature of 800°C or even lower in some circumstances, as discussed in Chapter 6). Metamorphism refers only to changes that happen in solid rock, not to changes associated with rock melting.

Low-grade metamorphism refers to metamorphic processes that occur at temperatures from about 150°C to about 550°C and at relatively low pressures (**FIG. 7.20**). (Note that diagenesis, the set of processes that happens when a pile of sediment undergoes burial and compaction, occurs below 150°C.) *High-grade metamorphism* refers to metamorphic processes at higher temperatures (above about 550°C) and, generally, high pressures.

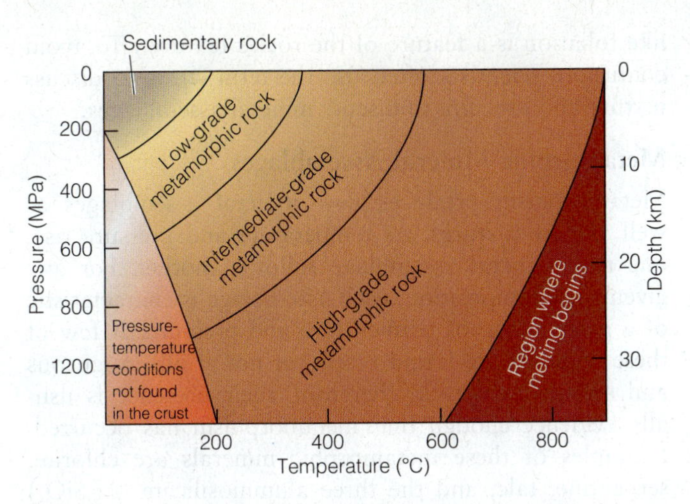

FIGURE 7.20 Temperature and pressure conditions for metamorphism
These are the ranges of temperature and pressure under which metamorphism occurs in the crust. At the lowest temperatures and pressures, sediment is converted to sedimentary rock by diagenesis. At the highest temperatures and pressures, melting commences and the result is magma. Between these conditions, metamorphic changes happen to rock in the solid state.

In a simplistic way, you can think of metamorphism as analogous to cooking. When you cook, what you get to eat depends on what you start with and on the cooking conditions. So too with rock; the end product is controlled by the initial composition of the rock and by the metamorphic (or cooking) conditions. The chemical composition of rock undergoing metamorphism plays a controlling role in the new mineral assemblage; so do the conditions of temperature and pressure. When a mixture of flour, salt, sugar, yeast, and water is baked, the high temperature causes a series of chemical reactions—new compounds are formed, and the result is a loaf of bread. When igneous or sedimentary rock is heated and squeezed, new minerals grow and the result is metamorphic rock.

In the case of rock, the "cooking" is brought about by Earth's internal heat, usually aided by the pressure associated with depth, and by tectonic forces. Other factors also play an important role in metamorphism. These include the presence or absence of fluids, how long the rock is subjected to high pressure or high temperature, and whether the rock is simply compressed or is also twisted and broken during metamorphism.

The Role of Fluids

The innumerable open spaces between grains in sedimentary rock and the tiny cracks and openings along grain boundaries in igneous rock are called *pores*. All rock pores are filled by a watery fluid. The fluid is never pure water, for it always has dissolved in it small amounts of gases such as CO_2 and salts such as NaCl and $CaCl_2$, as well as traces of all the mineral constituents present in the enclosing rock. At high temperature the fluid is more likely to be a vapor than a liquid. Regardless of its composition or state, the *pore fluid* (for that is its best designation) plays a vital role in metamorphism.

When the temperature and pressure of rock undergoing metamorphism change, so does the composition of the pore fluid. Some of the dissolved constituents move from the pore fluid into the new minerals growing in the metamorphic rock. Other constituents move in the other direction, from the minerals to the pore fluid. In this way the pore fluid serves as a transporting medium that speeds up chemical reactions in much the same way that water in a stew pot speeds up the cooking of a tough piece of meat.

When very abundant pore fluids are involved in metamorphism, the metamorphic process is so fundamentally altered that it has a different name altogether: **metasomatism**. This term comes from the Greek words *meta*, meaning "change," and *soma*, meaning "body." Metasomatism is often accompanied by a thorough redistribution of the chemical elements of the rock, resulting in a dramatic change in the overall composition and appearance of the rock.

Temperature, Pressure, and Stress

Rock can be heated by burial (starting with diagenesis, as discussed above), by exposure to the heat of a nearby igneous intrusion, or by thickening of the crust owing to collision. But burial, collision, and intrusion can also be associated with elevated pressures. Therefore, whatever the cause of the heating, metamorphism can rarely be considered to be entirely due to the rise in temperature. The effects attributable to changing temperature and pressure must be considered together.

When discussing metamorphic rock, scientists use the term **stress** instead of pressure. They do so because stress has the connotation of direction. Rock is solid, and solids can be squeezed more strongly in one direction than another; that is, stress in a solid, unlike stress in a liquid, can be different in different directions. The textures in many metamorphic rocks record *differential stress* (meaning not equal in all directions) during metamorphism. In contrast, igneous rocks have textures formed under *uniform stress* (meaning equal in all directions) because igneous rocks crystallize from liquids.

The most visible effect of metamorphism in a differential stress field is the parallel alignment of certain silicate minerals, such as micas and chlorites, that have flat, platy structures because they contain polymerized $(Si_4O_{10})^{4-}$ sheets. For example, the rock in FIGURE 7.21A is granite, an igneous rock, with typical randomly oriented mineral grains that grew in a uniform stress field. In contrast, FIGURE 7.21B shows a metamorphic rock (called *gneiss*) that contains the same minerals as in A, but was modified by metamorphism in a differential stress field. In metamorphic rock that contains sheet-structure minerals, such as the black mica (biotite) grains in Figure 7.21B, the sheets are oriented perpendicular to the direction of maximum stress. The parallel sheets produce a planar

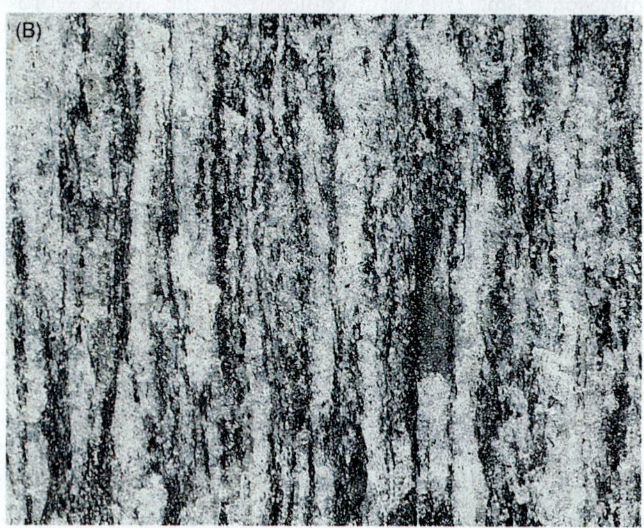

FIGURE 7.21 Effects of uniform and differential stress The two rocks in these photos have similar mineral assemblages but look very different because of their different stress histories. (A) This granite consists of quartz (glassy), feldspar (white), and biotite (dark), which crystallized from magma (a liquid) under conditions of uniform stress. Note that the biotite grains are randomly oriented. (B) This gneiss, a high-grade metamorphic rock, contains the same minerals as the granite, but they developed entirely in the solid state and under differential stress. The biotite grains are aligned, giving the rock a foliated texture.

fabric called **foliation**, named from the Latin word *folium*, meaning leaf.

Stress can be high or low. The magnitude of the stress, in combination with the temperature and the starting composition of the rock, determines the assemblage of minerals that will develop in a given metamorphic rock. Metamorphic rock *texture*, on the other hand, is controlled not by the magnitude of the stress but by the influence of differential versus uniform stress. (*Texture*, as discussed in Chapter 3, refers to the overall appearance of the rock as a result of the size, shape, and arrangement of its constituent mineral grains. Thus, a metamorphic fabric like foliation is a feature of the rock's texture.) To avoid confusion, scientists often use the term *stress* to discuss texture and *pressure* to discuss mineral assemblages.

Metamorphic Mineral Assemblages

Metamorphism produces new mineral assemblages as well as new textures. As temperature and pressure rise, one new mineral assemblage follows another. For any given rock composition, each assemblage is characteristic of a given range of temperature and pressure. A few of these minerals are found rarely (or not at all) in igneous and sedimentary rock; therefore, their presence is usually evidence enough that metamorphism has occurred. Examples of these metamorphic minerals are chlorite, serpentine, talc, and the three alumniosilicate (Al_2SiO_5) minerals andalusite, kyanite, and sillimanite. FIGURE 7.22 shows how mineral assemblages change with the grade of metamorphism as shale is metamorphosed.

Metamorphic Processes

The processes that result from changing temperature and pressure, and that cause the metamorphic changes, can be grouped under the terms *mechanical deformation* and *chemical recrystallization*. Mechanical deformation includes grinding, crushing, and the development of foliation. Chemical recrystallization includes all the changes in mineral composition, the growth of new minerals, and the losses of H_2O and CO_2 that occur as the rock is heated and squeezed. Different kinds of metamorphism reflect the different levels of importance of the two processes, and they can be dramatically influenced by the presence of abundant pore fluids. The three most important kinds of metamorphism are contact metamorphism, burial metamorphism, and regional metamorphism.

Contact Metamorphism

Where hot magma intrudes cooler rock, high temperatures cause chemical reactions and recrystallization in the surrounding rock. Cooling plutons may also release pore fluids (mainly heated water) into the surrounding rock, which can accelerate the growth of new minerals. This type of metamorphism, called **contact metamorphism**, is primarily temperature-driven; mechanical deformation and therefore the development of foliation are minimal. An igneous intrusion is commonly surrounded by a zone, or *aureole*, of contact metamorphosed rock (FIG. 7.23, on page 206). If the intrusion is very large, if it releases abundant pore fluid while cooling, or if it is surrounded by highly porous or chemically reactive rock such as limestone, the contact metamorphic aureole may extend for hundreds of meters. Unlike other kinds of metamorphism, contact metamorphism is not associated with any particular tectonic setting; it may occur wherever magmatic intrusions occur (labeled (4) in FIG. 7.24, on page 206).

Burial Metamorphism

Sediment, sometimes with interlayered pyroclastic material, may attain temperatures in excess of 150°C or more

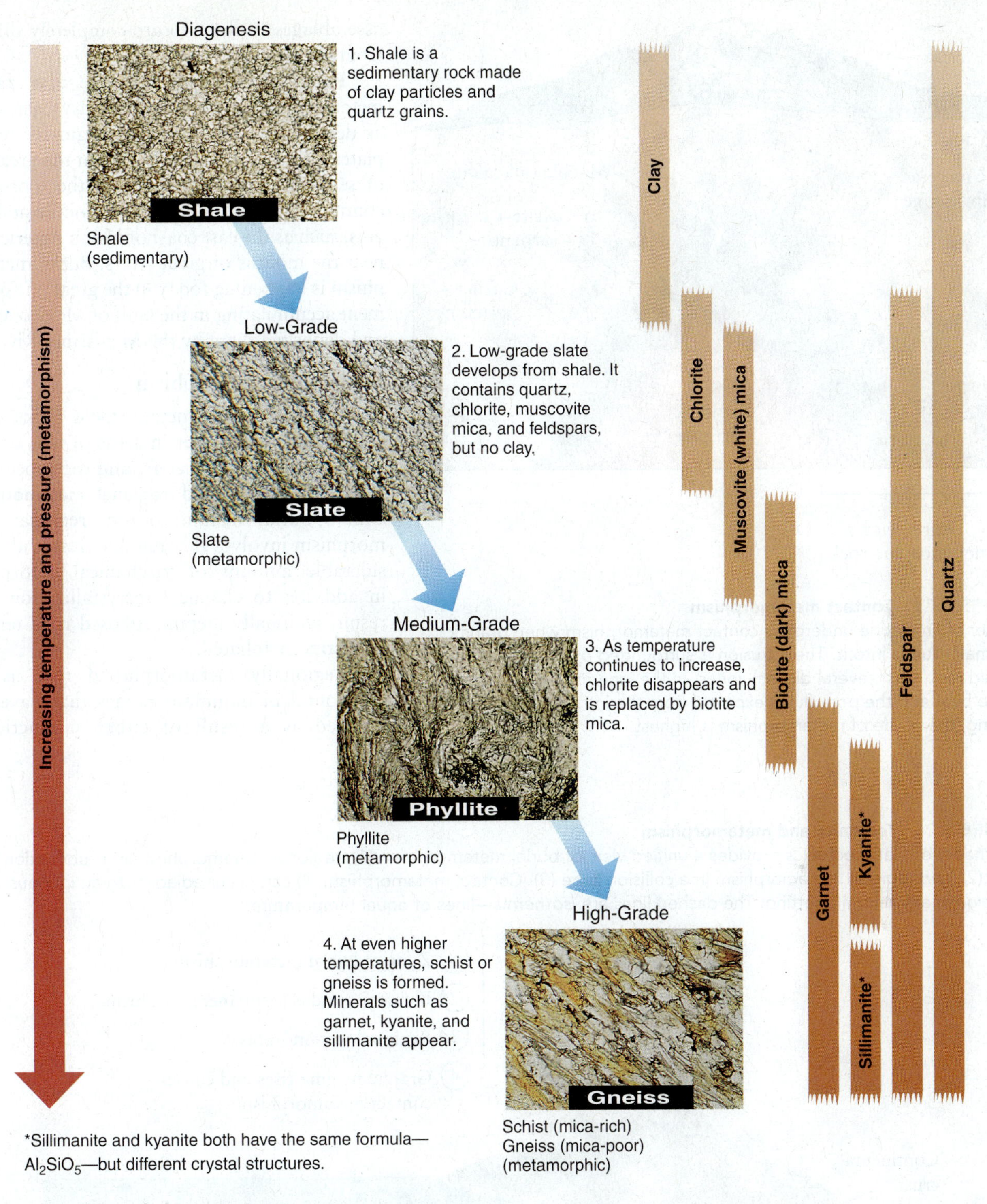

Diagenesis
1. Shale is a sedimentary rock made of clay particles and quartz grains.
Shale
Shale (sedimentary)
Low-Grade
2. Low-grade slate develops from shale. It contains quartz, chlorite, muscovite mica, and feldspars, but no clay.
Slate
Slate (metamorphic)
Medium-Grade
3. As temperature continues to increase, chlorite disappears and is replaced by biotite mica.
Phyllite
Phyllite (metamorphic)
High-Grade
4. At even higher temperatures, schist or gneiss is formed. Minerals such as garnet, kyanite, and sillimanite appear.
Gneiss
Schist (mica-rich)
Gneiss (mica-poor)
(metamorphic)
Increasing temperature and pressure (metamorphism)
Clay
Chlorite
Muscovite (white) mica
Biotite (dark) mica
Garnet
Kyanite*
Sillimanite*
Feldspar
Quartz
*Sillimanite and kyanite both have the same formula—Al_2SiO_5—but different crystal structures.

FIGURE 7.22 From shale to gneiss
As shale is subjected to higher and higher temperatures and pressures, it develops into a sequence of metamorphic rocks that have different mineral assemblages. The photos were all taken under a microscope with a 3-mm field of view, so they are greatly magnified.

when buried deeply in a sedimentary basin. This can initiate **burial metamorphism,** the first stage of metamorphism to occur in sedimentary rock after diagenesis. Abundant pore water is present in buried sediment, and this water speeds up chemical recrystallization and helps new minerals to grow. Because water-filled sediment is weak and acts more like a liquid than a solid, however, the stress during burial metamorphism tends to be uniform. As a result, burial metamorphism involves little mechanical deformation, and the metamorphic rock that results lacks foliation. The texture may look like that of an essentially unaltered sedimentary rock, even though the mineral

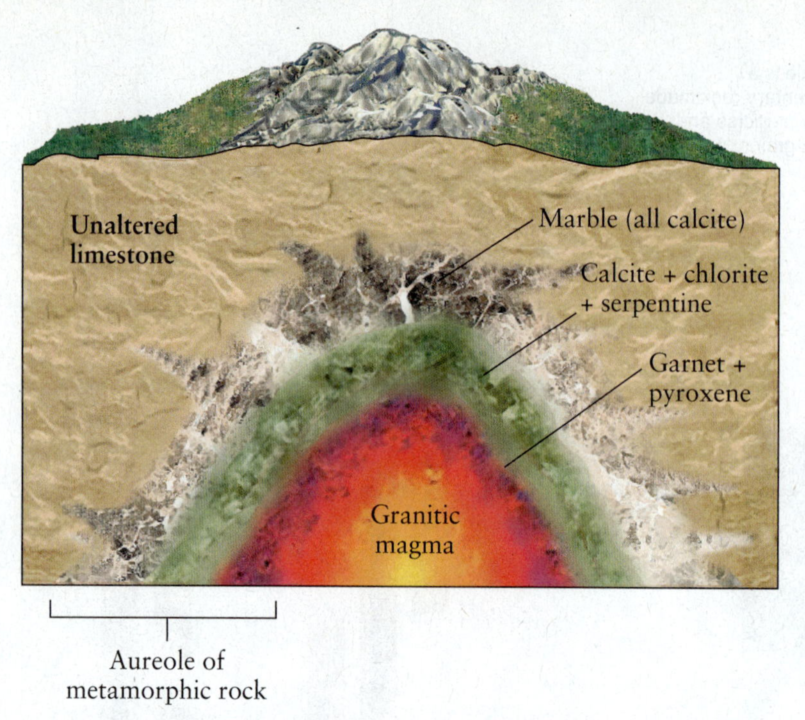

FIGURE 7.23 Contact metamorphism
A layer of limestone undergoes contact metamorphism when granitic magma intrudes into it. The intrusion is surrounded by an aureole of altered rock, with several distinct zones. In the innermost zone, closest to the heat and the pore fluids expelled from the intrusion during cooling, the grade of metamorphism is highest.

assemblages in the two are completely different from one another.

Burial metamorphism is the first stage of metamorphism in deep sedimentary basins, such as deep-sea trenches on the margins of tectonic plates (labeled (1) in Fig. 7.24), in the great piles of sediment that accumulate at the foot of the continental shelf along passive continental margins, such as the east coast of North America, and near the mouths of great rivers. Burial metamorphism is happening today in the great pile of sediment accumulating in the Gulf of Mexico, carried and deposited there by the Mississippi River.

Regional Metamorphism

The most common metamorphic rocks of the continental crust occur in areas of tens of thousands of square kilometers, and the process that forms them is called **regional metamorphism.** Unlike burial metamorphism, regional metamorphism involves differential stress and a considerable amount of mechanical deformation in addition to chemical recrystallization. As a result, regionally metamorphosed rock tends to be distinctly foliated.

Regionally metamorphosed rock is usually found in mountain ranges that have been uplifted as a result of either subduction or

FIGURE 7.24 Tectonics and metamorphism
The theory of plate tectonics provides a unified view of burial metamorphism (1), regional metamorphism in a subduction zone (2), and regional metamorphism in a collision zone (3). Contact metamorphism (4) can occur adjacent to an igneous intrusion in any tectonic setting. The dashed lines are *isotherms*—lines of equal temperature.

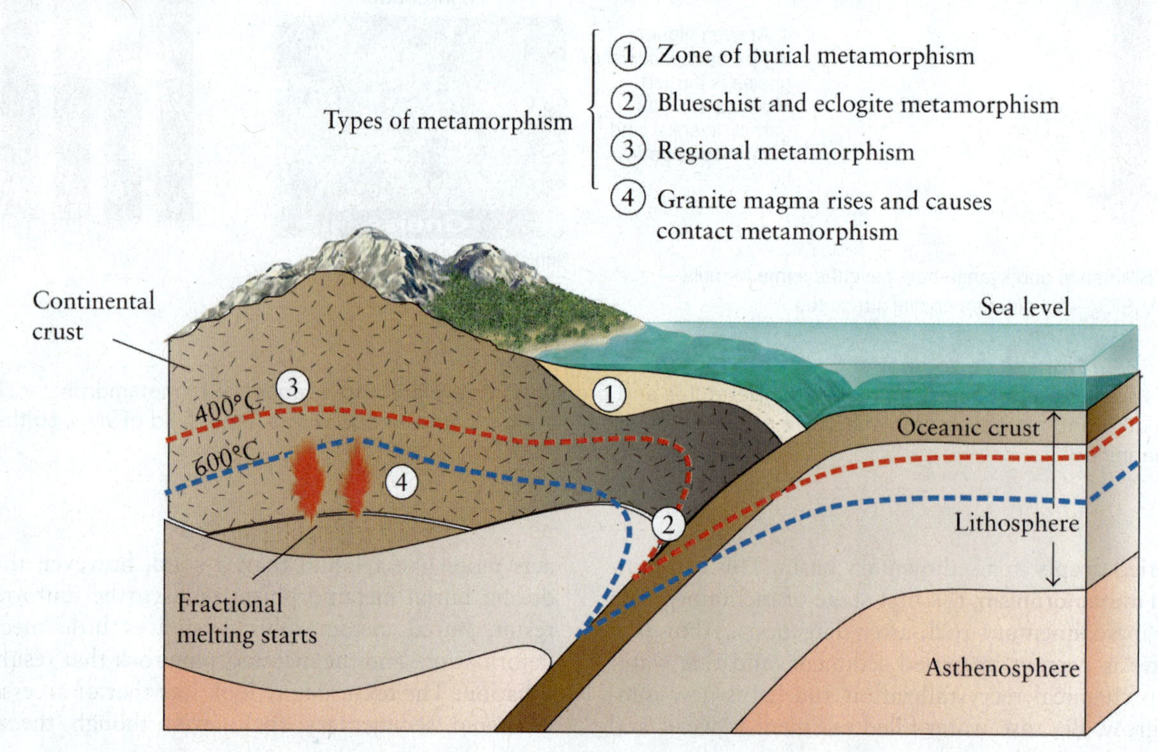

collision between fragments of continental crust (labeled (3) in Fig. 7.24). During both subduction and collision between continents, sedimentary rock along the margin of a continent is subjected to intense differential stresses. The foliation that characterizes these rocks is a consequence of the intense stresses; regional metamorphism is therefore a consequence of plate tectonics. Both ancient and modern orogens, such as the Appalachian Mountains, the Alps, the Himalaya, and all other mountain ranges formed by continental collisions, are composed of regionally metamorphosed rock.

A special kind of regional metamorphism occurs at some subduction margins (labeled (2) in Fig. 7.24). The subducting plate remains rigid and cool to great depths—that is why subduction zone megathrust earthquakes can have focal depths as great as 700 km, as discussed in Chapter 6. This occurs because the solid, cold oceanic crust dives rapidly (by geologic standards) into the hot, weak asthenosphere. The pressure on the subducting rock rises quickly, but the temperature cannot rise at the same rate. This is because of a basic property of rock that has been discovered through laboratory experiments: rock can transmit pressure very quickly, but transmits heat very slowly. Therefore, metamorphism in subduction zones, called *blueschist metamorphism*, typically occurs at very high pressures but low temperatures.

Metamorphic Rock

Metamorphic rock classification and naming are based partly on rock texture, and partly on mineral assemblage. The most widely used names are those applied to the metamorphic derivatives of shales, basalts, limestones, and sandstones. This is because shales, sandstones, and limestones are the most abundant sedimentary rock types, whereas basalt is by far the most abundant igneous rock.

Metamorphism of Shale

The low-grade metamorphic product of shale is called *slate* (as shown in Fig. 7.22). The minerals usually present in shale include quartz, clays, calcite, and feldspar. Slate contains quartz, feldspar, and mica or chlorite. Although slate may still look like shale, the tiny mica and chlorite grains formed under differential stress give slate a distinctive foliation called *slaty cleavage*. The presence of slaty cleavage is clear proof of the change from sedimentary rock (shale) to metamorphic rock (slate).

Continued metamorphism of slate produces both larger grains of mica and a changing mineral assemblage; the rock develops a pronounced foliation (Fig. 7.22) and is called *phyllite* (from the Greek *phyllon*, meaning leaf). In slate it is not possible to see the new grains of mica with the unaided eye, but in phyllite they are just large enough to be visible.

Still further metamorphism beyond that which produces phyllite leads to a coarse-grained rock with pronounced alignment of platy minerals. This texture is called *schistosity*, and the rock itself is called *schist* (Fig. 7.22). The most obvious differences between slate, phyllite, and schist are in their grain sizes. At the high grade of metamorphism characteristic of schist, minerals may start to segregate into separate bands. High-grade rock with coarse grains and pronounced foliation, and with bands of micaceous minerals segregated from bands of minerals such as quartz and feldspar, is called *gneiss* (pronounced nice, from the German *gneisto*, meaning to sparkle) (Fig. 7.22).

Metamorphism of Basalt

The main minerals in basalt are olivine, pyroxene, and feldspar, each of which is anhydrous (that is, lacking in water). If you heat basalt, nothing will happen until it reaches about 1000°C, at which point it will begin to melt. However, when basalt is subjected to metamorphism under conditions where H_2O can enter the rock and form hydrous minerals, distinctive metamorphic mineral assemblages develop. At low grades of metamorphism, mineral assemblages such as chlorite + feldspar + epidote + calcite form. The resulting rock is equivalent in metamorphic grade to slate but has a very different appearance. It has pronounced foliation, as phyllite does, but it also has a very distinctive green color because of its chlorite content: it is termed *greenschist*.

When greenschist is subjected to a slightly higher grade of metamorphism, chlorite is replaced by amphibole; the resulting rock is generally coarse-grained and is called *amphibolite*. Foliation is present in amphibolites but is not pronounced because micas and chlorite are usually absent. At the highest grade of metamorphism, amphibole is replaced by the mineral pyroxene, and an indistinctly foliated rock called *granulite* develops.

Metamorphism of Limestone and Sandstone

The metamorphic derivatives of limestone and sandstone are marble and quartzite, respectively (FIG. 7.25). Like the sedimentary rocks from which they are derived, marble and quartzite are typically monomineralic, or nearly so. *Marble* consists of a coarsely crystalline, interlocking network of calcite grains. During the recrystallization of limestone, the bedding planes, fossils, and other features of sedimentary rocks are largely obliterated. The end result, as shown in FIGURE 7.25A, is an even-grained rock with a distinctive, somewhat sugary texture. Pure marble is snow white and consists entirely of pure grains of calcite. Such marbles are favored for marble gravestones and statues in cemeteries, perhaps because white is considered to be a symbol of purity. Many marbles contain impurities such as organic matter, pyrite, goethite, and small quantities of silicate minerals, which impart various colors to the rock.

Quartzite is the metamorphic rock that is derived from sandstone by the filling in of the spaces between the original grains with silica, and by recrystallization of the entire mass (FIG. 7.25B). In both quartzite and marble, recrystallization greatly modifies the original rock textures. The recrystallizing mineral grains grow outward and crowd against each other, filling all spaces and forming an intricate, three-dimensional jigsaw puzzle that resembles an igneous rock more closely than the sedimentary rock from which

FIGURE 7.25 Marble and quartzite
(A) Marble is metamorphic rock composed mainly of calcite (seen through a microscope in the inset photo). Pure marble is snow white. The pink color of this sample (from Tate, Georgia) comes from impurities in the marble. (B) Quartz-rich sandstone becomes monomineralic quartzite when its pore spaces are filled with silica, and the entire mass recrystallizes as a result of metamorphism, as seen in the inset photo taken through a microscope. The sample shown here is from Minnesota.

they were derived. Sometimes the ghostlike outlines of the original sedimentary grains can still be seen in quartzite, or there may be the vague suggestion of fossil shapes in marble, even though recrystallization may have modified the original textures dramatically. Neither limestone nor quartz sandstone (when pure) contains the necessary ingredients to form platy minerals; as a result, marble and quartzite generally lack foliation.

Metamorphic Facies and Plate Tectonics

Finnish geologist Pennti Eskola pointed out in 1915 that the same metamorphic mineral assemblages are observed again and again. This led him to propose the concept of metamorphic **facies** (from the Latin for *face*, or appearance), which says that for a given range of temperature and pressure, and for a given rock composition, the assemblage of minerals formed during metamorphism is always the same. Based on mineral assemblages, Eskola defined a series of pressure and temperature conditions that he called metamorphic facies.

To help you understand Eskola's idea, another analogy with cooking is appropriate; think of a large roast of beef in which the center is rare and the outside is well done. The differences occur because the temperature was not uniform throughout. The center, where the meat is rare, is a low-temperature facies; the outside, well-done meat is a high-temperature facies. The composition of the beef does not vary from the outside to the center, so the development of the different facies was caused by a variation in temperature, not by a variation in composition.

The same is true for rock. In any rock of given composition, temperature (and pressure) determine the mineral assemblage. The same mineral assemblage will always form in rock of that composition under those conditions. Conversely, under given conditions of temperature and pressure, rocks of different compositions will develop characteristic mineral assemblages. By analogy, if you cooked a roast chicken under exactly the same temperature conditions as you cooked the roast beef, the results would be different.

One of the triumphs of plate tectonics is that, for the first time, it explains the distribution of metamorphic facies and regionally metamorphosed rock. For example, the conditions that lead to *blueschist facies* and (at even higher grades) *eclogite facies* metamorphism are reached when crustal rock is dragged down by a rapidly subducting plate; this setting is labeled (2) in Figure 7.24. As discussed earlier, under such conditions, pressure increases much more rapidly than temperature, and as a result metamorphism occurs under conditions of high pressure and relatively low temperature. The facies concept tells us that rock of a given composition, if subjected to blueschist-facies conditions, will always develop basically the same mineral assemblage. Blueschist metamorphism is happening today along the subducting margin where the Pacific Plate plunges under the coast of Alaska and the Aleutian Islands.

The metamorphic conditions characteristic of *greenschist* and *amphibolite facies* metamorphism occur where the crust is thickened by continental collision or heated by rising magma (labeled (3) in Fig. 7.24). Continental collision is the most common setting for regional metamorphism, and rock formed in this way occurs throughout the Appalachians and the Alps. Such metamorphism is known to be occurring today beneath the Himalaya, where the continental crust is thickened by collision, and beneath the Andes, where it is both thickened by subduction and heated by rising magma. If the crust is sufficiently thick, rock subjected to high-grade metamorphism can reach temperatures at which partial melting commences and metamorphism passes into magma formation.

FROM ROCK TO MAGMA AND BACK AGAIN

What happens if rock is heated to temperatures beyond the highest grades of metamorphism? In Chapter 6 you learned about the processes that can lead to rock melting, which occurs only at very high temperatures. The rock cycle includes processes leading to the development of magma, either through melting of rock deep within the crust or mantle, or melting of surface rock that has been returned to the mantle by subduction. Magma can then become rock again through the processes of cooling and crystallization, either underground or at the surface. Let's first briefly review how rock melting occurs, and then we will consider how crystallization leads to the formation and controls the properties of igneous rock.

Melting and Magma: Review

As you learned in Chapter 6, *magma* is molten rock that may also contain dissolved gases, crystals, and rock fragments. When magma reaches the surface, it is called *lava*. A lot of magma never reaches the surface, but instead remains trapped underground, where it eventually cools and solidifies.

Magma is generated by rock melting inside Earth. Scientists know from laboratory experiments and from observations of lava that rock begins to liquefy when heated, at atmospheric pressure, to a temperature between about 800°C and 1000°C. Because rocks contain many different constituents, melting proceeds over a range of temperatures rather than at a single temperature. The high pressures of Earth's interior have the effect of increasing the melting temperature of rock; in other words, rock that would melt at 1000°C at surface conditions would likely still be solid at 100 km depth, where the pressure is higher. However, if water is present in the magma, it has the opposite effect on rock-melting temperatures, causing them to decrease. Thus, in appropriate conditions, rock melting can be induced or sustained either by the addition of water or by the release of pressure, even if the temperature does not increase (as shown in Fig. 6.16).

Because rock melts over a range of temperatures, it is possible (even common) for melting to be incomplete (as shown in Fig. 6.17). This important process is called *partial* or *fractional melting*. If a rock melts only partially, the melted portion (which contains components that melt at lower temperatures) will have a different composition from the residual unmelted portion of the rock (which contains components that will only begin to melt at higher temperatures). Tectonic forces may cause the molten portion, which is less dense, to become separated from the denser rock residue. When the melted portion eventually solidifies, it will form an igneous rock with a very different composition from the unmelted residual rock. Partial melting thus contributes to the great diversity of igneous rock types known on Earth.

Magmas are mainly composed of Earth's abundant elements, but silica (SiO_2) is dominant and usually accounts for 45 to 75 percent of magma composition, by weight. A small amount of dissolved gas is typically present, most commonly water vapor or carbon dioxide. The dissolved gases, temperature, and proportion of silica strongly influence the properties of magma, as well as the eruptive style of the volcano from which it emerges. For example, high-temperature, low-silica magma with low dissolved gas would be expected to erupt relatively calmly, in fluid flows of lava. Lower-temperature, high-silica magma with higher gas content would be expected to erupt explosively, generating abundant pyroclastic material.

There are three common magma types: *basaltic magma* of low-silica composition (by far the most common on Earth), *andesitic magma* of intermediate-silica composition, and rhyolitic magma of *high-silica* composition. The processes that lead to the formation of these magma types and the settings in which they occur are influenced by plate tectonics. Basaltic magma is formed by partial melting of the mantle. The magma rises to the surface in areas where the crust is splitting, such as along midocean spreading centers, and above mantle hotspots. Andesitic magma is formed by wet partial melting of the wedge of mantle immediately overlying a subducting oceanic plate. Both the location in which andesitic magma occurs and the typically high water content are strongly controlled by the subduction zone setting. Rhyolitic magma forms as a result of melting at the base of the continental crust. The magma is silica-rich because it rises through the continental crust, assimilating silica-rich crustal material as it goes.

Crystallization and Igneous Rock

Whereas melting determines the properties of magma, cooling and crystallization determine the properties of **igneous rock**, that is, rock that forms as a result of the cooling and solidification of magma. **Crystallization** refers to the set of processes whereby crystals of the individual mineral components nucleate and grow in a cooling magma. In igneous rock, the crystals typically grow into an interlocking texture that holds the individual mineral grains together very firmly. Crystalline igneous and metamorphic rock is thus usually harder, with less pore space than sedimentary rock, whose grains may not be held together as tightly. The rate of cooling determines how large the individual mineral grains in the rock may eventually grow, and the grain size, in turn, affects the appearance or *texture* of the rock.

As discussed in Chapter 6, *extrusive* or *volcanic rock* is formed as a result of the rapid cooling and solidification of lava, and *intrusive* or *plutonic rock* is formed when magma cools and solidifies slowly within the crust or mantle. Both extrusive and intrusive igneous rocks are classified and named on the basis of rock texture and mineral assemblage (FIG. 7.26). Rocks that contain a lot of silica (≈70 percent) usually have a high proportion of light-colored minerals; they are referred to as *felsic* (from the words "feldspar" and "silica," two abundant constituents of such rocks). Rocks that contain little silica (≈50 percent) typically contain lots of darker-colored minerals rich in magnesium and iron; they are referred to as mafic (from the words "magnesium" and "ferric," or iron-rich). Rocks with about 60 percent silica are said to be *intermediate*.

When magma or lava of a given composition solidifies, the mineral assemblage that forms is essentially the

FIGURE 7.26 Volcanic and plutonic rock

Grain Size →

Silica Content of Magma	*Resulting Volcanic Rocks*	*Resulting Plutonic Rocks*
High (≈ 70%–75%)	**Rhyolite** lies at the felsic, high-silica end of the scale and consists largely of quartz and feldspars. It is usually pale, ranging from nearly white to shades of gray, yellow, red, or lavender.	 **Granite**, the plutonic equivalent of rhyolite, is common because felsic magmas usually crystallize before they reach the surface. It is found most often in continental crust, especially in the cores of mountain ranges. 
Intermediate (≈ 60%)	**Andesite** is an intermediate-silica rocks, with lots of feldspar mixed with darker mafic minerals, such as amphibole or pyroxene. It is usually light to dark gray, purple, or green.	 **Diorite** is the plutonic equivalent of andesite, an intermediate-silica rock. 
Low (≈ 45%–50%)	**Basalt,** a mafic rock, is dominant in oceanic crust and the most common igneous rock on Earth. Large, low-viscosity lava flows from shield volcanoes and fissures are usually basaltic. Dark-colored pyroxene and olivine give it a dark gray, dark green, or black color.	 **Gabbro** is the plutonic equivalent of basalt, a low-silica rock. 

↑ Silica Content

same for both intrusive and extrusive rock; the only differences, therefore, are textural. Once the texture of an igneous rock has been determined, specimens are further named on the basis of mineral assemblage.

Rapid Cooling: Volcanic Rock

The characteristic texture of volcanic rock is fine grain size. Lava cools so rapidly that mineral grains do not have sufficient time to become large. As a result, extrusive igneous rocks generally have individual grains less than 2 mm in diameter. Some lava cools so rapidly that the rock that forms is *glassy* and almost completely lacks individual mineral grains (FIG. 7.27). Although there are many compositional variations of volcanic rock, the three main classes are named for the magmas from which they solidify (as shown in Fig. 7.26): the mafic volcanic rock, *basalt*, crystallizes from basaltic magma; the intermediate volcanic rock, *andesite*, crystallizes from andesitic magma; and the felsic volcanic rock, *rhyolite*, crystallizes from rhyolitic magma.

Pyroclastic rock is particularly interesting because it is transitional between igneous and sedimentary. There is an old saying that "pyroclasts are igneous on the way up and sedimentary on the way down." *Pyroclasts* form directly from magmatic materials that have been erupted from a volcano. However, they are deposited gravitationally, like sediment, accumulating in deposits called *tephra*. Tephra can be converted to pyroclastic rock in two ways. The first, and most common, way is through the addition of a cementing agent, such as quartz or calcite introduced by groundwater. The second way is through the welding of hot, glassy, ash particles. When ash is very hot and plastic, the individual particles can fuse together to form a glassy pyroclastic rock called *welded tuff*. Pyroclastic rock is called *agglomerate* when the tephra is bomb-sized, and *tuff* when the pieces are smaller, either lapilli- or ash-sized. (The terms describing different sizes of tephra are defined in Chapter 6; refer to Fig. 6.19.) The magmatic origin of a pyroclastic rock is also indicated by the name for the mineral assemblage; for example, we would refer to rock of high-silica composition and corresponding mineral assemblage and texture as *rhyolite tuff*.

FIGURE 7.27 Volcanic glass
Obsidian is volcanic rock that has cooled so quickly it was not able to grow any individual crystals, so it has a glassy texture.

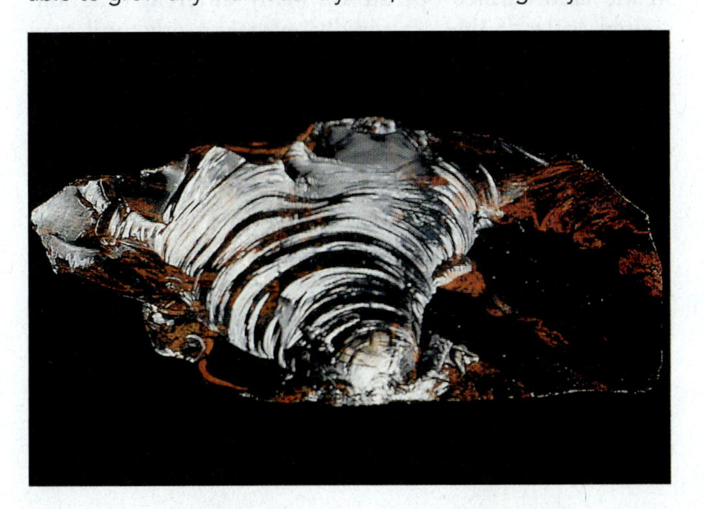

Slow Cooling: Plutonic Rock

Intrusive igneous rock tends to be coarse-grained because magma that solidifies in the crust cools slowly and has sufficient time to form large mineral grains. If an intrusive rock cools very slowly and has abundant pore fluid to facilitate the growth of crystals, it may become an extremely coarse-grained rock called *pegmatite*. Another special texture involves a distinctive mixture of large and small grains. Rock with this texture is called *porphyry*, meaning an intrusive igneous rock consisting of coarse mineral grains scattered through a mixture of fine mineral grains. The large grains in porphyry are formed in the same way that those of any other coarse-grained igneous rock are formed—by slow cooling of magma in the crust or mantle; the fine-grained mass that encloses the coarse grains provides evidence that partly solidified magma moved quickly upward. In the new setting, the magma cooled rapidly, and as a result, the later mineral grains are all tiny.

As discussed in Chapter 6, the three main magma types—basalt, andesite, and rhyolite—produce the corresponding coarse-grained plutonic rock equivalents gabbro, diorite, and granite, respectively (as shown in Fig. 7.26). The names of plutonic rocks, like the names of volcanic rocks, can reflect both their texture and their composition. For example, we might refer to rock of appropriate texture and mineral assemblage as *granite porphyry*.

Igneous Rock Diversification

This all seems fairly neat and simple—basaltic magma yields basalt and gabbro; andesitic magma yields andesite and diorite; rhyolitic magma yields rhyolite and granite. In fact, however, there is an *enormous* diversity of igneous rock types on Earth—there are hundreds of variations. How do all these different rock types arise from the three principal magma compositions? Part of the variation results from differences in cooling rate and texture, as discussed earlier; so rhyolitic magma might end up as rhyolite tuff, granite porphyry, or granite pegmatite, for example. Part of the variation also comes from partial melting, also discussed earlier, which can lead to the separation of melt from residual solid rock of a different composition.

Another important process, called *fractional crystallization*, also contributes to the diversification of igneous rock types, and it happens during cooling and crystallization. We know that rock melts over a range of temperatures. Similarly, crystallization—which is essentially the reverse of melting—occurs over a range of temperatures during the cooling of magma. Mineral grains will begin to form in cooling magma at about 1200°C, resulting in grains of high-temperature minerals floating in the remaining melt. Note that these first-formed minerals are the ones that would melt *last*, at the highest temperatures, during the melting process. As cooling proceeds, crystals of other mineral components will form at successively

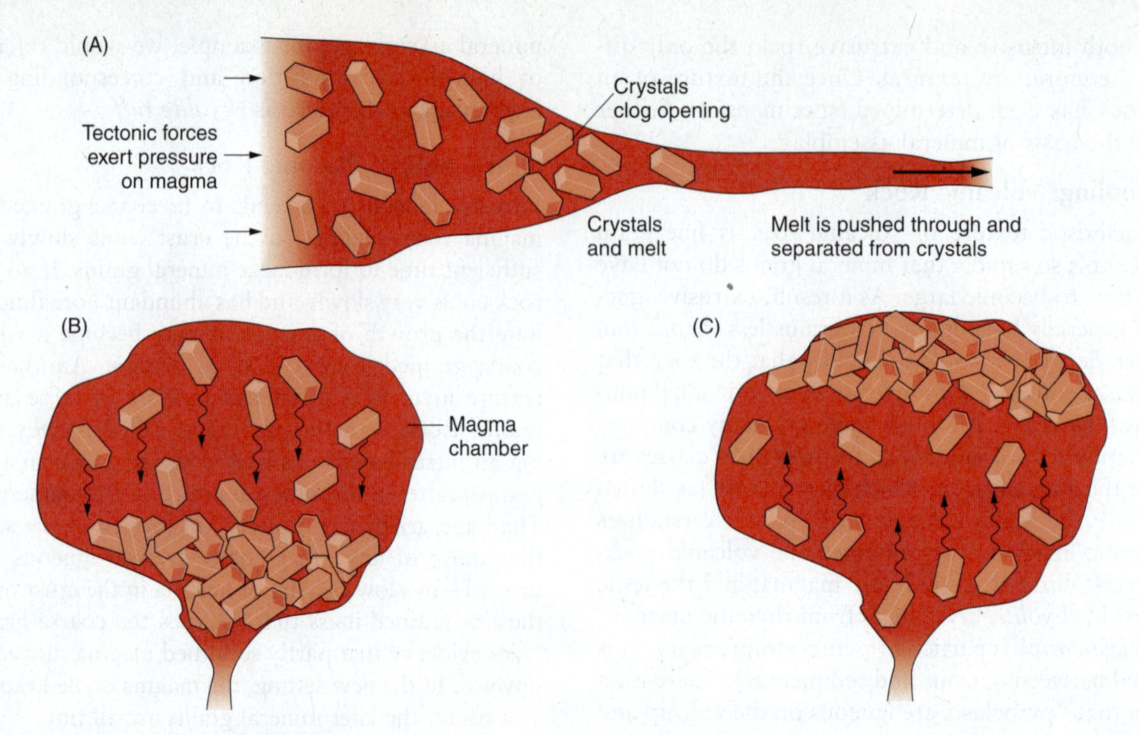

FIGURE 7.28 Fractional crystallization
Several processes can separate crystals from melt in a partially crystallized magma. (A) Filter pressing: Magma is squeezed through a small opening by tectonic forces. Only the liquid gets through, and the newly formed crystals are left behind. (B) Crystal settling: The first minerals to crystallize are denser than the melt and may sink to the bottom. (C) Crystal flotation: If the first crystals to crystallize are less dense than the melt, they may float and collect at the top of the magma chamber.

lower temperatures. For most rock types, crystallization will be complete, and the rock will be solid by about 800°C. (However, the solid rock still has a lot of cooling left to do at that point; it may expel pore fluids as it cools, and it may heat up the surrounding rock to cause contact metamorphism, as already discussed.)

What would happen if the crystallization process were halted midway, at some temperature between the onset and completion of crystallization? There will be a combination of high-temperature crystals floating in melt that is concentrated in the lower-temperature components of the magma. Now consider what might happen if the crystals were to become separated from the melt, of different composition. This can happen via several different processes, as shown in FIGURE 7.28. The separated melt and the higher-temperature crystals may now solidify into igneous rocks of very different compositions. Analogous to fractional melting, this is the process of fractional crystallization, and it is a very important contributor to the diverse array of igneous rock types found on Earth.

THE ROCK CYCLE, THE TECTONIC CYCLE, AND EARTH'S LANDSCAPES

The major components of the Earth system meet at the land surface. This is where the rock cycle and the tectonic cycle interact with the other cycles that we will consider in subsequent chapters, including the hydrologic and climate cycles. The evolution of Earth's surface and its landscapes involves a complex set of processes related to the geosphere, atmosphere, hydrosphere, and biosphere. Even a casual look at the land surface can raise some basic questions: How can Earth's varied landscapes be explained? If landscapes have changed through time, what processes have controlled the changes? Is the landscape still changing? What are the energy sources that drive these landscape-modifying processes? What clues do landscapes hold about the history of Earth's past environments?

To answer these questions we turn to the rock cycle and the tectonic cycle. The processes that produce changes in the land surface operate on many different time and spatial scales. Some act rapidly, even abruptly, and may cause only local changes. Sometimes these changes have a direct and adverse impact on people (including earthquakes, volcanic eruptions, storms, floods, landslides, and many other natural hazards). Other processes operate far more slowly—on geologic timescales—and are difficult or impossible to observe directly. Nevertheless, their effects can be seen both in surviving ancient landscapes and in the stratigraphic record.

Not only can surface processes affect people going about their daily lives, but everywhere they live people, too, are changing the face of the land. The construction of dams across streams, excavations for buildings or waste disposal sites, and the building of highways, cities, and airports all modify the landscape. Although such individual actions may seem insignificant at the global scale, they can quickly add up until, over the span of several human generations, their effect on the landscape is substantial. Thus, the anthroposphere also plays a significant role in the modification and evolution of Earth's surface. We will look more closely at some of these influences in Chapter 19.

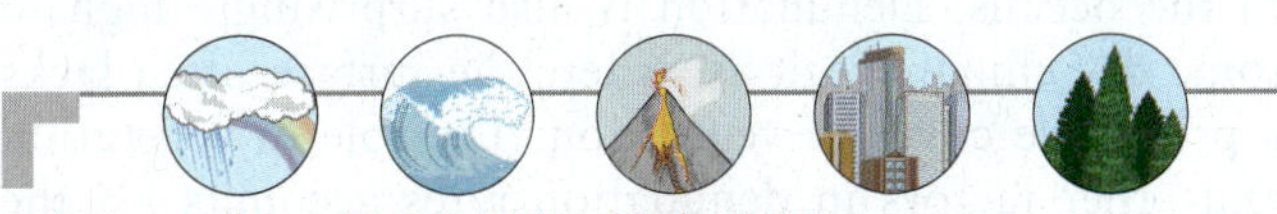

Make the CONNECTION

Looking around the area where you live, can you find some significant ways that people have modified the landscape? Look for major road cuts, quarries, tunnels, or dams. Are there some other, more subtle modifications that you hadn't noticed before?

Competing Geologic Forces

The constant changes of Earth's surface reflect an ongoing contest between internal forces that raise the lithosphere and external forces that wear it down. The uplifting, mountain-building processes of plate tectonics, isostasy (Chapter 5), and volcanism are driven by Earth's internal heat energy, controlled by gravity. Gravity, combined with energy from the Sun, also drives the external processes of weathering and erosion, causing mountains to be worn away and rock debris to be transferred from high places to low places. The net result is the progressive sculpture of the land into a surface of varied topography and *relief*, the difference in altitude between the highest and lowest points on a landscape. Relief varies regionally because geologic processes vary from place to place and because rocks of different type and structure have differing resistances to erosion.

Obtaining reliable geologic information about long-term rates of landscape change is not simple. If a landmass is rising tectonically, we must ask how rapidly it is rising and whether rates of uplift have changed through time. Second, we must calculate changing rates of **denudation**, the combined destructive effects of weathering, erosion, and mass wasting. Knowing uplift and denudation rates for a region, we can then infer something about how the landscape may have evolved and anticipate how it may continue to change in the future.

Uplift Rates

Measured and calculated uplift rates in active tectonic belts are quite variable—around 1 to 10 mm/yr, averaged over intervals of several thousand to several million years. For any given region, the rate of uplift changes through time, just as rates of seafloor spreading and plate convergence vary. Each such change is likely to lead to compensating isostatic adjustment in landscapes as they begin to evolve toward a new condition of equilibrium. The current local uplift rate for one high region of the western Himalaya is as high as 5 mm/year. If sustained for only 2 million years, the total uplift would be 10 km. Such a high rate of uplift is generally associated with steep slopes and high relief; this is certainly true in the Himalaya (FIG. 7.29). In these steep mountainous slopes, rapid rates of uplift are balanced by erosion at rates of about 3 to 4 mm/year—close to the rate of uplift.

FIGURE 7.29 Steep slopes, rapid erosion
A stream cutting a deep gorge in the Hindu Kush, Himalaya Mountains of northern Pakistan transports coarse sediment from the steep adjacent cliffs. In actively uplifting orogens like this, the rate of erosion almost matches the rate of uplift.

Denudation Rates

Denudation begins as soon as a mountain range is uplifted, and continues long after the active tectonic uplift has ceased. Ancient mountain ranges like the Appalachians were once high, steep, and craggy like the Himalaya, but have ceased to be actively uplifted and consequently have been worn down by erosion that continues even today.

The calculation of long-term denudation rates requires knowledge of how much rock debris has been removed from an area during a specified length of time. The total sediment removed from a mountain range, for example, will include sediment currently in transit, sediment temporarily stored on land on its way to the sea, and sediment deposited in the adjacent ocean basin, mostly on or near the continental shelf. If the volume of all this sediment can be measured, its solid rock equivalent and the average thickness of rock eroded from the source region can be calculated. Finally, if the duration of the erosional interval can be determined, the average denudation rate can be calculated.

For areas drained by major rivers, the volume of sediment reaching the ocean each year is a measure of the modern erosion rate. The volume of sediment deposited during a specific time interval can be estimated using drill-core and seismic records of the seafloor. Calculating the equivalent rock volume of this sediment and averaging it over the area from which it was derived gives an average denudation rate for the source region. The highest measured sediment yields are from the humid regions of southern Asia and Oceania, and from basins that drain steep, high-relief mountains of young orogenic belts, such as the Himalaya, the Andes, and the Alps (FIG. 7.30). As one might expect, rates are high on steep slopes, and much higher in areas underlain by soft, erodible, or highly jointed rock. The continent that discharges the most sediment to the ocean is Asia; rivers entering the sea between Korea and Pakistan contribute nearly half of the total world sediment input to the oceans. Denudation is also surprisingly high in some dry climate regions, where the surface often lacks a protective cover of vegetation. The role of vegetation and other factors in denudation rates accounts for the fact that the areas with the highest rates of denudation and the areas with the highest sediment yield are similar, but not identical.

One important factor that influences denudation rates is human activity, especially the clearing of forests, development of cultivated land, damming of streams, and construction of cities. Each of these activities has affected erosion rates and sediment yields in areas where they have occurred. Sometimes the results are dramatic: areas cleared for construction or mining may produce between 10 and 100 times more sediment than comparable rural areas or natural areas that are vegetated. On the other hand, in urbanized areas sediment yield tends to be low because the land is almost completely covered by buildings, sidewalks, and roads that protect the underlying rock and sediment from erosion.

FIGURE 7.30 Sediment yield from continents
The areas with the highest sediment yields are in Southeast Asia, which receives high precipitation, and in major high-relief mountain belts, including the Himalaya, the Alps, the Andes, and the coastal ranges of Alaska and British Columbia. Areas with the highest denudation rates are similar, but not exactly the same, because other factors such as vegetation cover also play an important role.

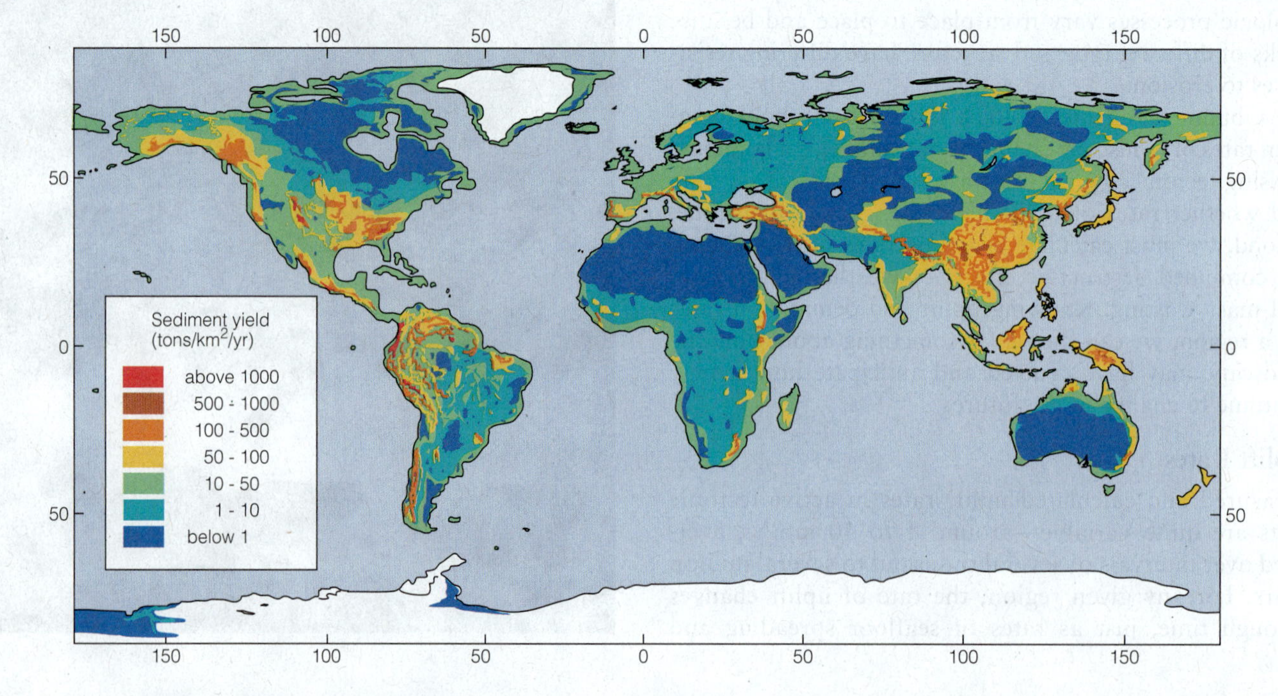

the BASICS

Factors Controlling Landscape Development

Five main factors interact to determine the character of landforms in a given location: *process*, *climate*, *lithology*, *relief*, and *time*.

Process. Distinctive landscape elements result from the activity of various surface processes. For example, a sand dune has a form that is different from that of a mud flat. The active process and depositional environment combine to result in a unique end product, or landform. Process, then, is one factor that helps dictate the character of landforms.

Climate. Climate, in turn, helps determine which processes are active in any area. In humid climates, streams may be the primary agent that moves and deposits sediment, whereas in an arid region wind may locally assume the dominant role. Glacial phenomena are largely restricted to high latitudes and high altitudes where frigid climates prevail. Because climate also controls vegetation, it further controls the effectiveness of some important processes; for example, a hillslope stabilized by plant roots that anchor the soil may become prone to mass wasting if the plant cover is destroyed. Thus, hillslope erosion is linked to vegetation cover, which in turn is determined by climate—a complex linkage in which the atmosphere, the hydrosphere, and the biosphere all play roles in determining how surface processes affect the land surface.

Lithology. Surface processes interact with exposed rock in various ways, depending on **lithology** or rock type. Some rock types are less erodible than others and will produce greater relief and more prominent landforms than those more susceptible to erosion (FIG. B7.3). Any single rock type, however, may behave differently under different climatic conditions. For example, limestone may underlie valleys in areas of moist climate where the rock dissolved easily, but in dry desert regions the same kind of rock may form bold cliffs. Structures such as faults and joints also play roles. Because of differential erosion, folded or faulted beds may stand in relief or control the drainage in such a way that they impart a distinctive pattern to the landscape.

Relief. Topographic relief, another factor in landscape development, is related to tectonic environment. Tectonically active regions have high rates of uplift, leading to high summit altitudes and steep slopes. Such landscapes tend to be extremely dynamic, with high erosion rates. In areas far away from active tectonism, relief typically is low, erosion rates are much lower, and landscape changes take place more gradually. Even in tectonically stable areas, however, a rapid rise or fall of sea level or regional isostatic movements related to changing ice or water loads may produce significant changes in landscapes.

Time. Finally, the concept of landscape evolution necessarily involves the element of time. Although some landscape features can develop rapidly, even catastrophically, others develop only over long geologic intervals. We know this, or at least we infer it, by measuring the present rates of surface processes and by dating deposits that place limits on the ages of specific landforms or land surfaces.

FIGURE B7.3 Differential erosion
Uluru, or Ayers Rock, in Australia is an *inselberg*, a rock mass that is more resistant to erosion than the surrounding highly erodible rock, and hence stands in high relief compared to the eroded, flat plains around it. Uluru consists of sandstone that has resisted erosion because it is remarkably unjointed.

Landscape Equilibrium

The major landscape features of Earth have developed over long intervals of time as the lithosphere has evolved and continents and ocean basins have continually been reorganized. The lateral movements and resulting collisions of lithospheric plates, leading to the generation of orogenic belts, have provided much of the driving force for landscape change over hundreds of millions of years. The study of landscape evolution focuses on the relative importance of uplift rates and erosion rates, and investigations of the complex relationships and feedbacks between the atmosphere,

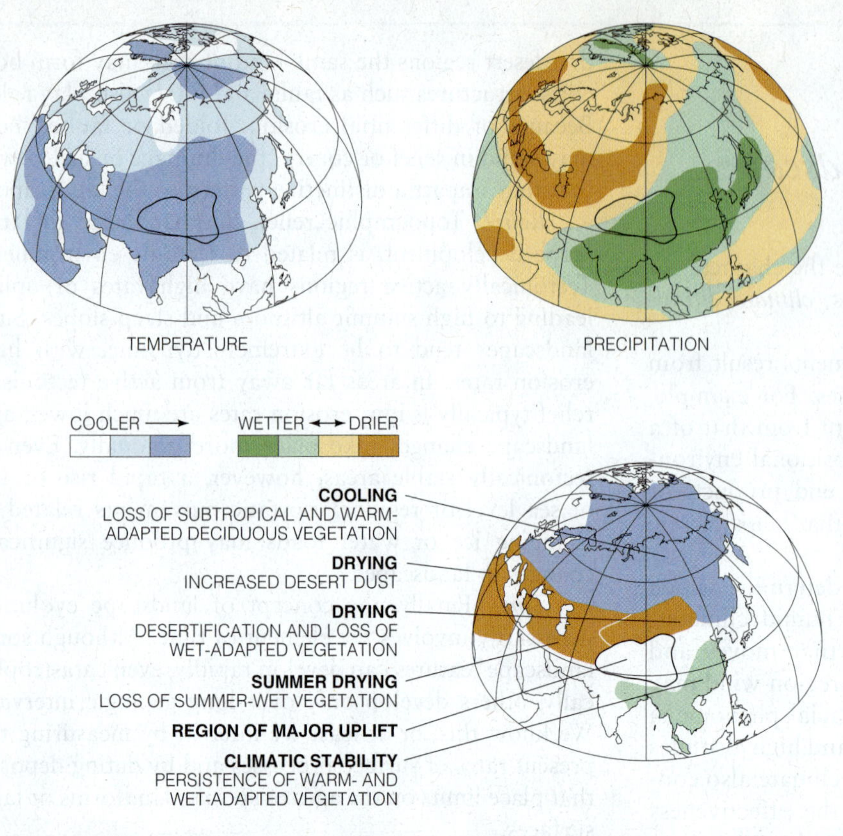

FIGURE 7.31 Computer simulation of environmental changes
Changes in the geosphere, such as the uplift of the Himalayan Mountains and the Tibetan Plateau, can lead not only to changes in relief but to changes in temperature, precipitation, and vegetation cover, as shown in this computer simulation.

biosphere, and lithosphere in dictating the character and evolution of the major landscape elements of the planet. Change is implicit in landscape evolution, and landscapes evolve if a change takes place in any of the controlling variables.

What initiates landscape change? Change may be started by a tectonic event that causes a landmass to be uplifted, or by a substantial fall of sea level that causes streams to assume new gradients. It can begin with a shift in climate that may modify the relative effectiveness of different surface processes. Over short intervals of time, rates of change may vary because of natural fluctuations in the magnitude and intensity of surface processes. Over longer intervals, the average rate of change may increase if the rate of tectonic uplift increases, or may gradually decrease as a land surface is progressively worn down toward sea level.

An interesting example of how a change in the geosphere can lead to environmental changes is illustrated in FIGURE 7.31. These results of a computer simulation show how uplift of the Himalaya and the Tibetan Plateau has led not only to changes in relief but also to significant changes in temperature and precipitation. These, in turn, have led to changes in major geographic and vegetation zones, illustrating some of the complex interconnections between the geosphere, hydrosphere, atmosphere, and the biosphere.

Does a landscape ever achieve a state of complete equilibrium in which no change takes place? The answer, apparently, is no. There is abundant evidence that Earth's surface is now, and very likely always has been, a dynamic surface. Landscapes are constantly changing in response to natural processes in the lithosphere, hydrosphere, atmosphere, biosphere, and even—in the most recent part of Earth history—the anthroposphere. In subsequent chapters we will consider more of these processes and interactions, which make Earth's surface a dynamic place as well as a hazardous place, but also give it the ability to support life.

SUMMARY

1. The rock cycle involves the cycling of material through the geosphere through the breakdown, rearrangement, and formation of rock.

2. Rock is transformed into regolith through physical and chemical weathering, which work together. Chemical weathering is the decomposition of rocks and minerals through chemical reactions; the principal agent is a weak solution of carbonic acid (H_2CO_3) that is formed when rainwater and groundwater dissolve carbon dioxide. Physical (mechanical) weathering is the physical breakup of rocks by processes such as jointing, frost wedging, precipitation of salts in fractures, fire breakage, and wedging by plant roots.

3. Erosion is the set of processes whereby weathered particles are transported under the influence of gravity, often carried or pushed by water, wind, or ice to a location where deposition occurs. Sediment is regolith that has undergone erosional transport and deposition.

4. Clastic sediment consists of broken rock and mineral particles. The four main size classes are gravel, sand, silt, and clay. Chemical sediment is formed when substances that have been dissolved are transported in solution and then precipitated by chemical processes such as evaporation. Biogenic sediment is formed as a result of biochemical processes.

5. Mass wasting is the downslope movement of regolith by gravity, with or without a transporting medium. The composition and texture of the sediment involved, the amount of water and air involved, and the steepness of the slope influence the type and velocity of movement. Landslides can be triggered by natural processes such as earthquakes, or by human activities that oversteepen or destablize slopes.

6. The locations where clastic sediment is deposited are largely controlled by plate tectonics. They include low-lying rift valleys, structural basins in mountain ranges, and deep oceanic trenches in subduction zones near continents. The deposition of chemical and biogenic sediment is not as strongly influenced by plate tectonics because it occurs in open aquatic environments.

7. Sediment or regolith becomes sedimentary rock through lithification, the first stage of which is burial and compaction. Loose grains of sediment adhere to one another as a result of cementation or recrystallization. Collectively, the changes that occur to sediment after deposition are referred to as diagenesis. Bedding and the presence of fossils are typical features of sedimentary rock. Sedimentary rock contains information about the environments in which the sediment was formed, transported, deposited, and lithified.

8. The four classes of clastic sedimentary rock are conglomerate, sandstone, siltstone (or mudstone), and shale, which form from gravel, sand, silt, and clay, respectively. Chemical sedimentary rock results from the lithification of chemical sediment. Limestone, which forms from lithified shells and other skeletal material from marine organisms, is the most important rock that forms from the lithification of biogenic sediment.

9. Sedimentary stratification refers to the layered arrangement of sedimentary deposits. Each stratum or bed differs from adjacent beds because of changes that occurred in the environment as the sediment accumulated. Stratigraphy is based on the principles of original horizontality, superposition, and lateral continuity. Stratigraphic correlation is the determination of equivalence in age of the succession of strata found in two or more different areas.

10. The principles of stratigraphy, stratigraphic correlation, and cross-cutting relationships have allowed scientists to determine the relative ages of rock units and geologic events around the world. The geologic column summarizes in chronological order the succession of known rock units. Using radiometric dating to establish the numerical ages of certain rock units, scientists have been able to fit a scale of numerical time to the geologic column.

11. Unconformities are breaks in the stratigraphic record that represent periods of nondeposition, erosion, or both. The three important kinds of unconformities are nonconformities, angular unconformities, and disconformities.

12. Metamorphic rock is igneous or sedimentary rock that has been changed in mineralogy, texture, or both, as a result of being subjected to elevated pressures and temperatures. Metamorphism describes changes in mineral assemblage and texture that occur in the solid state, at temperatures above 150°C and pressures in excess of 300 MPa. The upper limit to metamorphism is reached when rock melting begins.

13. Igneous rock textures form under uniform stress because igneous rocks crystallize from liquids, which cannot sustain differential stress. In contrast, the textures of many metamorphic rocks record evidence of differential stress, most commonly as foliation, a layered texture that results from the alignment of platy or elongate minerals. The starting composition of a rock, in combination with the temperature and pressure of metamorphism, determines the assemblage of minerals that will develop.

14. The three most important kinds of metamorphism are contact, burial, and regional metamorphism. They differ in conditions of temperature, pressure, and differential stress, and in the relative importance of mechanical deformation and chemical recrystallization. Mechanical deformation includes grinding, crushing, and the development of foliation. Chemical recrystallization includes changes in mineral composition, the growth of new minerals, and losses of H_2O and CO_2 that occur with metamorphism.

15. Metamorphic rocks are named on the basis of texture and mineral assemblage. The products of the metamorphism of shale are slate, phyllite, schist, and gneiss, in order of increasing metamorphic grade. Basalt undergoes metamorphism under conditions where H_2O can enter the rock and form hydrous minerals, resulting in the metamorphic rocks greenschist, amphibolite, and granulite. The metamorphic derivatives of limestone and sandstone are marble and quartzite, respectively; they are typically monomineralic and nonfoliated.

16. The metamorphic facies concept holds that for a given range of temperature and pressure, and a given rock composition, the assemblage of minerals formed during metamorphism is always the same. The conditions that lead to blueschist and eclogite facies metamorphism are reached when crustal rock is dragged down by a rapidly subducting plate, such that pressure increases much more rapidly than temperature. The metamorphic conditions characteristic of greenschist and amphibolite facies metamorphism occur where the crust is thickened by continental collision or heated by rising magma.

17. Melting leads to the development of three main magma types, basaltic, andesitic, and rhyolitic, the properties of which are mainly determined by their compositions. Partial and fractional melting can lead to variations in magma types. Cooling and crystallization, in which crystals nucleate and grow in a cooling magma, combine with composition to determine the properties of igneous rock. The rate of cooling influences how large the individual mineral grains eventually grow.

18. When magma or lava of a given composition solidifies, the mineral assemblage that forms is the same for intrusive and extrusive rock; the differences are textural. Volcanic rock is characterized by fine grain size or glassy texture, whereas plutonic rock tends to be coarse-grained because the magma cools slowly. Basaltic magma yields the mafic volcanic rock basalt and its plutonic equivalent, gabbro. Andesitic magma yields the intermediate volcanic rock andesite and its plutonic equivalent, diorite. Rhyolitic magma yields the felsic volcanic rock rhyolite and its plutonic equivalent, granite.

19. The enormous diversity of igneous rock types on Earth results partly from differences in texture that develop during cooling and crystallization, and partly from partial or fractional melting, which can lead to the separation of melt from residual solid rock of a different composition. Fractional crystallization also contributes to the diversification of igneous rock types.

20. Five main factors interact to determine the character of landforms: process, climate, lithology, relief, and time. The constant changes reflect ongoing interactions between the internal forces of plate tectonics, isostasy, and volcanism that uplift the lithosphere, and the external forces of denudation that wear it down. Denudation begins as soon as a mountain range is uplifted, and continues long after active tectonic uplift has ceased. The net result is the progressive sculpture of the land into a surface of varied topography and relief.

IMPORTANT TERMS TO REMEMBER

bedding *195*
biogenic sediment *192*
burial metamorphism *205*
cementation *195*
chemical sediment *192*
chemical weathering *186*
clastic sediment *190*
clay *190*
compaction *195*
contact metamorphism *204*
correlation *199*
crystallization *209*
denudation *213*
deposition *190*
diagenesis *195*
erosion *190*
facies *208*
foliation *204*
geologic time scale *200*
gravel *190*
igneous rock *209*
joint *187*
landslide *192*
lithification *195*
lithology *215*
mass wasting *192*
metamorphic rock *202*
metamorphism *202*
metasomatism *203*
physical (mechanical) weathering *187*
recrystallization *195*
regional metamorphism *206*
regolith *186*
rock cycle *186*
sand *190*
sediment *190*
sedimentary rock *195*
silt *190*
stratigraphy *198*
stratum (pl. strata) *198*
stress *203*
unconformity *199*
weathering *186*

QUESTIONS FOR REVIEW

1. What is the rock cycle? Summarize the main processes that comprise the rock cycle.
2. Why does the physical breakup of rock increase the effectiveness of chemical weathering?
3. What are the three main classes of sediment? Name them and describe how they are formed and transported.
4. What is erosion? Is it different from mass wasting? What are the main driving forces for erosion and mass wasting?
5. Describe two ways by which loose aggregates of sediment are transformed into rock.
6. Name and briefly explain the basic principles of stratigraphy.
7. How and why do breaks occur in stratigraphic sequences?
8. What is stress? How does it differ from pressure?
9. What is foliation? How does it form, and why does it provide a valuable clue that rock is of metamorphic origin?
10. What are the three main types of metamorphism? Name and briefly describe them.
11. What is crystallization? How is it related to melting? How does it differ from recrystallization?
12. What are the main types of igneous rock? How can we account for the enormous variability and diversity of igneous rock on Earth, given a small number of magma starting compositions?
13. What holds together the mineral grains in igneous and metamorphic rock? What about sedimentary rock?
14. Can a rock be uniquely defined on the basis of its mineral assemblage? If not, what additional information is needed? Consider sedimentary, metamorphic, and igneous rocks.
15. How can human activity influence erosion and sediment yield from an area? What influence might this have on changes in the landscape?

QUESTIONS FOR RESEARCH AND DISCUSSION

1. Discuss the role of plate tectonic settings in controlling the occurrence of different types of magma and igneous rock; sediment and sedimentary rock; and metamorphism and metamorphic rock.

2. During the first half of the twentieth century, erosional landscapes were generally believed to reflect successive cycles of erosion and landscape evolution, leading ultimately to a state of landscape equilibrium. This concept, called the *geographic cycle*, was proposed by William Morris Davis; it predated the theory of plate tectonics. Find and read some of Davis's work and discuss which aspects of his geographic cycle are consistent or inconsistent with our present understanding of plate tectonics.

3. Landscapes evolve through time as a balance between the forces of uplift and denudation. Do you live in an area where active uplift is occurring? Or are the wearing-away forces of denudation dominant at this point in the landscape evolution of the area?

QUESTIONS FOR *THE BASICS*

1. How are the principles of stratigraphy applied to the determination of the relative ages of geologic materials and events?

2. What are the main divisions of the geologic time scale?

3. What are the five factors that control landscape development?

QUESTIONS FOR *A CLOSER LOOK*

1. Why are banded iron formations all older than 1.8 billion years?

2. What comprises the alternating red, black, and white bands in banded iron formations?

3. How do banded iron formations reveal the chemical evolution of Earth's atmosphere and hydrosphere, and the progress toward an oxygenated atmosphere like the one we have today?

PART THREE

The Hydrosphere: Earth's Blanket *of* WATER *and* ICE

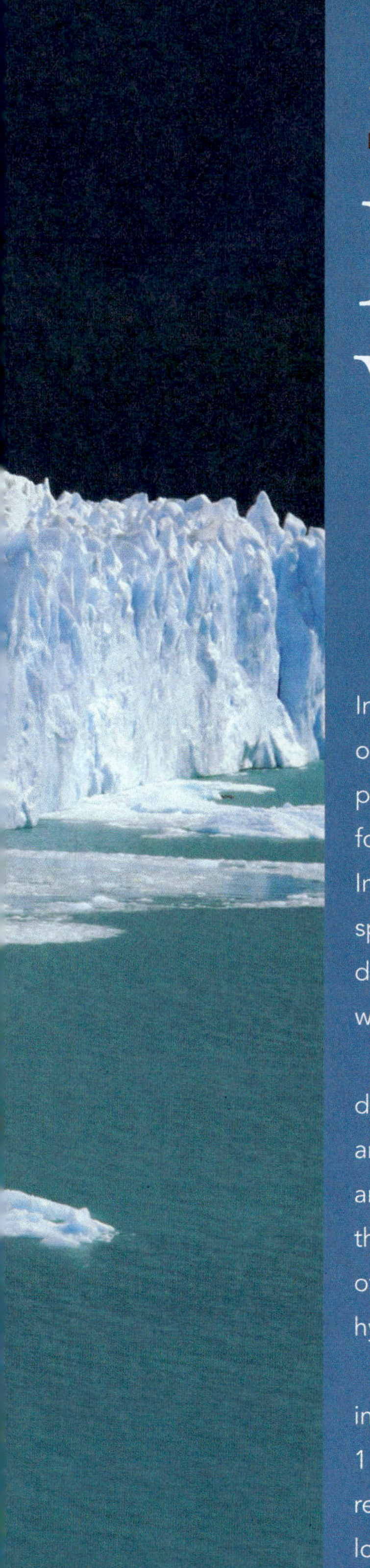

Hydrosphere: *from the Greek words* hydor, *meaning water, and* sphaira, *meaning globe, ball, or sphere.*

Cryosphere: *from the Greek words* kryos, *meaning icy cold, and* sphaira, *meaning globe, ball, or sphere.*

In Chapter 1, we defined the **hydrosphere** as the totality of Earth's water, including the ocean, lakes, streams, underground water, and snow and ice, but exclusive of atmospheric water vapor. Although this definition is straightforward, it is not strictly accurate, for water is present in *all* parts of the Earth system, even in the rock of Earth's interior. In other words, the hydrosphere overlaps with the geosphere, the atmosphere, the biosphere, and even the anthroposphere. It is mainly for convenience of discussion that we distinguish the hydrosphere as the "water sphere," for in it resides the bulk of Earth's water in the surface and near-surface environment.

Most of Earth's water, more than 97 percent, resides in the ocean. The ocean dominates the surface of our planet, and plays many crucial roles in influencing climate and supporting life. We examine the "world ocean" in Chapter 10. Next in volume are the myriad bodies of frozen water, held on a long-term basis (in human terms) in the form of snow and ice, mainly occupying the high mountains and two great ice caps of the polar latitudes. This is the **cryosphere**, the largest reservoir of fresh water in the hydrosphere, which we explore in detail in Chapter 9.

The rest of the water of the hydrosphere—in lakes and streams, in the atmosphere, in soil moisture, in living organisms, and under the ground—amounts to only about 1 percent of the total. Yet this is the water we are most conscious of and on which we rely in our daily lives. We begin our exploration of the hydrosphere in Chapter 8, with a look at the movement and storage of water on and under the land.

The chapters of Part III are as follows.

- Chapter 8. The Hydrologic Cycle
- Chapter 9. The Cryosphere
- Chapter 10. The World Ocean

Where land, water, and ice meet
In Patagonia, Argentina, the great ice cap of the Andes Mountains feeds numerous glaciers. This one, Perito Moreno Glacier, flows into Lago Argentino, and it is one of just a few Andean glaciers that are growing rather than shrinking in volume. The interrelationships among land, water, and ice have an enormous influence on Earth's climate system.

CHAPTER 8

The Hydrologic CYCLE

OVERVIEW

In this chapter we:

- Look at the reservoirs and pathways of the hydrologic cycle and how the hydrosphere interacts with other parts of the Earth system
- Describe how water flows in channels and is stored on the surface
- Examine flooding, its prediction, and its impacts on human interests
- Describe the occurrence and movement of water under the ground
- Consider society's dependence on water as a resource

Water on Earth's surface
The energy of flowing water is evident in this torrent pouring over the Atatürk Dam on the Euphrates River near Adiyaman, Turkey.

The most familiar cycle of the Earth system is surely the **hydrologic cycle**, which describes the fluxes of water between the various reservoirs of the **hydrosphere**, the totality of Earth's water on and just below the surface. We are familiar with these fluxes because we experience them as rain and snow or as a wet pavement drying by evaporation, and we see water moving and stored on the surface in streams, lakes, and wetlands. The movements of water through the hydrologic cycle and the important roles of water in the Earth system are the focus of this chapter.

WATER AND THE HYDROLOGIC CYCLE

Water in the atmosphere and large bodies of surface water play a central role in moderating temperature and controlling climate. They are the source of much of the water vapor in the atmosphere, and they store heat energy, exchanging it with the atmosphere. Another important consequence of the hydrologic cycle is the great diversity of Earth's landscapes. The erosional and depositional effects of streams, waves, and glaciers, coupled with tectonic movements, have produced landscapes that make Earth's surface unlike that of any other planet in the solar system. Through its effects on erosion and sedimentation, the hydrologic cycle is intimately related to the rock cycle (**FIG. 8.1**). Finally, water is a key component of an array of biogeochemical cycles that control the composition of the atmosphere and influence all living creatures on Earth.

FIGURE 8.1 The hydrologic cycle
The hydrologic cycle, shown here, interacts with the rock cycle, tectonic cycle, and biogeochemical cycles, and is interconnected with virtually all aspects of the Earth system.

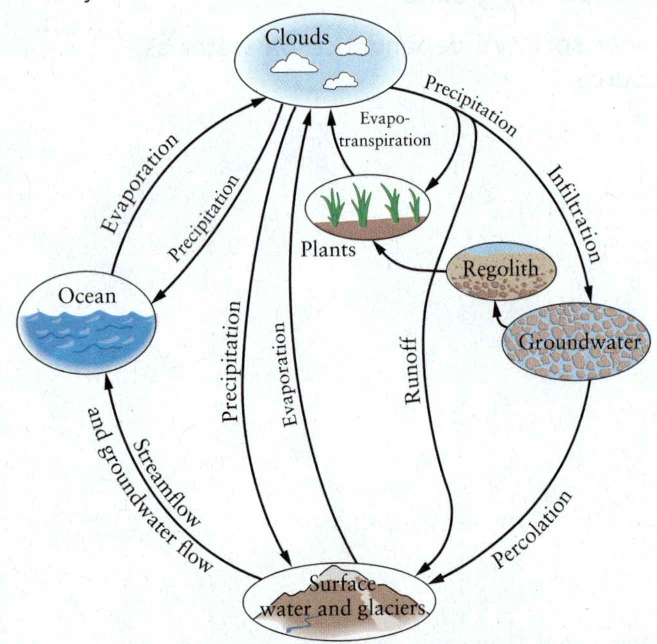

The unique properties of water as a chemical compound make life possible on this planet.

Like all of the cycles in the Earth system, the hydrologic cycle is composed of pathways and reservoirs. The pathways are the means by which water cycles between reservoirs. The reservoirs are the "storage tanks," where water may be held for varying lengths of time. (To review the basic concepts related to reservoirs, pathways, systems, and fluxes, see Chapter 1.) The total amount of water in the hydrologic system is fixed, but there can be quite short-term large fluctuations in local reservoirs. For example, a river may flood in one area, while a drought occurs in an adjacent area. On a global scale, however, these local fluctuations do not change the total volume of water in the Earth system; the hydrologic cycle therefore can be said to maintain a mass balance on a global scale. Overall it is a closed cycle, but within this closed cycle water is constantly shifting from one reservoir to another through a network of open subsystems.

Although water is continuously moving from one reservoir to another, the volume of water in each reservoir is approximately constant over short time intervals. Over lengthy intervals, however, the volume of water in the different reservoirs can change dramatically. During glacial ages, for example, vast quantities of water were evaporated from the ocean and precipitated on land as snow. The snow slowly accumulated to build ice sheets that were thousands of meters thick and covered vast areas where none exists today. At the culmination of the most recent glacial age, the amount of water removed from the ocean was so large that sea level was about 120 m lower, and the expanded glaciers increased the ice-covered area of Earth by more than 300 percent.

Reservoirs in the Hydrologic Cycle

The largest reservoir for water in the hydrologic cycle is the ocean, which contains more than 97.5 percent of all the water in the Earth system (**FIG. 8.2**). This means that most of the water in the hydrologic cycle is *saline*, not fresh. This has important implications for humans because we are dependent on fresh water as a resource for drinking, agriculture, and industrial uses. Surprisingly, the largest reservoir of fresh water is the great polar ice sheets, which contain almost 74 percent of all fresh water. The ice sheets are a long-term reservoir; water may be stored there for hundreds of thousands of years before it is recycled. Of the remaining unfrozen fresh water, almost 98.5 percent resides underground in the next largest reservoir, **groundwater**. Only a very small fraction of the water passing through the hydrologic cycle resides in surface freshwater bodies, such as streams and lakes. A smaller amount resides in pore spaces in soils, and an even smaller amount resides for short periods in the atmosphere and the bodies of living organisms.

In general, a correlation exists between the size of a reservoir and the average time that water stays in that reservoir, its *residence time* (Chapter 1). Residence time in the large-volume reservoirs, such as the ocean and ice

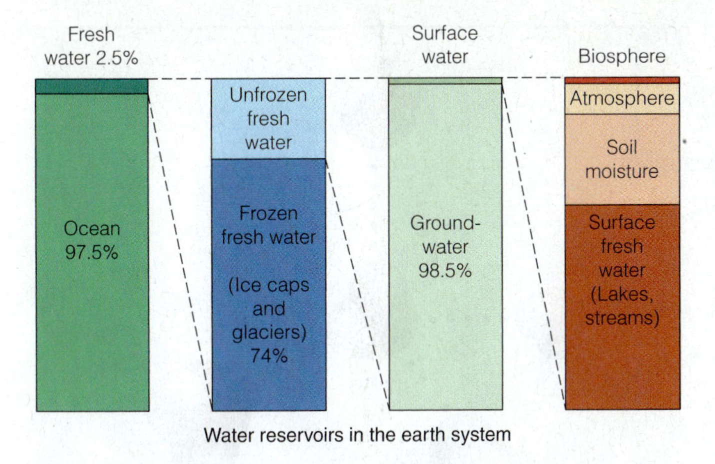
Fresh water 2.5%
Surface water
Biosphere
Ocean 97.5%
Unfrozen fresh water
Frozen fresh water
(Ice caps and glaciers) 74%
Ground-water 98.5%
Atmosphere
Soil moisture
Surface fresh water (Lakes, streams)
Water reservoirs in the earth system

FIGURE 8.2 Reservoirs in the hydrologic cycle The vast majority of Earth's water resides in the ocean. The perennially frozen cryosphere is the largest freshwater reservoir. Groundwater is the largest reservoir of unfrozen fresh water. Surface water, soil moisture, and water held in the atmosphere and biosphere are small in comparison.

FIGURE 8.3 Pathways in the hydrologic cycle Water moves around the Earth system in the hydrologic cycle, via the pathways shown here.

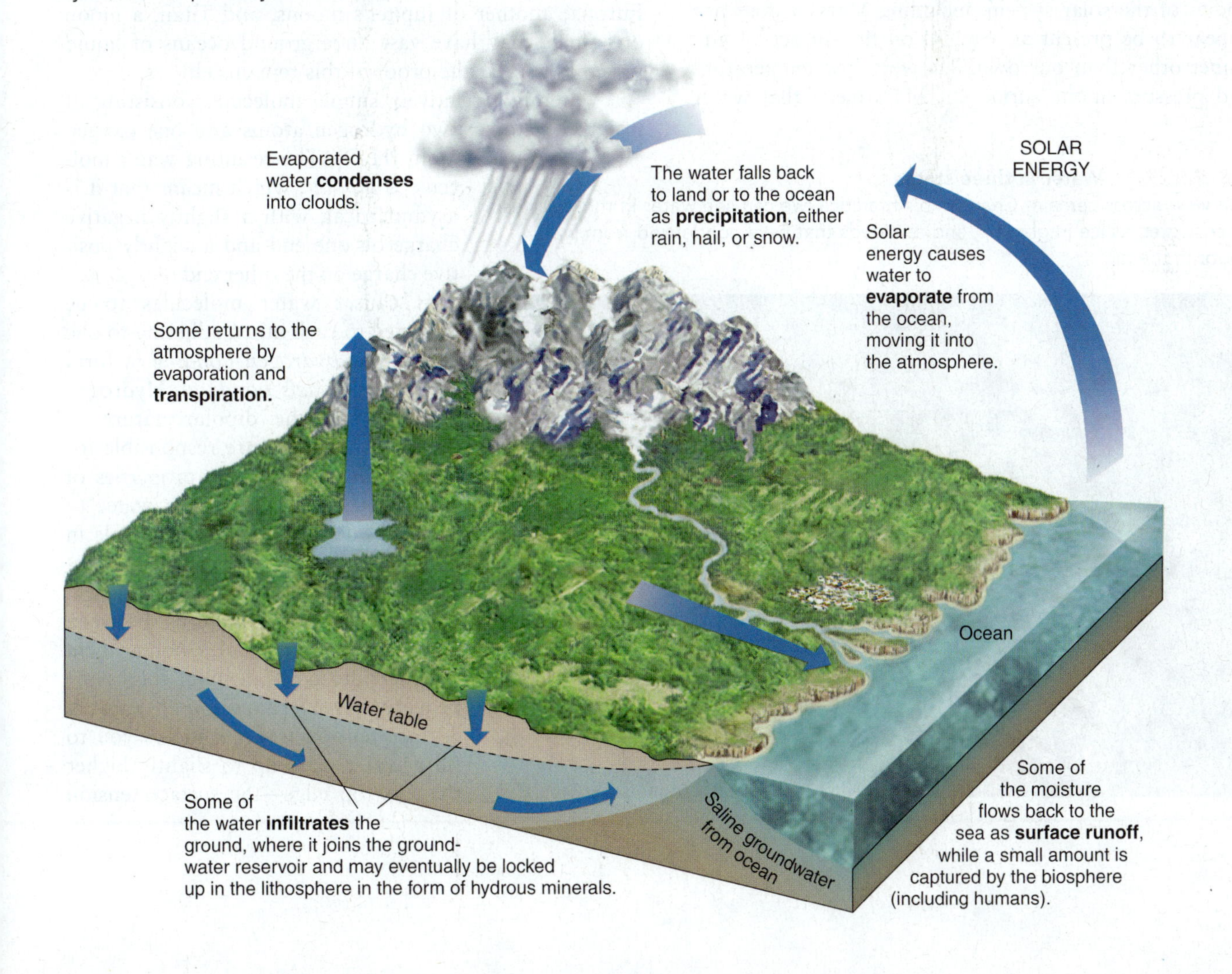
Evaporated water **condenses** into clouds.
The water falls back to land or to the ocean as **precipitation**, either rain, hail, or snow.
SOLAR ENERGY
Solar energy causes water to **evaporate** from the ocean, moving it into the atmosphere.
Some returns to the atmosphere by evaporation and **transpiration.**
Ocean
Water table
Saline groundwater from ocean
Some of the water **infiltrates** the ground, where it joins the ground-water reservoir and may eventually be locked up in the lithosphere in the form of hydrous minerals.
Some of the moisture flows back to the sea as **surface runoff**, while a small amount is captured by the biosphere (including humans).

caps, is many thousands of years, whereas in the small-volume reservoirs it is short—a few days in the atmosphere, a few weeks in streams and rivers, a few days or hours in the bodies of living organisms.

Pathways in the Hydrologic Cycle

The movement of water among Earth's reservoirs in the hydrologic cycle is powered by the Sun (FIG. 8.3). Heat from the Sun causes **evaporation** of water from the ocean and land surfaces, in which water is converted from its liquid form to its vapor (gaseous) form. The water vapor thus produced enters the atmosphere and moves with the flowing air. With changing atmospheric conditions, some of the water vapor in the atmosphere undergoes **condensation**, changing from a vapor back into a liquid or solid state. The condensed water, gathering into droplets or particles, falls under the influence of gravity as **precipitation** (such as rain or snow) on the land or ocean. In Chapter 11 we will look more closely at the processes of evaporation, condensation, and precipitation, and their role in weather.

Rain falling on land may evaporate, or it may be intercepted by vegetation, subsequently returning to the atmosphere by evaporation from leaf surfaces. Plants also can return water to the atmosphere by **transpiration**, in

the BASICS

Water, the Universal Solvent

We use the word "water" to refer to all three of the common states of water at Earth's surface—liquid, solid, and vapor—even though they differ physically. The physical state of water or H_2O is controlled by temperature and pressure. At high temperatures or low pressures, water vapor (or steam) is the stable state for H_2O, whereas water ice forms at low temperatures or high pressures. The air pressure at sea level and that at the top of Mount Everest represent the extremes of air pressures at Earth's surface. Surface temperatures range from about −100°C to +50°C. Within these limits of temperature and pressure, water can exist naturally in all three states of matter—solid (ice), liquid (water), or gas (water vapor) (**FIG. B8.1**).

The presence of water in solid, liquid, and vapor forms in the near-surface environment makes Earth unique among known planets and moons. Seen from space, Earth appears mostly blue and white because of its cover of water, snow, ice, and clouds. Although water in some form has been detected on the surfaces of other bodies of the solar system, including Mars, it does not appear to be present as a liquid on the surface of any planet other than our own. The very low temperature and pressure at the surface of Mars mean that water can exist there only as vapor or as ice, while the surface of Venus is so hot that water exists only as vapor. The surface of Ganymede, the largest of Jupiter's moons, is so frigid that it is covered by a thick "lithosphere" of ice. Europa, another of Jupiter's moons, and Titan, a moon of Saturn, may have vast underground oceans of liquid water, but so far the proof of this remains elusive.

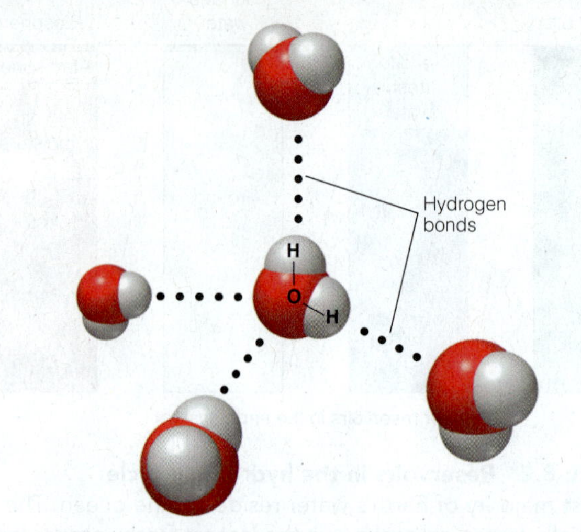

FIGURE B8.2 Water, a dipolar molecule
Water is a dipolar molecule, which means that it has a positive and a negative end. This causes it to adhere to other water molecules via hydrogen bonds.

Water is a relatively simple molecule, consisting of just three atoms: two hydrogen atoms and one oxygen atom (H_2O). The resulting water molecule is *dipolar*, which means that it is asymmetrical, with a slightly negative charge on one end and a slightly positive charge on the other end (**FIG. B8.2**). This causes water molecules to be attracted to each other, adhering to one another by *hydrogen bonding* (a form of *Van der Waals bonding*). Hydrogen bonding and the dipolar nature of the water molecule are responsible for determining many of the properties of water. These, in turn, allow water to play a mediating or facilitating role in countless physical, chemical, and biological processes in the Earth system.

The polar attraction between adjacent water molecules bonds them together into a strong film; this is the property of *surface tension* (**FIG. B8.3**). Surface tension is what allows you to fill a water glass up to slightly higher than its top edge—the surface tension

FIGURE B8.1 Water in three states
In a view across Lemaire Channel in Antarctica, we can see water in three states: as seawater, as ice in glaciers, and as clouds that have condensed from water vapor in the air.

FIGURE B8.3 Surface tension
Surface tension is one of water's most important properties. (A) Water forms droplets as a result of surface tension. (B) The water strider takes advantage of surface tension to move about on the undisturbed surface of the water.

of the water prevents it from immediately flowing over the edge. In conditions of surface temperature and pressure, the only common liquid with a higher surface tension than water is liquid mercury. Surface tension facilitates a variety of natural processes. It causes water to form into droplets, and it causes immiscible liquids, like oil and water, to separate from one another. Even the surface of a lake or pond is a manifestation of surface tension at the interface between water and air. Surface tension permits water to hold up substances denser than itself; some aquatic insects, such as water striders, rely on this property when they move about on the water's surface. Surface tension even plays a role in the formation of wind-driven waves and ripples; it is the force within the water that resists the dispersing force of the wind.

Water's dipolar character also results in *capillary attraction*, or *capillarity*, in which water molecules moving through a small tube are strongly attracted to the sides of the tube. As the water molecules are pulled into the tube, they pull or *wick* other water molecules along with them. Without this property, plants and trees would be unable to absorb nutrients through their roots, groundwater percolation would slow down, soil moisture would quickly dry up, and people would be unable to cry, because capillarity even plays a role in tears being able to flow out of the tear ducts.

In addition to binding easily to each other, water molecules bind readily to many other substances. This allows water to easily dissolve and transport a very wide variety of materials, and it is why water is sometimes called the "universal solvent" (although, in fact, many substances do not dissolve easily in water; oil is an example). The solvent property of water has many implications for life on Earth. For example, water dissolves plant nutrients from rock and soil, transporting them to the roots of plants. Water droplets in the atmosphere dissolve carbon dioxide, carrying it to Earth's surface where the resulting mildly acidic solution plays the central role in chemical weathering.

Hydrogen bonding also keeps water in a liquid form over a wider range of temperature than any other molecule of similar size. This is because a lot of energy is required to break the multiple hydrogen bonds present in liquid water. Thus hydrogen bonding gives water its high *heat of vaporization* (the energy required to transform it into a vapor), as well as its high *heat capacity* (the energy required to raise its temperature by a given amount). Of known materials, only ammonia (NH_3), which also has hydrogen bonding, has a higher heat capacity than water. These properties are responsible for the moderating effect that water bodies have on Earth's climate.

Water has many other fundamentally important properties. It is one of only a very few substances (elemental bismuth is another) in which the solid form is less dense than the liquid form. This is why ice floats, and why lakes don't freeze solid in the wintertime. Water is also transparent; without this property, which allows sunlight to penetrate, most aquatic life forms would be unable to exist. All life on Earth is fundamentally dependent on water and its unique properties.

which moisture taken up by plant roots passes through the plant and eventually evaporates from the leaf surfaces. Water that falls on the ground and doesn't evaporate may drain off the land surface, becoming **surface runoff**. Some of it may be stored for a time in a surface water body, such as a lake or wetland. Some may slowly penetrate into the soil by **infiltration**, eventually becoming part of the vast reservoir of groundwater. Snow may remain on the ground for one or more seasons until it melts and the meltwater flows away into soils or streams. Snow that nourishes glaciers remains locked up much longer, perhaps for thousands of years, but eventually it too melts or evaporates and returns to the ocean.

WATER ON THE GROUND

If you stand outside during a heavy rain, you can see that, initially, water tends to move downhill in a process called *overland flow* (or *sheet flow*, because the flowing water often takes the form of a thin, broad sheet). After traveling a short distance, overland flow begins to be concentrated into well-defined conduits, thereby becoming streamflow. **Streamflow** consists of *storm flow*, which comes from overland flow into the channel as a result of precipitation, and *base flow*, which is fed by groundwater directly into the channel. Streams that have no base flow dry up seasonally and are said to be *ephemeral*. Streams that have base flow do not dry up and are said to be *perennial*. Overland flow and streamflow together constitute surface runoff, one of the most important pathways in the hydrologic cycle (FIG. 8.4).

The Stream as a Natural System

All streams are part of a complex natural system with many interacting components. A **stream** consists of water flowing downslope in a clearly defined natural passageway; we commonly apply the term *river* to streams of significant size. As it moves, the water in a stream transports particles of sediment (*clasts*; see Chapter 7) and dissolved substances. The passageway is the stream's **channel**, an efficient conduit for carrying water. The stream's **load** is the total of all the sediment and dissolved matter it is transporting. Geologists refer to the sediment load as **alluvium**.

Every stream is surrounded by its **drainage basin** (sometimes called a *watershed* or a *catchment*), the total area that contributes water to the stream (FIG. 8.5A). The line that separates adjacent drainage basins is a topographic "high," or ridge, called a **divide**. Drainage basins range in size from less than a square kilometer to vast areas of subcontinental dimension. The area of any drainage basin is proportional to both the length of the stream that drains the basin and the average annual volume of water that moves through the drainage system. In North America, the huge drainage basin of the Mississippi River encompasses an area that exceeds 40 percent of the area of the contiguous United States and part of southern Canada (FIG. 8.5B).

Streams and drainage systems play a fundamental role in both the hydrologic cycle and the rock cycle. Water evaporates from the ocean into the atmosphere; a portion is precipitated on the land surface, and part of this travels across the land via overland flow and channeled streamflow on its way back to the sea. En route, streams erode the land, transport and deposit sediment, and support complex ecosystems that rely on a dependable supply of water. Earth's streams and drainage systems are constantly evolving in response to changing conditions of topography and relief (influenced by processes in the tectonic cycle), as well as climate and vegetation.

Stream Channels and Streamflow

The size and shape of a stream channel reflect several important controlling factors. They include the erodibility of the rock or sediment across which the stream flows, the steepness of its descent, and the average volume of water

FIGURE 8.4 Where precipitation goes
(A) When precipitation falls, much of it is returned directly to the atmosphere via precipitation—58 percent globally, as shown here, although this proportion varies substantially from one location to another. A small amount infiltrates and percolates to join the groundwater; the rest flows overland and in stream channels, becoming surface runoff as shown in (B), a photo of Mosquito Creek in Sequoia National Park, California.

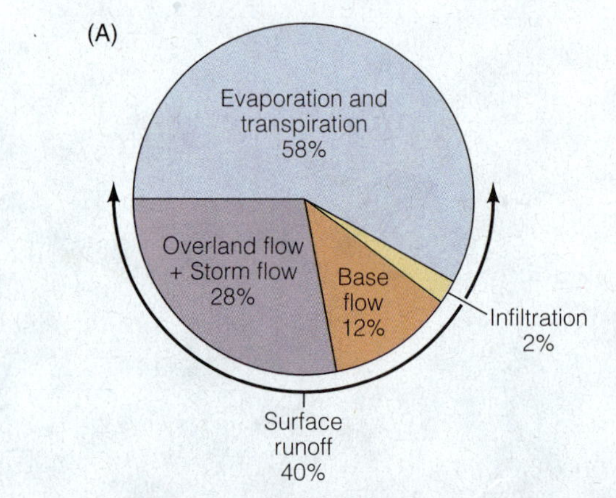

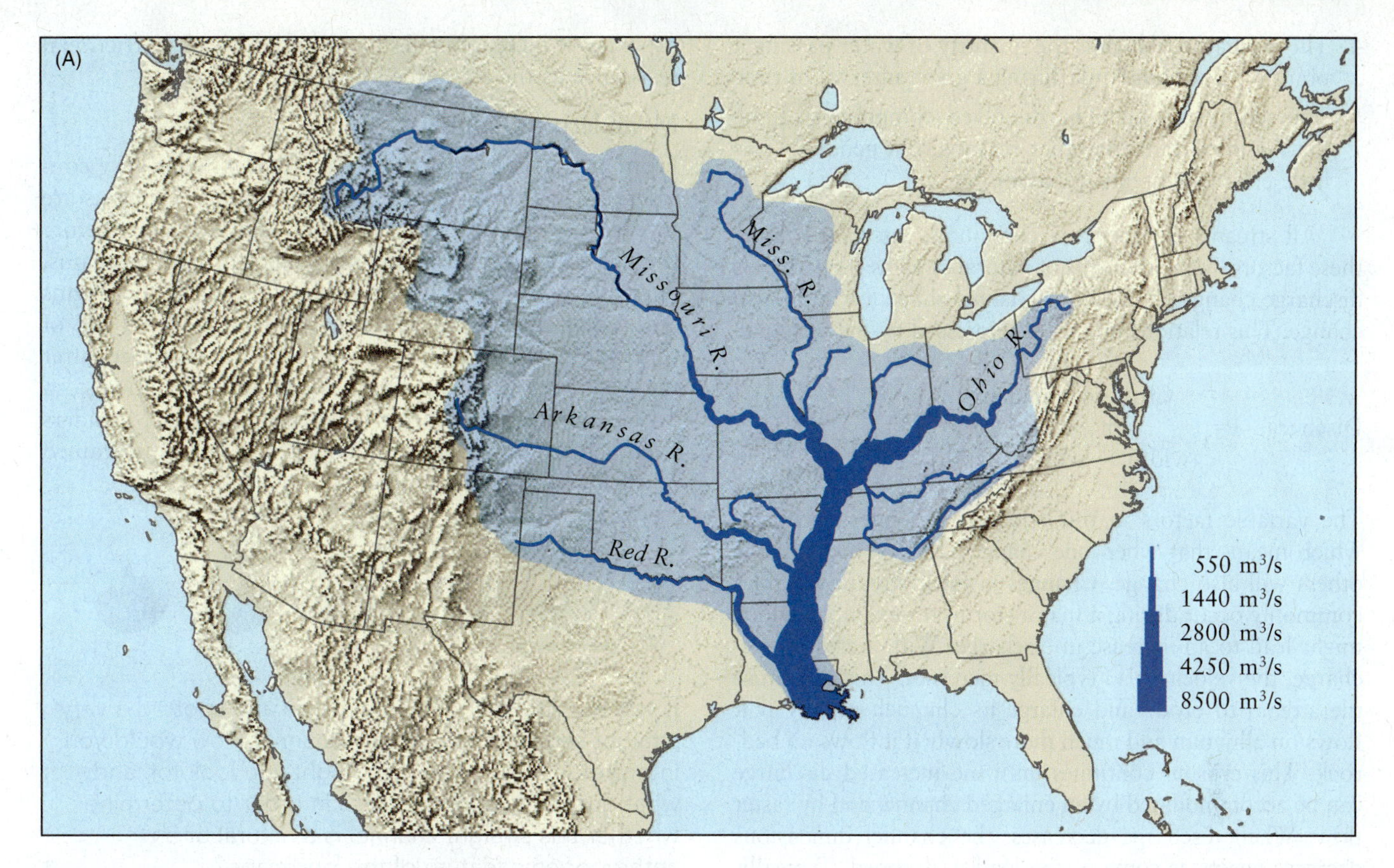

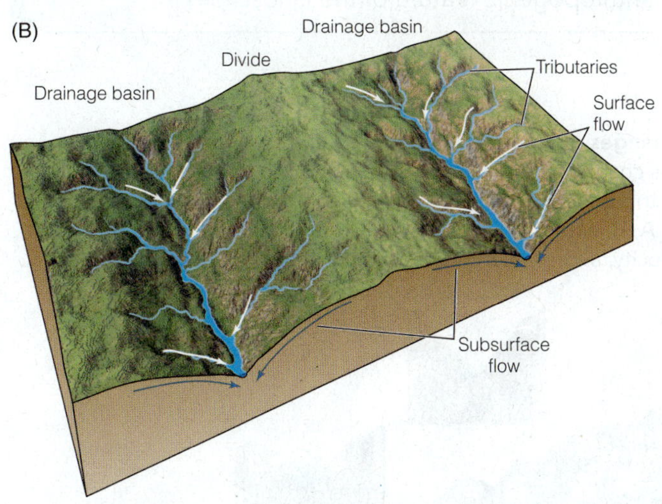

FIGURE 8.5 Drainage basins
(A) A drainage basin is the entire area drained by a given stream. The drainage basin of the Mississippi River encompasses most of the midwestern United States and extends into southern Canada. In this diagram, the widths of the rivers are exaggerated to represent the discharge in cubic meters per second. (B) Adjacent drainage basins are separated by divides, as shown here. Each tributary stream has its own smaller drainage basin.

passing through the channel. Some very small streams are about as deep as they are wide, whereas very large streams often have widths many times greater than their depths. Some channels are relatively simple and straight; others are twisted and complicated, with many branching tributaries.

If we measure the vertical distance that a stream channel descends between two points along its course, we obtain a measure of the stream's **gradient** between those points. The average gradient of a steep mountain stream may reach or exceed 60 m/km, whereas near the mouth of a large river the gradient may be less than 0.1 m/km. Overall, the gradient of a river decreases downstream but not always smoothly, as any white-water rafter or kayaker can attest. A local change in gradient may occur, for example, where a channel passes from resistant rock into more erodible rock, or where a landslide or lava flow forms a natural dam across the channel. At such places, water may tumble rapidly through a stretch of rapids or form a waterfall where it plunges over a steep drop.

Factors that Control Stream Behavior

A stream is thus a complex natural system, the behavior of which is controlled by five basic factors:

1. The average width and depth of the channel
2. The channel gradient
3. The average velocity of the water

4. The **discharge**, which is the quantity of water passing a point on a stream bank during a given interval of time
5. The sediment load (The dissolved component of the load is important for other reasons, but generally has little effect on stream behavior.)

All streams experience a continuous interplay among these factors. Measurements of natural streams show that as discharge changes, velocity or channel shape, or both, also change. This relationship can be expressed by the formula:

$$\text{Discharge} = \begin{array}{c}\text{Cross-sectional area} \\ \text{of channel} \\ (\text{width} \times \text{average depth})\end{array} \times \begin{array}{c}\text{Average} \\ \text{flow velocity}\end{array}$$

The variable factors in this equation are interdependent, which means that when one changes, one or more of the others will also change. Changes in these stream variables commonly occur during a major storm, for example, which might lead to an increase in discharge. With increased discharge, the velocity also typically increases. This can cause the stream to erode and enlarge its channel, rapidly if it flows on alluvium and much more slowly if it flows on bedrock. This erosion continues until the increased discharge can be accommodated by an enlarged channel and by faster flow. When discharge decreases, the channel dimensions decrease again, as some of the load is dropped. Typically, the velocity also decreases. In these ways channel width, channel depth, and velocity continuously adjust to changing discharge, and an approximate balance among the various factors is maintained.

Traveling along a stream from its head (or *source*) to its mouth, we can see orderly adjustments occurring along the channel. For example, the width and depth of the channel increase, the gradient decreases, flow velocity increases, and discharge increases (FIG. 8.6). The fact that velocity increases downstream seems to contradict the common observation that water rushes down steep mountain slopes and flows smoothly over nearly flat lowlands, but the physical appearance of a stream may not be a true indication of its velocity. Discharge is low in the headward reaches of a stream, and average velocity is low because of frictional resistance caused by the water passing over a very rough streambed. Here, where the flow is *turbulent* (agitated or disorderly), the water moves in many directions rather than uniformly downstream. Discharge increases downstream as each **tributary** (a stream joining a larger stream) introduces more water. The cross-sectional area of the channel increases to accommodate the increased volume. Despite a progressive decrease in slope, velocity also increases downstream as a result of the progressive increase in discharge and decrease in frictional resistance as the streambed becomes smoother.

Meandering Channels

From an airplane, it is easy to see that streams vary considerably in size and shape. Straight channel segments are rare and generally occur for only brief stretches before the channel assumes a sinuous shape. In many streams, particularly those with low gradients, the channel forms a series of smooth, looplike bends (FIG. 8.7). A bend of this type in a stream channel is called a **meander**, after the Menderes River (in Latin, *Meander*) in southwestern Turkey, which is noted for its winding course. Meanders occur most commonly in channels cut into fine-grained alluvium, with gentle gradients.

Make the CONNECTION

If you were traveling in an airplane and spotted a very straight segment of a stream channel, how would you interpret it? What evidence might you look for, and what questions might you ask in order to determine whether this straight channel is a natural or an anthropogenic feature of the landscape?

FIGURE 8.6 Changes in stream behavior
Stream properties change as the water moves along the channel. Discharge increases as new tributaries join the stream. Channel width and depth are shown by cross sections A, B, and C. The graphs show the relationship of discharge to width, depth, velocity, and gradient at the same three points along the channel.

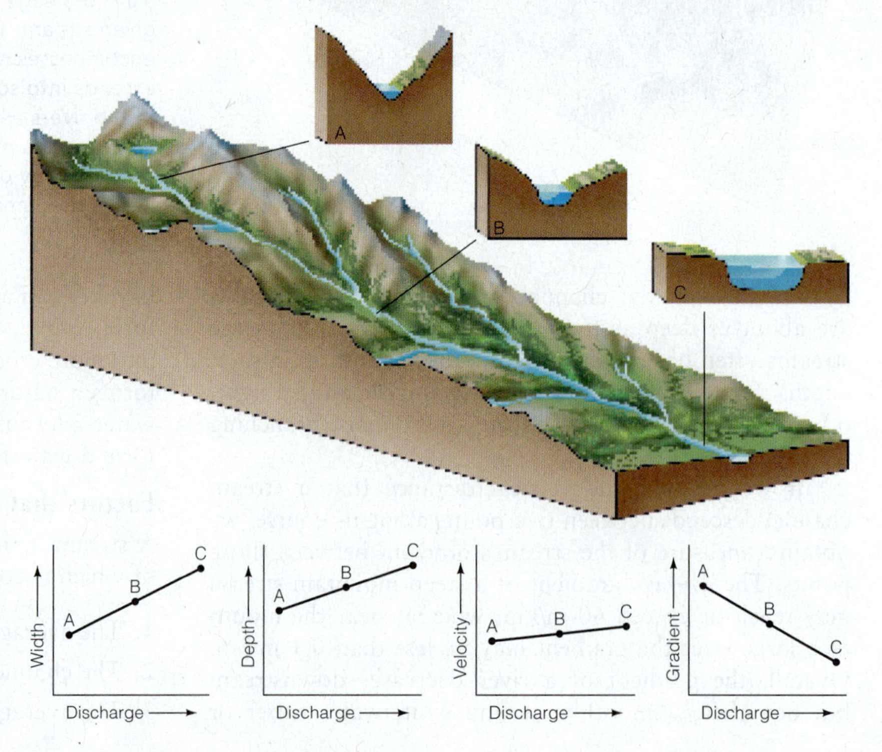

FIGURE 8.7 Meanders
In this meandering stream near Phnom Penh, Cambodia, light-colored point bars, composed of gravelly alluvium, lie opposite the steep banks on the outside of the meander loops. Two oxbow lakes, the product of past meander cutoffs, lie adjacent to the present channel.

The looping pattern of a meandering channel reflects the way in which a stream minimizes resistance to flow and dissipates energy as uniformly as possible along its course. If you try to wade or swim across a stream, it will quickly become apparent that the velocity of the flowing water is not uniform. Velocity is lowest along the perimeter of the channel because this is where the water encounters the greatest frictional resistance to flow. On the inside of a bend the velocity is least, but where the water rounds a bend the zone of highest velocity swings toward the outside of the channel.

The nearly continuous shift, or migration, of a meander is accomplished by erosion on the outer banks of the meander loops, where the velocity is greatest. Along the inner side of each meander loop, the water is shallow and the velocity and energy of flow are lowest. Sediment is therefore deposited on the inside of a bend and accumulates to form a *point bar* (Fig. 8.7). Collapse of the stream banks occurs most frequently along the outer side of a meander bend; there, the high current velocity causes outward erosion and undercutting of the stream bank. In this way, meanders tend to migrate slowly down a valley, progressively removing and adding land along the banks.

Sometimes a stream bypasses the channel loop between the upstream and downstream segments of a meander (**FIG. 8.8A**), cutting it off and converting it into an arc-shaped *oxbow lake*. Because the new course is shorter than the older course, the channel gradient is steeper there and the overall stream length is shortened (**FIG. 8.8B**). Nearly 600 km of the Mississippi River channel has been abandoned through such cutoffs since 1776. However, the river has not been shortened appreciably because the shortening due to cutoffs was balanced by channel lengthening as other meanders were enlarged.

Braided Channels

A stream that is unable to transport the entire available sediment load tends to deposit the coarsest and densest sediment to form a bar, which locally divides and concentrates the flow in the deeper segments of the channel to either side of the bar. As a bar builds up, it may emerge above the stream surface as an island and become stabilized by vegetation that anchors the sediment and inhibits erosion. A stream with many interlacing channels and bars is called a **braided stream.** The intricate geometry of a braided stream resembles the pattern of braided hair, for the water repeatedly divides and reunites as it flows through two or more adjacent, interconnected channels (**FIG. 8.9**).

Large braided rivers typically have numerous shallow channels that change size and shift position as the stream erodes laterally and deposits sediment. Although at any moment the active channels may cover no more than 10 percent of the width of the entire channel system, within a single season all or most of the surface sediment may be reworked by the laterally shifting channels.

A braided pattern tends to form in a stream that has a highly variable discharge and easily erodible banks, and that therefore can supply abundant sediment to the system. Streams of meltwater issuing from glaciers typically have braided patterns because discharges vary both daily and seasonally, and the glacier supplies the stream with large quantities of sediment. The braided pattern, therefore, seems to represent an adjustment by which a stream becomes more efficient in transporting an overabundance of sediment.

A Stream's Load

The energy of flowing water in a stream can be used to erode the rock across which it flows and to transport the

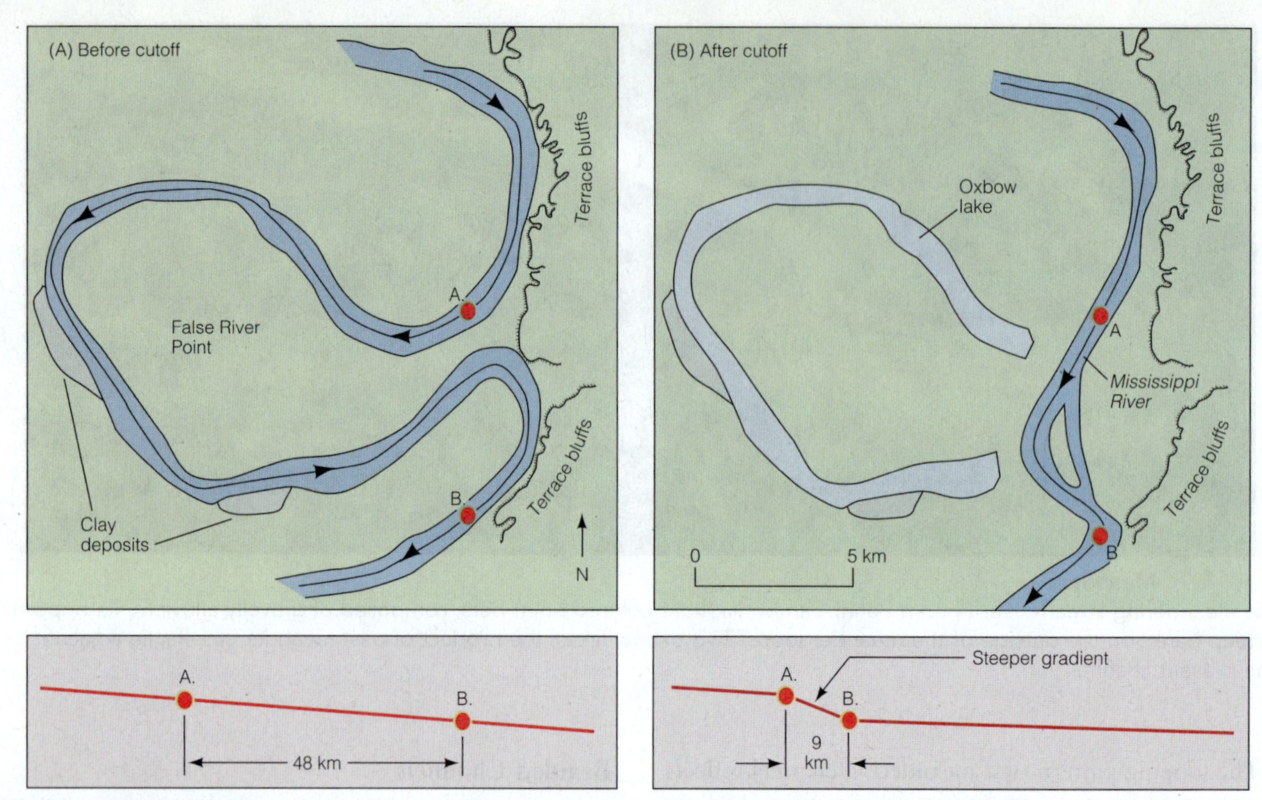

FIGURE 8.8 Oxbow lakes
This diagram shows the cutoff of a meander loop in the Mississippi River in Louisiana. (A) The downstream migration of the loop was halted when the channel segment on the south side of the loop encountered resistant clay deposits. Upstream, the channel continued to advance until it broke through, creating a more direct flow pathway. (B) Over its new, shorter path, the stream has a steeper gradient than in the abandoned course.

FIGURE 8.9 Braided stream
Braided streams typically develop in areas with a low gradient and a high or variable sediment load. This is the Waimakariri River and the Torlesse Range in New Zealand.

resulting sediment downstream. The size of clasts a stream can transport is related mainly to velocity; therefore, we might expect the average clast size of sediment to increase in the downstream direction as velocity increases. In fact, the opposite is true—sediment normally decreases in coarseness downstream. In mountainous headwaters of large rivers, tributary streams mostly flow through channels floored with coarse sediment that may include boulders a meter or more in diameter. Fine sediment is easily moved, however, so it is readily carried downstream, leaving the coarser sediment behind. When a large stream eventually reaches the sea, its load may consist mainly of sediment no coarser than sand.

A stream's load consists of three parts. Two of these, coarse particles that move along the stream's bed (the *bed load*) and fine particles that are suspended in the water (the *suspended load*), comprise the sediment load. In addition, streams carry dissolved substances (the *dissolved load*) that are chiefly a product of chemical decomposition of exposed rock, as well as suspended or floating organic debris. Streams accomplish the work of transporting sediment in these three ways, which are shown in Figure 7.9.

Bed Load

The bed load generally amounts to between 5 and 50 percent of the total sediment load of most streams. The average rate at which bed-load particles move is less than

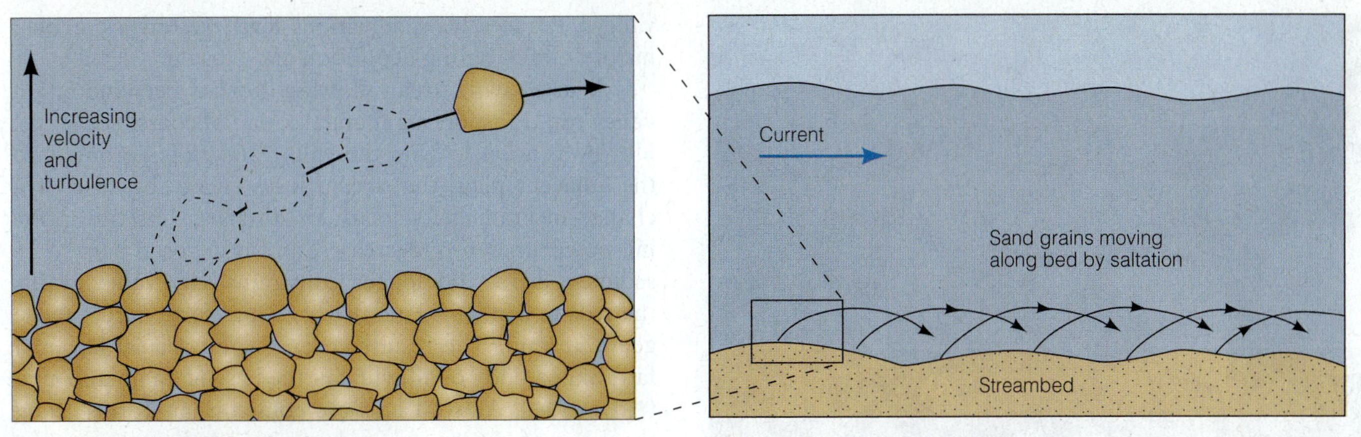

FIGURE 8.10 Bed-load saltation
A stream's bed load moves by saltation. Clasts are carried up into a stream at places where turbulence locally reaches the bottom or where suspended grains impact other grains on the bed. Once raised into the flowing water, the clasts are transported along arc-shaped trajectories as gravity pulls them toward the streambed, where they impact other particles which, in turn, are set in motion.

that of the water, for the particles are not in constant motion. Instead, they move discontinuously by rolling or sliding. Where forces are sufficient to lift a particle of the bed load off the streambed, it may move short distances by *saltation*, the progressive forward movement of a particle via short, intermittent jumps along arc-shaped paths (FIG. 8.10; compare to Fig. 7.9). Saltation continues as long as currents are sufficiently rapid or turbulent to lift particles and carry them downstream.

The distribution of bed-load sediment in a stream is generally related to the distribution of water velocity within the channel (FIG. 8.11). Coarse-grained sediment is concentrated where the velocity is high, whereas finer grained sediment is relegated to zones of progressively lower velocity.

Suspended Load

The muddy character of many streams results from particles of silt and clay carried in suspension. Most of the suspended load is derived from fine-grained regolith washed from areas unprotected by vegetation and from sediment eroded and reworked by the stream from its own banks. China's Huang He (or "Yellow" River) derives its color from the great load of yellowish silt it erodes from the thick, unconsolidated deposits of wind-blown dust that cover much of its catchment basin (FIG. 8.12).

Because upward-moving currents within a turbulent stream exceed the velocity at which particles of silt and clay can settle toward the bed under the pull of gravity, such particles tend to remain in suspension longer than they would in non-turbulent waters. They settle and are deposited only where velocity decreases and turbulence ceases, as in a lake or in the sea.

Dissolved Load

Even the clearest stream water contains dissolved substances that constitute part of its load. Seven ions comprise the bulk of the dissolved content of most rivers:

FIGURE 8.11 Bed-load grain size
The grain size of clasts in a stream's bed load varies with the velocity of flow. In this meandering channel, the coarsest sediment is associated with the zone of highest velocity, on the outside of the bend. The finest sediment is associated with the zone of lowest velocity, on the inside of the bend.

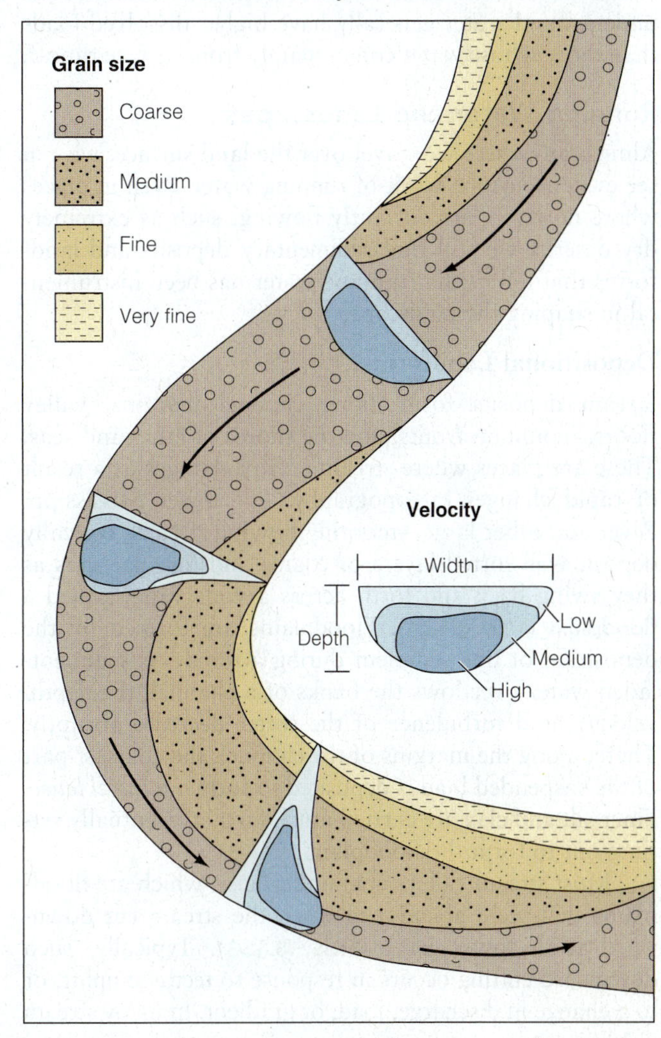

FIGURE 8.12 Suspended load
A large suspended load, eroded from extensive deposits of loose, buff-colored wind-deposited silt, gives the Huang He its characteristic yellowish color.

bicarbonate [$(HCO_3)^-$], calcium (Ca^{2+}), sulfate [$(SO_4)^{2-}$], chloride (Cl^-), sodium (Na^+), magnesium (Mg^{2+}), and potassium (K^+). In some streams the dissolved load may represent only a small percentage of the total load. However, streams that receive large contributions of underground water generally have higher dissolved loads than those whose water comes mainly from surface runoff.

Running Water and Landscapes

Almost anywhere we travel over the land surface, we can see evidence of the work of running water. Even in places where no rivers are currently flowing, such as extremely dry deserts, we still find sedimentary deposits and landforms that tell us that running water has been instrumental in shaping the landscape.

Depositional Landforms

Stream deposits form along channel margins, valley floors, mountain fronts, and the shores of lakes and seas. These are places where stream energy changes as a result of rapid changes in topography. The lower Mississippi River and other large, smoothly flowing streams typically deposit well-sorted layers of coarse and fine particles as they swing back and forth across a wide valley called a **floodplain** (FIG. 8.13A). Floodplains are formed by the deposition of fine sediment during a flood. As sediment-laden water overflows the banks of a channel, the depth, velocity, and turbulence of the water decrease abruptly. There, along the margins of the channel, the coarsest part of the suspended load is deposited to form a *natural levee*. Finer silt and clay are carried much farther, eventually settling out to form the floodplain.

Most stream valleys contain terraces, which are floodplains that were abandoned when the stream cut downward to a lower level (FIGS. 8.13B). Typically, such downward cutting occurs in response to tectonic uplift, or to a change in discharge, load, or gradient. In many stream valleys, terraces lying at various levels record a complex history of alternating deposition and erosion.

A large, swift stream flowing down a steep mountain valley can transport an abundant load of coarse sediment. However, upon leaving the valley and flowing out onto the adjacent plain the stream loses energy because of a change in gradient, velocity, or discharge. Its transporting power therefore decreases, and it deposits part of its sediment load. No longer constrained by valley walls, the stream can shift laterally back and forth across more gentle terrain. The resulting deposit, an **alluvial fan**, is a fan-shaped body of alluvium at the base of an upland area (FIG. 8.13C). Alluvial fans are common landforms along the base of most arid and semiarid mountain ranges. Some fans are so large and closely spaced that they merge to form a broad piedmont surface that slopes away from the base of the mountains.

When stream water enters the standing water of the sea or a lake, its speed diminishes rapidly, decreasing its ability to transport sediment. The water deposits its load of suspended sediment in the form of a **delta**, so named because the deposit may develop a crudely triangular shape that resembles the Greek letter delta (Δ) (FIG. 8.13D). Most of the world's largest rivers, among them the Nile, the Ganges-Brahmaputra, the Huang He, the Amazon, and the Mississippi, have built massive deltas at their mouths. Each delta has its own peculiarities, determined by such factors as the stream's discharge, the character and volume of its sediment load, the shape of the adjacent bedrock coastline, the offshore topography, and the strength and direction of currents and waves.

Tectonic and Climatic Controls on Divides

All continents except ice-covered Antarctica can be divided into large regions from which major through-flowing rivers enter one of the world's major oceans. The line separating any two such regions is a **continental divide**. In North America, continental divides lie at the head of major streams that drain into the Pacific, Atlantic, and Arctic oceans (FIG. 8.14, on page 236). In South America, a single continental divide extends along the crest of the Andes and divides the continent into two regions of unequal size. Streams draining the western (Pacific) slope of the Andes are steep and short, whereas to the east the streams take much longer routes along more gentle gradients to reach the Atlantic shore.

Because continental divides often coincide with the crests of mountain ranges, and because mountain ranges are the result of tectonic uplift, a close relationship must exist between plate tectonics and the locations of stream divides and drainage basins. The location of a divide plays an important role in determining the climate of a region, but—perhaps surprisingly—climate, in turn, can influence the location of the divide and the evolution of the landscape itself. When a divide lies close to a continental margin, there is typically a strong climatic gradient across it. An unequal distribution of precipitation can lead to a marked landscape asymmetry across the divide. Streams

(A)

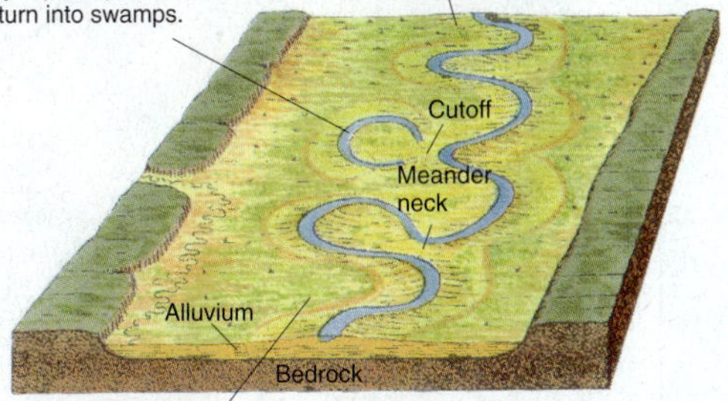
Oxbow lake After the cutoff, silt and sand are deposited across the ends of the abandoned channel, producing an oxbow lake. The oxbows fill with fine sediment and organic matter produced by aquatic plants and eventually turn into swamps.
Floodplain The meandering river channel dominates the floodplain, along with stretches of abandoned channels. These were abandoned after cutoff events, when the river cuts across a meander loop to create a shorter, more direct path.
Cutoff
Meander neck
Alluvium
Bedrock
Natural levees are created during overbank flooding, when sand and silt are deposited next to the channel creating belts of higher land on either side of the channel. Deposition is heavier closest to the channel, so the levee surface slopes away from the channel.

(B)

(C)

(D)

FIGURE 8.13 Streams and landforms
(A) Some of the landforms created by sediment in stream valleys are floodplains and natural levees. (B) Stream terraces adjacent to Cave Stream, New Zealand, record former floodplains that were abandoned when the stream incised its channel and reached a new level. (C) Where this stream emerges from the mountains into Death Valley, California, it abruptly slows and deposits its sediment load. This has created a symmetrical alluvial fan, covered by a braided system of channels. The stream was dry at the time the photograph was taken. (D) Where the Nile River empties into the Mediterranean Sea, the sediment it deposits has formed a triangle-shaped delta that supports the green vegetation seen in this satellite image.

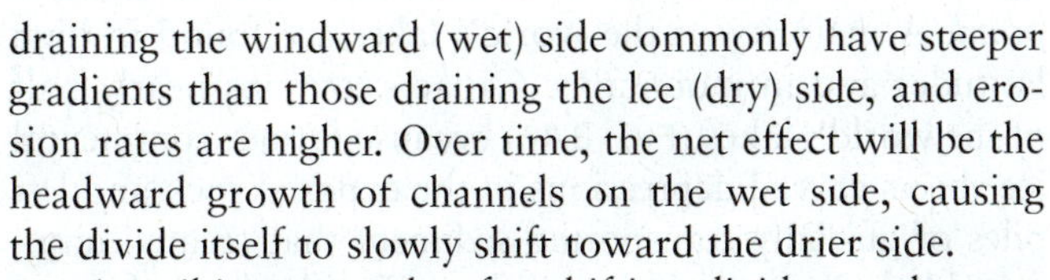

draining the windward (wet) side commonly have steeper gradients than those draining the lee (dry) side, and erosion rates are higher. Over time, the net effect will be the headward growth of channels on the wet side, causing the divide itself to slowly shift toward the drier side.

A striking example of a shifting divide can be seen in the Andes of southern South America (**FIG. 8.15**). Through nearly 60 degrees of latitude, from the northern tip of the continent to well south of Santiago, Chile, the drainage divide coincides with the topographic crest of the Andes. However, south of 45°, where easterly winds give way to westerlies flowing off the Pacific, the continental divide has shifted eastward in several locations. During glaciations, the mountain ice cap of the southern Andes apparently was thickest over the fjord region of southern Chile, which received heavy snowfall from the moist westerly winds flowing onshore. Outlet glaciers of a large ice cap flowed eastward across saddles in the main Andean

FIGURE 8.14 Continental divides
This map of North and South America shows the location of continental drainage basins and divides.

FIGURE 8.15 Climate divide
In southern Chile, the continental divide coincides with the topographic crest of the Andes. In some locations, however, the drainage divide (red dashed line) lies well east of the range crest, where it has migrated partly as a result of climatic influences.

crest, progressively deepening and ultimately eliminating these topographic barriers. In several sectors, the continental divide between the Pacific and Atlantic oceans now lies along the crest of the *moraines*—glacial deposits that mark the limits of the glaciers that terminated near or beyond the eastern front of the mountains.

Surface Water Reservoirs

Water moves on the surface via channeled streamflow and overland flow; it also can remain stored for a time in any of several different types of natural surface water reservoirs. The greatest of these reservoirs are the perennially frozen ice caps and glaciers of the cryosphere, and the saltwater ocean; we will consider these reservoirs in detail in Chapters 9 and 10, respectively. Although they are of much lesser volume, surface water reservoirs like lakes and wetlands are of crucial importance as a freshwater resource and for their support of ecosystems.

Lakes

A **lake** is a body of inland water of appreciable size with an open surface, occupying a depression in Earth's surface. A smaller body of fresh standing water is called a *pond*. A majority of the world's lakes are found in high latitudes and in mountains. Canada contains nearly half of the world's lakes (FIG. 8.16) because former continental ice sheets carved depressions in the exposed rock and left piles of glacially transported sediment that created innumerable hollows and natural dams. Besides being formed by glaciation, lakes can be formed by volcanism (such as crater and caldera lakes); tectonism (by the formation of basins, rift valleys, and other down-dropped terrain);

FIGURE 8.16 **Canadian lakes**
A multitude of lakes occupies glacially scoured basins in the frozen tundra of Canada's Northwest Territories. Some long, narrow lakes are aligned along faults in the underlying bedrock.

streamflow and sediment deposition (including oxbow lakes and tributary alluvial fans that dam stream valleys); formation of dams as result of landslides; collapse of cave systems; collapse of ice dams or thawing of permafrost; and various coastal processes.

Lakes are predominantly fed by runoff and direct precipitation, and their level and salinity commonly reflect a balance between freshwater input from streams and direct precipitation, outflow, and evaporation. In a small basin, the level of a lake may be controlled by the water table, with water table fluctuations controlling lake-level fluctuations. Most lakes contain fresh water. However, many lakes in arid and semiarid regions, where evaporation rates are high, have a large dissolved salt content (saline lakes). The salty Caspian Sea in Central Asia is the lake with the largest area (144,000 km^2), whereas Siberia's freshwater Lake Baikal is the deepest (1742 m). The five Great Lakes of Canada and the United States contain approximately one-fifth of the world's fresh surface water.

Lakes tend to be transitory features on the landscape. Few are more than a million years old, and most are no older than the end of the last glacial age, about 12,000 years ago (Chapter 13). A lake may drain away if its outlet becomes deeply eroded, or it may disappear if the climate leads to a negative water balance, in which more water is lost by evaporation and outflow than is added by inflow and direct precipitation. Small lakes may disappear with a drop in the water table. A lake may also gradually become shallow and disappear through the filling of its basin with organic and inorganic sediment, forming a wetland or bog. Scientists decipher the history of lakes by studying sediment cores. These contain a record of changing water chemistry, sediment input, and biologic productivity, typically related to changes in local or regional climate and human activity over time.

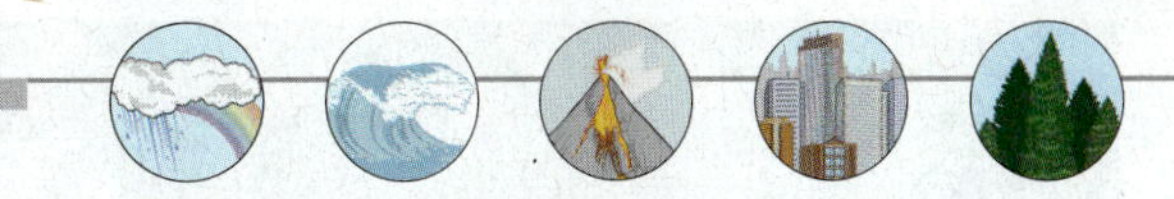

Make the CONNECTION

Are the reservoirs that form behind large hydroelectric dams similar to natural lakes? In what ways are they different?

Wetlands

A freshwater **wetland** is an area that is either permanently or intermittently moist. Wetlands, including swamps, marshes, and bogs, may or may not contain open, standing water, although most wetlands are covered by shallow water for at least part of the year. Wetlands tend to be highly biologically productive, with dense vegetation, migratory birds, reptiles, amphibians, and fish (FIG. 8.17).

Some wetlands are formed as a result of the natural development of a surface freshwater body. For example, a lake whose source of incoming water has been cut off will become shallower over time, gradually filling with sediment and organic material to become a peat bog or wetland, as mentioned above. This infilling and shallowing process is called **eutrophication**. Other wetlands occur in coastal areas, where the ground is water-saturated. Wetlands in coastal regions are transitional between freshwater and saltwater environments. These transitional environments include estuaries, salt marshes, and mangroves, which we will consider in greater detail in Chapter 10.

Wetlands were once considered to be dirty, mosquito-infested quagmires, prime targets to be drained and developed. More recently, wetlands have been recognized as natural storehouses for a great diversity of plant and animal species. Wetlands also perform many important

FIGURE 8.17 **Wetland**
Wetlands are surface water bodies that are intermittently or seasonally wet. They include a wide variety of bogs, marshes, and coastal wetlands, like this cypress swamp in Charleston, South Carolina.

FIGURE 8.18 **Mississippi River flood**
A pair of satellite images shows the region where the Missouri River joins the Mississippi River at St. Louis, Missouri. The photo on the left shows a dry summer with low flows (July 1998). The photo on the right shows the same region in July 1993. Weeks of rain caused the rivers to overflow their levees. Numerous towns, along with 44,000 km^2 of farmland in nine states, were flooded by an estimated 3 km^3 of floodwater.

environmental services, including storing groundwater and removing toxins from the soil.

Floods: When There's Too Much Water

The uneven distribution of rainfall through the year results in imbalances among the reservoirs of the hydrosphere. The result is that many streams rise seasonally in flood. A **flood** occurs when a stream's discharge becomes so great that it exceeds the capacity of the channel, so that water overflows the banks (FIG. 8.18). Lakes flood, too, mainly in storm conditions with high winds or when a natural or constructed dam fails. In the case of coastal flooding, the inflow of water from the ocean, rather than the runoff of water from the land, does most of the damage. The *storm surge* associated with Hurricane Katrina in 2005 temporarily raised the water level near New Orleans by 6 m or more and breached the levees even before the main force of the storm hit the city (FIG. 8.19).

Impacts of Flooding

People affected by floods are often surprised and even outraged at what the rampaging water has done to them. However, floods are normal and expected events that have been occurring throughout Earth's history—as long as there has been a hydrosphere, about 4.4 billion years. Major floods occur infrequently, perhaps only once in a century or more, but they can be devastating for people living in the area. The Huang He has a long history of catastrophic floods. In 1887, the river inundated 130,000 sq km and swept away many villages in the heavily populated floodplain. In 1931, another Huang He flood killed a staggering 3.7 million people. Yet these same floods transport and deposit abundant sediment, replenishing the mineral content of soil in the floodplain. This results in highly fertile soils, which explains why people keep moving back to the area.

Although major floods are thankfully rare, there is evidence of ancient floods of much greater volume and extent than any known historic floods. An example is seen in the Channeled Scabland in eastern Washington State, where the landscape consists of steep cliffs and plunge pools marking the sites of former huge waterfalls, deep basins carved in the basalt bedrock, massive gravel bars with enormous boulders, gravel deposits in the form of gigantic ripples, and scoured land that extends for hundreds of meters. The source of the enormous volume of floodwater that left these features was an ice-impounded lake that contained between 2000 and 2500 km^3 of meltwater from the ice sheet that covered western Canada during the last glaciation. When the ice dam failed, water was released rapidly from the basin and surged down the Columbia River to the Pacific Ocean.

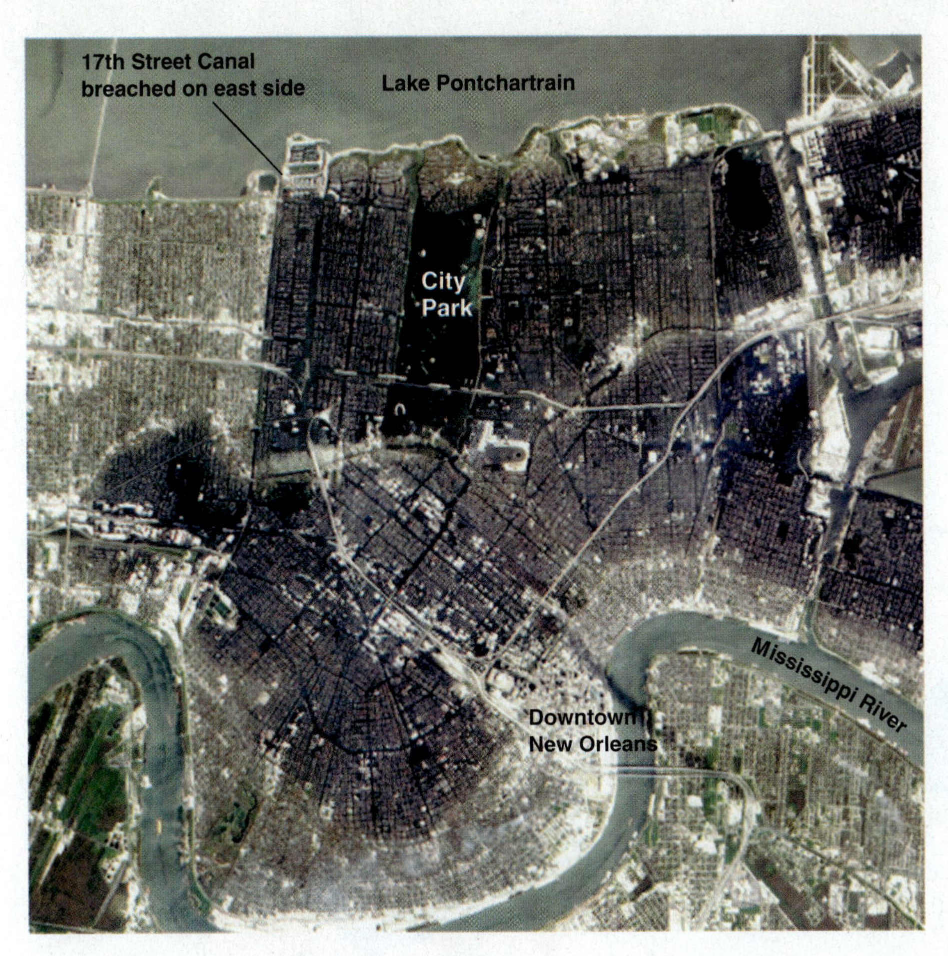

FIGURE 8.19 Hurricane Katrina coastal flooding After Hurricane Katrina hit the Louisiana coast of the Gulf of Mexico in August 2005, high waters breached protective levees in the city of New Orleans. In this satellite view, a week after the hurricane, dark areas are flooded, and Lake Pontchartrain, from which much of the water came, is at the top. The 17th Street Canal, where one of the levee breaks occurred, is clearly visible as the western edge of the flooded region.

During a flood, scientists record the passage of water on a *hydrograph*, a graph that indicates the stream discharge as a function of time. The highest or *peak* discharge in a stream flood usually comes well after the rains that produced it. After an interval of rainfall (FIG. 8.20), the surface runoff moves into stream channels, with the result that the channel discharge quickly increases. The *crest* of the resulting flood—the time when the peak flow passes the location where the measurements are being made—occurs later. This may be days or weeks later, in the case of a regional flood of long duration, or hours or even minutes later, in the case of a shorter but more intense *flash flood*. Eventually, the flood runoff passes through the channel, and the discharge returns to its normal level.

As a stream's discharge increases during a flood, so does the flow velocity. This enables the stream to carry a greater load, as well as larger particles. The collapse of the large Saint Francis Dam in Southern California in 1928 provides an extreme example of the exceptional force of floodwaters. When the dam gave way, the impounded water was released and rushed down the valley as a gigantic flood, moving blocks of concrete weighing as much as 9000 metric tons through distances of more than 750 m. Streams in flood can make dangerous projectiles out of trees, cars, and other debris. During the Great Flood of the Mississippi River in 1993 a propane tanker was ripped from its mooring by the surging river, putting parts of St. Louis at great risk.

Flood Prediction

Because floods can be so damaging and dangerous, it has become essential to predict them them. Scientists plot the frequency of past floods of different sizes on a graph, producing a *flood-frequency curve* (FIG. 8.21, on page 241). The average time interval between two floods of the same magnitude is called the *recurrence interval*. For example, a "10-year flood" has a recurrence interval of 10 years, which means that there is a 1-in-10 (or 10 percent) chance that such a flood will occur in any given year. A flood with an even greater discharge having a recurrence interval of 50 years would be termed a "50-year flood" for this particular stream and would have a 1-in-50 (2 percent) chance of occurring in any given year.

Regional planners should (and do) keep recurrence intervals in mind when planning development on or near a flood plain. However, the recurrence interval is strictly statistical and exerts no real influence on the timing of the next big flood. An area could experience a "50-year flood" one season, and then another "50-year flood" in the following rainy season, depending on the weather conditions. In fact, once a flood of great magnitude has occurred in an area, the widespread saturation of the ground may make it more difficult for water to infiltrate,

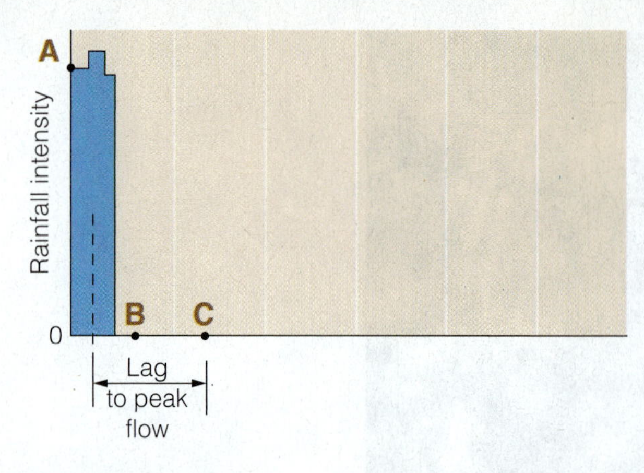

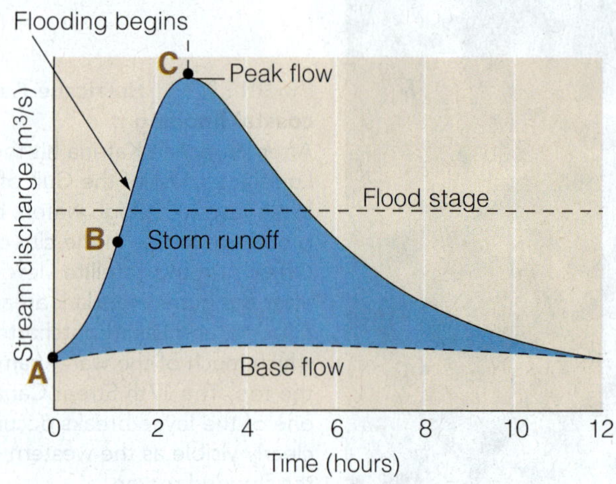

*Base flow is the "normal" flow of water in a stream, contributed by groundwater.

A Onset of storm (0 hours):
The peak discharge is delayed as the runoff collects and runs down the stream channel.

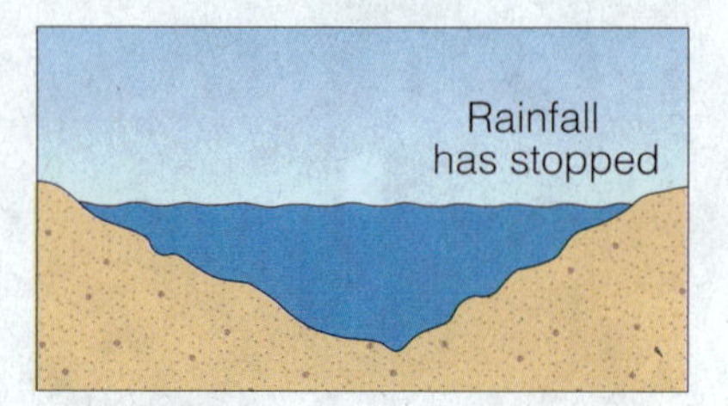

B 1 hour:
One hour after the cloudburst, the stream can still contain the increased volume.

C 2 hours:
After two hours, the stream reaches its peak flow and cannot be contained by its banks anymore.

FIGURE 8.20 Flood hydrographs
Flood hydrographs allow scientists to record and predict the progress of a flood by showing the change in discharge over time.

increasing the probability that a flood will occur in the next flood season.

Another aspect of flood prediction is the real-time monitoring of storms and water levels. Scientists can combine information about the weather with their knowledge of a river basin's geology to forecast the peak height of a flood and the time when the crest will pass a particular location. Such forecasts, which are made with the aid of remote sensing data, computer models, and integrated mapping and data management technologies known as *Geographic Information Systems (GIS)*, can be very useful for planning evacuation or defensive measures.

Flood Prevention and Channelization

Throughout history, people have been unsatisfied with simply predicting floods and have attempted to prevent them. River channels are often modified or "engineered" for the purpose of flood control and protection, as well as to increase access to floodplain lands, facilitate transportation, enhance drainage, and control erosion. The modifications usually consist of some combination of widening, deepening, straightening, clearing, or lining of the natural channel, or building flood walls or levees that effectively increase the height of the stream banks. These approaches are collectively called **channelization.**

Like dams, channelization projects undeniably contribute to our economic well-being, but at a price. Many opponents believe that channel modifications interfere with natural habitats and ecosystems. The aesthetic value of the river can be degraded and water pollution aggravated. A well-known example of poorly designed channelization is the Kissimmee River in south Florida. The straightening and deepening of the channel and the installation of concrete liners, undertaken for the purposes of flood control and irrigation, began in 1962 and

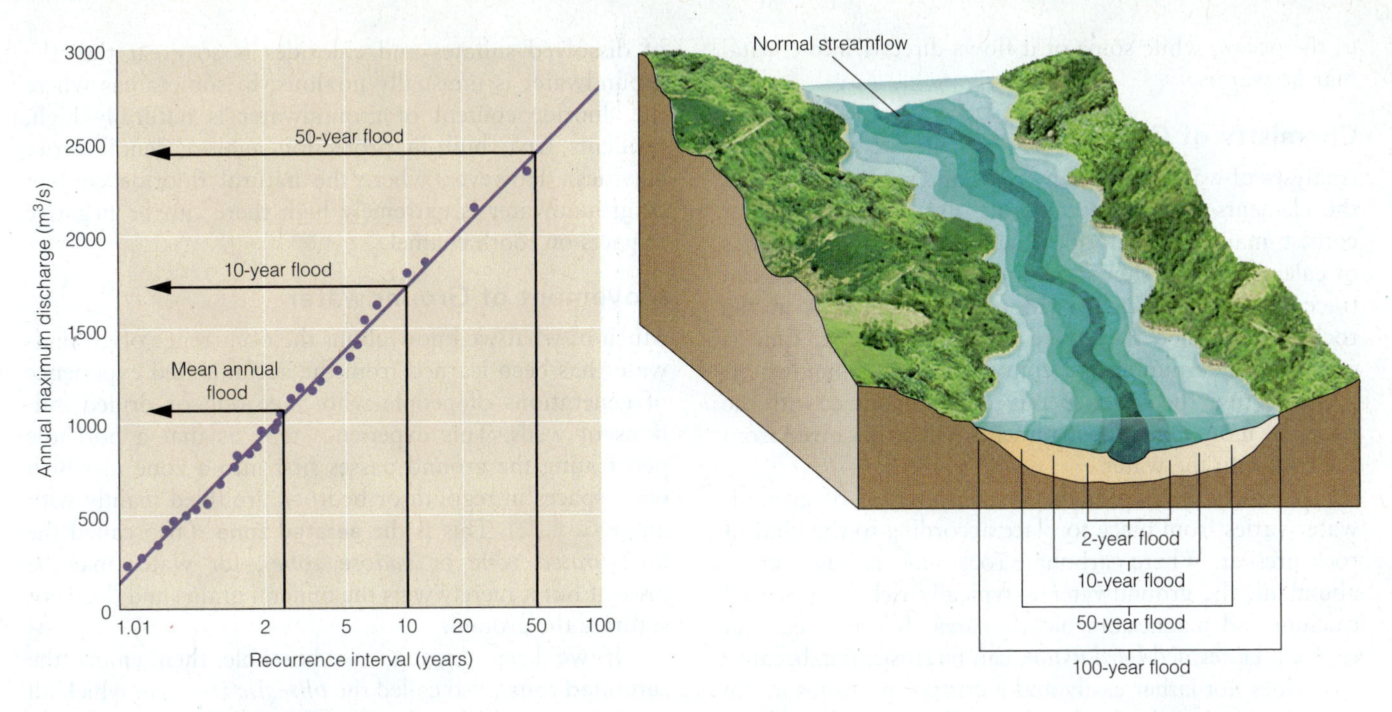

FIGURE 8.21 Predicting floods

took nine years and $25 million to complete. The channelization did not provide the anticipated flood control, but it did cause extensive damage to aquatic habitats and water quality. In the 1990s the state of Florida began an ambitious project to restore and rehabilitate portions of the Kissimmee River channel. In the end, the restoration will cost more than the original channelization project.

Both river engineering and other kinds of human activity may actually increase the chance that flooding will occur (instead of decreasing it). For example, urban construction on compressible sediments, often accompanied by withdrawal of groundwater, can lead to *subsidence*, which increases the danger of flooding (as it did in New Orleans in 2005, for example). The installation of storm sewers and the increase in impermeable ground cover associated with urbanization can contribute to flooding in urban areas because they allow the runoff from paved areas to reach the river channel more quickly. Floods in urbanized basins often have higher peak discharges and reach their crest more quickly than floods in undeveloped basins. This quicker peak time means that people living in a flood-prone area must be able to move very quickly, or be able to predict when a flood might strike. Even projects specifically undertaken for the purpose of flood control, such as the installation of flood walls and levees, may control flooding in the immediate area but contribute to more intense flooding downstream.

A final problem with channelization is that any modification of a channel's course or cross section renders invalid the hydrologic data collected there in the past. During the Mississippi River floods of 1973 and 1993, experts could not account for water levels that were higher than predicted by the historical data; the likely cause was extensive upstream modifications of the river channel by humans.

WATER UNDER THE GROUND

Less than 1 percent of the liquid water in the hydrosphere lies beneath the land surface, occupying openings in bedrock and regolith in the water-saturated portion of the upper geosphere. All of this subsurface water is called *groundwater*. Although the amount of groundwater as a percentage of the entire hydrosphere is small, it is still 35 times larger than the volume of all the water in freshwater lakes and flowing in streams, and nearly a third as large as the water frozen in all the world's glaciers and polar sea ice. More than half of all groundwater, including most of the water that is usable, occurs within about 750 m of Earth's surface. The volume of water in this zone is estimated to be equivalent to a layer of water approximately 55 m thick, if spread over the world's land areas.

Groundwater operates continuously as an integral part of Earth's water cycle (Figs. 8.1 and 8.3). Part of the water evaporated from the ocean falls on the land as rain, seeps into the ground, and enters the groundwater system, flowing under the influence of gravity. Some of this slowly moving underground water reaches stream channels and contributes to the water carried by streams to the ocean, while some of it flows directly into coastal marine waters.

Chemistry of Groundwater

Analyses of water from many wells and springs show that the elements and compounds dissolved in groundwater consist mainly of chlorides, sulfates, and bicarbonates of calcium, magnesium, sodium, and potassium. We can trace these substances to the common minerals in the rocks from which they were dissolved. Contact time is crucial in determining the composition of groundwater; the more time the water spends in direct contact with the rock, the more mineral constituents will be dissolved from the rock into the water.

As might be expected, the composition of groundwater varies from place to place according to the kind of rock present. Where carbonate rock such as limestone is abundant, the groundwater is typically rich in dissolved calcium and magnesium bicarbonates. Taking a bath in such water, termed *hard water*, can be frustrating because soap does not lather easily and a crusty ring forms in the tub. Hard water also leads to deposition of scaly carbonate crusts in water pipes, eventually restricting flow. By contrast, water that contains little dissolved matter and no appreciable calcium is called *soft water*. With it, we can easily get a nice soapy lather in the shower.

Groundwater in rock that has high arsenic or lead content may dissolve these toxic elements, making it dangerous to drink. Water circulating through sulfur-rich rock may contain dissolved hydrogen sulfide (H_2S), which, though harmless to drink, has the disagreeable odor of rotten eggs. In some arid regions, the concentration of dissolved sulfates and chlorides is so great that the groundwater is unusually noxious. In some areas where the fluoride content of groundwater is naturally high, residents have built-in protection against dental caries (cavities). However, where the natural fluoride content of groundwater is extremely high there can be negative impacts on tooth enamel.

Movement of Groundwater

Much of what we know about the occurrence of groundwater has been learned from the accumulated experience of generations of people who have dug or drilled millions of wells. This experience tells us that a borehole penetrating the ground passes first into a zone in which open spaces in regolith or bedrock are filled mainly with air (FIG. 8.22). This is the **aerated zone** (also called the *unsaturated zone* or *vadose zone*), for water may be present but it merely wets the mineral grains and does not saturate the ground.

If we keep digging, the borehole then enters the **saturated zone** (also called the *phreatic zone*), in which all openings are filled with water. We call the upper surface of the saturated zone the **water table.** Alternatively, you can think of the water table as the level below which the ground is water-saturated; therefore, if you drill a well, the top surface of the water that rises in the well is the water table.

Whatever its depth, the water table is a significant surface, for it represents the upper limit of all readily usable groundwater. In humid regions, the water table is a subdued imitation of the ground surface above it—higher beneath hills and lower under valleys. If all rainfall were to cease, the water table would slowly flatten and gradually approach the levels of the valleys; seepage of water into the ground would diminish, then cease, and the streams in the valleys would dry up. In times of drought, when rain may not fall for several weeks or even months, wells can dry up as the water table falls below their bottoms. Repeated rainfall, dousing the ground with fresh supplies of water, maintains the water table at a normal level.

Most groundwater within a few hundred meters of the surface is flowing. Groundwater flow, called **percolation**, includes the vertical flow of water from the surface down to the water table, and the lateral flow of water in the saturated zone. Unlike the swift flow of rivers, groundwater moves very slowly. The reason for the difference is that the water of a stream flows unimpeded through an open channel, whereas groundwater must move through small, constricted

FIGURE 8.22 Water under the ground
A well first passes through the zone of aeration, where pores in the soil are filled with both air and water. Eventually, it reaches the zone of saturation, where the pore spaces are completely filled with water. Underground water exists everywhere, and surface water occurs wherever the ground intersects the water table, the top of the saturated zone.

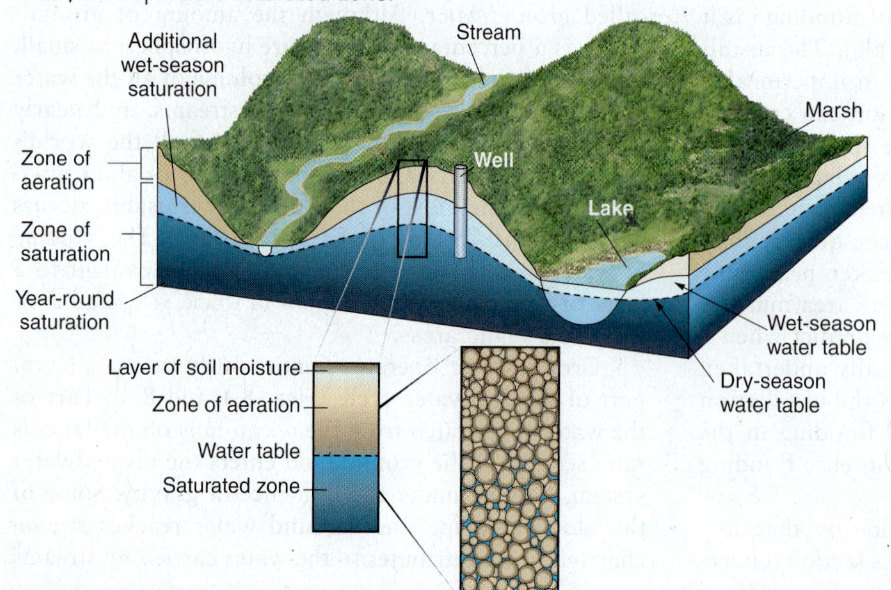

passages along innumerable curving, threadlike paths. Normally, groundwater flow velocities range from half a meter per day to several meters per year, rarely ranging up to hundreds of meters per year in exceptionally permeable material; in contrast, surface water flow may be measurable in terms of kilometers per hour. The flow of groundwater to a large degree depends on the nature of the rock or sediment through which it moves, and the most important characteristics of the surrounding medium are its *porosity* and *permeability* (see *The Basics: Porosity and Permeability*).

Recharge and Discharge

Responding to gravity, groundwater flows from areas where the water table is high toward areas where it is lowest (**FIG. 8.23**). In other words, it flows toward surface streams flowing in valleys, toward lakes and wetlands in low-lying areas, or toward the ocean. Some groundwater flows along paths that go very deep into the ground, responding to *hydrostatic pressure* (pressure due to the weight of overlying water at higher levels in the zone of saturation), but some travels along shallow paths not far beneath the water table.

The replenishment of groundwater occurs when rainfall and snowmelt enter the ground in areas of **recharge**, where precipitation seeps into the ground and reaches the saturated zone. The water then moves through the groundwater system to areas of **discharge**, where subsurface water emerges on the ground surface or is discharged to streams, lakes, ponds, or wetlands. The extent of a recharge area is invariably larger than that of the discharge area (Fig. 8.23). In humid regions, recharge areas encompass nearly all the landscape except streams. In more arid regions, recharge occurs mainly in mountains and along the channels of major streams that are intermittently dry and underlain by permeable alluvium. The time it takes for water to move through the ground from a recharge area to the nearest discharge area depends on rates of flow and the travel distance. It may take only a few days, or it may take many thousands of years in cases where water moves through the deeper parts of the groundwater system.

Like surface water, groundwater flows under the influence of gravity toward the lowest spots on Earth, the great ocean basins. Along the world's coastlines, groundwater can flow directly into the ocean through porous rock and regolith by means of submarine groundwater discharge. Examples of very rapid discharge through porous limestone are known in the Mediterranean and Red seas, and upwelling plumes of discharged groundwater were a source of fresh water for ancient mariners in the Persian Gulf.

Aquifers, Wells, and Springs

If we wish to find a reliable supply of groundwater, we search for an **aquifer** (Latin for water carrier), a body of rock or regolith sufficiently porous and permeable to store and conduct significant quantities of groundwater (**FIG. 8.24**, on page 245). Surface water becomes groundwater by infiltrating the ground and percolating down through porous regolith and rock to the water table, joining the water in the aquifer. Groundwater, in turn, becomes surface water by emerging from the ground in springs or by being pumped to the surface from wells for human use.

Aquifers and Wells

A good aquifer must be both porous, to store water, and permeable, to allow the water to flow through. Bodies of gravel and sand generally are good aquifers, for they tend to be both porous and permeable, and they often are very extensive. Many sandstones are also good aquifers. However, in some sandstone bodies, a cementing agent between the grains reduces the diameter of the openings, thereby reducing permeability and decreasing their potential as aquifers.

An aquifer that has a water table is called an *unconfined aquifer* (Fig. 8.24). If a well is drilled into such an aquifer, the water table is the surface of the water that you see when you look down into the well. The water in an unconfined aquifer is in contact with the atmosphere, through the porosity and permeability of the overlying rock or sediment. When water is pumped from a well in an unconfined aquifer, the rate of withdrawal may exceed the rate of local groundwater flow. This imbalance creates a *cone of depression* in the water table immediately surrounding the well. The locally steepened slope of the

FIGURE 8.23 Groundwater flow paths
The purple lines show possible pathways that groundwater may take, driven by gravity and influenced by hydraulic pressure, from a recharge area to a discharge area. The time required along various pathways depends on the permeability of the rock or regolith and the distance traveled.

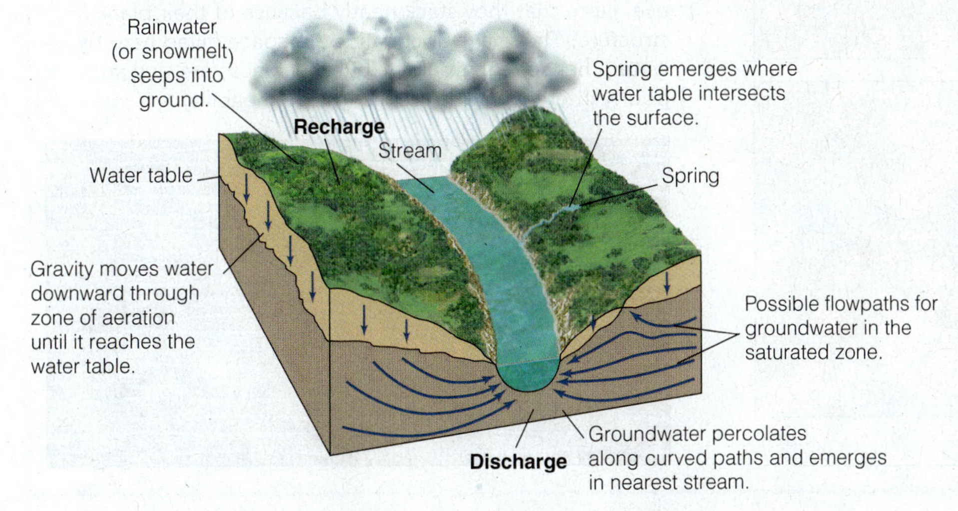
Rainwater (or snowmelt) seeps into ground.
Recharge
Stream
Spring emerges where water table intersects the surface.
Spring
Water table
Gravity moves water downward through zone of aeration until it reaches the water table.
Possible flowpaths for groundwater in the saturated zone.
Discharge
Groundwater percolates along curved paths and emerges in nearest stream.

the BASICS

Porosity and Permeability

The amount of water (or any other liquid, such as oil) that can be contained or stored within a given volume of rock or sediment depends on the **porosity**, the percentage of the total volume of a body of rock or regolith that consists of open spaces, called *pores* (FIG. B8.4). The porosity of sediment and sedimentary rock is affected by the sizes and shapes of the mineral particles. It is also controlled by the compactness of arrangement of the particles, and the degree to which the pores have been filled by natural mineral cement. In some well-sorted sands and gravels, the porosity may exceed 20 percent, while some very porous clays have a porosity of more than 50 percent. Sandstones typically have high porosities, because sand grains tend to be roughly spherical, and it is impossible to stack spheres tightly without pore spaces in between. However, sandstones often have mineral cement holding the grains together, which reduces the porosity. Non-fractured igneous and metamorphic rocks, which consist of tightly interlocked crystals, generally have lower porosities than sediments and sedimentary rocks; where fractures are present, the porosity of increases dramatically.

FIGURE B8.4 Porosity
In these examples, all of the pore spaces are filled with water, as they would be in the saturated zone. (A) The porosity is about 30 percent in this sediment with particles of uniform size. (B) This sediment, in which fine grains fill the spaces between the larger grains, has a lower porosity, around 15 percent. (C) In sedimentary rock, the porosity may be reduced by cement that binds the grains together and fills the pores.

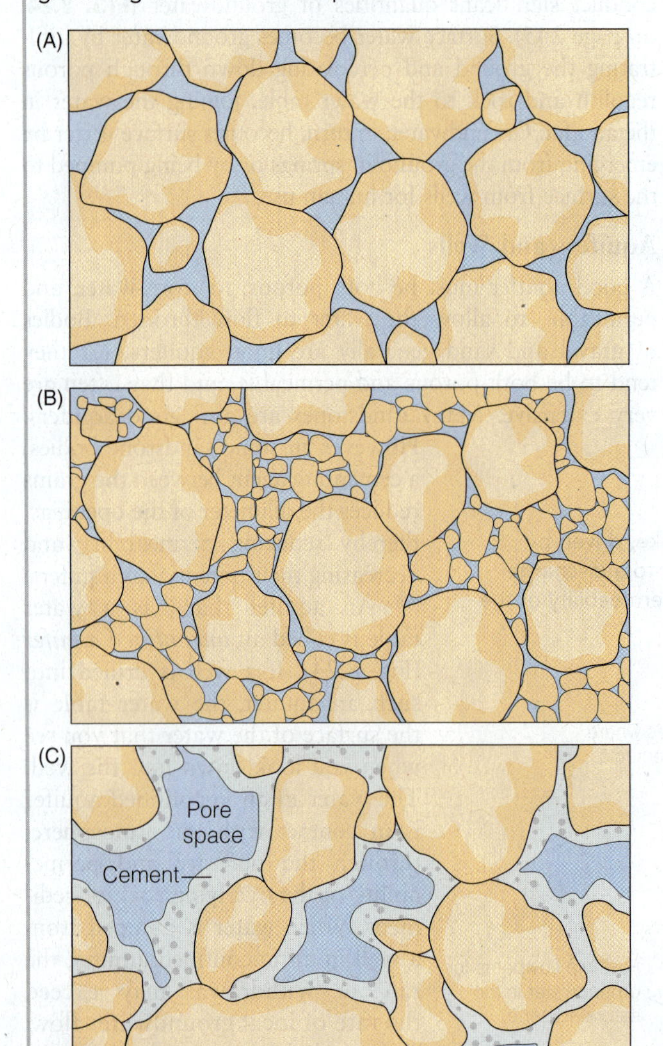

Permeability is a measure of how easily a solid allows fluids to pass through it. In a sense, it is a measure of the interconnectedness of the pores. Well-sorted gravel, with large clasts and large, interconnected openings between the clasts, is typically both porous and permeable, and can yield large volumes of water. A rock or sediment with low porosity will probably have a correspondingly low permeability. However, high porosity does not necessarily mean a corresponding high permeability. This is because the size and continuity of the openings influence permeability in an important way. For example, clay-rich rock and sediment often have high porosity but very low permeability. This is because the tiny, flat clay particles stack neatly on top of one another in a compact arrangement (FIG. B8.5). There can be many pore spaces between the particles, but the spaces are very tiny and poorly interconnected, reducing the permeability. Gravels, sands, sandstones, and limestones tend to be both porous and permeable, making them amenable to the storage and transmission of groundwater.

FIGURE B8.5 Clay
Clay is an example of a material that can have high porosity but low permeability. The individual tiny clay particles shown here would be too fine-grained to distinguish by eye. Note that they stack neatly because of their platy structure. They have abundant pore space (filled here by water), hence high porosity, but the spaces are tiny and poorly interconnected, hence low permeability.

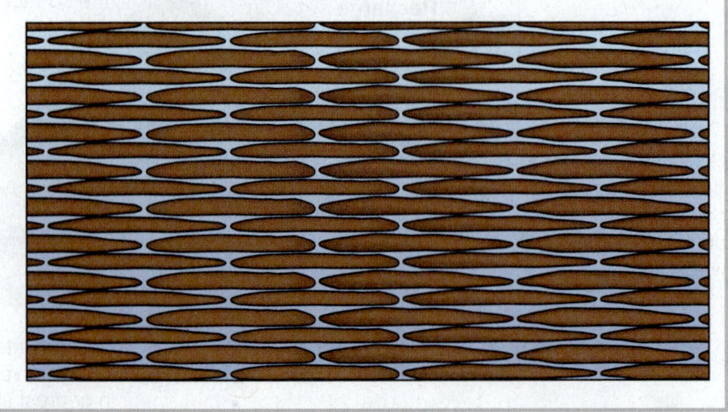

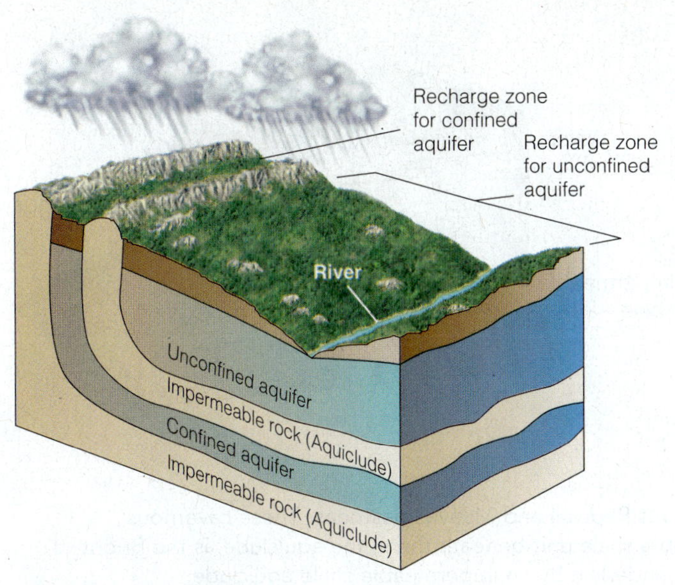

FIGURE 8.24 Aquifers
An unconfined aquifer is open to the atmosphere through pores in the rock and soil above the aquifer. In contrast, the water in a confined aquifer is trapped between impermeable rock layers.

water table will eventually increase the flow of water to the well. In most small domestic wells, the cone of depression is hardly discernible. Wells pumped for irrigation and industrial uses, however, withdraw so much water that such a cone can become very wide and deep and can lower the water table regionally, especially if the cones of depression from a number of pumped wells overlap.

In contrast to an unconfined aquifer, a *confined aquifer* is bounded above and below by bodies of impermeable rock or sediment, or by distinctly less permeable rock or sediment than that of the aquifer (Fig. 8.24). This prevents the water in the confined aquifer from rising to its natural water table. A rock or sediment unit that slows or prohibits the passage of groundwater is called an **aquiclude**. Clay-rich soil, sediment, and rock are common aquicludes. Their constituent particles are tiny and platy, so that they lie flat on top of one another, and the spaces between them are extremely small. Thus, the pore spaces are often too small and the interconnectedness of the pores is too poor to conduct water effectively. Igneous and metamorphic rocks are typically nonporous and impermeable, unless they are heavily fractured, so they may function as aquicludes.

Water that enters a confined aquifer in an upland recharge area flows downward under the pull of gravity. As it reaches greater depths, the water comes under increasing hydrostatic pressure. If a well is drilled into the confined aquifer, the difference in pressure between the water table in the recharge area and the level of the well intake will cause water to rise in the well. Such an aquifer is called an *artesian aquifer*, and the well is called an *artesian well*. Similarly, a freely flowing spring supplied by an artesian aquifer is an *artesian spring*. The term *artesian* comes from a French town, Artois (called Artesium by the Romans), where artesian flow was first studied. Under unusually favorable conditions, water pressure can be great enough to create fountains that rise as much as 60 m above ground level.

About 30 percent of the groundwater used for irrigation in the United States is obtained from the High Plains aquifer, an unconfined groundwater system that lies beneath the Great Plains, to the east of the Rocky Mountains. The aquifer is tapped by about 170,000 wells and is the principal source of water for a major agricultural region. It consists of a number of sandy and gravelly rock units in which groundwater can readily flow, and its saturated thickness averages about 65 m. The water table slopes gently from west to east, and water flows through the aquifer at an average rate of about 30 cm/day. The amount of water being withdrawn from the High Plains aquifer is currently much greater than the amount of recharge coming in from precipitation, so the inevitable result has been a long-term fall of the water table. The resulting decreased water yields and increased pumping costs have led to major concerns about the future of irrigated farming on the High Plains.

Springs

A **spring** is a flow of groundwater emerging naturally at the ground surface. The simplest kind of spring is one occurring where the land surface intersects the water table. Small springs are found in all kinds of rock, but almost all large springs flow from volcanic rock, limestone, or gravel. A change in permeability is a common reason for the localization of springs, and this often involves the presence of an aquiclude (FIG. 8.25). If a porous sand or volcanic rock overlies a relatively impermeable clay-rich aquiclude, water percolating downward will flow laterally when it reaches the aquiclude and will emerge as a spring where the boundary intersects the land surface.

Groundwater and Landscapes

Slowly moving groundwater has the capacity to perform an enormous amount of geologic work. In regions underlain by rock that is highly susceptible to dissolution and chemical weathering, groundwater has created distinctive landscapes that are among the most unusual on our planet. Most of these landscapes are created when groundwater dissolves the rock that it passes through.

As soon as rainwater infiltrates the ground, it begins to react with minerals in regolith and bedrock and weathers them chemically. An important part of chemical weathering involves mineral and rock constituents passing directly into solution, a process known as **dissolution**. Limestone, dolostone, and marble—the common carbonate rocks—are readily attacked by dissolution. Although carbonate minerals are nearly insoluble in pure water, they are readily dissolved by rainwater charged with CO_2, which becomes a dilute solution of carbonic acid. (Recall from Chapter 7 that natural waters are slightly acidic because they dissolve small quantities of atmospheric CO_2.) When carbonate rock weathers, it is slowly dissolved and carried away by the moving groundwater.

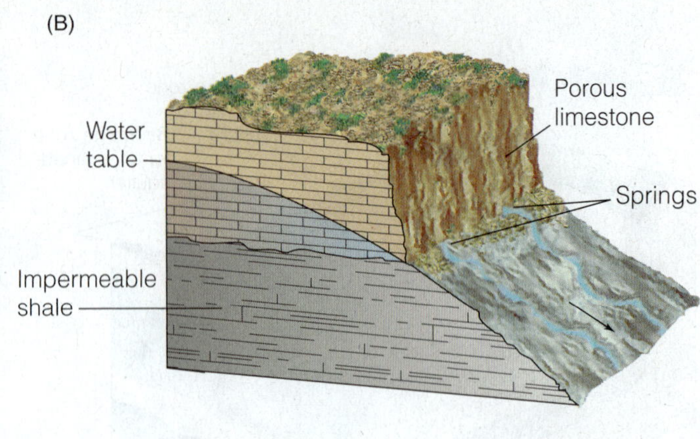

FIGURE 8.25 Springs
(A) This spring in the Grand Canyon is fed by water from the porous Redwall and Muav Limestones. These cavernous limestones are the water source for many springs. The impermeable shale unit beneath them, the aquiclude, is the Bright Angel Shale. (B) Water flows from a spring in a limestone aquifer underlain by an impermeable shale aquiclude.

Caves and Sinkholes

A **cave** will form when circulating groundwater slowly dissolves and removes carbonate rock, leaving an underground void with either a small opening or no opening to the surface. The process is thought to begin with dissolution along a system of interconnected open joints and bedding planes by percolating groundwater. The passage is progressively enlarged along the most favorable flow route by water that fully occupies the opening. This enlargement process can take thousands of years and perhaps up to a million years for full development of the passageway. It appears that most caves are excavated in the shallowest part of the saturated zone, along a seasonally fluctuating water table.

Carbonate caves come in many sizes and shapes, and they often contain spectacular formations on their walls, ceilings, and floors, including stalactites, stalagmites, and columns (FIG. 8.26), which are deposited by the precipitation of materials from the groundwater. Although most caves are small, some are of exceptional size. The Son Doong cave in Vietnam, possibly the world's largest, includes one chamber that is more than 5 km in length, 200 m high, and 150 m wide. Mammoth Cave in Kentucky is the longest cave system, with a series of interconnected chambers that have a combined known length of almost 600 km.

In contrast to a cave, a **sinkhole** is a large dissolution cavity that is open to the sky. Some sinkholes were caves whose roofs collapsed, whereas others are formed at the surface. Those produced by cave collapse can form abruptly. As a result, they pose a potential hazard for people whose houses or property may suddenly disappear into a widening conical depression tens of meters across.

FIGURE 8.26 Cave
An explorer in Lechuguilla Cave, a limestone cave in the Carlsbad Caverns region of New Mexico, examines the rock formations produced as carbonate precipitated from dripping and flowing water during past millennia.

Karst Terrains

In some regions of exceptionally soluble rock, sinkholes and caves are so numerous that they combine to form a distinctive topography characterized by many small, closed basins and intervening ridges or pinnacles. In this kind of landscape, streams disappear into the ground and eventually reappear elsewhere as large springs. Such terrain is called **karst** (German for bare, stony ground) after the classic karst region of Slovenia (where it is called *carso*). This remarkable landscape of closely spaced sinkholes and subsurface drainage has resulted from the dissolution of pure limestone.

FIGURE 8.27 **Karst**
Steep limestone pinnacles up to 200 m high, surrounded by flat expanses of alluvium, are part of a spectacular karst landscape around the Li River near Guilin, China.

Several factors control the development of karst landscapes. The topography must permit the flow of groundwater through soluble rock under the pull of gravity. Rainfall must be adequate to maintain the groundwater system, and soil and plant cover must supply an adequate amount of carbon dioxide. A warm and moist climate promotes carbon dioxide production, and therefore dissolution. Although karst terrain is found throughout a wide range of latitudes and at various altitudes, it is best developed in moist, temperate to tropical regions underlain by thick and widespread soluble rock. One of the most famous and distinctive of the world's karst regions lies in southeastern China, where vertical-sided limestone peaks stand up to 200 m high (FIG. 8.27). This dramatic landscape has inspired both classical Chinese painters and present-day photographers.

WATER AND SOCIETY

A reliable water supply is critical—not only for human survival and health, but also for the role it plays in industry, agriculture, and other economic activities, and for the environmental services it performs, such as supporting ecosystems, carrying away contaminants, and moderating climate. Water is under threat almost everywhere in the world, in terms of both quality and quantity.

Laws and policies relating to water are confusing and complicated—even more so for groundwater than for surface water. Because groundwater is hidden from view, it is difficult to monitor its flow and regulate its use. If you drill a well into an aquifer underlying your property, are you entitled to withdraw as much water as you want? Should you be allowed to withdraw water only for your own purposes, or is it acceptable to transport it and sell it elsewhere? What happens if withdrawing the water depletes the aquifer and causes your neighbor's well or lake to run dry? Similar problems arise when a landowner's actions cause a stream or an aquifer to become contaminated. What happens if the streamflow or groundwater percolation carries the contaminant beyond the borders of the property? Understanding the natural processes that control the movement of water in the hydrologic cycle is crucial to finding the answers to these and other challenging questions in water management.

Water Quantity

Globally as of 2009, according to the United Nations, crop irrigation accounts for about 75 percent of the demand for fresh water, industry for about 20 percent, and domestic use for the remaining 5 percent, though the proportions vary greatly from one region to another. The demand for water has led to dramatic increases in water withdrawal (that is, diversion of water from rivers, lakes, and groundwater) over the past few decades. Population growth is partly responsible for the increasing demand, but while the world's population tripled over the course of the twentieth century, demand for water increased sixfold. This means that development and improvements in standards of living have contributed to the large increase in water use per capita over the past few decades—we are using more water to produce the goods we consume and to carry out the activities of modern life. The total amount of water being withdrawn for worldwide human use is now about eight times the annual streamflow of the Mississippi River.

Water Shortages

Today twenty-nine countries worldwide, with a total population of 450 million people, suffer from significant water shortages; they are said to be experiencing *water stress* (FIG. 8.28). The lack of water in these countries places constraints on agricultural production, economic development, health, and environmental protection. The United Nations estimates that by 2050 as many as 54 countries with a combined population of 4 billion people could face water stress, partly as a result of the impacts of climate change on water supplies.

Make the CONNECTION

How will global climate change affect water resources around the world? How will it affect water resources where you live?

Sometimes regions with the greatest demand for water do not have an abundant and readily available supply of surface water. A classic example in North America is seen in California and the Southwest, where intensive agriculture and high population have combined to generate ovewhelming water demand in an area of naturally dry climate. The result is that surface water is moved from one drainage basin to another to meet the demand, sometimes over long distances; this is called *interbasin transfer*. Much of the water used for drinking and irrigation in

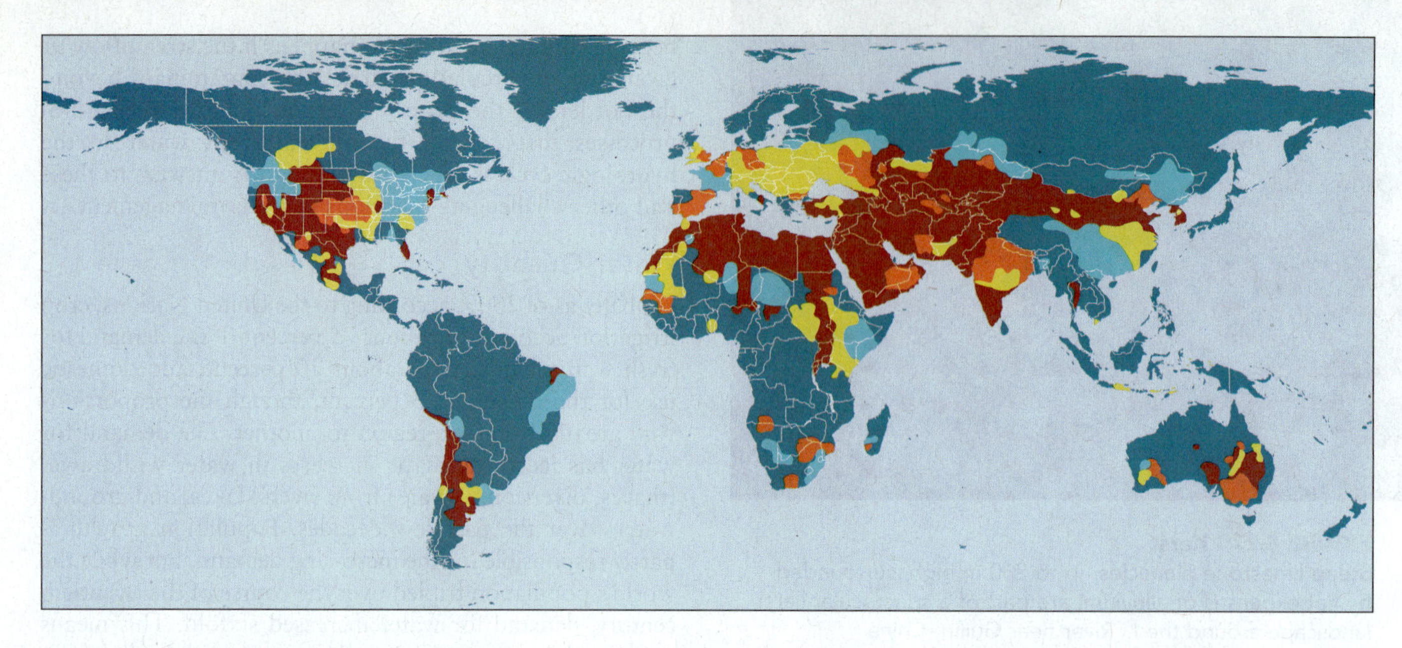

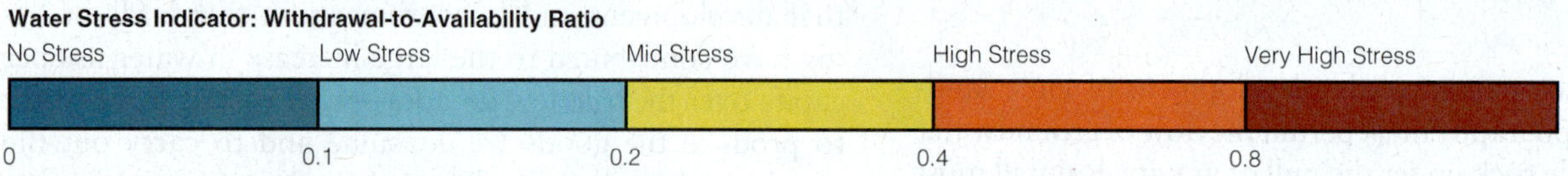

FIGURE 8.28 Water stress
This map shows areas of the world that are experiencing water stresses as a result of demand outstripping the available supply of water.

California and other states of the Southwest, for example, is transferred from the Colorado River.

Diversions and Withdrawals

Besides raising political issues related to water rights, interbasin transfer can have negative environmental impacts. The Colorado River, for example, has suffered from low flow, low sediment load, and high salinity as a result of the diversion of water from its channel. This has led to some tense political discussions between the United States and Mexico, where the mouth of the Colorado River is located. The classic example of the impacts of diversion is the Aral Sea, on the border between Kazakhstan and Uzbekistan. Once the world's fourth-largest inland water body, the Aral Sea has decreased dramatically in volume, depth, and area as a result of the diversion of water from its two major incoming rivers (see *A Closer Look: The Case of the Aral Sea*).

Groundwater, too, can suffer negative impacts as a result of excessive withdrawal. If the rate of withdrawal of groundwater regularly exceeds the rate of natural recharge, the volume of stored water steadily decreases; this is called *groundwater mining*. It may take hundreds or even thousands of years for a depleted aquifer to be replenished. The results of excessive withdrawal include lowering of the water table, drying up of springs and streams, compaction of the aquifer, and subsidence. Sometimes it is possible to recharge an aquifer by pumping water into it, but often the effects of depletion are permanent. When an aquifer suffers *compaction*—that is, when its mineral grains collapse on one another because the pore water that held them apart has been removed—it is permanently damaged and may never be able to hold as much water as it originally held.

Water Quality

In North America we are fortunate to have domestic water supplies that are generally safe and tightly regulated. Even so, the use of home filtration systems and bottled spring water has increased, as people have become wary of potential contaminants in tap water. (Note, however, that spring water and artesian water are essentially the same as other groundwater and that the purity of bottled water, in general, is not as closely monitored as tap water.) Elsewhere in the world, people are not so lucky with their drinking water. About 1.2 billion people, mainly in developing countries, do not have access to drinking water that is monitored for quality or improved in any way.

The water that comes to our homes and offices typically is drawn from relatively clean surface water bodies or groundwater aquifers, but may still require treatment to meet drinking-water standards. Water quality improvements can include treatments for hardness, the addition of fluoride (*fluoridation*), and, where necessary, *desalination* to remove salts from the water. Desalination is energy-intensive and very expensive, but it is worth the cost in some countries where water is particularly scarce. Most importantly, water for domestic use must be filtered to remove sediment and other solids, and treated by *chlorination* to kill harmful microorganisms.

Thousands of potentially harmful contaminants, or *toxins*, may enter a water supply, at any number of points in the hydrologic cycle. In addition to innumerable naturally occurring chemicals (some of which are highly toxic), there are millions of manufactured synthetic compounds, tens of thousands of which are in common use. Scientists don't have complete information about the human health impacts of all of these compounds, their impacts on other organisms in the biosphere, or their potential to interact with other substances in the hydrosphere. Of particular concern is that some contaminants have long residence times in natural reservoirs; they are said to be *persistent*. The residence time of a substance in a reservoir is determined by the chemical and physical characteristics of the substance, the nature of the medium through which or by which it is being transported, and the presence of other substances with which it may interact.

Ensuring the quality of water involves determining the health effects of toxins, and then setting and enforcing limits. An important aspect of this is to understand the pathways and processes by which toxic substances move through the various reservoirs of the Earth system and into organisms, including people.

Surface Water Contamination

The accessibility of surface water bodies makes them useful as resources, but renders them highly susceptible to contamination. Contaminants in surface water come mainly from urban, suburban, and agricultural runoff. Industrial *effluent* (contaminated runoff) is a significant contributor to surface water pollution. So are discharges related to resource extraction, among which mining, logging, and the petroleum industry are important. Effluents from poorly engineered landfills are another source of surface water contamination.

Not all contaminants in surface water bodies come from runoff. Airborne acid-forming substances contribute to the deposition of acidic precipitation on land and water surfaces. This can damage lakes and aquatic ecosystems, as well as soils and forests. Surface water reservoirs in industrial areas are also susceptible to *thermal pollution*, the release of heat or (more commonly) heated water, which is harmful to aquatic organisms and ecosystems. In this case, the heat itself is the contaminant.

Many wastes that end up in surface water bodies contain organic material. Lots of cities, even in wealthy countries, still discharge sewage directly into nearby water bodies with little or no treatment. When the organic matter in sewage breaks down, oxygen is used up in the process. Oxygen dissolved in the water is critical for the maintenance of most aquatic life forms. Effluents that cause the depletion of dissolved oxygen are said to place a *biochemical oxygen demand* on the system. The greater the amount of organic matter discharged into the system, the higher the biochemical oxygen demand. If sewage is discharged directly into a water body, the dissolved oxygen in the water downstream from the source of the sewage responds by becoming depleted.

One of the most common forms of surface water contamination results from an excess of plant nutrients, especially phosphorus and nitrogen, from fertilizer runoff and phosphorus-containing detergents. When an aquatic system is overloaded with nutrients, the growth of algae, plankton, and aquatic weeds can get out of control; this is called an *algal bloom*. When the algae and other aquatic plants die, their breakdown causes oxygen depletion. Other organisms in the water begin to die, and the water turns mucky and sickly green (FIG. 8.29). This is similar to eutrophication (discussed above), the natural process whereby a lake fills with plant matter and becomes swamplike. When eutrophication is accelerated by the addition of anthropogenic pollutants, it is called *cultural eutrophication*. If the water becomes completely oxygen-depleted, it may turn crystal-clear. This happened to parts of Lake Erie in the 1970s, and it is not a sign of a healthy water body—it is a sign of a water body in which nothing is able to survive.

The presence of infectious agents such as bacteria, viruses, and other disease-causing organisms is another form of surface water contamination associated with human and animal wastes. In many parts of the world, consumption of water contaminated with infectious agents is linked with outbreaks of diseases such as cholera. In North America, cholera and other waterborne diseases are less of a problem than they were 100 years

FIGURE 8.29 Eutrophication: A dying lake
This lake in Ontario has turned green and mucky because of the growth of algae stimulated by excessive nutrients. Eventually, the algae may use up all of the dissolved oxygen in the water, making it impossible for other aquatic life to survive. This process, called eutrophication, can occur naturally, but in this case it has been accelerated by human activities.

A Closer LOOK

THE CASE OF THE ARAL SEA

The Aral Sea, located along the border between Kazakhstan and Uzbekistan, was once the world's fourth-largest inland water body and a major resource for fishing, farming, and other local uses. Starting in the 1950s, significant quantities of water were diverted from the Aral Sea's two inflowing rivers, the Amu Darya and the Syr Darya, mainly to be used for agricultural purposes. For a few decades, the resulting irrigation allowed for large increases in cotton production in the former Soviet Union. Before long, however, the Aral Sea began to show the negative impacts of reductions to its supply of incoming fresh water.

In the 1960s, the sea began to shrink (FIG. C8.1). Over the next four decades its depth and surface area

FIGURE C8.1 The shrinking Aral Sea

(A) The Aral Sea, shown in 1973, was once the world's fourth-largest lake. However, it has been shrinking since the 1960s as a result of the withdrawal of water from its two incoming rivers, to irrigate cotton crops. It has now split into three much smaller water bodies, as shown in a satellite image taken in 2009 (B). Today restoration efforts are beginning to reverse the decline in the northern portion of the sea.

(A)

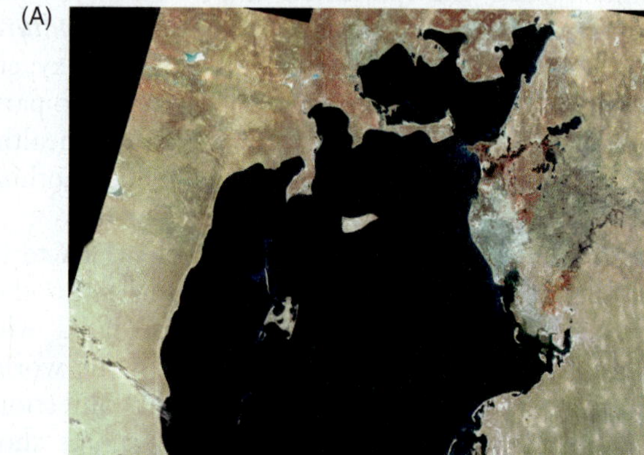

(B)

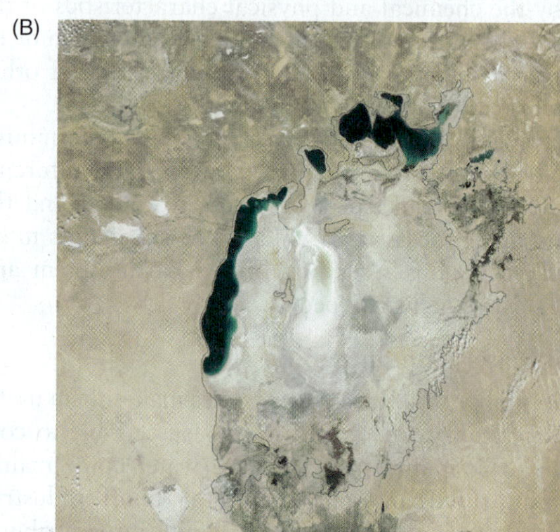

ago; still, high levels of *E. coli*, an infectious organism commonly found in feces, routinely cause beach closures in many coastal areas. In Walkerton, Ontario in 2000, the deaths of seven people and many serious illnesses were caused by a deadly strain of *E. coli* that entered the municipal groundwater supply through a cracked well-head. The source of the bacterium is thought to have been fecal-contaminated runoff from a nearby farm.

Some of the most significant toxic contaminants in surface water bodies are persistent chlorinated organic compounds, such as DDT and Mirex (both pesticides), polychlorinated biphenyls (PCBs), and dioxins and furans (industrial by-products). Note that in this particular usage, "organic" doesn't mean "environmentally friendly." It refers to the presence of organic molecules (Chapter 3), that is, molecules that have carbon and hydrogen as their structural building blocks, many of which are toxic. Another problematic category of contaminant is the heavy metals, such as cadmium, lead, arsenic, and mercury. Heavy metals can be released during mining and ore processing, metallurgical procedures, paper manufacture, petroleum refining, and other industrial processes. Long-term exposure to even very low levels of these substances can cause neurologic damage (especially in children), kidney damage, and a variety of other problems.

Contaminated surface water bodies can often be cleaned up, or *remediated*, where sufficient political will and finances are available. Some problems have responded well. For example, Lake Erie—which effectively "died" as a result of oxygen depletion in the 1970s—has since undergone substantial recovery, thanks to agreements between the United States and Canada that have limited the availability of phosphate in surface runoff into the Great Lakes. Other problems, notably the acidification of lakes as a result of acidic precipitation, have proven to be more difficult to resolve. Still others have been partially addressed by changes in human behavior, but await the response of the natural system. The most iconic example is DDT. This toxic organochloride pesticide was largely banned from North American manufacture and use in the 1970s; however, it is so persistent that it can still be detected in natural reservoirs such as the tissues and breast milk of animals, including humans. It will take many more years before natural processes can eliminate DDT from the hydrosphere and biosphere.

decreased dramatically, average temperature and salinity increased, and salt deposits began to form around its newly exposed banks. The lake eventually split into three much smaller water bodies, separated by a new desert called the Aralkum. By 2007 the lake had lost more than 80 percent of its original volume and had shrunk to about 10 percent of its original surface area. The lake's aquatic ecosystems have been irrevocably altered, and the once-vibrant local fishing industry has been decimated. As many as 40,000 fishing jobs have been lost, and fishing villages that were once located at the water's edge now lie many kilometers from the shoreline (FIG. C8.2). Winds blow pesticide- and salt-laden dust from the dry, exposed lake bed, which is not suitable for agriculture. The exposed and blowing salts and dust have caused health problems for nearby residents. The loss of the moderating influence of such a large body of water has even affected the local climate, contributing to decreased rainfall, colder winters, and hotter summers. Even some of the soil irrigated by the diverted water has become problematic; it is now waterlogged and salinized.

Starting in 2000, significant efforts began in an attempt to reverse the decline and to rescue at least the northern water body of the Aral Sea by restricting flow between the three arms of the lake. Programs sponsored by the United Nations brought together scientists and local people in this effort. By 2006 these programs were beginning to show some small positive signs, including a rise in lake level and the recovery of some fish populations. The eastern portion of the southern Aral Sea, however, may well be beyond rescue; satellite images show that it had all but disappeared by 2009. The western portion may last another 50 years or possibly longer, according to some current estimates. The Aral Sea, once a vital local resource, is an environmental catastrophe, an unfortunate case study of the impacts of water diversion, and an example of how a change in one part can lead to changes in many other parts of the Earth system.

FIGURE C8.2 Stranded boats
Fishing vessels lie stranded far from the shore because the waters of the Aral Sea receded so far and so quickly.

Groundwater Contamination

Many of the pollutants that affect surface water and soils also cause groundwater contamination. Because of its hidden nature, groundwater contamination can be more difficult to detect, control, and clean up. Harmful chemicals from leaking underground gasoline storage tanks and pipelines, poorly designed or abandoned landfills, and toxic waste disposal facilities can contaminate groundwater reservoirs.

Twenty of the top 25 toxic groundwater contaminants are volatile organic compounds, like benzene, toluene, ethylene, and xylene (collectively called *BTEX*), all of which are constituents of gasoline. Chlorinated organic solvents such as trichloroethylene (TCE), commonly known as dry-cleaning fluid, are highly toxic and pose a significant threat to groundwater quality in many sites. The most common pollutants in groundwater are untreated sewage and nitrate from agricultural chemicals; they are less acutely toxic than other contaminants, but no less problematic, because they are so widespread. An additional problem is saline contamination of groundwater, usually caused by the intrusion of seawater into coastal aquifers.

Once contaminated, groundwater can be extremely difficult to remediate because of its inaccessibility. There are two basic approaches to the remediation of polluted sites. The first is *passive remediation*; this involves relying on natural environmental processes to clean up the site. This may sound "lazy" or ineffective, but it is sometimes a viable approach. The natural environment can be very effective at filtering and removing contaminants from water and soil. If there is no immediate risk to people living or working nearby, no urgent need to prepare the land for new uses, and the environment itself is appropriate, it can save a lot of money and effort simply to wait and let nature take its course. However, passive remediation of even a mildly contaminated site can take years, or even decades. If the need for the land is urgent, or the risk to people or adjacent environments is severe, it is advisable to undertake *active remediation*, that is, to actively intervene in the cleanup of the site. This can include approaches such as injecting oxygen or other chemicals into the groundwater to speed the breakdown of the contaminants, or pumping the contaminated groundwater to the surface for treatment and eventual reinjection into the aquifer.

You have now learned about the characteristics and processes of the hydrosphere, the hydrologic cycle, and the freshwater reservoirs that are so important for all life on Earth. The presence of water in solid, liquid, and vapor form in the near-surface environment makes Earth unique among all known planets and moons. The unique chemical properties of water and its movement throughout the reservoirs of the Earth system support an almost endless variety of physical, chemical, and biological processes. In Chapter 19 we will revisit some of the human impacts on this most important subsystem, the hydrosphere.

In the next chapter we will look at the cryosphere—the perennially frozen part of Earth's hydrosphere.

SUMMARY

1. The hydrologic cycle describes the fluxes of water among the reservoirs of the hydrosphere, which includes the ocean, lakes, streams, underground water, and snow and ice. The hydrologic cycle is intimately related to the rock cycle and plays a central role in climatic and biogeochemical cycling.

2. The presence of water in its solid, liquid, and vapor states in the near-surface environment makes Earth a unique body in our solar system. Water (H_2O) has many unique properties, including very high surface tension, capillarity, ability to act as a solvent, very high heat of vaporization, very high heat capacity, and decrease in density upon freezing. These properties allow water to play a central role in countless physical, chemical, and biological processes in the Earth system.

3. The global hydrologic cycle maintains a mass balance, with large local fluctuations. The largest reservoir in the hydrosphere is the saline ocean. The largest freshwater reservoir is the perennially frozen polar ice sheets. Of the unfrozen fresh water, most is groundwater. A very small fraction of the water in the hydrosphere resides in surface freshwater bodies, soil moisture, and living organisms.

4. The hydrologic cycle is powered by the Sun's heat, and flow within the cycle is controlled by gravity. Heat from the Sun causes evaporation from the ocean and land surfaces. Water vapor that condenses in the atmosphere falls as precipitation. Some of this evaporates directly back to the atmosphere; the rest is intercepted by vegetation or flows downhill as surface runoff. Water collects in surface water bodies, returns to the atmosphere via transpiration, or infiltrates into the ground. Snow may remain on the ground until it melts, or it may be locked up for much longer in the cryosphere.

5. Overland flow and streamflow (storm flow + base flow) constitute surface runoff. A stream is water that flows downslope through a clearly defined natural channel, transporting particles of sediment and dissolved substances. Every stream is surrounded by its drainage basin, which is separated from adjacent drainage basins by topographic divides. Streams and drainage systems are constantly evolving in response to changing conditions of topography, climate, and vegetation.

6. Streams and their channels are part of a complex natural system, controlled by the interplay between the stream's cross-sectional area, gradient, average flow velocity, discharge, and sediment load. As discharge changes, the velocity or the channel cross-sectional area, or both, also change. From upstream to downstream, the channel's width and depth.

7. The maximum velocity of water flowing along a straight channel segment is midstream, away from the frictional resistance of the banks. Where the water rounds a bend, the zone of highest velocity swings toward the outside of the channel, facilitating the formation of meanders and point bars. A braided pattern with numerous shallow, shifting channels and bars is typical of streams that have highly variable discharge and an abundant supply of sediment.

8. A stream's load includes the coarse bed load, which moves along the bottom by saltation, and the finer suspended load; together these constitute the sediment load. The dissolved load consists of materials carried in solution.

9. Flowing water shapes our landscape. Stream deposits form wherever the stream energy changes, leading to deposition of the sediment load. Floodplains result from the deposition of fine sediment as a stream overflows its banks during a flood. An alluvial fan is a fan-shaped deposit at the base of an upland, where a stream emerges onto lowland and abruptly loses energy. When a stream enters the standing water of the sea or a lake, the water deposits its load in the form of a delta.

10. Continental divides separate all continents except Antarctica into large regions from which major through-flowing rivers enter one of the world's major oceans. They usually coincide with the crests of great mountain ranges, although their location can be influenced by other factors that control erosion and topography. Major divides exert a strong influence on climate.

11. A lake is a body of inland water with an open surface, occupying a depression. The level and salinity of a lake reflect a balance between freshwater input from streams and direct precipitation, outflow, and evaporation. Lakes can disappear as a result of changes in freshwater inputs, climate, or topography. Wetlands, including swamps, marshes, and bogs, are permanently or intermittently moist areas; they host significant biodiversity and perform many ecologic services. Wetlands in coastal regions, such as estuaries, salt marshes, and mangroves, are transitional between freshwater and saltwater environments.

12. A flood occurs when a stream's discharge exceeds the capacity of the channel and water overflows the banks. Floods are normal and expected events; they transport and deposit sediment and replenish the mineral content of soils. Scientists use flood-frequency curves to calculate statistical recurrence intervals, giving a basis for planning and preparation. Computer models and GIS are used during real-time monitoring of storms to forecast the peak of a flood. Channels can be engineered for flood control, but channelization can have unexpected negative side effects, including degradation of the aquatic environment.

13. Less than 1 percent of the fresh water in the hydrosphere is groundwater, which occupies pore spaces in rock and regolith, but this is still many times larger than the volume of surface fresh water. A portion of precipitation that falls on land infiltrates, flows downward under the influence of gravity, and joins the groundwater system. The composition of groundwater varies from place to place; the more time the water spends in direct contact with rock, the more mineral constituents will be dissolved into the water.

14. Immediately beneath the surface is the aerated zone, in which pore spaces contain some water and some air. This is underlain by the saturated zone, in which all openings are filled with water. The upper surface of the saturated zone is the water table. Groundwater flow, or percolation, is slow because groundwater must move through small, constricted passages. Percolation includes both the vertical flow of water from the surface down to the water table and the lateral flow of water in the saturated zone.

15. The amount of water that can be contained or stored within a given volume of rock or sediment depends on its porosity, the percentage of the total volume of a body of rock or regolith that consists of pore spaces. Permeability is a measure of the interconnectedness of pore spaces and thus controls how readily a rock or sediment will transmit fluids.

16. Groundwater flow is complicated, but basically water flows from areas where the water table is high toward areas where it is lowest. Replenishment occurs when rainfall and snowmelt enter the ground in areas of recharge. Water moves through the groundwater system to areas of discharge, where it is discharged to streams, lakes, ponds, or wetlands. Along coastlines, groundwater flows directly into the ocean through porous rock and regolith by means of submarine discharge.

17. An aquifer is a body of rock or regolith sufficiently porous and permeable to store and conduct significant quantities of groundwater. An aquifer that can rise to its natural water table is called an unconfined aquifer. In contrast, a confined aquifer is bounded above and below by an aquiclude. Springs can occur in circumstances where the ground surface intersects the water table.

18. Caves form when circulating groundwater slowly dissolves and carries away carbonate rock, leaving underground voids. The process begins with dissolution along interconnected joints and bedding planes, and continues with the passages being slowly enlarged by percolating groundwater. A sinkhole is a large dissolution cavity that is open to the sky. In some regions of exceptionally soluble rock, sinkholes and caves are so numerous that they form karst terrain, characterized by many small, closed basins and intervening ridges or pinnacles.

19. A reliable supply of fresh water is critical for human health, for industry, agriculture, and other economic activities, and for the diverse environmental services it performs. Water is under threat almost everywhere in the world, in both quality and quantity. The number of countries experiencing water stress is expected to increase in the next few decades, partly as a result of climate change. Both surface water and groundwater can suffer degradation and depletion as a result of withdrawals that exceed the rate of recharge.

20. Thousands of potentially harmful contaminants can enter water via the processes of the hydrologic cycle. Surface water, by virtue of its accessibility, is vulnerable to contamination from urban, suburban, and agricultural runoff, industrial and landfill effluents, and atmospheric deposition of acids. Sewage and excess fertilizers are among the most common surface water contaminants; they can lead to eutrophication, among other problems. Groundwater is susceptible to many of the same contaminants as surface water, but groundwater contamination is harder to detect and remediate.

IMPORTANT TERMS TO REMEMBER

aerated zone *242*
alluvial fan *234*
alluvium *228*
aquiclude *245*
aquifer *243*
braided stream *231*
cave *246*
channel *228*
channelization *240*
condensation *225*
continental divide *234*
delta *234*
discharge *230, 243*
dissolution *245*
divide *228*
drainage basin *228*
eutrophication *237*
evaporation *225*
flood *238*
floodplain *234*
gradient *229*
groundwater *224*
hydrologic cycle *224*
hydrosphere *224*
infiltration *228*
karst *246*
lake *236*
load *228*
meander *230*
percolation *242*
permeability *244*
porosity *244*
precipitation *225*
recharge *243*
saturated zone *242*
sinkhole *246*
spring *245*
stream *228*
streamflow *228*
surface runoff *228*
transpiration *225*
tributary *230*
water table *242*
wetland *237*

QUESTIONS FOR REVIEW

1. Describe the main pathways and reservoirs in the hydrologic cycle.

2. How does streamflow differ from overland flow, and how are these two kinds of surface flow related?

3. How do a stream's channel dimensions and velocity adjust to changes in discharge?

4. What is the difference between alluvium, sediment, and regolith? (Refer to Chapter 7 to review these terms.)

5. Why does stream velocity usually increase downstream, in spite of a decrease in stream gradient? Why, if velocity increases downstream, does sediment particle size usually decrease downstream?

6. How are floods predicted, and what are some of the ways in which people have tried to prevent floods?

7. In Chapter 3 you learned about the difference between a mixture and a solution. Recall that the components of a mixture can be separated from one another by physical means, whereas the components of a solution can only be separated by chemical means. Discuss the differences between bed load, suspended load, and dissolved load in the context of your understanding of mixtures and solutions.

8. What is the difference between a lake and a wetland? How do lakes form, and how do they disappear?

9. In what ways does the flow of groundwater differ from the flow of water in stream channels?

10. What determines how long it will take for water to move from a recharge zone to an aquifer and then to a discharge zone?

11. What is the difference between an aquifer and an aquiclude? What is the difference between a confined aquifer and an unconfined aquifer? Does groundwater differ from artesian water or spring water?

12. What is dissolution? What role does it play in the formation of caves and sinkholes?

13. What is interbasin transfer, and what are some of the problems associated with it?

14. What are some kinds of treatment that can be undertaken to ensure the quality of drinking water?

15. Both groundwater and surface water are susceptible to contamination, but with some differences; explain.

QUESTIONS FOR RESEARCH AND DISCUSSION

1. Let's assume you are considering the purchase of a 30-year-old house adjacent to a large river that experiences seasonal flooding. What information might you seek to evaluate the possibility that an unusually large flood could inundate the house sometime during the next half century?

2. How does groundwater differ from juvenile or primordial water, which also resides underground (Chapter 1)?

3. Find out where your community or city derives its supply of fresh water. Is the supply vulnerable to any types of contamination? Will the existing source of supply prove adequate if the population doubles during the next several decades? What might be the impacts of global climatic change on this water source?

4. Do you drink bottled water? Where does the water come from? How sure are you of its purity and healthfulness compared to the municipal water in your community?

QUESTIONS FOR *THE BASICS*

1. What are some of water's unusual chemical and physical properties, and how do they influence processes in the Earth system?

2. What is the difference between porosity and permeability?

3. Low porosity usually means low permeability, but high porosity doesn't necessarily mean high permeability; explain.

QUESTIONS FOR *A CLOSER LOOK*

1. What happened to cause the Aral Sea to shrink in volume, depth, and surface area?

2. What were some of the unexpected side-effects of the diversion of water from the Aral Sea?

3. Find out what kinds of programs have been undertaken to try to rescue at least the northern portion of the Aral Sea.

CHAPTER 9

The Cryosphere

OVERVIEW

In this chapter we:

- Discover the main features of the cryosphere
- Learn how glaciers form and why they move
- Describe how glaciers and ice sheets shape the landscape
- Examine how glaciers function as climatic and environmental archives
- Consider the causes of sea-ice formation and its important role in the Earth system

Ice from the land to the sea
This is the tidewater front where Shoup Glacier calves into the water at Shoup Bay State Marine Park, Southcentral Alaska.

Among the planets of the solar system, only Earth has an overall bluish color, imparted by the vast expanse of world ocean. Nevertheless, the polar regions of the northern and southern hemispheres appear largely white because of an extensive, perennial cover of snow and ice. In the northern hemisphere, much of the ice floats as a thin sheet on the Arctic Ocean, whereas in the southern hemisphere it consists of a vast glacier system that overlies the continent of Antarctica and adjacent islands and seas, as well as floating sea ice that extends far beyond the coastline. The vast mantles of polar ice are linked in important ways to the ocean and the atmosphere. They are involved in the generation of cold, dense water that drives deep-ocean circulation. They influence the albedo (reflectivity) of the planet, drive major wind systems, and play an important role in world climate. They are extremely dynamic features, which fluctuate seasonally as well as on geologic timescales. Earth's great reservoirs of frozen water are the subject of this chapter.

EARTH'S COVER OF SNOW AND ICE

The part of Earth's surface that remains perennially frozen constitutes the **cryosphere** (from the Greek words for "*icy cold sphere*") (FIG. 9.1). (*Perennial* in this context means that it does not melt on a seasonal basis but remains frozen year-round.) The cryosphere includes not only glaciers and sea ice, but also vast areas of frozen ground that lie beyond the limits of glaciers. Glaciers cover about 10 percent of Earth's land surface, while perennially frozen ground covers an additional 20 percent. Thus, nearly a third of Earth's land area currently belongs to the cryosphere. Some of the frozen material in the cryosphere crystallizes in the form of ice on the surface or in the near subsurface, in conditions of low temperature; some falls as precipitation and accumulates on the ground, where it may undergo subsequent changes (see *The Basics: Snow and Ice*, on page 263).

Snow

Earth's perennially frozen ice caps, glaciers, and sea ice constitute the major part of the cryosphere, but winter snow and seasonally frozen ground also play an important role in Earth's climate system. In the northern hemisphere, almost one-fourth of land area is covered by snow and frozen ground during the winter (FIG. 9.2), although its volume is still much less than the volume of perennial snow and ice that covers Antarctica. As you will learn, the accumulation of snow, which occurs when the amount that falls seasonally is greater than the amount that melts, is the main contributor to the growth of glaciers and ice caps. The highly reflective surface of snow bounces sunlight back into space, thereby reducing surface air temperature. When snow melts, it becomes a major source of water for rivers and moisture for agricultural soils. However, the timing or amount of snowfall can affect people adversely. For example, a heavy late-winter snowfall can generate widespread flooding, while a snowfall deficit can lead to water rationing during the summer.

FIGURE 9.1 The cryosphere
The major components of the cryosphere, including their extent (A) and the relative timescales of their processes (B), are seasonal snow cover, sea ice, ice shelves, ice sheets, and permafrost.

(A)

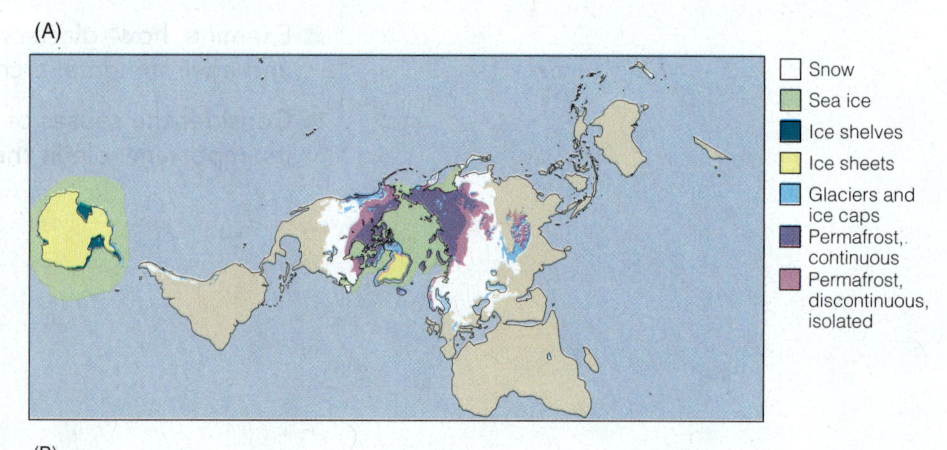

(B)

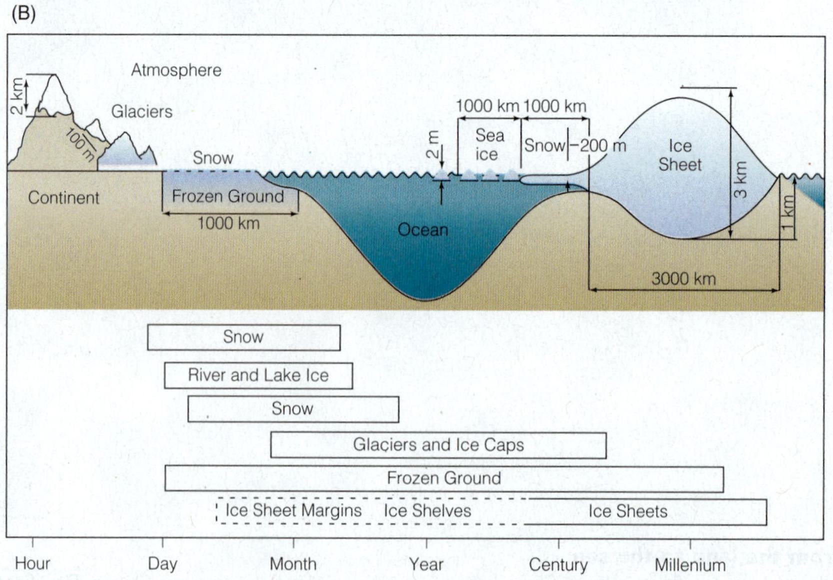

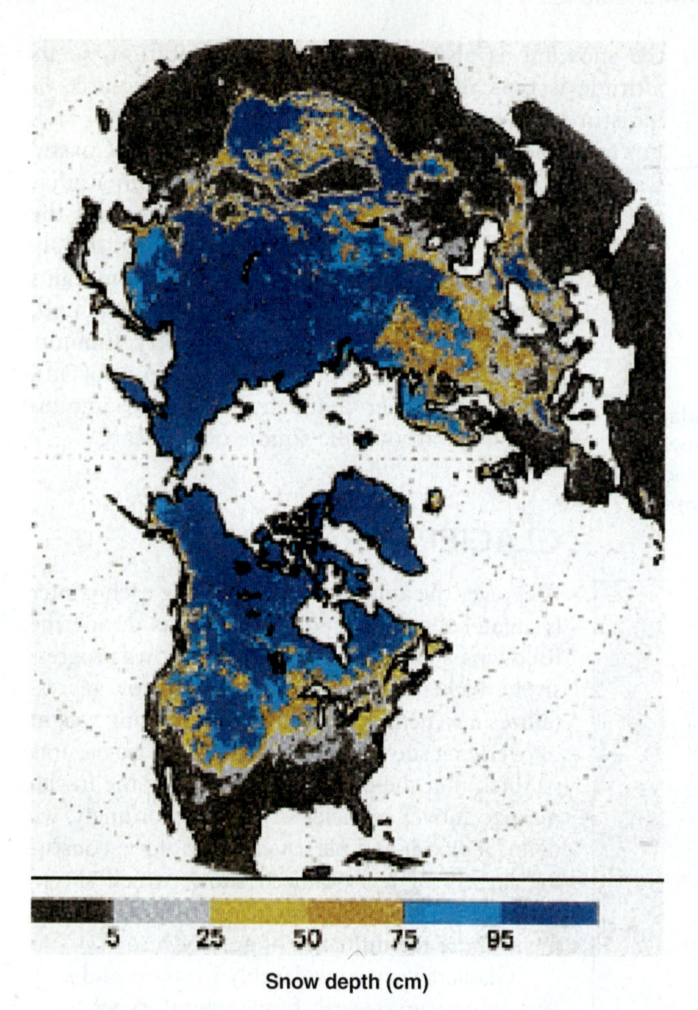

FIGURE 9.2 Snow cover
This map, based on a satellite image, shows the average snow cover in the northern hemisphere during December. The greatest snow cover lies in regions of continental climate in middle to high latitudes (northern North America and northeastern Asia) and high-latitude regions such as the Tibetan Plateau of Central Asia.

Annual Snow Cycle

Prior to the mid-1960s, estimates of variations in continental snow cover were obtained from limited ground-based measurements. Today, variations in snow depth and area are monitored by satellites. During a typical year, northern hemisphere snow cover first appears in northern Alaska and northeastern Siberia in mid-September to mid-October. Through November, the snow cover expands southward and begins to thicken. By December, the expanding snow cover reaches southern Russia, central Europe, and the northern United States, and snow blankets nearly all of the high Tibetan Plateau. From December through March, the snowpack thickens in continental interiors, but its southern limit begins to retreat as air temperature rises. The snowpack then recedes rapidly northward during late spring, and by mid-June the remaining snow is confined mainly to high mountains and to lands bordering the cold Arctic Ocean.

The Snowline

If you view a high mountain at the end of the summer, just before the earliest autumn snowfall, you commonly will see a snowy zone on its upper slopes. The lower boundary of this zone is the **snowline**, which is defined as the lower limit of perennial snow (FIG. 9.3). Above the snowline, part of the past winter's snow has survived the warm temperatures of summer, along with snow from earlier winters that has persisted through previous summers. In detail, the snowline is an irregular surface, its shape controlled both by variations in the thickness of the winter snowpack and by local topography. When viewed from a distance, however, the snowline appears as a line delimiting snow-covered land from snow-free land.

The altitude of the snowline and its horizontal position on the landscape typically change from year to year depending on the weather. Although a number of climatic factors are involved, the two principal ones are winter snowfall, which affects total snow accumulation, and summer temperature, which influences melting.

In the polar regions, the amount of annual snowfall is generally very low because the air is too cold to hold much moisture. This is why polar ice caps are sometimes referred to as *polar deserts*. At McMurdo Station, Antarctica, for example, an average of only 7.8 mm of precipitation falls each year, which easily fits the technical definition of a desert (a region that experiences less than 250 mm of precipitation per year). Nevertheless, because

FIGURE 9.3 The snowline
The lower limit of snow in late spring forms an irregular line across the flank of Mount Cook, the highest peak in New Zealand's Southern Alps. As the weather warms and the snow melts, the snow limit rises to its highest level at the end of the summer. This late-summer limit marks the annual snowline. Above the snowline, most of the ground remains snow-covered all year.

summer temperatures are also low near the poles, there is little melting of the snow that does fall. The result is that the snowline generally lies at low altitude.

Because mean summer temperatures increase toward the equator, the altitude of the snowline also rises toward the equator, but not uniformly. The position of the snowline is also controlled by precipitation, so its altitude is typically lower near the ocean (a source of moisture for precipitation) and higher inland (FIG. 9.4). For example, winter snowfall is very high in the Coastal Ranges of southern Alaska and British Columbia adjacent to the Pacific Ocean source of moisture, and the snowline lies as low as 600 m. However, it rises steeply inland to 2600 m in the Rocky Mountains as precipitation decreases by more than half. The snowline is highest on the lofty summits of Central Asia (in places more than 6000 m), which lie farther than any other high mountains from an oceanic source of moisture.

FIGURE 9.4 Influence of climate on the snowline
The map contours (in meters) show the regional altitude of the snowline throughout northwestern United States, western Canada, and southern Alaska for a representative year. The altitude of the snowline rises steeply inland from the Pacific coast in response to increasingly drier climates, and more gradually from north to south in response to progressively higher mean annual temperatures.

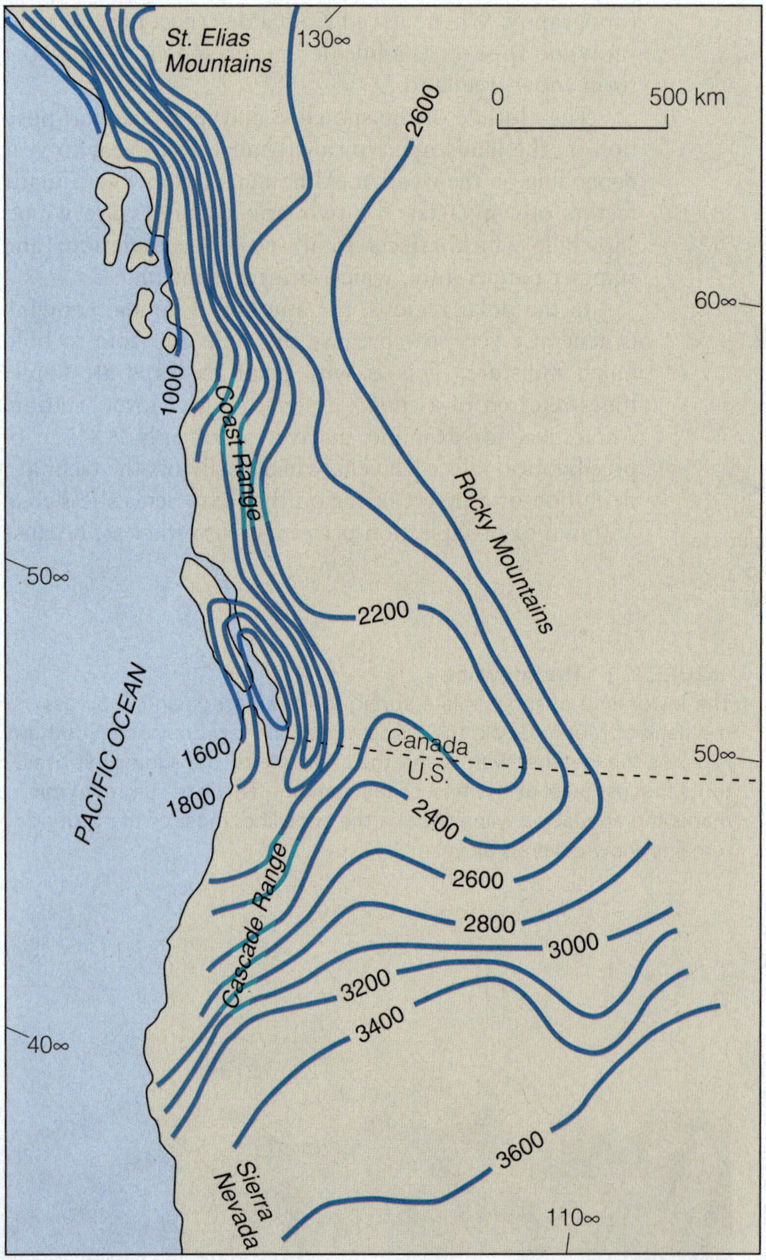

GLACIERS

Wherever the amount of snow falling each winter is greater than the amount that melts during the following summer, the snow pack grows progressively thicker. Gradually, the deeper snow recrystallizes into denser ice under the increasing weight of overlying snow. When the snow and ice become so thick that the pull of gravity causes the frozen mass to move, a glacier is born. Accordingly, we define a **glacier** as a persistent body of ice, consisting largely of recrystallized snow, which shows evidence of slow downslope or outward movement under the influence of gravity.

Glaciers vary considerably in shape and size. We recognize several fundamental types, and they are illustrated in FIGURE 9.5:

1. The smallest, a *cirque glacier* (FIG. 9.5A), occupies a protected, bowl-shaped depression (a **cirque**) on a mountainside that is produced by glacial erosion.
2. A cirque glacier that expands outward and downward into a valley becomes a *valley glacier* (FIG. 9.5B).
3. **Ice caps** cover mountain highlands or low-lying land at high latitudes and generally flow radially outward from their center (FIG. 9.5C).
4. Along some high-latitude seacoasts, nearly every large valley glacier occupies a deep **fjord**, the drowned seaward end of a glacier-carved bedrock trough. Such glaciers are called *fjord glaciers* (FIG. 9.5D).
5. When a valley glacier flows all the way out of the mountains and spreads out onto the surrounding lowlands, it is called a *piedmont glacier* (FIG. 9.5E).

Many of Earth's high mountain ranges contain glacier systems that include glaciers

FIGURE 9.5 Glaciers
(A) A *cirque glacier*, such as this one in Montana's Glacier National Park, occupies a bowl-shaped depression on a mountainside and often serves as the source for a valley glacier. (B) This *valley glacier* is the Tiedemann Glacier at Mt. Waddington, British Columbia. (C) An *ice cap* covers a mountaintop (or low-lying land in the polar regions) completely and usually displays a radial flow pattern. In this aerial photograph, the Greenland Ice Cap surrounds the Nunatak Mountains. (D) When a glacial valley is partly filled by an arm of the sea, the valley is called a *fjord*, and the glacier is a *fjord glacier*. Such glaciers often give rise to icebergs that break off and float away. (E) When a glacier flows all the way out of the mountains and onto the surrounding lowlands, it is called a *piedmont glacier*. The Malaspina Glacier in Alaska, shown here, starts as a valley glacier and then spreads out as a piedmont glacier.

tens of kilometers long, each heading in one or several cirques (FIG. 9.6). Huge continent-sized **ice sheets** overwhelm nearly all the land surface within their margins (FIG. 9.7). Earth's large ice sheets, which are now confined to Greenland and Antarctica, include about 95 percent of the ice in existing glaciers and reach thicknesses of 3000 m or more. Several small ice sheets are found among the Canadian Arctic islands. Floating **ice shelves** hundreds of meters thick occupy large embayments along the coasts of Antarctica; some of these, such as the Ross Ice Shelf, are labeled on Figure 9.7. Ice shelves are a form of sea ice, which we will discuss in greater detail later in the chapter.

How Glaciers Form

As discussed in *The Basics: Snow and Ice*, newly fallen snow is open and porous. As a result of compaction over the course of a year or more, accumulated snow gradually becomes denser and denser until it is no longer permeable to air, at which point it becomes glacier ice. Although now a rock, glacier ice has a far lower melting temperature than any other naturally occurring rock,

FIGURE 9.6 Valley glaciers
This is a satellite view of the valley-glacier complex that covers much of Denali National Park in south-central Alaska. Mount McKinley, the highest peak in North America, lies near the center of the glacier-covered region.

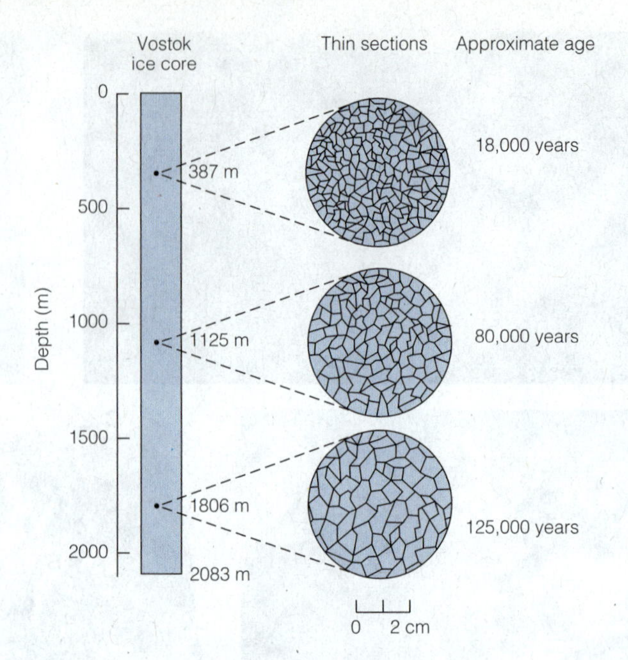

FIGURE 9.8 Recrystallization of glacial ice
This deep ice core drilled at Russia's Vostok Station penetrated through the East Antarctic Ice Sheet to a depth of 2083 m. Magnified sections of samples taken from different depths show a progressive increase in the size of ice crystals, the result of slow recrystallization as the thickness and weight of overlying ice slowly increase with time.

and its low density—about 0.9 g/cm^3—means it will float in water.

Further changes take place as the ice becomes buried deeper and deeper within a glacier. FIGURE 9.8 shows a core obtained by Russian glaciologists drilling deep in the East Antarctic Ice Sheet at Vostok Station. As snowfall adds to the glacier's thickness, the increasing pressure causes initially small grains of glacier ice to grow until, near the base of the ice sheet, they reach a diameter of 1 cm or more. This increase in grain size is similar to what happens in a fine-grained sedimentary rock that is buried deep within Earth's crust. As the rock is subjected to high pressure over a long time, recrystallization occurs and larger, interlocking mineral grains slowly develop (Chapter 7). The accumulation of layer upon layer of glacial ice eventually becomes the perennially frozen mass that we know as a glacier or an ice sheet.

FIGURE 9.7 Antarctic Ice Sheet
The East Antarctic Ice Sheet covers most of the continent of Antarctica, and the West Antarctic Ice Sheet overlies a volcanic island arc and the surrounding seafloor. In this satellite image you can also see four ice shelves that occupy large bays. Glaciers that flow down from the mainland feed these floating sea ice shelves.

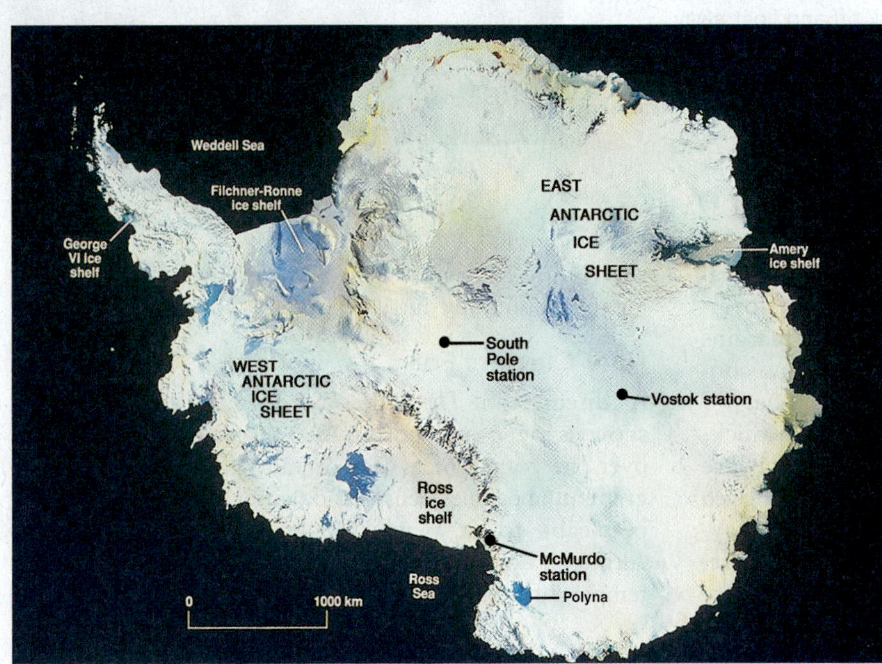

Distribution of Glaciers

A glacier can develop anywhere that the snowline intersects the topography and snow and ice are able to accumulate. This explains why glaciers are found not only at sea level in the polar regions, where the snowline is low, but also near the equator, where some lofty peaks in New Guinea, East Africa, and the Andes rise above the snowline. As we might expect, most glaciers are found in the high-latitude polar regions, the coldest regions of our planet. However, because low temperatures also occur at high altitudes, glaciers also can exist in lower latitudes on high mountains.

the BASICS

Snow and Ice

Snow is precipitation that consists of solid H_2O in crystalline form. It results from the condensation and crystallization of tiny water droplets into feathery ice crystals at very low temperatures in clouds. Snowflakes form from *supercooled* water at temperatures of −15°C or even −18°C—well below the normal freezing temperature of water. The resulting ice crystals are so tiny that several of them have to gather at a *nucleation site* before they can form a more substantial grain of snow.

Snow, like other forms of precipitation, falls when the grains gain sufficient mass that they settle under the influence of gravity. Solid H_2O can precipitate in other forms, too, such as *hail*. The crystal structure of snow has a *hexagonal* symmetry, which means that it typically forms six-sided or (occasionally) three-sided crystals (**FIG. B9.1**). The crystals are controlled by temperature and moisture in the cloud, and by the specific cooling history of the grain, so snow crystals vary widely.

The crystal structure of most snow grains is open and airy, with a density less than a tenth that of water. When the snow falls and is deposited on the ground, it settles under its own weight. The lower layers of snow may become compacted as a result of the weight of overlying snow accumulating, layer upon layer, like sediment.

Eventually, the seasonal snowpack melts and flows away. In polar regions and mountains above the snowline, however, the entire annual snowfall may not melt in a given season. As more snow accumulates on top, the underlying snow becomes more and more compacted. Air penetrates the pore spaces, and the delicate points of each snowflake gradually evaporate. The resulting water vapor condenses near the snowflake's center. The fragile crystals slowly become smaller, rounder, and denser, and the pore spaces between them disappear (**FIG. B9.2**). The open, feathery structure of the snow progressively recrystallizes into a much denser, interlocking crystalline texture that is characteristic of glacier ice.

Ice is also a solid form of H_2O. **Glacier ice** forms by the accumulation, compression, and recrystallization of snow, but some ice crystallizes directly from water in the atmosphere and falls as icy precipitation. Other types of ice form as the result of the crystallization of water, with no direct link to precipitation. For example, the ice floating on top of a frozen pond in the winter has mostly solidified in place, as a result of the crystallization of fresh water at temperatures less than 0°C. Similarly, sea ice—one of the most important components of the cryosphere—forms as a direct result of the crystallization of ocean water. Sea ice formation involves the crystallization of solid H_2O from a liquid that contains other components, notably salts, which are excluded from the solid during the crystallization process.

Ice is one of the few solids in nature that has a density less than that of its counterpart liquid. This is why ice floats and why most lakes don't freeze completely solid in the winter—the ice that forms as a layer floating on top of the water insulates the underlying water, allowing it to remain liquid. Ice, like snow, is hexagonal in crystal form, but more than a dozen other forms of ice have been observed in laboratory experiments. Although none of the other forms of ice is known to occur on Earth, it is likely that some of them do occur on other planets and moons in the solar system.

FIGURE B9.1 Snow crystals
Snow crystals are widely varied, although most are open and porous, and either hexagonal or triangular in shape. Although it is said that no two snowflakes are alike, snow crystals grown in exactly the same conditions would certainly be very similar.

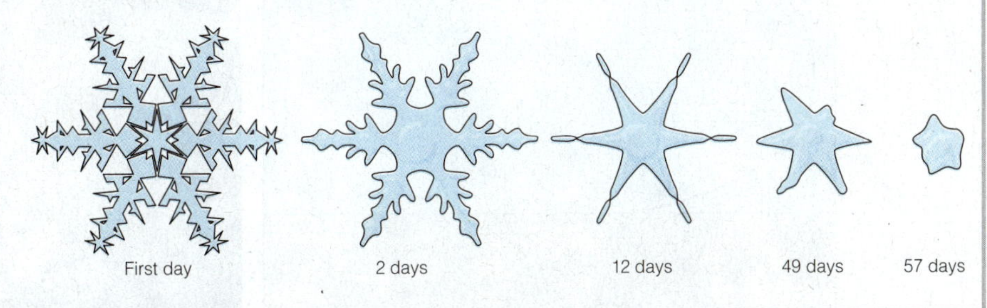

FIGURE B9.2 From snow to ice
As a snowflake is slowly converted into a granule of old snow and eventually into glacial ice, melting and evaporation cause its delicate points to disappear. The resulting meltwater refreezes, and vapor condenses near the center of the crystal, making it denser and less porous.

Low temperature is not the only factor that determines where glaciers can form, for a glacier must receive a continuing input of snow. Proximity to a moisture source is therefore another requirement. The abundance of glaciers in the coastal mountains of northwestern North America, for example, is related mainly to the abundant precipitation received from air masses moving landward from the Gulf of Alaska. Farther inland in the same latitude zone, the Rocky Mountains contain fewer and smaller glaciers because the climate there is much drier. Thus, the existence of glaciers is linked to the interaction of several Earth systems: tectonic forces that have produced high mountains; the adjacent ocean, which provides an abundant source of moisture; and the atmosphere, which delivers the moisture to the land as snow.

Warm and Cold Glaciers

Glaciers obviously are cold, for they consist of ice and snow. However, when we drill holes through glaciers in a variety of geographic environments, we find a large range in ice temperatures. This temperature range allows us to divide glaciers into warm and cold types. The difference between them is important, for ice temperature is a major factor controlling how glaciers behave.

Ice throughout a warm glacier, more commonly called a **temperate glacier**, can coexist in equilibrium with water. This ice is at its *pressure melting point*, the temperature at which ice can melt at a particular pressure (FIG. 9.9A). Pressure increases downward in a glacier, but temperate glaciers are rarely thick enough for the pressure melting effect to be more than a few tenths of a degree Celsius. Such glaciers are restricted mainly to low and middle latitudes.

In contrast, at high latitudes and high altitudes where the mean annual air temperature is below freezing, the temperature in a glacier remains below the pressure melting point, and little or no seasonal melting occurs (FIG. 9.9B). Such a cold glacier is commonly called a **polar glacier**.

If the temperature of snow crystals falling to the surface of a temperate glacier is below freezing, how

FIGURE 9.9 Temperate and polar glaciers
These graphs show temperature profiles through temperate and polar glaciers. (A) Ice in a temperate glacier is at the pressure melting point from surface to bed. The actual temperature depression due to the weight of the overlying ice is small—only a few tenths of a degree in a small glacier. The terminus of the glacier is rounded, as illustrated by Pré de Bar Glacier in the Italian Alps, because melting occurs at the surface. (B) Ice in a polar glacier remains below freezing, and the ice is frozen to its bed. Subfreezing temperatures inhibit melting at the terminus, which forms a steep cliff of ice, as illustrated by Commonwealth Glacier in Antarctica.

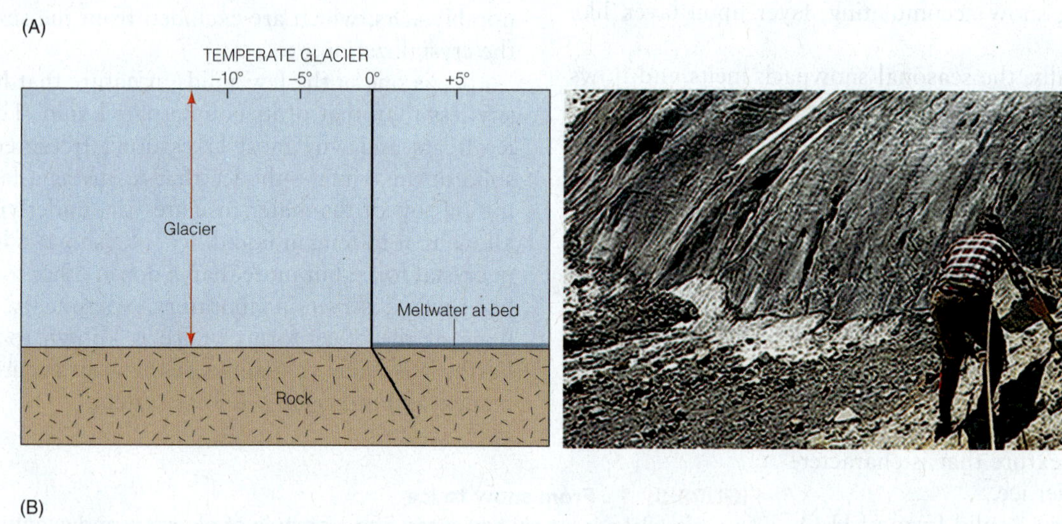

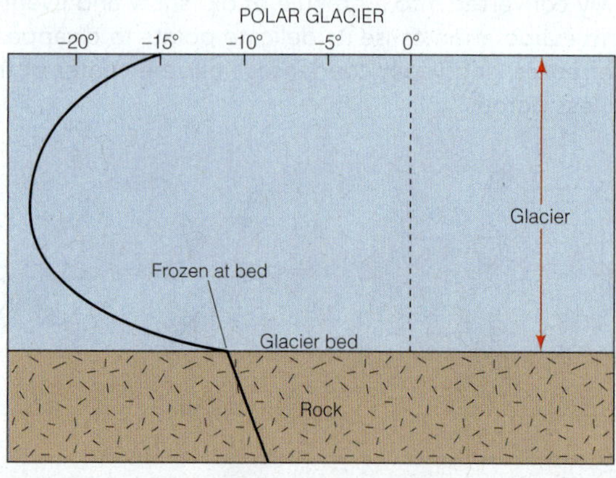

does ice throughout the glacier reach the pressure melting point? The answer lies in the seasonal fluctuation of air temperature and in what happens when water freezes to form ice. In summer, when air temperature rises above freezing, solar radiation melts snow and ice at the glacier's surface. The meltwater percolates downward, where it encounters freezing temperatures and it therefore freezes. When changing state from liquid to solid, each gram of water releases 335 J of heat. This released heat warms the surrounding ice and, together with geothermal heat flowing upward from the solid earth beneath the glacier, keeps the temperature of the ice at the pressure melting point.

Why Glaciers Change

Nearly all high-mountain glaciers have shrunk substantially in recent decades, exposing extensive areas of valley floor that only a century ago were buried beneath thick ice. Other glaciers have remained relatively unchanged, however, and a few have even expanded. To understand why glaciers advance and retreat, and why glaciers in the same region can show dissimilar behavior, we need to examine how a glacier responds to a gain or loss of mass.

The mass of a glacier is constantly changing as the weather varies from season to season and, on longer time scales, as local and global climates change. We can think of a glacier as being analogous to a checking account in a bank. The balance in the account at the end of the year is the difference between the amount of money added during the year and the amount removed. The balance of a glacier's account is measured in terms of the amount of snow added, mainly in the winter, and the amount of snow (and ice) lost, mainly during the summer. The additions are called **accumulation**, and the losses are **ablation**. The total in the account at the end of a year—in other words, the difference between accumulation and ablation—is a measure of the glacier's mass balance. The account may have a surplus (a positive balance) or a deficit (a negative balance), or it may have exactly the same amount at the end as at the beginning of a year.

If a glacier is viewed at the end of the summer ablation season, two zones are generally visible on its surface (FIG. 9.10). An upper zone, the *accumulation area*, is the part of the glacier covered by remnants of the previous winter's snowfall and is an area of net gain in mass. Below it lies the *ablation area*, a region of net loss where bare ice

FIGURE 9.10 **Glacial features**
This is a valley glacier. The accumulation area can be seen at the top of the glacier, above the equilibrium line, and the ablation area is lower, near the terminus. The glacier has been cut away along its center line to show its internal features. Arrows show the local directions of ice flow. A band of rock debris forms a medial moraine that marks the boundary between the main glacier and a tributary glacier joining it from a side valley. Crevasses form where the glacier flows over an abruptly steep slope.

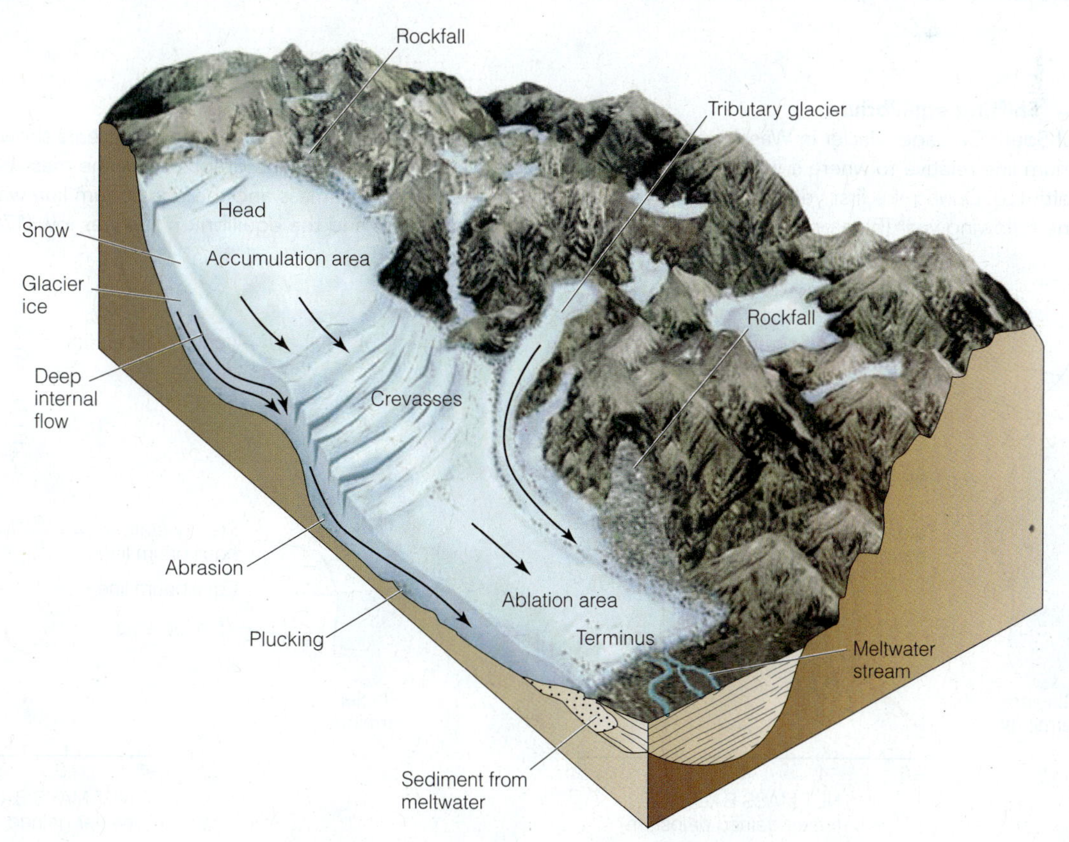

and old snow are exposed because the previous winter's snow cover has melted away.

The **equilibrium line** marks the boundary between the accumulation area and the ablation area (FIG. 9.11). It lies at the level on the glacier where net mass loss equals net mass gain. The equilibrium line on temperate glaciers coincides with the local snowline. Being very sensitive to climate, the equilibrium line fluctuates in altitude from year to year and is higher in warm, dry years than in cold, wet years. Because of this sensitivity, fluctuations in the altitude of the equilibrium line over time can provide us with a measure of climate change and its impacts on the glacier's mass balance.

When, over a period of years, a glacier gains more mass than it loses, its volume increases. The front, or **terminus**, of the glacier is then likely to advance as the glacier grows. Conversely, a succession of years in which negative mass balance predominates will lead to retreat of the terminus. If no net change in mass occurs, the terminus is likely to remain relatively stationary.

How Glaciers Move

The terminus of a glacier shifts in response to changes in the mass balance, but the entire glacier itself is also moving—flowing—under the influence of gravity. One way to prove that glaciers move is to walk out onto a glacier near the end of the summer and carefully measure the position of a surface boulder with respect to some fixed point beyond the glacier margin. Measure the position of the same boulder again a year later and you will likely find that the boulder has moved up to several meters in the down-glacier direction. Actually, it is the ice that has moved, carrying the boulder along for the ride.

What causes a glacier to move may not be immediately obvious, but we can find clues by examining the ice and the terrain on which it lies. These clues tell us that ice moves in two primary ways: by internal flow and by sliding of the basal ice across rock or sediment.

Internal Flow

When an accumulating mass of snow and ice on a mountainside reaches a critical thickness, the mass begins to deform and flow downslope under the pull of gravity. Flow takes place through movement within individual ice crystals, which are subjected to higher and higher stress as the weight of the overlying snow and ice increases. Under this stress, ice crystals are deformed by slow displacement (termed *creep*) along internal crystal planes in much the same way that cards in a deck of playing cards slide past one another if the deck is pushed from one end (FIG. 9.12). As the compacted, frozen mass begins to move, stresses between adjacent ice crystals cause some to grow at the expense of others, and the resulting larger crystals end up with their internal planes oriented in the same direction. This alignment of crystals leads to increased efficiency of flow because the internal creep planes of all crystals now are parallel.

In contrast to deeper parts, where the ice flows as a result of internal creep, the surface portion of a glacier has relatively little weight on it and is brittle. Where a glacier passes over an abrupt change in slope, such as a bedrock cliff, the surface ice cracks as tension pulls it apart. When a crack opens up, it forms a **crevasse**, a deep, gaping fissure

FIGURE 9.11 Shifting equilibrium
These maps of South Cascade Glacier in Washington's Cascade Range at the end of two successive balance years show the position of the equilibrium line relative to where it would be in a balanced (steady-state) condition. The graphs show the mass balance as a function of altitude. During the first year (A), a negative-balance year, the glacier lost mass and the equilibrium line was high (2025 m). In the following year (B), a positive-balance year, the glacier gained mass and the equilibrium line was low (1795 m).

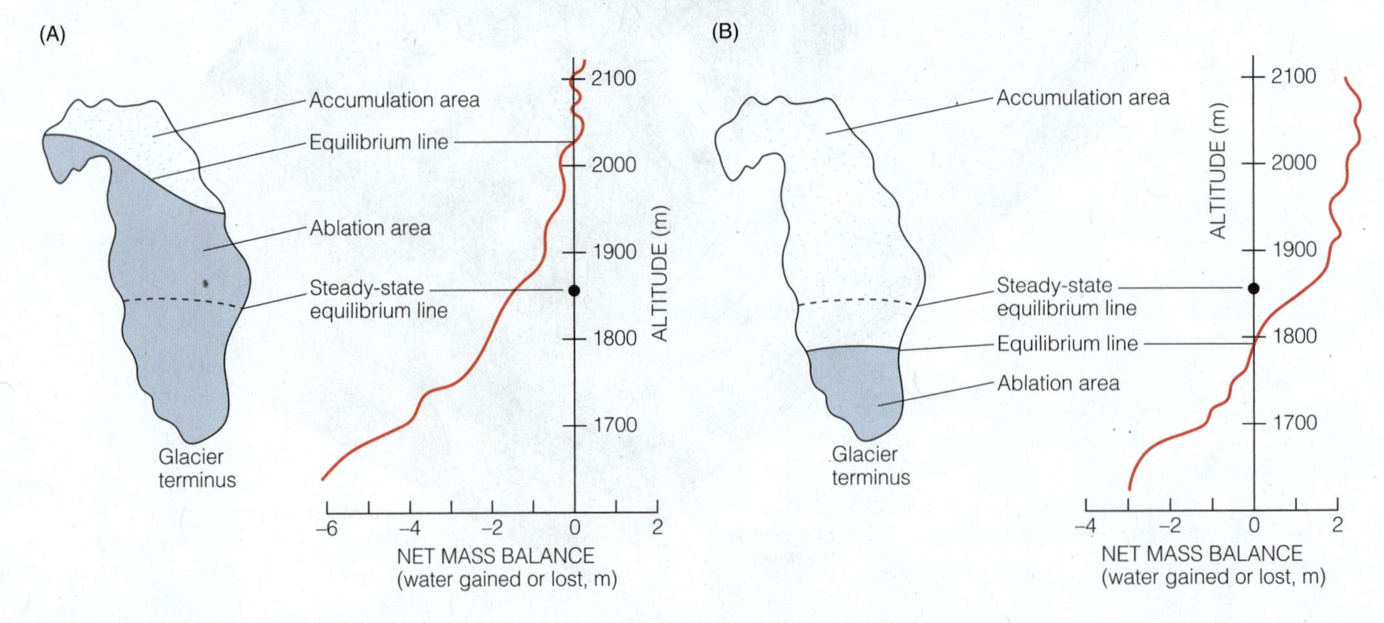

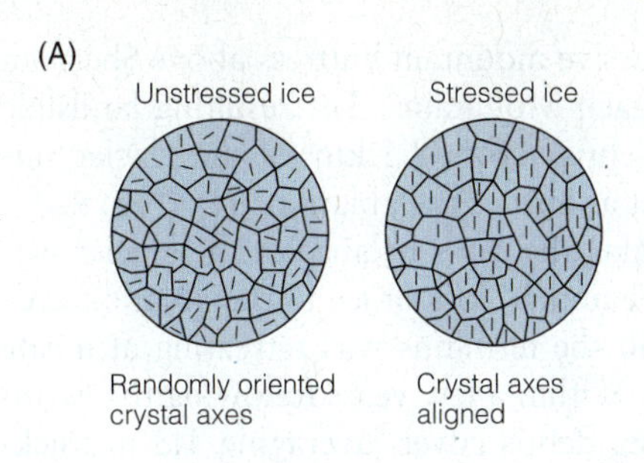

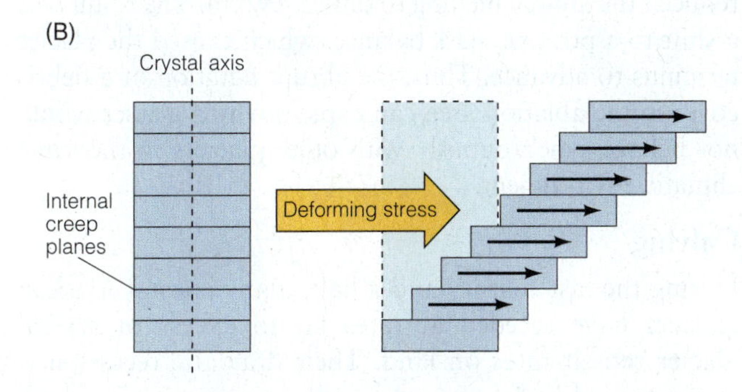

FIGURE 9.12 **Internal creep in a glacier**
(A) Randomly oriented ice crystals in the upper layers of a glacier are reorganized under stress so that their axes are aligned. (B) When stress is applied to an ice crystal, creep along internal planes causes slow deformation and movement of the glacial ice mass.

in the upper surface of a glacier, generally less than 50 m deep (FIG. 9.13 and shown in Fig. 9.10). Continuous flow of ice prevents crevasses from forming at depths greater than about 50 m. Because it cracks at the surface and yet flows at depth, a glacier is analogous to the outer part of Earth's geosphere, which includes a surface zone that cracks and fractures (the brittle lithosphere) and a deeper zone (in the upper mantle) that can flow slowly.

Basal Sliding

Ice temperature is very important in controlling the way a glacier moves and its rate of movement. Meltwater at the base of a temperate glacier acts as a lubricant and permits the ice to slide across its **bed** (the rock or sediment on which the glacier rests). In some temperate glaciers, such sliding accounts for up to 90 percent of the total observed movement. By contrast, polar glaciers are so cold that they are frozen to their bed. Their motion largely involves internal deformation rather than basal sliding, and so their rate of movement is greatly reduced.

Ice Velocity

Measurements of the surface velocity across a valley glacier show that the uppermost ice in the central part of the glacier moves faster than ice at the sides; this is similar to the velocity distribution in a river (Fig. 8.11). The reduced rates of flow toward the glacier margins are due to frictional drag of the ice against the valley walls. A similar reduction in flow rate toward the bed is observed in a vertical profile of velocity (FIG. 9.14). In most glaciers, flow velocities range from only a few centimeters to a few meters a day, or about

FIGURE 9.13 **Crevasse**
Deep fissures, called *crevasses*, like this one in Alaska's Mendenhall Glacier, open up as a result of stresses in the brittle surface layer of a glacier. The glacier flows in a direction perpendicular to the crevasse.

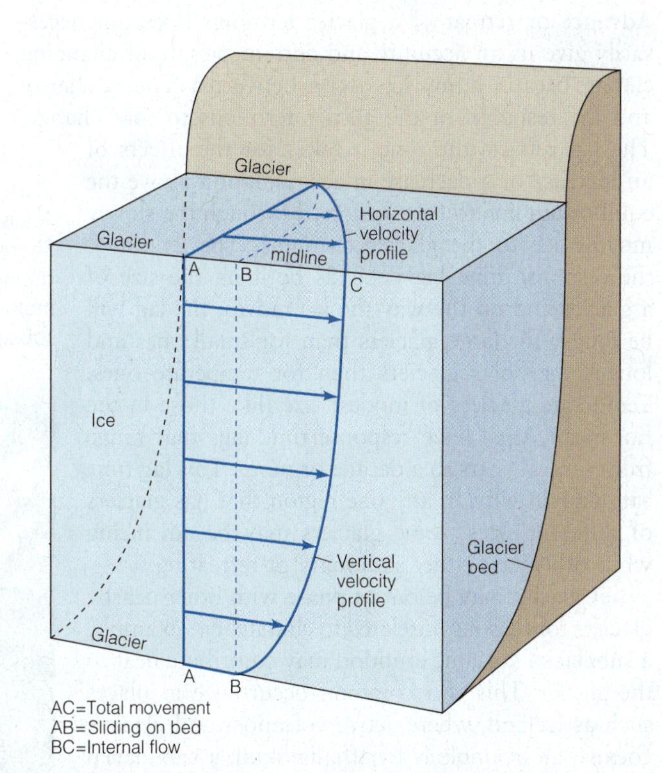

FIGURE 9.14 **Movement in a temperate glacier**
This three-dimensional view through half of a temperate glacier shows horizontal and vertical velocity profiles. Glacier movement is due partly to internal flow and partly to sliding of the glacier across its bed.

as fast as the rate at which groundwater percolates through crustal rock. Hundreds of years have elapsed since ice now exposed at the terminus of a very long glacier fell as snow near the top of its accumulation area.

Although snow piles up in the accumulation area each year, and melting removes snow and ice from the ablation area, a glacier's surface profile does not change much because ice is continuously being transferred from the accumulation area to the ablation area. In the accumulation area, the mass of accumulating snow and ice is pulled downward by gravity, so the dominant direction of movement is toward the glacier bed. However, the ice does not build up to ever-greater thickness because a down-glacier component of flow is also present. Ice flowing down-glacier replaces ice being lost from the glacier's surface in the ablation area, so in the ablation area the flow is upward toward the surface (as shown in Fig. 9.10). Ice crystals falling as snowflakes on the glacier near its head therefore have a long path to follow before they emerge near the terminus. Those falling close to the equilibrium line, on the other hand, travel only a short distance through the glacier before reaching the surface again.

Even if the mass balance of a glacier is negative and the terminus is retreating, the down-glacier flow of ice is maintained. Retreat does not mean that the ice-flow direction reverses; instead, it means that the rate of flow down-glacier is insufficient to offset the loss of ice at the terminus.

Response Lags

Advance or retreat of a glacier terminus does not necessarily give us an accurate and current picture of changing climate because a time lag occurs between a climatic change and the response of the glacier terminus to that change. The lag reflects the time it takes for the effects of an increase or a decrease in accumulation above the equilibrium line to be transferred through the slowly moving ice to the glacier terminus. The length of the response time lag depends both on the size of a glacier and on the way the ice moves; the lag will be longer for large glaciers than for small ones and longer for polar glaciers than for temperate ones. Temperate glaciers of modest size (like those in the European Alps) have response time lags that range from several years to a decade or more. This lag time can explain why, in any one region that has glaciers of different sizes, some glaciers may be advancing while others are either stationary or retreating.

A glacier may be out of phase with other nearby glaciers for reasons unrelated to climate. For example, a subglacial volcanic eruption may contribute heat to the glacier. This is a common occurrence in places such as Iceland, where active volcanoes and glaciers coexist; an example is Eyjaffjallajökull, a sub-glacial volcano that disrupted air travel by erupting in the spring of 2010. Major earthquakes can lead to massive landslides that affect the temperature balance of a glacier. For example, the Great Alaska earthquake of March 27, 1964 in Prince William Sound triggered the collapse of a massive mountain buttress above Sherman Glacier in the Chugach Mountains. The resulting landslide spread a layer of debris across 8.5 km^2 of the glacier surface, covering about a third of the ablation area (FIG. 9.15). Before the earthquake, the mass balance of the glacier was slightly negative, the annual loss of ice in the ablation area was about 4 m, and the terminus was retreating at a rate of about 25 m/year. Within a few years following the earthquake, the insulating debris cover (averaging 1.3 m thick) reduced the annual melting to only a few cm. The result was a shift to a positive mass balance, which caused the glacier terminus to advance. Thus, the abrupt addition of a debris cover in the ablation area can explain why a glacier would not behave synchronously with other glaciers in the same climatic environment.

Calving

During the last century and a half, many coastal Alaskan glaciers have receded at rates far in excess of typical glacier retreat rates on land. Their dramatic recession is characterized by frontal **calving**, the progressive breaking off of icebergs from the front of a glacier that terminates in deep water (FIG. 9.16). Although the base of a fjord glacier may lie far below sea level along much of its length, its terminus can remain stable as long as it is resting (or "grounded") against a shoal (a shallow submarine ridge) (FIG. 9.17). However, if the terminus retreats off the shoal, water will replace the space that had been occupied by ice. With the glacier now terminating in water, conditions are right for calving. Once started, calving will continue rapidly and irreversibly until the glacier front recedes into water too shallow for much calving to occur, generally near the head of the fjord.

FIGURE 9.15 Debris flow on a glacier
A vast sheet of dark, rocky debris covers the lower ablation zone of Sherman Glacier following the collapse of a large mountain buttress during the 1964 Alaska earthquake. The debris cover impeded melting of the underlying ice, leading to a positive mass balance and subsequent advance of the glacier terminus.

FIGURE 9.16 Calving
Chunks of ice calving from Alaska's Hubbard Glacier will become floating icebergs.

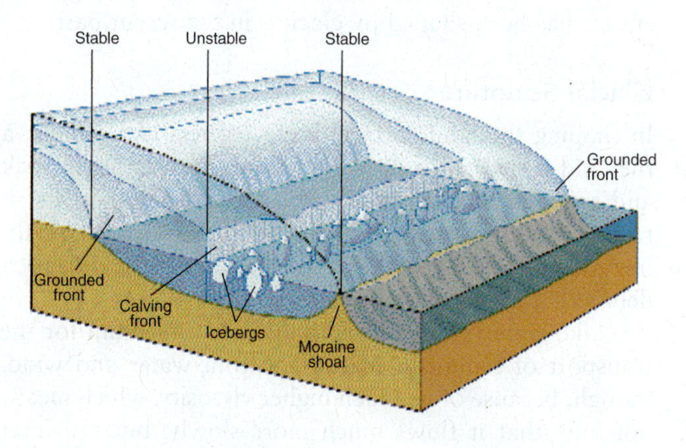

FIGURE 9.17 Fjord glacier
The terminus of a fjord glacier remains stable if it is grounded against a shoal (a submarine ridge), but if the glacier retreats into deeper water, calving will take place.

Icebergs produced by calving glaciers constitute an ever-present hazard to ships in subpolar seas. In 1912, when the S.S. *Titanic* sank after striking an iceberg in the North Atlantic Ocean, the detection of approaching bergs relied on the sharpness of sailors' vision. Today, with sophisticated electronic equipment, large bergs can generally be identified well before an encounter. Nevertheless, ice has a density of 0.9 gm/cm^3 (compared to 1.0 gm/cm^3 for water), which means that 90 percent of an iceberg lies under water, making it difficult to detect (**FIG. 9.18**). In coastal Alaska, where calving glaciers are commonplace, icebergs pose a potential threat to huge oil tankers. For this reason, Columbia Glacier, which lies adjacent to the main shipping lanes from Valdez at the southern end of the Alaska Pipeline, is being closely monitored as its terminus pulls steadily back and multitudes of bergs are released.

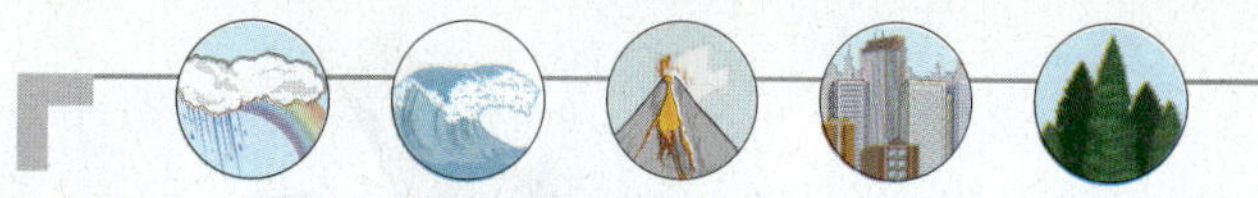

Make the CONNECTION

Some people have suggested that icebergs might make a good source for drinking water, and have proposed towing a large iceberg to an area of water shortage to harvest the fresh water from it. What do you think would be some of the potential impacts on the Earth system and marine ecosystems?

Glacial Surges

Although most glaciers flow slowly and grow or shrink as the climate fluctuates, some experience episodes of unusual behavior marked by rapid movement and dramatic changes in size and form. Such an event, called a **surge,** is unrelated, or only secondarily related, to a change in climate. When a surge occurs, a glacier seems to go crazy. Ice in the accumulation area begins to move rapidly down-glacier, producing a chaos of crevasses and broken pinnacles of ice in the ablation area. *Medial moraines,*

FIGURE 9.18 Iceberg
This is a composite photo of an iceberg, (that is, a digitally manipulated photo) showing that 90 percent of its mass lies below sea level. This makes it difficult to detect iceberg hazards, but it explains why the melting of sea ice wouldn't contribute very much to a rise in global sea level.

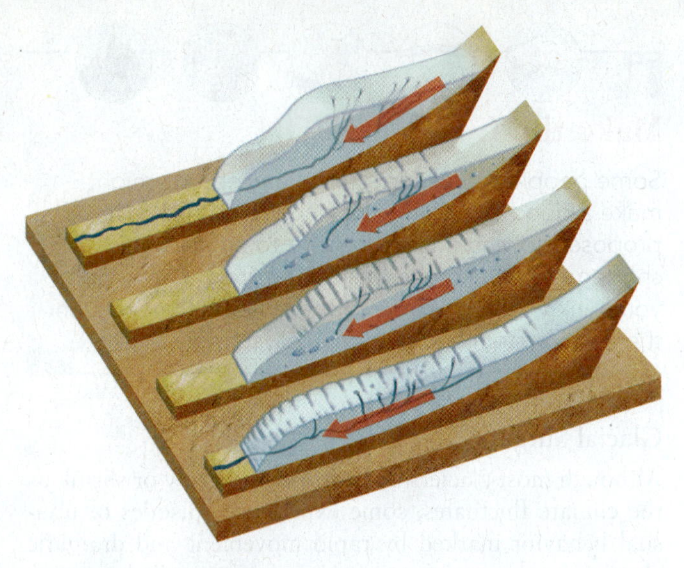

FIGURE 9.19 Glacial surge
According to one hypothesis, glacial surge begins when water at the base of a glacier gets blocked from flowing out. The buildup of pressure lubricates the base and allows the glacier to flow very rapidly, until the water finds a way out again.

which are bands of rocky debris marking the boundaries between adjacent tributary glaciers, are deformed into intricate patterns. In some cases, a glacier terminus has advanced up to several kilometers during a surge. Rates of movement as great as 100 times those of nonsurging glaciers and averaging as much as 6 km a year have been measured.

The cause of surges is still not fully understood, but available evidence supports a reasonable hypothesis (FIG. 9.19). The weight of the ice can produce high pressure in water trapped at the base of a glacier. Over a period of years, this steadily increasing pressure may cause the glacier to separate from its bed. The resulting effect is similar to the way a car hydroplanes on a wet road surface. According to this hypothesis, as the ice is floated off its bed, its forward mobility is greatly increased and it moves rapidly forward before the water escapes and the surge stops.

GLACIATION

Significant growth of glaciers and ice sheets requires that the average surface temperature remain low for an extended period, not just for one or two seasons. Periods during which the average temperature at Earth's surface drops by several degrees and stays low long enough for ice sheets to grow larger (and for new ones to form) are called **glaciations** (also *ice ages*, *glacial periods*, *glacial stages*, or *glacial epochs*). Periods between glaciations, when ice sheets retreat and sea levels rise, are called **interglacials** (also *interglacial stages* or *interglacial periods*). We are currently living in an interglacial period.

In Chapter 13 we will look closely at the history of climatic variation and the factors that cause Earth's climate to change and initiate glaciations. For now, let's consider the impacts of glaciation on Earth's landforms and inhabitants.

Glaciated Landscapes

Skiers racing down the steep slopes at Alta, Mammoth, or Whistler and rock climbers inching their way up the cliffs of Yosemite Valley, the granite spires of Mont Blanc, or the icy monoliths of the southern Andes owe a debt to the ancient glaciers that carved these mountain playgrounds. The scenic splendor of these and most of the world's other high mountains is the direct result of glacial sculpturing. Over other vast areas of central North America and northern Europe, farmers gain their livelihood from productive soils developed on widespread glacial sediment left by former continental ice sheets. In all, about 30 percent of Earth's land area not currently covered by persistent ice has been shaped by glaciers in the recent past.

Glacial Sculpture

In shaping the land surface, a glacier acts like a plow, a file, and a sled. As a plow, it scrapes up weathered rock and soil and plucks out blocks of bedrock. As a file, it rasps away firm rock. As a sled, it carries away the sediment acquired by plowing and filing, together with rock debris that falls from adjacent slopes.

Like water and wind, glacial ice is a medium for the transport of sediment. Ice differs from water and wind, though, because of its much higher viscosity, which means not only that it flows much more slowly, but also that part of its coarse load can be carried at its sides and even on its surface. A glacier can carry very large boulders and transports large and small pieces side by side without segregating them according to size and density. Thus, sediment deposited directly by a glacier is neither sorted nor stratified.

The load of a glacier typically is concentrated at its base and sides because these are the areas where glacier and bedrock are in contact. The coarse fraction of the load is derived partly from fragments of rock plucked from the lee (down-glacier) side of outcrops over which the ice flows, a process called *glacial plucking*. A significant component of the basal load of a glacier consists of very fine sand and silt grains informally called *rock flour*. If we examine such particles under a microscope, we find that they have sharp, angular surfaces that are produced by crushing and grinding.

Small rock fragments embedded in the basal ice scrape away at the underlying bedrock and produce long, nearly parallel scratches called *striations*. Larger rock fragments that the ice drags across a bedrock surface cut deep *glacial grooves*, aligned in the direction of ice flow (FIG. 9.20A). Rock flour in the basal ice acts like fine sandpaper and can polish the rock until it has a smooth, reflective surface.

FIGURE 9.20 **Glacial sculpting**
(A) These *glacial grooves* in Ohio were etched into limestone by the Wisconsin glacier during the most recent Ice Age, about 35,000 years ago. (B) The Western Cwm, a deep cirque on the west side of Mt. Everest, is flanked by sharp-crested *arêtes.* (C) The Lauterbrunnen Valley in Switzerland has the classic U-shape of a glacial valley. The glacier that formed it no longer exists.

In mountainous regions, cirques are among the most common and distinctive landforms produced by glacial erosion. The characteristic bowl-like shape of a cirque is the result of frost-wedging (Chapter 7), combined with plucking and abrasion at the glacier bed. As cirques on opposite sides of a mountain grow larger, they intersect to produce sharp-crested ridges. Where three or more cirques have carved through a mountain mass, the result can be a high, sharp-pointed peak called an *arête*, a classic example of which is the Matterhorn in the Swiss/Italian Alps (FIG. 9.20B).

A valley that has been shaped by glaciers differs from ordinary stream valleys in having a distinctive U-shaped cross profile and a floor that often lies below the floors of tributary valleys (FIG. 9.20C). The long profile of a glaciated valley floor may possess step-like irregularities and shallow basins related to the spacing of joints in the rock, which influences the ease of glacial plucking, or to changes in rock type along the valley. The valley typically heads in a cirque or group of cirques.

Glacially carved *fjords* deeply indent the mountainous, west-facing coasts of Norway, Alaska, British Columbia, Chile, and New Zealand. Typically shallow at their seaward end, fjords become deeper inland, implying deep glacial erosion. Sognefjord in Norway, for example, reaches a depth of 1300 m, yet near its seaward end the water depth is only about 150 m. Glacial erosion is also responsible for countless lakes that lie inside the limit of the last glaciation. Among the largest are the huge lakes that form an arc across southern and western Canada and include the Great Lakes, Lake Winnipeg, Lake Athabasca, Great Slave Lake, and Great Bear Lake.

Eroding ice sheets sometimes mold smooth, nearly parallel ridges of sediment or bedrock, called *drumlins,* which are elongated parallel to the direction of ice flow (FIG. 9.21). The streamlined drumlins offer minimum resistance to glacier ice flowing over and around them.

Glacial Deposition

A moving glacier carries with it rock debris eroded from the land over which it is passing, or rock dropped on the glacier surface from adjacent cliffs. As the debris is transported past the equilibrium line and ablation reduces ice thickness, the debris begins to be deposited. Some of the basal debris is plastered directly onto the ground as **till**, an unsorted sediment (FIG. 9.22A). Some also reaches the glacier margin, where it is released by the melting ice. The debris either accumulates there or is reworked by meltwater that transports it beyond the terminus, where it is deposited as *outwash*.

Ridge-like accumulations of sediment called **moraines** form as sediment is bulldozed by a glacier advancing across the land. Loose surface debris slides off and piles up along the glacier margin, or embedded debris melts out of the ice and accumulates along the edge of a glacier. If debris falls on the accumulation area, the flow paths of the ice will carry the debris downward into the glacier and then upward to the surface in the ablation area. If rock particles fall onto the ablation area, the debris will remain at the surface and be carried along by the moving ice. Moraines range in height from a few meters to hundreds

FIGURE 9.21 Drumlins
Drumlins in eastern Washington, each shaped like the inverted hull of a ship, are aligned parallel to the flow direction of the continental ice sheet that shaped them during the last glaciation.

of meters. The great thickness of some moraines results from the repeated addition of sediment from debris-covered glaciers during successive ice advances.

A moraine that is built up along the margin of a glacier is called an *end moraine*. An end moraine built at the terminus of a glacier is a *terminal moraine* (FIG. 9.22B), and one constructed along the side of a mountain glacier is a *lateral moraine*. Where two glaciers join, rocky debris at their margins merges to form a *medial moraine* (FIG. 9.22C). Moraines are important tools that scientists can use to determine the extent of ice coverage during an ancient glaciation.

When rapid melting greatly reduces a glacier's thickness in its ablation area, ice flow may virtually cease. Sediment deposited by meltwater streams flowing over or beside such immobile ice will slump and collapse as the supporting ice slowly melts away, leaving a hilly, often chaotic surface topography. Among the landforms associated with such terrain are *eskers*, curved ridges of unsorted sand and gravel (FIG. 9.22D). *Kettles* are closed basins

FIGURE 9.22 Glacial deposits
(A) Glacial till can sometimes include very large boulders, such as these boulders in Yellowstone National Park. When they are different from the bedrock, such boulders are called *erratics*. (B) This *terminal moraine* near Mt. Robson in British Columbia marks the farthest advance of the glacier at left in the nineteenth and twentieth centuries. (C) The dark stripes running down the center of Kaskawulsh Glacier, in the Yukon, are a *medial moraine*. (D) The curving ridge of sand and gravel in this photo is an *esker* in Kettle-Moraine State Park in Wisconsin.

the BASICS

Glacial and Interglacial Periods

During the past 1.8 million years, Earth has experienced more than 30 major glacial–interglacial cycles. The timing of these cycles has varied, with extreme temperature *minima* (low points) occurring roughly every 100,000 years over the past million years, and every 20,000 to 40,000 years before that. Glaciations have been especially prevalent during the Pleistocene Epoch, but the rock record contains evidence of glacial ages that occurred as long ago as 2.4 billion years.

The most recent ice age began about 70,000 years ago. It was a time when great woolly mammoths, mastodons, longhorn bison, and saber-tooth tigers roamed North America. Early humans migrated into North America from Asia, walking across exposed continental shelf in today's Bering Strait in Alaska. This land "bridge" was exposed during the ice age because a large volume of water was locked up in glacial ice, causing sea levels to drop globally. Huge floating ice sheets like those in present-day Arctic and Antarctic seas occupied large areas of the Atlantic Ocean.

Scientists can determine the timing and extent of glaciations from a variety of evidence. Glacial landforms, especially rock and sediment deposited by the glacier along its edges and at its terminus, reveal the geographic extent of land ice sheets. Scratches carved in bedrock by advancing ice indicate the direction of movement of the glaciers. Tree rings and radiocarbon ages of trees that were felled by advancing ice tell scientists when the ice arrived in a given region.

The ice sheets of the most recent glaciation reached their maximum extent in North America about 24,000 years ago (FIG. B9.3). Approximately 12,000 years ago the climate began to warm, and by 10,000 years ago Earth had emerged from the ice age and entered the present interglacial period. We have now passed the time of maximum warmth in the glacial–interglacial cycle. Temperatures peaked in a warm period about 6000–7000 years ago, known as the *Holocene optimum*. Since then temperatures have been cooling gradually, with some distinctly colder fluctuations, such as the *Little Ice Age*, which lasted from about AD 1300 to about AD 1900.

Glaciations are initiated by climatic change that is global in extent and long-lasting in duration. Several mechanisms cause natural climatic change; they involve the atmosphere, the geosphere, the ocean, and the biosphere interacting in complex ways. During the past decade or so, the role played by anthropogenic emissions in adding to the radiatively active gas content of the atmosphere, and the possibility of accelerated global climatic change occurring as a result, have been subjects of scientific and political debate. Climatologists and other scientists are working hard to understand how different parts of the Earth system work together to initiate or end glacial periods, and how the system can be affected by human actions. However, many sources of uncertainty in this complex system remain. We will look more closely at the natural causes of climatic change in Chapter 13 and then at the human influence on climate in Chapter 19.

FIGURE B9.3 Pleistocene glaciation
During the last glaciation, humans migrated into North America from Asia across a land bridge over the Bering Sea. The land was exposed because sea levels were lower, as a result of so much ice being locked up in glaciers, as shown here about 20,000 years ago at the height of the Pleistocene glaciation.

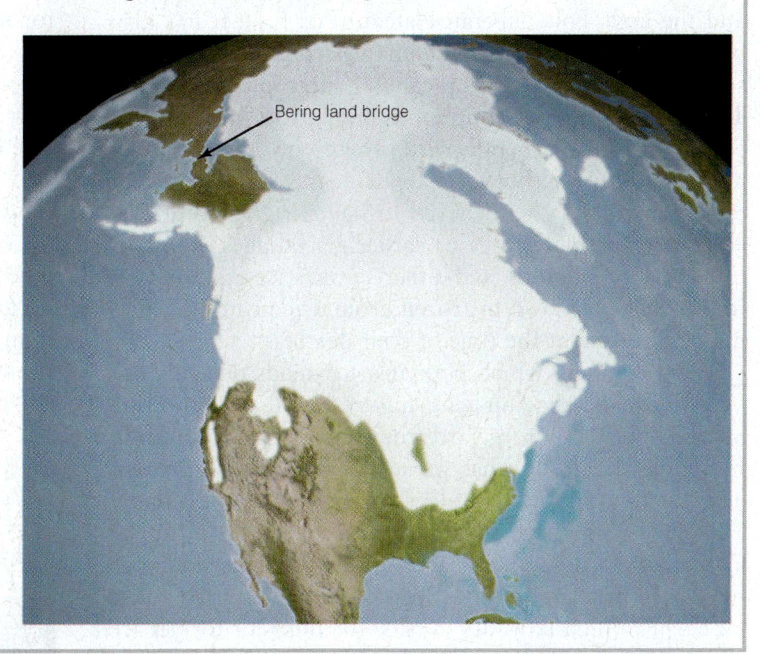

created by the melting away of a mass of underlying glacier ice. Landscapes marked by numerous kettles, now typically occupied by lakes, ponds, or wetlands, are clear evidence of previous glaciation and stagnant-ice conditions (FIG. 9.23).

Periglacial Landscapes and Permafrost

Land areas beyond the limit of glaciers also have an important influence on the landscape. In these **periglacial** zones, low temperature and frost action are important factors in determining landscape characteristics. Periglacial conditions are found over more than 25 percent of Earth's land areas, primarily in the circumpolar zones of each hemisphere and at high altitudes.

Permafrost

The most characteristic feature of periglacial regions is perennially frozen ground, also known as **permafrost**—sediment, soil, or even bedrock that remains continuously

FIGURE 9.23 Kettles
Lake-filled kettles are scattered over the surface of an end-moraine complex in the lake district of central Chile, which formed at the end of the last glaciation when debris-covered stagnant ice slowly melted away.

below freezing for an extended time (from two years to tens of thousands of years). The largest areas of permafrost occur in northern North America, northern Asia, and the high, cold Tibetan Plateau (Fig. 9.1). It has also been found in many high mountain ranges, even including some lofty summits in tropical and subtropical latitudes. The southern limit of continuous permafrost in the northern hemisphere generally lies where the mean annual air temperature is between about 5 and 10°C. Most of today's permafrost is believed to have originated during either the last glacial age or earlier glacial ages. Remains of woolly mammoth and other extinct ice-age animals found well preserved in frozen ground indicate that permafrost existed at the time of their death.

The depth to which permafrost extends depends not only on the average air temperature but also on the rate at which heat flows upward from Earth's interior and on how long the ground has remained continuously frozen. The maximum reported depth of permafrost is about 1500 m in Siberia. Thicknesses of about 1000 m in the Canadian Arctic and at least 600 m in northern Alaska have been measured. These areas of very thick permafrost all occur in high latitudes outside the limits of former ice sheets. The ice sheets would have insulated the ground surface and, where thick enough, caused ground temperatures beneath them to rise to the pressure melting point. On the other hand, open ground unprotected from subfreezing air temperatures by an overlying ice sheet could have become frozen to great depths during prolonged cold periods.

Living with Permafrost

In permafrost terrain, a thin surface layer of ground that thaws in summer and refreezes in winter is known as the *active layer*. In summer this thawed layer tends to become very unstable. The permafrost beneath, however, is capable of supporting large loads without deforming.

Permafrost presents unique problems for people living on it. If a building is constructed directly on the surface, the warm temperature developed when the building is heated is likely to thaw the underlying permafrost, making the ground unstable (FIG. 9.24). Arctic inhabitants learned long ago that they must place the floors of their buildings above the land surface on pilings or open foundations so that cold air can circulate freely beneath, thereby keeping the ground frozen.

Wherever a continuous cover of low vegetation on a permafrost landscape is ruptured, melting can begin. As the permafrost melts, the ground collapses to form impermeable basins containing ponds and lakes. Thawing can also be caused by human activity, and the results can be environmentally disastrous. Large wheeled or tracked vehicles crossing Arctic tundra can quickly rupture it. The water-filled linear depressions that result from thawing can remain as features of the landscape for many decades.

The discovery of a commercial oil field on the North Slope of Alaska in the 1960s generated the need to transport the oil southward by pipeline to an ice-free port. The company formed to construct the pipeline was faced with some unique problems. In order for the sticky oil to flow through a pipeline in the frigid Arctic environment, the oil had to be heated. However, an uninsulated, heated pipe in the frozen ground could melt the surrounding permafrost. Even if the pipe were insulated before placing it underground, the surface vegetation cover would be disrupted, likely leading to melting and instability. For these

FIGURE 9.24 Living with permafrost
This cabin in central Alaska settled more than a meter in eight years as permafrost beneath its foundation thawed.

reasons, along much of its course the Alaska Pipeline was constructed on piers above ground, thereby greatly reducing the possibility of ground collapse.

Periglacial Landforms

Many of the landscape features we associate with periglacial regions reflect movement of regolith within the active layer during annual freeze/thaw cycles (FIG. 9.25A, B). One of the most typical is a feature called *patterned ground* (also called *ice-wedge polygons*) in which a regular pattern of circles, polygons, stripes, or steps is formed in the active layer as a result of cyclical freezing and thawing of water in the pore spaces, combined with frost heaving (FIG. 9.25C). Another landform associated with frost heaving, which occurs as a result of the expansion of water upon freezing, is the pingo. *Pingoes* are small hills that form only in permafrost environments, by the expulsion of pore water through the active layer as a result of hydrostatic pressure from the expanding permafrost underneath.

Solifluction is a common mass-wasting process in periglacial regions in which waterlogged regolith in a thawed active layer moves downslope under the influence of gravity, forming large, soft lobes of sediment. (Solifluction can occur anywhere that waterlogged regolith overlies a harder subsurface layer; when the hard underlying layer consists of ice or permafrost, the process is technically referred to as *gelifluction*.)

FIGURE 9.25 Periglacial landforms
(A) An ice wedge forms when water seeps into an open crack in the ground and freezes. (B) In summer, the crack opens or partially melts, allowing more water to enter. In winter, the ice freezes and expands again. The ice wedge continues to grow as the melting and refreezing cycle repeats hundreds of times. (C) The result of many years of freezing and thawing can result in the development of patterned ground, ice-wedge polygons, and pingoes, as shown here near Tuktoyaktuk in Canada.

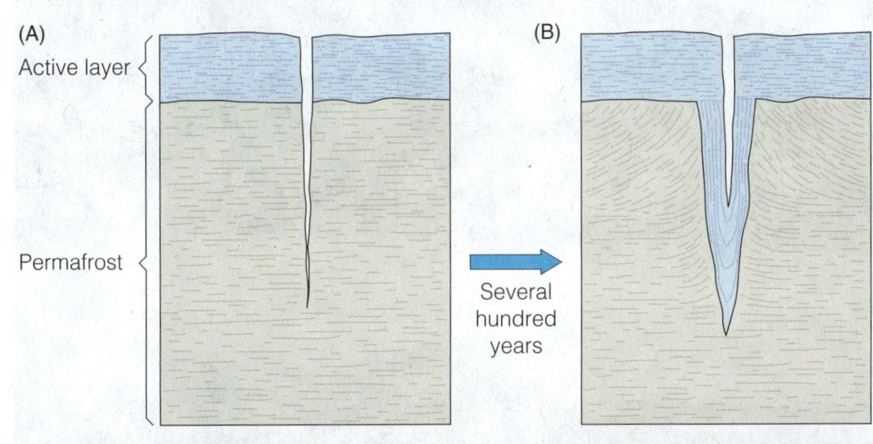

Periglacial regions are often associated with thick deposits of fine-grained, buff-colored sediment called **loess.** Loess is a windblown (or *aeolian*) sediment derived from rock flour and is generally indicative of deposition during a dry, cold climate. Although not all loess is glacial in origin, it occurs in thick deposits in some former periglacial regions such as northern China, the Rhine Valley in Europe, and the Mississippi River Valley, where it provides extremely valuable and fertile farmland.

The challenges of living in periglacial areas extend to nonhuman inhabitants, too. Ecosystems that form in areas underlain by permafrost are called *tundra*. The plants and animals of the tundra are specifically adapted to the extremes of temperature and moisture and, in near-polar regions, to the limitations of light that characterize these geographic settings. We will look more closely at tundra ecosystems and their characteristics in Chapter 16.

Glaciers and People

Permafrost is a challenge for modern inhabitants, but glaciation has been a central factor throughout much of human history. Modern humans evolved during the glacial ages under conditions that fluctuated between those like the present and those of the last ice age when vast areas of the northern hemisphere continents were covered with thick ice sheets, climates were cooler, and the skies were dustier. When the latest ice sheets retreated, people invaded the formerly glaciated landscapes and colonized them. Today, a substantial fraction of the world's population lies within or near the limit of former ice sheets, and this population is directly affected by the legacy of ice-age glaciers. The rich agricultural lands of the American Midwest and central Europe are a gift of the glaciers and of strong winds that eroded and deposited mineral-rich loess across the landscape.

Important subsurface groundwater supplies are mined from glacial gravels deposited along former meltwater stream systems. In the northern United States and parts of Europe, glacial sands and gravels are among the most valuable economic mineral resources, and in Alaska and Russia, such sediment is mined for placer gold, silver, and platinum. Vacationers, mountain climbers, and skiers are attracted to glaciated mountains the world over for their spectacular scenery and recreational opportunities. Where glaciers contribute to local or regional water supplies, they help moderate the flow of water throughout the year by releasing meltwater to streams during warm, dry summer months. By contrast, nearby streams that flow from basins that lack glaciers may experience stronger seasonal fluctuations in streamflow, including serious summer droughts.

Glaciers as Environmental Archives

Trapped in the snow that piles up each year in the accumulation area of a glacier is evidence of both local and global environmental conditions. The evidence includes physical, chemical, and biological components that can be extracted in a laboratory and studied as a record of the changing natural environment. The oldest ice in most cirque and valley glaciers is several hundred to several thousand years old, but large ice caps and ice sheets contain ice that dates far back into the ice ages. The record they contain is often unique, and it is of critical importance for understanding how the atmosphere, ocean,

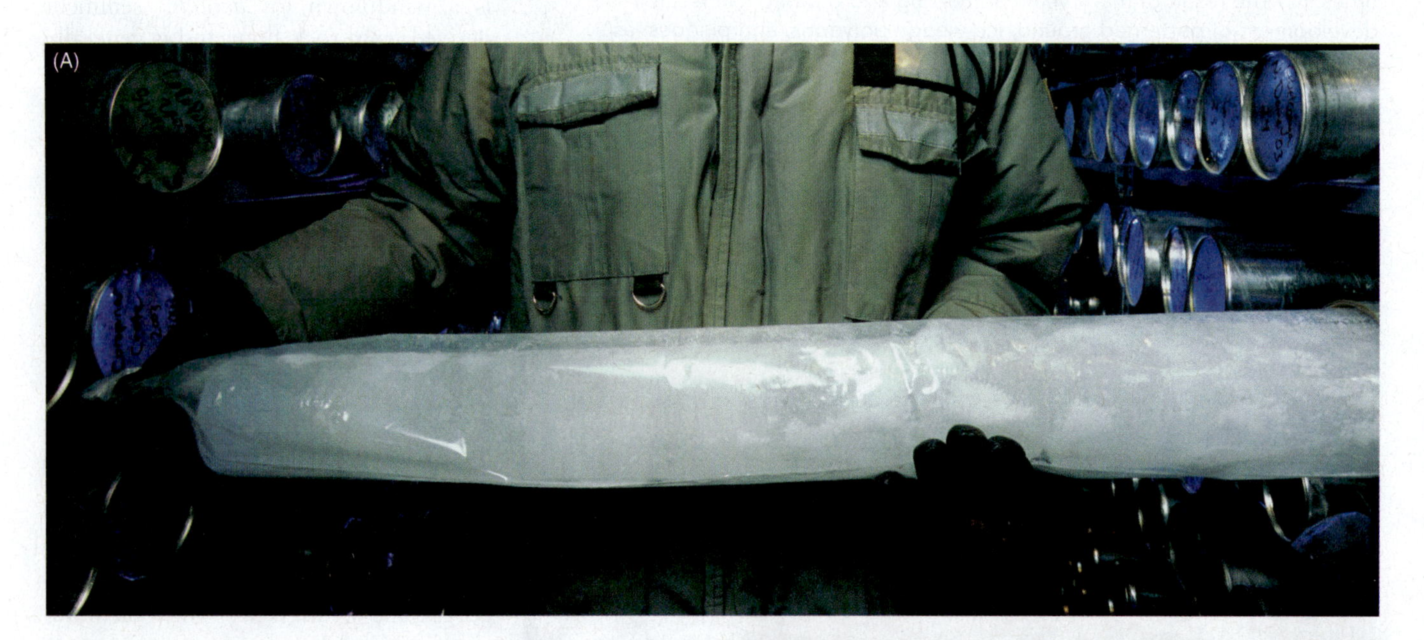

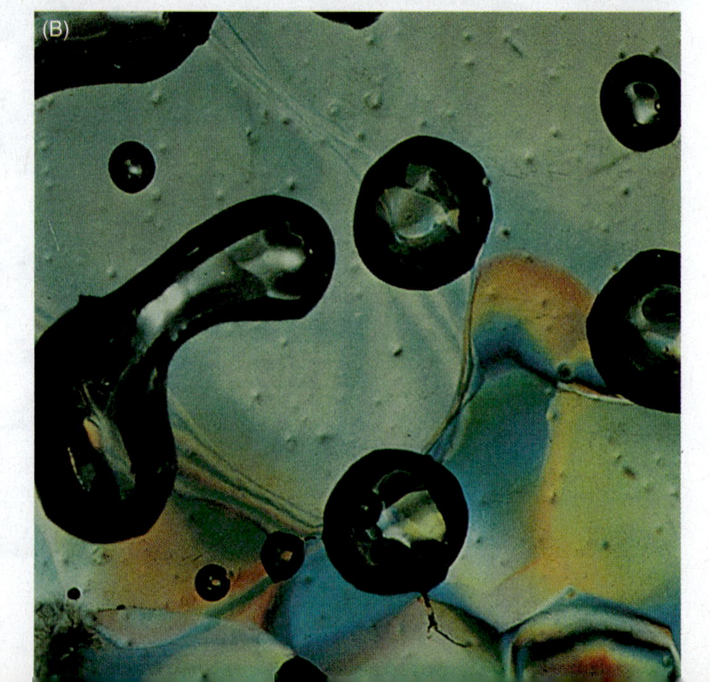

FIGURE 9.26 Ice as an environmental archive
(A) This ice core was extracted from the Quelccaya ice cap in Peru, which is rapidly retreating. (B) As the ice in a glacier recrystallizes, ice fills up the pore spaces between the crystals. But some tiny bubbles of air fail to escape and are locked permanently inside the ice, as seen here in a photomicrograph. When the ice is melted under controlled conditions in a laboratory, the chemical composition of the "fossil air" can be measured. Scientists are particularly interested in the concentration of carbon dioxide, a greenhouse gas. Samples from Antarctica and Greenland indicate that the atmosphere contained far less carbon dioxide during glacial ages than during interglacial periods.

and biosphere have changed over hundreds of thousands of years.

If we dig a pit many meters deep in the accumulation area of a glacier and look closely at the snow, layering can be seen. In each layer, the relatively clean snow that records a succession of winter snowfalls passes upward to a darker layer that contains dust and refrozen meltwater, a record of relatively dry, warmer summer weather. Depending on the local accumulation rate, one or several such annual layers may be exposed. Because digging a pit into a glacier is an inefficient way to examine the stratigraphy, glaciologists drill cores of ice that can be extracted and returned to a laboratory for analysis (**FIG. 9.26A**). Some drilling operations have penetrated to the base of the thick Greenland and Antarctic ice sheets, while others have focused on high-latitude and high-altitude ice.

Ice cores have proved a boon for atmospheric scientists who would like to know whether the concentrations of important atmospheric greenhouse gases like carbon dioxide and methane, which are trapped in air bubbles in the ice (**FIG. 9.26B**), have fluctuated as the climate changes (Chapter 13). Measurements of the chemistry of the ice itself can tell us the air temperature when the snow accumulated on the glacier surface. Ice cores also provide a record of major volcanic eruptions that generate sulfur dioxide gas which, combined with water, accumulates as a layer of acid snowfall on glaciers. High concentrations of loess in ice layers that date back to the last ice age provide evidence that the windy climate of glacial times was also extremely dusty. Tiny fragments of organic matter and fossil pollen grains trapped in the ice layers can tell us about the composition of local vegetation near a glacier and can be radiocarbon-dated to provide ages for the enclosing ice. Because these natural historical archives are trapped in annual ice layers that can be read like the pages of a book, they offer an unparalleled, detailed look at past surface conditions on our planet.

SEA ICE

Snow and glacial ice originate as precipitation, but some ice forms by the solidification of water at the surface, without falling as precipitation. Approximately two-thirds of the area of Earth's persistent ice cover floats as a thin veneer of **sea ice** on polar oceans (**FIG. 9.27**). It is vast in its extent, but sea ice comprises only about 1/1000 of Earth's total volume of ice.

An interesting feature of sea ice is that it consists of fresh water; the salt that is characteristic of seawater is excluded from the ice crystals as they form. Thus the formation of sea ice is one of the important Earth processes that control the salinity of ocean water. In periods of global coolness, when more sea ice forms, the ocean becomes saltier; in periods of global warmth, when sea ice melts, fresh water is released and the ocean becomes less salty.

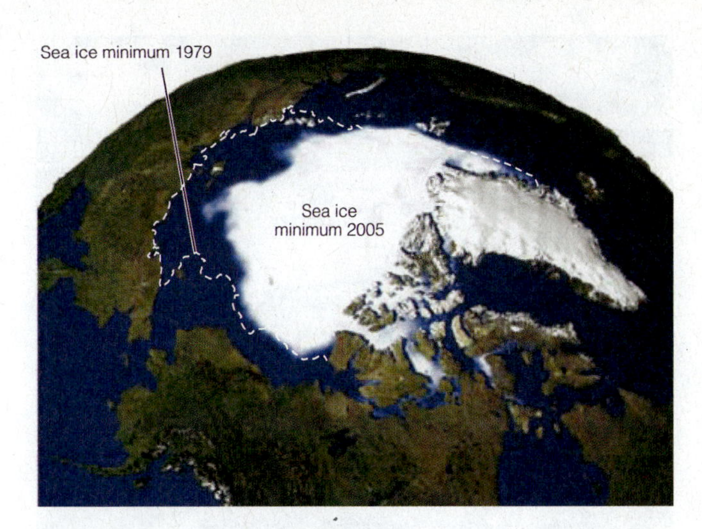

FIGURE 9.27 Arctic sea ice
Arctic sea ice typically reaches its minimum in September, at the end of the summer melt season, and then recovers over the winter. This satellite image shows the extent of sea ice at its minimum point for the year 2005, compared to the minimum for 1979 (dashed line), as determined from satellite images.

Make the CONNECTION

The formation of sea ice makes seawater saltier, and the melting of sea ice makes seawater fresher. What other processes control the salinity of seawater?

How Sea Ice Forms

Once the ocean surface cools to the freezing point of seawater, slight additional cooling leads to ice formation. The first ice to form consists of small crystalline platelets and needles up to 3 or 4 mm in diameter that collectively are termed *frazil ice*. As more ice crystals form, they produce a soupy mixture at the ocean surface. In the absence of waves or turbulence, the crystals freeze together to form a continuous cover of ice 1 to 10 cm thick. If waves are present, the crystals form rounded, pancake-like masses up to 3 m in diameter that eventually weld together into a continuous sheet of sea ice (**FIG. 9.28**).

Once a continuous cover of ice forms, the cold atmosphere is no longer in contact with the seawater, and sea-ice growth then proceeds by the addition of ice to the sea-ice base. In the Arctic, over the course of a yearly cycle, about 45 cm of ice is lost from the ice surface, but an equal amount is added to the base. As a result, an ice crystal added to the sea ice at its base will move upward through the ice column with an average velocity of about 45 cm/yr until it reaches the surface and melts away.

FIGURE 9.28 Sea ice
A nearly solid cover of seasonal ice forms a pavement on the Weddell Sea of Antarctica. Pressure ridges between plates of ice form when winds blow the fragments of ice together as the ice slowly melts in the Antarctic spring.

Sea-Ice Distribution and Zonation

The contrasting geography of Earth's north and south polar regions leads to important differences in the distribution of sea ice in the two hemispheres. The South Pole lies near the middle of the Antarctic continent, which is covered by a vast, thick ice sheet, whereas the North Pole lies in the middle of the deep Arctic Ocean basin. The open Southern Ocean adjacent to Antarctica contrasts with the largely landlocked Arctic Ocean, which is connected to the world ocean only by relatively narrow straits. In the Antarctic region, sea ice forms a broad ring around the continent and adjacent ice-covered archipelago, a ring that varies in width with the seasons. At its greatest northward extent in winter it covers 20 million km^2, but it shrinks to only 4 million km^2 in summer. By contrast, the Arctic Ocean is ice-covered most of the year, and several marginal seas (i.e., the Sea of Japan, the Sea of Ohkotsk, the Bering Sea, Davis Strait, Hudson Bay) are partially or wholly ice-covered during the winter. At its minimum extent in September, Arctic sea ice has typically covered about 6 million km^2, whereas during its winter maximum in March it expands to 14 million km^2. However, the maximum extent of sea ice in the Arctic has been declining; 2007 and 2008 were among the worst years yet recorded, with the summer ice extent shrinking to just over 4 million km^2 in both years (FIG. 9.29).

Scientists commonly categorize sea-ice zones as being either perennial or seasonal. The *perennial ice zone* contains sea ice that persists for at least several years (multiyear ice). In the Arctic, this zone lies north of 75° latitude and contains about two-thirds of all perennial sea ice. Near the center of the basin, the ice has an average thickness of 3 to 4 m and an age of several decades. In the Antarctic, multiyear ice is restricted to semienclosed seas (such as the Ross and Weddell seas, which can be seen in Fig. 9.7), where it reaches a thickness of up to 5 m but an age of less than five years. In the *seasonal ice zone,* the ice cover varies annually. In the Arctic, ice of the seasonal zone is less than 2 m thick where it has not been deformed by wind and ocean currents, but deformation within the pack ice often increases thickness substantially. In the Southern Ocean, the limit of seasonal ice shifts, on average, through 10° of latitude. Here, the ice front retreats poleward in summer largely in response to heat derived from the ocean water, whereas in the Arctic, surface melting that results from warmer air temperature in the summer is a major factor in the retreat of the ice margin.

FIGURE 9.29 Changing sea-ice
Sea-ice cover in the Arctic Ocean, shown here in graph form from the International Arctic Research Center, varies seasonally. The summer minimum was at its lowest point in recent years during 2007 and 2008.

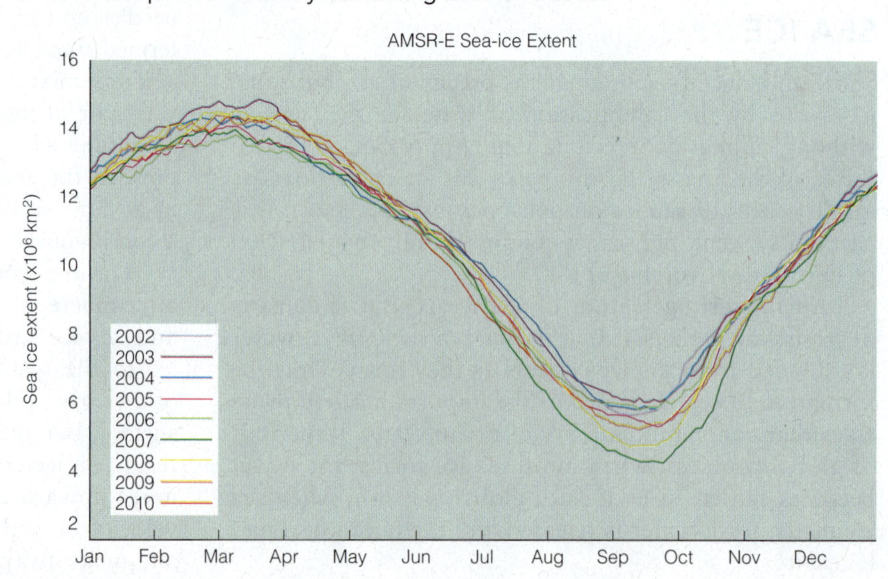

Sea-Ice Motion

Sea ice is in constant motion, driven by winds and ocean currents. Annual changes in the extent, character, and motion of sea ice can now be studied using radar imaging systems carried on orbiting satellites. Average drift rates in the Arctic Ocean are about 7 km/day, whereas in the Greenland and Bering seas velocities reach 15 km/day. Each year, about 10 percent of the Arctic Sea ice moves south into the Greenland Sea, where it eventually breaks up and melts away. Sea ice generally moves clockwise around Antarctica, but a large circulation system called a *gyre* in the Weddell Sea, east of the Antarctic Peninsula, causes the drifting ice to pile up to form a large region of multiyear ice.

Stresses resulting from diverging movement of the thin ice cover cause it to break, exposing the underlying water. Such a linear opening, called a *lead,* tends to be long and narrow and may extend for many kilometers. An exceptionally large lead may grow to become a huge area of open water called a *polynya* (several can be seen in Fig. 9.7; one is just below and to the left of McMurdo Station). Because of the large temperature gradient between the air and seawater in a lead, the water loses heat rapidly, causing a new, thin cover of ice to form quickly. As a result, the fractured ice pack becomes a changing complex mosaic of new ice and older ice. Although the exposure of surface water to the atmosphere permits substantial amounts of solar energy to reach the upper ocean, such open water commonly comprises less than 1 percent of the total area of the winter sea-ice cover.

Early explorers who tried to reach the North Pole by crossing the Arctic ice pack quickly found it rough going (see *A Closer Look: An Ice-Free Northwest Passage?*). The ice is not a vast smooth surface; rather, it is broken by numerous *pressure ridges*, formed when the shifting, fractured ice converges, shears, and piles up, in much the same way that converging lithospheric plates produce mountain chains on the continents. Beneath each pressure ridge is a submerged *keel* of deformed ice, much like the keel of a sailboat, up to five times as thick as the overlying ridge. Estimates suggest that as much as 40 percent of the mass of Arctic sea ice is contained in such deformation features. In the Antarctic, pressure ridges are far less common because prevailing winds and currents tend to disperse the pack ice, shifting it away from the continent at rates as high as 65 km/day.

ICE IN THE EARTH SYSTEM

Both sea ice and land ice play important roles in the Earth system, influencing processes that extend far beyond their own geographic limits. Circulation in the atmosphere and ocean, which in turn drive wind, weather, and global climate systems, are directly influenced by the presence of ice.

Influence on Ocean Salinity and Circulation

Interactions among ice, ocean water, and atmosphere in the seasonal ice zone influence ocean structure, salinity, and circulation. Because it is only a few meters thick, sea ice in particular is very sensitive to temperature changes in the overlying atmosphere and in the ocean water below. In turn, the ice cover affects both the atmosphere and ocean in important ways. As mentioned earlier, the growth of sea ice increases salinity at the top of the mixed layer of the ocean (Chapter 10) because salt is excluded when seawater freezes. Conversely, when the ice melts, the salinity is decreased as fresh water is added to the ocean surface.

The exclusion of salt as seawater freezes leads to the production of cold, saline (and therefore dense) water on the continental shelves. This dense water spills downward off the shelves into the ocean basins to produce deep water and bottom water. The process is enhanced in the marginal Antarctic seas, where offshore winds generate polynyas. Here, rapid ice growth under extremely cold conditions produces large quantities of dense water that sinks and is likely the source of much of the very cold, deep water in the Southern Ocean, called Antarctic Bottom Water. Similar processes operating in the Greenland and Norwegian seas are responsible for producing North Atlantic Deep Water. Both of these dense cold-water masses are crucial to maintaining the global oceanic circulation system, called the *thermohaline circulation* (Chapter 10).

Influence on Atmospheric Circulation and Climate

Both sea ice and land ice are important components of Earth's climate system. The floating cover of sea ice effectively isolates the ocean surface from the atmosphere, thereby cutting off the exchange of heat between these two reservoirs; the more extensive the ice cover, the stronger the effect. Ice also has a high *albedo* (Chapter 2), which means that an ice-covered surface reflects incoming solar radiation rather than absorbing it. This makes the ice-covered polar regions far colder than if the same areas were covered with water or land, both of which have lower albedo than ice.

The resulting steep temperature gradient between low (equatorial) latitudes and ice-covered polar regions is of major importance to atmospheric circulation. This temperature difference drives our global atmospheric circulation and wind systems. If the temperature gradient between the equator and the poles were to become less steep, such as by a warming of the polar regions, it would have a major disruptive influence on atmospheric circulation.

Ice Cover and Environmental Change

The fact that ice cover influences global climate leads to some important questions: How stable is the ice pack in the landlocked Arctic basin? What would it take to remove the thin ice cover completely, and how long would it take? In a time of interglacial climatic warming, might we expect the ice cover to disappear suddenly, thereby

A Closer LOOK

AN ICE-FREE NORTHWEST PASSAGE?

Bartolomeu Dias, a Portuguese explorer, sailed around the southern tip of Africa in 1488 and discovered a sailing route from Europe to Asia. In 1492, Columbus, sailing for Spain, set out to find a shorter way to Asia by sailing westward from Europe. Instead he discovered America. To stop trade disputes, Pope Alexander VI divided the then-known world into two parts, splitting it between Spain and Portugal. All trade and new lands to the east of a north–south line located 100 leagues (about 500 km) west of the Cape Verde Islands belonged to Portugal; everything to the west belonged to Spain. In 1494, Spain and Portugal reached an agreement that the line should be located at longitude 46°37'W. The countries of northern Europe, principally England, France, and Holland, were left out of the deal, and that started their long search for an alternative passage to Asia.

Initially, they explored both the east and west coasts of North America, hoping that a sea passage might cut through from the Atlantic Ocean to the Pacific Ocean. (The Panama Canal did not exist at that time.) When that hope failed, attention was turned to the search for a route across the northern coast of North America, through the icy Arctic Ocean. This fabled route, about 4000 km shorter than the passage that would eventually be made available by the opening of the Panama Canal, soon came to be known as the *Northwest Passage* (**FIG. C9.1**).

As the early explorers quickly realized, the northern coast of North America approximately coincides with the Arctic Circle (66°33'37"N), and north of the Circle there lies a vast archipelago of islands. John Cabot, an Englishman, was the first to seek a passage through the islands, in 1497. His attempt and all successive attempts, including the famous and ill-fated Franklin expedition of 1845, were unsuccessful, until the famous Norwegian explorer, Roald Amundsen, managed the feat between 1903 and 1906 (**FIG. C9.2**). The problem that plagued the explorers was persistent sea ice. Winds and ocean currents flowing between the

FIGURE C9.1 The Northwest Passage
This map shows some of the routes explored in the search for the fabled Northwest Passage from the Atlantic to the Pacific.

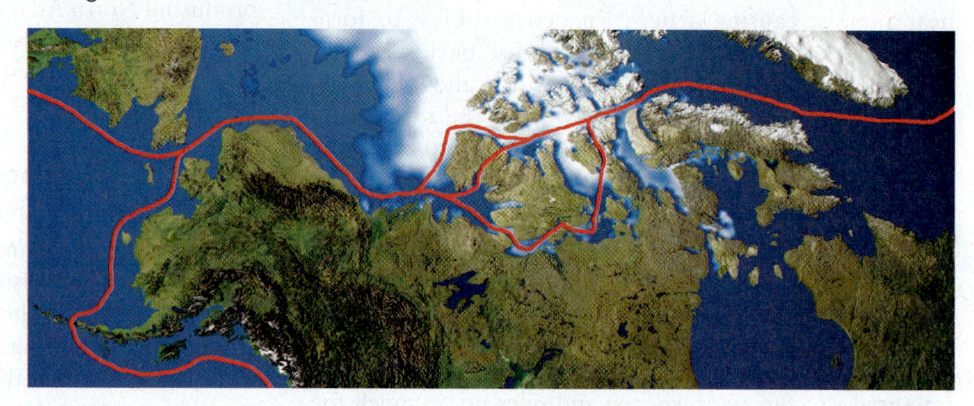

causing abrupt changes in climate in some of Earth's most densely populated regions? Would there be an impact on sea level or salinity if sea ice were to melt completely? What about land ice? What organisms and ecosystems would be affected if the distribution of land or sea ice were to change? We can begin to address some of these questions through climate modeling, which we will examine in greater detail in Chapter 13.

If the climate were to become colder, causing the ice cover to expand, the result would be a positive feedback (Chapter 1). The increased area of ice would raise the total planetary albedo, leading to further cooling. The resulting growth of ice cover would cause a further increase in albedo, and so on. On the other hand, if the ice cover shrinks or disappears, significant disruption in the pattern of atmospheric circulation could occur. If polar ice were melted to reveal land or water, or even if the ice surface were covered with a dark material such as dust, soot, or ash from a volcanic eruption, a similar effect could happen. The darkening of the ice surface would cause the overall albedo to decrease, such that more incoming solar radiation would be absorbed. This would lead to further melting and shrinking of the ice cover, revealing additional dark, low-albedo surfaces (water or land), and a positive feedback cycle would ensue.

Climate models suggest that in a warming climate, the warming at high latitudes will be several times that at low and middle latitudes. A possible scenario indicates that, as the Arctic warms up, we can expect a gradual shrinking of the ice pack followed by a discontinuous transition to ice-free conditions. We can easily calculate how much of a change would lead to the disappearance of perennial Arctic ice. It could occur, for example, with a 3° to 5°C increase in annual temperature, a 25 to 30 percent increase in solar radiation reaching the ice surface, a 15 to 20 percent decrease in summer albedo (brought about by increased surface melting), or a significant change in

FIGURE C9.2 **Early exploration of the Northwest Passage**
This engraving was based on an original drawing made during William Edward Parry's third voyage of exploration for the Northwest Passage, which took place in 1824–1825.

islands piled the ice into thick, impenetrable masses. Amundsen took three years to get through because he was trapped for three winters in the ice. The second successful passage was made in 1940, by the Canadian explorer Henry Larsen.

The Northwest Passage is still not an easy route, but as time has passed it has become easier to navigate and more ships have made it through. Icebreaker technology also has made advancements over the vessels used by the early explorers. Moreover, it has become easier to find passage through the Arctic archipelago because of changes due to global warming. Arctic sea ice is becoming thinner, and its extent is declining. A Canadian expedition in the summer of 2008 reported that the passage is not yet continuously navigable, even to an icebreaker. However, continuous monitoring of sea-ice conditions by satellite shows that soon an ice-free passage will be available and open to commercial shipping, research, and oil exploitation.

Some experts claim that as much as 25 to 30 percent of the world's undiscovered oil reserves lie beneath water, ice, and rock in the Arctic. If melting of sea ice were to open the Northwest Passage, the pressure to exploit this oil would be enormous. However, Russia, Canada, and the United States disagree deeply about how this region should be managed, and particularly about the locations of national boundaries and international waters. As the ice melts and the landscape changes, this disagreement is sure to escalate. A further concern is that oil spills in cold water are extremely difficult to clean up, leading many people to think that it would be prudent to leave Arctic oil where it currently lies, deep underground.

cloudiness. There are various ways in which a change in albedo could take place, but modeling suggests that once the albedo reaches a sufficiently low value, the shift to an ice-free Arctic Ocean would occur rapidly, measurable in years rather than in decades.

Interestingly, the melting of sea ice would have much less impact on sea levels than the melting of land ice, such as the continental ice sheets that cover the landmasses of Antarctica and Greenland. This is because sea ice is floating directly in the ocean. You may recall that the *Archimedes principle* tells us that a floating object is supported or *buoyed* by a force equal to the weight of the fluid displaced by the object. An object will float at the level where it displaces a volume of water equivalent to its own weight (as seen in Fig. 9.18). If the object—an iceberg, for example—were to melt, the water would simply fill up the displaced volume formerly occupied by the iceberg. The compensation is not exact because the density of freshwater ice is not exactly the same as the density of salty seawater; therefore, the melting of sea ice would cause only a small increase in sea level, even if *all* sea ice were to melt.

For land ice, however, the situation is completely different. Landmasses that are covered by ice sheets are basically mountains that stick up above sea level, with a thick coating of ice on top. Unlike sea ice, if this land ice were to melt, *all* of its volume would be added into the ocean. Depending on the extent of the melting, it could lead to a significant increase in sea level. The loss of either sea ice or land ice, and the resulting changes in ocean water temperature and salinity, would have significant disruptive effects on species and ecosystems that are adapted to the current distribution of ice cover, as well as the indigenous people who depend on these resources. In Chapter 13 we will consider in greater detail some of the possible global warming scenarios that might lead to a melting of sea or land ice, with a resulting increase in sea level and other ecosystem changes.

SUMMARY

1. The cryosphere is the part of Earth's surface that is perennially frozen. In the northern hemisphere, much of the ice floats as a thin layer on the Arctic Ocean; in the southern hemisphere, it is dominated by a vast ice sheet covering the continent of Antarctica and adjacent islands and seas. The cryosphere also includes large areas of permafrost, as well as seasonal snow cover, and glaciers in alpine areas.

2. Snow forms as a result of the crystallization of tiny water droplets into feathery ice crystals in clouds, at very low temperatures. The crystal structure of snow is open and porous, but when deposited as precipitation it settles and becomes compacted. Snow that accumulates from one season to the next becomes more compacted, and the feathery snow recrystallizes into a denser texture characteristic of glacier ice.

3. In the northern hemisphere, almost one-quarter of the land area is covered by seasonal snow and frozen ground during the winter. Above the snowline, part of the past winter's snow survives the summer and accumulates, along with snow that has persisted from earlier winters. The altitude and position of the snowline change from year to year, controlled by temperature and precipitation.

4. A glacier is a persistent body of ice, consisting largely of recrystallized snow, which moves slowly downslope under the influence of gravity. Important glacier types include cirque glaciers, valley glaciers, ice caps, fjord glaciers, and piedmont glaciers. The huge continental ice sheets of Greenland and Antarctica contain about 95 percent of existing glacial ice. Floating ice shelves occupy large embayments along the coasts of Antarctica and among the Arctic islands.

5. A glacier can develop anywhere that snow and ice accumulate and persist. Glaciers occur at sea level in the polar regions and above the snow line in high mountains near the equator. Low temperature and proximity to a moisture source are requirements for glacier formation. The existence of glaciers thus depends on tectonic forces to produce high mountains, the ocean to provide moisture, and the atmosphere to deliver the moisture to the land as snow.

6. In the temperate ("warm") glaciers of low and middle latitudes, ice is at its pressure melting point from surface to bed, so it can coexist with water. The temperature in polar ("cold") glaciers, which occur at high latitudes and high altitudes, is below the pressure melting point, so little or no seasonal melting occurs.

7. The mass balance of a glacier depends on the balance between accumulation (snow added over the winter) and ablation (snow and ice lost over the summer). When a glacier gains more mass than it loses over a period of years, its volume increases and the terminus advances. A succession of years in which negative mass balance predominates will lead to retreat of the terminus. A time lag occurs between a climatic change and the response of the glacier terminus because it takes time for the effects of changes above the equilibrium line to be transferred through the slowly moving ice to the terminus.

8. Glacial ice flows downslope under the influence of gravity. Deep within a glacier, individual ice crystals are subjected to stress from the weight of overlying snow and ice. Their internal crystal planes become aligned, facilitating deformation and flow. In contrast, the surface portion of a glacier is brittle; if it passes over an abrupt change in slope, it will develop deep crevasses. Meltwater at the base of a glacier acts as a lubricant, leading to basal sliding along the bed. In most glaciers, flow velocities range from a few centimeters to a few meters a day but glacial surges, up to 100 times faster, may be facilitated by water trapped at the base.

9. Recession of coastal glaciers is characterized by calving, the progressive breaking off of icebergs from the front of a glacier that terminates in deep water. Icebergs float with most of their volume below sea level.

10. Glaciations have been prevalent during the Pleistocene Epoch, but there is evidence of glacial–interglacial cycles as long ago as 2.4 billion years. During the most recent ice age, early humans migrated into North America from Asia across a land "bridge" that was exposed because such a large volume of water was locked up in glacial ice, causing sea levels to decline. The ice sheets of the most recent glaciation reached their maximum extent about 24,000 years ago; by 10,000 years ago Earth had emerged from the ice age and entered the present interglacial period.

11. Much of Earth's land area has been shaped by glaciers. As a medium for transporting sediment, ice can transport large and small pieces side by side without sorting them by size and density. The load of a glacier typically is concentrated at its base and sides, where glacier and bedrock are in contact. A significant component of the basal load of a glacier consists of very fine rock flour.

12. Glacial landforms can result from the sculpting action of glacial ice. Small rock fragments embedded in basal ice produce long, parallel striations and grooves, aligned in the direction of ice flow. In mountainous regions, cirques are produced by frost-wedging, combined with plucking and abrasion at the glacier bed. A valley that has been shaped by glaciers has a distinctive U-shaped cross section and a floor that lies below its tributary valleys. Smooth, parallel, streamlined drumlins are elongated parallel to the direction of ice flow.

13. Some glacial landforms are produced by the deposition of rock debris eroded by the glacier or dropped onto the glacier's surface from adjacent cliffs. Till is a glacial sediment that occurs in unsorted and unstratified deposits. Debris that reaches the margins or terminus of the glacier may be redistributed by meltwater and deposited as outwash. Moraine deposits form along the edges or terminus of a glacier, or where two glaciers merge.

14. The most characteristic feature of periglacial regions is permafrost. The largest areas of permafrost occur in northern North America, northern Asia, and the Tibetan Plateau. Most of today's permafrost is believed to have originated during the last glacial age or earlier glacial ages. Instability due to seasonal thawing of the active layer can make it challenging to live with permafrost.

15. Landscapes in periglacial regions reflect the movement of regolith in the active layer during annual freeze/thaw cycles. Patterned ground and pingoes form as a result of cyclical freezing and thawing in the active layer. Solifluction is a common mass-wasting process in which waterlogged regolith in a thawed active layer moves slowly downslope.

Periglacial regions are often associated with thick deposits of fine-grained, buff-colored loess, an aeolian sediment derived from rock flour.

16. Glacial ice contains abundant physical, chemical, and biological evidence of past changes in local and global environmental conditions. The retrieval of deep-ice cores for both polar and alpine glaciers has allowed atmospheric scientists to determine how the concentrations of atmospheric greenhouse gases have fluctuated as the climate changes. The chemistry of the ice itself can reveal the air temperature when the snow first accumulated. Ice cores also provide a record of major volcanic eruptions, dustfalls, and other evidence of changes in atmosphere and climate.

17. Approximately two-thirds of Earth's permanent ice cover is sea ice. The formation of sea ice has an important influence on the salinity of ocean water because salt is excluded from ice crystals as they form. Once the ocean surface cools to the freezing point of seawater, slight additional cooling leads to ice formation. It first forms small platelets and needles of frazil ice, eventually developing pancake-like masses that merge into a continuous sheet. Subsequent growth occurs by the addition of ice to the base.

18. The South Pole is located in the Antarctic continent, covered by a vast, thick ice sheet. The North Pole lies in the middle of the deep Arctic Ocean basin. The open Southern Ocean surrounding Antarctica contrasts with the largely landlocked Arctic Ocean. These differences affect ice distribution. In the Antarctic region, sea ice forms a broad ring around the continent, which varies seasonally; in contrast, the Arctic Ocean is ice-covered most of the year. Perennial ice zones contain sea ice that persists for at least several years; in seasonal ice zones, the ice cover varies annually.

19. Sea ice is in constant motion, driven by winds and currents. Stresses resulting from movement cause the ice to break, exposing the underlying water. A linear opening, or lead, may grow to become a large area of open water, or polynya. Because of the large temperature gradient between the air and seawater in a polynya, the water loses heat rapidly, causing a new, thin cover of ice to form. A fractured ice pack is thus a complex mosaic of new ice and older ice.

20. Both sea ice and land ice play important roles in the Earth system. The salinity and temperature, and therefore the global circulation of the world ocean, are directly influenced by the formation and melting of ice. Ice has a high albedo, which makes the ice-covered polar regions far colder than if the same areas were covered with water or land. In a warming climate, the melting of sea ice would have much less impact on sea level than the melting of land ice. The loss of sea ice or land ice would have significant disruptive effects on species, ecosystems, and people that are adapted to the current distribution of ice cover.

IMPORTANT TERMS TO REMEMBER

ablation *265*
accumulation *265*
bed *267*
calving *268*
cirque *260*
crevasse *266*
cryosphere *258*
equilibrium line *266*
fjord *260*
glaciation *270*
glacier *260*
glacier ice *263*
ice *263*
ice cap *260*
ice sheet *261*
ice shelf *261*
interglacial *270*
loess *275*
moraine *271*
periglacial *273*
permafrost *273*
polar glacier *264*
sea ice *277*
snow *263*
snowline *259*
surge *269*
temperate glacier *264*
terminus *266*
till *271*

QUESTIONS FOR REVIEW

1. How are glaciers related to the snowline?
2. Describe the steps in the conversion of snow to glacier ice.
3. What characteristics distinguish temperate glaciers from polar glaciers?
4. Why does the position of the equilibrium line provide a rough estimate of a glacier's mass balance?
5. Why is there commonly a time lag between a change of climate and the response of a glacier's terminus to the change?
6. How do glaciers move? In what ways does ice temperature influence the way a glacier moves?
7. What is a glacial surge? What causes glaciers to surge?
8. Summarize the main erosional (sculpting) landforms shaped by glaciers. Summarize the main depositional landforms shaped by glaciers.
9. What is the active layer in permafrost terrain, and how does it form?
10. What are some of the main periglacial landforms and how what processes lead to their formation?
11. What is permafrost? Where does it occur, and why is it challenging to live with permafrost?
12. How does sea ice form?
13. How does sea ice influence global climate? How does sea ice influence ocean circulation?
14. What is a glacial period? What is an interglacial period? What causes glaciations?
15. What would be the impact on global sea level if sea ice were to melt? How and why does it differ from the impact on sea level if land ice were to melt?

QUESTIONS FOR RESEARCH AND DISCUSSION

1. What are some of the ways in which you might be able to distinguish a landform shaped by a stream from a landform shaped by a glacier?
2. Suggest some ways in which changes in the geosphere, such as uplift, earthquakes, and volcanism, might affect the distribution and volume of glaciers.
3. Huge pools of petroleum likely underlie the Arctic continental shelves of Alaska, Canada, and Russia. In exploiting such resources, what problems might be encountered that are related to the presence of sea ice?

QUESTIONS FOR *THE BASICS*

1. Is snow a mineral? What about ice? Refer to Chapter 3 to check the definition of a mineral, and see if snow and ice fit all of the criteria.
2. What are the main differences between snow and ice?
3. Why was the most recent Pleistocene glaciation so important in human history?

QUESTIONS FOR *A CLOSER LOOK*

1. What is the Northwest Passage, and why is it important?
2. How is global warming changing the landscape of the Arctic, affecting the Northwest Passage?
3. When the Northwest Passage becomes ice-free, how should the area be governed or managed? Do you think oil exploration should be allowed? If so, who should regulate it?

CHAPTER 10

The World OCEAN

OVERVIEW

In this chapter we:

- Look at the distribution, size, and origin of the ocean and ocean basins
- Describe the deep and shallow circulation patterns in the ocean
- Consider the causes of ocean waves, surface currents, and tsunamis, and their effects on shorelines
- Explain the forces that produce tides
- Discover why sea level is slowly changing

Moonset over the Pacific
The Moon sets over the Pacific Ocean at sunrise on Northern California's Sonoma Coast. The San Andreas Fault runs through this section of coast.

Imagine, if you can, a dry Earth, devoid of water. Earth's surface would appear far different from the one familiar to us. Viewed from an orbiting spacecraft, a dry Earth would no longer look blue. On an Earth with no water, no clouds of condensed water vapor would obscure the surface, and there would be no vegetation cover on the land. Earth would appear as desolate as the rocky surfaces of the Moon or Mars. We would see the high-standing continents ending where their bordering continental slopes meet a great expanse of empty sea. As the spacecraft circled the planet, we would see several vast interconnected basins, each floored with oceanic crust and rimmed by thick continental crust.

If these huge basins were now slowly filled with water, the scene would be transformed. The rising water would initially fill the deepest parts of the basins, creating a number of shallow seas, but as the water level continued to rise, these seas would merge to form a larger and larger ocean that eventually would creep up the slopes of the continental margins. With the ocean basins filled to capacity, more than two-thirds of Earth's surface would now be covered by water and Earth would be a unique planet in the solar system, the only planet with water in all three forms—liquid, solid, and vapor—coexisting at the surface. In this chapter we consider the features and processes that characterize Earth's vast world ocean.

OCEAN BASINS AND OCEAN WATER

Under the ocean water, beyond the continental slopes, is the remote world of the deep-ocean floor. With devices for sounding the sea bottom and for sampling its sediment, teams of physical oceanographers and marine geologists have explored the ocean floor and greatly expanded our knowledge of the submarine regions. Scuba-diving geologists have visited, photographed, and mapped areas of seafloor at depths as great as 70 m, and observers in specially designed submersible crafts have descended more than 6 km to visit the greatest depths of the ocean floor. In recent years, available depth soundings have been combined with high-resolution marine gravity measurements from Earth-orbiting satellites to produce new, accurate, and more detailed topographic maps of the ocean floors (FIG. 10.1). These maps are based on very small differences in the surface of ocean water, caused by differences in the underlying topography.

Through all of this intensive research, which involves not just satellites, but ships and scientists from many nations, we are gradually coming to understand the ocean. The romanticist in each of us may regret that beliefs and legends built up through thousands of years of human history—monsters, mermaids, strange and threatening sea gods, fabled cities, and castles that sank into watery deeps—have faded away as scientific knowledge has steadily increased. In return, however, that knowledge has helped us appreciate the fragile environment of the ocean, an environment that is responsible for a large part of Earth's biologic heritage.

Ocean Geography

Seawater covers 70.8 percent of Earth's surface. The land comprising the remaining 29.2 percent is unevenly distributed. This uneven distribution is especially striking when we compare two views of the globe: one from a point directly above Great Britain and the other from a point directly above New Zealand (FIG. 10.2). In the first view, more than 46 percent of the viewed hemisphere is land, whereas in the second view less than 12 percent is land and more than 88 percent is water. The uneven distribution of land and water plays an important role in determining the paths along which water circulates in the open ocean and its marginal seas.

Most of the water on our planet is contained in three huge interconnected basins—the Pacific, Atlantic, and Indian oceans (seen in Fig. 10.1). All three are connected with the Southern Ocean, a body of water south of 50° S latitude that completely encircles Antarctica. Collectively, these four vast interconnected bodies of water, together with a number of smaller ones, are often referred to as the **world ocean**. The smaller water bodies connected with the Atlantic Ocean include the Mediterranean, Black, North, Baltic, Norwegian, Arctic, and Caribbean seas, the Gulf of Mexico, and Baffin and Hudson bays. The Persian Gulf, Red Sea, and Arabian Sea are part of the Indian Ocean, while the numerous marginal seas of the Pacific Ocean include the Gulf of California, Bering Sea, Sea of Okhotsk, Sea of Japan, and the East China, South China, Coral, and Tasman seas. All these seas and gulfs vary considerably in shape and size; some are almost completely surrounded by land, whereas others are only partly enclosed. Each owes its distinctive geography to plate tectonics, for this ongoing process has led to the creation of numerous small basins both in and adjacent to the major ocean basins.

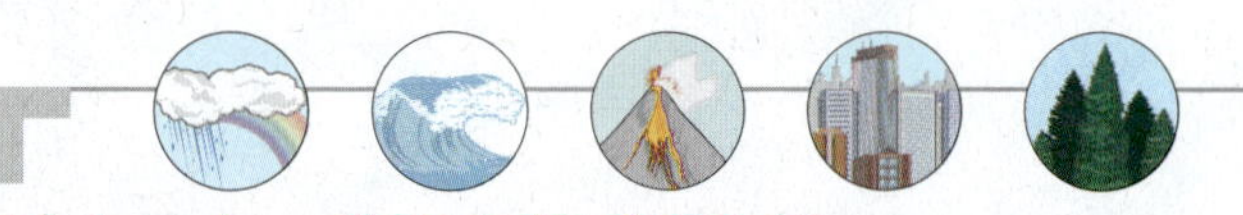

Make the CONNECTION

How would this planet differ if the ocean basins were distributed differently? Try to envision an Earth with one enormous ocean basin, and all the continents gathered together in a single giant continent. Then try to imagine an Earth with many smaller ocean basins surrounding smaller, islandlike continents. What would be the impacts on climate and on ecosystems? What would it be like to live on such an Earth?

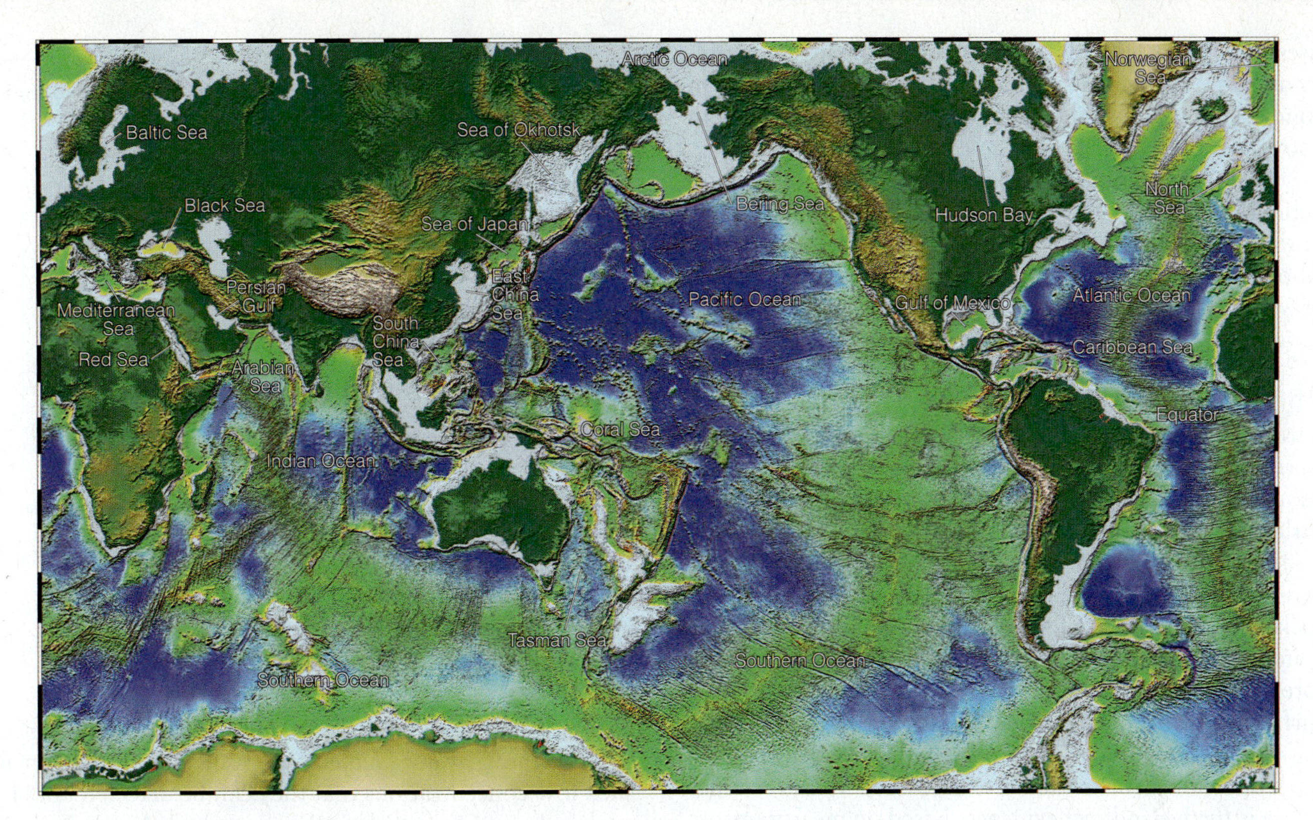

FIGURE 10.1 **The world ocean**
This shaded relief map of the ocean is based on high-resolution marine gravity data obtained by satellite altimetry and on ship depth soundings, and shows the major and minor basins of the world ocean.

Depth and Volume of the Ocean

Before the twentieth century, little was known about the depth of the ocean. Water depths were determined from soundings made with either a weighted hemp line or a strong wire lowered from a ship. Although this technique proved satisfactory and relatively rapid in shallow water, it could take 8 to 10 hours to recover a weighted wire in water thousands of meters deep. By the close of the nineteenth century, about 7000 measurements had been made in water more than 2000 m deep, and fewer than 600 in water deeper than 9000 m. In the 1920s, ship-borne acoustical instruments called *echo sounders* were developed to measure ocean depths. An echo sounder generates a pulse of sound and accurately measures the time it takes for the echo bouncing off the seafloor to return to the instrument. Because the speed of sound traveling through water is known, the water depth beneath a ship can be calculated.

Over the past 70 years, the world's oceans have been crossed many thousands of times by ships carrying echo sounders. As a result, the topography of the seafloor and the depth of the overlying water column are known in considerable detail for all but the most remote parts of the ocean basins. The greatest ocean depth yet measured (10,924 m) lies in the Mariana Trench near the island of Guam in the western Pacific. This is more than 2 km farther below sea level than Mount Everest

FIGURE 10.2 **Ocean and land**
The unequal distribution of land and ocean can be seen if we view Earth from above Britain and above New Zealand. In the first view, land covers nearly half the hemisphere, whereas in the other nearly 90 percent of the hemisphere is water.

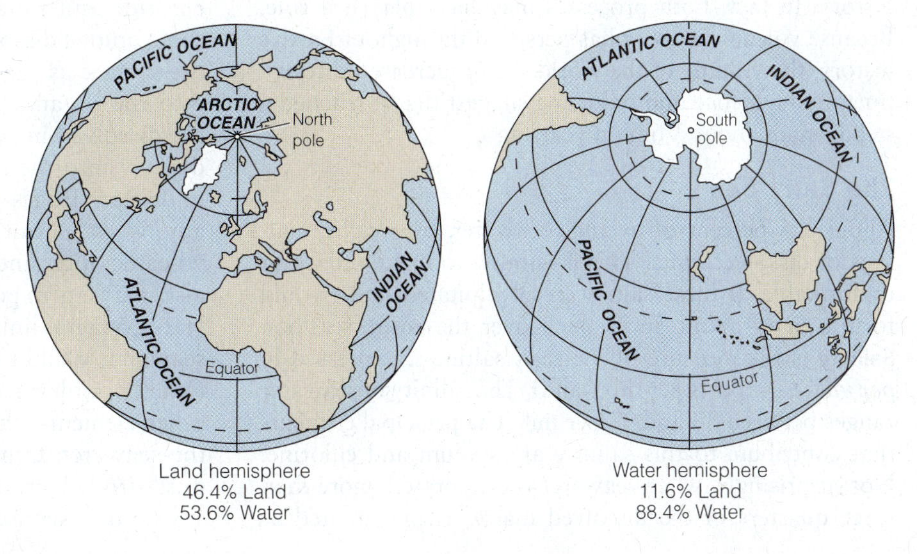

rises above sea level. Based on recent satellite measurements, the average depth of the sea is about 3970 m compared to an average height of the land of only 840 m above sea level.

The present volume of seawater is about 1.35 billion cubic kilometers; more than half of this volume resides in the Pacific Ocean. We say *present* volume because the amount of water in the ocean fluctuates over thousands of years, mainly because of the growth and melting of continental glaciers and ice sheets (Chapters 9 and 13).

Age and Origin of the Ocean

Earth's oldest rocks include sedimentary strata that were deposited by water and are similar to strata we see being deposited today. These rocks originated from still earlier igneous rock, through weathering, erosion, and re-deposition by water. From such direct evidence we are sure that, as far back as the oldest known sedimentary rocks—more than 4 billion years—Earth has had liquid water on its surface. We can be reasonably certain, therefore, that ocean water began to accumulate on Earth's surface sometime between 4.56 billion years ago, when the planet formed, and 4 billion years ago, when the oldest sedimentary rocks thus far discovered were forming. There is further indirect evidence, based on measurements of oxygen isotopes from individual grains of zircon even older than the sedimentary rocks that contain them, that the ocean may be at least 4.4 billion years old.

Where the water to create the ocean came from is still an open question. Primitive meteorites called *carbonaceous chondrites* (Chapter 4) are considered to be samples of the same material that accreted to form Earth, and they contain several percent of water combined in the form of hydrous minerals. If Earth formed by the accretion of this type of material, this water would have been released and vented to the surface when the planet heated up and partially melted early in its history. The most likely origin for Earth's surface water, therefore, is that it condensed from steam produced during primordial volcanic eruptions. Another possible answer is that the water came from icy comets that collided with Earth early in the planet's history. In fact, both processes may have played a role. Because volcanic activity has persisted throughout Earth's history, the volume of the world ocean increased through time, but evidence and modeling suggest that it reached a steady state about 3 billion years ago.

The Salty Sea

About 3.5 percent of average seawater, by weight, consists of dissolved salts. This is enough to make the water undrinkable. If these salts were precipitated, they would form a layer about 56 m thick over the entire seafloor. **Salinity** is the measure of the sea's saltiness, expressed in *per mil* (‰ = parts per thousand). The salinity of seawater ranges between 33 and 37 per mil. The principal elements that contribute to this salinity are sodium and chlorine. Not surprisingly, when seawater is evaporated, more than three quarters of the dissolved matter is precipitated as common salt (NaCl). However, seawater contains most of the other natural elements as well, many of them in such low concentrations that they can be detected only by extremely sensitive analytical instruments. As can be seen in **FIGURE 10.3**, more than 99.9 percent of the salinity is caused by only eight ions.

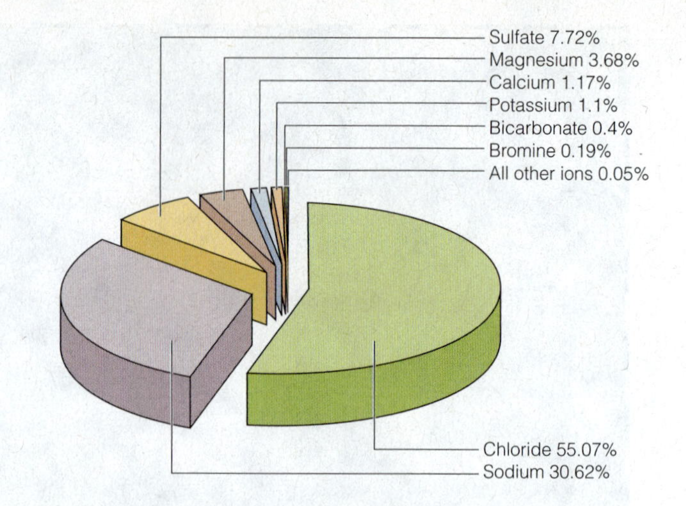

FIGURE 10.3 Principal ions in seawater
More than 99.9 percent of the salinity of seawater is due to eight ions, the two most important of which (Na^+ and Cl^-) are the constituents of common salt.

The Origins of Ocean Salinity

Where do the ions in seawater come from? Each year streams carry an estimated 2.5 billion tons of dissolved substances to the sea. As exposed crustal rock interacts with the atmosphere and the hydrosphere (rainwater), cations such as sodium and potassium are leached out and become part of the dissolved load of streams flowing to the sea. Chloride (Cl^-) and sulfate (SO_4^{2-}), the principal anions in seawater, are believed to have come primarily from the mantle. Chemical analyses of gases released during volcanic eruptions show that the two most important volatiles are water vapor and carbon dioxide (CO_2), and the two most important anions are chloride and sulfate. When released to the atmosphere, these anions dissolve in atmospheric water and return to the surface as precipitation, much of which falls directly into the ocean. Part of the remainder is carried to the sea dissolved in river water. Volcanic gas is also released directly into the ocean from submarine eruptions along midocean ridges and is another major source of the anions in ocean water. Interactions between heated volcanic rock and seawater along the midocean ridges also play an important role in the composition of seawater. Magnesium (Mg^{2+}) and sulfate are removed from seawater, while calcium (Ca^{2+}), iron (Fe^{2+}), manganese (Mn^{2+}), and trace amounts of copper, lead, zinc, and other elements are removed from the rock and added to the seawater. Other sources of ions in seawater include dust eroded from desert regions and blown out to sea, and gaseous, liquid, and solid pollutants released through human activity either directly into the ocean or carried there by streams or by polluted air.

The quantity of dissolved ions added by rivers and submarine volcanism over the billions of years of Earth history far exceeds the amount now dissolved in the sea. Why, then, doesn't the sea have a higher salinity? The reason is that chemical substances are being removed at the same time they are being added. Some elements, such as silicon, calcium, and phosphorus, are withdrawn from seawater by aquatic plants and animals to build their shells or skeletons. Potassium and sodium are absorbed and removed by clay particles and other minerals as they settle slowly to the seafloor. Still others, such as copper and lead, are precipitated to form sulfide minerals in oceanic crustal rock and bottom sediment. Ions also are an important component of sea spray, which can be carried landward by ocean winds. Because these and other processes of extraction are essentially equal to the combined inputs, the composition of seawater remains virtually unchanged.

Has the ocean always been salty? The best evidence of the sea's past saltiness is the presence, in marine strata, of salts precipitated by the evaporation of seawater. Marine strata containing salts that were concentrated by the evaporation of seawater are common in young sedimentary basins, but they are not present in rock older than about a billion years. Possibly this is because ancient marine deposits consisting of soluble salts have been completely removed from the geologic record through the slow dissolving action of percolating groundwater, or possibly by metamorphic processes. However, there is evidence in many ancient strata, older than one billion years, that evaporites were once present.

Controls on Ocean Salinity

The salinity of surface ocean water is closely related to latitude (FIG. 10.4). The most important factors affecting salinity are:

1. evaporation, which removes fresh water and leaves the remaining water saltier;
2. precipitation of rain and snow, which adds fresh water, thereby diluting the seawater and making it less salty;
3. inflow of fresh (river) water, which makes the seawater less salty; and
4. freezing of sea ice because when seawater freezes salts are excluded from the ice, leaving the unfrozen seawater saltier.

As one might expect, salinity is high in the latitudes where Earth's great deserts lie. In these zones evaporation exceeds precipitation, both on land and at sea. In a restricted sea, like the Mediterranean, where there is little inflow of fresh water, surface salinity exceeds the normal range; in the Red Sea, which is surrounded by desert, salinity reaches 41 per mil. Salinity is lower near

FIGURE 10.4 Ocean salinity
High salinity values are found in tropical and subtropical water where evaporation exceeds precipitation. The highest salinity has been measured in enclosed seas like the Persian Gulf, the Red Sea, and the Mediterranean Sea. Salinity values (shown here in per mil, or ‰) generally decrease poleward, but low values also are found off the mouths of large rivers.

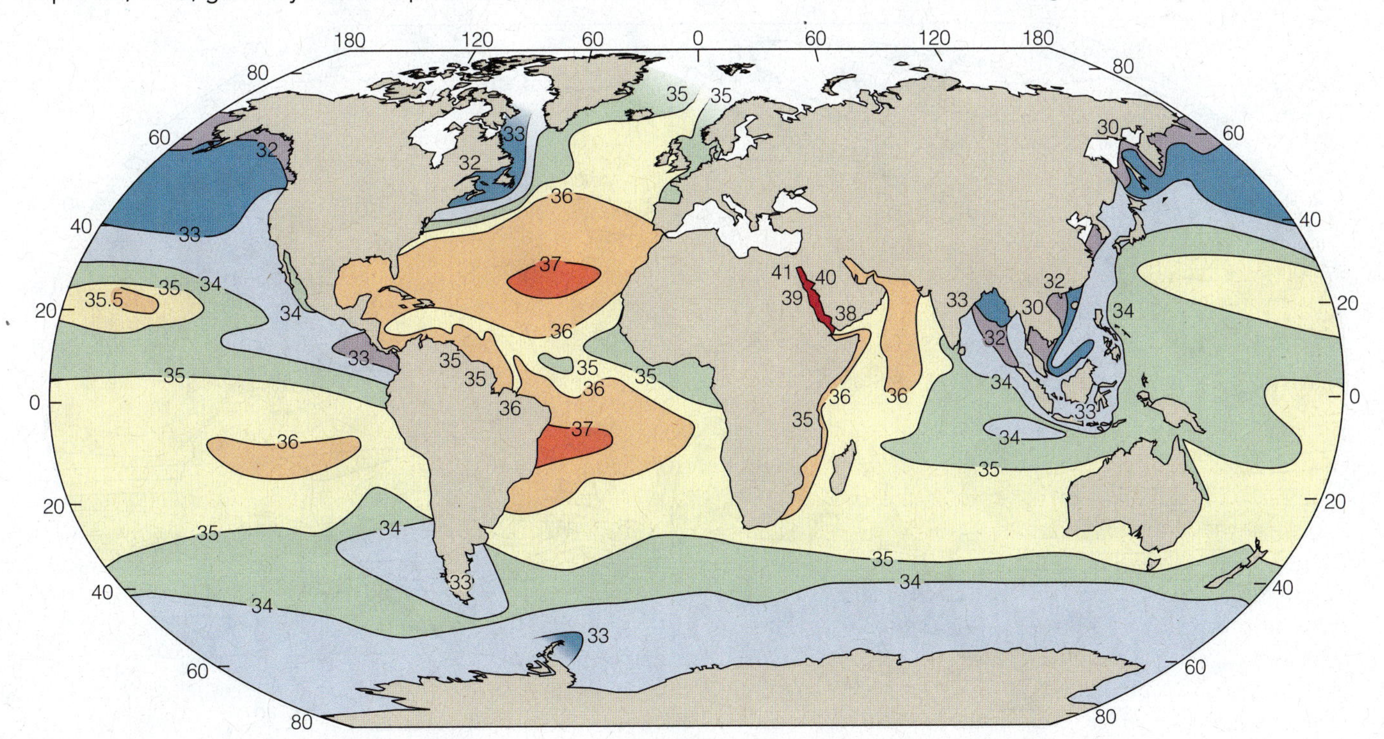

the equator because precipitation is high, and cool water, which rises from the deep sea and sweeps westward in the tropical eastern Pacific and eastern Atlantic oceans, reduces evaporation. Evaporation is also low at high latitudes that are rainy and cool. Up to 100 km offshore from the mouths of large rivers, the surface ocean water can be fresh enough to drink.

Temperature and Heat Capacity of Ocean Water

An unsuspecting tourist from Florida who decides to take a swim on the northern coast of Britain quickly learns how varied the surface temperature of the ocean can be. A map of global summer sea-surface temperature displays pronounced east–west temperature belts, with *isotherms* (lines connecting points of equal temperature) approximately paralleling the equator (FIG. 10.5). The warmest water during August exceeds 28°C, and occurs in a discontinuous belt between about 30° N and 10° S latitude in the zone where received solar radiation is at a maximum. In winter, when the zone of maximum received solar radiation shifts southward, the belt of warm water also moves south until it is largely below the equator. Water becomes progressively cooler both north and south of this belt, and reaches temperatures of less than 10°C poleward of 50° N and S latitude. The average surface temperature of the ocean is about 17°C, while the highest temperatures (>30°C) have been recorded in restricted tropical seas, such as the Red Sea and the Persian Gulf.

The ocean differs from the land in the amount of heat it can store. For a given amount of heat absorbed, water has a lower rise in temperature than nearly all other substances; in other words, it has a high *heat capacity*. Because of water's ability to absorb and release large amounts of heat with very little change in temperature, both the total range and the seasonal changes in ocean temperatures are much less than what we observe on land. For example, the highest recorded land temperature is 58°C, measured in the Libyan Desert, and the lowest, measured at Vostok Station in central Antarctica, is −88°C. The range, therefore, is 146°C. By contrast, the highest recorded ocean temperature is 36°C, measured in the Persian Gulf, and the coldest, measured in the polar seas, is −2°C, a range of only 38°C.

The annual change in sea-surface temperatures is 0–2°C in the tropics, 5–8°C in middle latitudes, and 2–4°C in the polar regions. Corresponding seasonal temperature ranges on the continents can exceed 50°C. Coastal inhabitants benefit from the mild climates resulting from this natural ocean thermostat. Along the Pacific coast of Washington and British Columbia, for example, winter air temperatures seldom drop to freezing. A short distance away, just to the east of the coastal mountain ranges, temperatures can plunge to −30°C or lower. In the interior

FIGURE 10.5 Sea-surface temperatures
This map shows typical sea-surface temperatures (in °C) in the world ocean during August. The warmest temperatures (≥28°C) are found in the tropical Indian and Pacific oceans. Temperatures decrease poleward from this zone, reaching values close to freezing in the north and south polar seas.

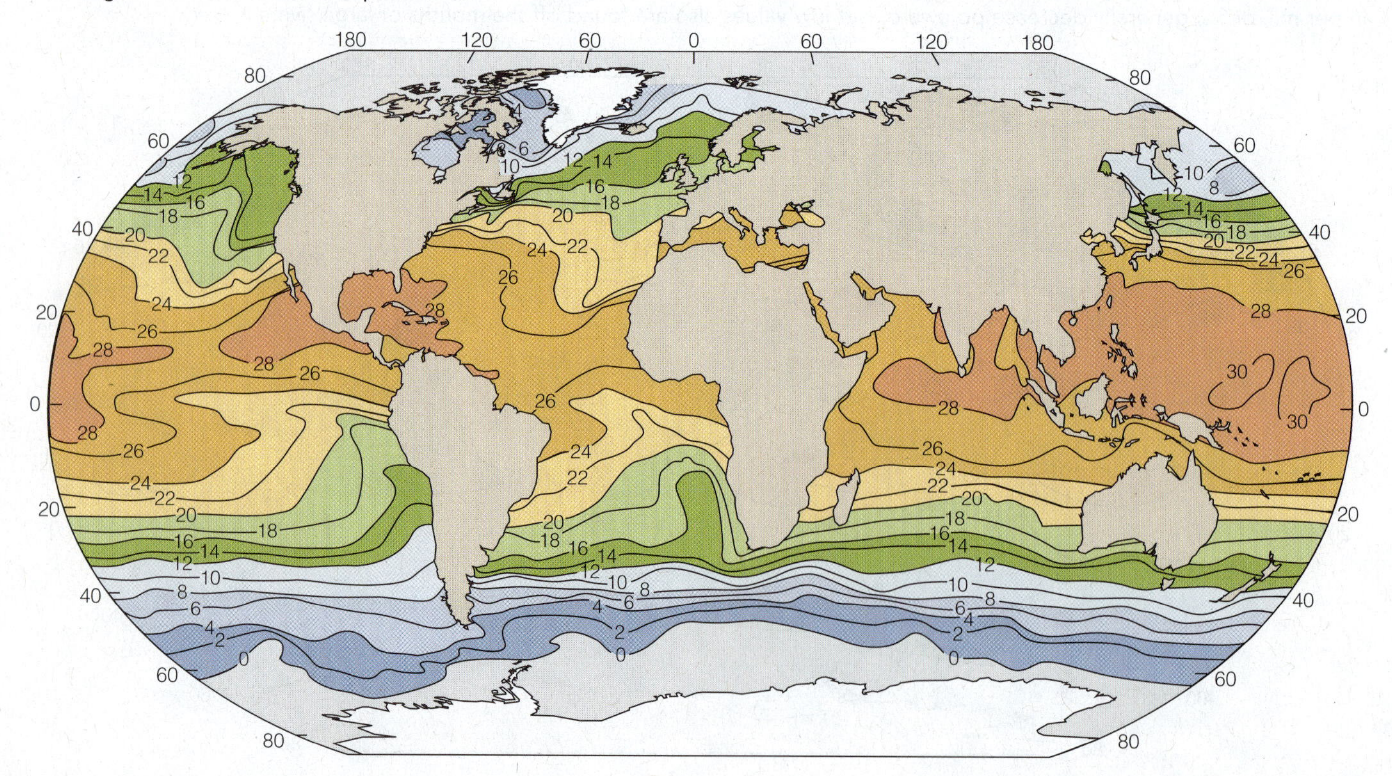

of a continent, summer temperatures may exceed 40°C, whereas along the ocean margin they typically remain below 25°C. Here, then, is a good example of the interaction of the hydrosphere, atmosphere, land surface, and biosphere: Ocean temperatures affect the climate, both over the ocean and over the land, and climate ultimately is a major factor in controlling the distribution of plants and animals.

Vertical Stratification of the Ocean

The physical properties of seawater vary with depth. The ocean is vertically stratified (layered) as a result of variations in the density of seawater. Seawater becomes denser as its temperature decreases and as its salinity increases, which means that colder, saltier water is denser than warmer, fresher water. Gravity pulls dense water downward until it reaches a level where the surrounding water has the same density. These density-driven movements lead to stratification of seawater, and drive circulation in the deep ocean.

Oceanographers recognize three major depth zones in the ocean (FIG. 10.6). The *surface zone*, typically extending to a depth of 100 to 500 m, consists of relatively warm water (except in polar latitudes, where the surface zone is absent). This zone is also referred to as the *mixed layer* because winds, waves, and temperature changes cause extensive mixing within it.

Below the surface zone lies a layer in which the ocean-water properties of temperature, salinity, and density experience significant changes with increasing depth. This zone goes by three different names, one for each property. In the open ocean, temperature commonly decreases markedly downward through the **thermocline** (FIG. 10.6B), and then more slowly with greater depth. The *halocline*, marked by a substantial increase of salinity with depth, is found over much of the North Pacific Ocean and in other high-latitude water where solar heating of the ocean surface is diminished and rainfall is relatively high (FIG. 10.6C). The rapid increase in water density with depth that defines the *pycnocline* (FIG. 10.6A) may result from a decrease in temperature, an increase in salinity, or both.

Below the zone that encompasses the thermocline, halocline, and pycnocline lays the *deep zone*, which contains about 80 percent of the ocean's water volume. In low and middle latitudes, the pycnocline effectively isolates water of the deep zone from the atmosphere, but in high latitudes, where the surface zone is absent, water of the deep zone lies in direct contact with the atmosphere.

Biotic Zones

Plants and animals living in the uppermost water of the ocean occupy the *pelagic* zone of the ocean and are called pelagic organisms. Animals that swim freely under their own locomotion include reptiles, squids, fish, and marine mammals. Benthonic organisms live on the bottom or within bottom sediment (the *benthonic* zone). Floating or drifting (*planktonic*) organisms include **phytoplankton**, which are mainly single-celled plants, and **zooplankton**, which are tiny animals. Among the most important zooplankton are single-celled foraminifera and radiolarians. Foraminifera have a calcareous shell, whereas radiolarian remains consist of silica, an important distinction in determining the composition of deep-sea sediment.

Plant life is restricted to the upper 200 m of the ocean because in this zone (the *photic* zone) sufficient light energy is available for the process of photosynthesis. However, plants also require nutrients, which are available mainly along coasts and shallow continental margins. Over much of the deep ocean, plant life is limited because of a lack of nutrients; therefore primary productivity (total organic material produced by photosynthesis) is low. We will look more closely at the characteristics of life zones in the ocean in Chapter 16.

Oceanic Sediment

Over vast areas, the deep ocean floor is mantled with sediment consisting largely of the skeletal remains of single-celled planktonic (free-swimming) and benthonic (bottom-dwelling) animals (FIG. 10.7). When more than 30 percent of the bottom sediment consists of such remains, it is called a *calcareous ooze* or *siliceous ooze*, depending on the chemical composition of the major component. Calcareous ooze covers broad areas of low to

FIGURE 10.6 Depth zones in the ocean Below the surface zone there is a zone in which the ocean-water properties experience a significant change with increasing depth. This zone is variously known as the *pycnocline* (A), a zone in which density increases with depth; the *thermocline* (B), a zone in which temperature decreases with depth; and the *halocline* (C), a zone in which salinity increases with depth. Still lower lays the *deep zone*, where water is dense as a result of low temperature and high salinity.

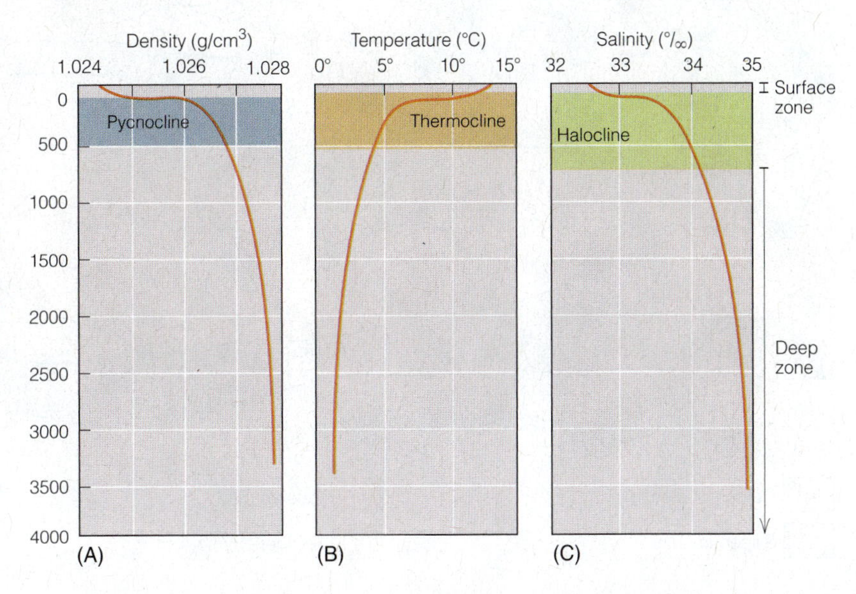

FIGURE 10.7 Marine single-celled organisms
These skeletons of calcareous foraminifera (smooth globular objects), siliceous radiolarians (delicate meshed objects), and siliceous rod-shaped sponge spicules from deep-sea ooze were photographed using a scanning electron microscope. The fossils are from a sediment core collected in the western Indian Ocean during a Deep Sea Drilling Project cruise.

middle latitudes where warm surface water favors the growth of carbonate-secreting organisms (FIG. 10.8). Their tiny shells settle to the seafloor in vast numbers, but they accumulate at an average rate of only about 1–3 cm per thousand years.

Calcareous ooze, however, is not found at these same latitudes in locations where the water is unusually deep. Cold, deep ocean water is under high pressure and contains more dissolved carbon dioxide than shallower water. As a result, deep water is more acidic and can easily dissolve carbonate particles. In the Pacific Ocean the depth at which this occurs is about 4000–5000 m, whereas in the Atlantic it is somewhat shallower. This explains why over large portions of the deep north and south Pacific Ocean and some marginal parts of the Atlantic Ocean calcareous ooze is absent.

In broad belts across the equatorial and far northern Pacific Ocean, sectors of the Indian Ocean, and a belt around the Southern Ocean biologic productivity is high. In these regions, siliceous organisms predominate and become the major component of the bottom sediments.

Lithic sediment, which consists of rock fragments derived from continental sources, mantles the continental shelves and slopes. Such sediment is important where

FIGURE 10.8 Ocean sediment
This map of the world ocean shows the generalized distribution of the principal kinds of sediment on the ocean floor.

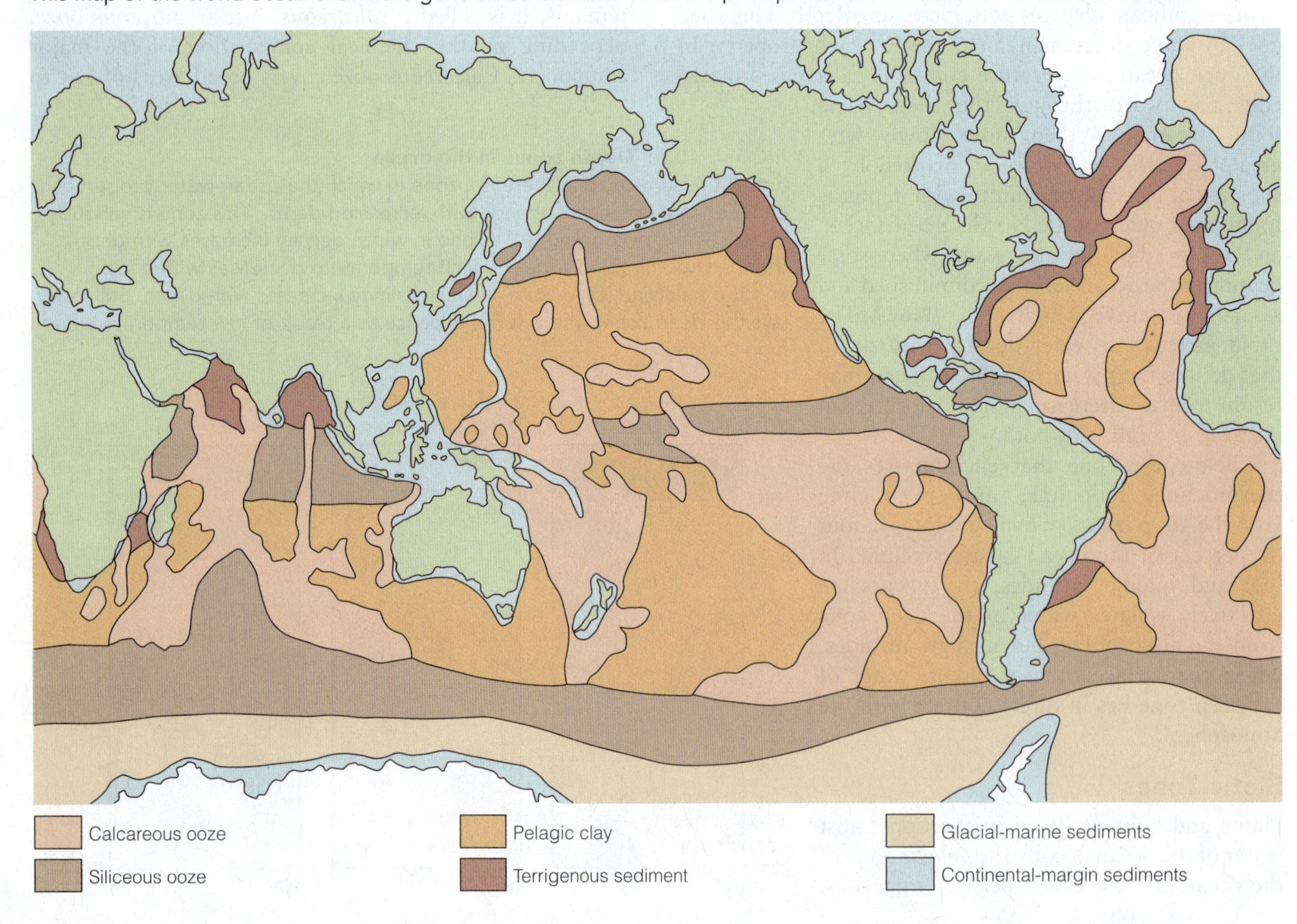

debris-laden glaciers generate icebergs that raft sediment seaward of glacier margins. Still other vast areas of the deep-sea floor, mostly far from land and in regions of low productivity, are dominated by very fine-grained reddish or brownish clay, generally called *red clay*. Much of the clay consists of fine windblown dust, the color of which is imparted by the oxidation of iron-rich minerals in the sediment.

OCEAN CIRCULATION

When Christopher Columbus set sail from Spain in 1492 to cross the Atlantic Ocean in search of China, he took an indirect route. Instead of sailing due west, which would have made his voyage shorter, he took a longer route southwest toward the Canary Islands, and then west on a course that carried him to the Caribbean islands where he first sighted land. In choosing this course, he was following the path not only of the prevailing winds but also of surface ocean currents. Instead of fighting the westerly winds and currents at 40° N latitude (westerly means "coming *from* the west"), he drifted with the easterly Canary Current and North Equatorial Current, as the trade winds filled the sails of his three small ships.

Factors That Drive Currents

Surface ocean currents, like those Columbus followed, are broad, slow drifts of surface water set in motion by the prevailing surface winds. There is friction between the surface of the ocean and the air that is flowing over it. Consequently, air that flows across the sea surface drags the water slowly forward, creating a current of water as broad as the current of air, but rarely more than 50 to 100 m deep. The ultimate source of this motion is the Sun, which heats the surface unequally, thereby setting in motion the planetary wind system. Thus, surface ocean currents result from the interplay of several key elements of the Earth system:

1. Radiation from the Sun provides heat energy to the atmosphere.
2. Nonuniform heating generates winds.
3. Winds, in turn, drive the movement of the ocean's surface water.

The forces that drive deep-ocean water circulation are different, as you will see, because this water—below the surface layer, at depths of 100 m and greater—is too deep to be influenced by wind and frictional drag. The movement of deep-ocean water is fundamentally driven by the characteristics of the water itself, principally temperature and salinity, which cause the water to be more or less dense and thus to sink or float more readily.

Factors That Influence Current Direction

In addition to the forces that set surface and deep water in motion, there are some important forces that influence the direction of these movements. Two important influences to be considered are the Coriolis force and Ekman transport.

The Coriolis Force

Wind sets the surface currents in motion, but the direction taken by ocean currents is also influenced by the **Coriolis force**, which causes all moving bodies to be deflected to the right in the northern hemisphere and to the left in the southern hemisphere. (See *The Basics: The Coriolis Force.*)

Although the Coriolis force does not cause ocean currents, it deflects them once they are in motion. Water that is flowing away from the equator and toward the North Pole in the northern hemisphere will be deflected toward the east (to the right) by the Coriolis force; water that is flowing away from the North Pole and toward the equator in the northern hemisphere will be deflected toward the west (still to the right; check this by looking at a map or a globe). In the southern hemisphere, water that is flowing away from the equator and toward the South Pole will be deflected toward the east (to the left); water that is flowing toward the equator and away from the South Pole will be deflected toward the west (still to the left).

The magnitude of the Coriolis force varies with latitude. This arises from the variation in angular velocity, which, as discussed in *The Basics*, reaches a maximum at the poles and a minimum at the equator. The Coriolis force therefore has its maximum influence on ocean currents at the poles, and minimal influence near the equator. This will become particularly important in Chapters 11 and 12, when we consider wind and wind systems, and the influence of the Coriolis force on their strength and directionality.

Ekman Transport

In 1893, Norwegian explorer Fridtjof Nansen (1861–1930) began an epic voyage across the frozen Arctic Ocean in his vessel, the *Fram*. Frozen into the shifting pack ice, the ship slowly drifted poleward, and then southward, eventually emerging into navigable water nearly three years later. Nansen's observations disclosed something totally unexpected: the floating pack ice moved in a direction 20° to 40° to the right of the prevailing wind. The reason is that wind affects the uppermost layers of the water column more than it affects the underlying layers, producing a net water flow that is at an angle to the wind. The surface layer, set in motion by the wind flowing across the surface, drags on the water immediately beneath. This sets the underlying layer in motion, and the process continues downward. Internal friction causes a decrease in the effectiveness of the wind with depth, however, and the result is a decrease in the velocity of the current in the deeper layers.

As the effectiveness of the wind and the velocity of a wind-driven current decrease with depth, the Coriolis force becomes relatively stronger and stronger, deflecting each successive, slower-moving layer farther to the right.

the BASICS

The Coriolis Force

The Coriolis effect is named for the nineteenth-century French scientist, Gustav Gaspard de Coriolis, who in 1835 first explained how Earth's rotation influences the movement of fluids. (*Note:* It will be *much* easier to understand the Coriolis force and its influence on oceanic currents if you read the following sections with a globe or a map at hand, and refer to it frequently. **FIGURE B10.1** will serve the purpose.)

An object on a rotating planetary body has an angular velocity (velocity due to rotation). The angular velocity is always in the direction of rotation and is a minimum at the equator and a maximum at the poles. It may seem odd to say that angular velocity is a minimum at the equator, but consider the following: Imagine a stone tower built exactly at the North Pole; every 24 hours the tower will rotate completely around as a result of Earth's rotation (**FIG. B10.1A**). A similar tower on the equator, however, would not rotate at all; rather, it would describe an end-over-end motion. At any latitude between the equator and the poles, some rotation and some end-over-end motion occur. For this reason the Coriolis force, which is due to rotation, is latitude-dependent and reaches a maximum at the poles and a minimum (zero) at the equator.

A body that is moving freely on Earth has two velocity components—the velocity of forward motion and the angular velocity of rotation. In order for a moving body to maintain the same velocity as its location on Earth, its angular velocity would have to change continually (unless it is sitting exactly on the equator, in which case its angular velocity would be zero). A change in velocity is an acceleration. The *Coriolis acceleration* is the angular acceleration that would be needed for a moving object to stay on track with respect to the rotating frame of reference, in this case the planet.

Because the angular acceleration is usually absent or insufficient, the *Coriolis force* occurs, and this force, as mentioned earlier, causes a deflection in the path of a moving object toward the right in the northern hemisphere and to the left in the southern hemisphere. Here's why it occurs: If a freely floating object moves away from a pole (**FIG. B10.1B**, *a*), its angular velocity will be slower than that of the surface of the planet. This causes the object to lag behind the rotation, and so it will be deflected in a clockwise direction (to the right). If such an object is moving toward the pole (*b*), its angular velocity will be faster than that of the surface, resulting in a counterclockwise deflection (again toward the right). Regardless of the direction of movement, an object in the northern hemisphere will be deflected to the right. In the southern hemisphere, the deflection is to the left (*c, d*), while at the equator the effect disappears (*e, f*).

Every moving body is subject to the Coriolis force; this includes water, air, and objects moving freely in water and air. One of the most dramatic demonstrations of the effect happened during World War I when the German army bombarded Paris from a distance of 120 km with a gun called "Big Bertha." The gunners discovered that their shots were falling 1 to 2 km to the right of the target. The reason for their poor shooting was their failure to account and correct for a deflection in the trajectory due to the Coriolis force.

FIGURE B10.1 Coriolis force
The Coriolis force results from the influence of angular momentum on the movement of bodies on the surface of a rotating planet. (A) A body at the pole rotates completely around every 24 hours, whereas a body on the equator goes end-over-end but does not rotate. The face on the tower at the pole rotates with respect to an external observer, whereas a tower on the equator always faces the same direction. This demonstrates that angular momentum is highest at the poles, decreasing to zero for objects at the equator. (B) On the rotating Earth, an object freely moving in the northern hemisphere (*a, b*) is deflected by the Coriolis force to the right, whereas in the southern hemisphere (*c, d*) it is deflected to the left. A moving object at the equator (*e, f*) is not deflected.

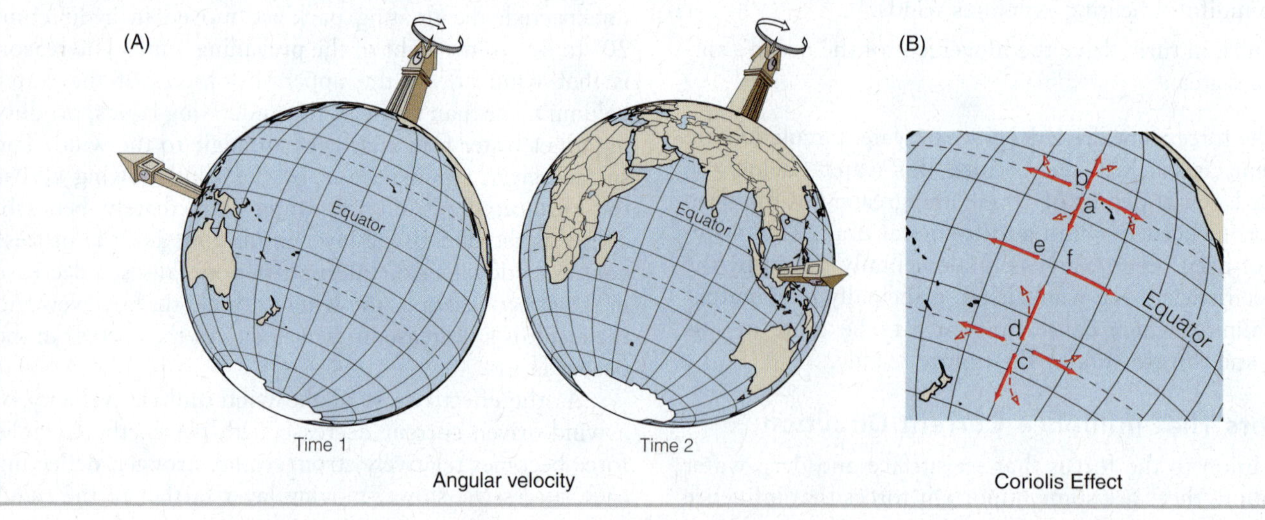

The balance of the two forces—the effect of the wind stronger near the surface and the Coriolis force stronger with depth, away from the frictional drag of the wind—generates a spiraling current pattern, called an *Ekman spiral* (FIG. 10.9). By a depth of about 100 m (the bottom of the surface layer), the wind has lost its effectiveness altogether (since it depends on surface frictional drag); the Coriolis force has taken over, and the wind-driven current speed drops to only a small percentage of the surface current velocity. (Remember that the Coriolis force doesn't *drive* currents, it only influences their direction; that's why the current velocity decreases with depth, even as the Coriolis force itself becomes more influential.) The average flow over the full depth of the spiral, called **Ekman transport**, results in a net direction of water movement at about 90° to the wind direction.

Near coasts, Ekman transport can lead to vertical movement of ocean water. Winds blowing parallel to the coast can drag a layer of surface water tens of meters thick toward or away from land, depending on wind direction (FIG. 10.10). If the net transport is away from land, subsurface water flows upward and replaces the water that is moving away, a process called **upwelling**. If the net Ekman transport is toward the coast, the surface water thickens and sinks in a process known as **downwelling**. Important areas of upwelling occur along west-facing low-latitude continental coasts (e.g., Oregon–California and Ecuador–Peru), where cold water, rich in nutrients and originating at depths of 100 to 200 m, supports productive fisheries.

FIGURE 10.9 Ekman transport
Wind blowing across the ocean in the northern hemisphere affects the surface water, which drags on the water immediately beneath, setting it in motion, and so on down the water column. Internal friction steadily reduces the wind's effect and the current's velocity with depth, and the Coriolis force begins to take over with depth, shifting each successively deeper layer farther to the right. The net result is an Ekman spiral. The average flow over the full depth of the spiral is called the Ekman transport, and is directed at 90° to the wind direction.

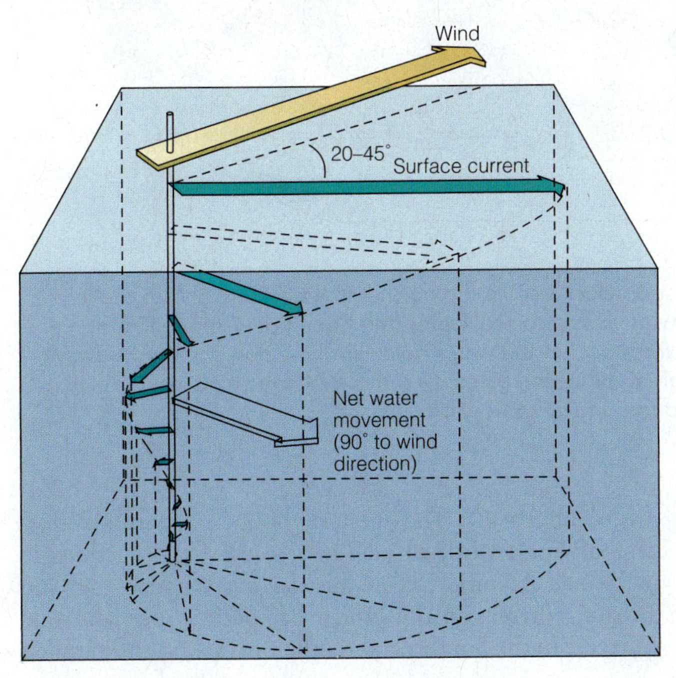

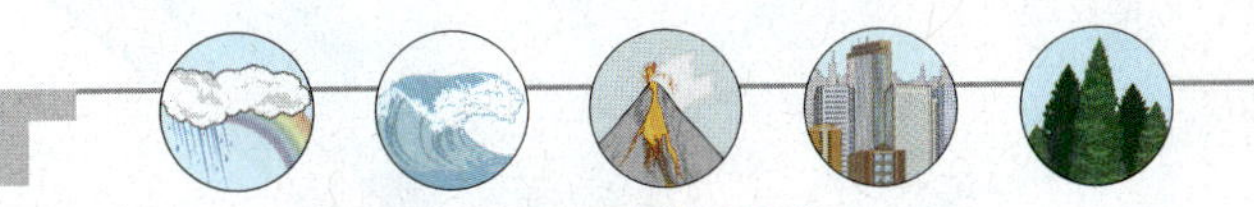

Make the CONNECTION

What would oceanic circulation be like if there were no Coriolis force because Earth did not rotate on its axis? What impacts might this have on ocean currents, weather, and climate?

Surface Current Systems

Now let's look at how these influences and drivers combine with geography to set up the major surface current systems in the world ocean. Each major ocean current is part of a large subcircular current system called a **gyre**. Within each gyre different names are used for different segments of the current system. FIGURE 10.11 shows Earth's five major ocean gyres, two each in the Pacific and Atlantic oceans and one in the Indian Ocean. Currents in the northern hemisphere gyres circulate in a clockwise direction, whereas those in the southern hemisphere circulate counterclockwise.

On either side of the equator, the low-latitude ocean regions are dominated by the warm, westward-flowing North and South Equatorial currents (Fig. 10.11). These are the regions, just to the north and south of the equator, where the easterly (*from* the east, *toward* the west) trade winds blow. (Why? Pause for a moment to think about winds moving away from the poles and toward the equator in the southern and northern hemispheres; how would their movement be influenced by the Coriolis force?) The trade winds blow toward the west on either side of the equator (just north and just south), dragging surface ocean water along with them. This was a significant factor for early explorers, such as Columbus, as they planned their routes, and it still influences transoceanic voyages by air and sea today.

Between the North and South Equatorial currents, flowing approximately along the equator is the eastward-flowing Equatorial Countercurrent. This current (called a countercurrent because it flows toward the east, sandwiched between the more vigorous North and South Equatorial currents, which are westward-flowing) is associated with the doldrums, an equatorial belt of light, variable winds. The trade winds, the doldrums, and other wind systems that are responsible for generating surface ocean currents are discussed in greater detail in Chapter 12.

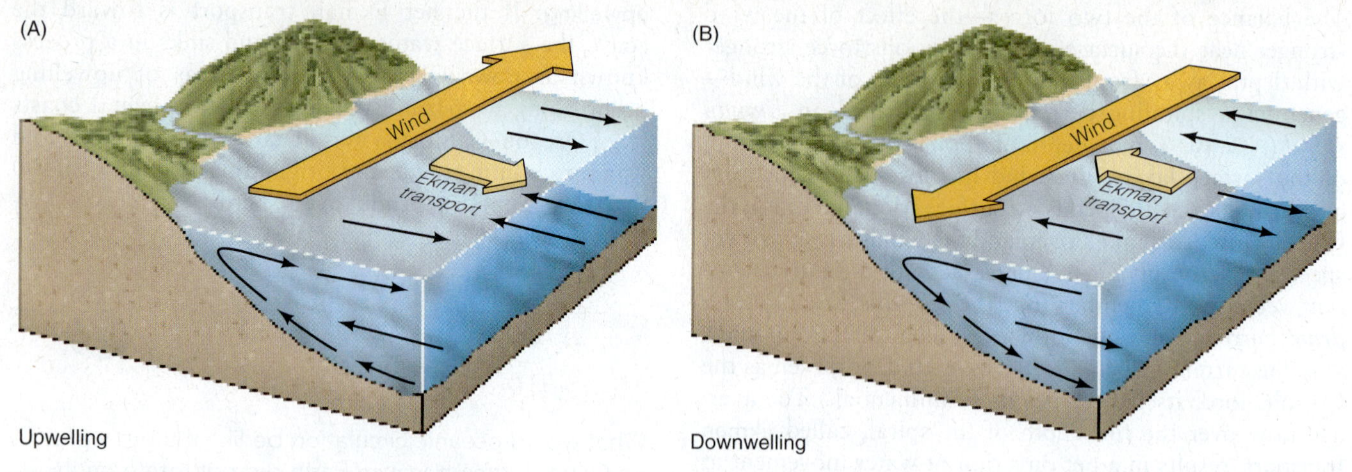

FIGURE 10.10 Upwelling and downwelling
Winds blowing parallel to a coast exert a drag on the surface water, forcing it away from (A) or toward the land (B), depending on wind direction. If the net Ekman transport is away from the land, rising deeper water replaces the surface water moving offshore, which produces upwelling. If the net Ekman transport is toward the shore, the surface water thickens and sinks, which produces downwelling.

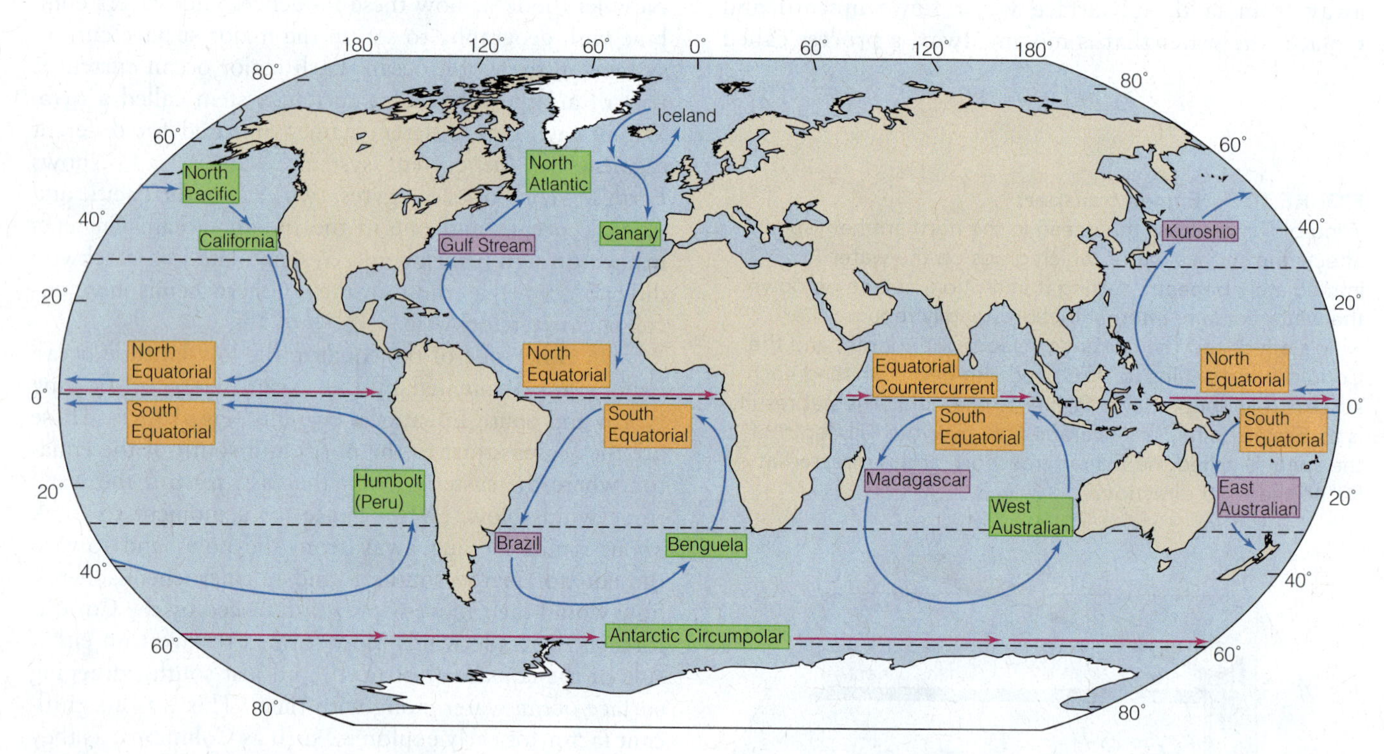

FIGURE 10.11 Surface ocean currents
Surface ocean currents form a distinctive pattern, curving to the right (clockwise) in the northern hemisphere and to the left (counterclockwise) in the southern hemisphere. The westward flow of warm, tropical water (the Equatorial currents) in the Atlantic, Pacific, and Indian oceans is interrupted by continents, which deflect the water poleward. The flow then turns away from the poles to define the middle-latitude margins of the five great midocean gyres: two in the Atlantic, two in the Pacific, and one in the Indian Ocean. Blue arrows show cold currents, and red arrows show warm currents.

In each major ocean basin, the westward-flowing North and South Equatorial currents are deflected poleward as they encounter land. Each current is thereby transformed into a *western boundary current* (because in each case they are flowing along the western edge of the ocean basin). The western boundary currents flow generally poleward, parallel to a continental coastline. In the North Atlantic Ocean this current is called the Gulf Stream, whereas in the North Pacific it is the Kuroshio Current. In the South Atlantic the Brazil Current parallels the South American coast. In the Pacific and Indian oceans the corresponding currents are the East Australian Current and the Mozambique Current. These current systems are all shown in Figure 10.11.

As you also can see from Figure 10.11, upon reaching the belt of westerly winds (again, westerly means *from* the west) around 40° N latitude, the Kuroshio Current changes direction to form the North Pacific Current, the poleward edge of the North Pacific gyre, eventually turning to flow southward along the west coast of North America as the California Current. In the Atlantic, the Gulf Stream similarly passes eastward into the northeast-flowing North Atlantic Current, eventually turning southward to form the Canary Current. In the southern hemisphere, the poleward-moving water of the Brazil, East Australian, and Mozambique currents join the Antarctic Circumpolar Current that circles the globe near latitude 60° S. Then, at the southeastern ends of the southern hemisphere gyres, cool southern water moves toward the equator along the western continental coasts, forming the West Australian Current in the eastern Indian Ocean, the Humboldt (or Peru) Current along the southwestern coast of South America, and the Benguela Current off the southwestern coast of Africa.

The northern Indian Ocean exhibits a unique circulation pattern in which the direction of flow changes seasonally with the monsoons (Chapter 12). During the summer, strong and persistent monsoon winds blow the surface water eastward, whereas in winter, winds from Asia blow the water westward. Regional surface current systems can be greatly influenced by seasonal variations, such as the ones that cause the monsoons, as well as by changes in the strength of wind systems over time. This happens every few years, for example, when the trade winds weaken or even reverse, causing the North and South Equatorial currents to weaken. This phenomenon, called El Niño, can cause dramatic changes in sea-surface temperatures across the Pacific, affecting upwelling along the western coast of South America and influencing weather patterns in many parts of the globe. We will discuss the El Niño phenomenon in greater detail in Chapter 12, in the context of regional wind and weather systems. Both the monsoon and El Niño are good examples of the complex interactions that are constantly occurring among the atmosphere, the ocean, and the land.

From Surface to Depth and Back Again: Major Water Masses

Ocean water is organized vertically into major water masses, stratified according to density. The identity and sources of these masses have been determined by studying the salinity and temperature structure of the water column at many places. (Recall from our discussion of vertical stratification that the density of ocean water is largely determined by its temperature and salinity; cold, salty water is denser than warmer, fresher water.) The Atlantic Ocean provides a good example of the major water masses (**FIG. 10.12**).

FIGURE 10.12 Water masses and ocean circulation
This transect along the western Atlantic Ocean shows the major water masses and general circulation pattern. North Atlantic Deep Water (NADW) originates near the surface in the North Atlantic where northward-flowing surface water cools, becomes increasingly saline, and plunges to depths of several kilometers. As NADW moves into the South Atlantic, it rises over denser Antarctic Bottom Water (AABW), which forms adjacent to the Antarctic continent and flows into the North Atlantic as Antarctic Intermediate Water (AAIW) at a mean depth of about 1 km.

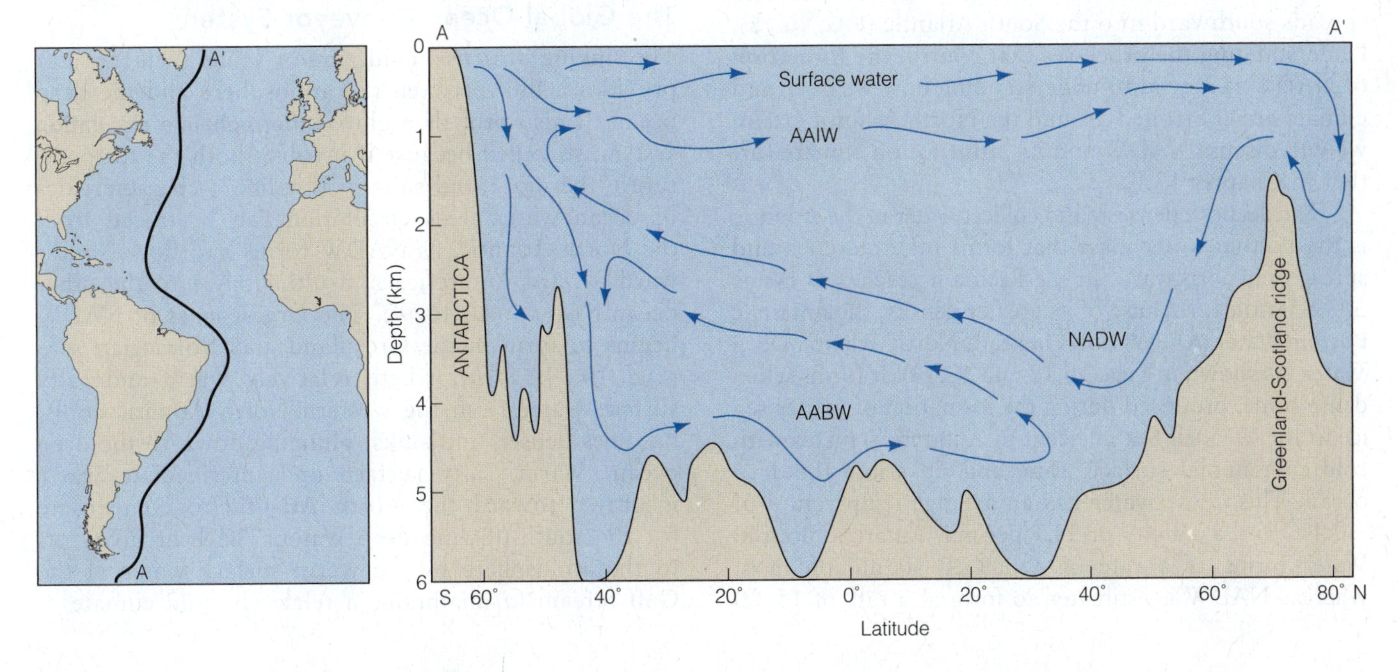

In the Atlantic, water in the surface zone forms a central water mass north and south of the equator to about 35° latitude. The temperature of this water typically ranges from 6° to 19°C, and the salinity ranges from 34 to 36.5 per mil. Cooler sub-Arctic and sub-Antarctic surface water masses are found at high latitudes where cool temperatures and high rainfall give rise to colder, less saline water. The largest mass of polar surface water is the Antarctic Circumpolar Current (shown in Fig. 10.11), which moves clockwise around Antarctica. Its temperature is 0–2°C, and its salinity is 34.6–34.7 per mil.

The central water mass of the Atlantic overlies an intermediate water mass that extends to a depth of about 1500 m. The Antarctic Intermediate Water mass (AAIW), the most extensive such body of water, originates as cold sub-Antarctic surface water that sinks and spreads northward across the equator to about 20° N latitude. Its temperature ranges from 3 to 7°C, and its salinity lies within the range of 33.8–34.7 per mil. Water entering the Atlantic from the Mediterranean Sea is so saline (37–38 per mil) that it flows over a shallow sill at Gibraltar (2400 m) and plunges downward beneath intermediate water to spread laterally over much of the ocean basin.

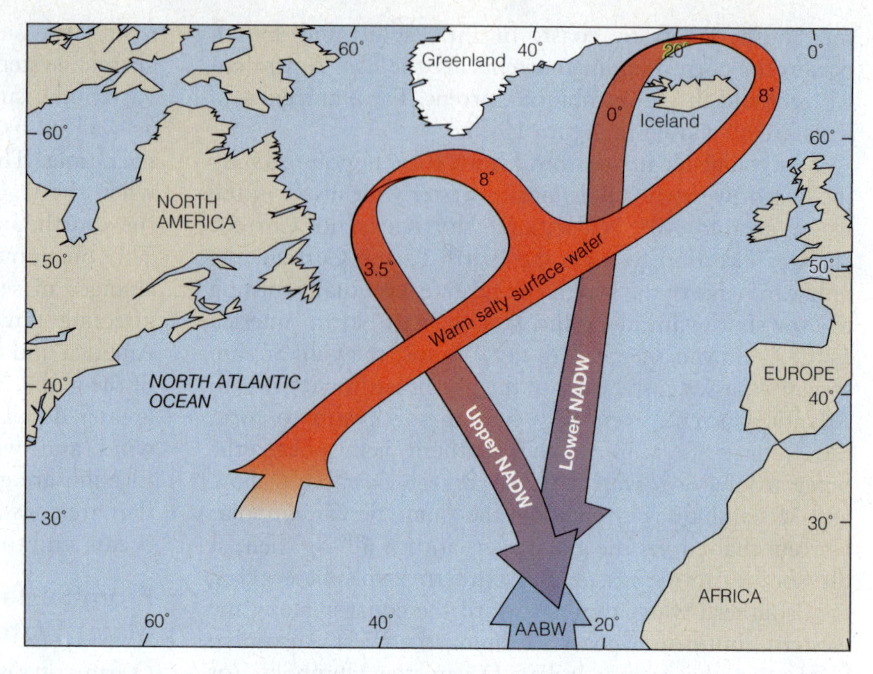

FIGURE 10.13 **North Atlantic Deep Water**
North Atlantic Deep Water (NADW) forms when the warm, salty water of the Gulf Stream/North Atlantic Current cools, becomes increasingly saline due to evaporation, and plunges downward to the ocean floor. The densest water then spills over the Greenland–Scotland ridge and flows southward as Lower NADW. Less dense water forming between Greenland and North America moves southeastward as Upper NADW and overrides denser Lower NADW. Because both water masses are less dense than northward-flowing Antarctic Bottom Water (AABW), they pass over it on their southward journey (as shown in a side view in Fig. 10.12).

In the North Atlantic, the deep ocean consists of a deep-water mass that extends from the intermediate water to the ocean floor. This dense, cold (2–4°C), saline (34.8–35.1 per mil) **North Atlantic Deep Water (NADW)** originates at several sites at the surface of the North Atlantic, flows downward, and spreads southward into the South Atlantic (FIG. 10.13). Understanding the processes that control the formation of NADW is crucial to understanding how weather and climate are controlled around the North Atlantic basin; we will discuss NADW and its influence on climate further in Chapter 13.

The deepest, densest, and coldest water in the Atlantic is the bottom water mass that forms off Antarctica and spreads far northward. In the Pacific it reaches as far as 30° N latitude. Because of its greater density, the **Antarctic Bottom Water (AABW)** flows beneath North Atlantic Deep Water (as shown in Figs. 10.12 and 10.13). It forms when dense brine, produced during the formation of winter sea ice in the Weddell Sea adjacent to Antarctica, mixes with cold circumpolar surface water and sinks into the deep ocean. This dense water has an average temperature of −0.4°C and a salinity of 34.7 per mil. Antarctic Bottom Water forms at a rate of about 20–30 million m^3/s, whereas NADW is estimated to form at a rate of 15–20 million m^3/s (equal to about 100 times the rate of outflow of the Amazon River). Together these water masses could replace all the deep water of the world ocean in about 1000 years.

The Global Ocean Conveyor System

The sinking of dense, cold, and/or saline surface water provides a link between the atmosphere and the deep ocean. It also propels a global **thermohaline circulation** system, so called because it involves both the temperature ("thermo") and salinity ("haline") characteristics of ocean water. This circulation can be traced from the North Atlantic, as NADW forms and flows southward toward Antarctica and ultimately into the other ocean basins (FIG. 10.14). The largest mass of NADW begins to form in the Greenland and Norwegian seas (Fig. 10.14, at *b*), where relatively warm and salty surface water from the western North Atlantic cools, becomes denser, and sinks, plunging down to the deep ocean. Warm, salty surface and intermediate water is drawn toward the North Atlantic to compensate for the south-flowing deep water. The heat given off to the atmosphere by the warm surface water of the Gulf Stream (*a*) maintains a relatively mild climate in

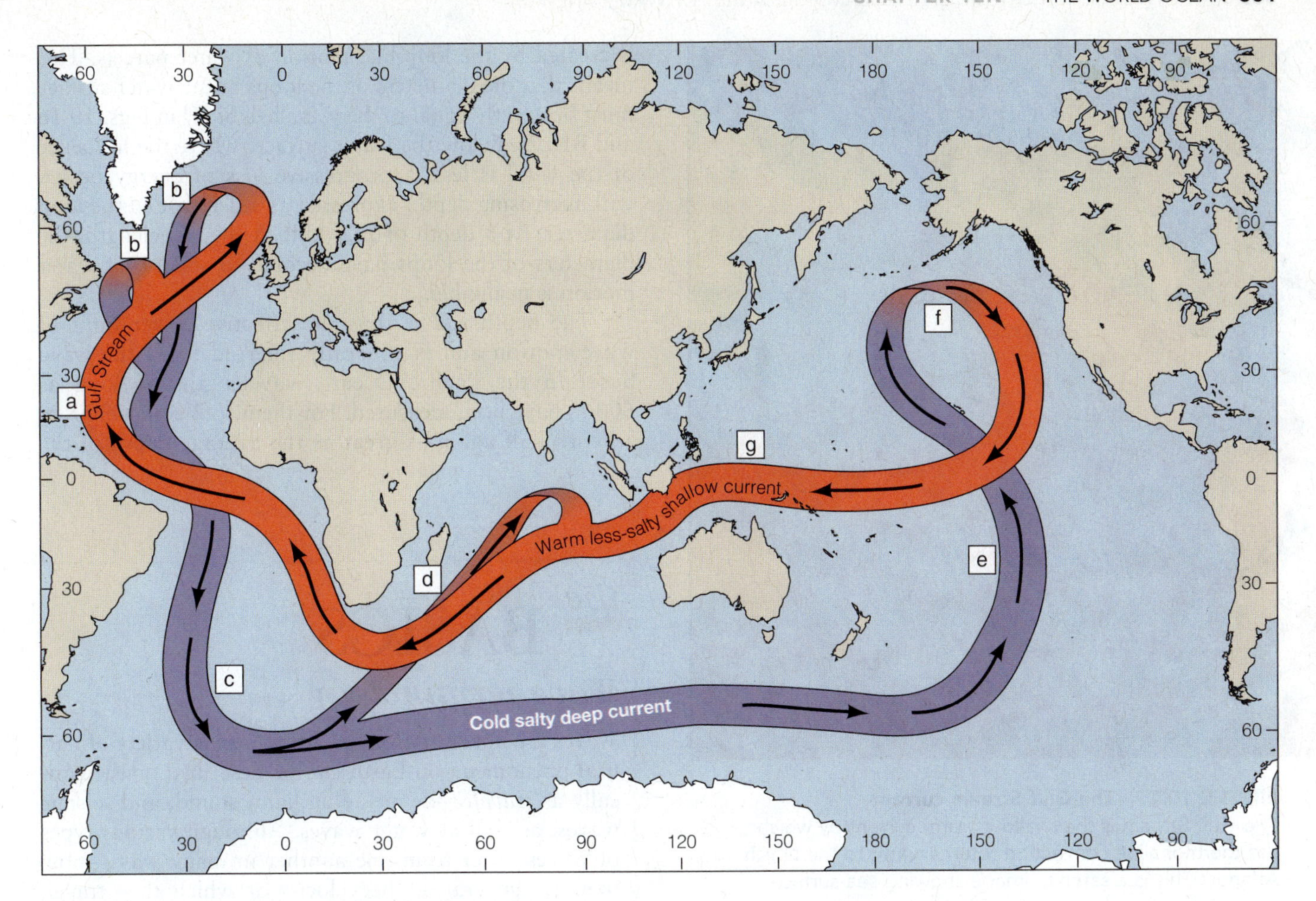

FIGURE 10.14 The thermohaline circulation
The major thermohaline circulation cells that make up the global ocean conveyor system are driven by the density of ocean water, which is in turn driven by the exchange of heat and moisture between the atmosphere and ocean. Warm water brought in by the Gulf Stream (*a*) cools and sinks at a number of sites in the North Atlantic (*b*). The North Atlantic Deep Water (NADW) spreads slowly along the ocean floor to the South Atlantic (*c*), eventually to enter both the Indian (*d*) and Pacific (*e*) oceans before slowly upwelling (*f*) and entering shallower parts of the thermohaline circulation cells. Meanwhile, Antarctic Bottom Water (AABW) forms adjacent to Antarctica (near *c*) and flows northward in fresher, colder circulation cells beneath warmer, more saline water in the South Atlantic (see Fig. 10.12) and South Pacific. Surface water warmed by solar energy flows into the western Pacific and Atlantic basins (*g*) to close the loop of the great global thermohaline cells.

northeastern North American and northwestern Europe (**FIG. 10.15**).

Reaching the South Atlantic (*c*), the southward-flowing NADW enters the Antarctic Circumpolar Current of the Southern Ocean, from which it spreads into the Indian and Pacific oceans. Surface and intermediate water flowing into the South Atlantic via the southern tip of Africa replaces the deep water moving out of the Atlantic basin. Meanwhile, Antarctic Bottom Water flowing into the southernmost Atlantic (near *c*) moves northward, slowly wells up, mixes with overlying NADW, and flows back toward Antarctica (as seen in a side view in Fig. 10.12B). These movements complete an important segment of the global system of ocean circulation, the Atlantic thermohaline circulation cell.

The rest of the global thermohaline circulation is linked to the Atlantic via the Antarctic Circumpolar Current, from which cold, deep water is drawn into the Indian (*d*) and Pacific (*e*) oceans (Fig. 10.14). Once there, the water moves northward, begins to warm, and rises (*f*), ultimately flowing back to the western Pacific (*g*) and Atlantic oceans as shallower, warmer currents. Together, these flows constitute the global thermohaline circulation system, which operates like an enormous conveyor belt to move ocean water around the globe.

OCEAN WAVES

Major currents and gyres are large-scale features of the ocean's surface layer. Finer-scale motions of surface water are a response to the interaction of the atmosphere and ocean surface and, in special cases, to movements of the geosphere. These forces result in the formation of water waves. Scientists use the same basic terminology

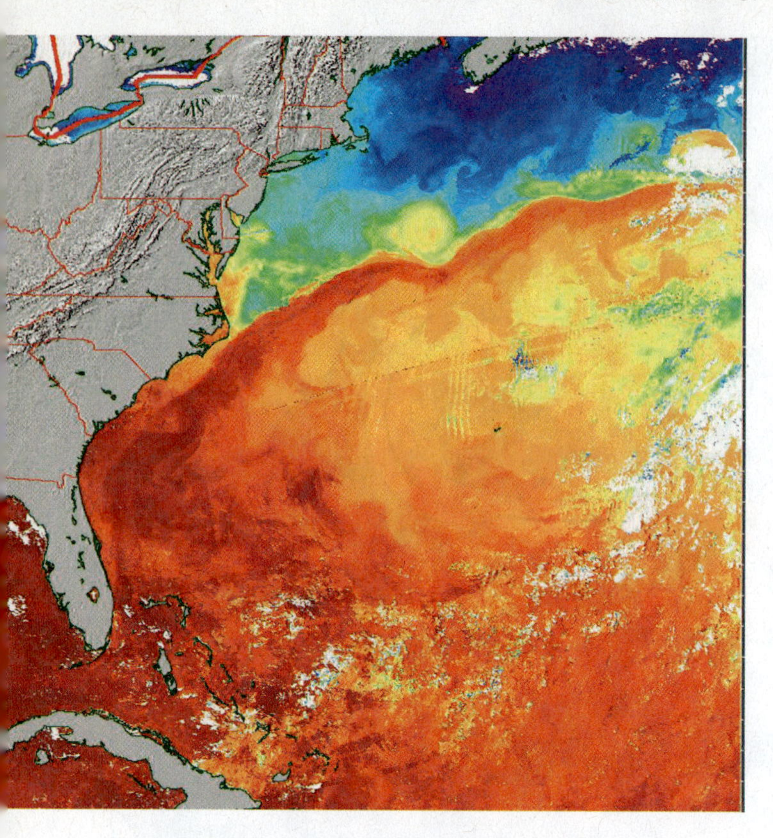

FIGURE 10.15 The Gulf Stream current
The Gulf Stream is the shallow, warm current by which the thermohaline convection returns water to the North Atlantic. This is a satellite image showing sea-surface temperatures, with the reds and oranges representing warm temperatures and blues and greens representing cool temperatures. The warm Gulf Stream current can be seen swirling from the Caribbean past the eastern coast of North America (at the left) and into the cooler water of the North Atlantic.

to describe all kinds of waves—water waves, light waves, sound waves, and seismic waves.

Surface ocean waves receive their energy from winds that blow across the water surface. The size of a wave is determined by how fast the wind blows (*wind speed*), the length of time it blows (*duration*), and the distance across which it blows (*fetch*). A gentle breeze blowing across a bay may ripple the water or form low waves less than a meter high. At the opposite extreme, storm waves produced by hurricane-force winds (>115 km/h) blowing for days across hundreds or thousands of kilometers of open water may become so high that they tower over ships unfortunate enough to be caught in them.

Wave Motion and Wave Base

FIGURE 10.16, on page 304 shows the significant dimensions of a wave traveling in deep water, where it is unaffected by the bottom far below. As the wave moves forward, each small parcel of water (shown as dots in Fig. 10.16) revolves in a loop, returning very nearly to its former position once the wave has passed. The waveform is created by the loop-like motion of water parcels. This means that the diameters of the loops at the water surface must be exactly equal to the wave height (H in Figs. 10.16 and B10.2). Below the water surface, where the influence of the wind is less, a progressive loss of energy occurs with increasing depth, expressed as a decrease in the loop diameter. At a depth of $L/2$, half of the wavelength, the diameters of the loops have become so small that water motion is negligible.

The depth $L/2$ is thus the effective lower limit of wave motion and is generally referred to as the **wave base.** In the Pacific Ocean, wavelengths as long as 600 m have been measured. For them, $L/2$ equals 300 m, a depth half again as great as the average depth of the

the BASICS

Wave Terminology

Waves are not restricted to water. A wide variety of natural phenomena on Earth can be described mathematically as *waveforms*, including light, sound, and seismic waves, as well as water waves. Although various types of waves differ from one another in many ways—how they are generated, the velocity at which they travel, and the medium through which the wave's motion is expressed, among other things—we use the same fundamental terms to describe all waves (**FIG. B10.2**).

The distance between two equivalent points on a wave—crest-to-crest or trough-to-trough are convenient distances, but any two equivalent points will do—is called the **wavelength**, denoted by the capital letter L (or sometimes the Greek letter λ). The distance from the bottom of the wave's *trough* to the top of the *crest* or *peak* is the wave's height, H. The *amplitude* of the wave, denoted A, is half of its height. This is typically measured from an "undisturbed" baseline—the horizontal

FIGURE B10.2 Wave terminology
Although different kinds of waves vary considerably in their properties and how they are generated, wave terminology is fundamentally the same whether it is used in reference to water waves, light waves, sound waves, or seismic waves.

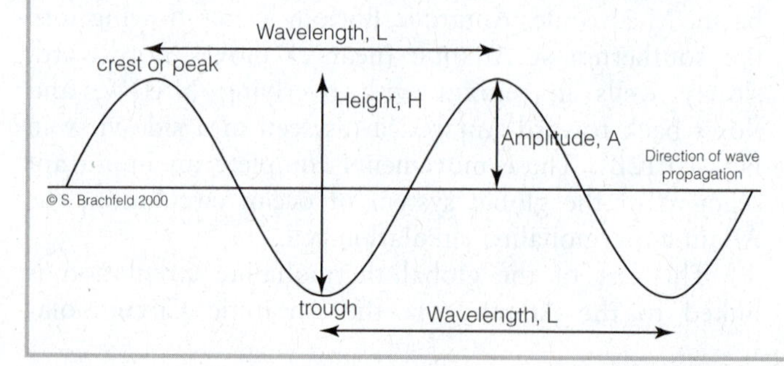

outer edge of the continental shelves (about 200 m). Although the wavelengths of most ocean waves are far shorter than 600 m, it nevertheless is possible for very large waves approaching these dimensions to move seafloor sediment even on the outer parts of the continental shelves.

Breaking Waves

Toward the land, as water depth becomes less than $L/2$, the circular motion of the deepest water parcels is influenced by the increasingly shallow seafloor, which restricts movement in the vertical direction. As the water depth decreases, the loops of the water parcels become progressively flatter until, in the shallow water zone, the movement of water at the seafloor is limited to a rapid back and forth motion. As depth decreases, the wave's shape is distorted; its height increases, and the wavelength shortens. At the same time, the wave's front grows steeper. Eventually, the steep front is unable to support the advancing wave, and as the rear part continues to move forward, the wave collapses or *breaks* (**FIGS. 10.17** and **10.18**).

When a wave breaks, the motion of its water instantly becomes turbulent **surf**, defined as wave activity between the line of breaking waves and the shore. In surf, each wave finally dashes against rock or rushes up a sloping beach until its kinetic energy is converted to potential energy; then, under the influence of gravity,

FIGURE B10.3 Wave fronts
When a set of waves is being propagated in the same direction (A), a plane passing through the crests of all advancing waves is called the wavefront. See if you can identify the locations of the wavefronts in this photograph of water waves (B).

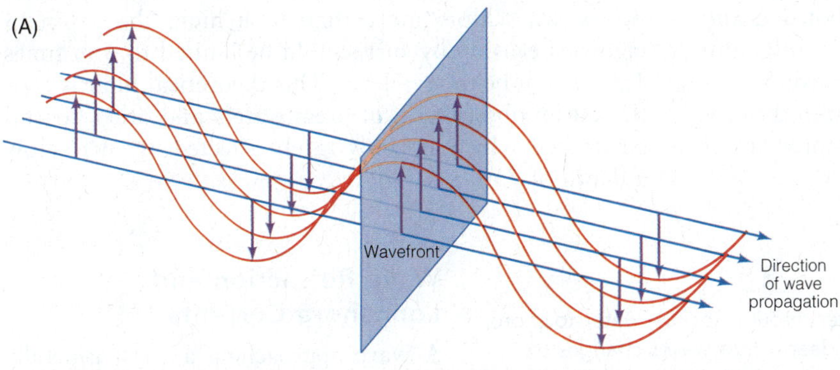

line in Figure B10.2—to either the crest or the trough.

Water waves, light waves, and seismic S waves are all examples of *transverse waves*. In a transverse wave the displacement that creates the wave is oriented at an angle to the *direction of propagation*, the direction in which the wave is traveling. This is shown in **FIGURE B10.3A**, in which the small arrows describe the displacement that is creating the waveform. Note that these arrows are oriented at a right angle to the direction of wave propagation. You may recall from Chapter 6 that seismic P waves are different; so are sound waves—both are *longitudinal waves*, in which the displacement that creates the wave is back-and-forth in a direction that is parallel to the direction of wave propagation.

When multiple waves are moving in the same direction, or moving forward in concert with one another, a plane that connects the crests of the advancing waves is called the *wavefront*, as shown in Figure B10.3. If a wavefront hits an obstacle that blocks its forward motion, the wavefront will be *reflected*. If the wavefront passes from one medium into another that causes a change in the velocity of propagation of the wave, *refraction* (bending) will occur. These processes are common to waves of all types. In Chapter 6, for example, we discussed the refraction of seismic waves as they pass through the various layers of Earth's interior.

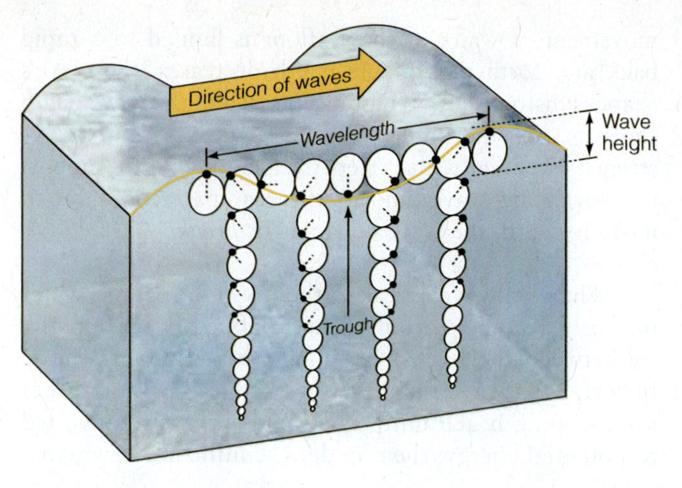

FIGURE 10.16 Waves
Water in a wave in deep water makes a looplike motion as the crest of the wave passes by. The dots mark the position of a particle of water as a wave passes through. Beneath the surface, where the influence of the wind is less, parcels of water travel in smaller loops.

FIGURE 10.18 Catching a wave
A surfer catches a breaking wave off the coast of Oahu, Hawaii.

it flows back. Surf possesses most of the original energy of each wave that created it. This energy is quickly consumed in turbulence, in friction at the bottom, and in moving the sediment that is thrown violently into suspension from the bottom. Although fine sediment is transported seaward from the surf zone, most of the erosive work of waves is accomplished by surf shoreward of the line of breakers.

Ocean waves typically break at depths that range between wave height and 1.5 times wave height. Because waves are seldom more than 6 m high, the depth of vigorous erosion by surf should be limited to 6 m times 1.5, or 9 m below sea level. This theoretical limit is confirmed by observation of breakwaters and other coastal structures, which are only rarely affected by surf below a depth of 7 m.

FIGURE 10.17 Waves moving ashore
Waves change form as they travel from deep water through shallow water to shore. The circular motion of the water parcels found in deep-water waves changes to elliptical motion as the bottom becomes shallower than the wave base (at a depth of $L/2$), and the wave begins to encounter frictional resistance from the bottom. Vertical scale is exaggerated, as is the size of loops relative to the scale of the waves.

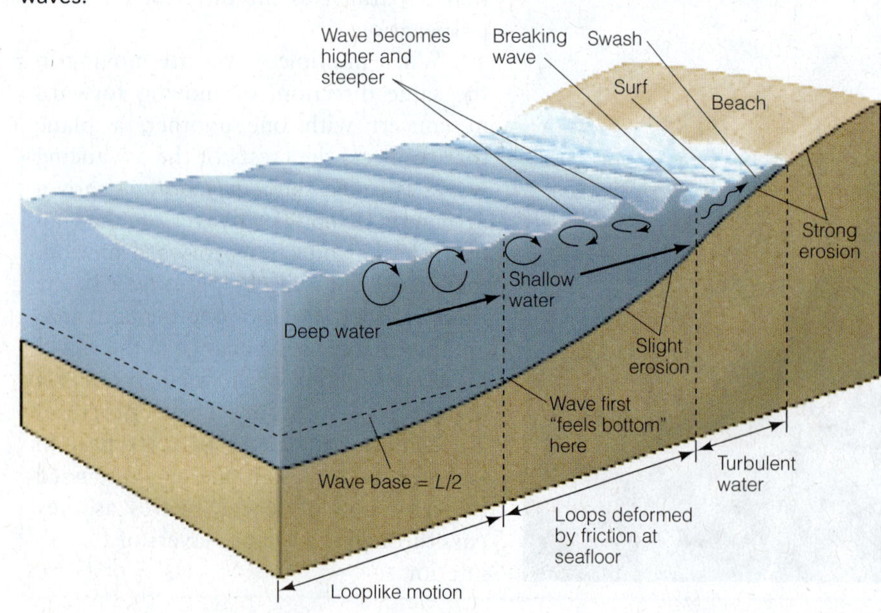

Wave Refraction and Longshore Currents

A wave approaching a coast generally does not encounter the bottom simultaneously all along its length. As any segment of the wave touches the seafloor, that part slows down, the wavelength begins to decrease, and the wave height increases. Gradually, the trend of the wave becomes realigned to parallel the bottom contours. Known as **wave refraction**, this process changes the direction of a series of waves moving in shallow water at an angle to the shoreline (**FIG. 10.19**). In this way, waves approaching the margin of a deep-water bay at an angle of 40° or 50° may, after refraction, reach the shore at an angle of 5° or less.

The path of an incoming wave can be resolved into two directional components, one oriented perpendicular to the shore and the other parallel to it. Whereas the perpendicular component produces the crashing surf, the parallel component sets up a **longshore current** within the surf zone, a current that flows parallel to the shore (**FIG. 10.20**). The direction of longshore currents may change seasonally if the prevailing wind directions change, thereby causing changes in the direction of the arriving waves.

Tsunami: A Different Type of Wave

A **tsunami** (technically, *seismic sea wave*) is a type of water wave that has a completely different origin and characteristics from normal water waves. Tsunamis are often erroneously called "tidal" waves; in fact, they have nothing to do with tides but are generated by sudden movements of the seafloor. The most common cause of such movements is submarine earthquakes, although submarine and coastal landslides and large volcanic eruptions also can cause sudden shifting of the ocean floor. When a sudden displacement of the seafloor occurs—but only if there is a significant vertical component to the movement—the overlying ocean water will be shoved upward or downward. When the displaced water falls back down, it splits into two components that travel in opposite directions away from the site of tsunami generation, as shown in **FIGURE 10.21**.

Seismic sea waves have physical characteristics that are very different from those of normal ocean waves. They travel at speeds up to 950 km/h, compared to typical velocities of 30 to 100 km/hr for normal ocean waves. Tsunamis typically have wavelengths measured in kilometers, compared to wavelengths measured in tens of meters or less for normal ocean waves. In contrast, the *wave height* (or *amplitude*, the height from the bottom of a trough to the crest) of a tsunami is typically only 1 to 2 m. (The wave heights of normal ocean waves range from very small to over 30 m, for extremely large storm-generated waves.) The oddly low wave height relative to the very long wavelength of the typical tsunami means that the enormous volume of water involved in the wave is spread out over an extensive distance; as a result, tsunami wave heights in the open ocean are so low that they are generally not seen or felt. Instrumentation specifically designed to measure water levels is required for the reliable detection of tsunamis in the open ocean; such instrumentation is now an integral part of all tsunami early-warning systems.

Another characteristic that differs greatly between normal waves and tsunamis is *periodicity*—that is, the

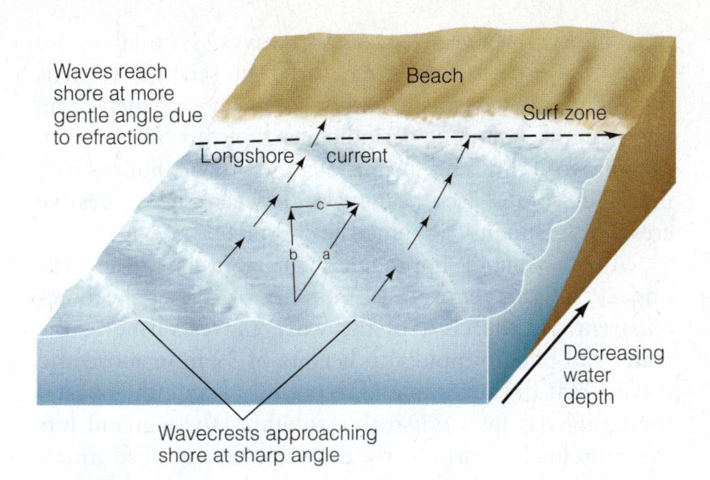

FIGURE 10.20 Longshore current
A longshore current develops parallel to the shore as waves approach a beach at a right angle and are refracted. A line drawn perpendicular to the front of each approaching wave (*a*) can be resolved into two components. The component oriented perpendicular to the shore (*b*) produces surf, whereas the component oriented parallel to the shore (*c*) is responsible for the longshore current. Such a current can transport considerable amounts of sediment along a coast.

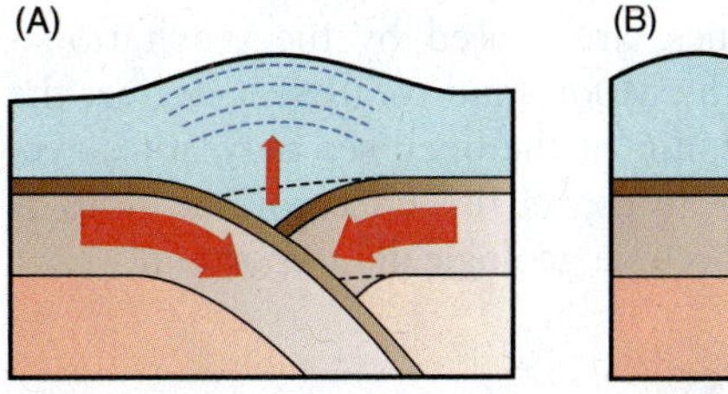

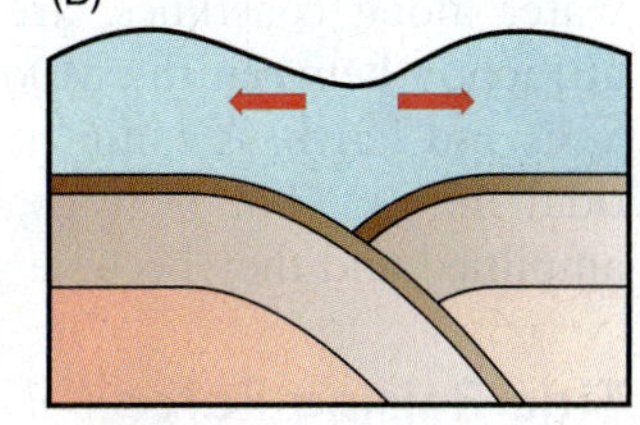

FIGURE 10.21 Tsunami
A tsunami can form any time the ocean floor undergoes a sudden movement that has a significant vertical component (A), causing the overlying water to be displaced upward. (Horizontal displacement will not cause the overlying water to be shoved upward and therefore will not generate a tsunami.) (B) When the displaced volume of water falls back down, it splits into two components that move away from the site of wave generation in opposite directions.

FIGURE 10.19 Breaking waves
Waves arriving obliquely along a coast near Oceanside, California, change orientation as they encounter the bottom and begin to slow down. As a result, each wavefront is refracted so that it more closely parallels the bottom contours. The arriving waves develop a longshore current that moves from right to left in this view.

time interval between successive waves. Normal ocean waves have periodicities measured in seconds to tens of seconds. In contrast, because of their extremely long wavelengths, tsunami periodicities typically range from 20 min to 1 hr. This means that up to an hour—even more, in some cases—can pass between the successive crests of a tsunami reaching the shoreline.

As a tsunami approaches the shoreline, the true impact of this enormous volume of water becomes apparent. As the crest of the wave moves onshore, the water can pile up rapidly to heights of 30 m or more and travel great distances inland, especially where the coastal topography is low. When the trough of the tsunami hits the shoreline, a particularly dramatic effect called *draw-down* can occur. During drawdown, the sea level falls as the trough of the wave passes through, causing the edge of the water to retreat. Many lives have been lost when curious onlookers have gone to the edge of the water to investigate this unusual phenomenon during the hour or so between successive crests of a tsunami, only to be caught by the extremely rapid onrush of water when the next crest arrives.

OCEAN TIDES

Tides, the rhythmic, twice-daily rise and fall of ocean water along coastlines, are caused by the gravitational attraction between the Moon (and to a lesser degree, the Sun) and Earth. A sailor in the open sea may not detect tidal motion, but near coasts the effect of the tides is amplified and they become geologically important.

Tide-Raising Force

As discussed in Chapter 2, the gravitational pull between the Moon and the solid Earth is balanced by an equal and opposite inertial force created by Earth's movement with respect to the center of mass of the Earth–Moon system (FIG. 10.22). At the center of Earth, the gravitational and inertial forces are balanced (equal but opposite), but they are not balanced from place to place on the surface. A water particle in the ocean on the side of Earth that is facing the Moon is attracted more strongly by the Moon's gravitation than it would be if it were at Earth's center, which is farther away. Liquid water is easily deformed, so each water particle on the facing side of the planet is pulled toward a point directly beneath the Moon. This *tide-raising force* creates a bulge on the ocean surface. On the opposite side of the planet, a particle on the surface of the ocean would be farther away from the Moon and would thus experience its gravitational pull even less.

Meanwhile, the inertial force keeps objects (including ocean water) moving straight ahead in a direction away from the center of mass of the Earth–Moon system. It has the same strength at any point on the surface. (Luckily for us, gravity prevents Earth's water from simply flying off into outer space as a result of this same inertial force.) On the side of Earth that is nearest the Moon, the magnitude of the gravitational attraction exceeds the magnitude of the inertial force, and the excess tide-raising force is in the direction of the Moon. On the opposite side of Earth, the magnitude of the inertial force exceeds the Moon's gravitational pull, and the resulting tide-raising force is directed away from the Moon (Fig. 10.22). These unbalanced forces generate the two bulges of water that account for daily ocean tides.

Tidal Bulges

The tidal bulges created by the tide-raising force on opposite sides of Earth appear to move continually around the planet as it rotates. In fact, the bulges remain essentially stationary beneath the tide-producing body (the Moon), while Earth rotates past them. At most places on the ocean margins, two high tides and two low tides are observed each day as a coast encounters each tidal bulge in turn. In effect, at every high tide, the coastline runs into a mass of water, which piles up against it. This water then flows back to the ocean basin as the coastline passes beyond each tidal bulge. The continual battering of the coastlines of the world by tides is slowly reducing Earth's rate of rotation. The effect of any one tidal cycle is tiny, but over long periods the total effect is large. It is estimated that 500 million years ago, Earth rotated much faster than it does today, so that a day was only 22 hours in length.

Earth–Sun gravitational forces also affect the tides, sometimes opposing the Moon by pulling in the opposite direction and sometimes aiding by pulling in the same direction. Twice during each lunar month, Earth is directly aligned with the Sun and the Moon, whose gravitational effects are thereby reinforced, producing higher

FIGURE 10.22 Tide-raising forces
Tide-raising forces are produced by the Moon's gravitational attraction and by inertial force. On the side of Earth that faces toward the Moon, gravitational attraction distorts the water level from that of a sphere and raises a tidal bulge. On the opposite side of Earth, a tidal bulge is created by inertia.

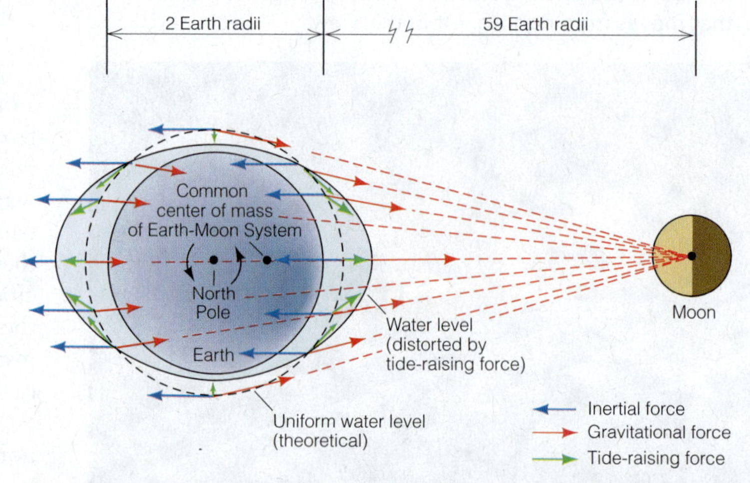

high tides and lower low tides (FIG. 10.23). At positions halfway between these extremes, the gravitational pull of the Sun partially cancels that of the Moon, thus reducing the tidal range. However, the Sun is only 46 percent as effective as the Moon in producing tides, so the two tidal effects never entirely cancel each other.

In the open sea, the effect of the tides is small (less than 1 m), and along most coasts the tidal range commonly is no more than 2 m. However, in bays, straits, estuaries, and other narrow places along coasts, tidal fluctuations are amplified and may reach 16 m or more (FIG. 10.24). Associated tidal currents are often rapid and may approach 25 km/h.

WHERE LAND AND OCEAN MEET

The ocean meets the land in a zone of dynamic activity marked by erosion and the creation, transport, and deposition of sediment. At a coast, waves that may have traveled unimpeded across hundreds or thousands of kilometers of the open ocean encounter an obstruction to further progress. They crash against the shore, erode rock and sediment, and move the resulting particles about. In the surf zone, blocks of bedrock are plucked out and carried away, and the continuous rubbings and grindings of moving rock particles wear away solid rock. In effect, the surf acts like a saw or file, cutting horizontally into the land. Over time, the net effect is substantial. Coasts change, often slowly, but at times very rapidly. At any given moment, the geometry of the shoreline represents an approximate equilibrium between constructive (depositional) and destructive (erosional) forces.

Beaches and Other Coastal Deposits

Beaches are a primary landform of most coasts (FIG. 10.25). Even coasts dominated by steep, rocky cliffs generally have beaches interspersed with rocky headlands. A **beach** consists of wave-washed sediment along a coast, including sediment in the surf zone that is in constant motion. Although some beach sediment is derived from erosion of adjacent cliffs or older beach

FIGURE 10.23 Tidal bulges
When Earth, Moon, and Sun are aligned (positions 1 and 3), tides of highest amplitude are observed. When the Moon and Sun are pulling at right angles to each other (positions 2 and 4), tides of lowest amplitude are experienced.

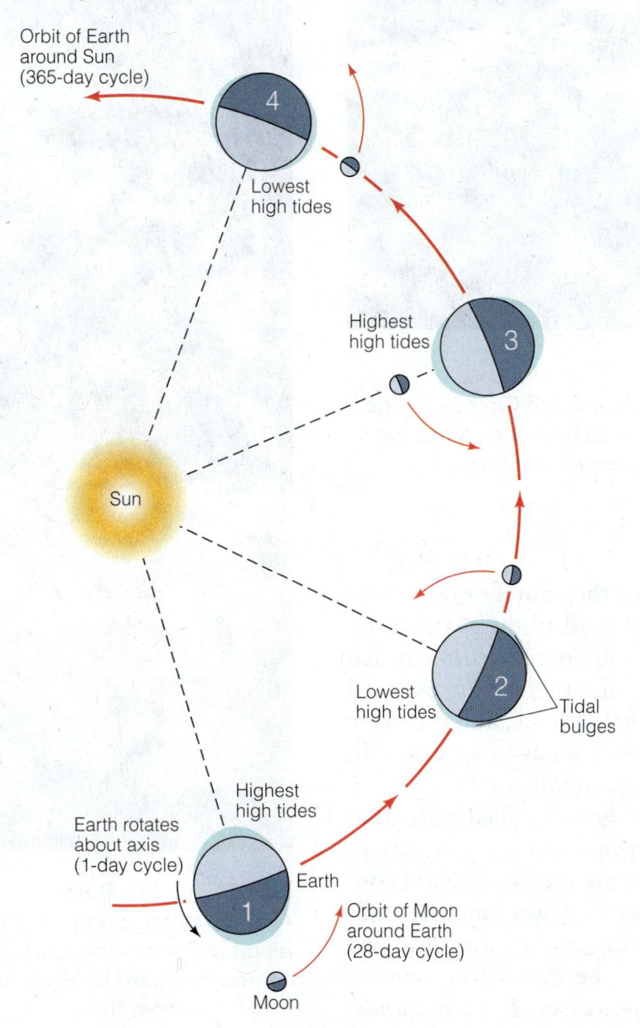

FIGURE 10.24 Tidal range
The tidal range in the Bay of Fundy, eastern Canada, is one of the largest in the world. (A) Coastal harbor of Alma, New Brunswick, at high tide. (B) Same view at low tide.

deposits, most sediment reaches the shoreline via rivers that are carrying their sediment load to the sea. During storms, powerful surf erodes the exposed part of a beach and makes it narrower. In calm weather, an exposed beach receives more sediment than it loses, and therefore becomes wider. Because storminess tends to be seasonal, beaches also change character seasonally.

Elongated ridges of sand or gravel, called *spits,* that project from land and end in open water are a conspicuous coastal landform. Most spits are merely seaward continuations of beaches (FIG. 10.26A). Many spits are built of sediment moved by longshore currents and dropped at the mouth of a bay, where the current encounters deeper water and its velocity decreases. *Barrier islands,* which are long, narrow sandy islands lying parallel to a coast and separated from it by a lagoon, are found along most of the world's lowland coasts (FIG. 10.26B). Some barrier islands, like those off the North Carolina coast, occasionally receive the full fury of destructive hurricanes, which erode and reshape these ephemeral landforms. These coastal landforms are typical of passive or trailing continental margins (Chapter 5), like the eastern coast of North America.

Marine Deltas

Where surf and currents are inadequate to erode all new sediment carried to the sea by a large stream, the sediment builds outward as a *marine delta* (FIG. 10.27, on page 310). The size and shape of a delta reflect the balance reached between sedimentation and erosion at the coast. Some deltas, such as that of the Mississippi River, consist of a complex system of deltas of different ages, indicating a long and complicated history.

FIGURE 10.25 Beach
A sandy beach along the shore of Bora Bora, a volcanic island in French Polynesia, consists of coral and shell debris carried landward by wave action and mixed with lava fragments from the eroding volcano.

FIGURE 10.26 Coastal landforms
(A) The long, curved spit of Cape Cod, Massachusetts has been built by longshore currents that rework glacial deposits forming the peninsula southeast of Cape Cod Bay. (B) These barrier islands off Corpus Christi, Texas (along the south side of a large bay) are seen from an orbiting satellite. To the right is the Gulf of Mexico. Padre Island National Seashore occupies the barrier island extending south from Corpus Christi Bay.

Deltas tend to be particularly fertile regions, both on land and in the adjacent marine coastal areas. Some of the world's largest deltas constitute prime agricultural lands and, collectively, are home to millions of people. Famous examples include Egypt's Nile Delta, China's Huang He and Yangtze deltas, the Mississippi Delta on the Gulf coast of the United States, and the Ganges Delta of Bangladesh. Each of these deltas is a geologically young feature, having been built during the latest postglacial rise of world sea level. The surface of each also lies within a few meters of present sea level and is therefore especially vulnerable to destructive coastal storms that can inundate farmland, destroy towns and villages, and devastate local populations.

Estuaries

An **estuary** is a semienclosed marine embayment that is diluted with fresh water, normally entering it by one or more streams. Several types of estuaries are recognized, distinguished on the basis of their internal circulation and salinity distribution. The simplest type is found where a river enters directly into seawater. The fresh water, being less dense, flows directly out to sea over denser saltwater. The rapid flow of the river retards saltwater incursion, producing a boundary tilted downward in the upstream direction that is between a wedge of saltwater moving upstream and an overlying thinner wedge of stream water. The turbulent stream water entrains saltwater as it flows seaward, increasing its salinity. However, very little river water is mixed downward, and so such estuaries are poorly mixed. In contrast, estuaries characterized by strong tidal turbulence and low river outflow tend to be well mixed, whereas those marked by both strong river and tidal inflow tend to be partially mixed.

Estuaries and associated coastal wetlands offer important habitats for an array of plants and animals.

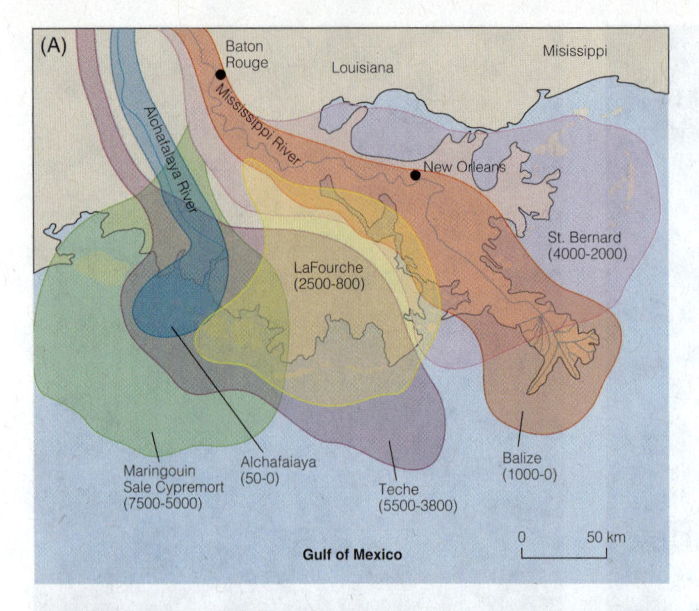

FIGURE 10.27 Deltas
The Mississippi River has built a series of overlapping deltas (A) as it has continually dumped sediment into the Gulf of Mexico. The ages of deltas are given in years before the present, determined by radiocarbon dating. Swirling clouds of sediment can be seen in the water of the current Mississippi Delta (Balize), in this satellite image (B).

They are spawning grounds for many species of commercial fish and shellfish, and often support large bird populations. The establishment of industrial cities, commercial ports, and fisheries in estuarine settings has led to their increasing modification and pollution (FIG. 10.28).

Reefs

Many of the world's warm-water coastlines are characterized by limestone **reefs,** vast colonies built by corals and other carbonate-secreting organisms. Three principal reef types are recognized: a *fringing reef* is either attached to or closely borders the adjacent land and lacks a lagoon (FIG. 10.29A); a *barrier reef* is a reef separated from the land by a lagoon and may be of considerable length and width (FIG. 10.29B); and an *atoll* is a roughly circular coral reef enclosing a shallow lagoon (FIG. 10.29C) that forms when a tropical volcanic island with a fringing reef slowly subsides. Charles Darwin was the first to deduce, during his voyage on the H.M.S. *Beagle* in the 1830s, that slow subsidence forces reef organisms to grow upward so that they can survive near sea level. As a volcanic island subsides, the fringing reef is transformed into an offshore barrier reef and eventually into an atoll. Atolls generally lie in deep water in the open ocean and are as large as 130 km in diameter. Darwin's hypothesis was confirmed a century after he proposed it, by drill holes on atolls that reached volcanic rock after penetrating thick sections of ancient reef rock.

Coral reefs are important environments because the varied habitats they offer generate intense biologic productivity and diversity. They are also highly sensitive environments, reacting quickly to even minor changes in water temperature and clarity. This makes them extremely useful as indicators of environmental change and degradation: Like a canary in a mine, when coral reefs suffer it can be a sign that worse is yet to come.

Coastal Erosion

Coastal zones are classic examples of systems in a state of dynamic equilibrium. Sediment coming into the shoreline system via rivers or from the erosion of adjacent cliffs is balanced by sediment being taken out of the system by waves and the longshore current. As soon as the local system becomes unbalanced—that is, more sediment is lost than gained, or vice versa—a change in the shoreline will become apparent. Sometimes a shoreline system exists in a balanced state for quite a long time, until an extraordinarily energetic set of storm waves arrives, carrying massive volumes of sediment away from the beach. Some beaches recover from such a loss very quickly; in other cases the change is permanent (on a human time scale, that is), and the beach system establishes a new state of balanced input and output of sediment.

Because shorelines are important to so many aspects of human society—transportation, fishing, and tourism, for example—coastal erosion tends to have significant impacts on human interests. When beaches are seen to be eroding, people tend to react in an attempt to protect the shoreline. However, if such intervention is carried out without regard for the natural erosional-depositional balance of sediment in the system, it can be disastrous, failing to protect the beach or even accelerating the rate

FIGURE 10.28 Estuary
This is a satellite image of the shrinking San Francisco Bay estuary. Filling and diking of tidal marshes to create farmland, evaporation ponds, and residential and industrial developments has reduced 2200 km^2 of wetland marshes that existed before 1850 to less than 130 km^2 today.

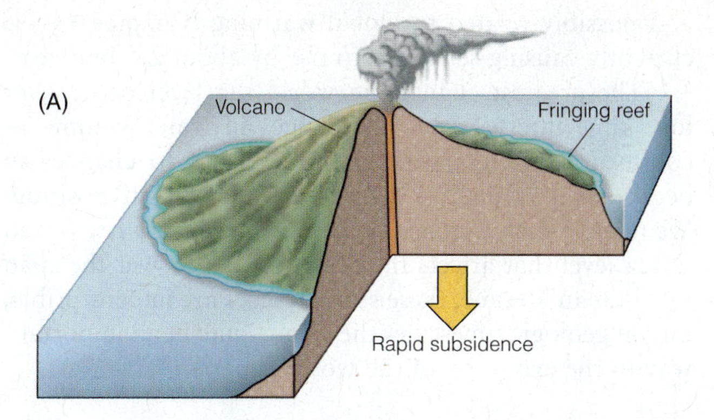

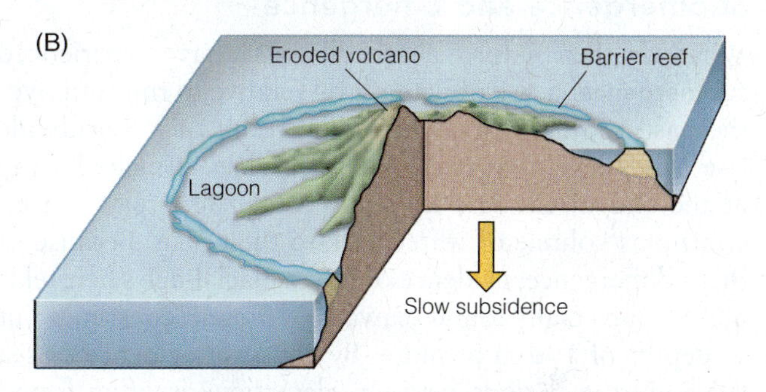

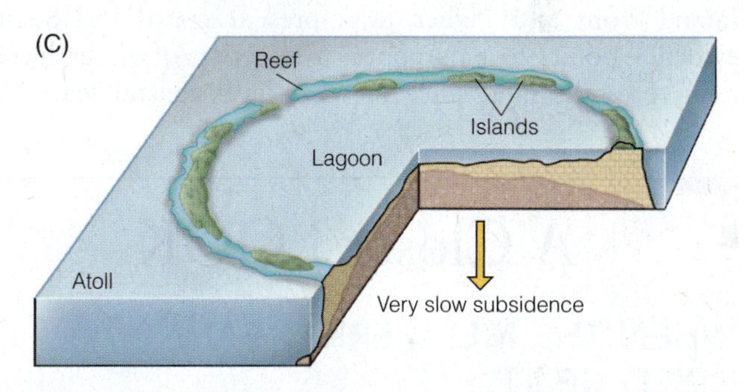

FIGURE 10.29 Island to atoll
Evolution of an atoll from a subsiding volcanic island. (A) Rapid extrusion of lava builds a volcano that begins to subside as the oceanic crust is loaded by the growing volcanic pile. A fringing reef grows upward, keeping pace with subsidence. (B) As the volcano becomes inactive, subsidence continues and the fringing reef becomes a barrier reef, separated from the eroded volcano by a lagoon. (C) With continuing subsidence and upward reef growth, the last remnants of volcanic rock are submerged, leaving an atoll reef surrounding a central lagoon.

of erosion in some cases. The responses to coastal erosion can be divided into three main categories:

1. *Hard stabilization* includes structural responses such as sea walls, jetties, and breakwaters; such structures typically slow erosion in some parts of the system (by breaking the energy of the incoming waves), but may accelerate erosion in the immediately adjacent areas.
2. *Soft stabilization* includes nonstructural approaches such as plantings to stabilize shifting sands, and beach replenishment or beach nourishment, the replacement of sand onto the shoreline by pumping it in from an offshore location.
3. *Retreat* operates on the premise that it is best to limit human interference in coastal systems and that areas where coastal erosion and destruction have been severe are best abandoned altogether.

Today most scientists who deal with coastal erosion prefer retreat or soft stabilization approaches rather than hard stabilization, which can be very expensive, as well as disruptive to the larger shoreline system.

CHANGING SEA LEVELS

On most coasts, the level of the sea is changing with respect to the land. Rapid changes can result from tectonic or isostatic movements of the crust, both local and regional. A much slower change, apparently worldwide and possibly related to global warming (Chapter 13), is currently causing sea levels to rise by about 2.4 mm/year.

The greatest changes in world sea level occur over long time intervals due to changes in water volume as continental glaciers wax and wane, and to changes in ocean-basin volume as lithospheric plates shift position. We refer to such changes as *eustatic*, meaning a rise or fall in sea level that affects the ocean globally. Over the span of a human lifetime, these slow changes are imperceptible, but on geologic timescales they contribute in an important way to the evolution of the world's coasts.

Submergence and Emergence

Whatever their nature, nearly all coasts have experienced **submergence**, a rise of water level relative to the land over the past 10,000 years. This is the result of a worldwide rise of sea level that occurred when glaciers melted away at the end of the last ice age (FIG. 10.30), releasing an enormous volume of water back to the ocean. Because of this submergence, evidence of lower glacial-age sea level is almost universally found seaward of present coastlines out to depths of 100 m or more. By contrast, evidence of sea levels during periods between glaciations is often found inland from and higher than present coastlines. Such evidence points to **emergence**, a lowering of water level relative to the land since the high-level coastal features formed. Repeated cycles of emergence and submergence along the world's coasts are related to the buildup and decay of vast ice-age glacier systems. Three components of the Earth system interact to produce such changes. The global climate system (atmosphere) controls the volume of glaciers (cryosphere) on land, which in turn determines the amount of water residing in the world ocean (hydrosphere). Thus changing ocean-water volume, related to changing climate, will cause sea level to rise or fall.

A change of global sea level that causes coastal submergence or emergence affects all parts of the world ocean at the same time. At the same time, uplift and subsidence of the land may cause local submergence or emergence. For example, vertical tectonic movements along the margins of converging plates have uplifted beaches and tropical reefs to positions far above sea level. Because tectonic movements and eustatic sea-level changes may occur simultaneously, either in the same or opposite directions and at different rates, unraveling the history of sea-level fluctuations along a coast can be a challenging exercise. (See *A Closer Look: When the Mediterranean Dried Up*.)

Sea Ice, Land Ice, and Sea Level

Ice on land has a significant impact on global sea level (as discussed in Chapter 9). The ice sheets covering Antarctica and Greenland comprise a volume of about 28 million km^3

A Closer LOOK

WHEN THE MEDITERRANEAN DRIED UP

The Mediterranean Sea is almost entirely surrounded by land—the name Mediterranean means "in the middle of the land." The only connection with the world ocean is a shallow, 14-km-wide passage called the Strait of Gibraltar that joins the Mediterranean with the Atlantic Ocean (FIG. C10.1). The Mediterranean, bounded to the south by the hot, dry countries of North Africa and to the north by the sunny countries of southern Europe, is famous for its warm water and sunny beaches.

The warmth of the region's climate means that water is constantly removed from the Mediterranean by evaporation. The amount removed by evaporation is greater, by far, than the inflow of water from all of the great inflowing rivers, including the Nile, Rhone, Po, Arno, and (via the outflow of the Black Sea) Danube. A strong current of Atlantic Ocean water flows in through the Strait of Gibraltar, replacing water lost to evaporation. As the cool Atlantic water flows toward the east it heats up, evaporates, and becomes saltier. Eventually, in the eastern Mediterranean, it becomes so dense that it sinks, setting up a counterflow at depth back toward Gibraltar. At the strait itself, the countercurrent of dense, saline water flows out to the Atlantic beneath the inward flow of cool Atlantic water.

What would happen if the tectonic movements of Africa or Europe closed the Strait of Gibraltar? We know the answer to this question because it happened about 5.9 million years ago. The strait closed, Atlantic Ocean water stopped flowing in, and the Mediterranean

FIGURE C10.1 The Mediterranean Sea and the Strait of Gibraltar
Water from the Atlantic Ocean flows through the Strait of Gibraltar into the Mediterranean Sea, returning to the ocean as a deep, dense, saline countercurrent.

(A)

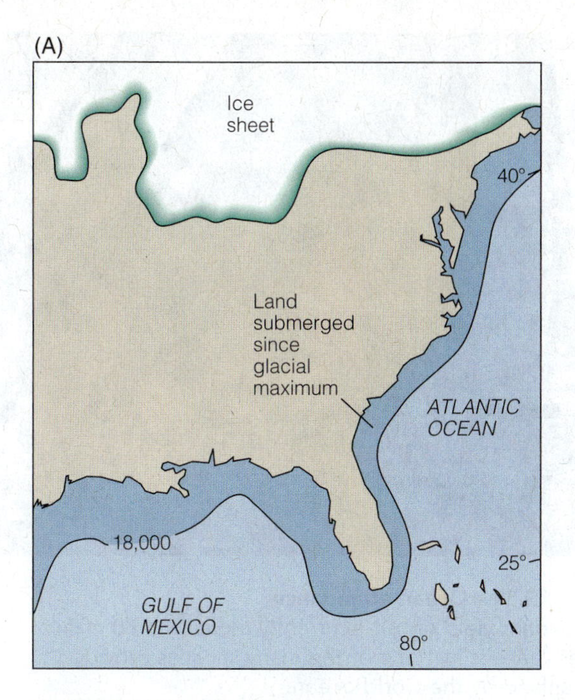

(B)

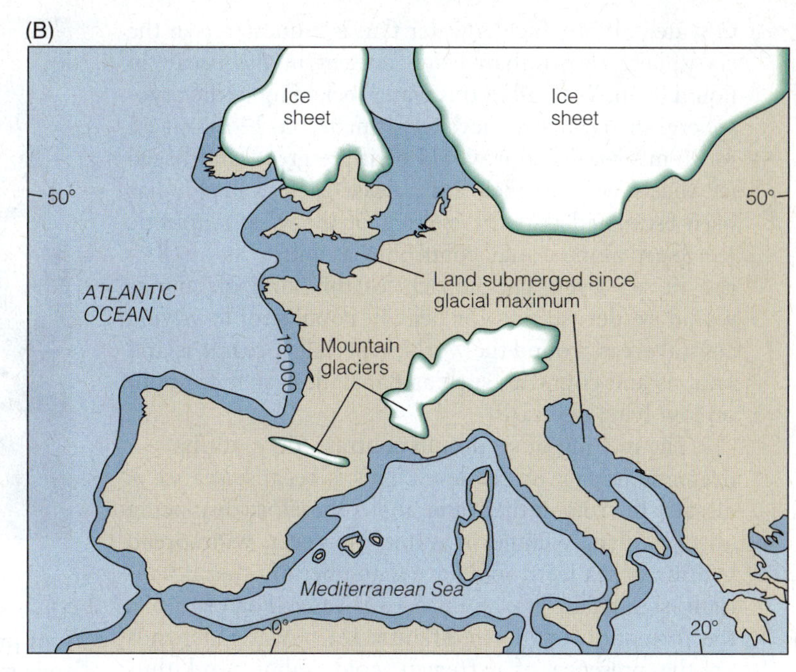

FIGURE 10.30 Coastal submergence
Coastal submergence of eastern North America and western Europe resulted when meltwater from wasting ice sheets returned to the ocean basins at the close of the last glaciation. (A) This is the area of northeastern North America covered by glacier ice during the last glacial maximum, the approximate position of the shoreline at the glacial maximum (18,000 years ago), and coastal areas submerged by the postglacial rise of sea level. (B) This is the area covered by ice sheets in western Europe at the last glacial maximum and by land areas that have been submerged during the postglacial rise of sea level.

became a desert rather than a sea. Within a thousand years, the water evaporated down to three shallow basins located in the east, west, and central portions of the sea (FIG. C10.2). In each basin, the evaporite mineral gypsum ($CaSO_4{\cdot}2H_2O$) was deposited, followed by halite (NaCl). Drilling by a scientific research vessel has revealed that the resulting evaporite salt deposits are up to 3 km thick in places. This is much too thick for a single cycle of evaporation, and the drilling results indicate that the basins were filled by seawater several times, followed each time by a period of drying out and evaporation. There is even evidence of ancient river valleys coming into the basin at levels far below the present sea level; these valleys are now submerged in the Mediterranean, but they point to a time when the water level was much lower.

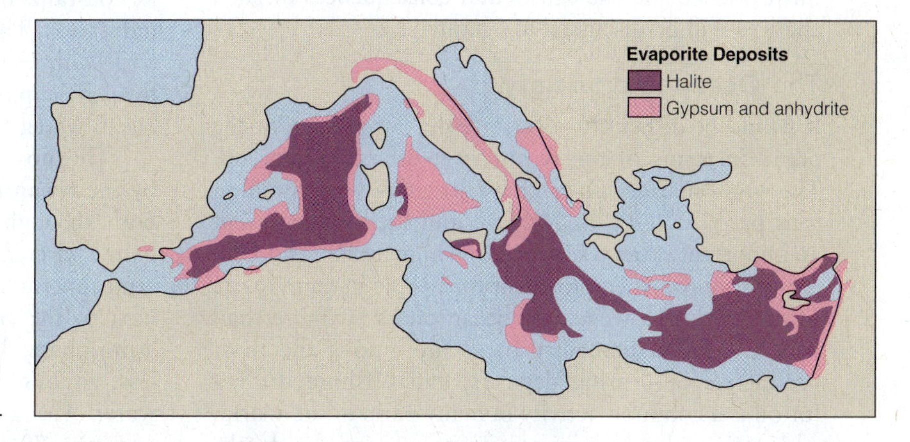

FIGURE C10.2 Mediterranean evaporite deposits
Evaporite deposits formed in three small basins during the Miocene Period when the Mediterranean repeatedly dried up and refilled.

What has not yet been discovered is why the water in the Mediterranean basin repeatedly flooded in and then was shut off. One possibility is that small tectonic movements caused a successive rise and fall of land levels and, therefore, a rise and fall of the land barrier at the Strait of Gibraltar. Perhaps the most likely scenario is that sea level was fluctuating as a result of the waxing and waning of the great Antarctic ice sheet, which began to form about 14 million years ago. The last refilling of the Mediterranean Sea was 5.33 million years ago, at the boundary between the Miocene and Pliocene epochs. The flood of water that came across the barrier and through the narrow strait would have lasted many hundreds of years and would have been larger than Victoria, Niagara, and Iguaçu Falls combined.

Could the Mediterranean dry up again? Both Africa and Europe are tectonically active. Scientists who have studied the case say it is highly likely that in the near-geologic future the Strait of Gibraltar could be closed once again.

of water, all of which is water that is sequestered in the cryosphere rather than being present in the ocean in liquid form. With all of this water locked up in the cryosphere, the global sea level is estimated to be about 65 to 80 m lower than it would be if the great land-based ice sheets were to melt and release their water. It has been estimated that the melting of the West Antarctic Ice Sheet alone could contribute as much as an 8-m rise in sea level. The impacts of this kind of change would be devastating for heavily populated, low-lying coastal areas around the world, especially oceanic island nations and countries such as Bangladesh that are built on low-lying deltas.

The melting of sea ice, in contrast, does not have a dramatic impact on sea level. This is because sea ice is already floating in the water and is therefore displacing an equivalent volume of water. However, widespread melting of sea ice would have a number of other potentially significant effects on ocean water. For example, the formation of North Atlantic Deep Water depends on the presence of extremely cold, saline (and thus dense) water. If NADW formation were suppressed as a result of an influx of warmer, fresher water from melted Arctic sea ice, the thermohaline circulation in the North Atlantic basin could shut down altogether. This would be disastrous for the climate in countries bordering the North Atlantic. There is evidence that dramatic, rapid changes of this type have occurred in the geologic past; this evidence and the causes and consequences of such changes will be discussed in Chapter 13.

FIGURE 10.31 Ocean abundance
The abundant yield of fish—currently more than 80 million tons per year—is just one of the many services provided to humankind by the world ocean.

The Ocean and Society

It would be difficult to overstate the importance of the ocean in terms of the resources it offers humankind. The world marine fish catch is currently over 80 million tons per year (FIG. 10.31). Fish and shellfish represent an important protein source in the human diet, and fishing is a source of employment for people in many parts of the world. However, the ocean offers far more than fish. Coastal sedimentary rocks are among the most productive oil-bearing deposits, and offshore drilling for oil is a common activity in many parts of the world. Although undersea mining is in its infancy, the depths of the ocean are already mined or explored for a variety of mineral resources, including salt, sand, gravel, manganese, copper, iron, tin, nickel, cobalt, and even diamonds. Throughout much of human history, the ocean has provided the means for travel and exploration of the globe. Goods are still transported worldwide by massive oceangoing freighters, and the ocean and its coastal zones are central to the travel and recreation industries. The ocean also has the potential—as yet largely unrealized—to provide a source of usable energy through exploitation of the natural thermal energy of ocean water, tidal energy, and wave energy.

The most important service rendered to humankind by the ocean is the role it plays in climate moderation both through the functioning of the global thermohaline circulation, and by acting as a reservoir for atmospheric carbon dioxide. We turn our attention next to the circulation of air in the atmosphere. The atmosphere, like the ocean, is a fluid, though much less viscous and hence more turbulent than ocean water. The atmosphere and the ocean are in contact over the 70 percent of Earth that is covered by the ocean, and the interactions between them are some of the most important interactions of the Earth system.

SUMMARY

1. Seawater covers 70.8 percent of Earth's surface. Most of the water is contained in three huge interconnected basins—the Pacific, Atlantic, and Indian oceans, which are all connected through the Southern Ocean. Together they comprise the world ocean. The greatest ocean depth is more than 11 km; the average is about 4.5 km. The present volume of seawater is about 1.35 billion km^3, more than half of which resides in the Pacific Ocean.

2. Evidence from sedimentary strata shows that Earth has had liquid water on its surface for at least 4 billion years. Most likely, the world ocean condensed from steam produced during primordial volcanic eruptions.

3. The salinity of seawater normally ranges between 33 and 37 per mil, or about 3.5 percent by weight. The principal elements that contribute to this salinity are sodium and chlorine. These and the other ions responsible for oceanic salinity are derived from the weathering of rock on land, as well as from volcanism, airborne dust, and anthropogenic pollutants.

4. Sea-surface temperatures are strongly related to latitude, with the warmest temperatures in the equatorial latitudes. Surface salinity is also strongly latitude-dependent, affected primarily by evaporation, precipitation, inflow of fresh (river) water, and the freezing and melting of sea ice.

5. Ocean water is vertically stratified as a result of variations in density (colder, saltier water is denser than warmer, fresher water). A thin surface zone consisting of relatively warm, fresh water is separated from the much colder deep zone by the thermocline, in which temperature decreases rapidly with increasing depth. The thermocline approximately coincides with the pycnocline and the halocline, in which density and salinity, respectively, increase rapidly with increasing depth.

6. The uppermost water of the ocean constitutes the pelagic zone; the lowermost water and bottom sediment comprise the benthic zone. Plant life in the ocean is restricted to the photic zone, where sufficient light energy is available for photosynthesis.

7. Calcareous and siliceous ooze are oceanic sediments that consist of the skeletal remains of planktonic (floating or drifting) organisms and benthonic animals and plants. Other important oceanic sediments are lithic sediment and red clay.

8. Oceanic currents are influenced by the Coriolis force, which causes all moving bodies to be deflected to the right in the northern hemisphere and to the left in the southern hemisphere, and by Ekman transport, which results from the interplay between wind-related frictional drag at the surface and the Coriolis force at depth. Near coasts, Ekman transport can lead to upwelling if the net transport is away from the coast, or downwelling if the net transport is toward the coast.

9. Surface ocean currents are broad, slow drifts of surface water set in motion by the frictional drag of the prevailing surface winds. Each major surface current is part of a large subcircular gyre, two each in the Pacific and Atlantic oceans and one in the Indian Ocean. Currents in northern hemisphere gyres circulate in a clockwise direction, whereas those in the southern hemisphere circulate counterclockwise.

10. The northern Indian Ocean exhibits a unique circulation pattern in which the direction of flow changes seasonally with the monsoon. Another important regional wind-current system is El Niño, which occurs every few years when the trade winds and their associated currents weaken or even reverse.

11. Ocean water is organized into major water masses, stratified according to density. The Antarctic Intermediate Water mass (AAIW) is the most extensive such mass. Dense, cold, saline North Atlantic Deep Water (NADW) originates at several sites near the surface of the North Atlantic, flows downward, and spreads southward into the South Atlantic, propelling the global thermohaline circulation, the great "conveyor belt" of the deep-ocean circulation system.

12. Surface ocean waves receive their energy from winds that blow across the water surface. Below the surface, where the influence of the wind is less, a progressive loss of energy occurs with increasing depth. By the wave base, at a depth of $L/2$, water motion is negligible. Near land, as the water becomes shallower, the wave base begins to interact with the bottom. This interference eventually distorts the wave until it collapses and breaks, becoming turbulent, energetic surf, which is responsible for most of the erosional work of waves.

13. Wave refraction occurs when the direction of an incoming wave realigns as it enters shallow water near a coastline. The component of the incoming wave that is perpendicular to the shoreline produces surf; the component that is parallel to the shoreline sets up a longshore current.

14. A tsunami, or seismic sea wave, is a water wave that is generated by a sudden vertical movement of the seafloor as a result of a submarine earthquake, submarine or coastal landslide, or large volcanic eruption.

15. Tides are caused by the gravitational attraction between Earth, Moon, and (to a lesser extent) the Sun. The oceanic tidal bulges created by the tide-raising force on opposite sides of Earth remain essentially stationary, while Earth rotates past them. At most places on the ocean margins, two high tides and two low tides are observed each day as a coast encounters each tidal bulge in turn.

16. Beaches, spits, barrier islands, and marine deltas are common depositional landforms in coastal zones. Estuaries and associated salty to brackish marine wetlands occur where fresh and saltwater mix in the coastal environment. Tropical limestone reefs and atolls are important and varied ecologic environments that consist of vast colonies built by corals and other carbonate-secreting organisms.

17. Sediment inputs and outputs in shoreline systems are locally balanced; if the supply of sediment varies, a change in the shoreline (either erosion or deposition) will become apparent. Human responses to coastal erosion may take the form of hard stabilization, soft stabilization, or retreat.

18. Geologically rapid changes in sea level can result from local or regional tectonic or isostatic movements of the crust. Eustatic (global) changes in sea level are caused by changes in water volume as continental glaciers wax and wane and by changes in ocean-basin volume as lithospheric plates shift position. Nearly all coasts show some evidence of past submergence or emergence.

19. Ice on land has a significant impact on global sea level; if all land ice were to melt, sea levels would likely increase by about 65 to 80 m. The melting of sea ice would not have a dramatic impact on sea levels, but could have other potentially significant effects, such as suppressing the formation of NADW.

20. Throughout human history the ocean has provided abundant resources, including fish and shellfish, transportation, and recreational opportunities, and, more recently, the opportunity to extract oil, minerals resources, and energy. The most important contribution of the ocean, though, is its moderating influence on climate.

IMPORTANT TERMS TO REMEMBER

Antarctic Bottom Water (AABW) *300*
beach *307*
Coriolis force *295*
downwelling *297*
Ekman transport *297*
emergence *312*
estuary *309*
gyre *297*
longshore current *305*
North Atlantic Deep Water (NADW) *300*
phytoplankton *293*
reef *310*
salinity *290*
submergence *312*
surf *303*
thermocline *293*
thermohaline circulation *300*
tide *306*
tsunami *305*
upwelling *297*
wave base *302*
wavelength *302*
wave refraction *304*
world ocean *288*
zooplankton *293*

QUESTIONS FOR REVIEW

1. In terms of plate tectonics, explain why the Pacific Ocean basin is the largest of Earth's three major ocean basins.
2. What geologic evidence indicates that liquid water has been present at Earth's surface for at least 4 billion years?
3. If the quantity of dissolved ions carried to the ocean by streams throughout geologic history far exceeds the known quantity of these substances in modern seawater, why is seawater not far saltier than it is?
4. How and why are the temperature and salinity of surface ocean water related to latitude?
5. What factors cause ocean water to be vertically stratified?
6. Explain why the seasonal temperature range is less at the western coast of North America than it is several hundred kilometers inland.
7. What processes cause the emergence or submergence of coastlines?
8. Explain the existence of large midocean gyres both north and south of the equator.
9. Describe three different types of coastal landforms.
10. What two interacting forces cause Ekman spirals?
11. What contrasting ocean conditions give rise to upwelling and downwelling along continental margins?
12. Why do ships tied to docks experience two high tides and two low tides each day?
13. Describe how North Atlantic Deep Water is produced and explain its role in the global ocean conveyor system.
14. What are the major biotic zones of the ocean?
15. How do coral reefs form and evolve, and why are they areas of such enormous biologic productivity and diversity?

QUESTIONS FOR RESEARCH AND DISCUSSION

1. Imagine yourself a navigational assistant on the H.M.S. *Beagle* as it sets sail, with Charles Darwin aboard as naturalist, from England toward the southeastern coast of South America. Suggest a course that Captain Fitzroy could follow in order to take advantage of winds and currents as you travel from Portsmouth (50.8° N) to Montevideo (35° S). Do some research—what course did they follow?
2. Consider the following hypothesis: The thermohaline circulation system of the ocean has not operated continuously but from time to time has shut down. What might cause the system to shut down, and what evidence might you look for to see if this had happened in the past?
3. Do you live or attend school near a coastline? If possible, arrange for a visit or a field trip to the coastal area. What are the sources of sediment into the shoreline system? Observe the incoming waves. What is their periodicity? How far from shore are they when they first begin to form breakers? From what direction are they approaching the shoreline? Are they refracted upon reaching the shore? Take note of any human interventions designed to control coastal erosion; are they hard stabilization or soft stabilization approaches?

QUESTIONS FOR *A CLOSER LOOK*

1. What is the current geographic setting of the Mediterranean Sea?

2. What happened 5.9 million years ago that caused the Mediterranean Sea to dry up and refill repeatedly? Could it happen again?

3. In *A Closer Look: When the Mediterranean Dried Up* we discuss two main types of evidence showing that the Mediterranean underwent repeated drying and refilling events during the Miocene Period. What are they? What other kinds of evidence might you look for if you were investigating the causes and impacts of this event? Do some research to discover what other evidence scientists have found. (The technical name for this event is the *Messinian Salinity Crisis* or *Messinian Event*.)

QUESTIONS FOR *THE BASICS*

1. What is the Coriolis force, and how does it influence the movement of oceanic currents? Would the Coriolis force cause the water in a sink or toilet to drain out in opposite directions in the southern and northern hemispheres, as Bart Simpson's research once investigated?

2. Why does the Coriolis force vary with latitude? Where is it strongest, and where is it weakest?

3. Figure B10.3A shows a set of transverse waves, in which the displacement (shown by the small arrows) is perpendicular to the direction of propagation. How would you need to change this diagram to illustrate a longitudinal wave? (*Hint:* You may wish to refer back to the discussion of seismic P waves in Chapter 6.)

PART FOUR

The Atmosphere: Earth's Gaseous ENVELOPE

***Atmosphere:** from the Greek words* atmos, *meaning vapor, steam, and* sphaira, *meaning globe, ball, or sphere.*

So far, we have considered the characteristics of Earth as a planet; the processes of the solid Earth; and the unique contributions of water to the Earth system. If a planet is to be habitable, however, yet another important characteristic must be in place: The atmosphere must be breathable, and the most essential ingredient of a breathable atmosphere is oxygen. Like many (though not all) organisms, the human body relies on oxygen, and where oxygen is concerned we rely on the **atmosphere**—the envelope of gas that surrounds our planet—to supply our needs. The atmosphere provides just the right amount of oxygen for life, as we know it, to exist on this planet. This characteristic, along with the dynamic nature of our geosphere, the existence of the biosphere, and the presence of liquid water at the surface, makes Earth unique in the solar system. We consider the history and basic properties of the atmosphere in Chapter 11.

In addition to supporting life by virtue of its chemistry, the atmosphere is an important medium for the storage of solar energy in the Earth system. Stored solar energy is expressed in the atmosphere in the form of winds and atmospheric circulation, which transport heat from the equator to the poles, redistributing it, and in the process driving global and regional weather systems. We look at wind- and weather-related processes in Chapter 12.

The atmosphere serves as an interface between the Earth system and outer space, and both incoming solar energy and outgoing terrestrial energy are modified as they pass through. Thus, the atmosphere is influenced both by processes that are internal to the Earth system and by external processes, including the orbital characteristics of Earth as a planet. These forces combine to shape our climate system, the focus of Chapter 13.

The chapters of Part IV are as follows.

Above the clouds
Earth's atmosphere provides a thin but vitally important protective blanket. Water, seen here in the form of clouds, is an important atmospheric constituent.

CHAPTER 11

The Atmosphere

OVERVIEW

In this chapter we:

- Learn how Earth's atmosphere has changed over the course of Earth history
- Introduce the composition and structure of the atmosphere
- Examine the role and behavior of moisture in the atmosphere
- Describe cloud types and cloud-forming processes
- Consider the causes and impacts of recent atmospheric changes

Earth's atmosphere
In this spectacular photograph, taken from the International Space Station as it orbited over the Pacific Ocean on July 21, 2003, you can clearly see the troposphere (the lower layer of the atmosphere) as a thin blue band above the curved horizon. Anvil-shaped clouds in the foreground are thunderstorms. Their unique shape is caused by the fact that warm air in the clouds cannot rise past the upper boundary of the troposphere.

Earth is surrounded by air. There is a difference between *atmosphere* and *air*, although we commonly use the words interchangeably. An **atmosphere** is a gaseous envelope that surrounds a planet or any other celestial body; Earth is one planet among many that have an atmosphere. **Air**, by contrast, is the invisible, odorless mixture of gases and suspended particles that surrounds one special planet—Earth. In other words, Earth's atmosphere is made of air.

As organisms that depend on oxygen for our survival, we can count ourselves lucky that the particular composition of air makes oxygen readily available to us. Although the human body has some capacity to adjust for a change in the amount of oxygen available in air, the range of adjustment is limited. We know about the lower limit of this range from studies of people who live, work, or visit at high altitude, and we know about the higher limit from studies on deep-sea divers and people in hospitals who have been given extra oxygen. Such studies show that to be habitable by humans, a planet's atmosphere must have an oxygen level ranging from 40 percent above to 44 percent below the level found in today's air. Other organisms are more tolerant of variations in the oxygen content of the atmosphere, or even the absence of oxygen; others are less tolerant.

In addition to supporting life chemically, the atmosphere provides many other important, life-supporting and Earth-modifying services. It stores moisture and solar energy, transporting it globally in the form of heat, and creating wind and weather systems in the process. It moves Earth materials from place to place; it generates ocean currents. It also protects humans and other organisms from life-threatening radiation. In this chapter we look at all of these basic characteristics and functions of Earth's atmosphere.

THE HABITABLE PLANET

Earth's atmosphere hasn't always had the same chemical composition that it has today, and it hasn't always supported organisms that, like humans, rely on breathable oxygen for their survival. As far as we can tell from the geologic record, the oxygen content of Earth's atmosphere has varied, but it has not moved outside the habitable range for at least the past several hundred million years. Let's begin our study of the atmosphere with a brief look at how it has changed over the course of Earth history, and how it has evolved into the composition that it has today.

Past Atmospheres of Earth

As you learned in Chapter 4, Earth (along with the other inner planets Mercury, Venus, and Mars) accreted from material in the solar nebula that included dense, rocky and metallic constituents, as well as lighter-weight or *volatile* constituents. Some of the volatile materials surrounded the planet, clinging to it and forming a gaseous envelope that was Earth's *primordial* or *primary atmosphere*. Most of this earliest atmosphere was stripped away by strong solar winds early in solar system history. Luckily for us, Earth still contained abundant volatile constituents in its interior, which were released little by little to the surface by volcanic outgassing. Volcanic venting of gases continues today, although it was more active early in Earth history. Over time, volcanic gases collected to form Earth's *secondary atmosphere*.

To retain an atmosphere and prevent the lightweight, gaseous elements from simply floating away into outer space, a planet must be either massive (so that its gravitational attraction is strong) or cold (so that the molecules in the gases are not moving energetically enough to escape), or both. The giant outer planets, Jupiter, Saturn, Uranus, and Neptune, are massive as well as cold; hence they have retained very thick atmospheres (see Figure 4.12B). In comparison, Earth is not a very massive planet, and early Earth was hot. It would have been difficult for the planet to retain an atmosphere under those conditions. However, Earth had cooled somewhat by the time volcanoes had emitted enough volatile material to form Earth's secondary atmosphere; this made it possible to retain the gaseous envelope that became our atmosphere.

Chemical Evolution of the Atmosphere

The early secondary atmosphere was not like the atmosphere we have today. For one thing, Earth was still quite hot during the Hadean Eon (the period from the origin of the planet to 3.8 billion years ago; the name itself comes from the Greek *Hades*, meaning hell). The composition of the Hadean atmosphere reflected the composition of volcanic emissions, which were (and still are) dominated by water vapor—as much as 97 percent by volume—with varying amounts of methane (CH_4), hydrogen (H_2), nitrogen (N_2), and water vapor (H_2O), carbon dioxide (CO_2), and noble gases, such as argon (Ar).

Many of these are *radiatively active* or *greenhouse gases*, so they contributed to even greater warming of the surface. It was too hot for liquid water to be sustained, but water vapor in the atmosphere combined chemically with carbon, sulfur, nitrogen, and hydrogen compounds in the atmosphere to form acids. The result was that Earth's earliest precipitation was acidic rain. There was no "free" oxygen available; that is, the small amount of oxygen that was present in Earth's early secondary atmosphere was immediately bound up into compounds such as carbon dioxide (CO_2), leaving no molecular oxygen (O_2) or ozone (O_3) available. The early atmosphere also was dense, and atmospheric pressure at the surface was very high.

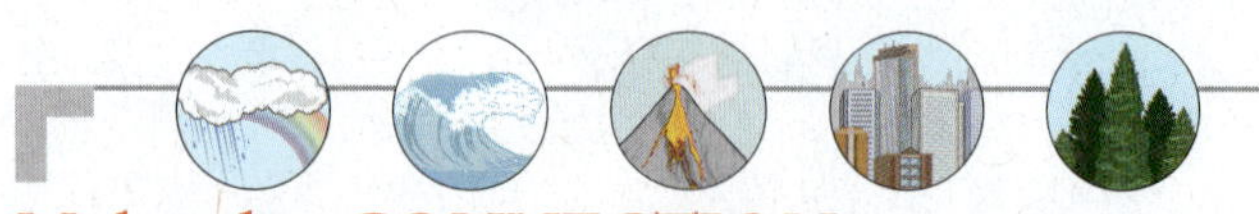

Make the CONNECTION

If the earliest rain on Earth was acidic, what impact might this have had on the first patches of crustal rock to solidify at the surface? What impacts might it have had on the first pools of water to accumulate at the surface, or the first life forms?

FIGURE 11.1 Titan, or Hadean Earth? This painting is an artist's version of the surface of Saturn's largest moon, Titan, with pools of hydrocarbons, an icy and rocky landscape, and a dense atmosphere that may be similar to the Hadean atmosphere of Earth.

In short, we would have found Earth's early atmosphere to be entirely inhospitable to life, as we know it today. In the Hadean Eon, Earth's surface may have resembled that of Titan, the largest moon of Saturn, which is one important reason why scientists are interested in exploring Titan (FIG. 11.1). The atmosphere and, in concert, the hydrosphere of Earth have changed dramatically over geologic time (FIG. 11.2). Perhaps surprisingly, life itself was the main contributor to the chemical changes that have occurred in Earth's atmosphere since then. The two main contributions of life to the chemical evolution of the atmosphere were the addition of oxygen and the removal of carbon dioxide.

Addition of Oxygen

The total volume of volcanic gases released over the past 4 billion years or so accounts for the present composition of the atmosphere, with one extremely important exception: oxygen. Earth had virtually no oxygen in its atmosphere more than 4 billion years ago, but the atmosphere is now approximately 21 percent oxygen. What caused this change?

Traces of oxygen were probably generated in the early atmosphere through the breakdown of water molecules into oxygen and hydrogen by ultraviolet light (a process called *photodissociation*). This is an important process, but it doesn't even come close to accounting for the present high levels of oxygen in the atmosphere. Almost all of the free oxygen now in the atmosphere originated through *photosynthesis*, the process whereby plants, the *primary producers* of Earth's biosphere, utilize light energy to produce carbohydrates from carbon dioxide and water, releasing oxygen as a by-product. We will explore photosynthesis and the role of primary producers in Chapter 15.

The oxygen produced by early photosynthetic organisms immediately combined chemically with other elements, especially iron, to make oxygen-bearing compounds. It took a long time for plants to produce sufficient oxygen that would allow free molecular oxygen (O_2) to begin to accumulate in the atmosphere. The transition to an oxygenated atmosphere happened around 2.5 to 1.8 billion

FIGURE 11.2 Composition of Earth's atmosphere over time The composition of Earth's atmosphere has changed over time. Changes in the oxygen and carbon dioxide contents, in particular, were driven by photosynthesis and other life processes.

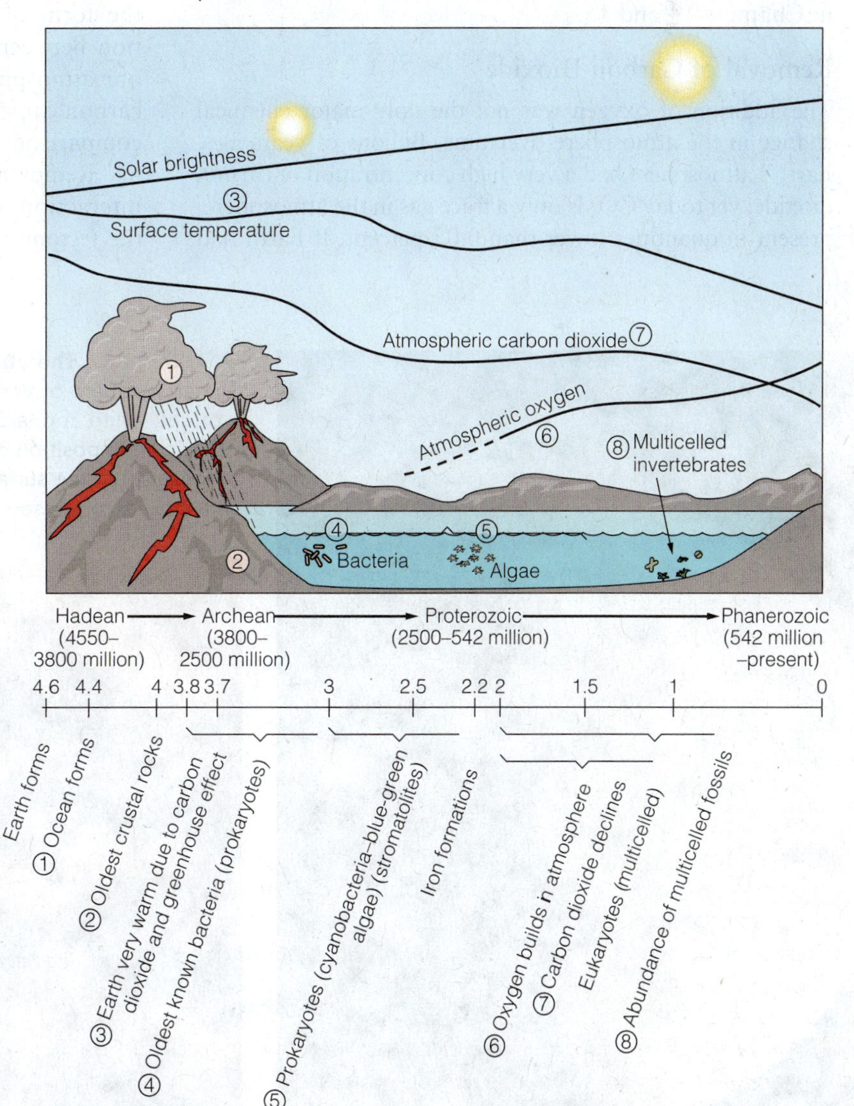

years ago, and this transition is recorded in the alternating black (*reduced*, oxygen-poor) and red (*oxidized*, oxygen-rich) *banded iron formation* sediments that accumulated in the ocean basins. (See *A Closer Look: Banded Iron Formations*, in Chapter 7.) Along with the buildup of molecular oxygen came an eventual increase in ozone (O_3) levels in the atmosphere. Because ozone absorbs harmful ultraviolet radiation, its buildup made it possible for life to flourish in shallow water and finally on land. This critical stage in the evolution of the atmosphere was reached between 1100 and 542 million years ago. Interestingly, the fossil record shows an explosive diversification of life forms 542 million years ago, at the beginning of the Phanerozoic Eon.

Since then, oxygen levels in the atmosphere have fluctuated somewhat. Over the past 200 million years, the concentration of oxygen has risen from 10 percent to as much as 25 percent, before settling (probably not permanently) at the current value of 21 percent. These fluctuations have had a major influence on life. For example, the overall increase in oxygen has benefited humans and other mammals, which are voracious consumers of oxygen to fuel their high-energy, warm-blooded metabolism. We will explore this and other impacts of the atmosphere on life in Chapters 14 and 15.

Removal of Carbon Dioxide

The addition of oxygen was not the only major chemical change in the atmosphere over time. Billions of years ago, Earth's atmosphere had a very high concentration of carbon dioxide; yet today CO_2 is only a trace gas in the atmosphere, present in quantities lower than 0.05 percent. If Earth had retained its CO_2-rich atmosphere, the planet would have ended up much more like Venus, which today has an atmosphere that is 97 percent carbon dioxide (FIG. 11.3). Earth and Venus have many characteristics in common, including size, overall composition, and location in the solar system (inner planets, with adjacent orbits, though Venus is closer to the Sun than Earth). But Venus has a very different atmosphere from Earth's, dominated by CO_2. This contributes to a runaway greenhouse effect on Venus and a surface temperature hot enough to melt lead, leaving no possibility for liquid water to persist at the surface. What process was responsible for removing most of the carbon from Earth's atmosphere, but not from that of Venus? Again, life played the central role, in concert with the tectonic cycle and the rock cycle.

In addition to generating oxygen through photosynthesis in the early atmosphere, the emerging biosphere had a profound impact on the carbon cycle. This is because the oxygen cycle and the carbon cycle are interconnected: Plants produce carbon-based organic matter, removing carbon from the atmosphere and releasing oxygen; in turn, the decay of carbon-based organic matter uses up oxygen and releases carbon back to the atmosphere in the form of carbon dioxide. This explains the connection between life and the oxygen and carbon dioxide in our atmosphere, but it still doesn't explain why so much carbon came to be "missing" from Earth's atmosphere, in comparison to our sister planet, Venus.

Removing carbon from the atmosphere requires the interaction of life processes with the rock cycle and the tectonic cycle. Recall that marine organisms construct

FIGURE 11.3 The atmosphere of Venus
The atmosphere of Venus is thick and cloudy (A), giving the planet a bright, white appearance when viewed through a telescope. The CO_2-rich composition causes a runaway greenhouse effect, with the result that the surface temperature is hot enough to melt lead, with clouds composed not of water but of sulfuric acid (B).

their shells primarily of calcium carbonate ($CaCO_3$), providing a reservoir for carbon dioxide (because $CaCO_3 = CaO + CO_2$). When these organisms die, their shells are buried by seafloor sediments and are eventually transformed into limestone by the process of lithification. Limestone acts as a long-term reservoir for carbon in the carbon cycle. Carbon-based organic matter also accumulates as sediment along continental margins, where it may be buried before it has fully decomposed. When limestone, other carbon-bearing sedimentary rocks, and organic sediments are buried, the carbon they contain is effectively isolated from further contact with the atmosphere (and hydrosphere) on a long-term basis. This process is called *carbon sequestration*. In contrast to Earth, on Venus there is no mechanism by which carbon sequestration occurs, so it remains in the atmosphere.

Still, how do the carbon-bearing sediments and rocks become buried? That is where the tectonic cycle enters the picture. When new ocean basins open up as a result of plate divergence and seafloor spreading, new continental margins are created. This happened, for example, as a result of the breakup of the supercontinent Pangaea at the boundary between the Triassic and Jurassic Periods, which started the opening of the Atlantic Ocean. Vast amounts of sediment accumulated on the new continental margins, accelerating the burial rate of organic matter. Through the process of subduction, as well, the tectonic cycle returns limestone and seafloor sediment to the mantle, removing its carbon from near-surface biogeochemical cycling, and isolating it from further contact with the atmosphere and hydrosphere.

The incorporation of carbon into organic matter, the formation of carbon-bearing rock and sediment, and the burial of organic matter have continued throughout Earth's long history. Thus, the geosphere, hydrosphere, and biosphere all interacted to produce and modify the atmosphere that we breathe today. If all the carbon dioxide currently stored in limestone, buried organic sediment, and other carbon-bearing sedimentary rocks were released, there would be as much CO_2 in Earth's atmosphere as in the atmosphere of Venus, and life as we know it would be impossible.

COMPOSITION AND STRUCTURE OF OUR ATMOSPHERE

We now turn our attention to the composition and structure of Earth's present-day atmosphere (**FIG. 11.4**). It may seem strange to say "structure" when we discuss a gaseous medium, but measurements show that, with increasing altitude, there are distinct variations in such things as the composition, temperature, pressure, and moisture levels in the atmosphere. We humans live at the bottom of the atmosphere, so most of the variations take place far above our heads. What is known about the upper reaches of the atmosphere, like our knowledge of the geosphere beneath our feet, comes mainly from instrument probes of one kind or another.

FIGURE 11.4 Earth and sky
Earth's atmosphere is the interface between our planet and outer space. Atmospheric processes are closely linked to processes in the geosphere and biosphere. Here trees in the Swiss Alps can be seen poking through a low-lying cloud layer.

Composition

Air—the particular material that makes up Earth's atmosphere—is a complex mixture of gases and tiny suspended particles. Because air pressure decreases with altitude, the amount of air per unit volume (that is, the density) also varies with altitude. In order to separate changes due to composition from those due to density, the composition of air is always discussed in terms of the *relative* rather than the absolute amounts of the different constituents present. The relative composition, it turns out, varies from place to place on the surface of Earth and even from time to time in the same place (**FIG. 11.5**). The two main reasons for the variations are the presence of aerosols and the presence of water vapor, both of which vary widely:

1. **Aerosols** are extremely tiny liquid droplets or solid particles, so small that they remain suspended in air. Aerosols are everywhere in the atmosphere, particularly in air that is nearest to the ground (as discussed in *A Closer Look: Aerosols in the Atmosphere*).
2. Water vapor is always present in the atmosphere because Earth has a hydrosphere from which water evaporates, moving freely into the atmosphere. The amount of water vapor in air, for which we use the term **humidity**, is quite variable. On a hot, humid day in the

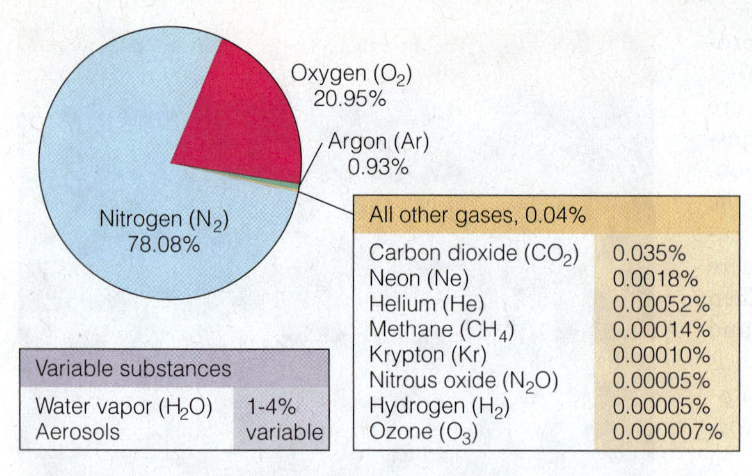

Oxygen (O_2) 20.95%
Argon (Ar) 0.93%
Nitrogen (N_2) 78.08%
All other gases, 0.04%
Carbon dioxide (CO_2) 0.035%
Neon (Ne) 0.0018%
Helium (He) 0.00052%
Methane (CH_4) 0.00014%
Krypton (Kr) 0.00010%
Nitrous oxide (N_2O) 0.00005%
Hydrogen (H_2) 0.00005%
Ozone (O_3) 0.000007%
Variable substances
Water vapor (H_2O) 1-4%
Aerosols variable

FIGURE 11.5 What air is made of
Air contains two substances whose concentration varies from place to place and day to day: water vapor and aerosols. The rest of the atmosphere consists primarily of nitrogen and oxygen, with small amounts of other gases.

tropics, as much as 4 percent of the air by volume may be water vapor, whereas on a crisp, cold day, less than 0.3 percent water vapor may be present.

Because the water vapor and aerosol contents of air vary widely, the relative amounts of the remaining gases in air are reported on a dry (meaning "free of water vapor"), aerosol-free basis. Once these two variable constituents are removed, the relative proportions of the remaining gases in air turn out to be essentially constant regardless of altitude. As shown in Figure 11.5 three gases—nitrogen, oxygen, and argon—make up 99.96 percent of dry air by volume.

The relative amounts of the remaining gases are very small, but these minor gases are profoundly important for life on Earth because they act both as a warming blanket and as a shield from deadly ultraviolet radiation. Carbon dioxide, water vapor, methane, ozone, and nitrous oxide are the minor gases that create Earth's life-maintaining blanket.

A Closer LOOK

AEROSOLS IN THE ATMOSPHERE

Aerosols are everywhere in the atmosphere. They have important impacts on human health, on a number of environmental processes, and even on the energy balance of the planet. Aerosols can consist of either liquid droplets or solid particles; their main characteristic is that they are so tiny that they remain aloft in the atmosphere very easily. By definition, aerosol particles are tiny—most are no more than 1 micrometer (one millionth of a meter, or 0.000001 m) in diameter, and some are considerably smaller than this. (Technically, the term *aerosol* refers to the suspension or solution itself, including the tiny particles or droplets of the aerosol, as well as the gaseous medium in which they are suspended.) These particles are too tiny to be seen by the eye (FIG. C11.1A), although we do see the results of the suspension of billions of tiny particles and droplets as smoke, fog, or clouds.

Aerosols come from both natural and anthropogenic sources. Common liquid aerosols, such as fog and clouds, are familiar to all of us. Solid aerosol particles, sometimes called *particulates*, include volcanic ash, smoke from forest fires, blown sea salt, blown dust and loess, and pollen. The tiny ice crystals that nucleate in high-level clouds at very cold temperatures are also solid aerosol particles. Haze is typically a combination of fog and smoke, often exacerbated by pollutants. Most anthropogenic particulates are pollutants that originate from industrial activities, transportation, the burning of fossil fuels, and the generation of electric power (FIG. C11.1B). Waste incinerators and the burning of wood and agricultural wastes also contribute particulates to the atmosphere. When particulate air pollution combines with sunlight in the atmosphere, it is known by the familiar name *smog*.

Because aerosols are so common and widespread, we breathe them in all the time. Fortunately, the human respiratory system is designed to prevent them from causing harm to our lungs in most common circumstances. Aerosols, both natural and anthropogenic, sometimes cause diseases or immune reactions in humans because they pass so easily into the lung where they can interact directly with the lung's sensitive tissues. For example, if you inhale air that contains pollen, your body's immune system may react by generating mucus and causing sneezing; this is an *allergic reaction*, and the pollen is called an *allergen*.

Aerosols play a variety of roles in the Earth system. For example, sea spray is one of the main ways that sodium is transferred to the land from the ocean, where it resides in the form of sodium chloride (NaCl, or salt, the mineral halite). Aerosols are extremely important in the atmosphere because they provide nucleation sites for water droplets and ice crystals. They are also very effective at scattering light, which means that they have an overall cooling effect on the climate. The concentration of aerosol droplets in clouds determines their ability to reflect sunlight. The greater the number of aerosol particles, the more water droplets (or ice crystals) will nucleate, and the smaller the individual

These five greenhouse gases absorb infrared radiation, creating a near-surface environment that is warmer than it would be if they were absent from the atmosphere (FIG. 11.6). Changes in the amounts of greenhouse gases in the atmosphere lead to changes in the heat-absorbing capacity of the atmosphere, and therefore to changes in temperature, and eventually changes in climate. In Chapter 13 there is a more detailed discussion of greenhouse gases and the role they play in global climate.

Minor gases also protect humans and other living creatures from ultraviolet radiation, the short-wavelength, high-energy part of the solar spectrum. Most of this deadly radiation is prevented from reaching Earth's surface as a result of absorption by three forms of oxygen gas: O, O_2, and O_3 (ozone) (FIG. 11.7). The most important of the three shielding gases is ozone. Even though it occurs in only trace amounts in the atmosphere, it is able to absorb the most lethal kind of ultraviolet rays. Absorption is one of several processes that can happen when sunlight passes through the atmosphere and interacts with molecules of gas and aerosols.

At very great altitudes, 80 km and higher, the composition of dry, aerosol-free air differs a little from what it is at Earth's surface; it is depleted in the heavier gases, such as argon and neon, and enriched in the lighter gases such as hydrogen and helium. For most purposes of discussion where the Earth system is concerned, however, we don't have to be worried about such high-altitude changes in the composition because weather- and climate-related processes happen in the lower atmosphere, where the relative compositions do not vary.

Temperature

Two things energize the atmosphere: the Sun's energy and Earth's rotation, which operate in concert. The Sun's energy reaches Earth in the form of electromagnetic radiation. Solar radiation warms the atmosphere and is the energy source responsible for clouds, rain, snowstorms, wind, and much of the local weather. The tilt of Earth's axis of rotation is responsible for the annual seasons because it allows more or less of the Sun's radiation to reach and warm the

droplets will be (because the total amount of water in the cloud does not differ). Clouds with more and smaller water droplets are whiter, and they reflect sunlight more effectively, promoting cooling.

Clouds of sulfur-bearing aerosols from large, explosive volcanic eruptions also can have a noticeable cooling effect on the global climate, sometimes lasting for years.

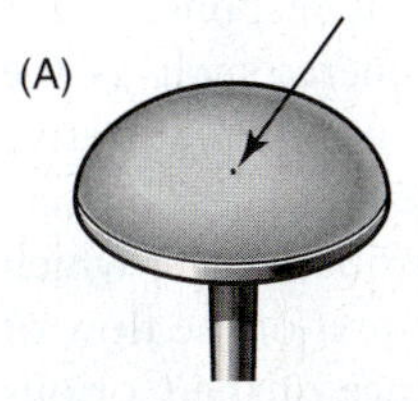

FIGURE C11.1 Aerosols
(A) Aerosol particles are extremely fine—less than 1 micrometer in diameter by definition—and are thus too small to be seen by the eye. (B) Although we can't discern the individual particles, we can see the results when billions of aerosol particles gather together to form fog, clouds, haze, smog, or smoke, as seen here in a photo of smoke from a forest fire in California in 1991. (C) This is a false-color electron micrograph of fly ash, a common type of solid aerosol particle. Fly ash comes from power plants and other sources of combustion. Individual spheres range in size from about 0.5 to 100 micrometers.

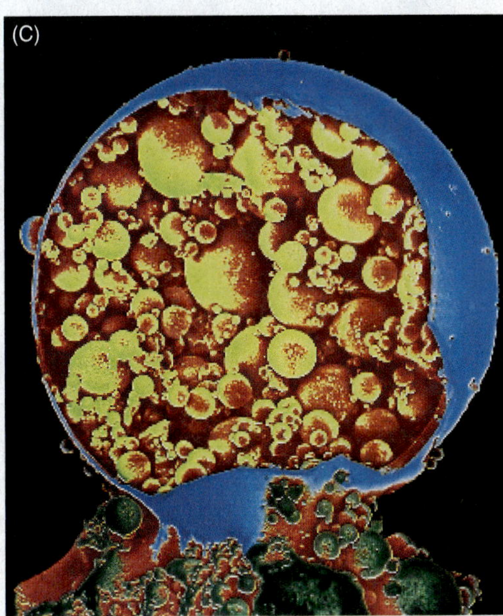

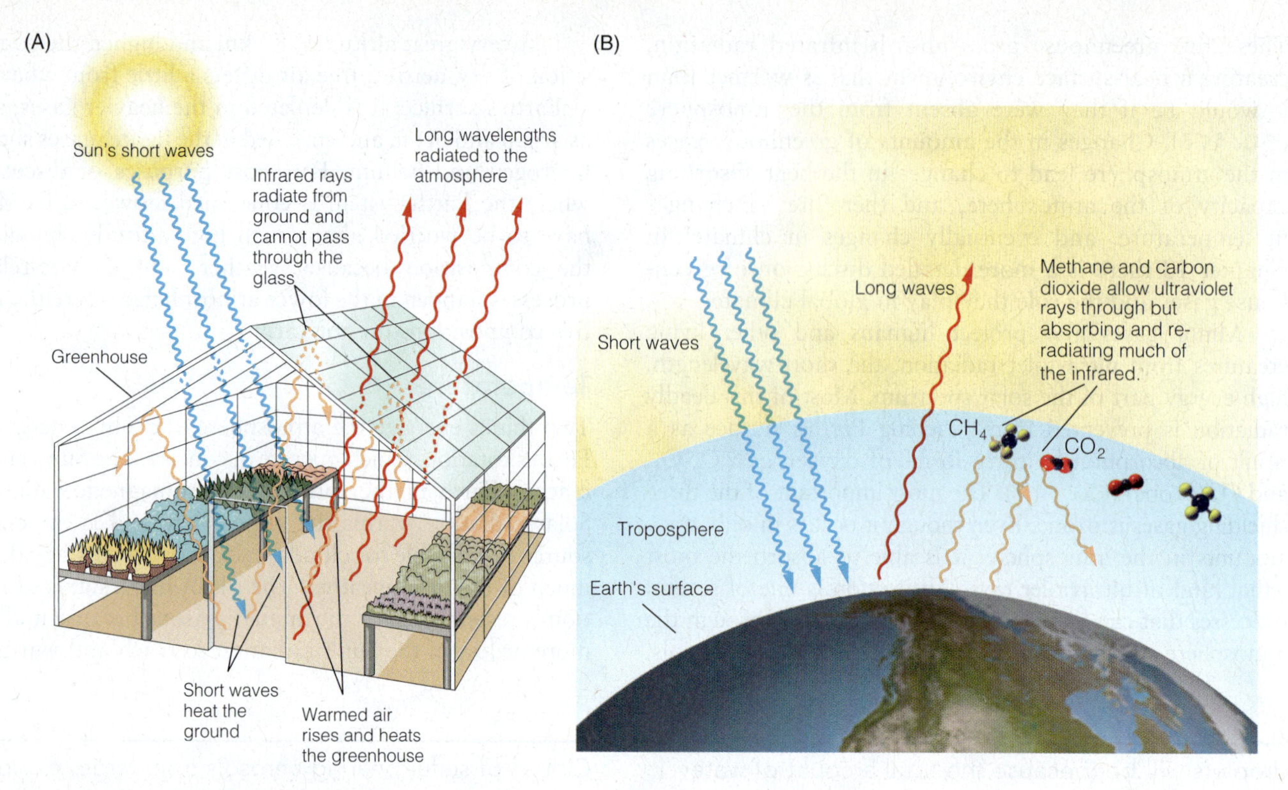

FIGURE 11.6 How a greenhouse works
The glass in a real greenhouse (A) and certain "greenhouse gases" in the troposphere (B) work in approximately the same way.

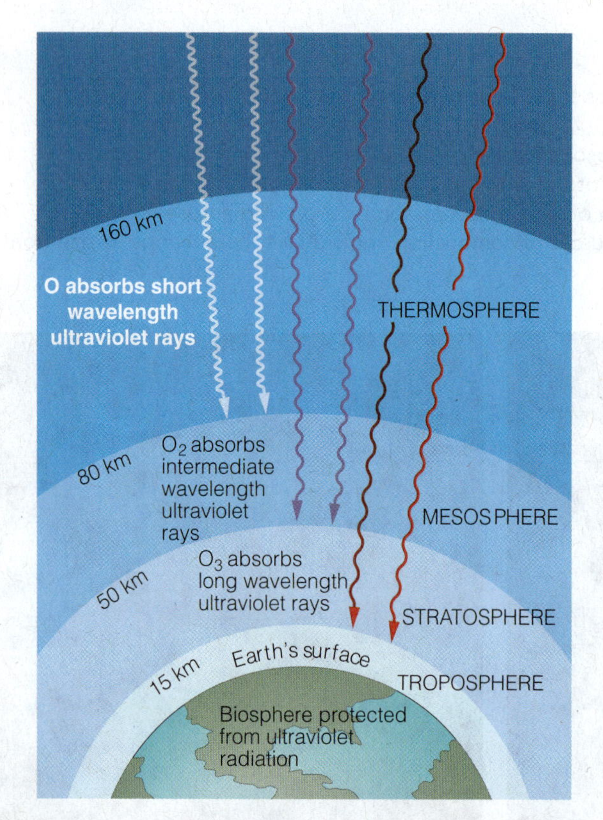

FIGURE 11.7 A shield against harmful radiation
Ultraviolet radiation coming from the Sun can be harmful or lethal; generally speaking, the shorter the wavelength, the more harmful the radiation. Fortunately, the atmosphere protects us from almost all these rays because they are absorbed by three kinds of oxygen—O, O_2, and O_3 (ozone).

atmosphere. Earth's rotation on its axis is responsible for large-scale flow patterns in the atmosphere, such as the midlatitude west-to-east movement of weather patterns, the jet stream, and global wind systems. These effects happen principally as a result of the Coriolis force, which was introduced in Chapter 10 in the context of the flow of ocean water. We will explore the influence of the Coriolis force on wind and weather systems in Chapter 13.

Temperature Versus Heat

Temperature is the most important and most familiar variable used to define the state of the atmosphere. The human body is sensitive to changes in temperature as small as 1°C; as a result, everyone is aware that temperature varies from hour to hour, from place to place, from day to night, and from season to season.

Before we discuss temperature variations in the atmosphere, it is important to establish the difference between heat and temperature. As discussed in Chapter 2, the definition of *heat* and *heat energy* (the two terms mean exactly the same thing) is the total kinetic energy (energy of motion) of all the atoms in a substance. Not all the atoms in a given sample move with the same speed, so there is a range of kinetic energies among them. *Temperature*, on the other hand, is a measure of the *average* kinetic energy of all the atoms in a body. Two bodies of the same substance, such as a cup of water and a pail of water, may have the same temperature, say 25°C, but there are so many more water molecules in the pail than in the cup that the pail has far more heat energy. If the temperature of the cup of water were raised from 25°C to boiling (100°C), the average speed of the atoms in the water molecules would increase; however, the heat energy in the lower-temperature but much larger pail would probably still exceed the heat energy in the smaller, higher-temperature cup. The reason is that the total heat energy of a large number of slow-moving atoms may well exceed the total heat energy of a small number of fast-moving atoms.

Insolation

The atmosphere gets its heat energy from the Sun. As discussed in Chapter 2, the flux of energy coming in from the Sun is 1370 W/m². This is the energy flux that would be measured by a satellite orbiting Earth outside the atmosphere. However, the *insolation*, which is the energy that actually reaches a surface (the surface of Earth, in this case), is considerably less than 1370 W/m². Insolation is less than 1370 W/m² for three principal reasons. The first two reasons concern the atmosphere, which reflects some of the incoming radiation back into space, and then acts as a filter by absorbing some more of the radiation as it passes through. The third reason is the shape of Earth. Even if Earth were devoid of an atmosphere, there is only one place on it that would receive 1370 W/m², and that is the spot where the Sun is directly overhead (as shown in Fig. 2.5). At all other places, because of Earth's curvature, the incoming 1370 W would be spread over an area larger than a square meter.

Temperature Profile of the Atmosphere

The way the temperature of the atmosphere changes with altitude is shown in FIGURE 11.8. There are four principal thermal layers in the atmosphere, separated by boundaries called **pauses**. The four layers, from bottom to top, are the troposphere, stratosphere, mesosphere, and thermosphere.

Troposphere

The bottommost layer of the atmosphere is the **troposphere**, which extends from the surface to variable altitudes of 10 to 16 km. The troposphere is named from the Greek, *tropos*, meaning to change or to turn, because it is endlessly convecting, with warm ground-level air rising upward and colder air from above sinking downward to take its place. Temperature in the troposphere decreases with altitude. This is because absorption of reradiated long-wavelength infrared rays is most effective at the bottom of the atmosphere where the air is most dense, and because air at the bottom is warmed by energy that is reradiated by the ground and the ocean. Most of our weather is a consequence of thermal motion of air in the troposphere.

The upper boundary of the troposphere is the *tropopause*. Because more heat per unit area reaches Earth's surface in the tropics than at the poles, the tropopause is 16 km high at the equator but only 10 km or less at the poles. The tropopause height does not change smoothly from the equator to the poles; rather, it declines very gently from the equator to about latitude 40° N and S, then drops sharply to about 10 km and continues near that height to the poles (FIG. 11.9).

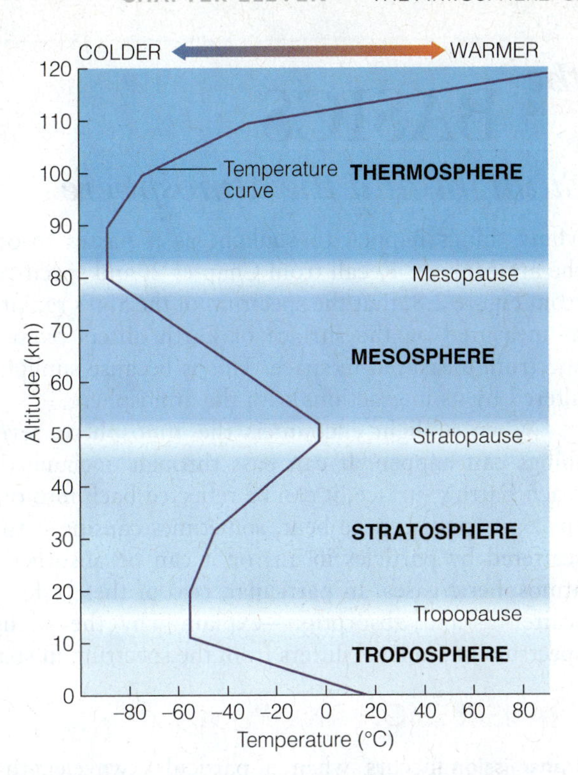

FIGURE 11.8 Temperature profile of the atmosphere Temperature varies with altitude in the atmosphere. In the lowest level, the troposphere, temperature drops rapidly with increasing altitude. In the next layer, the stratosphere, the reverse is true. Two more reversals occur in the mesosphere and the thermosphere. Upwards in the thermosphere the atmosphere becomes more and more tenuous, until it eventually merges into outer space in the exosphere.

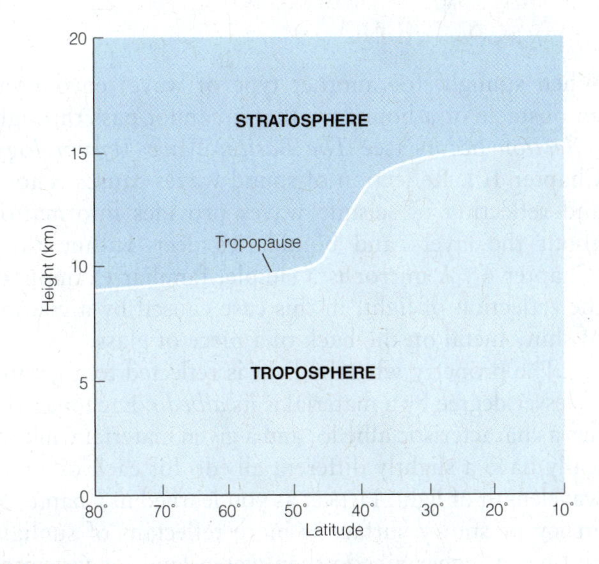

FIGURE 11.9 Altitude of the tropopause The altitude of the tropopause varies with latitude. It is high from the equator to about 40° latitude, where it drops precipitously, continuing at this lower level and declining gently toward the poles. The precipitous drop at 40° latitude facilitates the development of jet streams.

the BASICS

Sunlight and the Atmosphere

Many things happen to sunlight as it passes through the atmosphere. Recall from Chapter 2, and specifically from Figure 2.8, that the spectrum of the Sun's radiation as measured on the surface of Earth differs from the spectrum measured in space. This is because sunlight is altered by its interactions with the atmosphere.

When sunlight encounters the atmosphere, several things can happen: It can pass through unchanged to reach Earth's surface; it can be reflected back into outer space; its path can be bent, sometimes causing it to be scattered by particles in air; or it can be absorbed by atmospheric gases. In particular, two of these effects—scattering and absorption—explain why the sunlight spectrum at sea level differs from the spectrum in space.

TRANSMISSION AND WINDOWS

Transmission occurs when a particular wavelength of sunlight passes through the atmosphere unimpeded and unaltered, eventually reaching Earth's surface. Sunlight that is transmitted through Earth's atmosphere is shown as the sea level spectrum in Figure 2.8. This spectrum has a significant peak in the visible-light portion of the spectrum; in other words, visible light is transmitted easily through the atmosphere because there are no atmospheric gases that absorb light in these particular wavelengths. We therefore say that there is an **atmospheric window** for visible light to pass through (**FIG. B11.1**).

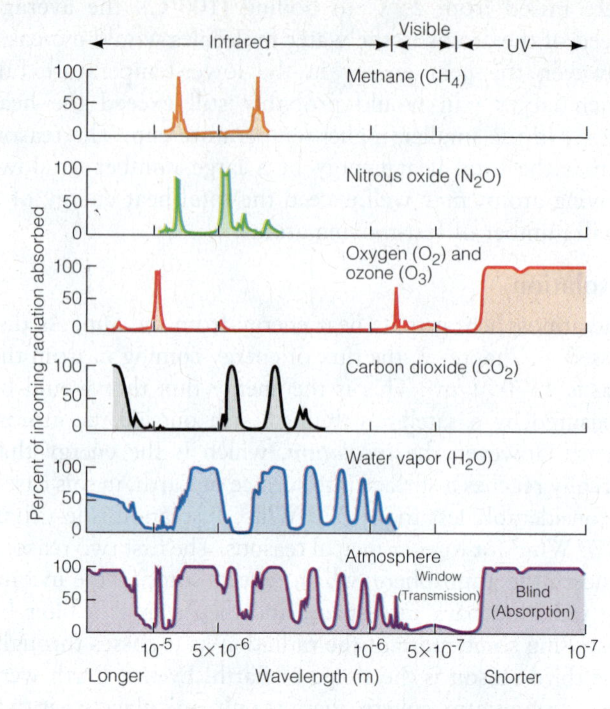

FIGURE B11.1 Atmospheric windows and blinds
Some wavelengths of electromagnetic radiation are transmitted easily through the atmosphere; there is an atmospheric window for those wavelengths of light. For example, there is an atmospheric window for light in the visible portion of the spectrum. Other wavelengths of electromagnetic radiation do not pass through the atmosphere because they are absorbed by atmospheric gases; there is an atmospheric window for those wavelengths of light. For example, ozone (along with some other gases) absorbs electromagnetic radiation in the ultraviolet part of the spectrum, creating an atmospheric blind.

REFLECTION, REFRACTION, AND SCATTERING

When sunlight (or another type of wave) encounters an obstacle or a boundary that it cannot pass through, *reflection* occurs (see *The Basics: Wave Terminology*, Chapter 10). Reflection of sound waves causes echoes, and reflection of seismic waves provides information about the layers and boundaries deep within Earth (Chapter 6). A mirror is a simple, familiar example of the reflection of light, in this case caused by a coating of shiny metal on the back of a piece of glass.

The property whereby light is reflected to a greater or lesser degree by a material is its *albedo*. Each material has a characteristic albedo, and a given material will typically have a slightly different albedo for each different wavelength of light. Hence, as you learned in Chapter 9, an icy or snowy surface is more reflectant of sunlight and has a higher albedo than water, land, or vegetated surfaces. Vegetated surfaces tend to be particularly reflectant in the near infrared portion of the solar spectrum; in other words, the infrared albedo of a forested surface is typically high. In comparison, water reflects infrared radiation poorly; its albedo for infrared wavelengths of radiation is close to zero.

If oncoming light encounters an obstacle but is able to pass through the boundary, some reflection generally occurs. In addition, the pathway of the reduced light that passes through the boundary will likely be bent upon entering the new material. In Chapter 6 you learned that seismic wave paths are bent when they pass from one medium into another medium of different seismic velocity. Similarly, light-wave paths are bent when they pass through a boundary from one medium to another. The term *refraction* is applied to describe the bending, whether it refers to light waves, seismic waves, water waves, or sound waves.

The reflection and refraction of light in the atmosphere is very complex because there are so many obstacles and boundaries in the atmosphere—every tiny particle, droplet, or gas molecule represents an obstacle. **Scattering** is a reflection phenomenon that involves the

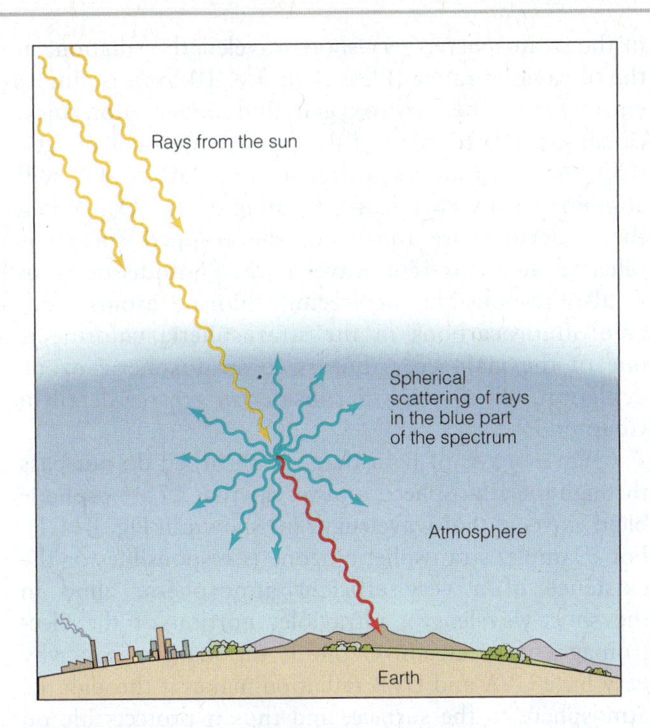

FIGURE B11.2 Scattering
Light coming from the Sun is scattered. Air molecules are so small that they scatter shorter blue wavelengths more easily then longer red wavelengths of light. The scattered blue light makes the sky appear blue in all directions. If there were no scattering by air molecules, the sky would appear pitch black and the stars would be visible all day.

FIGURE B11.3 Rainbow
Rainbows result from complex refraction and reflection of light by water droplets in the atmosphere.

dispersal of radiation in all directions, as **FIGURE B11.2** shows. (It is often called *spherical scattering* to emphasize that radiation moving in a straight line is scattered equally in all directions.) Radiation comes in from the Sun in a straight line; if some of the radiation is scattered, an observer will notice a reduction in the amount of radiation reaching the surface of Earth. We are all familiar with this effect because clouds, which are simply masses of suspended water droplets, ice crystals, aerosols, and other fine particles, scatter sunlight in all directions. When a cloud passes overhead, the intensity of sunlight drops.

Aerosols and gas molecules also cause scattering, but there is an important difference: aerosols scatter *all* wavelengths of visible light; gas molecules do not. In general, if the diameters of the particles that cause scattering are less than one-tenth the wavelength (λ) of the incoming radiation, as they are in gas molecules and many aerosols, then it is called *Rayleigh scattering*. The Rayleigh scattering relationship, which was discovered in 1881 by an English physicist, Lord Rayleigh, says that the amount of scattering is proportional to $\frac{1}{(\lambda)^4}$. The smaller λ is, the larger $\frac{1}{(\lambda)^4}$ will be, and therefore the greater the scattering will be for the short-wavelength end of the spectrum. If the scattering particles are much larger in comparison to the wavelength of the scattered light—a less common situation in the atmosphere—then a related process called *Mie scattering* pertains.

In the visible portion of the solar spectrum, the predominant Rayleigh scattering is at the blue end, because that is the short-wavelength end of the spectrum. The sky appears blue to us because white, unscattered light comes straight through, while blue radiation in sunlight is scattered in all directions. What we see when we look at the sky is this scattered blue radiation. Similarly, because the blue end of the spectrum is reduced in intensity by scattering, the Sun appears a little more yellow to an Earth-bound observer than it does to an observer in space.

A complicated combination of refraction and reflection of light by the surfaces of water droplets is responsible for the familiar phenomenon we call a *rainbow* (**FIG. B11.3**). When sunlight shines through droplets of rain, refraction, reflection, and differential scattering by the surfaces of the raindrops create an arc-shaped spectrum of light from red (in the inner part of the rainbow) to blue (in the outer part).

ABSORPTION AND BLINDS

Absorption occurs when the energy contained in electromagnetic radiation is taken up by the material through which it is passing. When light is absorbed as it passes through a material it is said to be *attenuated*, which means that it loses some of its intensity, depending on how absorbent the material is. Absorption, combined with reflection, is actually responsible for the ability of

(continued)

the BASICS *(Continued)*

our eyes to discern colors. If light shines on a material that is absorbent in every wavelength except red, then only the red light will be reflected (the other wavelengths are absorbed), and our eyes will perceive the material as being red.

Absorption can be of two kinds. In the first, certain specific wavelengths of solar radiation make atoms or molecules vibrate with the same frequency as the wavelength. Such an atom or molecule absorbs the radiation and then reemits it at the same wavelength. However, the radiation is reemitted equally in all directions, and so the observer of the incoming radiation sees a diminution of that wavelength. Most of the absorption of sunlight by atmospheric H_2O, CO_2, N_2O and CH_4 shown in Figure B11.1 happens in this way.

In the second type of absorption, molecules absorb the radiation and break apart into atoms or smaller molecules as a result. This process called, *photodissociation*, is part of the process by which ozone is formed in the stratosphere. Very-short-wavelength radiation, in the ultraviolet range (l less than 3×10^{-7}m), is almost entirely absorbed by oxygen and ozone molecules, causing them to photodissociate. Molecular oxygen (O_2) that encounters a free oxygen atom (O) will combine with it chemically, forming ozone (O_3). When the molecules are reformed, the trapped energy is released at a different wavelength. Photodissociation is also responsible for freeing chlorine atoms from chlorofluorocarbons in the stratosphere; chlorine is one of the main contributors to stratospheric ozone depletion, which we will consider in greater detail in Chapter 19.

Wavelengths of light that are absorbed do not pass through the atmosphere, and we say that an **atmospheric blind** exists in that wavelength (as shown in Fig. B11.1). For example, stratospheric ozone is responsible for the existence of a very efficient atmospheric blind in the short-wavelength, ultraviolet portion of the electromagnetic spectrum. Ozone is the main reason why very little UVA and UVB radiation makes it through the atmosphere to the surface, and thus it protects life on Earth from these damaging rays.

As will be discussed in Chapter 12, this sharp change in incline in the tropopause has important consequences for weather because it gives rise to the phenomenon known as the *jet stream*.

Stratosphere

The **stratosphere** is the layer of the atmosphere that lies above the tropopause. *Strato* means layer and is derived from *stratum*, the same Latin word used to describe layers of sediment. Temperature in the stratosphere increases with altitude, reaching a maximum at its upper boundary, the *stratopause*, at about 50 km in altitude. In other words, the stratosphere is stratified, with the densest, coldest air at the bottom and the warmest, least-dense air at the top. Temperature increases with altitude in this layer because of the presence of ozone. Most of the ozone in the atmosphere resides in the stratosphere, and ozone efficiently absorbs ultraviolet radiation coming from the Sun. Absorption converts the energy of ultraviolet rays into longer wavelength radiation, and this longer-wavelength radiation heats the air. The absorption of ultraviolet rays is at its maximum at the top of the stratosphere, so this is where the highest temperatures are found. As the Sun's rays pass through the stratosphere, less and less ultraviolet radiation remains to be absorbed, so the lowest temperatures in the stratosphere are found at the bottom.

Mesosphere

In the next layer of the atmosphere, called the **mesosphere** (from the Greek *mesos*, meaning middle), temperature again decreases with increasing altitude. The mesosphere is the coldest layer of Earth's atmosphere, reaching a minimum of about −100°C at its upper boundary, the *mesopause*, at about 85 km in altitude. The mesosphere does not contain ozone, so solar radiation passes through without being absorbed and, therefore, without causing warming.

Thermosphere

The **thermosphere** (from the Greek *thermos*, meaning warm), which reaches out to about 500 km from Earth's surface, is a layer in which temperature increases with altitude. The thermosphere reaches the highest temperatures of any layer of Earth's atmosphere. This may seem odd—we normally think of air as becoming colder and colder with altitude—but remember that the thermosphere is directly exposed to the Sun's radiation. The temperature increase in the thermosphere arises partly from the absorption of solar radiation and partly from the bombardment of gas molecules by protons and electrons given off by the Sun. During periods of strong sunspot activity, when the flux of protons and electrons is at a maximum, the bombardment is so great that the temperature at the top of the thermosphere may reach as high as 1500°C. Despite the high temperatures reached in the thermosphere, very few molecules of gas are present, so there is very little total heat energy. Strange as it may seem, we would feel very cold if we were exposed to a 1500°C atmosphere as thin as that in the thermosphere.

FIGURE 11.10 **Aurora**
When protons flowing from the Sun hit the ionosphere, they create the beautiful electrical phenomenon known as an aurora. This aurora was seen in a far-northern latitude, and at the moment it was photographed a meteor flashed across the sky.

One of the most spectacular sights on Earth, an aurora, occurs in the thermosphere (FIG. 11.10). When solar radiation is absorbed by molecules of gas in the thermosphere, some of the molecules are broken apart to form electrically charged ions. The region of ionized gases, from 100 to 400 km in altitude, is called the *ionosphere*. Auroras occur when electrons streaming in from the Sun combine with the ionized gases, form neutral atoms, and give off light rays in the process.

At the outermost fringes of the thermosphere, Earth's atmosphere becomes so thin and tenuous that it merges into outer space. This outermost part of the atmosphere is called the *exosphere*. The few gas atoms present in the exosphere are mostly atoms of the lightest elements, hydrogen and helium.

Air Pressure

Everyone knows that the higher you go above sea level, the less oxygen there is and the harder it becomes to breathe. That is why planes have emergency oxygen masks in case cabin pressures should fail and why mountain climbers often carry tanks of oxygen. The oxygen supply becomes short not because of a change in the composition of the atmosphere, but rather because of a reduction in **air pressure**, the force exerted by the weight of overlying air. The reduction in air pressure results in a reduction in air density—essentially, a smaller number of molecules of atmospheric constituents are contained in a given volume of air. Because the size of a human lung is fixed, we get less oxygen when we breathe less dense air. Air at sea level and air at 9000 m, near the summit of Mount Everest, have the same relative amount of oxygen—20.9 percent by volume in each case. However, a lungful of air at the top of Mount Everest has only 38 percent of the pressure that a lungful of air has at sea level, and therefore only 38 percent of the amount of oxygen.

Measuring Air Pressure

Air pressure is measured with a device called a **barometer**. Air pressure at the surface is closely monitored because, as you will learn in Chapter 12, changes in air pressure can indicate imminent changes in weather. When weather forecasters talk about a "high" or a "low," they are referring to air pressure that is higher or lower than the average air pressure at the surface.

There are two main types of barometers: mercury and aneroid. The *mercury barometer* was invented in 1644 by the Italian physicist Evangelista Torricelli (1608–1647), who performed the following experiment. He sealed a 1-m-long glass tube at one end and then filled it with mercury. Then, with his finger over the open end in order to prevent the mercury from running out, he inverted the tube and put the open end into a bowl of mercury (FIG. 11.11). When he removed his finger, some of the mercury flowed from the tube into the bowl, but most

FIGURE 11.11 **Mercury barometer**
This is a sketch of a simple mercury barometer. Air pressure on the surface of the open bowl of liquid mercury holds up the column of mercury inside the glass tube. The downward pressure exerted by the air exactly balances the downward pressure exerted by the column of mercury on the bowl. When the air pressure changes, the height of the column adjusts in response.

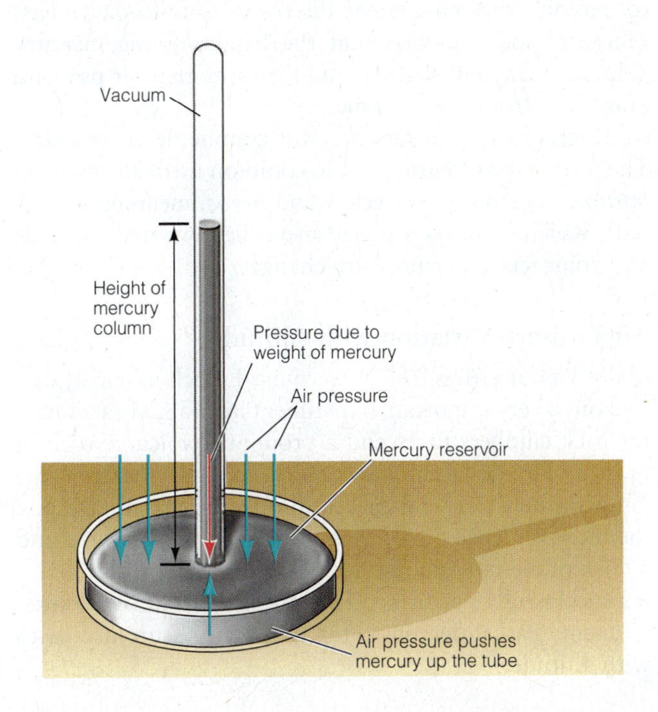

FIGURE 11.12 Measuring air pressure
(A) Puy-en-Velay in France is where Pascal showed that air pressure decreases with altitude. Mountaineers carried a barometer up the Puy, an ancient volcanic neck, making air pressure measurements along the way. (B) Today, scientists send up radiosondes to make such measurements. The helium-filled balloon will burst in the upper atmosphere, allowing the scientific instruments to parachute back to the surface with their measurements.

of it stayed up in the tube. Torricelli reasoned that air pressing on mercury in the bowl must be holding up the column of mercury in the glass tube. Scientists were quick to exploit Torricelli's great discovery. Day-to-day measurements soon showed that the height of the mercury column fluctuated slightly, and therefore that air pressure must vary from time to time.

Mercury barometers are still commonly used today. The other type of barometer in common use is the *aneroid barometer* (from the Greek *a* and *neros*, meaning no liquid), which employs a sealed metal bellows that expands and contracts as air pressure changes.

Air Pressure Variation with Altitude

Blaise Pascal (1623–1662), a young French scientist, carried out a very important experiment in 1658. He arranged for rock climbers to ascend a prominent volcanic rock in France, Puy-de-Dôme (FIG. 11.12A), and measure the air pressure at several places during the ascent. Despite the inconvenience of carrying a meter-long glass tube and a flask of mercury up the steep slope of the Puy, the climbers successfully performed the task, and from their measurements Pascal demonstrated that air pressure decreases with altitude.

Today, meteorologists use balloons filled with helium to carry pressure-recording instruments aloft. Such balloons, which carry an instrumental probe called a *radiosonde* (FIG. 11.12B), can reach an altitude of 30 km, at which point they burst and the recording instruments are parachuted back to the ground. To make measurements above 30 km, meteorologists resort to rockets and, most recently, to orbiting weather satellites.

At any given altitude, air pressure is caused by the weight of the air above. The average air pressure at sea level is about 10^5 Pa,[1] equivalent to the pressure produced by a column of water about 10 m high. Why doesn't this pressure crush us, along with the houses we live in? It doesn't, because air pressure is the same in all directions—up, down, and sideways, inside and outside. The outward air pressure from inside a house is exactly the same as the

[1] Because the earliest pressure measurements were made with mercury barometers, air pressures are still sometimes reported as the height of a mercury column. A model, or average pressure, called the **standard atmosphere** at sea level, is 760 mm of mercury. Another commonly employed unit is the bar, which is a pressure of 1 kg/cm^2. The standard atmosphere is 1013.25 millibars (mbar). In this book, SI units are used throughout (see Appendix A). The SI pressure unit is the pascal (Pa). The standard atmosphere is 101,325 Pa or 101.325 kilopascals (kPa).

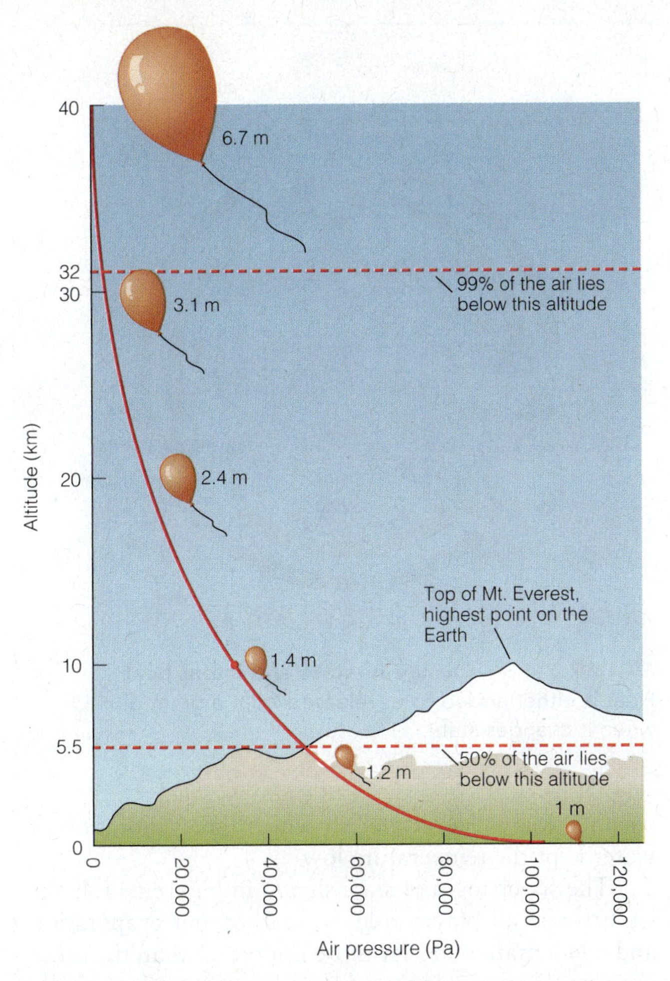

FIGURE 11.13 Change in air pressure with altitude
Air pressure decreases smoothly with altitude. If a helium balloon 1 m in diameter is released at sea level, it expands as it floats upward because of the pressure decrease. If the balloon did not burst, it would be 6.7 m in diameter at a height of 40 km.

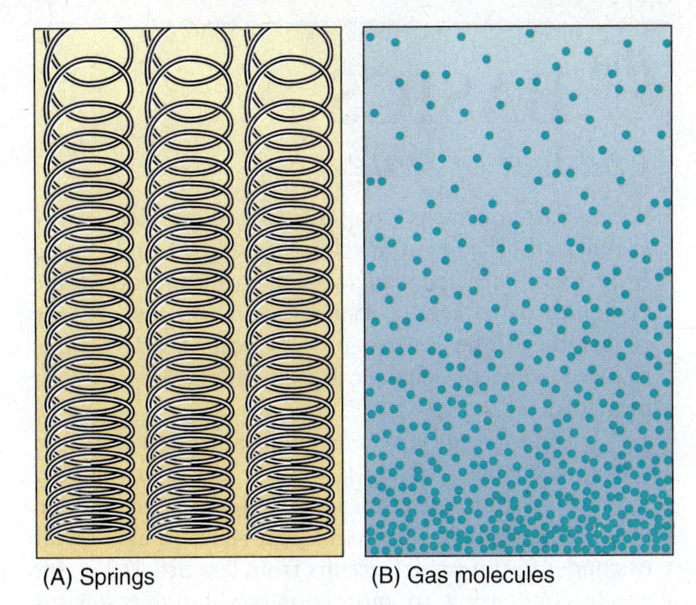

FIGURE 11.14 Air compressibility
Air is compressible and behaves like a pile of springs. (A) The springs near the base are compressed by the weight of the springs above. (B) Air, like the springs, is compressed by the weight of the air above. Molecules of the gases nearest the ground are squeezed closer together than molecules higher up. Compression is also the explanation for the shape of the curve in Figure 11.13.

inward air pressure from outside, so the net pressure on the house is zero.

As shown in FIGURE 11.13 air pressure decreases smoothly with altitude. The air pressure curve is not a straight line, however, because gases are highly *compressible*. This means that air near the ground is compressed by the weight of air above (FIG. 11.14). If air were not compressible (if it were more like water, for example), the pressure-versus-altitude curve in Figure 11.13 would be a straight line.

As a result of the compressibility of air, half of the mass of the atmosphere lies below an altitude of 5.5 km (within the troposphere) and 99 percent lies below 32 km (the troposphere plus the bottommost portion of the stratosphere). At a height of 32 km, about the middle of the stratosphere, the air is so thin that it is like a laboratory vacuum. The 1 percent of the atmosphere that lies above 32 km continues out to an altitude of about 500 km, with the air getting thinner and thinner until the exosphere simply merges into the vacuum of space.

MOISTURE IN THE ATMOSPHERE

Moisture is such an important component of the atmosphere, for so many reasons, that we need to consider its properties in detail. H_2O is the most remarkable substance around, and in no small measure the Earth system works the way it does because H_2O has the properties it does (Chapter 8). One of the most important properties of H_2O is its existence in three physical states at Earth's surface—as a solid (ice), a liquid (water), and a gas (water vapor). Among naturally occurring compounds only H_2O can form solid, a liquid, and a gas phase in the temperature and pressure conditions that exist at Earth's surface. The change of H_2O from one of these states to another is fundamentally important in weather-related processes and in the fundamental processes of the hydrologic cycle, such as *evaporation*, *condensation*, and *precipitation*.

Relative Humidity

Water vapor gets into air by **evaporation**, a process in which fast-moving liquid molecules manage to escape from the liquid and pass into the vapor above. Because molecules in a vapor move randomly in all directions, some of the gas molecules in the vapor will also move back into the liquid. When the number of molecules that evaporate (going from liquid to gas) equals the number of molecules that condense (going from gas to liquid), the vapor is referred to as *saturated*. Saturation determines the maximum concentration of

the BASICS

Changes of State

When matter changes from one state to another, energy is either absorbed by it or released (FIG. B11.4). In going from a more ordered state (a solid) to a less ordered one (a liquid) or to a fully disordered one (a gas), energy is absorbed. The reverse process occurs, and heat is released when the change is from a less ordered to a more ordered state.

The amount of heat released or absorbed per gram during a change of state is known as the *latent heat* (from the Latin *latens*, meaning hidden, hence hidden heat). For example, the latent heat released as a result of condensation (which results from less ordered water vapor condensing to more ordered liquid water) is 2260 J/g. The latent heat released as a result of freezing (again, a change from a less ordered to a more ordered state) is 330 J/g. The latent heat absorbed as a result of evaporation (more ordered liquid water vaporizing to less ordered water vapor) is 2260 J/g, while the latent heat absorbed as a result of melting (more ordered solid ice melting to less ordered fluid water) is 330 J/g.

One familiar phenomenon involving a change of state is evaporation. The 2260 J needed to evaporate a gram of water has to come from somewhere. The reason you feel cool after you wet yourself down on a hot day is that some of the heat needed for evaporation is absorbed from your skin, and as a result your body temperature drops. Before the invention of ice chests and refrigerators, the best way to keep food cool in hot weather was a "cool safe" in which the evaporation of water kept the temperature low.

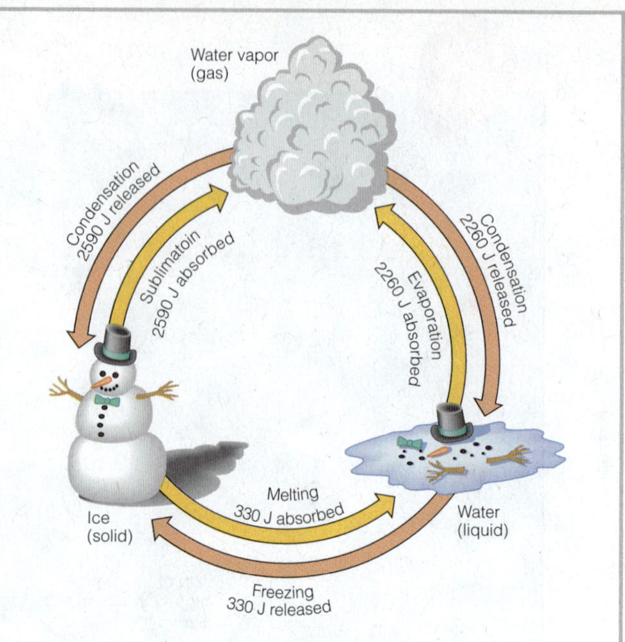

FIGURE B11.4 Change of state and latent heat Heat is either added to or released from a gram of H_2O when it changes state.

The six changes of state shown in Figure B11.4 (the six arrows) all play a role in weather, but evaporation and condensation are far more important than the other changes. Evaporation and condensation play vitally important roles in the weather because they give rise to clouds, fogs, and rain, and because they are the means by which enormous amounts of heat are moved from equatorial regions toward the poles.

H_2O molecules that can reside in the vapor phase at any specified temperature.

It is common practice to report the properties of a vapor in terms of its *vapor pressure*. One of the important properties of gases is that pressures in a mixture of gases are additive; this property is known as *Dalton's law of partial pressures*, and it means that the total pressure of a mixture of gases is the sum of the partial pressures exerted by all the individual gases present. Partial pressure, in turn, is a measure of the volume percent of a gas in a mixture. For example, the content of oxygen in dry air is 20.9 percent by volume. This means that the fraction of the pressure of standard air (101.325 kPa) that is attributable to oxygen is $101.325 \times 0.209 = 21.2$ kPa.

The additivity of gas pressures is why the water vapor content of air is often reported as a pressure rather than as a percentage. The *saturation vapor pressure*, which is also known as the *water vapor capacity* of air at any given temperature, cannot be exceeded. If the water vapor pressure exceeds the capacity, it will be reduced to the saturation vapor pressure by **condensation**, in which the molecules move from the gas into the liquid phase. The vapor pressure cannot be *higher* than the saturation value, but it can be *lower*. For example, if saturated air is removed from contact with water and then heated, the vapor pressure will fall below saturation level and the air will then be undersaturated. The saturation vapor pressure of water at various temperatures is shown in FIGURE 11.15.

When discussing the amount of water vapor in undersaturated air, meteorologists use the term **relative humidity**, which is the ratio of the vapor pressure in a sample of air to the saturation vapor pressure at the same temperature, expressed as a percentage. For example, saturated air at 20°C has a water vapor pressure of 2.338 kPa. Air at 20°C with a water vapor pressure of 1.403 kPa will therefore have a relative humidity of $(1.403 \div 2.338) \times 100\%$, or 60%.

Note that the relative humidity does not refer to a specific *amount* of water vapor in the air; it refers to the *ratio* of the water vapor that is present at a given temperature to the maximum possible amount of water vapor that the air could hold at the same temperature. The fact that relative humidity is a ratio sometimes confuses people and leads to misconceptions. One misconception is that if air feels damp and humid, it must contain more H_2O than air that feels dry. The confusion arises because temperature exerts a strong control on the water vapor capacity of air. For example, desert air at 30°C and a relative humidity of 25 percent feels very dry even though it contains 6.62 g of H_2O per kg of air. In contrast, air at 10°C and relative humidity of 80 percent, which feels damp and humid, contains only 5.60 g of H_2O per kg of air.

Relative humidity can be changed in two ways—by the addition of water vapor or by a change of temperature. When the relative humidity is below 100 percent and air is in contact with water, evaporation will occur, raising the relative humidity. This is why air in contact with the ocean usually has a high relative humidity and why so much of the water vapor that enters the atmosphere does so over the ocean (FIG. 11.16). Temperature changes also affect relative humidity, whether or not H_2O is added. If the amount of water vapor in air is kept constant and the temperature drops, the relative humidity will rise.

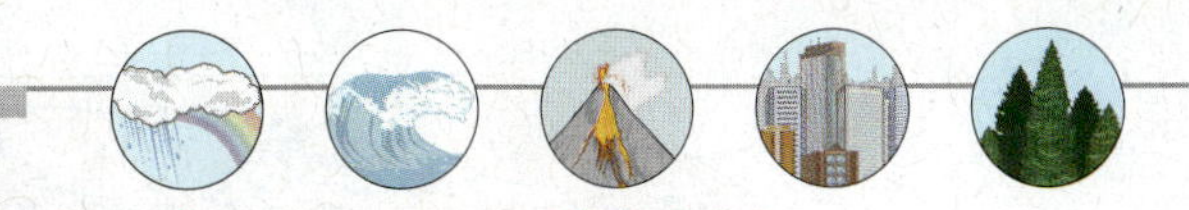

Make the CONNECTION

If the temperature of the ocean were to increase as a result of global warming, what impact would that have on evaporation rates and the relative humidity of air? What kinds of weather-related changes might this lead to?

The temperature at which the relative humidity of air reaches 100 percent and condensation starts is called the **dew point**. When the ground is cold and the air is warm, the layer of air in contact with the ground may cool sufficiently for the dew point to be reached, so that *dew*, a film of water coating the ground, condenses. If the ground temperature is below freezing, *frost* forms instead of dew. In contrast, if the temperature rises, the relative humidity drops. People who live in centrally heated houses are familiar with this effect. In winter, as air is drawn in from outside and heated by a furnace, the relative humidity drops and the air inside the house feels dry. For example, consider what happens when the outside air temperature is −10°C and the relative humidity is a comfortable 60 percent. If air is now drawn into the house and heated to 25°C without the addition of any water vapor, the relative humidity drops to 6 percent, a level at which many people feel that the air is uncomfortably dry. To counteract the effects of low humidity, many people use a humidifier to add water vapor to the heated air.

FIGURE 11.15 Saturation pressure of water vapor
This graph shows the variation in the saturation pressure of water vapor with temperature. The water shown in the small measuring cylinders is equivalent to the amount of H_2O present, if the vapor in a kg of saturated air at sea level were condensed at each temperature. This demonstrates that the water content in a kg of air increases with increasing temperature.

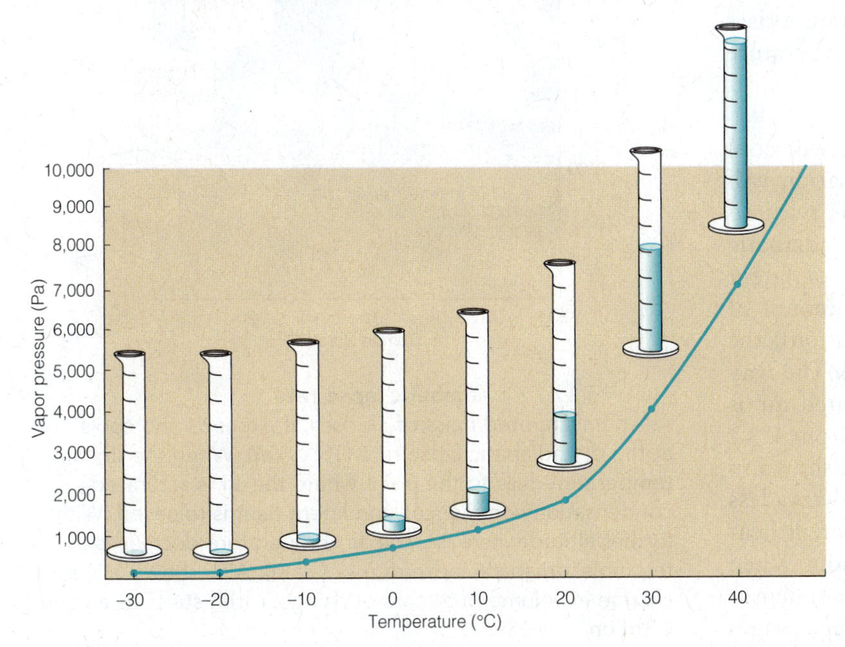

Adiabatic Lapse Rate

If you have ever pumped up a bicycle tire, you will have noticed that the pump became hot when the air was compressed. Similarly, if the compressed air in a tire is allowed to escape, the air is noticeably cool as it expands. These two effects, compressional warming and expansional cooling, are examples of what are called **adiabatic** processes, from the Greek *adiabatos*, meaning no passage. Adiabatic processes are so named because they are processes that occur without the addition or subtraction of heat from an external source. When air is compressed, the mechanical energy of pumping is converted to heat, and the temperature rises as a result. When compressed air expands, the energy required for the expansion comes from the heat energy of the gas; consequently, a temperature drop occurs.

Adiabatic processes have a profound effect on the behavior of air masses. Warm air is less dense than cold air and therefore rises, creating a convection cell. Warmed from below, air in the troposphere rises. Air pressure decreases with increasing altitude, so the

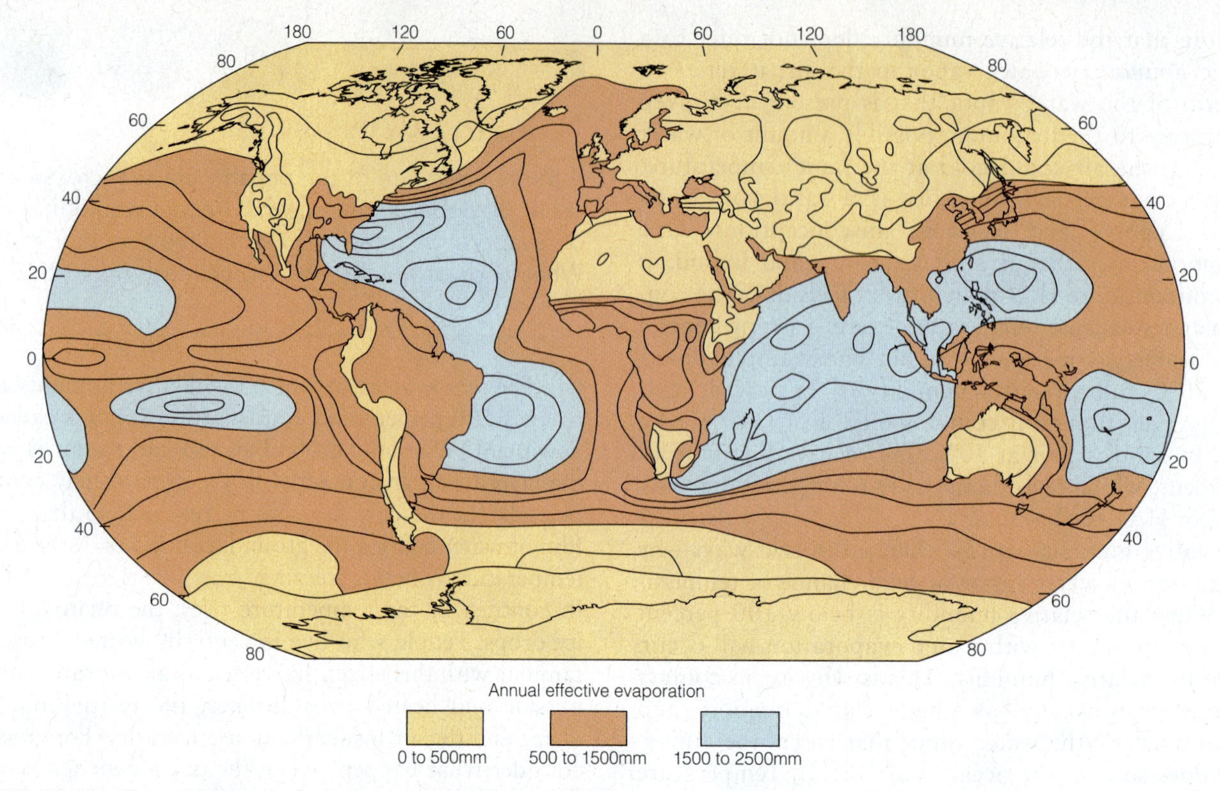

FIGURE 11.16 Evaporation and water vapor in the atmosphere
This map shows the annual addition of water vapor to the atmosphere as a function of geography. The amount evaporated per year is measured in millimeters of water. Areas of highest evaporation (blue) are over the ocean in equatorial and midlatitudes. Evaporation is low in the deserts (gold) because deserts have little water available for evaporation.

rising air expands. There is no heat source above ground level in the troposphere, so the expansion is adiabatic and the air temperature falls. Once cooled, air in the troposphere will sink; in this case, the reverse happens—the air is adiabatically compressed, and its temperature rises.

When a parcel of unsaturated air rises and expands adiabatically, the temperature drops at a constant rate of 10°C/km. Conversely, if cool, unsaturated air sinks toward Earth's surface, it is compressed and the temperature rises at a rate of 10°C/km (FIG. 11.17). The way temperature changes with altitude in rising or falling unsaturated air is called the *dry adiabatic lapse rate*.

Eventually, when air rises far enough, it will cool sufficiently to become saturated and condensation will start. The latent heat of condensation that is released as the vapor condenses works against the adiabatic cooling process; in other words, the release of latent heat slows the cooling rate. The greater the amount of latent heat released (the greater the condensation), the smaller the temperature increase with altitude. The way temperature drops in a rising mass of saturated air is called the *moist adiabatic lapse rate*; it ranges from 4° C/km to 9°C/km, with an average of 6°C/km (as shown in Fig. 11.17). The moist adiabatic lapse rate is always less than the dry rate because of the addition of latent heat to the rising air above the level of condensation. When a parcel of air has reached saturation and condensation has begun, the dry adiabatic lapse rate changes

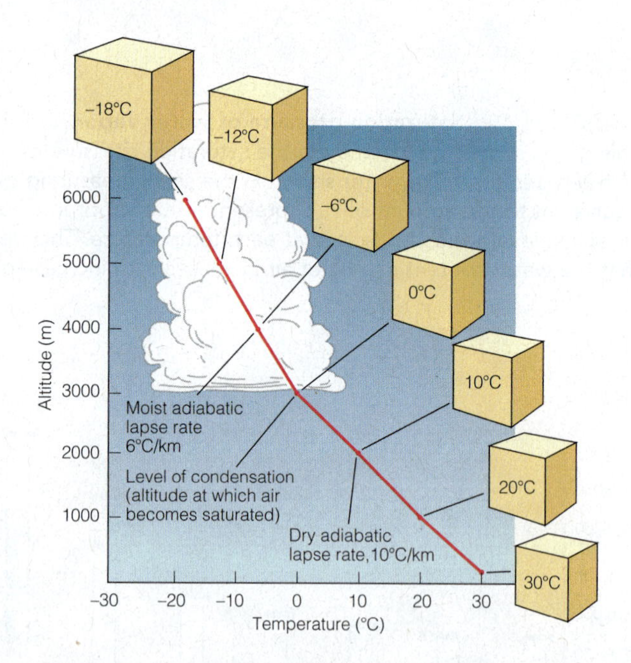

FIGURE 11.17 Adiabatic lapse rate
As an unsaturated mass of air rises, it expands and cools at the dry adiabatic lapse rate (10°C/km). When the dry air temperature falls to the point where the air is saturated, condensation commences and latent heat is released. With further altitude increase, the air temperature decreases at the moist adiabatic lapse rate (6°C/km). Also shown is the change in volume of a mass of rising air that starts as a cube 1 km on an edge.

immediately to the moist rate; this is the setting in which cloud formation occurs.

Cloud Formation

Clouds are visible aggregations of minute water droplets, tiny ice crystals, or both. Clouds form when air rises and becomes saturated with moisture in response to adiabatic cooling, leading to condensation.

Lifting Forces

There are four principal reasons for the upward movement of air. Any one of them can lead to the formation of clouds, although it is common for more than one lifting force to be at work in any individual circumstance. The four lifting forces are:

1. *Density lifting*; this occurs when warm, low-density air rises convectively and displaces cooler, denser air (FIG. 11.18A).
2. *Frontal lifting*; this occurs when two flowing air masses of different density meet. The boundaries between air masses of different temperature and humidity, and therefore different density, are called **fronts**. Fronts are between 10 and 150 km in width and mark the advance of one air mass into another. When warm, humid air advances over cold air (a *warm front*), the warm air rises up and flows over the cold air, forming clouds and possibly rain as a result (FIG. 11.18B). A similar process occurs when denser, cold, air flows in and displaces warm air by pushing it upward (a *cold front*), again producing clouds and possibly rain (FIG. 11.18C). When a cold front overtakes a warm front and two cooler air masses meet, the result is an *occluded front*.
3. *Orographic lifting*; this occurs when flowing air is forced upward as a result of passing over a sloping terrain, such as a mountain range (FIG. 11.18D). Some of the highest rainfall spots in the world—such as the western coast of Tasmania in Australia, the

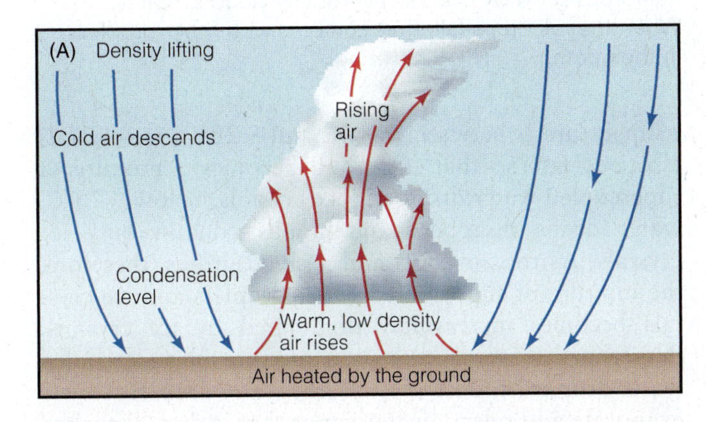

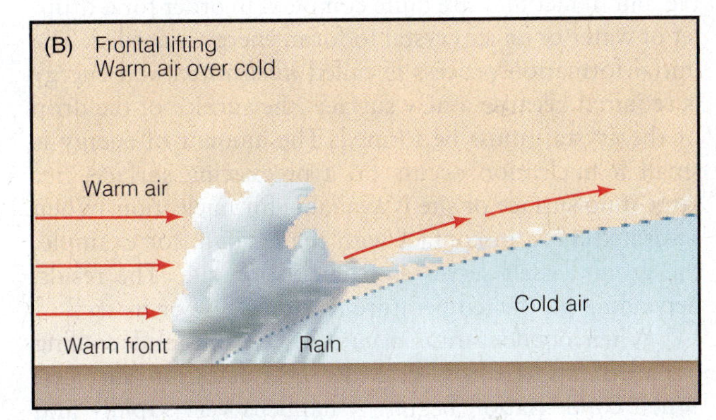

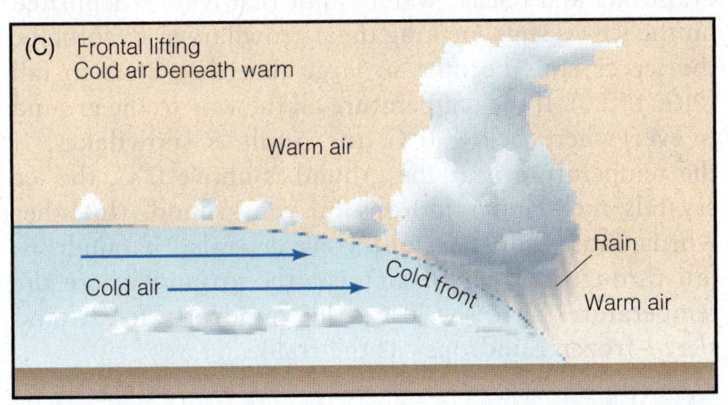

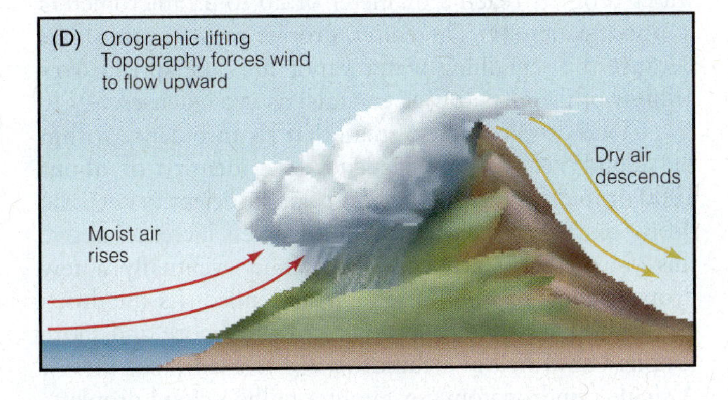

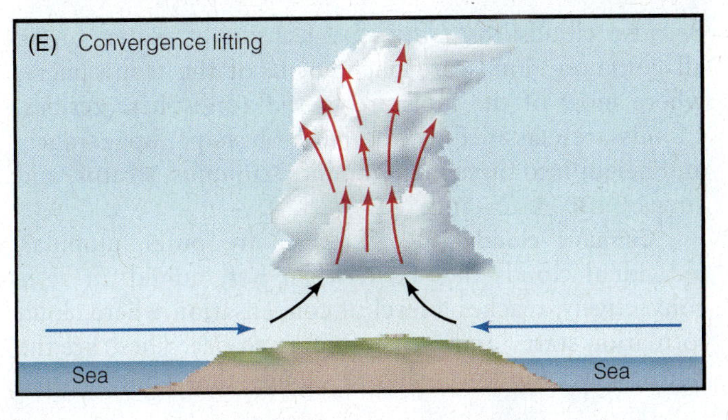

FIGURE 11.18 Lifting forces
These are the main lifting forces that lead to cloud formation. Under most circumstances two or more lifting forces operate at the same time. (A) Density lifting causes a convection cell as warm, low-density air rises and cold, higher-density air sinks. (B) Frontal lifting and a warm front occur when flowing warm air overrides cold air and is forced upward. (C) Frontal lifting and a cold front occur when a wedge of forward-moving cold air slides under a warm air mass and forces it upward. (D) Orographic lifting occurs when flowing air is forced upward by mountains or other sloping ground. (E) Convergence lifting occurs when masses of air collide and are forced upward.

Owen Stanley Range in New Guinea, and the Olympic Peninsula in Washington—result from orographic lifting. Deserts often form in the area to the landward side of such mountains, called a *rainshadow*, because the air loses its moisture as it traverses the mountains.

4. *Convergence lifting*; this occurs when flowing air masses converge and are forced upward (FIG. 11.18E). The Florida peninsula provides an example; air flows landward off the ocean from both east and west; the two flowing air masses collide and force some of air to rise. Clouds therefore form, and the result is the familiar frequent afternoon thunderstorms.

Condensation, Nucleation, and Precipitation

When air becomes saturated with water vapor (that is, the dew point is reached), one of two things happens: either water condenses or, if the temperature is low enough, ice crystals form. We see these at the surface in the form of **precipitation** if the droplets of water or crystals of ice grow large enough to settle by gravity. The processes seem simple, but in fact they are quite complex. In order for a droplet of water or an ice crystal to form, energy is needed. The initial formation process is called *nucleation*, and energy is required because a new surface (the surface of the drop or the crystal) must be formed. The amount of energy is small if nucleation occurs on a preexisting surface, but large if no surface or site is available for nucleation. When saturated air is in contact with the ground, for example, the ground itself serves as a nucleation site. The result, depending on the temperature, is either dew or frost.

When condensation occurs in a rising and cooling parcel of air, aerosols provide the nucleation surfaces. When condensation begins, it happens very rapidly and water droplets reach a diameter of 20 to 25 micrometers in about a minute. Thereafter, droplet growth rate slows because the remaining water vapor must be spread over billions of droplets (or ice crystals) as it condenses.

Cloud droplets are so small that air turbulence within the cloud keeps them suspended. A density of about 1000 droplets/cm^3 (or 1 drop/mm^3) is sufficient to keep the drops apart. When the density of droplets increases above this value, they start to coalesce, and eventually a few drops become too big to remain suspended. As the drops fall, they bump into, and coalesce with, more and more droplets until finally a raindrop has formed (FIG. 11.19). A single raindrop contains about a million cloud droplets.

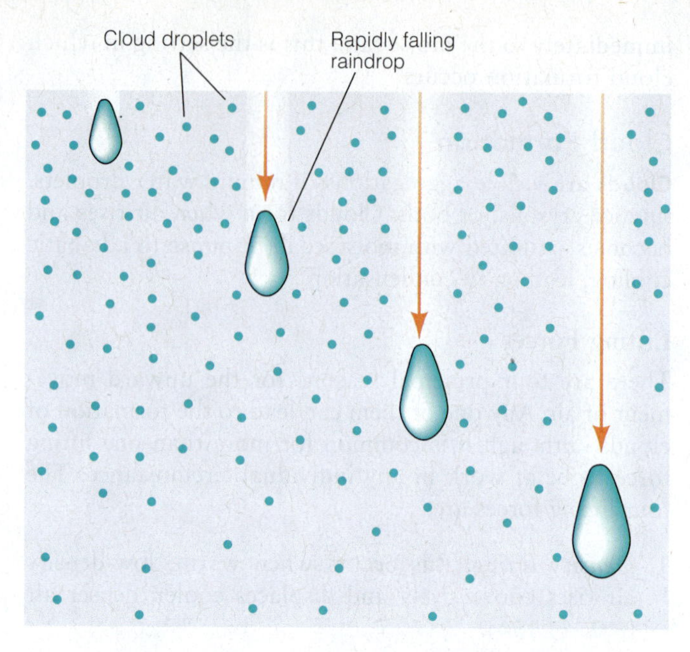

FIGURE 11.19 **Raindrops**
Raindrops grow by coalescence of tiny cloud droplets. When the raindrops fall, they combine with other droplets in their path.

When the cloud temperature is below 0°C, the process of precipitation is more complex because water droplets can be supercooled below 0°C without freezing to ice. The person who first recognized the role of supercooled water in cloud precipitation was a Scandinavian scientist, Tor Bergeron, and the process is now named the *Bergeron process* in honor of his discovery. What Bergeron discovered is that ice crystals grow at the expense of supercooled water droplets in clouds.

Clouds with temperatures between 0°C and −9°C contain only supercooled water droplets. When the temperature is between −10°C and −20°C, ice crystals also nucleate so that the cloud becomes a mixture of supercooled water drops and ice crystals. Below −20°C, water drops disappear and clouds contain only ice crystals. As the air cools and a cloud grows, therefore, the mixture of supercooled water droplets and ice crystals becomes increasingly dominated by ice crystals. Bergeron discovered that, in a mixture of supercooled water droplets and ice crystals, the water droplets slowly evaporate and release water vapor that is then deposited on the ice crystals, making them grow larger. Eventually, the ice crystals become so large that they start to fall (FIG. 11.20). If the temperature all the way to the ground is everywhere below 0°C, the result is snowflakes. If the temperature near the ground is above 0°C, the ice crystals melt and raindrops hit the ground. (In other words, most rain starts out as ice crystals.) If raindrops fall through a layer of air near the ground where the temperature is below 0°C, the raindrops freeze and *sleet*—frozen raindrops—is the result.

Cloud Types

Because clouds form by condensation of water vapor, all common clouds are phenomena of the troposphere, where most of the moisture in the atmosphere resides. Clouds are classified on the basis of shape, appearance, and height into three main families: cumulus, stratus, and cirrus (FIGS. 11.21 and 11.22).

Cumulus clouds (FIG. 11.22A) are puffy, globular, individual clouds that form when hot, humid air rises convectively, reaches a level of condensation where cloud formation starts, and then continues to rise. These are the flat-based, cauliflower-shaped clouds that children like

FIGURE 11.20 Bergeron process
Ice crystals grow by the Bergeron process, in which supercooled water droplets in clouds evaporate and ice crystals grow by incorporating the newly formed water vapor.

to draw—the flat base marks the level of condensation. When cumulus clouds coalesce to form a puffy layer, the term *stratocumulus* is applied (FIG. 11.22B). When large cumulus clouds rise to the top of the troposphere, they expand horizontally and form *cumulonimbus* clouds (FIG. 11.22C). These are the familiar anvil-shaped thunderstorm clouds or "thunderheads" of summer. There is a great deal of energy and turbulence in a cumulonimbus cloud, and some of the energy causes thunder and lightning within a cloud, between adjacent clouds, and between clouds and the ground.

Cumulus clouds that form at altitudes between 2 and 6 km are given the modifying name *altocumulus* clouds); those that form between 6 and 15 km are called *cirrocumulus* clouds. Frequently, altocumulus and stratocumulus clouds are arranged in regular rows or clumps separated by clear sky. Convection cells up to several hundred meters across give rise to these patterns.

Stratus clouds (FIG. 11.22D) are sheets of cloud cover that form at altitudes from 2 km to about 15 km and cover the entire sky. Stratus clouds form when air rises as a result of frontal lifting, reaches its level of condensation, and then spreads laterally but not vertically. If the cloud blanket is several kilometers thick, the day is dark and dreary and the cloud is called *nimbostratus* (FIG. 11.22E). Depending on their altitude, stratus clouds are given modifying names analogous to those given to cumulus clouds—between 6 and 8 km, *altostratus*, and between 8 and 12 km, *cirrostratus*.

Cirrus clouds (FIG. 11.22F) are the highest of the clouds in the troposphere. With the appearance of fine, wispy filaments or feathers, cirrus clouds form only above 6 km in altitude and are composed entirely of ice crystals.

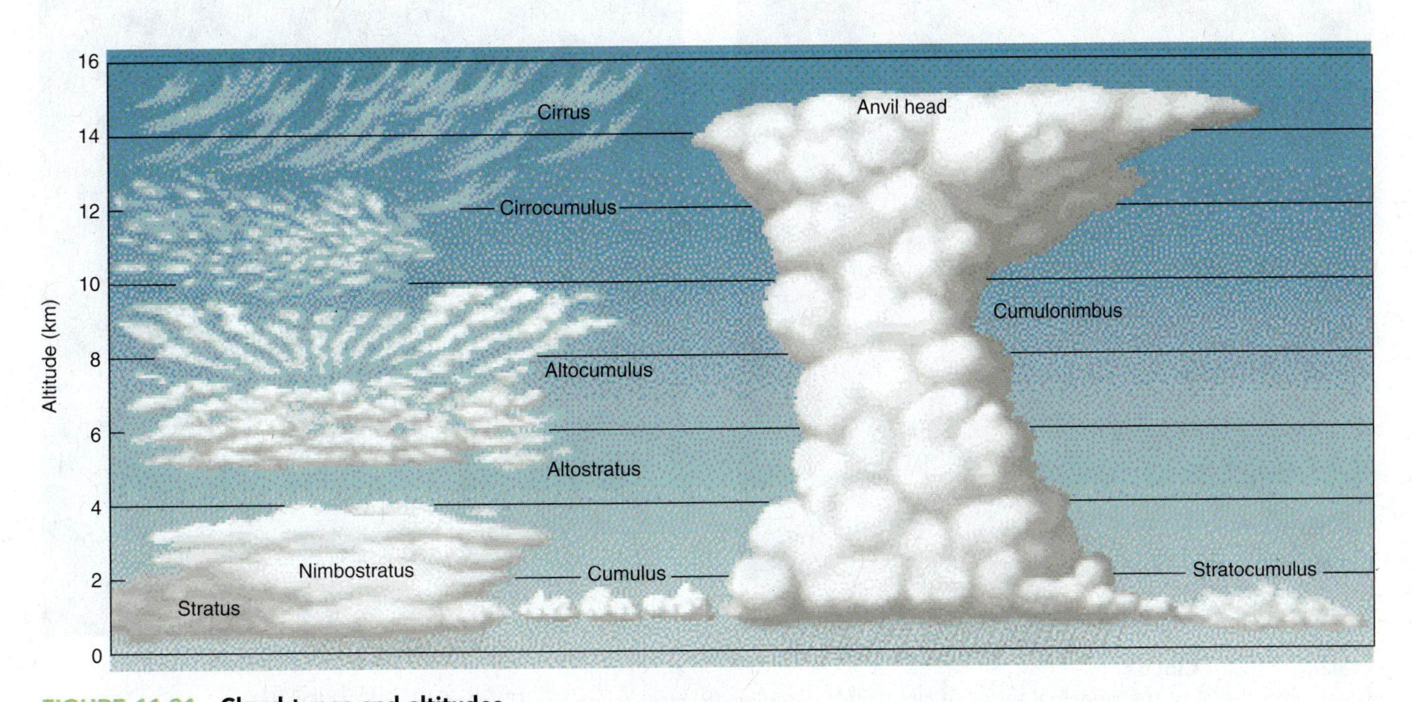

FIGURE 11.21 Cloud types and altitudes
This drawing shows the various types of clouds, their shapes, and the altitudes at which they characteristically occur.

FIGURE 11.22 Clouds
These are several of the principal types of clouds: (A) cumulus; (B) stratocumulus; (C) cumulonimbus; (D) stratus; (E) nimbostratus; (F) cirrus.

Two rare kinds of cloud are known to form in the stratosphere. *Nacreous clouds* are beautiful translucent sheets (meaning light gets through, like a frosted window pane) of minute ice crystals that form at altitudes between 20 and 30 km. Even less common are *noctilucent clouds*, which are so thin they look like gossamer veils. They are composed entirely of minute ice crystals and form at altitudes as high as 90 km. No clouds have ever been reported above 90 km, and none are likely because water vapor is too scarce to form clouds at such altitudes.

THE ATMOSPHERE IN THE EARTH SYSTEM

In Chapter 1 we introduced Earth's *life zone*, the place where Earth's four reservoirs interact in a narrow zone that allows and supports the existence of life on this planet. In this narrow zone all known forms of life exist because it is only here that conditions favorable for life are created by interactions between the geosphere, hydrosphere, and atmosphere, and modified by the biosphere. The atmosphere plays a central role in maintaining conditions within the life zone at the appropriate temperature and composition to support life.

The Atmosphere and the Life Zone

You learned at the beginning of this chapter that oxygen must be present within strictly defined limits in an atmosphere if a planet is to be habitable. In addition to the presence of oxygen in the atmosphere, two other essential criteria make a planet habitable: water vapor must be present, and the temperature must be neither too high nor too low.

Humans cannot live for long where the air is completely dry because water vapor is needed for our lungs to work properly. A habitable planet must therefore have a hydrosphere so that there is some way of getting water vapor into the atmosphere.

Where temperature is concerned, the limits of a habitable planet vary somewhat from one organism to another. Some organisms are adapted to extreme temperatures, but most—including humans—must live in conditions of fairly stable and moderate temperatures in order to survive. The parts of the human body that must be protected from temperature fluctuation are the core organs—the brain, heart, lung, liver, and digestive system. The temperature of these core organs, which is 37°C, cannot vary safely by more than ± 2°C. To maintain a stable core-organ temperature, the human body has two cooling mechanisms—perspiration and dilation of blood vessels in the skin, both of which cool because they get rid of heat—and two heat-conserving mechanisms—shivering and contraction of blood vessels in the skin, which retain heat.

The most effective way to handle the body's temperature requirements, of course, is to live where the annual mean temperature is comfortable. Over 90 percent of the human population of Earth lives in places where the annual mean temperature is between 6°C and 27°C. A planet that is habitable for people must therefore have regions that fall within this temperature range. Much of Earth enjoys annual mean temperatures in the equitable 6°C to 27°C range largely because of the atmosphere, which acts as a blanket that both warms and stabilizes the surface temperature.

Recent Atmospheric Changes

Considering our total dependence on air, and on the functioning of our atmospheric blanket in maintaining conditions supportive of life, you might think that everyone would know and care a lot about the atmosphere. Unfortunately, that does not seem to be the case. All too often, we take this most precious commodity for granted and carelessly treat air as if it had an endless capacity to absorb pollutants. A wide variety of anthropogenic contaminants cause pollution at low levels in the atmosphere, especially in urban centers. Some air pollutants remain local, but others are distributed regionally or even globally by atmospheric transport. Many of the common air pollutants are known to have negative impacts on human and ecosystem health.

The behavior of an airborne contaminant is controlled by a variety of interacting factors, including its own chemical and physical characteristics, the local topography, and prevailing atmospheric conditions. The characteristics of the pollutant itself (such as whether it occurs in aerosol form) determine how long it can remain in suspension in the atmosphere and, therefore, how far it will be transported. Wind systems and precipitation play important roles in the transport of airborne contaminants. When the air is very still, pollutants are not transported as far, and they tend to become concentrated over the source area. When it rains, pollutants wash out of the atmosphere by adhering to raindrops.

Today there is growing concern about anthropogenic pollutants causing changes in the concentration of radiatively active gases in the atmosphere, which, it is thought, will lead to changes in the greenhouse effect, and ultimately to global climatic change. Clouds are particularly important in this context, because they play a dual role in the climate system: they can warm the surface by holding heat underneath them, and they can cool the surface by reflecting, scattering, or absorbing incoming solar radiation. If a warming climate leads to more evaporation and thence to more cloud cover, will the resulting clouds have more of a warming influence, or more of a cooling influence on the climate? This is one of the great sources of uncertainty in our scientific understanding of the global climate system.

In Chapter 12 we will look more closely at the wind and weather systems that arise from the atmospheric characteristics that we have discussed in this chapter. Then, in Chapter 13, we will consider the causes and impacts of longer-term climatic change.

SUMMARY

1. A habitable planet must have a breathable atmosphere, and the most essential ingredient (for humans and many other organisms) is oxygen. Earth's atmosphere, which was formed as a result of volcanic outgassing of volatiles from the planet's interior, provides just the right amount of oxygen for life (as we know it) to survive.

2. Earth's early atmosphere was hot and dominated by water vapor and greenhouse gases. The earliest precipitation was acidic, and there was no "free" oxygen. The atmosphere was dense, with very high atmospheric pressure at the surface. The atmosphere was inhospitable to life, as we know it today.

3. Volcanic gas emissions over the past 4 billion years account for the present composition of the atmosphere—except the oxygen, which came principally from photosynthesis. The transition to an oxygenated atmosphere occurred slowly, around 2 billion years ago. With the buildup of molecular oxygen came an increase in ozone. The concentration of oxygen in the atmosphere continued to increase to its present value of 20.9 percent, which may have facilitated the rise of mammals.

4. Earth's carbon cycle has removed carbon from the atmosphere over time. The formation of limestone and the burial of organic sediments provide long-term reservoirs. The opening of new ocean basins accelerates the burial of organic sediments, and subduction returns limestone and seafloor sediment to the mantle. Both processes remove carbon from near-surface biogeochemical cycling and sequester it from further contact with the atmosphere and hydrosphere.

5. Air is Earth's atmosphere, a complex mixture of gases and suspended particles. The relative composition of air varies from place to place and from time to time in a given location. The two main reasons are aerosols and water vapor, both of which vary widely.

6. Aerosols are suspensions of extremely tiny liquid droplets or solid particles. They can be naturally occurring, in fog, clouds, volcanic ash, smoke from forest fires, blown sea salt, blown dust and loess, and pollen. They also can be anthropogenic, originating from industrial activities, transportation, the burning of fossil fuels, and the generation of electric power.

7. Nitrogen, oxygen, and argon comprise 99.96 percent of dry, aerosol-free air by volume. The remaining minor gases are profoundly important for life on Earth. Carbon dioxide, water vapor, methane, ozone, and nitrous oxide are the minor gases that create Earth's life-maintaining greenhouse effect. Ozone is the most important of the gases that shield life from harmful ultraviolet radiation.

8. Sunlight can be transmitted, refracted, reflected, scattered, or absorbed by the atmosphere. Atmospheric windows allow certain wavelengths of radiation to pass through, whereas atmospheric blinds result from the absorption of some wavelengths of radiation by gases. The reflection and refraction of light in the atmosphere is complex; any tiny particle, droplet, or gas molecule may cause scattering.

9. The atmosphere gets its heat energy from the Sun. Solar radiation is the energy source in the generation of clouds, rain, snowstorms, wind, and local weather. The tilt of Earth's axis is responsible for the annual seasons, and Earth's rotation drives large-scale flow patterns and global wind systems, as a result of the Coriolis force.

10. The principal thermal layers in the atmosphere, from bottom to top, are the troposphere (from the surface to an altitude of 10–16 km, bounded by the tropopause); the stratosphere (to 50 km, bounded by the stratopause); the mesosphere (to 85 km, bounded by the mesopause); and the thermosphere (to about 500 km). Beyond this altitude, the atmosphere merges into outer space. The troposphere is heated from the bottom, so temperature decreases with height; the stratosphere is heated from within, by the absorption of radiation by ozone, so temperature decreases with height.

11. Air pressure is measured using a barometer. Air pressure at high altitude can be measured using a radiosonde, carried aloft by a helium-filled balloon.

12. At any given altitude, air pressure is caused by the weight of the air above. The pressure of air at sea level doesn't crush us because it is the same in all directions. Air pressure decreases smoothly with height, but not linearly, because gases are compressible. As a result of the compressibility of air, half of the mass of the atmosphere lies below an altitude of 5.5 km and 99 percent lies below 32 km.

13. Humidity, the amount of water vapor in air, is quite variable. The change of H_2O from one state to another (solid, liquid, and vapor) is fundamentally important in weather-related processes such as evaporation and condensation, because the transitions release or absorb latent heat.

14. In evaporation, fast-moving molecules escape from the liquid phase and pass into the vapor above. Saturation determines the maximum concentration of H_2O molecules that can reside in the vapor phase at any specified temperature. If the saturation vapor pressure or water vapor capacity of air at a given temperature is exceeded, condensation will occur. Relative humidity is the ratio of the vapor pressure in a sample of air to the saturation vapor pressure at the same temperature; it can be changed by the addition of water vapor, or by a change of temperature.

15. Adiabatic processes occur without the addition or subtraction of heat from an external source. When warm, unsaturated air rises, expands, and cools adiabatically (or when cool, unsaturated air sinks, becomes compressed, and warms adiabatically), the temperature changes at a constant rate, the dry adiabatic lapse rate. When air rises and cools sufficiently to become saturated, condensation begins.

16. Clouds are visible aggregations of minute water droplets, tiny ice crystals, or both. Clouds form when air rises and becomes saturated with moisture in response to adiabatic cooling, leading to condensation. The four forces principally responsible for the upward movement of air are density lifting, frontal lifting, orographic lifting, and convergence lifting.

17. At the dew point, water condenses or, at low temperatures, ice crystals precipitate. For a droplet of water or an ice crystal to nucleate, energy is required because a new surface must be formed. When cloud droplets reach a certain density they coalesce, forming raindrops that can fall as precipitation once they reach sufficient density. If the temperature all the way to the ground is below 0°C, the result is snowflakes. If the temperature near the ground is above 0°C, the ice crystals melt and raindrops hit the ground.

18. All common clouds occur in the troposphere, where most of the moisture in the atmosphere resides. Clouds are classified into three main families: cumulus clouds, stratus clouds, and cirrus clouds. In addition, clouds are given modifying names, such as altostratus and cirrocumulus, which identify the altitude of formation. Some rare types of clouds form above the troposphere.

19. The atmosphere plays a central role in maintaining conditions in the life zone at an appropriate temperature and composition to support life. For a planet to be habitable, in addition to the presence of oxygen, water vapor must be present, and the temperature must be neither too high nor too low. Much of Earth enjoys moderate temperatures because of the atmosphere, which acts as a blanket that both warms and stabilizes the surface temperature.

20. Anthropogenic contaminants cause pollution at low levels in the atmosphere. Some air pollutants are local, but some are distributed regionally or globally by atmospheric transport. Many common air pollutants have negative impacts on human and ecosystem health. The behavior of an airborne contaminant is controlled by its own chemical and physical characteristics, the local topography, and prevailing atmospheric conditions.

IMPORTANT TERMS TO REMEMBER

absorption *331*
adiabatic (lapse rate) *337*
aerosol *325*
air *322*
air pressure *333*
atmosphere *322*
atmospheric blind *332*
atmospheric window *330*
barometer *333*
cirrus clouds *341*
cloud *339*
condensation *336*
cumulus clouds *340*
dew point *337*
evaporation *335*
front *339*
humidity *325*
mesosphere *332*
pause *329*
precipitation *340*
relative humidity *336*
scattering *330*
stratosphere *332*
stratus clouds *341*
thermosphere *332*
transmission *330*
troposphere *329*

QUESTIONS FOR REVIEW

1. What are the two main energizing forces in the atmosphere?
2. Mars and Venus have atmospheres, but are their atmospheres air? Why, or why not?
3. The main ingredient of air is nitrogen. What is the second most abundant ingredient? What percentage of air is made up by this second most abundant ingredient?
4. What is the difference between humidity and relative humidity?
5. Why is it possible for dry air in a desert to contain more water vapor than moist air in the Arctic?
6. Explain how it is possible for humans to live on Earth's surface without being harmed by lethal ultraviolet radiation from the Sun.
7. Explain why there is less oxygen to breathe at the top of Mount Everest than at sea level, given that the composition of the air is the same.
8. The way air pressure decreases with increasing altitude is smooth, but nonlinear. Explain why.
9. Name the four temperature layers of Earth's atmosphere in order of altitude, starting with the lowest. Give the approximate altitudes where one temperature zone passes into the next, and the name of the boundary in each case.
10. Why does temperature decrease with altitude in the troposphere, but increase with altitude in the stratosphere?
11. When air is rapidly compressed, as in a bicycle pump, it becomes heated. Explain why this is so. What is the name given to processes such as the heating or cooling of a gas as a result of compression or expansion?
12. What is the Bergeron process, and what role does it play in the formation of raindrops?
13. Describe four ways in which a mass of air can be lifted. Why are these lifting processes important for the formation of clouds?
14. What is the difference between a cold front and a warm front? Rain is commonly associated with both kinds of front. Why would that be so?
15. Name the three major families of clouds and describe their general differences.

QUESTIONS FOR RESEARCH AND DISCUSSION

1. Would you expect Earth's cloud cover 25,000 years ago, during the last glaciation, to have been more extensive, less extensive, or about the same as it is today? Did precipitation during the ice age have to be different from what it is today, or could precipitation have been the same and temperature the only thing that changed?

2. It is anticipated that global warming will enhance evaporation of ocean water, leading to an increase in cloud cover. Clouds play a dual role in climate; an increase in cloud cover could lead either to cooling or to additional warming. Do some research to find out more about the role of clouds in the climate system, what types of clouds might result from global warming, and whether they are anticipated to contribute to cooling or to warming.

3. Air pressure is usually the same in all directions. However, during the passage of a tornado or hurricane all of the windows in a house may break if they are left closed. Why might this be so?

QUESTIONS FOR *THE BASICS*

1. What is scattering? How does it differ from refraction and reflection?

2. What is the difference between an atmospheric blind and an atmospheric window?

3. What is latent heat? Give two examples of a change of state in which latent heat is released. Give two examples of a change in which latent heat is absorbed.

QUESTIONS FOR *A CLOSER LOOK*

1. What is an aerosol? How big are aerosol particles (or droplets)?

2. Name two common natural sources and two common anthropogenic sources for aerosols.

3. Describe one way in which aerosols can affect the global climate.

CHAPTER 12

Wind *and* Weather SYSTEMS

OVERVIEW

In this chapter we:

- Learn how weather differs from climate and how both are controlled by changes in the properties of the atmosphere
- Examine the forces that cause air to move and control its direction of flow
- Describe the major global and regional circulation systems of the atmosphere
- Describe some important local wind and weather systems
- Consider the causes and impacts of severe weather events

Quenching rain
Severe weather affects our lives in a variety of ways. Here, an approaching thunderstorm offers hope for a parched landscape at Lake Thunderbird, Oklahoma, during a severe drought in 2006.

The weather is said to be the most popular topic of conversation in all cultures. It deserves this high ranking because it plays such an important role in our daily lives—from the clothes we wear to the food we eat to the activities we pursue.

Even though we talk a lot about weather, there is sometimes a bit of confusion about what is actually being discussed. To avoid confusion, we follow the lead of *meteorologists* (weather scientists) who define **weather** as the state of the atmosphere at a given time and place. The five variables used to determine the state of the atmosphere, and therefore to describe weather, are:

1. Temperature
2. Air pressure
3. Humidity
4. Cloudiness
5. Wind speed and direction

In Chapter 11 you learned about the properties of the atmosphere, including the first four of these weather variables. Now, in Chapter 12, we will consider how various factors interact to influence circulation in the atmosphere, thereby creating wind systems and determining the weather.

These five weather variables are also used to measure and describe climate, but there is an important difference between *weather* and *climate*. Weather is short-term, whereas climate is long-term. Weather can change over a short time span; for example, it can be cold and wet in the morning but warm and dry in the afternoon in the same location. A rapid or dramatic change in the atmosphere can quickly cause changes in the weather on a regional or even a global scale. For example, a major volcanic eruption can affect weather around the world in a matter of days.

Climate, on the other hand, must be measured over a period of years, because **climate** is the *average* weather condition of a place. The climate of northern Canada is cold and wet, for example, even though there are many warm, dry days. Over a period of years, cold, wet days are more numerous than warm, dry days, so the average weather (that is, the climate) of northern Canada is cold and wet, regardless of what the weather may be on a given day or even during a given week or month. The opposite condition is found in the Sahara Desert region of northern Africa; there, hot, dry days are far more common than cold, wet ones, so the climate of the Sahara is classified as hot and dry.

The controls on climate are affected in the longer term by changes in the geosphere, hydrosphere, biosphere, atmosphere, and (most recently) the anthroposphere; these will be addressed separately in Chapter 13.

Make the CONNECTION

Climate is a longer-term phenomenon than weather, but not necessarily large in scale. *Microclimates* are small areas where the climate (not just the weather) differs from that of surrounding areas as a result of different conditions, such as the angle of the Sun on a hillslope or the moisture content of prevailing winds. See if you can identify some different microclimates in your neighborhood. What factors work together to control these microclimates?

WHY AIR MOVES

Wind is air movement that arises from differences in air pressure. Nature always moves to eliminate a pressure difference, and wind is the result when air flows from a place of high pressure to one of low pressure. As you learned in Chapter 11, a vertical gradient in air pressure is always present—air pressure is highest near the surface and decreases with altitude. Furthermore, air pressure is related to density: High pressure means the air is more compressed and therefore denser; this happens when cool air descends and collects near the surface. Low pressure means less compression and lower density; this happens when warm air rises away from the surface.

Therefore, the movement of air as a result of pressure differences is closely associated with both temperature and density, and horizontal movement is always associated with at least some vertical movement of air. The magnitude of the pressure difference in a particular location controls the wind speed, interacting with other factors to determine the direction of airflow.

Wind Speed

A wind speed of 20 km/h[1] is a pleasant breeze that rustles and moves all the leaves in a tree and produces small, white-capped waves on a lake. At 45 km/h, all the branches in a tree start to sway, and spray forms on open water. When wind speeds reach 65 km/h, twigs and small branches break off trees, waves are high, and foam forms on wave crests. At 90 km/h, trees are uprooted and the wind will knock you down; at 180 km/h, it can pick you

[1] Because wind was the power source for sailors for so many centuries, wind speeds are sometimes still reported in nautical units. If the weather reports from your local newspaper or radio or TV station use nautical units, here is how to convert them: The unit of distance at sea is the *nautical mile*, and the unit of speed is the *knot*. A knot is 1 nautical mile/hour. A nautical mile is equal to 1.1508 land miles, or 1.852 km. Thus, 30 knots is 30 × 1.852 or 55.56 km/hr.

up. Fortunately, such high-speed winds are rare. When they do occur, they tend to be associated with tornadoes or hurricanes, and the damage—although it can be severe—is localized.

The greatest wind speed ever recorded on the surface of Earth is 372 km/h on Mount Washington, New Hampshire, in April 1934. Speeds of 325 km/h have been recorded in hurricanes, and winds up to 335 km/h have been reported during severe storms in Greenland. Such high-speed winds can do remarkable things (FIG. 12.1). High-speed tornado winds, for example, have been reported to move houses from their foundations, carry away tractor-trailers, drive objects through tree trunks, and pluck all the feathers from chickens and ducks.

Most places around the world have wind speeds that average between 10 and 30 km/h. The windiest place on Earth is at Cape Dennison, in Antarctica, where the average speed is 70 km/h. Mount Washington, where the highest speed was recorded, averages only 55 km/h year round.

FIGURE 12.1 Wind
High-speed winds can be devastating, as shown here in a photo of damage caused by Hurricane Andrew in Florida, 1992.

the BASICS

Windchill Factor

On a typical winter day, you will hear the weather forecaster report on *windchill*, which combines temperature and wind speed to provide an indication of how cold you are likely to actually feel if you are outside on that day. Here's how windchill is calculated: Immediately adjacent to the human body (or any other solid surface) there is a thin layer of still air called a *boundary layer* (FIG. B12.1). This layer is still because friction prevents it from moving. Heat escaping from the body must pass through the boundary layer by conduction. Because air is a poor conductor, the boundary layer serves as an effective insulator. As wind speed increases, the thickness of the boundary layer decreases, thereby reducing its effectiveness as an insulator and increasing the rate at which heat is lost from the body. It usually *feels* colder when the wind is blowing, but the air temperature itself does not drop as a result of the wind. However, the higher the wind speed, the thinner the boundary layer and the faster heat is lost from the skin. If the skin reaches freezing temperature, frostbite ensues.

The windchill factor should correctly be called the *windchill equivalent temperature* because for a given air temperature and given wind speed, the windchill factor is reported as the air temperature at which exposed parts of the body would lose heat at the same rate if there were no wind. For example, if the air temperature is −3°C and the wind speed is 32 km/h, the windchill equivalent temperature is −18°C. Windchill is not the only factor that affects our comfort in cold weather, but it is certainly one of the most important as far as safety is concerned.

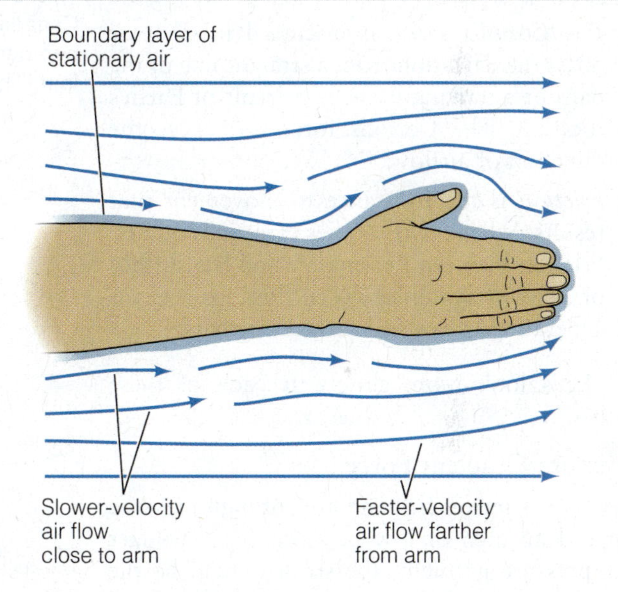

FIGURE B12.1 Windchill
Adjacent to any solid body, such as a human arm, there is a thin layer of air that is held stationary by friction. Away from the body, wind speed, indicated by the length of the arrows, increases as the effect of frictions weakens with distance. Higher wind speeds cause the thickness of the boundary layer to decrease. This lessens its insulating effect and allows the body to lose heat to the atmosphere more easily.

In places where temperatures drop below freezing, it has become customary for weather forecasters to report an additional parameter called **windchill**, which provides a measure of the heat loss from exposed skin as a result of the combined effects of low temperature and wind speed. (See *The Basics: Windchill Factor*.)

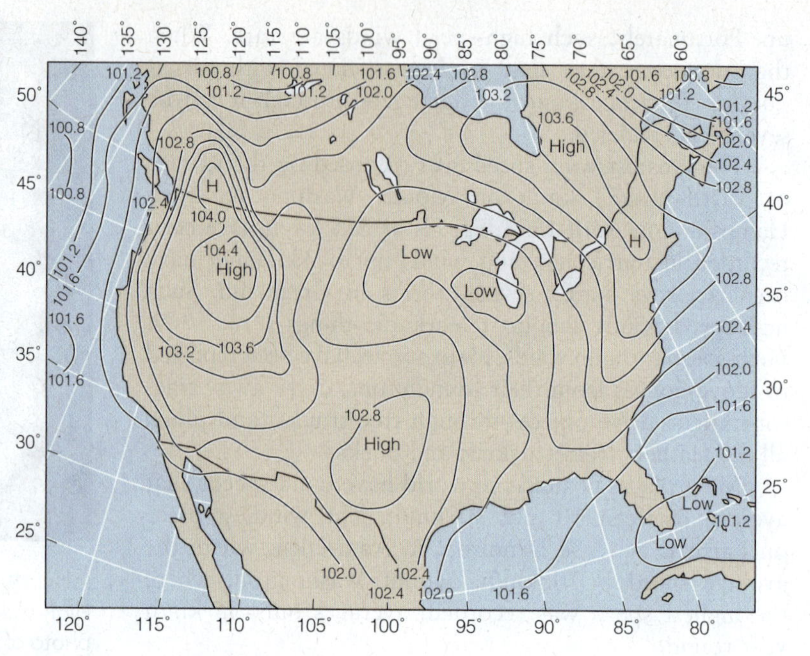

FIGURE 12.2 Isobars
This is a typical air pressure map of the United States and part of Canada. The blue lines are isobars, lines of equal air pressure, and the numbers are air pressure in kPa. (Weather maps often report air pressure in millibars; to convert kPa to mb, multiply by 10.)

Factors Affecting Wind Speed and Direction

Differences in pressure cause air to flow from one location to another, thereby generating wind. If Earth did not rotate, wind would blow in a straight line. If Earth had a frictionless surface, the wind would flow longer and harder than it does. Neither of these "ifs" applies, and wind is therefore controlled by the following three factors:

1. The **pressure-gradient force** results from the drop in air pressure per unit of distance. The pressure-gradient force drives the flow of air.
2. The *Coriolis force* as discussed in Chapter 10, is the deviation from a straight line of the path of a moving body as a result of Earth's rotation. The Coriolis force influences the direction of airflow.
3. *Friction* is the resistance to movement that results when two bodies are in contact. Friction between flowing air and the surface of land or water slows airflow, as well as generating waves and surface currents.

Let's look more closely at each of these forces.

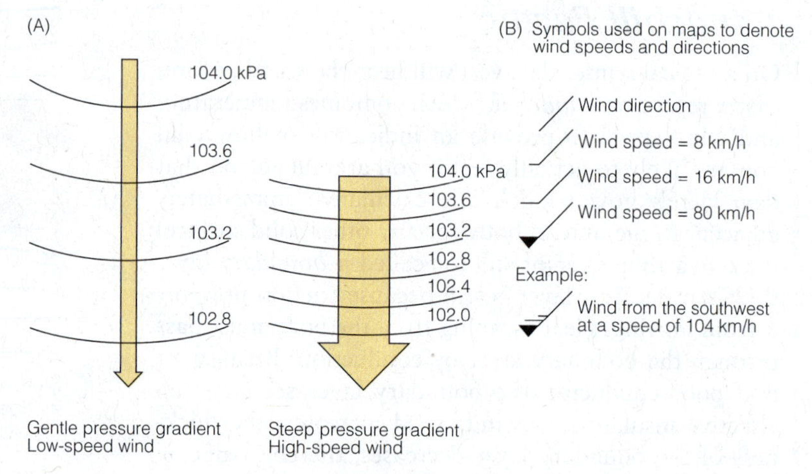

FIGURE 12.3 Wind and pressure gradients
(A) Widely spaced isobars indicate a slow pressure drop over a long distance and thus a low-pressure gradient; low-speed winds will result. Closely spaced isobars indicate a steep pressure gradient; high-speed winds will result. (B) Symbols like these are used to portray wind direction and speed on weather maps. The orientation of the stem indicates the direction, and the barbs indicate speed. If more than one barb is on the stem, add them together to get the wind speed.

Pressure-Gradient Force

Air always moves from an area of high pressure toward an area of low pressure. The stronger the pressure gradient, the stronger will be the resulting flow of air. Air pressure is measured using a barometer (Chapter 11), and the air pressure gradient can be determined from air pressure measurements portrayed on a weather map. On weather maps, places of equal air pressure[2] are connected by lines called **isobars** (FIG. 12.2). Isobars are analogous to contour lines that connect places of equal elevation on a topographic map. The spacing between topographic contour lines indicates the steepness and location of hills; similarly, the spacing between isobars on a weather map indicates the steepness of the air pressure gradient. When the isobars are close together, the gradient is steep; when they are far apart, the gradient is low. When isobars are close together, air flows rapidly down the steep pressure gradient, and a high-speed wind is the result (FIG. 12.3).

As mentioned earlier in this chapter, air pressure differences develop both horizontally and vertically in the atmosphere. Since air pressure decreases generally with altitude, and air of different density is constantly rising and falling in the troposphere, a vertical pressure gradient

[2] Although the SI unit for pressure is the pascal, weather maps usually report air pressure in an older unit, the millibar (abbreviated mb or, technically, mbar). A millibar is a pressure of $0.001 kg/cm^2$. To convert mb to kilopascals (kPa), divide the mb by 10.

is everywhere present. In order to assess the effects caused by horizontal pressure gradients, vertical pressure differences must be avoided. This requires measurements to be taken at a constant altitude, and mean sea level is the elevation generally chosen. In order to draw sea-level isobars on maps, air pressure measurements made at elevations higher than sea level are corrected to the sea-level value. Once the corrections are made, isobars are drawn, generally at a contour interval of 0.4 kPa.

Coriolis Force

The Coriolis force, introduced in Chapter 10 in the context of ocean currents, influences all freely moving objects on the surface of a rotating planet. Because wind is freely moving air, the directions of all winds—like all ocean currents—are subject to this effect. The Coriolis force causes a deflection of the path of the moving object toward the right in the northern hemisphere and toward the left in the southern hemisphere (see Fig. B10.1).

The speed of a moving object influences the magnitude of the Coriolis effect because a fast-moving object covers a greater distance in a given time than a slow-moving object. The longer the trajectory, the greater the change in angular velocity and therefore the greater the Coriolis deflection. Where airflow is concerned, the Coriolis effect is of greatest importance in large-scale wind systems such as the trade winds, but of only minor importance in small-scale, local wind systems such as thunderstorms.

Friction

When wind blows across the ground, through trees, or over solid objects of any kind, friction slows its speed. Friction is important for small-scale air motions and for the layer of air that is in contact with the surface of land or water. In other words, friction is most influential for winds that blow within 1 km of Earth's surface; it is much less important for winds higher than 1 km above Earth's surface, such as the winds of the jet stream.

Friction also can influence the direction of airflow because it influences the Coriolis effect. Remember that the magnitude of the Coriolis effect is proportional to the speed of the moving body. A reduction in speed as a result of friction will therefore reduce the Coriolis deflection. This, in effect, causes northern hemisphere winds to turn a little to the *left* and southern hemisphere winds to turn a little to the *right*.

Geostrophic Winds

Winds are always subject to more than one factor. Consider the least complicated example: a high-altitude wind that is not in contact with the ground and therefore is not affected by friction. Such a wind starts to flow because an air pressure gradient exists, and the direction of flow is down the gradient perpendicular to the isobars—the steeper the gradient, the greater the wind speed. Once flow starts, the Coriolis effect becomes important, so the flowing air is deflected. Deflection means that the wind direction is no longer perpendicular to the isobars; instead, the wind crosses them at an oblique angle (FIG. 12.4). Eventually, when the pressure-gradient flow and the Coriolis deflection are in balance, the wind flows parallel to the isobars.

Winds that result from a balance between pressure-gradient flow and the Coriolis deflection are called **geostrophic winds.** (In chapter 10 you learned about

FIGURE 12.4 Geostrophic wind
(A) A parcel of air is subjected to a pressure-gradient force and a Coriolis force; the resultant vector determines the direction of movement of the air. (B) The parcel of air moves in response to a pressure gradient. At the same time, it is turned progressively sideways until the pressure-gradient force and the Coriolis force balance, producing a geostrophic wind, whose flow is parallel to the isobars.

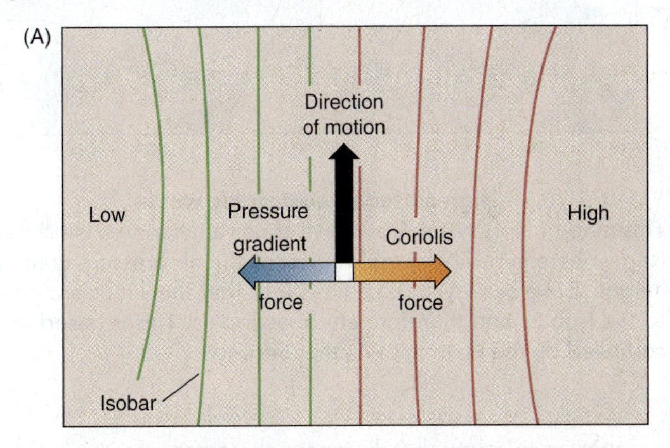

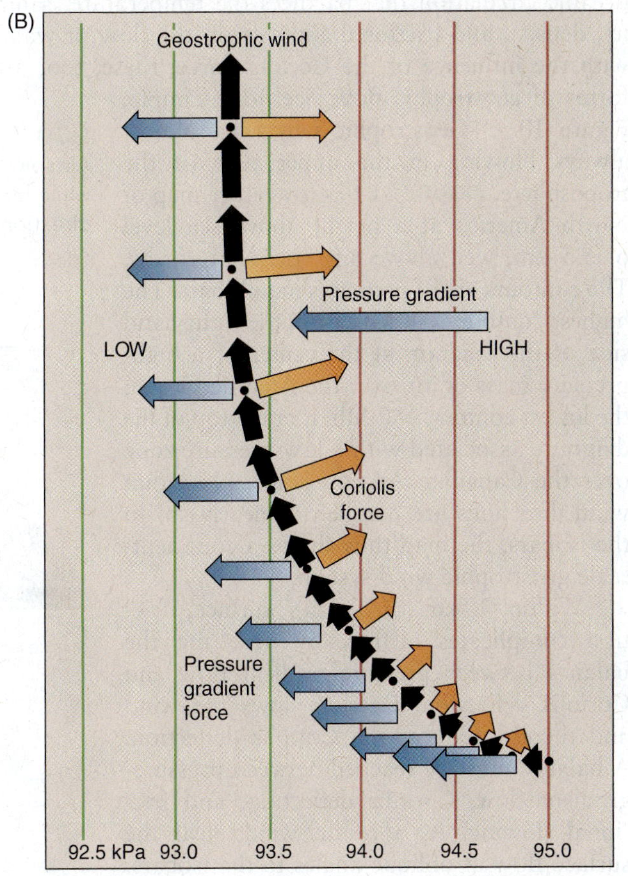

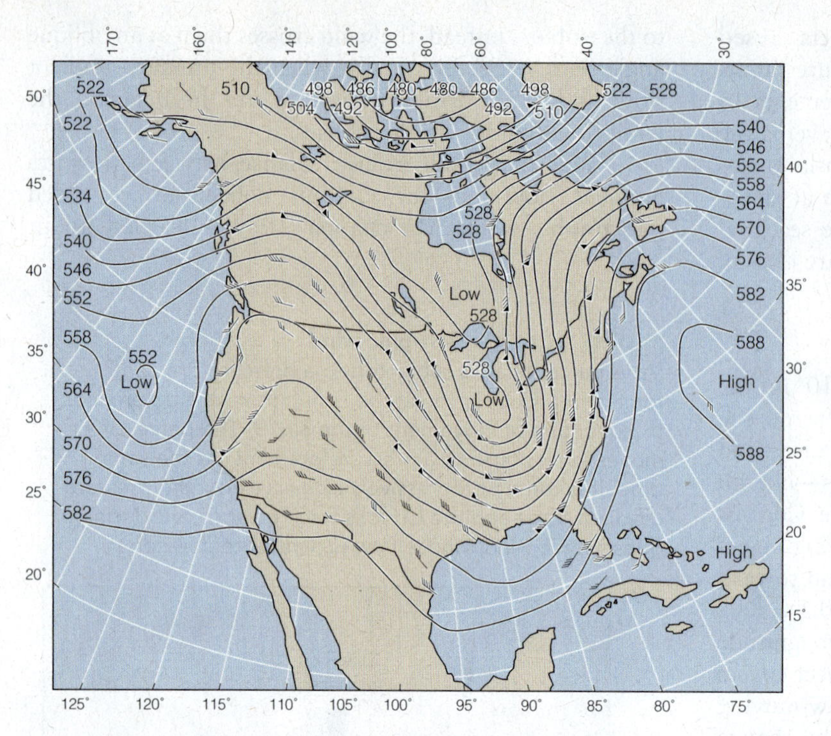

FIGURE 12.5 **High-altitude geostrophic winds**
This map of North America shows upper-atmosphere wind flow. The lines (shown here in millibars, mb) represent the air pressure contours at a height above sea level of 5.5 km. Note that the winds are nearly all parallel to the isobars and therefore are geostrophic. This is based on a map compiled by the National Weather Service.

oceanic circulation that balances the temperature, salinity, density, and frictional controls on the flow of water with the influence of the Coriolis force; these, too, are forms of geostrophic flow. See, for example, Figure 10.9.) Geostrophic winds are almost always blowing in the upper part of the troposphere. FIGURE 12.5 is a weather map of North America at a height above sea level of 5.5 km, well above any frictional effects. The contours are of pressures in millibars. The highest contour, 588 Mb, is on the right-hand side of the diagram at the center of a high-pressure mass of air over the Atlantic Ocean; the lowest contour, 480 Mb, is at the top of the diagram, associated with a low-pressure zone over the Canadian Arctic islands. Note that wind directions are parallel (or nearly so) to the isobars; the map thus shows a continent-scale geostrophic wind system.

Within 1 km of Earth's surface, friction complicates airflow by upsetting the balance between pressure-gradient flow and Coriolis deflection. Friction slows the wind and thereby reduces the Coriolis deflection. A balance must be reached between pressure-gradient flow, Coriolis deflection, and frictional slowing. As a result, winds near the surface flow at oblique angles to the isobars. The angle is a function of the roughness of the terrain. If the surface is very rough, the friction effect will be large, and the angle between the airflow and the isobars can be as great as 50°. If the surface is smooth, such as the surface of the sea, the angle will be closer to 10° or 20°.

Convergent and Divergent Flow

The three counterbalancing effects—the pressure-gradient force and the Coriolis effect, influenced near the surface by friction—interact such that winds around a low-pressure center develop an inward spiral motion (FIG. 12.6). By the same process, airflow spirals outward from a high-pressure area. In the northern hemisphere, the inward-flowing low-pressure spirals rotate counterclockwise, and the high-pressure spirals rotate clockwise. In the southern hemisphere, the reverse is true. Spiral flow was first explained by a Swedish scientist, Valfrid Ekman (1867–1954), and for this reason the spirals are sometimes called *Ekman spirals*. Ekman actually explained the spirals from his study of oceanography, as mentioned in Chapter 10, but the phenomenon is basically the same in the atmosphere as it is in the ocean.

The spiral pattern of airflow in the lower atmosphere can be seen almost daily on weather maps and is dramatically seen in satellite images of cloud patterns (FIG. 12.7). Air spiraling inward around a low-pressure center, designated **L** for **Low** on weather maps, is called a **cyclone**. Air spiraling outward, away

FIGURE 12.6 **Convergence and divergence**
Air spirals into a low and out from a high. Lows are centers of convergence, while highs are centers of divergence. In both lows and highs the flow direction is oblique to the isobars because of friction.

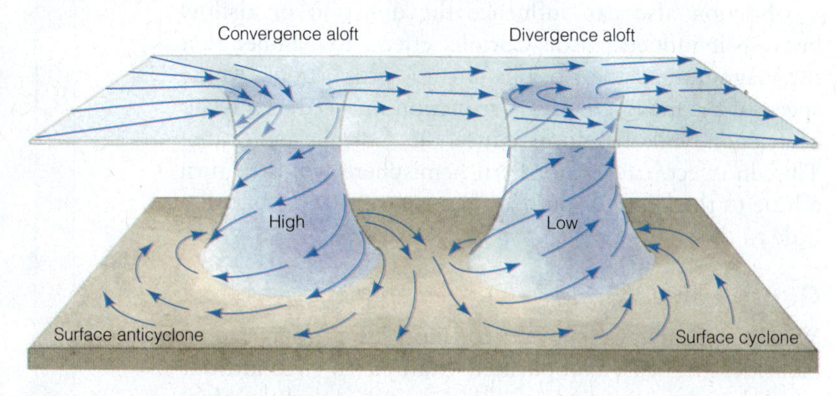

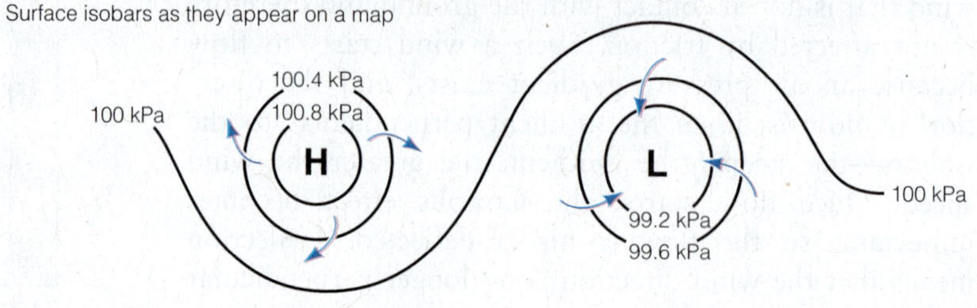

FIGURE 12.7 Low-pressure convergence
This low-pressure center (a cyclonic system) is centered over Ireland and moving eastward over Europe. The counterclockwise winds characteristic of a Northern Hemisphere low are clearly shown by the spiral cloud pattern.

from a high-pressure center, designated **H** for **High** on maps, is called an **anticyclone.**

The inward spiral flow in a cyclone causes **convergence,** which leads to an upward flow of air at the center of the low. As discussed in Chapter 11, upward flow leads to adiabatic cooling, saturation, and condensation in the air as it rises, with resulting cloud cover and rain (FIG. 12.8A). The outward spiral flow in an anticyclone, in contrast, causes **divergence,** which leads to an outward flow of air from the center. This outward flow means that high-altitude air will be drawn downward into the center (Fig. 13.9B). Cold air that is drawn downward will be compressed and heated adiabatically, causing the relative humidity to decrease and leading to clear, cloudless skies.

So low-pressure centers tend to be associated with cloudy, unsettled weather, and high-pressure centers with clear, dry weather. This is why weather forecasters always emphasize the location and movement of high- and low-pressure zones. It also explains why barometers are good predictors of weather changes. When the barometer is falling, the air pressure is dropping and a low is approaching, so cloudy weather can be forecast. When the barometer is rising, air pressure is increasing because a high is approaching, and dry, sunny weather is on the way.

GLOBAL AIR CIRCULATION

The pressure-gradient force and the Coriolis force thus combine in a planetary-scale circulatory system in which enormous volumes of air move, converge, diverge, rise, and fall, influenced by interactions with the surfaces of land and water. These enormous volumes of air, which can cover hundreds or even thousands of square kilometers of the surface, have consistent internal characteristics of temperature and humidity and, therefore, density. They are referred to as *air masses* (see *The Basics: Air Masses*), and they are responsible for our global wind and weather systems.

Mariners have long known about and used the global-scale wind systems that result from the movement of large air masses. Early Polynesian navigators, for example, were intrepid sailors who used the northeast and southeast trade winds (the word *trade* once meant a direction or course) to discover and then

FIGURE 12.8 Cyclone and anticyclone
(A) Convergence in a cyclone causes a rising updraft of air and with it clouds and probably precipitation. (B) Divergence in an anticyclone draws in dry, high-altitude air, creating a downdraft; clear skies and fair weather are the result.

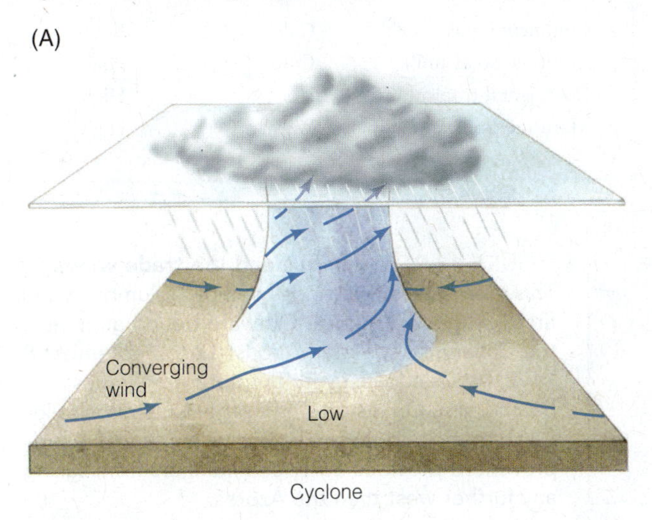

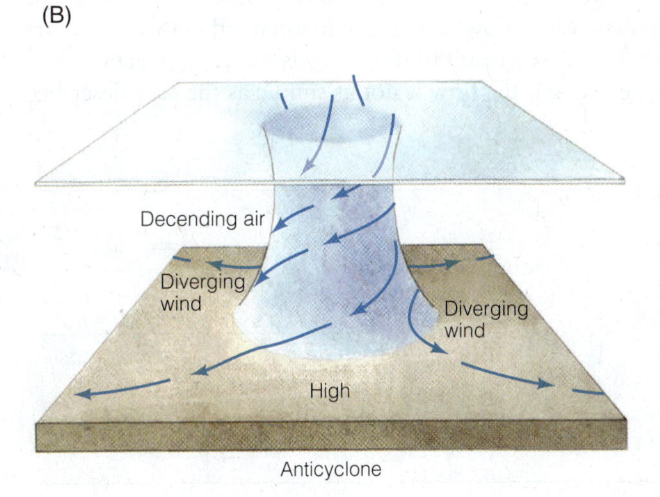

settle the Hawaiian Islands more than 1300 years ago. Christopher Columbus also relied on his knowledge of global winds when he set forth in 1492. He had previously sailed to the Azores Islands, which lie at latitude 37° N about one-third of the way from Spain to America. Persistent westerly winds slowed the trip to the Azores and prevented Columbus (and earlier sailors as well) from getting any farther west. (Remember that *westerly winds* blow from the west toward the east, and *easterly winds* blow from the east toward the west.) Columbus knew, however, from reports of Portuguese sailors that, if he sailed south down the coast of Africa, he would find easterly winds. When he left Spain on August 3, 1492, he therefore sailed south as far as the Canary Islands (FIG. 12.9), picked up the easterly-blowing trade winds, and the rest is history. With the trade winds behind him, he crossed the Atlantic Ocean and blazed the trail used by Europeans to invade the Americas. On his return voyage to Europe, Columbus sailed northward and picked up the westerly blowing winds in the North Atlantic.

The person who first offered an explanation for the persistent easterly and westerly winds reported by mariners was George Hadley (1685–1768), an English mathematician. Hadley pointed out in 1735 that the underlying cause of global winds is that more of the Sun's heat reaches the surface at the equator than at the poles. The reasons for the disparity of heat reaching the surface, as explained in Chapter 2, are that Earth is round and the axis of rotation is tilted. The solar heat imbalance, Hadley pointed out, means that warm equatorial air must flow toward the poles and cold air must flow toward the equator, creating huge convection cells.

If Earth were a nonrotating sphere, one simple convection cell in each hemisphere would carry heat from the equator all the way to the poles (FIG. 12.10, on page 358). Warm, low-density air rising above the equator would flow poleward, and cool polar air would flow back across the surface toward the equator. The equatorial region would be a zone of convergence, and therefore a low surface-pressure region, while the two polar regions would be zones of divergence, and hence high surface-pressure zones.

However, Earth is a rotating sphere, so both the poleward airflow and the equatorward return flow are deflected as a result of the *Coriolis force*. Convection does operate, but the flow is not as simple as the case described

the BASICS

Air Masses

People who dwell in the middle latitudes know that weather patterns generally last several days. The reason is that weather is controlled by huge air masses, up to 2000 km across and several kilometers high, which require several days to cross a continent. An **air mass** is a large volume of air that is characterized by a fairly homogeneous internal temperature and humidity, and, therefore, homogeneous density.

Within an air mass there are only small contrasts of temperature and humidity because any given air mass forms over a surface that has roughly uniform properties. Air masses interact with the surface over which they form and, subsequently, the surfaces over which they travel, taking on the temperature and moisture characteristics of the underlying land or water. This is why hurricanes develop strength as they move over warm ocean water—the circulating air mass of the hurricane system draws heat and moisture from the surface water.

Therefore, four variables tend to affect the major characteristics of air masses: whether a mass forms over a continent (c) or over a maritime region (m), and whether it forms in the tropics (T) or in the polar regions (P). The resulting characteristics of the four basic air-mass types are listed in TABLE B12.1. Warm fronts tend to be associated with mT air masses, and with anticyclones. Cold fronts are generally associated with cP or mP air masses, and with cyclones. The kinds of air masses of greatest importance as far as the weather of North America is concerned are cP and mT. The cP air masses originate in Canada, in the Arctic, and to a lesser extent in Alaska; the mT air masses originate in the Gulf of Mexico, the Atlantic Ocean, the Caribbean Sea, and the Pacific Ocean (FIG. B12.2).

TABLE B12.1 Characteristics of Air Masses

Origin of Air	Temperature	Humidity
Continental polar (cP)	Cold	Low
Maritime polar (mP)	Cool	High
Continental tropical (cT)	Hot	High
Maritime tropical (mT)	Warm	High

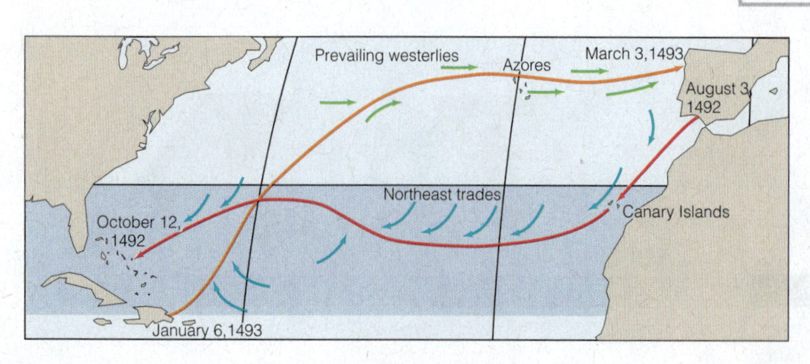

FIGURE 12.9 **Columbus and the trade winds** This map shows the winds used by Columbus on his first voyage to America. Outward bound after visiting the Canary Islands from August 12 to September 8, 1492, he sailed west with the northeast trades behind him. On his return voyage, Columbus sailed north to pick up the prevailing westerlies that had prevented previous European mariners from sailing any further west than the Azores.

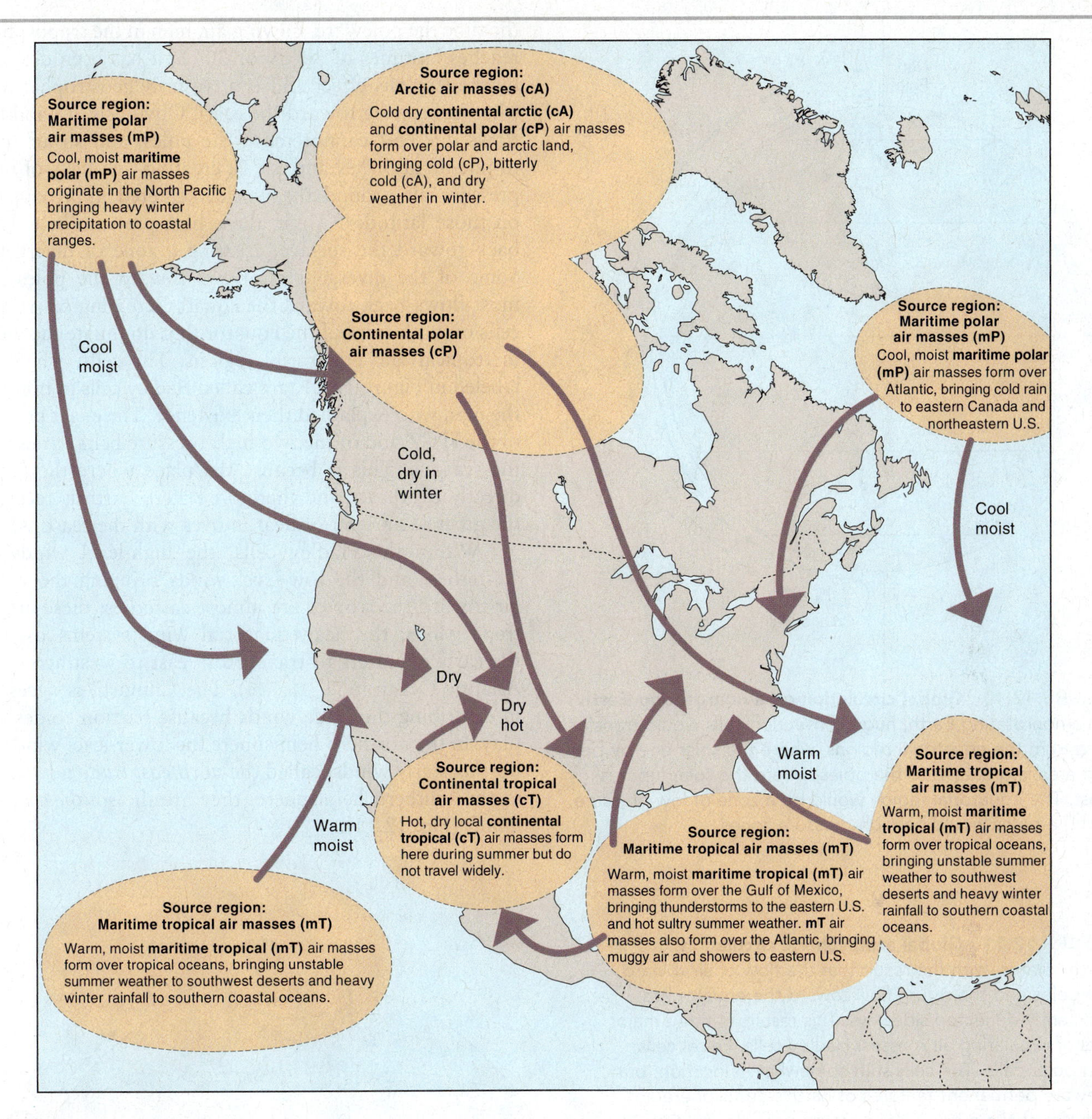

FIGURE B12.2 Air masses
This map shows the sources of the major air masses that typically control the weather in North America.

As discussed in Chapter 11, the boundaries between air masses of different temperature and humidity, and therefore different density, are called *fronts*. Fronts are generally between 10 and 150 km in width, and mark the advance of one air mass into another. Because they are locations where air masses with very different properties meet, fronts are commonly associated with active weather, including wind, rain or snow, and rapid changes in temperature.

for a nonrotating Earth. Instead, the *Coriolis effect* causes the global convection cells in the atmosphere to break into several global-scale circulation cells (FIG. 12.11).

Hadley Cells and the ITCZ

The first major circulatory cell stretches from the equator to approximately 30° N and 30° S latitude. On a rotating Earth, as on a nonrotating one, warm air rises in the tropics and creates a low-pressure zone of convergence, called the **intertropical convergence zone** (or **ITCZ**) (FIG. 12.12A). The convergence, heating, and rise of warm air in the ITCZ results in consistently low air pressure readings.[3] By

[3] Sea-level air pressure readings in equatorial regions are generally below the standard sea-level pressure of 101.3 kPa. They fall in the range 100.0 to 101.1 kPa, while pressures over the poles can be as high as 103.0 kPa.

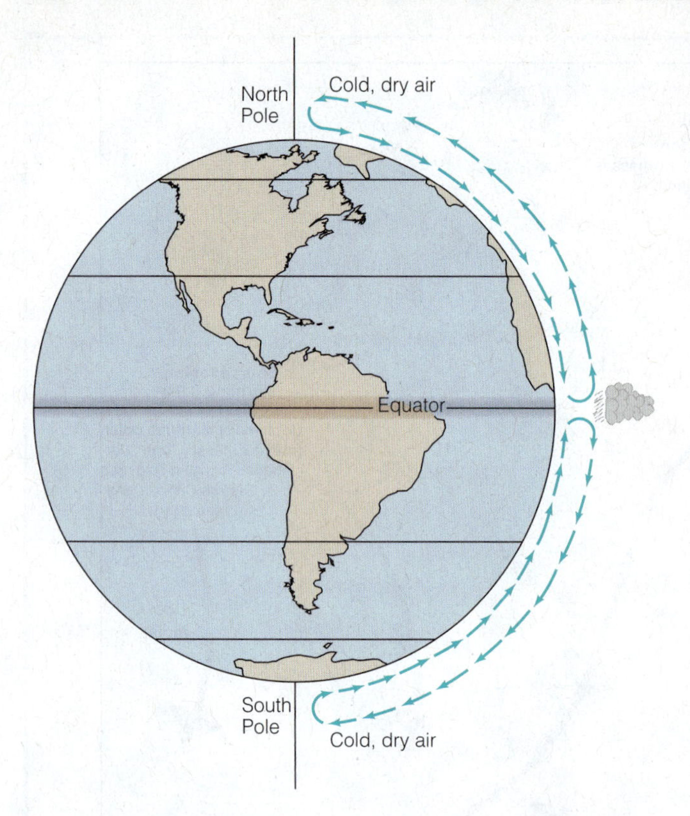

FIGURE 12.10 **Global circulation on a nonrotating Earth** On a nonrotating Earth, huge convection cells would transfer heat from the equatorial regions, where the solar energy per unit area is greatest, to the poles, where the solar input is least. The equatorial region would be a zone of low pressure, and the poles would be high-pressure zones.

the time the poleward-flowing air, high in the troposphere, reaches latitudes of 30° N or 30° S, it has been deflected by the Coriolis effect and is a westerly geostrophic wind (that is, blowing toward the east). Obviously, a wind that flows due east cannot reach the poles, and so air tends to pile up at 30° N and 30° S, creating two belts of high-pressure air around the world, centered approximately on those latitudes. Air in these high-pressure belts sinks back toward the surface, creating a zone of divergence. Some of the divergent air flows toward the poles, but most flows back toward the equator, creating convection cells on both sides of the equator that dominate the winds in tropical and equatorial regions. The cells, which are labeled in Figure 12.11, are called **Hadley cells** in honor of the man who explained their existence. The exact position of the ITCZ and of the two high-pressure belts varies with the seasons. This is because the place where the Sun is directly overhead, and therefore where Earth is receiving the greatest amount of heat, moves with the seasons.

Within the Hadley cells, the high-level winds are westerlies, and the low-level winds bringing the return air toward the tropics are almost easterlies; these are the **trade winds**, the near-equatorial wind systems used by the early explorers to travel from east to west across the Atlantic Ocean (FIG. 12.12B). The "almost" is necessary in describing the trade winds because friction comes into play. In the northern hemisphere the lower-level winds are northeasterly winds, called the *northeast trade winds*, and in the southern hemisphere, they are the *southeast trade winds* (Fig. 12.11).

FIGURE 12.11 **Global atmospheric circulation** Earth rotates, with the result that the flow of air toward the poles and the return flow toward the equator are constantly deflected sideways. This results in three major sets of circulating air masses: Hadley cells, Ferrel cells, and polar cells. The cells shift somewhat in location, but they are permanent features of Earth's atmosphere and therefore have a great influence on both day-to-day weather and long-term climate.

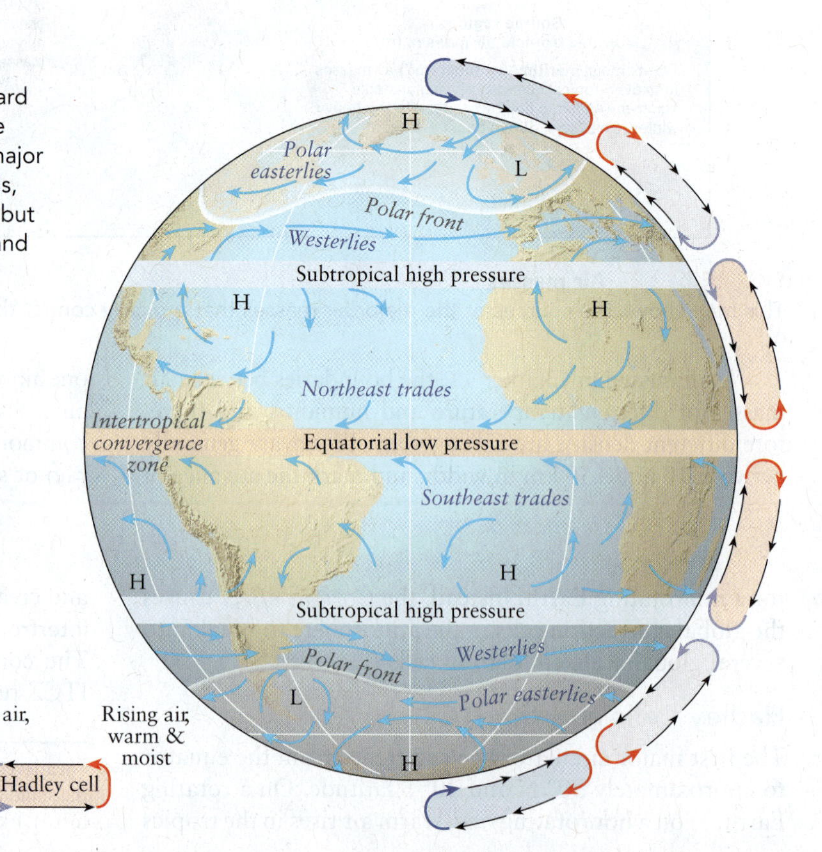

FIGURE 12.12 Trade winds and the ITCZ
(A) The Intertropical Convergence Zone, where the trade winds of the northern and southern hemispheres merge, shows up very clearly in this satellite photograph. Warm, rising air near the equator causes ocean water to evaporate and form a nearly perpetual band of storm clouds. (B) The continual blowing of the east-to-west trade winds has caused this tree, near the southern tip of Hawaii, to lean far to one side.

Ferrel Cells

In each hemisphere, on the poleward side of the Hadley cells, there is a second, middle-latitude circulation system (Fig. 12.11). The midlatitude cells are called **Ferrel cells** after the American meteorologist, William Ferrel (1817–1891). In Ferrel cells, the surface winds are westerlies because they are created, in part, by poleward flows of air from the high-pressure divergence regions at 30° N and 30° S latitude. These westerlies were the winds that prevented mariners before Columbus from sailing any farther west than the Azores Islands. Ferrel cells are sometimes described as acting like a "ball-bearing" between two great circulatory systems, the Hadley cells on the equator side and the polar cells on the other side.

Polar Fronts and Jet Streams

The third major region of atmospheric circulation cells, called the **polar cells**, lies on the poleward side of the Ferrel cells, extending to the polar regions (Fig. 12.11). In each polar cell, cold, dry, high-altitude air descends near the pole, creating a high-pressure area of divergence. Then air from this area moves toward the equator in a surface wind system called the polar easterlies. As this air moves slowly toward the equator, it encounters the middle-latitude belt of surface westerlies in the Ferrel cells. The two wind systems meet along a zone called the **polar front** and create a low-pressure zone of convergence that is analogous to the intertropical convergence zone. The polar front is a region of unstable air along which severe atmospheric disturbances occur.

The high-level winds in the polar cells are westerly. Indeed, because some of the high-level air in the Hadley cells spills over into the midlatitude Ferrel cells, the prevailing high-level winds poleward of 30° N and 30° S are all westerlies. Flow in the upper atmosphere is not uniform, however. Rather, the winds flow in great undulating streams called *Rossby waves*; these undulations resemble the meanders of streams and rivers.

Recall from Chapter 11 that the unequal heating of Earth's surface causes the top of the troposphere (the tropopause) to be much lower at high latitudes than at low latitudes (16 km at the equator and 10 km at the poles). The region where the height of the tropopause changes most rapidly is over the polar front (see Fig. 11.10). A large body of cold, polar air fills the troposphere poleward of the polar front, while warmer, subtropical air fills the troposphere on the equatorial side, but the top of the troposphere is everywhere at the same pressure. In other words, the tropopause is an isobar. This means that, in the stratosphere, there is a very steep pressure gradient over the polar front; high pressure is on the poleward side, and low pressure on the equatorial side of the stratosphere. Steep pressure gradients mean high-speed winds, and because friction is not involved, the winds are

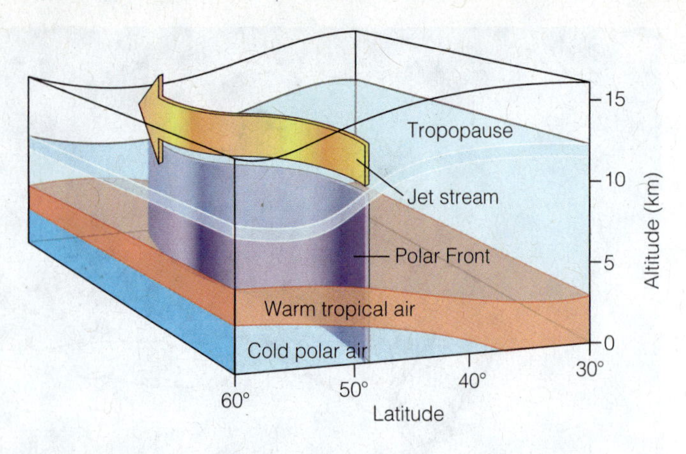

FIGURE 12.13 Jet stream
The jet stream is a high-speed westerly geostrophic wind that occurs at the top of the troposphere over the polar front, where a steep pressure gradient exists between cold polar air and warm subtropical air. At this location, the altitude of the tropopause declines precipitously.

geostrophic. Upper-atmosphere westerlies associated with this steep pressure gradient, called a **jet stream** (in this case, the *polar front jet stream*), can develop exceptionally high speeds (FIG. 12.13). Wind speeds up to 460 km/h have been reported by high-flying planes in jet streams.

Rossby waves distort the polar-front jet stream into great undulations (FIG. 12.14). As the jet stream undulates, it pushes and pulls the polar front with it and thus plays a major role in weather patterns between 45° and 60° north and south latitudes.

A second jet stream, also a geostrophic westerly, called the *subtropical jet stream*, forms above the tropopause over the Hadley cell, between latitudes 20° and 30° north and south. Speeds of westerly winds in the subtropical jet stream reach 380 km/h, but because the troposphere is higher at low latitudes, the subtropical jet is at a higher altitude than the polar-front jet and does not play the dominant weather role exerted by the polar-front jet stream.

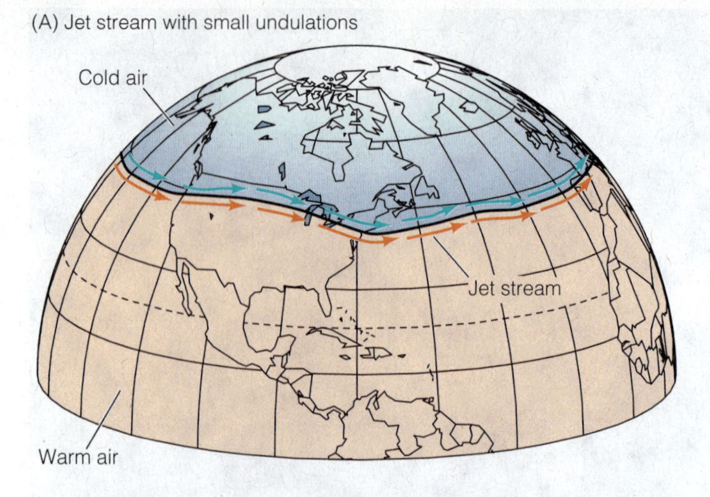

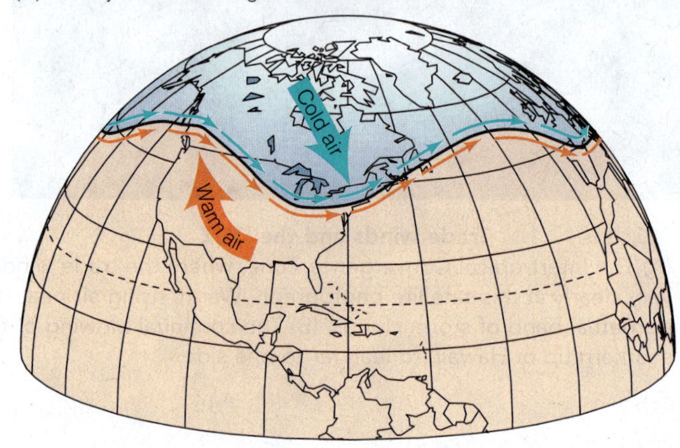

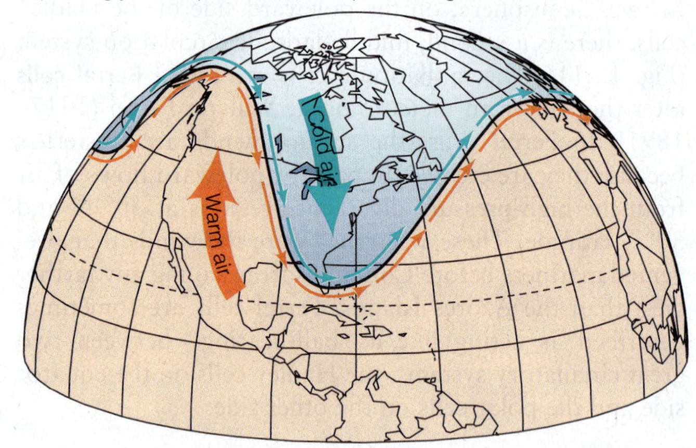

FIGURE 12.14 Rossby waves
Rossby waves distort the polar-front jet stream. (A) The axis of the jet stream starts out flowing to the east in a nearly straight line. Undulations grow into gigantic meanders that pull masses of cold polar air toward the south (B, C).

REGIONAL WIND AND WEATHER SYSTEMS

Our discussion of global atmospheric circulation systems has said little about the uneven distribution of land and sea around the world. In fact, both land distribution and land elevation play important roles in regional and local wind patterns and weather systems. Two of the most important regional weather systems are monsoons and the El Niño-Southern Oscillation. Let's examine these two regional systems in more detail.

Monsoons

A **monsoon** is a seasonally reversing wind system. The places on Earth where this phenomenon is most distinct are in Asia and Africa, although a weak monsoon develops over eastern and central North America, too.

The Asian monsoon is the most pronounced. Because the equator lies just south of the tip of India, the normal surface-wind pattern is a northeasterly trade wind blowing offshore from India into the Indian Ocean. For half a year during the winter months, the expected northeasterly

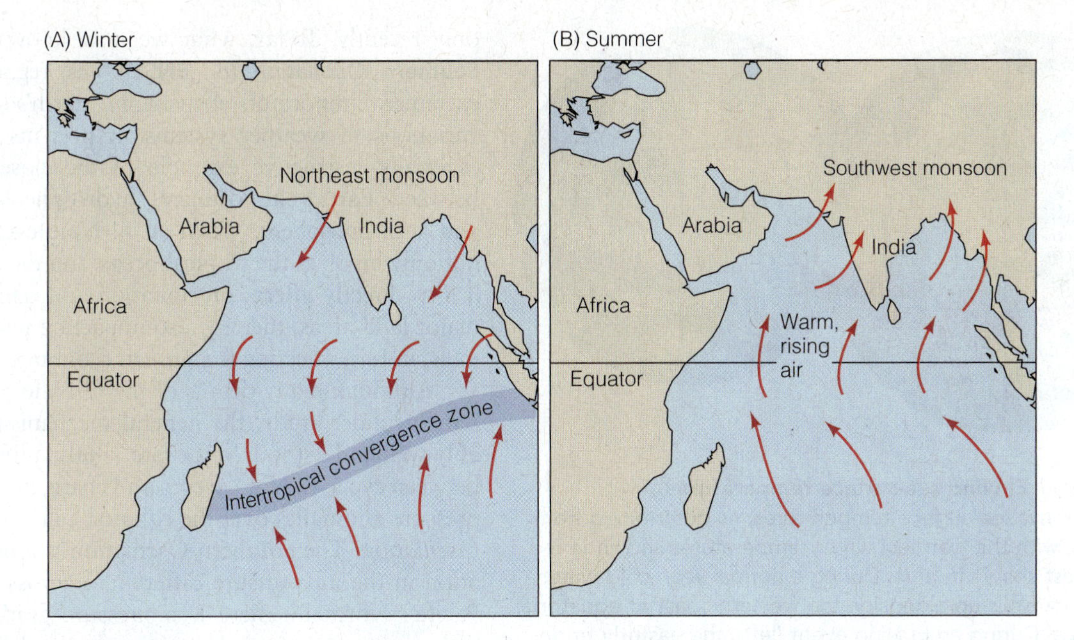

FIGURE 12.15 Monsoon
A monsoon is a regional weather system characterized by seasonally reversing winds. (A) During the winter when the Sun is overhead in the southern hemisphere, winds flow offshore from the northeast toward the intertropical convergence zone. Note how the winds curve toward the east as they cross the equator. (B) During the summer, the land heats up and winds flow from the southwest across Asia. When the Sun is overhead on land, the intertropical convergence zone moves farther to the north; in Asia, it becomes less pronounced and is not a distinct band of low pressure.

wind pattern is observed because a high-pressure anticyclone sits over the high, cold plateau of Central Asia, while the low-pressure intertropical convergence zone lies south of the equator, where the Sun is overhead (FIG. 12.15A). Winter months on the Indian subcontinent are therefore a time of cool, dry, cloudless days and northeast winds. For more than 2000 years, Arab sailors have used these northeasterly winds to sail home from India.

During the summer months, however, the Asian wind pattern is reversed (FIG. 12.15B). With the Sun overhead in the northern hemisphere, the intertropical convergence zone shifts north of the equator. As a consequence, the landmass of Asia heats up and is covered by low-pressure cyclones. The winds are now southwesterlies, blowing warm, moisture-laden air from the Indian Ocean onto the land. Summer months are therefore a time of hot, humid weather and torrential rains. The summer monsoons start in southern India and Sri Lanka in late May, progress to central India by mid-June, and reach China by late July. Arab sailors made use of this period to sail from Arabia to India. They referred to the change in wind direction as *mausim*, Arabic for change, and from this comes our word *monsoon*.

A monsoon system like the one that affects India also occurs in North and West Africa. There are local differences, but the main controlling factor in both cases is the seasonal movement of the intertropical convergence zone. A weak monsoon system also occurs in North America. During the summer months, there is a tendency for surface winds to bring warm, moisture-laden air from the Gulf of Mexico into the central and eastern United States. Humid weather and summer rains are the result. In the winter months, the winds reverse and there is a tendency for cold air to move southward from Canada toward the Gulf.

Make the CONNECTION

Regional weather systems such as the Asian monsoon and El Niño have obvious impacts on human interests, but what are some of their impacts on the plants, animals, soils, water bodies, and ecosystems of the regions in which they occur?

El Niño and the Southern Oscillation

The fishing grounds off the coast of Peru, among the richest in the world, are sustained by upwelling cold waters filled with nutrients. Periodically, a mass of unusually warm water appears off the coast in the Pacific Ocean (FIG. 12.16). Peruvians refer to this event as El Niño because it commonly appears at Christmas time, the season of the Christ Child (El Niño). During El Niño years, the trade winds slacken, upwelling is markedly reduced, and the fish population declines, accompanied by a great die-off of the coastal bird population, which depends on the fish for food. The Peruvian fishery is among the most important in the world, so the occurrence of a major El Niño event can constitute an economic disaster. Coincident with the Peruvian El Niño conditions,

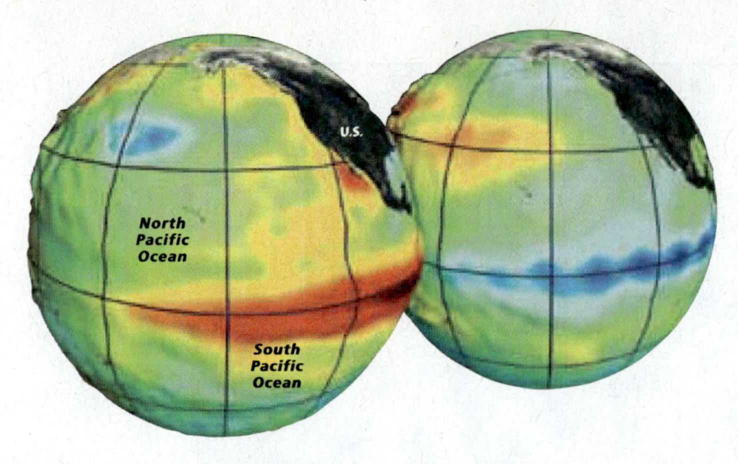

FIGURE 12.16 El Niño sea-surface temperatures
These maps show sea-surface temperatures, as determined from satellite data, with the warmest water temperatures shown in red and the coolest shown in blue. During a normal year, cold water from Antarctica wells up alongside the western coast of equatorial South America. During an El Niño event (left), the easterly trade winds fail to push warm surface waters away from South America, with the result that a tongue of warm water builds up in the eastern equatorial Pacific Ocean. The opposite phase, called *La Niña* (right), is characterized by cold ocean temperatures across the equatorial Pacific.

very heavy rains fall in normally arid parts of Peru and Ecuador, Australia experiences drought conditions, anomalous cyclones appear in Hawaii and French Polynesia, the seasonal rains of northeastern Brazil are disrupted, and the Indian monsoon may fail to appear. During exceptional El Niño years, weather patterns over much of Africa, eastern Asia, and North America are affected. In North America, unusually cold winters can result in the northern United States, while the Southeast becomes wetter; in California, abnormally high rainfall can produce major flooding and widespread landslides.

A Coupled Atmosphere-Ocean Phenomenon

El Niño has been experienced by generations of Peruvians, but its broader significance as a coupled atmosphere-hydrosphere phenomenon was recognized only recently. Today, what we now know as **El Niño/ Southern Oscillation (or ENSO)** is regarded as an extremely important element in Earth's year-to-year variations in weather systems. It presents us with an especially instructive example of the close interaction between Earth's atmosphere, hydrosphere, biosphere, and anthroposphere. When an El Niño event occurs, it not only involves the tropical ocean and the atmosphere, it also directly affects precipitation and temperature on major land areas, thereby also impacting plants and animals, in turn affecting the human economy.

Although many details of the El Niño phenomenon remain under study, the general mechanism is reasonably well understood. In the late 1960s, a link was made between cyclic El Niño events and changing atmospheric pressure anomalies over the equator, called the *Southern Oscillation*. The Southern Oscillation is a periodic variation in the air pressure differential across the tropical Pacific. Normally there is a pressure gradient from a high-pressure zone in the eastern tropical Pacific (typically measured near Tahiti) to a low-pressure zone in the western Pacific (typically measured near Darwin, Australia); this is what causes the trade winds to blow toward the west. In the western Pacific this warm, moist air rises, causing cloudiness and rain in the area of Indonesia. Once aloft, cooler and drier, the air returns to the eastern Pacific at high altitude (FIG. 12.17). This wind pattern, which is specific to the equatorial Pacific, is called the *Walker circulation*. The difference in air pressure

(A)

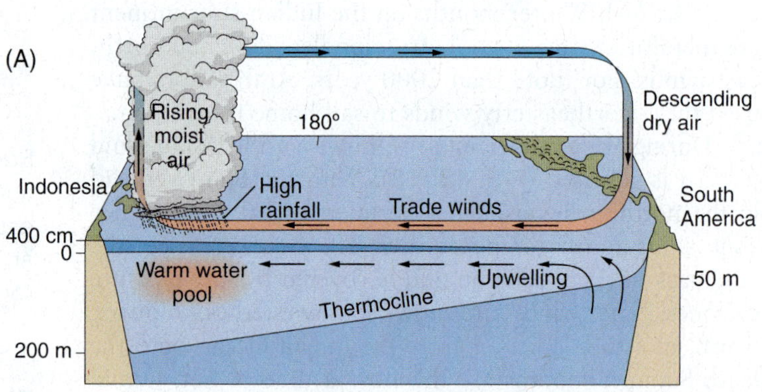

(B)

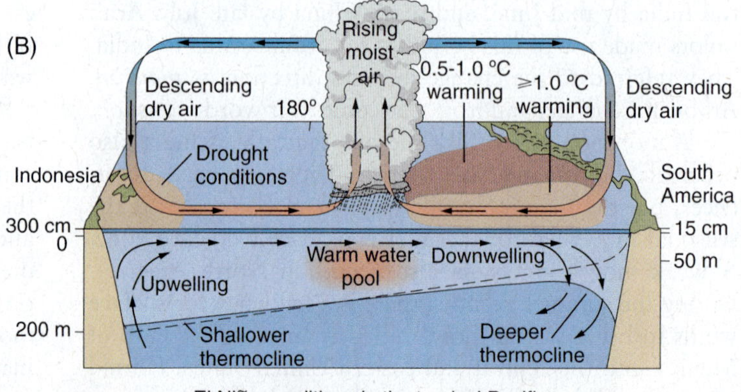

FIGURE 12.17 Walker circulation and El Niño
In a normal year (A), the air pressure difference between the eastern and western Pacific causes the trade winds to blow to the west across the tropical Pacific, pushing the warm water away from the coast of South America. This allows cold water from Antarctica to well up in the eastern Pacific. In the western Pacific, the warm water causes the moist air to rise and cool, bringing abundant rainfall in Indonesia. The drier, cooler air returns at high altitude to the eastern Pacific, in the Walker circulation. (B) During an El Niño year, the trade winds slacken or even reverse, and warm water accumulates in the eastern and central Pacific. A strong El Niño is disruptive to weather over much of the planet.

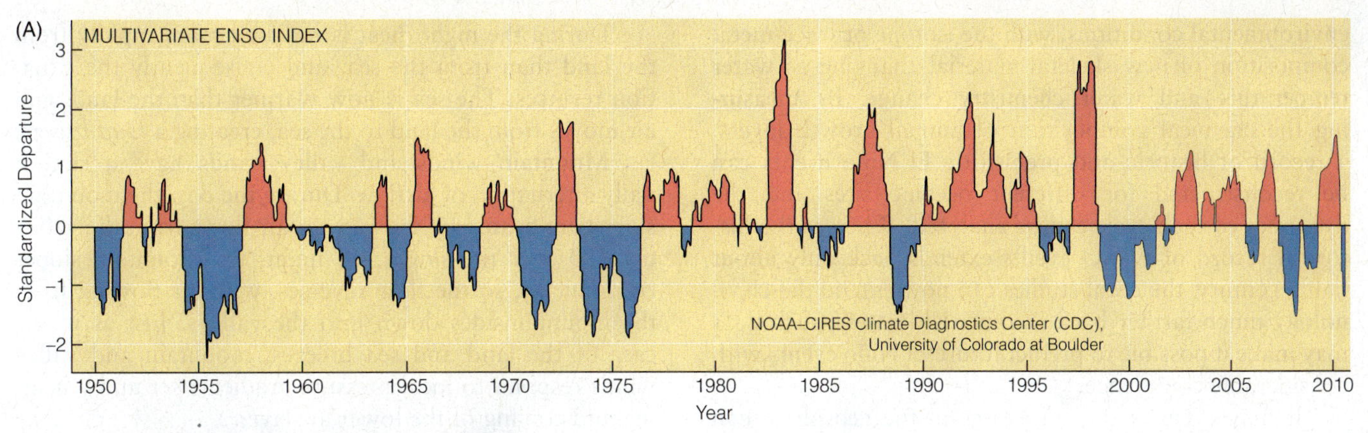

(B)

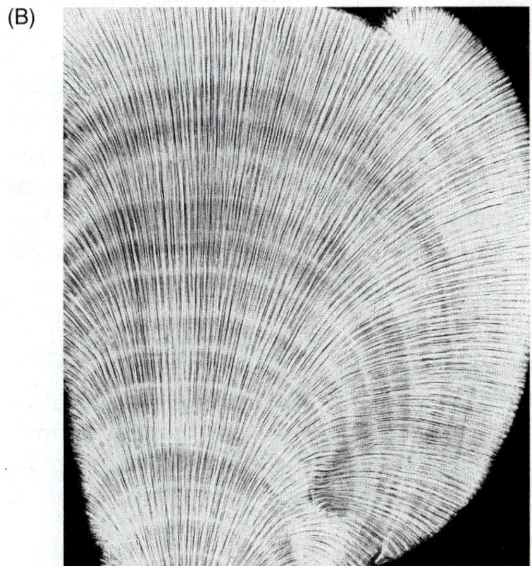

FIGURE 12.18 Past El Niño events
(A) This record of more than a half-century of El Niño events (1950–2010) shows the irregular frequency and magnitude of the events. Plotted on the vertical axis is the Multivariate ENSO Index, or MEI, which includes measures of sea-level pressure, surface winds, sea-surface temperature, surface air temperature, and total cloudiness. The zero line is a "normal" year, red represents El Niño, the warm phase of the cycle, and blue represents La Niña, the cold phase. The strongest El Niños on record were those of 1982–1983 and 1997–1998. (B) A slice through a coral from the Galápagos Islands shows annual layering (alternating light and dark bands). This layering preserves a record of changing surface water conditions, and therefore of sea-surface temperature and El Niño events.

from east to west in the tropical Pacific is referred to as the *Southern Oscillation Index*, or *SOI*. When the pressure difference is large, then the SOI is high, and the Walker circulation and trade winds will be strong; when the pressure difference is small, then the SOI is low, and the Walker circulation and trade winds will be weak.

During normal years, then, the Walker circulation causes the easterly trade winds to blow across the Pacific Ocean, pushing warm surface water toward the western Pacific, where it accumulates. This allows deep, cold water from the Southern Ocean to well up to the surface in the eastern tropical Pacific along the coast of South America (FIG. 12.17A). Recall that we discussed *upwelling* of ocean water in Chapter 10; in order for cold Antarctic water to well up to the surface in the eastern Pacific, the trade winds must move the warm surface water away from the equatorial coast of South America; if this doesn't occur, the upwelling will be suppressed.

An El Niño event begins with a weakening of the pressure differential between the eastern and western Pacific; in other words, the SOI decreases. This causes the Walker circulation to weaken, and the trade winds to weaken or even reverse. Without the trade winds, anomalously warm surface water accumulates in the central and eastern Pacific near South America, and this stops the upwelling of cold, nutrient-rich water (FIG. 12.17B). The zone of high rainfall that is normally situated near Indonesia shifts toward the central Pacific, simultaneously bringing drought conditions to Indonesia. At the peak of an event, equatorial surface water moves from west to east and also poleward. This flow gradually reduces the equatorial pool of warm water, leading to intensification of the trade winds and eventual return to normal conditions.

Understanding ENSO

We now recognize that El Niño recurs erratically, but on average about every four years (FIG. 12.18A). Its effect on weather is felt over at least half of Earth; therefore, a major area of research interest is to identify the factors that trigger El Niño events, or at least to be able to detect their onset as early as possible.

In an attempt to extend the detailed record of El Niño events farther back in time and improve our ability to predict future occurrences, scientists have examined the growth rings of living corals because the rings record annual variations in seawater conditions (FIG. 12.18B). Corals are abundant and widespread throughout the region most strongly affected by El Niño, and individual colonies can live as long as 800 years. Their skeletal chemistry closely reflects surrounding environmental conditions, with the isotope or trace-metal composition of new skeletal material changing as water temperature and water chemistry change. By measuring the chemical composition of annual growth layers, a record of historic and prehistoric El Niño events can be reconstructed for different oceanic sites and the dynamics of each cycle can be analyzed. Whereas the historical record of ENSO events extends back only about half a century, the coral studies can now extend the chronology much farther back in time. Ultimately, such data may make it possible to predict future El Niño events with considerable confidence.

It may even be that processes in the geosphere can trigger an El Niño event. The temperature of ocean water plays a role in determining the temperature, and thus the density and air pressure, of an overlying air mass. Some evidence suggests that submarine volcanic eruptions may introduce enough heat, concentrated in a limited area, to produce a perturbation in ocean temperature. If this temperature anomaly is then translated into the overlying air mass, it could influence the characteristics of the air masses that are normally involved in the tropical Walker circulation. If this does happen, then El Niño is truly a phenomenon in which the different spheres of the Earth system are interacting and influencing each other.

LOCAL WIND AND WEATHER SYSTEMS

In many localities, local winds are more important than global winds in influencing the weather on a daily basis. Local winds, which may flow for tens or hundreds of kilometers, rather than the thousands of kilometers involved in global winds, are the result of the interaction of local terrain with the forces that cause air to move.

Coupled Local Wind Systems

The least complicated example of a local wind system is the coupled land breeze and sea breeze that is familiar to anyone who lives on or near a coast. The origin of these breezes is illustrated in FIGURE 12.19. During the day the land heats up more rapidly than the sea, and the heated land causes the air in contact with it to heat up and expand. A pressure gradient develops, and the lower air layer flows toward the land, creating a *sea breeze*. Higher in the atmosphere, an upper-level reverse flow sets in; the coupled flows—rising air over the land and sinking air over the sea—form a convection cell.

During the night, heat is radiated more rapidly from the land than from the sea, and consequently the situation reverses. The sea is now warmer than the land, and air moves from the land to the sea, creating a *land breeze*.

Mountain winds and valley winds have a similar daily alternation of airflow. During the day, the mountain slopes are heated by the Sun, so air flows from the valley upward over the slopes. At night, the mountain slopes cool quickly, so the flow reverses, with air flowing from the mountainsides down into the valleys. Just as in the case of the land and sea breezes, mountain and valley winds respond to local pressure gradients set up by heating and cooling of the lower air layer.

Katabatic and Chinook Winds

The flow of cold, dense air under the influence of gravity is called a *katabatic* wind. Such winds occur in places where a mass of cold air accumulates over a high plateau or in a high valley in the interior of a mountain range. As the cold air accumulates, some eventually spills over a low pass or divide and flows down valleys on to the adjacent lowlands as a high-speed, cold wind.

FIGURE 12.19 Land and sea breezes
(A) During the day, the land heats up more rapidly than does the sea. Air rises over the land, creating a low-pressure area. Cooler air flows in to this area from the sea, creating a sea breeze. (B) During the night, the land cools more rapidly than the sea, and the reverse flow, a land breeze, occurs.

(A) Day

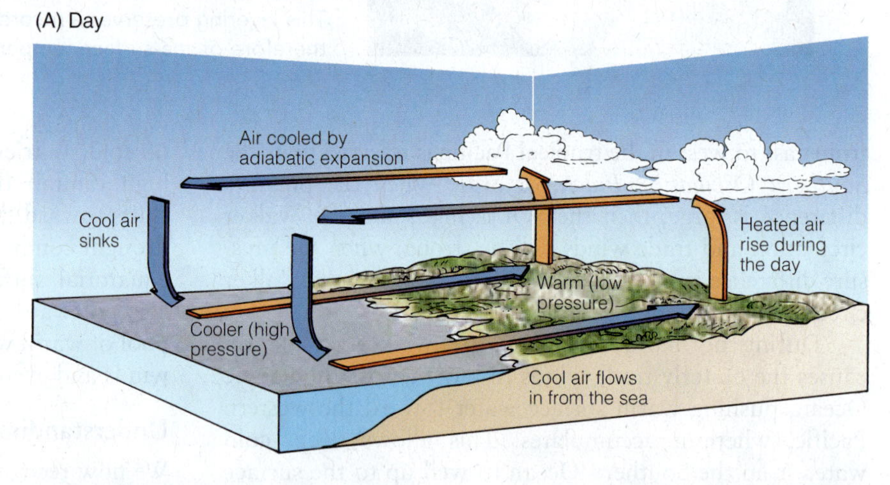

(B) Night

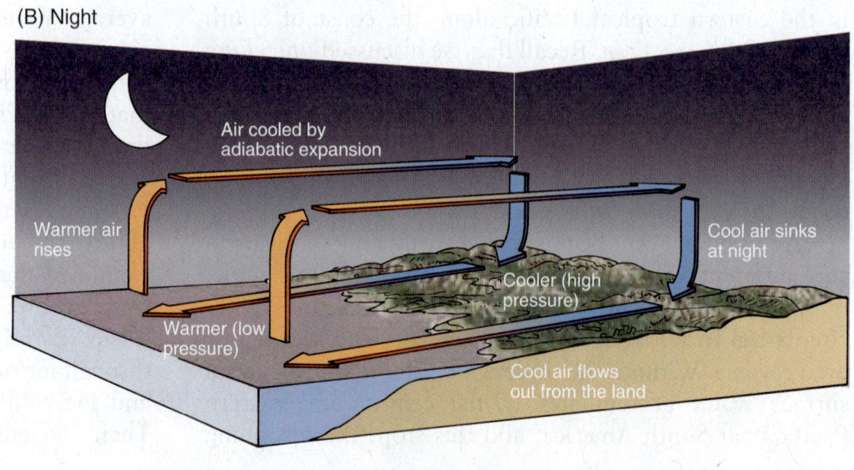

Katabatic winds occur in most mountainous regions around the world and commonly have local names. The *mistral* is a notable example; it is a cold, dry wind that flows down the Rhône Valley in France, past Marseilles, and out onto the Mediterranean Sea. Another notable example is the *bora*, a northeasterly wind that rushes down from the cold highlands of Croatia, Bosnia and Herzegovina, and Montenegro to the Adriatic Sea near Trieste. Wind gusts in Trieste during a bora can reach speeds of 150 km/h.

The most striking examples of katabatic winds are those that occur around the edges of Greenland and Antarctica, where the frigid, high-pressure air masses that accumulate above the continental ice sheets pour down the sloping margins of the ice and out onto the adjacent ocean waters. When the ice slope is steep, the katabatic wind speed can be terrifyingly high. It is because of a katabatic wind that Cape Dennison in Antarctica has a higher annual average wind speed than any other place on Earth.

Related to katabatic winds is another class of downslope land winds known by various local names, including *chinook* along the eastern slopes of the Rocky Mountains, *föhn* in Germany, and *Santa Ana* in Southern California. For simplicity, we speak now only of chinooks, but these wind systems are similar, regardless of the name being used. Chinooks are warm, dry winds. Because warm, dry air has a low density and so does not sink naturally, a chinook must be forced downward by large-scale wind and air pressure patterns. This forcing occurs when strong regional winds, commonly associated with anticyclones, rise and compress higher-level air masses as they pass over a mountain range and then are forced to flow down on the downwind side by the pressure of higher-level air. The result is that the downward-flowing air is heated adiabatically and is therefore dry—in short, a chinook.

SEVERE WEATHER

"Severe weather" generally refers to a violent disturbance of the atmosphere attended by strong winds and commonly rain, snow, hail, sleet, thunder, and lightning. Severe weather can have many causes, but most occurs along cold fronts. Severe weather causes much damage, economic disruption, injury, and loss of life worldwide every year. Blizzards and ice storms can paralyze local communities, but on a global scale the most damaging type of severe weather events are tropical cyclones and droughts.

Cyclones

The term *cyclone* refers to any cyclonic circulating wind system, that is, a wind system that is circulating around a low-pressure center, counterclockwise in the northern hemisphere or clockwise in the southern hemisphere, as dictated by the twin forces of air pressure gradient and Coriolis effect (FIG. 12.20).

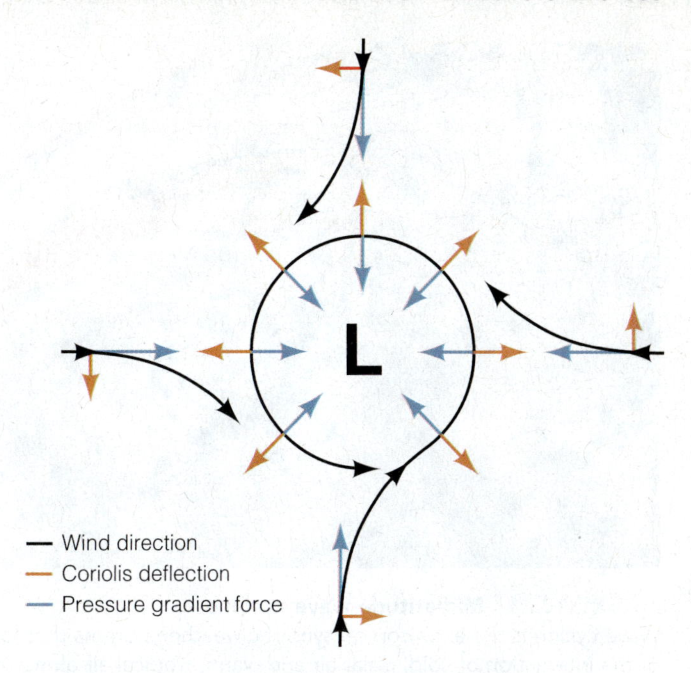

FIGURE 12.20 Rotation in a northern hemisphere cyclone
This is a map view showing how the Coriolis force and the pressure-gradient force combine to cause cyclonic wind systems to rotate in a counterclockwise direction in the northern hemisphere. The black arrows show the wind directions. The blue arrows are the pressure-gradient force, which tends to make the wind blow in toward the low-pressure center. The red arrows represent the Coriolis deflection, toward the right. The combination yields a counterclockwise-rotating wind—a cyclone.

Types of Cyclones

There are many different types of cyclones, distinguished by where they form, how they develop, and their characteristics of temperature, air pressure, wind speed, and geographic extent. The most intense and typically the most damaging are *tropical cyclones*, sometimes called *warm-core cyclones*. They are generally referred to as cyclones if they form over the Pacific Ocean, *typhoons* if they form over the western Pacific or Indian Ocean, and *hurricanes* if they form over the Atlantic Ocean.

Wave cyclones, also called *extratropical* or *midlatitude cyclones*, typically form between 30° and 60° N and S, as a result of the interaction between warm and cold air along the polar fronts. Wave cyclones are large in geographic extent, up to 2000 km across (FIG. 12.21). (In meteorology, systems that span several hundred to several thousand kilometers are called *synoptic weather systems*; synoptic comes from the Greek words meaning "view together".) They tend to last many days, and they are responsible for most of the everyday weather events in the midlatitude regions of the world, including rain showers, snowstorms, and other forms of precipitation, and they can spawn more local, intense weather systems such as thunderstorms.

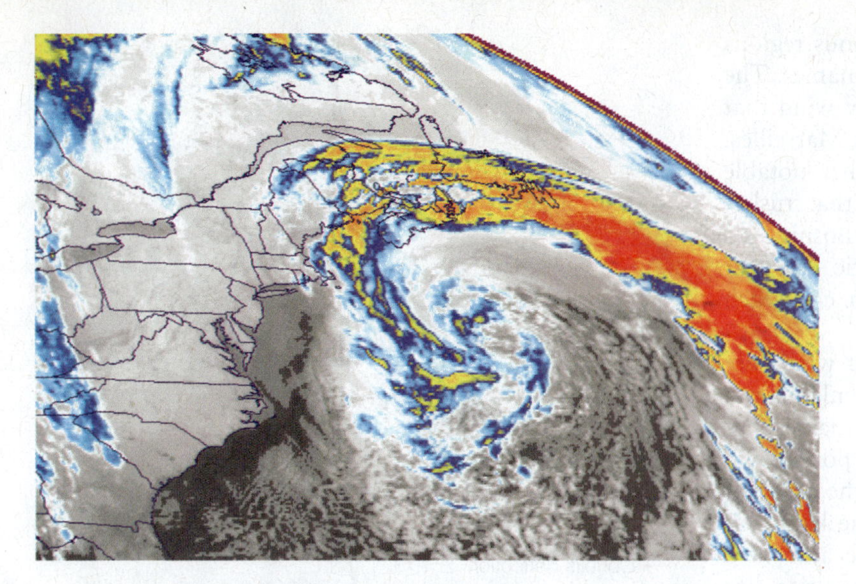

FIGURE 12.21 Midlatitude wave cyclone
Wave cyclones are extratropical synoptic weather systems that form as a result of the interaction of cold, polar air and warm, tropical air along the polar front. In this infrared image, collected by instruments on the GOES-7 satellite, shows a developing midlatitude cyclone off the eastern coast of North America.

Tropical Cyclones

A **hurricane**, by definition, is a tropical cyclonic storm with maximum wind speeds that exceed 119 km/h (FIG. 12.22A). They start as wind systems gently circulating over warm ocean water in the eastern Atlantic, near Africa (FIG. 12.22B), and they require a sea-surface temperature of at least 26.5°C to develop. Drawn upward into the low-pressure center, called a *depression*, the warm water evaporates and subsequently condenses. This releases latent heat and fuels the storm, which may develop into a *tropical storm* and eventually into a hurricane. Hurricanes and other cyclonic storms can be generated only in latitudes where the Coriolis effect is strong enough for cyclonic circulation to develop—that is, higher than about latitude 5° N or S. Recall that the Coriolis effect is zero at the equator, which means that hurricanes cannot form along the equator.

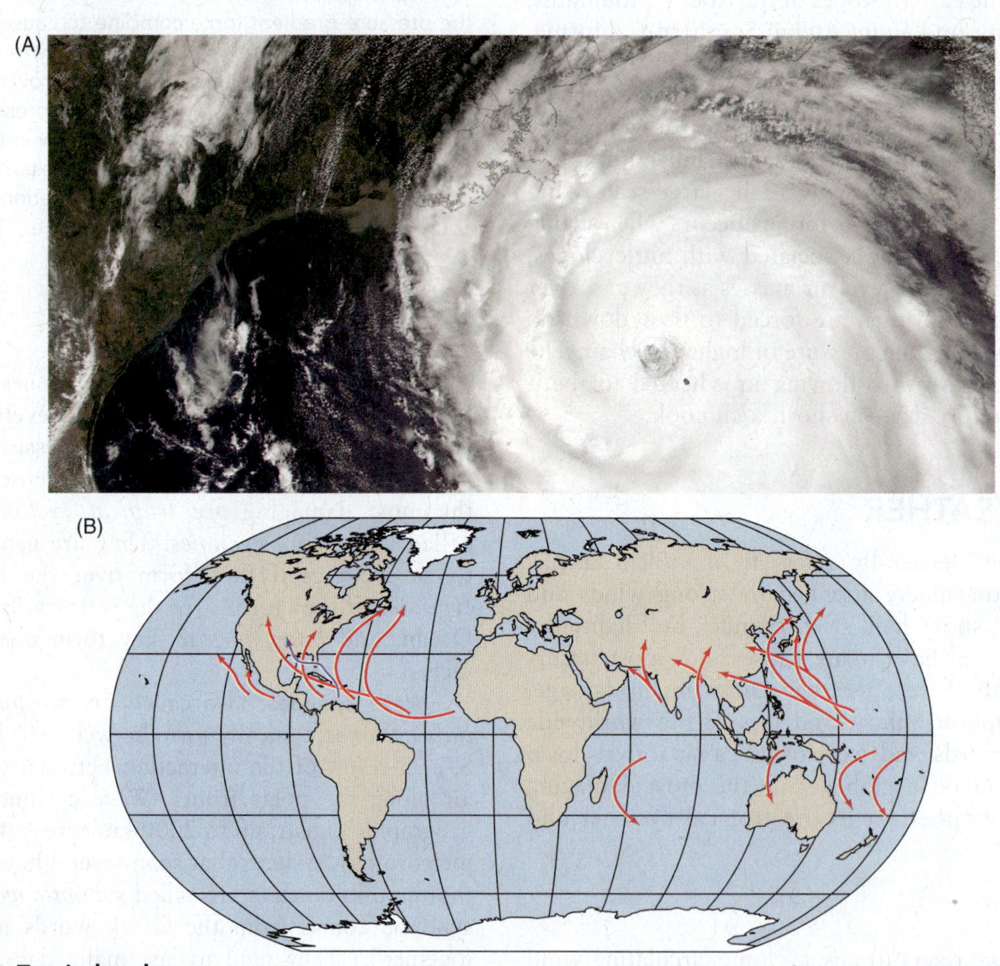

FIGURE 12.22 Tropical cyclones
(A) Hurricane Katrina slammed into the Gulf coast of Louisiana in August of 2005. Unusually warm sea-surface temperatures in the Gulf of Mexico that summer may have contributed to the very active hurricane season. (B) Cyclones and hurricanes form in places where the right conditions of ocean-water temperature and the Coriolis effect occur. Arrows show the usual directions followed by hurricanes once they form. In the Atlantic Ocean, hurricanes typically originate as tropical depressions off the coast of Africa. The path followed by Hurricane Katrina is shown in blue.

Because a hurricane draws its energy from warm ocean water, its wind speeds will diminish and the hurricane will quickly dissipate once it moves onshore, away from its energy source. For this reason, most hurricane wind damage occurs within 250 km of the coast. Besides wind damage, two other hurricane effects can be devastating. One is a *storm surge*, a local, exceptional flood of ocean water. The center of a hurricane is a region of very low air pressure—below 92 kPa in the greatest hurricanes—and this drop in air pressure raises local sea level, causing the surface to actually bulge upward. In the eye of a great hurricane, sea level may be 8 or 9 m above normal, and when hurricane-force winds drive such high seas onshore, extensive flooding results. Storm surge, combined with failure of the levee system, is what led to such massive devastation following Hurricane Katrina in New Orleans in 2005. Another damaging effect associated with hurricanes is rain. Torrential rains and consequential flooding accompany most hurricanes—falls of 25 cm are not uncommon—and even after wind speeds have dropped below hurricane force, violent rainstorms can continue.

Thunderstorms and Tornadoes

Unlike hurricanes and related tropical cyclonic storms, tornadoes and thunderstorms form over land rather than over the ocean.

Thunderstorms

Thunderstorms develop when an updraft of warm, humid air (called a *cell*) releases a lot of latent heat very quickly and becomes unstable. Most thunderstorms in North America form along cold fronts and are associated with mT air masses formed over the Gulf of Mexico. The released heat causes stronger updrafts, which pull in more warm, moist air, which in turn releases more latent heat, and so the process grows and the updraft intensifies. Cumulonimbus clouds form and heavy rainfall, hail, thunder, and lightning are the result (FIG. 12.23).

The towering masses of cumulonimbus clouds associated with thunderstorms can reach as high as 18 km, and winds can exceed 100 km/h. Large *hailstones* can form in the updrafts as a result of the layering and freezing of supercooled water on an initial tiny ice crystal, remaining aloft until a sudden downdraft deposits them on the ground. *Lightning* and *thunder* are due to electrical charges being released in the thunderstorm. The electrical charges form during the growth of a cumulonimbus cloud. The turbulent movement of precipitation inside the cloud causes particles in the upper part to become positively charged and particles in the lower part to become negatively charged. Exactly how the charge builds up is not clearly understood, but the buildup can reach hundreds of millions of volts. The charges can be released by a lightning strike either to the ground or to another cloud. As the lightning strike passes, it heats the surrounding air so rapidly that the air expands explosively and we hear the effect as thunder.

(A)

(B)

FIGURE 12.23 Thunderstorms
(A) A thunderstorm over Tucson, Arizona shows the classic dark, anvil-shaped cumulonimbus clouds, dense, rain, and lightning. (B) Lightning strikes farm fields in Oklahoma during a severe thunderstorm.

Tornadoes

Tornadoes are violent windstorms produced by a spiraling column of air that extends downward from a cumulonimbus cloud (FIG. 12.24). They commonly develop from large thunderstorms that have multiple updrafts, which are called *supercell* thunderstorms. Tornadoes are approximately funnel-shaped, and they are made visible

FIGURE 12.24 Tornado
A classic funnel-shaped tornado crosses the plains of Nebraska in 2003.

by clouds, dust, and debris sucked into the funnel. By convention, a tornado funnel is called a *funnel cloud* if it stays aloft and a tornado if it reaches the ground.

Tornadoes are small features relative to the thunderstorms with which they are associated. Because they are so violent, many details of their formation are still unresolved. The funnel develops as a result of a spiraling updraft in a thunderstorm. Such updrafts are commonly 10 to 20 km in diameter. For reasons not clearly understood, a spiraling updraft in certain thunderstorms will narrow and spiral down to a tornado funnel from 0.1 to 1.5 km in diameter. Interestingly, the rotation within a tornado, while typically cyclonic, is not directly caused by the Coriolis force but by other factors within the storm system; the scale of both thunderstorms and tornadoes is too small for the Coriolis force to have a direct impact.

Fortunately, most tornadoes are not especially strong and do not cause much damage. At the other end of the strength scale, there are some tornadoes that completely destroy any object in their path. The strength of a tornado and associated wind speeds can be estimated from the damage it causes by referring to a scale that correlates wind speeds with typical damage (TABLE 12.1). The *F-scale* (now called the *Enhanced Fujita Scale* or *EF Scale*) was named for Professor T. Theodore Fujita of the University of Chicago, who originally devised it.

Hundreds of tornadoes take place each year in the United States; they occur in all states, during any month of the year. However, there is a distinctly intense period of tornado activity from April to August, with a peak in May. Because the severe thunderstorms that are the parents of tornadoes form along cold fronts and because the most violent cold fronts are those associated with cP air from the Canadian Arctic and mT air from the Gulf of Mexico, most tornadoes occur in the midcontinent states because that is where the two air masses are most likely to meet.

Drought and Dust Storms

Deserts, or *arid* lands, where the average annual rainfall is less than 250 mm, represent one of the great terrestrial biomes of this planet. We will consider deserts as ecosystems in Chapter 16. However, it is worth noting here that the distribution of deserts is controlled principally by the global atmospheric circulation system, influenced by local topography and proximity to sources of moisture. There is also an important distinction to be made between deserts and *drought*, an extreme weather phenomenon in which a region experiences below-average precipitation for an extended period. Because drought requires *below-average* precipitation, an area that normally receives extremely low precipitation would not be said to be experiencing drought conditions.

TABLE 12.1 Enhanced Fujita Scale for Tornado Intensity

EF-Scale	Estimated Wind Speed, km/h	Damage
EF0	105–137	Minor. Peels surface off some roofs; some damage to gutters or siding; branches broken off trees; shallow-rooted trees pushed over.
EF1	138–178	Moderate damage. Roofs severely stripped; mobile homes overturned or badly damaged; loss of exterior doors; windows and other glass broken.
EF2	179–218	Considerable damage. Roofs torn off well-constructed houses; foundations of frame homes shifted; mobile homes completely destroyed; large trees snapped or uprooted; light-object missiles generated; cars lifted off ground.
EF3	219–266	Severe damage. Entire stories of well-constructed houses destroyed; severe damage to large buildings such as shopping malls; trains overturned; trees debarked; heavy cars lifted off the ground and thrown; structures with weak foundations blown away some distance.
EF4	267–322	Devastating damage. Well-constructed houses and whole frame houses completely leveled; cars thrown and small missiles generated.
EF5	>322	Extreme damage. Strong frame houses leveled off foundations and swept away; automobile-sized missiles fly through the air in excess of 100 m; steel- reinforced concrete structure badly damaged; high-rise buildings have significant structural deformation.

Drylands and Atmospheric Circulation

Deserts make up about 25 percent of the land area of the world outside the polar regions. A smaller, though still large, percentage of *semiarid* land exists, in which the annual rainfall ranges between 250 and 500 mm. Together, these desert and semiarid regions form a distinctive pattern on the world map (FIG. 12.25), which is directly related to the global atmospheric circulation and to local features of the Earth's geography.

When we compare Figure 12.24 with Figure 12.11, a relationship is immediately apparent. There are three global belts of high rainfall and four of low rainfall. The high-rainfall belts are the three regions of global convergence—the intertropical convergence zone and the two polar fronts. The four belts of low rainfall are the regions of divergence—the two belts of subtropical highs centered on latitudes 30° N and 30° S, and the two polar regions. The distribution of the low-rainfall belts exerts significant controls on the locations of deserts. In all, five types of desert are recognized (FIG. 12.26 and TABLE 12.2, on page 371).

The most extensive deserts are the Sahara, Kalahari, Great Australian, and Rub-al-Khali (FIG. 12.26A). These great deserts are associated with the two circumglobal belts of divergence, where dry air descends on the downward-flowing limbs of the Hadley cells, centered between latitudes 20° and 30° N and S. These and other subtropical deserts are associated with anticyclonic regions of high pressure.

A second type of desert is found in continental interiors, far from sources of moisture, where hot summers and cold winters prevail. The Gobi and Takla Makan deserts of Central Asia fall into this category (FIG. 12.26B). These deserts form because wind that travels a very long distance over land, especially land that rises up to high plateaus, eventually contains so little water vapor that hardly any is left for precipitation.

A third kind of desert is found where a mountain range creates a barrier to the flow of moist air, causing orographic lifting and heavy rains on the windward side along with a zone of low precipitation called a *rainshadow* on the downwind side (see Fig. 11.19D). The Cascade Range and Sierra Nevada of the western United States form such barriers and are responsible for desert regions lying immediately east of these mountains (FIG. 12.26C).

Coastal deserts, which constitute a fourth category, occur locally along western margins of certain continents. The flows of surface ocean currents can cause the local

FIGURE 12.25 Deserts and atmospheric circulation
This map shows the distribution of arid and semiarid climates and the major deserts associated with them. Many of the world's great deserts are located where belts of dry air descend near the 30° N and 30° S latitudes. Notice also that regions of cold, descending air also surround both of the poles. Despite being covered by ice, the polar regions receive little precipitation and are considered to be frozen deserts.

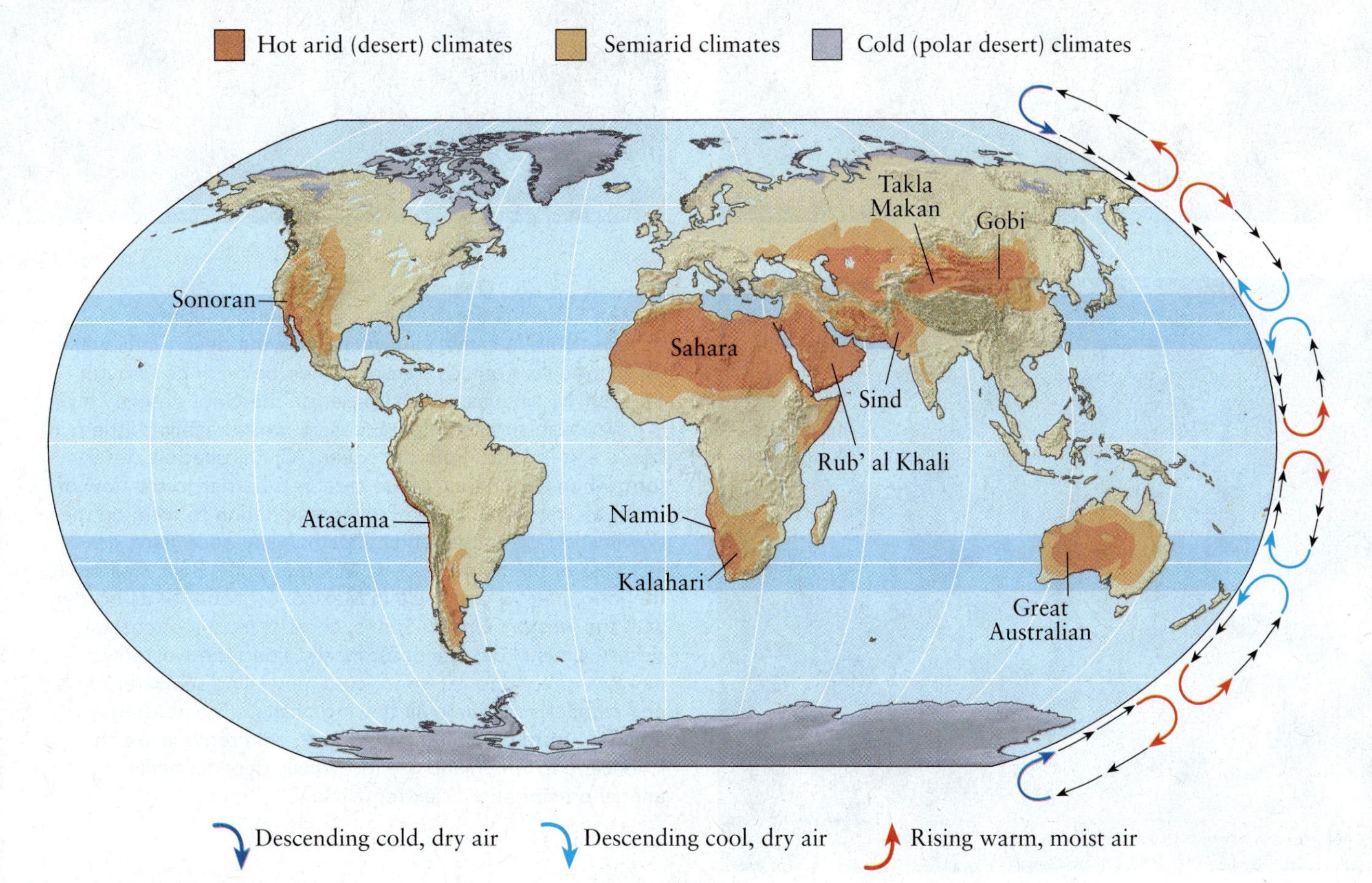

FIGURE 12.26 Deserts
(A) The Sahara is the greatest of the world's subtropical deserts. Here, a camel caravan crosses the desert in Libya. (B) Mongolian nomads transport their belongings through the Altai Mountains, which border on the Gobi desert. This is a typical inland continental desert, which receives little rain because it is so far from the ocean. (C) Rainshadow deserts form when a mountain range creates a barrier to the flow of moist air, causing a zone of low precipitation to form on the downwind side of the range. Death Valley, seen here, lies just east of the Sierra Nevada Mountains. (D) Baja, California, the long, narrow peninsula in Mexico just south of its border with the western United States, consists mostly of coastal desert. Coastal deserts occur locally along the western margins of continents, where cold, upwelling seawater cools and stabilizes maritime air flowing onshore, decreasing its ability to form precipitation. (E) Even cold regions like this landscape in Greenland are technically deserts, because annual precipitation is extremely low.

TABLE 12.2 Main Types of Deserts and Their Origins

Desert Type	Origin	Examples
Subtropical	Centered in belts of descending dry air at 20–30° north and south latitude	Sahara, Sind, Kalahari, Great Australian
Continental	In continental interiors, far from moisture sources	Gobi, Takla Makan
Rainshadow	On the sheltered side of mountain barriers that trap moist air flowing from oceans	Deserts on the sheltered sides of Sierra Nevada, Cascades, and Andes
Coastal	Continental margins where cold, upwelling marine water cools maritime air flowing onshore	Coastal Peru and southwestern Africa
Polar	In regions where cold, dry air descends, creating very little precipitation	Northern Greenland, ice-free areas of Antarctica

upwelling of cold bottom waters. The cold, upwelling seawater cools maritime air flowing onshore, thereby decreasing its ability to hold moisture. As the air encounters the land, the small amount of moisture it holds condenses, giving rise to coastal fogs. Nevertheless, in spite of the fog, the air contains too little moisture to generate much precipitation, and so the coastal region remains a desert (FIG. 12.26D). Coastal deserts of this type in Peru and southwestern Africa are among the driest places on the Earth.

The four kinds of desert mentioned thus far are all hot deserts, where rainfall is low and summer temperatures are high. In the fifth category are vast deserts of the polar regions where precipitation is also extremely low due to the sinking of cold, dry air (FIG. 12.26E). Remember that polar regions are high-pressure areas where cold, high-altitude air descends from the upper troposphere. However, cold deserts differ from hot deserts in one important respect: The surface of a polar desert, unlike the surfaces of warmer latitudes, is often underlain by abundant H_2O, nearly all in the form of ice. This ice accumulates, even though precipitation is exceedingly low, because the precipitation is always as snow and the snow doesn't melt. Even in midsummer, with the Sun above the horizon 24 hours a day, the air temperature may remain below freezing. Polar deserts are found in Greenland, Arctic Canada, and Antarctica (Fig. 12.26E).

Drought

As mentioned above, **drought** is an extreme-weather phenomenon in which a region experiences below-average rainfall for an extended period. There is no universal definition for drought because rainfall that would seem extremely low in a location that is normally rainy, such as Hawaii or Indonesia, would seem like monsoon-scale rains in a location that is normally dry, such as Phoenix or Jerusalem. Therefore, drought is typically defined with reference to a specific location, and with emphasis on the functional aspects of water supply, such as whether the lack of rainfall is affecting harvests.

Deserts typically receive little to no rainfall; inhabitants (both human and other kinds) are adapted to these conditions, so drought is not particularly problematic. However, the semiarid areas adjacent to deserts are highly susceptible to drought. In the region south of the Sahara, for example, there is a drought-prone belt of grassland known as the Sahel. There the annual rainfall is normally only 10 to 30 cm, most of it falling during a single rainy season. In the early 1970s, the Sahel experienced the worst drought of the twentieth century. For several years in a row, the annual rains failed to appear, causing the adjacent desert to spread southward by as much as 150 km. The drought-stricken region extended from the Atlantic to the Indian Ocean, and it affected a population of at least 20 million.

The fringes of deserts naturally migrate back and forth as a result of climate changes, and this has been true of the Sahara Desert and the Sahel for thousands of years. However, the results of the drought were intensified by the fact that between about 1935 and 1970 the human population of the region had doubled, and the number of domestic livestock had increased dramatically. This resulted in severe overgrazing that, in combination with the severe drought conditions, devastated the grass cover (FIG. 12.27). Livestock died, crops failed, and millions of people suffered from thirst and starvation. Today the Sahel region remains heavily populated, is still used intensively for livestock and agriculture, and continues to be at great risk from drought.

The expansion of desert conditions into adjacent areas, such as that described above in the case of the Sahel, is called *desertification*. It can happen as a result of natural causes (climate change), or human causes (such as the overgrazing mentioned above). Areas that are most susceptible to desertification are the semiarid fringe lands adjacent to the world's great deserts. One of the best-known examples of desertification is the Dust Bowl of the 1930s in North America, when dust storms swept across the Great Plains and drove many farm families from their land. Like the Sahel famine, the Dust Bowl was triggered by a multiyear drought, but exacerbated by decades of poor land-use practices. The Dust Bowl was responsible for initiating many of the soil-protection laws and practices in effect today in the United States and Canada.

Dust Storms

One of the most striking features associated with deserts and droughts is dust storms. As discussed in

FIGURE 12.27 Desertification
In the Sahel region of Niger, a herd of goats grazes on pasture at the edge of the desert. As the goats consume the remaining grass and bushes, the dunes of the desert will likely advance.

The Basics: Windchill Factor, at the surface of an object across which wind is flowing there is a boundary layer of still air. Above the boundary layer is a thin zone in which airflow is smooth, or *laminar*, and above the laminar air the flow is disordered, or *turbulent* (FIG. 12.28A). Large grains that protrude above the quiet, laminar-flow zone can be rolled and bounced along or may be swept aloft by rising turbulent winds. The larger grains that roll and bounce along mobilize fine sediment, which is then carried upward by the turbulent air. In this manner dust storms start (FIG. 12.28B).

Once in the air, dust constitutes the wind's suspended load (analogous to the suspended load of a stream; see Chapter 8). The grains of dust are continuously tossed about by eddies, like particles in a stream of turbulent water, while gravity tends to pull them toward the ground. Meanwhile, the wind carries them forward. In most cases grains suspended in the wind are deposited fairly near the place of origin, but the strong winds associated with large dust storms can carry very fine dust into the upper troposphere. There, it can be transported horizontally by geostrophic winds for thousands of kilometers.

FIGURE 12.28 Dust storms
(A) Particles of fine sand and silt at the ground lie within the boundary layer where wind speed is extremely low. As a result, it is difficult for the wind to dislodge and erode these grains. Turbulent air picks up the finest grains of sediment and carries it as a suspended load. (B) During major dust storms, like this one in Gao, Mali, visibility is greatly reduced.

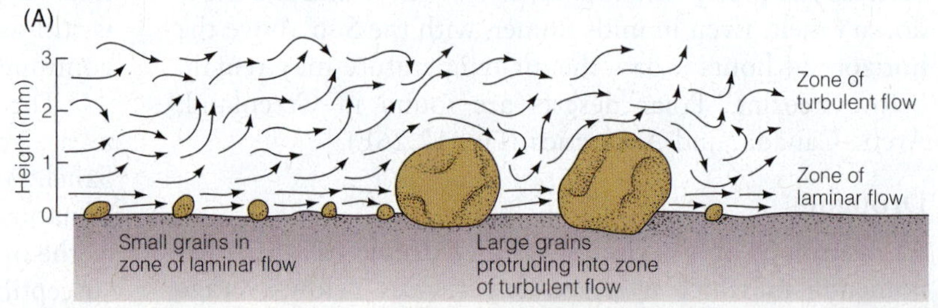

Dust storms are most frequent in the vast arid and semiarid regions of central Australia, western China, Russian Central Asia, Kazakhstan, the Middle East, and North Africa. In North America, blowing dust is especially common in the southern Great Plains and in the desert regions of California and Arizona. The frequency of dust storms is related to cycles of drought, with a marked rise in atmospheric dust concentration coinciding with severe drought. The frequency also has risen with increasing agricultural activity, especially in semiarid lands.

WEATHER AND THE EARTH SYSTEM

All life on Earth's surface is dependent on the atmosphere, and not just for its supply of oxygen. Global atmospheric circulation brings the moisture we need to survive. It also evens out the distribution of heat in the atmosphere. Thus, global atmospheric circulation is responsible for delivering oxygen and moisture, and ensuring a moderate surface temperature—the three fundamental requirements for a habitable planet.

By controlling temperature and moisture levels, atmospheric circulation also controls the distribution of Earth's major ecosystems. In turn, changes in the biosphere can have significant impacts on the atmosphere. For example, seasonal variations in plant growth cause the carbon dioxide levels of the atmosphere to fluctuate because carbon dioxide is utilized in photosynthesis. This causes seasonal variations in the capacity of the atmosphere to modify surface temperatures through the greenhouse effect.

Weather and climate are sensitive indicators of changes in the atmosphere, the biosphere, and the Earth system as a whole. In a system as complex as the global atmospheric circulation system, even seemingly small factors can dramatically influence the outcomes of weather-forming processes (see *A Closer Look: The Butterfly Effect: Chaos Theory and Weather Forecasting*).

Feedbacks

The reason that effects arising from small differences can propagate so quickly, affecting far-flung parts of the system, is that the Earth system and its component parts, including the atmosphere, are complex, dynamic systems with multiple feedbacks. As discussed in Chapter 1, a *positive feedback* is an interaction in which a change in the system tends to become magnified, whereas a *negative feedback* tends to be self-limiting, such that a change in the system tends to be minimized.

Earth's weather system is full of feedbacks, both positive and negative (FIG. 12.29). For example, a prolonged drought may cause trees and other ground-covering vegetation to die, revealing the bare, dry soil underneath. Dry soil is susceptible to erosion by water and wind, which may lead to dust storms and further erosion. Erosion and the loss of topsoil can make it difficult for anything to grow, leading to even less vegetation, more erosion, and so on—the positive feedback cycle continues and the initial change worsens. Drought and other climatic conditions may exacerbate the situation by creating conditions that are conducive to fires or pest outbreaks, for example. Human actions complicate things by interacting with, and sometimes reinforcing, the natural drivers of environmental change, as discussed earlier in the example of overgrazing in the Sahel.

On the other hand, an intense drought may cause animals (or people) living in the area to die or migrate away, leading to less pressure for food resources and providing an opportunity for plants to grow and the area to recover. This is an example of a self-limiting, or negative, feedback process.

FIGURE 12.29 Feedbacks in the climate system
Earth's weather and climate are affected by innumerable interconnected feedbacks, only a few of which are illustrated here. Human actions play a major role in these feedbacks. For example, as shown on the diagram, human laws, policies, institutions, technologies, and economic systems affect the composition of the atmosphere. This, in turn, affects weather and climate, which in turn affects ecological systems.

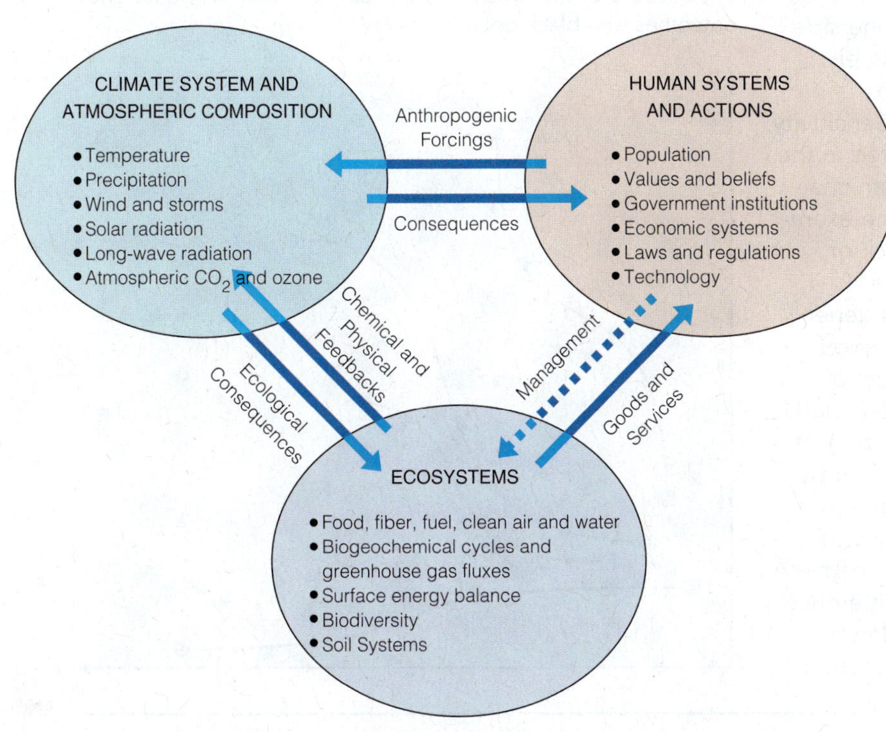

Thresholds

In a positive feedback situation, a prolonged drought might eventually

A Closer LOOK

THE BUTTERFLY EFFECT: CHAOS THEORY AND WEATHER FORECASTING

Weather is so very important to our lives—indeed, to *all* life on this planet. Why, then, is it so difficult to predict the weather accurately, particularly if we try to predict more than a couple of days in advance? You have learned that the atmosphere is a highly complex, dynamic system, in which phenomena grow, evolve, change, and interact, drawing energy from and losing energy to other parts of the Earth system, especially the hydrosphere and the geosphere. *Complex systems* are systems that have many interacting parts and in which processes are often *coupled* (linked to one another) and *nonlinear*. In other words, in a complex system A does not always, necessarily, neatly lead to B, and then to C, and then to D. Weather is a classic example of a complex system.

One of the most successful approaches to complex systems has been the application of *chaos theory*, which comes from a branch of mathematics that focuses on the study of complex, dynamic systems. In particular, chaos theory is applicable to systems that are extremely sensitive to small changes in initial conditions (FIG. C12.1). For example, a ball rolling down a hill seems like a fairly simple process, but its outcome is strongly influenced by many factors, including all of the tiny topographic variations in the hillslope. A tiny pebble at the top of the hill or the smallest breath of wind just as the ball starts on its downslope path could make the difference between the ball rolling down one side of the hill or the other side—two completely different outcomes from initial conditions that differed by only the tiniest factor. (Sensitivity to initial conditions was famously illustrated in the movie *Jurassic Park* by the mathematician character, played by Jeff Goldblum, using the example of a drop of water rolling off the back of his hand.)

The global atmospheric system that generates our weather is an example of a complex system that is highly sensitive to initial conditions. Sensitivity to initial conditions is popularly referred to as the *butterfly effect*, a phrase that came originally from the title of a talk given by Edward Lorenz, a widely recognized pioneer of chaos theory. Lorenz was experimenting with the application of computer models to weather forecasting. He discovered that a tiny difference in the original numerical inputs resulted in the model predicting completely different weather scenarios. Lorenz later gave a talk on this work, which was entitled, "Does the flap of a butterfly's wings in Brazil set off a tornado in Texas?" The idea is that a butterfly's wings can generate a tiny gust of wind, causing a slight variation in initial conditions. In a complex system like the global weather system, a flap or no-flap in the initial conditions could result in very different outcomes—a clear, sunny day versus a tornado in Texas, for example. Note that the flap of the butterfly's wings does not *cause* the tornado, but it is a crucial factor in the initial conditions, setting in motion a complex chain of events that eventually lead to the occurrence of the tornado.

Weather forecasting has improved dramatically over the past few decades, with the advent of advanced computer models and improvements in the scientific understanding of the global atmospheric circulation system. Chaos theory has been an important part of this. However, one thing that chaos theory has shown is that even if one could devise a perfect numerical model of the global weather system, the tiniest variation in initial conditions could still cause the system to diverge wildly from its predicted pathway. For this reason, forecasting the weather will likely remain a challenge for many years to come.

FIGURE C12.1 Sensitivity to initial conditions The small squares indicate four different starting conditions for a complex process that is sensitively dependent on initial conditions. The starting conditions differ from one another by just a tiny bit, but because of the sensitivity this results in four very different outcomes (the black dots).

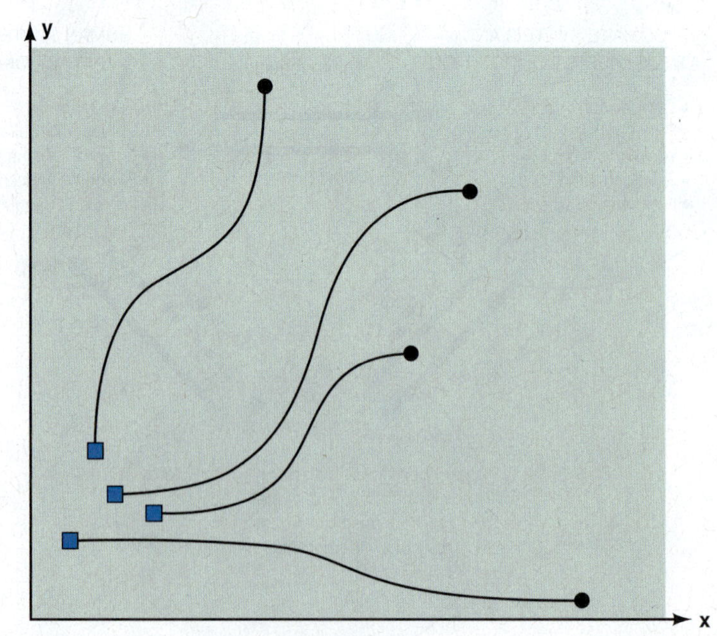

reach a point where it becomes impossible for the area to recover to its former state. Soils would wash away, lakes would dry up, and trees would die, to be replaced by sand dunes, dry riverbeds, sparse grasses, and other drought-tolerant species. A new, drier "normal" would have been established for the area. Scientists refer to this as a *threshold* situation. This implies that a system can handle and respond to changes by returning to its starting characteristics, but only up to a certain point. If pushed beyond that point—the threshold—the system will be unable to return to its earlier state, and a new, different set of equilibrium conditions will be established. Desertification can be an example of a threshold process; once desertification has progressed past a certain point, it becomes extremely difficult for non-drought-tolerant species to survive.

Environmental change is a natural part of the functioning of atmosphere and of the Earth system as a whole. However, today there is significant concern among atmospheric scientists that human actions have caused changes in the Earth system, particularly the atmosphere, that may push the system past a threshold—or, more likely, a number of thresholds—beyond which it will be impossible for the system to return to its former state. In Chapter 13 we will look more closely at Earth's climate system, the impacts of human actions on the system, and how close we may be to reaching one of these thresholds.

SUMMARY

1. The five variables that describe the state of the atmosphere and the weather are temperature, air pressure, humidity, cloudiness, and wind speed and direction. These variables are also used to measure and describe climate, but weather is short-term, whereas climate is long-term.

2. Wind is air movement that arises from differences in air pressure that cause air to flow from areas of high pressure to areas of low pressure. The movement of air is closely associated with temperature and density, and horizontal movement is always associated with at least some vertical movement. The magnitude of the pressure difference in a particular location controls the wind speed, and interacts with other forces to determine the direction of airflow.

3. Windchill provides a measure of heat loss from exposed skin as a result of the combined effects of low temperature and wind speed. The higher the wind speed, the thinner the boundary layer of still air and the faster heat is lost from the skin.

4. Wind is controlled by three factors: The pressure-gradient force drives the flow of air from areas of high pressure toward areas of low pressure. The Coriolis force, which results from Earth's rotation, deflects the pathway of flowing air; it is most important in large-scale wind systems. Friction slows airflow over the rough surface of land or water; it is most important for small-scale air motions.

5. Winds that result from a balance between pressure-gradient flow and the Coriolis deflection without the influence of friction, such that airflow is nearly parallel to the isobars, are called geostrophic winds.

6. Winds around a low-pressure center spiral inward, forming a cyclone. In the northern hemisphere, cyclones rotate counterclockwise; in the southern hemisphere, the reverse is true. Inward flow causes convergence; this leads to an upward flow of air and results in cooling, condensation, cloud cover, and rain. Winds around a high-pressure area spiral outward, forming an anticyclone. In the northern hemisphere, anticyclones rotate clockwise; in the southern hemisphere, the reverse is true. Outward flow causes divergence; this leads to high-altitude air being drawn downward, compressed, and heated adiabatically, leading to clear, cloudless skies.

7. An air mass is a large volume of air, characterized by homogeneous internal temperature, humidity, and density. The major characteristics of an air mass are determined by whether it forms over a continent (c) or a maritime region (m), and whether it forms in the tropics (T) or in the polar regions (P). The locations where air masses meet, called fronts, are associated with active weather and rapid changes in temperature.

8. The heat imbalance between the equator and the poles means that warm equatorial air flows toward the poles and cold air flows toward the equator. If Earth were a nonrotating sphere, one simple convection cell in each hemisphere would carry heat from the equator all the way to the poles. Instead, the Coriolis effect causes the convection cells in the atmosphere to break into several bands.

9. Warm air that rises in the tropics creates the intertropical convergence zone. At high altitude, the air turns poleward. By 30° N and 30° S latitude, it has been deflected by the Coriolis effect and is a westerly geostrophic wind. This creates two belts of high-pressure air that sink back toward the surface and return to the equator. This convective circulation creates the Hadley cells on either side of the equator. The positions of the ITCZ and the two high-pressure belts vary seasonally. The low-level winds that bring returning air back to the tropics in the Hadley cells, becoming easterly winds that flow along the equator, are the trade winds.

10. In each hemisphere, on the poleward side of the Hadley cells, there is a midlatitude circulation system. In these Ferrel cells, the surface winds are westerlies because they are created, in part, by poleward flows of air from the high-pressure divergence regions at 30° N and 30° S latitude. Ferrel cells act like a "ball-bearing" between two great circulatory systems, the Hadley cells on the equator side and the polar cells on the other side.

11. In the polar cells, cold, dry air descends near the poles, creating high-pressure divergence zones. Air then moves toward the equator via the polar easterlies, which meet the Ferrel cells along the polar fronts. Where the tropopause, which is an isobar, passes over the polar fronts, it undergoes a very steep gradient in altitude. The resulting high-speed geostrophic westerlies are called jet streams. Rossby waves distort the polar-front jet streams, pulling them into lower latitudes and thus influencing midlatitude weather patterns.

12. A monsoon is a seasonally reversing wind system; the Asian monsoon is the most pronounced. During the winter, a high-pressure anticyclone sits over the high, cold plateau of Central Asia, while the low-pressure intertropical convergence zone lies south of the equator over the Indian Ocean, where the Sun is overhead. The resulting northeasterlies bring cool, dry, cloudless days to the Indian subcontinent. During the summer, the wind pattern is reversed. The intertropical convergence zone shifts north of the equator, where the Sun is overhead, and the landmass of Asia heats up and is covered by low-pressure cyclones. The resulting winds are southwesterlies, blowing warm, moisture-laden air from the Indian Ocean onto the land and bringing hot, humid weather and torrential rain.

13. El Niño is a regional-scale, anomalous weather event that results from coupled atmosphere–ocean processes in the tropical Pacific. In a normal year, the Walker circulation causes the easterly trade winds to push warm surface water toward the western Pacific. This allows cold water from the Southern Ocean to well up in the eastern tropical Pacific, and generates rainy conditions in the western Pacific and Indonesia. An El Niño event begins with a weakening of the Southern Oscillation Index, a measure of the pressure differential between the eastern and western Pacific. This causes the Walker circulation to weaken, and the trade winds to weaken or reverse. Anomalously warm surface water accumulates in the central and eastern Pacific, which stops the upwelling of cold water. The zone of high rainfall normally situated near Indonesia shifts toward the central Pacific, bringing dry conditions to Indonesia and anomalous weather to many parts of the globe.

14. Local winds result from the interaction of terrain with the forces that cause air to move. Important processes include differential heating and cooling of land and water, and alternating daytime and nighttime airflow on slopes. Coupled land and sea breezes, katabatic winds, and chinook-type winds are examples of local wind systems.

15. Different types of cyclones are distinguished by where they form, how they develop, and their characteristics of temperature, air pressure, wind speed, and geographic extent. Extratropical or midlatitude cyclones typically form as a result of the interaction between warm and cold air along the polar fronts. They are synoptic weather systems that last many days and are responsible for most everyday weather events in the midlatitude regions.

16. Hurricanes originate as low-pressure centers over the Atlantic Ocean and develop into tropical cyclonic storms characterized by very high wind speeds. Hurricanes (also called cyclones or typhoons) draw warm water upward from the ocean surface. The water evaporates and subsequently condenses, which releases latent heat, fueling the storm. Cyclonic rotation requires Coriolis deflection, so tropical cyclones cannot form along the equator where the Coriolis effect is zero. A tropical cyclone draws energy from warm ocean water, so its wind speeds will quickly diminish once it moves onshore. High winds, storm surge, and torrential rains are common causes of damage from tropical cyclones.

17. Thunderstorms develop over land when an updraft of warm, humid air releases latent heat very quickly. The released heat causes stronger updrafts, which pull in more warm, moist air, releasing more latent heat. Towering, anvil-shaped cumulonimbus clouds form, resulting in heavy rainfall, hail, thunder, and lightning. Tornadoes are violent, funnel-shaped windstorms produced by a spiraling column of air extending downward from a cumulonimbus cloud. They commonly develop from large supercell thunderstorms with multiple updrafts.

18. The distribution of deserts is controlled principally by global atmospheric circulation. The great subtropical deserts are associated with the two belts of descending dry air at the downward-flowing limbs of the Hadley cells. Continental interior deserts form inland, far from sources of moisture. Rainshadow deserts form where mountain ranges create a barrier to the flow of moist air, causing orographic lifting, heavy rain on the windward side, and desert conditions on the leeward side. Coastal deserts occur along continental margins where cold-water upwelling occurs, cooling the air and reducing its ability to hold moisture. Polar deserts experience extremely low precipitation as a result of the sinking of cold, dry air.

19. Drought is an extreme-weather phenomenon in which a region experiences below-average rainfall for an extended period. Semiarid lands adjacent to deserts are highly susceptible to drought and desertification, which result from natural causes but can be exacerbated by overly intensive land use practices. Major dust storms occur in deserts as a result of fine sediment being lifted by turbulent airflow and carried along as the wind's suspended load.

20. Global atmospheric circulation delivers oxygen and moisture, and ensures a moderate surface temperature—the three requirements for a habitable planet. The atmosphere is a complex, dynamic system with multiple feedbacks, both positive and negative.

IMPORTANT TERMS TO REMEMBER

QUESTIONS FOR REVIEW

1. Explain why air density and air pressure are related.
2. What are the five factors that are used to describe weather (and climate)?
3. Name the three factors that control the speed and direction in which the wind blows, and briefly explain how each factor works.
4. On a weather map on which the isobar contour interval is 0.4 kPa, one region has the contours 20 km apart, while in another region they are 200 km apart. In which region would you experience the stronger winds?
5. What are geostrophic winds, and how do they arise? Name a well-known example of a geostrophic wind.
6. Describe the relationship between Hadley cells and the intertropical convergence zone.
7. How do the trade winds arise?
8. What are Ferrel cells, and where do they occur?
9. What is the polar front? How does the polar front jet stream form?
10. What distinguishes an El Niño year from a normal year? Describe the ENSO phenomenon.
11. What is a monsoon? Describe how the Asian monsoon occurs.
12. What is an extratropical cyclone? Some midlatitude, extratropical cyclones are synoptic; what does that mean?
13. Describe how the Coriolis force causes winds to rotate in cyclonic storm systems.
14. What is a katabatic wind? Give an example of a well-known katabatic wind.
15. What are the five main types of deserts? How is the formation of deserts related to global atmospheric circulation? What is the difference between desert and drought?

QUESTIONS FOR RESEARCH AND DISCUSSION

1. What would the weather be like where you live, if Earth was a nonrotating planet? Remember that although atmospheric convection would be much simpler, topography and the distribution of land and water bodies would still have an influence on regional and local wind and weather systems.
2. Investigate the current status of drought in the Sahel, or elsewhere. Can you find any information about weather changes that have led to the drought? What are residents doing to try to adjust to drought conditions and minimize the damage to farms and ecosystems?
3. What kinds of severe weather events are most common where you live? Find out about the broader weather systems that contribute to their occurrence.

QUESTIONS FOR *THE BASICS*

1. Why do weather reporters give the windchill factor during cold weather, but not during warm weather?
2. What is a boundary layer, and what role does it play in windchill?
3. List the four main categories of air masses and their characteristics.

QUESTIONS FOR *A CLOSER LOOK*

1. What are some of the characteristics of complex systems?
2. How can chaos theory be applied to weather forecasting?
3. What is the so-called butterfly effect?

CHAPTER 13

The Climate SYSTEM

OVERVIEW

In this chapter we:

- Introduce the components of Earth's climate system
- Consider the evidence that Earth's climate has changed
- Learn how scientists study past climate changes
- Describe the history of Earth's climatic variation
- Investigate the natural causes of climate change

Climate proxy
Each ring in this cut section of a cedar trunk represents one growth season. The rings vary in thickness and density, depending on the conditions in which the tree grew during that season. Natural archives like tree rings can provide valuable "proxy" information about climates of long ago.

Earth's climate system is dynamic. It has so many interacting components and feedbacks that it is characterized by change. In this chapter we will investigate the natural causes of climate change—some external to the Earth system, and some internal—and we will examine the evidence demonstrating that climates have changed throughout Earth history. We will also look at some of the tools and techniques that scientists use to study climates of the past. In Chapter 19, we will look more closely at how human activities are changing the atmosphere in ways that might lead to a significant change of climate during our lifetime.

EARTH'S CLIMATE SYSTEM

In Chapter 12 you learned that **climate** is a measure of the average weather conditions of any place on Earth, over time. As shown in **FIGURE 13.1**, Earth's climate system is complex and is driven by interactions between several major subsystems—the atmosphere, the hydrosphere (mainly the ocean), the cryosphere, the geosphere, and the biosphere (and, most recently, the anthroposphere). The subsystems interact so closely that a change in one of them causes changes in one or more of the others. All but the geosphere are driven by solar energy. Some incoming solar radiation is reflected back into space by clouds, atmospheric pollutants, ice, snow, and other reflective surfaces, as you learned in our discussion of Earth's energy cycle (Chapter 2). The remainder of the incoming solar energy is absorbed and subsequently reemitted at longer wavelengths by three great reservoirs—atmosphere, ocean, and land.

When changes happen, the atmosphere is the subsystem that responds most rapidly, commonly within a month or less. The surface layer of the ocean responds more slowly (generally over months or years), whereas changes involving the deep ocean may take centuries. Although the land may respond rapidly to local changes in the other components of the climate system, the geosphere is especially significant on timescales of millions of years. On these long timescales, processes in the geosphere affect the distribution of continents and ocean basins, and the location and height of mountain ranges. These influence circulation in the atmosphere and ocean, which in turn influence the location and extent of glaciers and sea ice, and the character, extent, and distribution of vegetation. Vegetation is important in the climate equation because it helps determine the reflectivity of the land surface. It also influences the composition of air by absorbing carbon dioxide, and affects humidity and therefore local cloud cover. Its absence can increase wind erosion, which in turn can

FIGURE 13.1 Earth's climate system
Earth's climate system has five major interacting components: the geosphere (including land relief); the atmosphere (including the role of gases and particulates); the hydrosphere (especially ocean basins and circulation); the cryosphere (including the reflectivity of the polar ice caps); and the biosphere (including vegetation, soils, and photosynthesis). The anthroposphere—human activities—has become a recent contributor to the climate system.

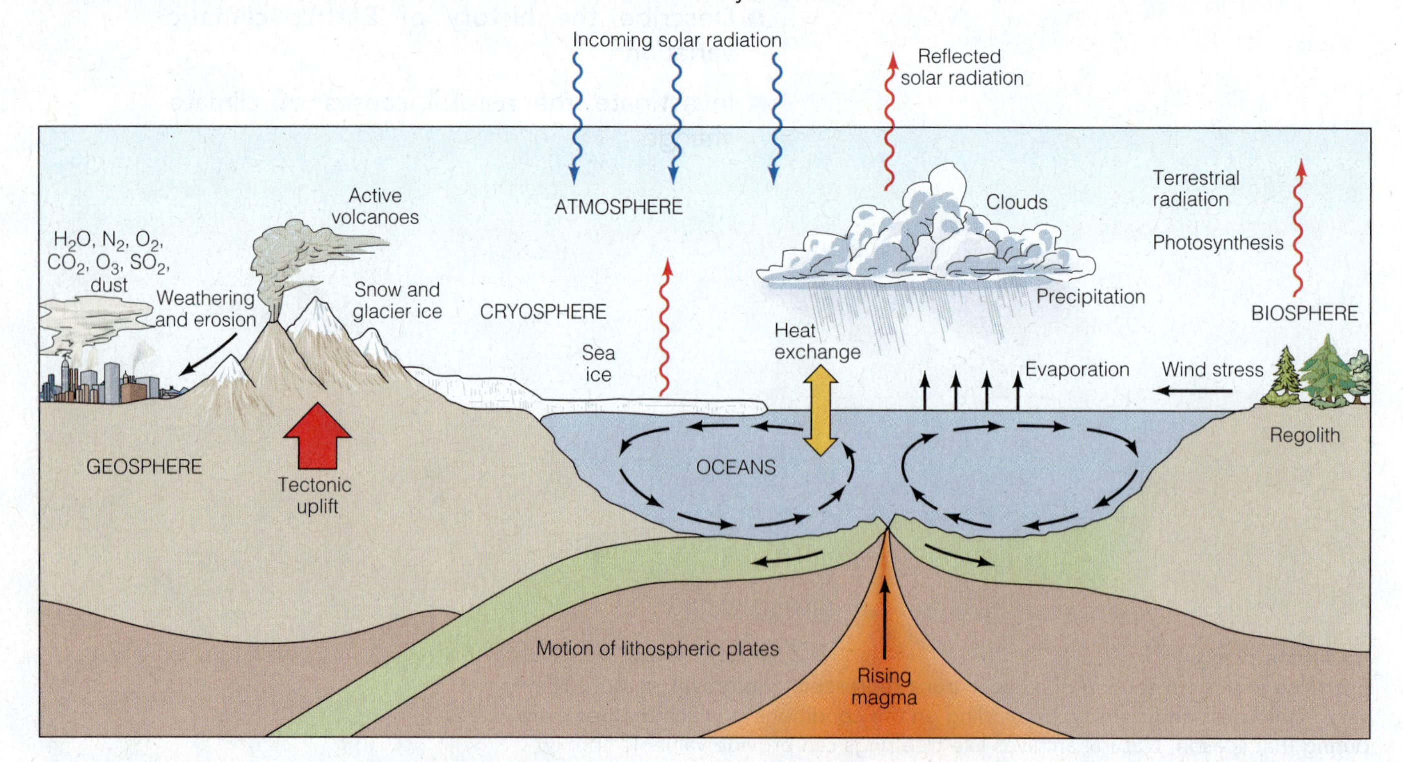

influence climate by affecting the dustiness of the atmosphere.

Because of its ability to absorb energy and retain it in the form of heat, the ocean serves as a great reservoir of thermal energy that helps moderate climate. The ocean's effect is illustrated by the contrast between a coastal region that has a maritime climate (little contrast between seasons) and an area farther inland that has a continental climate (strong seasonal contrast). The world ocean also is extremely important in controlling atmospheric composition, for the ocean contains a large volume of dissolved carbon dioxide. If the balance between oceanic and atmospheric carbon dioxide reservoirs were to change by even a small amount, the radiation balance of the atmosphere would be affected, thereby bringing about a change in world climates.

In *The Basics: Köppen System of Climate Classification* we provide an overview of Earth's present-day climatic zones. We will revisit these climatic zones in Chapter 16 when we look more closely at the organisms and ecosystems they support.

EVIDENCE OF CLIMATIC CHANGE

Last winter may have been colder than the winter before, and last summer may have been wetter than the previous summer, but such observations do not mean that the climate is changing. The identification of a climate change must be based on a shift in average conditions over a span of years. Several years of abnormal weather may not mean that a change is occurring, but trends that persist for a decade or more probably do signal a shift to a new climatic regime.

Climate commonly is expressed in terms of mean temperature and mean precipitation. Other parameters, including humidity, windiness, and cloudiness, are also important in characterizing climate, even though measurements of them are not routinely recorded in all places. Today, changes in the atmospheric conditions that define weather and climate are carefully tracked and measured on an almost continuous basis, at many locations around the world. However, instrumental tracking has only been available for the past century. Let's have a look at the historical record of climate, and then consider how scientists have managed to extend that record back into prehistoric times.

Historical Records of Climate

Our experience tells us that weather changes from year to year, but because climate is based on average conditions over many years, we may not find it easy to tell if the climate is changing. Your grandparents may recall that winters seemed colder half a century ago, but do such recollections actually point to a change of climate? In some places, instrumental records of climate have been maintained for a century or more, which is long enough to see if average conditions have really shifted.

One of the longest continuous climatic records available to us comes from Great St. Bernard Hospice, located at the crest of the Alps on the border between Switzerland and Italy. The Augustinian friars have recorded temperatures in that location since the 1820s and snowfall since the 1850s. This record shows that temperature and snowfall fluctuated approximately in phase, with times of cool temperature corresponding to times of above-average snowfall. Some short-term anomalies persisted for a decade or two, but over the entire period of the record the general trend has been toward warmer temperatures.

The temperature pattern demonstrated by this instrumental record from the Alps is representative of that in other parts of the northern hemisphere, where average temperatures experienced a fluctuating rise after the 1880s to reach a peak in the 1940s (FIG. 13.2). Thereafter, average temperatures declined slightly until the 1970s when they again began to rise. In the 1990s and then, again, in the 2000s they reached the highest values yet recorded.

FIGURE 13.2. Variations of Earth's surface temperature This graph, from the National Climatic Data Centre of NOAA's National Environmental Satellite, Data, and Information Service, shows global mean surface temperature anomalies from 1880 to 2010. A temperature anomaly is a departure from a long-term average – in this case, the average surface temperature for the 20th century. A positive anomaly indicates that the observed temperature was warmer than the reference temperature, whereas a negative anomaly indicates that the observed temperature was cooler than the reference value.

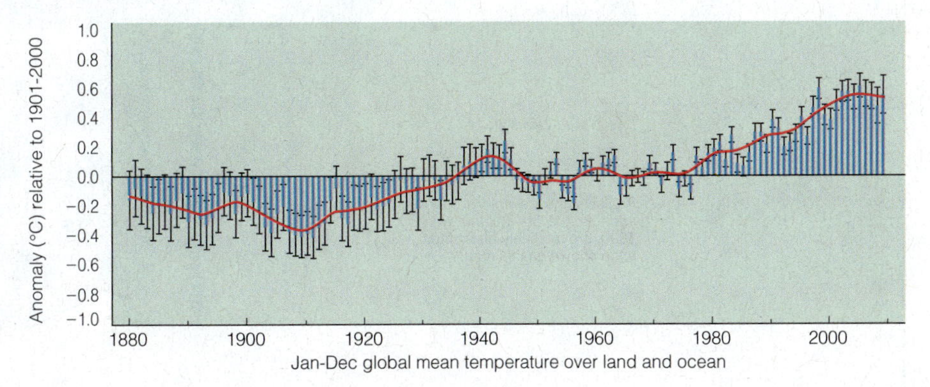

the BASICS

Köppen System of Climate Classification

The **Köppen system of climate classification**, originally developed by an Austrian climatologist named Vladimir Köppen, is one of the most widely used climate classification systems. It was most recently modified by Rudolf Geiger and therefore is often called the *Köppen-Geiger climate system*. This classification is based on the distribution of native vegetation types, on the premise that vegetation is the best indicator of climate. The system combines measurements of average temperature and precipitation, as well as seasonal variations, to define five major climate categories.

The most notable feature of a map showing the world distribution of climate zones (**FIG. B13.1**) is the close link between climate types and the prevailing wind currents in the global atmospheric circulation system, as shown in Figure 12.11. For example, the world's deserts are mostly concentrated along the 30°N and 30°S parallels, where the barometric pressure is high and the descending air mass is dry. The tropical jungles of Africa and South America lie near the equator, where we expect low barometric pressure and lots of warm moist air.

THE MAJOR CLIMATE CATEGORIES

There are five basic climate categories in the Köppen system, symbolized as A, B, C, D, and E, as well as an additional Highland category (H). These categories are further modified by lower-case letters that indicate variations of

FIGURE B13.1 Köppen climate classification
This world map shows the locations of some of the major climatic zones as classified by the Köppen system.

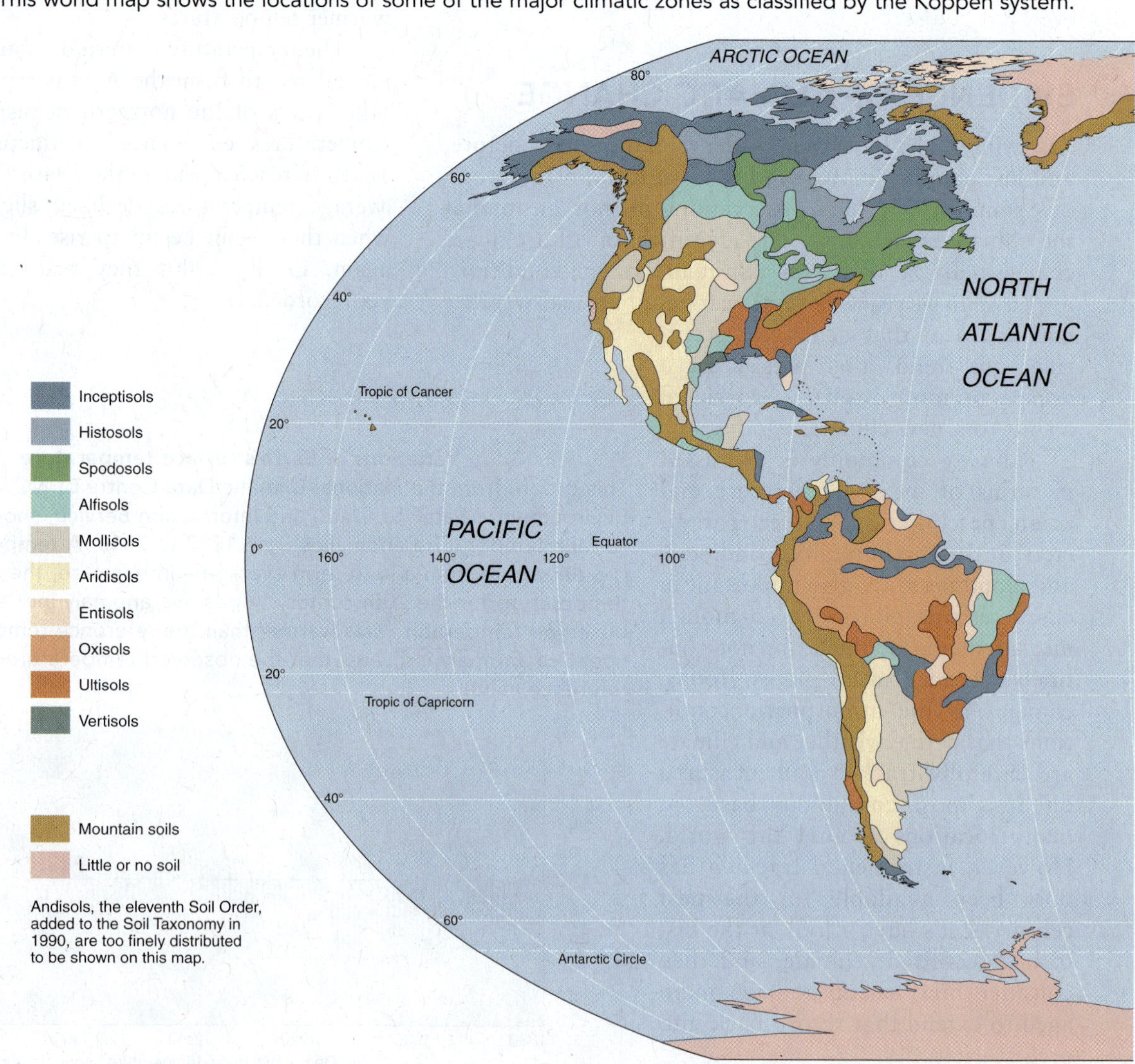

precipitation and temperature and, in some cases, seasonality with the major groupings.

A. EQUATORIAL CLIMATES

These are tropical humid climates, also called *megathermal* climates. The coolest month must be above 18°C. Examples include Hilo, Hawaii; Darwin, Australia; and Rio de Janeiro, Brazil.

Variations:

- Af—Tropical rain forest; no dry season; driest month must attain at least 6 cm of rainfall.
- Aw—Tropical wet or dry savanna; winter dry season; at least one month must attain less than 6 cm of rainfall.
- Am—Tropical monsoon; seasonally reversing monsoon winds; driest month with rainfall less than 6 cm.

B. DRY CLIMATES

No specific amount of moisture makes a climate dry; rather, the rate of evaporation relative to the amount of precipitation dictates how dry a climate is in terms of its ability to support plant growth. A modifier may be added to describe temperature. Examples include Yuma, Arizona; and Medicine Hat, Alberta.

Variations:

- BW—Desert, arid.
- BS—Semiarid or steppe.

C. WARM TEMPERATE CLIMATES

In this varied category, the warm temperature humid or *mesothermal* climates, the coldest month average must be below 18°C but above −3°C. The warmest month average must be above 10°C. One of the criticisms of

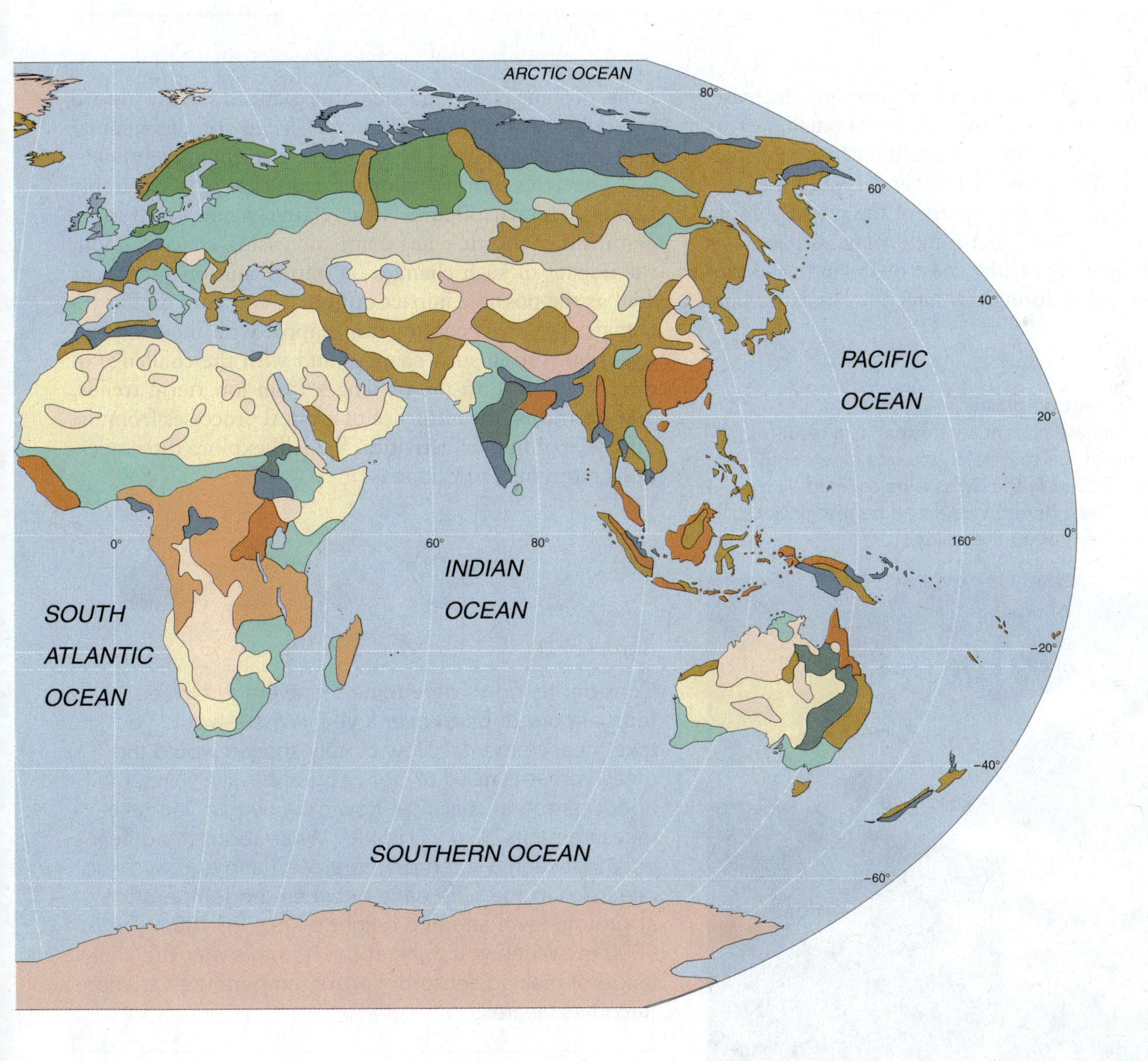

(continued)

the BASICS *(Continued)*

the Köppen system is that this category is too broad. Examples include Hong Kong, China; Madrid, Spain; Bergen, Norway; Prince Rupert, British Columbia.

Variations include:

Cf—No dry season; year-round weather dominated by the polar front.

Cw—Winter dry season; at least 10 times as much rain in the wettest month as in the driest.

Cs—Mediterranean climates; summer dry season; at least three times as much rain in the wettest month as in the driest; driest month must receive less than 3 cm of rainfall.

D. CONTINENTAL CLIMATES

These *microthermal* climates are characterized by winter snowfall and typically occur in continental interiors. The coldest month average must be below −3°C, and the warmest month average must be above 10°C. Examples include Chicago, Illinois; Anchorage, Alaska; and Fargo, North Dakota.

Df—Hot or warm summer continental climate; no dry season.

Dw—Continental subarctic or boreal (taiga) climates; mild or severe winters; winter dry season.

E. POLAR CLIMATES

In these cold climates the warmest month average must be below 10°C. Examples include Iqaluit, Nunavut; and Scott Base, Antarctica.

Variations include:

ET—Tundra; warmest month has an average temperature between 0°C and 10°C.

EF—Icecap; all 12 months below 0°C.

The amplitude of this recent temperature increase, amounting to less than 1°C overall, seems small, yet its effects have been seen widely, especially in high latitudes. During the six decades between 1880 and 1940, for example, mountain glaciers in most parts of the world shrank, some conspicuously (FIG. 13.3), and Arctic sea ice was observed less frequently off the coast of Iceland. The biosphere also responded during this interval. The latitudinal limits of some plants and animals expanded slightly toward the poles, and an increase in the length of the summer growing season led to a general improvement in crop yields.

That the world's climates can change measurably within a human lifetime is a relatively new realization. With this realization has come increasing concern about the impact of such changes on nature and on society, as well as the possible impact of human activities on Earth's climate. Scientists continue to grapple with the complexity of measuring such changes and with the challenge of disentangling short-term changes from long-term trends, and separating the influence of natural processes from the influence of human activities. We will explore these questions more fully in Chapter 19.

FIGURE 13.3 Retreating glacier
Many alpine glaciers are currently in retreat as a result of a general warming of the climate. In the late nineteenth century, Findelen Glacier in the Swiss Alps covered all the bare, rocky terrain seen here. Since that time the glacier's terminus has retreated far up the valley.

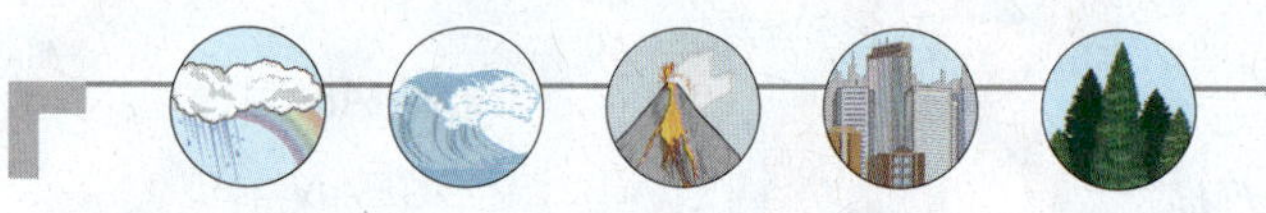

Make the CONNECTION

Consider trying to determine the average surface temperature of Earth over a year. Where would you take measurements? How closely spaced would the measurements need to be, geographically? What about temporal spacing—how often would you take measurements? Daily? Hourly? What about the difference between the temperature over land and over the ocean? And the difference between the temperature at ground level and at a height of 1 m, or 100 m, or 1000 m? When you think about it in this way, the challenge of making sense of Earth's temperature variations becomes clearer.

H. HIGHLAND CLIMATES
High-altitude climate variations.

TEMPERATURE, PRECIPITATION, AND SEASONALITY MODIFIERS

The following lower-case letters are added to the major climatic zone classifications for clarification, in special situations. For example, Dfa refers to a hot-summer continental climate; examples would be Chicago, Illinois or Seoul, South Korea. Dfb refers to a warm-summer continental climate; examples would be Helsinki, Finland or Moncton, New Brunswick.

a—hot summer. Warmest month must average above 22°C.
b—cool summer. Warmest month must average below 22°C. At least 4 months above 10°C.
c—short, cool summer. Less than four months over 10°C.
d—coldest month average must be below −38°C.
i—annual temperature range must be less than 5°C.
g—hottest month occurs prior to summer solstice.
h—hot; average annual temperature must be above 18°C.
k—cold; average annual temperature must be under 18°C.
k'—temperature of warmest month must be under 18°C.
m—monsoon; despite a dry season, total rainfall is so heavy that rain forest vegetation is not impeded.
n—frequent fog.
s—summer dry season. At least three times as much precipitation in the wettest month as in the driest.
w—winter dry season. At least 10 times as much precipitation in the wettest month as in the driest.
w'—autumn rainfall maximum.
w"—two dry seasons during a single year.
x—maximum precipitation in late spring or early summer.

The Geologic Record of Climatic Change

Historical records represent only a tiny fraction of Earth's evidence of climatic change. The majority of the evidence of climatic change comes from the geologic record. Scientists have long puzzled over the occurrence of geologic features that seem out of place in their present climatic environment. For example, abundant fossil bones and teeth of hippopotamus—the same kind that lives in East Africa today—have been recovered from sediments in southeastern England. These sediments were deposited about 100,000 years ago, and the evidence indicates that at the time southeastern England had a tropical climate like that in some parts of Africa today. On the other hand, fossilized plant remains in the temperate north-central United States show that this region formerly resembled Arctic landscapes like those now seen in far northern Canada. In each case, a significant change in local climate apparently has taken place, so that the organisms living in these areas today are very different from the fossil forms preserved in the geologic record (FIG. 13.4).

We refer to the climates of ancient times as **paleoclimates**. Besides fossils, other anomalous features tell us that paleoclimates in many localities differed dramatically from present-day climates. These include evidence of glaciation in lands that are now temperate, desert sand dunes now covered by lush vegetation, beaches of extensive former lakes in dry desert basins, channel systems of now-dry streams, remains of dead trees above the present treeline in mountainous areas, and soils with profiles that are incompatible with the present climate. The distribution of periglacial features indicates the extent of former permafrost conditions. At present, permafrost exists mainly in areas where the mean annual air temperature is below −5°C. If

FIGURE 13.4 **Tropical fossils in the Arctic**
The discovery of tropical organisms in areas that are now subarctic is a strong indication that global climates have changed. This 125-million-year-old fossilized turtle shell was found in northern British Columbia, Canada.

evidence of former permafrost is found at a place where the annual temperature is now 4°C, for example, then the former periglacial climate is inferred to have been at least 9°C colder in that location. All of these types of evidence, preserved in the geologic record, show that change has been the norm in the climate system throughout Earth history, rather than the exception.

Climate Proxy Records

Scientists attempting to reconstruct paleoclimates rely on records of natural events that are influenced by, and closely mimic, climate. Scientists call these **climate proxy records.** Where the fossil record provides a broad overview of climatic changes over great ranges of time, climate proxy records often can provide evidence of the year-to-year or season-to-season variability of weather in a specific location. Although lacking the precision of instrumental data, climate proxy records from different localities can add up to a detailed picture of local, regional, and global climatic trends.

We can access information through climate proxies in two basic ways. One way is to look at historical records maintained by people of events and processes that were controlled by climate; this works as far back as people were keeping records of the climate-controlled events. Another way is to examine natural records of climatic variations, recorded by processes like the deposition of annual or seasonal growth rings. In the first approach, dates are conveniently attached to dated historical records. In the latter approach, there must also be a mechanism available for determining the age of the natural record; without this mechanism, we would end up with a record of climate changes, but would have no way to determine when the changes actually occurred.

Human Records of Climate Proxies

Many human activities are controlled and influenced by climate, including just about everything related to farming, fishing, and harvesting. These activities are of obvious importance for human well-being, so people have been keeping track of such influences for a long time—more than 1000 years, in some cases. Note that this is different from the instrumental records of weather, discussed earlier, in which temperature and other weather-related factors are measured directly. In the case of climate proxies, the record-keeping doesn't track the weather directly, but tracks a process or event that was influenced by the weather.

Four of the most informative climate proxy series recorded in human historical records are shown in FIGURE 13.5. One important example is the frequency of dust storms in China, a record of which has been kept by Chinese scribes for over a thousand years. (Recall from Chapter 9 that abundant windblown dust, called *loess*, is indicative of a cold, glacier-dominated climate.) Other useful historical climate

FIGURE 13.5 Human records of climate proxies
Some human records of climate proxies span more than 1000 years. (A) The frequency of major dust-fall events in China is an indication of a cold, dusty atmosphere. (*Source:* After Zhang, 2982) (B) The severity of winters in England is based on the number of mild or severe months experienced. (*Source:* After Lamb, 1977) (C) The number of weeks per year during which sea ice reached the coast of Iceland is a record that has been kept by fishers for almost 1000 years. (*Source:* After Lamb, 1977) (D) The freezing date of Lake Suwa in Japan is represented here as being early or late relative to the long-term average date. (*Source:* After Lamb, 1966)

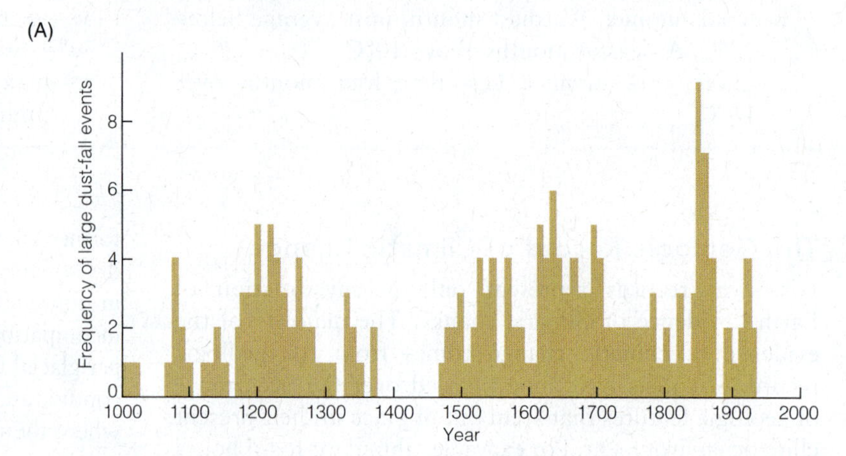

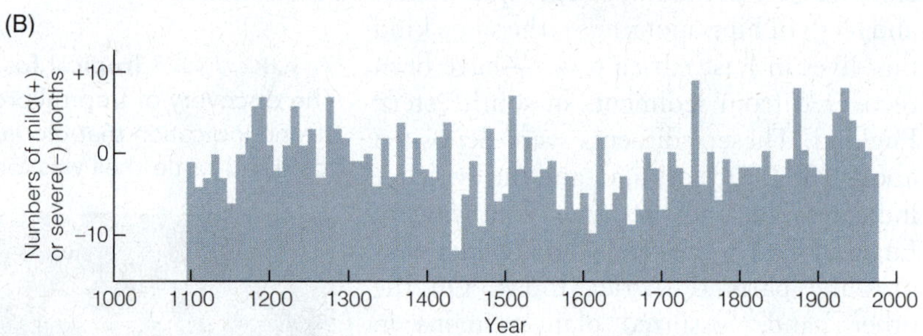

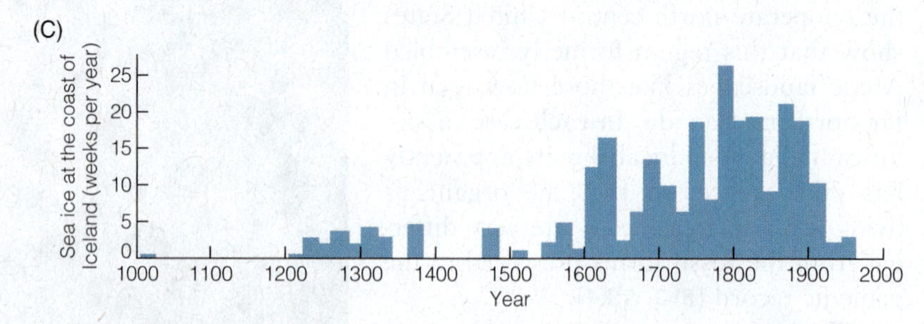

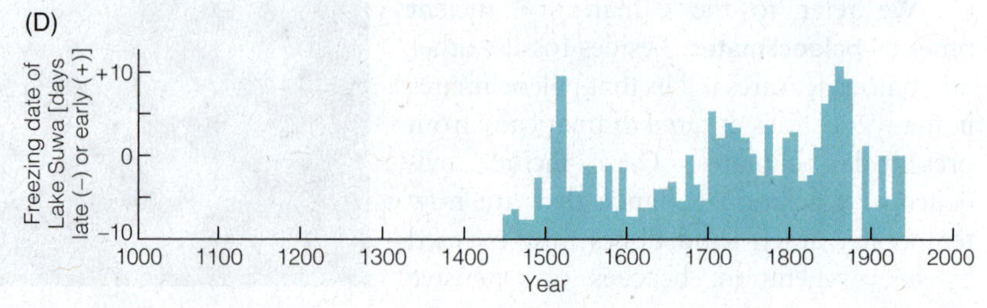

proxy records include the number of severe winters in England; the height of the Nile River at Cairo; the number of weeks that sea ice accumulates off the coast of Iceland (a record kept by Icelandic fishers since about 1200 CE); the quality of wine harvests in Germany; dates for the blooming of cherry trees in Kyoto, Japan; and variations in wheat prices (a reflection of climatic adversity) in England, France, the Netherlands, and northern Italy.

Each of these phenomena bears a relationship to some aspect of the prevailing climate and has been tracked in historical records for hundreds of years, making them useful proxies for climatic variability. However, even the longest historical records only go back 1000 years or so. To extend our use of climate proxies farther into the past, we must rely on nature's own record-keeping processes.

Ice Cores and Isotopic Studies

As you learned in Chapter 9, an important source of information about paleoclimates comes from ice cores collected from both polar ice sheets and alpine glaciers (**FIG. 13.6A**). Cores obtained from the Greenland and Antarctic Ice Sheets, as well as from several smaller ice

FIGURE 13.6 Reconstructing temperature records from climate proxies
(A) Chemical analysis of ice cores taken from glaciers and ice caps can provide a record of temperature at the time the ice was formed. (B) Glacier ice forms from snow, which is deposited in annual layers, as seen here in glacier ice from Glacier Bay National Park, Alaska. (C) Direct measurements of the temperature in the northern hemisphere from the middle of the nineteenth century onward are shown by the black curve. The other curves are based on indirect reconstructions of temperature from tree rings (green), glacier lengths (dark blue), and ice cores (light blue). The remaining curves (yellow, red, and purple) were reconstructed using a combination of different proxy data sources. All curves show that temperatures during the last few decades of the twentieth century were higher than during any comparable period in the last thousand years.

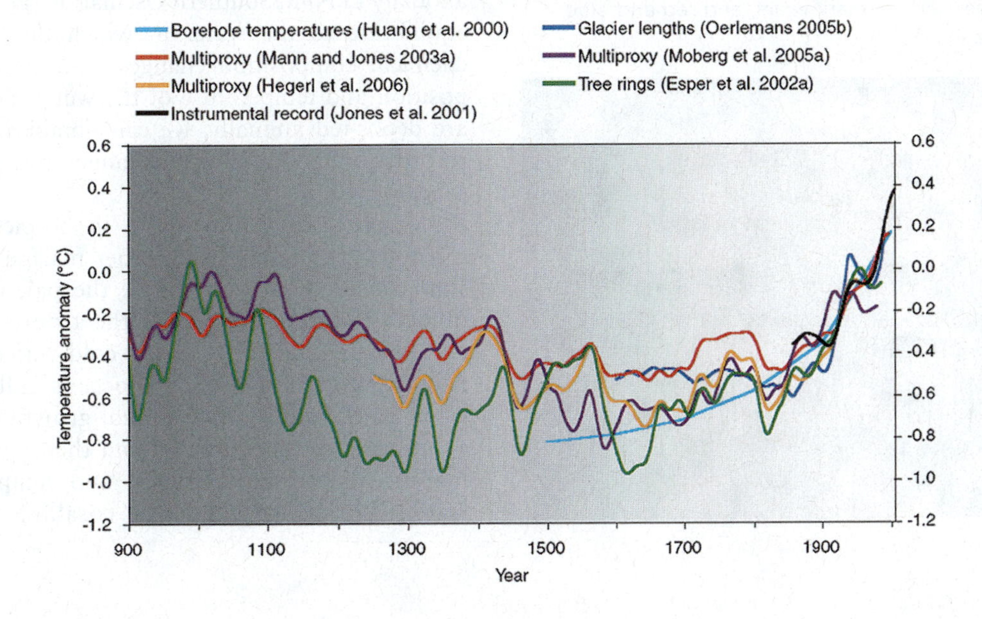

caps at lower latitudes, provide continuous records of weather conditions near the surface of these glaciers. Some of these records extend to hundreds of thousands of years before the present.

Measurements of the ratio of two isotopes of oxygen (^{18}O and ^{16}O) in glacier ice enable scientists to estimate the air temperature when the snow that later was transformed into that ice accumulated at the glacier surface. Other isotopic measurements carried out on glacier ice cores can reveal how much of Earth's water was locked up in glaciers at various times in the past (see *A Closer Look: Using Isotopes to Measure Past Climates*). A particularly useful aspect of ice-core analysis is that small bubbles of air are trapped in the glacier ice during its transformation from snow to ice, as discussed in Chapter 9 (see Fig. 9.26C). The trapped bubbles provide a sample of the ambient air at the time the snow fell, which can be extracted and chemically analyzed to determine the proportions of the various gases in the atmosphere at that time. In combination with the temperature information yielded by isotopic studies and the date provided by ice layering, this makes an extremely powerful data set; we can decipher what the climate was like in the past, and we can also find out what the chemistry of the atmosphere was at that time.

An important feature of ice-core analysis is that glacier ice is laid down in annual layers (**FIG. 13.6B**). By counting the annual layers, scientists can count back in time to attach dates to the temperature determinations (**FIG. 13.6C**). Without the layering, ice-core data would be of less use, since there would be no precise timescale attached to the temperature changes. The establishment of an ice-core chronology has enabled scientists to correlate data from several different locations around the world, thereby contributing to our understanding of paleoclimates on a global scale.

Ice is not the only Earth material that is amenable to isotopic study. Marine microorganisms that float in the surface layer of the ocean record the temperature of the water in which they lived because the oxygen isotopic composition of their shells change in response to changes in the temperature of the water. When the organisms die, they fall to the bottom of the sea and accumulate in the sediment, preserving a record of changes in the sea-surface temperature. Groundwater, soils, lake sediments, cave deposits—basically any material that equilibrates with its surroundings on a temperature-dependent basis, and preserves a chronologic sequence—can be used for paleoclimate reconstruction.

Annual Growth Rings

Many organisms lay down annual growth rings and are therefore potentially useful for providing paleoclimate information. For example, a tree living in middle latitudes typically adds a growth ring each year (chapter-opening photo), the width and density of which reflects the local climate. Many species live for hundreds of years; a few, like the Giant Sequoia and Bristlecone pine of the California mountains, live for thousands of years. Specialists in tree-ring analysis are able to reconstruct temperature and precipitation patterns from tree rings over broad geographic areas for any specific year in the past. The rings themselves provide the timescale. (The technique of establishing a date using tree rings is referred to as *dendrochronology*.) These reconstructions provide pictures of both changing regional weather patterns and long-term climatic trends.

Living corals also deposit annual growth rings (**FIG. 13.7**). In Chapter 12 you learned how scientists use the growth rings of corals to reconstruct water temperature oscillations related to the cyclic regional weather anomaly El Niño/Southern Oscillation (Fig. 12.18). Corals equilibrate to the water in which they grow, so their chemical composition changes in response to the composition and temperature of the water. Because the rings are deposited annually, we can obtain a high-resolution record of changes in water temperature spanning thousands of years.

Corals grow principally in tropical waters, but researchers are now using other marine organisms in a similar manner to investigate the paleoclimate history of ocean waters outside of the tropics. For example, coralline red algae live in the cold waters of the North Atlantic and North Pacific oceans. Like corals, they grow by the addition of annual growth rings, and they equilibrate to the water around them, preserving information about the chemistry and temperature of the water. Retrieving samples of coralline red algae from

FIGURE 13.7 Growth rings as climate proxies
The growth of some corals produces structures similar to the annual rings of trees. Analyzing the chemical composition of the coral can provide a temperature record. Black lines drawn on the lower section mark years, and red and blue lines mark quarters.

the bottom of the ocean in the far north is difficult because the water is very cold, and wind and waves are high. However, it is proving worthwhile because it provides an additional source of information about paleoclimates.

Sedimentary Evidence

Some lake sediments display a distinctive alternation of parallel layers with different *clast* sizes (Chapter 7). A pair of such layers deposited over the cycle of a single year is termed a **varve** (Swedish for cycle; FIG. 13.8). The layering results from a naturally occurring climatic cycle that influences sedimentation. The sediment preserves a record of the climatic variations. Because of the seasonal layering, a chronology can be established and attached to the data set.

FIGURE 13.8 Varves: seasonal layering in lake sediment These varves are seasonal layers deposited in a glacial-age lake near Seattle, Washington. Each pair of layers in a sequence of varves represents an annual deposit.

Varves are most common in deposits of high-latitude or high-altitude lakes, where there is a strong contrast in seasonal conditions. In spring, as the ice of nearby glaciers starts to melt, the inflow of sediment-laden water in a lake increases and coarse sediment is deposited throughout the spring and summer. With the onset of colder conditions in the autumn, streamflow decreases and ice forms over the lake surface. During autumn and winter, very fine sediment that has remained suspended in the water column slowly settles to form a thinner, darker layer above the coarse, lighter-colored summer layer. Varved lake sediments are common in Scandinavia and New England where they formed beyond the retreating margins of Ice Age glaciers.

Deep-sea sediment cores provide some of the best indirect evidence we have of past climatic changes. Seafloor sediments contain abundant fossils of tiny sea creatures called *foraminifera*, which once lived in surface waters. These microfossils (FIG. 13.9) equilibrate with the water around them, thus preserving a chemical record of past climatic changes.

Fossil Pollen Studies

Much of our knowledge of climatic conditions outside the great ice sheets during glacial times is based on interpretation of plant pollen fossils. Large plant fragments permit

FIGURE 13.9 Deep-sea microfossils as climate proxies Seafloor sediment obtained by drilling contains fossils of tiny sea creatures, foraminifera, which once lived in surface waters. These microfossils contain abundant information about the chemistry and temperature of the ocean. Foraminifera have calcium carbonate shells that contain oxygen. The ratio of oxygen-16 (the lighter isotope of oxygen) to oxygen-18 is a measure of the temperature of the seawater in which the microscopic creatures lived.

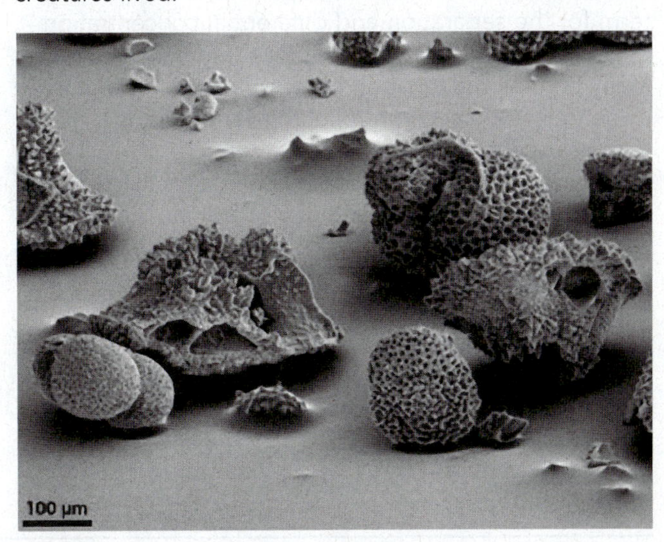

A Closer LOOK

USING ISOTOPES TO MEASURE PAST CLIMATES

Information about paleoclimates—especially *paleotemperatures*—can be partly uncovered through the analysis of stable isotopes in the H_2O molecules of ice extracted from polar and alpine glaciers. *Isotopes* (Chapter 3) are naturally occurring variations of elements, which vary just slightly from one another in mass, but not in other chemical characteristics.

For example, deuterium (^2H, or "heavy hydrogen")—a naturally occurring isotope of hydrogen (H) that is slightly heavier than hydrogen—behaves chemically like hydrogen. However, because of its slight difference in mass, deuterium is separated from hydrogen and becomes differently concentrated in Earth materials (such as snow) when acted upon by a process that is influenced by mass. Oxygen has three naturally occurring isotopes: ^{16}O, ^{17}O, and ^{18}O. ^{18}O is the heaviest; it behaves chemically like the others but is preferentially concentrated during processes influenced by mass. Other stable isotopes and materials are also useful in paleoclimate analysis, but water is so ubiquitous and so closely tied up with climatic processes that oxygen isotopes are the most widely used for this purpose.

ISOTOPIC FRACTIONATION

Consider, for example, the process of precipitation. Water that condenses from vapor in clouds, eventually falling as precipitation, is preferentially enriched in the heavier isotope of its component oxygen, that is, the ^{18}O isotope. Water that remains behind in the atmosphere as vapor in clouds, on the other hand, will be enriched in the lighter isotope, ^{16}O. The scientific term for the separation and differential concentration of isotopes of slightly different mass is **fractionation**. Many natural fractionation processes are temperature-dependent—that is, controlled by variations in temperature.

Many different Earth materials—not just ice, but groundwater, sediment, shells, bones, and others—equilibrate chemically with the water and other components of the environment around them, and therefore preserve the distinctive *isotopic signature* of their surroundings. Sampling and analysis of any material that was affected by a temperature-dependent fractionation process, that equilibrated with its surroundings, and that has preserved the isotopic signature will reveal the past temperature history of the material.

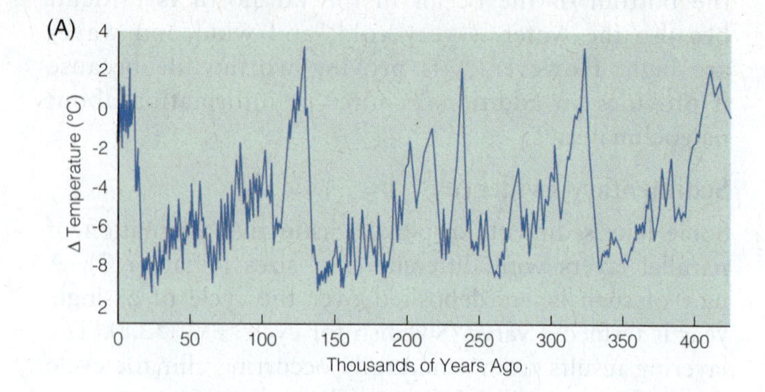

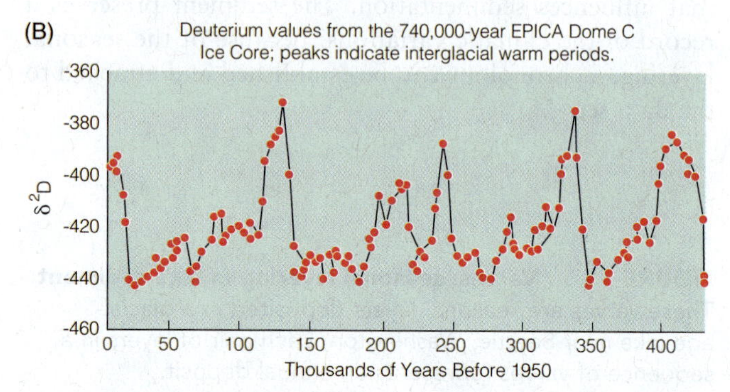

FIGURE C13.1 Temperature reconstruction from ice-core data
(A) Ice-core data from Antarctica shows a reconstruction of temperature, in terms of the difference (Δ) from present-day temperature, going back 420,000 years. This temperature reconstruction is based on analysis of oxygen isotopes from the ice. (B) The temperature reconstruction shown here, also from Antarctica, is based on Deuterium isotopic analyses and extends to 740,000 years before the present.

ISOTOPES AND GLOBAL CLIMATE

Some isotopic variations in glacier ice are believed to represent fluctuations in the air temperature near the glacier surface, exerting an influence on oxygen and hydrogen-deuterium fractionations in snow (FIG. C13.1). Other variations in ice cores, as well as variations seen in isotopic studies of marine sediments and corals, are thought to reveal changes in ice volume on a global scale.

During glacial ages, when water is evaporated from the ocean and precipitated on land to form glaciers, water containing the light isotope ^{16}O is more easily evaporated than water containing the heavier ^{18}O. (Evaporation requires the input of energy to separate a water molecule from the surface of the liquid, so it is more difficult for heavier water to evaporate when the environment is cold and less thermal energy is available.) As a result, Pleistocene glaciers received more of the light isotope, whereas

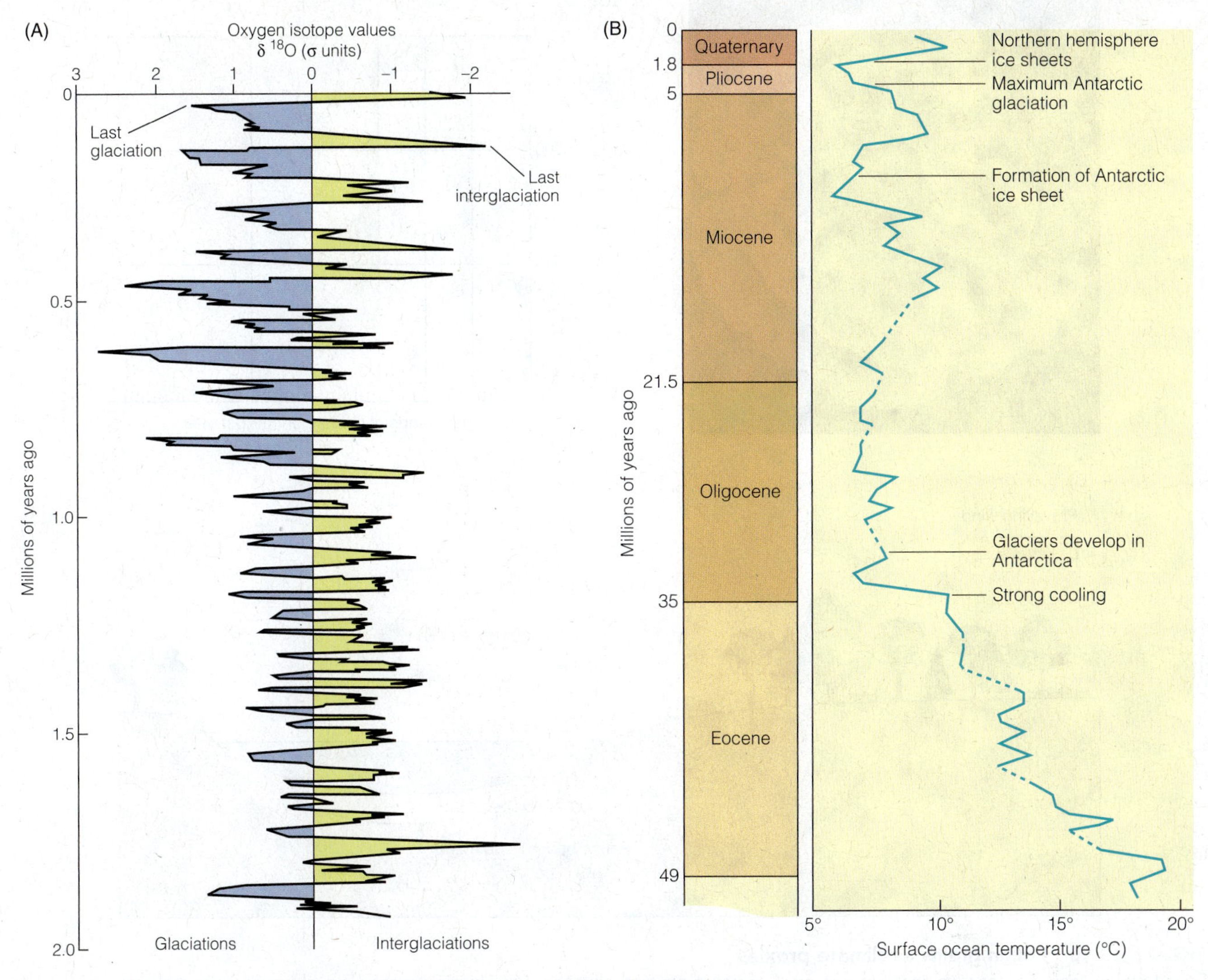

FIGURE C13.2 Temperature reconstruction from deep-sea sediment data
(A) This curve of average oxygen-isotope variations during the last 2 million years is based on analyses of deep-sea sediment cores. The curve illustrates changing global ice volume during successive glacial-interglacial cycles. (B) This record of surface ocean temperatures is based on oxygen-isotope ratios measured in a sediment core from the western Pacific Ocean. Relatively warm surface waters cooled abruptly about 35 million years ago, reflecting a dramatic change that led to the buildup of glaciers in Antarctica. With further cooling, an ice sheet developed over Antarctica, and by 2.5 million years ago, northern hemisphere ice sheets had formed.

the ocean water left behind became enriched in the heavy isotope. Marine organisms that grew and equilibrated in this water also preserve this isotopic signature. Isotope curves derived from marine sediments and organisms therefore give us a continuous reading of changing ice volume on the planet (FIG. C13.2). Because glaciers wax and wane in response to changes of climate, the isotopes also give a generalized view of global climatic change.

To get a truly global perspective, scientists need to combine multiple records from various areas. Traditionally, Antarctica and Greenland have been the principal sources for ice-core data, but more attention has been focused lately on retrieving the records from temperate (high-altitude) glaciers, many of which are disappearing. One of the most interesting challenges in stable isotope studies of paleoclimate has been to explain slight mismatches in timing (*asynchronicity*) between major climatic events recorded in ice-core records from Antarctic and Greenland ice cores. Because the number of available indicators decreases the further back in time we go, estimates of global climate conditions for the recent past tend to be more reliable than those for the distant past.

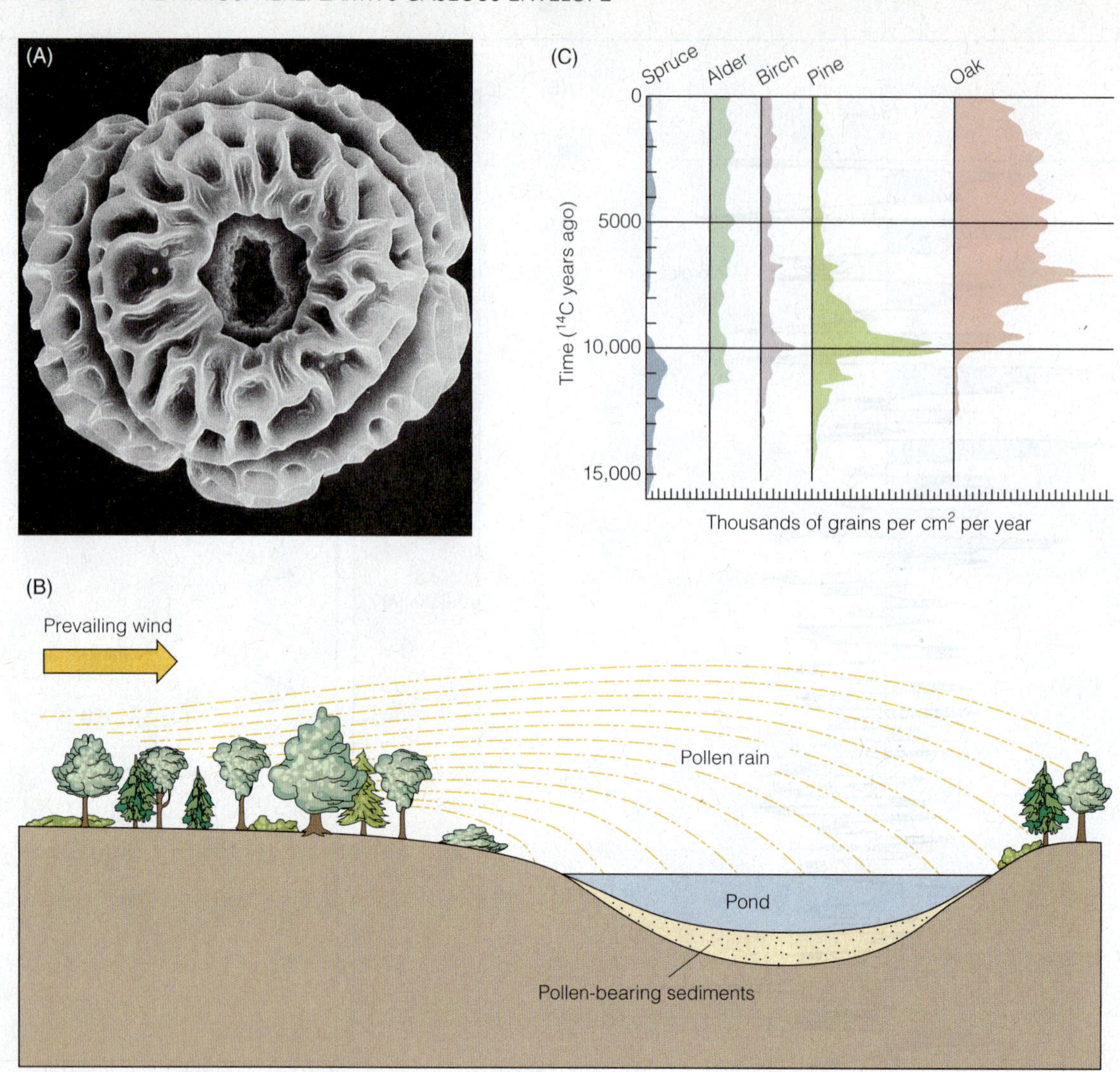

FIGURE 13.10 Pollen grains as climate proxies
Fossil pollen can be used to reconstruct past vegetation and climate. (A) This is a scanning electron microscope photograph of a grain of *Drymis winterii* pollen, diameter 42 micrometers. Its waxy coating protects it from degradation. (B) Windborne pollen grains from trees and shrubs fall into a nearby pond, where they are incorporated as part of the accumulating sediment. (C) This simplified diagram, based on data collected from Rogers Lake, Connecticut, shows variation in pollen counts as a function of time. A major change in forest composition occurred about 10,000 years ago at the end of the last glaciation, when the spruce/pine forest was replaced by a forest dominated by deciduous trees, mainly oak.

identification of individual species, but they are far less numerous than fossil pollen grains (FIG. 13.10A), which possess a hard, waxy coating that resists destruction by chemical weathering. Pollen can be identified and assigned to specific plants by trained experts, called *palynologists*. Most pollen is transported by the wind and settles into lakes, ponds, and bogs, where, protected from the wet environment by its waxy coating, it slowly accumulates (FIG. 13.10B). A sample of bog or lake sediment typically yields a vast number of pollen grains that can be identified by type, counted, and analyzed statistically. At any given level in a sediment core, the pollen grains reveal the assemblage of plants that flourished near the site when the enclosing sediment layers were deposited (FIG. 13.10C). When a modern vegetation assemblage can be found that has a composition like that implied by the fossil pollen, then the precipitation and temperature at the site of the modern assemblage can be used to estimate climatic conditions represented by the fossil assemblage.

EARTH'S PAST CLIMATES

Through the study of climate proxies and other evidence of past climates preserved in the geologic record, scientists have established a detailed chronology of climatic changes over Earth's long history (FIG. 13.11). Let's look more closely at these changes, starting with the most recent millennium and moving backwards in time.

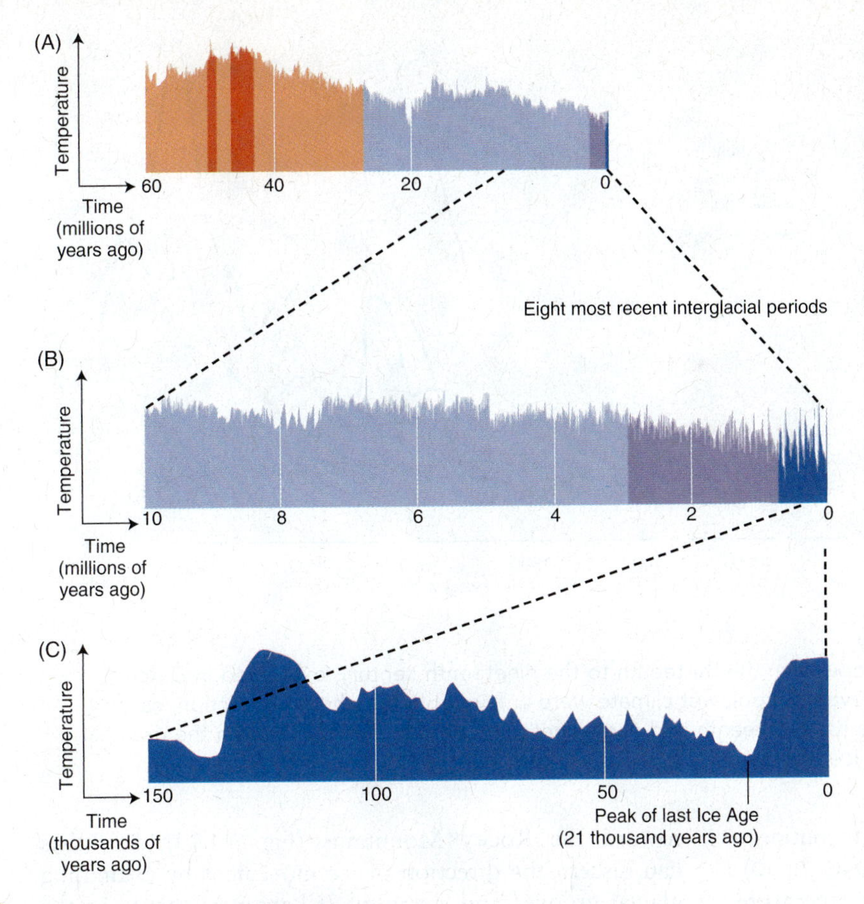

FIGURE 13.11 Past climatic change
These graphs show an estimate of global temperatures, based on deep-ocean sediments, over (A) the last 60 million years ago to present. At the beginning of the Cenozoic Era, Earth's surface was largely free of ice. Sea levels were higher and seawater could circulate freely, (B) As plate motions moved the major landmasses near their present locations, temperatures fell and glaciers appeared at the poles. In the last 800,000 years (the blue band in the temperature graph), the climate has fluctuated eight times between ice ages and warm interglacial periods. (C) At the peak of the last ice age, glaciers blanketed most of North America. Earth is now warmer than it has been at any time in the last 100,000 years, and roughly at the same temperature as it was in the last interglacial period 120,000 years ago.

Climate of the Last Millennium

A wealth of historical and climate proxy records have provided us with a comprehensive picture of climatic variations during the last thousand years. The varied evidence from the northern hemisphere shows that an episode of relatively mild climate during the Middle Ages, called the **Medieval Warm Period**, gave way about 800 years ago to a colder period, when temperatures in western Europe averaged 1 to 2°C lower (it is discernible in the temperature reconstructions in Fig. 13.6C). This cooler climate caused a lowering of the snowline by about 100 m in the world's high mountains, thereby causing glaciers to advance. Scientists refer to this interval of cooler climate and glacier advance as the **Little Ice Age**.

Throughout much of western Europe and adjacent islands, the Little Ice Age climate was characterized by unusually harsh conditions, marked by snowy winters and cool, wet summers, expansion of sea ice in the North Atlantic, and an increase in the frequency of violent wind storms and sea floods in mainland Europe. As summers became cooler and wetter, grain failed to ripen, wheat prices rose (**FIG. 13.12**), and famine became pervasive. In England the life expectancy fell by 10 years within a century.

By the early seventeenth century, advancing glaciers were overrunning farms in the Alps, Iceland, and Scandinavia. During the worst years of that century, sea ice completely surrounded Iceland, and the cod fishery in the Faeroe Islands failed because of increasing ice cover. The 1810–1819 decade witnessed renewed advances of glaciers in the Alps. Erratic weather led to further crop failures, rising grain prices, epidemics, and famines that resulted in large-scale emigrations of Europeans, especially to North America. Thus, many Canadians and Americans owe their present nationality to the Little Ice Age climate.

Little Ice Age conditions persisted until the middle of the nineteenth century, when a general warming trend caused mountain glaciers to retreat and the edge of the North Atlantic sea ice to retreat northward. Although minor fluctuations of climate have continued to take place since then (Figs. 13.2 and 13.6C), the overall trend of increasing warmth in the middle latitudes brought conditions that were increasingly favorable for crop production at a time when the human population was expanding rapidly and entering the industrial age.

The Last Glaciation

The last time Earth's climate was dramatically different from what it is now was during the last **glacial period** or *glaciation*, an interval when Earth's global ice cover greatly exceeded that of today. The last glaciation, which started about 70,000 years ago and ended about 10,000 years ago, was the most recent of a long succession of glaciations, or *ice ages*, that characterized the Pleistocene Epoch (Fig. 13.11B,C). To reconstruct the climate of this latest ice age, scientists have relied on sediments and glacier ice that contain fossil and isotopic evidence of ice-age conditions.

Temperature

In the popular imagination, glacial ages were times when temperatures were very cold, perhaps rivaling those in the middle of Antarctica today. Although such extreme cold did exist in some regions, in other places the *average* temperatures at the culmination of the last glaciation were not very different from what they are now. In the tropics, temperatures were about the same as they are today. In midlatitude coastal regions, temperatures on land were

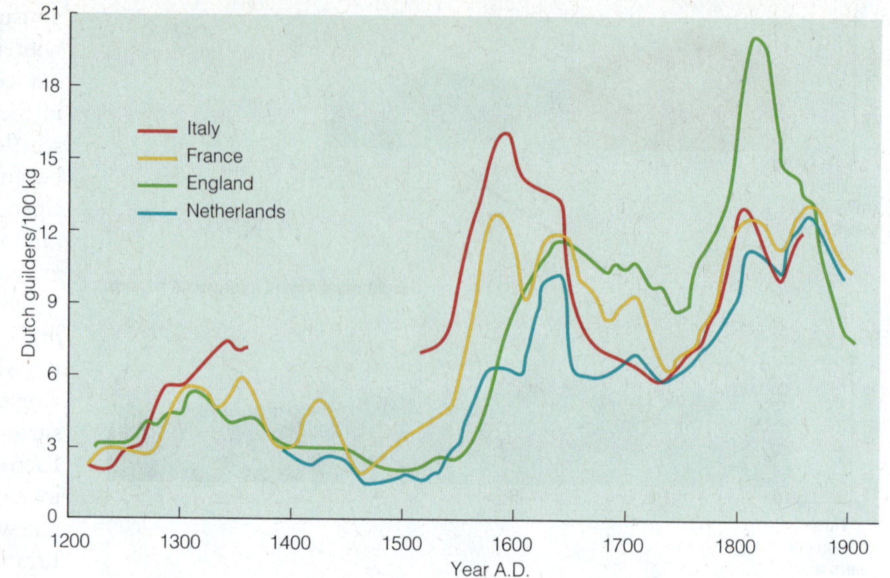

FIGURE 13.12 Wheat price as a climate proxy
Fluctuations in the price of wheat in western Europe from the thirteenth to the nineteenth century, expressed in Dutch guilders, track the course of climate change. Intervals of cool, wet climate were unfavorable for wheat production, causing the price to rise. The two largest peaks, in the early seventeenth and early nineteenth centuries, coincide with the greatest advances of glaciers in the Alps during the Little Ice Age.

generally reduced by about 5 to 8°C, whereas in continental interiors reductions of 10 to 15°C occurred (FIG. 13.13).

One way that scientists have derived temperature estimates for the last glaciation is by comparing the snowlines of ice-age glaciers with those of modern glaciers, to obtain a value for snowline lowering (FIG. 13.14, on page 396). An estimate of temperature can then be determined by comparison with the present average decrease of temperature with altitude (6°C/km). When this average rate is applied to the calculated snowline difference, the resulting depression in temperature can be found.

Other approaches to temperature reconstruction of the Pleistocene glaciation have included examination of periglacial features, fossil pollen studies, and analysis of microorganisms from deep-sea sediments. Isotopic measurements from sediment and ice cores show a marked change in isotope values at a level coinciding with the transition from a mild interglacial climate, recorded in the uppermost, recent parts of the cores, to cold ice-age temperatures in the older, Pleistocene-aged ice below (as shown in Fig. C13.1).

From these and other types of evidence obtained on land and from the ocean, an important fact has emerged: The changes accompanying a shift from interglacial to glacial conditions did not affect the whole world equally. The environments of some regions apparently changed little if at all, whereas others experienced profound changes.

Ice Extent

During the last glaciation, the climate of the northern middle and high latitudes became so cold that a vast ice sheet formed over central and eastern Canada and expanded southward toward the United States and westward toward the Rocky Mountains (Fig. 13.13). Scientists can discern the direction of ice movement by examining glacial grooves and striations (Chapter 9) carved in the bedrock by the advancing ice.

As the growing ice sheet moved across the Great Lakes region, it overwhelmed spruce trees growing in scattered groves beyond the ice margin. Ancient logs of that period, now exposed in the sides of stream valleys, are bent and twisted, indicating that they were alive when the glacier destroyed them. Some retain their bark, and some lie pointing in the direction of ice flow, like large aligned arrows. A radiocarbon date for the outermost wood or bark of such a log tells us the approximate time when the ice arrived and the tree was killed. The ages of buried trees discovered near the southern limit of the ice sheet indicate that the ice reached its greatest extent about 24,000 years ago. Additional evidence from tree ages suggests that the ice was advancing at an average rate of 25 to 100 m per year, a rate that is comparable to the movement of some large modern glaciers.

Simultaneously, other great ice sheets formed over the mountains of western Canada and over northern Europe and northwestern Asia. As ocean water was evaporated and then deposited as snow on these growing ice sheets, world sea level fell. The falling sea level allowed the great ice sheets of Greenland and Antarctica to grow larger as they spread across the adjacent, exposed continental shelves. Large glacier systems also formed in the Alps, Andes, Himalayas, and Rockies, and smaller glaciers developed on numerous other ranges and isolated peaks scattered widely through all latitudes.

We assume that sea-ice shelves also existed under full-glacial conditions, but their size and distribution are

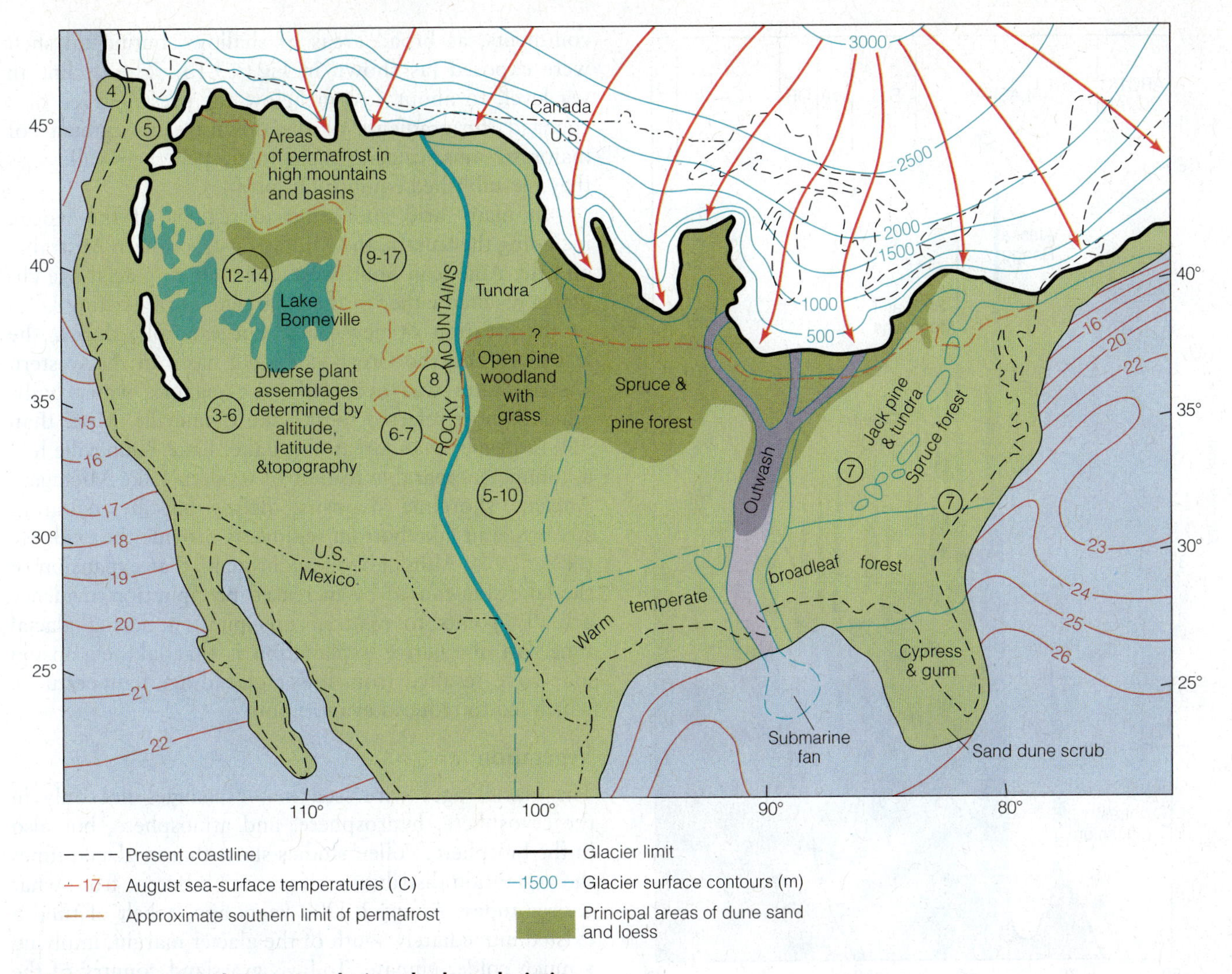

FIGURE 13.13 North America during the last glaciation
This map shows some aspects of the geography of North America about 20,000 years ago, during the last glaciation. Coastlines lie farther seaward, owing to a fall in sea level of about 120 m. Sea-surface temperatures are based on analysis of microfossils in deep-sea cores. Circled numbers show the estimated temperature lowering, relative to present temperatures, at selected sites, based on climate-proxy evidence.

not easy to determine. Some scientists postulate that an ice shelf may have covered all of the Arctic Ocean and extended south into the northern reaches of the Atlantic Ocean, thereby linking the major northern ice sheets into a continuous glacier system that covered nearly all of the arctic and much of the subarctic.

With the southward spread of ice sheets on the northern continents, periglacial zones were displaced to lower latitudes and lower altitudes. In Russia, permafrost extended 1000 km or more south of the ice margin. However, in North America evidence of full-glacial permafrost is mainly restricted to Alaska, to a narrow belt adjacent to the southernmost limit of the ice sheet in the northern Great Plains and Great Lakes regions, and to the high mountains of the American West, especially the Rockies. The contrast may reflect the fact that, whereas the massive Eurasian glacier lay north of 50° latitude, the ice sheet over central North America extended south of 40° into more temperate latitudes. The periglacial zone was therefore much narrower in the United States because the north-to-south gradient of climate there was far steeper.

The Dusty Ice-Age Atmosphere

At the height of the glacial age, the middle latitudes were both windier and dustier than they are today. Glacial-age loess is found south of the ice limit in the midwestern United States, where cold katabatic winds (Chapter 12) flowing south from the ice mass picked up fine sediment from the floodplains of glacial meltwater streams and carried it further south. The thick loess deposits of central China lie near desert basins in Central Asia that were also swept by cold, dry winds during glacial times, and loess deposits in eastern Europe lie downwind from extensive meltwater sediment near the southern limit of the great European ice sheet. These deposits contain fossil plants and animals consistent with cold, dry conditions. In each of these regions, successive sheets of loess are separated by soils, each formed during an

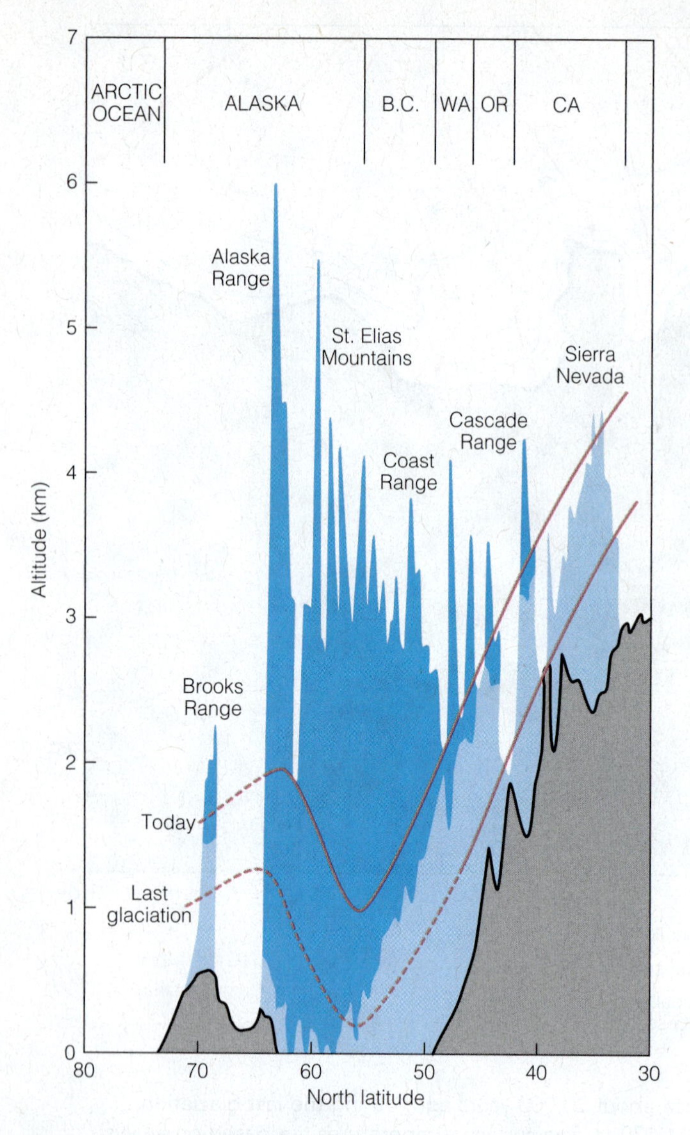

FIGURE 13.14 Snowline altitude
A transect along the coastal mountains of western North America shows the relationship of the present snowline to existing glaciers (blue) and of the ice age snowline to expanded glaciers during the last glaciation (light blue). The difference between present and ice-age snowlines was about 900 to 1000 m along the southern part of the transect and about 600 m in northern Alaska. The depression in altitude of the snowline gives an indication of the lowering of temperature during the glacial period. The change in slope of the two snowlines at about 55° latitude occurs where the transect passes inland across the Alaska Range and then northward across the Brooks Range to the Arctic Ocean.

interglacial period, a time when both the climate and the global ice cover were similar to those of today. That glacial times were both windy and dusty is also shown by studies of fine dust found in ice cores from the Greenland Ice Sheet; the percentage of windblown dust rises significantly in the part of the cores that corresponds to the last glaciation.

Water Levels and Precipitation

The fall of sea level that accompanied the buildup of glaciers on land changed the shapes and coastlines of the continents, as broad areas of shallow continental shelf were exposed (as shown in Fig. 13.13). The decline in sea level, to about 120 m below the present level, also changed the gradients of the downstream segments of major streams, causing them to deepen their valleys as they reestablished equilibrium profiles.

In many arid and semiarid regions of the world, including the Sahara, the Middle East, southern Australia, and the American Southwest, the shift to glacial-age climates resulted either in the enlargement of existing lakes or the creation of new ones. For example, during the last glaciation the Great Salt Lake basin in the western United States was occupied by a gigantic water body, which is now referred to as Lake Bonneville. More than 300 m deeper than Great Salt Lake, Lake Bonneville had a volume comparable to that of modern Lake Michigan. Ancient shoreline deposits, deltas left by tributary streams, and lake-bottom sediments provide the evidence (**FIG. 13.15**). Although we might guess that expansion of the lake was caused by increased precipitation, evidence actually points to *reduced* precipitation during glacial time. An alternative explanation is that lake expansion may have resulted from lower glacial age temperatures, which led to reduced evaporation.

Vegetation

The glacial ages witnessed major changes not only in the cryosphere, hydrosphere, and atmosphere, but also in the biosphere. Pollen studies show that in glacial times the vegetation distribution was quite different from what we see today. About 20,000 years ago, a belt of tundra existed immediately south of the glacier margin, implying a much colder climate. Today's grassland country of the Great Plains was then mostly open pine woodland. Pollen studies reveal that vegetation changes accompanying the advance and retreat of the great ice sheets were dynamic

FIGURE 13.15 Glacial Lake Bonneville
Horizontal beaches at several levels above the surface of Great Salt Lake, Utah, mark shorelines of Lake Bonneville, a vast Pleistocene lake. At its maximum extent and depth during the last glaciation, the surface of Lake Bonneville stood more than 300 m above that of the present lake.

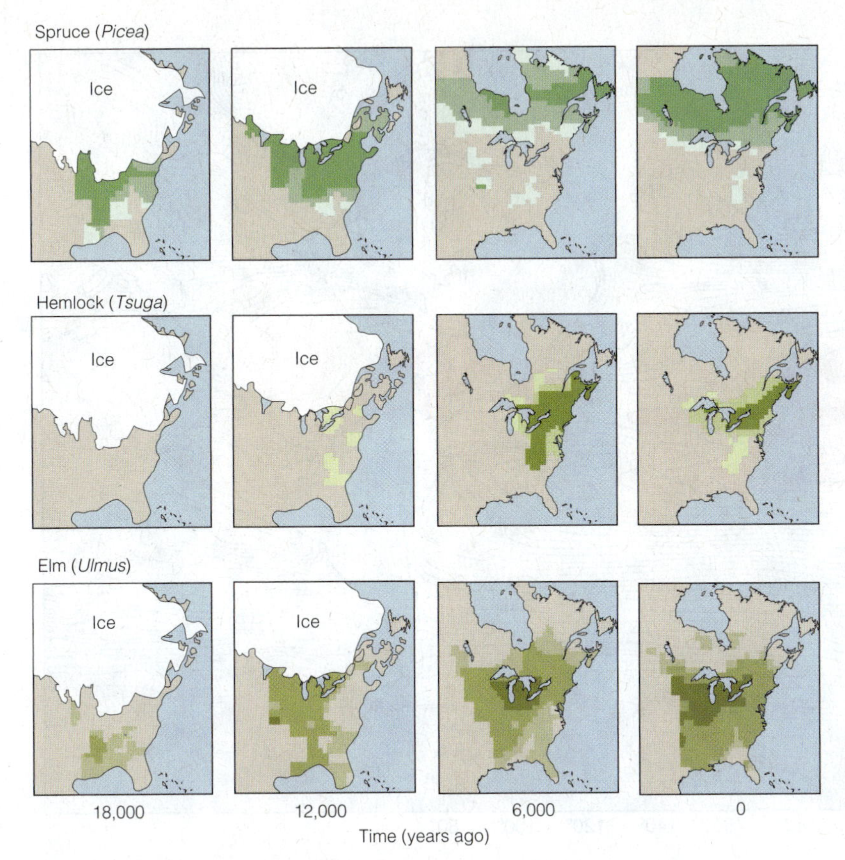

FIGURE 13.16 Changing forests
These maps, based on fossil pollen data, show the changing distribution of spruce, hemlock, and elm trees in North America at 6000-year intervals between 18,000 years ago and the present day. The color intensities indicate the relative abundance for each species, with the darkest shade of green being the highest and the lightest shade the lowest.

and complicated (FIG. 13.16). Species were displaced in various directions, forming new plant communities that are unknown on the present landscape. A similar approach can be taken using animal fossils, such as beetles, which give comparable results.

In Europe the response to glaciation was similar, but with one major difference. In North America, plant species forced southward by the advancing ice could inhabit the relatively warm lowlands that extended to the Gulf of Mexico. But in Europe the glacier-clad Alps—800 km long and 150 km wide—constituted a high, cold barrier north of the Mediterranean Sea. Many species were trapped between the large ice sheet to the north and the Alpine glaciers to the south, and were driven to extinction. Thus, western Europe, which before the glacial ages had an abundance of tree types, now has only 30 naturally occurring species. By contrast, North America, with no mountain barrier standing between the Great Lakes and the Gulf of Mexico, has 130 tree species.

Pleistocene and Older Glacial Ages

As recently as a few decades ago, it was thought that Earth had experienced only four glacial ages during the Pleistocene Epoch. This traditional view was discarded when studies of deep-sea sediments disclosed evidence of a long succession of glaciations during the Pleistocene. Paleomagnetic dating (Chapter 4) of deep-sea cores shows that during the last 800,000 years alone there have been about eight such episodes. Seafloor sediments provide a continuous historical record of climatic change, whereas evidence of glaciation on land generally is incomplete and interrupted by many unconformities. When all of the evidence is accounted for, it shows that for the Pleistocene Epoch as a whole (the last 1.8 million years), about 30 glacial ages are recorded.

The Glacial Epoch

Much of the evidence of repeated glaciation in the Pleistocene comes from extracted cores of seafloor sediment and their microfossils. With increasing depth in the sediment cores, the biologic component shows repeated shifts from warm interglacial biota to cold glacial biota. The ^{18}O to ^{16}O isotope ratios in the shells of deep-sea microorganisms fluctuate with a similar pattern, indicating high global ice volumes during the cold periods indicated by the biota. Whereas the isotopic variations in ice cores represent fluctuations in air temperature near the glacier surface, in marine sediments they are thought to reflect changes in global ice volume. (See *A Closer Look: Using Isotopes to Measure Past Climates*, Figs. C13.1 and C13.2.)

Isotopic analyses suggest that the amount of ice on land was about the same during each glaciation, for most of the last 800,000 years, and the average length of a glacial-interglacial cycle was about 100,000 years (Fig. C13.2A). Prior to that time, however, the ice volume on land during each glaciation appears to have been smaller, and the duration of each cycle averaged only about 40,000 years. Why the cycle length changed is not yet known with certainty, but it clearly represented a fundamental shift in Earth's climate system.

A record of ocean-surface temperatures, based on oxygen isotope values in deep-sea sediment cores, shows that the ocean has grown colder over the last 50 million years (Fig. C13.2B). During one pronounced cooling event about 35 million years ago, surface ocean temperatures declined by nearly 8°C in less than 100,000 years. Glaciers spread from highlands in Antarctica and reached the sea. As temperatures continued to fall, an ice sheet formed over Antarctica about 10 to 12 million years ago in the most recent glaciation. The presence of such a large polar ice mass reduced average temperatures still further and caused a substantial drop in sea level (FIG. 13.17). From that time onward, large glaciers have occupied the mountain valleys of Alaska and the southern Andes.

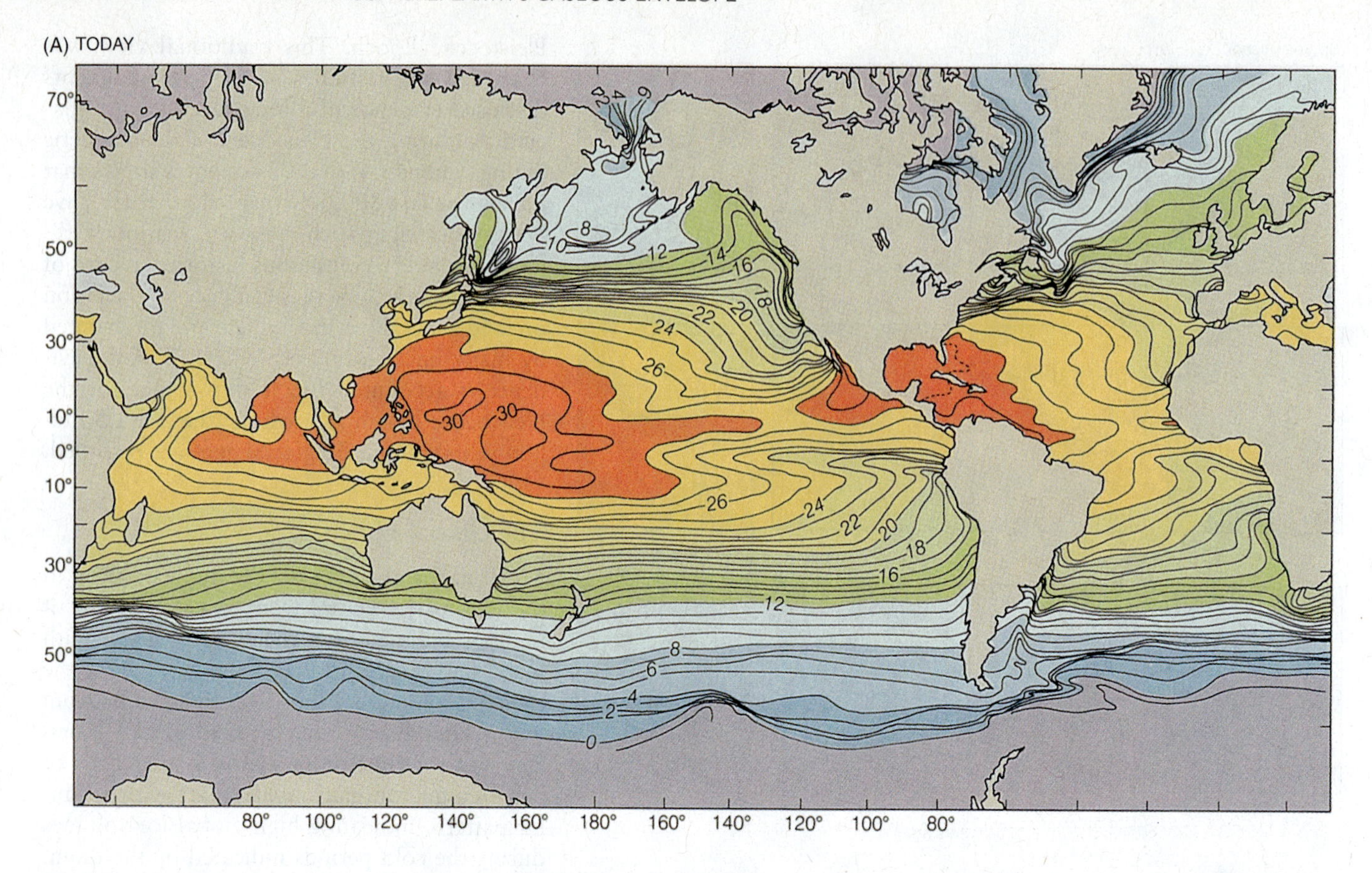

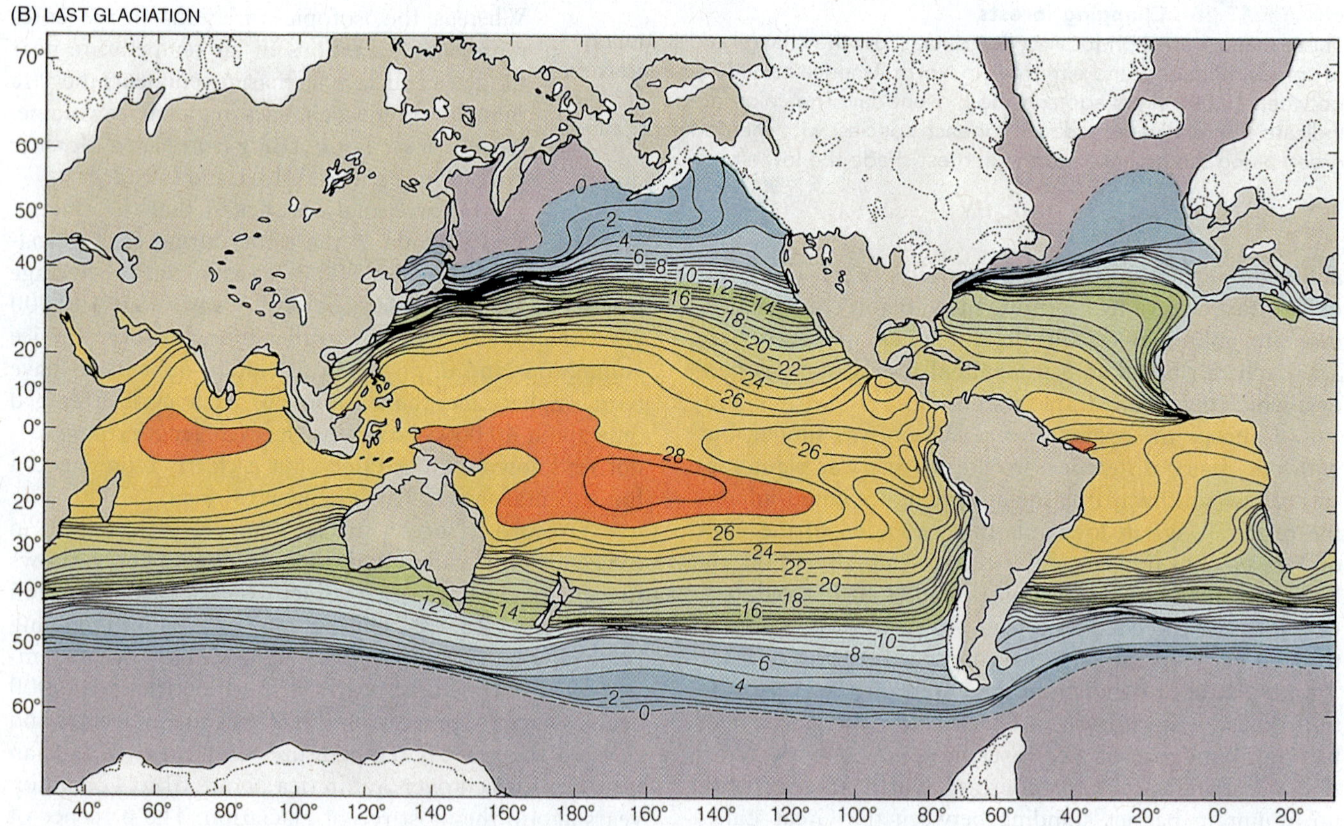

FIGURE 13.17 Present and past ocean temperatures
(A) This map shows the modern August sea-surface temperatures (in °C). (B) The other map is a reconstruction, showing the August sea-surface temperatures during the last glaciation, about 18,000 years ago. Cold polar water extended far south of its present limit in the North Atlantic, and plumes of cool water extended westward from South America in the equatorial Pacific and from Africa in the Atlantic.

Ancient Glaciations

Ancient glaciations, identified mainly by rocks of glacial origin and associated polished and striated rock surfaces, are known from far back in Earth's history as well. The earliest recorded glacial episode dates to about 2.4 billion years ago, in the early Proterozoic. Evidence of other glacial episodes has been found in rocks of late Proterozoic, early Paleozoic, and late Paleozoic age (refer to the geologic time scale, Fig. B7.2). During the latest of these intervals, 50 or more glacial advances and retreats are believed to have occurred. The geologic record is fragmentary and not always easy to interpret, but evidence from today's low-latitude regions, such as South America, Africa, Australia, and India, as well as from Antarctica, suggests that Earth's land areas must have had a very different relationship to one another during the late Paleozoic glaciation than they do today. In the Mesozoic Era, glaciation of similar magnitude apparently did not occur, consistent with geologic evidence that points to a long interval of mild temperatures both on land and in the ocean.

The Warm Middle Cretaceous

It's probably a good thing we did not live 100 million years ago during the Middle Cretaceous Period in the late Mesozoic Era. Not only was the world inhabited by huge carnivorous dinosaurs, but also the climate was one of the warmest in Earth's history. Evidence that the world was much warmer than it is today is compelling (**FIG. 13.18A**).

Warm-water marine faunas were widespread, coral reefs grew 5° to 15° closer to the poles than they do now, and vegetation zones were displaced about 15° poleward of their present positions. Peat deposits, indicative of tropical swamp environments, formed at high latitudes, and dinosaurs, which are generally thought to have preferred warm climates, ranged north of the Arctic Circle. Sea level was 100 to 200 m higher than today, implying the absence of polar ice sheets, and isotopic measurements of deep-sea sediments indicate that intermediate and deep water in the ocean was 15 to 20°C warmer than now. Based on such evidence, average global temperature is estimated to have been at least 6°C milder than today and possibly as much as 14°C (**FIG. 13.18B**), with the greatest difference being in the polar regions. Whereas today the difference in temperature between the poles and the equator is 41°C, during the Middle Cretaceous it may have been no more than 26°C and possibly as little as 17°C.

Computer simulations of past climates provide insights into the Middle Cretaceous world and suggest that interactions involving geography, ocean circulation, and atmospheric composition were likely involved in producing such warm conditions. However, geography alone is inadequate to explain warmer year-round temperatures at high latitudes; an additional factor is required. This additional factor was probably related to the composition of the atmosphere, and specifically to the concentration of greenhouse gases in the atmosphere (as shown in Fig. 13.18B). As we are considering what may have contributed to the warm climate of the Middle Cretaceous, let's look in general at the factors that cause climatic change.

FIGURE 13.18 The warm Middle Cretaceous
(A) During the Middle Cretaceous Period, sea level was 100 to 200 m higher than now, and ocean waters flooded large areas of the continents, producing shallow seas. Warm-water animal assemblages (labeled "W") and evaporite deposits (labeled "E") were present at low to middle latitudes. Coal deposits (labeled "C") developed from tropical swamps in northern latitudes, implying warm year-round temperatures. (B) This reconstruction shows changing atmospheric carbon dioxide levels and average global temperature over the past 100 million years. High CO_2 values and high temperatures in the Middle Cretaceous contrast with much lower modern values. Other intervals of higher temperature and CO_2 occurred during the Eocene and Middle Pliocene Epochs.

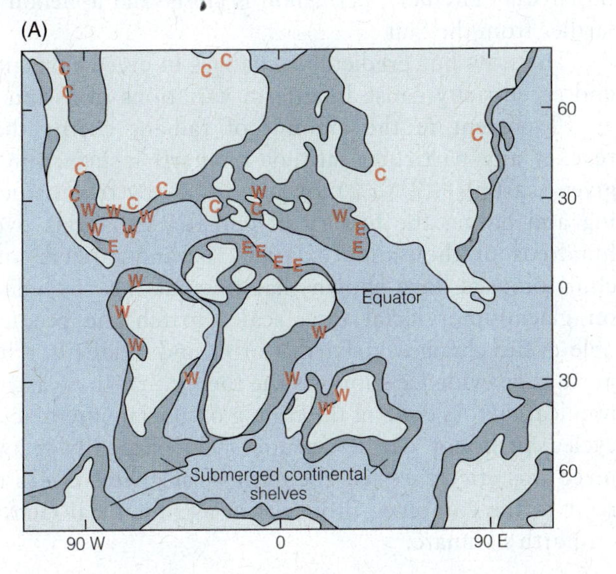

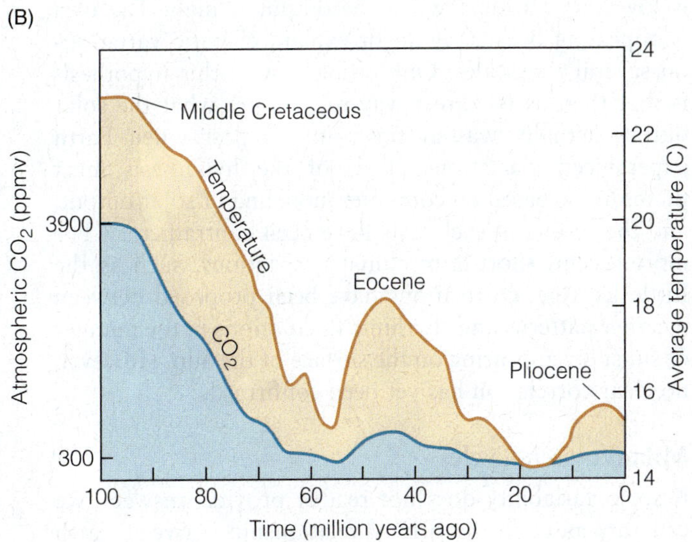

WHY CLIMATES CHANGE

What factors cause the climate to warm and cool, bringing about great changes in Earth's surface processes and environments? The search for answers has proved difficult because climate fluctuates on different timescales, ranging from decades to many millions of years, and several quite different mechanisms appear to be responsible for these changes. Furthermore, these mechanisms involve not only the atmosphere, but also the geosphere, the ocean, the biosphere, and extraterrestrial factors, all interacting in a complex way. The search for causes of climatic variability is therefore a challenging one.

External Causes of Climatic Change

Determining the cause of the cyclic pattern of glacial and interglacial ages has long been a fundamental challenge to the development of a comprehensive theory of climate. To begin our exploration of the causes of climatic change and the cyclicity of glaciations, we will look first at external causes—that is, causes from outside of the Earth system. Such causes result in **climate forcing**, by which we mean changes in the amount of solar radiation reaching the top of the atmosphere (i.e., insolation). *Positive forcing* causes an increase in insolation and would be expected to lead to warming; *negative forcing* causes a decrease in insolation and would be expected to lead to cooling.

Solar Variation

The first logical place to look for external forcing is the Sun itself. We know that the Sun varies in its radiative output, on a number of timescales. The principal cycle of solar variability is the 11-year sunspot cycle. (A *sunspot* is a temporary cool, dark spot on the Sun's photosphere—see Chapter 4—which affects the overall solar luminosity.) Longer- and shorter-term variations in solar output are also known.

Therefore, one hypothesis regarding the cause of glacial events is that fluctuations in the energy output of the Sun result in cooling of Earth's climate when the output is low and warming when the output is high. The idea is appealing because it might explain climatic variations on several timescales. One problem with this hypothesis is that there is no direct way to measure what the solar output actually was at times in the past when Earth experienced glaciations. Tests of the hypothesis must therefore be based on computer modeling of solar output, and the results of such tests have been contradictory. For more recent, short-term climatic variations, such as the Little Ice Age, correlations have been proposed between weather patterns and rhythmic fluctuations in the number of sunspots appearing on the surface of the Sun. However, no clear correlation has yet been confirmed.

Milankovitch Cycles

If solar variability does not readily provide answers, we can turn next to the orbital relationship between Earth and the Sun. Nineteenth-century Scottish geologist John Croll and early twentieth-century Serbian astronomer Milutin Milankovitch recognized that minor variations in Earth's orbit around the Sun, and in the tilt of Earth's axis, cause variations in the amount of radiant energy reaching any given latitude. The variations are slight but important, and Milankovitch proposed that the effects might combine to influence the timing of glacial-interglacial cycles.

Three movements are involved (FIG. 13.19). First, the axis of rotation, which now points in the direction of the North Star, wobbles like the axis of a spinning top (FIG. 13.19A). The wobbling movement causes the North Pole to trace a cone in space, completing one full revolution every 26,000 years. At the same time, the axis of Earth's elliptical orbit around the Sun is also rotating, but much more slowly, in the opposite direction. These two motions together cause a progressive shift in the position of the four cardinal points of Earth's orbit (spring and autumn equinoxes and winter and summer solstices). As the equinoxes move slowly around the orbital path, a motion called *precession of the equinoxes*, they complete one full cycle in about 23,000 years.

Second, the *tilt* of Earth's axis varies, which influences the extent to which a given portion of Earth's surface points away from or toward the Sun, at any given time. The tilt of Earth's axis would be 0° if the axis were oriented exactly perpendicular to the plane of the solar system, called the *ecliptic*. However, the axis is not perpendicular; it tilts at an average angle of about 23.5°, and it shifts about 1.5° to either side during a cycle of about 41,000 years (FIG. 13.19B).

Finally, the *eccentricity* of Earth's orbit, which is a measure of its departure from circularity, changes over a period of 100,000 years. About 50,000 years ago, the orbit was more circular (lower eccentricity) than it has been for the last 10,000 years (FIG. 13.19C). Eccentricity influences *perihelion* and *aphelion*, the points at which Earth is (respectively) nearest and farthest from the Sun. When Earth's orbit is more eccentric (that is, elongated rather than circular), perihelion is closer and aphelion is farther from the Sun.

The slow but predictable changes in precession, tilt, and eccentricity cause long-term variations of as much as 10 percent in the amount of radiant energy that reaches any particular latitude on Earth's surface in a given season (FIG. 13.20, on page 402). By reconstructing and dating the history of climatic variations over hundreds of thousands of years, oceanographers and climatologists have shown that fluctuations of climate on glacial-interglacial time scales match the predictable cyclic changes in Earth's orbit and axial tilt. This persuasive evidence supports the theory that these astronomical factors control the timing of glacial–interglacial cycles. In honor of the scientist who originally recognized this effect, we use the term **Milankovitch cycles** to refer to the combined influence of astronomical factors on Earth's climate.

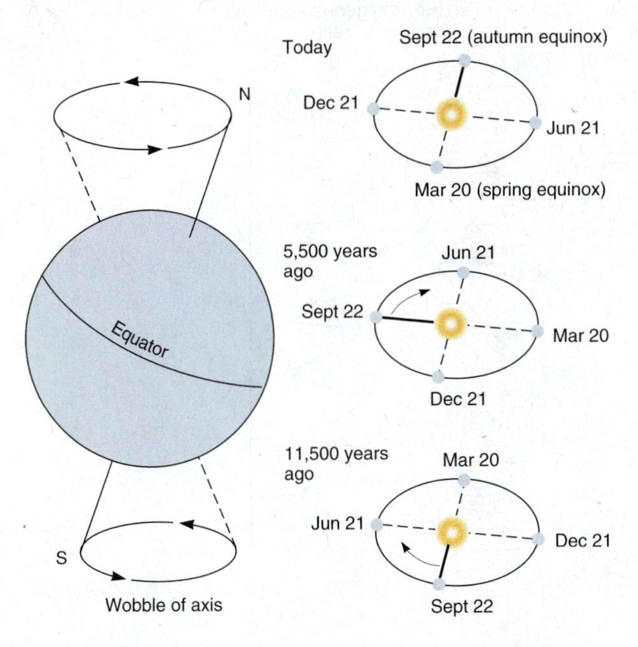

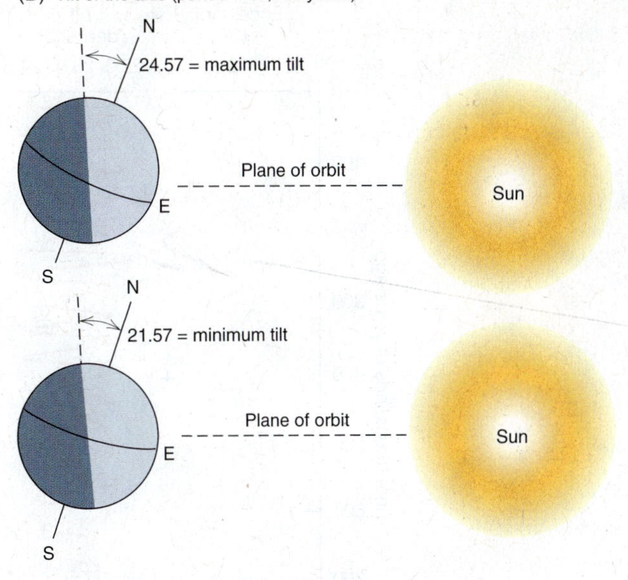

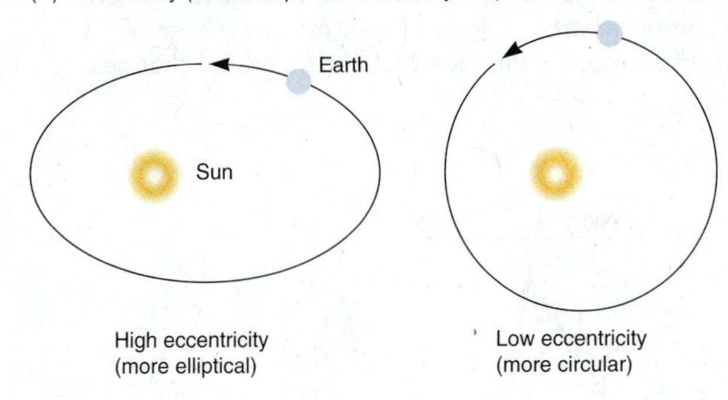

FIGURE 13.19 Milankovitch cycles
The geometry of Earth's orbit and axial tilt influence insolation, which in turn influences climate and glacial cycles. (A) Earth wobbles on its axis like a spinning top, making one revolution every 26,000 years; this is called precession. The axis of Earth's elliptical orbit also rotates, though more slowly, in the opposite direction. These motions together cause a progressive shift, or precession, of the spring and autumn equinoxes, with each cycle lasting about 23,000 years. (B) The tilt of Earth's axis, which now is about 23.5°, varies from 21.5° to 24.5°. Each cycle lasts about 41,000 years. Increasing the tilt means a greater difference, for each hemisphere, between the amount of solar radiation received in summer and that received in winter. (C) The shape of Earth's orbit is an ellipse with the Sun at one focus. Over 100,000 years, the shape of the orbit changes from almost circular (low eccentricity) to more elliptical (high eccentricity). The higher the eccentricity, the greater the seasonal variation in radiation received at any point on Earth's surface.

Internal Causes of Climatic Change

Although orbital factors can explain the timing of the glacial-interglacial cycles, the variations in solar radiation reaching Earth's surface are too small to account for the average global temperature changes of 4° to 10°C implied by paleoclimatic evidence. Somehow, the slight temperature decreases caused by orbital changes must have been amplified into temperature changes sufficiently large to generate and maintain the huge Pleistocene ice sheets. There must be some other factors involved, and the candidates for these factors are internal to the Earth system.

Atmospheric Filtering

The first obvious candidate for an internal influence on climate is the atmosphere. Solar radiation must pass through the atmosphere before it reaches Earth's surface and again on its way back to outer space. Gases in the atmosphere interact with solar radiation as it passes through, reflecting, refracting, and scattering some of it, and absorbing some of it in particular wavelengths (Chapter 11). Radiatively active gases, mainly in the troposphere, selectively absorb infrared radiation, which causes a layer of warmed air to accumulate close to the surface.

Earth's atmosphere thus acts like a selective filter for solar radiation in different wavelengths, and this causes greenhouse warming. Of interest, then, as we attempt to sort out the influences on glacial-interglacial cycles, is how effective the atmospheric filter was at any given time in Earth history. To determine this, we must figure out the abundance of greenhouse or radiatively active gases in the atmosphere at that time.

Luckily, we have samples of ancient air available for analysis, trapped in glacier ice (Fig. 9.26C). The chemical composition of air bubbles trapped in ice indicates that during glacial times the atmosphere contained less carbon dioxide and methane than it does today (**FIG. 13.21**). Carbon dioxide and methane are both important greenhouse gases. If their concentration is low, as it was during

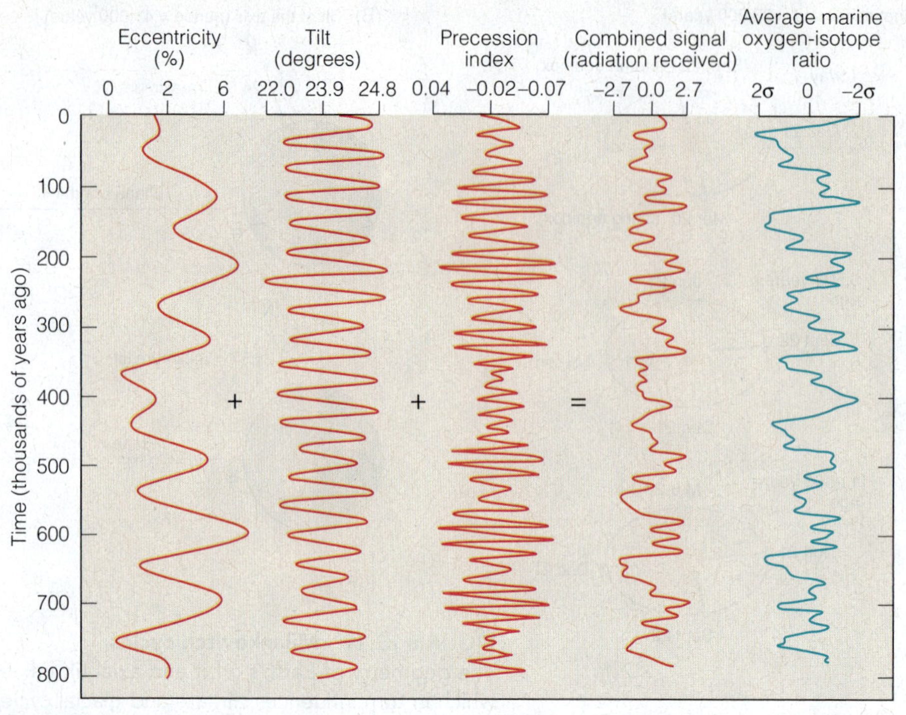

FIGURE 13.20 Orbital influences on glacial cycles
These curves show variations in eccentricity, tilt, and precession during the last 800,000 years. Summing these factors produces a combined signal that shows the amount of radiation received on Earth at a given latitude, through time (the curve labeled "Combined signal"). The magnitude and frequency of oscillations in the combined orbital signal closely matches those of the marine oxygen isotope curve, which is a temperature proxy (at right). This supports the idea that Earth's orbital changes influence the timing of the glacial-interglacial cycles.

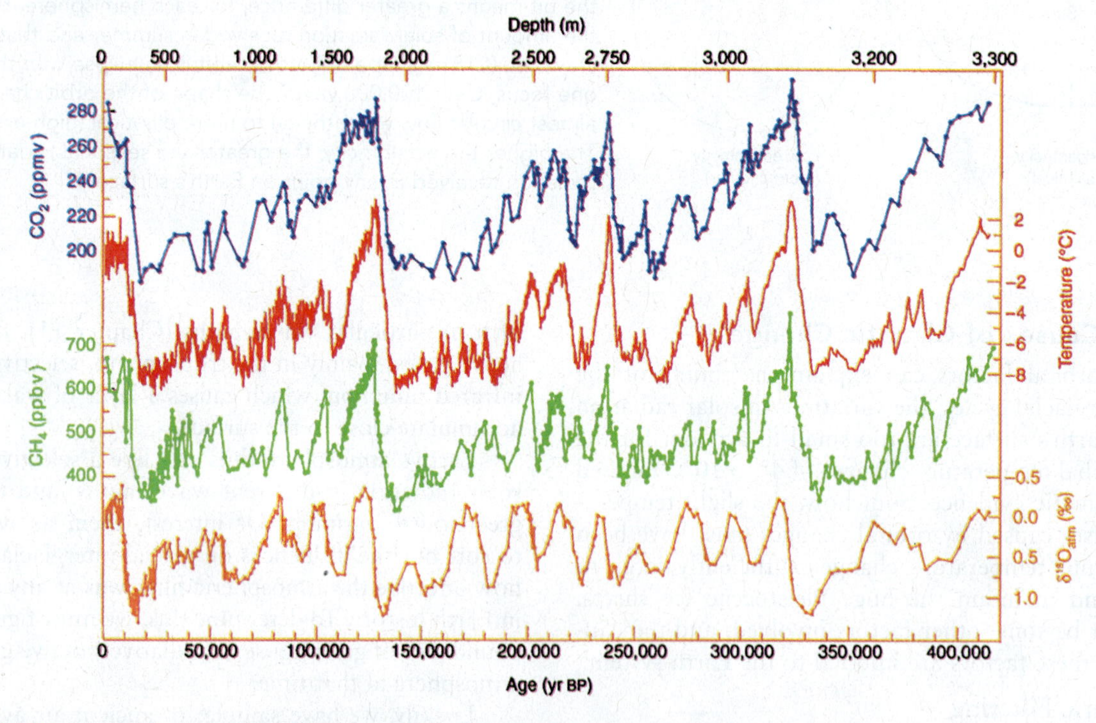

FIGURE 13.21 Atmospheric carbon dioxide and methane over time
These curves show changes in atmospheric carbon dioxide and methane (based on chemical analysis of trapped air samples) compared to changes in temperature (based on oxygen–isotope values from ice) in samples from deep ice cores drilled at Vostok Station, Antarctica. Concentrations of the greenhouse gases were high during the early part of the last interglaciation, just as they are during the present interglaciation, but they were lower during glacial times. The curves are consistent with the hypothesis that the atmospheric concentration of these gases contributed to warm interglacial climates and cold glacial climates. This remarkable record goes back 420,000 years.

glacial times, the effectiveness of the greenhouse effect would be reduced and surface air temperatures would be lower. Calculations suggest that the low levels of these two important atmospheric gases during glacial times can account for nearly half of the total ice-age temperature lowering.

Although we know that the atmospheric concentration of greenhouse gases was lower during glacial times, we do not yet know for certain what caused them to fall. It is a chicken-and-egg question: Was the temperature low because the greenhouse gas concentrations were low, or were the greenhouse gas concentrations low because the temperature was low? Scientists do not yet have the answer to either question, and it is possible that both are true to a certain extent. This is because of the existence of positive feedbacks within Earth's climate system, which we will explore in more detail shortly.

Changes in Albedo

Another factor that controls how much solar radiation reaches Earth's surface is the albedo, or reflectivity, of the atmosphere. Aerosols—extremely fine, suspended particles of dust, volcanic ash, and other substances—can cause dramatic changes in atmospheric albedo. As you learned earlier, ice-core studies have shown that the amount of dust in the atmosphere was unusually high during glacial times, when midlatitude climates were generally drier and windier. The fine atmospheric dust scattered incoming radiation back into space, which would have further cooled Earth's surface.

Large explosive volcanic eruptions can eject huge quantities of fine ash into the atmosphere, creating a veil of fine dust that encircles the globe (FIG. 13.22). The fine ash particles scatter incoming solar radiation, resulting in the cooling of Earth's surface. The volcanic dust settles out rather quickly, generally within a few months to a year. However, tiny droplets of sulfuric acid, produced by the interaction of volcanically emitted SO_2 gas and water vapor, remain in the upper atmosphere for several years. They scatter the Sun's rays and increase the reflectivity of clouds; therefore, explosive, sulfur-rich volcanic eruptions are strongly connected with global cooling. The major eruptions of Krakatau (1883) and Tambora (1815) in the East Indies, and Pinatubo (1991) in the Philippines caused average surface temperatures in the northern hemisphere to decline between 0.3 and 2.0°C over a period of years. A far greater eruption of Toba volcano about 74,000 years ago, the largest known prehistoric explosive eruption, may have lowered surface temperatures in the northern hemisphere by 3 to 5°C.

The albedo of land surfaces also plays a role in the climate system. Whenever the world enters a glacial age, large areas of land are progressively covered by snow and glacier ice, and the extent of high-latitude sea ice increases. Instead of radiation being absorbed and heating the land surface, the highly reflective surfaces of snow and ice scatter incoming radiation back into space, leading to further cooling of the lower atmosphere. Together with lower greenhouse gas concentrations and increased atmospheric dust, this additional cooling would favor the expansion of glaciers.

Volcanic CO_2

Interspersed with ancient glacial intervals there have been episodes of exceptionally warm climate, like that of the Middle Cretaceous. Increased atmospheric CO_2 was likely an important factor in Middle Cretaceous warming; however, we still are faced with explaining how the atmospheric concentration of this greenhouse gas increased so substantially. Although explosive, sulfur-rich volcanic eruptions can cause cooling, volcanism is also a major natural source of CO_2. Could volcanism have been responsible for the high atmospheric CO_2 and Middle Cretaceous warming?

Model simulations of the Middle Cretaceous climate show that, by rearranging the geography of continents and ocean basins, and increasing carbon dioxide six to eight times above the present concentration, the warmer temperatures can be explained. Reconstructions of changing atmospheric CO_2 levels over the past 100 million years point to at least a tenfold increase in CO_2 during the Middle Cretaceous (as shown in Fig. 13.18B). Under such greenhouse conditions, it is easy to see why ice volume on Earth was unusually low and world sea level was so high.

Where did the CO_2 come from? Geologic evidence points to an unusually high rate of volcanic activity in the Middle Cretaceous. Vast outpourings of lava created a succession of great undersea volcanic plateaus across the southern Pacific Ocean. One of these—the Ontong-Java Plateau in the southwestern Pacific—has more than twice the area of Alaska and reaches a thickness of 40 km. Such a massive outpouring of lava likely released a huge volume of CO_2. By one calculation, the eruptions could have released enough CO_2 to raise the atmospheric concentration to 20 times its modern natural value, in the process raising average global temperature as much as 10°C.

Geologists have hypothesized that such vast lava outpourings are associated with *superplumes*—vast masses of unusually hot rock that rise by convection all the way from the core-mantle boundary, spreading outward at the base of the lithosphere. A superplume would be a highly efficient mechanism for allowing heat to escape from Earth's core. If this hypothesis is correct, then the core and the atmosphere are linked dynamically, and the warm Middle Cretaceous climate was a direct consequence of the cooling of Earth's deep interior.

Shifting Continents

The positions, shapes, and altitudes of landmasses have changed with time over the course of Earth history (Chapter 5) and have had major influences on both regional and world climates. The changes include the movement of

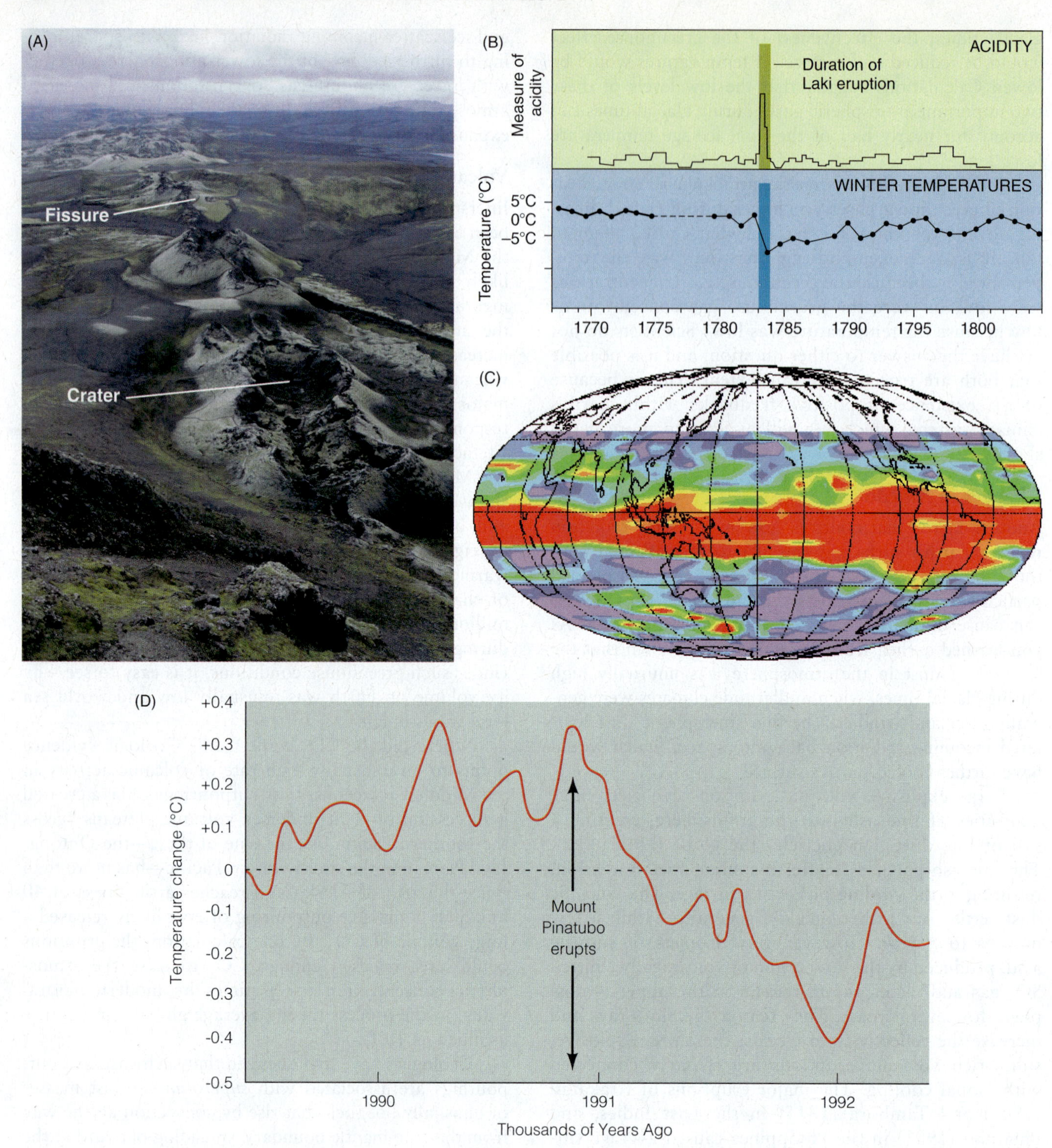

FIGURE 13.22 Volcanoes and climate
Major volcanic eruptions can cause global cooling. (A) The fissure eruption of Laki, a volcano in Iceland, lasted from 1783 to 1784 and was the largest flow of lava in recorded history. (B) In the winter after Laki's eruption, the average temperature in the northern hemisphere was about 1°C below normal. In the eastern United States, the decrease was closer to 2.5°C. At the same time, ice cores from Greenland record a dramatic spike in acidity, due to acid precipitation caused by the volcanic emissions reacting with water vapor in the atmosphere. (C) Mount Pinatubo, a stratovolcano in the Philippines, erupted violently in June, 1991, producing a sulfur-rich aerosol haze that encircled the globe. The color scale in this diagram shows atmospheric sulfur dioxide in ppb, as measured in the upper atmosphere by the Microwave Limb Sounder (MLS) in September, 1991. The warmer colors indicate higher levels of sulfur dioxide. (D) Sulfate aerosols from the eruption of Pinatubo caused a global decrease in temperature of about 0.4°C, and more in the northern hemisphere.

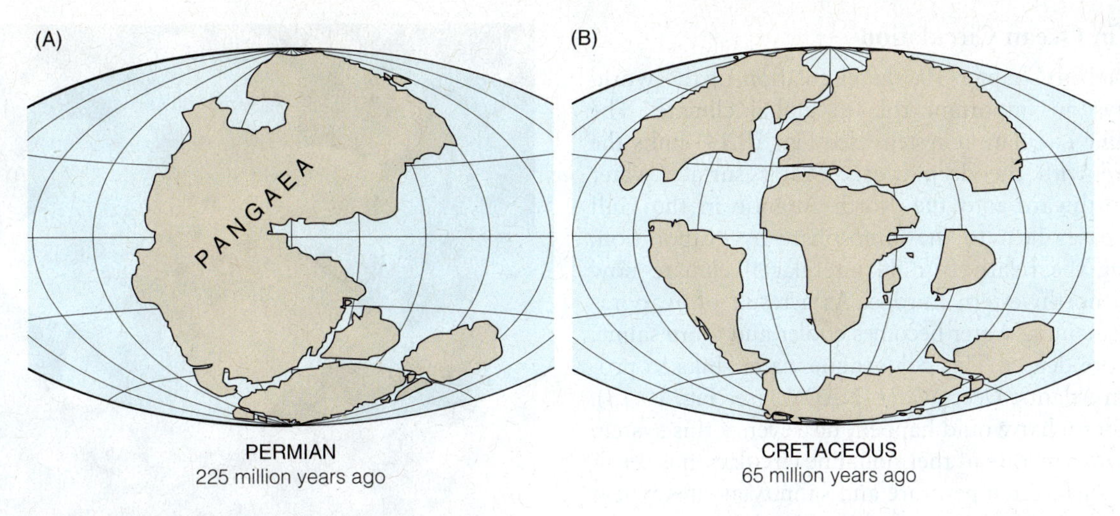

FIGURE 13.23 Pangaea and Panthalassa
(A) Around 225 million years ago the continents were still gathered together in one supercontinent, Pangaea, with vast areas of land located far from the temperature-moderating influence and moisture source of the global ocean, Panthalassa. (B) Early in the Cretaceous Period, however, Pangaea began to split apart, which would have brought much more land in closer contact with the ocean, leading to warmer temperatures, higher precipitation, and lesser extremes in temperature.

continents as they are carried along with shifting lithospheric plates, the creation of high mountain chains and plateaus where plates collide, and the opening or closing of ocean basins and seaways between moving landmasses. In the process, the paths of ocean currents and atmospheric circulation have been altered. Part of the effect results from landmasses and mountain chains standing in the way of oceanic and atmospheric circulation. Another part is due to the fact that masses of water and land store and release heat differently, affecting the characteristics and therefore the movement of overlying air masses.

Recall from Chapter 5 that 225 million years ago the continents were gathered together in one supercontinent, called *Pangaea* (FIG. 13.23A). This left vast areas of land in the center of the supercontinent located far from the temperature-moderating influence and moisture source of the one great, global ocean (which scientists refer to as *Panthalassa*). It is probable, therefore, that much of the interior land of Pangaea experienced extreme variations in temperature and desert-like conditions. Early in the Cretaceous Period, however, Pangaea began to split apart (FIG. 13.23B), accompanied by rapid seafloor spreading and extreme volcanic activity. This would have brought much more land in closer contact with the ocean, leading to warmer temperatures, higher precipitation, and lesser extremes in temperature.

Today, Earth's largest existing glacier is centered on the South Pole, where temperatures are constantly below freezing. The only glaciers found at or close to the equator lie at extremely high altitudes. As landmasses and ocean basins have shifted position, occasionally they have assumed an arrangement that was optimal for widespread ice-sheet development in high latitudes. Where evidence of ancient ice sheet glaciation is now found in low latitudes, we invariably find evidence that such lands formerly were located in higher latitudes. For example, glacial deposits that formed in the south polar environment of Pangaea are now located at latitudes far from the pole, in South America, Africa, Madagascar, Arabia, India, and Australia.

Of singular importance to the world's present climate system has been the uplift of the Himalaya and the Tibetan Plateau in the lower middle latitudes of Central Asia over the last 50 million years, caused by the collision of the Indian Plate with Asia. This vast uplifted region has had a major effect on Asia's climate and on global circulation patterns. To explore the effects of large-scale uplift on Earth's climate system, global climate models have been used to compare conditions with a Tibetan Plateau (like today) and without (prior to uplift). The simulations, supported by geologic evidence, point to some major global changes, including cooling of midlatitude high plateaus, strong westerlies, and enhancement of the jet stream, bringing cold air south, changes in precipitation, strengthening of the Asian monsoon, and changes in ocean salinity and temperature.

Central Asia, though clearly the most important, is not the only continental region to have experienced major regional uplift in the recent geologic past. Other areas include the high plateaus and mountains of the American West, the high Andes and Altiplano of South America, and the extensive plateaus of eastern and southern Africa that have mean altitudes of more than 1000 m. The variety and magnitude of environmental changes linked to the uplift of mountains point to the complexity of the interlinked components of the Earth system. A major tectonic event resulting from the collision of two continents can affect not only the geosphere in and beyond the belt of collision, but also the global atmosphere, hydrosphere, cryosphere, and biosphere.

Changes in Ocean Circulation

As discussed in Chapter 10, the circulation of the world ocean plays an important role in global climate. The thermohaline circulation system (see Fig. 10.14) links the atmosphere with the deep ocean. Warm surface water moving northward into the North Atlantic in the Gulf Stream releases heat to the atmosphere by evaporation, maintaining the relatively mild interglacial climate now enjoyed in northwestern Europe. As a result of evaporation, the remaining water becomes cooler and more saline, and therefore denser. This cold, saline water sinks to produce North Atlantic Deep Water (NADW) (see Fig. 10.13).

Consider what would happen, however, if this system closed down. The rate of thermohaline circulation is sensitive to the surface temperature and salinity at sites where deep water forms. If the water of the North Atlantic were to become fresher, for example, such that NADW failed to form, the thermohaline circulation could be disrupted, bypassing the northern portion of the Atlantic altogether. This would effectively shut down the warm Gulf Stream Current, plunging the coastal areas of North America and western Europe into a deep freeze. You might recognize this as the scenario explored in the 2004 movie *The Day After Tomorrow*—which is a *science fiction* movie, it is important to point out.

Once the Gulf Stream portion of the thermohaline circulation is cut off, the expanding sea-ice cover in the North Atlantic and extensive ice sheets on the adjacent continents would cause the climate to become increasingly cold. Thus, changes in the thermohaline circulation system may help explain why Earth's climate system appears to fluctuate between two relatively stable modes: one in which the ocean conveyor system is operational (interglaciation) and one in which it has shut down (glaciation). In *The Day After Tomorrow*, this change happens over the course of about 4 days, an impossible timescale in the real world. However, over the past decade or so, considerable interest has been focused on fluctuations of climate that start and end very abruptly, last only a few hundred to a thousand years, and recur at intervals much shorter than the tens of thousands of years that we commonly associate with major glaciations. The shift from one climate state to another may take no more than a few decades, or less than a human life span.

A rapid cooling interval of this type appears to have happened between 11,000 and 10,000 years ago, when Earth was in the process of emerging from the last glaciation. The climate had been steadily warming, ice sheets had been retreating, and plants and animals had begun to reoccupy land emerging from beneath the melting ice mass. Then temperatures in North America fell abruptly; the ice sheets stopped retreating and began to advance once more. This cold episode lasted about 1300 years, and scientists refer to it as the **Younger Dryas event** (FIGS. 13.24 and 13.25).

The Younger Dryas was not unique; it now appears that the climate fluctuated rapidly in and out of full-glacial conditions during much of the last glacial age.

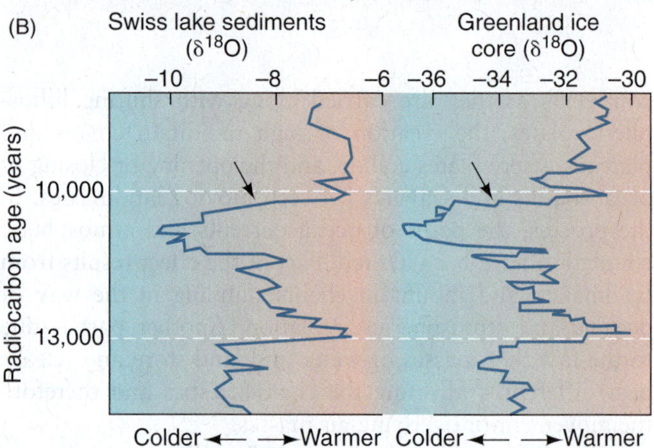

FIGURE 13.24 The Younger Dryas event
(A) Under full-glacial conditions, plants that are currently limited to polar and high-altitude regions could move into forests in northwestern Europe. Among these plants is *Dryas octopetala*, shown here. A large amount of *Dryas* pollen was found in deposits dating to the cold period now known as the Younger Dryas event. (B) Measurements of oxygen isotopes in sediment taken from a Swiss lake (left) and an ice core from the Greenland Ice Sheet (right) show that both the onset and the end of the Younger Dryas event were rapid. At the end of the event, the climate over Greenland warmed by about 7°C in only 40 years.

These warm-cool fluctuations are now referred to as **Dansgaard-Oeschger events** (or *D/O events*), and they were discovered in the ice-core records. D/O events are also clearly displayed in isotope records from deep-sea sediment cores, and they have been detected in pollen, loess, lake-sediment, and glacial records on land. The evidence reveals that D/O events recur approximately every 1500 years, and there were about 50 events during the last glaciation. The Younger Dryas may have been the first in a series of such fluctuations that accompanied Earth's emergence from the last major glaciation. Interestingly, about

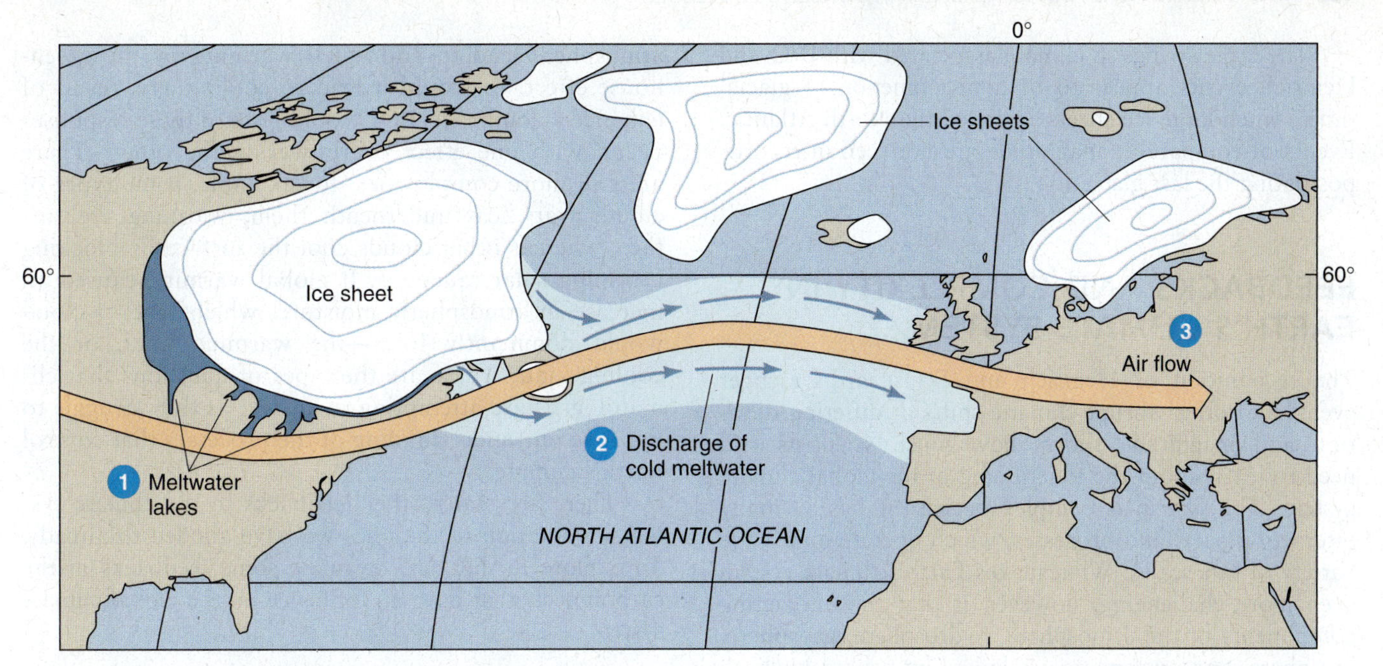

FIGURE 13.25 The Younger Dryas and Heinrich events
(1) As the ice sheet over eastern North America retreated, vast meltwater lakes holding icy water were created. At the same time, the ocean conveyor belt was at work in the North Atlantic. Wind-driven warm surface currents, such as the Gulf Stream, headed toward the poles, cooling and eventually sinking at high latitudes., having transferred energy around the globe. (2) As the ice shrank further, it uncovered a natural drainageway between the meltwater lakes and the North Atlantic. The meltwater flooded rapidly into the ocean, forming a freshwater lid over the denser salty seawater. The cold surface meltwater in turn reduced the salinity of the water and the rate of evaporation from the ocean surface, shutting down the normal pattern of ocean thermohaline circulation. (3) Without the thermohaline circulation system, air passing over the cold North Atlantic brought colder conditions to northwestern Europe that led to the growth of glaciers and a major change in vegetation.

65,000 to 70,000 years ago a similar series of climatic fluctuations may have accompanied the *onset* of the last glaciation as well. The cause of the rapid fluctuations is still uncertain, but clearly, we need to learn more about these changes if we are to understand the global climate system.

Make the CONNECTION

Glaciation may have been the cause of the demise of the Neandertals, the closest relative to modern humans. Neandertals disappeared about 28,000 years ago, after coexisting with modern humans for some 15,000 years. But Neandertals were well suited to the cold and had survived glacial conditions before. An alternative hypothesis suggests that it was climatic *variability* during the onset of the last glaciation, rather than the cold itself that led to the end of the Neandertal. Why would climatic variability have made it difficult for Neandertals to survive? What would have been the impacts of rapid fluctuations of temperature on their lifestyle and ability to access resources?

An important clue about the nature of rapid warming and cooling fluctuations came with the recognition of layers of ice-rafted sediment in the upper parts of deep-sea cores from the North Atlantic Ocean. Referred to as *Heinrich layers*, after the marine geologist who first described them, they record sudden massive releases of icebergs and ice-rafted debris into the North Atlantic from the surrounding continental ice sheets during the emergence from the last glaciation. The melting of the continental ice sheets also contributed massive volumes of cold, fresh meltwater to the oceans at this time (as shown in Fig. 13.25). The climatic effects of these events must have been profound and must have reverberated through key elements of the Earth's interacting atmosphere, hydrosphere, cryosphere, and biosphere systems.

During such a **Heinrich event**, a "lid" of cold, fresh meltwater laden with icebergs would have interrupted the generation of saline North Atlantic Deep Water, thereby shutting down the ocean conveyor system. Because that system affects the oceans worldwide, world climate would also be affected. Air passing across the vast field of icebergs and cold, low-salinity surface water associated with them was cooled, causing temperatures in adjacent land areas to plummet. Similar climatic shifts at the same times are found in lake sediments from western Europe, pollen records from Florida, loess deposits of China, and alpine glacial deposits of western North America and South America, which supports the suggestion that Heinrich events were global in extent.

The large swings in climate associated with D/O and Heinrich events appear to be a phenomenon of glacial times, when large ice sheets encircled the North Atlantic. Events of comparable magnitude are not seen in records postdating the last glaciation.

FEEDBACKS AND COMPLEXITY IN EARTH'S CLIMATE SYSTEM

The recognition of Heinrich and Dansgaard-Oeschger events and other abrupt climatic shifts of differing duration and magnitude shows how much scientists still need to learn about the functioning of the global climate system. The system is complex because it has so many interacting parts and processes, which operate on a wide variety of timescales. What makes Earth's climate system even more challenging, however, is that the interacting components of the atmosphere, hydrosphere, geosphere, biosphere, and cryosphere are linked by feedbacks, both positive and negative. This means that a change in one subsystem will not only cause a change in another subsystem, but the change may become amplified as it moves through the positive feedbacks of the climate system. The opposite can also occur: A change in one subsystem may be immediately diminished if acted upon by negative feedbacks in the climate system.

Feedbacks

Recall from Chapter 1 that a *negative feedback* occurs when a system is stabilizing, self-limiting, or *homeostatic*. In this case, the system's response to a change is in the opposite direction from the initial input. For example, if the global surface temperature warms, evaporation would be expected to increase. This would lead to more water vapor in the atmosphere and thus to more clouds. Clouds can block incoming solar radiation; hence, the surface temperature will cool down again. The response of the system is in the opposite direction to the initial change.

In contrast, a *positive feedback* is self-perpetuating and self-reinforcing. In this case, the system's response to a change is in the same direction as the initial input. For example, if the global surface temperature warms, evaporation would be expected to increase. This would lead to more water vapor in the atmosphere. Water vapor is a highly effective greenhouse gas—it is the most important naturally occurring greenhouse gas. More water vapor in the atmosphere would lead to further warming, which would lead to more evaporation, more water vapor in the atmosphere, and more warming. This is thought to be the most significant positive feedback in the climate system.

Feedbacks make Earth's climatic processes very challenging to disentangle, as you can tell from the preceding example. Would additional water vapor in the atmosphere lead to additional warming by the greenhouse effect, or would it lead to cooling as a result of enhanced cloud cover? Or would both of these responses occur, with one effect overpowering the other? There are still more complexities; for example, some types of clouds trap heat underneath them, warming the surface, whereas other clouds cool the surface by blocking incoming solar radiation. If global warming caused an increase in atmospheric moisture, which type of cloud would dominantly form—the warming kind, or the cooling kind? These are the types of questions that climatologists are attempting to answer as they struggle to improve our understanding of the processes that control Earth's climate.

There are many other feedbacks in the climate system, in addition to the ones we have alluded to already. To explore further, let's examine some feedbacks in the carbon cycle that have an influence on the global climate system.

Feedbacks in Carbon Cycling

The long-term trend of climate during the last 100 million years bears a general relationship to the trend in atmospheric CO_2 concentration (Figs. 13.18B and 13.21). Explaining this long-term trend is important in trying to understand how Earth's climate system has evolved. Of special interest is explaining how the climate maintains a degree of equilibrium. Why doesn't atmospheric CO_2 rise to such high values that Earth becomes a "hothouse" like Venus? Why doesn't it fall so low that Earth cools down to become a frigid planet like Mars?

Negative Feedbacks in the Carbon Cycle

Controlling the amount of carbon dioxide in the atmosphere requires negative feedbacks in the carbon cycle. Without self-limiting negative feedbacks, or if positive feedbacks were to overwhelm the self-limiting capacity of the system, Earth would have a runaway greenhouse effect, as there is on Venus. The geosphere, atmosphere, and biosphere interact in interesting ways to produce negative, self-limiting feedbacks on Earth.

The tectonic cycle plays a large role in controlling the rate of CO_2 buildup in the atmosphere through the processes of seafloor spreading and subduction. Seafloor spreading regulates atmospheric CO_2 by controlling the rate of volcanism, and therefore the production of volcanic CO_2, along midocean ridges and in subduction zones. The metamorphism of carbon-rich sediments carried downward in subduction zones is an additional and possibly larger source of atmospheric CO_2 that also is controlled by the tectonic cycle. Carbon is released from carbon-rich sediments, such as limestones and deep-sea oozes, when they are heated and broken down by metamorphism. During periods of intense tectonic activity, volcanic emissions contribute large quantities of CO_2 to the atmosphere; during periods of tectonic quiescence, volcanic emissions are less.

At the same time, CO_2 is being removed from the atmosphere by the weathering of surface rock. As you learned in Chapter 7, rainwater combines with CO_2 to form carbonic acid (H_2CO_3), and this acid causes chemical weathering of silicate rocks, as in the following reaction:

$$\underset{\text{(silicate rock)}}{CaSiO_3} + \underset{\text{(carbonic acid)}}{H_2CO_3} \rightarrow \underset{\text{(weathering products)}}{CaCO_3 + SiO_2} + \underset{\text{(water)}}{H_2O}$$

The weathering products (carbonate and silica) are carried to the ocean by streams. There, marine plankton use the carbonate and silica to build their shells or skeletons. When they die, their remains accumulate on the seafloor and are ultimately buried and stored as sediment. Weathering of silicate rocks, therefore, is a negative feedback in the climate system, which can help move global climate toward an equilibrium condition.

If seafloor spreading speeds up, more CO_2 enters the atmosphere. Earth warms, the water vapor content of the atmosphere rises, and vegetation density increases (because carbon dioxide has a fertilizing effect on plants, which utilize it for photosynthesis). The increase of atmospheric CO_2 speeds up the rate of chemical weathering of silicate rocks, which removes CO_2 from the atmosphere, thereby keeping the system in a more balanced state. If seafloor spreading slows, less CO_2 enters the atmosphere. The climate cools, weathering rates decrease, and the system adjusts to a near-balanced condition.

A further important negative feedback occurs with the burial of organic carbon on continental shelves. High erosion rates can lead to rapid sedimentation in sedimentary basins and submarine fans along the margins of continents. Isolated from the weathering environment at the land surface, the carbon is quickly buried, stored, and sequestered, rather than being returned to the atmosphere by surface weathering. Recent studies suggest that carbon burial may be even more important than silicate weathering in long-term carbon storage.

Another hypothesis has been proposed in which tectonic uplift is the driving force behind changes in atmospheric CO_2 levels. Uplift could increase the rate of removal of atmospheric CO_2 because faulting and landslides expose fresh rock to weathering. Glacial (physical) weathering is dominant on unvegetated, high-altitude slopes. The products of physical weathering are transferred from high slopes to lower slopes by stream runoff. There, in the flat terrain and warm, moist climates of floodplains and deltas, chemical weathering becomes dominant. Instead of being a relatively weak negative feedback, chemical weathering could be a major factor in controlling the long-term concentration of CO_2 in the atmosphere, and areas of rapid uplift may be sites of primary importance in Earth's carbon budget.

Positive Feedbacks in the Carbon Cycle

We know from ice-core data that a strong positive correlation exists between paleotemperatures and the concentration of greenhouse gases, especially carbon dioxide, in the atmosphere, as shown in Figure 13.21. A crucial question, though, is the extent to which one causes or reinforces the other. In a positive feedback situation, high temperatures would lead to increases in atmospheric CO_2, which would drive additional warming, in a self-reinforcing loop. But does this actually happen?

One major source of uncertainty in the carbon cycle lies in the response of terrestrial ecosystems to a warmer global climate. As mentioned earlier, increased CO_2 concentration in the atmosphere leads to enhanced growth of vegetation, which removes CO_2 from the atmosphere—a negative feedback. At the same time, however, warmer conditions can lead to an increase in soil respiration, resulting in an increased flux of CO_2 and CH_4 to the atmosphere. (This is one of the reasons tropical soils are very poor in carbon; the warmer the soil, the more easily carbon-rich soil gases escape.) In balance, it appears that a warmer climate could decrease the ability of the land to act as a sink for carbon, therefore contributing to a positive feedback.

The ocean, too, is crucial as a sink for carbon dioxide. Ocean-surface water is in chemical equilibrium with the atmosphere; if the atmospheric concentration of CO_2 increases, the uptake of CO_2 by ocean water should also increase. However, warmer conditions decrease the ability of the ocean to absorb atmospheric CO_2. As warming occurs, the nutrient content of ocean water declines. This limits the growth of the larger marine microorganisms that are the most effective at taking up carbon. (By the same token, fertilization of oceanic phytoplankton has been suggested as one way to potentially enhance the uptake of CO_2 by ocean water.)

An additional complexity is that as atmospheric CO_2 concentrations increase and the ocean takes in CO_2, ocean water becomes more acidic ($CO_2 + H_2O = H_2CO_3$, carbonic acid). A more acidic ocean would be less capable of absorbing atmospheric CO_2, perhaps eventually becoming a source rather than a sink.

The presence of vast reservoirs of hydrocarbons frozen into seafloor sediments is also of concern. Hydrocarbon molecules (mainly methane, CH_4) can become trapped by molecules of ice in sediment that is near the freezing temperature of seawater. The gas-ice combination takes a chemical form in which the hydrocarbon is trapped inside a tiny "cage" constructed of molecules of ice; this is called a **hydrocarbon gas hydrate** or *clathrate hydrate*. An enormous volume—perhaps more than 6 trillion tonnes—of methane is trapped in this form in seafloor sediment. If ocean water warms, even by a small amount, the frozen clathrate that traps these hydrocarbon molecules could melt, allowing this vast quantity of methane to bubble out. This could lead to a significant positive feedback—methane is a potent greenhouse gas, and its release into the atmosphere would cause warming, potentially leading to further warming of ocean water.

Anthropogenic Causes of Climate Change

The entire Earth system is involved in the interrelated processes that control our climate system. Not only are the basic mechanisms involved, but complex positive and negative feedbacks play a role too, leaving us with difficult questions to answer and challenging research problems to pursue. To this challenge of understanding the natural causes of climatic change, we must now add human causes.

Within the past two centuries, since the Industrial Revolution, atmospheric carbon dioxide has risen at a rate faster than at any time in the past 100,000 years. Something altogether new is happening, and it is difficult to escape the conclusion that the new ingredient is human consumption of fossil fuels. Humans pump roughly 8 billion metric tons of carbon, most of it as carbon dioxide, into the atmosphere every year from fossil fuel use *alone*. This is actually more than enough to explain the observed increase, without accounting for the other significant anthropogenic sources of carbon.

Because we understand how carbon dioxide and other greenhouse gases act to slow the escape of heat from the lower atmosphere, both logic and past climate records suggest that an increase in atmospheric carbon dioxide should be accompanied by an increase in average global temperature. The evidence for this is becoming more conclusive year by year. Worldwide temperature records show an overall increase of at least 0.6°C in the past century, and each of the past 10 years has been one of the 25 warmest years on record.

Understanding how Earth's climate system works is a challenging task, and we are far from having all the answers. Important insights have been gained through the study of past climates, evidence of which is preserved in the geologic record. Such evidence offers important clues that can help tell us what causes climate to change, and how the different physical and biological systems of Earth respond to changes of climate on different timescales. Every one of these clues is important as scientists race to understand how human actions may be contributing to atmospheric change, how (and how fast) the climate system will react to these changes, and what we can do, as a global society, to halt, mitigate, or adjust to the changes. The answers will not come from any single field, but rather from collaborative investigations by scientists in many different fields of Earth science. We will explore the human role in climatic change, and other human impacts on the Earth system, in Chapter 19.

SUMMARY

1. Earth's climate system is dynamic, with many interacting components. All but the geosphere are driven by solar energy.

2. The Köppen system is a widely used approach to climate classification, based on the distribution of native vegetation types. The system combines measurements of average temperature, precipitation, and seasonal variation to define five major climate categories: equatorial, dry, warm temperate, continental, and polar, with highland variations.

3. Climate is expressed in terms of mean temperature and precipitation, as well as humidity, windiness, and cloudiness. Instrumental tracking has only been available for the past century or so. These records show that the average temperatures in the northern hemisphere peaked in the 1940s, then declined slightly until the 1970s, when they again began to rise. The 1990s and the 2000s were the warmest decades on record.

4. That fossils of organisms adjusted to climatic conditions very different from those of today can allow us to infer that the climate of a location has changed. Other paleoclimatic evidence includes signs of glaciation and periglaciation in temperate zones; desert sand dunes now covered by vegetation; beaches of extensive former lakes in dry desert basins; channel systems of now-dry streams; remains of dead trees above the present treeline; and surface soils with profiles that are incompatible with the present climate.

5. Climate proxies are records of natural events that are controlled by, and closely mimic, climate. They can be historical records, or they can be natural records of climatic variations archived by processes like the deposition of annual or seasonal growth rings. For a climate proxy to be useful, there must be a mechanism for determining the age of climatic evidence in the record. The longest historical records of climate-controlled events, typically associated with farming, fishing, and harvesting, date back 1000 years or so.

6. Measurements of oxygen isotopes (^{18}O and ^{16}O) in glacier ice enable scientists to estimate the air temperature when the snow was accumulating at the glacier surface.

Some records extend to hundreds of thousands of years before the present. Other isotopic measurements reveal how much of Earth's water was locked up in glaciers in the past. Groundwater, soils, lake sediments, and cave deposits also equilibrate chemically with their surroundings and preserve a chronology, and therefore are amenable to isotopic analysis for paleoclimate reconstruction.

7. Trees, corals, and some other organisms lay down annual growth rings, which can provide paleoclimate information. Varved lake sediment and deep-sea sediment containing shelled microorganisms are also useful in paleoclimate reconstruction. Fossil pollen grains reveal the assemblage of plants that flourished when the enclosing sediment was deposited. The climate at the site of a similar modern assemblage can be used to estimate climatic conditions represented by the fossil assemblage.

8. A wealth of historical and climate proxy records provide a comprehensive picture of climatic variations during the last millennium. The Medieval Warm Period gave way about 700 years ago to the Little Ice Age, which persisted until the middle of the nineteenth century. Minor fluctuations have occurred since then, but the overall trend in the middle latitudes is one of increasing warmth.

9. The most recent Pleistocene glaciation started about 70,000 years ago and ended about 10,000 years ago. In mid-latitude coastal regions, temperatures on land were lower by about 5 to 8°C; in continental interiors, reductions of 10 to 15°C occurred. The changes that accompanied the shift from interglacial to glacial conditions did not affect the whole world equally.

10. During the last glaciation, a vast ice sheet formed over central and eastern Canada and northern United States, reaching its greatest extent about 24,000 years ago. World sea level was about 120 m lower than at present, and the atmosphere was windier and dustier. Communities of plant species were distributed quite differently from today.

11. Seafloor sediments provide a continuous historical record of climatic change, but the record of glaciation on land is incomplete. In the Pleistocene Epoch about 30 glacial ages are recorded. Evidence comes from extracted cores of seafloor sediment and their microfossils. These show repeated shifts from warm interglacial biota to cold glacial biota, and ^{18}O to ^{16}O isotope ratios indicate high global ice volumes during the cold periods indicated by the biota.

12. The average length of a glacial-interglacial cycle over the past 800,000 years was about 100,000 years. Prior to that time, the ice volume on land during each glaciation was smaller and the duration of each cycle averaged only 40,000 years. It is not clear exactly why the cycle changed.

13. The earliest recorded glacial episode dates to about 2.4 billion years ago. There is evidence of other glacial episodes in rocks of late Precambrian, early Paleozoic, and late Paleozoic age. In the Mesozoic Era, glaciation of similar magnitude did not occur, consistent with geologic evidence that points to a long interval of mild temperatures, accompanied by high atmospheric greenhouse gas concentrations.

14. Minor variations in the eccentricity of Earth's orbit and in the tilt and precession of Earth's orbital axis cause variations in the amount of solar radiant energy reaching any given latitude. The combined effects of these orbital changes, called Milankovitch cycles, cause climate forcing and influence the timing of glacial-interglacial cycles. Fluctuations in the energy output of the Sun, such as those related to sunspot cycles, may also influence insolation.

15. Earth's atmosphere acts as a selective filter for solar radiation in different wavelengths, causing greenhouse warming. The chemical compositions of ancient air samples trapped in glacier ice show that atmospheric concentrations of the greenhouse gases CO_2 and CH_4 have generally been higher during warm periods and lower during cool periods.

16. Aerosols such as fine dust and volcanic ash can cause dramatic changes in atmospheric albedo, causing global cooling. The albedo of land surfaces also plays a role in the climate system. When large areas of land are covered by snow and glacier ice, the highly reflective surfaces scatter incoming radiation back into space, further cooling the lower atmosphere. Volcanic emissions can also contribute to global warming, through the emission of CO_2, a greenhouse gas.

17. The positions, shapes, and altitudes of landmasses have a major influence on climate. Part of the effect results from landmasses and mountain chains standing in the way of oceanic and atmospheric circulation, and part is due to the fact that water and land store and release heat differently. Uplift of major land masses, like the Tibetan Plateau, can lead to enhancement of jet streams, changes in precipitation, strengthening of monsoons, and changes in ocean salinity and temperature.

18. The thermohaline circulation links the atmosphere with the deep ocean and plays a central role in climate. Earth's climate system fluctuates between a mode in which the thermohaline circulation is operational (interglaciation) and a mode in which it shuts down (glaciation). The Younger Dryas was the first in a series of fluctuations, called Dansgaard-Oeschger events, which accompanied Earth's emergence from the last major glaciation. Sudden massive releases of icebergs and cold, fresh meltwater into the Atlantic Ocean from the continental ice sheet interrupted the generation of saline North Atlantic Deep Water and shut down the thermohaline circulation, affecting climate worldwide.

19. Earth's climate system is characterized by feedbacks, both positive (self-reinforcing) and negative (self-limiting). Important controls in the carbon cycle include the rates of seafloor spreading, subduction, metamorphism, uplift, chemical weathering, burial of carbon-rich sediments, soil degassing, and oceanic uptake of carbon dioxide, in addition to the shorter-term mechanisms of photosynthesis and respiration by plants.

20. Human causes of climate change now contribute to the complexity of the climate system. Humans emit more than 8 billion tons of carbon into the atmosphere every year through fossil fuel burning alone, which may eventually cause an increase in average global temperature.

IMPORTANT TERMS TO REMEMBER

climate *380*
climate forcing *400*
climate proxy record *386*
Dansgaard-Oeschger events *406*
fractionation *390*
glacial period *393*
Heinrich event *407*
hydrocarbon gas hydrate *409*
interglacial period *396*
Köppen system of climate classification *382*
Little Ice Age *393*
Medieval Warm Period *393*
Milankovitch cycles *400*
paleoclimate *385*
varve *389*
Younger Dryas event *406*

QUESTIONS FOR REVIEW

1. Describe two types of records kept by humans, which can be used to study past climates.

2. Describe two types of natural climate proxy records and explain how they are related to changing climate.

3. Evidence indicates that climates during glacial times were drier than they are today. If true, how can you explain the expansion of lakes during glacial times? (Try to think of two possible explanations.)

4. How might we try to obtain an estimate of ice-age land-surface temperature by studying fossil pollen grains?

5. How are oxygen isotopes from glacier ice used in temperature reconstructions? What about oxygen isotopes from deep-sea sediment?

6. What evidence gained from the study of deep-sea sediment cores indicates that glacial-interglacial cycles occurred repeatedly during the Pleistocene Epoch?

7. Is there any evidence of glaciations older than the Pleistocene Epoch?

8. What factors may have contributed to making the Middle Cretaceous climate so much warmer than the present climate?

9. Describe the three orbital motions of Earth that contribute to variations in insolation, and therefore influence glacial-interglacial cycles.

10. What effect can large volcanic eruptions have on climate?

11. How do scientists obtain analyses of the greenhouse gas compositions of ancient atmospheres?

12. Describe the Younger Dryas event and its possible causes.

13. What is a Dansgaard-Oeschger event?

14. What is a Heinrich event? What must conditions have been like during a Heinrich event?

15. Describe some of the important positive and negative feedbacks in the carbon cycle that have an influence on climate.

QUESTIONS FOR RESEARCH AND DISCUSSION

1. How can sediment accumulating at the *floor* of the ocean provide information about *surface* water conditions?

2. Other isotopes, besides those of hydrogen and oxygen, can be used in studies of paleoclimate. Do some research to find out about sulfur isotopic analysis, and look into other materials that can be analyzed for paleoclimate information, such as cave deposits and groundwater.

3. How did sea-surface temperatures at the peak of the last glaciation differ from those of the present? Why do you think some regions of the ocean showed more change than others? What influence would these changes have had on atmospheric circulation and weather?

QUESTIONS FOR *THE BASICS*

1. On what characteristics is the Köppen system of climate classification based?

2. What are the five basic climate categories in the Köppen system?

3. How are variations in temperature, precipitation, and seasonality denoted?

QUESTIONS FOR *A CLOSER LOOK*

1. What is an isotope?

2. What are the three naturally occurring isotopes of oxygen? Which is the heaviest? What are the two naturally occurring isotopes of hydrogen? Which is the heaviest?

3. How can isotopic analyses of deep-sea sediments reveal changes in global ice volumes?

PART FIVE

The Biosphere: LIFE *on* EARTH

Biosphere: *from the Greek words* bios, *meaning life, way of living, lifetime, and* sphaira, *meaning globe, ball, or sphere.*

People are fascinated by life—by its complexities, and by the marvelous adaptations that plants and creatures make to meet their needs. Scientists have learned a tremendous amount about those complexities and adaptations. We understand the genetic code by which the shape, form, and function of organisms are transmitted from one generation to the next. We know a lot about how individual cells function, about the complex chemicals that make them up, and about the roles of these chemicals. We can create hybrids of crop plants and even animals that are designed for specific environmental conditions. We have learned to cure many diseases. But when it comes to understanding *what is necessary for life to persist* and *how life came to be on this planet* in the first place, we are still seeking answers.

In the next three chapters we explore what we know about life and its environment—the **biosphere**. We offer some answers as we understand them today, and we discuss some remaining unanswered questions. In Chapter 14, we focus on life at its most fundamental. We look at the characteristics of life itself and at the cell—a basic biologic building block. We consider how life might have originated and how it progressed from its beginnings on this planet, through the process of evolution, to the rich diversity that now surrounds us. In Chapter 15 we look more closely at life-supporting systems, especially *ecosystems*, the basic environmental units that sustain life, and we examine the role of biogeochemical cycling in ecosystems. In Chapter 16 we zoom in to consider some of the many ways in which individual organisms interact with one another, and how resource availability and environmental change affect individuals, species, populations, and communities. Finally, we take stock of Earth's biological diversity and discuss why some species are at risk.

The chapters of Part V are as follows.

Insect world
This unusual-looking creature is a net-winged beetle in the genus *Duliticola*, seen here in a tropical rain forest in Borneo. There are more beetles than any other insect and more insects than any other type of animal, to which evolutionary biologist J.B.S. Haldane is said to have remarked that God "must have an inordinate fondness for beetles."

CHAPTER 14

Life, Death, and EVOLUTION

OVERVIEW

In this chapter we:

- Introduce the basic processes and characteristics of life
- Describe the fundamental units in the hierarchy of life
- Identify the life-supporting characteristics of the Earth system
- Learn how life has both adapted to and altered the Earth system over the course of the planet's history
- Summarize the history of life, death, and evolution on this planet

Life is persistent
Life takes hold on Earth, even in environments that seem incredibly hostile. Here a plucky fern grows from a solidified lava flow in Volcanoes National Park, Hawaii.

Although life is a mystery in many respects, we do know that it is a planetary phenomenon—global in extent but unique to this particular planet, as far as we have discovered. Therefore, the Earth system is an appropriate and important context in which to consider fundamental questions about life. The discovery of a meteorite known to have originated on Mars and observed to contain structures that may be fossils of bacteria-like life forms makes this question all the more intriguing (FIG. 14.1).[1] Could life have existed at one time on Mars but died out as water and oxygen became scarce? Could it be that life—that is, a **biosphere**—is not just a phenomenon of Earth, but rather a more general phenomenon that requires just the right planetary size, complexity, and environmental conditions in order to originate and persist? Many complex issues are bound up in this question, and the first and most fundamental is the question "What is life?"

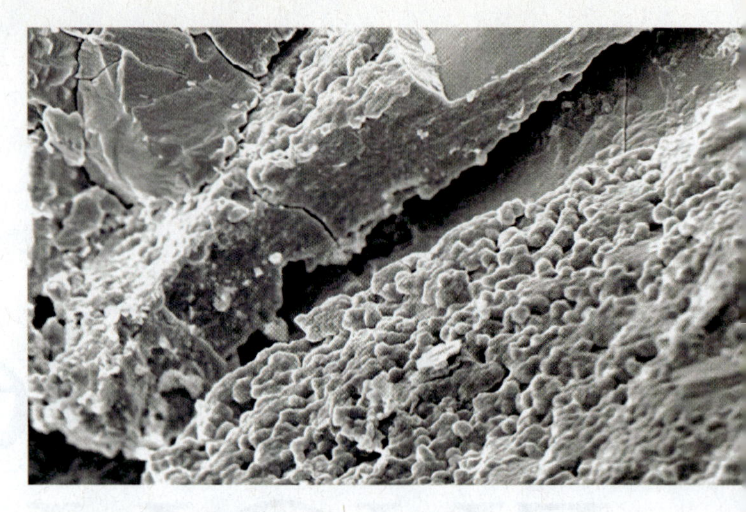

FIGURE 14.1 **Life on Mars?** Scientists suggest that this 2009 scanning electron microscope photo of the surface of Martian meteorite ALH84001 shows bumpy surfaces similar to deposits left by bacteria on Earth. The structures of possible biological origin are just 20 to 100 nm across (a nanometer is one-billionth of a meter).

WHAT IS LIFE? AN OVERVIEW OF BASIC BIOLOGICAL PROCESSES

What do we mean when we say that something is alive? What are the *essential* differences between living and nonliving matter? Most of us seldom stop to ask this question. A dog (alive) runs about and barks, while a stone (not alive) lies still and silent. But what about a dog and a potato? The differences must obviously be something else besides the ability to run and bark. How, then, do we know when something is alive?

The Necessities of Life

Four essential properties differentiate living organisms from nonliving matter. They are *metabolism*, *reproduction*, *growth*, and *evolution*. In certain circumstances, nonliving matter can do *some* of these things; for example, a crystal (not alive) can grow. The big difference, however, is that living organisms do *all* of these things. Let's look more closely at each of these essential processes.

Metabolism

One crucial characteristic of life is **metabolism**, which refers to the entire set of chemical reactions by which an organism derives energy for life processes. Some organisms create their own food energy from inorganic materials available to them in the environment; others derive food energy by eating other organisms.

[1] The meteorite itself was discovered in Antarctica in 1984. The paper that first described structures of possible biogenic origin was published in 1996 by NASA scientist Dr. David McKay and colleagues (David S. McKay, Everett K. Gibson Jr., Kathie L. Thomas-Keprta, Hojatollah Vali, Christopher S. Romanek, Simon J. Clemett, Xavier D. F. Chillier, Claude R. Maechling, and Richard N. Zare, 1996. *Search for Past Life on Mars: Possible Relic Biogenic Activity in Martian Meteorite AL84001*. Science 273:5277, 924–930). The evidence for biogenic activity in this meteorite is still controversial, although it is widely accepted that the meteorite itself did come from Mars.

AUTOTROPHS. Organisms that produce their own organic food energy from inorganic chemicals are called **autotrophs** ("self-feeders" or "self-energy-providers"). Most autotrophs produce food energy by the process of **photosynthesis**, in which inorganic carbon dioxide is combined with water in the presence of light energy to produce sugars (in the form of carbohydrates). Oxygen, which is generated as a by-product of photosynthesis, has been critically important for life on Earth. (Photosynthesis and autotrophs are discussed again in Chapter 15.)

Photosynthesis is by far the most important autotrophic process on Earth, but a few autotrophs produce their own food energy from inorganic chemicals in nonoxygenated or *anaerobic* environments, through a process called **chemosynthesis**. An example of an environment in which such organisms thrive is at seafloor spreading centers (Chapter 5), where the oceanic crust is actively rifting. In these tectonic environments, ocean water is in close contact with the underlying hot mantle. Water circulates through cracks at the rift, reemerging on the ocean floor as plumes of superheated water, called **black smokers**, which are loaded with dark mineral content. Chemoautotrophs at these locations are able to synthesize organic compounds from inorganic minerals, in the absence of sunlight.

HETEROTROPHS. Organisms that derive food energy by feeding on other organisms or on organic compounds produced by other organisms are called **heterotrophs** ("other-feeders"). When heterotrophs consume another organism, the energy stored in the organic compounds is released by one of two metabolic processes. Organisms that cannot tolerate oxygen obtain their energy through the anaerobic process of *fermentation*, in which carbohydrate molecules release energy as they are partially decomposed to form alcohol, carbon dioxide, and water. Heterotrophs that are oxygen-tolerant obtain their energy through the *aerobic* (oxygenated) process of **respiration** (or *cellular respiration*), which means that they use oxygen to break down carbohydrates, releasing carbon dioxide, water, and energy. Respiration is many times more efficient than fermentation as a metabolic process.

Reproduction

Another important feature that distinguishes living from nonliving things is that living organisms can reproduce themselves. The genetic plan of a living organism is encoded in its **DNA** (deoxyribonucleic acid), which consists of two chainlike molecules held together by organic molecules that store genetic information (see *The Basics: DNA and RNA*). The information and instructions stored in the DNA are decoded and executed by **RNA** (ribonucleic acid). A cell cannot reproduce without RNA because the RNA contains the information required to construct an exact duplicate of the proteins that make up the cell.

In single-celled organisms, reproduction occurs when cells divide (FIG. 14.2A). Reproduction in multicelled organisms occurs either through sexual or asexual reproduction. An example of *asexual reproduction* is live twigs of some trees that, when stuck in the ground, take root and produce a new tree (FIG. 14.2B). *Sexual reproduction* involves combining genetic information from two different individuals; this genetic information is then passed along to the resulting offspring (FIG. 14.2C), resulting in *genetic recombination*. Exactly how and why sexual reproduction began is one of the big questions remaining in our efforts to understand the history of life on this planet.

Growth

Growth is a fundamental characteristic of living organisms. Growth involves the ordering and organizing of atoms and small molecules to make larger molecules. At a microscopic level, this ordering of atoms and molecules can happen in two ways: polymerization and crystallization. As discussed in Chapter 3, *polymerization* is the stringing together of small molecules, like beads on a necklace, to make large chain- or sheetlike molecules. *Crystallization*, as discussed in Chapter 7, is the packing of atoms or molecules in ordered geometric arrays. Crystallization is what happens when the randomly ordered H_2O molecules in water vapor form ice crystals on a cold windowpane, or when atoms gather to form mineral grains within a cooling magma body.

In both polymerization and crystallization, an ordered pattern of atoms or molecules is replicated throughout the structure. There is an important difference, however: polymerization *absorbs* energy, whereas crystallization *releases* energy. All growth in living matter involves the polymerization of small organic molecules to form large organic molecules (*biopolymers*, Chapter 3, of which DNA is an example); therefore, growth in living organisms requires a source of energy, and this energy is acquired through metabolism.

Crystals that are not alive can grow, but they lack the other characteristics of life, including the ability to reproduce and to metabolize food energy. Viruses also can grow, and their atoms are ordered in biopolymers; however, viruses utilize the reproductive machinery of other organisms to replicate themselves, and they lack the ability to metabolize. Hence, viruses are somewhere between living and nonliving things.

Evolution

We have discussed three characteristics that distinguish living from nonliving things—metabolism, reproduction, and growth. Living organisms also possess a fourth defining characteristic; it is the potential to evolve and develop into new life forms, or *species*.

FIGURE 14.2 Reproduction
(A) Single-celled organisms reproduce by the process of mitosis, seen here at a magnification of almost 1000x in a type of human skin cell called a HaCaT. (B) Cholla cacti (*Cylindropuntia fulgida*) reproduce asexually when young plants take root and grow as individuals after falling off the parent plant. (C) In sexual reproduction, mating pairs like these Emperor penguins (*Aptenodytes forsteri*) pass their combined genetic information along to their offspring.

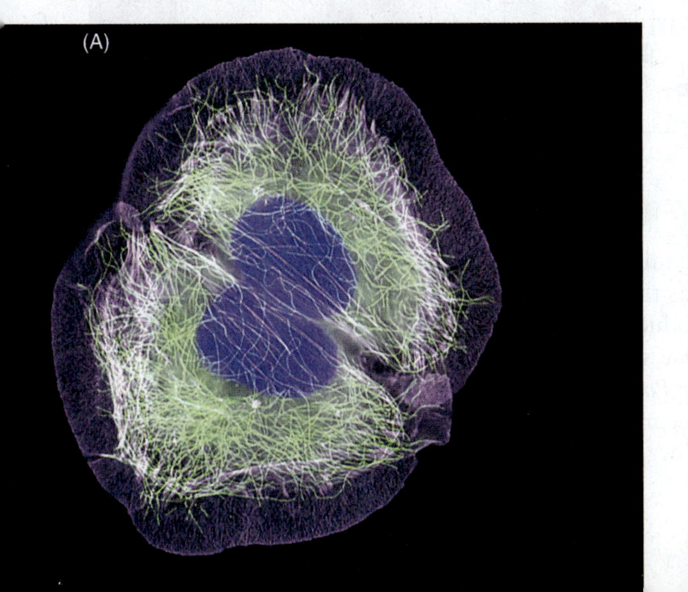

the BASICS

DNA and RNA

The genetic plan of a living organism is encoded in its **DNA** (*deoxyribonucleic acid*). DNA consists of two chainlike molecules, called *biopolymers*, held together by organic molecules that store the organism's genetic information (**FIG. B14.1**). The twisted, ladderlike structure of the DNA molecule is called a "double helix," and the structural units that are stacked together to make the ladder are *nucleotides*.

Growth, one of the characteristic features of living organisms, involves the construction of new cells, which in turn requires input from DNA. Cell growth follows specific plans; each kind of cell has its own special plan, and the full details of these plans are passed from cell to cell, generation after generation. The information for the growth plan is stored in the nucleotides, the organic molecules that form the structural basis of the DNA. Each nucleotide consists of a phosphate group, a sugar molecule, and a *nitrogenous base*. There are four bases—guanine (G), cytosine (C), thymine (T), and adenine (A)—that hook together to form *base pairs*; G always pairs with C, and T with A. Each of the two strands of DNA thus carries four kinds of base pairs, and the sequence of base pairs along the chain can be varied almost infinitely.

A **gene** is a nucleotide sequence—that is, a portion of the long DNA molecule—encoded with information or instructions about specific traits that can be passed along to the organism's offspring. Like the bar codes used in supermarkets, the sequence of bases in the gene is the code that stores the genetic information and instructions. Variants of genes that are encoded for the expression of specific inherited (or, more precisely, *heritable*) traits are called *alleles*. For example, a gene carries instructions for hair color; there are specific alleles, or variants of that gene, that carry instructions for black hair or red hair or blond hair. The specific combination of alleles that determine the expression of heritable traits in an individual organism is the organism's *genotype*.

The stored genetic information provides a cell with a reference library of how to carry out the activities of life, such as the details of reproduction, growth, and maintenance, including the polymerization of protein. The information and instructions encoded in the DNA are read and executed by the organism's RNA. RNA (or *ribonucleic acid*) is a single-strand molecule similar to one-half of a double DNA chain. A cell cannot reproduce without RNA because the RNA decodes and carries out the instructions for constructing an exact duplicate of the proteins that make up the cell.

However, carrying out the instructions for protein synthesis is not the final step in determining the physical appearance of the organism. That happens over the course of the organism's lifetime as a result of interactions between the organism's genotype and the environment. The combined outcome is the organism's overall physical appearance—its *phenotype*.

FIGURE B14.1 DNA and RNA
This is a diagrammatic representation of the twisted, ladderlike molecule of DNA. The two strands of DNA are joined by organic molecules, called bases, of which there are four types: adenine (A), cytosine (C), guanine (G), and thymine (T). The binding molecules always join as base pairs: A always joins to T, and C always joins to G. The sequence of bases is the code that directs the activities of a cell, and the segment of DNA that encodes a particular type of instruction is called a gene.

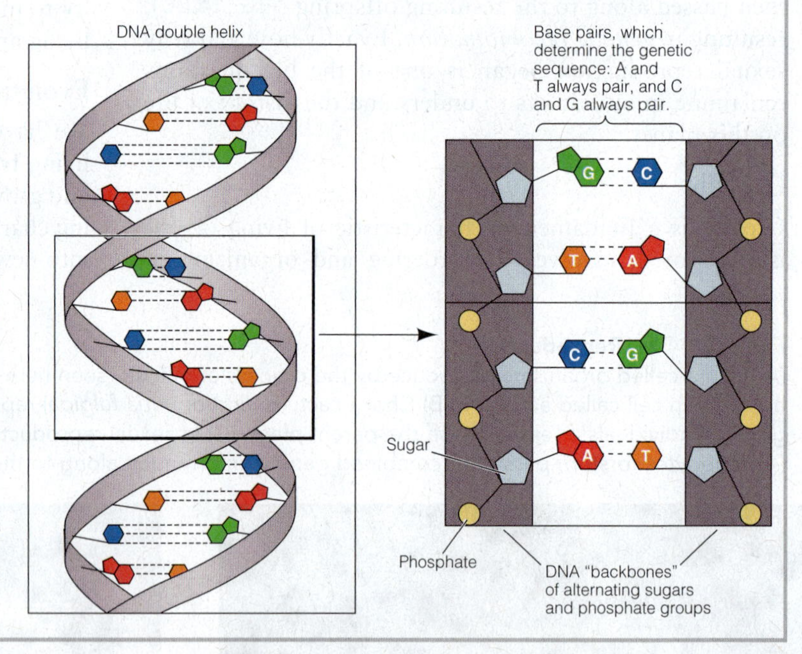

It is more difficult to define the term **species** than it might seem. Some scientists prefer a definition based on the organism's *morphology*, that is, its appearance (from the Greek *morph*, "form"). Thus, organisms that look similar to one another and share many of the same structural characteristics are said to belong to one species. Other scientists prefer a definition based on the organism's biological characteristics; specifically, organisms that are able to breed successfully with one another, producing fertile offspring, are said to belong to the same species.

Paleontologists, who study the fossilized remains of ancient plants and animals, have no choice but to rely on an organism's morphology to define it as a member of one or another species; there is no way to determine whether two organisms that lived millions of years ago would have been able to breed successfully. However, sometimes a morphological approach doesn't work very well. Consider domestic dogs (FIG. 14.3A), which all belong to the same species even though they vary widely in size and morphology. Although it might be logistically challenging, all dogs are biologically capable of breeding to produce fertile offspring; thus, they belong to a single species. Horses and zebras, which do look quite similar, can breed but their offspring are not fertile (FIG. 14.3B); hence, they belong to different species.

This potential to develop new forms and species is the fourth defining characteristic of life; it is called **evolution**.

FIGURE 14.3 Defining species
Species can be defined on the basis of either morphological similarity or their ability to breed successfully. (A) Dogs can look very different—as seen here—but they can breed successfully and hence all belong to the same species (*Canis lupus familiaris*). (B) Horses (*Equus ferus caballus*) and zebras (*Equus zebra, Equus grevyi*, and *Equus burchelli* are the three extant species) look very similar, but their offspring are infertile; they belong to different species. This little "zorse," named N'Soko, is the offspring of the male zebra and female horse seen here.

Evolution is the process whereby the genetic makeup of species changes over time and new species arise. The basis for evolution is the passing of genetic information from one generation to the next. This leads to the expression of heritable traits in offspring, such as a certain eye color, tail length, number of toes, or thickness of fur. Individuals in a population whose characteristics help them survive, reproduce successfully, and adapt to constraints and changes in their environment leave more offspring than others. Their heritable characteristics eventually come to dominate the population, gradually changing the overall genetic composition and, therefore, the characteristics of the population as a whole. If the characteristics of one portion of the population change to the extent that those organisms can no longer successfully breed with the rest of the population, then a new species has arisen.

While "change" is typical of all parts of the Earth system, the process of biological evolution *only* occurs in living organisms; it is therefore a fundamental characteristic of life. We will give much more consideration to evolution and related processes later in this chapter and in the remaining chapters of Part V. Just as the theory of plate tectonics (Chapter 5) was a unifying theory in geology, so has evolution proven to be a unifying theory in biology. It is the backdrop and central theme against which we will consider all other life processes, historical and current, in our examination of Earth's biosphere.

The Hierarchy of Life

Given the task of studying the enormous variety of life on this planet, biologists have sought to organize species into various classification systems. This has been a complicated task, made even more complicated by the explosion of discoveries in recent years by biologists and biochemists investigating the genetic linkages and relationships among organisms. Let's begin our study of the categorization of life by looking at *cells*, a basic structural unit and building block of life.

The Cell as a Structural Unit of Life

A **cell**, the structural and functional unit of all living organisms, is a complex grouping of chemical compounds enclosed by a membrane (FIG. 14.4). (Cells are constructed of organic molecules and biopolymers, including proteins and nucleic acids; these and other basic organic materials are discussed in Chapter 3.) The cell's *membrane* (a thin layer of lipids with proteins embedded in it) separates the materials inside from the environment outside and facilitates the controlled exchange of materials and energy between the cell and its environment. Many bacteria are *unicellular* (consisting of just one cell), but most organisms are *multicellular* and consist of hundreds to trillions of cells.

PROKARYOTES AND EUKARYOTES. Cells may be small (0.01 to 0.02 mm) or large (0.05 mm to a few centimeters or larger, in rare cases). Whether large or small, however, all cells are of two kinds: prokaryotic or eukaryotic. **Prokaryotic cells** (from the Greek *pro* = before and

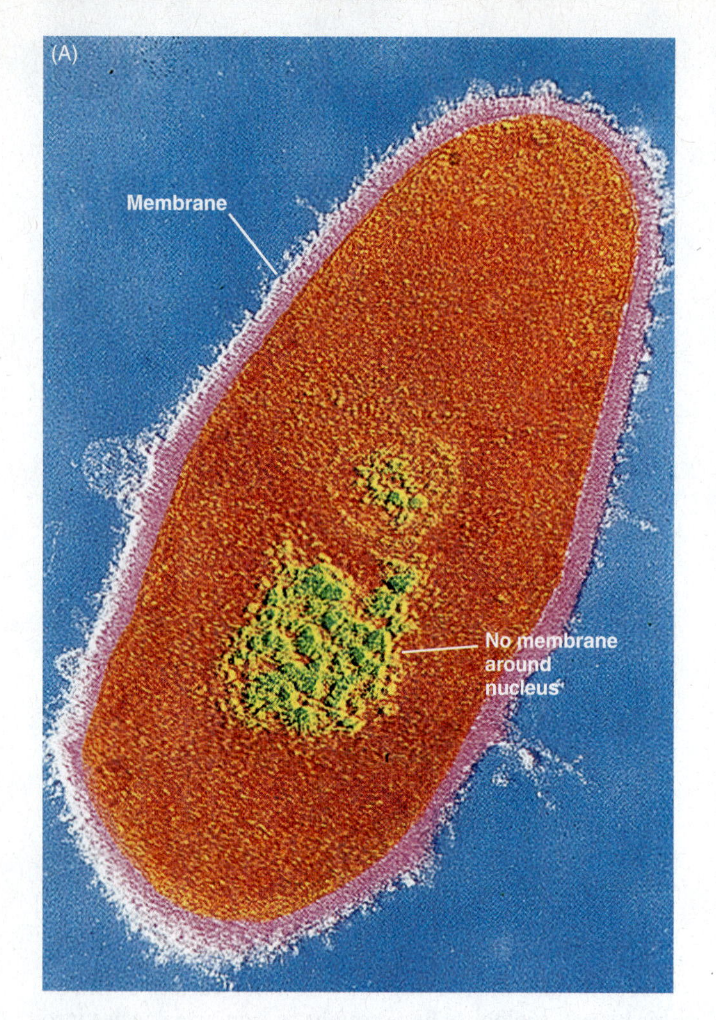

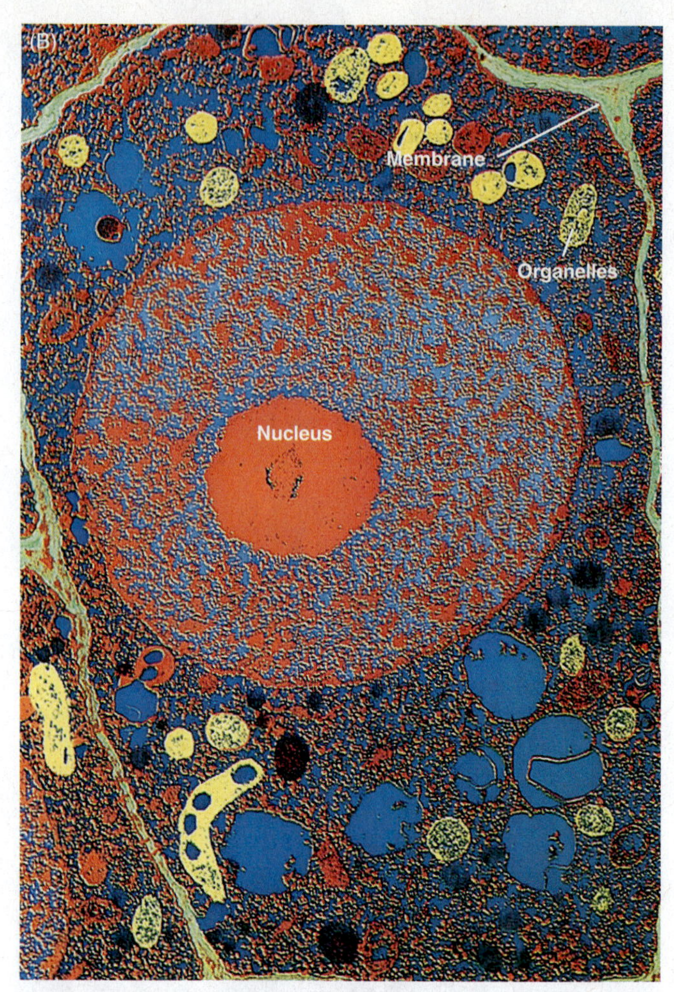

FIGURE 14.4 Cells: Building blocks of life
(A) Prokaryotic cells like this bacterium are devoid of visible organelles. Their DNA is concentrated in a poorly defined area that is not separated from the cytoplasm by a membrane. (B) Eukaryotic cells, such as this one from a plant root, have a well-defined, membrane-bound nucleus and varied organelles. These cells are colored because they have been stained for greater visibility.

karyote = nucleus, hence before a nucleus) are generally small and comparatively simple in structure. These cells house their DNA with all its genetic information in a poorly demarcated part of the cell (FIG. 14.4A). The main body of the cell, called the *cytoplasm,* lacks distinctly defined areas in which the various cell functions are carried out. Most importantly, the portion of the cell that houses the genetic information is not separated from the cytoplasm by a membrane. Prokaryotes occur as bacteria and related organisms called *archaebacteria* and *mycoplasmas.*

Eukaryotic cells (from *eu* = true, hence with a true nucleus) are larger and more complex than prokaryotic cells. Their genetic information is housed in a well-defined nucleus that is separated from the cytoplasm by a membrane (FIGURE 14.4B). The cytoplasm in a eukaryotic cell contains a variety of well-defined cell parts, called *organelles,* each having a particular function in the operation of the cell. Humans and all other animals, as well as plants and fungi, are eukaryotes.

From the Cell to the Biosphere

Life is organized in a hierarchy of many kinds of systems, starting from single cells and ranging up to the biosphere (FIG. 14.5). At a fundamental level, cells combine to form organs, and organs combine to form individual multi-celled organisms. A grouping of organisms of the same kind that shares genetic information is a *population.* Together, all populations of the same type of organism constitute a *species,* a group of individuals capable of interbreeding and exchanging genes.

On a local scale, populations of one species interact and are interdependent with populations of other species in *ecological communities.* An *ecosystem* comprises all of the interactions of an ecological community with the nonbiological components of its local environment. As discussed in Chapter 3, the *biotic* or living components of ecosystems are its animals, plants, fungi, and microorganisms; air, water, and rock are the principal *abiotic* or nonbiological components.

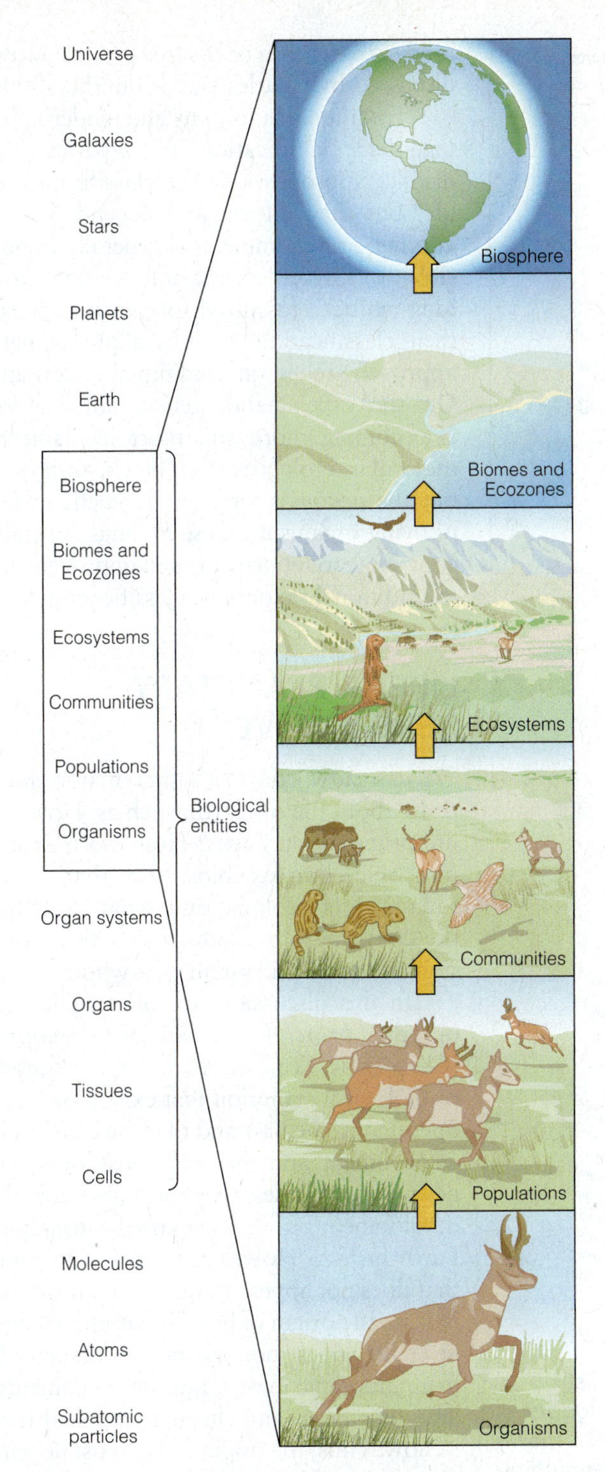

FIGURE 14.5 Hierarchical structure of life
Biological entities are organized in a hierarchical structure, ranging from cells to the biosphere.

On a regional scale, all ecosystems of one general type are collectively referred to as a *biome*. Biomes are classified according to their characteristic climate and vegetation suites. For example, the *tropical rain forest* biome includes individual examples of forests that occur in Asia, Africa, Central America, and South America; together, these individual forests with similar characteristics make up the tropical rain forest biome. All of the ecosystems that coexist in a particular geographic region are together referred to as a *bioregion* or, on an even broader geographic scale, an *ecozone*. A single ecozone may thus include examples of several different biomes.

On a planetary scale, all of Earth's ecozones together constitute the biosphere. The biosphere and the rest of the components of the global environment—the atmosphere, hydrosphere, and geosphere—form the Earth system as a whole. Each level within this hierarchy affects the levels above and below it. From a planetary perspective, this hierarchical character is an important feature of life.

The Kingdoms of Life

Given the task of studying the enormous variety of life on this planet, biologists have sought to organize species into a hierarchy of **taxonomic ranks** (the word *taxonomy* comes from the Greek word that means "order"). Traditional biological classification was based on the work of early biologists such as Carl Linnaeus (1701–1778), who identified two major classes or **kingdoms** of living things: *Animalia* (animals) and *Plantae* (plants). Refinement of this two-kingdom scheme led to the Linnaean System of taxonomic classification, which is still used today. Through decades and centuries of subsequent work on the classification of living organisms, this simple scheme has been greatly expanded and refined to better account for the diversity of living organisms and their complex evolutionary histories and genetic relationships.

Today, while there is still considerable discussion among biologists about the best way to categorize life, a classification scheme based on six kingdoms of life, rather than just the two identified by Linnaeus, is the most widely accepted taxonomic system. In the *six-kingdom system* (FIG. 14.6), all prokaryotes are said to belong to one of two kingdoms: *Archaea* (formerly known as *Archaebacteria*) or *Bacteria* (also known as "true bacteria" or *Eubacteria*). All living organisms that are not prokaryotes are eukaryotes, and all eukaryotes belong to one of four kingdoms: *Protista* (single-celled and simple multicellular eukaryotes); *Fungi* (mushrooms, yeasts, molds, and their relatives); *Animalia* (animals, which are multicellular eukaryotic organisms that obtain their food by consuming other organisms); or *Plantae* (multicellular, sexually reproducing eukaryotes that produce their own food and are more complicated than algae). The kingdoms are further subdivided, in a hierarchic manner, into successively narrower categories according to the traditional Linnaean classification scheme.

Many biologists group the six kingdoms into three **domains** or *superkingdoms*—the highest taxonomic ranking of organisms (FIG. 14.6A). The three domains of life are *not*, as you might expect, Plants, Animals, and everything else; rather, they are Archaea, Bacteria, and Eukarya. This gives us some perspective on the relative importance of microorganisms in general, and prokaryotes in particular, in the overall context of life on this planet. *All* animals, plants, fungi, and protists are placed together into a single taxonomic group (Eukarya) at the domain level, whereas it

(A)

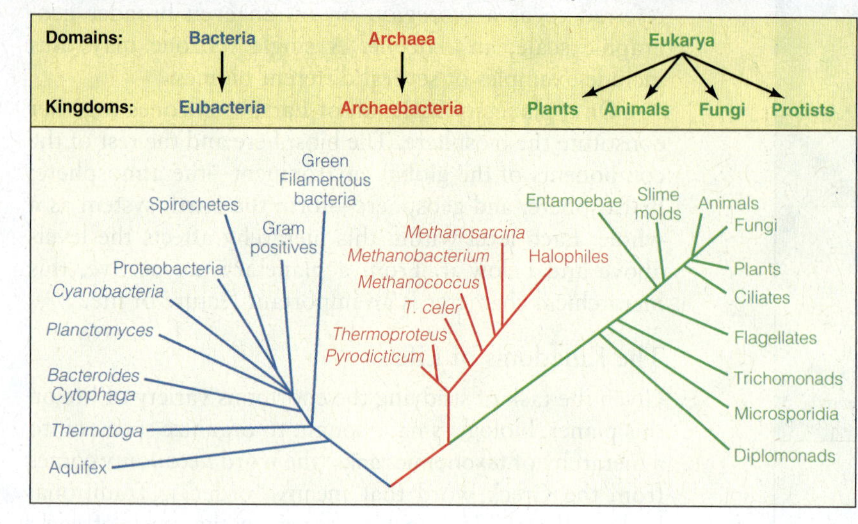

(B)

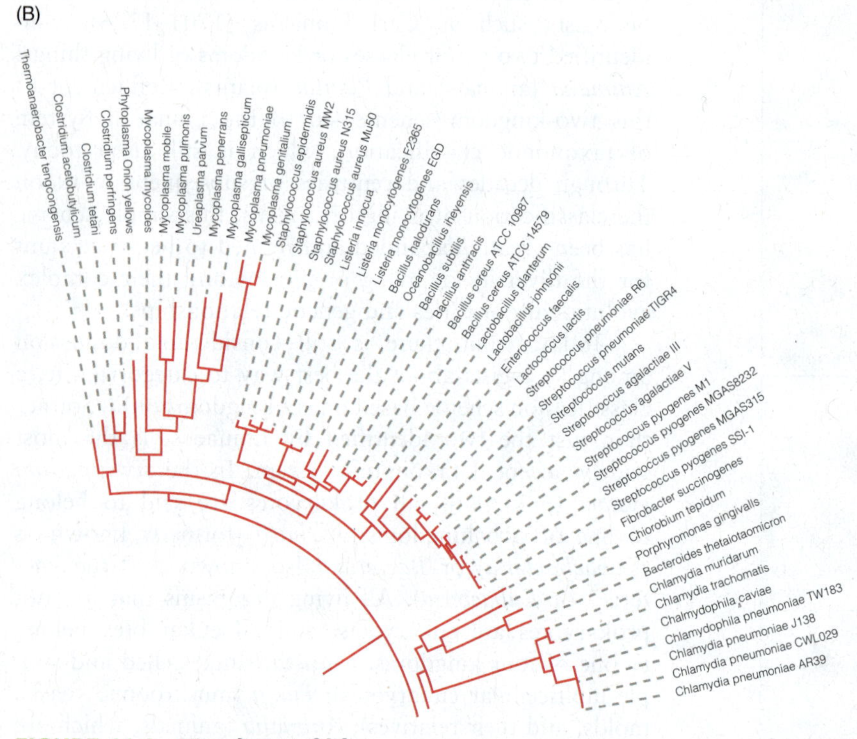

FIGURE 14.6 Kingdoms of life
(A) The six-kingdom classification system is now widely accepted but is still undergoing constant modifications and refinement by evolutionary biologists and taxonomists. (B) A phylogenetic tree, like the very detailed fragment shown here, displays organisms in terms of their genetic and evolutionary relationships.

requires *two* groups of equivalent taxonomic rank (Archaea and Bacteria) to account for the prokaryotes.

Many of the assignments of organisms to taxonomic rank based on traditional studies of morphological similarity are now being modified as new findings emerge concerning the complex genetic relationships among organisms. A taxonomy that organizes the forms of life into groupings according to their genetic and evolutionary relationships is called a **phylogenetic tree** (*phylogeny* means "evolutionary relatedness," and such groupings tend to take a branching form, hence "tree") (**FIG. 14.6B**). The phylogenetic tree is a modern extension of the *tree of life* concept discussed by Charles Darwin in the 1800s.

Traditional taxonomy and modern phylogenetic classification are separate and distinct approaches to the classification of life, but they inform and depend on one another. For example, it is generally impossible to extract genetic information from long-extinct, fossilized organisms; hence their classification, even in a phylogenetic approach, relies on traditional taxonomy. On the other hand, genetic information is becoming more and more available as molecular biologists and biochemists sort out the *genomes*, or overall genetic codes, of living organisms; these findings naturally contribute to refinements and improvements in traditional taxonomic classification.

LIFE: A PLANETARY PERSPECTIVE

We can now consider other major questions about life on Earth, such as: How did life originate on Earth? How did it evolve into the complex biosphere that exists today? Over geologic time what effect has the changing biosphere had on the evolution of the Earth system as a whole?

In this discussion, we once again rely on the *principle of uniformitarianism* (Chapter 4). We assume that the processes of biological evolution that exist today also operated in the past and that the changes in environment and the new constraints and new opportunities they provided for the development of life also existed throughout Earth history. However, there is one thing that does not appear to fit with uniformitarianism: the origin of life. To our knowledge, new life does not spring spontaneously from inorganic matter, but the fundamental laws of physics and chemistry applied then as now. Thus, the origin of life must lie with phenomena that exist today, even if the conditions that make this origin possible are no longer available on Earth.

Let's first examine the planetary properties of Earth that make it capable of supporting life. Then we will consider how life might have originated on this planet and how life has influenced the Earth system over the course of geologic time.

The Ecosphere and the Life Zone

As discussed in Chapter 1, Earth's four reservoirs (atmosphere, hydrosphere, geosphere, and biosphere) interact most intensively in a narrow zone that we might call the

the BASICS

The Linnaean System of Taxonomic Classification

In the traditional Linnaean System of classification, which is still widely used today, the kingdoms of life are subdivided, in a hierarchic manner, into successively narrower categories. These categories are phylum, class, order, family, genus, and species (**FIG. B14.2**).

Like the term *species*, which we have already considered briefly, *phylum* (pl. *phyla*)—the taxonomic ranking immediately below the kingdom—can be defined on the basis of the general morphological similarity of a group of organisms, or on the basis of the closeness of the organisms in genetic or evolutionary terms. An example of the former is the phylum Arthropoda, which encompasses both crabs and spiders—genetically unrelated but somewhat similar in their body plans and characteristics. An example of the latter is seen in worms, which are generally similar in body plan (invertebrates with long, cylindrical bodies and no legs) but can belong to any of a number of different phyla, including Nematoda, Platyhelminthes, Annelida, and several others. (Note that botanists commonly use the term *division* instead of phylum in the classification of plants.)

At the other end of the scale of taxonomic ranks is the *species*, which we have defined as a group of organisms capable of interbreeding successfully, or that has similar morphological or genetic characteristics. Species are sometimes subdivided even further into *subspecies*, *varieties*, or *types*. The taxonomic ranks of *class*, *order*, *family*, and *genus* are intermediate to phylum and species. For example, a modern human is classified as follows: kingdom: Animalia; phylum: Chordata; class: Mammalia; order: Primates; family: Hominidae; genus: *Homo*; species: *Homo sapiens*; subspecies: *Homo sapiens sapiens*. The subspecies of modern humans (*H. sapiens sapiens*) is the only member of the genus *Homo* that is still living, although other species of *Homo* lived in the geologic past. Another example is the common fruit fly, which is classified as follows: kingdom: Animalia; phylum: Arthropoda; class: Insecta; order: Diptera; family: Drosophilidae; genus: *Drosophila*; species: *funebris* (although there are several other species that belong to the genus *Drosophila*).

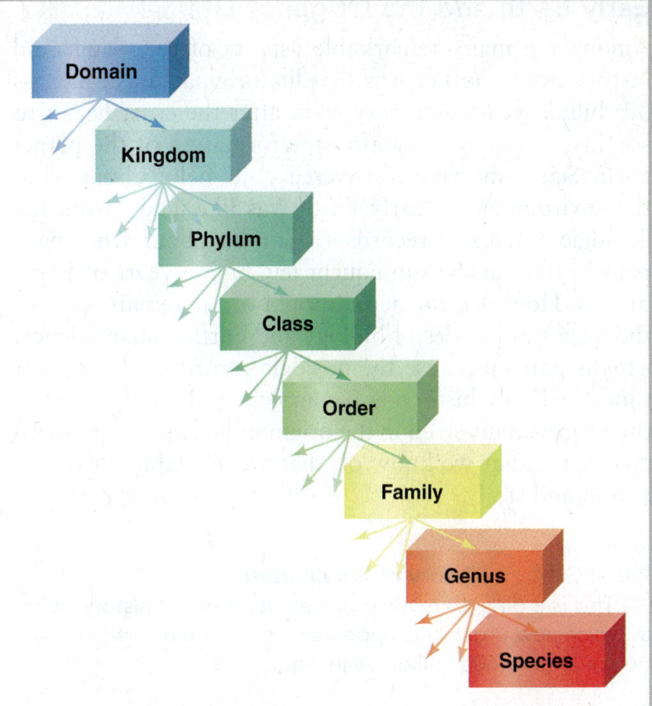

FIGURE B14.2 Linnaean System
The traditional Linnaean System of taxonomic classification is still widely used to classify individual organisms.

habitable zone or **life zone**, because its most important characteristic is that it supports life and allows it to exist on this planet. The life zone ranges from about 10 km above Earth's surface to about 10 km below the surface (see Fig. 1.8). In this narrow zone all known forms of life exist, and it is here that conditions favorable for maintenance of the biosphere are created by interactions between the lithosphere, hydrosphere, and atmosphere, and by energy from Earth's interior and from the Sun.

Earth is habitable and is able to offer a life-supporting zone by virtue of its particular relationship with the Sun. Our planet is just the right size and composition, and just the right distance from the Sun—not too near (where it would be too hot), and not too far (where it would be too cold). A zone around a star, within which orbiting planets and their moons would be just the right distance to allow for the existence of liquid water and thus, possibly, to support life, is called an **ecosphere.**[2] It now appears that there are stars in other solar systems that have planets orbiting within their ecospheres. We do know of other planets and moons that have water—even in our own solar system—but none other that we know of (yet) has exactly the right combination of temperatures and materials to provide a life zone near the surface. As mentioned at the beginning of this chapter, there is some evidence that Mars might have had suitable conditions for life a few billion years ago. However, Earth is the only planet on which we have so far confirmed the existence of life, either in the past or in the present.

[2] Some people use the term *ecosphere* as a synonym for biosphere. Others use it in reference to the sum of all ecosystems on Earth's surface. Astronomers commonly use the term *ecosphere* as it is defined here: the orbital zone around a star, within which a planet could potentially be supportive of life.

Early Earth and the Origin of Life

Among the many remarkable aspects of the origin and history of life on Earth is that life originated as much as 3.9 billion years ago, very soon after the crust began to solidify, and very early after the formation of the planet itself. Scientists have discovered quite a lot about what the environment of early Earth was like, and—from the geologic and fossil records—a fair bit about what happened to life in the subsequent few billion years of Earth history. However, the actual origin of life remains one of the biggest mysteries of biology and Earth system science. This is partly because the geologic record of the earliest times in Earth history is incomplete and partly because the processes involved in the origin of life have no modern analogues, that we know of, that are available for observation and study.

The Environment of Early Earth

During the **Hadean Eon** (from the origin of the planet 4.6 billion years ago to about 3.8 billion years—FIG. 14.7; see also Fig. 4.25), Earth and its atmosphere were very different from how they are today. There was no free oxygen (O_2) in the atmosphere, a necessity for most forms of life today. It was also very hot; the early atmosphere consisted primarily of greenhouse gases, which trapped heat near the surface. In the beginning, it was too hot for water to exist as a liquid, so there were no oceans, lakes, or rivers. Atmospheric pressure was much greater than it is today. The Hadean Earth was probably entirely inhospitable to life, as we know it (FIG. 14.8).

As the planet cooled, water vapor condensed and fell as rain. It collected in low-lying areas, forming bodies of water on the surface that became oceans, possibly

FIGURE 14.7 History of life on Earth
(A) This is a diagrammatic representation of the history of life on Earth from 4.6 billion years ago to the present. (B) The rates at which new organisms appeared and of the diversity of existing organisms both increase with time. The Cambrian Radiation began around 600 million years ago.

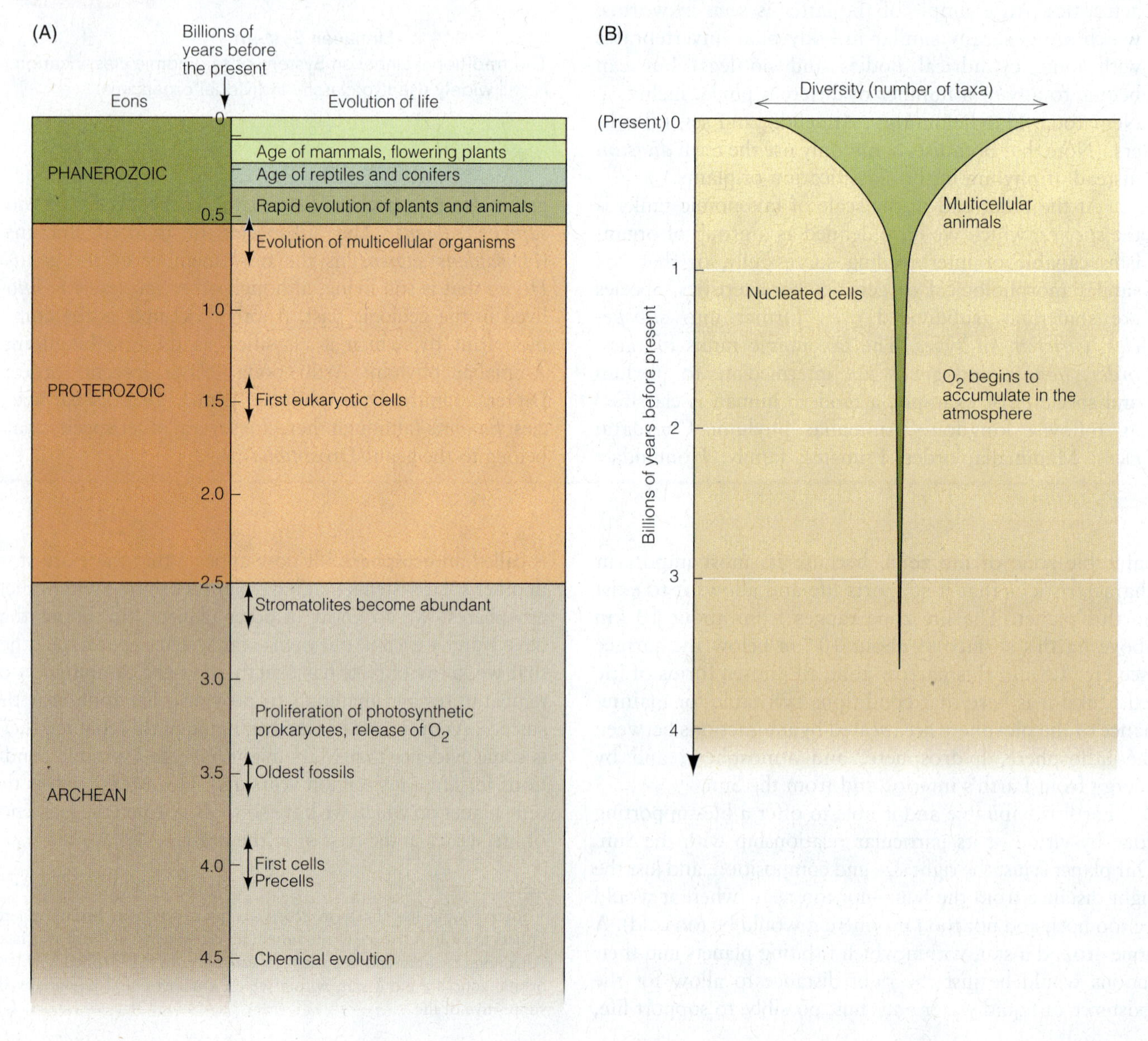

FIGURE 14.8 Early Earth
Early Earth had a high surface temperature and atmospheric pressure, with abundant greenhouse gases and acidic precipitation, as shown here in an artist's rendition.

starting as early as 4.4 billion years ago. The water in the atmosphere and oceans came mainly from volcanic gases, in which water vapor is still dominant today. There is also evidence of contributions of water and other volatiles from comets and meteorites—extraterrestrial materials that have fallen to Earth's surface. The early hydrosphere reacted with gases in the atmosphere to form acids. The acids, falling as rain, reacted with the rocks of the crust to cause chemical weathering. The compositions of the atmosphere, hydrosphere, and geosphere began to change dynamically as materials cycled between them.

As you know from Chapter 11, Earth's atmosphere now contains approximately 21 percent oxygen. Volcanic gases (which are typically oxygen-poor) and material from comets could not have supplied all of this oxygen. Where did it come from? Some oxygen was generated in the early atmosphere by the breakdown of water molecules (H_2O) into hydrogen and oxygen as a result of interactions with solar radiation—a process called *photodissociation*. This is an important process, but it doesn't come close to accounting for the present level of oxygen in the atmosphere. Another oxygen-producing process was required, and you will see that it came from life itself.

The Origin of Life

Although the details of how, when, and where life began are not known, scientists are getting closer to understanding the chemical and biological processes involved. The oldest fossil remains yet found are 3.5 billion years old (**FIG. 14.9**), although the "chemical signatures" of biological processes have been detected in rocks even older than this. These fossils come primarily from rocks of the **Archean Eon** of Earth history (which lasted from 3.8 to 2.5 billion years ago, as shown in Fig. 14.7). Some of the very earliest "fossils" are biochemical deposits left behind by simple, single-celled organisms similar to modern photosynthetic bacteria.

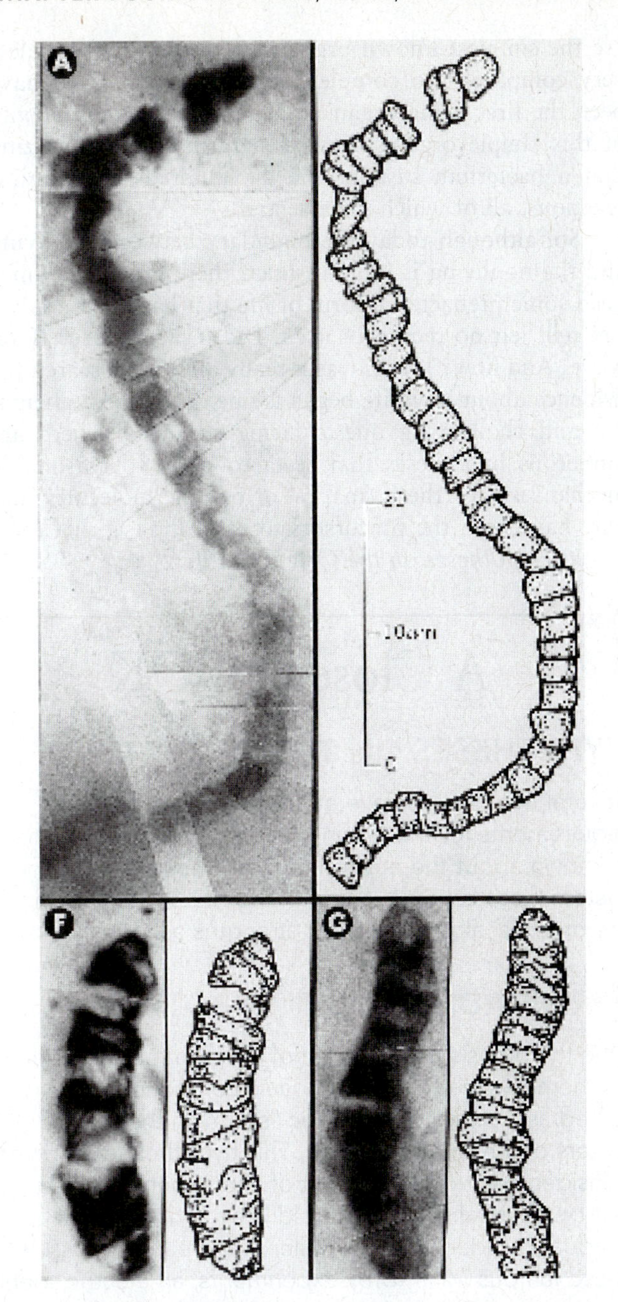

FIGURE 14.9 Ancient fossils
These are examples of some of the oldest fossils ever found. They are 3.5-billion-year-old microfossils from Western Australia. Adjacent to each photograph is a sketch. Magnification is indicated by the scale; 10 μm is equal to 0.1 mm.

How did these earliest life forms arise? A number of scientific hypotheses have been proposed, but none of them has been proven. Could either a virus or a bacterium—two of the first known manifestations of life—have been the first living life form? The answer seems to be "no." Life cannot have begun as a virus because a virus requires a living host. Nor is it likely that life began as a bacterium because, even though bacteria are the simplest known organisms, they are nevertheless very complex—too complex, in all probability, to have been the first living organism. The operating complexity of this simple organism can be grasped by recognizing that a bacterium sustains life by hundreds of chemical reactions, all of which are essential.

So, although today the boundary between the living and the nonliving is quite distinct, there must have once been some prebacterial form of life that has (so far as we can tell) left no record of itself. Did it metabolize? If so, what? And how? If not, was it really alive? The search for evidence about how life began is one of the most intriguing and challenging quests facing scientists. There are numerous hypotheses that seem to point to reasonable mechanisms for the formation of organic molecules that may have been the precursors to early life (see *A Closer Look: Hypotheses on the Origin of Life*).

Whatever may have been the initial state of matter destined to become alive, we can specify three steps that must have been accomplished on the way to the complex life forms we know today: (1) *chemosynthesis*, the synthesis of small organic molecules, such as amino acids, from inorganic precursors; (2) *biosynthesis*, the polymerization of small organic molecules to form biopolymers, especially proteins; and (3) the development of all the complex chemical machinery, including DNA and RNA, needed for metabolism and replication. It is possible to explain, and to test in a laboratory, the chemosynthesis of organic molecules (hydrocarbon compounds) from inorganic precursors, and even the biosynthesis of larger organic molecules (such as proteins and carbohydrates, Chapter 3). However, the discovery of the specific pathways that led to the development of the complex processes of life remains elusive.

A Closer LOOK

HYPOTHESES ON THE ORIGIN OF LIFE

Several hypotheses have attempted to address the requirements for the origin of life on Earth, given what is known about the environment of the early Earth system. Let's consider some of the more well-known hypotheses, along with their strengths and weaknesses.

"PRIMORDIAL SOUP" HYPOTHESIS

One hypothesis of the origin of life is that the products of chemosynthesis from inorganic compounds collected as a primordial organic "soup" in the surface waters of the primitive Earth. This hypothesis was given considerable weight by a set of experiments carried out by physical chemist Harold Urey and his (then) graduate student Stanley Miller in the early 1950s.[3] In these famous *Miller-Urey experiments*, an electric spark (to simulate lightning) was applied to a sterile flask containing water and a combination of gases (methane, ammonia, and hydrogen) thought to represent Earth's early atmosphere. The resulting compounds were extracted and analyzed. It was found that much of the carbon in the flask had been converted into organic forms, including amino acids, sugars, and lipids, which are among the most important basic building blocks of life. Although problems and shortcomings have been identified with this set of experiments (for example, the combination of gases used was probably not accurately representative of Earth's early atmosphere), they remain among the most powerful experiments ever carried out on the origin of life.

According to the hypothesis, then, once this primordial "soup" had formed, some of its less soluble organic compounds would clump together, like butterfat in milk. Proteinlike compounds also may have formed by the drying out of organic "soup" along ancient shorelines, and subsequent polymerization by solar or volcanic heat. Polymerization might also have occurred if organic molecules had adsorbed onto the surfaces of clay minerals and the reactions had taken place there.

Regardless of how polymerization may have occurred, these organic compounds had no obvious means of metabolism, self-replication, or evolution. In other words, it is possible to envision a mechanism whereby some of the important organic constituents of life would arise, but there is no clearly understood pathway whereby these compounds would have become truly "alive." Another problem is that when amino acids polymerize to make proteins, water is eliminated: In an aqueous environment, this is a difficult, if not an impossible, feat.

"BLACK SMOKER" HYPOTHESIS

Since the discovery of submarine hot springs and black smokers at oceanic spreading centers in 1977 (FIG. C14.1), a number of scientists have suggested these as possible sites for the origin of life. In this hypothesis, the first organic molecules were formed on the surfaces of pyrite grains that form around the vents, with pyrite serving as a concentrating mechanism. Prebiotic evolution could have begun with reactions that took place in organic layers as thin as one molecule thick. Reactions in the organic layer were akin to metabolism, and the products were the first cells. When completed, the cells became detached from the mineral substrate.

[3] They carried out these experiments to test a hypothesis put forward by Alexander Oparin and J.B.S. Haldane, which suggested that conditions on early Earth would have favored the synthesis of organic compounds from inorganic precursors.

The Influence of Life on Earth Systems

The history of the biosphere is closely intertwined with that of the atmosphere, hydrosphere, and geosphere. Without a hospitable atmosphere and hydrosphere, life as we know it could not have survived. (Indeed, one set of hypotheses about the origin of life holds that life may have originated and been extinguished in a number of different local environments on Earth, until it finally found a hospitable environment in which it was able to take hold and flourish.) In turn, without life, the atmosphere and ocean would not exist in their present forms. There are two major ways in which life, once formed, contributed to the chemical evolution of the Earth system over the planet's history—first, by contributing to the buildup of oxygen, and second, by providing a mechanism for sequestering carbon.

OXYGEN BUILDUP. As discussed above, one problem associated with the chemistry of our current atmosphere is the origin of its abundant "free" molecular oxygen (O_2). Oxygen is generated as a by-product of photosynthesis, and almost all of the free oxygen in the atmosphere originated through photosynthesis. However, oxygen itself is toxic to most organisms, unless specialized enzymes are present. Therefore, it is likely that early photosynthetic organisms needed to escape from the oxygen they generated as a waste produce. This means that the early buildup of oxygen in the atmosphere was probably fairly slow (as shown in Figure 11.3).

When there was very little oxygen being released, all of it was used up in combinations with other elements in the atmosphere, such as iron and sulfur. "Free" oxygen—that is, molecular oxygen that is not in a chemical

---

The "black smoker" hypothesis is highly speculative. These earliest "cells" are not to be confused with the unicellular, chemoautotrophic bacteria that live around black smokers today and that now serve as the base of the food chain in this strange ecosystem. A strong point of the hypothesis is that the hot sulfur-containing gases from black smokers provide an energy source that would have allowed for a rapid rate of production of organic compounds.

"PANSPERMIA" HYPOTHESIS

Arguing that the basic organic molecules of life would have been difficult to synthesize on Earth, some scientists hypothesize that organic molecules (or possibly even life, itself) may have arrived, ready made, from some other part of the solar system or even from the galaxy beyond the solar system. This is known as the *panspermia hypothesis* (from the Greek words for "universal" and "seeds"), and it has attracted attention for two reasons. First, astronomers have demonstrated that many small organic molecules occur in interstellar space. Second, among the many kinds of meteorites that fall to Earth, two of them—comets and carbonaceous chrondrites (Chapter 4)—have been found to contain a variety of small organic molecules. If interstellar dust, carbonaceous chrondrites, comets, or all of these extraterrestrial materials fell on the surface or into the oceans of early Earth, perhaps they provided the starting molecules for life. The big problem is the next step: How did the early molecules polymerize to become "life" molecules? It has even been hypothesized that biopolymers could have grown in space from small molecules and formed organisms there.

The panspermia hypothesis is attractive because it extends by a few billion years the time available for biosynthesis of the earliest biopolymers, for life could have originated before the solar system formed. It is hard to imagine, however, that conditions in interstellar space could have favored biosynthesis. How could these highly unlikely compounds form in high enough concentration over a short enough time for life to originate in space? What would be the energy source? Any location with a suitable temperature range would have been subject to concentrated short-wave electromagnetic radiation unless it was on a planet with an atmosphere that filtered out the radiation. Such radiation is deadly to biopolymers. Furthermore, the dispersion of matter in outer space is so thin that the formation of polymers is unlikely in any case. Therefore, this hypothesis seems highly speculative and unlikely.

FIGURE C14.1 Birthplace of life?
The "black smoker" hypothesis suggests that life might have originated near submarine hot springs like this one at Endeavour Hydrothermal Vents in the Pacific Ocean, off the west coast of Canada. In this photo, tubeworms bask in superheated (400°C) water issuing from a 5-m-high chimney of rock, 2250 m below the ocean surface.

compound bonded to other elements—could persist in the atmosphere only after oxygen was present in higher concentrations. Eventually, enough oxygen was generated through photosynthesis to permit molecular oxygen to build up. This change happened over a long period, but the onset of major oxygenation of the atmosphere coincides roughly with the beginning of the **Proterozoic Eon**, defined as starting around 2.5 billion years ago (Fig. 14.7). The history of oxygenation of the atmosphere is recorded in the rock record, most notably in *banded iron formations*—marine chemical sediments, mostly between 2.5 and 1.8 billion years old, that consist of alternating layers of reduced and oxidized iron-rich minerals (see Chapter 7, *A Closer Look: Banded Iron Formations*).

Along with the buildup of oxygen in the atmosphere came an increase in ozone (O_3). When sufficient ozone had accumulated, it began to function as a screen to filter out harmful ultraviolet radiation—we know this as the *stratospheric ozone layer* (Chapter 2). Once the ozone layer began to protect the surface from harmful, energetic shortwave radiation, organisms would have been able to survive and flourish in shallow waters and, eventually, on land. This critical stage was reached around 600 million years ago, just before the onset of the **Phanerozoic Eon** (542 million years ago, as shown in Fig. 14.7).

The oxygen content of the atmosphere has fluctuated through Earth history, which surely has had major consequences for the biosphere in ages past (**FIG. 14.10**). For example, samples of air trapped in amber (fossil tree resin) suggest that 100 million years ago, during the Cretaceous Period, the oxygen content of the atmosphere was 40 percent higher than in today's atmosphere. As the Cretaceous came to a close, the oxygen level started to decline. The dinosaurs died out suddenly at the end of the Cretaceous, probably as a result of a great meteorite impact, but they had been declining in numbers for a long time before then. One hypothesis for the decline of the dinosaurs is that they had small lungs because the air contained so much oxygen. As the oxygen level dropped, their small lungs could not adjust and so, like high-flying balloonists, they died out as a result of respiratory stress.

Carbon Sequestration and Other Effects. The emerging biosphere also had a profound impact on biogeochemical cycling, especially in the carbon cycle. The shells of marine organisms are composed primarily of calcium carbonate ($CaCO_3$), providing a storage reservoir for carbon dioxide ($CaCO_3 = CaO + CO_2$). When these organisms die, their shells are buried by seafloor sediments and are eventually transformed into limestone. In the carbon cycle, limestone is a long-term storage reservoir for carbon dioxide. Other sediments that contain organic matter, principally seafloor sediments, also store carbon, isolating it from contact with the atmosphere and hydrosphere; this is called *sequestration* (Chapter 1). If all the carbon dioxide currently stored in limestone and other sedimentary rocks were released, there would be as much CO_2 in Earth's atmosphere as in the atmosphere of Venus, where the greenhouse effect runs rampant and the surface temperature is 480° C.

Make the CONNECTION

Where do the mineral salts in the ocean come from? How are these salts used by marine organisms? How does this affect the chemistry of both ocean water and the atmosphere?

Today life continues to affect the physics and chemistry of the environment at a global level in a variety of ways. Photosynthetic organisms—especially land plants, but also algae and photosynthetic bacteria in the oceans—change the color of Earth's surface, and thereby change

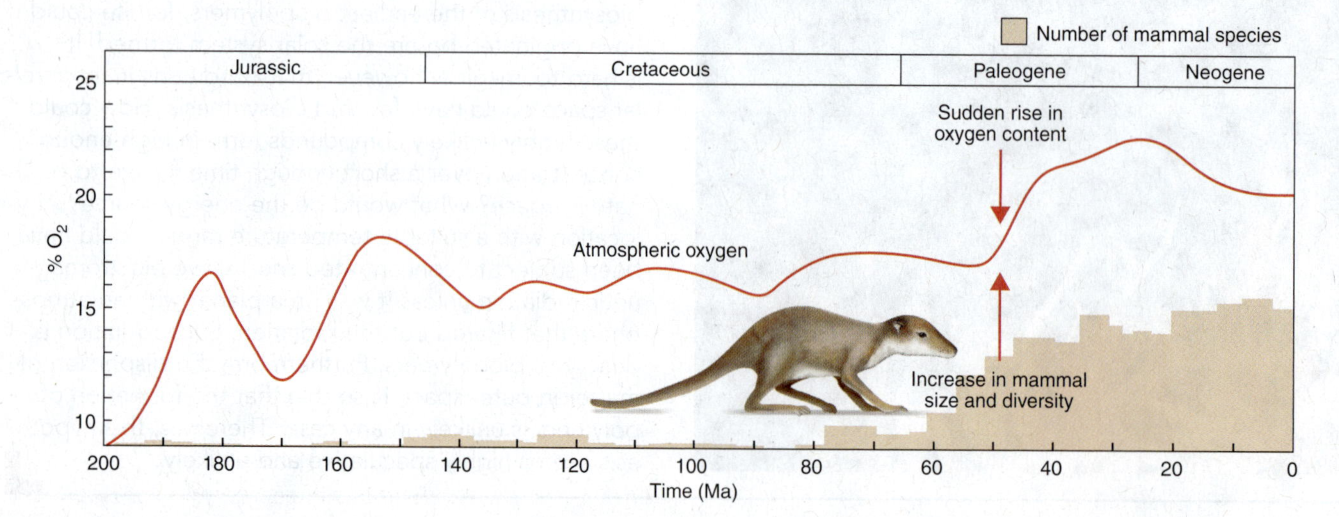

FIGURE 14.10 The changing Earth
Over the past 200 million years, oxygen levels in the atmosphere have increased. The rise of mammals, aided by the demise of the dinosaurs 65 million years ago, may also have resulted partly from the plentiful oxygen supply.

the reflectance properties and the amount of energy that is stored and reflected. The structure of vegetation on the land affects the roughness of the surface, which in turn reduces the flow of wind and thereby affects weather and climate. Life also affects the atmosphere through the exchange of water between land, plants, and the atmosphere. Plants take up water from the soil and release it through their leaves, a process that is necessary for plant survival; a result is that plants increase evaporation and therefore decrease runoff. This, together with the ability of plant roots to stabilize soil, affects rates of erosion, thereby influencing the rate of flux of dissolved and suspended particulate matter from the land to the oceans via streams and rivers. In turn, life in the oceans affects chemical transfers between the ocean and atmosphere, as well as between the ocean and marine sediments, many of which are biological products or influenced indirectly by life.

The Gaia Hypothesis

Several intriguing questions arise concerning the extent to which life modifies the global environment. For example, are these modifications necessary for the persistence of life? If not necessary, are they beneficial to the biosphere—is life able to persist more successfully or in greater abundance if these life-induced changes take place?

Some answers to these questions have been phrased as the *Gaia Hypothesis* (from the Greek word "Gaia" for Mother Earth), proposed by chemist James Lovelock and biologist Lynn Margules. This hypothesis proposes that life has altered the environment at a global scale throughout life's history on Earth and continues to do so; that these alterations contribute to biogeochemical stability, expressed as a condition of *homeostasis* or *steady state* (Chapter 1), on Earth; and that the alterations benefit life by increasing the probability of the persistence of life. Much evidence exists that life affects the global environment, so the first part of the Gaia Hypothesis is generally accepted. However, the other parts remain controversial. Well-known writer Lewis Thomas describes Lovelock's fundamental contribution as a view that the Earth system is "a coherent system of life, self-regulating, self-changing, a sort of immense organism."[4] This "superorganism" idea has made the Gaia Hypothesis popular, but controversial.

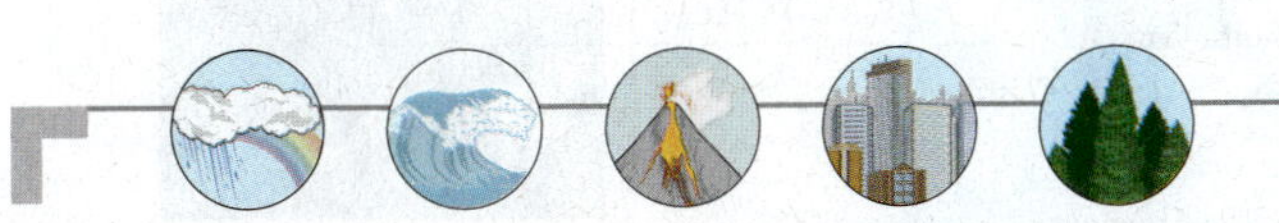

Make the CONNECTION

It is thought that the Earth system has evolved through time in response to changes in the biosphere. What kind of changes in the biosphere cause changes in the other parts of the Earth system?

[4] Thomas, Lewis (1995) in the *Foreword* to *The Ages of Gaia: A Biography of Our Living Earth*, by James Lovelock. W.W. Norton & Company, Inc., NY; p. x.

EVOLUTION: THE HISTORY OF LIFE

Evolution gives a quality to life that sets it apart from the nonliving world. A crystal of quartz that grows today is fundamentally the same as a crystal of quartz that grew 3 billion years ago. Living organisms, however, change over time. Today's organisms have arisen as a result of change. This change affects not only the individual organism in its life cycle of birth, metabolism, growth, reproduction, and death, but also the population to which it belongs, which is always responding to environmental fluctuations and occasionally undergoing more radical changes that can result in the generation of new species.

Through geologic time, the early, simple, single-celled organisms eventually diversified into the vast array of species that inhabit Earth today. The *theory of evolution* says that new species evolve from old species. Applied to the history of life on Earth, this means that all present-day organisms are descendants, through a gradual process of genetic change and adaptation to environmental conditions, of different kinds of organisms that existed in the past.

The Mechanisms of Evolution

Evolution is achieved through the process of **natural selection**, in which individuals that are well adapted to their environment survive and are reproductively successful, whereas individuals that are poorly adapted to their environment are reproductively unsuccessful and therefore tend to be eliminated from a population.

Natural selection occurs because the characteristics of individuals are passed from one generation to the next through their genes. In any given environment, some traits will be more advantageous than others. For example, certain characteristics might enable an individual to compete more effectively for scarce resources or to escape predators more easily. If environmental conditions change, possession of a particular trait may provide a new survival advantage to some individuals. When individuals possess characteristics that enable them to respond effectively to changes in environmental conditions, they will be more successful and therefore more likely to pass along their genes to offspring. Eventually, the characteristics of the entire population will change over time, as natural selection favors individuals that are particularly well suited to their environment and eliminates individuals that are poorly suited to their environment. This process of change in response to environmental pressure is called **adaptation.**

If the characteristics of a population, or part of a population, change so dramatically over time that the individuals would no longer be able to breed successfully with individuals from the original population, then a new species has arisen; this is called **speciation**. Speciation requires that a population become reproductively isolated from others of the same species, which can happen if part of a population becomes geographically isolated. For example, a rising mountain chain or an invasion by the sea might provide a physical barrier that separates two groups, or some individuals might migrate across a large river or to an island, thereby becoming isolated from the main population (FIG. 14.11). In such cases the separated group must become adapted to a new and different environment. Before the separation occurred, the two groups were part of a single species; after the separation, as a result of adaptation through natural selection, they may eventually become so different that they can no longer interbreed successfully. The process of speciation resulting from geographic isolation is called *allopatric speciation*. *Sympatric speciation* occurs when populations become segregated and reproductively isolated without the presence of a geographic barrier.

FIGURE 14.11 Natural selection by adaptation
When Charles Darwin visited the Galápagos Islands, he counted 13 species of apparently related birds, but with different beak shapes and diets. What could account for such a dramatic difference in diversity over such a short geographic distance? Finches from the mainland had colonized the islands and had subsequently changed as a result of having to adapt to their new environments. Different beak shapes developed because they gave specific survival and reproductive advantages to birds in different environments. This is an example of allopatric speciation and natural selection by adaptation.

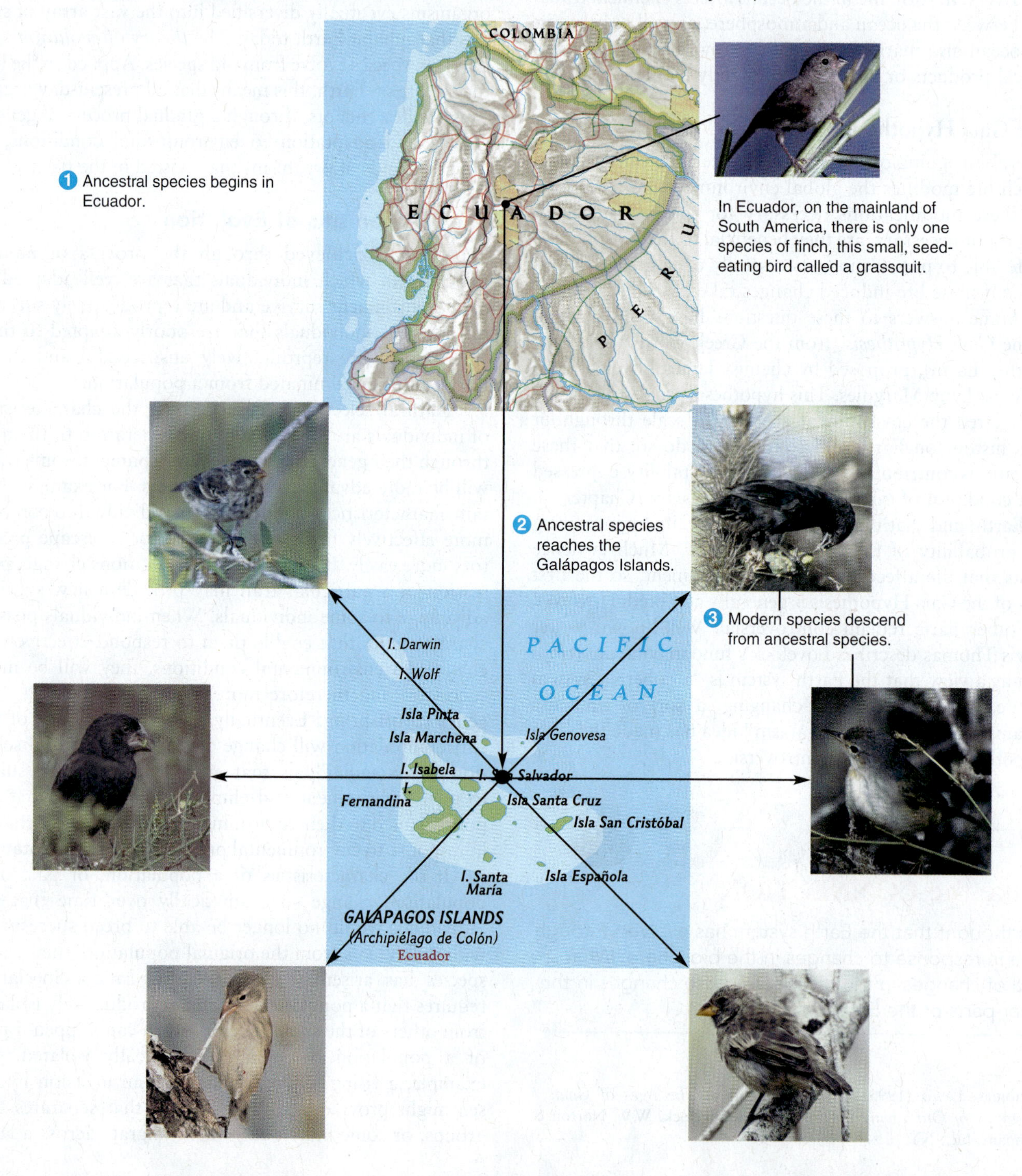

In natural selection, then, certain genetic variants (called *alleles*) become more or less common as a result of the survival-related usefulness of the heritable traits that they confer. Another mechanism whereby certain alleles can become more or less common within a population, or even disappear altogether, is **genetic drift**, which results from random sampling of genes from one generation to the next (FIG. 14.12). When reproduction samples the range of alleles available within a population, purely by chance the frequency (or statistical distribution) of alleles passed to offspring may not be exactly the same as in the parent population. The next generation will therefore be sampling from a slightly different genetic pool. Over time, this can have a significant effect, especially in small populations (where one variation would have a proportionately greater impact than in a larger population), and it is thought the genetic drift is an important contributor to evolution.

An organism's genetic sequence can also change through **mutation**, which happens spontaneously during the passage of genes from parents to offspring and can also be caused by environmental factors. Mutation is the main source for new genetic variants within a population.

Natural selection is not random; it is driven by environmental change, so alleles that are *beneficial* for adaptation and conducive to reproductive success are selected for and tend to dominate over time. However, mutation and genetic drift are largely random processes; therefore, the genetic variations they introduce may be *beneficial*, *harmful*, or *neutral* to the organism and its reproductive success. They are not entirely unrelated processes, though; for example, mutation provides the genetic variety on which natural selection can act.

FIGURE 14.12 Genetic drift
Genetic drift results from the random sampling of genes from one generation to the next. Over time, this can result in a shift in the genetic makeup of the population, even leading to the disappearance of an allele.

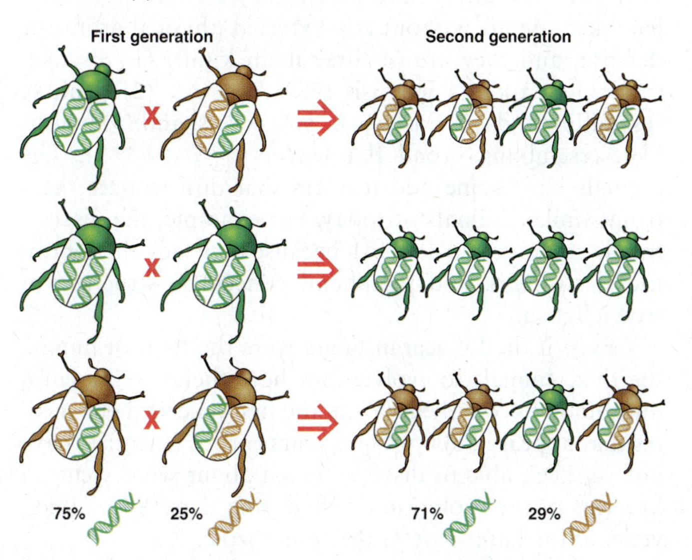

Early Life Forms

The most ancient fossils that have been found so far are 3.55 billion years old. Some of the earliest fossils from the Archean Eon are the remains of microscopic prokaryotes (see Fig. 14.9); others are structures made up, layer upon layer, of thin sheets of calcium carbonate that were precipitated as a result of certain bacteria (also prokaryotes) influencing the chemistry of seawater (FIG. 14.13A). The layered structures, called *stromatolites*, are not fossils of actual organisms, but they provide clear evidence of their presence because we can see and study similar structures being formed today by living organisms (FIG. 14.13B).

FIGURE 14.13 Stromatolites
Stromatolites are layered growths that form in warm, shallow seas when photosynthetic bacteria cause dissolved salts to precipitate. (A) These are fossil stromatolites more than 1.5 billion years old, from the northern Flinders Range, South Australia. (B) These modern stromatolites are forming in the intertidal zone of Shark Bay, Western Australia.

Prokaryote World

Whatever its origin, early life was prokaryotic. It must have been anaerobic (i.e., lived in the absence of oxygen), too, since we know that the early atmosphere contained little or no free oxygen. During the anaerobic phase of life's history, living organisms could acquire energy only by fermentation. The energy yield from CO_2 fermentation is low, as discussed above, and the by-products include CO_2 and alcohol. Alcohol is a high-energy compound and, for the cell, a waste product that must be eliminated. The inability to use energy contained in alcohol puts limitations on the anaerobic cell:

1. A large surface-to-volume ratio is required to allow rapid diffusion of food in and waste out. Anaerobic cells must therefore be small.
2. Anaerobic cells have enough trouble keeping themselves supplied with energy. They cannot afford to deploy energy resources on the maintenance of specialized organelles, including the nucleus.
3. Prokaryotes need free space around them; crowding interferes with the movement of nutrients and water into and out of the cell. They also require space for the elimination of toxic waste products—alcohol and, in the early photosynthetic organisms, oxygen. Therefore, they live singly or are strung end-to-end in chains. They cannot form three-dimensional structures.

It took about 2 billion years for Earth's oxygen sinks to be filled, and during this entire time—almost half of Earth history—the prokaryotes had the world to themselves. Several different kinds of prokaryotes evolved over their 2-billion-year supremacy, and some of them were photosynthetic. The first of these were probably **cyanobacteria** (also called *blue-green algae*, although this is a misnomer; cyanobacteria are prokaryotes, in the domain Bacteria, whereas algae are eukaryotes, in the domain Eukaryota). In a fundamental sense, we owe our current atmosphere to the development of photosynthesis by cyanobacteria.

Eventually, an oxygenated atmosphere formed. When it did, organisms seem to have wasted little time in turning the lethal waste product, oxygen, to an advantage. While some prokaryotes developed oxygen-tolerance, the major change really happened through the appearance of eukaryotic cells.

The Emergence of Eukaryotes

For at least 2 billion years the only life on Earth was prokaryotic. Then, about 1.4 billion years ago, during the Proterozoic Eon, a profound change occurred: eukaryotes emerged. How and where the first simple eukaryotes came into being is a matter for speculation. We can, however, be reasonably sure that eukaryotes arose from prokaryotes. The chemical pathways in the two classes of cells are so similar that it is clear they must be related. Furthermore, the organelles in eukaryotes so closely resemble some of the smaller prokaryotes that most authorities hypothesize that organelles were once prokaryotic bacteria and that eukaryotes arose by larger prokaryotes enclosing the smaller cells, thereby being able to use their chemical products.

Some of the characteristics of eukaryotic cells are as follows:

1. They are aerobic; that is, they use oxygen for respiration. Because oxidative respiration is much more efficient than fermentation, they do not require as large a surface-to-volume ratio as anaerobic cells do. Such cells are therefore typically larger.
2. Aerobic cells, because of their superior metabolic efficiency, can maintain a nucleus and organelles.
3. Aerobic eukaryotes are not inhibited by crowding. This means that, unlike prokaryotes, eukaryotes can form three-dimensional colonies of cells.

With the appearance of eukaryotes and the establishment of an oxygenated atmosphere, the biosphere started to change rapidly and to influence more processes on Earth. We can study the evolutionary history of the biosphere by examining the rock record. This history is mainly revealed through **fossils**, the altered remnants of ancient organisms preserved in rocks. Microorganisms such as the prokaryotes that dominated the first 2 billion years of life on Earth are not readily preserved as fossils; they are delicate and extremely tiny, which makes them particularly challenging to study. In the latter part of the Proterozoic Eon, however, larger and more complex eukaryotic organisms began to emerge, and some of these have been preserved as fossils in the rock record.

The Ediacaran Fauna

The earliest animal fossils were first discovered in 600-million-year-old rocks in the Ediacara Hills of South Australia and are known as the *Ediacaran fauna*. Nearly identical fossils have subsequently been discovered in similar-aged rocks in other parts of the world. The Ediacaran fauna were animals that lived in quiet marine bays and lacked any hard parts. They seem to have been jellylike animals without any external physical armor or defense, and they are of three main kinds: (1) disclike, resembling today's jellyfish (FIG. 14.14A); (2) penlike, resembling today's sea-pens or soft corals; and (3) wormlike, resembling broad flat worms (FIG. 14.14B). The animals have some odd features that differentiate them from similar animals of today. For example, the disclike fossils are not really jellyfish because they lack the central radial structure and peripheral concentric structure of true jellyfish.

Even if the Ediacaran fauna were the first, or among the first, animals to evolve, they nevertheless represent a huge jump in complexity from the first unicelled eukaryotes that appeared 800 million years earlier. Scientists have not yet been able to discover much about what went on in terms of the evolution of life during those 800 million years of the latter part of the Proterozoic.

FIGURE 14.14 Ediacara fauna
These are among the most ancient multicelled animals that have been found. (A) *Mawsonia spriggi* was a disc-shaped, floating animal, something like a jellyfish; this sample is 13 cm in diameter. (B) *Dickinsonia costata* was a curious flat, wormlike creature; this sample is about 7.5 cm in diameter.

Phanerozoic Life

The Phanerozoic Eon encompasses all of geologic history starting from 542 million years ago to the present (FIG. 14.15), thus including the Paleozoic, Mesozoic, and Cenozoic eras. The term *Phanerozoic* comes from Greek words meaning "visible life," and it is no accident that early geologists, seeking to establish a worldwide geologic column, placed a major boundary at this point in time. In earlier rocks, they could see little evidence of life (although now we have better tools with which to study earlier *microfossils*). From this point forward, however, clearly visible evidence of ancient life, in the form of *macrofossils*, was abundant.

The Cambrian Radiation

The Cambrian Period, beginning 542 million years ago, was the time of the introduction of skeletons, both internal and external. A famous shallow-marine fossil assemblage in the Burgess Shale of British Columbia (FIG. 14.16, on page 437) preserves many soft-bodied forms and a few skeletonized ones. The fossil assemblage in the Burgess Shale, which was deposited 505 million years ago, demonstrates a richness of forms that we no longer see in contemporary bottom-dwelling marine fauna.

From the Burgess Shale and other similar fossil assemblages of about the same age in other locations, we know that the early Cambrian Period was a time of almost unbridled growth and diversity in the marine environment. During a relatively short period of Earth history, termed the **Cambrian radiation** (or *Cambrian explosion*), the pace of evolution increased dramatically (see Fig. 14.7B). (An *evolutionary radiation* is a rapid increase in diversity of organisms evolving from a common ancestor.) Most of the major modern groups of organisms emerged at this time, and their basic body plans were established. This was accompanied by a rapid increase in the complexity of many organisms. Compact animals with skeletons evolved to replace the floppy ones of Ediacaran times: trilobites, mollusks (clams), echinoderms (sea-urchins), and sea-snails—all of them (except the trilobites) types that have persisted up to the present, and all equipped with gills, filters, efficient guts, a circulatory system, and other complex structures.

Why did the **biodiversity**, or overall variety of life forms, increase so dramatically in the early Cambrian? This is another of the great, unanswered questions about the biosphere. Many hypotheses have been offered, but none, as yet, is backed by hard evidence. One hypothesis is that the emergence of sexual reproduction allowed more rapid evolution. Prokaryotes are asexual and reproduce by a process called *mitosis*, in which the cell splits its DNA into two equal halves in order to create a new cell. The new cell is identical to the parent. Eukaryotes reproduce sexually, and two individuals contribute their genetic information equally; thus, any genetic variety that exists is quickly spread among the growing population. Another hypothesis is that the availability of oxygen allowed for the development of skeletons composed of calcium phosphate and calcium carbonate, making more complex forms possible. A final hypothesis is that the stratospheric ozone layer (which, you will recall, is thought to have begun functioning as an ultraviolet screen around this time) made it possible for life to emerge from the depths of the oceans into millions of new environments in shallow water and, eventually, on land.

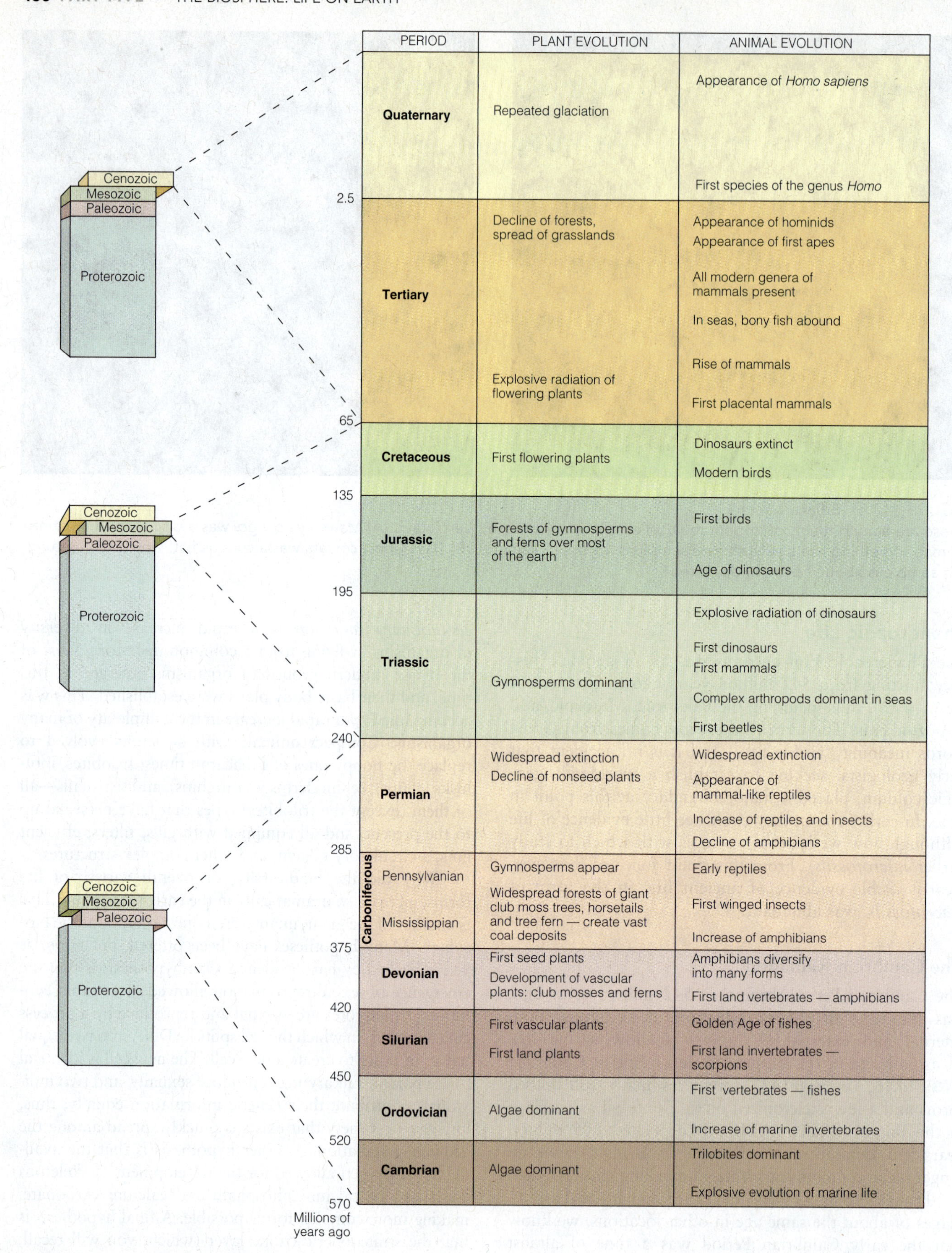

PERIOD
PLANT EVOLUTION
ANIMAL EVOLUTION
Quaternary
Repeated glaciation
Appearance of *Homo sapiens*
First species of the genus *Homo*
2.5
Tertiary
Decline of forests, spread of grasslands
Explosive radiation of flowering plants
Appearance of hominids
Appearance of first apes
All modern genera of mammals present
In seas, bony fish abound
Rise of mammals
First placental mammals
65
Cretaceous
First flowering plants
Dinosaurs extinct
Modern birds
135
Jurassic
Forests of gymnosperms and ferns over most of the earth
First birds
Age of dinosaurs
195
Triassic
Gymnosperms dominant
Explosive radiation of dinosaurs
First dinosaurs
First mammals
Complex arthropods dominant in seas
First beetles
240
Permian
Widespread extinction
Decline of nonseed plants
Widespread extinction
Appearance of mammal-like reptiles
Increase of reptiles and insects
Decline of amphibians
285
Carboniferous
Pennsylvanian
Mississippian
Gymnosperms appear
Widespread forests of giant club moss trees, horsetails and tree fern — create vast coal deposits
Early reptiles
First winged insects
Increase of amphibians
375
Devonian
First seed plants
Development of vascular plants: club mosses and ferns
Amphibians diversify into many forms
First land vertebrates — amphibians
420
Silurian
First vascular plants
First land plants
Golden Age of fishes
First land invertebrates — scorpions
450
Ordovician
Algae dominant
First vertebrates — fishes
Increase of marine invertebrates
520
Cambrian
Algae dominant
Trilobites dominant
Explosive evolution of marine life
570
Millions of years ago
Cenozoic
Mesozoic
Paleozoic
Proterozoic

FIGURE 14.15 Phanerozoic life

The onset of the Phanerozoic Eon, at the start of the Cambrian Period, is placed at 542 million years ago. This version of the geologic timescale shows the major plant and animal types that dominated each of the three eras of the Phanerozoic: the Paleozoic, Mesozoic, and Cenozoic.

(A)

(B)

FIGURE 14.16 Fossils from the Burgess Shale
Most of the fossils preserved in the 505-million-year-old Burgess Shale are of soft-bodied organisms. Of the 25 different species depicted in the drawing, only the five circled had hard parts. The Burgess Shale is the most complete assemblage of Cambrian fauna ever found, and it is presumed that the abundance of soft-bodied animals reflected the situation elsewhere in the ocean. The ancient chordate *Pikaia* (also shown in Fig. 14.20) resembles a small fish. (B) This trilobite, from the genus Olenoides, is from the Burgess Shale. They were up to 10 cm in length, although some types of trilobite were much larger.

Life on Land

The Cambrian Period was the first of the six periods in the Paleozoic Era, which lasted about 330 million years. The Cambrian Period was followed, in order, by the Ordovician, Silurian, Devonian, Carboniferous, and Permian Periods. The five periods of the Mesozoic and Cenozoic Eras followed. Each period is characterized by the appearance of major groups of plants and animals, as summarized in Figure 14.15. Early geologists established these time boundaries largely on the basis of these major fossil horizons, and the *stratigraphic correlation* (Chapter 7) of corresponding rock units around the world.

The great proliferation of life in the Cambrian Period was entirely confined to the sea, but by 500 million years ago the main plans for animal life had been established. The one big step that remained for organisms was to leave the sea and occupy the land and, finally, the atmosphere.

Eventually, plants, insects and other animals, bacteria, and fungi all evolved land-based forms. Here are some of the requirements for multicellular organisms to survive on land:

1. Structural support (such as a skeleton or a strong stem) is needed because aquatic organisms are buoyed by water, but gravity on land becomes a real force with which to contend.
2. An internal aquatic environment is required, with a plumbing system that gives it access to all parts of the organism and with devices for conserving the water against losses to the surrounding atmosphere.
3. A mechanism is required for exchanging gases with air (such as lungs) instead of with water (such as gills).
4. A moist environment for the reproductive system is essential for all sexually reproducing organisms.

Selection for these necessities is largely what has shaped terrestrial organisms into the familiar forms we know today.

PLANTS. Early land plants likely evolved from the earliest known eukaryotes, green algae (FIG. 14.17). Eventually,

FIGURE 14.17 Earliest life on land?
The first plants to colonize the land may have looked like the green algae shown in this photograph.

vascular plants (or *higher plants*) evolved. These had woody stems for structural support (requirement 1), as well as a vascular system (requirement 2), a system of channels by which water is transferred from the roots to the leaves, and the products of photosynthesis from leaves to roots. Requirement 3 (gas exchange) in vascular plants occurs by diffusion and is controlled by adjustable openings in the leaves called *stomata* (from the Greek for "mouth"). When carbon dioxide pressure inside the leaf is high, the stomata open; when low, they close. (The stomata also close when the plant is short of water, thereby protecting it against drying out.) Gas exchange was managed by plants in the Devonian Period in much the same way as it is managed in today's plants.

The earliest plants, of which mosses and leafy liverworts are modern examples, were seedless (FIG. 14.18A, B). Many of these *seedless plants* can tolerate some drought; for example, mosses survive dry spells by releasing spores that lie dormant until moisture returns. But these ancient plant forms still rely on moisture for the sexual phase of the reproductive cycles; without it the sex cells have no medium in which to reach each other and fuse. Consequently, the seedless plants have never been able to colonize habitats where moisture is not unfailingly present for at least part of the growing season. Seedless plants reached their peak in the Carboniferous Period, when they dominated vast forests on the tropical floodplains and deltas of North America, Europe, and Asia.

By the Middle Devonian Period, a few plants were already on the way to meeting requirement 4—that is, providing their own moist environment in order to facilitate sexual reproduction. The plants that evolved were the **gymnosperms**, or *naked-seed plants*, such as *Glossopteris* of Gondwana fame (see Fig. 5.4). The female cell is attached to the vascular system and thus has a supply of moisture. The male cell is carried in a pollen grain that has a hard coating. When the two fuse, a seed results. The seed is simply a supply of moisture and nutrients that will sustain the early growth of the young plant until it becomes self-supporting by photosynthesis. Naked-seed plants survive today; examples are the gingko (FIG. 14.18C) and conifers.

Gymnosperms had huge success. Freed from the swampy habitat, they did not have to compete with the great seedless trees of the coal forests, but could occupy the drier uplands of the newly forming supercontinent, Pangaea. By the end of the Carboniferous Period, they had spread over most of the world, and by the Triassic Period they were rivaling in size their former cousins of the swamps.

Life has one drawback for gymnosperms, however. The male cell-carrier, the pollen, is spread through the air.

FIGURE 14.18 Evolution of plant life
(A) This fossil fern (genus *Pecopteris*), preserved in shale, is about 300 million years old. It is an example of a seedless plant. (B) Compare the fossil fern to a modern fern. This photograph shows the spore-producing organs, the dark spots on the underside of the frond. (C) Naked-seed plants developed from the seedless plants late in the Devonian Period. These are modern (*Gingko biloba*, right) and fossilized (*Gingko adiantoides*, left) ginkgo leaves. (D) Flowers and fruits represented an advantageous adaptation for plants because they serve as an incentive for insects to do the work of distributing pollen. This 15-million-year-old fossil (left) found in Idaho is shown next to its modern equivalent, the sweet gum fruit (genus *Liquidambar*).

(A)

(B)

(C)

(D)

What are the chances that a pollen grain, loose in the air, will find a female cell? To ensure success, gymnosperms have to make huge amounts of pollen. Flowering plants (**angiosperms**, or *enclosed-seed plants*) solved the problem of the random distribution of pollen. Through a small incentive (nectar, or a share of the pollen), insects are enticed to deliver the pollen (FIG. 14.18D). It took longer for the angiosperms to evolve than it did for the gymnosperms, but by the end of the Cretaceous Period angiosperms had become the dominant land plants. Their life cycle is not significantly different from that of gymnosperms, but they have developed specialized relationships with animals: insects for pollination, and birds and mammals for seed dispersal.

The last frontier for plants—so far, at least—has been the dry steppes, savannas, and prairies. These areas were not colonized until the Tertiary Period, when the first grasses evolved. In arriving at this period, we have also reached the culmination of animal life on land, the great grazing faunas of the high plains of all continents except Antarctica.

INSECTS. Among the numerous creatures in the Cambrian seas, many belonged to the phylum Arthropoda, so-called because of the presence of jointed legs. They include crabs, spiders, centipedes, and insects, as well as trilobites. This remains the most diverse phylum on Earth. These animals—not the fish from which we humans are descended—were the first to make the change from sea to land.

Early arthropods, with a few exceptions, were quite small, had lightweight structures, and were covered with shell composed of a hard substance called *chitin* to provide structural support. Thus, they were admirably preadapted for life on land in regard to both structural support and water conservation. The earliest to go on land were probably Silurian centipedes and millipedes. By the Carboniferous Period, insects were abundant and included dragonflies with a wing span of up to 60 cm (FIG. 14.19). For all their success as land creatures, the arthropods have always had very primitive respiratory and vascular systems. They breathe through tiny tubes that penetrate the outer coating. Because the respiring mass of an aerobic organism increases as the cube of its length, while the area available for gas exchange increases only as the square, this mode of respiration severely limits the size of an organism. This is one reason that most insects are small.

Arthropods have an open vascular system. That is, their "blood" does not circulate in closed vessels, but is simply body fluid bathing the internal organs and generally kept in sluggish motion by a "heart" that is little more than a contractile tube. At first, it seems odd that the arthropods, with such a primitive arrangement, should have diversified into more than a million terrestrial species. The arthropods' vascular system isn't great, but it obviously works. And it's close to indestructible; whoever heard of a cockroach having a heart attack?

ANIMALS WITH BACKBONES. Inconspicuous among the fossils of the Burgess Shale is a small fossil called *Pikaia* (FIG. 14.20; see also Fig. 14.16). Pikaia is an early example of a *chordate*, a member of the phylum *Chordata*, to which humans also belong. Chordates possess a *notochord*, a cartilaginous rod running along the back of the body. (Humans and other *vertebrates*—animals with backbones—have a notochord as embryos, later replaced by the backbone.) Pikaia and other Cambrian fish were jawless, probably feeding on organic matter dredged from the seafloor. Jawed fish evolved afterward, and with their evolution a great burst of diversification took place. The original jawless fish, only a few centimeters in length, were quickly joined by much larger, armored fish. These included sharks and other boneless (cartilaginous) fish, as well as the huge order of ray-finned fish that are familiar as today's game and food fish.

FIGURE 14.19 Ancient insect
A single wing of the giant dragonfly *Megatypus schucherti* is 16 cm long. The largest dragonfly today is only 15 cm in total width. This specimen, Lower Permian in age, is from Dickinson County, Kansas.

FIGURE 14.20 *Pikaia*, the first chordate
This soft-bodied animal from the Burgess Shale, *Pikaia*, is one of the earliest known chordates (phylum Chordata).

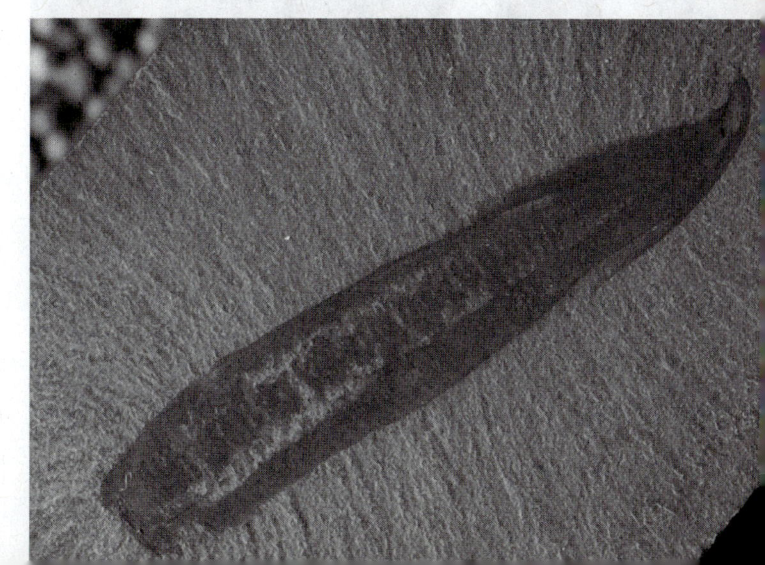

The first fish to venture on land, probably from an obscure group called the *crossopterygians*, did so in the Devonian Period (about 400 million years ago). They gave rise to the amphibians. The crossopterygians (now typically placed within the class Sarcopterygii) had several features that may have served to facilitate the transition to land (FIG. 14.21). Their lobelike fins were preadapted as limbs because the lobes contain the (foreshortened) elements of a quadruped limb, complete with small bones to form the extremity. They also had internal nostrils characteristic of air-breathing animals. Being fish, the crossopterygians already had a serviceable vascular system that was adequate to make a start on land. Water conservation, however, never became a strong point with amphibians: They retain permeable skins to this day, which is one reason they have never become independent of the aquatic environment.

Despite their limitations, the amphibians ruled the land for many millions of years during the Devonian Period. They had one difficulty that limited their expansion into many niches: They never met the reproductive requirement for life on land. In most species, the female amphibian lays her eggs in the water, the male fertilizes them there after a courtship ritual, and the young hatch as aquatic organisms. Like the seedless plants, the amphibians have remained tied to the water for breeding. Although some became quite large, the amphibians are not highly diverse. One branch evolved to become reptiles; the rest that survive are frogs, toads, newts, salamanders, and limbless water "snakes."

The reptiles freed themselves from the water by evolving a water-tight skin and an egg with a shell that could be incubated outside of the water. These two "inventions" gave them the versatility to occupy terrestrial niches that the amphibians had missed because of their bondage to the water. The amniotic egg did for reptilian diversity what jaws did for diversity in fishes. Reptiles originated in the Carboniferous Period swamps, about 375 million years ago. By the Jurassic Period, some 185 million years later, the reptiles had moved onto the land, up in the air, and back to the water. This resulted in the production of the two orders of dinosaurs, the largest quadrupeds ever to walk the surface of Earth, and gave rise to two new vertebrate classes—mammals and birds (FIG. 14.22).

FIGURE 14.21 Pioneer fish?
When a living coelacanth was caught in the Indian Ocean in 1938, it created a sensation because not only its species but its entire order had been believed to be extinct. The first fish to haul themselves onto land may have been relatives of the coelacanth. Today's coelacanths, like this one (*Latimeria chalumnae*) photographed off the coast of South Africa, are exclusively deep-sea creatures and are once again at risk of extinction.

FIGURE 14.22 Early birds and mammals
(A) The skeletons and teeth of *Archaeopteryx* were very similar to those of dinosaurs. However, the very detailed impressions of feathers in this fossil identify *Archaeopteryx* as an early bird. (B) Discovered in 2002 in China, this shrew-sized *Eomaia scansoria* specimen is the oldest-known fossil of a placental mammal (that is, a mammal that gives live birth to its offspring). Its body was about 8 cm in length, and it lived 125 million years ago, during the height of the dinosaur age.

Mammals are in many ways better equipped to occupy terrestrial niches than were the great reptiles. It is difficult to pick out a single mammalian "invention" comparable to the jaws of fish or the reptilian egg, for the mammal is a fine-tuned quadruped, better adapted to a faster and more versatile life than the reptiles. The placental uterus is sometimes regarded as the key to mammalian success. It is mandated by the delicate intricacy of the fetus that lives in it, especially the brain. The true mammalian "invention" is perhaps just that it possesses a set of interdependent improvements managed by a more capable brain, and supported by a faster metabolism.

THE HUMAN FAMILY. Charles Darwin was often accused of believing that humans are descended from the apes. In fact, the family of humans, *Hominidae*, and the family of apes, *Pongidae*, are both descended from an earlier common ancestor that was neither human nor ape. The emergence of humans during the Cenozoic Era is one of the most complex and controversial fields in paleontology, in part because of the paucity of the fossil record and the lack of transitional forms. But it is clear that *hominids*—humanlike organisms—are a very recent evolutionary development.

The first hominid that was clearly *bipedal* (walked upright routinely) was *Australopithecus*, of which the famous fossil "Lucy" is an example (**FIG. 14.23**). These hominids were only about 1.2 meters in height but had a brain capacity larger than that of chimpanzees. Their fossils range from about 3.9 to 3.0 million years in age. From the shape of its pelvis and from footprints left in soft volcanic mud (see Fig. 14.23), we know that *Australopithecus* walked upright. Lucy's descendants never spread beyond Africa, and they disappeared altogether about 1.1 million years ago.

Homo erectus, possibly the first species of our own genus (*Homo*), was more widely traveled than *Australopithecus*. Fossils of *Homo erectus*, dating back about 1.8 million years, have been found in Africa, Europe, China, and Java. Even earlier, a problematic species called *Homo habilis* used stone tools. Because toolmaking is a distinguishing feature of the genus *Homo*, some experts include *habilis* in this genus; others argue that the skill of this species is more like that of the Australopithecines.

Homo erectus disappeared and was replaced by *Homo neanderthalensis* ("Neanderthal man"), no later than 230,000 years ago. Unfortunately, the poor fossil record between 400,000 and 100,000 years ago has made it difficult for scientists to determine how this transition occurred. We hypothesize from burial sites that Neanderthals might have practiced some form of religion. Because of similarities in teeth and brain size (slightly larger than our own), some experts have argued that Neanderthal was part of our own species; however, recent DNA studies suggest that *Homo sapiens* is not

FIGURE 14.23 The human family
(A) This drawing by Michael Rothman depicts a mother and child of the species *Australopithecus afarensis*. These humanlike individuals lived together in small groups, formed lasting bonds with mates, and looked after their children through infancy. (B) This 3.3-million-year-old fossil is the skull of an *A. afarensis* baby, who probably looked much like the child in the drawing. (C) From footprints like these, preserved in soft volcanic mud, scientists know that australopithecines walked upright on two feet. This 70-m trail includes the footprints of two adults and possibly a child, stepping in the footprints of one of the adults. To the right are footprints of an extinct three-toed horse.

(A)

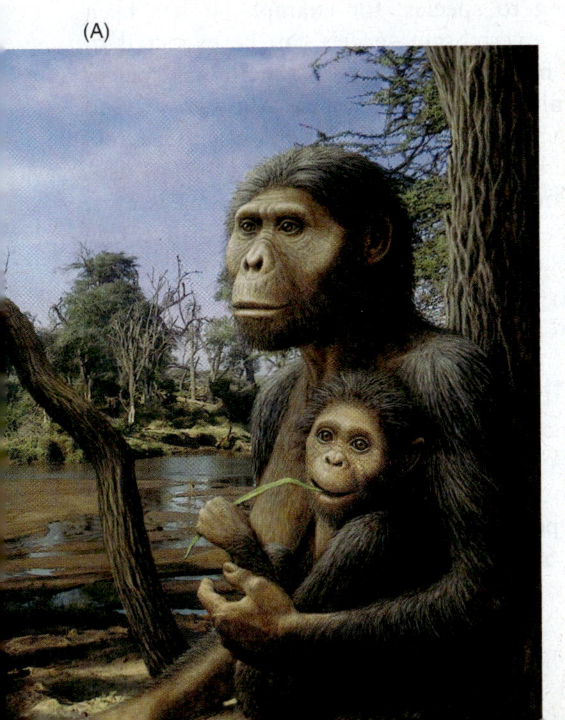

(B)

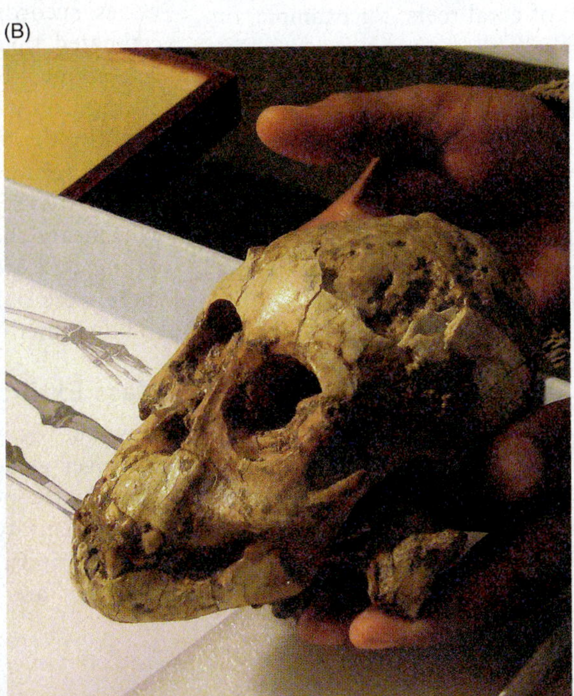

(C)

a direct descendant of Neanderthals. The Neanderthals disappeared about 30,000 years ago and were replaced rather suddenly by biologically modern people, the first indisputable members of our own species, *Homo sapiens*.

Did *H. sapiens* evolve from Neanderthals, or were they a distinct species? What happened during the 5000-year period when both kinds of humans were alive and overlapped geographically in Europe? The Neanderthals become extinct very rapidly—possibly within a timescale of just 10,000 years. Did the modern *Homo sapiens* kill the Neanderthals? Or, as some scientists have suggested, were the Neanderthals simply unable to adjust to the rapidly cooling climate of the time? These and many other questions await answers, as paleontologists continue to look for clues in the fossil record.

Environmental Change and Biodiversity

The history of life's biological diversification has been driven by the availability of new environments on the one hand, and evolutionary innovations by successful organisms on the other. Rapid diversification results when a new "invention" (jaws in fish, the amniotic egg in reptiles) provides an organism with adaptive advantages for a life in a new or changing environment.

Organisms not only respond to environmental change but also create it. The oxygenation of the atmosphere by early photosynthetic organisms is an example of a major environmental change that provided many new opportunities for diversification. An example at a smaller spatial scale is the vertical structural complexity provided by trees. In a tropical rain forest, for example, the top of the tree provides a sunlit environment that receives much rainfall, while down near the base of branches there can be shaded and comparatively dry habitats. This structural complexity provides opportunities for organisms to adapt to multiple, varied environments, and this has contributed fundamentally to the evolution of the immense biological diversity that characterizes tropical rain forests today. Other complex environments also tend to be biologically diverse; think of coral reefs, for example, or mangrove swamps.

However, a modification that adapts a species to a *niche* (functional role, Chapter 16) that is already occupied may confer little immediate advantage, even though the new species may be adaptively superior to the incumbent one. Such was the case with the mammals. Originating in the Triassic Period, mammals were from the start more attuned to the niches they now occupy than were the established occupants, the dinosaurs. With more capable brains, faster metabolism, a uterus to shelter the fetus, milk for postnatal nourishment, and parent–offspring bonding, the early mammals had greater potential to exploit dinosaur niches than had the dinosaurs themselves. The great dinosaurs had command of the food supply, however, and the mammals, instead of growing larger, became smaller—perhaps small enough not to interest carnivorous dinosaurs. So they remained for 150 million years, until the end of the Cretaceous Period. When dinosaurs became extinct, the mammals moved in to occupy vacated niches, with a burst of diversification that began in the Paleocene Epoch and continued through the Eocene Epoch. Mammals continued to adapt over that time. For example, by comparing brain-to-body weight ratios in archaic and modern reptiles and mammals, it can be shown that mammalian brain size has increased. This is a continuing process, whereas in reptiles such an increase has not occurred: The brain-to-body ratios in modern reptiles do not differ significantly from those in archaic ones.

EXTINCTION: THE HISTORY OF DEATH

Few of the species alive at the beginning of the Cambrian Period had living descendants at the end. Throughout the Cambrian, species died out and became extinct; new species evolved to occupy vacated niches. Why and how did great groups of animals suddenly die out and new ones arise to take their places? In the nineteenth century, when paleontologists started to study the fossil record in a systematic way, they recognized that evidence of **extinction**—the permanent disappearance of a species—is widespread, and that most of the species that have ever lived on Earth—probably 99 percent of them—are now extinct.

Background Rate of Extinction

Organisms have disappeared (by extinction), just as new types of organisms have emerged (by speciation) throughout the history of life; we refer to the ongoing disappearances as the *background rate of extinction*. There are several ways of measuring the background extinction rate; one is to calculate the normal length of time that a species would be expected to last. For example, the evidence of extinctions in the fossil record of marine animals is extremely detailed. From this record scientists have estimated that the average time span of an ocean-dwelling species is about 4 million years. This varies according to species; for example, it has been estimated that invertebrate species can be expected to last about 11 million years, whereas mammal species average only about 1 million years. Another way to quantify the background rate of extinction is to express an average number of extinctions per year, which is usually cited as about 1 extinction per million species-years. This means that if there were a million species alive, one would be expected to go extinct every million years; if there were only one species alive, it would be expected to go extinct in about a million years.

Mass Extinctions

However, extinctions have not happened at a constant rate over geologic time. Evidence from the fossil record also clearly points to a series of extinction events in which many types of organisms have died out over very short periods. Such events are called **mass extinctions**, and they are typically followed by radiations, periods

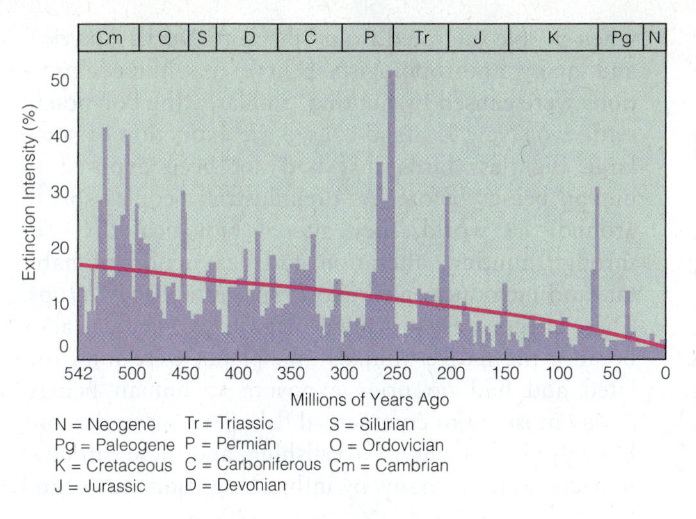

FIGURE 14.24 Mass extinctions
The background extinction of species has been punctuated with a number of major mass extinctions; several of the biggest are seen here, including those at the end of the Cretaceous and Permian periods.

of rapid diversification and the emergence of many new species. Analysis of the fossil record shows evidence for at least five major mass extinctions and possibly as many as 19 smaller ones (FIG. 14.24). Two of these major events occurred at the Cretaceous-Tertiary boundary, the end of the Mesozoic Era, and at the Permian-Triassic boundary, the end of the Paleozoic Era.

What caused these major extinction events? For many years, gradual environmental change and associated climatic change were thought to be the principal causes of mass extinctions. Then evidence began to emerge that there may have been extraordinary contributing factors in some mass extinctions.

For example, a popular hypothesis is that the impact of a gigantic meteorite at the end of the Cretaceous caused worldwide impacts that led to the extinction (FIG. 14.25). A huge impact crater of exactly the right age has been found in Mexico; the devastation caused by the impact must have been massive, and there is evidence indicative of global effects in rock units of this age, worldwide. An alternative explanation is that the Cretaceous-Tertiary mass extinction was caused by the global effects of a massive outpouring of lava. The Deccan Traps of India, one of the world's largest flood basalt deposits, were formed approximately 65 million years ago, coinciding with this extinction.

The greatest of all extinctions in Earth's history, when some 95 percent of known fossil species were lost, is the event that marks the Permian-Triassic boundary at the end of the Paleozoic Era. There is no convincing evidence that the Permian-Triassic boundary extinction had an extraterrestrial cause. Although the Cretaceous-Tertiary extinction severely affected both terrestrial and marine life, mainly marine species suffered in the Permian-Triassic one. This earlier extinction is associated with the assembly of Pangaea, the supercontinent that existed briefly from Late Permian to Late Triassic time. The extinction also was associated with a marked drop in sea level. Consider the effect of these developments—continents massed together and sea level lowered—on the continental shelves. If continents are stuck together to make a single supercontinent, there is a significant loss of coastline and, consequently, of continental shelf. A 100-m drop in sea level would approximately halve the remaining shelf area, reducing the total shelf area to about 25 percent of what was there before—a drastic reduction in living space for shelf organisms. For this and other reasons, many scientists conclude that plate tectonics and specifically supercontinent formation made a major contribution to the Permian-Triassic mass extinction.

FIGURE 14.25 The day the dinosaurs died?
There is evidence that the extinction of the dinosaurs 65 million years ago may have been caused by the impact of a meteorite approximately 10 km in diameter, possibly in the ocean near Mexico's Yucatán Peninsula, as shown in this artist's rendition. The impact would have caused a massive tsunami, scattering debris across the hemisphere and igniting continent-wide forest fires. Soot from the fires likely remained in the atmosphere for months, or even years, blocking sunlight and effectively shutting down photosynthesis worldwide.

The Sixth Great Extinction

Extinction is a natural process, and the ultimate fate of all Earth-restricted species is extinction. However, human beings have greatly accelerated the rate of extinction, especially since the Industrial Revolution, possibly thousands of times the background rate of extinctions. This very rapid rate of extinction may have global impact and has been termed by some the "sixth great extinction."

Ever since people learned how to use fire, we have had major effects on large areas of the land surface and have influenced the abundance of many species of animals and plants. Even preindustrial, hunter-gatherer peoples affected biological diversity through hunting, as well as through the use of fire for land clearing. The activities that have most affected biodiversity over the course of human history are hunting and overharvesting; deforestation, land-clearing, and desertification; and the introduction of new species that outcompete native species. Today, pollution has joined the other major causes of habitat destruction and biodiversity loss.

Extinctions of some of the largest mammals in North America occurred near the end of the ice age when people migrated from Siberia to North America, and many anthropologists believe that these extinctions were caused by hunting. Similarly, the Polynesian settlers of New Zealand caused the extinction of some large flightless birds that had not been exposed to human beings before. As preindustrial peoples spread around the world, they altered biological diversity through hunting, alteration and destruction of habitat, and introduction of exotic species and new crops. These effects are especially evident in the Pacific islands because the native animals and plants were long isolated and had no prior exposure to human beings. Today most major commercial fisheries have undergone large declines owing to overfishing, and these can have impacts on food chains by influencing competition and predation.

A rapid, human-induced decrease in genetic diversity could mean that the Earth system will lose many kinds of biological capabilities, only a few of which we understand at present, but which might be important to ecosystem dynamics or have practical uses to human beings. We will explore these ideas in greater detail in Chapters 16 and 19.

END OF CHAPTER 14

SUMMARY

1. Four essential characteristics differentiate living organisms from nonliving matter: metabolism, reproduction, growth, and evolution. Metabolism is the set of chemical reactions by which an organism derives energy for life processes. Autotrophs create their own food energy, most commonly by photosynthesis. Heterotrophs derive food energy by eating other organisms. Heterotrophs that are oxygen-tolerant obtain their energy through respiration, using oxygen to break down carbohydrates.

2. In single-celled organisms, reproduction occurs by cell division. Reproduction in multicelled organisms can occur either sexually or asexually. The genetic plan of a living organism is encoded in its DNA, which consists of two biopolymers in a double-helix structure, held together by bases. A gene is a sequence of DNA that encodes for a certain trait, and an allele is a specific variant of that trait. RNA decodes and carries out the instructions for duplicating the proteins that make up the cell.

3. Growth involves the ordering and organizing of atoms and small molecules to make larger molecules. This can happen by polymerization or by crystallization. In living matter, all growth involves polymerization of small organic molecules to form biopolymers. Polymerization absorbs energy; therefore, growth in living organisms requires a source of energy, which is acquired through metabolism.

4. Living organisms have the potential to adapt and to develop into new life forms, or species. A morphological definition of a species holds that organisms that look similar to one another and share many structural characteristics belong to one species. An alternative definition is that organisms that are able to breed successfully with one another, producing fertile offspring, belong to the same species. Species change genetically over time, through evolution.

5. A cell, the structural and functional unit of all living organisms, is a complex grouping of chemical compounds enclosed by a membrane. Prokaryotic cells are generally small and comparatively simple in structure; they occur as bacteria and related organisms. Eukaryotic cells are larger and more complex; their genetic information is housed in a well-defined nucleus, and they contain a variety of organelles that have specific functional roles. All animals (including humans), plants, and fungi are eukaryotes.

6. Life is hierarchical. Cells form organs, which combine to form multicelled organisms. A grouping of organisms of the same kind that interbreed is a population. All populations of a particular organism constitute a species. Different species interact with each other in ecological communities. Interactions within a community, and between the biotic and abiotic components of the environment, constitute an ecosystem. Biomes are major ecosystem types. All of the ecosystems in a particular geographic region form an ecozone, and Earth's ecozones together constitute the biosphere. The biosphere, atmosphere, hydrosphere, and geosphere form the Earth system.

7. In the six-kingdom system of taxonomic classification, all prokaryotes belong to one of two kingdoms: Archaea or Bacteria. All other living organisms are eukaryotes and belong to one of four kingdoms: Protista, Fungi, Animalia, or Plantae. The six kingdoms are grouped into three domains: Archaea, Bacteria, and Eukarya. In the Linnaean System, the kingdoms are subdivided into successively narrower categories: phylum, class, order, family, genus, and species. Traditional assignments of organisms to taxonomic rank based on morphological similarity are being modified as new findings emerge concerning complex genetic relationships among organisms. A phylogenetic tree is a taxonomy that organizes the forms of life according to their genetic and evolutionary relationships.

8. During the Hadean Eon there was no free oxygen in the atmosphere, a necessity for most forms of life on Earth today. It was too hot for water to exist as a liquid, and atmospheric pressure was much greater than it is today. As the planet cooled, water vapor condensed, fell as rain, and collected as surface water. The early hydrosphere reacted with gases in the atmosphere to form acids, which reacted with the rock of the crust. The compositions of the atmosphere, hydrosphere, and geosphere thus changed dynamically as materials cycled among them.

9. Earth is the only planet on which we have confirmed the existence of life, past or present. Life originated on Earth as much as 3.9 billion years ago. Three steps must have been accomplished on the way to the complex life forms we know today: (1) chemosynthesis of small organic molecules from inorganic precursors; (2) biosynthesis, the polymerization of small organic molecules to form biopolymers, especially proteins; and (3) the development of the complex chemical machinery needed for metabolism and replication. The specific mechanisms by which life originated are still hypothetical.

10. Life contributed to the chemical evolution of the Earth system through the buildup of oxygen and sequestration of carbon. Oxygenation happened over a very long time, but the onset of major oxygenation coincides roughly with the beginning of the Proterozoic Eon. Once the ozone layer began to absorb harmful shortwave radiation, organisms would have been able to survive in shallow waters; this critical stage was reached around 600 million years ago, just before the onset of the Phanerozoic Eon. The emerging biosphere had a profound impact on carbon cycling; limestone and organic sediment provide long-term storage reservoirs for carbon, sequestering it from the atmosphere and hydrosphere.

11. The theory of evolution states that all present-day organisms are descendants, through a gradual process of genetic change and adaptation, of different kinds of organisms that existed in the past. Evolution is achieved through natural selection, in which individuals that are well adapted to their environment survive and are reproductively successful, whereas individuals that are poorly adapted tend to be eliminated. If the genetic characteristics of a group change to the extent that individuals are no longer able to breed successfully with individuals from the original population, a new species has arisen. Speciation requires that a population be reproductively isolated from others of the same species.

12. The most ancient known fossils are 3.55 billion years old, found in rocks from the Archean Eon. Some early fossils are the remains of microscopic prokaryotes; others are stromatolites that precipitated as a result of bacteria influencing the chemistry of seawater. The earliest life, which dominated Earth for about 2 billion years, was prokaryotic and anaerobic. Anaerobic cells acquire energy by fermentation, which restricts them to a small size (to allow rapid diffusion of food in and waste out) and limits their structural complexity (to save energy). The first photosynthetic prokaryotes were probably cyanobacteria.

13. With the development of an oxygenated atmosphere, organisms turned a lethal waste product, oxygen, to an advantage. About 1.4 billion years ago, during the Proterozoic Eon, the aerobic eukaryotes emerged. They likely arose from prokaryotes but are typically larger, and can maintain a nucleus and organelles because oxidative respiration is a more efficient energy source. Eukaryotes are not inhibited by crowding, so they can form three-dimensional colonies.

14. The evolutionary history of the biosphere is revealed through fossils, the altered remnants of ancient organisms preserved in rocks. The 600-million-year-old Ediacaran fauna, among the oldest animal fossils, were jellylike animals that lived in quiet marine bays. The Ediacaran fauna represent a huge jump in complexity from the first unicelled eukaryotes, 800 million years earlier, but scientists have not yet discovered much about the evolution of life in the intervening period.

15. The Phanerozoic Eon includes the Paleozoic, Mesozoic, and Cenozoic eras. From the Burgess Shale and other fossil assemblages, we know that the early Phanerozoic was a time of rapid evolution in the marine environment. Most of the major modern groups of marine organisms emerged at this time. This was accompanied by a rapid increase in complexity, including the appearance of skeletons (internal and external), gills, filters, efficient guts, circulatory systems, and other complex structures. The Cambrian radiation may have occurred in response to the emergence of sexual reproduction; or the protection afforded by the ozone layer; or the availability of oxygen, which facilitated the construction of skeletons.

16. The requirements for multicellular organisms to survive on land include structural support; an internal aquatic environment; a means for exchanging gases with air; and a moist environment for the reproductive system. Selection for these necessities has shaped terrestrial organisms into the forms we know today.

17. The first land plants were seedless; these evolved into naked-seed plants (gymosperms), enclosed-seed plants (angiosperms), and grasses. Arthropods were probably the first animals to move from sea to land. Pikaia, the first known chordate, was jawless; jawed fish evolved later. The first fish to venture onto land were crossopterygians, which gave rise to amphibians. Amphibians have permeable skins and lack a moist environment for reproduction, and thus remain dependent on the aquatic environment. Reptiles became liberated from the water by evolving a water-tight skin and an egg with a shell; they evolved into two orders of dinosaurs and two new vertebrate classes—mammals and birds.

18. The first clearly bipedal hominid was *Australopithecus*, whose fossils range from about 3.9 to 3.0 million years in age. Fossils of *Homo erectus*, possibly the first species of our own genus, date to about 1.8 million years ago. *Homo erectus* disappeared 300,000 years ago and was replaced by *Homo neanderthalensis*. The poor fossil record between 400,000 and 100,000 years ago has made it difficult for scientists to determine how the transition occurred, but DNA studies suggest that *Homo sapiens* is not a direct descendant of

Neanderthals. The Neanderthals disappeared about 30,000 years ago and were replaced by the first indisputable members of our own species, *Homo sapiens*.

19. The diversification of life has been driven by the availability of new environments and by evolutionary modifications in successful organisms. Rapid diversification results when a new "invention" provides an organism with adaptive advantages for life in a new or changing environment. Organisms not only respond to environmental change, but also create it. The oxygenation of the atmosphere by early photosynthetic organisms was a major environmental change that provided many new opportunities for diversification. An example at a smaller spatial scale is the vertical structural complexity provided by trees, which provides opportunities for organisms to adapt to multiple, varied environments.

20. About 99 percent of the species that have lived on Earth are now extinct. Extinction is a natural process, but it has not happened at a constant rate over geologic time. Superimposed on background extinctions are at least five major mass extinction events, in which many types of organisms died out over short periods. These may have been caused by climatic and other environmental changes, possibly accelerated by major meteorite impacts, massive episodes of volcanism, or the formation of supercontinents. Humans have greatly accelerated the rate of extinction of species.

IMPORTANT TERMS TO REMEMBER

adaptation *431*
angiosperms *439*
Archean Eon *427*
autotroph *418*
biodiversity *435*
biosphere *418*
black smokers *418*
Cambrian radiation *435*
cell *421*
chemosynthesis *418*
cyanobacteria *434*
DNA *419*
domain *423*
ecosphere *425*
eukaryotic cell *422*
evolution *421*
extinction *442*
fossil *434*
gene *420*
genetic drift *433*
gymnosperms *438*
Hadean Eon *426*
heterotroph *418*
kingdom *423*
life zone *425*
mass extinction *442*
metabolism *418*
mutation *433*
natural selection *431*
Phanerozoic Eon *430*
photosynthesis *418*
phylogenetic tree *424*
prokaryotic cell *421*
Proterozoic Eon *430*
respiration *419*
RNA *419*
speciation *431*
species *420*
taxonomic rank *423*
vascular plants *438*

QUESTIONS FOR REVIEW

1. What are the essential characteristics of life?
2. Could life have started as a virus? Could it have started as a bacterium?
3. What is the difference between prokaryotic cells and eukaryotic cells?
4. What is the difference between autotrophs and heterotrophs? Give an example of each.
5. What is evolution?
6. How does natural selection work?
7. What is a species? Why is it difficult to define "species"?
8. What is speciation? Describe one mechanism whereby speciation can occur.
9. What are the four major eons of Earth history, and what environments and life forms characterized each of them (in general)?
10. What kinds of things have scientists learned from the Ediacaran fauna and the fossils of the Burgess Shale?
11. When did life start leaving the sea and inhabiting the land? In what order did animals, insects, and plants become established on the land?
12. What four requirements had to be met in order for organisms to leave the sea and live on land? Are the requirements the same for plants, insects, and animals?
13. In what ways have plants modified their reproductive cycle since moving on to the land?
14. Amphibians are tied to the water. What characteristics did the reptiles develop that allowed them to free themselves from a dependence on water?
15. What evidence in the fossil record suggests that rapid extinctions have occurred many times in the past? Describe two hypotheses for the extinctions.

QUESTIONS FOR RESEARCH AND DISCUSSION

1. Why did biodiversity explode during the Cambrian radiation?
2. How does life differ from other phenomena in the universe?
3. Which of the hypotheses for the origin of life on Earth do you think is the most supported by scientific evidence? On the basis of this hypothesis, what do you think are the chances that life originated and took hold on another planet?

QUESTIONS FOR *THE BASICS*

1. What is DNA? What is RNA? Why are they so fundamental to life?
2. What is the six-kingdom classification system, and how is it related to the traditional Linnaean System of taxonomic classification?
3. What is a phylogenetic tree?

QUESTIONS FOR *A CLOSER LOOK*

1. What is the "primordial soup" hypothesis? Where and how did chemosynthesis occur in the hypothesis, and why is this part of the hypothesis now in doubt?
2. What was the possible role of the "black smoker" environment in the origin of life?
3. What is the "panspermia" hypothesis? Describe some of the pros and cons of this hypothesis.

CHAPTER

15

Ecosystems, Biomes, and Cycles of LIFE

OVERVIEW

In this chapter we:

- Examine the functions and pathways of energy in ecosystems
- Consider the fundamental importance of material recycling in ecosystems
- Define the minimum characteristics of a life-supporting system
- Explore the major biogeochemical cycles and how they are affected by human activities
- Summarize the characteristics of Earth's major aquatic and terrestrial biomes

A marsh ecosystem
This green frog (*Rana clamitans*) is intimately connected to both the biotic and abiotic components of its shallow-water marsh home by innumerable biogeochemical and ecological processes and pathways.

In Chapter 14 we summarized the characteristics that differentiate living organisms from nonliving matter. We also looked at the long story of life on Earth through geologic time. Starting with the origin of life, we considered how evolution, natural selection, and extinction shaped the great diversity of life that now exists on this planet. In this chapter we consider the systems that maintain life and enable it to persist. We will examine the interactions and flows of energy and matter in life-supporting systems, and summarize the variety of such systems on Earth.

ENERGY AND MATTER IN ECOSYSTEMS

In the Arizona desert an exotic, futuristic structure rises above the landscape; it is Biosphere-II, the most ambitious human attempt ever to develop a closed, habitable system on Earth (FIG. 15.1). It cost $120 million to build and had the best material seal ever produced. Eight people—four men and four women—lived in that closed system for two years, along with 3800 species of plants, nonhuman animals, and an unknown number of microorganisms. The goal was to demonstrate that people could produce their own food, recycle all their own wastes, and maintain a breathable atmosphere and healthy environment in a system completely closed to the exchange of matter with the universe outside.

The inhabitants came close to achieving their goal. But during the second year the oxygen concentration in Biosphere-II began to decline. When it reached an oxygen pressure equivalent to that found at 4500 m above sea level, a medical decision was made to open the air locks and provide some more oxygen for the health of the inhabitants. Throughout this experiment—when it was proposed, while it was being planned, and during its early stages of operations—many scientists suspected that the system would fail, and each gave a reason. Some suggested that the carbon dioxide level would get too high; others suggested that nitrates would not be removed rapidly enough by bacteria and would pollute the waters; still others suggested that a crop disease or an insect outbreak would make it impossible for the people to feed themselves.

But nobody forecast an oxygen decline. The cause of that decline was found to be bacteria, which, in an atmosphere with elevated carbon dioxide, could break down compounds in the cement walls. In doing so, they used up oxygen as they respired. In terms of its stated goals, Biosphere-II failed, but it provided important insights and is the first example we have of a large-scale, human-supporting, materially closed system.

In the sealed container of Biosphere-II, the eight humans obtained food from the plants and animals they farmed. They depended on plants, algae, and bacteria to give off oxygen, which is necessary for animal life, and to remove carbon dioxide from the air they breathed out. They had to recycle water for drinking, washing, and irrigation, and all materials, including their own wastes. In short, their survival within the sealed structure depended on two primary processes: (1) flow of energy and (2) continual recycling of chemical elements. These are the two fundamental requirements for any life-supporting system. They provide a framework for our discussion of the **ecosystem**, the basic life-supporting system in which living organisms interact with each other and with the abiotic components of their environment.

FIGURE 15.1 Biosphere-II
Biosphere-II was an attempt to create a closed, life-sustaining system, like that of Earth. It failed because of an unanticipated depletion of oxygen, caused when bacteria digested the cement walls of the structure.

Energy Flow in Systems

In Chapter 2 we discussed some basic concepts about energy. Here we revisit that discussion to explain the role and importance of energy and energy flow in life-supporting systems. *Energy flow* refers to the movement of energy through an ecosystem, from the external environment (the Sun) through a series of organisms and back to the external environment (space). It is a fundamental process common to all ecosystems.

At first glance energy flow seems simple enough. But when we consider the role of energy in ecosystems, we discover that energy flow is what distinguishes life and life-supporting systems from the rest of the universe. All life requires energy. As anyone who has followed a diet knows, our body mass is a delicate balance between the energy we take in through food and the energy we use through activities; what we don't use, we store. This is true not only of people, but also of other organisms, and of populations, communities, ecosystems, biomes, and the biosphere.

Two aspects of energy and its role in ecosystems are essential to an understanding of energy and life. The first is the *functions* of energy in living systems, especially its involvement in the production of organic matter in ecosystems, which can tell us about the ultimate limits on the abundance of life. The second is the *pathways* through which energy flows, the efficiency with which it is used, and the role it plays in the various processes of life.

Functions of Energy: Biological Productivity

The basic function of energy in a life-supporting system is to make possible the production of organic matter. The amount of usable energy in a system provides an upper limit on the amount of organic matter, and thus the amount of life that can be sustained by any ecosystem, and ultimately by the Earth system as a whole.

BIOMASS AND PHOTOSYNTHESIS. The total amount of organic matter on Earth or in any particular ecosystem is called the **biomass** of that system. This includes all living things and all products of living things. (Plant biomass is called *phytomass*, and animal biomass is called *zoomass*.) Most of life is on or close to Earth's surface, so biomass is usually measured as an amount per unit surface area (for example, as grams per square meter).

Biomass increases as a result of **biological production**—that is, through the transformation of energy into matter by biological processes. In Chapter 14 you learned that

the BASICS

Thermodynamics Revisited

In Chapter 2 we introduced the basic concepts of energy and the laws of thermodynamics. Here we review some of these concepts to highlight their significance to life.

Matter and energy are both subject to the *first law of thermodynamics,* which addresses the observation that energy changes form, not amount. In any physical or chemical change, matter is neither created nor destroyed, but merely changed from one form to another. This first law of thermodynamics raises a seemingly simple question: If the total amount of energy is always constant, why can't we just recycle energy inside our bodies? Similarly, why can't energy be recycled in ecosystems and in the biosphere the way chemical elements are recycled?

To answer this question, first consider a simple, hypothetical ecosystem involving, say, frogs and mosquitoes. Frogs eat insects, including mosquitoes. Mosquitoes suck blood from vertebrates, including frogs. Consider an imaginary closed ecosystem consisting of water, air, a rock for frogs to sit on, frogs, and mosquitoes. In this system, the frogs get their energy from eating the mosquitoes, and the mosquitoes get their energy from biting the frogs (FIG. B15.1). Why can't this system maintain itself indefinitely? Such a closed system would be a biological perpetual motion machine: It could continue indefinitely without an input of any new material or energy. This sounds nice, but unfortunately it is impossible.

The general answer as to why this system could not persist is found in the *second law of thermodynamics,* which addresses the kind of change in form that energy undergoes. The second law tells us that energy always changes from a more useful, more organized form to a less useful, less organized form. Energy cannot be completely recycled to its original state of organized, high-quality usefulness because it takes energy to maintain order in any natural system. Life is no exception to this. Life makes orderly, complex carbohydrate molecules out of disorderly, simpler molecules, and this process takes energy. For this reason, without an energy source, the mosquito-frog system will eventually run down and stop; there will not be enough useful energy left to maintain the system.

So a life-supporting system must have a source of usable energy and a sink for degraded energy, which occurs in the form of heat. These three components: the energy source + the system in which life processes occur (the organism or, more broadly, the ecosystem) + the energy sink form a thermodynamic system. Earth is a closed thermodynamic system, meaning that matter is neither added nor lost, but energy flows freely into and out of the system. As we saw in Chapter 1, all closed and open systems respond to inputs, and all have outputs. The same principle holds for the Earth system that holds for smaller ecosystems: In order to support life, Earth must have a source of usable energy and a sink for heat. For Earth, the main usable energy source is the Sun, and the sink for degraded heat energy is the vastness of space.

FIGURE B15.1 An impossible ecosystem
The energy in this impossible ecosystem is being recycled again and again. In the real world, the energy would become degraded and would eventually be lost, and the system would run down.

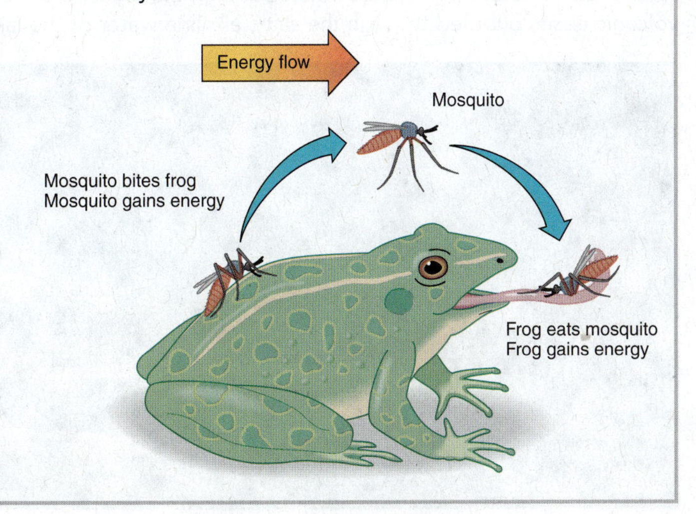

organisms that synthesize their own organic compounds are called *autotrophs*, and they accomplish this most commonly through photosynthesis. The conversion of sunlight into food energy by photosynthesis happens according to the following reaction:

$$CO_2 + H_2O + \text{energy} \rightarrow CH_2O + O_2$$

The product of this reaction, in addition to oxygen, is organic matter in the form of carbohydrates (Chapter 3). *Carbohydrates* are a group of chemical compounds, including sugars, starches, and cellulose, that are found in all organisms and have the general formula $C_x(H_2O)_y$. Carbohydrate production and the consequent building of body mass by autotrophs is called **primary production.** Chemosynthesis (Chapter 14), a much less common process of primary production, produces carbohydrates by combining CO_2 with H_2S, rather than with H_2O. The production of body mass by *heterotrophs*—organisms that derive their food energy by eating other organisms—is called **secondary production.**

NET PRIMARY PRODUCTION. Primary producers convert energy and inorganic compounds into biomass, in the form of carbon-based organic compounds. There are three steps in the production of biomass and its use as a source of energy by autotrophs. First, the autotrophic organism produces organic matter within its body (by photosynthesis or, less commonly, by chemosynthesis); then it uses some of this new organic matter as a fuel in *metabolism* and *respiration* (Chapter 14), releasing energy back to the surrounding environment in the form of heat; finally, it stores some of the newly produced organic matter for future use, in the form of carbon-based compounds.

The first step, production of organic matter before any use, is called **gross production.** The biomass that is left over from gross production after it has been used to fuel the processes of life (the second step) is called **net production.** In these terms,

$$\text{net production} = \text{gross production} - \text{respiration}$$

This is a fundamental production relationship. The difference between gross and net production is like the difference between a person's gross and net incomes. Gross income is the dollars you are paid, whereas net income is what you have left after money is deducted for taxes and other fixed costs. Respiration, which supports metabolism and allows the organism to "run" all of its life-maintaining processes, releases energy back to the environment; this is equivalent to taxes and other costs that must be paid in order for you to do your work and live your life.

Pathways of Energy: Trophic Dynamics

Autotrophs transform energy, use it, and store some of it in the form of organic compounds. But organisms are not even close to being 100 percent efficient in converting sunlight or chemical energy into energy stored in organic matter. The process is even more inefficient when it comes to passing this food energy from one organism to another, through feeding relationships. Let's have a closer look at some of the pathways by which energy flows through ecosystems.

FOOD CHAINS. Energy is transferred along **food chains** in which one organism eats another and is, in turn, is eaten by another organism. In a basic sense, a food chain is a pathway by which energy (in the form of food) moves through an ecosystem. A simple kind of natural ecosystem, and one that we can use to examine this idea of food chains, is found in salt lakes, examples of which include the Aral Sea in Uzbekistan, the Great Salt Lake in Utah, and Mono Lake in California.

Mono Lake covers 160 km^2 in the high desert just east of the Sierra Nevada Mountains (FIG. 15.2). It is famous

FIGURE 15.2 Mono Lake
Mono Lake in California, east of the Sierra Nevada Mountains, is a salt lake that has few species and a relatively simple food-chain structure (see Figure 15.3), but high biological productivity. The rock structures (called *tufa*) were formed when volcanic gases bubbled through the salty, alkaline water of the lake.

for its beauty, for its strange salty waters—twice as salty as the ocean—and for more than 1 million birds that feed there and depend on small animals that grow in the lake. **FIGURE 15.3** is a simplified diagram of food chains at Mono Lake. A number of autotrophs—species of algae and photosynthetic bacteria—grow in the salty, alkaline waters. Only two heterotrophs—a brine shrimp and the larvae of the brine fly—are able to eat the algae and live in the salty water. Although the lake's chemistry creates an environment of great stress, it also provides an abundance of the chemical elements required by these species. Five species of birds feed on the brine shrimp and brine fly. The birds feed on the brine shrimp and brine fly, which feed on the algae and bacteria—this is the basic food chain structure at Mono Lake.

Trophic Pyramids. Notice in Figure 15.3 that each set of organisms is a certain number of steps away from the original source of energy. The brine shrimp and brine fly are one step away from the autotrophic source of food energy, and the birds are two steps away. Each group of species that is the same number of steps away from the original source of energy is called a **trophic level.**

The autotrophs of Mono Lake—the algae and photosynthetic bacteria—form the first trophic level; they are the **primary producers** of the system. The brine shrimp and brine fly, both heterotrophs, form the second trophic level; they are **consumers** and *secondary producers*, meaning that they produce biomass by consuming other organisms rather than through autotrophic processes. Because the brine shrimp and brine fly directly consume the primary producers, they are referred to as *first-order consumers*. The birds of Mono Lake are also heterotrophs; they are consumers and secondary producers, too, and they form the third trophic level in the system. Because they are two steps removed from the primary producers, they are referred to as *second-order consumers*.

These food-energy levels together form a **trophic pyramid. FIGURE 15.4** shows the example of a trophic pyramid based on grass (the autotroph and primary producer, forming the first trophic level), eaten by deer (a heterotroph and first-order consumer, forming the second trophic level), eaten in turn by wolves (another heterotroph and second-order consumer, forming the third trophic level). The word "trophic" is generally interpreted to mean "energy," but in fact it comes from the Greek word meaning "food" or "nourishment," which demonstrates the close equivalency of food and energy in ecosystems.

Consider again the Mono Lake food chains. Energy enters the ecosystem as sunlight. It is converted into organic matter by the primary producers at the bottom of the trophic pyramid, and then flows to the higher trophic levels, being passed along in the form of food. However, photosynthetic organisms (the primary producers) are inefficient; they are able to convert and store only about 1 percent of the energy available in sunlight.

FIGURE 15.3 Mono Lake food chains
This diagram shows who feeds on whom in the Mono Lake ecosystem. Compared to most ecosystems, Mono Lake has few species and the food chains are relatively simple.

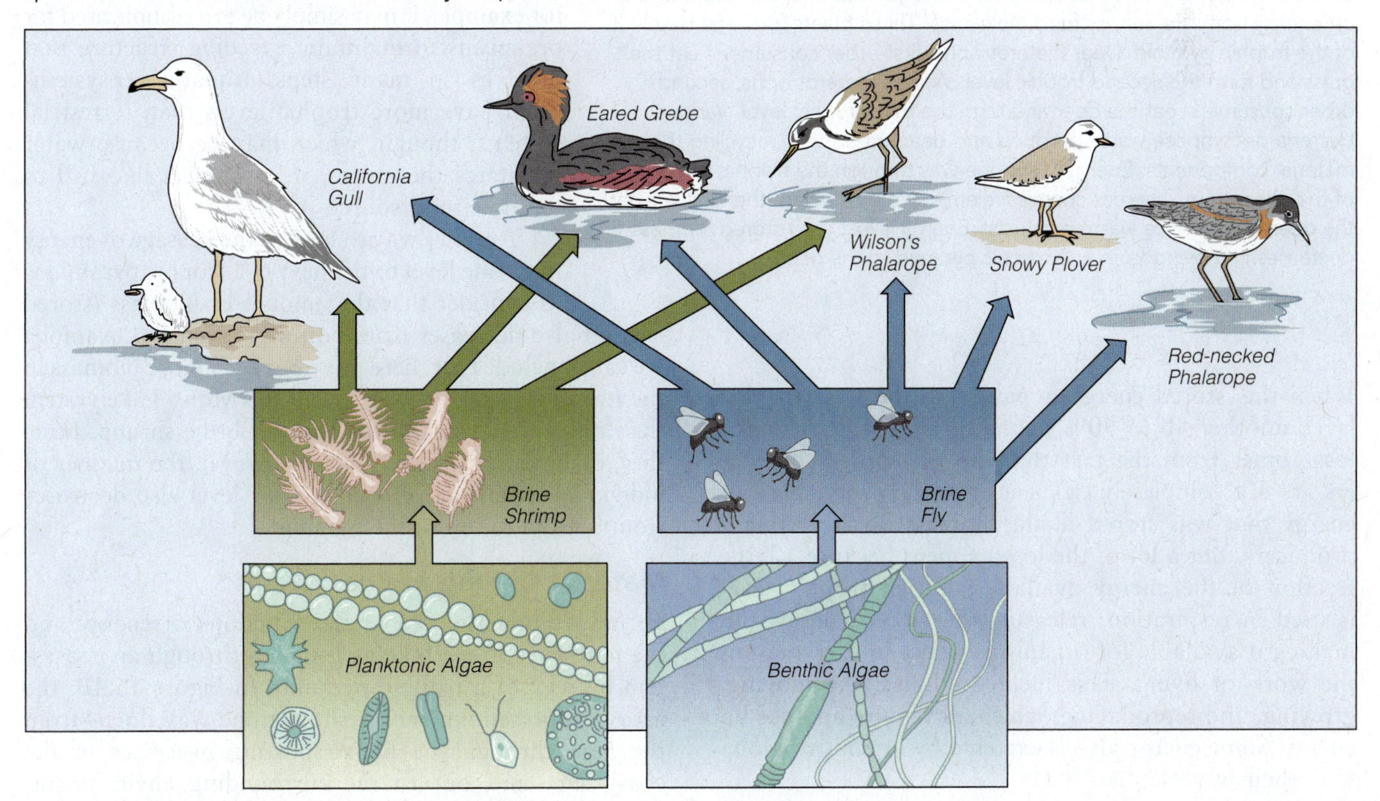

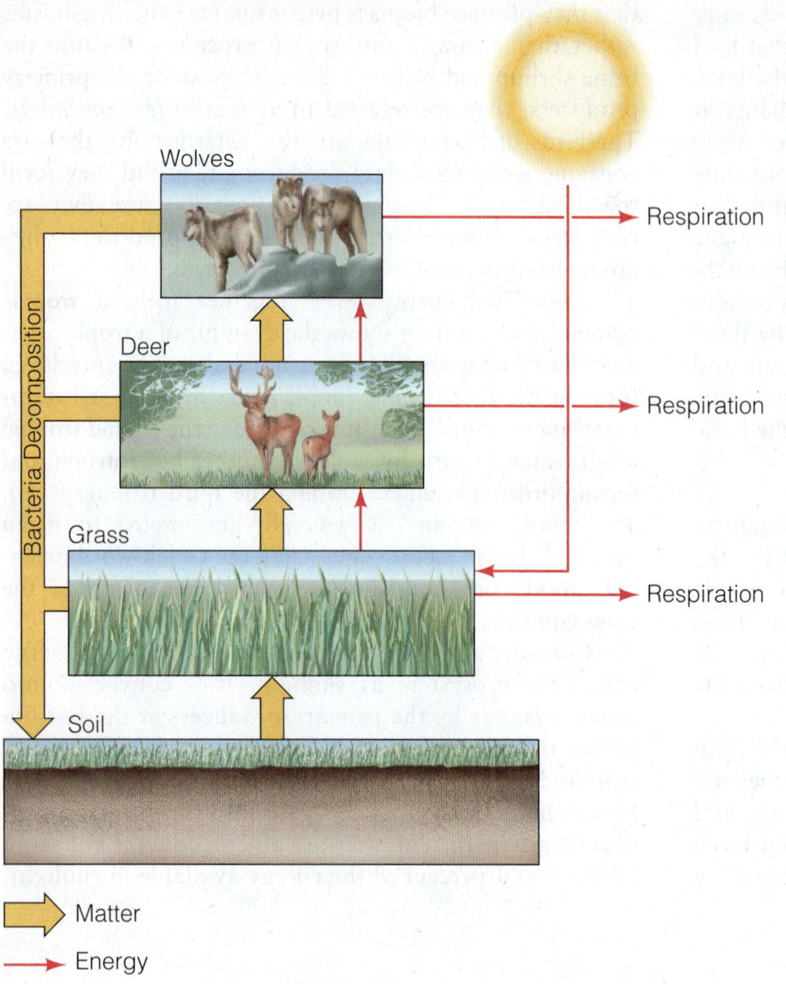

FIGURE 15.4 Trophic pyramid
In this ecosystem, grasses (autotrophs; primary producers) capture and lock up energy from the Sun, as food molecules. These plants form the base of the trophic pyramid. Deer (heterotrophs; first-order consumers) eat the grass and form the second trophic level. Wolves (heterotrophs; second-order consumers) eat the deer and form the third trophic level. Various bacteria decompose waste products and dead organisms, recycling their material components. Energy flows one-way through the trophic levels of the ecosystem, whereas chemical elements are recycled (the wolf will eat the deer, and excrete waste products). Energy is lost, and stored biomass decreases in moving from one trophic pyramid to the next.

When this stored energy is passed to the next trophic level, another 80 to 90% of it will be lost. Part of the loss comes from the fact that the first-order consumers are not completely efficient at eating every shred of energy that was stored in the biomass of the primary producers. But a lot of the loss happens because a large fraction of the energy available at each trophic level is used in respiration, releasing the stored energy and making it available for organisms to use in carrying out the work of living. This includes things like moving, growing, and reproducing, which all require and use up energy. Some energy also is excreted by organisms along with their wastes.

This used energy is not passed along to subsequent trophic levels in the food chain, and it can never be recovered or recycled. Once it has been used to run life-supporting processes, the used energy (now degraded and less ordered) is released to the surrounding environment in the form of heat. The movement of energy through the trophic pyramid and through the ecosystem is thus a one-way flow, as shown in Figure 15.4—it comes in, it is used, and it is released to the surrounding environment. A similar magnitude of energy loss occurs between the second and third levels in a trophic pyramid; that is, only about 10 to 20% of the available energy in the first-order consumers is passed along to the second-order consumers. An equivalent loss of energy would occur again if there were higher trophic levels in the system. (For example, if the Mono Lake ecosystem included coyotes that ate the birds—it doesn't—this would constitute a fourth tropic level).

This explains why trophic pyramids and food chains are limited. In principle, there is no limit to the number of levels that could exist in a trophic pyramid; however, the actual number of trophic levels in food chains is typically around four, and it is extremely rare to find a food chain with more than six trophic levels. So much energy is lost in moving from one level to the next that it would require an enormously large base of primary producers to support a trophic pyramid with more than six levels. (There are additional reasons for this limitation; for example, it may simply be too complicated for organisms to maintain a feeding structure that involves so many steps. Aquatic ecosystems often have more trophic levels than terrestrial systems, though, which may be because water facilitates the access of consumers like fish to multiple food sources.)

Another way to look at the passage of energy from one level to the next in a trophic pyramid is to consider that the amount of biomass (stored energy) also decreases from bottom to top. For example, we can conclude that there must be much more biomass in the form of algae and bacteria in the Mono Lake system than there is biomass in the form of brine shrimp, brine flies, or birds. Typically (but not always), the number of individual organisms at each trophic level also decreases from bottom to top of the pyramid.

Material Cycling in Ecosystems

So far we have looked at the functions of energy and the pathways it can take as it moves through an ecosystem by way of a trophic pyramid. In Figure 15.4B, the arrows representing energy show a one-way flow—from the Sun, through the life-supporting processes of the ecosystem, and out to the surrounding environment.

Notice that there is another set of arrows in the figure, representing the flow of chemical elements. These arrows are cyclical—they show that chemical elements, unlike energy, can be recycled again and again through an ecosystem. Let's examine the processes by which this occurs.

Decomposers

In the Mono Lake and grass–deer–wolf food chains (Figs. 15.3 and 15.4), there is an additional pathway that we haven't yet discussed. Organisms at each level in these trophic pyramids excrete waste products; they also die. Both of these processes leave behind material composed of organic matter—that is, biomass. (Note that biomass is not necessarily *currently alive*, but includes recently dead biological materials and the organic wastes of living organisms.) These leftover materials are an important part of the energy that is *not* passed along to the next level in the trophic pyramid. We know what happens to the energy; it is released to the surrounding environment, in the form of heat. But what happens to the chemical elements of which the excreted waste and dead organisms are composed? They are recycled, and you will recall that the recycling of chemical elements is one of the fundamental characteristics of life-supporting systems.

A cycling of chemical elements is necessary because no one organism can make all its own food from inorganic chemicals and then break down all of those complex organic chemicals, returning them to the original inorganic form that would allow them to be reused. At the minimum, therefore, there must be at least two kinds of organisms in a life-supporting system: The first type of organism (autotrophs) produces its own organic compounds from energy and inorganic chemicals available in the environment. The other type of organism decomposes complex organic compounds so that the chemical elements can be recycled; these are called **decomposers** (or *saprotrophs*), and they play a crucial role in life-supporting systems.

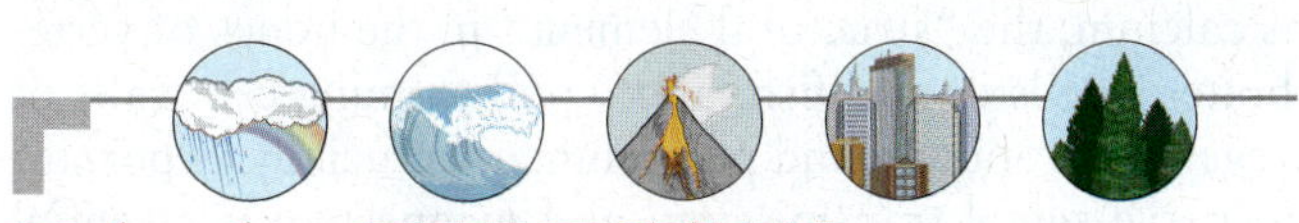

Make the CONNECTION

What would life be like on Earth if there were no decomposers? What are the connections between decomposers and the atmosphere, hydrosphere, biosphere, and geosphere?

Organisms that decompose complex organic compounds are heterotrophs because they obtain their energy from the organic matter of other organisms; they are mainly (though not exclusively) bacteria and fungi. At Mono Lake, for example, an unknown number of bacteria act to decompose the fecal material of the shrimps, flies, and birds, and the dead bodies of these organisms. To our knowledge, no single species of bacteria or fungus can decompose all of the compounds present in all autotrophs. Some of these compounds are not only complex, but also quite resistant to decomposition. A good example is *lignin* (Chapter 3), an important structural component of woody plants. Lignin is indigestible by animals, and its decomposition requires the secretion of specialized enzymes by certain fungi and bacteria.

Food Webs

So both energy and materials are cycled through ecosystems by way of feeding relationships. However, in most ecosystems, feeding relationships are more complicated than at Mono Lake, where the food chains are simple enough that each organism functions on just one trophic level. When there are several interconnected food chains in an ecosystem that involves complex eating relationships and interactions, we call it a **food web**.

Consider, for example, the food web of the harp seal (FIG. 15.5), in which five trophic levels are identifiable. *Phytoplankton*, microscopic floating plants, are the autotrophs; they carry out primary production at the first trophic level. Tiny floating animals, or *zooplankton*, are the first-order consumers; they are heterotrophs that feed on the phytoplankton at the second trophic level. At the third trophic level are fish, mollusks, and other organisms that consume the zooplankton. At the fourth level are several types of fish that consume other fish. The harp seal feeds at several trophic levels, from the second through the fourth, eating everything from tiny zooplankton to all types of fish. It even feeds on the predators of some of its prey and thus is a competitor with some of its own food. (Predator-prey and other interactions between individual organisms will be discussed in Chapter 16.) A species that feeds at several trophic levels is classified as belonging to the trophic level above the highest from which it feeds, so we consider the harp seal to be on the fifth trophic level.

In a more realistic manner than the simplified food chain, a food web illustrates the movement of both energy and matter through a complex set of feeding interactions. (Note, however, that for all its complexity, the harp seal food web still includes just five trophic levels, or six if you consider that a polar bear might come along and eat the seal.) Energy moves through the food web in a complicated, circuitous, but ultimately one-way flow, but materials are cycled over and over again to provide nutrients for organisms at all levels. This recycling of materials is not 100 percent leak-proof, though. One can imagine many circumstances whereby materials might be transferred from one part of an ecosystem, or from one local ecosystem to another—an eagle could carry a mouse to its nest far away, or a river might bring waste products and sediments to the ocean, for example. It is part of the fundamental nature of environmental systems to be open, to allow energy to flow freely through, and to recycle materials, while still allowing some passage of materials into and out of the local system.

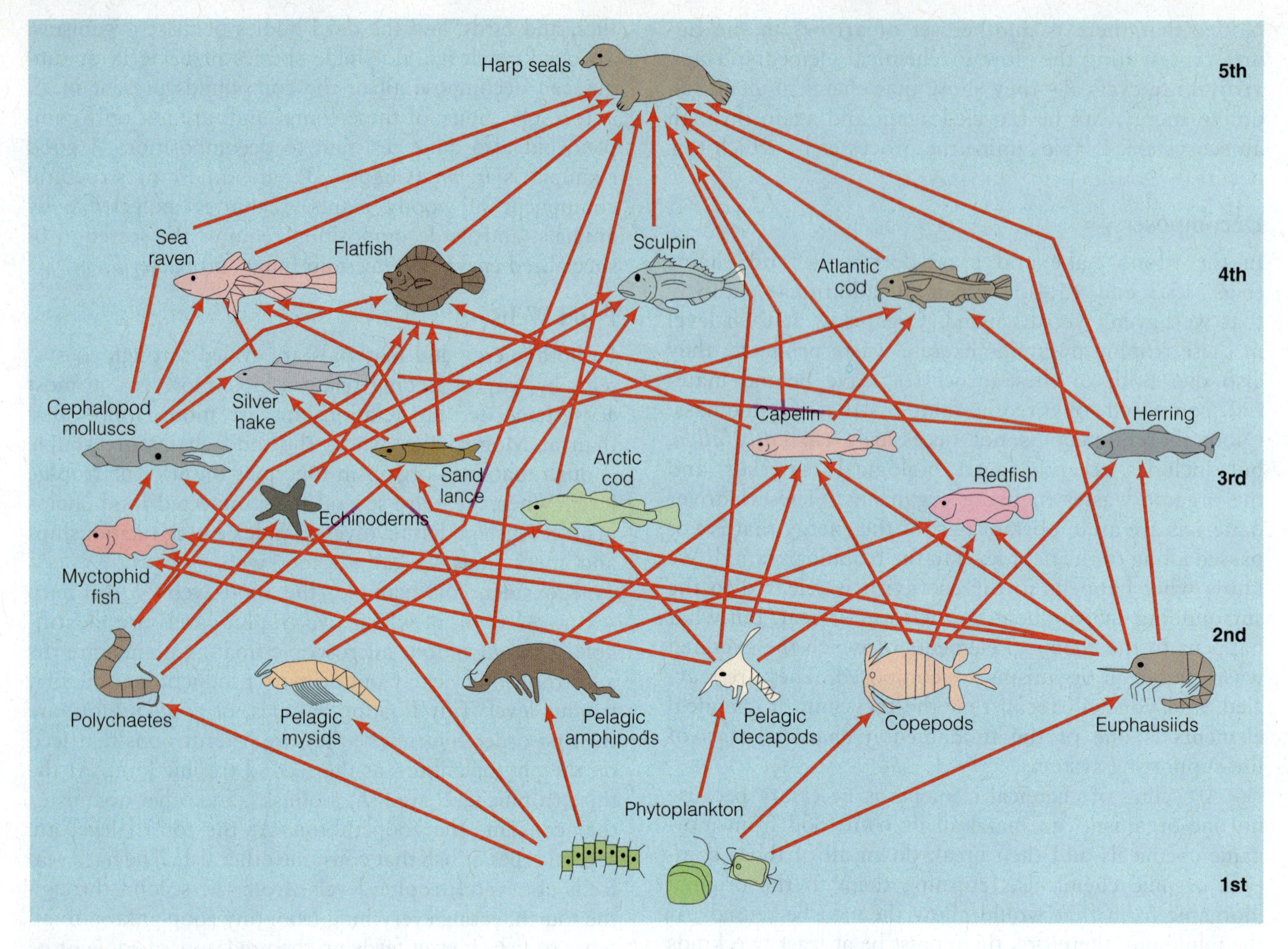

FIGURE 15.5 Food web
This is the food web of the harp seal. The arrows show who feeds on whom, indicating the pathways taken by both energy and matter. The harp seal feeds on more than one trophic level.

Nutrients, Toxins, and Limiting Factors

What are these life-supporting materials that cycle endlessly through ecosystems? Only 24 chemical elements are **essential nutrients**, that is, elements that are known to be required for life (FIG. 15.6). (Some listings of nutrients also include compounds, such as amino acids, fatty acids, and vitamins, but the 24 chemical elements are the basic components of these.) Nutrients are divided into two basic categories: the **micronutrients**, elements required either in small amounts by all life or in moderate amounts by some forms of life and not others; and the **macronutrients**, elements required in large amounts by all life. The macronutrients include the "Big Six"—the elements that form the fundamental building blocks of life. They are carbon, hydrogen, nitrogen, oxygen, phosphorus, and sulfur.

Each of the macronutrients plays a special, fundamental role in organisms. Carbon is the basic building block of organic compounds; along with oxygen and hydrogen, it forms carbohydrates. Nitrogen, along with oxygen, hydrogen, and carbon, makes amino acids, the basic structural units of proteins. Phosphorus is the "energy element" occurring in the compounds called ATP and ADP, which are important in the transfer and use of energy within cells. Sulfur is a crucial component of some amino acids and the proteins and enzymes that contain them, and it serves as a source of chemical energy for some chemosynthetic organisms. Another macronutrient is calcium, the "structural element" in the bones of vertebrates, shells of shellfish, and wood-forming cell walls of vegetation. Sodium and potassium are crucially important in nerve signal transmission, and magnesium is essential for plant growth and photosynthesis.

For any form of life to persist, the required chemical elements must be available at the right times, in the right amounts, and in the appropriate relative concentrations to each other. When this does not happen, an element can become a **limiting factor**, preventing the growth of an individual, population, or species, or even causing its local extinction. We can determine which chemical element is limiting in an ecosystem by comparing the ratios of the concentration of elements in the environment with the ratios of those same elements in living tissue. The element with the lowest concentration in the environment in comparison to the other required elements will be limiting to growth.

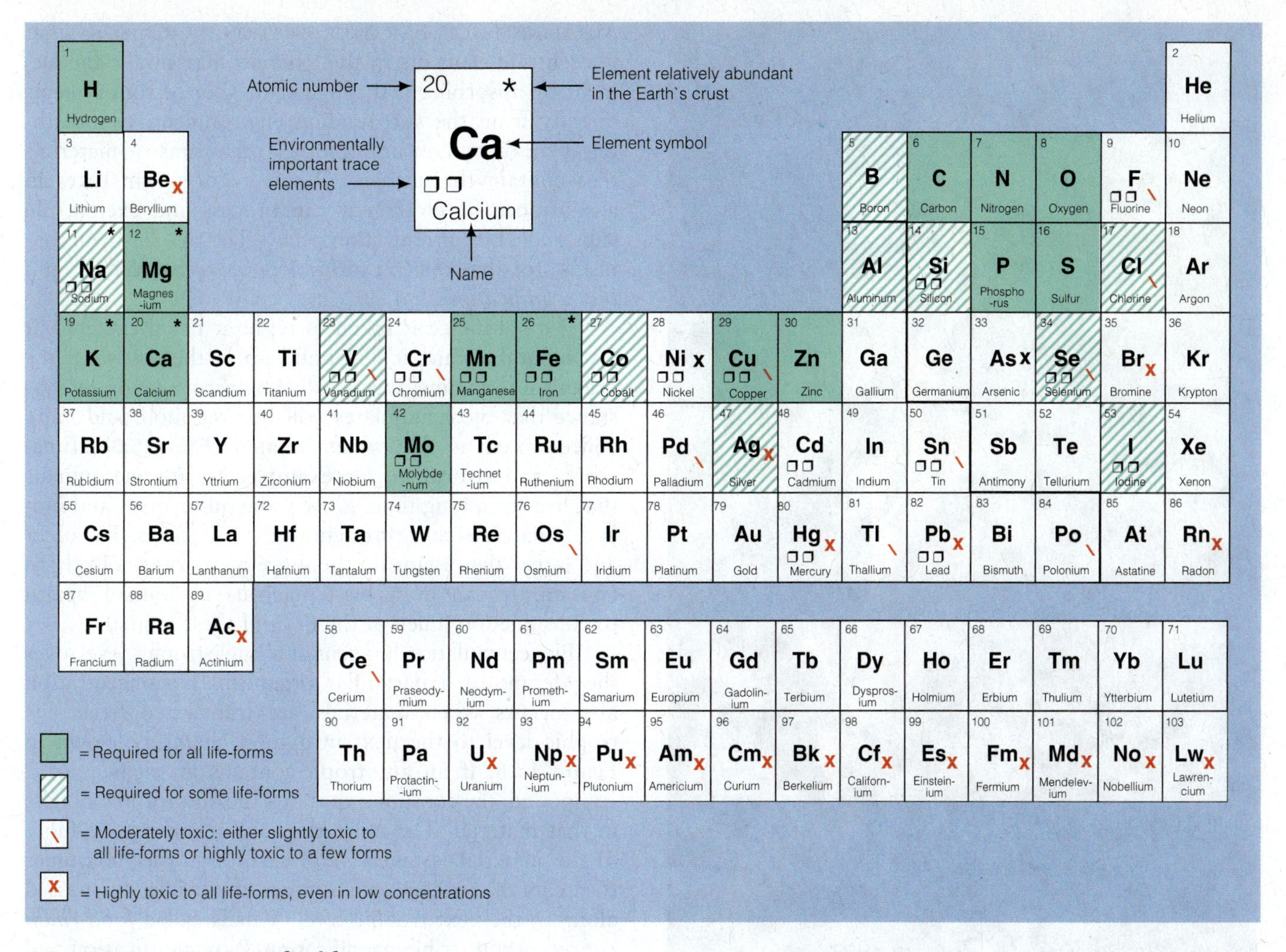

FIGURE 15.6 Nutrients for life
This version of the periodic table of the elements shows which elements are required for life and which are toxic to living things.

The idea that growth is limited not by the total amount of resources available but by the scarcest resource is sometimes referred to as *Liebig's Law*[1], or the *Law of the Minimum.*

Just as some elements are required for life, other elements are harmful, or *toxic*, to organisms. Some, like mercury, are toxic even in low concentrations. Others, like copper, are required in low concentrations but can be toxic if present in high concentrations. Every chemical element has a spectrum of possible effects on a particular organism; for example, selenium is required in small amounts by living things but may be toxic in cattle and wildlife when it is present in high concentration in soil. Paracelsus, a fifteenth-century physician who established the role of chemistry in medicine, wrote that, "Everything is poisonous, yet nothing is poisonous." By this he meant that any substance in too great amounts can be dangerous, yet anything in extremely small amounts can be relatively harmless.

Still other elements are neutral for life. Some of these are chemically inert elements, such the *noble gases* (helium, neon, etc.), which do not react with other elements. Others are present on Earth's surface in extremely low concentrations, and still others do not play any known role in life processes; gold and platinum are examples.

Biological Concentration of Elements

Organisms are selective in their uptake of chemical elements, allowing for the synthesis of highly specific compounds in their cells. This is an important and special ability of living things. As a result of this selectivity, chemical elements can become much more concentrated in organisms than they are in the local, surrounding environment. The amount by which an element is concentrated within an organism, compared to the concentration of the element in the surrounding medium, is called its **concentration factor**. For example, the concentration factor for iron in marine algae is about 100,000; that is, algae have an internal concentration of iron 100,000 times more than that found in ocean water (FIG. 15.7). Nitrogen, phosphorus, and manganese have concentration factors of 10,000 to 100,000 in marine phytoplankton, and zinc, nickel, copper, cadmium, and aluminum have concentration factors of 1000 to

[1] Justus von Liebig was a German chemist who made fundamental contributions to the development of organic chemistry in the 1800s.

FIGURE 15.7 Biological concentration
Marine algae like this example bioconcentrate chemicals by selectively taking them in from the environment.

10,000. Marine algae concentrate *all* elements from ocean water except chlorine and sodium (the two most abundant dissolved elements in ocean water), and fluorine and magnesium (whose concentration factor is about 1, or the same as the water).

Land plants also concentrate chemical elements, taking up the required elements through their roots. Concentration factors differ for different elements and plants. For example, the vegetable kale concentrates phosphorus by 1600 times compared to the level in a soil water solution, and potassium by 1100 times. Calcium and magnesium tend to have low concentration factors (less than 20), and some plants actively reject sodium. Some plants concentrate gold and have even been used as biochemical tracers to locate soils (and, by extension, underlying rocks) that contain ore-grade concentrations of gold. Plants that concentrate substances that are toxic to humans or animals can be used to extract those substances from contaminated soils; this is called *phytoremediation*.

MECHANISMS OF BIOCONCENTRATION. Organisms that concentrate elements in their tissues may intake the element by absorbing it through their skin or root systems directly from the surrounding environment; by breathing it in; or by consuming other organisms or materials that contain the element. When an organism takes in a substance faster than it can process and excrete the substance, **bioconcentration** occurs The two basic mechanisms for bioconcentration are *bioaccumulation* and *biomagnification*.

If a substance is taken in without being excreted at a comparable rate, it will build up in the tissues of the organism over time; this is called **bioaccumulation**. A substance that bioaccumulates will become more and more concentrated as the organism ages (FIG. 15.8). Tuna, shark, and swordfish, for example, are marine animals that live for a long time, grow to be quite large, and bioaccumulate mercury throughout their lifetimes. The older and larger the organism, the more mercury it is likely to contain. It can even be dangerous to human health to consume too much of the flesh of these animals.

Bioaccumulation happens at a single trophic level, over the lifetime of a particular organism. Bioconcentration also occurs when materials are transferred from one trophic level to the next in a food chain, as shown in Figure 15.8. If an autotroph contains an element or a compound, the consumer that eats the autotroph will take in that material. The more the consumer eats, the more of the material is taken in. If a second-order consumer then eats the first-order consumer, the material is passed along to the next trophic level. This is called *food-chain concentration*, or **biomagnification**. Biomagnification and bioaccumulation typically occur in tandem. The result is that organisms that are large, long-lived, and feed at the top of the food chain tend to end up with the highest concentrations of bioaccumulated and biomagnified materials in their tissues. This has implications

FIGURE 15.8 Biomagnification and bioaccumulation
Chemicals from the environment (including toxins) bioaccumulate in organisms over time, if they are taken in faster than they are excreted. Substances can also become concentrated through biomagnification, when consumers eat organisms from lower trophic levels that contain the material.

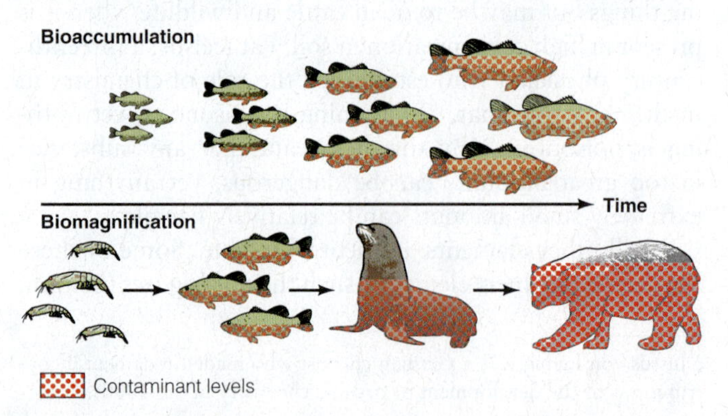

not only for the chemistry of the organism, but also for the chemistry of the surrounding environment and for human health (see *A Closer Look: The Human Body and Element Cycling*).

CONSEQUENCES FOR THE SURROUNDING ENVIRONMENT. Through uptakes of chemicals from the environment, living things change their own chemistry and the chemistry of the environment. In this way, life is an active chemical entity in any ecosystem. Bioconcentration happens locally, in individual organisms, but the combined impact means that organisms affect the concentration and distribution of elements in the global environment, through normal metabolic activity.

Organic matter has a fairly constant composition, and therefore the *ratios* (that is, the *relative concentrations*) of chemical elements in living things are fairly constant. For example, marine plankton have an average chemical makeup in which there are 106 atoms of carbon and 16 atoms of nitrogen for every atom of phosphorus. The ratio of carbon:nitrogen:phosphorus for marine plankton is therefore 106:16:1. This is called the *Redfield ratio* after Alfred Redfield, who first discovered this general property of living things. Through active uptake, plankton bring the concentration of carbon, nitrogen, and phosphorus into this same ratio in ocean water. If excess phosphorus is present, algae and bacteria add carbon through photosynthesis or fix additional nitrogen from the atmosphere; the organisms then take up the excess phosphorus. Only when the concentration of carbon:phosphorus ratio approaches 106:1 and the nitrogen:phosphorus ratio approaches 16:1 will the additional growth stop. In this way, marine organisms keep the chemical composition of ocean water essentially constant. The Redfield ratio is an important example of the mechanisms by which life affects the chemical environment.

Minimum Requirements of a Life-Supporting System

On the basis of this discussion of energy flow and material recycling, we can conceive of a system with the minimum requirements to support and sustain life. It is a system with a continually renewed source of energy and a sink for heat energy; in other words, energy passes through the system along a unidirectional pathway. Materials will be recycled again and again within the system, although it is difficult to imagine that the system would be leak-proof. Thus, this system could theoretically be closed but it is more likely an open system that allows energy to flow through freely, and recycles matter with some leakage across the system boundaries.

This minimum life-supporting system must have at least one autotroph (a primary producer) and one heterotroph (a decomposer), but will likely have a number of each, as well as other heterotrophs that feed on the autotrophs and on other heterotrophs. In addition to these organisms, there must be a fluid medium—a gas or a liquid or both—to facilitate the movement of chemical elements among the different kinds of organisms. And it must have a nonliving environment consisting at least of fluids—water and air—with mineral sources to provide inorganic chemical nutrients.

Looking back at the definition of an ecosystem—a system involving interactions among living organisms and interactions between organisms and the inorganic environment—we can see that what we have just described is a simple ecosystem. In other words, *the minimum system that can sustain life is an ecosystem*. Here we obtain a fundamental insight about life. Typically, we associate life with individual organisms, like ourselves, or a tree, a mushroom, or a one-celled bacterium because these are alive. But no individual is capable of life support. *The capacity to support life is a characteristic of ecosystems, not of individual organisms or populations.* Life can persist only in a system in which a number of kinds of organisms interact with each other and with the nonliving environment.

GLOBAL CYCLES OF LIFE

Because organisms evolve, the biosphere itself has changed through time and is still changing today. The Earth system as a whole has undergone changes, driven—at least in part—by life processes. One way these changes can be seen is in the cyclical transfer of chemical elements among Earth's four major reservoirs or subsystems—the geosphere, hydrosphere, atmosphere, and biosphere. The cycles of some elements have changed dramatically over Earth's history, but the most marked changes have been in the biologically important elements carbon, oxygen, nitrogen, and phosphorus. Elements that are essential to and strongly influenced by the biosphere, that are transformed from organic to inorganic forms, and that move among biological and nonbiological reservoirs are said to undergo *biogeochemical cycling*.

Basic Principles of Biogeochemical Cycling

In its most general form, a **biogeochemical cycle** is the complete pathway that a chemical element follows through the Earth system—from the geosphere to the biosphere, atmosphere, and hydrosphere, and back again. It is a *chemical* cycle because the cycling materials take various chemical forms and undergo chemical transformations. It is *bio-* because these are cycles that involve life processes and biological reservoirs. It is *geo-* because the cycles include rocks and soils. Although there are as many different biogeochemical cycles as there are elements, certain general concepts hold true for these cycles. The key that unifies biogeochemical cycles is the involvement of the four principal components of the Earth system: rock, air, water, and life.

Modeling Biogeochemical Cycles

The simplest way to view a biogeochemical cycle is as a *box model* (Chapter 1), in which the boxes represent the

A Closer LOOK

THE HUMAN BODY AND ELEMENT CYCLING

In 1956, doctors described curious symptoms in people who lived near Minamata Bay (FIG. C15.1). The symptoms were indicative of severe toxification of the central nervous system. It was soon discovered that patients afflicted with the disease were all fishers and their families, and that they regularly ate fish caught in Minamata Bay. Further study showed that mercury was involved and that a chemical plant on Minamata Bay, which produced basic chemicals for the plastics industry, used mercury salts in the processing and was the source of the problem. The amounts used were tiny, however, so the question then became, why did the people have so much mercury in their bodies? The answer turned out to be a case of food-chain concentration.

Organisms eliminate poisons at a rate that depends on the concentration of the poison in their tissues. If an organism receives a single dose M_0 of the poison, a graph of M (the amount of poison in the organism) against time looks like FIGURE C15.2A. M decreases in a regular manner, so that after 100 days one-half of the initial dose remains; after another 100 days M has again been halved, and so on. The halving period of 100 days is called the *half-life* of the poison in the organism (this term was introduced in Chapter 4).

Suppose now that instead of receiving a single dose, the organism has a diet in which a constant amount of poison is ingested every day. At the start, the organism's poison content, M, will rise fairly steeply. As it builds up, the rate of elimination increases. Eventually, a steady state is reached where the rate of elimination becomes equal to the rate of intake (FIG. C15.2B). Strictly speaking, the steady state is reached only after an infinite time, but it is near enough for practical purposes after about six half-lives, assuming that the poison has not killed the organism by then.

FIGURE C15.1 Minamata disease
This person suffers from a neurological syndrome called Minamata Disease, caused by extreme mercury poisoning incurred as a result of eating mercury-contaminated fish.

Imagine an ecosystem with the trophic levels and diets given by the first two columns (mass and daily diet of individual) in TABLE C15.1. Now suppose that a small amount of mercury is introduced into the system and is taken up by the algae, producing a mercury concentration in them of 10 parts per billion (1 part in 10^8). The daily diet of an herbivore, 10 grams of algae, contains 10×10^{-8} or 10^{-7} grams of mercury. At steady state, the herbivore will contain $10^{-7}/0.007$ or 1.4×10^{-5} grams mercury, and the mercury concentration in its tissues will be $1.4 \times 10^{-5}/100$, or 1.4×10^{-7} (140 parts per billion).

reservoirs where chemical elements are stored and the arrows represent the pathways of transfer (FIG. 15.9A, on page 462). A biogeochemical cycle is generally drawn for a single chemical element, but can be drawn for a compound; water (H_2O) is a common example. Figure 15.9B shows the basic elements of a biogeochemical cycle, using water as an example. Water is stored temporarily in clouds in the atmosphere (A). It moves to a lake (B) as precipitation or from land into the lake as runoff (C). It leaves the lake through evaporation to the atmosphere, or by runoff or subsurface flow.

We can view each individual reservoir in a biogeochemical cycle as an open system and the entire biogeochemical cycle as a set of open systems linked by the transfer of materials. Because Earth, as a whole, functions as an essentially closed system, these interconnected systems maintain a *mass balance*; that is, when material moves from one reservoir, it will show up in another one of the connected reservoirs. This is the material "budget" that is maintained by all global biogeochemical cycles.

A crucial aspect of any biogeochemical cycle, therefore, is the set of processes that controls the *flux*, that is, the rates of flow of substances from one reservoir to another. For example, in FIGURE 15.9B, on page 462, the rate of evaporation of water from the lake would be influenced by air temperature and wind velocity; the rate of runoff would be influenced by topography and the permeability of the soil; and the rate of precipitation would be influenced by moisture and temperature conditions in the atmosphere.

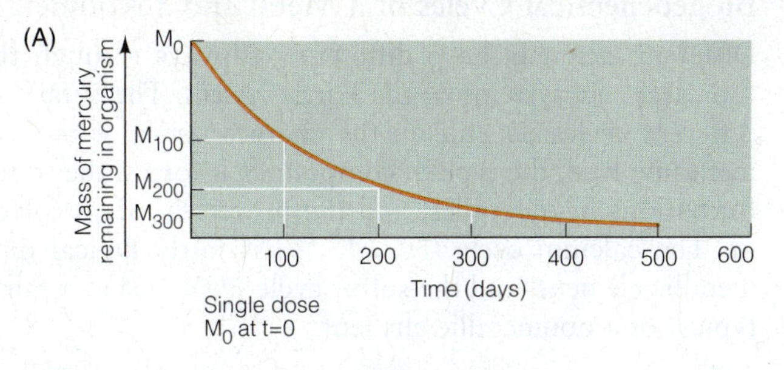

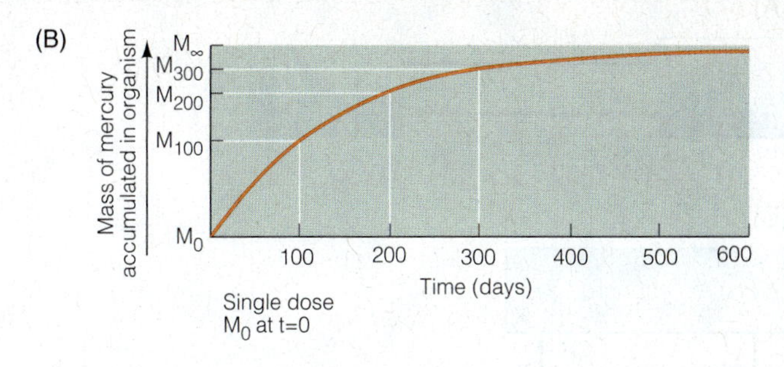

FIGURE C15.2 Mercury bioaccumulation
(A) This graph shows the decrease of mercury in the body of an organism that eliminated half of a single-exposure dose each 100 days. (B) This graph shows the buildup of mercury in the body of an organism that receives a steady intake of mercury each day. As the amount in the body builds up, the rate of elimination is increased. A balance is reached when the intake equals the elimination rate.

TABLE C15.1 Increase in Mercury Concentration Up the Food Web

Trophic Level	Mass of Individual	Daily Diet of Individual	Daily Mercury Intake	Steady-State Mercury Content	Mercury Concen./Tissues (parts per billion)
Algae	—	—	—	—	10
Herbivores	100 g	10 g algae	10^{-7} g	1.4×10^{-5} g	140
Primary carnivores	1000 g	1 herbivore	1.4×10^{-5} g	2×10^{-3} g	2000
Secondary carnivores	10,000 g	1 primary carnivores	2×10^{-3} g	0.29 g	29,000

We can now complete Table C15.1 by calculating the mercury concentrations in the species at successive trophic levels in this food web.

In moving three trophic levels up the food web from the algae, the mercury gets concentrated by a factor of nearly 3000 ($2.9 \times 10^{-5}/10^{-8}$), reaching a concentration of 29 parts per million in the secondary carnivores—many times the acceptable limit for foods in North America.

There are some weaknesses in this oversimplified model. For example, the half-life varies from one kind of tissue to another. Furthermore, it is unlikely that all of the prey species would survive the six half-lives necessary to reach the steady state; some would get killed and be eaten with lower concentrations of mercury in them. Finally, such an ecosystem might not be closed; fish that swam in after having fed elsewhere could be mercury-free. Nevertheless, in Minamata Bay top-level carnivores were found with as much as 50 ppm of mercury in them. A 60-kg person eating 200 grams of such fish a day would bioaccumulate 24 ppm of mercury in her tissue—enough to cause harm.

Another important aspect is the set of factors that control how long a particular substance will remain in a reservoir, its average *residence time*. Residence time is influenced by the properties of the substance itself and by the properties of the surrounding medium in the reservoir. By their basic characteristics, the four major reservoirs of the Earth system have different influences on average residence times. In general, the residence time of chemical elements is long in the geosphere, short in the atmosphere and biosphere, and intermediate in the hydrosphere.

One important factor that influences the residence time of a substance in a particular reservoir is **bioavailability,** which is a measure of how easily the substance can be absorbed and used by organisms. Bioavailability thus influences how readily a substance may be transferred into a biological reservoir. Elements that occur in forms that are not bioavailable must go through intermediate steps in transforming to a more bioavailable form, before they can be taken into biological reservoirs.

Make the CONNECTION

What is the role of volcanism in biogeochemical cycles?

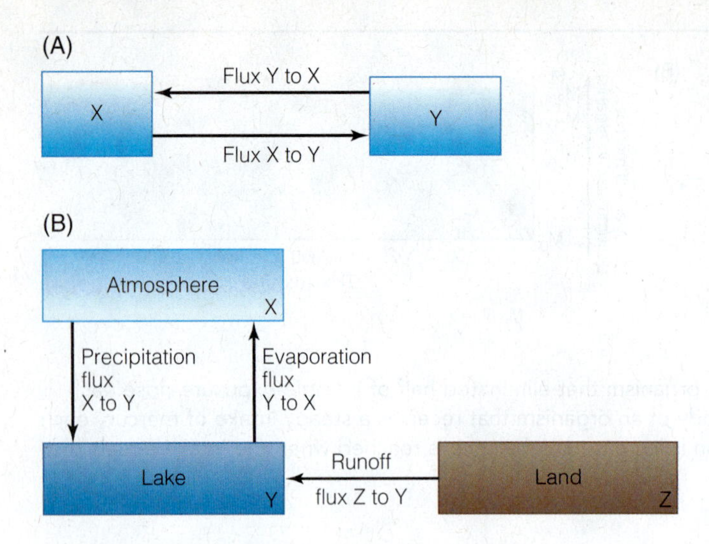

FIGURE 15.9 Biogeochemical box models
(A) Box models are a convenient way to represent the transfer of materials from reservoir to reservoir in biogeochemical cycles. (B) These are the basic elements of a biogeochemical cycle, using water as an example. X, Y, and Z are reservoirs, and the arrows represent transfer processes and fluxes of material between the reservoirs.

Biogeochemical Cycles of a Metal and a Nonmetal

Different elements have different pathways through the four great subsystems of the Earth system. The details of different cycles depend on the characteristics of the element involved, the type of compounds it forms, the transformations it undergoes, and the processes it is involved in. The calcium cycle (FIG. 15.10) is fairly typical of a metallic element, and the sulfur cycle (FIG. 15.11) is fairly typical of a nonmetallic element.

THE CALCIUM CYCLE. Like most metals, calcium does not form a gas and therefore has no major phase that occurs in the atmosphere; it cycles mainly between the geosphere, the hydrosphere, and the biosphere (Fig. 15.10). In the geosphere, its main reservoir, calcium, is a constituent of several common rock-forming minerals in igneous, metamorphic, and sedimentary rocks. Calcium ions (Ca^{++}) are released from minerals by weathering, typically becoming attached to water molecules. Water carries the calcium to the ocean, where it can reside for quite a long time (perhaps a million years).

In the ocean, calcium can occur as dissolved ions, or it can be precipitated in the form of inorganic minerals (such

FIGURE 15.10 Calcium cycle
This diagram shows the annual calcium cycle in a forest ecosystem. In the circles are the amounts transferred per unit time (the flux rates, in kilograms per hectare per year). The other numbers are the amounts stored (kilograms per hectare). Unlike sulfur, calcium does not have a gaseous phase. The information in this diagram was obtained from Hubbard Brook Ecosystem. (*Source:* G. E. Likens, F. H. Bormann, R. S. Pierce, J. S. Eaton, and N. M. Johnson, 1977, *The Biogeochemistry of a Forested Ecosystem*, Springer-Verlag, New York.)

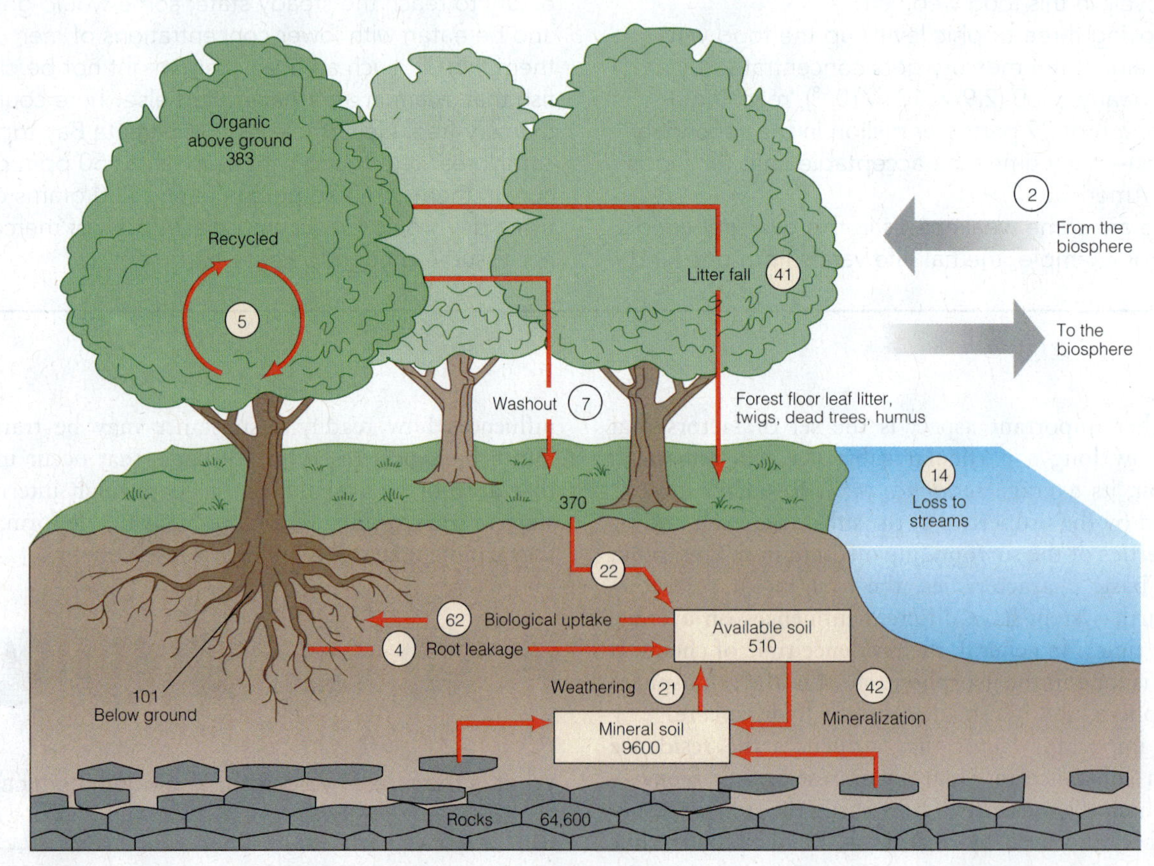

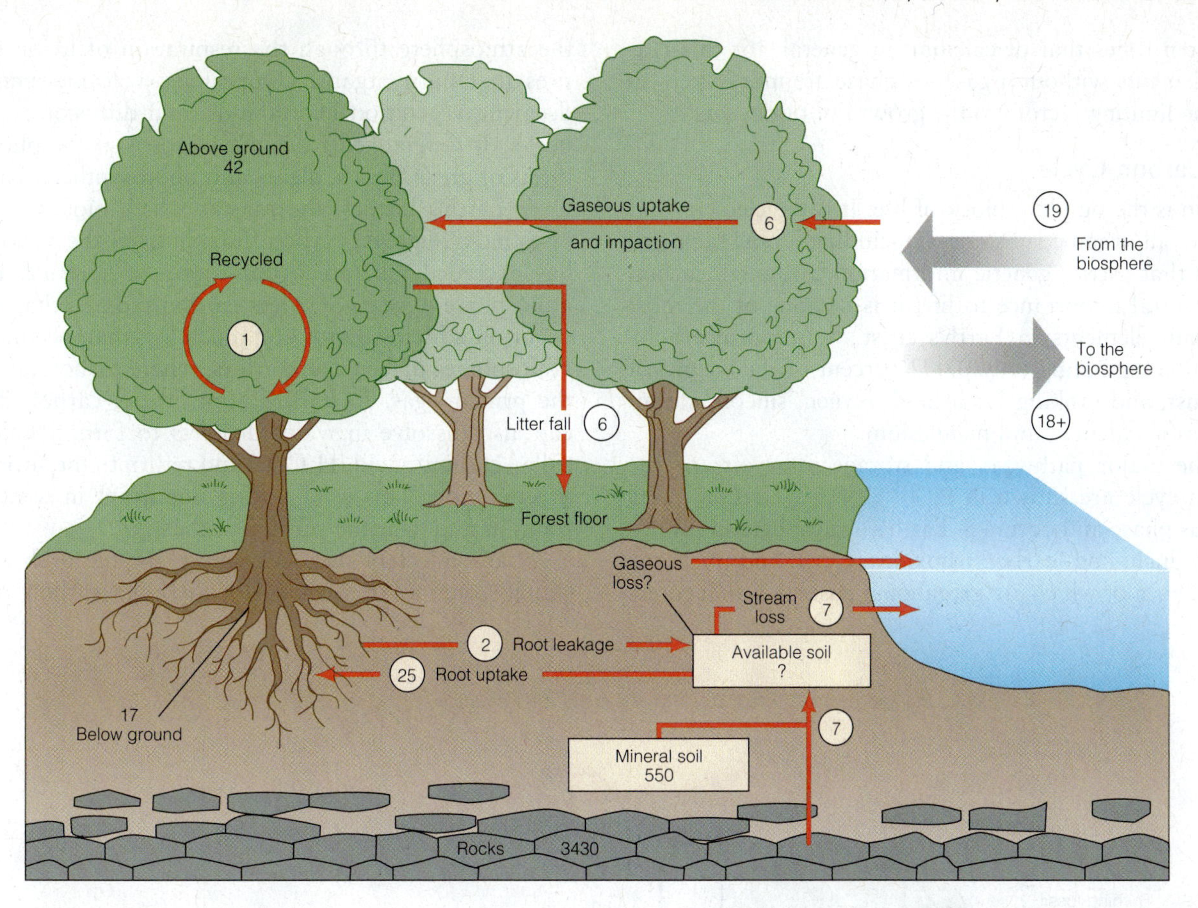

FIGURE 15.11 Sulfur cycle
This diagram shows the annual sulfur cycle in a forest ecosystem. The circles show the amounts transferred per unit time (the flux rates) (kilograms per hectare per year). The uncircled numbers are the amounts stored (kilograms per hectare). Sulfur has a gaseous phase as H_2S and SO_2. The diagram is based on studies of the Hubbard Brook Ecosystem. (*Source:* G. E. Likens, F. H. Bormann, R. S. Pierce, J. S. Eaton, and N. M. Johnson, 1977, *The Biogeochemistry of a Forested Ecosystem*, Springer-Verlag, New York.)

as gypsum, a calcium sulfate mineral). Calcium in the ocean also can be precipitated biologically through a process called *biomineralization*, in which marine organisms utilize calcium carbonate ($CaCO_3$) to build their shells and skeletons. When these organisms die, their shells collect in bottom sediments, some of which eventually become transformed into limestone. This provides a mechanism for sequestering both calcium and carbon dioxide.

From the ocean, the principal pathway for calcium back to land is via the geological processes of plate tectonics and continental uplift. These processes take a very long time to operate; that is why limestone makes such an effective long-term reservoir. A relatively small amount of calcium also returns to land by the harvesting of fish and shelled organisms by animals and humans. On land, calcium from weathered rocks is an important constituent of soil. From there, it is taken up by plants and then passed along to animals that eat the plants; humans, in turn, receive calcium through both plant and animal sources, such as milk. Calcium is an important component of many biological structures on land, including bones, teeth, and eggshells. It is also fundamentally important in many cell processes.

The Sulfur Cycle. Most of the sulfur in the Earth system is in long-term reservoirs in the geosphere, either in rocks or buried in seafloor sediments (Fig. 15.11). A significant difference between sulfur and calcium is that sulfur occurs in its elemental state and in a range of oxidation states; this means that sulfur can form a wide variety of both reduced and oxidized compounds, including a wide range of compounds of biological origin. Another significant difference is that sulfur forms several gaseous compounds, including sulfur dioxide (an inorganic, oxidized compound that is a major air pollutant and a component of acid rain) and hydrogen sulfide (or "rotten egg" gas, a reduced compound that is usually produced biogenically). Sulfur is released to the atmosphere from Earth's interior through volcanic emissions.

In the atmosphere, sulfur plays several important roles; for example, sulfate aerosols whiten the atmosphere, contributing to increased planetary *albedo* (reflectivity). Plants require sulfur for several fundamental processes; they can absorb the required sulfur directly from the atmosphere through their leaves (as sulfur dioxide, or from soil water through their roots (as dissolved sulfate). Animals, in turn, acquire the sulfur they need by eating plants. Because sulfur has gaseous forms, it can be returned to the biosphere via the atmosphere much more rapidly than can calcium; the annual input of sulfur from the atmosphere to a forest ecosystem has been measured to be ten times that of calcium. In general, for this reason, elements without a gaseous phase are more likely to become limiting factors to the growth of organisms.

The Carbon Cycle

Carbon is the building block of life. It is the element that anchors all organic substances, including DNA, the compound that carries genetic information. Although carbon is of central importance to life, it is not one of the most abundant elements in Earth's crust. It is fourteenth by weight, contributing only 0.032 percent of the weight of the crust, and ranking far behind oxygen, silicon, aluminum, iron, calcium, and magnesium.

The major pathways and storage reservoirs of the carbon cycle are shown in FIGURE 15.12. Carbon has a gaseous phase and occurs in Earth's atmosphere in several forms, including carbon dioxide (CO_2) and methane (CH_4), both of which are greenhouse gases. Carbon enters the atmosphere through the respiration of living things; fires that burn organic compounds; volcanic emissions; decaying vegetation in wetlands; and diffusion from the ocean. It is removed from the atmosphere by photosynthesis of green plants, algae, and photosynthetic bacteria. Over Earth's history, the rate at which biological processes have removed carbon dioxide from the atmosphere has exceeded the rate of addition. As a result, Earth's atmosphere now has far less carbon dioxide than would occur on a lifeless Earth and much less than occurs in the atmospheres of Venus and Mars, where carbon dioxide is the primary gas. In Earth's atmosphere, carbon dioxide can also dissolve in water droplets to form a mild acid called carbonic acid (H_2CO_3) and return to the surface via precipitation. This mild acid is important in weathering rocks at and near the surface of the land.

Carbon enters the ocean from the atmosphere as a simple solution of carbon dioxide. The carbon dioxide

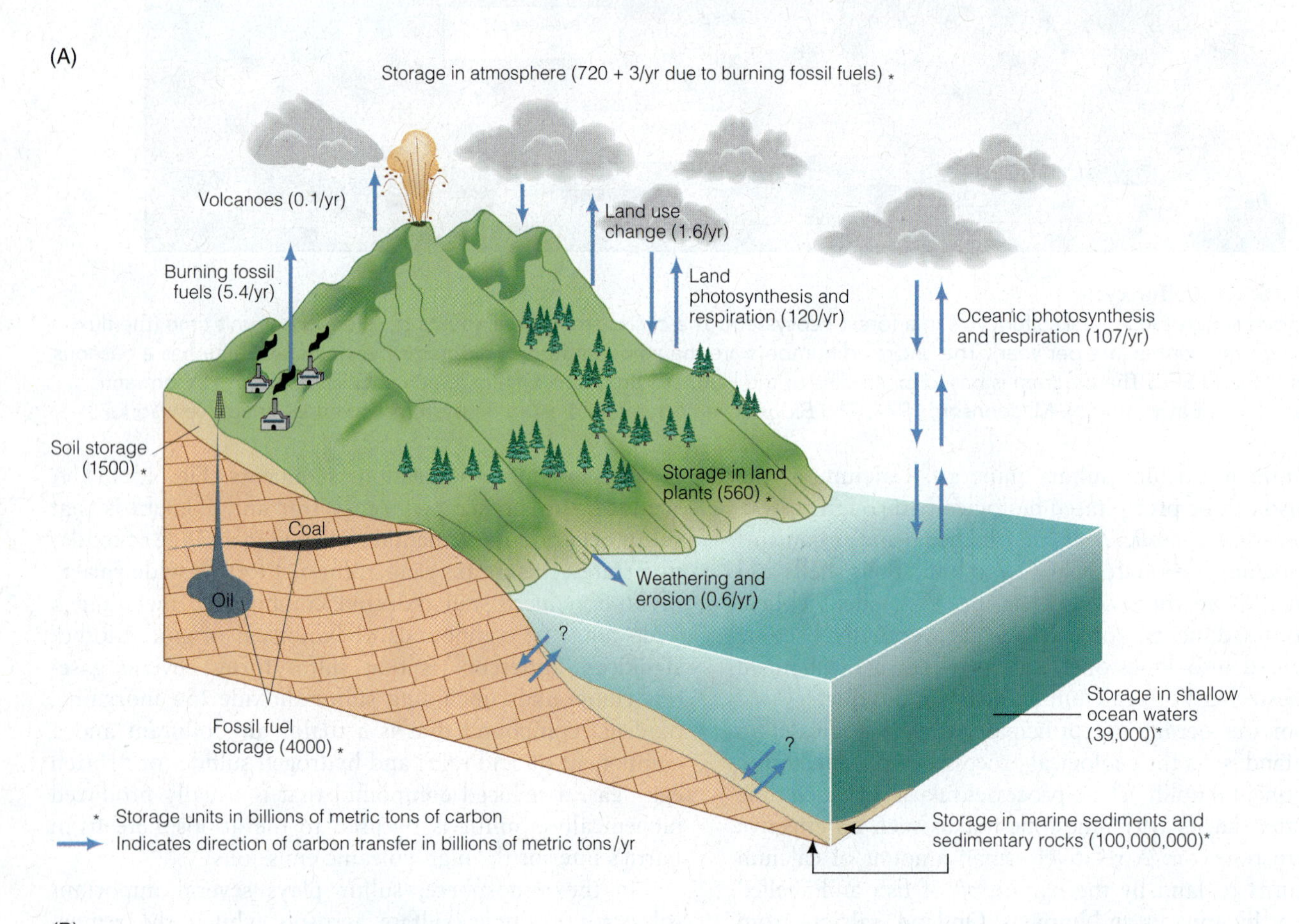

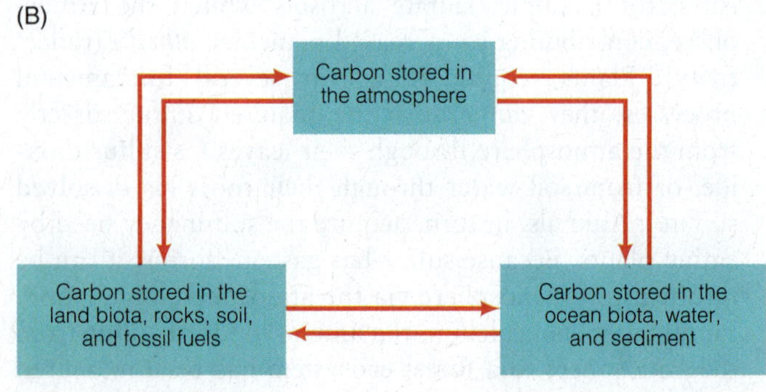

FIGURE 15.12 Carbon cycle
(A) This is a generalized version of the global carbon cycle. (B) Parts of the carbon cycle are simplified in a box model to illustrate the cyclic nature of the movement of carbon. (*Source:* Modified after G. Lambert, 1987, *La Recherche*, 18, pp. 782–783, with some data from R. Houghton, 1993, *Bulletin of the Ecological Society of America*, 74(4), pp. 355–356.)

dissolves and is converted to carbonate $(CO_3)^{2-}$ and bicarbonate $(HCO_3)^-$ ions. Marine algae and photosynthetic bacteria obtain the carbon dioxide they use from the water, in one of these forms. Carbon is also transferred from the land to the ocean by rivers and streams, in the form of dissolved carbon and particles of organic matter, and winds blow small organic particles from the land to the ocean. Locally and regionally the flux carried by rivers and streams is of great importance, influencing near-shore areas that are often highly productive biologically.

When an organism dies, most of its organic material decomposes into inorganic compounds, including carbon dioxide. Some of the organic matter may be buried, where there is no oxygen to make conversion possible or where the temperatures are too cold for decomposition. In this situation, the organic matter will be stored. Over years, decades, and centuries, such storage occurs in wetlands including parts of floodplains, lake basins, bogs, swamps, deep-sea sediments, and near polar regions. Over longer periods, some of this material is buried under other sediments, becomes incorporated in sedimentary rocks, and is eventually transformed into *fossil fuels* (Chapter 18). Large amounts of carbon are also sequestered (along with calcium) in the form of limestone, produced primarily by biomineralization in the shells of marine organisms. In fact, nearly all of the carbon stored in Earth's crust exists as sedimentary rocks. Carbon also occurs in a few inorganic forms, including graphite and diamond.

The carbon cycle can be understood in terms of timescales. Short-term cycling (up to tens of years) is dominated by biological processes, including photosynthesis by individual plants. Medium-term cycling (up to thousands of years) is dominated by the storage of organic chemicals in the woody tissue of trees, in forest soils, and in other organic sediments. Long-term cycling (up to millions of years) is dominated by geologic processes, and involves the production of carbonate rocks and the eventual return of CO_2 to the atmosphere via weathering, metamorphism, subduction, and volcanism.

Because carbon is the most important organic compound and because it forms two of the most important greenhouse gases, much research has been devoted to understanding the carbon cycle. However, at a global level some key issues remain unanswered. For example, scientists estimate that as much as 2 billion tonnes of carbon per year are unaccounted for in the global carbon budget. As you can see in Figure 15.12, human activities release roughly 7 billion tonnes of carbon per year, primarily through the burning of fossil fuels and through land-use changes involving deforestation. (Deforestation leads to the decomposition of trees and soils, thus converting organic carbon to inorganic carbon dioxide.) Of this amount, about 3 billion tonnes remain in the atmosphere and 2 billion tonnes are absorbed into the ocean. This leaves about 2 billion tonnes of carbon unaccounted for. Inorganic processes cannot account for the fate of this "missing carbon." Either marine or land photosynthesis, or both, must provide the additional flux, transferring carbon to terrestrial biosphere reservoirs (forests and soils) or aquatic reservoirs (the ocean) for storage. Scientists do not yet fully understand which processes dominate, or in what regions of Earth this missing flux occurs. The "missing carbon" problem illustrates the complexity of biogeochemical cycles, especially those where the biota play an important role, and particularly the carbon cycle.

The Nitrogen Cycle

Nitrogen has one of the most important and most complex biogeochemical cycles (**FIG. 15.13A**). Oxygen is life's gift to itself, but nitrogen is the atmosphere's main gift to life. Nitrogen is a relatively unreactive gas that seems to have been a minor constituent of the initial atmospheres of all the planets in our solar system. It is essential to life because it is a necessary component of all proteins, including DNA. The key to the nitrogen cycle is in understanding how the reduction and oxidation of nitrogen take place and how nitrogen moves between the four major reservoirs—the atmosphere, the ocean, the geosphere, and the biosphere (**FIG. 15.13B**).

Free nitrogen (N_2 uncombined with any other element) makes up approximately 78 percent of Earth's atmosphere by volume. However, except for a few bacteria, organisms cannot use free nitrogen directly; in other words, nitrogen in its common form in the atmosphere is not bioavailable. Plants, algae, and bacteria can take up nitrogen—a crucial plant nutrient—only in the form of the nitrate ion $(NO_3)^{2-}$ or the ammonium ion $(NH_4)^+$. Animals, in turn, can only intake nitrogen in the form of organic compounds made by primary producers.

However, nitrogen is a relatively unreactive element, and few processes convert molecular nitrogen into one of these useful componds. In nature, about 90 percent of the conversion of molecular nitrogen into biologically useful forms is conducted by bacteria; most of the rest is caused by lightning, which oxidizes atmospheric N_2. The process of converting N_2 into NO_3 is called **nitrogen fixation**. Some bacteria also can convert molecular nitrogen to the ammonium ion. Once in either of these forms, nitrate or ammonium, nitrogen can be taken up on the land by plants and in the ocean by algae. Bacteria, plants, and algae then convert these inorganic nitrogen compounds into organic ones, and the nitrogen becomes bioavailable, through food chains, in the form of organic compounds. When organisms die, other bacteria convert these organic compounds back into nitrate, ammonia, and eventually (by a series of chemical reactions) molecular nitrogen. In that form, it can be returned to the atmosphere. The process of converting fixed nitrogen back to molecular nitrogen and then releasing it to the atmosphere is called **denitrification**.

It follows that ultimately all other terrestrial organisms depend on nitrogen-converting bacteria. Some organisms have evolved highly specific, mutually dependent or *symbiotic* (Chapter 16) relationships with these bacteria. For example, plants of the clover and pea families have nodules in their roots that provide a habitat for such bacteria (Fig. 15.13C). The bacteria

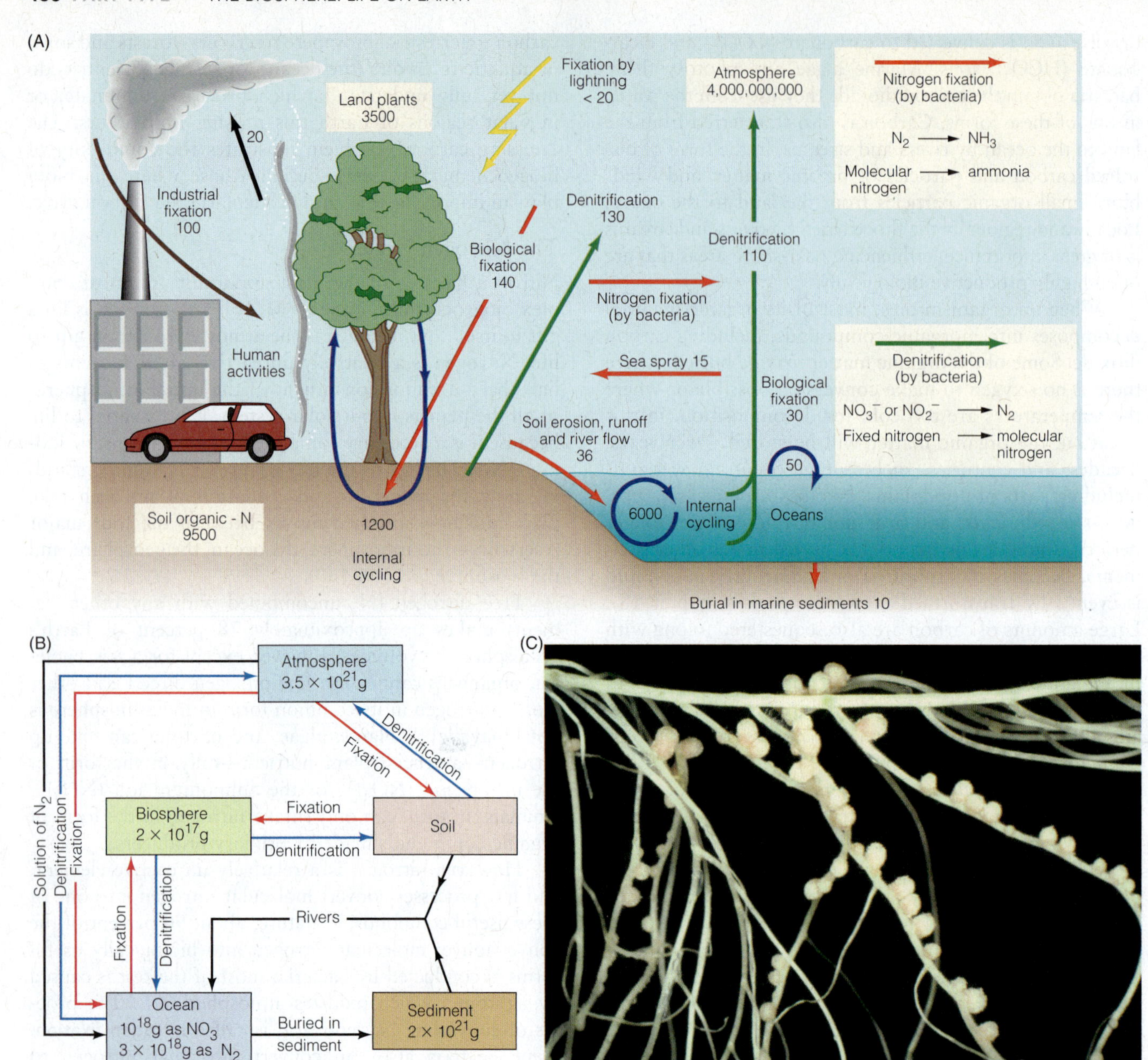

FIGURE 15.13 Nitrogen cycle
(A) This is a box model showing the basic processes in the global nitrogen cycle. (B) Some details of the global nitrogen cycle are shown here. Pools or reservoirs for nitrogen and their contents are shown in boxes; transfer processes are shown as arrows, with annual fluxes in 10^{12} g N_2. Note that industrial fixation of nitrogen is nearly equal to global biological fixation. (*Source:* Data from R. Söderlund and T. Rosswall, 1982, in O. Hutzinger (ed.), *The Handbook of Environmental Chemistry,* Vol. 1, Pt. B, Springer-Verlag, New York.) (C) These root nodules on white clover are produced by colonies of nitrogen-fixing bacteria.

obtain organic compounds for food from the plants, and the plants obtain usable nitrogen. Such plants can grow in otherwise nitrogen-poor environments. When these plants die, they contribute relatively nitrogen-rich organic matter to the soil, thereby improving the fertility of the soil. Alder trees also have nitrogen-fixing bacteria in their roots. These trees grow along streams, and their nitrogen-rich leaves fall into the streams and increase the supply of the element in a biologically usable form to freshwater animals.

The nitrogen cycle is interesting because of its complexity and because parts of the cycle have had to evolve as the atmosphere became oxygenated. Because organisms cannot use N_2 directly, either some reduced nitrogen must have been available when life arose or the earliest organisms had the ability to reduce N_2. Anaerobic nitrogen-fixing bacteria are ancient, and the fixation chemistry that evolved with them will not work in the presence of oxygen. Such bacteria must have evolved before the atmosphere contained oxygen. Today these bacteria live only in oxygen-free environments. A few nitrogen-fixing bacteria have developed an oxygen tolerance, even though they still use the old, anaerobic fixation chemistry. They perform this trick by making sure that the sites in their cells where fixation occurs are carefully guarded from oxygen. As the oxygen content of the atmosphere increased, the amount of nitrate that rained into the soil also increased. This opened new niches that were soon occupied by organisms that learned to reduce NO_3 to NH_3. Many of the higher plants have this ability.

Once reduced, nitrogen tends to stay reduced, remain in the biosphere, and be either reused by other organisms or changed back into N_2 and returned to the atmosphere. The main route by which nitrogen returns to the atmosphere is the reduction of nitrate. This route is kept open by bacteria that use the oxygen in nitrate to oxidize carbon compounds during metabolism. Denitrifying bacteria must therefore have evolved quite late in the history of the biosphere, after oxygen started to accumulate in the atmosphere. Thus, the simple nitrogen cycle of the early Earth has evolved into today's complex cycle in response to a changing atmosphere.

The Phosphorus Cycle

Phosphorus plays two essential roles in the biosphere: in the form of sugar-phosphate units, phosphorus forms the helical framework of the DNA molecule, and it facilitates all of life's energy transactions. It commonly occurs in its oxidized state as phosphate, which combines with calcium, potassium, magnesium, and iron to form minerals found in soils and in waters. Like calcium, phosphorus lacks a gaseous phase and exists in the atmosphere only in very small amounts in the form of dust particles. This lack of a gaseous phase exerts a major influence on the behavior of phosphorus in global biogeochemical cycling (FIG. 15.14).

Phosphorus in rocks on the continents is slowly eroded, used temporarily by life on the land, and then washed to the ocean via streams and rivers. In the ocean, phosphorus is temporarily available to plankton but is eventually deposited in the deep ocean or in marine sediments. There is no short-term nonbiological return of phosphorus from the ocean to the land; without life, the return of phosphorus could take place only through

FIGURE 15.14 Phosphorus cycle
Phosphorus is recycled to soil and land biota by geologic processes that uplift the land and erode rocks, by birds that produce guano, and by human beings. Although Earth's crust contains a very large amount of phosphorus, only a small fraction of it can be mined by conventional techniques. Phosphorus is therefore one of our most precious resources. Values of the amount of phosphorus stored or in flux are compiled from various sources. Estimates are approximate to the order of magnitude. (*Source:* Based primarily on C. C Delwiche and G. E. Likens, 1977, "Biological Response to Fossil Fuel Combustion Products," in W. Stumm, ed., *Global Chemical Cycles and Their Alterations by Man,* Abakon Verlagsgesellschaft, Berlin, pp. 73–88, and U. Pierrou, 1976, "The Global Phosphorus Cycle," in B. H. Svensson and R. Soderlund (eds.), "Nitrogen, Phosphorus and Sulfur—Global Cycles," *Ecological Bulletin,* Stockholm, pp. 75–88.)

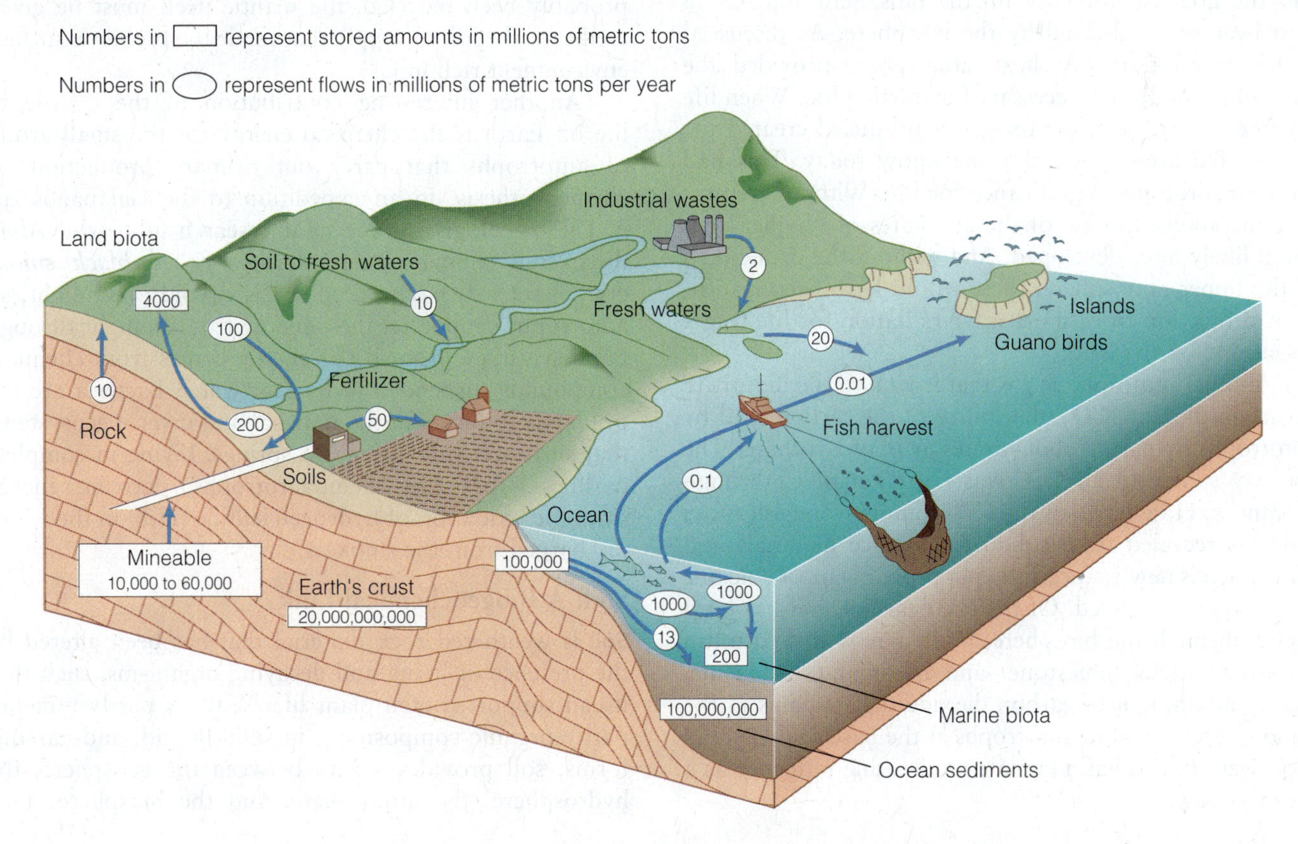

the very long-term geological process of the uplift of continents. Phosphorus also tends to form compounds that are relatively insoluble in water and therefore are not readily eroded as part of the hydrologic cycle. As a result, the rate of transfer of phosphorus tends to be slow in comparison to carbon and nitrogen.

Phosphorus is often a limiting element for plant and algal growth, which is typical of nutrients that do not occur in gaseous forms. Phosphorus enters the biosphere through uptake by plants, algae, and bacteria—autotrophic organisms. Plants can take up phosphorus in its oxidized form, as phosphate (PO_3^{-4}), a common ion. Phosphorus slowly becomes available to plants in this form through the weathering of rocks or rock particles in the soil. In a relatively stable ecosystem, much of the phosphorus that is taken up by vegetation is returned to the soil. Some, however, is lost to wind and water erosion, transported out of the soil in a water-soluble form and from there via rivers and streams to the ocean.

Biogeochemical Links among the Spheres

Based on the discussion in this chapter, we can abstract some of the principal ways that life affects each of the other major components of the Earth system through biogeochemical cycling.

The Atmosphere

We have seen that elements that occur in gaseous phases are much more readily transferred to biospheric reservoirs than elements that have no gaseous phase. Not surprisingly, among Earth's spheres the atmosphere has had the greatest influence on the biosphere and has in turn been most affected by the biosphere. As discussed in Chapter 14, the Archean atmosphere provided the anaerobic conditions necessary for starting life. When life became autotrophic, the oxygen it produced created the oxygenated atmosphere that we know today. This had the most profound significance for life. Without respiratory metabolism, none of the attributes of "higher" life could likely have developed. And without the ozone (O_3) of the upper atmosphere, there would have been insufficient protection from short-wave radiation for life to get a foothold on the land.

Another atmospheric gas that is vital to the biosphere is carbon dioxide, CO_2, the source of the carbon used by autotrophs to make carbohydrates by photosynthesis. The primary source of all CO_2 was Earth's mantle, through volcanoes. How much of the CO_2 coming from volcanoes today is recycled from subducted surface materials and how much is new is uncertain, but probably most volcanic CO_2 today is recycled. Of the total carbon that has been cycled through the biosphere, most is now locked up in carbonate rocks (limestone) and in organic sediments. The small amount of carbon dioxide that remains in the atmosphere is vital to autotrophs at the base of the global food web. It also has a climate-moderating influence as a greenhouse gas.

The Hydrosphere

Water is indispensable for the biosphere. As a chemical compound, water has many unique properties that are of fundamental importance to life (Chapter 3). The hydrosphere provides the medium for life in aquatic ecosystems. Globally, water either directly participates in or facilitates most environmental and biological interactions. Most of the gas exchange between aquatic ecosystems and the atmosphere is mediated by water. Water also provides the transfer medium for nutrients to move from soils into plant roots. Essential elements are transferred from the land (soil) to aquatic ecosystems by flowing water, either carried in solution or in suspension. In shallow aquatic environments (rivers, most lakes, and the continental shelves), essential material resources are recycled fairly rapidly; in the deep-sea environment, the cycles are longer and slower.

The Geosphere

Ultimately, the geosphere contributes all of the elements necessary to support life. The most important growth-limiting element it contributes is phosphorus, which is released from the crust by weathering. Since the growth of most ecosystems is limited by phosphorus availability, weathering is an important regulator of total biomass. The geosphere gives up its phosphorus rather grudgingly from relatively insoluble minerals such as apatite (calcium phosphate).

The geosphere, specifically the mantle, is also the source (through volcanism) of the carbon (as CO_2) that is the starting point for synthesis of organic compounds in the biosphere. Although today most of this carbon has probably been recycled, the mantle itself must be given credit for the initial supply that made the prebiotic surface environment rich in CO_2.

Another interesting contribution of the mantle to life on Earth is the chemical energy for the small group of autotrophs that carry out primary production by chemosynthesis. In an expedition to the Galápagos rift in 1977, geologists in the small research submarine *Alvin* discovered submarine hot springs called *black smokers* (Chapter 14). The deep sea is devoid of sunlight, so autotrophs cannot get their energy from sunlight through photosynthesis. Instead the energy comes from chemical compounds suspended in the hot water. Bacteria are the primary producers in the unusual and diverse ecosystems that surround submarine hot springs. Living in complete darkness, they are chemoautotrophs, deriving energy from the oxidation of hydrogen sulfide (H_2S) in the water discharged from the smokers.

Soil: A Biogeochemical Link

Soil is weathered rock material that has been altered by the presence of living and decaying organisms, such that it can support rooted plant life. With its partly mineral, partly organic composition, in solid, liquid, and gaseous forms, soil provides a link between the geosphere, the hydrosphere, the atmosphere, and the biosphere. Like

FIGURE 15.15 Soil: one of Earth's unique features
These microscopic views show Earth's soil (A) compared to regolith from the Moon (B). Earth's soil contains organic material and hydrous minerals such as clay, while lunar regolith contains neither.

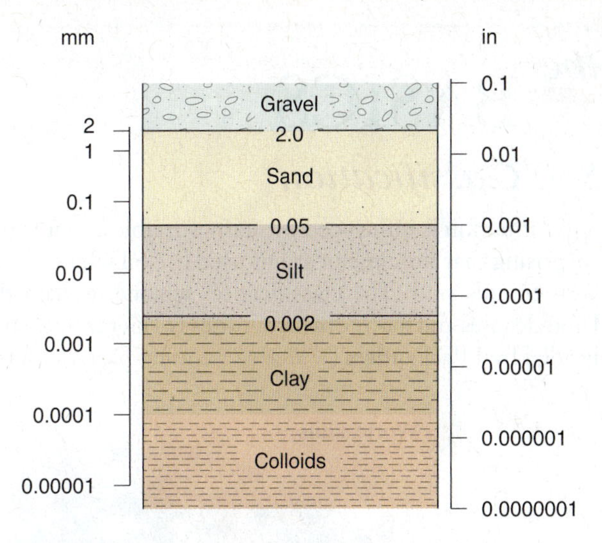

FIGURE 15.16 Soil particle sizes
Mineral particle sizes in soils are named sand, silt, and clay (which includes colloids). Gravel is typically not included when discussing soil texture. Size grades are defined using the metric system, and each unit on the scale represents a power of 10. English equivalents are also shown.

liquid water, soil is one of the features that make Earth a unique planet (FIG. 15.15). Soil is the base of almost all terrestrial ecosystems. The inorganic components of soil supply habitat to the organisms of the soil; together, they constitute an ecosystem of many species and great complexity. In addition, the inorganic constituents of soil act as a substrate, together with water, nutrients, and essential trace elements, for the larger ecosystem (forest, prairie, or whatever it may be).

SOIL PROPERTIES AND FORMATION. If you examine soil with a microscope or magnifying glass, you will see that it contains a variety of materials, including tiny fragments of rocks and minerals (FIG. 15.16). The smallest are *colloids* (Chapter 3), particles that are extremely fine (less than 0.0002 mm) and easily suspended in water. Next (and discussed in Chapter 7) are *clay* particles, tiny mineral fragments less than 0.002 mm in diameter. *Silt* particles are slightly larger (0.002 to 0.005 mm), and *sand* particles are the largest and most easily visible (0.05 to 2 mm). Larger rock or *gravel* fragments may also be present. Soil also contains **humus**, partially decayed organic matter, and a variety of small living organisms such as worms, spiders, mites, and microorganisms such as bacteria and fungi.

The relative proportions of these constituents, called **soil texture**, determine many of the properties of the soil. For example, *loam* is an organic-rich soil that is particularly well suited for growing plants. In addition to humus, the inorganic portion of loam consists of about 40 percent sand, 40 percent silt, and 20 percent clay particles. These constituents give the soil just the right texture to be able to hold moisture and supply nutrients to root systems. A soil that is richer in one or another of these constituents will have different physical, chemical, and biological properties.

Other soil properties that are important include the color of the soil, which is indicative of its chemistry; for example, a red soil likely contains abundant iron, whereas a dark soil likely contains abundant organic matter. **Soil structure** is another important property; it is a measure of the "clumpiness" of the soil. Other properties that are important in different contexts include the electrical resistivity, thermal properties, water content, and a variety of properties that determine the engineering behavior of the soil, such as its shear strength. Soils are classified on the basis of their properties, as well as the processes of formation, the environment in which they form, and the presence or absence of the different kinds of *soil horizons* (see below). There are about as many variations of soil as there are places on this planet, which makes them very complicated to classify.

It can take many thousands of years for soil to form from bare rock. The rate of soil formation depends on a variety of factors. Most important is the rate of weathering (Chapter 7), which, in turn, is influenced by the

the BASICS

Soil Classification

Soil formation is influenced by many variables, so it is not surprising that the classifications used by soil scientists are very complicated. The classification scheme used in the United States and many other countries is a hierarchical one headed by 11 **soil orders** (TABLE B15.1 and FIG. B15.2), the highest classification rank for soils. Each order is distinguished by specific characteristics, such as the presence or absence of well-developed horizons, accumulations of certain minerals, distinctive color, or high acidity. The orders are divided into suborders and then, in increasing detail, into groups, subgroups, families, series, and types. At least 17,000 types of soils are defined, and each of them has a name.

FIGURE B15.2 **World soils**

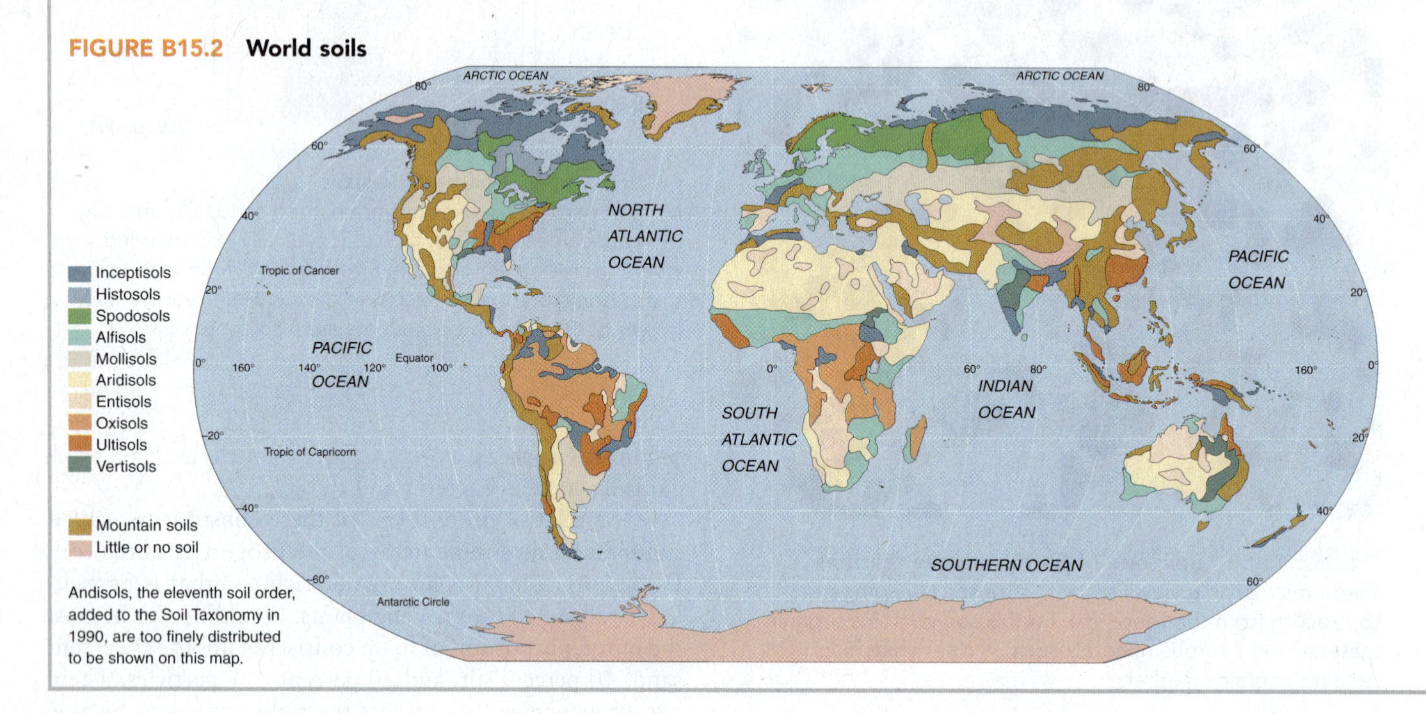

Andisols, the eleventh soil order, added to the Soil Taxonomy in 1990, are too finely distributed to be shown on this map.

composition of the rock from which the soil forms, called the **parent rock.** High temperatures and abundant rainfall promote rapid chemical weathering by enhancing the rate of chemical breakdown of the minerals. Weathering is more intense and extends to greater depths in warm, wet tropical climates than in cold, dry, arctic and alpine climates. In cold, dry regions and on high mountain peaks, weathering is shallow and proceeds slowly by mechanical weathering processes, such as cracking and breaking. Steep slopes also weather more rapidly than shallow, vegetated slopes.

As the parent rock weathers, soil gradually develops downward from the surface. A fully developed soil consists of a succession of roughly horizontal layers, or **soil horizons** (FIG. 15.17). Each horizon has distinct characteristics of color, texture, chemistry, organic content, and water content. The entire sequence of horizons from the surface to the underlying bedrock constitutes the **soil profile.** Soil profiles vary considerably from one location to another. The uppermost horizon in many soil profiles is an accumulation of decaying organic matter, commonly underlain by a dark, humus-rich horizon; these horizons together constitute **topsoil.** Any of a variety of mineral-rich horizons may underlie the topsoil, determined by the chemistry of the soil, the climate, and other physical, chemical, and biological conditions in which the soil formed. The deepest horizon typically contains large fragments of the original bedrock from which the soil formed.

Our definition of *soil* mentions "the presence of living organisms"; indeed, this is the main factor that distinguishes soil from sediment. Plants, animals, and a multitude of microorganisms reside in soil, and their importance in soil formation cannot be overstated. Plants are the main source of the organic matter in soil. Microorganisms such as bacteria and fungi break the organic matter down to humus, and animals such as worms, mice, and moles burrow in the soil, mixing the components, and providing passageways for water and air to enter.

SOIL FERTILITY. The soil of the continents serves as the base from which nutrition in all the terrestrial ecosystems (including those that support people) is derived. The availability of *arable* soils—those suited for growing

TABLE B15.1 Soil orders*

Group I Soils with well developed borizons or with fully weathered minerals, resulting from long-continued adjustment to prevailing soil temperature and soil-water conditions.	
Oxisols	Very old, highly weathered soils of low latitudes, with a subsurface horizon of accumulation of mineral oxides and very low base status.
Ultisols	Soils of equatorial, tropical, and subtropical latitude zones, with a subsurface horizon of clay accumulation and low base status.
Vertisols	Soils of subtropical and tropical zones with high clay content and high base status. Vertisols develop deep, wide cracks when dry, and the soil blocks formed by cracking move with respect to each other.
Alfisols	Soils of humid and subhumid climates with a subsurface horizon of clay accumulation and high base status. Alfisols range from equatorial to subarctic latitude zones.
Spodosols	Soils of cold, moist climates, with a well-developed B horizon of illuviation and low base status.
Mollisols	Soils of semiarid and subhumid midlatitude grasslands, with a dark, human-rich epipedon and very high base status.
Aridisols	Soils of dry climates, low in organic matter, and often having subsurface horizons of accumulation of carbonate minerals or soluble salts.
Group II Soils with a large proportion of organic matter.	
Histosols	Soils with a thick upper layer very rich in organic matter.
Group III Soils with poorly developed horizons or no horizons, and capable of further mineral altcration.	
Entisols	Soils lacking horizons, usually because their parent material has accumulated only recently.
Inceptisols	Soils with weakly developed horizons, having minerals capable of further alteration by weathering processes.
Andisols	Soils with weakly developed horizons, having a high proportion of glassy volcanic parent material produced by erupting volcanoes.

* *Base status* refers to degree of leaching. *Low* means highly leached; *high* means minimal leaching.

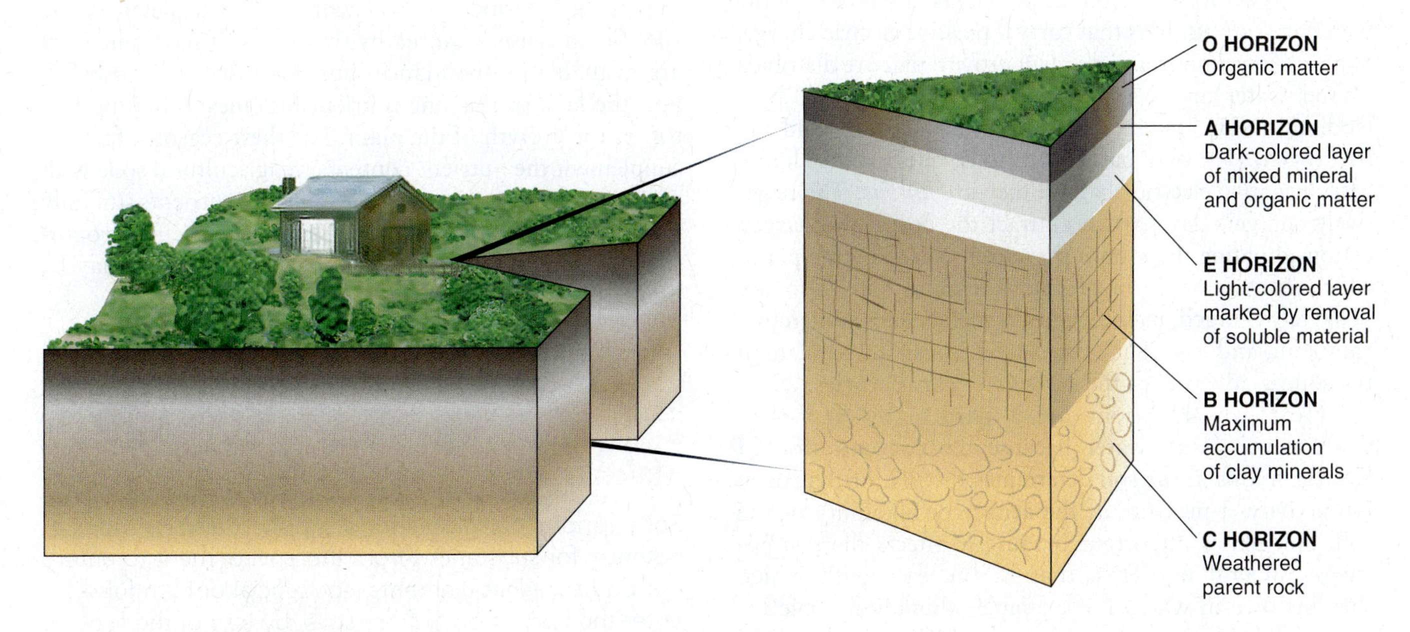

FIGURE 15.17 Soil profile
This is a typical sequence of soil horizons that would commonly develop in moist, temperate climates. The A horizon, which lies within reach of plant roots, is commonly called the topsoil.

FIGURE 15.18 **Arable soil**
Fertile, farmable soil is crucial for global food security. Modern farmers use contour plowing, as shown here, to slow erosion and preserve topsoil.

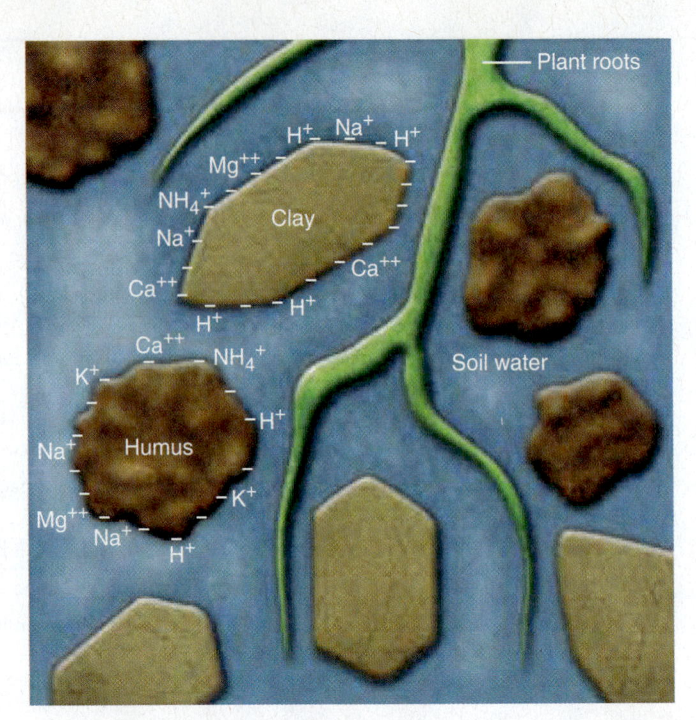

FIGURE 15.19 **Soil water**
Water that fills or partially fills the pore spaces between the mineral grains in the soil occurs as a film that adheres to the surfaces of mineral particles. Most of the components dissolved in the water are cations; they can include vital plant nutrients, as well as applied fertilizers and other agrochemicals, and some pollutants.

agricultural crops—is crucial for the maintenance of the human food supply (FIG. 15.18). **Soil fertility** is the ability of soil to provide nutrients, including phosphorus, nitrogen, and potassium, to growing plants. Humus is a crucial component in soil fertility because it stores some of the chemical nutrients released by decaying organisms and weathered minerals.

Soil moisture (or *soil water*) is another crucial component of soil fertility. Water that fills or partially fills the pore spaces between the mineral grains in the soil occurs as a film that adheres to the surfaces of mineral particles (FIG. 15.19). Most of the components dissolved in the water are *cations*, ions that carry a positive electric charge. Some cations important for plant growth that are dissolved in soil water are K^+ (potassium), Ca^{2+} (calcium), Na^+ (sodium), NH_4^+ (ammonia), Mg^{2+} (magnesium), and H^+ (hydrogen). The clay particles in soils, on the other hand, carry negative electric charges; they are *anions*. The negatively charged clay particles attract the positively charged cations, holding them until they are required by the plant. Many contaminants in groundwater and soils also are positively charged, including heavy metals such as mercury, cadmium, and lead. Such contaminants can accumulate in the soil by adhering to the surfaces of clay particles.

Hydrogen (H^+) ions are a particularly important component of soil water because the concentration of hydrogen ions in the soil determines the *pH*, or hydrogen ion activity, a measure of the acidity or alkalinity of the soil. Soil pH is important because it affects the solubility of mineral nutrients, that is, the ease with which they dissolve in water. For example, aluminum, which is potentially toxic to both plants and animals, becomes more water-soluble in soils with low pH (more acidic). Other mineral nutrients that are essential for plant growth become less soluble, and therefore less available, in soils with higher pH (more alkaline).

For optimal growth, as discussed earlier in this chapter, plants require nutrients in specific proportions. Therefore, if one type of plant is grown in the same soil over a long period, the soil will become depleted in the specific nutrients required by that plant. If most nutrients are available in abundance, but one essential nutrient is not, the lack of this one nutrient becomes a limiting factor on the growth of the plant. For these reasons, farmers supplement the nutrient contents of agricultural soils with plant fertilizers, mixtures of mineral nutrients (mainly nitrates, phosphates, and potassium in specific proportions) designed to supplement soil nutrients that may be limited in availability.

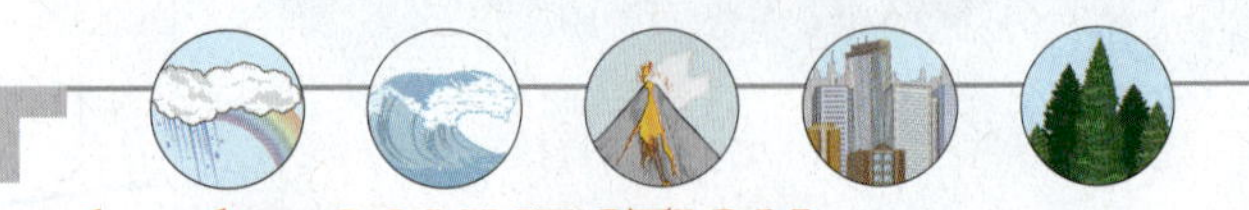

Make the CONNECTION

Soil supports our global food supply. It also provides a pathway for substances from the environment to enter our bodies. What elements—beneficial or harmful—enter the human body from soils, by way of the food crops that we consume? Trace the pathways of these substances.

Soil is a component of the geosphere that is vulnerable to damage and erosion. Before humans began intensive farming and grazing, the average rate of soil erosion was about 10 billion tonnes a year from all continents, and soil production and loss rates were in balance. The present rate of erosion is about 25 billion tonnes a year. With the average rate of soil formation being 10 billion tonnes a year, the current erosion rate could remove most of the world's topsoil in less than a century. This calculation is crude, but even if it's off by a factor of 5, a century from now there will be much less freedom to choose where to grow crops in the world's productive agricultural zones than there is today. The good news is that even though soil erosion and degradation are already severely impacting many countries, effective—and often affordable—control measures can substantially reduce these adverse effects.

The Anthroposphere: Impacts on Biogeochemical Cycles

Activities in the anthroposphere affect global biogeochemical cycles by causing changes in the *balance* among reservoirs; changes in the *fluxes* of materials between reservoirs; and changes in the *rates* at which natural processes occur. Let's have a closer look at some of these impacts.

Impacts on the Carbon Cycle. Probably the most obvious impact of human activity on the carbon cycle is the remobilization of carbon that results from the burning of fossil fuels. Coal, oil, natural gas, and other forms of fossil fuels (Chapter 18) are long-term storage reservoirs for carbon-based organic matter that has been chemically and thermally altered and compressed. When these fuels are burned, their carbon content is released from long-term sequestration. Cement manufacture causes a similar impact; limestone, also a storage reservoir for carbon dioxide, must be crushed and burned (or *coked*) to make cement, which releases the carbon dioxide from sequestration.

Deforestation, especially when trees are burned to clear the land, also releases carbon back to the atmosphere. Fires happen naturally, but this highlights one of the main features of human impacts on biogeochemical cycles: Even when human activity involves processes by which materials already move around in nature, *it almost always alters the balance among reservoirs by changing the rates at which those processes occur.* For example, when a tree dies and falls to the forest floor it will decompose, releasing its carbon to the atmosphere. This natural process can take anywhere from years to centuries, depending on the physical and chemical environment. When a forest is intentionally burned for land clearing, however, its carbon is released instantaneously. The disappearance of a forest also diminishes the available sinks for carbon moving from the atmosphere into the biosphere, and exposes soils to erosion, drying, and loss of soil moisture and other gases.

Impacts on the Sulfur Cycle. Human activity also modifies the biogeochemical cycling of sulfur. It is thought that more sulfur is now mobilized by human activity than by all natural processes combined. Earth's mineral resources (both metals and coal) exist underground in an oxygen-free environment. When exposed to the atmosphere or hydrosphere by mining, some compounds from this environment (mainly the sulfide mineral pyrite, FeS_2) become oxidized to form acids (such as sulfuric acid), which then acidify the surrounding environment (**FIG. 15.20A**). Stream ecosystems are especially damaged by this *acid-mine drainage*. In addition, the acidified water mobilizes toxic elements held in the rocks, such as lead and cadmium, releasing them into the aquatic environment.

The burning of coal affects the sulfur cycle. Coal, formed mostly in swamps on ancient floodplains near sea level, contains sulfur (derived from marine sulfate), which, on burning, is converted to sulfur dioxide. This latter compound forms sulfurous acid by combining with water. Eventually the acid is oxidized, forming sulfuric acid that rains out of the atmosphere, downwind of the coal-burning installations. Like acid mine drainage, acid rain traumatizes aquatic ecosystems by acidifying the water and can have deleterious effects on terrestrial ecosystems as well.

Impacts on the Nitrogen and Phosphorus Cycles. The nitrogen and phosphorus cycles also are affected by human activity. Nitrogen combines with oxygen when exposed to a high-temperature environment. Many modern industrial combustion processes produce oxides of nitrogen, including the burning of fossil fuels in gasoline and diesel engines. These oxides play a significant role in urban smog.

Human populations also produce large amounts of sewage. Whether raw or treated, this wastewater usually contains fairly high concentrations of nitrogen and phosphorus. The release of nutrient-carrying wastewaters into aquatic environments can result in *eutrophication* (Chapter 8). Eutrophic bodies of water have a high level of nutrients and consequently vigorous growth of algae, in contrast to water bodies with natural levels of phosphorus and nitrogen. As the algae die, their decay creates an oxygen demand that quickly makes the environment *anoxic* (lacking in oxygen) and asphyxiates aerobic organisms living in it.

Runoff from farms can cause eutrophication by transporting plant nutrients from agricultural fields to nearby aquatic systems. Modern agriculture depends on the widespread application of nitrogen, phosphorus, and potassium fertilizers. Phosphorus and potassium are mined from rocks and then made into fertilizers that are added as soil amendments (**FIG. 15.20B**). However, nitrogen normally comes to soils, and from soils to plants, from an atmospheric source; to make this transition, nitrogen-fixing bacteria must transform N_2 from the atmosphere into a bioavailable form. In the early part of the twentieth century, scientists discovered that electric sparks produced by industrial processes could convert molecular nitrogen into compounds usable by plants. This process, named the *Haber-Bosch process* after

FIGURE 15.20 **Human impacts on biogeochemical cycles**
Modern human activities have significant impacts on global biogeochemical cycles. (A) This forest is suffering from acid mine drainage associated with coal mining in East Germany. (B) Here phosphate is being mined near Jasper, Florida. Human industrial activities such as mining change the fluxes of naturally occurring substances like phosphate.

two of the scientists involved in its development, greatly increased the availability of nitrogen for use in fertilizers. Today *industrial fixation* is the major source of commercial nitrogen fertilizer and of fixed nitrogen in the nitrogen cycle. Like the sulfur cycle, the global nitrogen cycle is dominated by human activity, which has greatly altered the rates of processes that occur in nature.

BIOMES: EARTH'S MAJOR ECOSYSTEMS

So far in this chapter we have summarized the fundamental characteristics of life-supporting systems, defined the ecosystem as the fundamental life-supporting system, and examined the processes whereby materials and energy cycle around through these systems. Now let's look at the characteristics and geographic distribution of Earth's major ecosystems.

Biogeography

The geographic distribution of living organisms and the characteristics of their communities and ecosystems are the central focus of *biogeography*. The most important unit of biogeography is the **biome**, a large geographic area defined by its environmental attributes—mainly temperature and precipitation—as well as by the plants, animals, and soils that inhabit and characterize the area. Biomes are Earth's major ecosystem types (FIG. 15.21).

Biomes and Ecozones

In Chapter 14 we discussed the idea that a particular biome can have examples that occur in various geographic locations, with similar characteristics. For example, a desert is basically a desert. There are many variations and many deserts, but all deserts—wherever they occur—have certain characteristics in common. These include, notably, very low precipitation and (therefore) organisms that are adapted to dry conditions. For example, the Mohave Desert has low precipitation and dry soil; it features plants with waxy coatings and spines instead of leaves (to minimize water loss through the broad, unprotected surfaces of leaves). The Sahara and Gobi deserts have similar physical characteristics; they host plants that are unrelated to the plants of

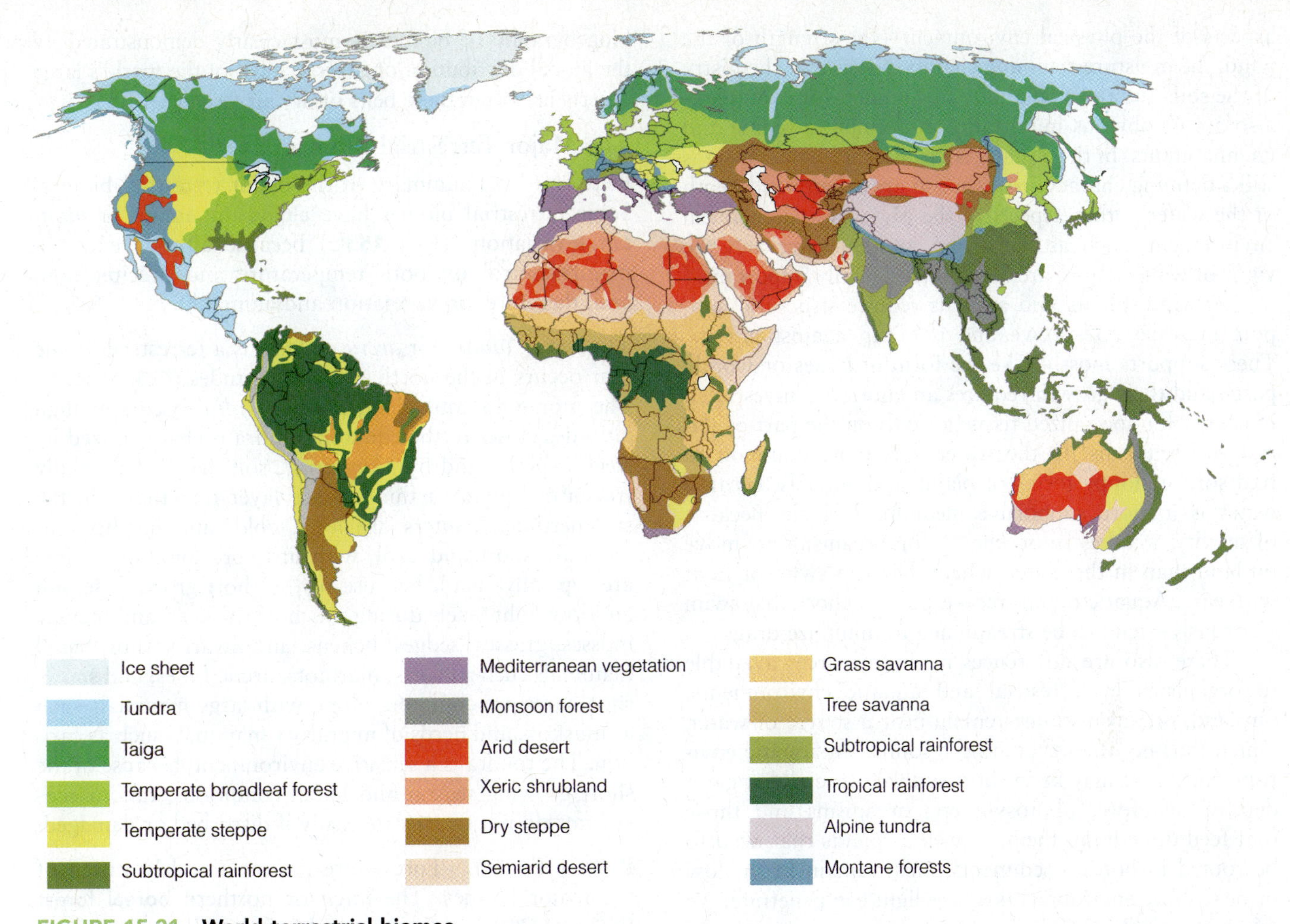

FIGURE 15.21 World terrestrial biomes
The locations of biomes are influenced by temperature, precipitation, atmospheric circulation, and other factors.

the Mohave Desert, but nevertheless share these dry-environment adaptive characteristics.

Different biomes that coexist in a particular geographic region constitute a *bioregion* or, on an even broader (roughly continental) geographic scale, an **ecozone.** A single ecozone thus includes several different biomes. Approximately eight ecozones constitute the biosphere, although different researchers categorize ecozones differently.

These terms may seem confusing; exactly what is the difference between a biome and an ecozone? The biome concept acknowledges that ecosystems with *similar* characteristics, and similar flora and fauna, can evolve separately, in locations that are widely separated from one another geographically. Thus, for example, tropical rain forests with similar characteristics are found in many different locations, on different continents.

The bioregion and ecozone concepts, on the other hand, acknowledge that ecosystems with *different* characteristics can share an evolutionary history because they coexist in a given geographic location or region. A tropical rain forest in Brazil may display similar characteristics to a tropical rain forest in Inodonesia, but these rain forests—both of them exemplars of the tropical rain forest biome—do not share a common geographic location or evolutionary history, and therefore they belong to distinct ecozones. On the other hand, the Afrotropic ecozone (as an example) includes the Sahara and Namib deserts, as well as the tropical forests of Madagascar and central Africa, and the grasslands and savannahs of the Sahel—a variety of biomes.

Differences between Terrestrial and Aquatic Biomes

Different biome types grade into one another; the boundaries between them are not sharp or distinct, and there are many variations (which makes the classification of biomes particularly challenging). However, there are two basic types of biome: *terrestrial* and *aquatic.* (The biome concept has been more widely applied to terrestrial ecosystems than to aquatic ecosystems, where the term *biotic zone* is more commonly used.)

Obviously, terrestrial and aquatic environments differ fundamentally in their physical attributes; this affects the properties of ecosystems and the organisms that characterize them. In a terrestrial environment, the most important attributes that define a biome are usually temperature and precipitation, which, in turn, influence soil type. Other aspects of the physical environment—the strength of the wind, the moisture-retaining characteristics and chemistry of the soil, and the availability of sunlight, for example—also are of obvious importance in defining a biome and its inhabitants. In the aquatic environment, temperature is still a defining characteristic, but so are salinity and depth of the water. Other aspects of the physical and chemical environment, such as the acidity of the water and the vigor of wave action, are also fundamentally important.

On land, plants and animals require structural support in order to survive and hold up against gravity. These supports mostly take the form of bones or woody parts, and their growth requires an enormous investment in energy for specialized tissue growth on the part of terrestrial organisms. In the water it is more common to find soft or floppy forms of plants and animals, because water is a more supportive medium than air. Because of gravity, it takes more energy for organisms to move on land than in the water, where they can swim or float with ease. Aquatic organisms—especially those that swim vigorously—tend to be streamlined to minimize drag.

There also are differences in the resources available to organisms in terrestrial and aquatic environments. On land, organisms must remain near a source of water, which thus becomes a limiting resource. In aquatic environments, light may be in short supply, especially at great depths; therefore, photosynthetic organisms and those that feed directly on them, as well as plants that need to be rooted in bottom sediments, must remain in shallow or near-shore environments where light can penetrate. An adequate supply of light is not a problem in most terrestrial environments, with the exception of a few unusual locations such as caves, or the floor of a dense rain forest. The basic types of food also differ from terrestrial to aquatic biomes. Land plants mainly produce food energy in the form of carbohydrates; in the ocean, proteins are the main form of organic matter. Organic matter and mineral nutrients also tend to be unevenly distributed and poorly circulated in deep aquatic environments.

Let's have a closer look at terrestrial and aquatic biomes, their defining characteristics, and some important examples of them.

Terrestrial Biomes

The distribution of terrestrial biomes is closely related to the geography of climate, which is fundamentally influenced by the air masses that move heat and moisture around the planet. Recall from Chapter 12 that convection in the atmosphere combines with the Coriolis effect to create huge cells of rising and falling air masses, forming three belts of high rainfall and four belts of low rainfall. The belts of high rainfall lie along the equator, resulting in moist tropical climates, and along the two polar fronts, resulting in moist temperate climates. The belts of low rainfall generate two zones of dry climate in the polar regions and two in the subtropical regions, along the 30° N and 30° S latitudes. The effect of the wet and dry belts on the biogeography of biomes is most clearly demonstrated by the global distribution of deserts; most of the world's large deserts lie within these belts of dry air (see Fig. 12.25).

The Major Terrestrial Biomes

Here is a brief summary of the major terrestrial biomes. Some terrestrial biomes have **alpine** (mountain or *montane*) variations (FIG. 15.22) because altitude exerts a major control on both temperature and precipitation, and, therefore, on vegetation and animal life.

TUNDRA. **Tundra**, or *arctic tundra*, is a terrestrial biome that occurs in the north, at high latitudes (FIG. 15.23A). The mountain equivalent, *alpine tundra*, occurs at high altitudes closer to the equator. Tundra is characterized by treeless plains and by *permafrost*, soil that is perennially frozen except for a thin surface layer that thaws in the summertime. Winters are long, cold, and harsh; summers are short and cool, with little precipitation. Plants are typically small, because of the short growing season and low light levels during much of the year, and include mosses, grasses, sedges, lichens, and dwarf shrubs. Small mammals such as voles, marmots, arctic foxes, and snowshoe hares are common, along with large mammals such as muskox, and herds of migratory mammals such as caribou. The tundra is a sensitive environment; because of the short growing season and harsh conditions, tundra ecosystems do not regenerate easily if disturbed or damaged.

BOREAL FOREST. Forests are the most widely varied of the major biomes. The *taiga* or northern **boreal forest** (FIG. 15.23B) lies to the south of the tundra in North America and Eurasia. There is no southern equivalent to either the tundra or the boreal forest because there isn't much land area in comparable latitudes in the Southern Hemisphere. The taiga is characterized by cold winters (though not as cold as the tundra), short growing seasons, and low levels of precipitation. The boreal forest mainly consists of *coniferous* trees (evergreens with cones), such

FIGURE 15.22 Alpine biomes
With increasing altitude, biomes change in ways that are similar to changes observed from the equator to the poles.

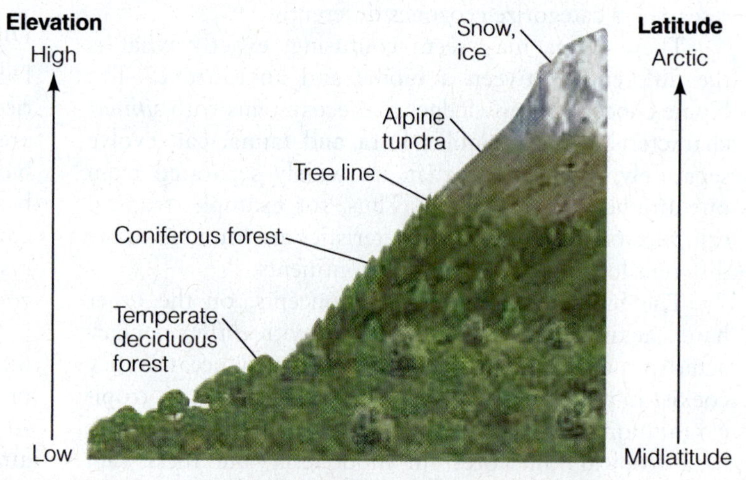

FIGURE 15.23 Some important terrestrial biomes
(A) Tundra (Yukon Territory, Canada) (B) Boreal forest (Minnesota, USA) (C) Temperate rain forest (Washington, USA) (D) Temperate deciduous forest (New England, USA) (E) Tropical rain forest (Georgia, USA) (F) Savanna (Kgaligadi, South Africa) (G) Grassland (Colorado, USA) (H) Desert (Utah, USA).

as spruce, fir, and pine trees, as well as mosses and lichens. Conifers are adapted to conditions of low precipitation. They have needlelike leaves with little surface area, which minimizes moisture evaporation. Animals of the taiga include large migratory mammals such as caribou and moose, and many smaller mammals such as rabbits, foxes, wolves, a wide variety of rodents, birds, and insects, but few reptiles or amphibians (because there is not enough water).

TEMPERATE RAIN FOREST. The **temperate rain forest** is a mainly coniferous forest (FIG. 15.23C), typical of the northwest coast of North America (Oregon, Washington, and British Columbia), but also found in other locations, such as southeastern Australia. The winters are milder than in the far north and precipitation is high, so the evergreen forests grow thick and tall. These rich wood-producing regions have hosted some bitter confrontations between loggers and environmentalists. Typical trees of the temperate rain forest include pine, fir, redwood, and cedar. *Epiphytes* (ferns, vines, and mosses that grow attached to the branches of trees) also grow in abundance. Bears, mountain lions, wolves, and elk are common, as well as smaller rodents, birds, reptiles, and amphibians.

TEMPERATE DECIDUOUS FOREST. The **temperate deciduous forest** occurs mainly in the northeastern United States, Europe, and eastern China, where the climate is characterized by seasonal changes from summer to winter (FIG.15.23D). These forests consist mainly of broad-leaved *deciduous* trees, that is, trees that shed their leaves each year, including maple, oak, birch, and elm trees. The soils are rich in organic material, and well suited for agriculture. Deer, bears, and wolves are common, though now limited to the localities where temperate deciduous forests have not been cleared for agriculture or urban development.

TROPICAL RAIN FOREST. The great **tropical rain forests** of the equatorial regions host an enormous diversity of organisms (FIG. 15.23E). Both temperature and precipitation are high, and the growing season lasts all year, so the vegetation is very tall and dense. These are *closed forests*, in which the *canopy*, or top layer of vegetation, forms an almost continuous cover (FIG. 15.24); consequently, the forest floor can be quite dark. There is an astonishing array of plant and animal life in these forests, including a wide range of birds, insects, mammals, reptiles, amphibians, epiphytes, and flowering plants, such as orchids and bromeliads. Most of the mammals, such as monkeys and lemurs, live in the canopy and rarely descend to the forest floor or *understory*. The soils in tropical rain forests tend to be highly weathered and low in organic matter. This is because most of the organic matter resides in the lush vegetation, rather than in the soils. With so much warmth and moisture available, there is an abundance of decomposers, including bacteria, fungi, and some types of insects, which break down and consume organic debris almost as quickly as it falls to the ground. Because tropical soils contain so little organic matter, when exposed by deforestation they dry out and are easily eroded. This is one of the reasons why it is problematic when tropical rain forests are cut down.

TROPICAL DECIDUOUS FOREST. The equatorial region also hosts large **tropical deciduous forests**, also called *tropical seasonal forests* or *monsoon forests*, in which the trees shed their broad leaves during the dry season. The main seasonal variation is in the amount of precipitation, although there is greater temperature variation than in the tropical rain forest. Tropical rain forests and tropical deciduous forests together are referred to as the *tropical moist forest system*. The largest of these is Amazonia, in South America, but other major tropical moist forests are in Southeast Asia (Indonesia) and equatorial Africa. In addition to hosting much of the planet's biodiversity, the tropical moist forests represent an enormous reservoir for carbon in the global carbon cycle.

SAVANNA. Tropical and subtropical **savanna** is a type of *open forest* (Fig. 15.24) consisting of broad, grassy plains with scattered trees and lacking a continuous canopy (FIG. 15.23F). Temperatures are high but rainfall is low, particularly during the long dry season. The best-known savanna is the African Sahel (from *sahel,* the Arabic word for "border"), which lies along the southern margin of the Sahara Desert. African savannas host huge herds of migratory mammals, such as antelope, zebras, and elephants, and large predators, such as lions and tigers. This is a fragile environment, primarily because of the low precipitation. When overstressed by intensive agriculture and grazing, savanna can quickly turn into desert.

CHAPARRAL. The biome that is typical of a Mediterranean climate—hot, dry summers and cool, wet winters—is the **chaparral**, characterized by low, scrubby evergreen bushes and short, drought-resistant trees. Chaparral occurs in the region surrounding the Mediterranean Sea, but is also found in the southwestern United States, as well as parts of Australia, Africa, and South America. Fires are common in chaparral regions, and some plants have fire-resistant adaptations, such as below-ground growth that is able to reemerge following a fire.

GRASSLAND. Temperate **grasslands** are the huge, extensive prairies typical of the midwestern United States and Canada, as well as the Ukraine (FIG. 15.23G). Grasses have extensive, interconnected root systems. Grasslands are particularly well suited to agriculture because of the rich organic content of the soils. There are two basic types of grasslands. *Temperate moist grasslands*, also known as *tallgrass prairies*, once hosted great herds of bison and elk, and predatory wolves. *Shortgrass prairies* are drier, and the grasses are shorter and more drought-resistant. Typical animals in shortgrass prairies include prairie dogs, snakes, and lizards.

DESERT. We finish off the low-precipitation extreme of the terrestrial biomes with deserts (FIG. 15.23H). Although

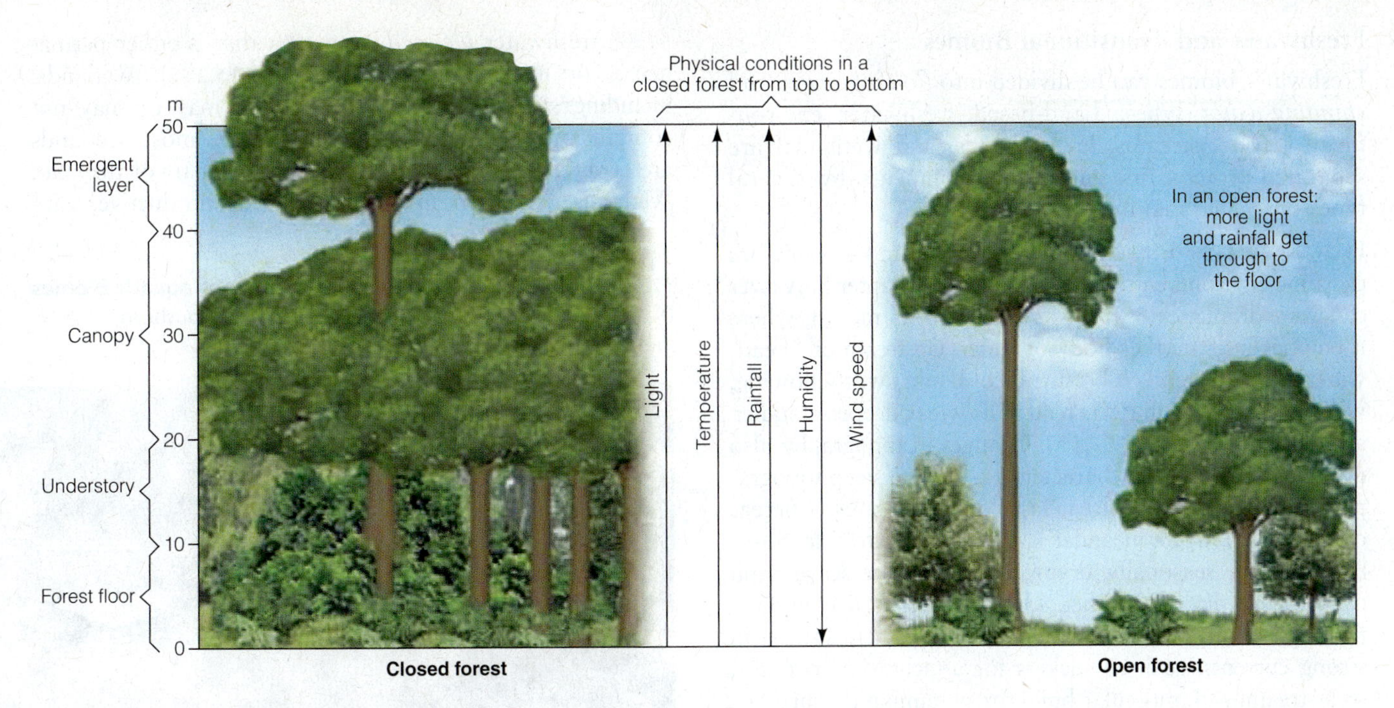

FIGURE 15.24 Closed and open forests
In a closed forest, the top layer of vegetation, called the canopy, provides essentially continuous cover. In an open forest, there are openings in the canopy where light can pass through to the forest floor.

we often think of deserts as being hot, lack of precipitation is actually the main defining characteristic. The term **desert** refers to *arid* lands, where annual precipitation is less than 250 mm. Four types of hot desert and one type of cold desert are recognized, as discussed in Chapter 12:

- *subtropical deserts*, in the two subtropical dry belts; examples include the Sahara, Kalahari, and Great Australian;
- *continental interior deserts*, far from any source of moisture, such as the Gobi and Takla Makan of Central Asia;
- *rainshadow deserts*, where a mountain range creates a barrier to the flow of moist air, causing a zone of low precipitation; examples are the deserts to the east of the Cascade Range and Sierra Nevada in the western United States;
- *coastal deserts*, such as the western coastal deserts of Peru, Chile, and southwestern Africa; and,
- the vast, cold, *polar deserts*, where precipitation is extremely low owing to the sinking of cold, dry air.

In all deserts, plant cover is sparse. Desert plants are adapted to the lack of water, including the many varieties of cacti and succulents, which have few or no leaves and are able to retain large quantities of water. Animals adapt to the extreme temperatures and dryness by storing water, remaining hidden during the day, hunting or foraging at night, or tunneling underground during extended dry periods.

Aquatic Biomes

Aquatic biomes are distinguished primarily on the basis of differences in salinity. Thus, there are three major groups of aquatic biomes—*freshwater*, *marine*, and *transitional*—with further subdivisions based on temperature, depth, penetration of light, and distance from the shore. Saltwater (marine) and freshwater biomes differ from one another in some fundamental ways, aside from the salinity of the water. For example, the ocean is much more strongly affected by tides than even the largest lakes. Waves and deep currents are generally stronger in the ocean than in lakes, and the depth of the water is considerably greater. This means that low light levels, cold temperatures, and even high pressures can place significant constraints on oceanic organisms.

Organisms in these various aquatic environments generally fall into one of three categories: *plankton*, *nekton*, and *benthos*. *Plankton* (Chapter 10) are extremely tiny, free-floating or swimming organisms. Of these, as mentioned previously, *phytoplankton* are aquatic plants and algae, the primary producers that form the base of the food chain in aquatic ecosystems, and *zooplankton* are aquatic animals, including the larvae of some larger organisms. **Nekton** are larger animals that are active swimmers, including fish, whales, turtles, and dolphins. Finally, **benthos** is the collective term for bottom-dwelling organisms. This includes organisms that attach themselves to rocks, such as barnacles and mussels; plants that are rooted in bottom sediments; organisms that burrow into bottom sediments, such as octopi; and some organisms that move along the bottom, such as crustaceans.

Freshwater and Transitional Biomes

Freshwater biomes can be divided into *flowing-water* and *standing-water* types. Land-based freshwater environments (rivers, streams, lakes, ponds, and wetlands) are separated from saline marine environments by several types of transitional biome.

FLOWING-WATER ENVIRONMENTS. Rivers and streams are dynamic environments, as you learned in Chapter 8. A river can vary dramatically from its source to its mouth, where it empties into another body of water. Upstream or "headwater" streams tend to be small, cold, and swiftly flowing, whereas downstream rivers tend to be wider, deeper, cloudier, and not as cold (FIG. 15.25A). Changes in topography also dramatically affect the character of a stream; steep topography leads to deeply incised, fast-flowing streams, whereas on flat plains, rivers meander lazily from side to side. Some rivers change seasonally, drying up during one season and overflowing during another. Organisms that live in fast-flowing streams have adaptations that help them survive in strong currents, such as suckers for attaching onto rocks, or a streamlined, muscular build for swimming up-current.

STANDING-WATER ENVIRONMENTS. Standing-water environments include lakes, ponds, and wetlands. A lake is a standing body of fresh water that occupies a large depression; ponds are smaller standing freshwater bodies (FIG. 15.25B). Lakes contain several zones, defined by a combination of depth, temperature, and distance from the shore. The warm, shallow, near-shore area is called the **littoral zone**. This is the most biologically productive part of a lake, with abundant plant life (cattails, sedge grasses), as well as birds (ducks and other waterfowl), insects, reptiles, and amphibians. The open-water environment is called the **limnetic zone**, which hosts zooplankton and phytoplankton, as well as the larger organisms, mainly fish, that eat them. The limnetic zone extends as far down as light can penetrate (the *photic zone*; see below). Below this is the **profundal zone**, which only occurs in the largest and deepest lakes. Organic material in the profundal zone consists mainly of dead organisms that float down and are consumed by bacteria.

The temperature of the water in a large lake can vary greatly from the surface, where sunlight heats the water, to deeper levels, where the water is typically much colder. This is called *thermal stratification*, and it greatly influences the biota that can occupy the various zones. Cool water sinks to the bottom of a deep lake because it is dense. As sunlight hits the lake, the surface waters become warmer and less dense. This is a stable thermal stratification—cold, dense water at the bottom, and warm, less dense water at the top. The depth where the temperature changes very rapidly from warm to cold is called the *thermocline* (defined in Chapter 10 in the context of ocean thermal stratification). In the tropics, the temperature difference between the surface and deep layers is less strong because there is no seasonal cooling of the surface waters; thus, the thermal stratification in tropical lakes is not as pronounced as in temperate lakes.

A freshwater *wetland* is an area that is either permanently or intermittently moist (FIG. 15.25C). Wetlands, including *swamps*, *marshes*, and *bogs*, may or may not contain open, standing water, although most wetlands are covered by shallow water for at least part of the year. Wetlands tend to be highly biologically productive, with

FIGURE 15.25 Some important freshwater aquatic biomes (A) River (Newfoundland, Canada) (B) Lake (Southern Germany) (C) Wetland (Mantecal, Venezuela).

dense vegetation, migratory birds, reptiles, amphibians, and fish. Some wetlands represent a developmental stage in the natural lifetime of a freshwater body. For example, a lake whose source of incoming water has been cut off will become shallower over time, gradually filling with sediment and organic material to become a peat bog or swamp—this is the natural process of eutrophication. Wetlands were once considered to be dirty, mosquito-infested quagmires, prime targets to be drained and developed. More recently, wetlands have been recognized as natural storehouses for a great diversity of plant and animal species. Wetlands also perform many important environmental services, including storing groundwater and removing toxins from the soil.

TRANSITIONAL ENVIRONMENTS. Aquatic environments in coastal regions are transitional between freshwater and saltwater environments. An example is the *estuary*, a body of water that is connected to the open ocean but has an incoming supply of fresh water from a river. Salt water and fresh water mix in estuaries, and water levels, salinity, and temperature fluctuate with the rise and fall of the tides. Organisms that inhabit estuaries must be adapted to tolerate these variations. A common feature of temperate estuaries is a *salt marsh*, a coastal wetland dominated by salt-tolerant grasses. Salt marshes typically host abundant shore birds. The tropical equivalent is *mangrove forests*, coastal wetlands that host some of the most productive fisheries in the world. Estuaries, salt marshes, and mangroves, like other wetlands, perform many important ecological functions, including cleansing the soil and providing habitats for many different species. Coastal wetlands also protect the shoreline against the battering energy of oceanic storms. Unfortunately, coastal development is threatening mangrove forests and other coastal wetlands in many localities.

Marine Biomes

We looked briefly at the zones of the ocean in Chapter 10, but here we revisit them from a biological perspective. The primary subdivision of oceanic biotic zones is based on the depth of the water, and thus on the penetration of light into the water (FIG. 15.26). The **photic zone** (or *euphotic zone*) is the topmost layer of ocean water, where light penetrates and photosynthesis can occur. The photic zone extends from the surface down to the maximum depth of sunlight penetration—about 150–200 m. Below this is the **aphotic zone**, where little to no light is able to penetrate. (Sometimes a transitional zone of low light levels, called the *disphotic zone*, is identified between these two zones.) Marine environments and the biotic zones they support are further categorized on the basis of distance from the shore, depth of water, and relationship to the bottom.

NEAR-SHORE ENVIRONMENTS. As in lakes, the parts of the ocean that are near the shoreline are referred to as the *littoral zone*. The oceanic littoral zone is sometimes called the *intertidal zone*, which captures its transitional nature. This is a dynamic environment, where the influence of tidal variations and the energy of breaking waves are strong. Rocky or cliffed shorelines typically host organisms that can attach themselves to rocky surfaces,

FIGURE 15.26 Oceanic zones
The zones or biomes of the ocean can be divided on the basis of depth, distance from shore, and light levels.

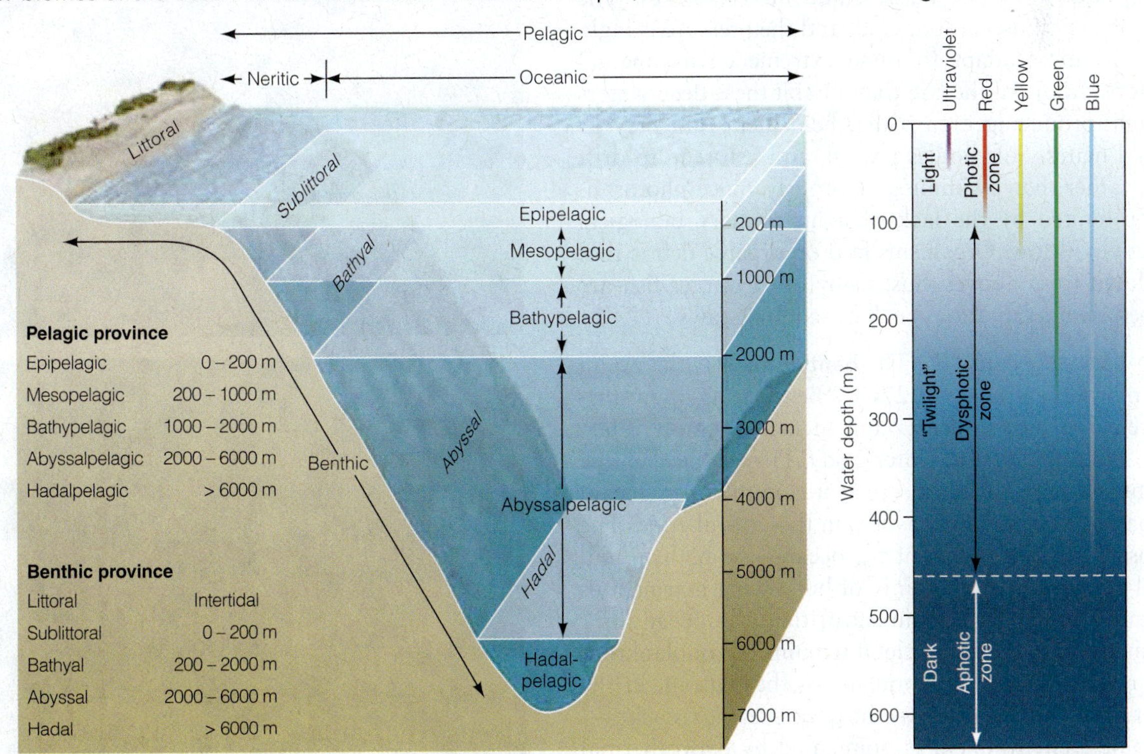

such as mussels and certain types of algae, and other organisms that can survive periodic drying out during low tide (FIG. 15.27A). Some residents of rocky shorelines survive by burrowing and hiding between the rocks. Another common type of shoreline is the sandy beach, where small animals like sand crabs burrow into the sand to escape the breaking waves and the ebb and flow of the tides.

OPEN-WATER ENVIRONMENTS. The **pelagic zone** is the open-water environment of the ocean (Figs. 15.26 and 15.27B). The pelagic zone consists of the *neritic province*, which extends from the shore out to where the water reaches a depth of 200 m, and the *oceanic province*, which encompasses the rest of the pelagic zone—that is, most of the open ocean. Further subdivisions are based on depth: the topmost layer is the *epipelagic zone*, extending to a depth of about 200 m (coincident with the photic zone); the *mesopelagic zone*, extending to a depth of 1000 m; the *bathyal* (or *bathypelagic*) *zone*, extending to a depth of 4000 m; and the *abyssal* (or *abyssopelagic*) *zone*, extending to a depth of 6000 m. The deepest part of the ocean, called the *hadal* (or *hadopelagic*) *zone*, exists only in the deepest oceanic trenches, more than 6500 m below the surface.

Organisms that inhabit the pelagic zone are floating or swimming organisms. The near-shore photic zone is the most biologically productive marine environment; the water is relatively warm, and light levels are high. This environment supports abundant phytoplankton and, therefore, zooplankton and the nekton that feed on them. Nekton include the rich diversity of marine life, including large and small fish, sharks, swimming mollusks, and marine mammals, such as whales.

Beneath its relatively warm, light surface layer, the oceanic province extends to great depths. No light penetrates the deep waters of the bathyal and abyssal zones. The temperature is constant, but cold, and the pressure is high. Organisms must adapt to these extreme environmental conditions. Many of the fish that inhabit these deep waters have light-producing organs that help them find prey or potential mates. Soft bodies prevail, and fish tend to drift slowly, rather than swimming vigorously. Since photosynthesis is limited by the lack of light, primary biological productivity is low. Organisms feed on organic debris that rains down from above; thus, many of the larger fish are equipped with adaptations such as scooping jaws.

BOTTOM ENVIRONMENTS. The **benthic zone** is the ocean floor environment (Fig. 15.27C). "Benthic" is often misinterpreted to mean "deep," but in fact it simply refers to the "bottom" of the water body. Hence, the benthic environment may be very deep, as it is in the open ocean, or it may be very shallow, as it is in the littoral zone. The benthos of the deep parts of the ocean—the bathyal and abyssal zones—consists mainly of burrowing marine animals and bacteria. The sediment of the deep ocean floor is dominated by the tiny skeletal remains of zooplankton, which rain down and accumulate on the bottom. In the deepest parts of the benthos—the deep oceanic trenches—the geologic environment is dominated by hydrothermal vents with black smokers. The organisms in this environment are adapted to a completely dark environment, and their primary biological productivity comes from chemosynthesis, rather than from photosynthesis.

In contrast to the extreme environment of the deep ocean bottom, the shallow-water benthic zone is a warm, light, and highly productive environment. In water up to

FIGURE 15.27 Some important marine biomes
(A) Intertidal zone (Monterey Bay, California) (B) Open ocean (Atlantic) (C) Benthic zone (Clayoquot Sound, British Columbia, Canada)

about 10 m deep, salt-tolerant sea grasses grow in bottom sediments. They provide food and protection for a wide variety of fish, crustaceans, reptiles, and bottom-dwelling detritivores, such as mud shrimp. In deeper, cooler water (up to about 25 m), huge beds of kelp—brown algae—form the bottom of the marine food chain for many organisms, including sponges, sea cucumbers, clams, crabs, fish, and mammals such as sea otters. A particularly productive benthic environment is the coral reef. Coral reefs are worthy of special mention, not only because they are a biologically productive biome, but also because they are sensitive indicators of environmental stress (as discussed in Chapter 19). They also perform an important role in the recycling of nutrients in shallow coastal environments and provide physical barriers that dissipate the force of waves.

SUMMARY

1. A flow of energy is a fundamental requirement for any life-supporting system. The basic function of energy in a life-supporting system is biological production, the transformation of energy into organic matter by biological processes. The amount of usable energy in a system provides an upper limit on the amount of biomass, and thus the amount of life that can be sustained by any ecosystem and by the Earth system as a whole.

2. Gross production occurs when an autotrophic organism synthesizes organic compounds, commonly through photosynthesis. The autotroph uses some of the organic matter as fuel in metabolism and respiration, releasing energy in the form of heat, and stores some of it for future use, in the form of carbon-based compounds. Net production is biomass that is left over from gross production after it has been used to fuel the processes of life.

3. Energy is transferred through an ecosystem along food chains. Autotrophs, the primary producers, are the first trophic level. Heterotrophs that are first-order consumers form the second trophic level; second-order consumers form the third trophic level; and so on. The food-energy levels together form a trophic pyramid. Stored energy and biomass decrease dramatically from one trophic level to the next, because organisms are not efficient at converting and using energy, and because a large fraction of the energy available at each trophic level is used in respiration. The movement of energy through a trophic pyramid is thus a one-way flow—it comes in, it is used, and it is released to the surrounding environment.

4. Unlike energy, which cannot be recycled, materials are cycled endlessly through ecosystems. No one organism can make all its own food from inorganic chemicals and then break down all of those complex organic chemicals, returning them to the original inorganic form. The chemical elements in excreted waste and dead organisms are broken down and recycled by decomposers, mainly fungi and bacteria.

5. Most ecosystems have several interconnected food chains involving complex eating relationships and interactions, which form a food web. Energy moves through the food web in a complicated, circuitous, but ultimately one-way flow, but materials are cycled over and over again to provide nutrients for organisms at all levels.

6. Some elements are essential for life, others are toxic, and others are neutral. Macronutrients are required in large amounts by all life; these include carbon, hydrogen, nitrogen, oxygen, phosphorus, and sulfur. Each macronutrient plays a special, fundamental role in organisms. For any form of life to persist, the required chemical elements must be available at the right times, in the right amounts, and in the appropriate relative concentrations to each other. The element in scarcest supply becomes a limiting factor.

7. Organisms are selective in their uptake of chemical elements; as a result, elements can become more concentrated in organisms than they are in the local, surrounding environment. The two basic mechanisms for concentration are bioaccumulation, in which an element builds up in the tissues of an organism as it ages, and biomagnification, or food-chain concentration. Through active uptake of chemicals from the environment, living things change their own chemistry and the chemistry of the environment.

8. A system that supports and sustains life must have a continually renewed source of energy and a sink for heat energy. It is likely an open system that allows energy to flow through, and it recycles matter with some leakage across system boundaries. It must have at least one autotroph (a primary producer) and one heterotroph (a decomposer). There must be a fluid medium to facilitate the movement of chemical elements among the different kinds of organisms, and a nonliving environment to provide inorganic chemical nutrients. The minimum system that can sustain life is thus an ecosystem, and the capacity to support life is a characteristic of ecosystems.

9. Elements that are essential to and strongly influenced by the biosphere, that are transformed from organic to inorganic forms, and that move among biological and nonbiological reservoirs are said to undergo biogeochemical cycling. In a box model of a biogeochemical cycle, the boxes represent reservoirs where chemical elements are stored, and arrows represent fluxes and pathways of transfer. Each reservoir is an open system, and the entire biogeochemical cycle is a set of open systems linked by the transfer of materials.

10. The residence time of a substance in a reservoir is influenced by the properties of the substance itself, and by the properties of the surrounding medium. In general, residence times are long in the geosphere, short in the atmosphere and biosphere, and intermediate in the hydrosphere.

11. Different elements take different pathways through the four great reservoirs of the Earth system, depending on the characteristics of the element. Most metals do not form gases and therefore have no major phase in the atmosphere; they typically cycle between the geosphere, hydrosphere, and biosphere. Nonmetals more commonly form gaseous compounds, which can be returned rapidly to the biosphere via the atmosphere.

12. The carbon, nitrogen, and phosphorus cycles are three of the most important biogeochemical cycles. Carbon cycling is dominated in the short term by biological processes, primarily photosynthesis and respiration; in the medium term by the storage of organic chemicals in trees, soils, and other organic sediments; and in the long term by geologic processes, including the production of carbonate rocks and return of CO_2 to the atmosphere via weathering, metamorphism, subduction, and volcanism. Nitrogen cycling involves the conversion of molecular nitrogen from the atmosphere into biologically useful forms, mostly by nitrogen-fixing bacteria in soils, and its release back to the atmosphere through denitrification. Phosphorus cycling occurs mainly in the geosphere, hydrosphere, and biosphere because it lacks a gaseous phase. Phosphorus is often a limiting element for plant and algal growth.

13. Earth's spheres interact through biogeochemical cycles. The atmosphere has had a great influence on the biosphere and has, in turn, been deeply affected by the biosphere over Earth's history. Water in the hydrosphere directly participates in or facilitates most environmental and biological interactions. The geosphere, specifically the mantle, is the source (through volcanism) of the carbon that is the starting point for synthesis of organic compounds in the biosphere. Activities in the anthroposphere affect global biogeochemical cycles by altering the balance among reservoirs and by changing the rates at which natural processes occur.

14. Soil, a biogeochemical link, is weathered rock material that has been altered by the presence of living and decaying organisms, such that it can support rooted plant life. Soil consists of tiny fragments of rocks and minerals mixed with water, soil gas, humus, and small organisms. The rate of soil formation depends on the rate of weathering, which, in turn, is influenced by temperature, rainfall, topography, and the composition of the parent rock. A fully developed soil profile has horizons with distinct color, texture, chemistry, organic content, and water content.

15. There are two basic types of biome: terrestrial and aquatic (although the biome concept has been more widely applied to terrestrial ecosystems). In the terrestrial environment, the attributes that define a biome are temperature and precipitation, which, in turn, influence soil type and vegetation. In the aquatic environment temperature, salinity, and depth of the water are defining characteristics.

16. Organisms that live on land differ fundamentally from those in aquatic environments. On land, plants and animals require structural support in order to hold up against gravity, and they must remain near a source of water, which thus becomes a limiting resource. In aquatic environments, light may be in short supply, especially at great depths; this is not a problem in most terrestrial environments.

17. The distribution of biomes is closely related to the geography of climate, which is fundamentally influenced by air masses that move heat and moisture around the planet. The major categories of terrestrial biome include tundra; forest (boreal forest, temperate coniferous forest, temperate deciduous forest, tropical rain forest, and tropical deciduous forest); savanna; chaparral; grassland; and desert, with many variations of each, including alpine (or montane) variations.

18. There are three major groups of aquatic biomes—freshwater, marine, and transitional—with further subdivisions based on temperature, depth, penetration of light, and distance from the shore or bottom of the water body. Aquatic organisms are plankton (extremely tiny, free-floating or swimming organisms); nekton (larger animals that are active swimmers); and benthos (bottom-dwelling organisms).

19. Freshwater biomes can be divided into flowing-water types (rivers and streams) and standing-water types (lakes, ponds, and wetlands). Land-based, freshwater environments are separated from saline marine environments by several types of transitional biome, including estuaries, salt marshes, and mangroves.

20. The photic zone is the topmost layer of ocean water, where light penetrates and photosynthesis can occur. Below is the aphotic zone, where little to no light is able to penetrate. Marine environments and the biotic zones they support are further categorized on the basis of distance from the shore, depth of water, and relationship to the bottom. The littoral or intertidal zone is near the shoreline, and the near-shore photic zone is the most biologically productive marine environment. The pelagic zone is the open-water environment, consisting of the neritic and oceanic provinces. The benthic zone is the ocean floor environment; it may be very deep, as in the open ocean, or very shallow, as in the littoral zone.

IMPORTANT TERMS TO REMEMBER

QUESTIONS FOR REVIEW

1. What are the two fundamental requirements for any long-term life-supporting system?
2. Describe the characteristics of a minimum life-supporting system.
3. What is biological productivity? What is the difference between gross productivity and net productivity?
4. What is the difference between a food chain and a food web? Where are humans on the food chain? Which trophic level?
5. How does energy move through an ecosystem? In what ways is this different from, or similar to the movement of matter through an ecosystem?
6. What is the role of decomposers in ecosystems?
7. What is the main difference between a macronutrient and a micronutrient, in terms of how they are used by organisms?
8. In what important ways does phosphorus differ from the other important biogeochemical cycles?
9. What are the main impacts of human activity on the global biogeochemical cycles?
10. Why is the residence time of chemical elements long in rocks and short in the atmosphere?
11. What is soil, and in what ways does it form a biogeochemical link between the spheres of the Earth system?
12. What are the fundamental differences between terrestrial and aquatic biomes? On the basis of which properties are terrestrial and aquatic biomes defined?
13. Choose one major terrestrial biome to describe.
14. Choose one major freshwater or transitional aquatic biome to describe.
15. Choose one major marine biome (or zone) to describe.

QUESTIONS FOR RESEARCH AND DISCUSSION

1. Take one of the "Big Six" macronutrients and describe its biogeochemical cycle in detail. Describe how the evolution of the Earth system would be affected if the cycling of this element had been dramatically different.
2. Why does the carbon cycle have such an important influence on the chemistry of the atmosphere?
3. In what ways it the water cycle (or hydrologic cycle) like the other biogeochemical cycles? In what ways is it different?

QUESTIONS FOR *THE BASICS*

1. Why can't we recycle energy within our bodies or between species?
2. On the basis of the map of world soils, what type of soil dominates the area where you live? What are the characteristics of this type of soil?
3. What are the 11 soil orders?

QUESTIONS FOR *A CLOSER LOOK*

1. Why do some toxic elements, such as mercury, become more concentrated in animal tissues as they move up through the trophic levels? Describe the processes whereby this can occur.
2. Do some research to find out the average levels of mercury in fish that you regularly consume. What are the recommended levels of consumption for this type of fish, in terms of kg per week? Are you eating more, or less than this amount?
3. What are some other substances, besides mercury, that have been known to bioaccumulate in humans?

CHAPTER 16

Populations, Communities, and CHANGE

OVERVIEW

In this chapter we:

- Examine the factors that influence the health of populations
- Consider carrying capacity and factors that limit the growth of populations
- Describe interactions between individual organisms in ecological communities
- Explore the concepts of niche and habitat
- Summarize the types and importance of biodiversity, and its current threats

Honeybee community
Populations of honeybees in many parts of the world have undergone a dramatic collapse in the past few years, for reasons that are not thoroughly understood. Here, a community of bees (*Apis mellifera*) cooperates to repair its honeycomb. Humans depend on honeybees for the pollination of many food crops.

In Chapter 14 we took a broad view of life on Earth—what life is, how it may have originated, how it has changed, and how it has influenced the Earth system over the course of geologic time. Then, in Chapter 15 we looked more closely at the systems that support life: their defining characteristics, and the mechanisms whereby energy and nutrients cycle among organic and inorganic reservoirs. In Chapter 16 we zoom in to examine the individual organisms, populations, and communities that constitute the living components of Earth's ecosystems. We consider the factors that cause populations to grow and shrink, and the interactions that occur between individuals of the same and different species in ecological communities. We will end Chapter 16 and our examination of the biosphere by considering how all of these processes have worked together to bring about the enormously complex, rich diversity of life on this planet.

POPULATIONS

A **population** is a group of individuals of the same species, which interbreed and share genetic information. We commonly think of populations as inhabiting the same physical space, such as a population of birds that live and nest in a particular forest, or a population of trees that grows in the forest, or a population of ants living in one colony. However, depending on the behaviors and life cycle of a particular species, the individual members of a population need not live in close proximity. Consider whales, for example; they typically travel many thousands of kilometers in small groups, coming together only rarely for mating with others from the breeding population. This type of geographic dispersal is a characteristic of many populations of migratory and territorial species.

Factors That Cause Changes in Populations

To understand populations, we need to understand what makes them healthy or unhealthy and what makes them grow or shrink. To accomplish this we need to consider factors that are *intrinsic* (or internal), that is, specific to the characteristics of a particular species or population. This can include characteristics such as a slow reproductive rate or specific requirements for food, living space, or light. Factors that are *extrinsic* or external to the population itself also exert an influence on its health and survival; examples are climate change, a forest fire (FIG. 16.1), or the seasonal abundance of a predator species.

We also need to consider how interactions between organisms at the individual level influence the health and growth of a population as a whole. These include things such as competing with a member of the same species for food or for mating rights. These are *intraspecific* interactions, that is, interactions between individuals of the same species within a population. *Interspecific* interactions, that is, interactions between individuals of different species, are also important because populations do not live in isolation; they live in communities composed of many interacting and interdependent species. Examples of interspecific interactions that have an impact on population growth include competition, predation, and parasitism. Interspecific and intraspecific interactions and the behaviors and characteristics of individual organisms influence rates of birth, death, and survival—the fundamental components of population growth.

Ultimately, the Earth system controls the physical availability of resources, including food, water, space, shelter, sunlight, and an appropriate reproductive environment, all of which have an obvious impact on the health of populations and the individuals of which they are composed. There are important differences between the physical environment where a particular species or population lives or could live—its *habitat*—and the ecological role or *niche* of the species within that environment.

The niche concept, which we will return to later in this chapter, opens up all sorts of questions about the possible interactions within populations and among populations in a particular habitat. For example, what roles can species play in their interactions with one another? Can two species within a community occupy the same niche? Do individuals from a single population compete with one another for resources? Do they compete with individuals from other species? Can species cooperate with one another to fulfill a particular role within a habitat? We will examine the possible answers to these and other questions about populations and communities.

FIGURE 16.1 Extrinsic factors can affect populations Here, fireweed (*Epilobium angustifolium*) is the first foliage to reemerge amid blackened spruce trees after a forest fire near Eagle Point, Yukon Territory.

Population Dynamics

Population growth and change is known as *population dynamics*. The basic question of population dynamics is: What controls the abundance and variety of life on Earth? We can ask this question at many different levels of biological organization, from a single population within a species, to all species in an ecosystem, to all life in a major region of Earth, or to all life on Earth. The question is important because a rapidly growing population can occupy large amounts of space, alter the physical and chemical environment, and affect other populations and the Earth system as a whole. It is also important because, as we are learning, equilibrium in the biosphere relies on delicate balances among species. The loss of a population or species may have unforeseen consequences, and we need to understand as much as possible about the factors that keep populations healthy and stable.

Let's begin by considering how populations change over time. The change in the number of individuals in a population over time is called its **growth rate**. Growth rate (or r) is simply a function of the rates at which individuals are added to or taken away from a population during a given period. It is often expressed as a ratio, in this way:

$$\text{growth rate } (r) = \frac{N \text{ (population at end of period} - \text{population at beginning of period)}}{N_0 \text{ (population at beginning of period)}}$$

If more individuals are added than are taken away, then the population is increasing and the growth rate is positive. If more individuals are lost from the population than are added, then the growth rate is negative and the population is decreasing.

Given sufficient energy and nutrients, life forms are capable of incredible rates of growth. For short periods, populations can undergo exponential growth. For example, the Northern elephant seal, once believed to be extinct but actually reduced to about 20 individuals at the turn of the twentieth century, grew to approximately 60,000 by the 1970s and more than 120,000 currently, increasing at growth rates of about 9 percent per year (FIG. 16.2). However, populations cannot grow exponentially forever. In a finite world, the growth of populations is limited by a variety of factors.

Limits to Growth

If real populations cannot grow exponentially indefinitely, then how do they grow? One of the earliest proposals of the modern scientific era was that under constant environmental conditions, including a constant food supply, a population would grow smoothly until further growth became limited by environmental factors. Such growth follows an S-shaped curve, in contrast to the J-shaped curve that is characteristic of unlimited exponential growth. This *S-curve* has become known as the **logistic growth curve** (FIG. 16.3).

The idea behind the logistic curve is simple. When a population is small in comparison to its resources, every individual has plenty to eat and an abundance of needed resources; the population is able to grow exponentially. At first, as is typical of exponential growth, the absolute increase in population will be small because there are so few individuals reproducing (point A on Fig. 16.3). As the numbers increase, more individuals are available to reproduce, but overcrowding is still minimal and resources are still abundant; as a result, population growth is rapid (point B on Fig. 16.3). Eventually, as the **population density** or number of individuals per unit area increases, the environment becomes more crowded. The amount of food and other resources available for each individual declines. As this decline takes place, individual growth and reproduction also decline (point C on Fig. 16.3); death rates may also increase as a result of overcrowding and resource scarcity. Population growth begins to slow. Eventually, a population level is reached at which there is just enough food for each individual. Birth and death rates are equal, and the population ceases to grow.

Biotic Potential and Limiting Factors

The growth of a population in optimal environmental conditions, and in the absence of resource limitations, is called the **biotic potential** of the population. But as we have seen,

FIGURE 16.2 History of the elephant seal population
The Northern elephant seal (A, *Mirounga angustirostris*) was brought to near-extinction by 1890, primarily as a result of hunting. Estimates of the number of individuals alive at that time range from 20 to 200. Subsequently, as a result of protection afforded by its status as an endangered species, the population of elephant seals increased exponentially, at a rate of up to 9 percent per year (B). The current population of about 127,000 individuals is no longer considered to be endangered, but may be susceptible to disease because all extant individuals are descended from the same few survivors. This is an example of a *genetic bottleneck*.

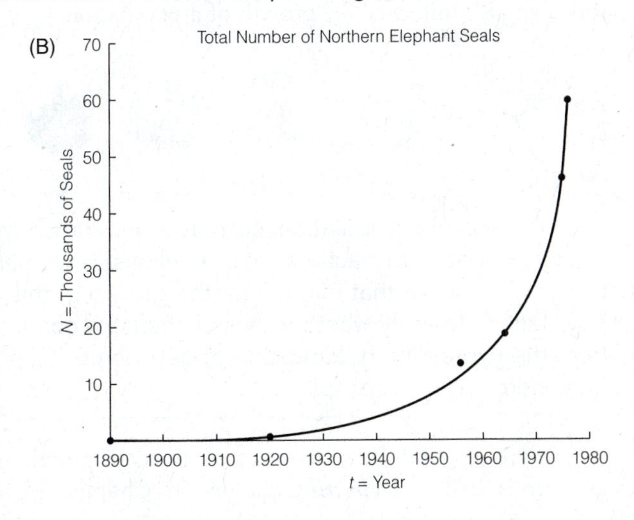

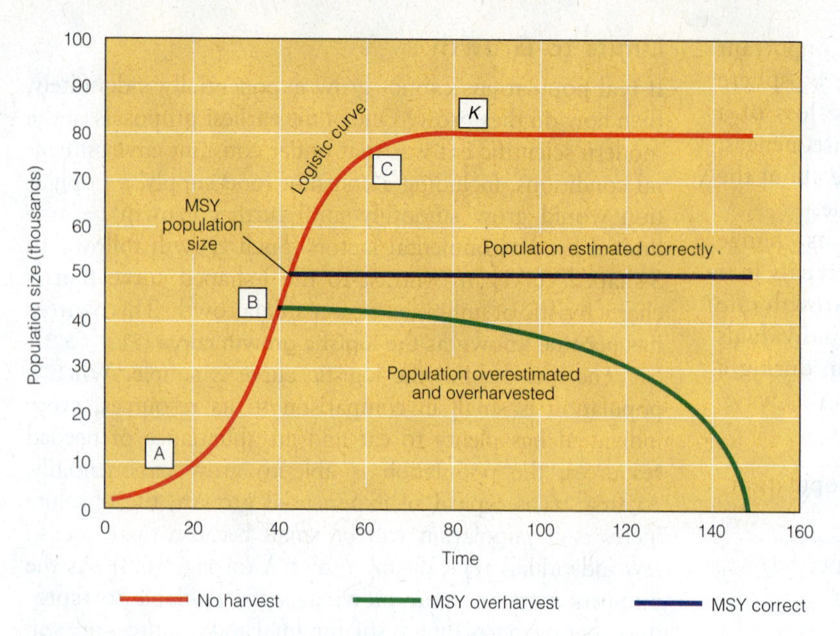

FIGURE 16.3 Logistic growth curve
The logistic curve is a model of population growth based on exponential growth that is limited by lack of resources and other factors. Growth rate is slow (A) when the population is small, even though resources and space are abundant, because few individuals are reproducing. Growth reaches a maximum at (B), because more individuals are reproducing but resources are still sufficient. Population growth slows at (C) in response to limiting factors such as lack of resources, overcrowding, or abundance of predators. The leveling-off point represents the carrying capacity (*K*).

the reproduction, survival, and growth of organisms in the real world—and therefore the growth of populations—are limited by the availability of resources. As a population grows, many factors begin to place limitations on its growth. If the system is densely populated, individuals may have a difficult time gaining access to the resources they need, such as light, water, food, shelter, or living space. Resources may become depleted or run out altogether. In a community with more than one species, additional factors come into play. For example, with increasing population density, an individual might be more likely to encounter a predator. Environmental risk factors, such as drought, a cold winter, an intense storm, or a population explosion among predators, can also influence the growth of a population.

Make the CONNECTION

Think of a natural population, such as a population of insects, or bears, or bacteria, or dandelions. How many things can you list that might limit the growth of this population? Identify whether these limitations come from the hydrosphere, atmosphere, geosphere, biosphere, or anthroposphere.

Anything that acts to control or cap population growth is called a *limiting factor*. In Chapter 15 we discussed the role of essential nutrients as limiting factors to biological productivity and to the growth of individual organisms, but chemical nutrients are not the only resources that can be limiting to population growth. For example, on the ground floor of a densely vegetated rain forest, light could be a factor that limits growth or survival. In a desert, water is usually the most important limiting factor. For pandas, which eat only bamboo, the availability of bamboo forest habitat is a limiting factor. For snowshoe hares, an abundance of lynx—an important predator—might be a limiting factor; and so on.

All of these factors—limitations to food, water, space, or light; barriers to reproductive success; competition, predation, disease; climate change; and many other environmental limitations and risk factors, both intrinsic and extrinsic—provide resistance to unlimited growth in populations. In other words, they prevent population growth from following an infinite exponential growth trajectory, limiting it instead to following a logistic growth curve.

Carrying Capacity

The leveling-off point of the logistic curve in Figure 16.3 represents the influence of limits, both intrinsic and extrinsic, on population growth. It also provides a theoretical maximum number of individuals in a population that can be supported by that particular ecosystem. This number is called the **carrying capacity**, denoted by the letter *K*.

Carrying capacity is an important concept both for populations and for the ecosystems that support them. There are two commonly used definitions: (1) the constant abundance reached by a hypothetical population growing according to the logistic curve; and, a more general definition, (2) the maximum number of individuals of a species (or population) that can persist in an area on a long-term basis, without affecting the future ability of the population to maintain that same abundance. Note that carrying capacity is species- and environment-specific. In other words, a particular forest ecosystem would have a certain carrying capacity for grizzly bears, but another—completely different—carrying capacity for, say, honeybees. The concept of carrying capacity also refers explicitly to the *future* or continuing ability of the ecosystem to support the population. A related concept, the ability to support a population at a particular level of abundance on a long-term basis without incurring serious or permanent damage to the ecosystem, is referred to as **sustainability**.

Population Stability

Carrying capacity represents a state of stability or equilibrium for a population. If the population is reduced below this level, births will (in principle) exceed deaths and the population will return to the carrying capacity. If the population overshoots and reaches a level above the carrying capacity, deaths will exceed births and the population will decline to the carrying capacity. A population that remains at carrying capacity is in a steady state and is called a *stable population*. A stable population has two attributes: First, it has a constant abundance unless disturbed; and second, when briefly disturbed and then released from the disturbing factor, it returns to the same abundance. This type of self-regulatory equilibrium, introduced in Chapter 1, is called *homeostasis*. A long-standing question in ecology has been whether populations can truly exhibit this kind of stability.

Evidence available today suggests that real populations undergo much more complex patterns of change over time; they do not grow according to a simple logistic curve, nor do they automatically return to a steady-state condition when disturbed. Instead, they are affected in complex ways by environmental change, by interactions with other species, by random events, by effects of individuals within the population on one another, and by the passage of time.

The rapid growth of a population is often used as a sign that a population is "healthy" in some sense. But rapid growth for a long period must inevitably be detrimental to a population. This is illustrated by the potential of the human population for growth, discussed in Chapter 19. One of the often-stated concerns about the growth of the human population is that, as it gets larger and larger, it faces the danger of running out of resources and polluting its habitat to an extent that might lead to extinction. Thus, a population or species that is likely to persist will not grow in a rapid, uncontrolled way over a long period.

"Boom-and-Bust" Cycles

A period of very rapid population increase is called a *population "boom."* Sometimes a booming population overshoots the carrying capacity of the ecosystem. This often results in a sudden dieback—a *population "bust"*—that can result from the rapid depletion of resources by the burgeoning population (FIG. 16.4A), or from an overabundance of predators, flourishing because their favorite prey is so available. In some cases the dieback returns the population to a sustainable level for the carrying capacity; in other cases, if the crash is too severe, the population may not be able to recover. Other populations approach the carrying capacity of the ecosystem more slowly, fluctuating but then leveling off without the dramatic "boom-and-bust" oscillation (FIG. 16.4B).

A well-known example of boom-and-bust oscillations was discussed by Maclulich (1937), in a classic study of variations in the populations of the snowshoe hare and lynx in northern Canada. The snowshoe hare, an r-strategist (see *A Closer Look: K-strategists and r-strategists*), reproduces quickly in the absence of predators (FIG. 16.5). The number of hares increases

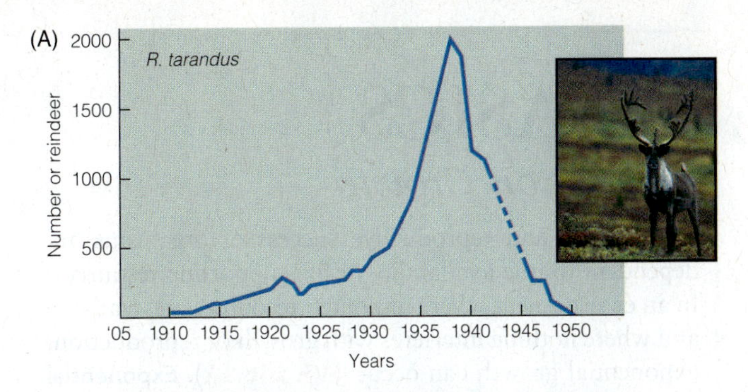

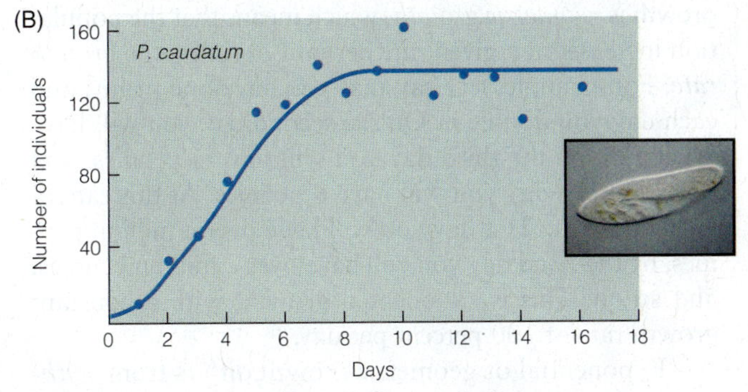

FIGURE 16.4 Growth in real populations
(A) In 1911, reindeer (*Rangifer tarandus*) were imported to St. Paul Island in the Bering Sea, where there were no natural predators. The population grew extremely rapidly, reaching more than 2000 by 1938. This exceeded the carrying capacity of the island, which became severely overgrazed. The winter of 1940–1941 was particularly cold and many reindeer starved, leading to a large dieback. Other fluctuations have since occurred, but the present reindeer population has stabilized at about 450–500. (B) In the 1930s, Gause carried out classic experiments on populations of *Paramecium*, single-celled organisms that feed on bacteria. The graph shows *P. caudatum* in a laboratory experiment in which food and other environmental conditions were held constant and no competitors were present. The population increases, overshoots slightly, fluctuates, and then levels off.

FIGURE 16.5 Boom-and-bust cycles
This graph, based on a classic study by MacLulich (1937), shows the abundance of the Canada lynx (*Lynx canadensis*) varying in response to the availability of the snowshoe hare (*Lepus americanus*), its main food source.

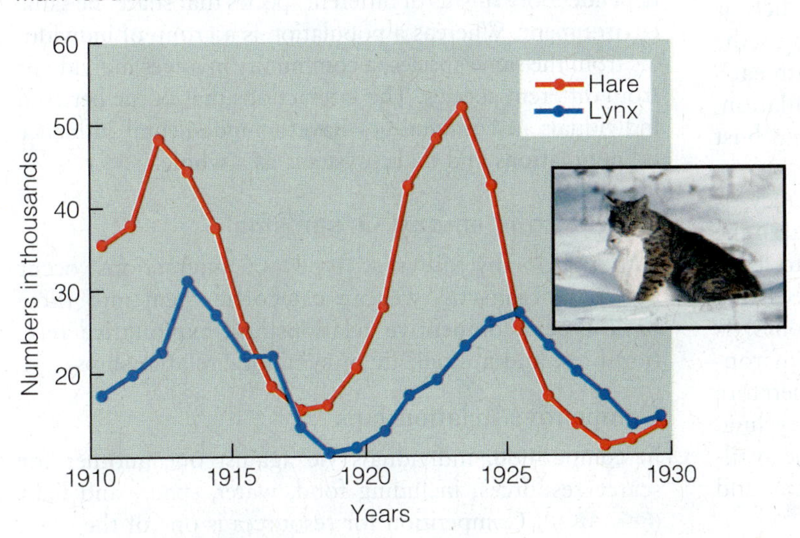

the BASICS

Population Growth

The growth and reproductive success of organisms are dependent on the availability of life-supporting resources. In an environment where unlimited resources are available and where nothing interferes with growth or reproduction, exponential growth can occur (FIG. B16.1A). **Exponential growth** is *geometric growth*, which means that the population increases at a given *rate* per unit of time—the *growth rate*. For example, let's say that you have one penny, and each day you double it. On the second day you will have 2 pennies, on the third day you will have 4 pennies, and on the fourth day you will have 8 pennies. At this rate of growth, by the 21st day you will have over a million pennies, by the 22nd day you will have over 2 million pennies, and so on. This is exponential growth, with a constant growth rate of 100 percent per day.

Exponential or geometric growth differs from *arithmetic* or **linear growth**, in which the *amount* of growth per unit of time, rather than the *rate* of growth, is a constant (FIG. B16.1B). If you again start with one penny, but this time simply *add* one penny each day, instead of doubling the number of pennies, the growth would be arithmetic. On the second day you would have 2 pennies, 3 pennies on the third day, 4 pennies on the fourth day, 21 pennies on the 21st day, and so on. You can see that the outcome would be very different.

The mathematical equation that describes exponential growth is:

$$dN/dt = rN$$

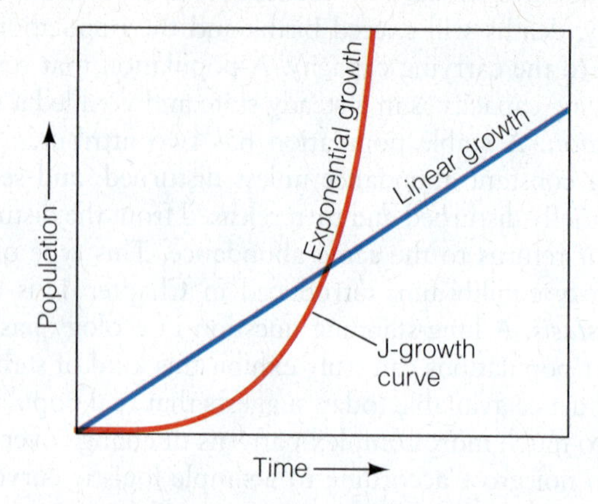

FIGURE B16.1 Exponential and linear growth Exponential growth is geometric growth, in which the *rate* of growth per unit of time is a constant. The J-curve is characteristic of exponential growth. Linear growth is arithmetic growth, in which the *amount* of growth per unit of time is a constant.

where N is the number of individuals, t is the unit of time, and r is the growth rate (like the interest rate in a bank account). The term d is a mathematical term that can be read as "change." Therefore, the equation can be read, "the change in the number of individuals per change in unit of time is equal to the number of individuals times the growth rate."

Try out this equation with a calculator, using the example of an organism that reproduces quickly, such as a hypothetical bacterium that reproduces itself by dividing in half every 30 minutes. Start with $N = 1$, $t = 0.5$ hr,

so rapidly that the population overshoots the carrying capacity of the ecosystem. After each population boom, the hare population encounters environmental resistance and experiences a dieback. The resistance comes from a lack of food, from environmental conditions such as cold, and from predation, as shown on the graph by the concomitant increases in the number of lynx. With each decrease in the hare population, the lynx population, lacking its prey, also decreases, and the boom-and-bust population cycle repeats itself.

It is clear that variability, rather than constancy, is the rule in nature. The question then is how to determine what factors cause the variations, and from this to learn to make forecasts about population change. Forecasting the fate of populations, especially endangered ones, is important to understanding how change in the environment might affect the diversity of life and might therefore affect other global factors, including chemical cycling, energy flow, the surface coloration of Earth, and the availability of resources that are important to humans and other species.

COMMUNITIES

Populations do not live in isolation; they live in ecological communities. A **community** is a group of interacting, interdependent organisms of different species that share the same environment. Whereas a population is a group of individuals from the same species, a community involves individuals from different species. The interactions that occur between individuals in communities have a fundamental influence on populations and on ecosystems as a whole.

Interactions among Organisms

Although many different types of interactions occur between organisms, we can categorize them into three basic types: competitive relationships, exploitative relationships, and mutualistic or symbiotic relationships.

Competitive Relationships

In **competition**, individuals vie against one another for scarce resources, including food, water, space, and light (FIG. 16.6). Competition for resources is one of the main and $r = 1.0$ (that is, 100 percent). After each calculation, the result gets recycled back into the equation as the new N. When does N reach one thousand? When does N surpass one million? One billion? (For a comparison, try the calculation again using a typical reproductive rate for another organism that reproduces more slowly, such as fruit flies, which might reproduce at a rate of 150 percent or $r = 1.5$ per day, or elephants, which might reproduce at a rate of 100 percent or $r = 1.0$ per decade.)

You should find that for the rapidly reproducing bacterium, N is greater than 1000 by the time 5 hours have passed, greater than 1,000,000 by 10 hours, and greater than 1,000,000,000 by 14.5 hours. If you plot the results of the penny example or the bacterium calculations on a graph, using N as the vertical axis and t as the horizontal axis, you will see that a graph of exponential growth is shaped sort of like the letter "J," increasing very slowly over time at first and then very rapidly. This *J-curve* (as seen in Fig. B16.1A) is characteristic of exponential growth. If you plot an arithmetic increase, on the other hand, you should get a straight-line graph, as shown in Fig. B16.1B.

Clearly, unlimited exponential growth would be problematic. If a bacterial population can grow from 1 to over a billion individuals in less than 15 hours, why isn't the whole world completely awash in bacteria (not to mention other organisms that reproduce rapidly, such as cockroaches, fruit flies, or mosquitoes)? The answer, of course, is that growth in biological populations in the real world is *not* unlimited; in fact, the reproduction of organisms in the real world is limited by a variety of factors.

One obvious oversight in the setup for our calculation is that we have only considered growth as a function of **birth rate**, the births in the population per unit of time (also called *natality*, and usually expressed as a ratio or percentage of the total population). However, organisms also die. The **death rate** in a population represents the individuals that have died during the same period (also called *mortality*, and usually expressed as a ratio or percentage of the total population), and must therefore be subtracted from the population. In the natural environment, organisms also can enter a population through *immigration*, and they can leave through *emigration*. These adjustments to population also must be made. The growth rate—a constant in simple exponential growth—is therefore the difference between the death (and emigration) rate and the birth (and immigration) rate.

Finally, and crucially, there are environmental limitations on population growth. At the beginning of this discussion we referred to a hypothetical situation in which unlimited resources are available to the organism. In the real world, this is never the case. Although populations can experience periods of explosive growth, limits to growth will eventually be imposed by the environment. Classically, ecologists have suggested three possible processes that limit populations: (1) a population is self-limiting; (2) interactions among species lead to competition for resources, therefore limiting the abundance of any one species; and (3) the nonbiological environment sets limitations. In the latter, the limitations can come in the form of shortages in the availability of food or nutrients, lack of water or sunlight, or an inhospitable climate.

FIGURE 16.6 Competition
Competition can be *interspecific* (between individuals of different species) or *intraspecific* (between individuals of the same species). Here, two American buffalo males (*B. bison*) compete over territory in Yellowstone National Park.

A Closer LOOK

K-STRATEGISTS AND *R*-STRATEGISTS

Different types of organisms employ different reproductive strategies. Examining these strategies can help us to understand the interplay between intrinsic and extrinsic factors and how they relate to population growth.

Some organisms invest their reproductive energy into producing a very large number of offspring, only a few of which are expected to survive. Such organisms are called ***r*-strategists** (FIG. C16.1A, B). The *r* stands for *reproduction*, and *r*-strategists are common in unstable or changeable environments where the ability to reproduce quickly is an asset to survival. Although there are exceptions, *r*-strategists tend to have small body sizes, early maturity, and short life spans. They invest heavily in the production of offspring, but then expend little or no energy caring for the offspring or raising them to adulthood. They typically produce large numbers of eggs, offspring, or seeds on a frequent basis, and have the capacity for offspring to be dispersed widely. Examples include many weeds—think of dandelions, for example, which broadcast their seeds via airborne fluff over great distances—and insects, some of which can produce thousands of individuals from a single egg.

***K*-strategists** (as in *K* for *carrying capacity*) rely on a very different reproductive strategy. What distinguishes *K*-strategists from *r*-strategists is their ability to compete for limited environmental resources. These organisms tend to be larger (although, again, there are exceptions), with longer life spans and later maturity. They produce a small number of offspring on an infrequent basis, but the expectation is that a higher proportion of the offspring will survive to adulthood; therefore, *K*-strategists devote considerable energy to the parental care and raising of their offspring (FIG. C16.1C, D). Examples include elephants, owls, whales, and humans. *K*-strategists (with the exception of humans) tend to be more vulnerable to extinction because of their low reproductive rate and susceptibility to environmental change. Where *r*-strategists can adapt to a changeable environment by reproducing in huge numbers—essentially playing a reproductive lottery game in the hope that some of their offspring will survive—*K*-strategists are heavily invested in their few offspring and their ability to compete for the resources they require, so stability and predictability of the environment are important.

FIGURE C16.1 *r*-strategists and *K*-strategists
Dandelions (A, *Taraxacum officinale*) and many other weeds are r-strategists. So are rabbits (B, *Lepus arcticus*). Gorillas (C, *G. gorilla gorilla*) and humans (D, *H. sapiens sapiens*) have few offspring but invest significant energy in ensuring their survival; they are *K*-strategists.

(A)

(B)

(C)

(D)

FIGURE 16.7 Parasitism and predation
(A) A blood-sucking river lamprey (*Lampetra fluviatilis*, which does inhabit marine areas in spite of its name) parasitizes a rainbow trout (*Oncorhynchus mykiss*). (B) Ladybugs (family Coccinellidae) are beneficial to gardeners because they feed on aphids (superfamily Aphidoidea), as seen here. (C) Many species develop features that protect them from predators. Here, a crab spider (family Thomisidae) confronts a moth caterpillar (undetermined species). The long, stiff hairs on the caterpillar act as a defense against predators.

factors that controls population in a community. In its most basic form, competition is related to the principle of *natural selection* (Chapter 14), which says that the organism that is best adapted to a particular environment will survive and prevail.

Competition can occur within a species (*intraspecific competition*), or between individuals of different species (*interspecific competition*). Intraspecific competition might take the form of territorial aggression, for example, in which an individual stakes and defends its territory against other individuals of the same species. This is common in many species of birds and among predatory animals such as wolves. Plants also compete for resources such as nutrients, water, and light. You have seen this if you have ever planted seeds too close together in a garden; at some point, the density of the foliage becomes so great that access to light will favor an individual plant. The plant that grows the tallest or broadest blocks the light, preventing the light from reaching adjacent seedlings.

Exploitative Relationships

In competitive relationships, the outcome is invariably negative for one species and often negative for both because each takes resources the other might have used. In exploitative interactions, one organism exploits another for its own gain, with the result that the relationship is decidedly beneficial for one and harmful for the other. The two most common exploitative relationships are *parasitism* and *predation*.

PARASITISM. A relationship in which one partner benefits while the other is harmed is called **parasitism** (FIG. 16.7A). The partner that benefits is the *parasite*; the other is the *host*. Parasites usually live in close physical proximity to the host. Some parasites live on the outside of the host's body; an example is a tick, which attaches itself to a mammalian host and feeds on its blood. Other parasites live inside the host's body; examples are human or canine intestinal parasites, such as tapeworms.

Many diseases, including many human diseases, are caused by parasites. Bubonic plague, for example, is caused by a bacterium that lives as a parasite in rats and the fleas that feed on them, and is then transferred to people. Another example is malaria, which spends part of its life cycle in the intestines of mosquitoes before being passed to a human host. A parasite that either causes or transmits disease or kills its host is called a *pathogen*. Killing the host is not a particularly good strategy, especially if the parasite depends exclusively on the availability of one particular type of host. In most cases, parasites just weaken their hosts by extracting nutrients, without actually killing them; alternatively, they may develop strategies whereby they are easily passed to a new host if the original host dies.

PREDATION. **Predation** is a type of exploitative interaction in which one organism, the *predator*, eats another organism, the *prey* (FIG. 16.7B). Technically, any organism that directly feeds on another—whether or not the prey is killed—is a predator. Thus, even parasites that derive nutrients from their hosts without killing them, or while killing them very slowly, are predators of a specialized type. However, we commonly think of predators as organisms that specifically hunt, kill, and then feed upon a certain type of prey.

Wolves, lions, owls, and alligators are examples of large predators; they are **carnivores**, consumers that only eat other animals. A specific type of carnivore is the **insectivore**, which eats only insects; examples include many types of birds, reptiles, and amphibians. Even some plants are insectivorous, deriving chemical nutrients from insects that they catch and dissolve (although their energy is still derived from photosynthesis). Some predators, such as caterpillars and pandas, are **herbivores**, or plant-eaters. Some predators are **omnivores**, which means they eat both plants and animals; common examples include pigs, crows, raccoons, and humans.

Populations of predators in the wild are dependent on the availability of their main prey species. You saw an example of this in Figure 16.5, showing the cyclic variation in abundance of the Canada lynx, a predator whose main prey is the snowshoe hare. MacLulich (1937) demonstrated that the abundance of the lynx was directly correlated to the availability of the snowshoe hare (and vice versa). This type of regularly alternating increase and decrease in population is called a *predator–prey cycle*. Many additional environmental factors contribute to population fluctuations; the relationship is rarely as straightforward as the one found by MacLulich in his classic study.

Through natural selection and evolution, predators develop characteristics that allow them to be efficient hunters, such as speed, a venomous attack, a disguise, or a clever means of attracting prey. Prey species also develop adaptations that help them evade capture or defend themselves. In plants, these include spines, thorns, or poisons that discourage animals from eating them. Animals display these and a host of other defenses, including warning colors, noises, postures, and other behaviors. Warning colors are intended to advertise the poisonous or otherwise unsavory nature of the prey. Camouflaging colors or patterns allow the animal to hide by blending in with the background. Some animals puff up, so they look fierce; others shoot barbed spines or smelly chemicals at potential predators (FIG. 16.7C). All of these tactics—both predatory and defensive—are genetic adaptations. Thus, in its own way, predation has led to the evolution of a greater diversity of species.

Mutualistic Relationships

Sometimes species interact in ways that are beneficial to both, or beneficial to one without harming the other.

SYMBIOSIS AND MUTUALISM. In its strict sense, the term **mutualism** refers to relationships that are beneficial to both of the interacting organisms. Many mutualistic relationships involve organisms of different species living in close physical contact, which is referred to as **symbiosis** (from the Greek words meaning "together life").[1] Symbiosis is closely related to parasitism because the parasite and the host commonly live in close physical proximity, too; the difference in mutualism is that both partners benefit, whereas in parasitism one partner is harmed.

Symbiosis (that is, symbiotic mutualism) is very common; most organisms, even humans, participate in symbiotic relationships. Humans, for example, host a wide range of microorganisms (about 100 common ones), some of which are of benefit to us, such as the organisms that reside in our intestines and help us digest our food. Another example is the relationship between photosynthetic microorganisms called *zooxanthellae* and the corals they inhabit (Chapter 19). The zooxanthellae provide food energy for the corals; the corals, in turn, provide nitrogen—an essential plant nutrient—for the zooxanthellae.

The mutual benefit in symbiotic relationships can come in the form of food or services provided. Mutualism is sometimes expressed by one species offering shelter or protection to another. For example, some kinds of ants "farm" aphids (FIG. 16.8A). They corral the tiny aphids, offering them protection from predators in exchange for being allowed to harvest a sugary liquid produced by the aphids.

In some cases, such as that of the coral–zooxanthellae relationship, the symbiotic partnership is mutually beneficial but the partners can still exist without one another. In other cases, one or both of the organisms is completely dependent on the symbiotic relationship and could not exist without it. An example occurs in ruminants, animals such as cattle, deer, moose, and reindeer, which have four-chamber stomachs. These animals are herbivores; they eat grasses that contain *cellulose*, a relatively indigestible plant material. Specialized bacteria that live in the guts of ruminants help them digest the cellulose, making the plant nutrition available; the animal, in turn, supplies the bacteria with food. This type of mutualism is an example of **coevolution**, in which two species evolve together into their dependency on one another.

COMMENSALISM. Another form of beneficial relationship is **commensalism**, in which one partner benefits without affecting the other, either negatively or positively. For example, some of the microorganisms that are commonly hosted by the human body appear to be neutral in their impact on us; they are just along for the ride, such as the mites that inhabit our eyelashes. Birds that ride along, eating the insects that are stirred up by a bull's hooves, benefit from a relationship that has little or no impact on the bull (FIG. 16.8B). Another example of commensalism is seen in epiphytes, a category of plant that is common in temperate and tropical rain forests. Epiphytes include mosses, vines, ferns, lichens, and other plants that grow by attaching themselves to the branches of trees. The epiphytes use the tree as an anchor, but they do not obtain any nutrients directly from the tree. The trees, in turn, apparently do not suffer any ill effects from the relationship, nor do they benefit from it. Desert plants that provide shade for other plants, shielding them from burning sun and providing a cool, moist environment for growth, provide another example of commensalism.

[1] Some scientists consider symbiosis to be limited to relationships in which both participants benefit—that is, mutualism. Others define symbiosis to include parasitic and commensal, as well as mutualistic, relationships. The defining characteristic in either definition is that the partners in a symbiotic interaction are different types of organisms that live in close proximity.

FIGURE 16.8 Symbiosis
(A) An ant (genus *Camponotus*) herds aphids (*Aphis nerii*) on a seed pod of milkweed (*Asclepias lanceolata*). The ant guards the aphids against predators and, in return for this service, feeds on aphid honeydew; this is symbiosis and mutualism. (B) Red-billed oxpeckers (*Buphagus erythrorhynchus*) catch a ride on a Cape buffalo (*Syncerus caffer*) in South Africa; this is symbiosis and commensalism.

Keystone Species

The examples used here to illustrate interactions among species, such as competition, symbiosis, commensalism, parasitism, and predation, seem simple and straightforward. In real communities and ecosystems, of course, the interrelationships among species are anything but simple. Ecosystems are open systems in which species interact in immeasurably complex ways with one another and with the physical environment. It may be tempting to try to disentangle one species from this complex web and to examine its role in isolation from the rest of the system. What is the ecosystem role of a particular species of toad, for example, or a certain type of insect, or a flowering plant?

In some cases, it seems as if it might be possible to remove a species from an ecosystem, with no discernible short-term effect on the rest of the system. However, ecologists do not fully understand how ecosystems respond to the removal of one or more species in the long term. This speaks to the central theme of **conservation**—that is, conserving and protecting species and the natural environment (sometimes with the implication that protection is occurring in order to ensure the availability of resources for later human use; we will return to this concept). If one or another species is lost from an ecosystem, we can't really be sure of the long-term impact on the rest of the ecosystem.

In some cases, a particular species is known to play a central role in maintaining the ecological balance in a given ecosystem (FIG. 16.9). For example, alligators burrow into the mud of subtropical wetlands, creating depressions that retain water during dry periods. These "gator holes" are crucial to the health of communities in subtropical wetlands—not just the health of the alligators themselves. Gator holes create habitat for aquatic organisms, where none would otherwise exist. Without alligators and gator holes, the entire ecological community and the food web of the ecosystem would be altered.

Other large predators, such as wolves, and some smaller, less conspicuous species, such as beavers, play similarly crucial roles in their particular ecosystems. A species that plays a fundamental role in an ecosystem, or whose influence is much greater than might be expected given its abundance, is called a **keystone species.** Not all keystone species are large, and not all of them are even animals; the crucial feature of a keystone species is its disproportionate impact on the functioning of the community and ecosystem as a whole.

We know that removing a keystone species from an ecosystem will be damaging or even disastrous for the ecosystem. Less well known is the ultimate or cumulative effect of removing other, less central species. Given

FIGURE 16.9 Keystone species
The sea otter (*Enhydra lutris*) is a keystone species because it controls the population of sea urchins (*Strongylocentrotus purpuratus*), which would otherwise inflict serious damage on kelp forest ecosystems in coastal marine environments. Here an otter eats a sea urchin in Monterey Bay, California.

this uncertainty it does seem wise, as suggested by famed ecologist Aldo Leopold, to mitigate our impacts on ecosystems by striving to "keep every cog and wheel."[2]

Habitat and Niche

The physical environment in which a population or an organism lives or could live is its **habitat.** In ecological communities, different species must share habitats. This leads necessarily to competition and other interspecific interactions of the types that we have discussed. An important aspect of community equilibrium is for organisms to sort out this competitive relationship and access the resources they require while sharing their physical habitat with others.

To understand how this works, we turn to the concept of **niche,** which refers to an organism's functional role and use of resources within the community. In his classic book *Animal Ecology,* published in 1927,[3] ecologist Charles Elton emphasized the role of the organism in the location that it inhabits. This was expanded upon by other scientists, including well-known ecologists (and brothers) Howard and Eugene Odum, who wrote that the "habitat is the organism's 'address,' and the niche is its 'profession,' biologically speaking."[4] For example, a squirrel's habitat—its address—is a forest or woodland, but its profession or niche is eating the seeds of certain trees. Over time, species evolve to avoid competition by adapting to different niches, allowing them to flourish while sharing the same habitat.

Fundamental and Realized Niche

The niche represents the set of all environmental conditions under which a species can persist and which it would be capable of occupying in the absence of any competition or crowding. An example is shown in FIGURE 16.10—a geometric representation of the niche of a reef-forming coral in terms of temperature, depth, and salinity. In such diagrams the niche always appears as a space, not a point, because organisms tolerate a *range* of values of each variable. The narrower the tolerable range, the smaller the niche and the more specialized the requirements of the organism.

We can illustrate what happens when niches overlap by considering one example of an experiment and one example of a study of organisms in their natural setting. In the first example, flour beetles were studied experimentally. One species of flour beetle prefers a warm, wet environment, and the other prefers a cool, dry environment (FIG. 16.11A). In a uniformly warm/wet or cool/dry environment, one species always dominated (FIG. 16.11B). In an environment with a range of conditions, however, the species were able to share the niche and coexist (FIG. 16.11C). Note that in both cases the beetles are not coexisting in exactly the same niche (FIG. 16.11D).

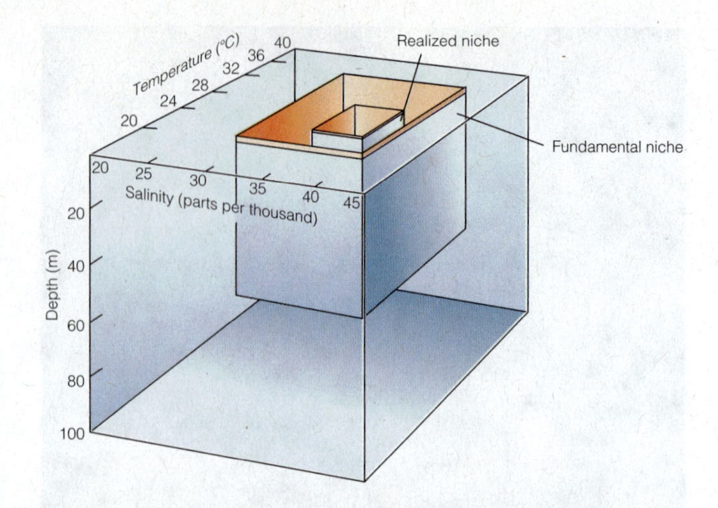

FIGURE 16.10 Geometric representation of a niche This diagram illustrates the temperature-salinity-depth tolerances that bound the niche of a marine organism. The graph tells us that the water temperature reaches from 16° to 40°C, but the organism thrives only between 20° and 36°C. The water salinity ranges from 20 to 45 parts per thousand, but the organism lives only when the salinity is between 29 and 41 parts per thousand (per mil, or ‰). The fundamental niche is the complete range of environmental conditions favored by the organism. If there is competition for this niche, the organism may end up only occupying a portion of that range, shown here as the realized niche (the most restricted set of conditions).

In the second example of overlapping niches, tiny flatworms called *Planaria* (family Planariidae) were studied in their natural environment, the bottom of freshwater streams. Studies in Great Britain showed that some streams contained one species, some streams another species, and still other streams contained both. The stream waters are cold at their source in the mountains and become progressively warmer as they flow downstream. Each species of flatworm occurs within a specific range of water temperatures. In streams where species A occurs alone, it is found from 6°C to 17°C (FIG. 16.12A). Where species B occurs alone, it is found from 6°C to 23°C (FIG. 16.12B). When they occur in the same stream, however, their temperature ranges are both reduced: Species A occurs in the upstream sections where the temperature ranges from 6°C to 14°C, whereas species B occurs in downstream areas where temperatures are warmer, from 14°C to 23° (FIG. 16.12C). The temperature conditions in which species A occurs when it has no competition from species B is called its **fundamental niche.** The set of conditions under which it persists in the presence of B is called its **realized niche.** The realized niche is thus a subset (a portion) of the fundamental niche. Species B is likewise occupying a realized niche in Figure 16.12C; its fundamental niche is larger, as shown in Figure 16.12B.

The flatworm example shows that species divide up resources along environmental gradients in nature. (You can imagine how the discussion might go if flatworms could talk: "Look, there are two of us living in this stream. Both of us could survive anywhere in the stream, but we

---

[2] "To keep every cog and wheel is the first precaution of intelligent tinkering." *Leopold, Aldo (1966). A Sand County Almanac.* Ballantine Books, 295 pp.

[3] Elton, C.S. 1927. *Animal Ecology*. Sidgwick and Jackson, London, UK.

[4] Odum, Howard, and Odum, Eugene, 1953. *Fundamentals of Ecology* (1st edition). W.B. Saunders, Philadelphia. This was probably the first textbook to be written on the subject of ecology.

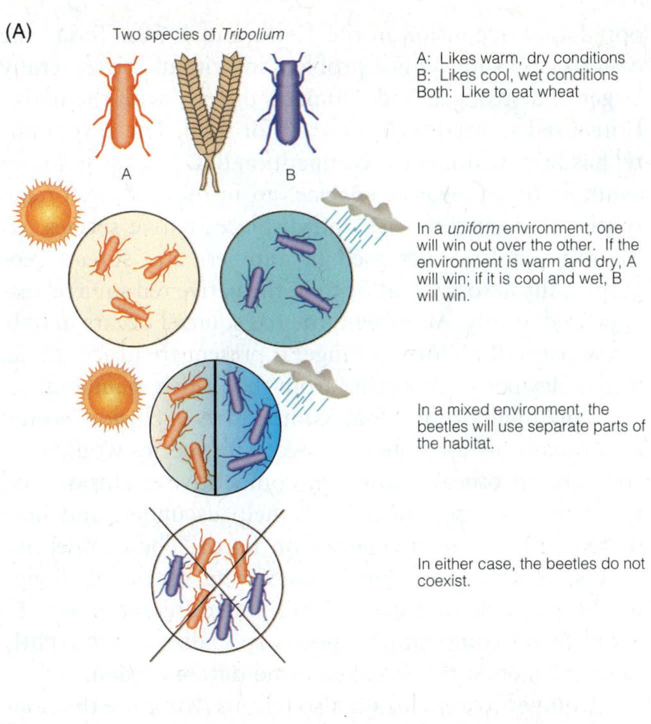

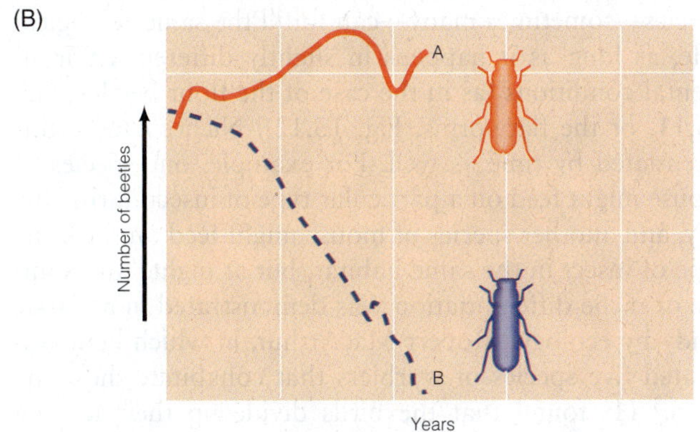

FIGURE 16.11 Overlapping niches
What happens when niches overlap? Experiments with flour beetles (*Tribolium*) demonstrate that one species will dominate if the environment is uniform, but they can split the niche if there are variations in environmental conditions.

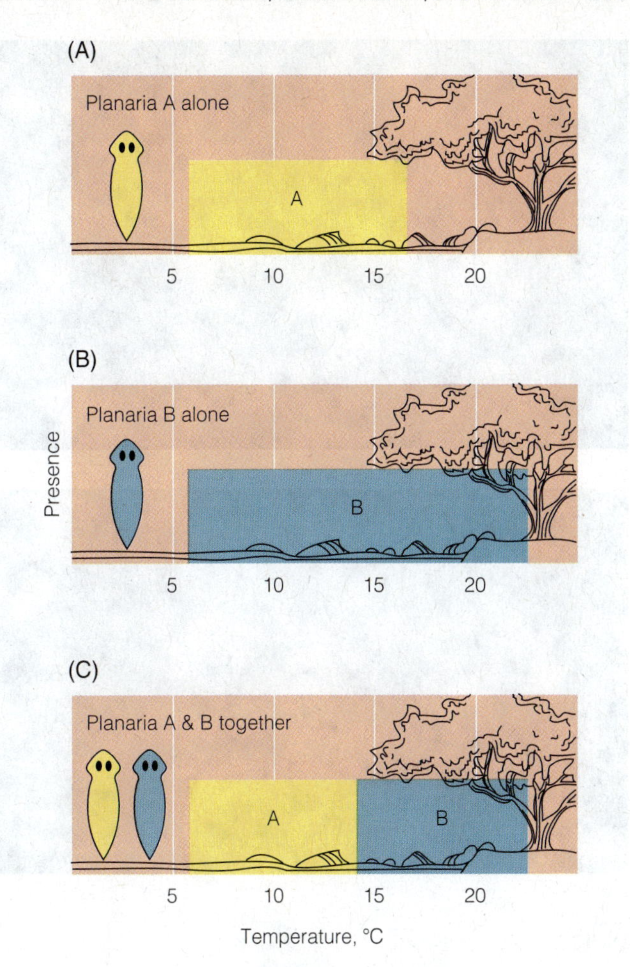

FIGURE 16.12 Fundamental and realized niche
Freshwater flatworms (*Planaria*) live in cold mountain streams in Great Britain. (A) Species A occupies a fairly restricted temperature range, even in streams where it occurs alone, preferring cold water. (B) Species B has a broader range of temperature in which it survives in streams where it occurs alone. (C) In streams where the two species occur together, both of their temperature ranges are limited; they have shared the niche according to the environmental gradient in the temperature. (A) and (B) represent the fundamental niches of the species, whereas (C) shows their more restricted realized niches.

can't both occupy the same niche. I'm more comfortable over here where it's warm, and you prefer the cold. So let's just stay out of each other's way, and everyone will be happy.") Of course, temperature is only one aspect of the environment. Flatworms also have requirements in terms of the acidity of the water and other factors. We could create graphs for each of these factors showing the range under which A and B occurred. The collection of all those graphs would constitute the complete description of the fundamental and realized niche of each species.

Competitive Exclusion

Another way to think about competition for resources within a given habitat is to consider the **competitive exclusion principle**, which states that two species that have exactly the same requirements cannot coexist in exactly the same habitat and niche. In that situation, one species will always win out over the other, either gaining the larger or richer niche, or driving the losing species out of the habitat altogether. Russian biologist G.F. Gause (mentioned in the context of Figure 16.4) demonstrated this principle in his classic 1934 study on *Paramecium caudatum* and *Paramecium aurelia*. In a series of experiments, *P. aurelia* drove *P. caudatum* to extinction in the laboratory environment because the two species were competing for exactly the same resources.

Here is an example of how competitive exclusion works. The eastern gray squirrel from North America (**FIG. 16.13A**) was introduced into Great Britain because people thought it was attractive and would add to the scenery in parks and towns. About a dozen different attempts were made at this introduction, the first as early as 1830. By the 1920s, the gray squirrel was well established in Great Britain, and it underwent a major

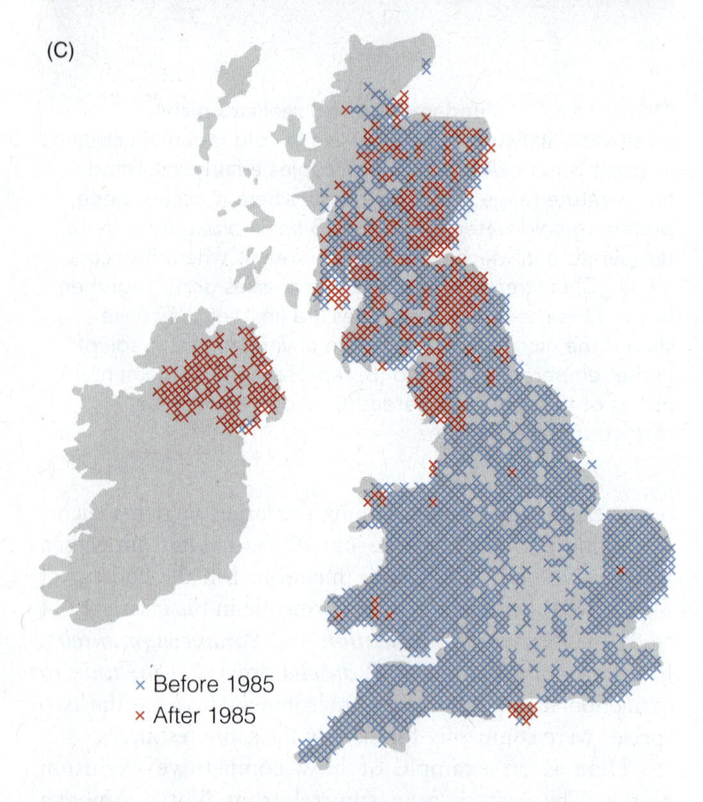

FIGURE 16.13 Competitive exclusion
The eastern gray squirrel (A, *Sciurus carolinensis*) was introduced into the British Isles and now outcompetes the red squirrel (B, *Sciurus vulgaris*) in Great Britain and Ireland, an example of the competitive exclusion principle in action. The British red squirrel is declining; its current range (C) is much smaller than its original range and is highly fragmented.

population expansion in the 1940s and 1950s. Today the eastern gray squirrel is a problem in Britain. It is generally larger and stronger, and competes directly with the native British red squirrel (**FIG. 16.13B**) for food. The gray squirrel has been winning the competition (**FIG. 16.13C**); it is an example of an **invasive species**, an introduced or foreign species that outcompetes and displaces native species. As the gray squirrel increased in numbers and spread geographically across Great Britain, the native red squirrel disappeared locally. At present, the red squirrel occurs in only a few parts of its former range; if present trends continue, it may disappear altogether from the British mainland.

You might think that competitive exclusion would mean that only a few highly successful species would outcompete all others. However, competitive exclusion and the niche concept can actually help us understand how there can be so *many* species on Earth. The competitive exclusion principle *requires* coexisting species to divide up the available resources by fulfilling different niches. To avoid direct competition, species specialize to fill slightly different niches; this is called **niche differentiation**.

Competitive exclusion also tells us that more than one species—sometimes many—can fulfill the same ecological role, as long as it happens in slightly different environmental conditions (as in the case of the flour beetles, Fig. 16.11, or the flatworms, Fig. 16.12). Niches can be differentiated by time as well. For example, one species of mouse might feed on a particular type of insect during the day, and another species of mouse might feed on the same type of insect in the same habitat, but at night. An example of niche differentiation was demonstrated in a classic study by ecologist Robert MacArthur, in which he investigated five species of warblers that cohabitate the same trees.[5] He found that the birds divide up their feeding both temporally and geographically, with different species focusing their feeding efforts on slightly different parts of the trees and at different times of the day (**FIG. 16.14**).

This sharing differentiates the species' use of the resource and avoids direct competition; it is an example of **resource partitioning.** Animals are not the only organisms that partition resources. For example, in a deciduous forest environment it is common for ground-covering plants to be the first new growth to emerge in the springtime. That way, the low, understory plants can take advantage of the available light to photosynthesize and grow, making the best possible use of the resource before the taller trees sprout new leaves and block the light.

Specialists and Generalists

All of this suggests that the greater the complexity of the environment, the more different niche opportunities there may be and the more opportunity for organisms to partition resources and to specialize. Thus, a planet with a great variety of environments can support many species, each adapted to different ranges of conditions. In an overall

[5] MacArthur, R.H. 1958. Population ecology of some warblers of northeastern coniferous forests. *Ecology* 39:599–619.

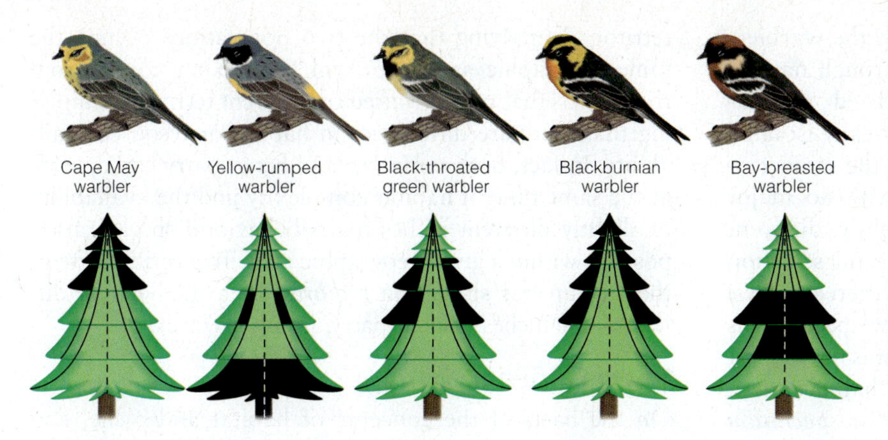

FIGURE 16.14 Niche differentiation
MacArthur discovered that five species of warbler occupying the same tree could forage for the same resources if they differentiated their niche both temporally and spatially. The parts of the tree shown in black represent the areas where each of the different species spent at least 50 percent of their foraging time. He also found that the birds moved differently from one feeding area to another, some moving concentrically around the tree and others moving vertically, thus splitting the resource geographically.

sense, this provides the biosphere with a kind of insurance against environmental change because the more species there are in different niches, the greater the likelihood that some of them will survive a large-scale catastrophe.

Species that have a wide range of tolerance occupy large, broadly defined niches; they are **generalists.** Species with a narrower range of tolerance occupy smaller, more narrowly defined niches, and their requirements for survival are more specific; they are **specialists.** An environment that varies rapidly over a wide range does not allow specialists to survive. For example, in central Alaska near Fairbanks, temperatures range in midsummer from the high 20°s C to the minus 40°s C in the winter. A species that can survive only when temperatures are between 20° and 25°C could not persist there. As a result, the boreal forests of the far north have few species of trees—only about 20 worldwide. Some occupy wetlands and some dry sandy soils, but all are generalists in terms of temperature tolerance.

We can refer to the niches in the far North as "large," meaning that they have a wide range of one or more environmental variables. The species that thrive in this type of environment cannot be too picky; they must be generalists, able to withstand wide variation in environmental conditions. Typically, the larger the niches in a broad geographic region, the more generalists there are, the fewer species overall, and the lower the diversity in that ecosystem.

In contrast, where smaller, more narrowly defined niches are available within a geographic region, the more specialization, and therefore the more different species there will be, the higher the diversity. In low latitudes, where climatic conditions are more or less constant year-round, one might expect narrower niches—that is, niches with a narrow range of variability, promoting higher diversity. In fact, such is the case both on land and in the oceans. The tropical rain forest, with optimal conditions of heat and moisture that permit year-round growth, is the most diverse terrestrial ecosystem. In the marine environment, the continental shelves of the intertropical region are the most diverse ecosystems.

Speciation

By examining species and their niches, we can see that natural selection and evolution will take a species along a pathway leading to one or the other of these survival strategies. We know (from Chapter 14) that *speciation*—the process whereby a new species arises—depends on a subpopulation of an existing species becoming reproductively isolated from the original population. We can relate this to the concepts of habitat and niche.

Perhaps the simplest way to begin thinking about populations becoming reproductively isolated is to think about it in a geographic sense. Consider the case of a population that becomes geographically split as a result of the habitat being segregated by a geologic event (such as a lava flow) or the influx of a major river. The two populations, now living separately, can still reproduce successfully—they still belong to the same species. However, over time it is likely that each population will become adapted, through natural selection, to the specific environmental conditions in which they are now living. Over time, they may become so differentiated genetically that they would be unable to breed successfully with each other, even if the barrier between their two geographic habitats were to be removed.

When a new species arises in this manner—through complete geographic isolation—it is called **allopatric speciation** (FIG. 16.15A). *Allopatry* comes from root words that mean "separate" and "territory," and the term specifically implies that the two populations (eventually two species) occupy completely separate habitats.

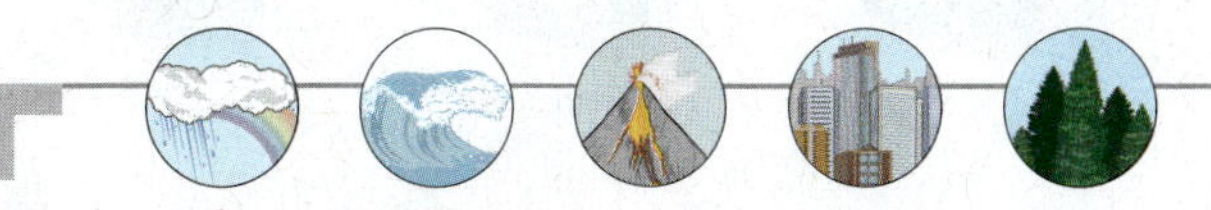

Make the CONNECTION

Allopatric speciation can occur when part of a population becomes geographically isolated. Causes of geographic isolation could be a river (the hydrosphere) or a lava flow (the geosphere). Can you think of some other ways that part of a population might become separated and isolated from the rest? Try to identify possible causes from each of Earth's subsystems.

We have also seen that niche differentiation within a given habitat can drive organisms to become specialized in their patterns of resource utilization (like the warblers that feed in different parts of the tree). Through natural selection, eventually, such small, specialized populations may become reproductively and genetically isolated (FIG. 16.15B). Even though they still live in the same geographic area as the original population, with no major geographic barrier, they may occupy slightly different habitats and fulfill different roles, and natural selection will therefore act upon them in slightly different ways. If this process eventually gives rise to a new species, it is called *peripatric speciation* if the new habitat is nearby but geographically separate from the original habitat (*peri-* means "around" or "beyond"), or *parapatric speciation* if the new habitat is immediately adjacent to the original habitat (*para-* means "near," "close," or "beside").

We can take this idea even further to propose the idea of **sympatric speciation** (and the related concept of *heteropatric speciation*), in which a new species arises in the absence of any geographic barrier—that is, in completely overlapping habitats (FIG. 16.15C). In this case, reproductive isolation and speciation may take place as a result of behavioral specialization and differentiation.

Let's look at this idea a bit more closely. The term *sympatry* comes from root words that mean "same territory," implying that the two populations occupy the same geographic area; the term *heteropatry* comes from root words that mean "varied or different territory," implying that there are differences in habitat or resource availability. In fact, both sympatry and heteropatry can operate at the same time, if habitat complexity and the availability of slightly different niches make behavioral specialization possible within a given geographic area. To a certain extent, then, sympatric speciation *requires* the availability of differentiable niches—heteropatry, in other words.

FIGURE 16.15 Speciation
In allopatric speciation, a portion of the population becomes reproductively isolated as a result of a geographic barrier. In peripatric and parapatric speciation, part of the population colonizes an adjacent but slightly different habitat, causing natural selection to have a slightly different outcome. In sympatric speciation, variation occurs within the population, in the absence of any barriers, perhaps as a result of genetic variation or behavioral differentiation. The final result is that the new species is distinct from the original species; even if all barriers were removed, the two species would not be able to interbreed successfully.

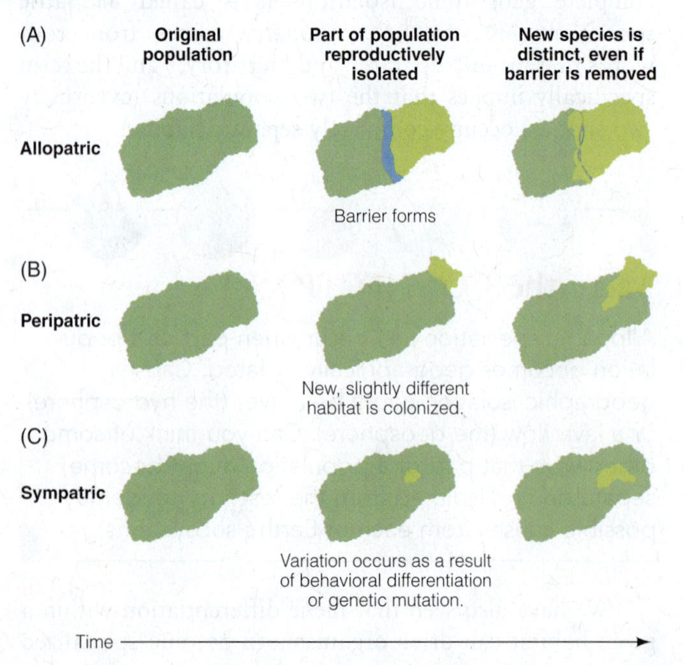

Species Vulnerability

On the basis of the concepts of habitat and niche, and the intrinsic and extrinsic factors that affect population growth and stability, we can begin to make some predictions about which species may be at greatest risk as a result of environmental change.

For example, consider the Giant Panda, a member of the bear family (*Ursidae*). Pandas are extreme specialists; they survive on a vegetarian diet that consists almost exclusively of bamboo (FIG. 16.16). They do not store much body fat (an intrinsic factor), and therefore they require a constant and predictable source of food. This puts them at great risk because protected bamboo forest habitat is very limited worldwide. Pandas are quite specialized in several other ways, too, such as the temperature, topography, and elevation of their habitat (extrinsic factors), mostly because these characteristics all affect the growth and availability of the types of bamboo they require.

Pandas are also extreme *K*-strategists, which means that they produce a small number of offspring infrequently (typically one cub every two years) and then invest considerable energy in the protection and raising of the offspring. Because their reproductive rate is so slow and their habitat needs are so restricted, it is very difficult for a panda population to adapt, adjust, and recover after a major environmental disruption.

BIODIVERSITY

As we have seen, competition, predation, and other types of interactions, resource partitioning, niche differentiation, habitat complexity, and specialization have led to increasing richness and variety of species over geologic time. We call this variety of life forms **biodiversity** (or *biological diversity*). At any particular time on Earth, some species have been very abundant and others have been very rare. Over the history of the Earth system, the kinds of organisms that have been abundant have changed greatly, as you learned in Chapter 14.

Defining and Measuring Biodiversity

High diversity makes communities more resilient, more able to adapt to change, and more likely to withstand major environmental upheavals, with at least some species emerging intact. It would be helpful if we could measure biodiversity quantitatively; this would allow us to monitor it and perhaps better protect it. But biodiversity is actually

FIGURE 16.16 Giant Panda: Specialist and *K*-strategist Many factors contribute to the vulnerability of the Giant Panda (*Ailuropoda melanoleuca*) from China. Its population is small and scattered. It is an extreme specialist, feeding only on certain types of bamboo, which are at risk because of habitat modification. It is also an extreme *K*-strategist, reaching sexual maturity late, bearing small litters, and investing significant energy in the rearing of offspring.

TABLE 16.1 Kinds of Biological Diversity

Biological diversity involves three different concepts:

(1) *Genetic diversity*—the total number of genetic characteristics, sometimes of a specific species, subspecies, or group of species.
(2) *Habitat diversity*—the diversity of habitats in a given unit area.
(3) *Species diversity*—which in turn involves three ideas:
  (a) *species richness*—the total number of species
  (b) *species evenness*—the relative abundance of species
  (c) *species dominance*—the most abundant species

a more complicated concept than it might seem to be at first glance. There are several kinds of biodiversity (see TABLE 16.1) and different ways of measuring each of them. We will consider *genetic diversity*, *habitat diversity*, and *species diversity*; each of these tells us something different about the variety of life and life-supporting environments on this planet.

Genetic Diversity

Genetic diversity refers to the variability or heterogeneity that is available among the DNA of individuals within a population or species. This is important, as you will recall from the discussion of natural selection in Chapter 14, because genetic variability gives a species a better chance to compete, adapt, and survive in a changing environment. The variety present in the gene pools of populations is the raw material for natural selection, making possible an almost infinite number of genetic recombinations and (therefore) evolutionary developments. If a population experiences a crisis, such that there are only a few individuals left alive, that population may be doomed—even if there are still breeding pairs alive—if their genetic diversity is too limited to allow them to adapt to changing environmental conditions.

We have seen an example of this in the elephant seal, reduced to a population of perhaps 20 individuals. The current population of more than 120,000 elephant seals is descended from just those few individuals. Their genetic diversity is extremely limited, and this makes the species vulnerable, even though there are many more individuals alive than there were before. This type of situation is referred to as a **genetic bottleneck**. The population has survived, but it has come through the bottleneck with a greatly reduced range of genetic variation with which to support natural selection and adaptation (FIG. 16.17).

There are many different ways of measuring genetic diversity, most of which are beyond the scope of this book. For example, genetic diversity is high in a population with individuals that have many different alleles. *Alleles* are alternative forms of the same *gene* (a portion of DNA) occupying a given position on a *chromosome* (the physical structural unit of DNA). Alleles encode for the production of proteins, so they control the expression of all kinds of inherited traits.

FIGURE 16.17 Genetic bottleneck
If a population undergoes a significant decline, it may rebound. However, a genetic bottleneck can cause the available genetic diversity in the new population to be significantly reduced.

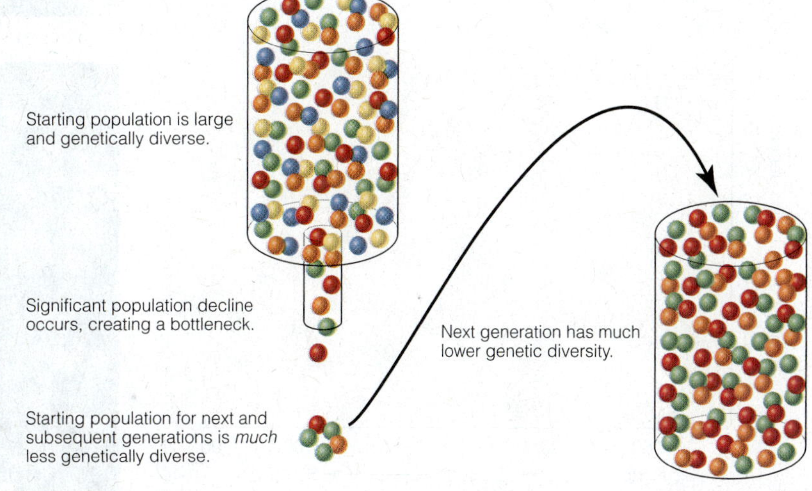
Starting population is large and genetically diverse.
Significant population decline occurs, creating a bottleneck.
Starting population for next and subsequent generations is *much* less genetically diverse.
Next generation has much lower genetic diversity.

Having a wide variety of alleles (a condition called *heterozygosity*) means that the expression of these traits will likely be more varied in the population. This increases the chances that some individuals will have characteristics that are favorable for survival in a particular environment. In contrast, the sameness or lack of variation in alleles (*homozygosity*) means that there will be less variation in inherited traits, and thus less capacity to adapt to a changing environment.

Habitat Diversity

Habitat diversity (or **ecosystem diversity**) refers to the variety of habitat types available in an ecosystem, and the biological richness and complexity of those habitats. Many experts believe that preserving habitat diversity is the key to preserving biodiversity in general, and may be even more important than focusing on the survival of individual species.

We have seen that narrower niches and greater niche diversity promote biodiversity by driving geographic and behavioral specialization, differentiation of resource utilization, and (ultimately) speciation. Only ecosystems that are stable and consistent for long periods of time can provide the complexity and subtle variety of niches that are required to support a large number of species. Diversity *depends on* ecosystem stability, but it also *promotes* ecosystem stability (a positive feedback cycle). In a general way, therefore, it makes sense that diversity is highest in low latitudes (in tropical regions), where the climate is stable and the environment has been unchanged for a very long time, and lowest in high latitudes (near the poles), where the climate is less predictable and ecosystems periodically undergo glaciation—a massive environmental change (FIG. 16.18).

It might be imagined that if the tropical climate extended over the whole Earth, global diversity would be greater than it is today. The high latitudes would be lush with vegetation and new tropical species. But this reasoning does not take into account the possibility that, in expanding poleward, the equatorial habitats might support not new species but simply more of the same species that live in the tropics at present. Species that now live in colder climates, such as polar bears, penguins, skuas, moose, caribou, and the rest, would not exist, and global diversity would be impoverished by their absence.

Here, then, another factor in global diversity comes into play. It is the extent to which the Earth system is divided into subsystems by barriers to the migration of organisms, sometimes referred to as *provinciality*. These barriers can take the form of climatic gradients, seaways, or mountains between land masses, or land between seas. Provinciality is high at present as a result of such barriers, and so is global diversity. For instance, several species of "anteater" (they really eat termites) are living in South America, South Africa, southwestern Asia, and Australia because they are kept apart by geographical barriers and cannot compete with each other (FIG. 16.19). Thus, the

FIGURE 16.18 Biodiversity and latitude
(A) The number of mosquito genera varies as a function of latitude, peaking in low-latitude regions. Comparable patterns are observed in clams, turtles, parrots, foraminifera, termites, snails, frogs, snakes, lizards, crocodiles, reef-forming corals, amphibians, butterflies, and palm trees.
(B) Low-latitude environments, where the climate is stable and predictable year-round, tend to offer complex habitat and many narrow niches, contributing to high biodiversity.
(C) High-latitude environments, like this northern boreal forest, have a wide variation in climatic and other environmental conditions. The niche is very broad, which tends to limit overall biodiversity.

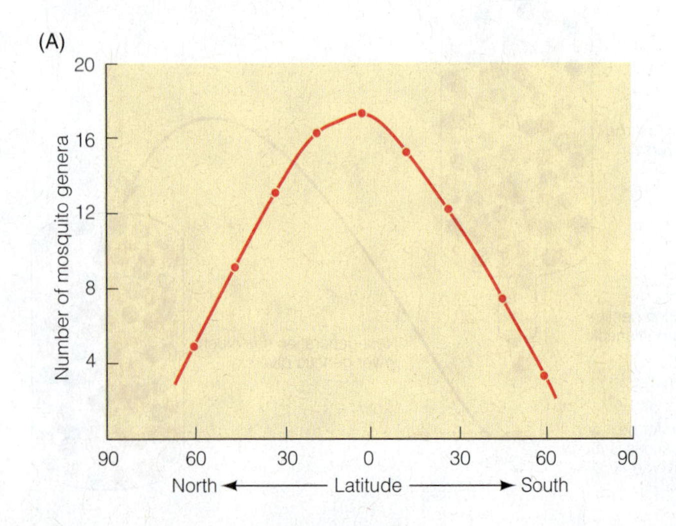

FIGURE 16.19 **Provinciality and biodiversity**
So-called anteaters (usually they eat termites) from different parts of the world are different species: They fill the same niches but are not competitive because they do not come in contact. The species shown here are (A) short-beaked echidna (*Tachyglossus aculeatus*) from Australia, (B) tamandua (*Tamandua mexicana*) from Central America, (C) giant anteater (*Myrmecophaga tridactyla*) from Venezuela, and (D) pangolin (*Manis javanica*) from Malaysia.

more Earth's ecosystems are separated from each other, the greater is the potential for allopatric speciation, and the greater the global biodiversity.

Habitat and ecosystem diversity depend on variations in physical aspects of the environment, such as differences in temperature, precipitation, and light levels, topography, altitude, soil type and chemistry, and other factors. But if a complex physical environment favors biodiversity, high biodiversity, in turn, contributes to the complexity of the habitat. In other words, the more species of trees that are available, the more complex is the available habitat for tree-dwelling species.

Species Diversity

To fully quantify biodiversity we will to approach the problem of measuring how many species live in the biosphere and in specific ecosystems; thus, **species diversity** is the third major component of biodiversity.

SPECIES RICHNESS. There are several different ways to think about species diversity, which can lead us to some possible quantitative measures. **Species richness** is a very basic measure that refers to the number of species in an ecosystem or community. Ecologists have been able to demonstrate some fundamental relationships between species richness and habitat diversity. For example, the larger the area of habitat, in general, the more species it can support; this is referred to as the *species-area relationship* (FIG. 16.20).

It is more challenging to measure the species richness of an ecosystem than you might think. For example, should you count the number of species, or the number of genera, or orders, or families, or some other taxonomic level? Counting species would be ideal, but it is much easier to be accurate at higher taxonomic levels. Measurements of species richness are also heavily dependent on the sampling procedure used; unless the sampling is broad and exhaustive *and* the ecosystem is extremely simple, you stand a good chance of missing many of the less well-represented species in the ecosystem.

FIGURE 16.20 Species-area relationship
In general, a larger area of habitat can support a larger number of species.

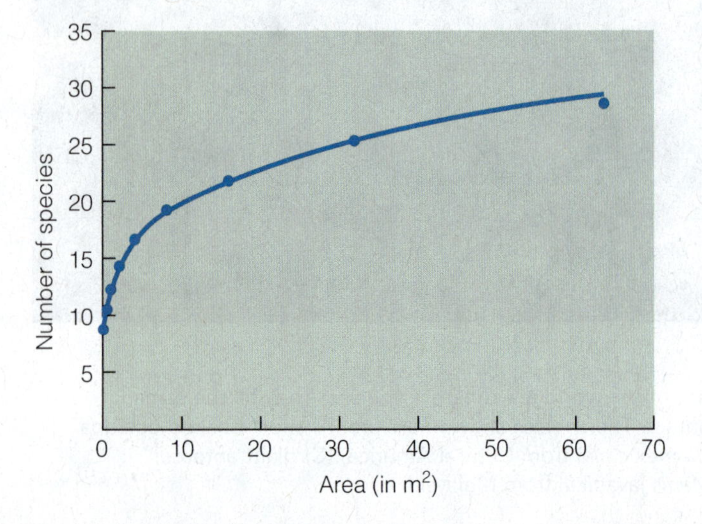

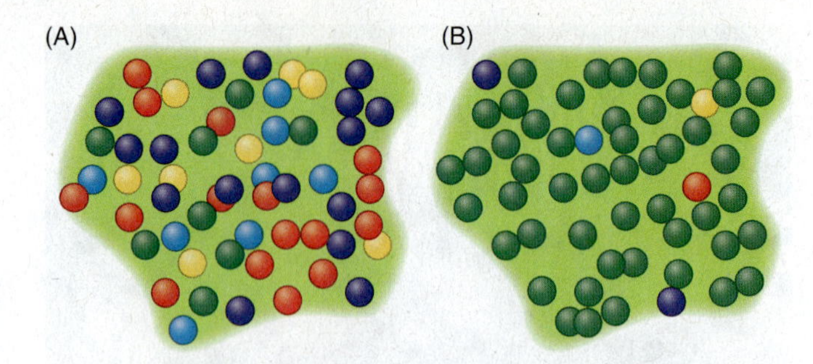

FIGURE 16.21 Richness, dominance, and evenness
Habitats (A) and (B) have the same number of organisms and the same number of species (*species richness*). However, one species is overwhelmingly dominant in habitat (B), so it has lower *species evenness* than habitat (A).

SPECIES EVENNESS. In general, high species richness (that is, a large number of species) is a good thing for ecosystem stability, but this number on its own is not a complete indicator of biodiversity. Think about two communities that have the same number of species (FIG. 16.21). Let's say that in one of the communities, 90 percent of the individuals belong to the same species. In the other community, however, each species is equally represented with individuals. Which community is more diverse? In the first, a single species is overwhelmingly dominant; all others are only marginally represented. This limits the overall species diversity of that community. If you were to take a walk through that ecosystem, there is a good chance that you would only encounter the dominant species.

In the second community, the species richness (overall number of species) is the same as in the first, but the **species evenness**—the relative abundance of individuals within each species—is much higher. This time, if you were to take a walk through the ecosystem, the chance of encountering a variety of species is much higher, and the overall diversity of the ecosystem is thus greater. A number of indexes have been devised to measure species evenness. One example is the *Shannon index*, which is at a maximum in a situation where all species are equally represented in the community.

Threats to Biodiversity

The extinction of species is not a new phenomenon, but today we are in the midst of an extinction caused by human actions, rather than by a natural environmental catastrophe. We will revisit this "sixth great extinction" and some of its human causes in Chapter 19.

Endangered Species

A species in imminent danger of extinction is called an **endangered species**. Current examples include the black rhino, Siberian tiger, giant panda, mountain gorilla, and

the BASICS

How Many Species?

There are so many kinds of organisms that even today no one knows the exact number of species on Earth. New species are being discovered all the time, especially in little-explored areas such as tropical rain forests and the deep ocean. Approximately 1.5 million species have been identified and named, but estimates of the total number that may be present on Earth are much higher, ranging from 30 million to as high as 100 million.

Insects and plants make up most of the known species: There are approximately 500,000 insect species and 400,000 plant species. Many of the insects are tropical beetles that inhabit local areas in rain forests; in fact, of all the insects, beetles have the largest number of known species (about 350,000, leading the famous evolutionary biologist J.B.S. Haldane, an atheist, to remark that, if there is a God, he has shown "an inordinate fondness for beetles").

As more explorations are carried out, especially in tropical areas, the number of identified invertebrates and plants will continue to increase. Mammals, the animal group to which people belong, comprise a comparatively small number of species—slightly more than 4000. However, even with their relative rarity, new mammalian species continue to be discovered. Over 400 new species of mammals—about 10 percent of the total inventory—have been identified and named since 1990. These include a new primate genus (*Callibella*, discovered in Brazilian forests in 1998), and numerous mice, voles, shrews, and other small rodents (FIG. B16.2).

FIGURE B16.2 Newly discovered species
New species are being discovered all the time. This black-headed dwarf marmoset (*Callibella humilis*) is the second-smallest primate species ever discovered; the adult is only 10 cm long. It was discovered in a Brazilian rainforest in 1998.

many others. Some are well known; others are obscure (FIG. 16.22). There are currently about 1500 endangered species in North America, and more than 16,000 species are listed as threatened or endangered worldwide. An especially large number of these are amphibians, some of which may be having trouble adapting to environmental changes that affect their access to water. Approximately 1 in 4 mammal species are currently in danger of extinction in the wild, according to the International Union for the Conservation of Nature's *Red List of Endangered Species*, which is the longest-lasting compilation of data on extinct, endangered, and vulnerable species (started in 1948).

A species that has shown a significant decrease in population or *range* (the natural area in which the species is found), or shows signs of imminent local extinction (called *extirpation*), is a **threatened species**. Examples include the gray wolf and sea otter, which are abundant in some parts of their natural range, but are locally endangered.

As discussed in the example of the giant panda, biological and lifestyle factors that can lead to species vulnerability include highly selective feeding habits and a limited natural range. A small population also renders a species more vulnerable, partly because of lack of genetic variability (as for the elephant seal), and partly because it makes it difficult to find a mate. An example is the California condor, which was brought back from the brink of extinction by a captive breeding program. The condor is still at risk because so few are alive in the wild, and none of them, as yet, is known to have bred successfully. Extremely selective breeding habits also cause vulnerability. The green sea turtle, for example, breeds only in a few isolated locations on a specific type of sandy beach, and salmon do not breed in the wild unless they can make their way upstream to their original spawning ground.

Human activity endangers species and biodiversity in many ways. Significant causes of biodiversity loss are conflicts between human interests and the needs of wild species, increasing as a result of human population growth; overexploitation of species; pollution and environmental change; and importation of invasive species that outcompete native or *endemic* species. We will revisit these causes

FIGURE 16.22 Endangered species
The African black rhinoceros (*Diceros bicornis*), osprey (*Pandion haliaetus*), and Mauna Kau silversword (*Argyroxiphium kauense*) are among many species now listed as endangered on the IUCN Red List.

in Chapter 19 when we consider some of the many ways that human activity influences the Earth system. Above all, the single most important cause of extinction and loss of biodiversity is the destruction of habitat. Because we have focused so much of our attention in this chapter on concepts related to habitat, let's take a moment to consider how habitat loss can affect biodiversity.

Habitat Loss and Fragmentation

It has been estimated that as many as 85 percent of endangered species have been affected by the destruction of their natural habitat. A significant part of the problem is **habitat fragmentation**, in which large tracts of natural area are broken up into smaller patches by roads and other disruptions. Fragmentation can irretrievably damage the quality of habitat, even if it does not dramatically reduce the spatial extent of the habitat (**FIG. 16.23**).

For example, some species require habitat in the interior or *core* of a forested area. Near the edge of a forest, the characteristics of the ecosystem grade into those of adjacent areas; this is called the *edge effect*. If a forested area is cut or otherwise disturbed, new edge is created, which damages interior habitat. Migratory species require habitat that allows them to move freely from one natural area to another. Some

FIGURE 16.23 Habitat fragmentation
Loss of habitat, particularly through fragmentation, is the main threat to biodiversity today. Fragmentation not only decreases the size and connectivity of habitat, but changes it character by modifying the proportion of edge to core, as shown here.

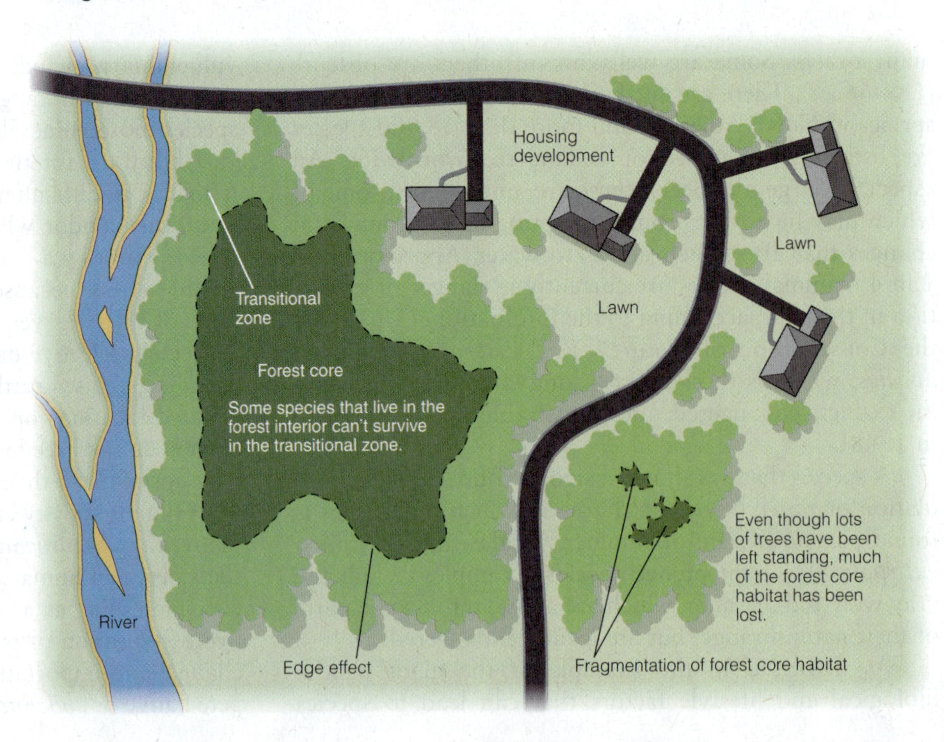

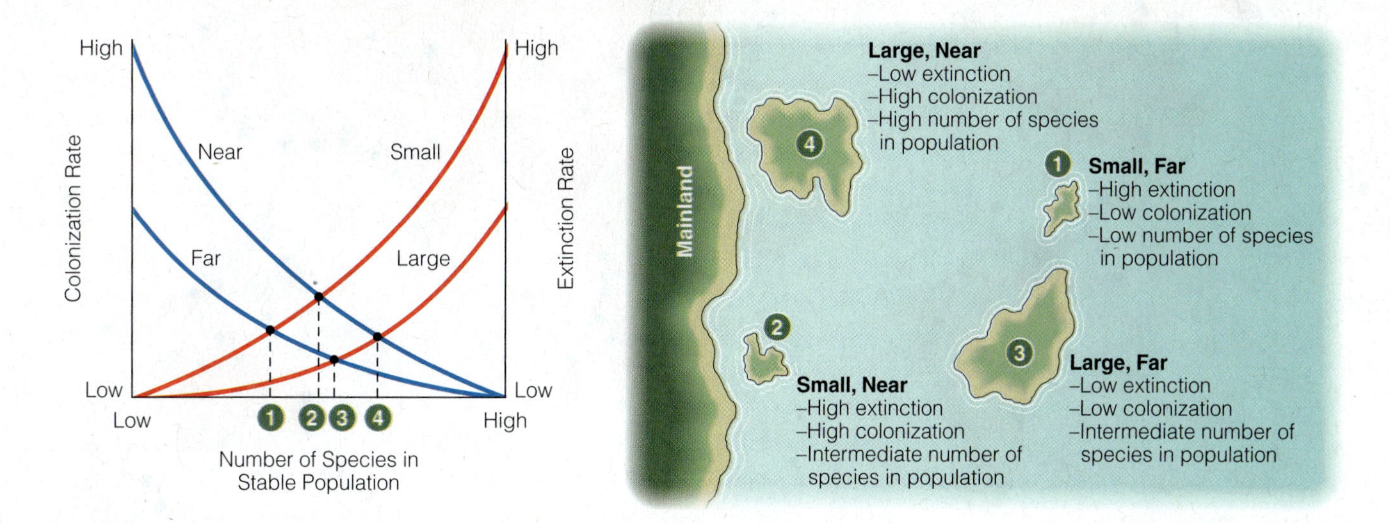

FIGURE 16.24 Island biogeography
The study of island biogeography can tell us some important things about the characteristics of habitat fragments, in terms of supporting species diversity. An island (or fragment of habitat) that is small and isolated will support the lowest number of species, whereas one that is large and not too far from the mainland (or connected to similar habitat) will support the largest number of species. The classic work on island biogeography was carried out by MacArthur and Wilson (1967), both referenced previously in this chapter.

migratory species, such as elephants, travel very long distances. This often puts elephants into direct conflict with human land development needs. Migratory wildlife can coexist with development if natural corridors are preserved that connect habitat areas together, allowing wildlife to migrate freely.

Much of what is known about habitat fragmentation comes from the study of biodiversity on islands, called **island biogeography.** The basic premise of island biogeography is that species diversity in an isolated area is a balance between the loss of species through extinction and the arrival of new species through colonization (immigration). An island that is small, or far from land, has less chance of being colonized by new species. A large island has greater habitat diversity and can support more species (we know this from the species–area relationship); one that is close to land has a better chance of being colonized. Thus, a large island that is close to land should have greater species diversity than a small island far from land (FIG. 16.24).

The basic concepts of island biogeography have been supported by observations of species diversity on actual islands. An extension of the concept considers small fragments of land-based habitat to be like "islands," in the sense that they are limited in size and are isolated from similar habitat areas. As with islands, the size and connectedness of terrestrial habitat fragments influence their ability to support species and habitat diversity. This leads to the conclusion that it is not only important to preserve habitat, but to preserve habitat in areas that are as large, undisturbed, and interconnected as possible.

Conserving Habitat and Biodiversity

Natural habitat and wilderness are under increasing pressure from economic and political forces that promote industrial development, urbanization, and the exploitation of natural lands and resources. Today only about 6.4 percent of the world's land area outside of Antarctica is protected (7.2 percent in developed countries, 5.8 percent in developing countries), and only about 2 percent is protected by international agreements (FIG. 16.25). Unfortunately, the protected areas are not always the same as the areas that host the greatest concentrations of biodiversity.

Luckily, protection can also occur under national jurisdictions, typically in the form of nature reserves, wilderness areas, national parks, natural monuments, habitat or species management areas, or protected landscapes. Protected seascapes and marine ecosystems are less well documented than terrestrial protected areas, and there are fewer of them. Protective legislation for natural areas and species, and the extent of enforcement, vary widely from country to country.

How can we do a better job of protecting species, habitats, and ecosystems? One way is to improve our scientific understanding of species and their needs. Species likely to become endangered through human activities

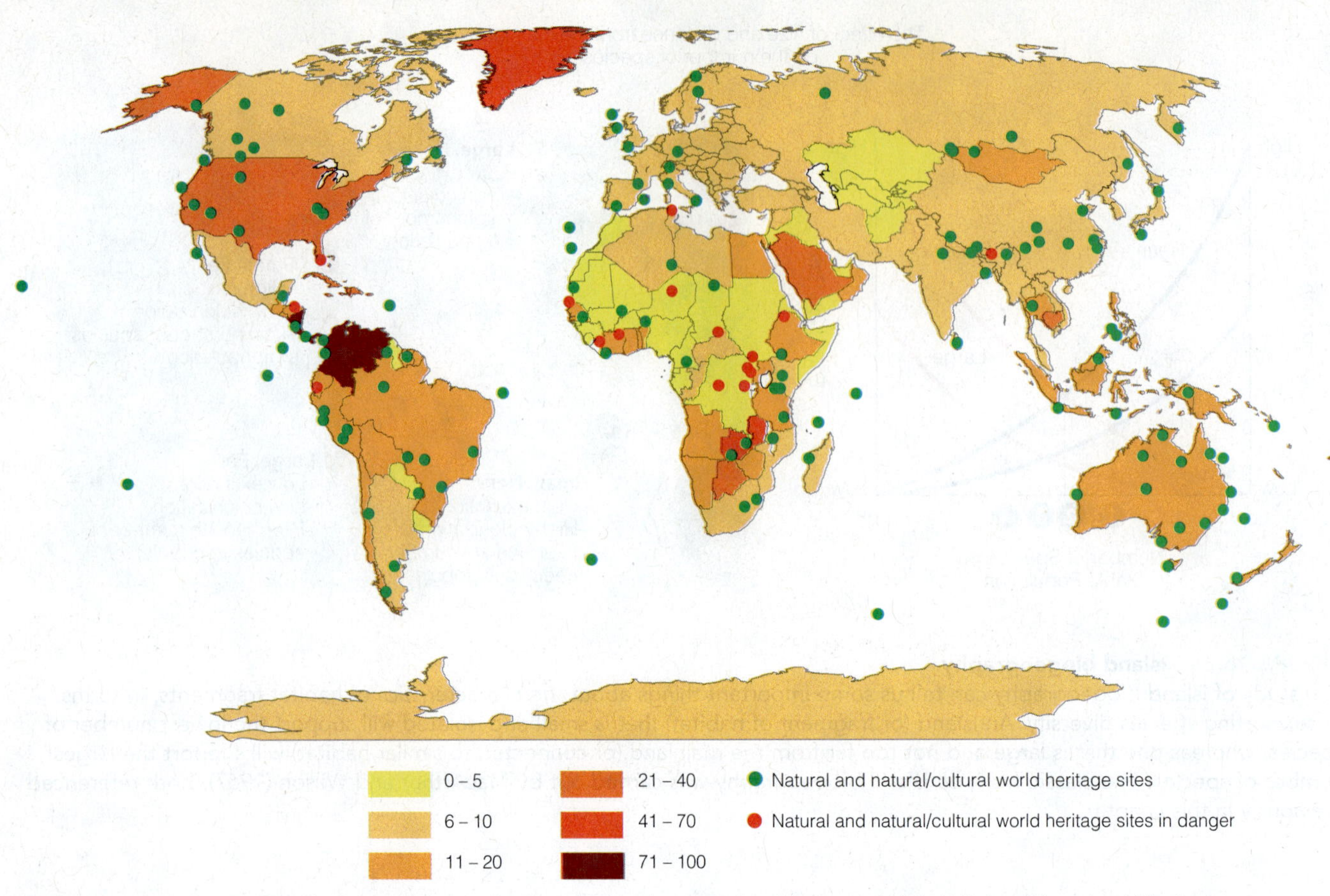
0 – 5
6 – 10
11 – 20
21 – 40
41 – 70
71 – 100
Natural and natural/cultural world heritage sites
Natural and natural/cultural world heritage sites in danger
The map shows the percentage of each country that is protected, and the location of World Heritage Natural and Mixed Cultural and Natural sites. Some 12.6 percent of the world's land area is protected.

FIGURE 16.25 World protected areas
Only a small fraction (about 2 percent) of the world's area is protected by international agreement. Luckily there are other national-level mechanisms that can also be used to protect habitat and biodiversity, as shown on this map from the United Nations Environment Program.

tend to share certain traits; knowledge of these common traits can help us protect and manage them. Easily endangered species, particularly vertebrates, are generally long-lived and large. Such species tend to have low reproductive rates and to recover slowly from lowered population levels. Another factor in their vulnerability is that the biggest and largest also require the largest territories and the most food per individual. Carnivores and large herbivores are particularly susceptible to extinction. Specialist species—those adapted to very narrow sets of conditions and having highly specific habitat and behavior requirements—are also vulnerable, as we saw in the case of the Giant Panda. As people clear land and modify the environment, the diversity of habitats is reduced and specialist species are impacted.

A key lesson to be learned is that a species will do better with a small population in an ecosystem that is in good condition, than with an abundant population in an ecosystem in poor condition. Another lesson from this discussion points to an irony in evolution. When the environment is constant, a way to "win" at evolution is to become more and more specialized. But the more specialized the species is, the more vulnerable it is to environmental changes or changes in its competitors, predators, or prey.

When the population of an endangered species drops too low, or its habitat has become too severely damaged or fragmented, it may be impossible for the species to successfully endure on its own *in situ* (or "in place") in the natural environment. In such cases, the only way to preserve the species may be through *ex situ* conservation ("out of place"), such as in a captive breeding program at a zoo or aquarium (FIG. 16.26). Breeding in captivity obviously is not as desirable as helping a species to survive in the wild. Some species are difficult to breed in captivity. Others that are bred in captivity can never be released back into the wild because they lack crucial survival skills, or because their natural habitat has been irretrievably damaged. If the population of a species is very small, its genetic diversity may be so low that the species will never be self-sustaining outside of the zoo or aquarium.

FIGURE 16.26 ***Ex situ* conservation**
Using a puppet, keepers feed an endangered California condor (*Gymnogyps californianus*) at the San Diego Wild Animal Park. There are only 384 condors left in the world, including only 188 known in the wild.

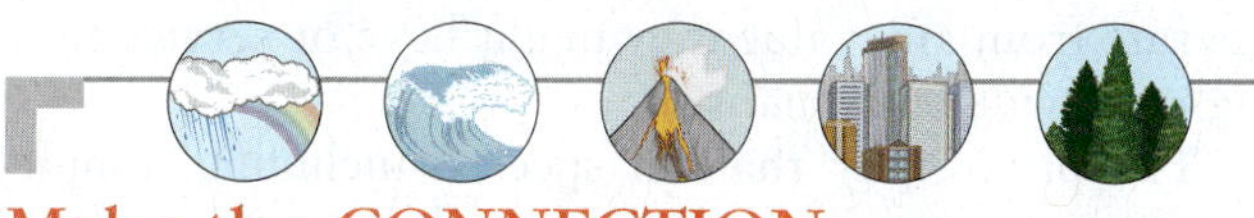

Make the CONNECTION

Can animals bred in captivity fulfill the same role in the biosphere as animals born in the wild? What do you think is the ultimate value of *ex situ* conservation?)?

Botanical gardens and seed banks play a similar role in the *ex situ* conservation of plant biodiversity. For many years, farmers and botanists interested in preserving genetic diversity in plants have maintained sample collections, both living and preserved. There are currently more than 100 seed banks in the world, where more than 3 million seed samples have been dried and preserved. Many of these are wild varieties of plants that were long ago domesticated as crops and have subsequently lost some of their genetic diversity.

All of these environments, including zoos, aquaria, and botanical gardens, are examples of **managed** or **domesticated ecosystems**. Additional examples are gardens, farms, aquaculture enclosures, municipal parks, and forest plantations. These environments are ecosystems—biotic communities that interact with an abiotic environment—but they rely on human intervention and management for their continued existence. A managed ecosystem is one way to keep species alive and available for human use and enjoyment. However, the biodiversity in a managed ecosystem will never match the biodiversity in a natural ecosystem. When a forest is clear-cut, it is certainly better to plant trees than to leave the soil bare and exposed. However, in doing this we are creating a plantation, not a forest, and we should never lose sight of the difference.

Why Is Biodiversity Important?

As we end this discussion, and close the part of the book that focuses on the biosphere, it is valid to revisit the question of why it is worthwhile to study biodiversity and, in particular, why it is important to concern ourselves with conserving and protecting biodiversity. There are two main sets of arguments concerning the value of biodiversity: arguments based on the usefulness of species and biodiversity for humans, and arguments based on the intrinsic value of species and biodiversity.

Utilitarian Arguments

Most people begin a list of reasons why biodiversity is important with something like, "Scientists might find new drugs to cure cancer from undiscovered species in the rain forest." Indeed, new substances are being tested and discovered every day—not just from the tropical rain forest species, but from other ecosystems as well. For example, it was recently discovered that a substance called *taxol*, derived from the Pacific yew tree, might be useful in the treatment of ovarian cancer. Penicillin—which completely revolutionized health care—comes from a natural fungus. What would the world be like today if that fungus had gone extinct before its medicinal use had been discovered?

These are fundamentally **utilitarian** arguments for the importance of biodiversity, that is, arguments based on the usefulness or potential usefulness of species to humans. Another utilitarian argument for the value of biodiversity, aside from medical and pharmaceutical purposes, is that many species have direct economic value. Virtually all of the food we eat, and many other products we depend on, come from the natural environment. For example, those concerned with preserving the tropical rain forests have begun to build upon the strength of economic arguments (FIG. 16.27). By marketing rain forest products without damaging the forest they are emphasizing the economic value of the intact rain forest, rather than its value in terms of cut lumber.

Other financial arguments are based on the recreational and tourism value of biodiversity. Tourism in the natural environment—including hunting, fishing, hiking, boating, enjoyment of scenery, and many other outdoor and wilderness activities—is a major contributor to the world economy. The natural environment also has aesthetic value. The concept of aesthetics—appreciation of the innate value of beauty—may seem somewhat vague. However, imagine a home overlooking a wild ravine, with songbirds and small animals in the yard and a mountain in the background. Now imagine a similar home overlooking a shopping mall. Which home would have greater real estate value? We appreciate the beauty of a mountain view, but it may be more important, for the long-term preservation of biodiversity, to be able to attach a specific dollar value to it.

Something that is more difficult to evaluate quantitatively is the spiritual and cultural importance of the natural environment, especially for aboriginal peoples. Many

FIGURE 16.27 Value of biodiversity
The berries of the acai palm (*Euterpe oleracea*), seen here in baskets waiting to be taken to a market in Abaetetuba, near the mouth of the Amazon River, are among many new products that can be harvested sustainably from Amazonian rain forests.

indigenous populations depend on the natural world for life support, as well as for their spiritual fulfillment and sense of identity. As aboriginal communities around the world lose access to their lands and natural resources, their traditional knowledge is lost.

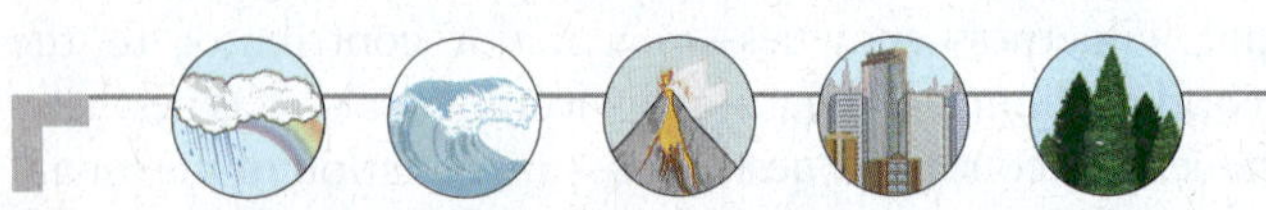

Make the CONNECTION

Some economists are concerned with the problem of how to quantify the value of the aesthetic components of the Earth system, such as the value of a beautiful view. See if you can identify aesthetically valuable components from each of Earth's subsystems. We have mentioned the aesthetic value of a mountain view (geosphere); how about a beautiful sunset (atmosphere)?

Biodiversity also provides ecological benefits and services. For example, wetlands perform a variety of environmental services, such as storing groundwater and cleansing toxins from soils and water. Coral reefs, in addition to hosting extremely productive biological communities, protect shorelines from storms. Rain forests provide an enormous storage reservoir for carbon, without which our global climate would suffer massive changes. Bees are crucially important in the pollination of many crops. Bacteria and other detritivores protect us from drowning in our own waste products. And so on—we cannot even begin to comprehend the value of the environmental services performed by species and ecosystems.

Intrinsic Value Arguments

Note that *every one* of the above utilitarian reasons for valuing biodiversity is related—directly or indirectly—to meeting the needs of humans. This is an **anthropocentric**, or human-centered, view of the value of biodiversity. The economic and utilitarian justifications are obviously anthropocentric, but even the spiritual, aesthetic, and ecological service arguments ultimately represent a human perspective. But many people believe that biodiversity should be protected and preserved for moral reasons because species have a value of their own—an **intrinsic value** that is wholly separate from any value they might have or service they might provide to humans.

The philosophy that all species, including nonhuman species, have intrinsic value and, thus, the right to continue to exist, and that humans have a moral obligation to protect other species, is called *deep ecology*. When pitted against human interests, this can lead to conflicts. For example, in the case of taxol, the cancer treatment made from the yew tree, several 100-year-old trees must be harvested to yield enough taxol for one treatment. Unfortunately, yew trees are an integral part of the old-growth forests of the Pacific Northwest, providing habitat for the northern spotted owl and other endangered species. This highlights the types of conflict that may arise when humans seek to benefit from biodiversity.

Overall, the concept of biodiversity speaks to a holistic or an integrated vision of the environment in which the cumulative value of species and ecosystems together is worth incalculably more than individual species considered separately. This is akin to the idea of "marginal value" in economics; the marginal value of a single, obscure species of salamander or insect, taken on its own, may be next to nothing. The true value of a species may lie in the role it plays within the whole. One way to think of this is that species are like individual cards in a house of cards, or like the rivets in an airplane; they are being removed, one by one, but the airplane is still flying. How many rivets can we remove without causing the structure to fail?

SUMMARY

1. A population is a group of individuals of the same species that interbreed and share genetic information. Both intrinsic and extrinsic factors exert an influence on health and survival, and thus influence the growth of a population.

2. Growth rate (r) is the change in number of individuals in a population over time, controlled by birth rate, death rate, immigration, and emigration. Theoretically, population growth will be exponential if unlimited resources are available and nothing interferes with growth or reproduction, but real populations cannot grow exponentially forever. A typical biological population will grow until it is limited by environmental factors. This more realistic growth model is represented by the logistic growth curve.

3. Factors that limit or control population growth in the real world include limitations to food, water, space, or light; barriers to reproductive success; competition, predation, disease; climate change; and many other environmental limitations and risk factors.

4. The leveling off of the logistic curve represents carrying capacity, the theoretical maximum number of individuals that can be supported by an ecosystem without sustaining damage. If the population is reduced below this level, births will exceed deaths and the population will rebound. If the population overshoots the carrying capacity, deaths will exceed births and the population will decline. Populations that grow rapidly, overshoot carrying capacity, and then experience diebacks are said to undergo boom-and-bust cycles.

5. Organisms that invest their reproductive energy into producing a large number of offspring but expend little or no energy caring for them are r-strategists. They are common in changeable environments where the ability to reproduce quickly is an asset to survival. K-strategists tend to be larger, with longer life spans and later maturity. They produce a small number of offspring infrequently, but devote considerable energy to their care and upbringing.

6. Populations live in communities—groups of interacting, interdependent organisms of different species that share the same environment. Interactions between individuals in communities have a fundamental influence on populations and on ecosystems as a whole.

7. In competition, individuals vie for scarce resources. Competition can be interspecific or intraspecific. The outcome of competition is invariably negative for one individual, and often for both, because each takes resources the other might have used.

8. In parasitism, one partner (the host) benefits while the other (the parasite) is harmed. In most cases, parasites weaken their hosts by extracting nutrients, without actually killing them. In predation, one organism (the predator) kills and eats the other (the prey). Carnivores are predators that only eat other animals, herbivores only eat plants, and omnivores eat both.

9. A mutualistic relationship is beneficial to both interacting organisms. The benefit can be in the form of food, protection, or other services provided. Many mutualistic relationships involve symbiosis, organisms living in close physical contact. In coevolution, two species evolve into their dependency on one another. In commensalism, one partner benefits without affecting the other.

10. A keystone species plays a central role in an ecosystem. Removing a keystone species is damaging, but the cumulative effect of removing less central species may be equally damaging to the ecosystem.

11. The physical environment in which a population or an organism lives or could live is its habitat. The niche is the organism's functional role and use of resources within the community. The fundamental niche is the set of all environmental conditions under which a species can persist and which it would be capable of occupying in the absence of competition or crowding. When niches overlap, competition ensues and the realized niche may only be a portion of the fundamental niche.

12. The competitive exclusion principle states that two species with the same requirements cannot coexist in exactly the same habitat and niche; one will gain the larger or richer niche, or drive the other species away. Species can avoid direct competition by differentiating their niches through specialization or resource partitioning. The greater the complexity of the environment, the more niche opportunities there may be, and the more opportunity for organisms to specialize.

13. Generalists have a wide range of tolerance and occupy large, broadly defined niches. Specialists have a narrower range of tolerance and occupy smaller, more narrowly defined niches. An environment that varies rapidly over a wide range does not allow specialists to survive. The larger the niches in a broad geographic region, the more generalists there are, the fewer species overall, and the lower the diversity. Where smaller, more narrowly defined niches are available, the more specialization and therefore the more different species there will be. Low-latitude areas, where climatic conditions are constant year-round, have niches with narrow ranges of variability, promoting higher diversity.

14. Allopatric speciation occurs when a new species arises as a result of a geographic barrier that leads to reproductive isolation. If populations occupy slightly different habitats and fulfill different roles, natural selection will act upon them in slightly different ways. Thus, through niche differentiation and specialization, a population may become reproductively isolated even without a major geographic barrier. In the case of sympatric speciation, reproductive isolation and speciation may take place as a result of behavioral specialization and differentiation.

15. Genetic diversity is the heterogeneity in DNA of individuals in a population or species. If genetic diversity is too limited, a species or population will be less adaptable to environmental change. Habitat diversity is a measure of the variety and complexity of available habitat in an ecosystem. Species richness is the number of species in an ecosystem or a community. The species–area relationship suggests that the larger the habitat, the more species it can support. Species evenness refers to the relative abundance of individuals of each species in an ecosystem; an ecosystem in which a single species is overwhelmingly dominant has low species evenness.

16. No one knows the exact number of species on Earth, and new species are being discovered all the time. Approximately 1.5 million species have been identified and named (mostly insects and plants), but the total number of species may be 30 million or even higher.

17. An endangered species is in imminent danger of extinction. A threatened species has undergone a significant decrease in population or range, or shows signs of imminent local extinction (extirpation). Factors that can lead to species vulnerability include highly selective feeding and breeding habits, and a limited natural range. A small population also renders a species more vulnerable, partly because of lack of genetic variability and partly because it is difficult to find a mate.

18. Human activity damages ecosystems and endangers biodiversity in many ways, but the most important are the destruction and fragmentation of habitat. An isolated fragment of habitat is like an island; its size and connectedness influence its ability to support species diversity. It is important to preserve habitat in areas that are as large, undisturbed, and interconnected as possible.

19. Natural habitat and wilderness are under increasing pressure. Protected areas are examples of *in situ* conservation. When the natural population of an endangered species drops too low, or its habitat becomes too severely damaged, it may require *ex situ* conservation, such as in a zoo, aquarium, botanical garden, or seed bank. These are examples of domesticated ecosystems that rely on human intervention for their continued existence.

20. Utilitarian arguments for the value of biodiversity are based on the usefulness of a species to humans; this is an anthropocentric perspective. Arguments based on intrinsic value hold that biodiversity should be protected and preserved for moral reasons, because species have a value of their own that is wholly separate from any value they might have or service they might provide to humans.

IMPORTANT TERMS TO REMEMBER

allopatric speciation *501*
anthropocentric *512*
biodiversity *502*
biotic potential *489*
birth rate *493*
carnivore *496*
carrying capacity *490*
coevolution *496*
commensalism *496*
community *492*
competition *492*
competitive exclusion principle *499*
conservation *497*
death rate *493*
ecosystem diversity *504*
endangered species *506*
exponential growth *492*
fundamental niche *498*
generalist *501*
genetic bottleneck *503*
genetic diversity *503*
growth rate *489*
habitat *498*
habitat diversity *504*
habitat fragmentation *508*
herbivore *496*
insectivore *496*
intrinsic (value of biodiversity) *512*
invasive species *500*
island biogeography *509*
keystone species *497*
K-strategist *494*
linear growth *492*
logistic growth curve *489*
managed (domesticated) ecosystem *511*
mutualism *496*
niche *498*
niche differentiation *500*
omnivore *496*
parasitism *495*
population *488*
population density *489*
predation *495*
realized niche *498*
resource partitioning *500*
r-strategist *494*
specialist *501*
species diversity *506*
species evenness *506*
species richness *506*
sustainability *490*
sympatric speciation *502*
symbiosis *496*
threatened species *507*
utilitarian (value of biodiversity) *511*

QUESTIONS FOR REVIEW

1. What is the difference between a population and a community?
2. What are some intrinsic factors that might affect population growth? What are some extrinsic factors that might affect population growth?
3. What is the equation that expresses the definition of growth rate?
4. Define carrying capacity.
5. What is the difference between intraspecific competition and interspecific competition?
6. What is parasitism? What is predation? How are they similar, and how are they different?
7. Define and give an example of an herbivore, an omnivore, a carnivore, and an insectivore.
8. What is mutualism? What is symbiosis? How are they similar, and how are they different? How are they similar to or different from commensalism?
9. What is the difference between an organism's habitat and its niche?
10. Describe the competitive exclusion principle, and discuss its relationship to resource partitioning and niche differentiation.
11. What is the difference between a specialist and a generalist? Give an example of each.
12. What kinds of intrinsic and extrinsic factors can make a species vulnerable to extinction?
13. Name and define the three basic types of biodiversity.
14. Describe how habitat fragmentation can threaten ecosystems and biodiversity.
15. What is the difference between *in situ* and *ex situ* conservation? Give examples of each.

QUESTIONS FOR RESEARCH AND DISCUSSION

1. Why is a logistic growth curve a more realistic model for the growth of a biological population than either a linear or an exponential model? Are there any problems or limitations with the logistic model?

2. Why is biodiversity valuable? Does biodiversity have value for reasons that are not anthropocentric?

3. We put a lot of effort into the protection of some species, but it is not uncommon for some populations to be intentionally reduced or brought to near-extinction by pest-eradication programs. For example, government animal control officers in the United States kill approximately 100,000 coyotes each year, to prevent them from preying on domestic animals. This brings up some interesting questions, such as "What constitutes a pest?" and "Is it ever justifiable for humans to intentionally cause the extinction of another species?" Before answering, consider the case of smallpox (a virus, therefore not quite alive, but a good example nonetheless). Through vigorous inoculation programs in the 1960s and 1970s, the smallpox virus was essentially eradicated; only a few populations remained, protected in laboratories. Some argued for its complete annihilation; others argued that if the virus were to reemerge, for whatever reason, scientists would need laboratory populations from which to manufacture a vaccine. Which do you think was the correct course to take?

QUESTIONS FOR *THE BASICS*

1. What is the difference between exponential growth and linear growth? Draw and label a graph to illustrate each of them.

2. Why is the logistic growth curve a more realistic model for a biological population than either exponential growth or linear growth?

3. How many species are known? How many might there be altogether?

QUESTIONS FOR *A CLOSER LOOK*

1. Describe the reproductive strategy of a *K*-strategist.

2. Describe the reproductive strategy of an *r*-strategist.

3. Name three examples of *K*-strategists and three examples of *r*-strategists.

PART SIX

The Anthroposphere: HUMANS *and the* EARTH SYSTEM

Anthroposphere: *from the Greek words* anthropos, *meaning human being, and* sphaira, *meaning globe, ball, or sphere.*

Our consideration of the pressures that threaten biodiversity (Chapter 16) provides a fitting transition into Part VI, in which we shift our focus to the **anthroposphere**—humans and our activities, technologies, and use of resources. The anthroposphere encompasses the human population, the built environment, and the parts of the Earth system that have been modified and manipulated by human actions.

Throughout this book we have stressed the life-supporting functions of the Earth system and the processes that influence the cycling of materials and energy. Earth materials and processes work together to create the *life zone* (Chapter 1), an environment that is—as far as we know—uniquely suited to support life. All of Earth's life forms depend on the natural environment for survival. But one species has a far greater impact on the planet, withdraws more resources, displaces more material, and generates more waste, than any other. That is, of course, the human species.

Given that Earth's human population is currently almost 7 billion, we must ask ourselves how long the Earth system can continue to support this growing population and at what level of resource consumption. In Chapter 17 we focus, in particular, on resources that are renewable or replenishable on a humanly accessible timescale. In Chapter 18 we examine mineral and energy resources, and then in Chapter 19 we look at some anthropogenic forces that are causing changes in the Earth system on a global scale.

The chapters of Part VI are as follows.

- Chapter 17. The Resource Cycle
- Chapter 18. Mineral and Energy Resources
- Chapter 19. The Changing Earth System

Desert palms
This is Palm Jumeirah in Dubai, one of a series of artificial islands, peninsulas, and breakwaters constructed from sand dredged from the bottom of the Persian Gulf. The palm-tree-shaped island measures 5 km by 5 km. This is an infrared image, and the red color shows areas that are vegetated.

CHAPTER 17

The Resource CYCLE

OVERVIEW

In this chapter we:

- Introduce the resource cycle and consider the significance of Earth resources for human society
- Summarize different management approaches for renewable and nonrenewable resources
- Investigate the characteristics, use, and status of forests, wilderness and wildlife, fisheries, soil, and water resources
- Describe some current challenges in the management of renewable resources
- Consider whether there are limits to population growth and resource extraction

Demolition derby
A pile of junked cars awaits recycling and final disposal in Los Angeles, California.

In this chapter we begin the final part of the book, and we embark on our consideration of the **anthroposphere**—the human population, the built environment (or *technosphere*), and the influence of human activities on the Earth system. Many thousands of years ago, when there were fewer than 10 million humans living on this planet, the Earth system could easily replenish resources of food and fuel. Resource supplies were vast in comparison to the level of demand placed on them, and the discarded wastes from resource use were easily absorbed, with little impact.

Today, with a population of almost 7 billion, our level of resource extraction is beginning to push the limits of the system's carrying capacity. We are using some resources faster than they are being replenished, and we are degrading the physical systems on which their replenishment depends. The huge quantity and harmful characteristics of some of the wastes generated by human activities threaten to overwhelm the systems into which they are emitted. Humans are by far the largest cause of displacement of Earth materials—we move more rock and sediment each year than all the great rivers of the world combined (FIG. 17.1). Let's begin by considering how we became such prodigious users of Earth resources.

FIGURE 17.1 **People and the built environment**
The Three Gorges Dam on China's Yangtze River, seen here under construction in 1999 and completed in 2006, is one of the biggest engineering projects ever undertaken, and the largest electricity-generating facility in the world.

RESOURCES FROM THE EARTH SYSTEM

Civilization and natural resources are inseparable—the former would not have been possible without the latter. Scholars have even marked the stages of civilization by the natural resources our ancestors learned to use: Stone Age, Bronze Age, Iron Age. Human history and the story of civilization are thus inextricably linked to the human innovations in the use of Earth resources.

History of Human Resource Use

The first resources that early humans used were the fruits, nuts, animals, and fish they hunted and gathered, a lifestyle that is still maintained by some traditional nomadic people today (FIG. 17.2). These materials are **renewable resources** because their supply is replenished by new growth each season.

FIGURE 17.2 **Hunter-gatherers**
Some modern people, including this Mbuti hunter from the Ituri rain forest in Zaire, still follow the hunter-gatherer lifestyle of our distant ancestors which requires a large area of land to support a small population.

Millions of years ago our ancestors started to use another class of natural resources, when they picked up suitably shaped stones for use as hunting aids (FIG. 17.3). Stones for tool use are a **nonrenewable resource**, so-called because even though they are replenished naturally, the rates of replenishment are measured in millions of years rather than in annual seasons.

Around the time when Earth was emerging from the last glaciation (about 12,000 years ago; Chapter 13), humans took a giant step forward when they began the cultivation and domestication of renewable resources, both plants and animals, and **agriculture** was born. Hunting and gathering requires a very large land area to support a given population. Agriculture removed this environmental constraint, allowing people to settle down and extract food resources more intensively from a smaller area of land. As a result, population increased dramatically (FIG. 17.4).

As a consequence of agriculture, the human diet shifted to a greater dependence on grains. Out of necessity, people started gathering and trading a nonrenewable mineral resource: salt. Originally, the dietary need for salt was satisfied by eating meat brought home by hunters. When farming started and diets became cereal-based,

FIGURE 17.3 Stone tools
Stones for tools, like this stone roller and metate used for grinding corn in the prehistoric Fremont culture of Utah, were among the first nonrenewable resources that were used and traded by our distant human ancestors.

extra salt was needed. We don't know when or where the mining of salt started, but trading routes for salt crisscrossed the globe long before history was recorded in writing.

Metals were first used about 17,000 years ago. Copper and gold were the first metals to be used; both occur naturally in their metallic form as pure or *native* metals. These native metals are rare, though, so eventually the most obvious sources became depleted and additional sources were needed. The extraction of copper by *smelting* was discovered about 6000 years ago (FIG. 17.5); this involves heating certain rocks to such extreme temperatures that the copper-bearing minerals break down, releasing the copper in a molten form. Before another thousand years had passed, people had discovered how to smelt minerals of lead, tin, zinc, silver, and other metals. The technique of mixing metals to make *alloys* came next, and bronze (copper + tin) and pewter (tin + lead + copper) came into use. The smelting of iron is more difficult than the smelting of copper, so development of an iron industry came much later—about 3300 years ago.

FIGURE 17.4 Human population and environmental resistance
Great technological advances, like the origin of agriculture and the Industrial Revolution, caused an easing of environmental resistance, effectively increasing Earth's carrying capacity and allowing major, rapid increases in human population.

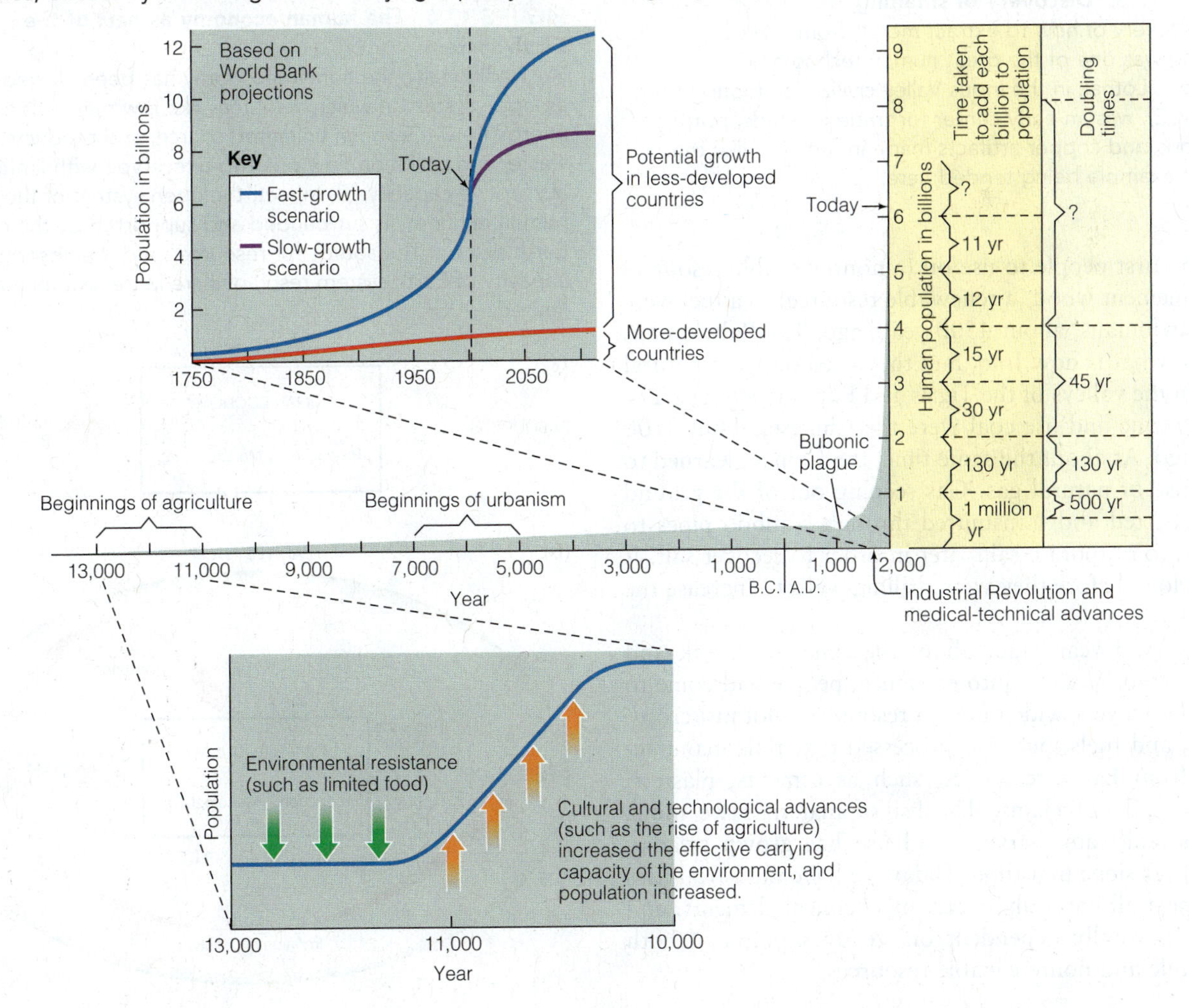

FIGURE 17.5 Discovery of smelting
The discovery of how to extract metals from rocks by smelting was one of the great human technological advances. Lothal, in the Indus Valley civilization (dating from 2400 BCE), was an early center for trade in beads, pottery, shellwork, and copper artifacts made in furnaces like the ancient example being tended here.

The first people to use oil (a nonrenewable resource) to supplement wood (a renewable resource) as a fuel were the Babylonians, about 4500 years ago. The Babylonians lived in what is now Iraq, and they used oil from natural seeps in the valleys of the Tigris and Euphrates rivers. The first to mine and use coal were the Chinese, about 3100 years ago. At about the same time, the Chinese learned to make use of natural gas. Gas seeping out of the ground was collected and transmitted through bamboo pipes to be used to evaporate saltwater in order to recover salt. It wasn't long before they were drilling wells to increase the flow of gas.

By 2500 years ago, about the time the Greek and Roman empires came into existence, people had come to depend on a very wide range of resources—not just crops, metals, and fuels, but also processed materials manufactured from Earth resources, such as cements, plasters, glasses, and porcelains. The list of materials we mine, process, cultivate, harvest, and use has grown steadily larger ever since that time. Today we have industrial uses for almost all naturally occurring chemical elements, and society is totally dependent on steady supplies of both renewable and nonrenewable resources.

Resources In, Wastes Out

Traditionally, the human economy has been considered as a reservoir of productive activity through which resources and energy (including human labor) flow (FIG. 17.6A). Resources and energy drive production within the economy; the outputs are finished products, as well as pollution and waste. In this traditional view, the Earth system is seen as a limitless source of raw materials, with a correspondingly boundless capacity to absorb harmful outputs.

However, because we are now pushing the limits of so many parts of the Earth system, this model no longer seems to work very well. We now understand that the human economy exists as part of the larger Earth system and that it depends on the Earth system in many important ways. The human economy is an *open system* (Chapter 1) into and out of which materials and energy flow. Earth, in contrast, is a *closed system*—its material resources are finite, and so is its waste-absorbing capacity.

This switch to a systems perspective of the human economy changes everything. It means that resources—once viewed as limitless—suddenly have inherent value, and there are good reasons for preserving them and maintaining the integrity of the natural systems from

FIGURE 17.6 The human economy as part of the Earth system
(A) Traditionally, the human economy has been viewed as an open system in which resources like raw materials and energy flow in from an unlimited source, and products, waste, and pollution flow out into a reservoir with limitless absorptive capacity. (B) In fact, the open system of the human economy is surrounded and supported by the closed Earth system. The source for resources and the absorptive capacity of Earth system reservoirs are large, but ultimately limited.

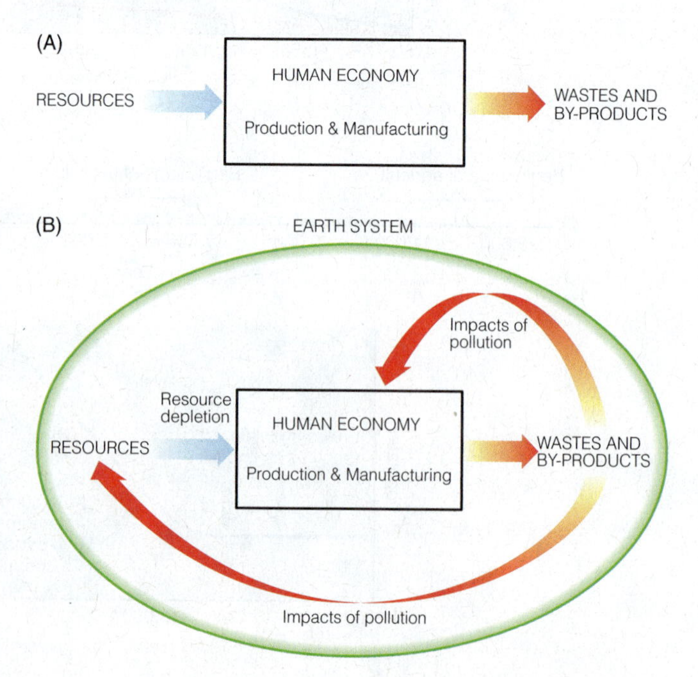

which they are derived. It also means that waste, byproducts, and pollution don't disappear into an endlessly receptive vacuum, but must be absorbed by the atmosphere, hydrosphere, geosphere, and biosphere, with all of the resulting negative impacts.

Consequently, we can now draw a wider boundary that surrounds and encompasses the human economy (FIG. 17.6B). Every time we use materials from within this boundary, we must think about the status of the remaining resources. Every time we emit wastes and pollution, we must think about their impacts on the rest of the system. In the **resource cycle** of human production, there is potential for such impacts at every stage, from the extraction of raw materials through transportation, production, packaging, delivery, use, and disposal of products (FIG. 17.7). If we want to ensure the survival and integrity of natural systems, maximize the availability and quality of resources, and minimize the negative impacts of waste and pollution, we can look at each stage in the resource cycle as a potential site for improving our management approach.

Basic Concepts in Resource Use and Management

Because they involve such different timescales, nonrenewable and renewable resources require fundamentally different management approaches. Let's consider some of the differences.

Managing Nonrenewable Resources

The key feature of nonrenewable resources is that the more we use, the less remains in the stock. A resource's **stock** is like the content of a reservoir (Chapter 1); it is the physical mass of the resource. When we use nonrenewable resources like copper and oil, they are not replenished on a humanly accessible timescale, so their use automatically leads to the depletion of the stock (FIG. 17.8). History demonstrates that new deposits can be found, but for some materials this is becoming increasingly difficult; we will discuss this challenge in greater detail in Chapter 18. The availability of known nonrenewable resources can only be extended through conservation, substitution, reuse, or recycling. A wise management strategy might look toward replacing reliance on nonrenewable resources with reliance on renewable or inexhaustible ones; for example, it might be a good management strategy to replace the use of nonrenewable fossil fuels with solar, geothermal, tidal, or biomass sources of energy, each of which is a renewable resource. In Chapter 18 we will look more closely at nonrenewable mineral resources, and at fossil fuels and the potential for their replacement by alternative energy sources for the future.

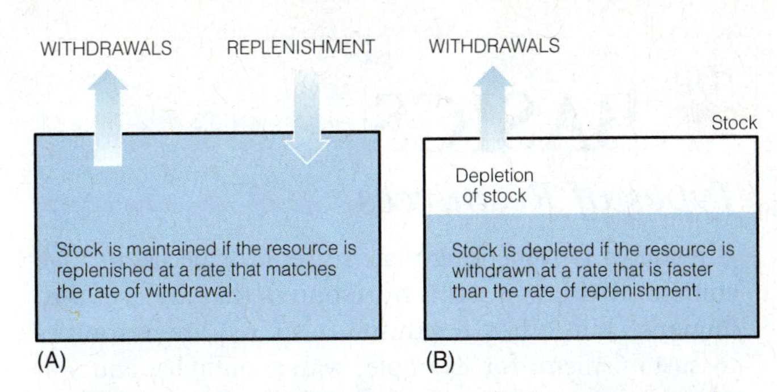

FIGURE 17.8 Depletion and renewal of resource stocks
(A) Renewable resources like trees, fish, and groundwater are, in principle, renewed or replenished by growth or addition to the stock on a seasonal or an ongoing basis. (B) Problems arise when renewable resources are harvested at a rate that is faster than the rate of renewal. This is a form of "mining," and it renders the resource effectively nonrenewable because the stock is being depleted with each withdrawal.

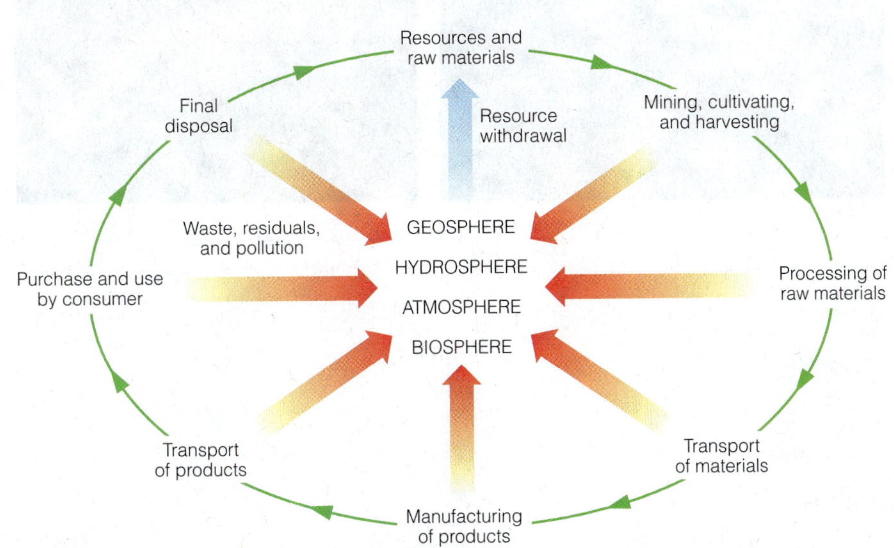

FIGURE 17.7 The resource cycle
Waste can be generated at every stage in the resource and production cycle, from extraction of raw materials through production, packaging, transportation, use, and disposal of products. Closing this cycle means taking care to minimize waste and maximize reuse and recycling of materials at every stage.

Managing Renewable Resources

Living resources like fish and trees are renewable, in principle, as long as they are managed properly. When resources of the biosphere are used at a rate that is faster than the rate at which they can reproduce and grow they can become depleted, just like nonrenewable resources. This happens when fish are harvested faster than new fish can hatch and grow, for example, or when trees are cut down faster than new trees can take root and grow. If the portion of the stock that is withdrawn from a living resource is exactly replaced by new growth each year, then the stock will remain constant; it is at *steady state* (Chapter 1) because what goes into the stock balances what comes out.

the BASICS

Types of Resources

A **resource** is something that is useful or necessary. We commonly think in terms of resources that are useful to humans, but other organisms also require resources to sustain them; for example, water, sunlight, and soil nutrients are vital resources for plants. The survival of populations and, indeed, of entire species depends on the success of its individuals in competing for scarce resources.

There is a whole branch of economics, well beyond the scope of this book, that deals with how humans value resources, both monetarily and otherwise. For example, would you consider a beautiful view to be a resource? What about a forest—is it of value only for the wood that it holds, or are there other values in a forest? What about resources that are of value to other organisms but not of direct use to humans? For example, oxygen in stream water is vital for the survival of fish, but most of us wouldn't immediately notice if the oxygen in the water of a particular stream became depleted. If the oxygen-depleted stream water became unfit to support fish, it might have a measurable impact on human interests; on the other hand, what if it only affected fish that are not consumed by people—does that mean the resource is of no value? These are challenging questions, and there has been much consideration lately in economics, environmental science, Earth science, and related fields, about the ramifications of various approaches to the valuation of resources.

Different types of resources require different management approaches (FIG. B17.1). Living resources, such as fish, crops, and trees, are *renewable*, in principle because they are replenished by the birth or seeding and subsequent growth of new individuals on a seasonal or continuous basis. As long as organisms are given the environment they need to sustain growth, and sufficient time and opportunity to reproduce, a biological resource should be indefinitely renewable.

Not all renewable resources are living resources; nonliving resources also can be renewable. For example, groundwater is replenished every time it rains, when water infiltrates and percolates downward to join an aquifer. Soil is another example of a (mostly) nonliving renewable resource; it can become depleted when it is washed away by erosion, or when overly intensive use strips it of its nutrients. It can be replenished by weathering and

FIGURE B17.1 Resources
Resources exist on a continuum, from those that are perpetual or inexhaustible (left) through those that are renewable on humanly accessible timescales (middle) to those that are nonrenewable because the timescale for their replenishment is much longer than a human lifetime. On the far right is a resource that is truly nonrenewable because it can never be replenished—a split atom.

disintegration of rock, and by the addition of organic material. But soil-forming and groundwater regeneration processes take a lot of time—from hundreds to tens of thousands of years. Renewable resources that have very long regeneration periods are extremely vulnerable, because they can so easily become effectively nonrenewable as a result of overly intensive use.

In contrast, *nonrenewable resources* are only replenished on a timescale that greatly exceeds the scale of a human lifetime—or even a human civilization. The geologic processes responsible for the formation of most types of mineral and energy resources are still operating today, but they may take hundreds of millions of years to complete. As a society, we can't wait that long to replenish a resource that has been used up. Therefore, it is more accurate to say that nonrenewable resources are not renewable or replenishable *on a humanly accessible timescale*. For example, the geologic processes that led to the formation of the great coal deposits that we currently use to generate much of our electricity happened, for the most part, 100 million years ago, during the Middle Cretaceous Period when tropical climates and swamps were widespread. Coal-forming processes are still happening today, but as a society we cannot afford to wait another 100 million years for those resources to be ready for use.

Some resources are *truly* nonrenewable. When an atom is split for the purpose of extracting nuclear energy, for example, that atom will *never* be put back together again. Similarly, some resources are renewed on such a continuous basis that they might be referred to as **inexhaustible** or **perpetual resources.** An example is solar energy, along with some of the resources that derive from it, such as wind and wave energy. As long as there is an Earth–Sun system, solar energy will arrive at the surface of the planet—it is an inexhaustible source.

Finally, it can be useful to categorize resources on the basis of how they are managed, rather than how they are formed. An important category is **common property resources**—resources that are commonly or communally owned, accessed, or managed. Parks, town squares, and national, state, or provincial forests are examples of common property resources. The Great Lakes are a common property resource; they are publicly owned and jointly managed by the governments of the United States, Canada, and aboriginal groups. The atmosphere and the ocean are common property resources; some people include Antarctica in this category as well. The scale of these resources and their importance to all people are so great that they are referred to by a special name: the *global commons*.

FIGURE B17.1 Resources (*continued*)

If the stock of a living renewable resource becomes severely depleted—even if the population does not fall to zero—it may never regenerate, no matter how long we wait. For example, fisheries in some parts of the world have collapsed (both economically and biologically) because the fish were harvested at a rate that was faster than their natural rate of regeneration. Populations of some species fell below a critical level, and reproduction and survival became too challenging. There are many reasons for the inability of species to recover quickly—for example, other species may move in and take over the environment once inhabited by the depleted species. Some depleted populations, though not extinct, will never recover to their former levels.

It is also possible for *nonliving* renewable Earth resources to become depleted to the point where they cannot regenerate naturally. For example, groundwater reservoirs are replenished by rainfall; the process can be slow, but it happens more or less continuously—every time rain falls and infiltrates. However, if water is withdrawn from an aquifer at a rate that exceeds the rate of recharge, the groundwater supply will become depleted. Ultimately, if too much water is withdrawn, the sediment grains that comprise the aquifer may become compacted. When this happens the aquifer can never be fully replenished, no matter how much it rains and no matter how long we wait.

RENEWABLE RESOURCES: SEEKING BALANCE

In this book you have learned about the systems of the biosphere, atmosphere, geosphere, and hydrosphere that support the growth and replenishment of forests, fish, soils, and other renewable Earth materials. Now let's look specifically at some current issues in the management of these resources.

Forest Resources

The world's forest resources encompass a great variety of forest types, as discussed in Chapter 15. About 95 percent of the world's forests are natural forests; the remaining 5 percent are forest **plantations**, that is, managed forests where trees have been planted in rows, usually for the eventual purpose of harvesting (FIG. 17.9A). Plantations can help maintain the stock of a forest resource by balancing the cutting and removal of trees with the seeding and growth of new trees. However, a plantation forest can never replace a natural forest in terms of biodiversity and habitat.

Natural forests with communities of different species, and of varied ages and maturities, typically host much wider varieties of plant life and more diverse habitats than plantations. "Natural" forests are not intentionally cultivated, but neither are they free of human intervention. Significant human impacts on forests date back tens of thousands of years; land clearing and burning of forests are among the earliest human impacts on the environment. Today, even pristine forests deep in wilderness areas are undeniably affected by human activity in the form of contaminants and pollutants transported and deposited by atmospheric and hydrospheric processes.

A forest that has endured for a long time—hundreds or thousands of years—without extensive intentional

FIGURE 17.9 Plantations and old-growth forests
Plantations (A) can replace harvested forest area, but their neat rows, lack of undergrowth, and single-aged trees demonstrate that they fall short of replacing the immense biodiversity and habitat diversity offered by old-growth forests (B).

human intervention is called an **old-growth forest.** Old-growth forests with very large, old trees and high biodiversity are extremely valuable ecosystems (**FIG. 17.9B**). They are also highly valued for their timber, which makes them controversial from an environmental and resource-management perspective.

The Value of Forests

Timber and fuel wood are chief among the economically important products harvested from forests, but there is also a great variety of nonwood products, including latex, nuts, citrus fruits, bananas, oils, and bush meat. The world's great forests host many species, as yet undiscovered, that may be useful in the development of new pharmaceuticals and other products. Forests also provide recreational opportunities, and they are of great cultural significance for aboriginal peoples whose communities and livelihood depend on them.

The environmental services provided by forests are also of great value (**FIG. 17.10**). Trees stabilize soil by holding it in place with root systems. They keep the soil replenished with organic material by returning nutrients to the forest floor in the form of dead leaves and branches, or *litter*. Trees play an important role in the hydrologic cycle, drawing water from the soil through their roots and returning it to the atmosphere by transpiration. They intercept precipitation, returning much of the water to the atmosphere by evaporation and transpiration from leaf surfaces. Trees also hold humidity under their canopies and prevent high winds from reaching the soil surface.

On a global scale, forests harbor extensive reserves of biodiversity. About half of all plant and animal species live in tropical rain forests. Forests help to regulate the global climate by serving as an enormous reservoir for carbon. Approximately 600 billion metric tons of carbon resides in above-ground biomass in forests, with another 1500 billion metric tons in the form of litter and soil organic matter. This is almost three times as much carbon as in the atmospheric reservoir. If this carbon were released, Earth's climate system would undergo an enormous upheaval.

Make the CONNECTION

Try to summarize all of the ways that forests and trees interact with the climate system, on a local scale and on a global scale.

Logging, Forest Management, and Agroforestry

Clear-cutting, the removal of all trees from an area, is generally the most economically efficient way to harvest wood from a forest, but also the most ecologically destructive. Removing trees damages the undergrowth and eliminates habitat, killing or driving away much of the animal life in the area. More subtle but equally harmful are the impacts of clear-cutting on soil and on the cycling of water and nutrients. Soil that is exposed to the effects of sun and wind quickly dries out, becoming susceptible to erosion. Clear-cutting accelerates the loss of soil nutrients, which are washed away by runoff. Depending on the circumstances, a forest that has been clear-cut may be able to regenerate only after many years, if at all.

In contrast, forests can recover and, in some cases, even benefit from natural disturbances such as forest fires and landslides. Boreal forests, for example, are adapted to forest fires; some boreal species depend on periodic fires to release their seeds or clear away competitors. *Selective cutting*—the harvesting of certain trees or selected small areas of trees, leaving the rest of the forest intact—is increasingly being designed to mimic the characteristics of clearings that result from natural disturbances. Timber harvesting that minimizes harm to the forest ecosystem in this and other ways is called *reduced-impact logging*.

Another important development in forest management is the increasing use of *agroforestry*, in which trees (typically but not always fruit or nut trees) and crops (typically but not always cash crops, such as coffee) are planted together. Trees typically have much deeper root systems than crops; the deep roots stabilize the soil and draw water from depth, making

FIGURE 17.10 Forest services
Forests link the biosphere to the hydrosphere, atmosphere, and geosphere, and provide many valuable environmental services.

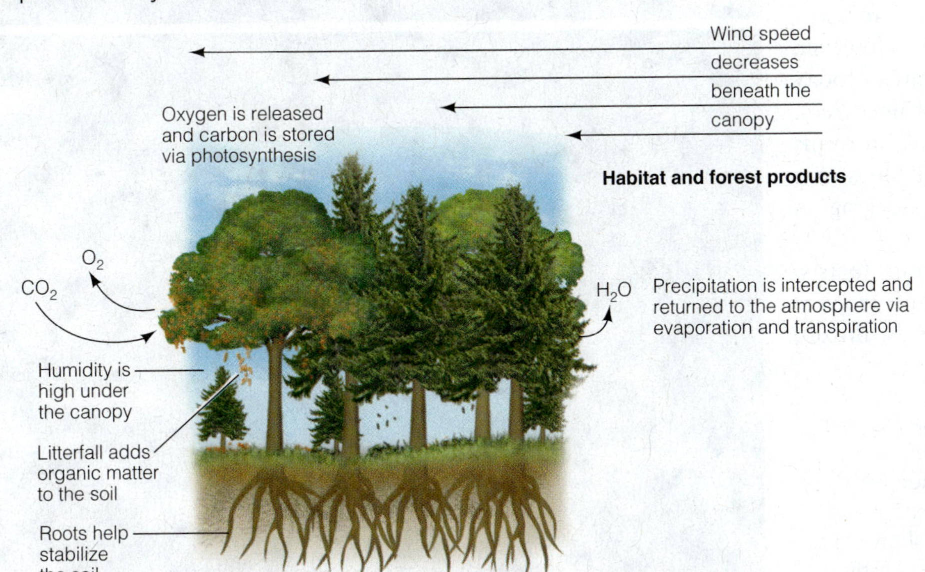

it available to the crops. Trees also contribute to the organic content of the soil, through litterfall. If the trees grow so tall that they begin to shade the crops, they can be selectively harvested, providing fuel-wood, timber, fruits, nuts, and leaves and other clippings to feed domesticated animals.

The management of forest plantations for the purpose of timber harvesting, or *silviculture*, focuses on maintaining a **sustainable yield**, that is, taking only as many trees as can be replaced by new, harvestable growth each year. The concept of sustainable yield is fundamental to the success of any resource that is renewed by annual or seasonal growth. As shown in Figure 17.8B, sustainable yield is based on a level of harvesting that ensures that the rate of withdrawal is matched by the rate of replenishment of the resource.

Deforestation

It is challenging to keep track of natural and anthropogenic changes in forest resources. The assessment process combines ground-based information sources with data from remote sensing. Satellite images are analyzed to determine the amount, type, and health of forest cover. By comparing images from year to year, and combining this with on-site mapping, scientists can determine the extent of changes in the amount and quality of forested land (**FIG. 17.11**).

Today forest cover in developed countries is stable or slightly increasing, while **deforestation**—the loss of forest cover—continues at a rapid pace in many of the poorer countries of the developing world. Of course, this does *not* mean that logging does not occur in wealthy countries. Extensive logging and clearing of land occurred historically throughout the developed world, particularly during the latter part of the nineteenth century and the early twentieth century, and continues in many areas. Only about 5 percent of original forested land remains intact in North America. However, logging in economically developed countries today is typically balanced by the planting of new trees. Reforestation can protect forested areas, in terms of hectares of coverage; however, it does not resolve the controversies surrounding logging in old-growth forests.

Of particular concern are tropical rain forests. The three remaining large expanses of tropical rain forest are in South America (mainly Brazil),

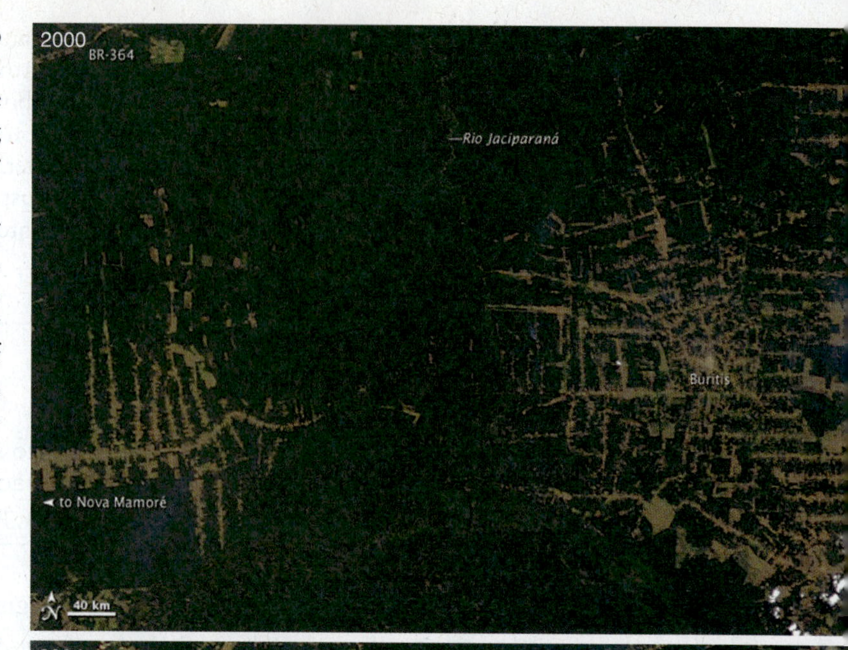

FIGURE 17.11 Forest changes
Human settlement and subsequent land-use changes have had substantial impacts on forest cover worldwide. Remote sensing can be combined with on-site mapping to assess changes in the extent and health of forest cover from year to year, as shown here in this series of satellite images showing progressive deforestation in the Rondonia region of the Amazonian rain forest in Brazil.

Central Africa (mainly Democratic Republic of Congo), and Southeast Asia (mainly Indonesia). In Brazil alone, the loss of forested area averages 5 million ha per year. If present rates of deforestation continue, the world's tropical rain forests are at risk of virtual extinction by the end of this century.

Deforestation is problematic in tropical rain forests not only because of the loss of biodiversity and habitat, but also because tropical soils are fragile. Although rain forests are lush and thick with vegetation, most of the nutrients and organic matter reside in the vegetation, not in the soil (**FIG. 17.12**). Deforestation removes most of the organic matter from the system, leaving the nutrient-poor soils exposed and susceptible to drying, compaction, and erosion. In temperate forests, deeper root systems stabilize the soil and litter decays more slowly, so organic matter has a chance to accumulate at the surface. This provides nutrients and holds moisture in the top layer of the soil.

Aside from clear-cut logging, the most important direct cause of deforestation is the clearing of land for agriculture and ranching. Road building can contribute by providing access by which other types of development, including industry, mining, hydroelectric dams, and the towns that accompany them, can move into an area. Hydroelectric dams also cause deforestation because of the extensive flooding behind the dam when a reservoir is created. Cutting trees for fuel wood is an important cause of deforestation in developing countries, where wood and charcoal account for a large proportion of household energy use.

Other causes of deforestation are indirect. For example, forests in northeastern North America, Europe, and Southeast Asia are threatened by air pollution, particularly acid precipitation, which weakens the trees and makes them more susceptible to pests and disease. Other types of environmental change, such as drought, severe weather, and climatic change, can weaken trees and allow pest populations to take hold. And finally, poverty and the debt loads of developing nations—currently almost 3 trillion dollars in total—are an underlying cause of deforestation in countries that sell off rich natural resources to make their loan payments.

Wilderness and Wildlife

Wilderness is any area where natural forces are more significant than human intervention, and where people do not live on a permanent basis. The protection of wilderness areas began with the national parks movement in the United States and the creation of Yosemite Park in 1864 and Yellowstone Park in 1872. Since then virtually every country has set aside some land and marine areas to be protected in one way or another, either as state, provincial, or national parks and forests, international preserves, or *wildlife refuges*—sanctuaries for the animals (and sometimes plants) that inhabit wilderness areas.

Many questions and controversies surround wilderness preservation. Why should we protect wilderness and wildlife? Are parks and refuges the only way to preserve wildlife? Is there a necessary or an optimal size and shape for a protected area? Do we set aside protected areas because of the inherent value of wilderness and wildlife? Or do we protect wilderness for the benefit of people, so that they can visit these special places to be rejuvenated, to experience nature, and to learn about wildlife and the natural world? In a world with many economic and political pressures, how can we ensure that wilderness and wildlife will continue to be protected? Are preservation and human use mutually exclusive?

FIGURE 17.12 Water and nutrient cycling by forests
Temperate and tropical forests differ in the way they cycle water and nutrients between the atmosphere, hydrosphere, and soils. Temperate forests (A) typically have deep root systems that draw up water and nutrients from depth. Tropical forests (B) have shallow root systems, and most of their organic matter resides in the vegetation, not in the soil. The trees return a large proportion of precipitation directly back to the atmosphere.

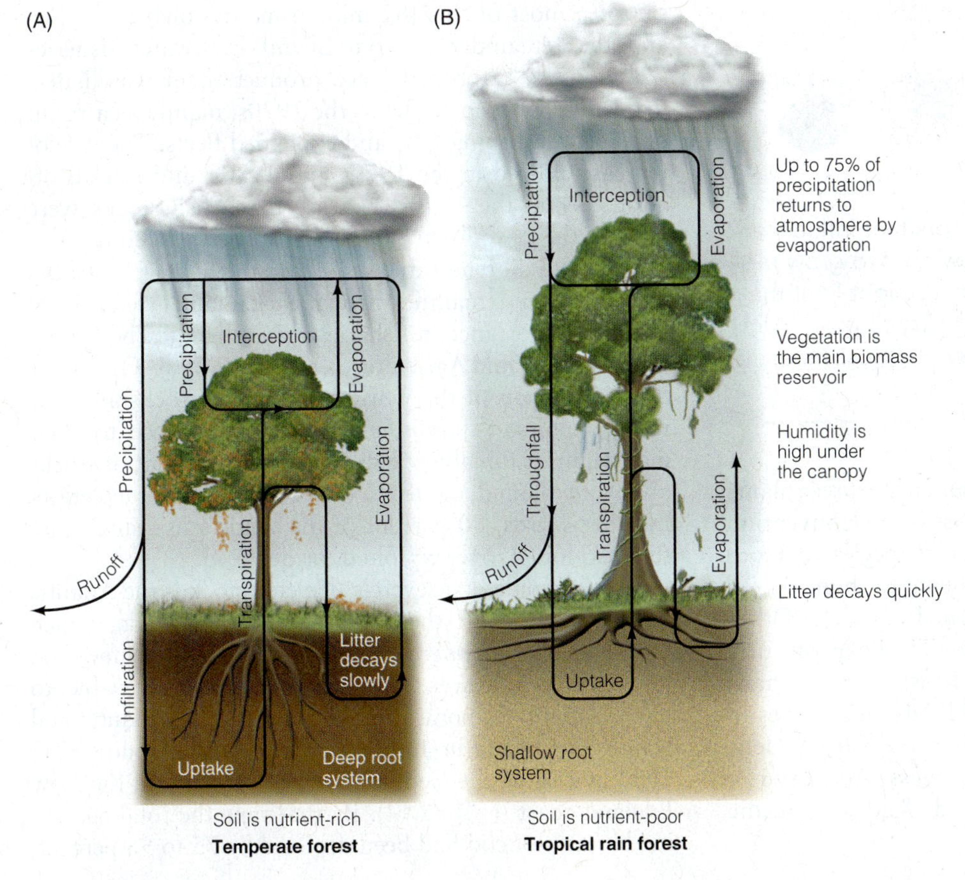

The question of why wilderness and wildlife should be protected speaks to the value of ecosystems and biodiversity, discussed in Chapter 16. In parks, forests, and wildlife refuges, however, there are almost always conflicts between protection and human use. It is important to distinguish between *conservation*—the conscientious management of natural heritage and resources—and *preservation*—the maintenance of natural wilderness areas in near-pristine condition. Both may be possible, and both may be desirable, but by definition they are not achievable at the same time in the same area.

Most parks and forests are designed to fulfill more than one purpose. Recreational use, logging, mineral exploration, oil drilling, hunting, and even agriculture and ranching are permitted in many protected areas. Some of these uses are incompatible, forcing us to examine our reasons for setting aside protected areas in the first place. For example, logging may not make sense in an area where the preservation of old-growth forest is of prime concern. Enormous controversy surrounds the use of wilderness areas for mineral and oil exploitation because of the potential for permanent impacts. Certain types of tourist use, such as off-road vehicles, dune buggies, mountain bikes, and even heavy trail use can be damaging to fragile environments. Hunting and ranching may be inconsistent with wildlife preservation; for example, controversies rage about the preservation of wolves near Yellowstone Park because the wolves—protected and supported within the park—defy park boundaries and prey on the sheep of nearby ranchers.

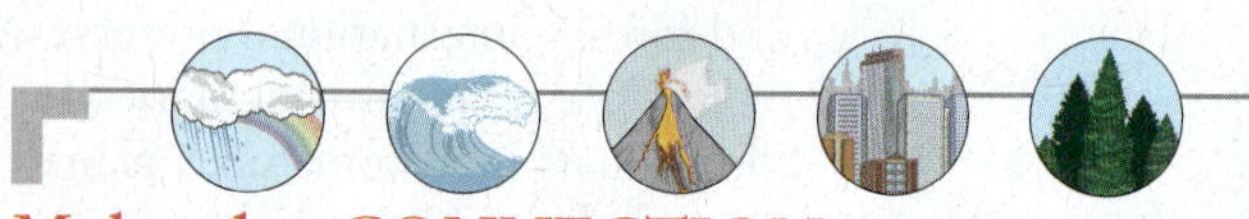

Make the CONNECTION

In Chapter 16 we discussed the factors that influence a wild population, limiting it or allowing it to grow rapidly. What are some of the factors that might limit the wolf population of Yellowstone Park? Try to give examples from the biosphere, hydrosphere, atmosphere, and anthroposphere.

A number of international agreements protect plants and animals; among these agreements is the Convention on International Trade in Endangered Species of Flora and Fauna (CITES, 1975), which regulates the hunting and trade of endangered and threatened species. Other international agreements are aimed at the protection of specific species, such as whales. Most countries have some form of protection of natural lands and species in their national legislation, although these vary widely. Examples include the U.S. Wilderness Act (1964), Endangered Species Act (1973), and Wild and Scenic Rivers Act (1968).

Fisheries Resources

Living resources in aquatic environments are under increasing pressure, both from overharvesting and from development and environmental change that affect the quality and quantity of available aquatic habitat. Fish are an important source of animal protein worldwide, especially in low-income countries. About one billion people in the world depend on fish as their main source of protein, and in developing countries fish accounts for more than 20 percent of animal protein consumption.

Status of the Resource

Assessing the world's fisheries is even more challenging than assessing forests because it relies on individual fishing vessels and countries to report their catch in a truthful, accurate, and timely manner. It is difficult to monitor the status of a resource that is so large, so widely dispersed, and hidden from direct view. Aquatic environments, particularly marine environments, also are less well protected by law than terrestrial environments. National jurisdiction over marine fisheries—the best means of controlling and monitoring their use—is subject to constant challenges and misuse by competing nations. The open ocean, stretching beyond national territorial waters, is particularly difficult to regulate.

Total world fish production is currently about 142 million metric tons per year. About 77 percent of this goes to human consumption; most of the rest is made into oils or fishmeal, mainly for animal consumption. **Capture fisheries**, where fish are caught in the wild, account for the majority of world production, about 92 million metric tons; most of this (82 million metric tons) is marine, and the remainder is from inland freshwater fisheries (FIG. 17.13). Capture fishery production increased dramatically from the 1950s to the 1970s, mainly as a result of improved fishing gear and expanded fleets. The rate of increase slowed in the 1970s and 1980s, and leveled off altogether in the 1990s, because many fish stocks were nearing their productive limits.

As with all renewable resources, the balance of stocks and flows and maintenance of sustainable yields is of primary importance in fisheries management. The United Nations Food and Agriculture Organization (FAO) reports that 52 percent of the world's major fish stocks are now fully exploited—that is, yielding catches close to their maximum sustainable yield. Another 28 percent of stocks are overexploited, depleted, or recovering from depletion. In other words, 80 percent of the world's fish stocks are fully exploited, overexploited, or depleted.

An example of a severely depleted stock is the Atlantic cod population off the eastern coast of Canada, which was fished to *commercial extinction*; that is, so few cod remained that it was no longer economically viable to harvest them. A moratorium was placed on Atlantic cod fishing in Canada in 1992; the United States followed in 1994 by closing the Georges Bank cod fishery off the New England coast (FIG. 17.14). By that time the total spawning biomass of cod had been depleted by 75 to 98 percent,

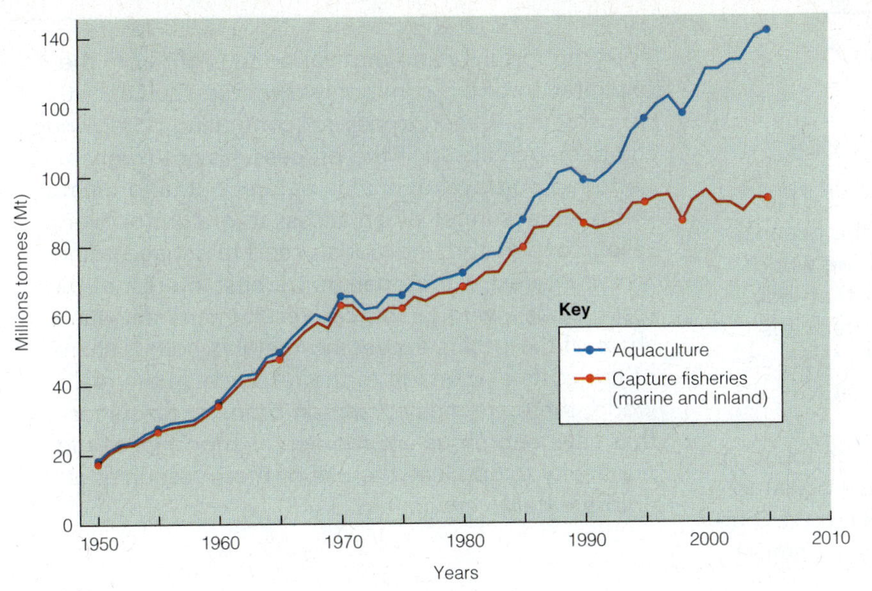

FIGURE 17.13 Global fisheries production
Global capture fishery production has increased dramatically over the past few decades, but the rate of increase began to decline in the 1970s and 1980s, probably because many fisheries are nearing their productive limits. The proportion of global production provided by aquaculture has surged, especially in the past decade, to the point where it is now equal to about half of the production from capture fisheries.

and some stocks have continued to decline since then. It is not known when the fishery will recover, if ever. Populations of other species have now largely replaced the cod, changing the whole balance of the ecosystem. As of 2010, the International Union for the Conservation of Nature's *Red List of Endangered Species* (Chapter 16) lists the Atlantic cod as "vulnerable"—one step short of being an endangered species.

Aquaculture

Production from marine and inland **aquaculture**, the raising of fish, shellfish, crustaceans, and aquatic plants in captivity (FIG. 17.15), has increased dramatically over the past 25 years as the world's natural fisheries have come under stress. Current production is about 52 million metric tons; this represents an increase of about 10 million metric tons in just five years, to a level that is now more than half of the total production from capture fisheries. Most of the growth in aquaculture activity has occurred in Asia, especially China.

FIGURE 17.14 Lost abundance
Fishermen in the 1950s off Canada's New Brunswick coast haul in their catch of cod. Large cod and abundant hauls like this are no longer available since the collapse of the Atlantic cod fishery off the eastern coast of North America.

FIGURE 17.15 Aquaculture
Aquaculture is increasing in importance as the world's capture fisheries come under greater stress. Seen here are pens for fish farming on Lake Sebu in Mindanao, Philippines.

A Closer LOOK

THE TRAGEDY OF THE COMMONS

Managing common property resources is tricky. No matter how much we may all want to see the oceans, atmosphere, forests, soils, and fisheries of the world properly preserved and cared for, it seems that their depletion and degradation continue unabated. It is very difficult to get the nations of the world to agree on rules for their use, and even more difficult to enforce such rules.

Part of the reason for this situation can be found in a deceptively simple concept that was first articulated by ecologist Garrett Hardin in a 1968 paper in the journal *Science*.[1] The concept relates to the management of common property resources, and the paper was called *The Tragedy of the Commons*. Although the paper is quite detailed (and worth reading in the original), the basic idea of the tragedy of the commons is the following:

Imagine a goat pasture, to which all the people of a certain town have equal access—in other words, a commons. The pasture has a specific *carrying capacity*—the maximum number of goats that can be sustained biologically by the pasture for an indefinite time, without causing permanent damage to the ecosystem. (Compare this to the definition of carrying capacity in Chapter 16.)

Let's say that the carrying capacity of the pasture allows for the grazing of five goats from each family (FIG. C17.1). No one polices the resource; it is taken on good faith that everyone will put only 5 goats out to graze. Then imagine that one person gets a bit greedy and decides to put 15 goats out to graze; after all, who will know? After a while, the pasture ecosystem starts to be degraded because there are too many goats grazing. The grass is eaten down to the ground, and the soil begins to erode. But in the meantime, the person with the 15 goats has become rich from selling goat's milk and cheese.

The point is that the individual with the 15 goats—indeed, every individual with access to the resource—had every opportunity and motivation to profit from the exploitation of the commonly owned resource, but no particular reason to pay for controlling or repairing the damage caused in the process. It's not a very flattering portrayal of human nature, but unfortunately it is often accurate: When access to or control over a resource is shared, individuals tend to act according to self-interest, rather than in the best interest of the resource as a whole. This can result in an individual's share of the profits exceeding that individual's share of the costs of damage to the resource. In the real world, there are many common property resources that have enormous environmental importance, and our ability to "police" the use of these resources is, unfortunately, very limited.

FIGURE C17.1 The Tragedy of the Commons
(A) This pasture appears to have sufficient carrying capacity to support the number of goats grazing here. (B) It seems that too many goats have been pastured here, leading to degradation of the resource. The Tragedy of the Commons suggests that it is difficult to regulate commonly-held resources; individuals may benefit from over-using the resource, but the entire community shares the cost of the resource degradation.

[1] Hardin, Garrett, 1968. "The Tragedy of the Commons" *Science*, vol. 162, no. 3859, p. 1243–1248.

Much aquaculture production focuses on products that are of high economic value, such as shrimp and salmon. The specific approach depends on the type of organism being farmed. For example, salmon are started (*spawned*) in captivity and then released to the wild, where they mature. Because of their natural breeding habits, the salmon eventually return to their original spawning ground to breed; this predictability of behavior allows them to be harvested. Other aquatic organisms, such as shrimp, are raised in enclosures in brackish-water coastal areas. Saltwater aquaculture is used mainly for aquatic plants, such as kelp, and mollusks, such as oysters. Inland or freshwater aquaculture is used mainly for finned fish, such as carp, and for recreational fishing purposes.

Aquaculture can have significant environmental costs if not properly managed. For example, if cultured organisms escape, they can contaminate the natural gene pool or spread diseases into wild populations. Depending on circumstances, aquaculture can require major inputs of fresh water, energy, and even wild fish (in the form of meal for fish food). In Southeast Asia, coastal aquaculture farms have contributed significantly to the loss of coastal wetlands and mangrove forests.

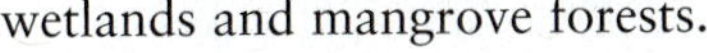

Make the CONNECTION

Do the same factors that limit natural populations in the wild apply to populations of organisms in managed ecosystems, like fish in aquaculture ponds? What are some of the similarities and differences?

Soil Resources

Soil is weathered rock material that has been altered by the presence of living and decaying organisms, such that it can support rooted plant life. We examined the science of soil, soil formation, and soil classification in Chapter 15; here we concern ourselves with soil as a resource.

Soil as a Critical Resource

The availability of **arable soil**—soil that is suited for growing agricultural crops—is crucial for the maintenance of our global food supply. Soil is the link between the atmosphere, geosphere, hydrosphere, and biosphere, and the processes by which soil is regenerated are constantly at work. Like all renewable resources, however, the rate of regeneration is crucial. It only takes a few uncultivated seasons for agricultural soil to replenish its nutrient content, but it can take thousands of years for an arable soil horizon to form from bare rock. In addition to physical loss through erosion, soil can be degraded through contamination, compaction, contamination, or loss of nutrients.

As discussed in Chapter 15, *soil fertility* is the ability of soil to provide nutrients such as phosphorus, nitrogen, and potassium to growing plants. If one type of plant is grown in the same soil over a long period, the soil will become depleted in the specific nutrients required by that plant. If most nutrients are available in abundance, but one essential nutrient is not, the lack of this one nutrient becomes a *limiting factor*; that is, it limits the growth potential of the plant. This is why farmers supplement the nutrient contents of agricultural soils with plant fertilizers. These mixtures of mineral nutrients (mainly nitrates, phosphates, and potassium in various proportions) are specifically designed to supplement soil nutrients that may be limited in availability.

Traditional and Modern Agriculture

Thousands of years ago, humans were nomadic hunter-gatherers; they extracted the foods they required from the environment around them. About 12,000 years ago, for reasons that are not entirely understood, people began to settle into a sedentary lifestyle that involved production, rather than just extraction of food resources. As people selectively cultivated and bred certain types of plants and animals, domestication occurred and agriculture was born.

The hunter-gatherer lifestyle requires a large land area to support a small human population. The development of agriculture involves a much more intensive use of resources. This effectively increased the carrying capacity of the land, allowing the human population to increase accordingly. Much later, during the Industrial Revolution, the invention of the steam engine and the resulting development of mechanized farming tools again spurred a huge increase in agricultural production. This was the beginning of modern agriculture.

The switch to modern agricultural practices entailed new impacts on soils and crops. Traditional agriculture involves the planting of diverse crop types in one field, interspersed with seasons in which the field is left *fallow*, or unplanted, allowing time for depleted nutrients, particularly nitrates, to be replenished by the action of soil bacteria. Modern agriculture has tended toward shorter fallow periods and the planting of large tracts of a single crop, or *monoculture*, which makes crops easier to harvest but can lead to depletion of the specific soil nutrients required by that crop. Monoculture has also contributed to a decrease in the genetic diversity of crop species (**FIG. 17.16**); about 5000 plant species are cultivated, but 95 percent of our nutrition comes from just 30 of these. Genetic uniformity renders species less able to respond to pests, diseases, and changes in the physical environment. Seed banks (a type of *ex situ* conservation, Chapter 16) have been set up in an effort to protect the vanishing genetic diversity of crops.

Over the past 40 or 50 years, modern farming methods have led to an enormous increase in agricultural productivity. For example, wheat and corn production almost tripled in the decades following the 1950s, a

FIGURE 17.16 **Monoculture**
One of the hallmarks of modern agriculture is monoculture, the sowing of large areas of a single crop type. This is a field of corn from Nebraska.

FIGURE 17.17 **Salt-affected soil**
Salinization can occur when soils become water-logged. Salts are drawn upward as the water evaporates and are deposited at the surface, as seen in this photo of a rangeland in Colorado.

period referred to as the **Green Revolution**. Rice and other grains more than doubled, and similar increases occurred for livestock production. The Green Revolution was brought about by the development of high-yield, disease-resistant seed types through bioengineering. There was also enormous expansion in the use of irrigation and in the amount of land under cultivation. The increase in productivity also required significant inputs of energy from *agrochemicals* such as fertilizers and pesticides, and from the fossil fuels required to run large farm machinery.

Impacts of the Green Revolution

There may be limits to the tremendous increases in productivity realized in the Green Revolution. In the mid-1980s, productivity in some agricultural sectors began to level off. When combined with global population growth, productivity per capita had begun to stagnate or even decline. Some crop yields per hectare also declined. The practices that led to the Green Revolution have had some impacts on environmental quality, which, in turn, may have affected agricultural productivity. For example, the increased use of chemical pesticides has led to the emergence of pesticide-resistant insect species, leading farmers to use more agrochemicals with less satisfactory results. The increased use of fertilizers, required as a consequence of monoculture and short fallow times, has led to surface and groundwater pollution from nitrate- and phosphate-bearing runoff.

Greatly expanded irrigation, another hallmark of the Green Revolution, also has drawbacks. When soils in arid and semiarid environments are overirrigated, *salinization* (also called *salination*) can occur. The top layer of the soil becomes soaked, or *waterlogged*. As this wet layer dries in the hot sun, salts are drawn up from deeper in the soil by capillary action, leaving a crust of dried salt on the surface (FIG. 17.17). Waterlogging and associated salinization can permanently damage agricultural land. Some extensive irrigation projects also have led to the overexploitation and depletion of groundwater resources. Agricultural researchers in arid regions are developing new, high-efficiency irrigation techniques in response to these problems.

Erosion and Loss of Agricultural Soil

The estimated global rate of soil loss through erosion now exceeds 25 billion metric tons each year; this may be the most serious limitation to agricultural productivity worldwide. "Soil loss" doesn't mean that the soil vanishes; it is carried away from agricultural lands by erosion. Eroded topsoil generally winds up in a body of water, where it may clog a stream channel or contribute contaminants that have adhered to the surfaces of clay particles in the soil. Thus, soil loss is problematic on farms and rangelands, but it also causes problems in streams, rivers, and lakes. Contaminants such as agrochemicals and farm waste can have a serious impact on water quality. For example, the deaths of seven people in Walkerton, Ontario in 2000 from *E. coli* infections have been attributed to the contamination of drinking water by livestock manure from nearby farms.

Soil scientists talk about soil loss in terms of a *tolerable annual rate of loss*, referring to a rate of erosional loss that is balanced by regeneration. In the United States, the amount of arable soil lost to erosion at the beginning of the 1990s exceeded the amount of newly formed soil by almost 2 billion metric tons per year. The rate of loss since then has been reduced substantially by the implementation of aggressive soil erosion control programs. However, soil loss in the other major food-producing countries, China and Russia, is at least as rapid, while in India the rate has been estimated to be more than twice as high.

Modern agriculture is probably capable of meeting the challenge of feeding a population of 8 to 10 billion by the middle of the twenty-first century. In attempting to meet this challenge, the maintenance of a fertile soil resource will continue to be of paramount concern.

Water Resources

A reliable supply of fresh water is crucial not only for the survival of people and ecosystems, but also for industry, agriculture, recreation, transportation, and fisheries. As discussed in Chapter 8, irrigation currently accounts for about 75 percent of the demand for water globally, industry for 20 percent, and domestic use for 5 percent, although the proportions vary greatly from one region to another. Global water use has more than tripled since 1950; both population growth and improved standards of living have contributed to this increase.

Although most large public water-supply systems draw from surface water sources, more than half of the people in North America get their drinking water from groundwater supplies. Aquifers are replenishable; however, if the rate of groundwater withdrawal exceeds the rate of recharge over time, the volume of stored water steadily decreases. It may take hundreds or even thousands of years for a depleted aquifer to recharge.

The results of excessive groundwater withdrawal include depression of the water table; drying up of springs, streams, and wells; compaction; and subsidence. When an aquifer suffers compaction—that is, when its mineral grains collapse on one another because the pore water that held them apart has been removed—it is permanently damaged and may never be able to hold as much water as it originally held. Urban development also contributes to groundwater depletion, not only by increasing the demand for water but also by increasing the amount of impermeable ground cover. In a recharge area that is covered by roads, parking lots, buildings, and sidewalks, the rate of groundwater replenishment is substantially reduced.

Consumption, Loss, and Supply

It is important to distinguish between water consumption or *use*, and "irretrievable" consumption, or *loss*. Water that is "lost" is not returned to the local hydrologic cycle. Water loss occurs mainly through evaporation; some water also is lost through leakage from underground pipes. Globally, irrigation is not only the largest consumer of water, but also the largest cause of water loss. Much of the water used for irrigation is lost through evaporation (FIG. 17.18A). Drip irrigation, in which a measured amount of water is delivered directly to the root of each plant, has been developed as a response to this problem (FIG. 17.18B); it can also help minimize soil salinization. Water stored behind dams is problematic because of the very large amount of water lost to evaporation from the surface of the reservoir. Industrial processes consume a significant amount of water, but a large proportion of the water is typically returned to the same surface water body after treatment and is therefore not lost from the local hydrologic cycle.

Hydrologists designate as **water-stressed** a country or region with annual renewable water supplies of 1000–2000 m^3 per person. This refers to water that is available

FIGURE 17.18 Irrigation and water loss
Irrigation is the biggest consumer of water, as well as the main source of irretrievable water loss worldwide. (A) Here, a standard irrigation approach loses a large proportion of water to evaporation. (B) In modern drip irrigation approaches (seen here in Australia), exactly the right amount of water is delivered directly to the roots of the crops, minimizing water loss to evaporation.

for all purposes, including industrial and agricultural, not just personal use. For comparison, the available water supply in most parts of Canada, a water-rich country, is about 50,000 m^3 per person per year. If the available supply drops below 1000 m^3, the area is considered to be **water-scarce**—that is, lack of water represents a serious constraint on agricultural production, economic development, environmental protection, and personal nutrition, health, and hygiene.

Today 29 countries worldwide, with a total population of 450 million people, suffer from significant water shortages (FIG. 17.19). Most of these are developing countries with rapidly increasing population rates, foreshadowing ever-increasing water problems. At least nine countries (Libya, Qatar, United Arab Emirates, Yemen, Jordan, Israel, Saudi Arabia, Kuwait, and Bahrain) are using more than 100 percent of their internally available water and have to make up the difference by importing fresh water or by desalination of seawater.

Even if a country is not designated as water-scarce, there may be shortages. In California, for example, there is a water surplus in the northern part of the state, but water scarcity in the southern part of the state, as a result of population density and intensive agricultural use. Much of the water used in Southern California is imported from northern California or from out-of-state. A mismatch between local supply and demand typically leads to the diversion and *interbasin transfer* of water. An example of this occurred with devastating environmental consequences, when water from the rivers that replenish the Aral Sea in Central Asia was diverted for agricultural use (see *A Closer Look: The Case of the Aral Sea*, Chapter 8).

Water Management

Issues around the allocation and regulation of water use can be very controversial. Conflicts over water rights have caused or intensified many international disputes, which continues to be true in areas where water is scarce, such as the Middle East. Water rights tend to be founded on one of two basic ideas. *Riparian rights* guarantee water access only to those who own or have legally established rights to the *riparian zone*, the land adjacent to the banks of a stream or lake. The doctrine of *prior appropriation*, on the other hand, assigns water rights to those who have established a history of water use from the source, whether or not they have legal ownership of riparian lands.

Water distribution rights become much more complicated when applied to groundwater, which is hidden from view, or where a surface water body crosses or straddles a political boundary. This situation pertains in many of the world's great aquifers and rivers, including the Rhine, Mekong, Ganges, Nile, Danube, and Colorado. Almost half of the world's land area lies in drainage basins shared by two or more countries. One of the most successful international collaborations on the management of trans-boundary surface water bodies is the International

FIGURE 17.19 Water stress
This map shows areas of the world that are experiencing water stresses as a result of demand outstripping the available supply of water.

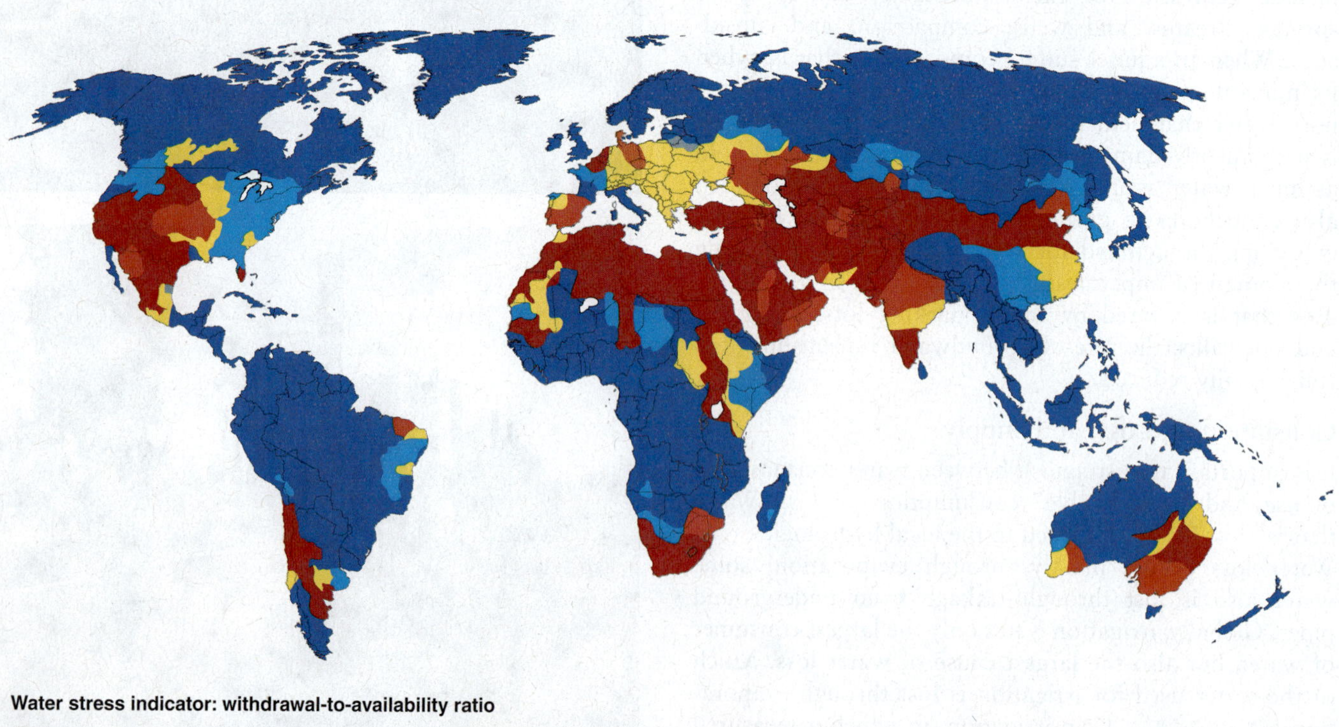

Joint Commission, set up by Canada and the United States to facilitate cooperation in the management of the Great Lakes, an important shared resource.

LIMITS TO GROWTH

It is difficult to know what Earth's carrying capacity may turn out to be and whether we have reached or even surpassed it. The human population continues to grow, and we continue to find, process, and make use of resources in a wide variety of ways. Are there enough renewable and nonrenewable resources available to support this growing population?

Perhaps the only firm conclusion we can reach is that while Earth's carrying capacity is very large, it is not infinite. The available resources can be extended through careful use and conservation practices, new technologies, and new discoveries. However, Earth is a closed system; in the long run, there must eventually be some limits to the growth[2] of the human population and resource use. It seems inevitable that our descendants will have to manage Earth resources very differently from how we and our parents and grandparents have managed them.

Increasingly, the guiding principle in the management of renewable resources is **sustainable development**—the cautious, planned utilization of Earth resources to meet current needs, without degrading ecosystems or jeopardizing the future availability of those resources. Is sustainable development a useful concept, or is it too vague and impossible to implement? This question is still being debated. It is not even clear that sustainable development is possible in the context of nonrenewable resources, which—by definition—become more depleted every time they are used.

The basic concepts involved in the management of renewable resources seem simple enough, especially when viewed from an Earth system science perspective—balance stocks and flows, monitor the status of natural systems, and maintain a sustainable yield. In practice, however, these are difficult concepts to implement, and renewable resources are subject to many uncertainties and pressures. As a global community, our record in the management of renewable resources has been less than commendable. We must improve if we hope to protect these precious resources.

[2] *Limits to Growth* (1972) is the title of a book by Dennis Meadows *et al.*, one of the first studies that used computer modeling to make projections about remaining supplies of resources.

END OF CHAPTER 17

SUMMARY

1. Humans have withdraw more resources, displace more material, and generate more waste than any other species. The anthroposphere comprises the built environment and the parts of the Earth system that have been modified by human actions.

2. Renewable resources, both living and nonliving, are replenished seasonally or continuously. Nonrenewable resources, such as mineral resources and fossil fuels, are replenished on timescales that greatly exceed the scale of a human lifetime. Some resources, such as an atom that has been split, are truly nonrenewable. Others, such as solar, tidal, and geothermal sources of energy, are inexhaustible.

3. Common property resources are commonly, jointly, or communally owned. When control over a resource is shared, individuals sometimes act according to self-interest rather than in the best interest of the resource as a whole. This is the tragedy of the commons; it can lead to a situation in which the carrying capacity of the resource is exceeded and the resource becomes degraded.

4. Civilization and natural resources are inseparable. The first resources used by early hunter-gatherers were fruits, nuts, animals, and fish. Stones for toolmaking, salt, and native metals were the first nonrenewable resources to be used by early humans. The emergence of agriculture about 12,000 years ago allowed people to extract resources more intensively. Today we have industrial uses for almost all naturally occurring chemical elements.

5. Traditionally, we have viewed the human economy as a reservoir into which resources and energy flow from a limitless source, and we have seen the Earth system as having limitless capacity to absorb waste and pollutants. We now understand that the human economy and the resource cycle exist as part of a larger Earth system in which material resources and waste-absorbing capacity are finite.

6. Nonrenewable and renewable resources require different management approaches. The more we use of a nonrenewable resource, the less remains in the stock. New deposits can be found, but otherwise the availability of nonrenewable resources can only be extended through conservation, substitution, reuse, or recycling. Living resources are renewable, but problems arise when they are harvested more quickly than the rate at which they reproduce and grow. If withdrawals are balanced by new growth or replenishment, the stock will remain constant; if the stock is drawn down faster than new stock can accumulate, the resource is being depleted.

7. Significant human impacts on forests date back tens of thousands of years. Managed forest plantations can help maintain the stock of forest resources, but they cannot replace the biodiversity and varied habitat of old-growth forests.

8. Trees stabilize soil by holding it in place with root systems and by replenishing its content of organic material. Trees also play an important role in water cycling. On a global scale, forests harbor extensive reserves of biodiversity and help to regulate the climate by serving as an enormous reservoir for carbon.

9. Clear-cutting is the most economically efficient way to harvest wood from a forest, but it is the most ecologically destructive. Selective cutting can reduce impacts by mimicking natural disturbances. Global forest cover is assessed using ground-based information and data from remote sensing. Of particular concern are tropical rain forests, because of the loss of biodiversity, habitat, and carbon reservoirs, and because tropical soils are fragile. Direct causes of deforestation are logging, and clearing of land for agriculture and ranching. Indirect and underlying causes include air pollution, environmental and climatic change, and socioeconomic forces.

10. In agroforestry, trees and crops are planted together. The deep roots of the trees stabilize the soil and draw water from depth, making it available to crops. Trees contribute to the organic content of the soil, and they can be selectively harvested, to provide fuel-wood, timber, fruits, nuts, and clippings to feed domesticated animals.

11. In wilderness areas, natural forces are more significant than human intervention, and people do not live on a permanent basis. Conservation is the conscientious management of natural heritage and resources, whereas preservation is the maintenance of wilderness areas in near-pristine condition. Most protected wilderness areas are designed to fulfill multiple purposes.

12. Fish is an important source of dietary protein and income, especially in low-income countries. It is difficult to monitor the status of fisheries because the resource is large, widely dispersed, and hidden from view. Marine capture fishery production increased dramatically from the 1950s to the 1970s as a result of greatly improved fishing gear and expanded fleets. About 80 percent of the world's major fish stocks are now fully exploited, overexploited, or depleted, mainly as a result of overharvesting.

13. Aquaculture is increasing in importance as the world's natural fisheries come under greater stress. Aquaculture can have environmental costs, including the potential for contamination of wild populations by genes or diseases from cultured organisms. Some types of aquaculture require major inputs of fresh water and energy, and contribute to the loss of coastal wetlands and mangroves.

14. Arable soil is crucial for the maintenance of our global food supply. It can take thousands of years for an arable soil horizon to form from bare rock. In addition to physical loss through erosion, soil can be degraded through contamination, compaction, or loss of fertility, the ability of the soil to provide nutrients to growing plants.

15. Since the 1950s, modern farming methods have led to an enormous increase in agricultural productivity. The Green Revolution was brought about by increased use of agrochemicals; expanded irrigation; monoculture, and the development of high-yield crops; enormous expansion of cultivated land; and significant inputs of energy.

16. Agricultural production per capita and some crop yields have begun to level off, suggesting that there may be environmental limits to the enormous increases in productivity realized in the Green Revolution. Increased use of agrochemicals has led to the emergence of pesticide-resistant insect species, as well as to surface and groundwater pollution. Poorly designed irrigation has caused waterlogging and salinization of soils and depletion of some groundwater resources. Monoculture has contributed to soil nutrient depletion and the loss of crop biodiversity.

17. The estimated global rate of soil loss through erosion now exceeds 25 billion metric tons each year. A tolerable annual rate of loss would be a rate of erosional loss that is balanced by regeneration. Soil loss is problematic on farms and rangelands, but also in streams, rivers, and lakes. Eroded topsoil may clog stream channels or contribute contaminants that have adhered to the surfaces of clay particles in the soil.

18. A reliable supply of fresh water is crucial for the survival and health of people and ecosystems, and for industry, agriculture, recreation, transportation, and fisheries. Aquifers are replenishable, but if the rate of groundwater withdrawal exceeds the rate of recharge, the volume of stored water steadily decreases. The results of excessive groundwater withdrawal can include depression of the water table; drying up of springs, streams, and wells; compaction of aquifers; and subsidence.

19. Water loss—that is, water that is consumed and not returned to the local hydrologic cycle—occurs mainly through evaporation. Irrigation is the largest cause of water loss, but new approaches in irrigation technology are addressing this issue. A country or region with annual renewable water supplies of 1000–2000 m^3 per person is water-stressed; if the available supply drops below 1000 m^3, the area is considered to be water-scarce. Twenty-nine countries suffer from significant water shortages; most are developing countries with rapidly increasing population rates.

20. Earth's carrying capacity is very large but not infinite. Nonrenewable resources can be extended through careful use and conservation practices, new technologies, and new discoveries. A guiding principle in the management of renewable resources is sustainable development—the cautious, planned utilization of Earth resources to meet current needs, without degrading ecosystems or jeopardizing the future availability of those resources.

IMPORTANT TERMS TO REMEMBER

agriculture *520*
anthroposphere *520*
aquaculture *531*
arable soil *533*
capture fisheries *530*
common property resource *525*
deforestation *528*
Green Revolution *534*
nonrenewable resource *520*
old-growth forest *527*
perpetual (inexhaustible) resource *525*
plantation (forest) *526*
renewable resource *520*
resource *524*
resource cycle *523*
stock *523*
sustainable development *537*
sustainable yield *528*
water-scarce *536*
water-stressed *535*
wilderness *529*

QUESTIONS FOR REVIEW

1. When did humans first develop agriculture? How does the timing of this activity compare to the times when other types of resource use first started?
2. Describe the human economy as part of the larger Earth system, using a diagram to illustrate your answer.
3. How do management approaches differ between renewable and nonrenewable resources, and why?
4. What is a "sustainable yield," and to what kinds of resources does it apply?
5. What is an old-growth forest? Are any forests today completely free of human influence? Explain.
6. What are some of the differences between a natural forest and a managed forest plantation?
7. What is the difference between conservation and preservation of wilderness?
8. Why is it so challenging to evaluate the state of fisheries?
9. What is aquaculture? Can it replace capture fisheries?
10. What are four practices that led to the enormous increases in agricultural productivity that characterized the Green Revolution?
11. What is arable soil, and why is it such an important resource?
12. Compare modern and traditional agriculture, and comment on their environmental impacts.
13. What is the difference between water use and water loss?
14. Is it possible for a country that is not water-scarce or water-stressed to nevertheless experience water shortages? Explain.
15. Define sustainable development.

QUESTIONS FOR RESEARCH AND DISCUSSION

1. Do you think it is important to preserve wilderness and protect wildlife? Why (or why not), and to what extent should human activity in wilderness areas be subject to limitations?
2. Do you think sustainable development is a useful concept or guide for resource management? Why (or why not)?
3. In what ways is the concept of a tolerable annual rate of soil loss similar to sustainable yield in a forest or fishery? In what ways are these concepts different?

QUESTIONS FOR *THE BASICS*

1. Give examples of renewable, nonrenewable, and inexhaustible resources.
2. What is a common property resource? Give an example.
3. How is it possible to transform a renewable resource into a nonrenewable or depletable resource?

QUESTIONS FOR *A CLOSER LOOK*

1. Explain the concept of the tragedy of the commons.
2. How is the tragedy of the commons expressed in the management of the world's ocean fisheries? What about the management of the atmosphere and the global climate system?
3. What is carrying capacity, and how does it relate to the idea of the tragedy of the commons?

CHAPTER 18

Mineral *and* ENERGY RESOURCES

OVERVIEW

In this chapter we:

- Consider how conservation, recycling, and waste management can help close the resource cycle for nonrenewable resources
- Examine the geologic processes that lead to the formation of ore deposits
- Summarize the environmental impacts of mining
- Examine the geologic processes that lead to the formation of fossil fuel deposits
- Learn about the impacts of fossil fuel use and the potential for alternative energy sources in the future

Molten ore
Molten gold is poured into containers at an ore refinery.

In Chapter 17 we introduced a key feature of nonrenewable resources—that the more we use, the less remains in the stock. Thus, management of nonrenewable mineral and energy resources differs fundamentally from management of renewable resources. We must concern ourselves with the effective management, efficient use, conservation, and recycling of nonrenewable raw materials because they are finite and depletable. We need to understand the resource itself: how nonrenewable resources are formed, where they occur, how to locate them, and how to extract them. Finally, we need to learn as much as possible about the environmental impacts of the extraction and use of nonrenewable resources, and how those impacts can be minimized.

In this chapter we will consider the formation, extraction, and use of nonrenewable mineral and energy resources. We will also have a look at the great variety of energy sources that have the potential to replace nonrenewable sources in the future.

NONRENEWABLE RESOURCES: CLOSING THE CYCLE

Conserving nonrenewable resources and minimizing the impacts of their use involves "closing" the resource cycle as much as possible by recycling useful materials and containing harmful wastes. As discussed in Chapter 17, waste is generated at every stage of the resource cycle, from extraction and processing of raw materials through production and packaging of goods, transportation, use, and post-consumer disposal (shown in Fig. 17.7). Each of these points represents a possibility for pollution or waste to occur, but each point can also represent an opportunity for more efficient management of the resource, through waste minimization and recycling.

Recycling refers to the extraction of usable raw materials, such as metal, glass, and pulp, from waste products such as discarded manufactured objects, paper and cardboard, cans, jars, and bottles (FIG. 18.1). Many industrial waste materials are recyclable, and industry has traditionally been relatively efficient at recycling useful waste materials from manufacturing processes, simply because it makes economic sense to do so. The real "leakage" in the resource cycle comes at the post-consumer stage. Curbside recycling programs are now commonplace in North America, but many of these programs remain uneconomic or only marginally economic. Demand for products that contain post-consumer recycled materials may be the best way to change this situation.

FIGURE 18.1 Recycling
Recycling of raw materials is one of the few ways to extend the lifetime of a nonrenewable resource. Seen here are bales of pop bottles waiting to be recycled; the plastic of which they are made is a petroleum product.

MINERAL RESOURCES

In our modern world of almost 7 billion people we use mineral resources to make clothes, build shelters, fertilize crops, provide transportation, distribute electric power, and communicate electronically (FIG. 18.2). The list of materials that we mine, process, and use has grown enormously since our ancestors first discovered how to extract useful minerals, thousands of years ago (Chapter 17). Today we have uses for almost all naturally occurring chemical elements, and more than 200 different kinds of minerals are mined.

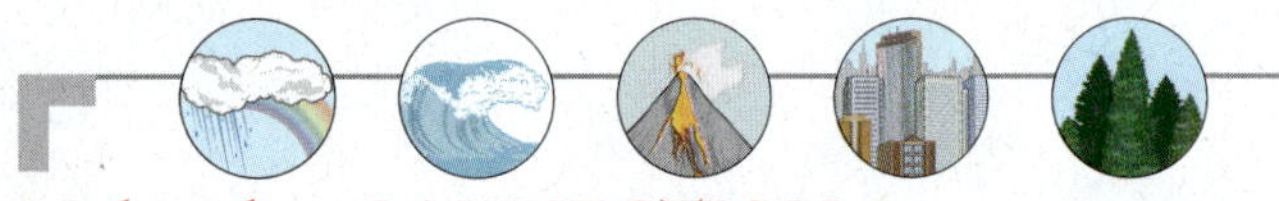

Make the CONNECTION

Look around. Even if you are reading this book under a tree in a meadow, you are surrounded by products made from nonrenewable resources. Your watch has metal parts; the face is glass (made from silica) or plastic. Your pen is metal or plastic (made from petroleum); so are your jewelry, your glasses, and even your clothes if you are wearing synthetic fabrics (also made from petroleum). The food in your lunch was grown in soil tilled by metal machinery and nourished by mineral fertilizers. Even the paper in this book has mineral additives to give it texture and color. The digital version of the book is available on a computer made entirely of nonrenewable mineral and petroleum resources. And virtually every part of the car, bicycle, or bus that transported you here was made from nonrenewable resources.

Nearly every kind of rock and mineral can be used for something, although those that are most valuable tend to be rare. Elements that make up more than 0.1 percent of Earth's crust are called *geochemically abundant elements*, and those that make up less than 0.1 percent of the crust are called *geochemically scarce elements*. There are 92 naturally occurring chemical elements; only 12 of them are geochemically abundant.

It is convenient to group mineral resources according to how they are used (see TABLE 18.1). **Metallic minerals**

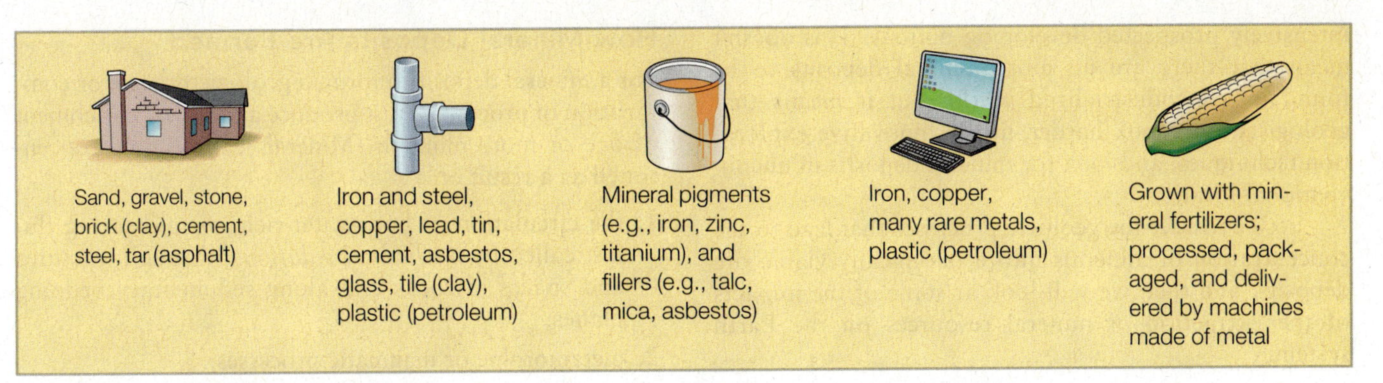

FIGURE 18.2 Minerals in everyday life
Our modern society is fundamentally dependent on nonrenewable mineral resources, as shown by these everyday objects made or grown using minerals.

TABLE 18.1 Mineral Resources and Their Uses

Metals	
Abundant metals	iron, aluminum, magnesium, manganese, titanium, silicon
Scarce and rare metals	copper, lead, zinc, nickel, chromium, gold, silver, tin, tungsten, mercury molybdenum, uranium, platinum, and many others
Nonmetals	
Used for chemicals	sodium chloride (halite), sodium carbonate, borax, calcium fluoride (fluorite)
Used for fertilizers	calcium phosphate (apatite), potassium chloride, sulfur, calcium carbonate (limestone), sodium nitrate
Used for building	gypsum (for plaster), limestone (for cement), clay (for brick and tile), asbestos, sand, gravel, crushed rock, shale (for brickmaking), decorative stone
Used for jewelry	diamond, corundum (ruby and sapphire), garnet, amethyst, beryl (emerald), and many others
Used for glass and ceramics	clays, feldspar, quartz (silica sand)
Used for abrasives	diamond, garnet, corundum, pumice, quartz

are mined specifically for the metals that can be extracted by *smelting*—that is, by heating and breaking the mineral apart so that the metal in it melts and separates from the other mineral constituents. Examples of metallic mineral resources include sphalerite (zinc sulfide, ZnS), from which zinc is recovered, and galena (lead sulfide, PbS), from which lead is recovered. In contrast, **nonmetallic minerals** are mined for their chemical or physical properties, not for the metals they contain. Examples of nonmetallic mineral resources include salt, clay, gravel, building stone, and gemstones.

Each of us relies directly or indirectly (through industry and public works) on a very large annual supply of these nonrenewable mineral resources, both metallic and nonmetallic. Without a steady supply of mineral resources, modern society, agriculture, and industry would collapse. Finding new mineral resources, managing and conserving their supply, and controlling the impacts of their extraction and use are among the great challenges of our time.

Locating and Assessing Mineral Resources

Metallic and nonmetallic minerals are mined from concentrations formed under suitable conditions by geologic processes such as weathering, sedimentation, and volcanism. The "suitable conditions" for concentration are not common, however, and for this reason mineral deposits are hard to find. The concentration processes also take many millions of years to form a deposit, which means that new deposits are not being generated quickly enough to replace those depleted by mining.

Economically exploitable mineral deposits are unevenly distributed and localized within Earth's crust. This uneven distribution is the main reason why no nation is self-sufficient in mineral supplies. Localization makes it difficult to assess the available quantity of a given material, and even more difficult to anticipate where and when new deposits might be discovered. A country that can meet its needs for a given mineral substance today may face a future in which it will become an importing nation. For example, a century and a half ago England was a great mining nation, producing and exporting such materials as tin, copper, tungsten, lead, and iron. Today, most of England's known mineral deposits have been exhausted. This pattern—intensive mining followed by depletion of the resource, declining production and exports, and increasing dependence on imports—can be applied on a local, regional, or global scale to estimate the remaining lifetime of a given mineral resource.

In the industrialized world today, the geologic locations most favorable for mineral exploration have mostly been prospected, assessed, and in some cases already mined and depleted. Over the past few decades, therefore, there has been a slow but steady shift in the emphasis of mineral exploration and production, away from the industrialized nations and toward the less intensively prospected developing nations. This doesn't mean that there are no more mineral deposits to be found in the industrialized world, but it means that geologists must look harder, utilize innovative exploration techniques, and look for mineral deposits in unconventional locations.

Let's consider the geologic processes that lead to the concentration of minerals into economically viable ore deposits, and then we will look at some of the impacts of the extraction of mineral resources on the Earth system.

How Mineral Deposits Are Formed

For a mineral deposit to form, a geologic process or combination of processes must produce a localized enrichment of one or more minerals. Minerals can become concentrated as a result of:

1. the circulation of hot, water-rich, metal-bearing fluids, called *hydrothermal solutions*, through fractures and spaces in rocks and along sedimentary bedding planes;
2. metamorphic or magmatic processes;

the BASICS

Deposits, Ores, and Reserves

A **mineral deposit** is a local concentration or enrichment of a given mineral. Deposits are sought that will yield the highest quality and quantity of materials at the lowest cost. The costs of extraction vary widely, depending on the location of the deposit, how concentrated it is, how deeply buried it is, how big it is, what kinds of technologies must be used to extract and process the material, and how far the processed material must be transported to get to a market, as well as the costs of labor, mine management, and environmental protection.

To distinguish between profitable and unprofitable mineral deposits we use the word **ore**, a deposit from which one or more minerals can be extracted profitably. Whether a given mineral deposit is an ore is determined by how much it costs to extract the mineral, and how much the market is prepared to pay for it. Note that "ore" is an economic term, whereas "mineral deposit" is a scientific term. All ores are mineral deposits because each of them is a local enrichment of one or more minerals; however, not all mineral deposits are ores. The degree of concentration, or **grade**, of ore is an important factor. In general, the more highly concentrated an ore is, the higher the grade and the more valuable the deposit. Ores that are less concentrated—that is, low-grade—contain a higher proportion of nonvaluable minerals, called **gangue** (pronounced "gang"), mixed with the sought-after ore minerals.

To assess how much of a particular nonrenewable resource remains in the stock, we use the concept of a **reserve**, which is that portion of a resource that has been identified and is economically extractable using current technologies (FIG. B18.1). "Identified" refers to deposits that have been found and studied, and for which scientists have relatively detailed information concerning the grade, location, and extent of the deposits. Deposits that are less well known are designated as speculative or hypothetical resources, not reserves. For a material to be part of a reserve it must also be economically feasible to extract the material, using known technologies, at its present market value.

FIGURE B18.1 Resources and reserves
A reserve, as shown here, is that portion of a resource that has been identified and is economically extractable using current technologies. Portions of a resource that are less well known and/or not economically extractable are designated as hypothetical, speculative, or subeconomic resources.

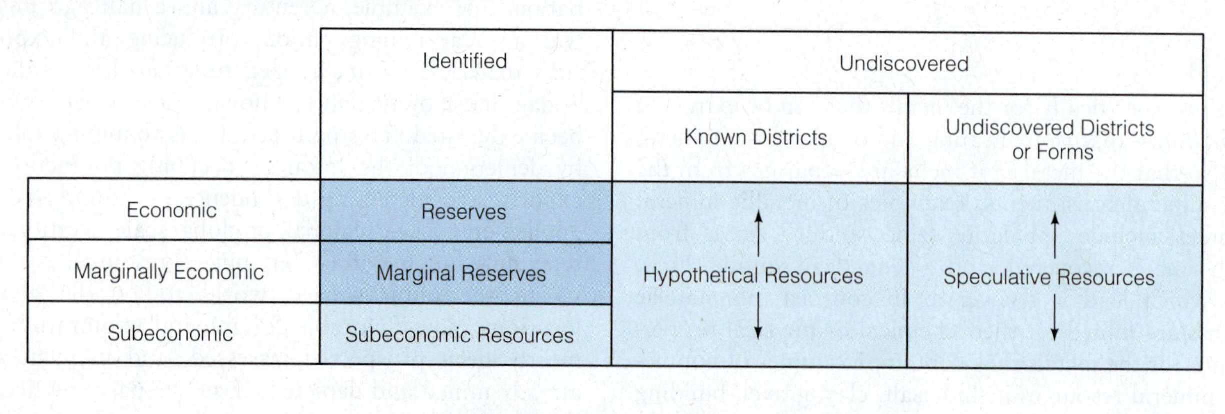

Identified
Undiscovered
Known Districts
Undiscovered Districts or Forms
Economic
Marginally Economic
Subeconomic
Reserves
Marginal Reserves
Subeconomic Resources
Hypothetical Resources
Speculative Resources
Reserve
Resource

3. chemical sedimentary processes that lead to precipitation of minerals from lake water or seawater;
4. the action of waves or currents in flowing surface water; or
5. weathering.

These concentrating processes are illustrated in FIGURE 18.3; let's examine each of them in turn.

1. *Hydrothermal Ore Deposits*

Many famous mines contain ores that formed when minerals were deposited from **hydrothermal solutions** (from the Greek words *hydro*, meaning water, and *therme*, meaning heat)—hot, aqueous, metal-saturated fluids. More mineral deposits have probably been formed by deposition from hydrothermal solutions than by any other mechanism (FIG. 18.3A). Hydrothermal deposits are the

FIGURE 18.3 Mineral deposits

There are many ways that mineral deposits can form and processes by which ore minerals can be concentrated. The mechanisms of formation of these deposits are described in the text. (A) **Hydrothermal:** A miner in Potosi, Bolivia, points to a rich vein containing chalcopyrite (a copper mineral), sphalerite (a zinc mineral), and galena (a lead mineral). (B) **Metamorphic:** This rock is from a metamorphic mineral deposit at the Tempiute Mine in Arizona. The ore minerals are sphalerite (brown, lower left), pyrite (gold-colored), and scheelite (pale grayish brown, lower right). (C) **Magmatic:** In this unusually fine magmatic ore outcrop at Dwars River in South Africa, layers of pure chromite (black) are sandwiched by layers of plagioclase. (D) **Sedimentary:** Evaporite deposits, such as this salt pan in Death Valley, California, form when lake water or seawater evaporates and leaves its dissolved mineral behind. (E) **Placer:** The world's richest known gold deposit, in Witwatersrand, South Africa, is an ancient placer deposit that was formed about 2.7 billion years ago. (F) **Residual:** In this bauxite sample from Queensland, Australia, rounded masses of aluminum hydroxide (gibbsite) are embedded in a matrix of iron and aluminum hydroxides.

primary sources of many metals, including copper, lead, zinc, mercury, tin, molybdenum, tungsten, gold, and silver.

Some hydrothermal solutions originate when water dissolved in magma is released as the magma rises and cools; others are formed from rainwater or seawater that circulates deep in the crust. The ore-mineral constituents in hydrothermal solutions originate from the rocks of the crust. Heated water reacts chemically with rock that it contacts, causing changes in the composition of both the rock and the solution. Trace metals such as copper, zinc, and gold are released from minerals in the rock, becoming concentrated in the slowly evolving hydrothermal solution.

As a metal-bearing hydrothermal solution moves through cracks and spaces in rocks, it cools and deposits its dissolved constituents. When a hydrothermal solution moves slowly upward, as with groundwater percolating through an aquifer, the solution cools very slowly. Dissolved minerals that precipitate from such a slow-moving solution are spread over great distances and do not become sufficiently concentrated to form an ore. But when a solution flows rapidly, such as through an open fracture or a layer of porous volcanic rock, where the flow is less restricted, cooling can be sudden and can happen over short distances. Rapid precipitation and a concentrated mineral deposit are the result. Other effects—such as boiling, rapid decrease in pressure, compositional changes caused by reactions with adjacent rock, and cooling as a result of mixing with seawater—can also cause rapid precipitation leading to the formation of concentrated deposits. When valuable minerals are present, an ore can be the result (FIG. 18.4B).

Irregular deposits called *veins* are formed when hydrothermal solutions deposit minerals in open fractures. Many such veins are found in regions of volcanic activity. The famous gold deposits at Cripple Creek, Colorado, and the huge tin and silver deposits of Bolivia are in fractures localized in volcanic rocks adjacent to volcanoes. In each case, the magma chambers that fed the volcanoes served as sources for the hydrothermal solutions that rose up through the shattered rock to form the ore-bearing veins.

A cooling plutonic magma body (Chapter 6) can also be a source for hydrothermal solutions. Solutions moving outward from a cooling pluton will flow through any available fracture or channel, altering the surrounding rock in the process, and commonly depositing valuable minerals. The tin deposits of Cornwall, England, and the copper deposits at Butte, Montana; Bingham, Utah; and

FIGURE 18.4 Hydrothermal ore deposits
(A) Hydrothermal ore deposits can form when groundwater or seawater is heated by nearby magma or when hot, aqueous solutions are expelled from a cooling plutonic body. (B) If a hydrothermal solution carries valuable metals, and if their precipitation is sufficiently sudden to concentrate them, an ore can result, like this gold ore hosted by a quartz vein at Burgin Hill Mine, California.

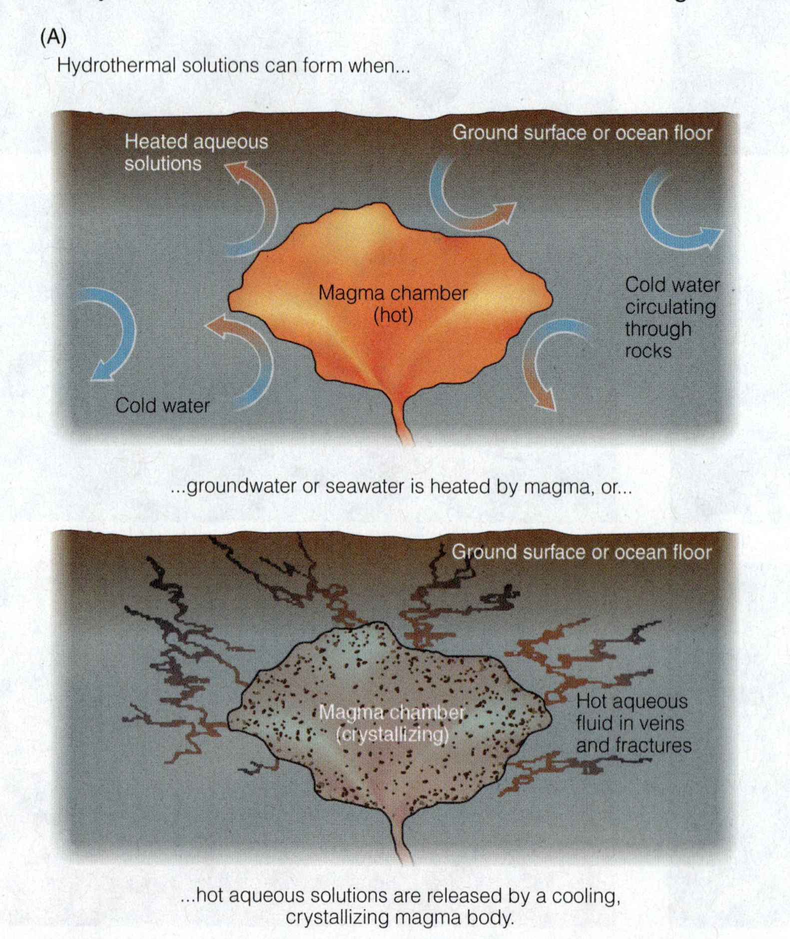

Bisbee, Arizona, are examples of ore deposits that are associated with plutons.

Some of the world's most important ores of lead, zinc, and copper occur in regular, fine layers that look like sedimentary layers. The ore mineral layers are enclosed by and parallel to the sedimentary strata in which they occur; for this reason, such deposits are called *stratabound mineral deposits*. They look sedimentary but are not sedimentary in the truest sense of the term. Stratabound deposits are formed when a hydrothermal solution invades and reacts with sediment. Chemical reactions between sediment grains and the solution cause deposition of the ore minerals, sometimes even before the sediment has become a sedimentary rock. The famous copper deposits of Zambia, in central Africa, are stratabound ores, as are the great Kupferschiefer deposits of Germany and Poland. Three of the world's largest and richest lead and zinc deposits ever discovered, at Broken Hill and Mount Isa in Australia, and at Kimberley in British Columbia, are stratabound ores.

2. Metamorphic and Magmatic Ore Deposits

The process of rock metamorphism—the alteration and recrystallization of rocks as a result of exposure to high temperatures and pressures (Chapter 7)—also can act to concentrate minerals. In response to the changes that occur during metamorphism, minerals may become segregated and concentrated into distinct bands or layers (FIG. 18.3B). Many valuable nonmetallic mineral resources occur as concentrations in metamorphic rocks, including mica, asbestos, graphite, marble, and some gemstones.

There are several ways by which ore minerals may become concentrated as a result of igneous or magmatic processes, that is, geologic processes involving magmas or lavas (Chapter 6). For example, when basaltic magma solidifies into rock, one of the first minerals to crystallize is sometimes chromite, the main ore mineral from which we derive chromium, an important constituent of steel. Chromite crystals are denser than the magma that surrounds them, and may settle and accumulate at the bottom of the magma chamber (FIG. 18.3C). This *fractional crystallization* process (Chapter 6) can produce almost pure layers of chromite. Other mineral resources that can be concentrated by fractional crystallization include iron, platinum, nickel, vanadium, and titanium.

Pegmatites (Chapter 7), coarse-grained igneous rocks formed by fractional crystallization and slow cooling of water-rich granitic magma, commonly contain concentrations of rare elements such as lithium, beryllium, cesium, and niobium. Much of the world's lithium is mined from pegmatites such as those at King's Mountain, North Carolina, and Bikita in Zimbabwe. The great Tanco pegmatite in Manitoba, Canada, produces cesium and other rare and valuable elements.

Another important kind of magmatic mineral deposit is *kimberlite*, a long, thin, pipelike body of rock derived from magma that originates deep in the mantle (FIG. 18.5). Kimberlite magma rises explosively, transporting broken fragments of mantle rock (called *xenoliths*) from great depths to the surface. One of the mineral constituents of kimberlite is diamond, a high-pressure mineral that forms only at depths greater than 150 km. The only way diamonds can reach Earth's surface is through kimberlite pipes.

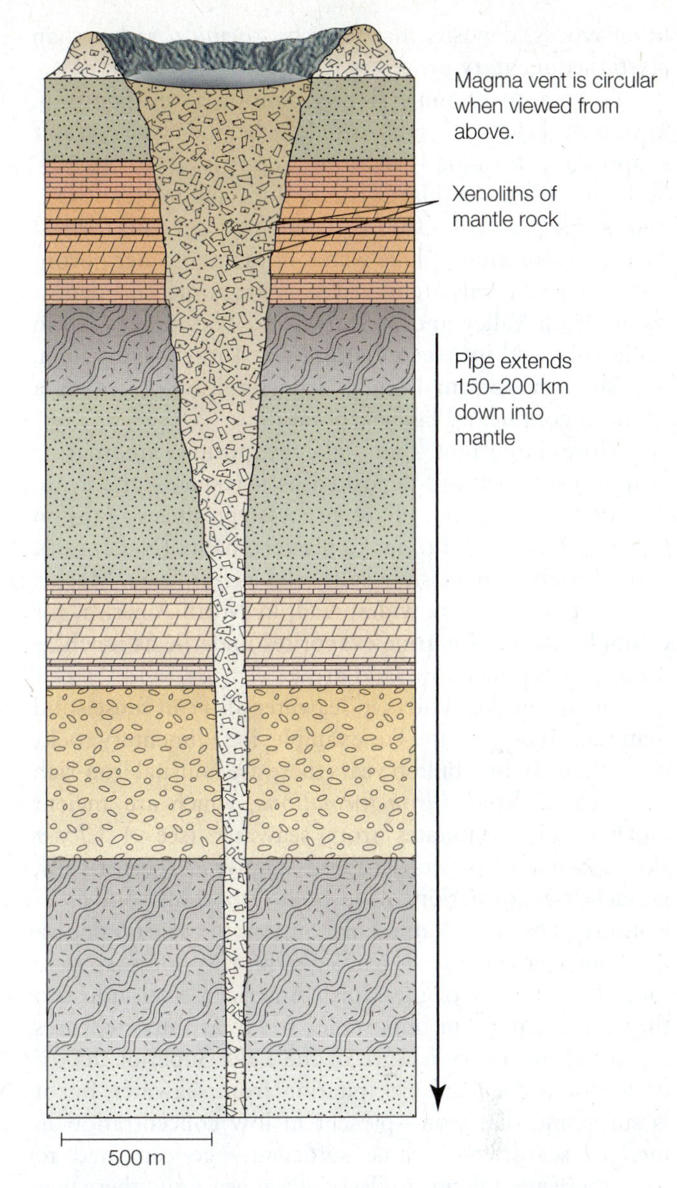

FIGURE 18.5 Kimberlite: a magmatic ore deposit
To reach the surface from their great depths, diamonds must be carried by an eruption of unusual ferocity. These eruptions leave behind a long, cone-shaped tube of solidified magma, called a *kimberlite pipe*. Although most people treasure diamonds for their beauty and luster, geologists treasure them also as "messengers"—samples from an otherwise inaccessible region of Earth's interior.

3. Sedimentary Ore Deposits

The term *sedimentary deposit* is applied to any local concentration of minerals formed through processes of sedimentation. The term is used mainly in reference to deposits that form near Earth's surface when dissolved substances precipitate from lake water or seawater—in other words, deposits that form by *chemical* rather than *clastic* sedimentary processes (Chapter 7).

One cause of mineral precipitation is evaporation, in which layers of salts are left behind when water evaporates, forming **evaporite deposits** (FIG. 18.3D). Sodium carbonate (Na_2CO_3), sodium sulfate (Na_2SO_4), and borax ($Na_2B_4O_7 \cdot 10H_2O$) come from deposits formed by the evaporation of lake water. Borax and other boron-containing minerals are mined from evaporite lake deposits in Death Valley and in Searles and Borax lakes, all in California, and in Argentina, Bolivia, Turkey, and China, and some important lithium deposits also formed as a result of continental lake evaporation.

More important than lake evaporites are marine evaporites, which are deposited as a result of the evaporation of seawater. This process produces gypsum ($CaSO_4 \cdot 2H_2O$, used to make plaster), halite (NaCl, salt), and a variety of potassium salts used as fertilizers. Marine evaporite deposits are widespread; in North America, for example, strata of marine evaporites underlie as much as 30 percent of the entire land area.

Biochemical reactions in seawater also can cause solid materials to precipitate. For example, in sedimentary rocks older than about 2 billion years there are unusual iron-rich rocks called *banded-iron formations*, which are ancient biochemical precipitates from seawater (see *A Closer Look: Banded Iron Formations*, Chapter 7). Historically, banded-iron formations have been important sources of iron ore. The remarkable Lake Superior-type iron deposits, mined principally in Michigan and Minnesota, were long the mainstay of the North American steel industry; they are declining in importance today as imported ores replace them. Every feature of the Lake Superior deposits indicates chemical precipitation from seawater, but it is surprising that iron—present in low concentration in modern seawater—became sufficiently concentrated to form these great deposits. Two billion years ago there was very little oxygen in the atmosphere, so iron in a reduced, soluble state could be easily dissolved and was therefore abundant in seawater. Today, in contrast, the oxygenated atmosphere keeps iron in an oxidized state that is much less soluble in seawater. Lake Superior-type iron deposits provide some of the most convincing evidence that Earth's atmosphere has changed with time.

4. Placer Ore Deposits

Heavy mineral grains can be concentrated as a result of the sifting or winnowing action of flowing water acting on clasts of differing density (FIG. 18.6). The flowing water may be in a stream or in waves along a shoreline. Recall from Chapter 7 that flowing water carries a sediment load. When the current slows down for one reason or another—if the stream goes around a bend or flows over a large boulder, for example—the stream will lose energy and drop the heaviest particles from its load. Along a shoreline, as the waves sift through the sand, the densest minerals become concentrated in layers while the less dense minerals are washed away. A deposit that forms in this way is called a **placer deposit.** Heavy minerals that typically become concentrated in placer deposits include gold, platinum, cassiterite (SnO_2), zircon ($ZrSiO_4$), rutile (TiO_2), and diamonds.

Gold is the most important mineral recovered from placers; more than half of the gold recovered throughout all of human history has come from placers. This is the result of the huge gold production from South Africa and Russia, almost all of which has come from placers (FIG. 18.3E). Most placers are found in stream gravels that are geologically young, but the placer gold deposits of South Africa's Witwatersrand Basin are gold-bearing conglomerates that were laid down about 2.7 billion years ago in the shallow marginal waters of a marine basin. As far as size and richness are concerned, nothing like the deposits in the Witwatersrand Basin has been discovered anywhere else; nor has the original source of all the placer gold been discovered. It is therefore not possible to say, with any degree of certainty, why so much of the world's mineable gold should be concentrated in this one sedimentary basin.

5. Residual Ore Deposits

As discussed in Chapter 7, all rocks are slowly altered or dissolved by naturally acidic rainwater and groundwater, causing chemical weathering. Chemical weathering

FIGURE 18.6 Placer deposits
Dense ore minerals, such as gold, platinum, and diamond, tend to sink and accumulate at the bottom of sediment layers deposited by moving water, such as streams or longshore currents. Such concentrations are called placers. This diagram shows three typical geologic locations where placer concentrations can be found.

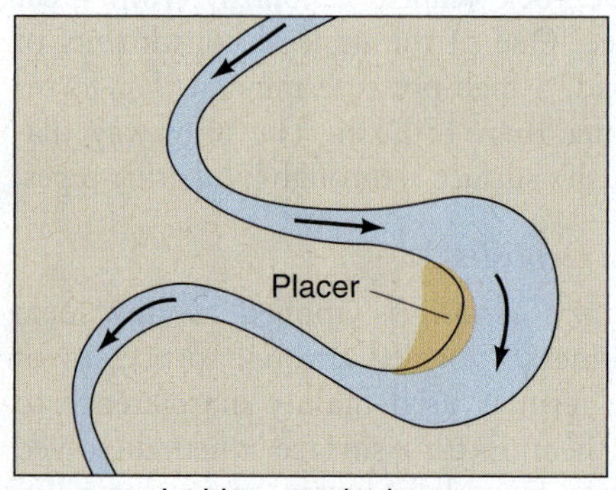

Inside meander loops

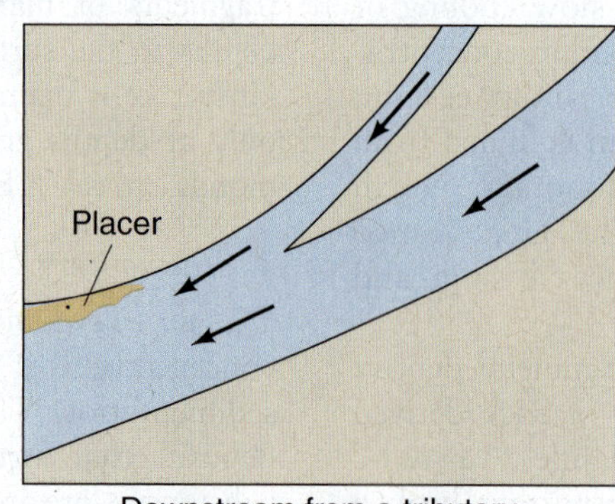

Downstream from a tributary

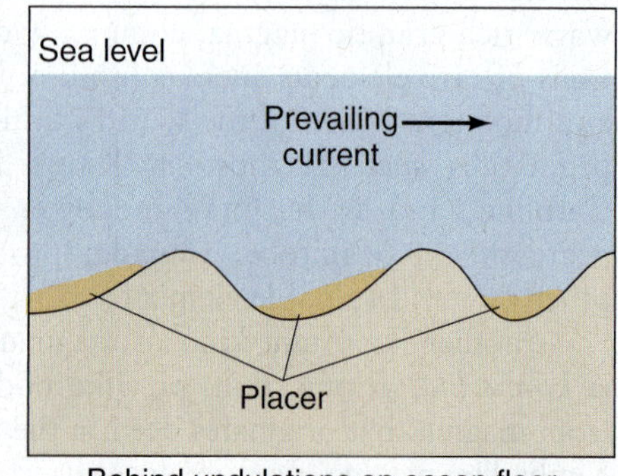

Behind undulations on ocean floor

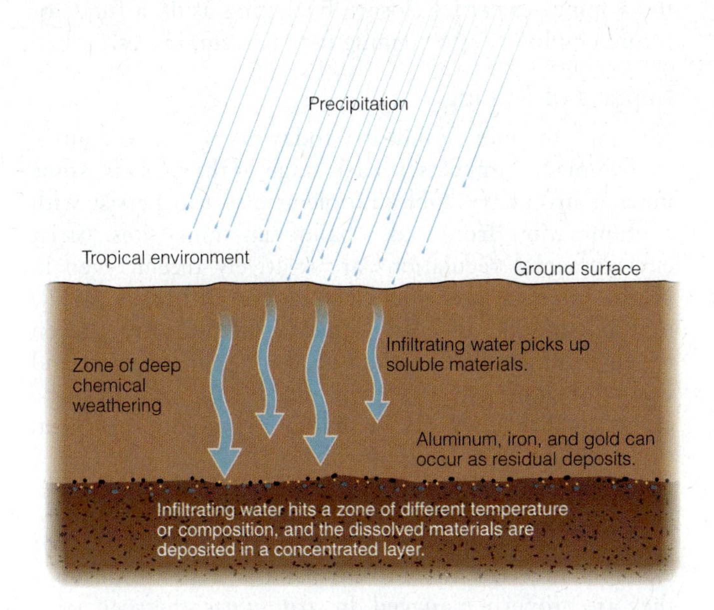

FIGURE 18.7 Residual ore deposits
Residual ore deposits form in tropical climates, when soluble materials are picked up by slightly acidic rainwater and carried downward and deposited at a deeper level. The insoluble, residual material and the soluble, transported material can both become concentrated in this manner.

removes the most soluble materials first, and leaves the less soluble minerals behind (FIG. 18.7). The dissolved materials are carried downward by the infiltrating water. If the water encounters a zone of sharply different temperature or composition, it may deposit all of its dissolved materials in one concentrated layer. Both the dissolved, transported, precipitated material and the insoluble, residual material become concentrated through this process, resulting in ores.

Weathering-related **residual deposits** form mainly in the tropics, where chemical weathering is most effective because of high rainfall and high temperatures. The most common type of residual deposit is *laterite*, a highly weathered material that consists mainly of the mineral limonite (FeO·OH). Under some circumstances laterite can be mined for iron and nickel, but the most important type of laterite is the aluminum-rich variety, *bauxite*, the source of aluminum ore (FIG. 18.3F). The principal mineral of bauxite ores is gibbsite ($Al(OH)_3$).

Where bauxite is found in present-day temperate conditions, such as France, China, Hungary, and Arkansas, we can infer that the climate was tropical at the time when the deposit was formed. Bauxites are vulnerable to erosion. They are not found in glaciated regions, for example, because glaciers scrape off the soft surface materials. The vulnerability of bauxite means that most bauxites are geologically young. More than 90 percent of all known deposits of bauxites formed during the last 60 million years, and all of the very large deposits formed less than 25 million years ago.

Mining

Mining is the set of processes whereby useful resources are withdrawn from the stock of any nonrenewable resource; the term is most commonly used in reference to the extraction of mineral resources. Mining is of critical importance to industrialized society; it is also a controversial industry, which suffers from a negative public image due, in part, to many years of neglectful practices. The mining industry has undergone an enormous change in practices, attitudes, and regulations since the 1970s. Modern mining companies—the largest of which are American, Canadian, British, Australian, and Brazilian—are, for the most part, committed to employing environmentally sound practices, even for operations in countries with less rigorous environmental regulations.

One issue that remains particularly controversial is whether mining should be permitted in parks, forests, and wilderness areas, as discussed in Chapter 17. In favor are those who believe that we cannot afford to forego development of these important mineral (and energy) resources. In some cases it may be possible to mine and then restore the land to its former state, to be used for recreational or other purposes. This multiple-use approach is called *sequential land use*. In very remote areas, however, land reclamation may not be economically feasible. In fragile environments such as the Arctic, unique ecosystems, or areas inhabited by threatened or endangered species, the environmental changes caused by mining can be permanent; it may not be possible to restore these damaged ecosystems after mining. For these reasons and others, environmentalists tend to be vehement opponents of mining and oil drilling in parks and wilderness areas.

Steps in the Mining Process

During the *prospecting* or *exploration* phase of the mining process, an area is assessed for ore potential. This phase is typically short-lived and involves little or no permanent environmental damage. Dirt roads or tracks may be constructed, and there may be drilling. The noise of helicopters used by exploration geologists has been an issue in some wilderness areas because of the potential impacts on animal migrations.

Extraction is the next step in the mining process (FIG. 18.8). The way a mineral or metal occurs in nature determines the type of mining operation needed and the amount of waste produced in the extraction, separation, and concentration of the ore. In the case of sand and gravel extraction, for example, very little material is discarded. Such deposits tend not to be economically viable unless they are quite high-grade; in other words, very little waste rock is present. In other cases, huge quantities of rock must be removed and discarded to obtain a relatively small amount of ore. For example, a typical copper grade of 2 percent will produce just 20 kg of pure metal from each ton of ore processed, leaving 980 kg of waste.

Once the ore has been extracted, it is crushed and concentrated; this is the *milling* phase. For many

FIGURE 18.8 Ore extraction
Extraction is the second phase of the mining process, following exploration. In this photo a miner drills into gold ore at a mine in Gardiner, Montana, just outside Yellowstone National Park.

nonmetallic minerals and construction materials (such as sand, limestone, and gravel), basic processing is all that is needed before the substance is transported and used. Other nonmetallic minerals (such as asbestos, talc, and potash) require further processing of the concentrate to separate the ore from the gangue. This may involve the use of chemical reagents such as kerosene, cyanide solutions, or mercury. The material that is discarded after the initial processing, usually at the mine site, is called *tailings*. After the concentrate leaves the mine site, metallic minerals, in particular, must be further processed at refineries, using high-temperature smelting and/or chemical processes.

Finally, the mine site itself must be managed in the *postoperational* phase, after the deposit has been exhausted and mining has ceased. When the mine is closed, steps are taken to ensure that discarded materials are properly contained and monitored on an ongoing basis; this closure process is called **minesite decommissioning.** Once a mine is closed permanently, laws require that the mine site be cleaned up and, if appropriate, reclaimed for other uses. In some cases, restoration of the land to its former status is a legal requirement that companies must agree to before they are even granted a permit to begin mining. Mining companies now also typically plan ahead for a mine's eventual closure by setting aside a fund for future employment retraining for mine employees.

Impacts of Mining

Although the mining industry today takes care to control environmental impacts at each stage of the process, some impacts are unavoidable. Problems also can persist with contamination from old, abandoned mine sites. Strict environmental regulations are relatively recent, even in North America, and mines abandoned before the 1970s may not have received the kind of attention now given to environmental management during production and decommissioning. Abandoned mines (FIG. 18.9A) can be especially difficult to deal with if there is no one with legal or financial responsibility for the site.

IMPACTS ON THE GEOSPHERE. The impacts of mining on land depend on the type of mine. *Open-pit mines*, for example, are very disruptive to land (FIG. 18.9B), even if they are properly managed. In *strip mines*, the vegetation and top layer of regolith are removed so that the underlying ore can be extracted. Some strip mining involves the removal of entire mountaintops. Underground or *subsurface mines*, with small openings but extensive underground works, are much less disruptive at the surface, although subsidence of the land overlying the underground works can be a problem. Other impacts on land include the construction of roads and buildings, which can be damaging in wilderness areas, and the storage of rock waste, tailings, and other discarded materials.

IMPACTS ON THE ATMOSPHERE. Smelting and refining can emit pollutants into the air, including particulates (smoke and fine particles), nitrogen oxides and sulfur oxides (which have a role in the production of acid precipitation), vaporized metals, and volatile organic compounds. Blowing dust from waste rock piles and tailings is another potential atmospheric problem, especially if there are hazardous materials such as chromium or asbestos in the tailings. Mining companies today generally seal their tailings and ore stockpiles either by spraying on a latex cover, maintaining a water cover, or re-vegetating the area.

IMPACTS ON THE HYDROSPHERE. Liquid wastes containing a wide range of contaminants can be generated during the milling and processing stages of mining. Flowing liquid wastes, or *effluents*, may carry metals like arsenic, cadmium, iron, and lead, or chemical reagents like cyanide, kerosene, activated carbon, and sulfuric acid. Some mine effluents can be saline or alkaline (caustic). **Acid mine drainage**—possibly the single most challenging environmental problem associated with mining today—results when water interacts chemically with sulfide minerals in waste rock and tailings piles, creating sulfuric acid (FIG. 18.9C). Mining companies deal with liquid waste by designing the operation to function, as much as possible, as a closed system. Wastes are generally stored and treated on-site, until they are clean enough to be released back into the environment.

FIGURE 18.9 Environmental impacts of mining
(A) Abandoned mine sites like this one in the San Juan Mountains of Colorado present significant management challenges; untended, they can continue to cause negative environmental impacts, and responsibility for them is often unclear. (B) Open-pit mines, even well-managed, are unavoidably disruptive to the natural environment. This is Bingham Canyon (Kennecott) Copper Mine in Utah, possibly the largest human excavation in the world. (C) Acid mine drainage, shown here at a mine site in Denali National Park, Alaska, can be extremely damaging to aquatic environments.

Impacts on the Biosphere and Human Health. All of the impacts described above can have negative impacts on plants and animals, including humans. Of particular concern for local ecosystems is acid mine drainage. Some human health hazards are directly related to the mining process, although health regulations for miners are now strictly enforced. The health risks of mining include black lung disease and coal dust explosions, associated with coal mining; mine collapse; and various cancers, such as lung cancer and mesothelioma, associated with the mining of asbestos and uranium.

ENERGY RESOURCES

Energy is fundamental to life. Just living entails the use of energy, as discussed in Chapter 2. The most fundamental characteristic of any life form is the nature of the processes whereby it acquires the energy it needs for survival, growth, and reproduction. For humans, this includes food energy derived from photosynthesizing plants and from the animals that eat them. For modern humans, it also includes the various sources of energy that we use to sustain the activities and structures of modern society and civilization.

Many sources of energy are available to us through the energy cycle, in a wide variety of forms. In the remainder of this chapter we will look at energy that occurs in the form of fossil fuels—the nonrenewable energy source that has dominated human use since the Industrial Revolution—and some of the unconventional hydrocarbon resources that are being developed to extend the availability of fossil fuels and improve their environmental performance. Then we will consider energy sources that may provide viable alternatives to fossil fuels in the future. Some of these, such as hydroelectric and nuclear power, have already been widely developed. Others are so-called **new renewables**—alternative energy sources based on new and developing technologies, some of which are poised to replace fossil fuels as supplies dwindle. We will end this chapter with a look at some of the environmental impacts of energy use and our continued dependence on fossil fuels, and the prospects for the future.

Energy from the Earth System

As discussed in Chapter 2, Earth's energy comes from three sources: solar radiation; geothermal energy from heat sources inside the planet; and tidal energy, from Earth's gravitational interactions with the Sun and Moon (**FIG. 18.10**; see also Fig. 2.14). Energy circulates through the many pathways and reservoirs of Earth's energy cycle, driving processes such as photosynthesis and atmospheric circulation. *All* energy for human use, regardless of its application, is derived from the energy circulates in the energy cycle.

Some of the energy sources available to us are renewable or inexhaustible; others are nonrenewable (see Chapter 17, *The Basics: Types of Resources*). Renewable energy sources include wood, charcoal, and other biomass fuels. Energy sources driven by the Sun and gravity are inexhaustible; as long as there is a Sun–Moon–Earth system with an atmosphere and hydrosphere, there will be winds, tides, flowing water, and solar energy. Earth's own internal energy is also inexhaustible, in principle, although individual concentrations of geothermal energy can become depleted. Energy sources that are nonrenewable on a humanly accessible time scale include coal, oil, and natural gas—the fossil fuels. Nuclear energy, though unimaginably vast, is also truly nonrenewable; once an atom has been split, there is no way to put it back together.

The energy available to us in various forms through the energy cycle is of a magnitude that is difficult to comprehend. We use energy technologies to convert this raw energy into power that we can use to run the machines of daily life. A healthy person, working hard and continuously all day, can do just enough muscle work so that if the work were converted to electricity, the electricity would keep a single 75-watt light bulb alight. It costs less than 10 cents to purchase the same amount of electrical energy from the local electrical utility.[1] Viewed strictly as energy-generating machines, therefore, humans aren't worth much. By comparison, the amount of commercially produced mechanical and electrical energy used each working day in North America could keep four hundred 75-watt bulbs burning for every person living there.

Commercially produced energy is used in every aspect of our lives, from food production to transportation, housing, manufacturing, and recreation, and North Americans are the world's biggest users of energy, per capita. To see where all the energy is used, it is necessary to sum up all the energy employed to grow and transport food, make clothes, cut lumber for new homes, light streets, heat and cool office buildings, and do myriad other things. The uses can be grouped into three categories: transportation,

[1] Assuming the cost of a KWh to be 15 cents: 75 watts × 8 hours = 600 watt-hours = 0.6KWh, and 0.6KWh × \$0.15/KWh = \$0.09

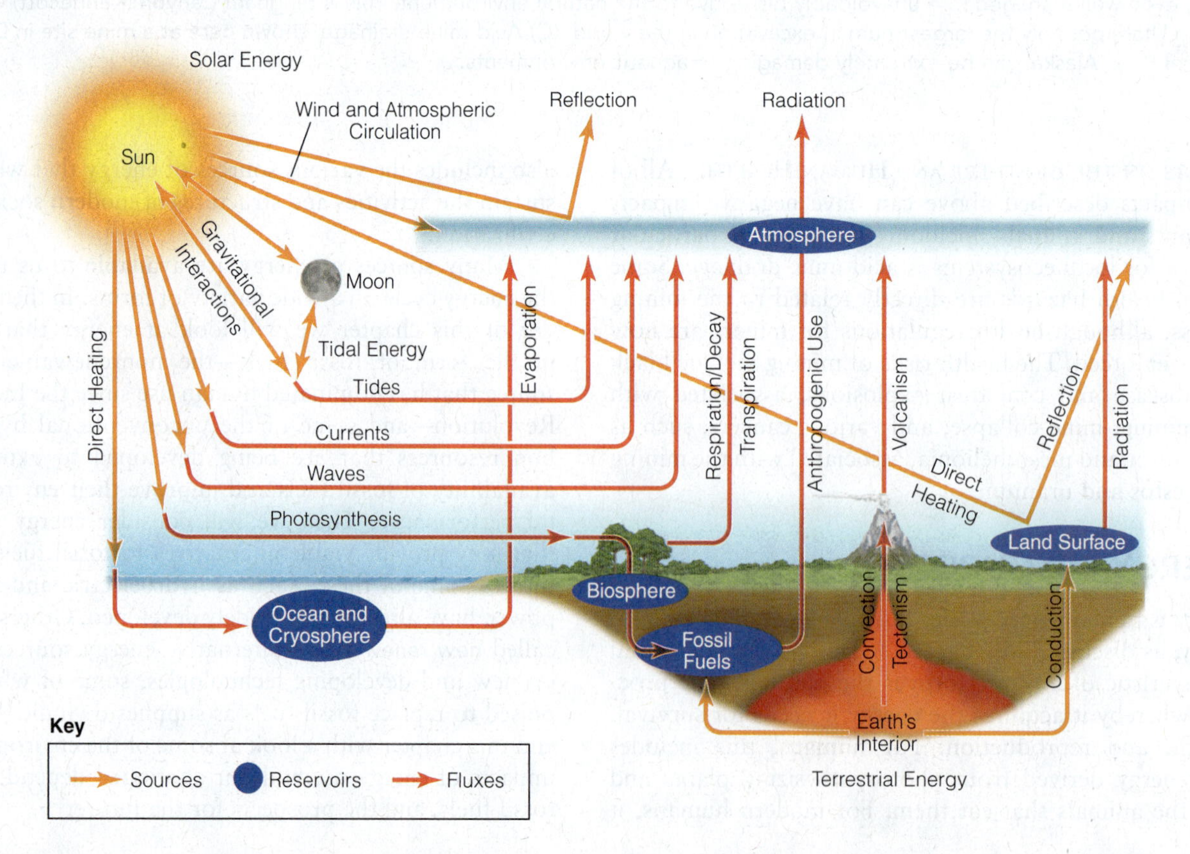

FIGURE 18.10 Earth's energy cycle
All of the energy that we use to power the activities of modern society comes from some part of Earth's energy cycle, shown here in a simple box model format.

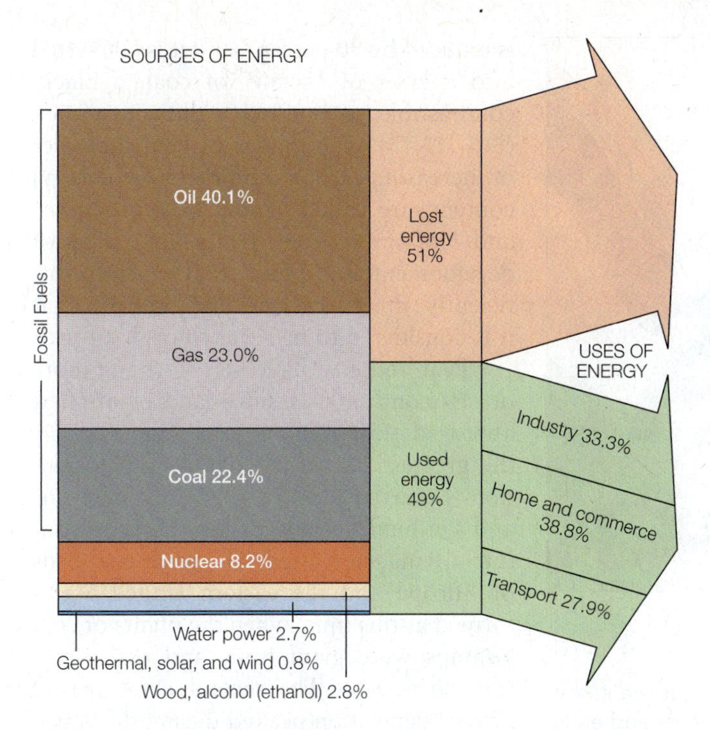

FIGURE 18.11 Energy sources and uses
In North America, fossil fuels (oil, natural gas, and coal) account for 85 percent of the energy used. The large amount of lost energy—nearly half, as shown in the upper-right arrow—arises both from inefficiencies in energy use and from the fundamental physical limits on the efficiency of any heat engine.

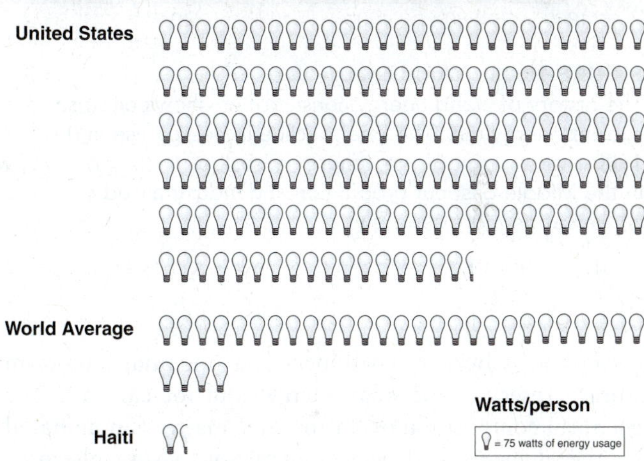

FIGURE 18.12 Energy consumption
An average American or Canadian uses energy, directly or indirectly, at a rate equivalent to burning more than 150 75-watt light bulbs every minute of the day, every day of the year. Here a highway sign admonishes Los Angeles commuters, "Don't be fuelish—be carpoolish."

home and commerce, and industry (meaning all manufacturing and raw material processing plus the growing of foodstuffs). These are summarized in FIGURE 18.11.

The energy drawn in 2008 from the major fuels—coal, oil, and natural gas—as well as that from nuclear power plants, was 5.0×10^{20} J. This was equivalent to the burning of 2.7 metric tons of coal or 12.2 barrels of oil for every living man, woman, and child each year! Energy consumption around the world is very uneven, however (FIG. 18.12). In less developed countries such as Haiti and Tanzania, energy use is equivalent to burning only 2 or 3 barrels of oil per person per year, while in a developed country such as the United States and Canada, energy use is equivalent to burning more than 50 barrels of oil per person per year.

Fossil Fuels

Everywhere in the world, even in the least developed countries, nonrenewable sources supply at least half of the energy used. **Fossil fuels**, which include coal, oil, natural gas, and other hydrocarbon-based fuels such as peat, are the main source of commercial energy worldwide today. Energy consumption and the character of the world's energy "mix" have changed over time (FIG. 18.13). During the Industrial Revolution, coal replaced wood as the main energy source in industrializing societies. Coal was burned to create steam to drive the newly invented steam engines. By the end of World War II, oil had risen to dominance in the world's energy mix. The shift from coal to oil was driven by new technologies, primarily the internal combustion engine, which requires fuel in liquid or gaseous form. Recently, the use of natural gas has increased, driven by price, as well as environmental concerns.

Fossil fuels consist of altered organic matter from the remains of plants or animals, trapped in sediment or sedimentary rock. This means that fossil fuels ultimately derive their energy from the Sun. Through photosynthesis, plants use the Sun's energy to combine water (H_2O) and carbon dioxide (CO_2) into oxygen and organic carbohydrates. When organic matter decays, it is converted back into water and carbon dioxide, and the solar energy stored in it is released. Any organic matter that escapes

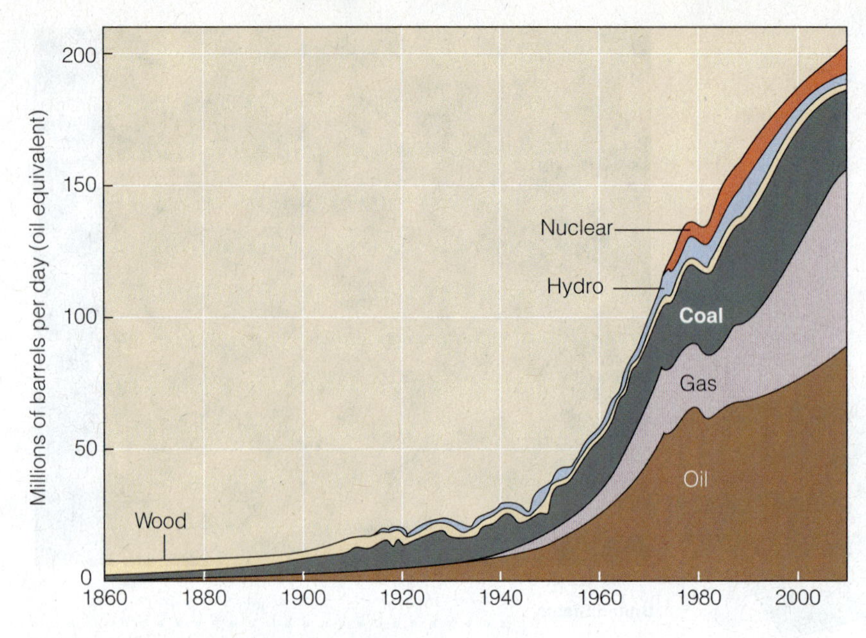

FIGURE 18.13 World energy use
The history of world energy consumption shows oil's rise to prominence in the current energy mix. The brief decline in energy use in the late 1970s and early 1980s was a result of the "energy crises" of 1973 and 1979, when countries in the Middle East cut production and industrialized countries began to take conservation more seriously.

decay and is buried in sediment is a potential long-term storage reservoir for solar energy (and for carbon). The chemical composition of this organic matter is dominated by the elements hydrogen and carbon, so an alternate name for fossil fuels is **hydrocarbons.**

Various changes occur during and subsequent to the burial of the organic remains. The kind of sediment, the kind of organic matter, and the nature and extent of post-burial changes determine which type of fossil fuel will form. The principal types are coal and petroleum (oil and natural gas). There are also abundant deposits of so-called "unconventional" hydrocarbons, which include tar sands and oil shales. We will look at the formation and occurrence of each of these.

Coal

Coal is a solid fossil fuel that forms from organic matter in a terrestrial environment. Organic matter on land comes mainly from trees, bushes, and grasses. These large land plants are rich in *resins*, *waxes*, and *lignins*, which tend to remain solid. In water-saturated places, such as swamps and bogs, the organic remains of plants accumulate to form **peat**, an unconsolidated deposit with a high carbon content (FIG. 18.14).

Peat is the initial stage in the formation of coal, in a process called **coalification**. Over millions of years, the peat is compressed by overlying sediments. Water is squeezed out, and gaseous (volatile) compounds such as carbon dioxide (CO_2) and methane (CH_4) escape. By compaction and gas escape, the thickness of the layer of peat is reduced by 90 percent, and it is converted into a layer or "seam" of coal, a black, combustible, carbon-rich sedimentary rock (FIG. 18.15). The grades of coal, in order of increasing carbon (and therefore energy) content, are called *lignite*, *subbituminous*, and *bituminous* coal. In its final stage of development, *anthracite*, coal has been sufficiently altered by heat and pressure that it is considered to be a metamorphic rock.

Peat and coal have been forming more or less continuously since land plants first appeared 450 million years ago. By far the greatest period of coal swamp formation occurred during the Carboniferous and Permian Periods, when the supercontinent Pangaea existed. The great coal beds of Europe and the eastern United States formed at this time, when the plants of coal swamps were giant ferns and scale trees (*gymnosperms*). The second great period of coal deposition peaked during the warm Cretaceous Period. The plants of the coal swamps during this period were flowering plants (*angiosperms*). Today, peat accumulation is occurring in vast wetlands such as the Okefenokee Swamp in southern Georgia, and the Great Dismal Swamp on the border between southeastern Virginia and North Carolina (FIG. 18.16).

Petroleum: Oil and Natural Gas

Petroleum (from the Latin words *petra,* meaning rock, and *oleum,* meaning oil) refers to naturally occurring gaseous, liquid, and semisolid substances that consist chiefly of hydrocarbon compounds and form in a marine environment. In the ocean, microscopic phytoplankton

FIGURE 18.14 Cutting peat
A peat cutter harvests peat from a bog in Ireland. When dried, peat provides fuel for heat and cooking. It is higher in energy content than firewood but lower than coal because it is in the process of changing from plant matter to coal. If the peat cutter could wait a few million years, he might be able to harvest much higher-energy coal.

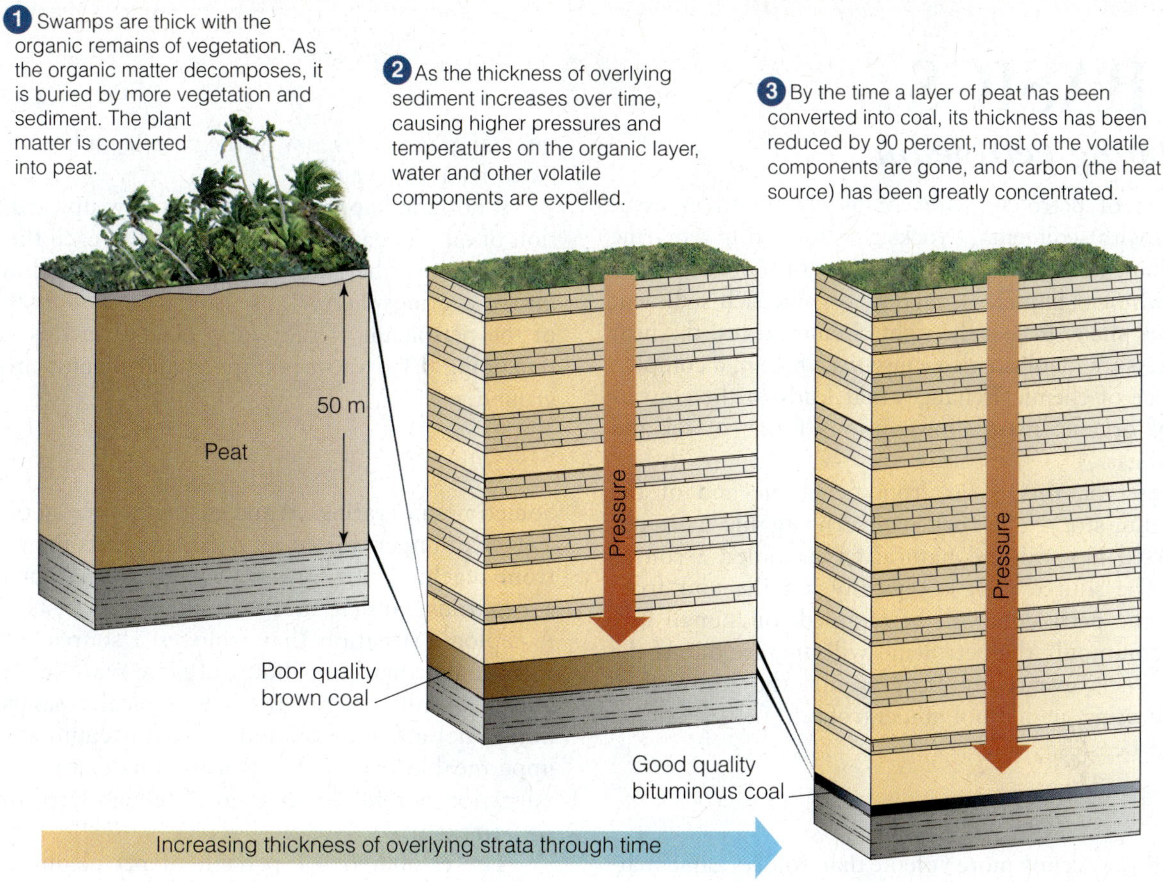

1 Swamps are thick with the organic remains of vegetation. As the organic matter decomposes, it is buried by more vegetation and sediment. The plant matter is converted into peat.

2 As the thickness of overlying sediment increases over time, causing higher pressures and temperatures on the organic layer, water and other volatile components are expelled.

3 By the time a layer of peat has been converted into coal, its thickness has been reduced by 90 percent, most of the volatile components are gone, and carbon (the heat source) has been greatly concentrated.

50 m

Peat

Pressure

Poor quality brown coal

Pressure

Good quality bituminous coal

Increasing thickness of overlying strata through time

FIGURE 18.15 From peat to coal
The conversion of plant matter to coal, or coalification, happens over a period of millions of years, as layers of peat are buried and compressed by overlying sediment.

and bacteria are the main sources of organic matter. When these marine microorganisms die, their remains settle to the bottom and collect in the fine seafloor mud, where they start to decay. The decay process quickly uses up any oxygen that is present. The remaining organic materials—mainly *proteins*, *lipids*, and *carbohydrates*—are preserved and covered with more layers of mud and decaying organisms.

Over time, with continued burial, the muddy sediment with the partially decayed organic material is subjected to heat and pressure, eventually being transformed into sedimentary rock (typically shale). During burial and the conversion into rock, the organic compounds are chemically transformed into petroleum. A complex sequence of chemical reactions is involved in converting the organic matter into petroleum, and additional chemical changes may occur even after it has formed. This explains why subtle chemical differences exist between one body of petroleum and another.

The processes by which sedimentary organic matter is transformed into various forms of petroleum are collectively referred to as **maturation**. The conditions of temperature and pressure under which maturation will occur are very specific—if the temperature increases too rapidly or too slowly with depth in a particular location, the conditions for maturation will not be reached and petroleum will not be formed.

Oil, the liquid form of petroleum, came into use in 1847 when oil from natural seeps was used as a lubricant.

FIGURE 18.16 Coal-forming environment
Today's major coal deposits formed millions of years ago, starting out in moist, heavily vegetated terrestrial environments like this one in the Great Dismal Swamp in North Carolina.

the BASICS

Trapping Petroleum

Deposits of petroleum are commonly found in association with sedimentary rocks that formed in a marine environment. Through the processes of *diagenesis* and *lithification* (Chapter 7), muddy, organic-rich sediment becomes shale. In just the right circumstances, the heat and pressure involved also cause *maturation*, a complex sequence of chemical changes that leads to the conversion of organic matter into various forms of oil and natural gas.

However, this is far from being the end of the petroleum story. The rock in which organic matter is converted into oil and natural gas is called a *source rock*. The source rock is typically not the unit from which the petroleum will be pumped for human use. Most commonly the petroleum will migrate out of the source rock, eventually escaping into the atmosphere or collecting in a pool underground. Let's examine these processes.

PETROLEUM MIGRATION

Oil and gas occupy more volume than solid organic matter, so the conversion process creates internal stresses that cause small fractures to form in the source rock, typically marine shale. The fractures allow the oil and gas to escape into adjacent, more porous rocks. Being light, the oil and gas slowly migrate upward toward the surface. When oil and gas are squeezed out of the shale in which they originated and enter a body of sandstone or limestone somewhere above, they migrate readily because sandstone and limestone are much more permeable than shale.

Water has a stronger molecular attraction to mineral grains than oil does (see *The Basics: Water, The Universal Solvent*, Chapter 8). Therefore, the salty water tends to become attached to the mineral grains, while the oil and natural gas occupy the central parts of the larger openings in the porous sandstone or limestone. Because oil and gas are lighter than water, they tend to glide upward past the water, which adheres to the mineral grains. In this way, the petroleum becomes segregated from the water.

If nothing happens to stop this slow upward percolation of oil and gas, they will eventually reach the surface, seep out onto the land or into the ocean, and evaporate into the atmosphere. It is estimated that 99.9 percent of the petroleum in the world escapes in this way. The remaining 0.1 percent is trapped and accumulates underground.

PETROLEUM TRAPS

Sometimes migrating oil and gas encounter an obstacle—a layer of impermeable rock that prevents the petroleum from reaching the surface. Such a formation is called a *cap rock*, and it most commonly consists of shale. A geologic situation that includes a source rock (typically shale) that contributes organic material, a porous and permeable *reservoir rock* (typically sandstone or limestone) in which oil and gas can accumulate, and an impermeable cap rock (typically shale) that stops the migration is referred to as a **petroleum trap** (or *hydrocarbon trap*).

The estimated 0.1 percent of petroleum that does not escape to the surface is trapped by a cap rock and accumulates in a reservoir rock, forming an underground *pool*. Although this may seem like a rather rare circumstance, several types of petroleum traps are known. Some of the most common types of petroleum traps are illustrated in **FIGURE B18.2**. Recognizing and finding the settings in which oil can become trapped is the job of petroleum geologists. Most of the obvious settings for petroleum accumulation have already been explored and identified, and many of them have been partially or fully exploited already. To find new ones, geologists must employ more sophisticated exploration techniques and look in unconventional locations where they might otherwise not expect to find deposits. Petroleum engineers are also working to develop new extraction technologies, so that petroleum deposits that are of lower grade or difficult to extract can be made economically viable.

The first oil wells were hand dug in Oil Springs, Ontario, after a Canadian chemist discovered how to produce *kerosene*, a distilled product that could be burned in lamps. In 1859, the first commercial well was drilled in Titusville, Pennsylvania. **Natural gas,** a naturally occurring hydrocarbon that is gaseous at ordinary temperatures and pressures, was discovered in 1821 at Fredonia, New York, when a water-well produced bubbles of a mysterious gas that burned. Before long, natural gas was being piped for the purpose of lighting lamps. Today, oil and natural gas dominate commercially available energy sources worldwide.

Oil and gas typically are found together, usually in association with salty water. Petroleum typically forms in one kind of rock (shale) and at some later time migrates to another (mainly sandstone or limestone). Oil and gas are less dense than the surrounding rock, so they migrate upward through the overlying rock and sediment. If the

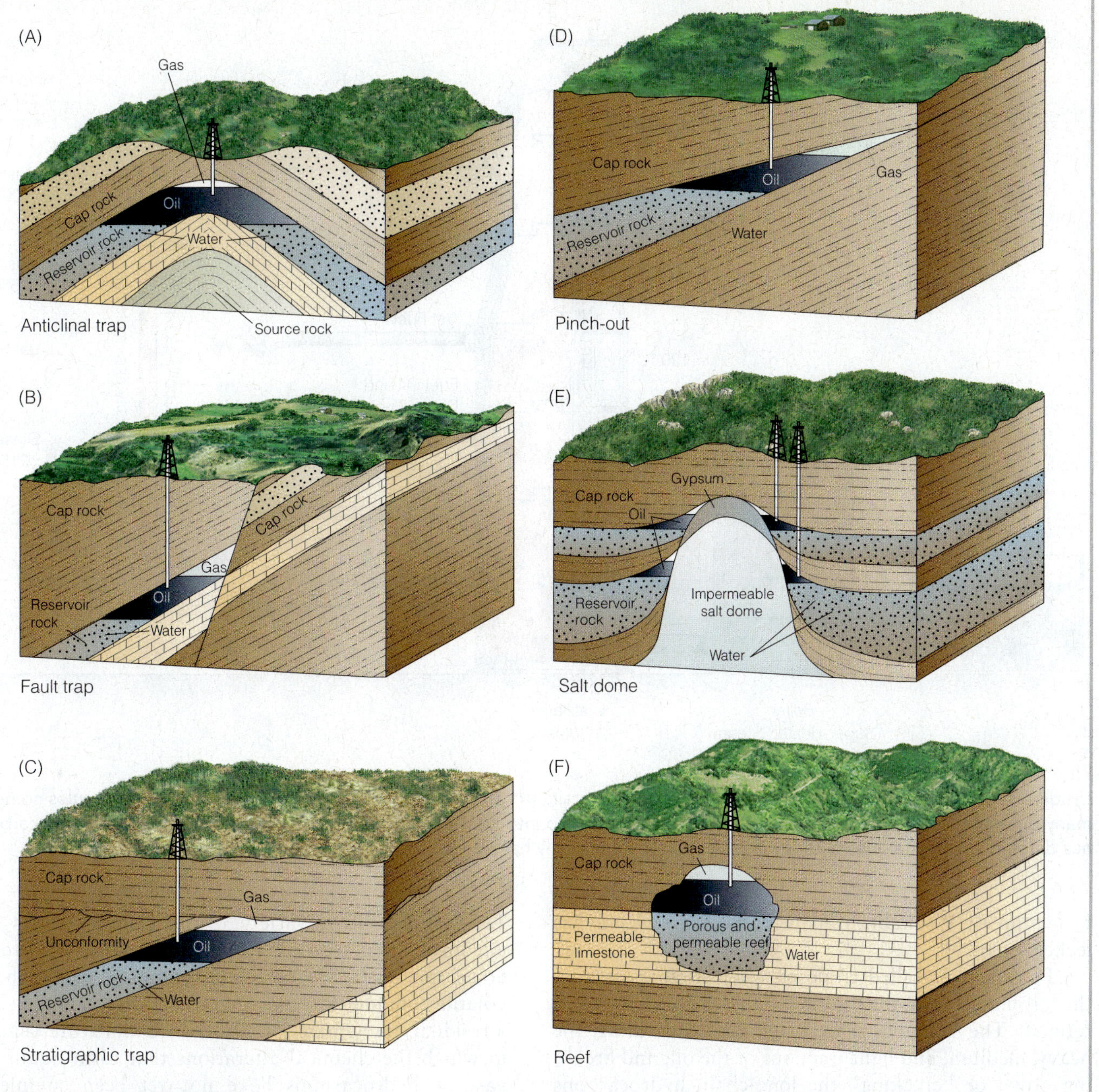

FIGURE B18.2 Petroleum traps
This figure illustrates six geologic circumstances in which oil can become trapped. Each kind of trap requires a source rock to provide organic material, a reservoir rock to store the oil or gas, and a cap rock to prevent its migration. Oil doesn't usually form in an underground pool; it is trapped and accumulates in the pores of a reservoir rock. Oil is usually found with natural gas and salty water. The gas lies on top because it is the least dense, and the water lies underneath because it is densest.

migrating petroleum encounters an impermeable rock unit, it can collect into a pool (see Fig. B18.2 in *The Basics: Trapping Petroleum*). These underground pools are the main targets of oil exploration.

Most of the petroleum that is formed does not find a suitable trap and eventually makes its way, along with groundwater, to the surface or the sea. It is not surprising, therefore, that the highest ratio of oil and gas pools to volume of sediment is found in rock no older than 2.5 million years—young enough so that little of the petroleum has leaked away. This does not mean that older rocks produced less petroleum; it simply means that oil in older rocks has had a longer time in which to leak away.

Petroleum deposits, like coal deposits, are distributed unevenly, even though suitable source sediments for petroleum are widespread. The critical controls seem to be a supply of heat to effect the conversion of solid organic matter trapped in the sediment to oil and gas, and the

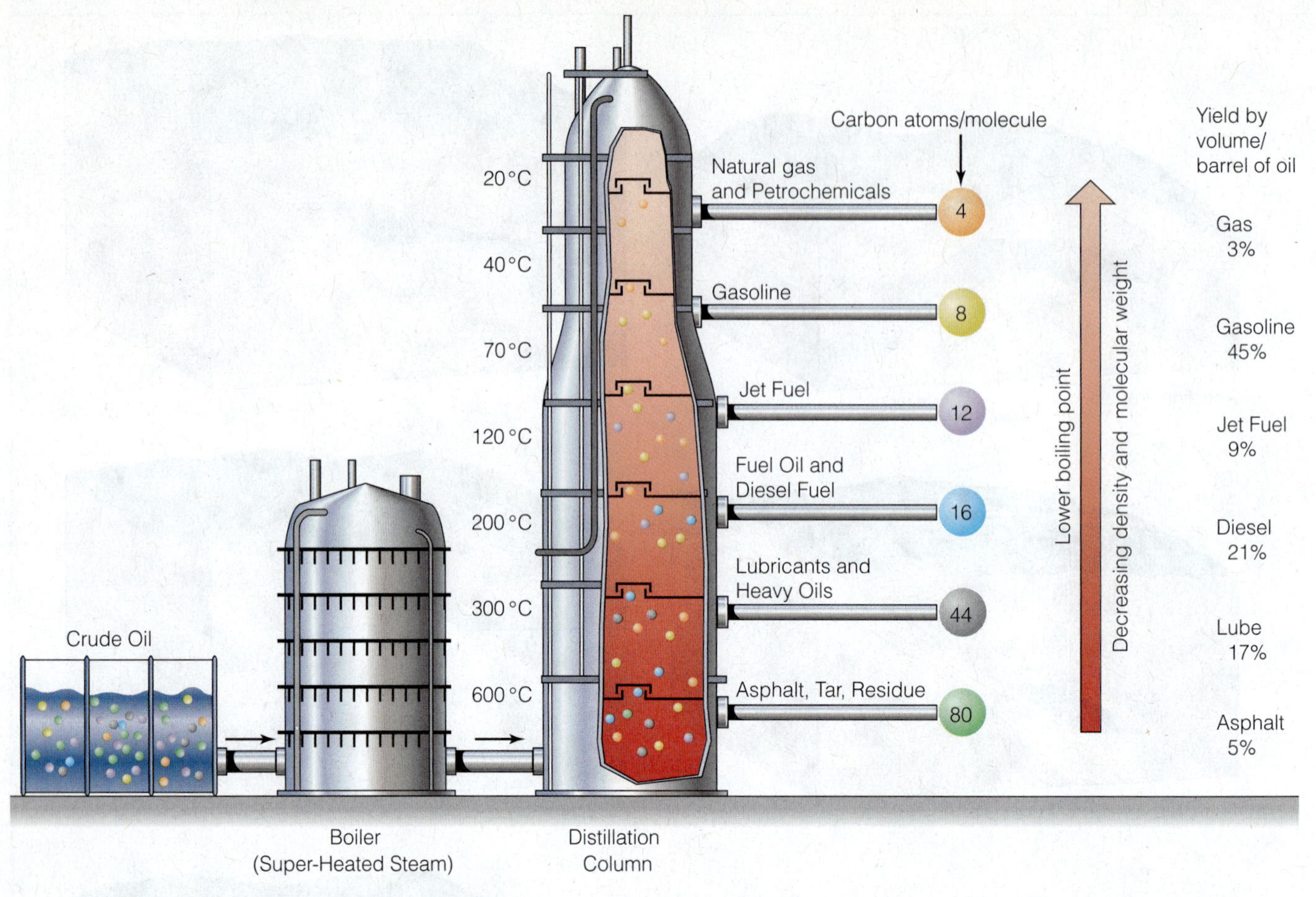

FIGURE 18.17 Petroleum as a raw material
Crude oil consists mostly of hydrocarbons—molecules containing carbon and hydrogen but no oxygen. The molecules come in many different sizes. At an oil refinery, the crude oil is distilled into heavier and lighter components. Each type of hydrocarbon has different uses. The ones with fewer carbon atoms generally have a lower boiling point and are more useful as fuels.

formation of a suitable trap before the petroleum has leaked away.

Petroleum is called **crude oil** when it emerges from the ground. From this state it must be distilled and refined. The refining process involves separating the heavy, medium, and light fractions of the oil, and breaking down or "cracking" the long-chain hydrocarbons to form compounds with lower boiling points, such as gasoline and kerosene. The hydrocarbon components of petroleum also are used to make fertilizers, lubricants, asphalt, and an array of synthetic materials, such as plastic, as well as fabrics, such as nylon and rayon (FIG. 18.17).

Unconventional Hydrocarbons

Hydrocarbon resources other than oil, natural gas, and coal are called *unconventional hydrocarbons*, *synthetic fuels*, or *synfuels* (even if they are naturally occurring). It is tempting to call synfuels "alternative" energy sources; however, they are not truly alternatives, in the sense that they represent an extension of—rather than a departure from—our dependency on fossil fuels.

Tar sands are deposits of dense, thick, asphalt-like oil called *tar*, which cannot be pumped easily. Tar is found in a variety of sedimentary rocks and unconsolidated sediments (not just sand or sandstone). Some tar sands are petroleum deposits in which the volatile components have migrated away, leaving behind a residual, tarry material. Others are immature deposits in which the chemical alterations that form liquid and gaseous hydrocarbons have not yet been completed. The largest known occurrence is in Alberta, Canada, where the Athabasca Tar Sand covers an area of 5000 km^2 and reaches a thickness of 60 m (FIG. 18.18A). The Athabasca deposit may contain as much as 600 billion barrels of oil from tar.

When organic material is buried, compacted, and cemented in very fine-grained sedimentary rocks, a wax-like compound called *kerogen* may be formed if burial temperatures are not high enough to initiate the chemical breakdowns that lead to the formation of oil and natural gas. If kerogen is heated, it breaks down into liquid and gaseous hydrocarbons, similar to those in oil and gas. All fine-grained sedimentary rocks, such as shale, typically contain some kerogen. To be considered an energy resource, however, the kerogen in an **oil shale** must yield more energy than is required to mine and heat it. The world's largest deposit of oil shale is located in

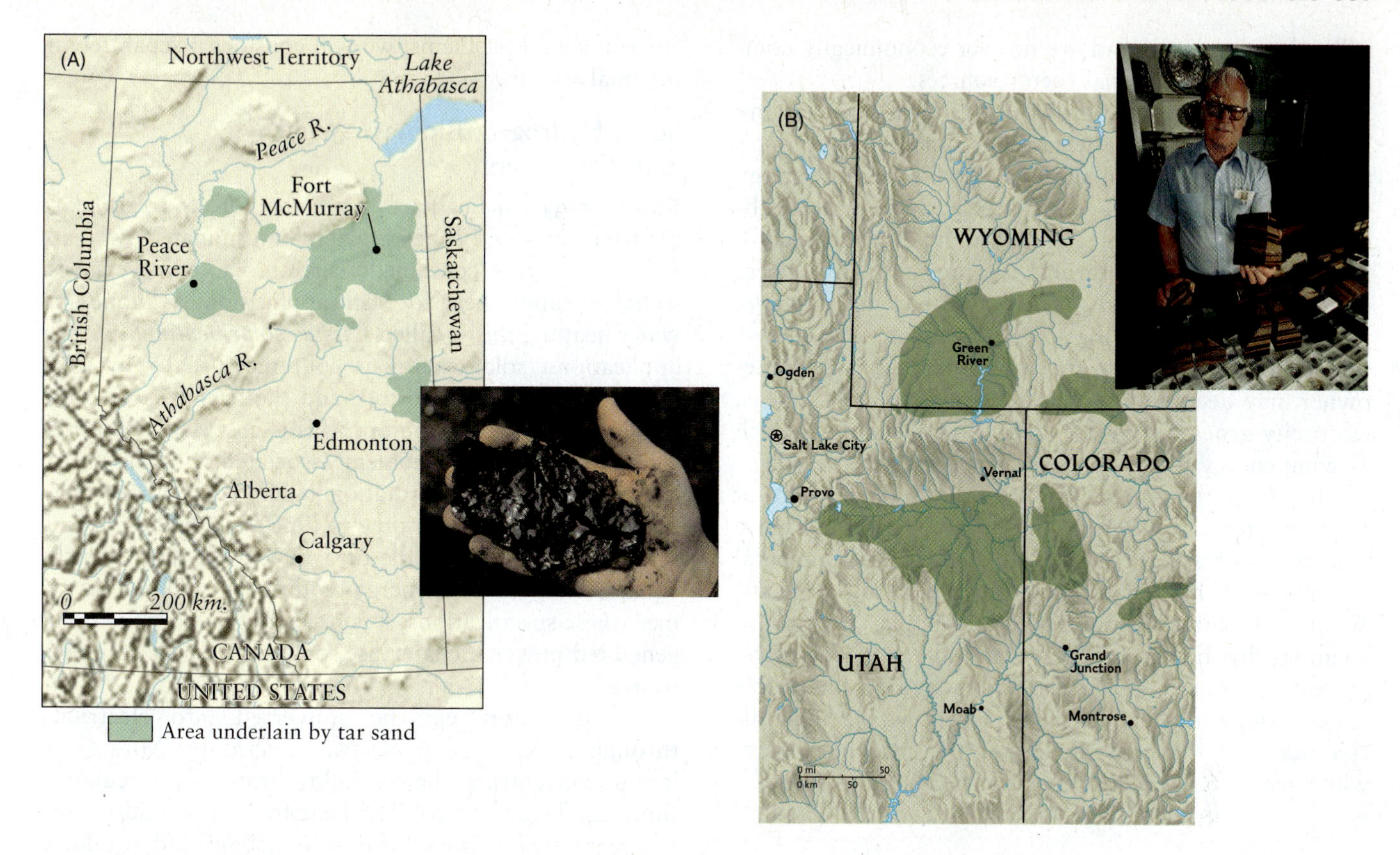

FIGURE 18.18 Unconventional hydrocarbons
(A) The Athabasca Tar Sand covers about 7500 square kilometers in the province of Alberta, Canada. The tar is easily seen in the rock shown here. Nonetheless, extracting it from the rock is a complicated operation that involves cooking the stone with hot water and steam. (B) Another hydrocarbon-rich rock is oil shale, found in huge amounts in the Green River formation in Colorado, Wyoming, and Utah. The shale bookends held by this gem dealer contain about half a pint of oil.

the United States, in the Green River Oil Shale deposits of Colorado, Wyoming, and Utah (FIG. 18.18B). These deposits ultimately could yield about 2000 billion barrels of oil.

Another unconventional hydrocarbon resource that has yet to be exploited is *gas hydrates*. These are deposits of methane (CH_4), frozen into ice in permafrost and in seafloor sediments (Chapter 13). The amount of methane trapped in gas hydrates in seafloor sediment may be greater than all known deposits of oil and natural gas. To date, however, it is still economically unfeasible to exploit methane gas hydrates. In the meantime, there is some concern that global climatic change may lead to the warming of ocean water, which, in turn, would melt the gas hydrates and release methane. This could result in a positive feedback cycle and greenhouse warming because methane is a highly effective greenhouse gas.

New hydrocarbon processing technologies are being developed that may extend existing hydrocarbon resources, or limit the negative environmental impacts of their use. Among these technologies is *fluidized-bed combustion*, in which coal is burned more efficiently but at lower temperatures than in a conventional power plant. The more efficient burning produces more heat and less CO_2, a greenhouse gas and the main gas emitted by the burning of fossil fuels. If carried out in a pressurized environment, the technology can also remove sulfur and nitrogen oxides (SO_2 and NO_x) from the emissions; these are problematic because they combine with water vapor in the atmosphere to form acidic precipitation.

Other technologies have been developed to turn solid coal into a liquid fuel that is similar to oil and less polluting than coal; this is called *coal liquefaction*. Technologies that use oil could be retrofitted to use liquefied coal relatively easily. Other technologies can produce methane gas from coal. An advantage of *coal gasification* is that impurities, notably sulfur compounds, can be removed in the pre-combustion stage, minimizing air pollution from coal burning.

Alternatives to Fossil Fuels

New hydrocarbon technologies and applications may extend the lifetime or enhance the environmental acceptability of some fossil fuel energy sources, but there are other energy sources that are true alternatives to fossil fuels. Some of these sources, including nuclear and hydroelectric power, are already widely used; others are still being developed and are not yet economically competitive with conventional energy sources.

Today there is a growing trend toward living "off the grid," that is, generating some or all of one's required electricity on-site, using alternative energy sources, rather than importing electricity from the established power grid. In some areas building owners receive financial credit for returning to the grid any excess electricity they generate. It has also become possible, through deregulation, for some homeowners to choose the source of the electricity they purchase; the homeowner may decide to purchase *"green" power*—that is, electricity generated by a nonpolluting technology, such as wind energy.

It is important to recognize that, although many of the new alternative energy sources are significantly more benign than fossil fuels in terms of their environmental impacts, each has limitations, drawbacks, and impacts. Wind and geothermal power, for example, are tied to locations that have specific favorable characteristics. And *all* energy sources—even renewable and inexhaustible ones—require the use of some nonrenewable minerals resources for the construction of the machinery with which we convert the raw energy into useable power.

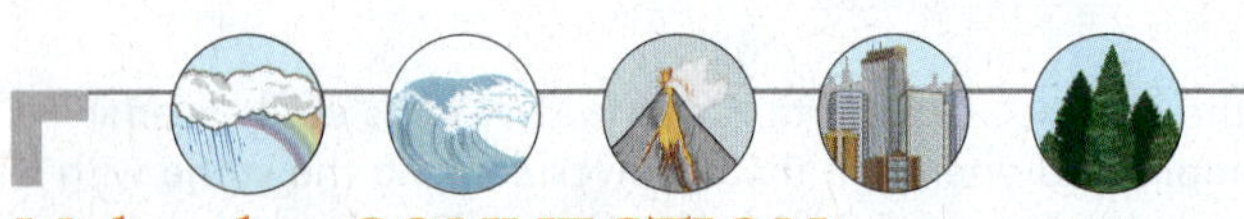

Make the CONNECTION

Consider the power sources in your home, school, or office. Chances are that a pipeline delivers natural gas to the building, or perhaps a truck comes periodically to fill an oil tank. These fuels are often used in homes and offices to run appliances such as furnaces, air conditioners, and stoves. Other appliances use batteries, a form of electrochemical energy. There is also a line that joins your home or school building to an extensive grid system through which electricity is distributed. Typically, a utility company owns and maintains the power plants that generate the electricity, and another company distributes it to homes, offices, and factories through the power grid. The power plants that generate the electricity are coal-fired, natural gas, hydroelectric, or nuclear generators. Take some time to determine the major forms of power that you use in your everyday activities, and consider the energy sources.

We will begin our look at alternative energy sources by considering the various ways in which we can utilize energy from the Sun directly. Then we will examine some indirect sources of solar energy, including biomass, wind, and wave energy, as well as hydrogen fuels. In the subsequent sections we will look at sources of energy that are derived from gravity (tidal and hydroelectric) and Earth's own internal heat. Nuclear energy comes with its own environmental problems; we will consider it separately in the final section.

Solar, Hydrogen, Biomass, Wind, and Wave Energy

Solar energy reaches Earth from the Sun at a rate more than ten thousand times greater than human power use from all sources combined. Direct solar energy is best suited to supply heat for such applications as home and water heating; this is called *passive solar heating*. In other applications, solar energy is collected, usually by solar panels located on a rooftop, and the heat is stored and distributed by fans or pumps; this is *active solar heating*. The most significant problem associated with the use of solar energy is that insolation (Chapter 2) varies from place to place and from time to time—it is highest during the summer, in the middle of a noncloudy day, near the equator. Because peak energy requirements do not always meet these specifications, it is necessary to store the solar-generated power for later use, or install a backup power source.

Solar energy can be converted into electricity through *solar thermal electric generation*. Mirrors or lenses concentrate the sunlight, heating a circulating fluid (see Figure 2.3A). The heat from this fluid is used to create steam, which turns a turbine and produces electricity. Solar energy can also be directly converted into electricity through the use of *photovoltaic cells*, thin wafers or films that are treated chemically so that they absorb solar energy, emitting a stream of electrons in response. The cost of photovoltaic cells is high and their efficiency too low for most uses. The energy they produce can be stored in batteries for later use, but an impediment has been the challenge of developing industrial strength batteries with sufficient storage capacity at an acceptably low (and competitive) cost. Consequently, they are used mainly in small appliances such as solar-powered calculators, watches, and radios. Photovoltaic technology is constantly improving, however, and the cost is decreasing rapidly.

Another way to convert solar energy into usable power is to use the electricity from a photovoltaic cell to split water into its component parts (hydrogen and oxygen), a process called *electrolysis*. Hydrogen released by electrolysis or derived from other sources can be used as a fuel. **Hydrogen fuel** can be stored or transported long distances via pipelines, an advantage over other alternative power sources. One of the potential uses of hydrogen is in **fuel cells**, which are similar, in some respects, to huge batteries. Normal batteries produce power from electrochemical reactions between the chemicals inside them. Fuel cells also run on electrochemical reactions, but the chemical fuel (such as hydrogen, although other fuels can be used) is imported into the fuel cell from an external source (FIG. 18.19). Some experts feel that our future energy economy may be based on a system of distributed, rather than centralized, energy production, provided by fuel cells of various types.

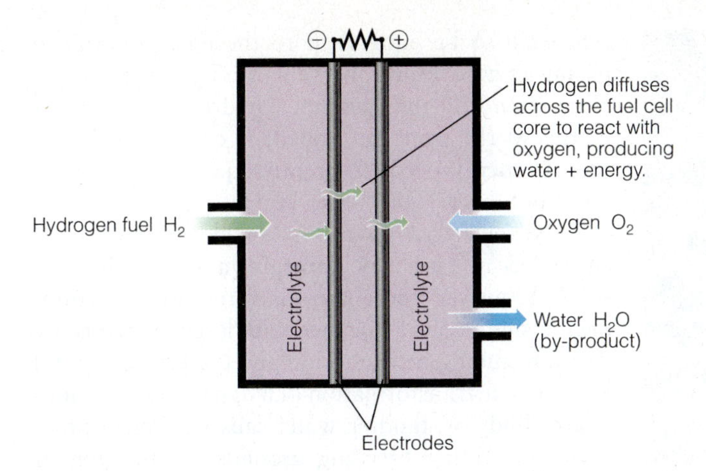

FIGURE 18.19 Energy from chemical reactions
There are many types of fuel cells, but they all derive energy from chemical reactions, in a manner similar to regular batteries. Here, hydrogen fuel is introduced on one side of the fuel cell and diffuses through the core, reacting with oxygen to produce water and energy as by-products of the reaction.

Biomass energy is energy derived from Earth's plant life. Biomass in the form of fuel wood was the dominant source of energy until the end of the nineteenth century, when it was displaced by coal. Biomass fuels, primarily wood, peat, animal dung, and agricultural wastes, are widely used throughout the world, especially in developing countries. Today there are still more than 1 billion people who use fuel wood for cooking and heating.

Organic materials such as animal dung also can be burned, or converted into methane for power generation. This converted material is called *biogas*. It has become an important fuel in developing countries, particularly China and India, where it is used for cooking and lighting purposes. Methane gas from the breakdown of organic garbage also is collected from some landfill sites. Biomass can be converted into a variety of liquid fuels, notably the alcohol fuels methanol and ethanol. This is most commonly done using crop residues. For example, Brazil has an extensive program to produce alcohol fuel from sugarcane residues, which would otherwise be agricultural waste (FIG. 18.20).

The total new plant growth on land each year would yield about 1.5×10^{11} metric tons of dry plant matter. If all of this were burned, or used in some other way as a biomass energy source, it would produce almost nine times more energy than the world uses each year. Obviously, this is a ridiculous suggestion because in order to do so all the forests would have to be destroyed, plants could not be eaten, and agricultural soils would be devastated. Nevertheless, controlled harvesting of fuel plants could probably increase the fraction of the biomass now used for fuel without serious disruption to forests or to food supplies. In several parts of the world, such as Brazil, China, and the United States, experiments are already under way to develop this energy source.

Biomass fuels are advantageous because they are widely available. They are particularly important in developing countries, where other sources of energy may be financially inaccessible. Biomass energy also makes use of materials that might otherwise go to waste, such as agricultural cuttings. A disadvantage, however, is that the removal of organic wastes prevents valuable nutrients from being recycled into the soil. Crucially, the use of agricultural crops for biomass fuels diverts them from use as a food source. Some types of biomass energy, notably fuel wood, have been associated with deforestation, soil depletion, and resulting land degradation. Biomass has a lower caloric value than hydrocarbons because of its high oxygen content, and trace elements in biomass (including alkalis, sulfur, and halogens) can produce pollutants.

Wind energy is another indirect expression of solar energy. For thousands of years, wind has been used as a source of power. Today, huge windmill "farms" are being erected in suitably windy places (FIG. 18.21). In Denmark, where tax incentives encourage the development of alternative energy sources, about 6000 wind turbines supply electricity throughout the country. Although there are some problems with windmill technologies, notably noise pollution, it is likely that windmills will soon be

FIGURE 18.20 Energy from biomass
At Araras in São Paulo State, Brazil, sugar cane is crushed before being used to produce gasohol, a biomass fuel that is an alternative to gasoline.

FIGURE 18.21 Energy from wind
Windmills at Tehachapi Pass, California, generate pollution-free electricity. Each fan's rotary motion turns an electric generator. Wind farms can operate economically only where steady surface winds prevail year-round.

cost-competitive with coal-burning power plants. Getting wind-generated electricity from the production site to the location where it will be used is a difficult challenge, and existing power grids are not sufficient for this task. At best, steady surface winds can provide only about 10 percent of the amount of energy now used by humans. Therefore, wind power may become locally important but probably will not become globally significant.

Waves are created by wind, so **wave energy** is also an indirect expression of solar energy. Waves contain an enormous amount of energy, but so far no one has discovered how to tap them as a source of power on a large scale. Locally, wave power stations produce electricity using a hollow, tubelike chamber. As a wave rises and crests, it pushes air into the tube. The air, in turn, spins a turbine, which generates electricity. Wave power stations have been plagued by problems related to exposure to weather and the battering power of waves.

Hydroelectric, Tidal, and Geothermal Energy

Hydroelectric power is generated from the energy of a stream of water flowing downhill; thus, it is primarily a form of gravitational energy (see Chapter 8 opening photo). Hydroelectric power is the only form of water-derived power that currently fulfills a significant portion of the world's energy needs. To convert the power of flowing water into electricity, it is necessary to access a waterfall or build a dam. The flowing water is used to run turbines, which convert the energy into electricity. Hydroelectric power is very important for countries like Canada with large rivers and suitable dam sites. However, the total recoverable energy from the water flowing in all of the world's streams has been estimated to be equivalent to the energy obtained by burning 15 billion barrels of oil per year. Thus, even if *all* of the potential hydroelectric power in the world were developed, it could not satisfy today's energy needs (currently equivalent to about 83 billion barrels of oil per year).

Hydroelectric power is a "clean" source of energy because it has no damaging atmospheric emissions. However, there are many negative environmental side effects associated with large hydroelectric dams, including deforestation and the loss of natural habitat. The transformation of a dynamic stream into a large body of standing water causes problems, too, such as creating breeding grounds for mosquitoes and other disease-bearing insects. Downstream areas are starved for sediments and mineral nutrients once delivered by rushing streams. Reservoirs fill with silt, limiting the productive lifetime of the dam. In some countries there have been significant problems associated with the loss of farmland and cultural or historic artifacts, and the displacement of people—entire towns, in some cases—to make way for the filling of major reservoirs.

Tidal energy is another water- and gravity-related power source (FIG. 18.22). The energy in tides comes from the rotation of Earth and its gravitational interaction with the Moon. The usefulness of tidal energy in any particular location depends on the configuration of the coastline. Most advantageous are coastlines with long, narrow bays and a large tidal range (the difference between water levels at low tide and high tide). A dam is constructed across the mouth of the narrow bay. With the gates open, water flows in during high tide and is trapped behind the dam. The water is released at low tide, driving a turbine as it rushes out. The use of tidal energy is not new; a mill in Britain dating from 1170 still runs on tidal power. Like wind power, tidal power is insufficient to satisfy more than a small fraction of human energy needs and thus can only be locally important.

Geothermal energy, Earth's internal energy, is used commercially in a number of countries, including New Zealand, Italy, Iceland, and the United States. People in Iceland use water warmed by hot volcanic rocks to heat their houses, grow plants in hothouses, and swim in naturally heated pools. Icelanders also use volcanically produced steam to generate most of their electricity. The most easily exploited geothermal deposits are hydrothermal reservoirs, underground systems of circulating hot water and/or steam in fractured or porous rocks near the surface. Most hydrothermal reservoirs are close to the margins of tectonic plates, where recent volcanic activity has occurred and hot rock or magma is close to the surface. To be used efficiently for geothermal power, hydrothermal reservoirs must be 200°C or hotter, and this temperature must be reached within 3 km of the surface. An example is The Geysers in California, the largest producer of geothermal power in the world. In principle, geothermal energy is inexhaustible. However, intensive exploitation

FIGURE 18.22 Energy from ocean currents
This is the Seaflow Marine Current Turbine off the coast of England, with its rotor raised for maintenance. These turbines work like submerged windmills, driven by flowing water rather than air. They can be installed at places with high tidal current velocities or swift, continuous ocean currents, to draw energy from these huge volumes of flowing water. This energy source is more predictable than wind or wave energy, which respond to the shorter-term variations of the weather system.

of a particular reservoir can lead to local depletion, to the point where the reservoir is no longer useful.

The newest geothermal technologies take advantage of small temperature differences between the ground surface and the shallow subsurface. Water or another fluid is circulated through an underground system of pipes that are deep enough so that the temperature is more or less constant. Geothermal *ground source heat pumps* exploit the thermal energy differences between surface and subsurface. The ground varies in temperature from season to season less than air does, so the pumps heat buildings in the winter by transferring heat from the ground into buildings, and they cool buildings in the summer by transferring heat from buildings into the ground. The heat transfer is accomplished using a network of underground plastic pipes that circulate water or another fluid. Electricity is required to drive the system, but because heat is simply moved from place to place rather than being produced using outside energy inputs, heat pumps can be highly energy-efficient.

Nuclear Energy

Nuclear energy comes from the heat energy produced during the induced transformation of a chemical element into other chemical elements. In theory, nuclear energy can be generated in two ways: by inducing a heavy atom to split into lighter atoms, or by causing two light atoms to combine to make a heavier atom. In practice, only one of these (splitting) is usable for power generation with existing technology.

Splitting heavy atoms into lighter atoms is called **fission**. Fission is induced by bombarding fissionable atoms with *neutrons*, electronically neutral particles from the nucleus of an atom (Chapter 2). When a fissionable atom is hit by a neutron, it splits into two different atoms, both of which are lighter than the original. Some matter is "lost" during the splitting process. However, because of the Law of Conservation of Matter and Energy (Chapter 2), this matter is not actually lost but is instead transformed into energy. Thus, when the atom splits, it creates two new (lighter) atoms and releases the leftover energy as heat. The split atom also ejects some neutrons from its own nucleus. These neutrons can be used to induce more atoms to split, creating a chain reaction (FIG. 18.23). The entire process is carried out inside a nuclear reactor.

FIGURE 18.23 Energy from nuclear fission
In a chain reaction, a neutron strikes a uranium nucleus and causes it to split into smaller nuclei. The process releases more neutrons, which can go on to collide with more uranium nuclei. Each time a uranium nucleus splits it releases heat. A nuclear power plant uses that heat to generate steam, which turns turbines that generate electricity.

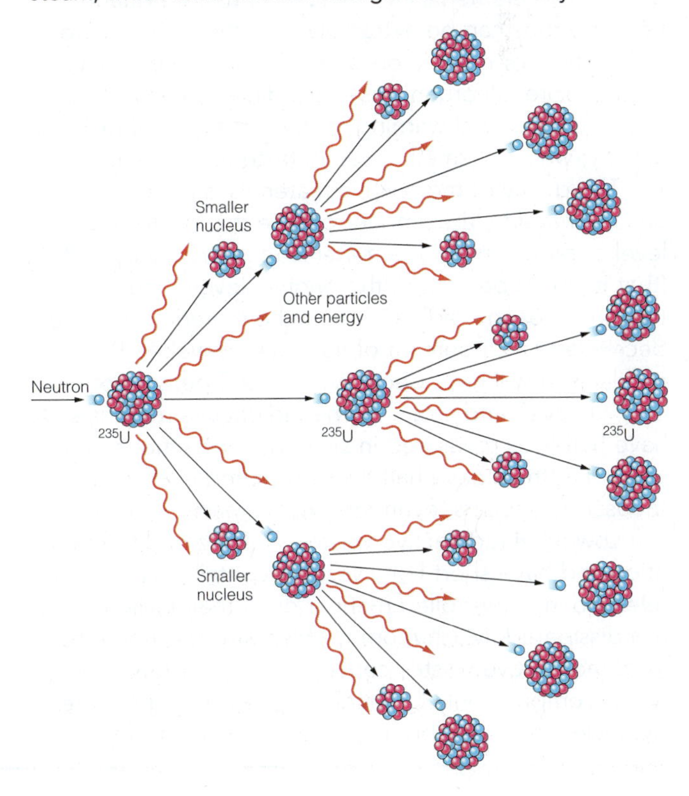

The rate of neutron bombardment is controlled, or moderated, usually by water. When a chain reaction proceeds without control, an atomic explosion occurs.

Uranium-235 is a naturally occurring fissionable material that is mined and used as fuel in nuclear reactors (in addition to other fissionable materials). The fissioning of just 1 gram of ^{235}U produces as much heat as the burning of 13.7 barrels of oil. The ^{235}U is processed and concentrated into fuel pellets, which are packed into a bundle of hollow tubes, called *fuel rods*. The fuel rods are loaded into the core of the reactor, where the fission process is induced. The heat generated by fission is carried away by water, which also moderates the chain reaction. The heated water makes steam, which turns a turbine, producing electricity. If the heat were not removed from the fuel bundle, it could get so hot that the reactor core would melt, releasing its radioactive contents; this is called a *meltdown*, and it is what happened at Chernobyl in Ukraine (in the former Soviet Union) in 1986. Current nuclear reactor technologies are designed to minimize or eliminate the possibility of a meltdown.

Nuclear power is considered to be a clean source of energy—even by some environmentalists—because it causes no harmful atmospheric emissions. Approximately 17 percent of the world's electricity is derived from nuclear power plants. In France, more than half of all electric power comes from nuclear plants; the proportion

A Closer LOOK

NUCLEAR POWER AND RADWASTE

Radioactive waste, or *radwaste*, is leftover radioactive materials or equipment from nuclear power plants, laboratories and medical procedures, uranium mining, and the decommissioning of nuclear weapons. *Radioactivity* is the energy and energetic particles that are emitted during the natural transformation of one element into another element, called *radioactive decay*. Radioactivity can damage organisms, even causing death if the dose is high enough. It also causes genetic damage, that is, damage to the organism's offspring. Radioactive elements can be toxic in minute quantities, and they can be extremely persistent, remaining radioactive for many thousands of years in some cases. Furthermore, all organisms—including humans—lack any kind of natural warning for the presence of radioactivity; we can't smell it, see it, taste it, or feel it.

The decay of radioactive materials is measured in units of *half-life*, the amount of time it takes for the level of radioactivity in a material to decrease by half (that is, to 50 percent of the original level). Thus, in two half-lives the radioactivity in the material will have decreased to 25 percent of its original level. In three half-lives, it will have decreased to 12.5 percent of its original level, and so on. Some radioactive materials have half-lives measured in seconds, or fractions of a second. Others have half-lives measured in days, years, thousands of years, even millions of years.

Low-level radioactive wastes, which emit little radiation and have short half-lives, are contained and then released in a controlled manner once their radioactivity has dissipated. Much more problematic are the high-level radioactive wastes contained in spent fuel rods, which comprise only 1 percent of the waste generated by nuclear reactors, but generate 99 percent of the radioactivity. So far, there are no permanent repositories for high-level radioactive waste (FIG. C18.1). Most high-level wastes are stored temporarily in concrete bunkers or water pools (because water dissipates the heat, as well as absorbing the radioactivity). There has been an intensive search, for the past two decades or more, for appropriate disposal sites and designs for permanent repositories for high-level nuclear waste.

All permanent disposal options for high-level waste involve concentration and immobilization of the waste, usually in inert glass or ceramic pellets; containment in specially engineered canisters; and subsequent isolation from the biosphere and hydrosphere. An appropriate disposal site also must be isolated from potential resources and human activities, such as mining. It must be engineered so that it cannot be easily entered, damaged, or sabotaged. It must be economically feasible and capable of holding a large quantity of waste for a long time—10,000 years is commonly proposed. Ideally, disposal will not require transportation over long distances because the transport of hazardous materials is risky. Finally, the medium that surrounds the waste must be geologically stable and nonfractured (because fractures provide pathways along which escaping contaminants could travel), and must provide good heat conductivity (to dissipate the heat generated by the radioactive materials) and excellent chemical absorption.

It's a tall order, but many scientists feel that it is technically possible to find an appropriate site and design a permanent disposal facility for nuclear waste that will meet these criteria. So far, all serious options for the permanent disposal of high-level nuclear waste involve land-based *geologic isolation*—that is, using the properties of natural rocks to isolate and contain the material. This is combined with a *multiple barrier concept*, in which the repository is engineered to place many physical and chemical barriers in the way of any escaping contaminants.

is rising sharply in some other European countries and in Japan. The reason for the increase is that Japan and most European countries do not have adequate supplies of fossil fuels to be self-sufficient. In North America, however, the growth of the nuclear industry appears to have stalled; this is partly a result of negative public opinion stemming from nuclear catastrophes like Chernobyl and from the intractable problems associated with nuclear waste disposal.

In normal operation, nuclear power plants generate very little environmental radiation; we are actually exposed to far more radiation per year from natural sources than from nuclear reactors. However, nuclear fission generates highly radioactive by-products—*nuclear waste*, which must be isolated from the biosphere and hydrosphere. This presents a technically difficult disposal problem that has not yet been resolved (see *A Closer Look: Nuclear Power and Radwaste*).

In principle, nuclear **fusion**—the joining together or fusing of two small atoms to create a single larger atom, with attendant release of heat energy—is another potential source of nuclear power. Nuclear fusion utilizes as its fuel a "heavy" isotope of hydrogen, called deuterium. Earth has a virtually endless supply of deuterium in the form of a very common chemical compound—water (hydrogen dioxide, H_2O). The primary byproduct of nuclear fusion would be helium, a non-toxic, chemically inert gas.

Among the proposed sites for land-based geologic isolation are old salt mines. Salt does not fracture easily. There is also a great deal of confidence that materials contained in salt will be isolated from the hydrosphere; if water were circulating, the salt would have dissolved long ago. Deep-well injection into shale also has been proposed. This is based on scientific studies of a naturally occurring nuclear fission site in Gabon, West Africa, where shale layers have naturally contained fission by-products for at least a billion years. Canada's model involves permanent disposal of high-level waste in an underground facility, deep in the stable granitic rock of the Canadian Shield. Until 2009, when it was rejected by Congress, the U.S. plan called for the permanent disposal of high-level radioactive wastes in dense volcanic rock at Yucca Mountain, Nevada. This proposal was plagued by controversy, stemming partly from the fact that Yucca Mountain has been volcanically active in the geologic past. Transportation of radioactive wastes to the site, the presence of nearby mineral resources, and fluctuating groundwater levels also raised questions. So far there are no clear alternative locations for the permanent disposal of high-level radioactive waste in the United States.

FIGURE C18.1 Radioactive waste management
(A) Nuclear reactors produce radioactive waste that must be isolated from the hydrosphere and biosphere, including human contact, for thousands of years. Here, a technician uses a radiation detector at a waste storage site in France where reprocessed waste from 10 reactors is stored for five years before disposal. (B) In the United States, nuclear waste is currently stored at 125 temporary sites in 39 states. No permanent storage site has been built yet. In 2009 the U.S. government finally rejected the proposed permanent disposal facility site at Yucca Mountain, Nevada, shown here.

This conjures up images of a cheap, clean, virtually inexhaustible power source. So why are we not using energy provided by nuclear fusion? Fusion is the nuclear process that occurs in the cores of stars, the process responsible for the tremendous heat energy generated by the Sun. But that, in a nutshell, is the problem. For two atomic nuclei to fuse, the ambient conditions must be similar to those at the core of a star—on the order of 100 million degrees. The possibility that nuclei could be induced to fuse at something close to room temperature (so-called cold fusion) has led many scientists to search for this "holy grail" of energy, but routine use of fusion power remains an unrealized goal.

Energy and Society

The energy flowing in the Earth system from all sources, all of which could theoretically be used, is $174{,}000 \times 10^{12}$ watts (174,000 terawatts[2]). The world's 6.8 billion people use power at an average rate of 1600 watts per capita, for a total of 10.9×10^{12} watts (10.9 terawatts). It is clear, therefore, that we are not on the verge of running out of energy in an absolute sense. However, we do need to find sources of energy for the future that will meet society's needs in a way that is socially, environmentally, and economically acceptable. Of primary concern today are the environmental impacts of our continuing dependence on fossil fuels as the main sources of energy for the running of our modern industrial society, and the issue of how reliable supplies of these nonrenewable resources are.

The Impacts of Fossil Fuel Use

Fossil fuel use causes environmental impacts at every stage, from extraction, refining, and transport to the emissions that result from its combustion. Coal is the most abundant of the fossil fuels, but its use entails the most damaging environmental impacts, principally air pollution. Most of the coal that is mined is eventually burned under boilers to make steam for electrical generators, or converted into coke, an essential ingredient in the smelting of iron ore and the making of steel. Coal is often strip mined, causing extensive land degradation. Underground coal mines can be particularly hazardous; coal dust is highly combustible, and many of history's great mine disasters, including underground fires, explosions, and collapses, have occurred in coal mines.

Natural gas and oil are fluids, so they are extracted from the ground by drilling and then transported by pipelines, trucks, and oil tankers. Their environmental impacts can occur at the point of drilling, during transport, or as a result of their use. Oil spills are the most visible impact of both drilling and transport of oil by pipeline or tankers. Crude oil consists of hundreds of chemical components, many of them toxic, and spills can be devastating to the local environment.

One of the best-known oil spills, though not the largest, happened in 1989 when the tanker *Exxon Valdez* ran aground off the coast of Prince William Sound, Alaska. Hundreds of kilometers of Alaska's shoreline were contaminated with crude oil, killing tens of thousands of birds, fish, and mammals, and it took more than 20 years for the resulting lawsuits to be settled and for the environment to recover. The *Exxon Valdez* spill was completely and dramatically eclipsed in April 2010, however, when BP's offshore drilling rig *Deepwater Horizon* underwent an explosion and fire. The rig ultimately sank, killing 11 platform workers and spewing millions of liters *per day* of thick, sticky, toxic crude oil into the Gulf of Mexico (FIG. 18.24). Efforts to cap the well, deep underwater, or collect the flowing oil were largely unsuccessful for over two months. This spill is by far the largest in North American waters, and one of the largest and most devastating oil spills ever.

The *Exxon Valdez*, *Deepwater Horizon*, and other major spills and the resulting cleanup efforts have revealed gaps in scientific knowledge about the natural processes by which oil is dispersed in oceanic and coastal environments, and the technologies that are most effective in cleaning up spilled oil, especially in cold or wind-driven environments. They have also raised many issues concerning the legal and financial responsibility for spills and cleanups, as well as the technological and procedural aspects of oil extraction and transport.

As visible as they are, however, oil spills are not the principal environmental problem associated with petroleum. Combustion of oil and natural gas (and coal) releases atmospheric emissions, including CO_2, SO_2, NO_x, and particulate matter, all of which cause extensive harm to the environment. Airborne particulates that result from the burning of fossil fuels are extremely tiny particles—soot, essentially—that cause respiratory problems and other human health impacts. Carbon dioxide (CO_2) contributes to the anthropogenic greenhouse effect, implicated in global warming. Sulfur dioxide (SO_2) and nitrogen oxides (NO_x) react with water in the atmosphere to form acids, which then fall to the ground as either dry acidic deposition or acid precipitation. Coal is particularly high in both CO_2 and SO_2 emissions. Coal-fired power plants are often installed with air-cleaning technologies that remove some of the sulfur from post-combustion emissions.

One of the biggest environmental controversies over the past few decades has centered on whether to allow oil drilling in environmentally sensitive areas, such as offshore areas, national parks, and wildlife refuges. This

[2] A watt is a unit of power rather than energy; it is a measure of the amount of energy used per unit of time. Specifically, 1 watt = 1 joule per second (J/s). Joules, calories, and British thermal units are all measures of the amount of energy. Watts and horsepower are measures of the *rates* at which energy is used. Thus a kilowatt-hour (Kwh), the unit by which we buy electricity from power companies, is an energy unit because it is a rate unit multiplied by a time unit. The prefix *tera-* means 10^{12}, or 1,000,000,000,000. For more information about units, conversions, and large numbers, see Appendix A.

FIGURE 18.24 Impacts of oil use
(A) Dark clouds of smoke and fire emerge as oil burns during a controlled fire in the Gulf of Mexico on May 6, 2010. The U.S. Coast Guard conducted the burn to aid in preventing the spread of oil following the April 20 explosion of BP's offshore drilling unit, the Deepwater Horizon. (B) A brown pelican covered with oil is cleaned by a team hired to rescue animals affected by the Deepwater Horizon oil spill.

debate is intensified by the fact that many domestic oil and natural gas deposits are located in wilderness areas, such as Alaska's Prudhoe Bay. Drilling can have local environmental consequences, such as the release of briny or heated fluids (used in extracting hydrocarbons from the ground).

Will We Run Out?

How much flowing oil is there in the world? This is an extremely controversial question. It has been estimated that about 950 billion barrels of oil have already been pumped out of the ground. A lot of additional oil has been located by drilling but is still waiting to be pumped out. Possibly a great deal more oil remains to be found. Unlike coal, for which the volume of strata in a basin of sediment can be accurately estimated, the volume of undiscovered oil can only be guessed at, based on accumulated experience from a century of drilling. Knowing how much oil has been found in an intensively drilled area, such as eastern Texas, experts make estimates of probable volumes in other regions where rock types and structures are similar to those in eastern Texas. Using this approach, and considering all the sedimentary basins of the world (FIG. 18.25), experts estimate that somewhere between 2100 and 3000 billion barrels of flowing oil will eventually be discovered.

Are supplies of fossil fuels adequate to meet future demands? If we use a barrel of oil as our unit of measurement, we can compare quantities of all fossil fuels directly. Approximately 0.22 ton of coal produces the same amount of heat energy as one barrel of oil. Thus, the estimated recoverable coal, $13,800 \times 10^9$ metric tons, is equivalent to about 63,000 billion barrels of oil. Considering the approximate world-use rate of barrels of oil (30 billion barrels a year), it is apparent that only coal seems to have the capacity to meet our long-term demands.

What happens when we reach the limit? We can hope that, as a result of increased energy efficiency and conservation, the demand for energy will level off. Realistically, however, it is unlikely that the demand for fossil fuels will decrease substantially or that another fuel source will replace oil in the near future. For the next two decades, the known deposits of natural gas and oil likely will be sufficient to meet demand. As supplies dwindle, however, prices will increase. Only coal—the most abundant fossil fuel, but also the most environmentally damaging—may be in sufficient supply to meet the world's demands beyond the middle of the twenty-first century.

Increasing concerns about the environmental impacts of fossil fuel use will inevitably lead to greater interest in alternative sources of energy. Further concerns arise from issues of national security and from the need to import oil from politically unstable regions. In the meantime, as dwindling supplies of fossil fuels drive prices up, more attention will be paid to enhancing *energy efficiency*, the amount of energy consumed per unit of productive output. The energy efficiency of technologies used in industrialized countries has been increasing (that is, less energy is being used to accomplish the same tasks) over the past decade or so.

If we can manage to move beyond our dependence on fossil fuels, it is clear that there are numerous sources of

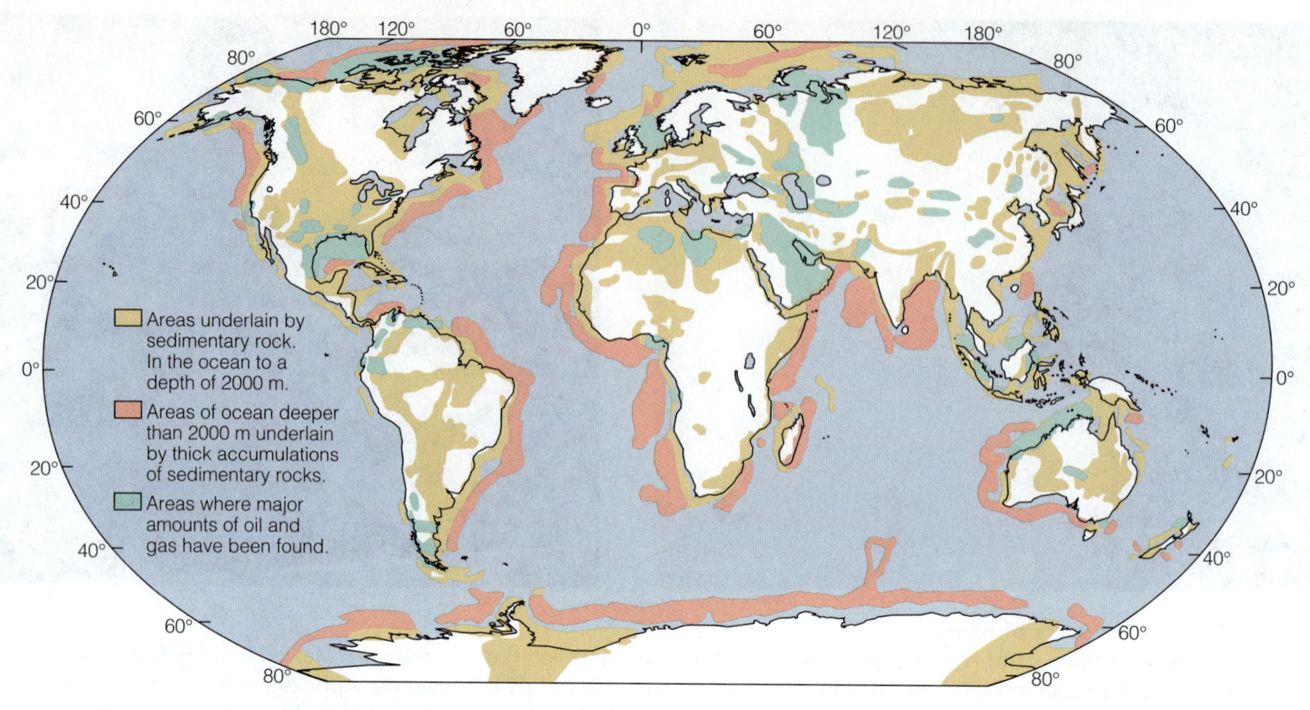

FIGURE 18.25 How much oil is left?
This map shows areas underlain by sedimentary rock and regions where large accumulations of oil and gas have been located. Where the ocean is deeper than 2000 m, sedimentary rock has yet to be tested for oil and gas potential.

energy available to us from various sources in the Earth system—far more energy than we can use. What is not yet clear is when, or even whether, we will be clever enough to learn how to tap these energy sources in ways that will not disrupt the environment and endanger the future availability of resources. It makes sense to look for alternative, renewable, or inexhaustible sources of energy to fill a greater proportion of our energy needs.

In the next and final chapter of the book we will look broadly at some of the changes that human activity is causing in Earth systems, including the scale, degree, and reversibility of these changes.

SUMMARY

1. Metallic minerals are mined for the metals that can be extracted from them by smelting; nonmetallic minerals are mined for their physical or chemical properties. Conserving nonrenewable mineral resources and minimizing the impacts of their use involves closing the resource cycle by recycling useful materials and containing harmful wastes.

2. A mineral deposit is a local enrichment of a given mineral. An ore is a deposit from which one or more minerals can be extracted profitably. The more highly concentrated an ore, the higher the grade and the more valuable the deposit. Lower-grade ores contain a higher proportion of nonvaluable gangue minerals. Economically exploitable mineral deposits are unevenly distributed. For a mineral deposit to form, a geologic process or combination of processes must produce a localized concentration of one or more minerals.

3. Hydrothermal deposits are the primary sources for many metals. Some metal-bearing hydrothermal solutions originate when water dissolved in magma is released as the magma rises and cools; others are formed from rainwater or seawater circulating near a source of heat.

4. Minerals may become segregated and concentrated into distinct bands or layers in response to metamorphism. Ore minerals also may be concentrated as a result of igneous or magmatic processes, including fractional crystallization. An important kind of magmatic mineral deposit is kimberlite, the source for mantle xenoliths and diamonds.

5. Sedimentary deposits are formed as a result of chemical sedimentation near Earth's surface. Evaporite deposits occur when salts are left behind during the evaporation of lake water or seawater. Biochemical reactions in seawater also can cause solid materials to precipitate; banded-iron formation, an important source of iron ore, is an example.

6. Heavy mineral grains can become concentrated as a result of the sifting or winnowing action of flowing water acting on clasts of differing density. A deposit that forms in this way is called a placer deposit.

7. Chemical weathering causes soluble materials to be picked up from regolith and rock, and transported downward with infiltrating rainwater, leaving the less soluble minerals behind. Both the dissolved, transported material and the insoluble, residual material can be concentrated into residual ore deposits in this manner.

8. Mining is the set of processes whereby useful resources are withdrawn from the stock of a nonrenewable resource. The stages are prospecting or exploration; extraction; milling, smelting, and other ore processing; and postoperational management. The mining industry today takes great care to control environmental impacts, although some are unavoidable. Mining can have impacts on the geosphere, the atmosphere, the hydrosphere, and the biosphere, including human health.

9. Energy is available in many forms through Earth's energy cycle. Renewable sources include wood, charcoal, and other biomass fuels. Energy derived from the Sun, gravity, and Earth's internal energy is inexhaustible, in principle. Sources that are nonrenewable on a human timescale include coal, oil, natural gas, and unconventional hydrocarbons. Nuclear energy is truly nonrenewable because it involves the splitting of atoms. Everywhere in the world, nonrenewable fossil fuels supply at least half of the energy used.

10. Coal is a solid fossil fuel that forms from organic matter in a terrestrial environment. In water-saturated swamps and bogs, the organic remains of plants accumulate and undergo compression to form peat, the initial stage of coalification. By compaction, gas escape, and chemical alteration, peat is transformed into coal. The greatest periods of coal formation occurred during the Carboniferous-Permian and Cretaceous Periods.

11. Petroleum is naturally occurring gaseous, liquid, and semisolid hydrocarbons that originate in a marine environment. A petroleum trap includes a source rock that contributes organic material; a porous and permeable reservoir rock in which oil and gas accumulate; and an impermeable cap rock that stops their upward migration. From the crude state, petroleum is distilled and refined into heavy, medium, and light fractions, which are used to make fuels, fertilizers, lubricants, asphalt, and many synthetic materials and fabrics, including plastics.

12. Unconventional hydrocarbons or synfuels include tar sands, oil shales, and methane gas hydrates. New hydrocarbon processing technologies such as coal liquefaction, coal gasification, and fluidized-bed combustion may help to extend existing hydrocarbon resources, or limit the environmental impacts of their use.

13. New alternative energy sources are more benign than fossil fuels in terms of their environmental impacts, but each has limitations, drawbacks, and impacts. Some are tied to locations that have specific characteristics, and all require the use of some nonrenewable minerals resources to construct the machinery with which to convert the raw energy into useable power.

14. Direct or passive solar power is useful for such applications as home and water heating. Active solar power can be stored and distributed by fans or pumps, or converted into electricity through solar thermal electric generation or photovoltaic cells. The most significant problem associated with the use of solar energy is that insolation varies spatially and temporally, and does not match peak energy requirements. Wind and wave energy are indirect expressions of solar energy. They can be cost-competitive, but are restricted to specific favorable locations.

15. Hydrogen fuel can be derived from the splitting of water by electrolysis. Hydrogen and other fuels can be used in fuel cells, which are similar to batteries in that they produce energy from electrochemical reactions.

16. Biomass energy is derived from Earth's plant life. Biomass fuels, primarily wood, peat, animal dung, and agricultural wastes, are widely used throughout the world, especially in developing countries. Conversion of organic matter to biogas, collection of methane gas from landfills, and conversion of crop residues into alcohol fuels are some important new technologies. A disadvantage of biomass energy is that crops are diverted from use as food, and the removal of organic wastes prevents nutrients from being recycled into the soil.

17. Hydroelectric power, the only water-derived power that currently fulfills a significant portion of the world's energy needs, is generated from the energy of water flowing downhill. Hydroelectric power is a clean source of energy, but there are negative environmental side effects associated with large hydroelectric dams and their reservoirs. Tidal energy is another water- and gravity-related power source. The usefulness of tidal energy in any particular location depends on the configuration of the coastline, which means that it can only be of local importance.

18. Geothermal energy is used commercially for space heating and water heating. The most easily exploited geothermal deposits are hydrothermal reservoirs, most of which are close to the margins of tectonic plates. In principle, geothermal energy is inexhaustible, but intensive exploitation of a particular reservoir can lead to local depletion. The newest geothermal technologies are geothermal ground source heat pumps, which take advantage of slight temperature gradients between surface and subsurface materials to extract energy.

19. Nuclear energy can be generated by inducing a heavy atom to split into lighter atoms. Fission is carried out inside a nuclear reactor, using fissionable materials such as Uranium-235 in a chain reaction. Nuclear power is a clean source of energy, but it generates radioactive waste that must be isolated from the biosphere and hydrosphere for thousands of years. Most proposals for the permanent disposal of high-level radioactive waste rely on geologic isolation. Nuclear fusion is a potential source of cheap, clean, virtually inexhaustible power, but room-temperature fusion remains an unrealized goal.

20. We need to find sources of energy that can meet society's needs in a way that is socially, environmentally, and economically acceptable. Of concern are the environmental impacts of fossil fuel use and the potential for a shortfall in supply. Fossil fuel use causes environmental impacts at every stage, from extraction, refining, and transport to combustion. Of the known and speculated fossil fuels resources, only coal—the most damaging environmentally—has the capacity to meet long-term demands. Concerns about the environmental impacts, dwindling supply, and increasing price of fossil fuel use will inevitably lead to greater interest in alternative sources of energy.

IMPORTANT TERMS TO REMEMBER

acid mine drainage *550*
biomass energy *561*
coal *554*
coalification *554*
crude oil *558*
evaporite deposit *548*
fission *563*
fossil fuels *553*
fuel cell *560*
fusion *565*
gangue *544*
geothermal energy *562*
grade *544*
hydrocarbons *554*
hydroelectric power *562*
hydrogen fuel *560*
hydrothermal solution *545*
maturation *555*
metallic mineral *542*
mineral deposit *544*
minesite decommissioning *550*
mining *549*
natural gas *556*
new renewables (energy sources) *551*
nonmetallic mineral *543*
nuclear energy *563*
oil *555*
oil shale *558*
ore *544*
peat *554*
petroleum *554*
petroleum trap *556*
placer deposit *548*
recycling *542*
reserve *544*
residual deposit *549*
solar energy *560*
tar sand *558*
tidal energy *562*
wave energy *562*
wind energy *561*

QUESTIONS FOR REVIEW

1. Describe five ways in which mineral deposits can form.
2. How do hydrothermal solutions form, and how do they lead to the formation of ore deposits?
3. Briefly describe the formation of three different kinds of sedimentary ore deposit.
4. How are placer deposits formed? Name four minerals that are commonly found in placer deposits.
5. How do residual mineral deposits form? In what kind of environment are they most likely to form?
6. What are the main phases of the mining process?
7. What causes acid mine drainage?
8. Summarize the main environmental impacts of mining on land, air, water, and the biosphere, including human health.
9. What is peat, and what are some examples of modern environments in which peat is forming today?
10. Explain the steps that occur as organic matter becomes coal.
11. What kinds of rocks serve as source rocks for petroleum? Is this the kind of rock in which petroleum is normally found by drilling? Why (or why not)?
12. What is the source of the organic matter that becomes petroleum? How does it differ from the organic matter that becomes coal?
13. Explain what this means, making reference to Earth's energy cycle: "Wind and wave power are expressions of solar energy."
14. What are "unconventional hydrocarbons"? Describe two major types of unconventional hydrocarbons and two new hydrocarbon technologies.
15. What limitations are there on the development of geothermal power? wind power? wave power?

QUESTIONS FOR RESEARCH AND DISCUSSION

1. Should mining be allowed in national parks, forests, and wilderness areas?

2. Are there any economic mineral deposits or significant energy sources in the area where you live or study? If so, what kind? How did they form, and how are they exploited?

3. Discuss the role recycling might play in making available metallic resources in the future. Can you imagine a scenario in which the mining of new metallic minerals might cease? Do you think there can be sustainable development of nonrenewable resources?

QUESTIONS FOR *THE BASICS*

1. What is the difference between a resource and a reserve?

2. Define ore and gangue. What is the difference between an ore and a deposit? What is the difference between gangue and tailings?

3. What circumstances are required for the formation of a petroleum trap?

QUESTIONS FOR *A CLOSER LOOK*

1. What is radioactivity?

2. Why is nuclear waste problematic?

3. Describe the most common approach taken to nuclear waste disposal today.

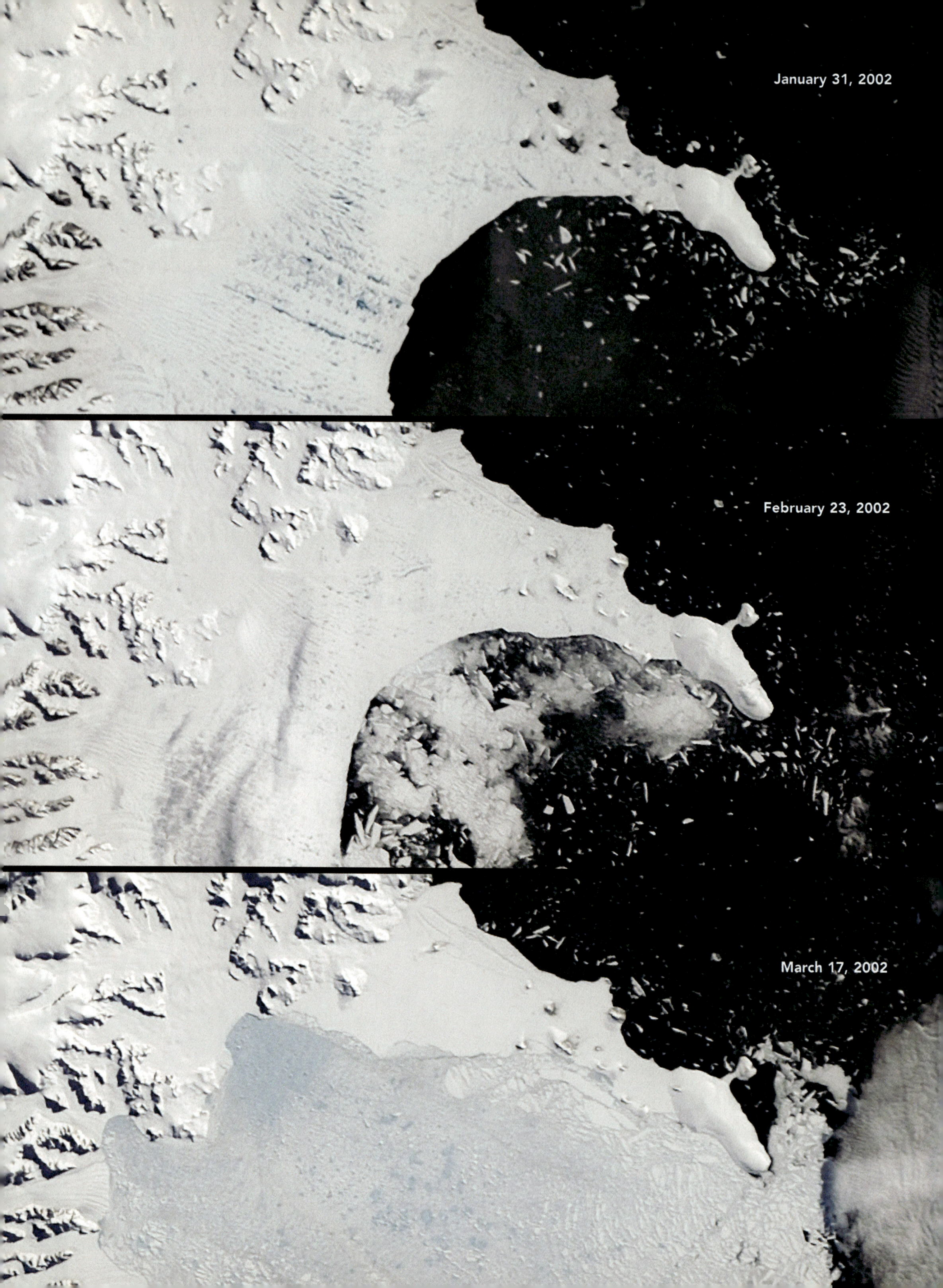
January 31, 2002
February 23, 2002
March 17, 2002

CHAPTER 19

The Changing Earth System

OVERVIEW

In this chapter we:

- Examine how population and lifestyle contribute to human impacts on the Earth system
- Consider the role of scientific uncertainty in our approaches to the management of Earth systems
- Summarize the global-scale impacts of human activity on the geosphere, hydrosphere, atmosphere, and biosphere
- Evaluate the possibility of anthropogenic climate destabilization and its potential impacts
- Contemplate human responses to global Earth system changes

Antarctic ice shelf collapse
A portion of the Larsen Ice Shelf off the coast of Antarctica, known as Larsen B, collapsed over a period of weeks during February and March, 2002, as shown in this series of satellite images. The collapse may have been accelerated by warm summer temperatures that caused meltwater to fill crevasses along the landward edge of the ice sheet, causing increased pressure. This portion of the shelf was more than 20 m thick and larger than the state of Rhode Island. This is one of a series of major Antarctic ice sheet collapses.

Throughout human history, people have asked important questions about Earth: How did the major features of our planet originate? Have the land, water, and air always been as they are now? What is the place of *Homo sapiens* among the multitude of living things that inhabit our planet? How can we ensure that the Earth system will continue to supply the resources we need for survival? In the context of a growing human population and rising levels of resource consumption, these questions have become increasingly important.

In this book you have learned that Earth functions as a closed system, so it has finite material resources and a finite waste-absorbing capacity. You have learned, too, that human actions are having significant impacts on the system—some long-lasting, some perhaps even permanent. We have touched on many such impacts in previous chapters, but in this chapter we will specifically consider the global-scale and cumulative impacts of human activity.

FIGURE 19.1 Early whalers
By the 1800s, activities like whaling (shown near the coast of New England in this woodcut from the nineteenth century) had already begun to produce impacts on ocean ecosystems.

UNDERSTANDING ANTHROPOGENIC CHANGE

In the 1840s, Henry David Thoreau traveled to Cape Cod and viewed the ocean from Provincetown at the very tip of the Cape. At that time the Cape was a rural countryside of fishers and farmers. To Thoreau, the ocean seemed to be the wildest part of Earth's surface, one that could never be affected by human actions. "The ocean is a wilderness reaching around the globe, wilder than a Bengal jungle, and fuller of monsters," he wrote. "Serpents, bears, hyenas, tigers, rapidly vanish as civilization advances, but the most populous and civilized city cannot scare a shark far from its wharves."[1] Little did he realize that events already underway would lead to human-induced changes to the ocean, and to the whole Earth system.

Two human activities in the ocean off Thoreau's home state of Massachusetts were already having an effect: direct harvesting of marine populations and chemical alterations of the ocean water. During one of his visits to Cape Cod, Thoreau saw fishers driving small whales ashore where they were harvested for blubber (FIG. 19.1). He reported this as an interesting economic activity, never suspecting that human actions might eventually threaten the whale population with extinction. But even then, whaling ships were departing from New Bedford, Massachusetts, armed with newly developed, specialized guns and harpoons, on a global hunt for Bowhead whales. In the early twentieth century, modern factory ships replaced whaling schooners, greatly increasing the killing power of the fleets. By 1914, at the beginning of World War I, the abundance of these huge mammals had decreased from 50,000 to about 3000. (Today, as a result of a whaling moratorium introduced in 1966, the global population of Bowhead whales has recovered to about 25,000—half of its pre-whaling level.)

Since Thoreau's time, humans have developed many technologies that affect the global environment. The use of technology and the impacts of our lifestyle are multiplied by the sheer number of people on the planet. Humans, as a collective, have altered the Earth system to a greater extent than any other organism throughout Earth history, aside from cyanobacteria.

Make the CONNECTION

What did cyanobacteria do to modify the Earth system so fundamentally?

The IPAT Equation

One way to quantify the human impact on Earth systems and to explore interactions among population, lifestyle,

[1] Thoreau, Henry David, *The Writings of Henry David Thoreau,* vol. 4, p. 188, Houghton Mifflin, 1906.

the BASICS

Human Population Growth

Scientists had little direct knowledge about the size of the human population until rather recently. The first attempts to estimate global population were made near the end of the seventeenth century, when world population was about 700 million. National censuses helped improve estimates during the eighteenth century, but only after World War II did estimates of global population become reasonably reliable. Estimates of earlier human population size are based, in part, on the way people obtained food. Hunter-gatherers can live only at a low density without overusing their food supply. The advent of agriculture therefore allowed a major increase in human population density, but far less than modern agriculture permits. Based on this kind of information, and the fossil and historic record of the distribution of our species, estimates of human population growth have been made (FIG. B19.1).

There have certainly been times in human history when population has surged as a result of technological advancements, as well as times when growth has slowed in response to environmental resistance and stresses. However, the long-term trend is clear: it took many millennia for the human species to number a billion (toward the beginning of the nineteenth century), but that number has doubled twice since then and is now approaching 7 billion (6.8 billion, as of early 2010).

At present, the human population is growing at the rate of approximately 1.2 percent per year. The rate is considerably higher in many developing nations, but lower in some industrialized nations. The overall rate of growth has been steadily decreasing over the past few decades, although it still represents the addition of over 80 million people to the population each year—over twice the population of Canada. The sheer growth in population has led to increasing interaction between people and natural systems, so that now even the most remote parts of the planet are affected by human activity.

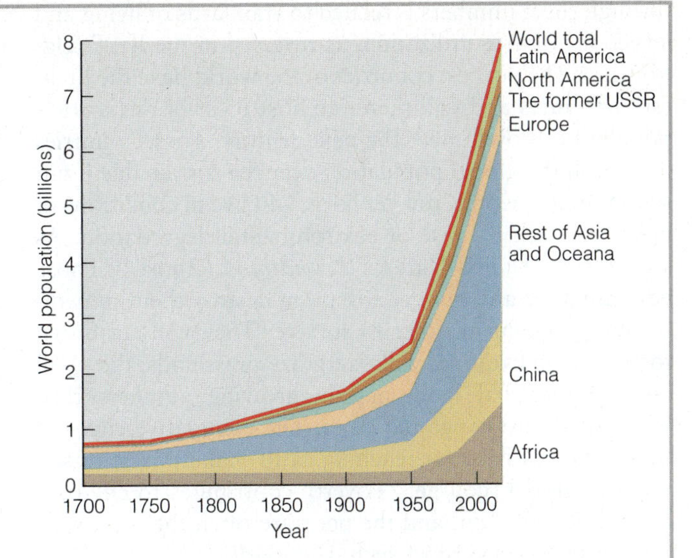

FIGURE B19.1 Human population growth
Extremely rapid increase of human population is a relatively recent phenomenon in the context of human history as a whole. Throughout history, there have been times when human population surged as a result of the development of new technologies, and times when population growth slowed as a result of environmental resistance.

technology, and the environment is the **IPAT equation.**[2] The equation is a conceptual relationship, rather than a mathematical equation, which expresses human impacts (I) as a function of population (P); consumption and waste generation, represented by affluence (A); and technology (T). An additional term is included to account for the sensitivity (S) of a particular environment to the impacts of human activity. The relationship is stated:

$$I = P \times A \times T \times S$$

It is difficult to insert actual numbers into this equation—for example, how can the impacts of technology be quantified in terms that are directly comparable to the impacts of lifestyle? Nevertheless, the IPAT relationship has provided a useful framework for discussions about the impacts of human activities on Earth systems. In the early 1970s, when the research that led to the IPAT equation was carried out, environmental impact was generally thought of in terms of pollution. But the relationship demonstrates that factors relating to other technological impacts, population, waste generation, and lifestyle are also important.

Poverty and Affluence

Is the *carrying capacity* (Chapter 16) of this planet sufficient to support a human population of 12 billion, or even 8 billion? Have we already overshot the planet's long-term carrying capacity as we approach a population of 7 billion? There is no simple answer to this question.

One of the interesting issues related to human population is the question of how our impact on the Earth system

[2] The first statements of the IPAT relationship and early explorations of these interactions, from which the IPAT equation first emerged, are generally credited to Ehrlich, Holden, and Commoner, as follows:

Ehrlich, Paul R., and John P. Holdren, 1971. Impact of Population Growth. *Science*, 171:1212–1217.

Commoner, Barry, 1972. The Environmental Cost of Economic Growth. In: *Population, Resources and the Environment*. U.S. Government Printing Office, Washington, DC. pp. 339–363.

through sheer numbers is related to standards of living and levels of resource utilization, as revealed in the IPAT relationship. The poorest countries of the world have the highest birth rates and will therefore host most of the world's population growth over the next century. These countries may push the global population over the sustainable limit, so that many people, not yet born, will live in conditions of poverty—itself a cause of environmental degradation. As stated in the United Nations' *Brundtland Report*,[3] "Those who are poor and hungry will often destroy their immediate environment in order to survive: They will cut down forests; their livestock will overgraze grasslands; they will overuse marginal land; and in growing numbers they will crowd into congested cities. The cumulative effect of these changes is so far-reaching as to make poverty itself a major global scourge." Poverty contributes to environmental degradation, and the poor are often the most seriously affected victims of such degradation.

On the other hand, what if we measure our impacts on the basis of intensity and rate of resource use? On a per capita basis, Americans use about 4 times as much steel and 23 times more aluminum than people in Mexico. The average Japanese consumes 9 times more steel than the average Chinese. Canadians routinely top the global list in per capita energy consumption. In overall use of nonreplenishable resources, the average Swiss consumes as much as 40 Somalis. The point is not that the lifestyle of the Swiss or Somalis or Americans or Canadians is right or wrong. Nor should everyone's living standard be lowered in order to fit within Earth's carrying capacity. Rather, the point is that there are many choices to be made, many ways to use resources wisely, and many ways to live more efficiently. That a Somali family could have 39 children and still not consume as much as a Swiss or North American family gives a somewhat different perspective on population and resource use.

Ecological Footprint

To support our modern lifestyle, we draw resources from an area much greater than the physical area that we actually occupy. Have you ever eaten fish from the ocean, or a hamburger made from beef raised in South America, or strawberries grown in California's Imperial Valley? If so, then you have made use of resources from far beyond your own backyard. This idea is encapsulated in the concept of the **ecological footprint**, which is a measure of the resources required to support a particular person or community, rendered in terms of land area. The calculation translates the requirements for six categories of activities that support our material lifestyle into land area equivalents (usually expressed in terms of hectares). The six categories are: growing crops for food, feed, and fiber; grazing animals for milk and meat; fishing; harvesting timber; building infrastructure (roads, houses, etc.); and burning fossil fuels (FIG. 19.2).

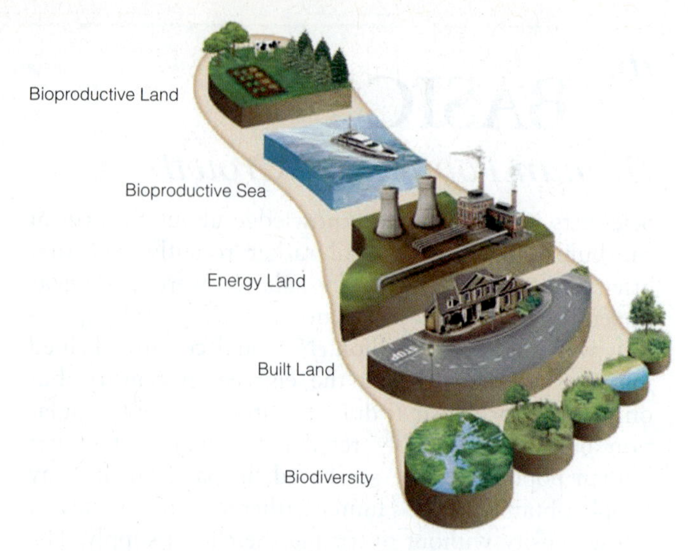

FIGURE 19.2 Ecological footprints
The ecological footprint is a calculation of the resources required to support a person or community, rendered in terms of land area. The calculation translates into terms of land area the requirements for six categories of activities that support our material lifestyle: growing crops for food, feed, and fiber; grazing animals; fishing; harvesting timber; building infrastructure; and burning fossil fuels

Most people in the industrialized world have ecological footprints much larger than the biologically productive area available to each person on this planet—about 2.1 ha. The ecological footprint of the average American is almost 10 ha. The average Canadian lives a bit more lightly, at 7 ha, and the average Italian's ecological footprint is 3.6 ha. According to ecological footprint calculations (which vary widely, depending on the specific assumptions made by the researcher), the current human population of almost 7 billion has already exceeded Earth's carrying capacity. It would take about 1.3 Earths to meet our current resource needs sustainably, and three or more Earths for everyone on the planet to live at the level of resource consumption of the average North American.

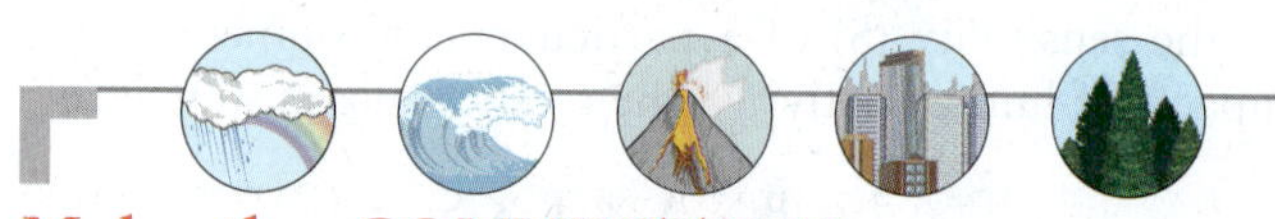

Make the CONNECTION

How is the concept of the ecological footprint related to that of carrying capacity? (*Hint:* Turn it upside down.) What are the components of your own ecological footprint; in other words, from what spheres of the Earth system are you drawing the resources needed to sustain your standard of living? Try visiting an online ecological footprint calculator, and calculate your own ecological footprint.

[3] *Our Common Future* (1987), commonly known as the *Brundtland Report*, was the report of the United Nations' World Commission on Environment and Development, in which the term *sustainable development* was first popularized.

A Closer LOOK

MALTHUS, POPULATION, AND RESOURCE SCARCITY

Traditionally, the debate about the growing human population has been divided into two camps: the Malthusians and those who disagree with the Malthusian model. About 200 years ago, English economist Thomas Malthus (1766–1834) stated the human population problem so eloquently that his argument remains central to the discussion, even today. Malthus argued that reproduction is a biological phenomenon, beyond our control, and that the human population will grow *exponentially* (Chapter 16), unlike the availability of food and other resources, which increase *linearly* (FIG. C19.1). As a result, he argued, the human population will eventually exceed Earth's capacity to provide necessary resources, and, he wrote, "The power of population is indefinitely greater than the power in the earth to produce subsistence for man." The consequences, according to Malthus, would be widespread famine, pestilence, disease, conflicts over scarce resources, and deaths.

Anti-Malthusians argue, on the other hand, that technology can support a greater density of people on Earth, and it will continue to do so into the indefinite future. Certainly Malthus was wrong on two central points: humans have found both the means and the will to limit reproduction, and technological development has allowed us to greatly expand Earth's capacity to provide food and other resources.

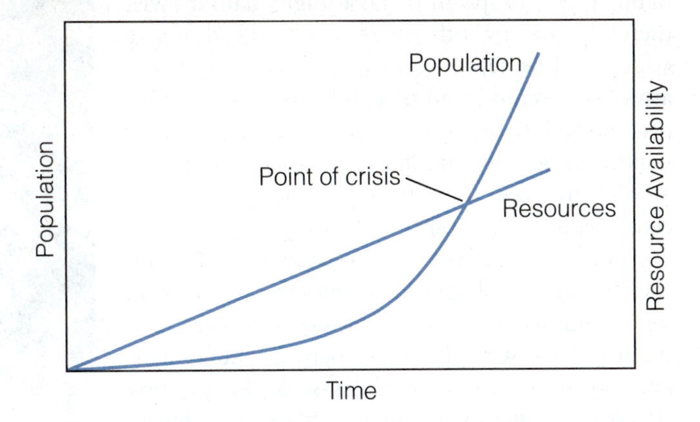

FIGURE C19.1 **Malthus, population, and resource scarcity** Thomas Malthus predicted that human population would grow exponentially, eventually exceeding the capacity of Earth to provide food and other resources.

However, Malthus' model seems relevant today, in a world where resource scarcity leads to conflicts and war, and where increasing numbers of refugees are leaving their homelands as a result of devastating environmental change. The global perspective of this book makes it clear that the Earth system is finite; there must eventually be a limit to human population growth and resource use. The exact limit depends not only on the availability of resources and the number of people on the planet, but also on the technologies we develop and the quality of life we want to maintain.

Scientific Uncertainty

As a society, we try to manage environmental impacts with laws, regulations, agreements, and economic instruments such as fines, fees, and taxes. A discussion of such tools is beyond the scientific scope of this book. But one aspect of the scientific endeavor has a particularly important influence on environmental management; it is *uncertainty* in science.

Scientists are always dealing with uncertainty. When scientists approach a problem they are testing *hypotheses* (Chapter 1), and they rarely expect to come up with a final, ironclad answer. Most commonly, scientific research provides tentative answers to some questions, while raising many new ones. In science, everything is open to questioning and testing. Uncertainty is part of what makes the complex world we live in exciting and challenging to scientists.

On the other hand, policymakers usually seek to avoid or at least reduce uncertainty. If a natural event is predicted, policymakers would like to know when and where it will happen, how big it will be, how widespread its impacts, and how to prepare for or mitigate the effects. The uncertainty involved in our scientific understanding of the Earth system makes it particularly challenging for decision makers and politicians to deal with things such as natural hazards and global climatic change.

One way for decision makers to deal with risk is to rely on the **precautionary principle**, the concept that if the potential consequences of an anticipated event are unacceptably severe, those in authority have a responsibility to take action to avoid or mitigate those consequences—even if the probability of occurrence is small and even if there is scientific uncertainty. Many people feel that a global problem such as climate destabilization, which is surrounded by scientific uncertainty but would have dire consequences in the worst-case scenarios, obliges us to invoke the precautionary principle when making decisions that might affect the future of the Earth system and our species.

HUMAN IMPACTS ON THE EARTH SYSTEM

Our study of the Earth system has made it abundantly clear that this planet is a complex, dynamic set of open systems and cycles that are in a constant state of change and shifting balances. The solid, liquid, gaseous, and organic realms of Earth are closely interlinked, and a change in one part of the system affects other parts. A massive earthquake might raise an extensive zone of submerged coast, exposing and destroying nearshore marine habitats. An eruption of lava might dam a river, thereby affecting other streams in the drainage system, while tephra and gases ejected into the atmosphere could lead to a global drop in air temperatures. Fluctuations in air pressure can alter wind and oceanic circulatory systems, leading to an El Niño event that affects weather and oceanic ecosystems throughout much of the world.

We can observe and measure such natural changes as they happen, or interpret them from the natural records of Earth history. However, only recently have we come to recognize that humans also can have profound, global-scale, long-lasting effects on the geosphere, hydrosphere, atmosphere, and biosphere. Some of the effects are global in their reach; stratospheric ozone depletion and global climatic change are two examples. Others are local or regional, but their collective or **cumulative effects** over time are such that the impacts have reached worldwide proportions; soil erosion, acid precipitation, and groundwater depletion are examples. In the remainder of this chapter we will look more closely at **global change**, the cumulative, long-term, global-scale changes that human activities are imposing on the Earth system.

FIGURE 19.3 Desertification
Desertification is the advance of desert conditions into once-productive land. It can occur naturally, but it can be accelerated by human activity. Here, a grid of fencing has been set up to slow the advance of sand dunes into the ancient oasis of Tekenket, Mauritania.

Geosphere: Impacts on Land

Human impacts on land tend to be local, but local impacts can add up to regional and global problems. Land-based impacts typically stem from overly intensive or inappropriate uses of land already sensitive to environmental change and from the disposal of hazardous materials on land.

Desertification

Desertification is the expansion of desert conditions into previously productive land areas (FIG. 19.3). The term was coined when the United Nations convened a conference in 1977 to study the problem of land degradation resulting from human impact. Although desert expansion can result from natural processes, it is now recognized that increasing numbers of people and livestock in drylands can lead to progressive deterioration of the land and ultimately to desertification. Sometimes the process is triggered or exacerbated by natural drought. The most obvious symptoms include crop failures or reduced yields; reduction in rangeland biomass available to livestock; reduction in fuelwood supplies; reduction in water supplies resulting from decreased streamflow or a depressed groundwater table; and advance of dune sand over agricultural lands.

Significant expansion of deserts in the Sahel region of Africa, the semiarid belt bordering the southern Sahara, has attracted world attention because of the resulting widespread famine (FIG. 19.4). A strong case can be made that human

FIGURE 19.4 Land degradation
Overgrazing during years of drought killed much of the vegetation in this part of the Sahel in the Gao region of Mali. Without vegetation, topsoil is eroded and the land becomes infertile.

activity has caused desertification in some areas, but scientific opinions are not always consistent. For example, some researchers have concluded that arid land areas have increased by more than 50 million ha since 1931 as a result of a 30 percent decline in rainfall during this interval.

The complex linkages between humans and the geosphere, biosphere, and atmosphere are illustrated by a hypothesis of progressive desertification that links grazing animals, the vegetation they consume, and the atmosphere. As animals consume vegetation cover in drylands, the *albedo* (or reflectivity, Chapter 2) of the land surface increases because sand and bare rocks have a higher albedo than grassland. This causes incoming solar radiation to be reflected back into space rather than being absorbed, leading to a cooler ground surface. The cooler ground is associated with descending dry air and therefore with reduced precipitation. In this way, the degraded area becomes more and more desert-like. In other words, a positive feedback is set in motion that promotes increasing desertification. Dust kicked up by the hooves of massive migrating herds of cattle may also play a role in this feedback cycle.

Soil Erosion

With world population approaching 7 billion, increasing demand on agricultural land is causing serious erosion of soils. Although soil erosion happens naturally as a result of changes in topography, climate, or vegetation cover, the effects of human activity contribute substantially. For example, accelerated erosion is often closely related to deforestation. Clear-cut logging leads to accelerated rates of surface water runoff and to the destabilization of soils as a result of the loss of anchoring roots. In the humid tropics, forests hold most of the organic matter in the live vegetation (unlike in temperate forests, where much organic matter is stored as dead plant debris in soils). Consequently, when tropical land is cleared and cultivated, the organic-poor soils quickly lose their fertility (FIG. 19.5).

Mining, road construction, and other earth-moving activities also contribute dramatically to exposure and ultimately erosion of soil. Other activities that cause damage to soils include overly intensive grazing, poorly designed irrigation, dams, and pollution, which can lead to flooding, deposition of salts, or chemical contamination of soils.

So widespread are the effects of soil erosion and degradation that the problem has become global. Agriculture is the foundation of the world economy, so a progressive loss of soil signals a growing crisis that could undermine the economic stability of many countries. It generally takes between 80 and 1000 years (and even longer under some circumstances) to form one centimeter of topsoil; therefore, soil erosion—for all practical purposes—is tantamount to mining the soil. Worldwide, the most productive soils are being depleted at the rate of 7 percent each decade. One estimate projected that, because of excessive soil erosion and increasing population, only two-thirds as much topsoil is available to support each person now as there was in 1980.

FIGURE 19.5 **Deforestation and erosion** Widespread deforestation in the Rio Branco area of the Amazon in Brazil has devastated a formerly luxuriant rain forest and led to accelerated runoff and deep gully erosion. Soils on the landscape quickly lose their natural fertility when forest is converted to crops or grazing land, leaving a degraded landscape with little value.

Although soil erosion and land degradation are already affecting many countries, effective control measures can substantially reduce these adverse trends. One method of reducing soil loss involves crop rotation and increased fallow periods, which helps avoid nutrient depletion and allows the soil time to be replenished. The most serious soil erosion problems occur on steep hillslopes, but even some steep slopes can be exploited sustainably through terracing and other accessible, affordable erosion control measures. Such measures are being applied in many countries today.

Waste Disposal and Toxins

Vast quantities of garbage and industrial wastes resulting from human activity are deposited each year in landfills. When a landfill reaches its capacity, it generally is covered with earth and re-vegetated. The buried waste products may be mobilized by rainwater that percolates through the site, carrying away soluble substances. In this way, harmful chemicals may slowly leach into surrounding soil and groundwater systems and contaminate them. Pollutants travel from landfill sites as contaminated plumes that follow regional groundwater flow patterns (FIG. 19.6). Modern landfills can be engineered to avoid problems like these, but older, abandoned sites are numerous and continue to leak. Some programs are now in place to clean up such sites and render them safe; however, problematic sites number in the tens of thousands, and significant time and money will be required to complete the task.

Another hazard is posed by pesticides and herbicides sprayed over agricultural fields to help improve the quality

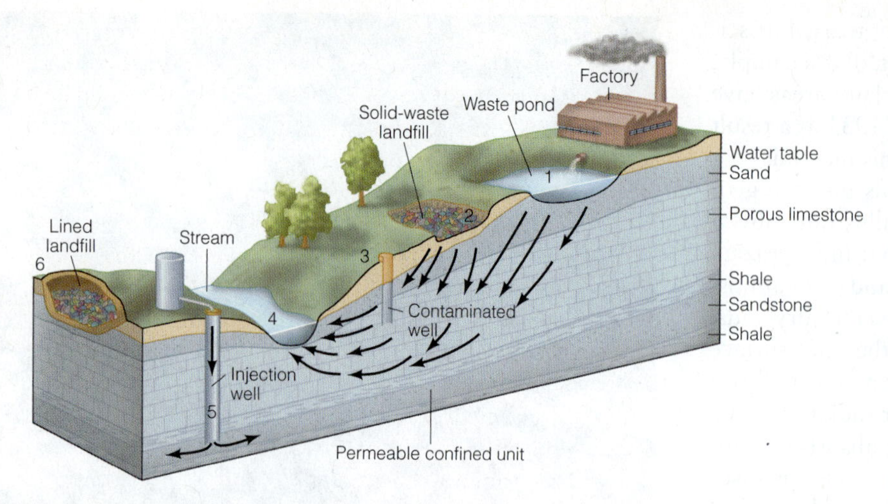

FIGURE 19.6 Soil and groundwater contamination
Toxic materials in improperly engineered disposal sites (1) or landfills (2) can percolate downward, contaminating both soil and groundwater. Also contaminated, in the scenario shown here, are a well downslope (3) and a stream (4) at the base of the hill. Alternative approaches to waste disposal that can be more secure include deep-well injection (5) and properly engineered secure landfills (6). Because neither of the latter approaches is foolproof, constant monitoring is required.

and productivity of crops. Some agrochemicals have been linked with cancers and birth defects in humans, and others have led to major declines of wild animal populations. If they are broadcast over large areas of land, excess chemicals inevitably contaminate soils and invade the groundwater system. They can also affect nontarget species. A current concern, for example, is the precipitous decline in both wild and managed bee populations, possibly as a result of exposure to pesticides (although there are several other possible causes). These risks are increasingly taken into consideration by modern farm managers, who seek to minimize chemical applications and to direct applications as specifically as possible.

Another cause for concern is radioactive waste. Countries with nuclear power plants have the special problem of disposing of radioactive waste products that are so highly toxic that exposure to even minute quantities can be fatal. Most studies concerning disposal of toxic and nuclear wastes have concluded that safe underground storage is possible, provided appropriate sites can be found. However, some nuclear wastes can remain dangerous for tens of thousands of years; therefore, a primary requirement of a safe disposal site is to ensure that radioactive materials remain isolated from the biosphere and hydrosphere. The waste products and their containers must be stored in a stable, secure location, such that they will not be affected chemically by groundwater, physically by natural deformation such as earthquakes, or accidentally by people.

Hydrosphere: Impacts on Water

As the population has grown and become more industrialized, people have altered the natural flow of rivers, withdrawn water from underground sources, and released ever-larger amounts of waste, much of which inevitably finds its way into the water that we rely on for our existence. In many places, water is dwindling in both quantity and quality, which raises important questions for the communities involved: Will there be enough clean water to sustain future needs? Is its quality adequate for the uses to which we put it? Is the water being used with a minimum of waste? We introduced some of these questions in Chapter 8, but here we consider the cumulative effects of human activities on the hydrosphere.

Dams and Diversions

Few of the world's large rivers in the densely populated regions of Earth now flow unrestricted to the sea; the estuaries of almost every major river of the world have been significantly altered by human actions, and diversion occurs to a greater or lesser extent in almost every major river channel. Today the ecological effects of dams, diversions, and stream-channel control are widely recognized. The worldwide alteration of major rivers is having an effect on biodiversity, through the loss of habitat for fish, amphibians, migrating waterfowl, and aquatic plants. An iconic example of the devastating impacts of diverting water from river channels is that of the Aral Sea (see Chapter 8, *A Closer Look: The Case of the Aral Sea*), where the diversion of river water has left a once-great water body in a state of ecological collapse. We now face a fundamental question: how can we meet the energy and water requirements of human beings, while sustaining the natural integrity of river ecosystems?

Dams are often constructed to extract hydroelectric power from the potential energy of water flowing downslope. Hydropower is "clean" in the sense that it does not directly contaminate Earth's atmosphere or waters. However, the major dams required for controlling the flow can have significant impacts on land and river ecosystems, as discussed in Chapter 18.

Another common objective of large dams is flood control. The results can be impressive, as illustrated by Egypt's Aswan Dam (FIG. 19.7A). This huge dam (actually a series of dams), with its vast reservoir (Lake Nasser), markedly reduces the seasonal variability in discharge that formerly produced annual floods in the Nile Valley (FIG. 19.7B). However, the Aswan Dam project brought unanticipated consequences. The Nile, like all other large streams, is a complex natural system, and its behavior reflects a delicate balance between water flow and sediment load. Prior to construction of the dam, an average of 125 million metric tons of sediment passed downstream each year, and most was deposited on the floodplain. The dam reduced this total to 2.5 million metric tons

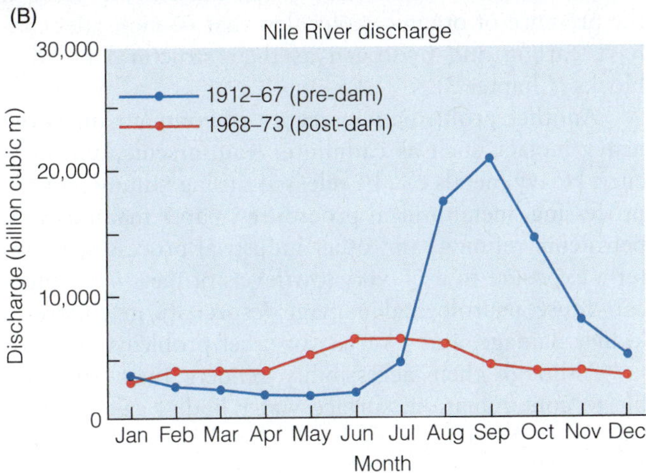

FIGURE 19.7 Dams and their impacts
Dams now exist on virtually every major waterway, and they have significant impacts on aquatic ecosystems. (A) The Aswan High Dam impounds the Nile River (top) to form Lake Nasser (behind the dam, at the bottom of photo). (B) Prior to construction of the Aswan Dam, the discharge of the Nile River varied seasonally, with peak discharge coming during the late summer and early fall interval of flooding. Controlled release of water after the dam was built greatly reduced the seasonal variability in discharge. Sediment formerly carried to the Mediterranean Sea is now settling in Lake Nasser; it will eventually fill the reservoir and make it unusable.

because nearly 98 percent of the suspended sediment is now deposited in Lake Nasser. Under pre-dam conditions, floodwater sediment replenished the rich agricultural soils of the Nile Valley and Delta. With this natural source of nourishment eliminated, farmers now resort to artificial fertilizers to keep the land productive. Some of the fertilizer seeps back into the river, causing pollution problems downstream. Furthermore, because the annual discharge of sediment to the Mediterranean Sea has now been cut off, the delta and coast have now become increasingly vulnerable to erosion. The creation of a very large body of standing water, Lake Nasser, has also contributed to the waterlogging of soils and increases in the incidence of mosquito-borne diseases such as schistosomiasis (also know as bilharzia, or snail fever).

Mining Groundwater

In many parts of the world, groundwater is the major source of water for human consumption. If withdrawal exceeds the natural rate of recharge of an aquifer, the volume of stored water will diminish and the water table will decline; this is referred to as *groundwater mining*. We regard petroleum, coal, and minerals as nonrenewable resources because they form over geologically long intervals of time and are depleted by extraction, but we don't often think of groundwater as a nonrenewable resource. In locations where precipitation is low, natural recharge takes so long to replenish depleted aquifers that formerly vast underground water supplies have essentially been lost to future generations.

Even where the problem has been recognized and measures have been taken to mitigate the impacts, centuries or millennia of natural recharge may be required to return aquifers to their original state.

In some cases, aquifer changes may be permanent. When groundwater is withdrawn, the water pressure is reduced, and the particles of the aquifer shift and settle slightly. As a result, the land surface subsides, and it may never again be possible for water to infiltrate into the aquifer. The interrelated impacts of groundwater withdrawal, compaction of soft sediments, ground subsidence, and resulting impacts on buildings are common to large coastal cities throughout the world, including Venice, Shanghai, Houston, Bangkok, and many others.

Surface Water Contamination

The accessibility of surface water bodies makes them useful as resources, but also renders them susceptible to contamination. Contaminants in surface water come mainly from urban, suburban, and agricultural runoff. Industrial *effluent* (contaminated runoff) is a significant contributor to surface water pollution; so are discharges related to resource extraction, among which mining, logging, and oil drilling are important. Effluents from poorly engineered landfills are another important source of surface water contamination. Airborne substances can contribute to the deposition of acids on land and water surface, which can damage lakes and aquatic ecosystems, as well as soils and forests. Surface water reservoirs in industrial areas are also susceptible to *thermal pollution*—the release of heat or (more commonly) heated water, which is harmful to aquatic organisms and ecosystems.

Wastes that end up in surface water bodies often contain organic matter. Many cities, even in wealthy countries, still discharge sewage directly into nearby water bodies with little or no treatment. When the organic matter in sewage breaks down, oxygen is used up in the process. Oxygen dissolved in the water is critical for the maintenance of most aquatic life forms, and effluents that cause the depletion of dissolved oxygen are said to place a *biochemical oxygen demand* on the system. The greater the amount of organic matter discharged into the system, the higher the biochemical oxygen demand. If sewage is discharged directly into a water body, the dissolved oxygen in the water downstream from the source of the sewage responds by becoming depleted.

One of the most common forms of surface water contamination results from an excess of plant nutrients, especially phosphorus and nitrogen, from fertilizer runoff and phosphorus-containing detergents. When an aquatic system is overloaded with nutrients, the growth of algae, plankton, and aquatic weeds can get out of control; this is called an *algal bloom*. When the algae and other aquatic plants die their breakdown causes oxygen depletion. Other organisms in the water begin to die, and the water turns mucky and a sickly green (FIG. 19.8). This is similar to *eutrophication* (Chapter 8), the natural process whereby a lake fills with plant matter and sediment to become a swamp. When eutrophication is accelerated by the addition of anthropogenic pollutants, it is called *cultural eutrophication*. If the water becomes completely oxygen-depleted, it may turn crystal-clear. This happened to parts of Lake Erie in the 1970s, and it is not a sign of a healthy water body—it is a sign of a water body in which nothing is able to survive.

The presence of infectious agents such as bacteria, viruses, and other disease-causing organisms is another form of surface water contamination associated with human and animal wastes. In many parts of the world, consumption of water contaminated with infectious agents is linked with outbreaks of diseases such as cholera. In North America, cholera and other waterborne diseases are less of a problem than they were 100 years ago; still, high levels of *E. coli*, an infectious organism commonly found in feces, routinely cause beach closures in many coastal areas. In Walkerton, Ontario in 2000, the deaths of seven people and many serious illnesses were caused by a deadly strain of *E. coli* that entered the municipal groundwater supply through a cracked wellhead. The source of the bacterium is thought to have been fecal-contaminated runoff from a nearby farm.

Some of the most significant toxic contaminants in surface water bodies are persistent chlorinated organic compounds, such as DDT and Mirex (both pesticides), polychlorinated biphenyls (PCBs), and dioxins and furans (industrial by-products). Note that "organic" doesn't mean "environmentally friendly"; in this case, it refers to the presence of organic molecules, that is, molecules that have carbon and hydrogen as their structural building blocks (Chapter 3).

Another problematic category of contaminant is the heavy metals, such as cadmium, lead, arsenic, and mercury. Heavy metals can be released during mining and ore processing, metallurgical procedures, paper manufacture, petroleum refining, and other industrial processes. Long-term exposure to even very low levels of these substances can cause neurological damage (especially in children), kidney damage, and a variety of other problems.

Although their accessibility renders them susceptible to contamination, surface water bodies can often be

FIGURE 19.8 Eutrophication: A dying lake
This lake in Florida has turned green and mucky because of the growth of algae stimulated by excessive nutrients (possibly from sewage or fertilizer runoff). Eventually, the algae will use up all of the dissolved oxygen in the water, making it impossible for other aquatic life to survive. This process, called eutrophication, can occur naturally, but it can be accelerated by human activities.

cleaned up, or *remediated*. Some have responded well; for example, Lake Erie, which effectively "died" as a result of oxygen depletion in the 1970s, has since undergone substantial recovery, thanks to agreements between the United States and Canada that have limited phosphate runoff into the Great Lakes. Other problems, notably the acidification of lakes as a result of acidic precipitation, have proven to be more difficult to resolve. Still others have been partially addressed by changes in human behavior, but await the response of the natural system; the most iconic example is DDT. This toxic organochloride pesticide was largely banned from North American manufacture and use in the 1970s; however, it is so persistent that it can still be detected in natural reservoirs such as the tissues and breast milk of animals, including humans. It will take many more years before natural processes can eliminate DDT from the hydrosphere and biosphere.

Groundwater Contamination

Many of the pollutants that affect surface water and soils also cause groundwater contamination. Because of its hidden nature, groundwater contamination can be very difficult to detect, control, and clean up. Harmful chemicals from leaking underground gasoline storage tanks and pipelines, poorly designed or abandoned landfills, and toxic waste disposal facilities can contaminate groundwater reservoirs. Many of the top groundwater contaminants are toxic organic compounds that are constituents of petroleum and other industrial chemicals, such as solvents.

Perhaps surprisingly, however, the most common pollutants in groundwater are untreated sewage and nitrate from agricultural chemicals. These are less acutely toxic than other contaminants, but no less problematic, because they are so widespread. Another common problem is saline contamination of groundwater, usually caused by the intrusion of seawater into coastal aquifers.

Once contaminated, groundwater can be extremely difficult to remediate because of inaccessibility. There are two basic approaches to the remediation of polluted groundwater. The first is *passive remediation*; this involves relying on natural processes to clean up the site. This may sound "lazy" or ineffective, but it is sometimes a viable approach. The natural system can be very effective at filtering and removing contaminants from water and soil. If there is no immediate risk to people living or working nearby, no urgent need to prepare the land for new uses, and the environment itself is appropriate, it can save a lot of money and effort simply to wait and let nature take its course.

Passive remediation of even a mildly contaminated site can take years, or even decades. If the need for the land is urgent, or the risk to people or ecosystems is severe, it is advisable to undertake *active remediation*, that is, to actively intervene in the cleanup of the site. This can include approaches such as injecting oxygen or other chemicals into the groundwater to speed the breakdown of the contaminants, or pumping the contaminated groundwater to the surface for treatment and eventual re-injection into the aquifer.

Oil Spills and Other Marine Impacts

Like surface freshwater bodies and aquifers, the marine system suffers from the impacts of human activity. The water of the open ocean is still relatively free of pollution, although deposition from atmospheric sources and litter in shipping lanes are becoming more problematic. The status of water in coastal zones, however, is considerably more worrisome. The proximity of coastal waters to human activities renders them vulnerable, and their relatively shallow, warm, biologically productive waters are particularly sensitive to environmental changes.

Oil spills are one of the most visible impacts of human activity in the oceanic environment, especially in coastal zones. Oil can enter the ocean by runoff from spills on land, or directly as a result of accidental spills from oil tankers and drilling rigs. Crude oil consists of hundreds of chemical components, many of them toxic, and spills can be devastating to local ecosystems. The water, coastal zones, ecosystems, and human inhabitants of the area surrounding the Gulf of Mexico will take decades to recover from the BP *Deepwater Horizon* oil spill of 2010 (shown in Fig. 18.24). Interestingly, though, significantly more oil enters the marine environment every year as a result of thousands of small, unreported spills than from the few big spills that receive massive media attention.

Many of the world's coastal waters also suffer from eutrophication, discussed previously in the context of freshwater bodies. Eutrophication can result from nutrient-laden runoff (that is, runoff carrying sewage, fertilizer, animal manure, and other plant nutrients), leading to an overgrowth of aquatic plants and algae. When the algae die, they decay, using up oxygen. In the extreme scenario, this can lead to a situation of oxygen depletion, or *hypoxia*, in which dissolved oxygen levels fall too low to support aquatic life. There is much concern that oceanic hypoxic zones, known as "dead zones," have been increasing in number, extent, and intensity of oxygen depletion over the past few decades, possibly enhanced by the warming of ocean waters.

Atmosphere: Impacts on Air

In Chapter 17 we introduced *common property resources* or *commons*—resources that are open and accessible to all people, or to a group of users. The world ocean and Antarctica are two examples of resources held in common by all people, but the atmosphere is the ultimate example of a **global commons**. Any common property resource can be difficult to monitor and regulate, and the atmosphere is certainly no exception. We all depend on it for life support, yet we use it almost without thought as a receptacle for potentially harmful emissions from our daily activities.

Long-Range Transport of Pollutants

Some of the impacts of air pollution are local, but many are regional; they result from the long-range transport of airborne pollutants. Atmospheric transport is the main reason why anthropogenic pollutants now touch virtually every part of the world—it is why chemicals from industry and agriculture end up in the fat of polar bears in the high Arctic, for example.

One of the best-known examples of long-range atmospheric transport occurred when a nuclear reactor in Chernobyl, Ukraine, experienced a meltdown and explosion in 1986. Radioactive particles from this disaster encircled the globe within two weeks (FIG. 19.9). Some of the contamination eventually settled out as *dry deposition* (that is, settling to the ground as tiny, solid particles). However, scientists who investigated the spread of the radioactive contamination discovered that the most significant contamination was experienced in localities that were near Chernobyl and in areas where it rained in the days following the explosion. Precipitation facilitated the settling out or *wet deposition* of the contaminants, by dissolving some components and adhering to others.

The behavior of an airborne contaminant is thus controlled by a variety of interacting factors, including its own chemical and physical characteristics, topography, and prevailing weather patterns. Wind systems and precipitation play an important role in the transport and eventual deposition of the contaminant, and atmospheric impacts that are generated locally by industrial activity can become regional or global problems through long-range transport.

Stratospheric Ozone Depletion

As discussed in Chapter 11, ozone (O_3) is present in the atmosphere in only trace amounts. Without this trace gas, life on Earth would be very different, for ozone provides living organisms with a protective shield against harmful ultraviolet radiation from the Sun. For humans, direct exposure to ultraviolet light damages the immune system, produces cataracts, substantially increases the frequency of skin cancer, and causes genetic mutations.

The average global concentration of ozone in the atmosphere is approximately 300 DU, which varies vertically, geographically, and temporally (seasonally, and over time). **DU** stands for the **Dobson Unit,** and 100 DU is defined to be the equivalent of a layer of pure ozone 1 mm thick at surface temperature and pressure. The 300-DU average means that if you took all of the ozone from the atmosphere and collected it at surface temperature and pressure it would yield a layer of pure ozone just 3 mm thick.

Maximum concentrations of ozone in the atmosphere occur in the stratosphere, between 25 and 35 km above Earth's surface. In this region of the stratosphere, ultraviolet radiation breaks down molecules of oxygen (O_2) into two oxygen atoms, which then combine with other O_2 molecules to form molecules of ozone (O_3). The ozone is in turn broken down by ultraviolet radiation, creating a natural balance among O, O_2, and O_3.

In 1985 British scientists reported a startling discovery: a vast area of depletion—an **"ozone hole"** about the size of Canada—had developed in the ozone layer above Antarctica (FIG. 19.10A). By 1987 measurements showed that concentrations of this life-protecting gas had dropped by more than 50 percent since 1979. Stratospheric ozone levels over Antarctica have dipped as low as 90 DU (1992 and 2006 were particularly low years). Record low values were subsequently measured over Australia and New Zealand, and continuing surveys showed that ozone values at all latitudes south of 60° had continued to decrease (FIG. 19.10B).

What happened to upset the natural balance among the three gaseous forms of oxygen? A decade before the ozone hole was discovered, it was recognized that a group of synthetic industrial gases called **chlorofluorocarbons (CFCs)** were entering the lower atmosphere and spreading rapidly. As CFCs rise into the upper atmosphere, ultraviolet

FIGURE 19.9 Atmospheric transport of pollutants
(A) In 1986 at the Chernobyl nuclear reactor in Ukraine, reactor number 4 (seen here from the roof of an adjacent reactor) experienced a core meltdown and explosion that caused a large release of radioactive material. (B) After the meltdown wind systems transported the emissions, leading to radioactive fallout throughout Europe within days. Radioactive emissions were eventually transported as far as North America.

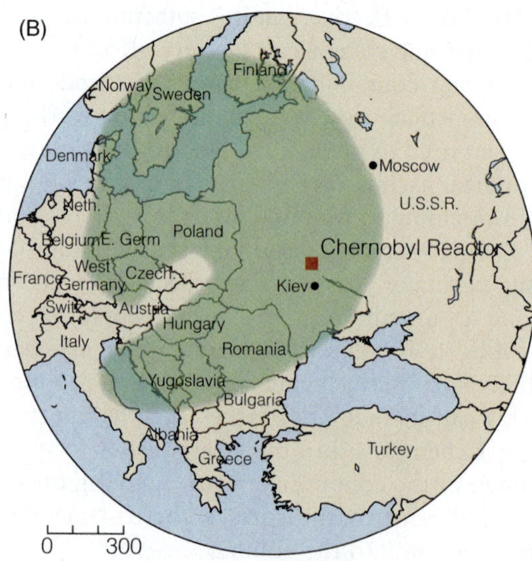

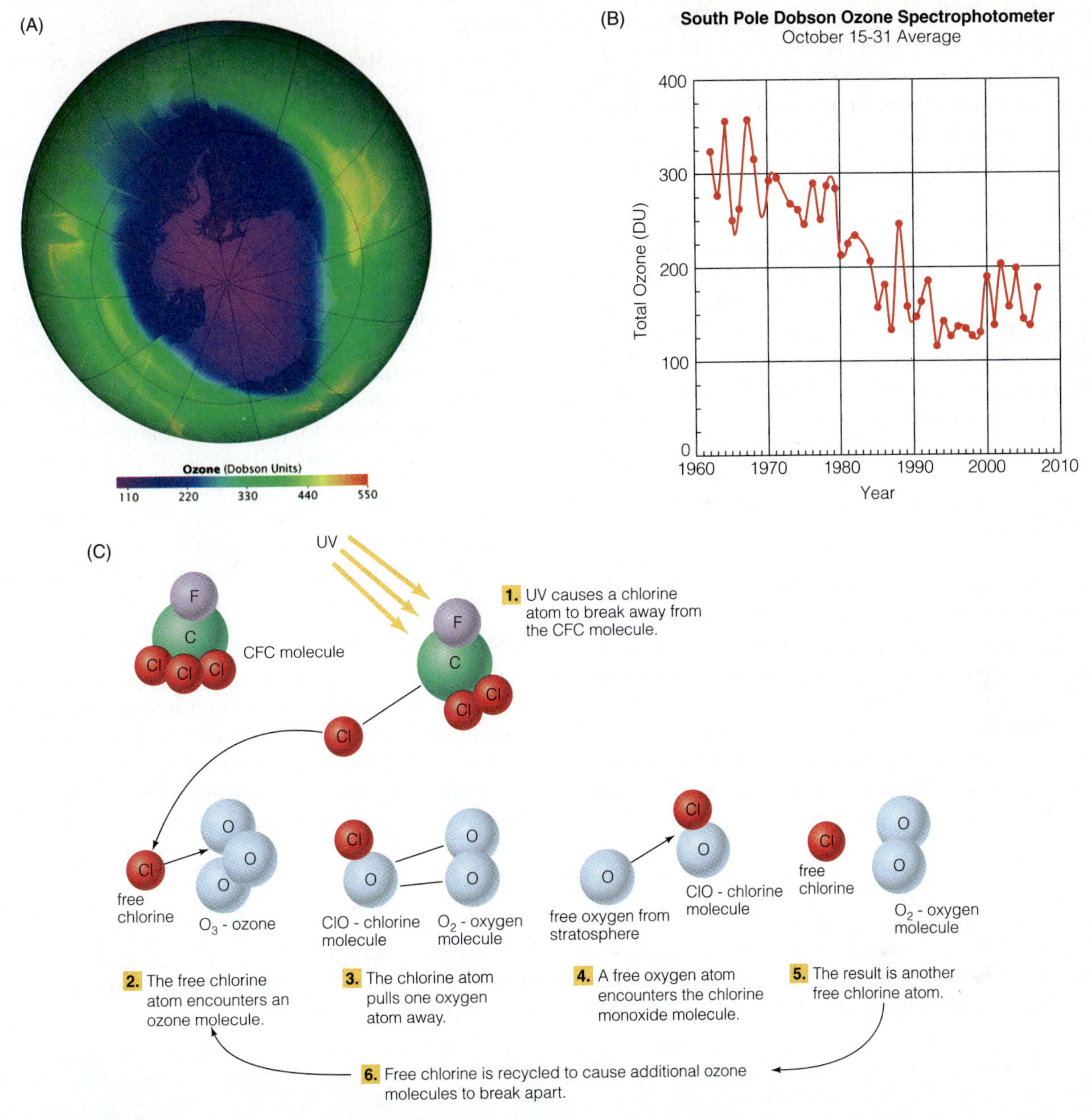

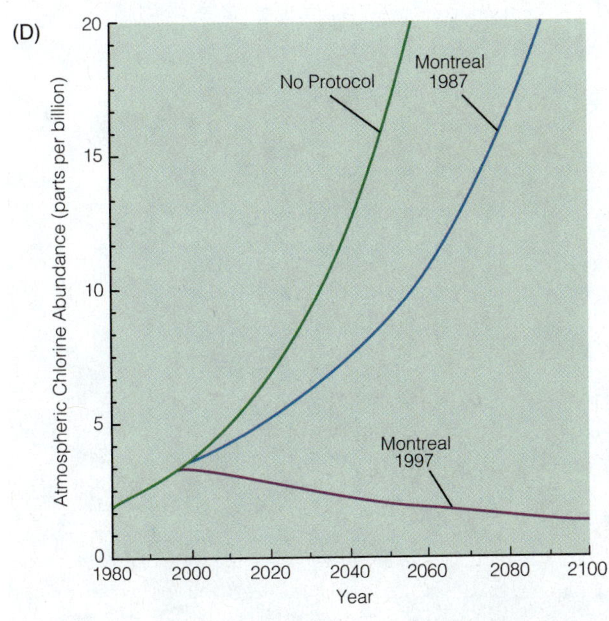

FIGURE 19.10 Stratospheric ozone depletion
The ozone "hole" is an area of depletion of stratospheric ozone. It is most severe during the springtime over the poles, especially Antarctic, as shown here (A) for the year 2006. The purple colors show the most strongly depleted areas. (B) Measurements from Antarctica show a steady depletion in ozone until the early 1990s. Regulations on ozone-depleting substances were implemented through the Montreal Protocol starting in 1987. (C) Chlorine atoms derived from chlorofluorocarbons can be recycled many times—perhaps tens of thousands of times—to cause depletion of ozone molecules in the stratosphere. (D) As a result of measures to cut back on emissions of ozone-depleting substances, stratospheric ozone levels are expected to return to 1980 levels (the date when significant ozone depletion began to be observed) by around 2045 in midlatitude regions and 2080 in Antarctica.

radiation breaks them down, releasing chlorine in the form of chlorine monoxide (ClO). Each chlorine atom is capable of destroying as many as 100,000 ozone molecules before other chemical reactions remove the chlorine from the atmosphere (FIG. 19.10C). Sunlight and very cold temperatures (−80°C or lower) in the upper atmosphere are critical to ozone destruction. These conditions are present in the South Pole region in the springtime, which is why the ozone hole is especially pronounced over Antarctica. In the Arctic, the period of the critical spring conditions is much shorter. With the documentation of ozone depletion in the upper atmosphere, scientists for the first time could show that human activity was having a detrimental global effect on one of Earth's natural systems.

Ongoing ground-based and satellite measurements show that atmospheric ozone concentration has continued to decrease at all latitudes outside the tropics. Following restrictions on use and production of CFCs and related materials that came into effect in 1987, the atmospheric concentration of some CFCs has begun to decline. However, because of the residence time of ozone-depleting substances in the stratosphere, the concentration of ozone itself will continue to decrease, albeit at a slower rate. The complete recovery of the ozone layer to natural conditions is not likely to occur before at least 2050 (FIG. 19.10D). The fact that the ozone layer is expected to recover, and that an international agreement was reached to take actions to achieve the recovery, make this a hopeful example of an environmental success story.

Smog

Ozone in the stratosphere is crucial for the survival of life on Earth, but ozone in the troposphere, the lowest layer of the atmosphere, is detrimental to human health. Ozone at ground level comes from chemical interactions of sunlight, heat, nitrous oxides, and carbon-bearing compounds, the latter two of which primarily result from the burning of fossil fuels. Together, these components comprise the type of air pollution that we know as **photochemical smog.** Smog is common in all cities of the industrialized world, and represents a major worldwide health hazard (FIG. 19.11).

Ozone is also a greenhouse gas. Ground-level ozone and nitrous oxide are increasing annually at rates of 0.5 to 2 percent and 0.3 percent, respectively, and together account for about 13 percent of the greenhouse effect. Nitrous oxide—released by microbial activity in soil, the burning of timber and fossil fuels, and the decay of agricultural residues—has a long lifetime in the atmosphere. Atmospheric concentrations are likely to remain well above preindustrial levels even if we are successful in stabilizing industrial emission rates.

Acid Precipitation

Acid precipitation, or *acid rain*, is rain that has a pH less than that of natural rain. All rain is somewhat acidic; the pH (hydrogen ion activity, a measure of acidity) of natural rain is around 5.6. Rain is naturally acidic because there are trace gases naturally present in the atmosphere, especially CO_2, that combine with water vapor to make acids.

FIGURE 19.11 Photochemical smog
Photochemical smog, which comes from the alteration of precursor pollutants in the atmosphere, has become a concern for human health throughout the industrialized and newly industrializing world. In Mexico City, shown here, the accumulation of smog is exacerbated by the location of the city in a geographic basin surrounded by mountains.

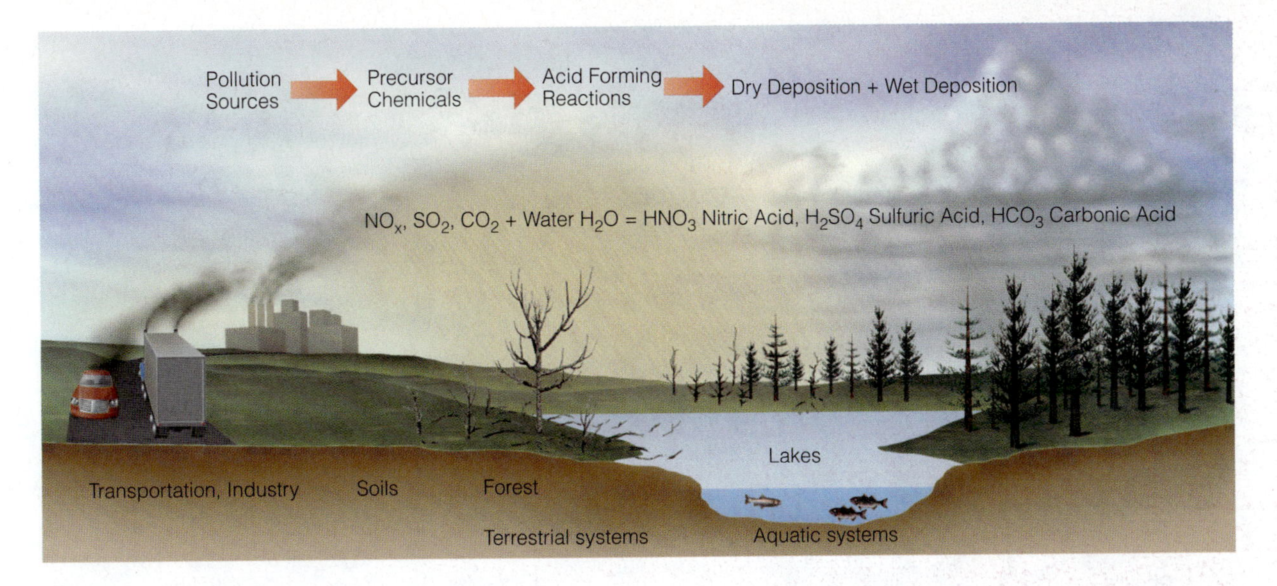

FIGURE 19.12 Acid deposition
Acid deposition in all forms (wet or dry) comes from the chemical interaction of natural chemicals and pollutants, especially nitrous oxides, sulfur dioxide, and carbon dioxide, with water vapor in the atmosphere. Both the precursor pollutants and the resulting acids can be transported far from their source by atmospheric processes.

(Groundwater is naturally slightly acidic for the same reason.) Rain also can become acidic if it interacts with smoke from a forest fire or ash from a volcanic eruption. Acid rain commonly has a pH in the range of 3 to 5, although rains with pH as low as 1.5 have been recorded. Acid rain is a form of *acid deposition*, which may be wet or dry; there have been incidents of acidic snow, and even acidic fogs.

Acids form in the atmosphere when precursor chemicals interact with water (FIG. 19.12). The most common acid precursors are CO_2, SO_x (sulfur oxides, especially SO_2 and SO_3), and NO_x (nitrous oxides). All of these are present in the atmosphere naturally, but their concentrations have increased as a result of anthropogenic emissions, principally from the burning of fossil fuels. Once in the atmosphere, the acid precursors are transported by the wind. They become oxidized, and some of the resulting sulfate and nitrate particles may fall as dry deposition close to the emission source, becoming acidic when they interact with moisture on the ground. The remainder of the airborne material interacts with water vapor in the atmosphere, reacting to become carbonic acid (HCO_3), sulfuric acid (H_2SO_4), or nitric acid (HNO_3). Depending on local and regional weather conditions and circulation patterns, gaseous pollutants and acids can stay in the atmosphere up to two weeks.

The areas that are hardest hit by acid precipitation are downwind of heavily industrialized areas, including northeastern United States and southeastern Canada, eastern and western Europe, and newly industrialized countries in Asia. Acid deposition has negative impacts on both natural and human systems. For example, it can speed the degradation of rubber, paint, and plastics, as well as building materials such as steel, cement, masonry, and building stone. Acid precipitation also has a negative effect on trees, rendering them more susceptible to the effects of cold and disease. Soils and surface waters can become acidified, particularly those underlain by bedrock such as granite, which lacks acid-buffering capability. In aquatic ecosystems, particularly lakes, the acidification of water can cause reproductive failure and death in fish. In some cases, it is possible to remediate an acidified water body, but the best solution is to eliminate the emission of acid precursors from industrial and other sources.

Biosphere: Impacts on Life and Ecosystems

In this chapter we have already touched on some of the global-scale impacts of human activity on the biosphere—it is, of course, impossible to separate them completely from impacts on the atmosphere, hydrosphere, and geosphere. But human activity is causing profound changes directly to the global biosphere; we will look specifically at impacts on forests, fisheries, corals, and global biodiversity.

Loss of Forests

Over the last 4000 years, humans have drastically changed the world's vegetation cover. In major areas of the world the natural vegetation has all but disappeared, replaced by an artificial, managed landscape dominated by agriculture and introduced species. It is estimated that forests still covered a quarter of Earth's land area in 1950. By 1980 the area had been reduced to only one-fifth. Today there are approximately 2.5 billion ha of closed forest and 1.2 billion ha of open woodlands and savannas in the world, virtually none of which is without signs of human intervention.

Although much deforestation has been related to commercial logging and exploitation of agricultural and forest products, in Africa nearly 60 percent of deforestation is due to fuelwood production. Both historically and in modern times, forests have been removed not so much for their resources but to make room for crops. This means that human population pressure plays a direct role

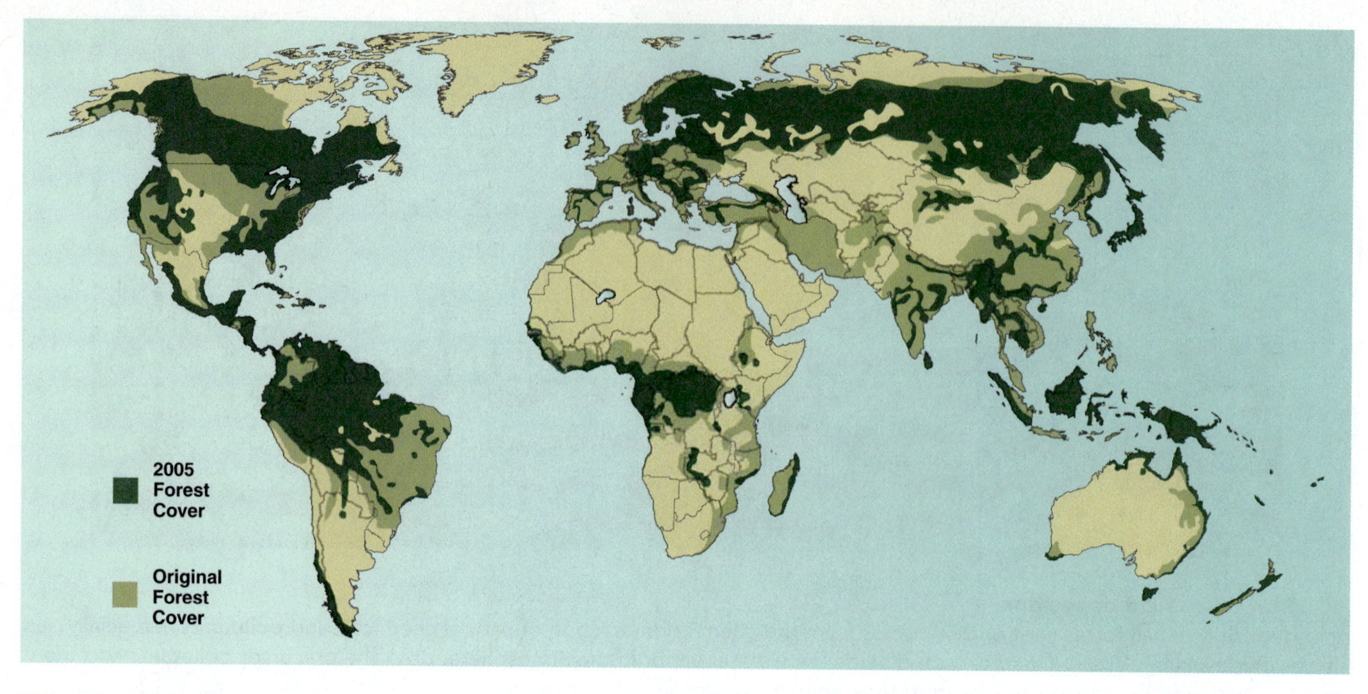

FIGURE 19.13 Changing forest cover
The global forest cover has decreased over time, as shown in this map comparing original forest cover to 2005. Whereas recent forest cover can be accurately assessed using satellite imagery, "original" forest cover is necessarily only an estimate.

in deforestation. Deforestation, in turn, has significant impacts on habitat and biodiversity, as well as on soils, water cycling, and biogeochemical cycling through the atmosphere and hydrosphere.

Forest monitoring is carried out by the Food and Agriculture Organization (FAO) of the United Nations through its Global Forest Assessment. The assessment process combines a variety of ground-based information sources with data from remote sensing to determine the amount, type, and health of forest cover, and detect changes over time (FIG. 19.13). Over the past decade the average amount of forest added each year has been approximately 5.2 million ha, compared to 14.6 million ha of forest area lost each year. In other words, the average net annual change in forested area over the past decade has been a loss of approximately 9.4 million ha.

At present, tropical forests are experiencing the greatest impacts from clear-cut logging and other forms of land clearing, such as slash-and-burn agriculture. In recent decades land clearing has been especially intense in Southeast Asia, parts of Africa, and the tropical Americas. If land clearing continues at current rates, a total of about 40 percent of the Amazonian rain forest may have been lost within two more decades.

When tropical rain forest is cleared, the hydrologic balance is upset because the forest plays a key role in maintaining the balance of the hydrologic cycle. Without forest cover to intercept the rain and promote infiltration, more rainfall runs off the land and far less is recycled through plants back to the atmosphere. As a result, the potential exists for a negative feedback cycle, whereby destruction of forest leads to reduction in the rainfall that the forest requires for its very existence.

Clear-cut logging also has a deleterious effect on soils. When trees are felled and slopes are laid bare, the natural protection provided by tree roots is lost. Hillslopes and logging roads can become sites of increased runoff and erosion, as shown in Figure 19.5. This can cause a dramatic increase in sediment supplied to streams, accompanied by a loss of soil nutrients in the forest itself. The end result is stream channels clogged with debris and disruption of natural ecosystems. To avoid this, new forestry practices, such as selective harvesting, are designed to permit commercial logging to operate with minimum disruption of forest ecosystems.

The great forests of this planet also act as enormous reservoirs for carbon. The impact of the photosynthetic "breathing" of forests is so great that atmospheric carbon dioxide increases and decreases with rhythmic regularity in response to the seasonal growth of trees, particularly in the northern hemisphere. If all the carbon currently held in forests and soils were released back to the atmosphere, the impacts on the atmosphere and the global climate system would be profoundly destabilizing—and yet, we are in the process of doing just that, through widespread clearing of the world's forested lands.

Empty Nets

Living resources in aquatic environments are under increasing pressure. Preserving species and habitat diversity and avoiding overexploitation of these resources present some unique challenges. Marine ecosystems and

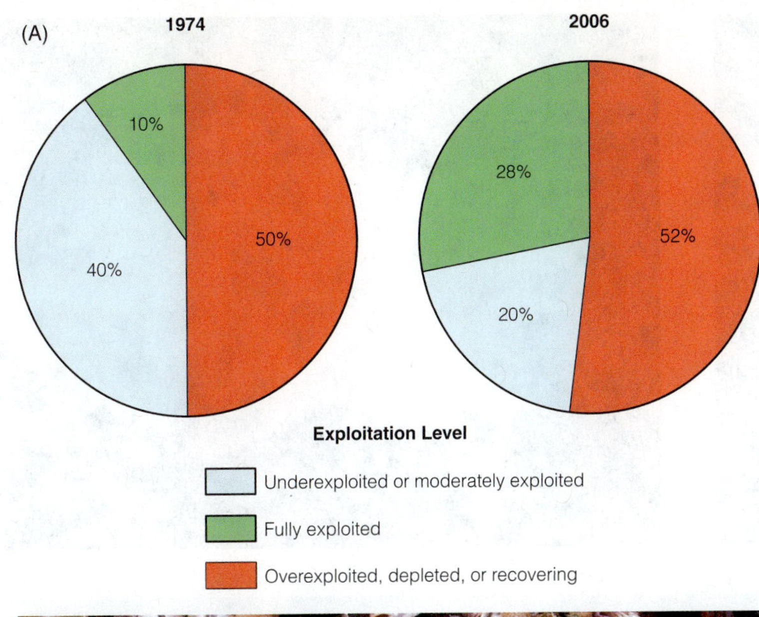

FIGURE 19.14 Depleted stocks
The United Nations Food and Agriculture Organization, which monitors the world's fishery resources, estimates that 80 percent of fish stocks worldwide are now fully exploited to overexploited or depleted, as shown in (A). One cause of the depletion of marine stocks is harmful fishing techniques. Shown here (B) is dead by-catch from shrimp harvesting.

populations are harder to study and monitor, and thus less well understood than most terrestrial ecosystems.

The FAO, which also monitors forest resources, reports that 80 percent of the world's fish stocks are now fully exploited, overexploited, or depleted (FIG. 19.14A). This is mainly due to overfishing, although environmental change and harmful fishing techniques and equipment can also have significant negative impacts on fish stocks. Overexploited stocks can be at risk of permanent damage if steps are not taken to limit overfishing. In most major fisheries today, the allowable catch is closely monitored in an effort to avoid overfishing, but regulation is challenging. Environmental change, both natural and anthropogenic, can also have a significant impact on fish stocks. In 2009 the FAO reported that marine fish stocks, inland fisheries, and aquaculture operations, both marine and fresh water, are all at risk from the impacts of global climatic change.

The impacts of some fishing practices are also of significant global concern. For example, drift nets are enormous nets that are dragged behind a boat, often entangling non-target fish and other marine organisms, such as dolphins. A large proportion of marine organisms that are inadvertently caught by fishers are discarded, dead, as by-catch (FIG. 19.14B). Trawls are huge baglike nets that are dragged along the bottom; they can cause considerable damage to sensitive seafloor habitats, such as coral reefs. To minimize such impacts, fishing gear and practices are more closely monitored and regulated than ever before, as are the size and capacity of fleets.

Coral Bleaching

One of the most productive and sensitive marine ecosystems is the coral reef. We touched briefly on the topic of corals in previous chapters; here we are concerned with their role as sensitive indicators of change in the marine environment.

Coral reefs are formed by colonies of tiny animals, which coexist in a symbiotic relationship with single-celled, photosynthetic algae called *zooxanthellae* (pronounced "zoe-zan-thell-ay"). The zooxanthellae require sunlight for photosynthesis, providing the corals with food energy. The corals, in turn, provide nitrogen—a plant nutrient—to the zooxanthellae. Although corals can exist without zooxanthellae, their presence stimulates the corals to build calcareous shells, which eventually accumulate into reefs; corals without zooxanthellae do not build reefs.

To remain healthy (FIG. 19.15A), corals require clear water in which the temperature remains above 18°C. Because of their very specific light and temperature requirements, corals are highly susceptible to damage from human activities, as well as from natural causes such as tropical storms and tsunamis. For example, industry and development in coastal zones can lead to soil erosion; the resulting sediment clouds coastal waters, inhibiting photosynthesis by the zooxanthellae. When this happens, the corals may either die or expel the zooxanthellae, turning ghostly white in the process (FIG. 19.15B); this is called **coral bleaching**. Coral bleaching is an important indicator of the status of ocean waters, and it is spreading dramatically (FIG. 19.15C).

Ocean acidification also threatens the health of corals worldwide. As the carbon dioxide content of the atmosphere increases, more CO_2 is dissolved by ocean water. In this way the ocean acts as a *sink* (Chapter 1) for carbon dioxide. The increasing carbon dioxide content of ocean water is leading to an increase in carbonic acid (H_2CO_3), which is formed when carbon dioxide reacts chemically

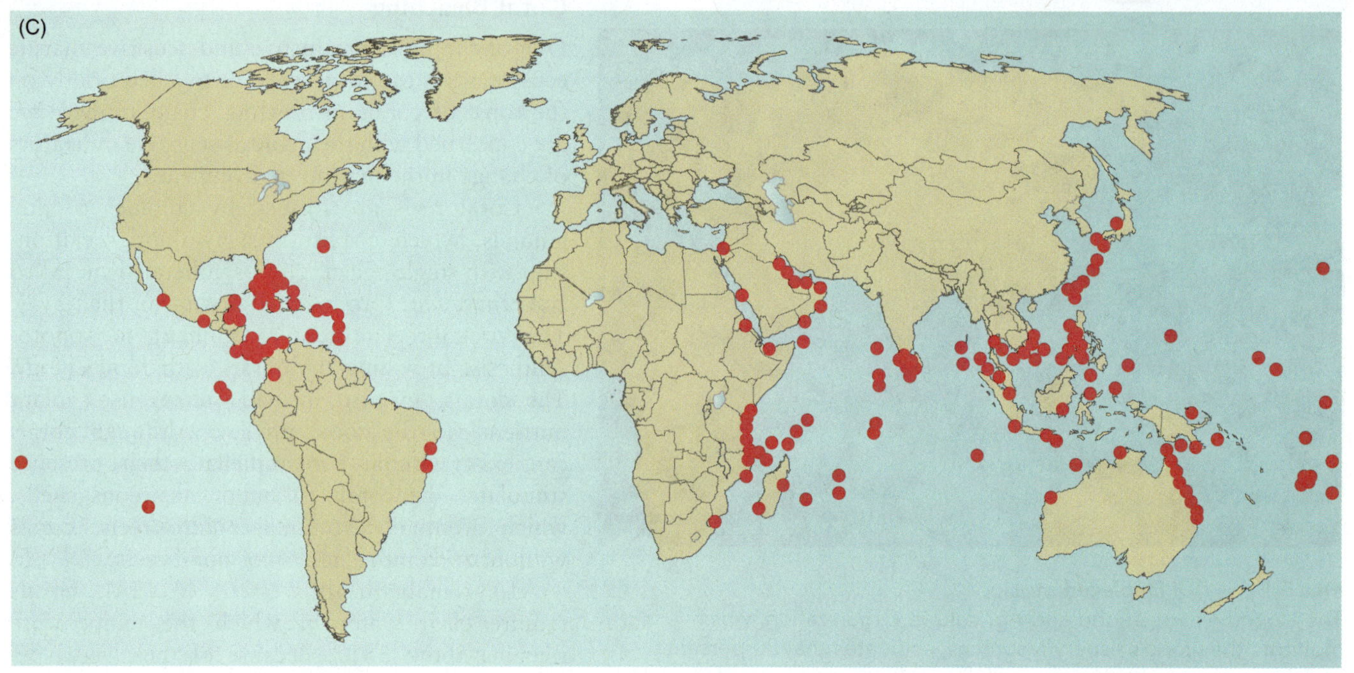

FIGURE 19.15 Coral bleaching
Coral reefs like this one from West Papua, Indonesia (A) are extremely productive and biologically diverse habitats. Corals are highly sensitive to environmental changes such as increasing water temperature, acidity, and cloudiness, and they may exhibit stress by undergoing bleaching, as shown here (B) in a sample from the South Male Atoll, Maldives. Coral bleaching is occurring in many locations throughout tropical and subtropical seas, as shown on this map (C).

with water. As a result, the upper layer of the ocean is acidifying. This has a direct impact on any marine organism that builds its shell or other hard body parts out of calcium carbonate ($CaCO_3$), including corals. Calcium carbonate becomes more soluble (more easily dissolved) as the acidity of the water increases, causing organic structures that are composed of this material to be vulnerable to dissolution.

Species at Risk

Biodiversity is increasingly under threat. New species arise as a result of *evolution*, *natural selection*, and *speciation*, and they can disappear forever through *extinction* (Chapter 14). Extinction is not a new phenomenon; throughout the history of life on this planet, species have gone extinct at a rate of roughly one per year. Overall, in the geologic history of life on this planet, the evolution of new species has outpaced extinction. Today, however, we are in the midst of an extinction that rivals or exceeds the great mass extinctions of our geologic past; in Chapter 14 we referred to this as the "sixth great extinction." This extinction is caused by human actions rather than by a natural catastrophe. Scientists estimate that in the past 200 years the rate of species extinctions has soared to thousands per decade—many times higher than the natural rate of extinction.

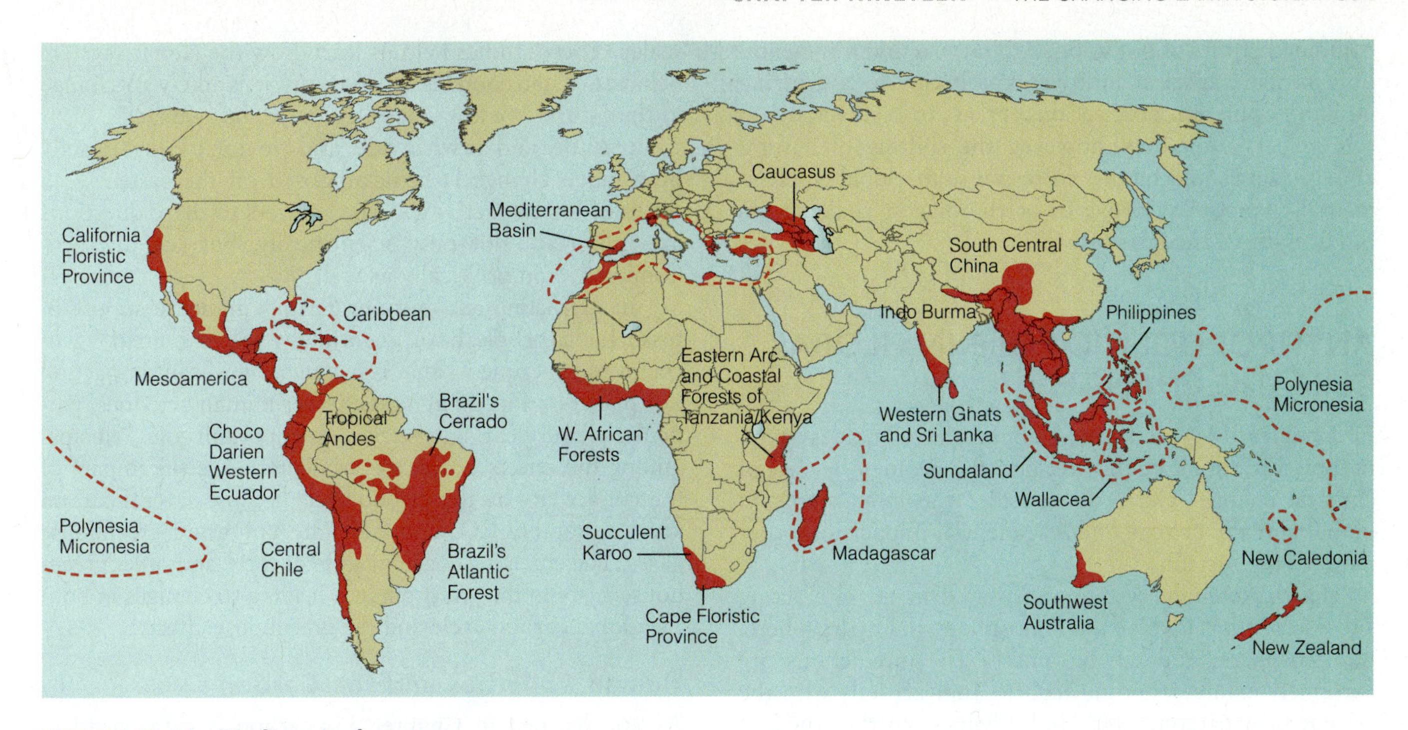

FIGURE 19.16 Biodiversity hotspots
Biodiversity clusters in certain "hotspots" around the world, as shown here. Unfortunately, these important areas don't always coincide with areas that are protected.

Endangered species can be protected by national laws, such as the Endangered Species Act in the United States, or by international agreements, such as the United Nations Convention on International Trade in Endangered Species (CITES). About 2 percent of the world's wilderness land area is protected by international agreements; unfortunately, it's not the same land area that hosts the greatest biodiversity. One of the interesting things about biodiversity is that it tends to cluster in certain highly productive areas—**biodiversity "hotspots."** These areas, mostly tropical, represent just 2 percent of the world's land surface, but collectively hold at least 44 percent of all vascular plant species and 38 percent of all vertebrate animal species (FIG. 19.16).

In Chapter 16 we looked at habitat loss and fragmentation as the single greatest threat to biodiversity. Another significant threat is *poaching*, the illegal harvesting of wild species for commercial exploitation. Animals are killed for everything from their fur to their horns, feathers, or body parts used as trophies or for reputed medicinal treatments. A rhinoceros or an elephant, for example, is likely to be killed by poachers who are only interested in the horn or tusk of the animal. Many exotic fish and birds are captured live and sold to collectors and pet owners. Often they are caught and transported illegally, and many die in the process. Plants, too, are harvested illegally, usually for medicinal or aesthetic purposes. For example, the popularity of cacti for arid landscaping (called *xeriscaping*) has led to the poaching of cacti from public lands; in some cases, the plants are hundreds of years old and may be rare or endangered. Poaching is difficult to control because it requires intensive policing of wilderness areas. Poachers have strong incentives because their harvest may bring a staggering price on the black market.

Another important threat to biodiversity is the introduction of exotic species into an ecosystem. Although this can happen naturally, accidental or purposeful importation is common and is increasing as a result of global transportation. Lacking its natural predators, an exotic species may take hold and flourish in the new environment, becoming an *invasive species* (Chapter 16) that outcompetes native or *endemic species*.

There are many unfortunate examples. The African so-called killer bee was imported to South America intentionally, in an effort to increase the honey production of domestic bees; the aggressive insect has now spread far into the United States. Purple loosestrife was originally grown as an ornamental flower, but it "escaped" from gardens and has spread into wetlands across the northern United States and southern Canada, choking and crowding out native plants. Zebra mussels are thought to have hitched a ride on a cargo vessel across the Atlantic Ocean and into the Great Lakes, where they have proliferated, covering all available solid surfaces and clogging intake pipes. The northern snakehead, a fish native to northern China, was imported for medicinal purposes. The fish is very aggressive and eats just about everything in its path; worse, it can survive and travel, pulling itself along on its fins, for up to three days on dry land. It is becoming increasingly widespread in streams of the eastern United States.

Pollution also contributes to the loss of biodiversity. Acid precipitation and ozone depletion, in particular, are known to have detrimental effects on organisms. Many animals have choked on discarded plastic and other debris, especially in the marine environment. The combined effects of pollution, habitat loss and fragmentation, invasive species, and overexploitation by a growing human population pose serious threats to biodiversity on this planet. Ultimately, however, the shifting of natural climatic zones and biomes in response to global environmental change may prove to be the biggest challenge to biodiversity in the near future.

ANTHROPOGENIC ROLE IN GLOBAL CLIMATIC CHANGE

In Chapter 13 you learned about the climate system, natural variations in climate, and the geologic record of climatic change. Here, we consider the human role in global climatic change and its potential impacts on other parts of the system.

Earth's climate system consists of many interacting subsystems that involve the atmosphere, the hydrosphere, the geosphere, and the biosphere. The interactions are extremely complex and difficult to analyze; only with the advent of supercomputers and advances in the study of paleoclimates have we begun to answer some of the basic questions about how the climate system works. By reconstructing past climates, scientists are trying to determine the range of natural climatic variability on different time scales. These studies help us learn how the climate system behaves, what controls it, and how it is likely to change in the future. We know it will change, but we lack a clear view of how and at what rate, and we don't know exactly how much change is being imposed on the system as a result of human activities. The answers to these questions are important not only scientifically, but socially, economically, and politically as well.

In extracting and burning Earth's immense supply of fossil fuels, people have unwittingly begun a great "geochemical experiment" that is having a significant impact on our planet and its inhabitants. Other human activities, particularly land clearing, also play an important role. Principal among the effects of this experimentation is the impact of human activity on the carbon cycle because several carbon species (notably CO_2 and CH_4) are important greenhouse gases. The changing concentration of these and other greenhouse gases in the atmosphere is leading to changes in both the global carbon cycle and the greenhouse effect.

Human Activities and the Carbon Cycle

As you learned in Chapter 15, carbon is transferred in biogeochemical cycles among the four major reservoirs of the Earth system. Human activities change these reservoirs and influence the exchanges of carbon between them (**FIG. 19.17**).

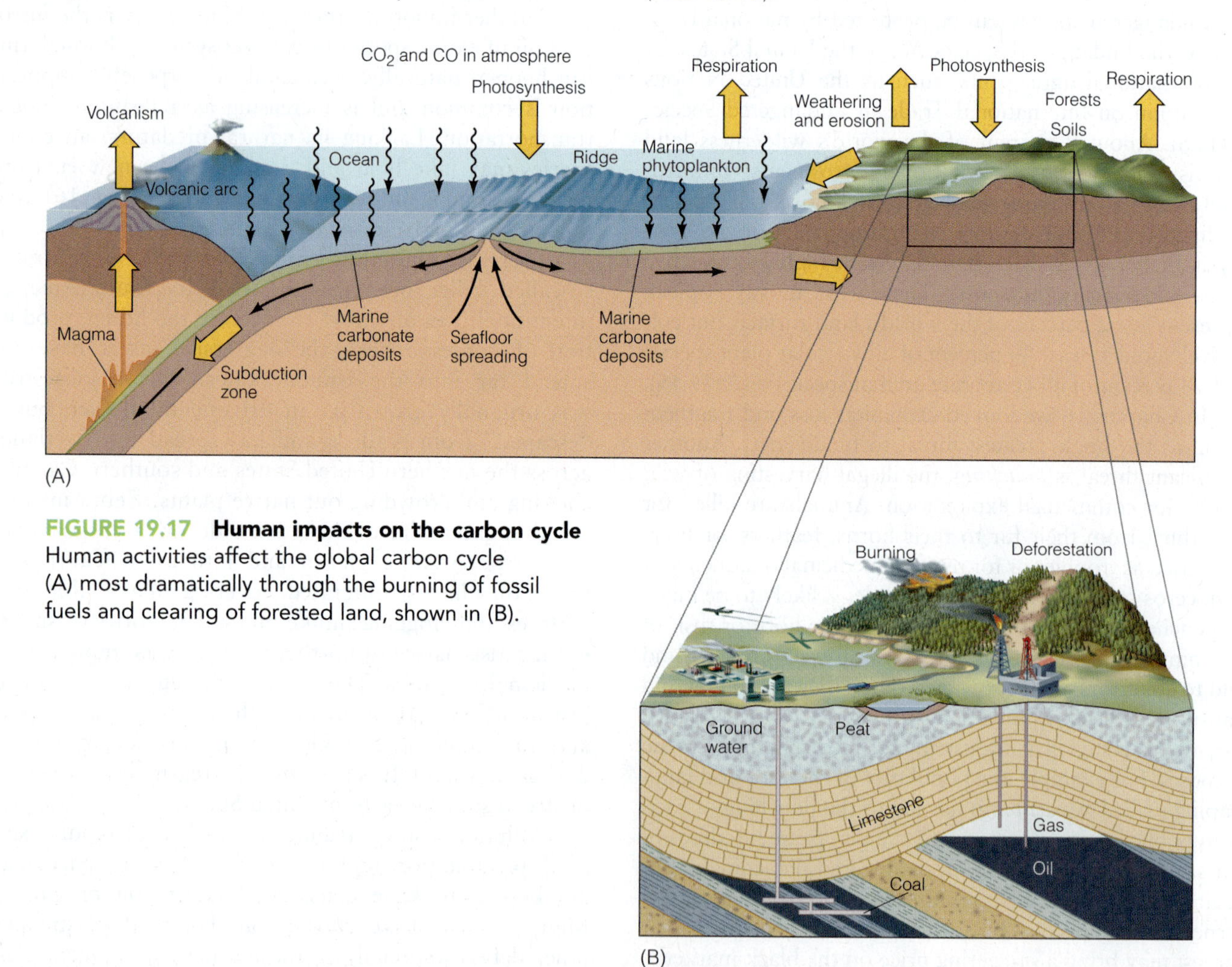

FIGURE 19.17 Human impacts on the carbon cycle Human activities affect the global carbon cycle (A) most dramatically through the burning of fossil fuels and clearing of forested land, shown in (B).

For example, the burning of fossil fuels and the clearing of forested land cause carbon (primarily in the form of CO_2) to move from the geosphere and the biosphere to the atmosphere, at rates much faster than they would move naturally. An individual action—the clearing of a small patch of forest, or the emissions from one coal-burning power plant—may appear insignificant; however, the cumulative effects of human activities are now so great as to be measurable on a global scale.

Unless this additional CO_2 is dissolved in the hydrosphere or is buried and sequestered in sediments as fast as it is generated, the CO_2 content of the atmosphere must inevitably increase. However, the rate at which natural processes are removing CO_2 from the atmosphere is slower than the rate at which human activities are adding it, which led scientists many years ago to conclude that the CO_2 content of the atmosphere might be increasing. Indeed, such a change has now been measured, and it has been accompanied by other important changes in atmospheric chemistry.

Changes in Atmospheric Chemistry

Earth's atmosphere has changed as a result of human activity, particularly since the Industrial Revolution (FIG. 19.18); this finding is not controversial. Some changes have resulted from the emission of gases that are of wholly anthropogenic origin, such as increases in chlorofluorocarbon compounds that contribute to the breakdown of ozone in the stratosphere. Other changes have resulted from human activities that cause the mobilization of naturally occurring compounds, such as the release of sulfur and carbon compounds by the burning of fossil fuels. Among these, greenhouse gases have received increasing scientific and public attention over the past two decades, as it has become clear that their atmospheric concentrations are rising.

While it may be troubling, in itself, that human action has had a measurable effect on the chemistry of the atmosphere, our real concern in this context is the impact of these changes on the stability of Earth's climate system. As discussed in Chapter 2, the greenhouse effect makes Earth habitable. Without it, the surface of our planet would be as inhospitable as the surfaces of the other planets in the solar system. The greenhouse effect results from the absorption of outgoing heat from Earth's surface by *radiatively active* gases in the atmosphere. When changes occur in the atmospheric concentrations of greenhouse gases, whether from natural or anthropogenic causes, there is a resulting change in the radiative balance maintained by the greenhouse effect.

Water vapor is the most important gas in the natural greenhouse effect. Human activity affects the water content of the atmosphere in complex ways that are not thoroughly understood. When considering the human contribution to the greenhouse effect, also known as the **anthropogenic** (or *accelerated* or *enhanced*) **greenhouse effect**, we are primarily concerned with carbon dioxide (CO_2) because of the magnitude of our emissions, and methane (CH_4), because of its extreme efficiency as a greenhouse gas. Let's consider the cause of recent changes in atmospheric concentration of these two important gases.

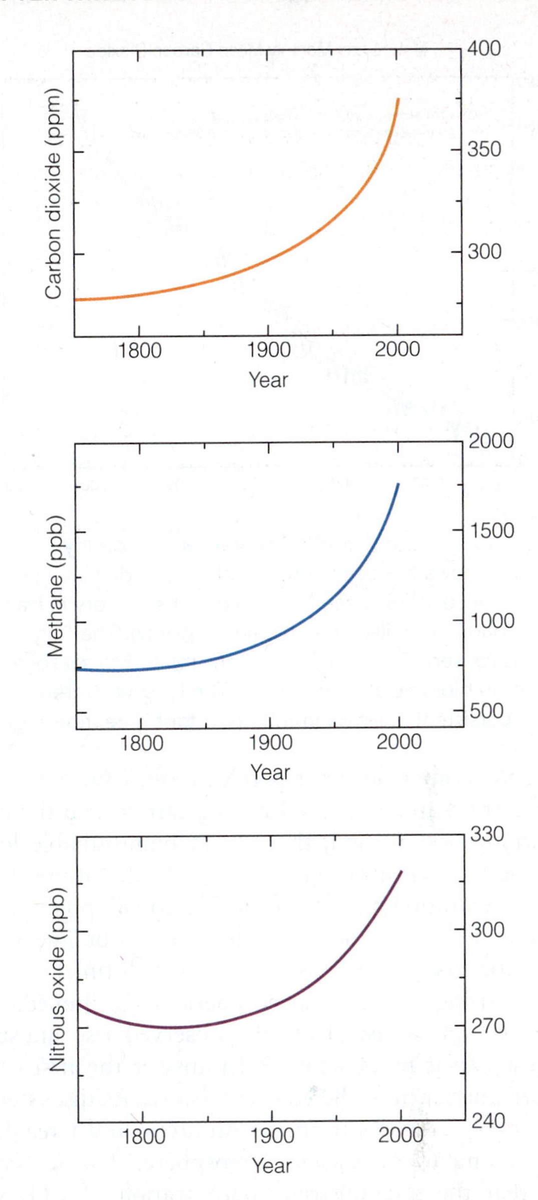

FIGURE 19.18 Changing atmospheric chemistry Atmospheric concentrations of carbon dioxide, methane, nitrous oxides, and several other key constituents have increased dramatically over the past 200 or so years, since the Industrial Revolution.

CARBON DIOXIDE. Beginning in 1958, the carbon dioxide concentration in the atmosphere has been measured at a site near the top of Mauna Loa, a dormant volcano in Hawaii. This site was chosen because of its altitude and remote location, far from ground-level sources of atmospheric pollution. The measurements show two remarkable things. First, the amount of CO_2 fluctuates regularly with an annual rhythm (FIG. 19.19). In effect, Earth (or, more correctly, the biosphere) is "breathing." During the northern hemisphere growing season, CO_2 is absorbed by vegetation, and the atmospheric concentration falls; then,

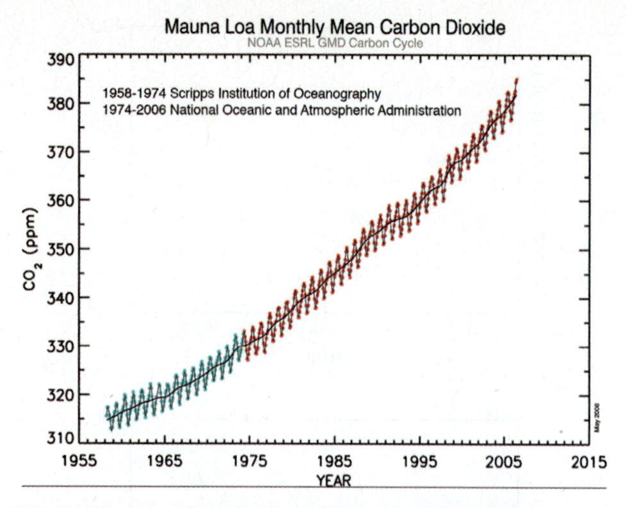

FIGURE 19.19 Carbon dioxide varies seasonally This graph shows the concentration of carbon dioxide in dry air, measured since 1958 at the Mauna Loa Observatory in Hawaii (given in parts per million by volume, or ppmv). The "zig-zag" pattern is caused by annual fluctuations related to seasonal variations in biologic uptake of CO_2. The long-term trend shows a persistent increase in this important greenhouse gas.

during the winter dormant period, more CO_2 enters the atmosphere than is removed by vegetation, and the concentration rises. Second, there is an unmistakable long-term rise in concentration. Since 1958, the atmospheric CO_2 concentration has risen from 315 to 387 ppmv (parts per million, by volume), and the rise is not linear but exponential (i.e., the rate is increasing with time).

The rising curve of atmospheric CO_2 immediately raises two questions: (1) Is the observed rise unusual? (2) How can it be explained? To answer the first question, we must turn to the geologic record. As discussed in Chapter 13, ice cores from the Antarctic and Greenland contain samples of ancient atmosphere. The ice cores show that the preindustrial concentration of CO_2 was close to 280 ppmv, the typical value for an interglacial age. The subsequent rapid increase to 387 ppmv during the last 200 years is unprecedented in the ice-core record and implies that something very unusual is taking place.

This increase in CO_2 closely resembles the increase in the rate at which carbon has been released to the atmosphere by the burning of fossil fuels. No known natural mechanism can explain such a rapid increase; the burning of fossil fuels must be a primary reason for the observed increase in atmospheric CO_2. Additional contributing factors are widespread deforestation, with its attendant burning and decay of cleared vegetation, and the use of wood as a primary fuel in many underdeveloped countries that have rapidly growing populations.

METHANE. Methane gas (CH_4) absorbs infrared radiation 25 times more effectively than CO_2, making methane an important greenhouse gas despite its relatively low atmospheric concentration. The concentration of atmospheric methane has increased by about 150 percent since the Industrial Revolution (from 700 ppb to about 1745 ppb), at an average rate of about 0.8 percent per year. Methane levels for earlier times obtained from ice-core studies show an increase that essentially parallels the rise in the human population. This is not surprising, for much of the methane now entering the atmosphere is generated either by biologic activity related to rice cultivation or as a by-product of the digestive processes of domestic livestock, especially cattle. The global livestock population has increased greatly in the past two centuries, and the total acreage under rice cultivation has increased more than 40 percent since 1950. In prehistoric times, methane levels, like CO_2 levels, increased and decreased along with processes involved in the onset and end of glacial cycles.

Assessing Climatic Change

These changes in atmospheric chemistry are well documented, and the links to human activity since the Industrial Revolution seem indisputable. What is much less certain, however, are the potential future impacts of the changes. If the atmospheric concentration of greenhouse gases is rising, what does this portend for future climate? Will there be a significant anthropogenic contribution to climatic change in the near future? Does it mean that Earth's surface temperature is warming, and, if so, by how much and at what rate? To try to find answers to these questions, we first look at the recent historical record of global temperatures, and then we see how forecasts of the future can be made.

Historical Temperature Trends

As discussed in Chapter 13, assessing global changes in temperature is a very difficult task because few instrumental measurements were made before 1850, and the majority date to the time since World War II. The earliest records are from western Europe and eastern North America. Data for oceanic areas, which encompass 70 percent of the globe, are sparse, especially prior to 1945. Therefore, most "global" temperature curves are reconstructed primarily from land stations located mainly in the northern hemisphere.

Numerous graphs of average annual temperature variations since the middle- to late-nineteenth century have been published. Although they differ in detail, they all show one characteristic feature: a rise in temperature over the course of the twentieth century (FIG. 19.20). Although short-term departures from this trend are evident, the overall increase from 1860, when reliable hemispheric-wide records began, to 2010 appears to be about 0.5 to 0.8°C. Because the interval of rising temperatures coincides with the time of rapidly increasing greenhouse gas emissions, it is tempting to draw the immediate conclusion that the two trends are causally related. However, temperature reconstructions prior to 1950 are based on relatively few data, unequally distributed across the globe, so it is very difficult to make a conclusive scientific case on this basis alone.

Future Climatic Change

It is of obvious importance to human society to be able to apply the scientific understanding of Earth's climate

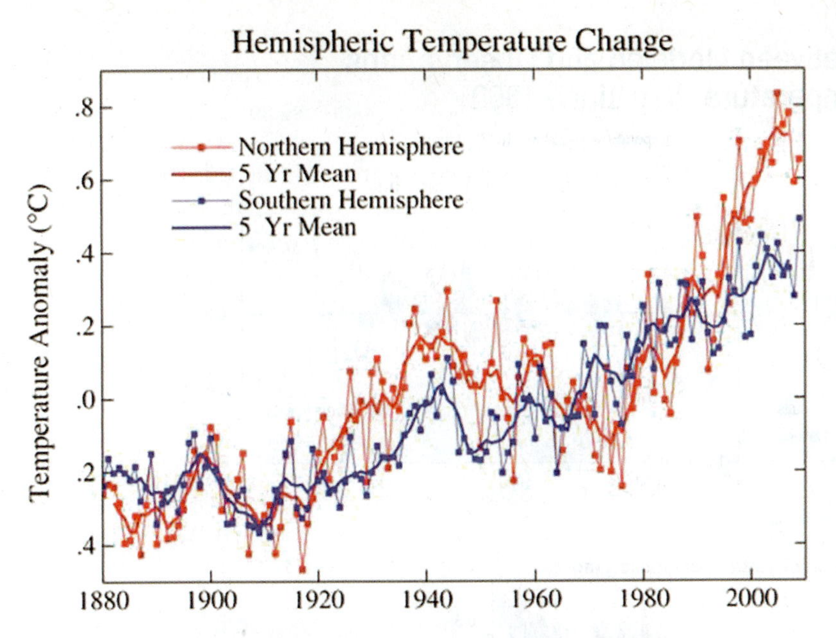

FIGURE 19.20 Surface temperature anomalies since 1880 It is extremely challenging to reconstruct average surface temperatures, but most data suggest an increase of at least 0.6°C in the long-term trend. Here data showing average surface temperatures in the Northern and Southern Hemispheres are compared to a baseline (the 1951-1980 average). The Southern Hemisphere shows an overall increase of about 0.6°C; the Northern Hemisphere increase is slightly greater.

system to the prediction of future climatic changes. The projection of future rates of warming is dependent on a number of basic uncertainties: How rapidly will concentrations of the greenhouse gases increase? How, and how rapidly, will the oceans, a major reservoir of heat and a fundamental element in the climate system, respond to changing climate? How will changing climate affect ice sheets and cloud cover? What is the range of natural variations in the climate system on the century timescale?

The potential complexity inherent in such questions is well illustrated by clouds. If the temperature of the lower atmosphere increases, more water will evaporate from the oceans. Water is a greenhouse gas, and this may contribute to further warming. The increased atmospheric moisture will create more clouds, but clouds reflect solar energy back into space; this may have a cooling effect on the surface air and consequently a result opposite to that of the greenhouse effect.

Because of such uncertainties and complexities, scientists are reluctant to make firm forecasts, and they tend to be cautious in their predictions. They qualify their conclusions with adjectives like "possible," "probable," and "uncertain." To the scientists who carry out this research, these terms have very specific, clearly defined meanings, and they represent an appropriate level of acknowledgement of the scientific uncertainty involved in the study of Earth's climate system. However, their understandably cautious approach has complicated the public discourse on climate change; it emphasizes the gap between what we know about the Earth system and what we would like to know, and it points to the many challenges that face both the scientists who study global change and the policymakers who must manage it.

CLIMATE MODELS. If the historical temperature record is judged to be inconclusive, we can still explore the linkages between greenhouse gas emissions and present and future climate by turning to models of the climate system. Three-dimensional mathematical models of Earth's climate system are an outgrowth of efforts to forecast the weather. The most sophisticated are **general circulation models (GCMs)** that link atmospheric, hydrospheric, and biospheric processes. The sheer complexity of these natural systems means that such models, of necessity, are greatly simplified representations of the real world. Furthermore, as mentioned, many of the linkages and processes in the climate system are still poorly understood and therefore difficult to model. For instance, the models do not yet adequately portray the dynamics of ocean circulation or cloud formation, two of the most important elements of the climate system. Also missing are many of the complex biogeochemical processes that link climate to the biosphere.

Despite these limitations, GCMs have been successful in simulating the general character of present-day climates and have greatly improved weather forecasting. One of the most powerful tests of climate models has been their use in "predicting the present." The idea is to input into the model conditions of past climates as we know them to have existed, then run the model to see if it successfully predicts the climate conditions of the present day (FIG. 19.21). The success of GCMs in "predicting the present" enhances the confidence of scientists in their ability to predict climatic change. Scientists have applied the results of climate modeling to the development of predictions about future changes that we may expect in our climate system.

SCIENTIFIC CONSENSUS AND THE IPCC. Despite the uncertainties inherent in climate modeling and data analysis, the strong scientific consensus is that (1) human activities have led to increasing atmospheric concentrations of carbon dioxide and other trace gases that have enhanced the greenhouse effect; (2) global mean surface air temperature has increased by up to 0.8° during the last 100 to 150 years, an increase that appears to be the direct result of the enhanced greenhouse effect; and (3) during the next century global average temperature will likely continue to increase.

The most thoroughly reviewed and widely accepted synthesis of scientific information concerning climate change is a series of reports issued by the **Intergovernmental Panel on Climate Change (IPCC)**. This international panel of scientists and government officials was established

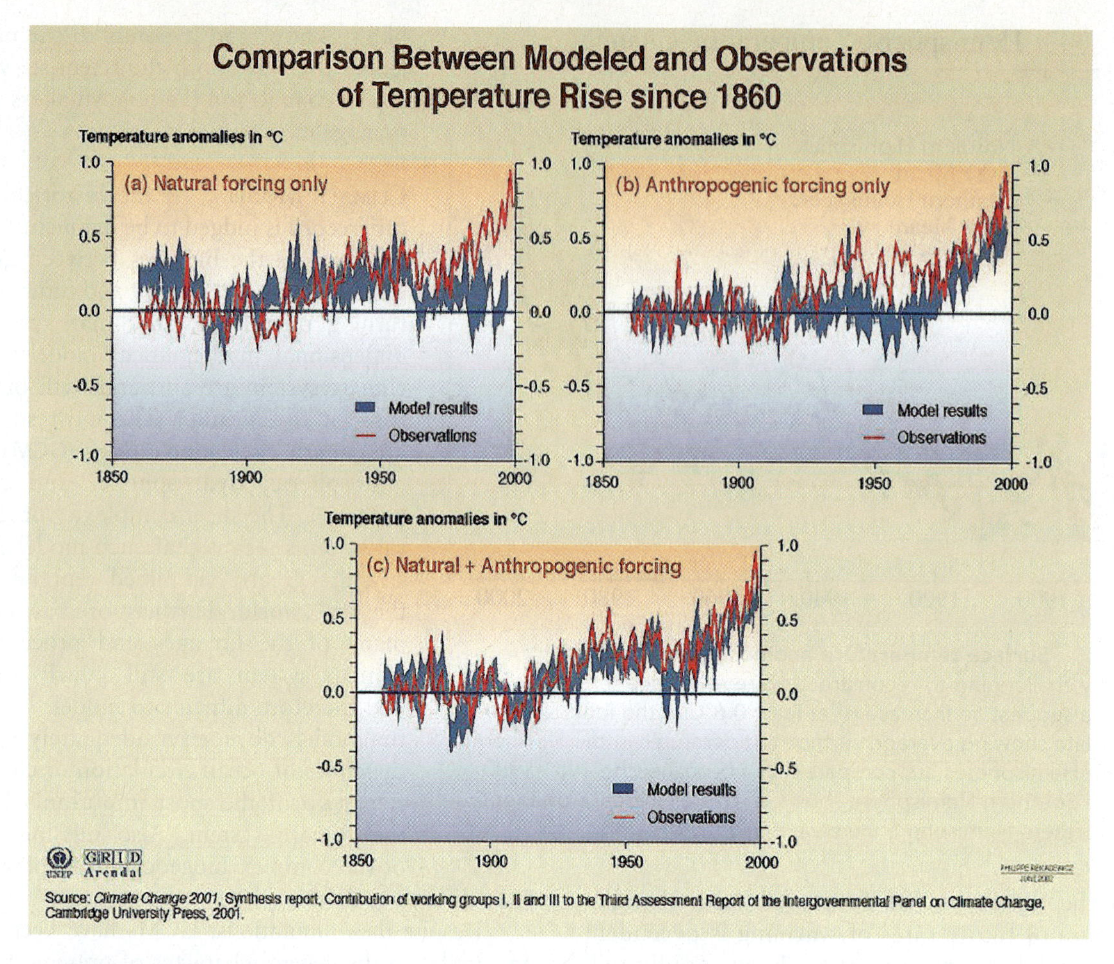

FIGURE 19.21 Predicting the present
Scientists test climate models by entering climate data from past years and comparing the model predictions (blue areas) with actual observations (red lines). Models that incorporate only natural factors (A) or only anthropogenic factors (B) do not predict real climate trends as closely as models that incorporate both (C). (*Source:* IPCC 3rd Assessment Report, 2001).

in 1988 by the United Nations Environment Program (UNEP) and the World Meteorological Organization.

In 2007 the IPCC released its *Fourth Assessment Report*, which represents the consensus of scientific climate research from around the world. This report summarizes many thousands of scientific studies, and it documents observed trends in surface temperature, precipitation patterns, snow and ice cover, sea levels, storm intensity, and other factors (TABLE 19.1). It also predicts future changes in these phenomena after considering a range of potential scenarios for future greenhouse gas emissions. The report also addresses the impacts of current and future climate change on wildlife, ecosystems, and human societies, as well as strategies that we might pursue in response to climate change.

As with all scientific endeavors, the IPCC must deal in uncertainties; its authors have therefore assigned statistical probabilities to all conclusions and predictions. In addition, estimates regarding impacts of change on human societies are conservative because its scientific conclusions had to be approved by representatives of the world's national governments, some of which are reluctant to move away from a fossil-fuel-based economy.

The IPCC report concludes that average surface temperatures on Earth increased by an estimated 0.74°C in the century from 1906 to 2005, with most of this increase occurring in the last few decades. Eleven of the years from 1995 to 2006 were among the 12 warmest on record since global measurements began 150 years earlier. In the future, we can expect average surface temperatures on Earth to rise by at least 0.2° C per decade for the next two decades, according to IPCC analysis.

The temperature increase related to the continued release of greenhouse gases will be larger and more rapid than any experienced in human history—indeed, larger and more rapid than anything in the geologic record. If we were to cease greenhouse gas emissions today, temperatures would still rise 0.1°C per decade because of the time lag from gases already in the atmosphere that have yet to exert their full influence. The IPCC predicts that by the end of the twenty-first century global temperatures will be at least 1.8–4.0°C higher than today's, depending on the emission scenario. Temperature change will not be uniform throughout the world, but is predicted to vary from region to region in ways that parallel existing regional differences. Projected temperature increases are

TABLE 19.1 Projected impacts of global warming
This table provides examples of impacts associated with projected global average surface warming. Examples of global impacts are projected for climate changes (and sea level and atmospheric CO_2 where relevant) associated with different amounts of increase in global average surface temperature in the 21st century. The black lines link impacts; broken-line arrows indicate impacts continuing with increasing temperature. The left-hand side of text indicates the approximate level of warming that is associated with the onset of a given impact. Adaptation to climate change is not included in these estimations. Confidence levels for all statements are high. (*Source:* Figure SPM-7 from IPCC 4th Assessment Report, *Climate Change 2007: Synthesis Report*. http://www.ipcc.ch/publications_and_data/ar4/syr/en/spms3.html).

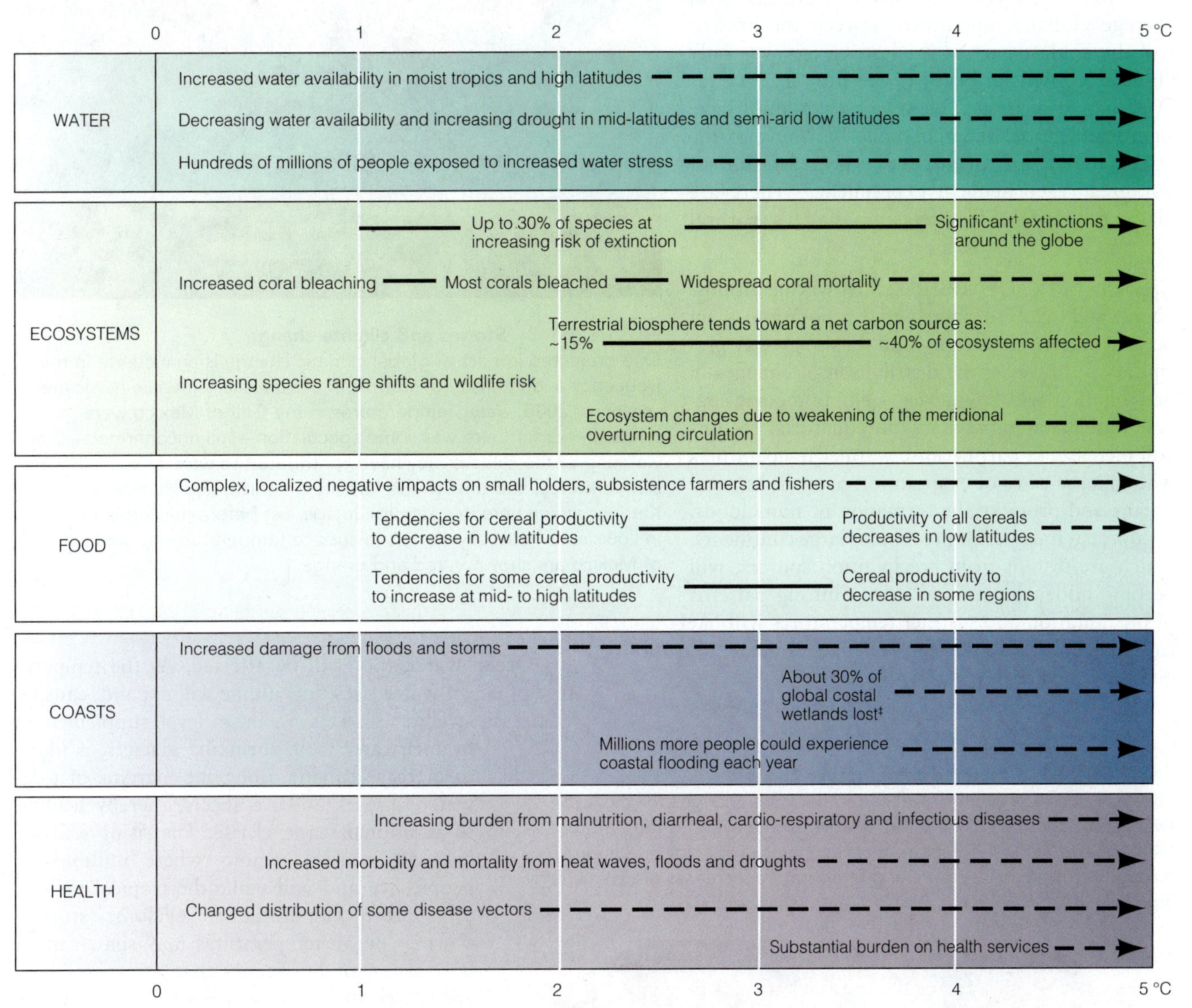

† Significant is defined here as more than 40%. ‡ Based on average rate of sea level rise of 4.2 mm/year from 2000 to 2080.

as great as 7.5°C in some areas, notably the Arctic (FIG. 19.22). It is likely that both average and extreme temperatures will increase.

Potential Impacts of Climatic Change

An increase in global surface air temperature by a few degrees does not sound like much; surely we can put up with this rather insignificant change. However, if we consider that the difference in average global temperature between the present and the coldest part of the last ice age was only about 5°C, we can begin to see how a temperature change of even a degree or two could have significant global repercussions. Warming of the atmosphere is just one of the projected results of our great "geochemical experiment." There are many potential side effects on Earth's natural systems.

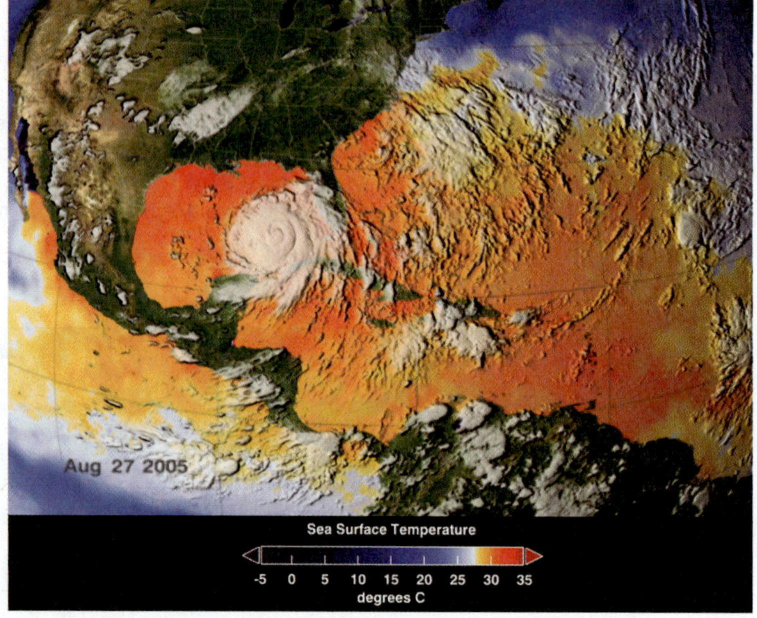

FIGURE 19.23 Storms and climate change
One projected impact of global climatic change is an increase in the frequency and intensity of storms. In the unusually intense hurricane season of 2005, water temperatures in the Gulf of Mexico were unusually warm; there was some speculation—still unconfirmed—that warming of the climate may have contributed to warmer sea-surface temperatures, and thus to the ferocity of storms like Hurricane Katrina, shown here in a satellite image just before making landfall in Louisiana. Unusually warm sea-surface temperatures in the Gulf of Mexico are shown in red and orange.

CHANGES IN THE HYDROSPHERE. A warmer atmosphere will lead to increased evaporation from oceans, lakes, and streams, and thus to greater precipitation; however, the distribution of changes in precipitation will be uneven. Climate models suggest that precipitation rates in equatorial regions will increase, in part because warmer temperatures will increase rates of evaporation over the tropical oceans and promote the formation of rain clouds. In contrast, the interior portions of large continents, which are distant from precipitation sources, will become both warmer and drier. Shifting patterns of precipitation and warmer temperatures will likely lead to significant local and regional changes in stream runoff and groundwater levels.

FIGURE 19.22 Projected temperature changes
This map shows projected surface temperature changes for the late twenty-first century (2090–2099), relative to the average temperature in the period 1980–1999. Note that the changes will not be uniform throughout the Earth system, and temperature increases are particularly high in the Arctic. (*Source:* IPCC Climate Change 2007).

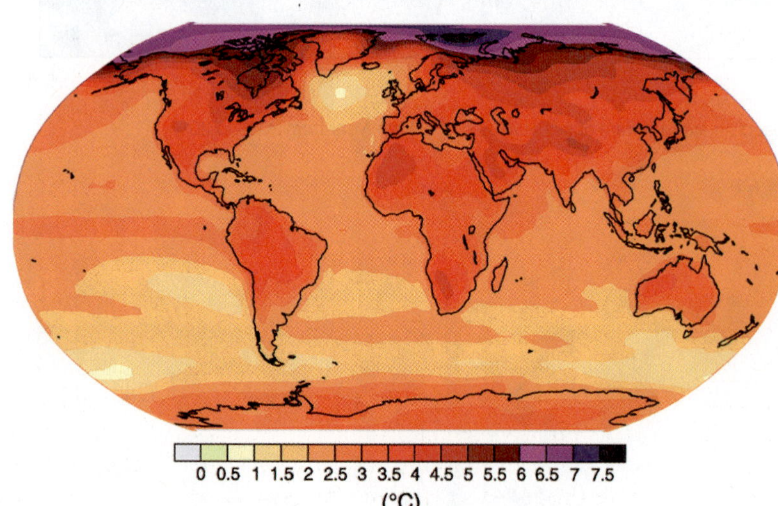

Ocean water, too, will be affected. As the temperature of ocean water rises, its volume will expand, causing world sea level to rise. This rise in sea level, supplemented by meltwater from shrinking glaciers, is likely to increase calving along the margins of tidewater glaciers and ice sheets, thereby leading to additional sea-level rise. The rising sea will inundate coastal regions where millions of people live and will make the tropical regions even more vulnerable to cyclonic storms. Warmer ocean temperatures may spawn more hurricanes, or hurricanes that are more powerful. Recent analyses of storm data suggest that warmer seas may not yet be increasing the number of storms, but may be increasing their power, and possibly their duration. The record number of hurricanes and tropical storms in 2005—Katrina and 27 others—left many people wondering whether global warming was to blame (FIG. 19.23).

An additional concern is the potential for impacts of climatic warming on the global thermohaline circulation. The addition of warmer, fresher water to the oceans could inhibit the formation of the cold, deep, saline waters that drive the global oceanic conveyor-belt circulation system. If the water of the North Atlantic were to become fresher, for example, such that North Atlantic Deep Water failed to form, the thermohaline circulation could be disrupted, causing it to bypass the northern portion of the Atlantic altogether. This would effectively shut down the warm Gulf Stream Current, destabilizing the regional climate system and plunging the coastal areas of North America and western Europe into a deep freeze. (This process, for which there is abundant evidence in the paleoclimate record, is discussed in detail in Chapter 13.)

Changes in the Biosphere. Shifting precipitation patterns are likely to upset ecosystems, causing vegetation communities and the animals dependent on them to migrate or adjust to new conditions. Forest boundaries may shift during coming centuries in response to altered temperature and precipitation patterns. Some prime midcontinental agricultural regions are likely to face increased droughts and substantially reduced soil moisture that will negatively impact crops. Higher-latitude regions with short, cool growing seasons may see increased agricultural production as summer temperatures increase, although this benefit may be tempered by the lack of arable soil in regions such as the Arctic. Fisheries and other aquatic ecosystems, such as coral reefs, may already be changing as a result of changes in water temperature.

Changes in the Cryosphere. Because warmer summers favor greater ablation (loss of ice volume in glaciers), worldwide recession of low- and middle-latitude mountain glaciers is likely in a warmer world. On the other hand, warmer air in high latitudes can evaporate and transport more moisture from the oceans to adjacent ice sheets, leading to more precipitation, which may cause the ice sheets to grow larger. This is a source of great uncertainty in climate modeling.

Temperature changes to date have been greatest in the Arctic, and this will likely continue to be the case throughout the twenty-first century. The warming projected for high northern latitudes (as shown in Fig. 19.22) favors the shrinkage of sea ice. A reduction in polar sea ice, which has a high albedo, should reduce the amount of shortwave solar radiation reflected back into space, thereby increasing the heat absorbed by the ocean. Models show much less heating in the high-latitude southern hemisphere, suggesting little change in sea–ice cover there.

Rising summer air temperatures will also thaw perennially frozen ground at high latitudes, which will affect ecosystems as well as cities and engineering works built on frozen ground. In Arctic areas, altered conditions are already posing challenges for people and wildlife. As sea ice melts earlier, freezes later, and recedes from shore, it becomes harder for Inuit people and polar bears to hunt the seals they rely on for food.

Changes in the Geosphere. As temperature rises, the rate of decomposition of organic matter in soil will increase. Soil decomposition releases CO_2 to the atmosphere, thereby further enhancing the greenhouse effect. If world temperature continues to rise at its present apparent rate of increase, during the next 50 years soils may release an amount of CO_2 equal to nearly 20 percent of the projected CO_2 release due to combustion of fossil fuels, assuming the present rate of fuel consumption continues.

Sediments also host abundant natural deposits of methane in the form of gas hydrates, which are icelike solids in which gas molecules, mainly methane, are trapped along with water ice in ocean sediments and beneath frozen ground. By one estimate, worldwide gas hydrates may hold 10,000 billion metric tons of carbon, twice the amount in all the known coal, gas, and oil reserves on land. They accumulate in ocean sediments beneath a water depth of 500 m, where the temperature is low enough and the pressure high enough to permit their formation. They also accumulate beneath permafrost, which acts as a seal to prevent upward migration and escape of the gas. When gas hydrates break down, they release methane. Warming of temperatures at high latitudes will thaw frozen ground, and this thawing may destabilize the hydrates there, releasing large volumes of methane, amplifying the greenhouse effect.

Feedbacks. A number of the potential changes associated with global climatic warming may further enhance the warming trend, and therefore constitute a positive feedback; that is, they may tend to move the system further in the direction of change. These include the reduction in global albedo as sea-ice cover contracts in the polar oceans, release of additional CO_2 to the atmosphere as soil organic matter decomposes, and release of methane from gas hydrates as frozen ground thaws. Any of these changes could feed back positively to contribute to further warming.

However, there are also some potential negative (that is, self-limiting) feedbacks associated with climatic change. For example, plants grow faster in air enriched with carbon dioxide, so as the atmospheric concentration of CO_2 rises, plant growth may increase, thereby removing this greenhouse gas from the atmosphere at an increasing rate. The result could then be a reduction in the rate that CO_2 is added to the atmosphere (a negative feedback). Determining which of these feedbacks will dominate future climatic changes is one of the most daunting tasks facing climate scientists today.

ANTHROPOSPHERE: HUMANS AND EARTH SYSTEM CHANGE

How do we, as a global society, respond to the changes our activities have brought upon the natural Earth system? In many respects, environmental quality has changed for the better in the decades since the 1970s, when the world's first national-level environmental legislation was passed (the U.S. Environmental Protection Act) and the first federal-level environmental agency was created (the U.S. Environmental Protection Agency). In the older industrialized nations, at least, air and water quality have improved, environmental protection and monitoring technologies have advanced, and we have greatly developed our general understanding of the impacts of our lifestyle on the Earth system.

However, as this chapter's quick overview of human impacts has revealed, we have a long way to go to ensure the integrity and even the survival of natural Earth systems. In particular, if the predicted impacts of global climatic warming and destabilization do occur, the consequences will be very severe indeed. At the beginning of this chapter we discussed the precautionary principle, which holds that in a risk management situation in which the potential consequences are unacceptably severe, measures should be taken to mitigate the risk—even if there is some scientific uncertainty, and even if the probability of occurrence of the worst-case scenario is low. The world community is currently in a risk management situation to which the precautionary principle might reasonably apply. We must determine whether the risks we face are acceptable or unacceptable, and then decide upon a course of action.

Mitigation, Adaptation, and Intervention

There are three basic options for dealing with the potential impacts of global climatic change. We can act now to minimize predicted future impacts; we can wait and prepare to adapt to the changes that occur; or we can intervene in Earth's climate system on a grand scale.

Mitigation: Minimizing Our Impacts

Our first option is to *migitate* the anthropogenic impacts on global systems, principally by cutting back on our emissions of greenhouse gases and other environment-altering materials. This is the most cautious option and perhaps the most achievable. It has received the most serious consideration by the international community, and it has been formalized through international agreements such as the *Montreal Protocol* on ozone-depleting substances and the *Kyoto Protocol* (with its follow-up in Copenhagen in 2009), designed to limit emissions of greenhouse gases.

However, such agreements, particularly the Kyoto Protocol, have had a controversial history and therefore limited success, so far. Several important nations, including the United States, have not joined in, partly because of fears about economic disruption and partly because of questions about shared or differentiated responsibility for the problems and their solutions. Some scientists now fear that our window of opportunity for taking this pathway may be closing, forcing us to consider other alternatives.

Adaptation: Preparing for Change

A second option is that we can decide to take no decisive action to mitigate global climatic change, for the time being, but instead wait and prepare to *adapt* to whatever changes may occur. Adaptations could include building seawalls around low-lying areas like Florida to protect against the impacts of rising sea level, for example, or improving the management of vital water resources, or educating the public about tropical diseases that may migrate to more northern areas.

The risk inherent in this approach is that we may wait too long, until it becomes too late to prevent or mitigate the worst impacts. Even today, some small island nations are at risk of being flooded by rising seas, and would have neither the time available nor the funding to attempt such adaptations.

Intervention: Engineering the Earth System

A final option, and the one that has been the least discussed, would be to undertake a large-scale *intervention* to modify the functioning of the Earth's atmosphere and climate system. Such interventions commonly come under the label of **geo-engineering**. All geo-engineering options carry huge price tags—probably in the trillions of dollars—as well as the risk of unforeseen consequences for Earth systems. Scientists have long discussed geo-engineering as an option for dealing with global change, but they have been reluctant to open up the discussion to the broader community. This is partly because the potential impacts are risky and poorly understood, and partly because there is a general reluctance to consider interfering with the Earth system on such a grand scale. Many scientists also fear that the public and policymakers would cease to take seriously the need to minimize our environmental impacts, if a comprehensive, "easy" solution were seen to exist. Nevertheless, it is important to consider all options.

Since Earth's climate system is a balance between incoming solar radiation and outgoing terrestrial radiation, two fundamental approaches could be taken in the context of geo-engineering: limit incoming solar radiation or prevent heat from being trapped near Earth's surface.

BLOCKING INCOMING SOLAR RADIATION. One possible geo-engineering approach would be to take steps to limit incoming radiation, thus cooling the planet. For example, we could launch enormous orbiting mirrors or reflective solar shields. Another possibility would be to inject a fine mist of water or sulfate aerosols into the atmosphere. This would have the effect of whitening the atmosphere, thereby increasing its albedo reflecting more of the incoming solar radiation and cooling the planet. The model for this is great volcanic eruptions, like that of Pinatubo (Chapter 6), which cause global cooling by increasing the albedo of the atmosphere with volcanic aerosol particles.

PREVENTING GREENHOUSE TRAPPING OF TERRESTRIAL HEAT. The other basic approach of geo-engineering the climate system would be to decrease the effectiveness of the greenhouse effect, which is responsible for trapping and holding terrestrial energy close to the planet. One approach would be to enhance the effectiveness of the ocean in taking up atmospheric carbon dioxide, perhaps by fertilizing oceanic phytoplankton so that they become more active photosynthetically. Another option would be to prevent society's carbon dioxide emissions from entering the atmosphere; rather than ceasing the emissions, this would involve collecting and storing emitted carbon dioxide, either by injecting it into the ocean or by storing it underground. This group of developing technologies is known by the name of *carbon capture and storage* (or *CCS*), and there is considerable interest in the potential of this approach.

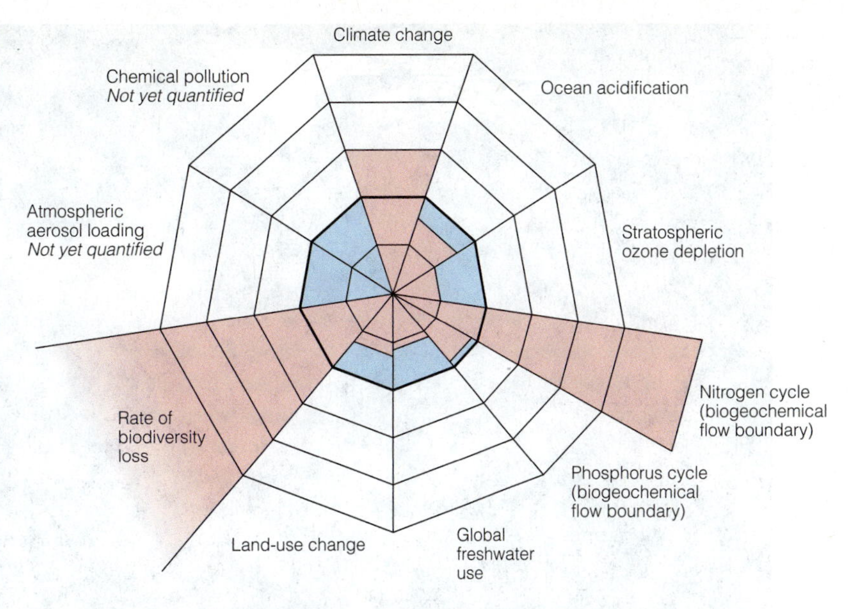

FIGURE 19.24 Crossing planetary boundaries
According to scientists from the International Geosphere-Biosphere Program, the inner (blue) circle of this diagram represents the safe operating space for the key planetary systems. The red wedges indicate the best estimate of the current situation. Three boundaries have already been crossed: climate change, the nitrogen cycle, and biodiversity loss.

Our Past, Our Future

Over many millennia prior to the Industrial Revolution, people slowly changed Earth's natural landscapes as they built villages and cities, converted forests to agricultural land, and locally dammed and diverted streams. With the development of industrial technology, mineral and energy resources were needed to fuel an increasingly populous and demanding society. The exploitation of fossil fuels helped raise the standard of living for most people well beyond that of their forebears.

In spite of the obvious benefits involved, the exploitation of our planet's rich natural resources has not been without cost. In many parts of the world, environmental deterioration is epidemic. In addition to scarring and poisoning Earth's land surface, we have also, unwittingly, polluted the oceans and groundwater and changed the composition of the atmosphere. Even in places long considered to be the most remote on the planet—the frigid ice sheets of Antarctica, the vast Amazon rain forest, the trackless desert of Saudi Arabia, the lofty summits of the Himalaya—the impact of human activities is being felt.

Today, human and natural geologic activities are inextricably intertwined. As we approach a global population of 7 billion, it is increasingly apparent that people have become a major factor—a global factor—in environmental change. There is good evidence that human activity may already have driven several key Earth systems beyond the boundaries of their "safe" operating level (FIG. 19.24).

Despite considerable research, we do not yet have a clear vision of our future on this planet, and uncertainty about global climatic change plays a major role. At present, the best we can conclude is that the force of scientific evidence and theory makes it very probable that the climate is warming due to human actions and will continue to warm as we add greenhouse gases to the atmosphere. As you have learned, there also is a high probability that average global temperatures ultimately will increase by at least 2 to 4°C—more in some places, less in others—leading to widespread environmental changes that will likely affect the everyday lives of all people on the planet.

Although the short-term prospect (on the scale of a few human generations) is for a warmer world, if we stand back and look at our great geochemical experiment from a geological perspective, we will perceive that it is only a brief, very rapid, yet nonrepeatable perturbation in Earth's climatic history. It is nonrepeatable because once Earth's store of easily extractable fossil fuels is used up, most likely within the next several hundred years, the human impact on the atmosphere will inevitably decline. The greenhouse perturbation may well last a thousand years, and perhaps more, but ultimately the changing geometry of Earth's orbit will propel the climate system into the next glacial age.

The Earth system will endure, long after the demise of humans. But as the current stewards of this planet it is our solemn responsibility to manage our impacts and maintain the integrity of all Earth systems to the greatest possible extent. It is in our own best interests to do so because all life on Earth, including our own, depends on the functioning of these delicately balanced interacting systems. The central challenge now facing the world community is to find pathways through which to achieve this vision of a sustainable future. Humanity does have a future on this planet; it is not too late for us to ensure that our children will inherit the Earth in a reasonably healthy state. Achieving this will require creative solutions to a variety of complex problems on a wide range of scales, from local to regional to global. Developing those solutions will depend, in turn, on how successful we are at furthering our scientific understanding of this unique planet (FIG. 19.25).

To Our Readers

We hope this book has provided you with a useful, enjoyable, and perhaps eye-opening introduction to the science of the Earth system. Scientific understanding of this

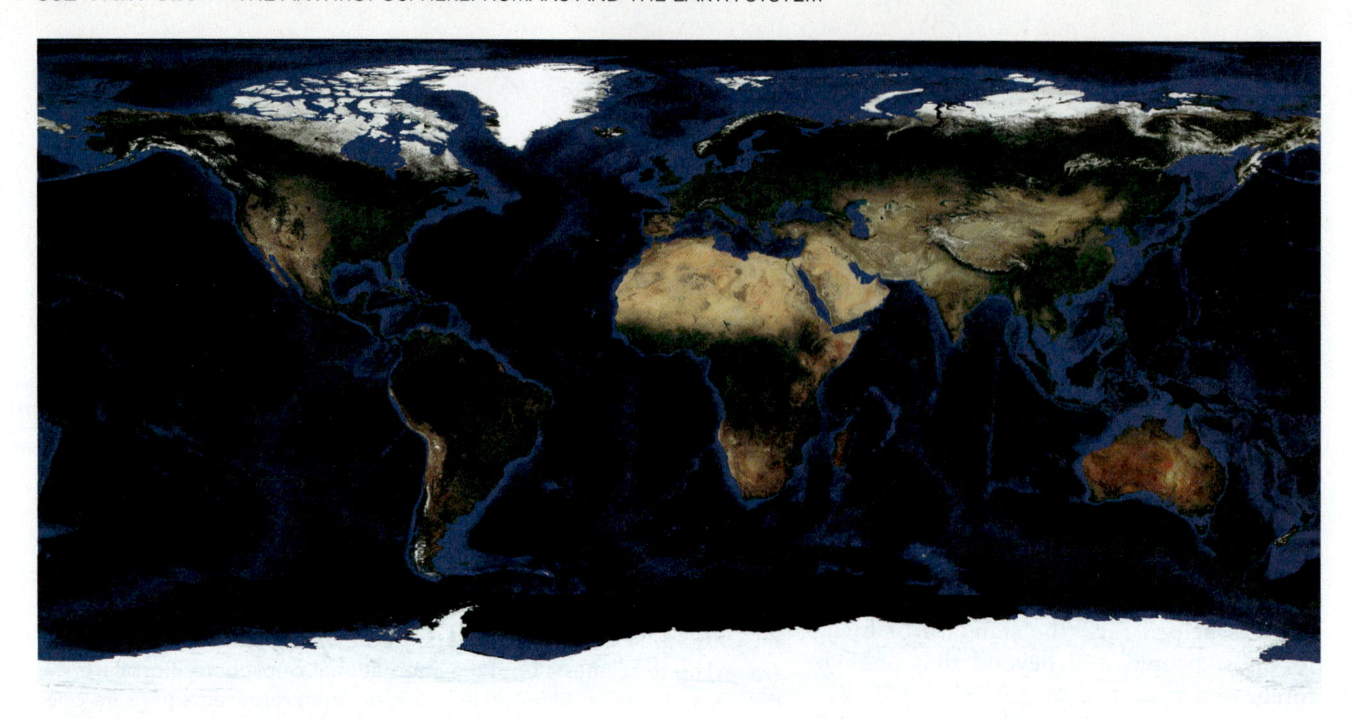

FIGURE 19.25 A delicate balance
Humans and other organisms depend fundamentally on the continued functioning and integrity of interacting Earth systems. This composite, cloud-free satellite image gives a unique perspective on Earth's land, ice caps, and vast oceans.

complex and beautiful world grows with each passing day. Through this understanding of the natural world, our impacts on it, and its influence, in turn, on human lives and activities, we can begin to design policies, invent technologies, and forge agreements that will allow the world community to chart a course toward a sustainable future. We wish you many years of discovery and learning about the Earth system.

> *"We travel together, passengers on a little space ship . . ."*
>
> —Adlai Stevenson, 1965

SUMMARY

1. Human actions are having significant, long-lasting impacts on the Earth system. The human population is now 6.8 billion, growing at 1.2 percent per year. Human use of resources already outstrips carrying capacity in some local areas. Population, lifestyle, and technology interact to drive human impacts on natural Earth systems.

2. The ecological footprint is a measure of the resources required to support a particular person or community, by translating into land area the requirements for growing crops, grazing animals, fishing, harvesting timber, building infrastructure, and burning fossil fuels. It would take about 1.3 Earths to meet current resource needs sustainably.

3. Malthus argued that human population grows exponentially while resource availability grows linearly, so the population must eventually exceed Earth's capacity to provide food and other necessary resources. However, humans have been able to limit reproduction, and technology has greatly expanded Earth's capacity to provide food and other resources. Malthus's model still has relevance today, in a world where resource scarcity often leads to conflict.

4. Scientific uncertainty in our understanding of the natural world makes global change particularly challenging. One approach to risk management is the precautionary principle, which states that if a consequence would be unacceptably severe—even if there is uncertainty or the probability of occurrence is low—we must take action to avoid or mitigate the consequence.

5. Human activity affects the geosphere. Land-based impacts tend to be local, but cumulatively they can become regional or global problems. The natural process of desertification is accelerated by intensive human use of drylands. Productive agricultural soils are being depleted by erosion. Deforestation, overly intensive agriculture, and overgrazing contribute substantially to the problem. Improperly managed garbage, agrochemicals, radioactive materials, and toxic industrial wastes can contaminate soil, posing risks for humans and other organisms.

6. Human activity affects the hydrosphere. Most of the world's large rivers have been altered by diversion, dams, or contamination. The quality and quantity of freshwater supplies are declining in many parts of the world. The open ocean is relatively free of pollution, but marine coastal zones are vulnerable to the impacts of human activity. Their shallow, warm, productive waters are sensitive to environmental changes, and many coastal waters suffer from eutrophication and hypoxia.

7. Human activity affects the atmosphere. Local air pollution can become regional or global as a result of long-range transport. Anthropogenic pollutants, especially CFCs, have depleted the stratospheric ozone layer, leading to the possibility of an increased flux of harmful ultraviolet radiation to the surface. Limitations on ozone-depleting substances may lead to the recovery of the ozone layer by about 2050. Tropospheric ozone is detrimental to human health and, with nitrous oxides, the main constituent of photochemical smog. Acid precipitation forms when anthropogenic emissions, principally from industrial sources, combine with water vapor in the atmosphere.

8. Human activity affects the biosphere. Deforestation, especially in tropical forests, affects habitat, biodiversity, soils, water cycling, biogeochemical cycling, and carbon sequestration. Living resources in aquatic ecosystems, both freshwater and marine, are under increasing pressure. Eighty percent of the world's major fish stocks are fully exploited, overexploited, or depleted. Corals are highly sensitive environmental indicators; worldwide they are showing signs of stress, exhibited as coral bleaching. Biodiversity is increasingly under threat. We are in the midst of an anthropogenic extinction that rivals the great natural mass extinctions of the geologic past. Highly productive biodiversity hotpots hold much of the world's biodiversity, but they are not all protected by law or agreement.

9. The climate system is complex, involving all major components of the Earth system. We know Earth's climate will change, but we lack a clear view of how and at what rate, and we don't know exactly how much change is being caused by human activities. Human activity has changed the composition of the atmosphere since the Industrial Revolution. Some changes have resulted from the emission of gases of wholly anthropogenic origin, such as chlorofluorocarbons; other changes have resulted from anthropogenic mobilization of naturally occurring compounds, such as sulfur and carbon compounds released by the burning of fossil fuels.

10. Changes in the concentration of atmospheric greenhouses gases have the potential to affect the stability of Earth's climate system. In the enhanced or anthropogenic greenhouse effect, we are primarily concerned with carbon dioxide (CO_2), because of the magnitude of emissions, and methane (CH_4), because of its efficiency as a greenhouse gas. The rate at which natural processes are removing CO_2 from the atmosphere is slower than the rate at which human activities are adding it, and the CO_2 content of the atmosphere has therefore been increasing.

11. The first step in predicting future climatic change is to understand past climatic change by applying the tools of paleoclimatology and by analyzing measured and observed climatic indicators. Measured average annual temperature curves all show a rise in temperature over the past century, of approximately 0.5° to 0.8°C.

12. General circulation models allow scientists to explore the linkages among Earth's component systems, human greenhouse gas emissions, and present and future climatic change. GCMs have been successful in simulating present-day climates, which increases the confidence of scientists in their ability to predict future climatic change.

13. The most thoroughly reviewed and widely accepted synthesis of scientific information on global climatic change is the series of reports issued by the Intergovernmental Panel on Climate Change. These reports reflect the strong scientific consensus that (1) human activities have led to increasing atmospheric concentrations of carbon dioxide and other trace gases that have enhanced the greenhouse effect; (2) global mean surface air temperature has increased by up to 0.8° during the last 100 to 150 years as a direct result of the anthropogenic greenhouse effect; and (3) during the next century global average temperatures will continue to increase, such that by the end of the twenty-first century, global temperatures will be at least 1.8–4.0°C higher than today.

14. Global climatic change will have impacts on the hydrosphere. Precipitation in equatorial regions may increase, but the interior portions of large continents will become both warmer and drier. Shifting patterns of precipitation and warmer temperatures will likely lead to changes in stream runoff and groundwater levels. As the temperature of ocean water rises, world sea level will rise, causing flooding of coastal regions. Warmer ocean temperatures may spawn increased storm activity. The addition of warmer, fresher water to the oceans could inhibit the formation of the cold, deep, saline waters that drive the global thermohaline circulation.

15. Global climatic change will have impacts on the biosphere. Shifting precipitation will force ecosystems and communities to migrate or adjust to new conditions if they are to survive. Midcontinental agricultural regions are likely to face increased droughts and substantially reduced soil moisture that will negatively impact crops. Fisheries and other aquatic ecosystems, such as coral reefs, may already be changing as a result of changes in water temperature.

16. Global climatic change will have impacts on the cryosphere. Worldwide recession of low- and middle-latitude mountain glaciers is likely in a warmer world, although high-latitude continental ice sheets may grow larger as a result of greater precipitation. Melting of high-albedo polar sea ice would contribute to further climatic change by reducing the amount of short-wave solar radiation reflected back into space. Rising summer air temperatures in the Arctic, where predicted changes are the greatest, will thaw permanently frozen ground.

17. Global climatic change will have impacts on the geosphere. As temperature rises, the rate of decomposition of organic matter in soil will increase, potentially releasing additional greenhouse gases to the atmosphere. Warming of temperatures at high latitudes may destabilize frozen gas hydrates and release large volumes of methane.

18. Some of the responses of natural Earth systems to global climatic warming may further enhance the warming trend, leading to positive feedbacks; there are also some potential negative or self-limiting feedbacks. Determining which of these feedbacks will dominate future climatic changes is one of the most daunting tasks facing climate scientists today.

19. There are three basic options for dealing with global change. (1) We can act to mitigate predicted impacts, principally by cutting back on emissions of greenhouse gases. (2) We can wait and prepare to adapt to the changes that occur. (3) We can intervene to modify Earth's climate system on a grand scale. Such geo-engineering would likely involve either limiting incoming solar radiation through technologies intended to increase the planet's albedo, or preventing heat from being trapped near Earth's surface through technologies intended to enhance the sequestration of greenhouse gases.

20. Over the past few decades, air and water quality have improved, environmental protection and monitoring technologies have advanced, and we have greatly advanced our understanding of human impacts on the Earth system. However, we have a long way to go to ensure the survival of natural Earth systems and, in particular, the human species. It is our responsibility to manage our impacts and maintain the integrity of all Earth systems to the greatest possible extent. It is in our own best interests to do so because all life on Earth depends on the functioning of these delicately balanced interacting systems.

IMPORTANT TERMS TO REMEMBER

acid precipitation 586
anthropogenic or greenhouse effect 593
biodiversity "hotspots" 591
chlorofluorocarbons (CFCs) 584
coral bleaching 589
cumulative effects 578
desertification 578
Dobson Unit (DU) 584
ecological footprint 576
general circulation models (GCMs) 595
geo-engineering 600
global change 578
global commons 583
Intergovernmental Panel on Climate Change (IPCC) 595
IPAT equation 575
"ozone hole" 584
photochemical smog 586
precautionary principle 577

QUESTIONS FOR REVIEW

1. In what ways does human population growth place stress on the Earth system?
2. How do population, technology, and resource use interact to determine the impact of human activities on Earth systems?
3. What is the ecological footprint concept? How is an ecological footprint calculated, and how does it reveal the impacts of our resource use on the Earth system?
4. What role does scientific uncertainty play in our management of Earth systems? How does the precautionary principle come into it?
5. What are cumulative effects? What are some of the ways that local or regional impacts of human activity can become global in extent?
6. Summarize the major global impacts of human activity on the geosphere.
7. Summarize the major global impacts of human activity on the hydrosphere.
8. Summarize the major global impacts of human activity on the atmosphere.
9. Summarize the major global impacts of human activity on the biosphere.
10. What are the main human activities that disrupt the global carbon cycle, and what are the resulting effects?
11. What are General Circulation Models? How are they constructed, how are they tested, and what are some of their strengths and weaknesses?
12. What does the IPCC predict will be the major changes in our global climate over the next century?
13. What will likely be the impacts of these changes on the hydrosphere, biosphere, cryosphere, and geosphere?
14. What is the difference between mitigation, adaptation, and intervention? Describe how each of these might be used in constructing our response to global environmental change.
15. What is geo-engineering? How might it help us to deal with the effects of global climatic change?

QUESTIONS FOR RESEARCH AND DISCUSSION

1. Why is there uncertainty about the potential for future climatic change or destabilization?
2. Why was the world community apparently successful at finding a solution for stratospheric ozone depletion (through the Montreal Protocol), but apparently unsuccessful—so far, at least—at solving the problem of global climatic change?
3. Among the available strategies for mitigation, adaptation, or intervention to address global environmental change, which do you think provides humanity with its best hope for the future?

QUESTIONS FOR *THE BASICS*

1. How have researchers arrived at estimates of very early human populations?
2. What is the current human population?
3. What is the current rate of increase of the human population?

QUESTIONS FOR *A CLOSER LOOK*

1. Summarize the position of Thomas Malthus on human population growth and resource availability. Use a diagram to illustrate your answer.
2. On what points was Malthus wrong in his predictions? In what ways does the Malthusian position continue to be relevant in today's world?
3. What is the Intergovernmental Panel on Climate Change, and what is its role in the international dialogue about global climatic change?

APPENDIX A

Units and Their Conversions

ABOUT SI UNITS

Regardless of the field of specialization, all scientists use the same units and scales of measurement. They do so to avoid confusion and the possibility that mistakes can creep in when data are converted from one system of units, or one scale, to another. By international agreement the SI units are used by all, and they are the units used in this text. SI is the abbreviation of Système International d'Unités (in English, the International System of Units).

Some of the SI units are likely to be familiar, some unfamiliar. The SI unit of length is the meter (m), of area the square meter (m^2), and of volume the cubic meter (m^3). The SI unit of mass is the kilogram (kg), and of time the second (s). The other SI units used in this book can be defined in terms of these basic units. Three important ones are:

1. The newton (N), a unit of force defined as that force needed to accelerate a mass of 1 kg by 1 m/s^2; hence 1 N = 1 $kg \cdot m/s^2$. (The period between kg and m indicates multiplication.)
2. The joule (J), a unit of energy or work, defined as the work done when a force of 1 newton is displaced a distance of 1 meter; hence 1 J = 1 N.m. One important form of energy so far as Earth is concerned is heat. The outward flow of Earth's internal heat is measured in terms of the number of joules flowing outward from each square centimeter each second; thus, the unit of heat flow is $J/cm^2/s$.
3. The pascal (Pa), a unit of pressure defined as a force of 1 newton applied across an area of 1 square meter; hence 1 Pa = 1 N/m^2. The pascal is a numerically small unit. Atmospheric pressure, for example (15 lb/in^2), is 101,300 Pa. Pressure within Earth reaches millions or billions of pascals. For convenience, earth scientists sometimes use 1 million pascals (megapascal, or MPa) as a unit.

Temperature is a measure of the internal kinetic energy (expressed as movement) of the atoms and molecules in a body. In the SI system, temperature is measured on the Kelvin scale (K). The temperature intervals on the Kelvin scale are arbitrary, and they are the same as the intervals on the more familiar Celsius scale (°C). The difference between the two scales is that the Celsius scale selects 100°C as the temperature at which water boils at sea level, and 0°C as the freezing temperature of water at sea level. Zero degrees Kelvin, on the other hand, is absolute zero, the temperature at which all atomic and molecular motions cease. Thus, 0°C is equal to 273.15 K, and 100°C is 373.15 K. The temperatures of processes on and within Earth tend to be at or above 273.15 K. Despite the inconsistency, earth scientists still use the Celsius scale.

Appendix A provides a table of conversion from older units to Standard International (SI) units.

COMMONLY USED UNITS OF MEASURE

Length

Metric Measure

1 kilometer (km)	= 1000 meters (m)
1 meter (m)	= 100 centimeters (cm)
1 centimeter (cm)	= 10 millimeters (mm)
1 millimeter (mm)	= 1000 micrometers (μm) (formerly called microns)
1 micrometer (μm)	= 0.001 millimeter (mm)
1 nanometer (nm)	= 10^{-7} centimeters (cm)
1 angstrom (Å)	= 10^{-8} centimeters (cm)

Nonmetric Measure

1 mile (mi)	= 5280 feet (ft) = 1760 yards (yd)
1 yard (yd)	= 3 feet (ft)
1 fathom (fath)	= 6 feet (ft)

Conversions

1 kilometer (km)	= 0.6214 mile (mi)
1 meter (m)	= 1.094 yards (yd) = 3.281 feet (ft)
1 centimeter (cm)	= 0.3937 inch (in)
1 millimeter (mm)	= 0.0394 inch (in)
1 mile (mi)	= 1.609 kilometers (km)
1 yard (yd)	= 0.9144 meter (m)
1 foot (ft)	= 0.3048 meter (m)
1 inch (in)	= 2.54 centimeters (cm)
1 fathom (fath)	= 1.8288 meters (m)

Area

Metric Measure

1 square kilometer (km^2) = 1,000,000 square meters (m^2) = 100 hectares (ha)
1 square meter (m^2) = 10,000 square centimeters (cm^2)
1 hectare (ha) = 10,000 square meters (m^2)

Nonmetric Measure

1 square mile (mi^2) = 640 acres (ac)
1 acre (ac) = 4840 square yards (yd^2)
1 square foot (ft^2) = 144 square inches (in^2)

Conversions

1 square kilometer (km^2) = 0.386 square mile (mi^2)
1 hectare (ha) = 2.471 acres (ac)
1 square meter (m^2) = 1.196 square yards (yd^2) = 10.764 square feet (ft^2)
1 square centimeter (cm^2) = 0.155 square inch (in^2)
1 square mile (mi^2) = 2.59 square kilometers (km^2)
1 acre (ac) = 0.4047 hectare (ha)
1 square yard (yd^2) = 0.836 square meter (m^2)
1 square foot (ft^2) = 0.0929 square meter (m^2)
1 square inch (in^2) = 6.4516 square centimeter (cm^2)

Volume

Metric Measure

1 cubic meter (m^3) = 1,000,000 cubic centimeters (cm^3)
1 liter (l) = 1000 milliliters (ml) = 0.001 cubic meter (m^3)
1 centiliter (cl) = 10 milliliters (ml)
1 milliliter (ml) = 1 cubic centimeter (cm^2)

Nonmetric Measure

1 cubic yard (yd^3) = 27 cubic feet (ft^3)
1 cubic foot (ft^3) = 1728 cubic inches (in^3)
1 barrel (oil) (bbl) = 42 gallons (U.S.) (gal)

Conversions

1 cubic kilometer (km^3) = 0.24 cubic miles (mi^3)
1 cubic meter (m^3) = 264.2 gallons (U.S.) (gal) = 35.314 cubic feet (ft^3)
1 liter (l) = 1.057 quarts (U.S.) (qt) = 33.815 ounces (U.S. fluid) (fl. oz.)
1 cubic centimeter = 0.0610 cubic inch (in^3) (cm^3)
1 cubic mile (mi^3) = 4.168 cubic kilometers (km^3)
1 acre-foot (ac-ft) = 1233.46 cubic meters (m^3)
1 cubic yard (yd^3) = 0.7646 cubic meter (m^3)
1 cubic foot (ft^3) = 0.0283 cubic meter (m^3)
1 cubic inch (in^3) = 16.39 cubic centimeters (cm^3)
1 gallon (gal) = 3.784 liters (l)

Mass

Metric Measure

1000 kilograms (kg) = 1 metric ton (also called a tonne) (m.t)
1 kilogram (kg) = 1000 grams (g)

Nonmetric Measure

1 short ton (sh.t) = 2000 pounds (lb)
1 long ton (l.t) = 2240 pounds (lb)
1 pound (avoirdupois) (lb) = 16 ounces (avoirdupois) (oz) = 7000 grains (gr)
1 ounce (avoirdupois) (oz) = 437.5 grains (gr)
1 pound (Troy) (Tr. lb) = 12 ounces (Troy) (Tr. oz)
1 ounce (Troy) (Tr. oz) = 20 pennyweight (dwt)

Conversions

1 metric ton (m.t) = 2205 pounds (avoirdupois) (lb)
1 kilogram (kg) = 2.205 pounds (avoirdupois) (lb)
1 gram (g) = 0.03527 ounce (avoirdupois) (oz) = 0.03215 ounce (Troy) (Tr. oz) = 15,432 grains (gr)
1 pound (lb) = 0.4536 kilogram (kg)
1 ounce (avoirdupois) (oz) = 28.35 grams (g)
1 ounce (avoirdupois) (oz) = 1.097 ounces (Troy) (Tr.oz)

Pressure

1 pascal (Pa) = 1 newton/square meter (N/m^2)
1 kilogram.force/square centimeter (kg/cm^2) = 0.96784 atmosphere (atm) = 14.2233 pounds/square inch (lb/in^2) = 0.098067 bar
1 bar = 0.98692 atmosphere (atm) = 10^5 pascals (Pa) = 1.02 kilograms.force/square centimeter (kg/cm^2)

Energy and Power

Energy

1 joule (J) = 1 newton meter (N.m) = 2.390×10^{-1} calorie (cal) = 9.47×10^{-4}

British thermal unit (Btu)	=	2.78×10^{-7} kilowatt-hour (kWh)
1 calorie (cal)	=	4.184 joule (J) = 3.968×10^{-3} British thermal unit (Btu) = 1.16×10^{-6} kilowatt-hour (kWh)
1 British thermal unit (Btu)	=	1055.87 joules (J) = 252.19 calories (cal) = 2.928×10^{-4} kilowatt-hour (kWh)
1 kilowatt hour	=	3.6×10^{6} joules (J) = 8.60×10^{5} calories (cal) = 3.41×10^{3} British thermal units (Btu)

Power (energy per unit time)

1 watt (W)	=	1 joule per second (J/s) = 3.4129 Btu/h = 1.341×10^{-3} horsepower (hp) = 14.34 calories/minute (cal/min)
1 horsepower (hp)	=	7.46×10^{2} watts (W)

Temperature

Fahrenheit (°F)	=	(K.9/5) − 459.67
Fahrenheit (°F)	=	(°C.9/5) + 32
degrees Celcius (°C)	=	K+ 273.15
degrees Kelvin (K)	=	°C −273.15
degrees Fahrenheit (°F)	=	(°C.9/5) + 32
degrees Celcius (°C)	=	(°F − 32).5/9

APPENDIX B

Tables of the Chemical Elements and Naturally Occurring Isotopes

TABLE B.1 Alphabetical List of the Elements

Element	Symbol	Atomic Number	Crustal Abundance, Weight Percent
Actinium	Ac	89	Human-made
Aluminum*	Al	13	8.00
Americium	Am	95	Human-made
Antimony	Sb	51	0.00002
Argon	Ar	18	Not known
Arsenic	As	33	0.00025
Astatine	At	85	Human-made
Barium	Ba	56	0.0456
Berkelium	Bk	97	Human-made
Beryllium	Be	4	0.00019
Bismuth	Bi	83	0.000018
Boron	B	5	0.0011
Bromine	Br	35	0.000088
Cadmium	Cd	48	0.000008
Calcium*	Ca	20	5.06
Californium	Cf	98	Human-made
Carbon[a]	C	6	0.02
Cerium	Ce	58	0.0043
Cesium	Cs	55	0.0002
Chlorine	Cl	17	0.0244
Chromium	Cr	24	0.0135
Cobalt	Co	27	0.0266
Copper	Cu	29	0.0027
Curium	Cm	96	Human-made
Dysprosium	Dy	66	0.00036
Einsteinium	Es	99	Human-made
Erbium	Er	68	0.00021
Europium	Eu	63	0.00011
Fermium	Fm	100	Human-made
Fluorine	F	9	0.0553
Francium	Fr	87	Human-made
Gadolinium	Gd	64	0.00037
Gallium	Ga	31	0.0016
Germanium	Ge	32	0.00013
Gold	Au	79	0.00000013
Hafnium	Hf	72	0.00037
Hahnium	Ha	105	Human-made
Hassium	Hs	108	Human-made
Helium	He	2	Not known
Holmium	Ho	67	0.000077
Hydrogen[b]	H	1	0.14
Indium	In	49	0.0000052
Iodine	I	53	0.000071
Iridium	Ir	77	0.000000037
Iron*	Fe	26	5.80
Krypton	Kr	36	Not known
Lanthanum	La	57	0.0020
Lawrencium	Lw	103	Human-made
Lead	Pb	82	0.0011
Lithium	Li	3	0.0016
Lutetium	Lu	71	0.000030
Magnesium*	Mg	12	2.77
Manganese*	Mn	25	0.100
Meitnerium	Mt	109	Human-made
Mendelevium	Md	101	Human-made
Mercury	Hg	80	0.000003
Molybdenum	Mo	42	0.00008
Neodymium	Nd	60	0.0020
Neon	Ne	10	Not known
Neptunium	Np	93	Human-made
Nickel	Ni	28	0.0059
Nielsbohrium	Ns	107	Human-made
Niobium	Nb	41	0.0008
Nitrogen	N	7	0.0056
Nobelium	No	102	Human-made
Osmium	Os	76	0.000000041
Oxygen[b]	O	8	45.2
Palladium	Pd	46	0.00000015
Phosphorus*	P	15	0.1010
Platinum	Pt	78	0.00000015
Plutonium	Pu	94	Human-made
Polonium	Po	84	Footnote[d]
Potassium*	K	19	1.68
Praseodymium	Pr	59	0.0049
Promethium	Pm	61	Human-made
Protactinium	Pa	91	Footnote[d]

TABLE B.1 *(Continued)*

Element	Symbol	Atomic Number	Crustal Abundance, Weight Percent	Element	Symbol	Atomic Number	Crustal Abundance, Weight Percent
Radium	Ra	88	Footnote[d]	Technetium	Tc	43	Human-made
Radon	Rn	86	Footnote[d]	Tellurium[c]	Te	52	0.000001
Rhenium	Re	75	0.0000000188	Terbium	Tb	65	0.00006
Rhodium[c]	Rh	45	0.00000001	Thallium	Tl	81	0.00005
Rubidium	Rb	37	0.0049	Thorium	Th	90	0.00056
Ruthenium	Ru	44	0.000000057	Thulium	Tm	69	0.000028
Samarium	Sm	62	0.00039	Tin	Sn	50	0.00017
Scandium	Sc	21	0.0022	Titanium*	Ti	22	0.86
Seaborgium	Sg	106	Human-made	Tungsten	W	74	0.00010
Selenium	Se	34	0.000013	Uranium	U	92	0.00013
Silicon*	Si	14	27.20	Vanadium	V	23	0.0138
Silver	Ag	47	0.0000056	Xenon	Xe	54	Not known
Sodium*	Na	11	2.32	Ytterbium	Yb	70	0.00019
Strontium	Sr	38	0.0320	Yttrium	Y	39	0.0019
Sulfur	S	16	0.0404	Zinc	Zn	30	0.0072
Tantalum	Ta	73	0.00007	Zirconium	Zr	40	0.0132

Source: R.L. Rudnick and S. Gao (2004)

*K. K. Turekian (1969).

[a]Estimate from S. R. Taylor (1964).

[b]Analyses of crustal rocks do not usually include separate determinations for hydrogen and oxygen. Both combine in essentially constant proportions with other elements, so abundances can be calculated.

[c]Estimates are uncertain and have a very low reliability.

[d]Elements formed by decay of uranium and thorium. The daughter products are radioactive with such short half-lives that crustal accumulations are too low to be measured accurately.

TABLE B.2 **Naturally Occurring Elements Listed in Order of Atomic Numbers, Together with the Naturally Occurring Isotopes of Each Element, Listed in Order of Mass Numbers**

Atomic Number[a]	Name	Symbol	Mass Numbers[b] of Natural Isotopes	Atomic Number[a]	Name	Symbol	Mass Numbers[b] of Natural Isotopes
1	Hydrogen	H	1, 2, [3][c]	19	Potassium	K	39, [40], 41
2	Helium	He	3, 4	20	Calcium	Ca	40, 42, 43, 44, 46, [48]
3	Lithium	Li	6, 7	21	Scandium	Sc	45
4	Beryllium	Be	9, [10]	22	Titanium	Ti	46, 47, 48, 49 50
5	Boron	B	10, 11	23	Vanadium	V	[50], 51
6	Carbon	C	12, 13, [14]	24	Chromium	Cr	50, 52, 53, 54
7	Nitrogen	N	14, 15	25	Manganese	Mn	55
8	Oxygen	O	16, 17, 18	26	Iron	Fe	54, 56, 57, 58
9	Fluorine	F	19	27	Cobalt	Co	59
10	Neon	Ne	20, 21, 22	28	Nickel	Ni	58, 60, 61, 62, 64
11	Sodium	Na	23	29	Copper	Cu	63, 65
12	Magnesium	Mg	24, 25, 26	30	Zinc	Zn	64; 66, 67, 68, 70
13	Aluminum	Al	27	31	Gallium	Ga	69, 71
14	Silicon	Si	28, 29 30	32	Germanium	Ge	70, 72, 73, 74, 76
15	Phosphorus	P	31	33	Arsenic	As	75
16	Sulfur	S	32, 33, 34, 36	34	Selenium	Se	74, 76, 77, 80, 82
17	Chlorine	Cl	35, 37	35	Bromine	Br	79, 81
18	Argon	Ar	36, 38, 40	36	Krypton	Kr	78, 80, 82, 83, 84, 86

Atomic Number[a]	Name	Symbol	Mass Numbers[b] of Natural Isotopes
37	Rubidium	Rb	85, [87]
38	Strontium	Sr	84, 86, 87, 88
39	Yttrium	Y	89
40	Zirconium	Zr	90, 91, 92, 94, 96
41	Niobium	Nb	93
42	Molybdenum	Mo	92, 94, 95, 96, 97, 98, 100
44	Ruthenium	Ru	96, 98, 99, 100, 101, 102, 104
45	Rhodium	Rh	103
46	Palladium	Pd	102, 104, 105, 106, 108, 110
47	Silver	Ag	107, 109
48	Cadmium	Cd	106, 108, 110, 111, 112, 113, 114, 116
49	Indium	In	113, [115]
50	Tin	Sn	112, 114, 115, 116, 117, 118, 119, 120, 122, 124
51	Antimony	Sb	121, 123
52	Tellurium	Te	120, 122, 123, 124, 125, 126, 128, 130
53	Iodine	I	127
54	Xenon	Xe	124, 126, 128, 129, 130, 131, 132, 134, 136
55	Cesium	Cs	133
56	Barium	Ba	130, 132, 134, 135, 137, 138
57	Lanthanum	La	[138], 139
58	Cerium	Ce	136, 138, 140, [142]
59	Praseodymium	Pr	141
60	Neodymium	Nd	142, 143, [144], 145, 146, 148, 150
62	Samarium	Sm	144, [147], [148], [149], 150, 152, 154
63	Europium	Eu	151, 153
64	Gadolinium	Gd	[152], 154, 155, 156, 157, 158, 160
65	Terbium	Tb	159
66	Dysprosium	Dy	156, 158, 160, 161, 162, 163, 164
67	Holmium	Ho	165
68	Erbium	Er	162, 166, 167, 168, 170
69	Thulium	Tm	169
70	Ytterbium	Yb	168, 170, 171, 172, 173, 174, 176
71	Lutetium	Lu	175, [176]
72	Hafnium	Hf	174, 176, 177, 178, 179, 180
73	Tantalum	Ta	180, 181
74	Tungsten	W	180, 182, 183, 184, 186
75	Rhenium	Re	185, [187]
76	Osmium	Os	184, 186, 187, 188, 189, 190, 192
77	Iridium	Ir	191, 193
78	Platinum	Pt	190, 192, 195, 196, 198
79	Gold	Au	197
80	Mercury	Hg	196, 198, 199, 200, 201, 202, 204
81	Thallium	Tl	203, 205
82	Lead	Pb	204, 206, 207, 208
83	Bismuth	Bi	209
84	Polonium	Po	[210]
86	Radon	Rn	[222]
88	Radium	Ra	[226]
90	Thorium	Th	[232]
91	Protactinium	Pa	[231]
92	Uranium	U	[234], [235], [238]

[a]Atomic number = number of protons.

[b]Mass number = protons + neutrons.

[c]☐ indicates isotope is radioactive.

Tables of the Properties of Selected Common Minerals

TABLE C.1 Properties of the Common Minerals with Metallic Luster

Mineral	Chemical Composition	Form and Habit	Cleavage	Hardness / Specific Gravity	Other Properties	Most Distinctive Properties
Chalcopyrite	$CuFeS_2$	Massive or granular.	None. Uneven fracture.	3.5–4 / 4.2	Golden yellow to brassy yellow. Dark green to black streak.	Streak. Hardness distinguishes from pyrite.
Copper	Cu	Massive, twisted leaves and wires.	None. Can be cut with a knife.	2.5–3 / 9	Copper color but commonly stained green.	Color, specific gravity, malleable.
Galena	PbS	Cubic crystals, coarse or fine-grained granular masses.	Perfect in three directions at right angles.	2.5 / 7.6	Lead-gray color. Gray to gray-black streak.	Cleavage and streak.
Hematite	Fe_2O_3	Massive, granular, micaceous.	Uneven fracture.	5–6 / 5	Reddish-brown, gray to black. Reddish-brown streak.	Streak, hardness.
Limonite (*Goethite* is most common.)	A complex mixture of minerals, mainly hydrous iron oxides.	Massive, coatings, botryoidal crusts, earthy masses.	None.	1–5.5 / 3.5–4	Yellow, brown, black, yellowish-brown streak.	Streak.
Magnetite	Fe_3O_4	Massive, granular. Crystals have octahedral shape.	None. Uneven fracture.	5.5–6.5 / 5	Black. Black streak. Strongly attracted to a magnet.	Streak, magnetism.
Pyrite ("Fool's gold")	FeS_2	Cubic crystals with striated faces. Massive.	None. Uneven fracture.	6–6.5 / 5.2	Pale brass-yellow, darker if tarnisned. Greenish-black streak.	Streak. Hardness distinguishes from chalcopyrite. Not malleable, which distinguishes from gold.
Sphalerite	ZnS	Fine to coarse granular masses. Tetrahedron shaped crystals.	Perfect in six directions.	3.5–4 / 4	Yellowish-brown to black. White to yellowish-brown streak. Resinous luster.	Cleavage, hardness, luster.

TABLE C.2 Properties of Rock-forming Minerals with Nonmetallic Luster

Mineral	Chemical Composition	Form and Habit	Cleavage	Hardness / Specific Gravity	Other Properties	Most Distinctive Properties
Amphiboles. (A complex family of minerals, *Hornblende* is most common.)	$X_2Y_5Si_8O_{22}(OH)_2$ where X = Ca, Na; Y = Mg, Fe, Al.	Long, six-sided crystals; also fibers and irregular grains.	Two; intersecting at 56° and 124°.	5–6 / 2.9–3.8	Common in metamorphic and igneous rocks. *Hornblende* is dark green to black; *actinolite*, green; *tremolite*, white.	Cleavage, habit.
Apatite	$Ca_5(PO_4)_3$ (F, OH, Cl)	Granular masses. Perfect six-sided crystals.	Poor. One direction.	5 / 3.2	Green, brown, blue, or white. Common in many kinds of rocks in small amounts.	Hardness, form.
Aragonite	$CaCO_3$	Massive, or slender, needle-like crystals.	Poor. Two directions.	3.5 / 2.9	Colorless or white. Effervesces with dilute HCl.	Effervescence with acid. Poor cleavage distinguishes from calcite.
Calcite	$CaCO_3$	Tapering crystals and granular masses.	Three perfect; at oblique angles to give a rhomb-shaped fragment.	3 / 2.7	Colorless or white. Effervesces with dilute HCl.	Cleavage, effervescence with acid.
Dolomite	$CaMg(CO_3)_2$	Crystals with rhomb-shaped faces. Granular masses.	Perfect in three directions as in calcite.	3.5 / 2.8	White or gray. Does not effervesce in cold, dilute HCl unless powdered. Pearly luster.	Cleavage. Lack of effervescence with acid.
Feldspars: Potassium feldspar (*orthoclase* is a common variety)	$KAlSi_3O_8$	Prism-shaped crystals, granular masses.	Two perfect, at right angles.	6 / 2.6	Common mineral. Pink, white, or gray in color.	Color, cleavage.
Garnets	$X_3Y_2(SiO_4)_3$; X = Ca, Mg, Fe, Mn; Y = Al, Fe, Ti, Cr.	Perfect crystals with 12 or 24 sides. Granular masses.	None. Uneven fracture.	6.5–7.5 / 3.5–4.3	Common in metamorphic rocks. Red, brown, yellowish-green, black.	Crystals, hardness, no cleavage.
Graphite	C	Scaly masses.	One, perfect. Forms slippery flakes.	1–2 / 2.2	Metamorphic rocks. Black with metallic to dull luster.	Cleavage, color. Marks paper.
Gypsum	$CaSO_4.2H_2O$	Elongate or tabular crystals. Fibrous and earthy masses.	One, perfect. Flakes bend but are not elastic.	2 / 2.3	Vitreous to pearly luster. Colorless.	Hardness, cleavage.
Halite	NaCl	Cubic crystals.	Perfect to give cubes.	2.5 / 2.2	Tastes salty. Colorless, blue.	Taste, cleavage.

Mineral	Chemical Composition	Form and Habit	Cleavage	Hardness / Specific Gravity	Other Properties	Most Distinctive Properties
Kaolinite	$Al_2Si_2O_5(OH)_4$	Soft, earthy masses. Sub-microscopic crystals.	One, perfect.	2–2.5 / 2.6	White, yellowish. Plastic when wet; emits clayey odor. Dull luster.	Feel, plasticity, odor.
Mica: *Biotite*	$K(Mg, Fe)_3 AlSi_3O_{10}(OH)_2$	Irregular masses of flakes.	One, perfect.	2.5–3 / 2.8–3.2	Common in igneous and metamorphic rocks. Black, brown, dark green.	Cleavage, color. Flakes are elastic.
Mica: *Muscovite*	$KAl_3Si_3O_{10}(OH)_2$	Thin flakes.	One, perfect.	2–2.5 / 2.7	Common in igneous, and metamorphic rocks. Colorless, pale-green or brown.	Cleavage, color. Flakes are elastic.
Olivine	$(Mg,Fe)_2SiO_4$	Small grains. granular masses.	None Conchoidal fracture.	6.5–7 / 3.2–4.3	Igneous rocks. Olive green to yellowish-green.	Color, fracture, habit.
Pyroxene (A complex family of minerals. *Augite* is most common.)	$XY(SiO_3)_2$ $X = Y = Ca, Mg, Fe$	8-sided stubby crystals. Granular masses.	Two, perfect, nearly at right angles.	5–6 / 3.2–3.9	Igneous and metamorphic rocks. *Augite*, dark green to black; other varieties white to green.	Cleavage
Quartz	SiO_2	6-sided crystals, granular masses.	None. Conchoidal fracture.	7 / 2.6	Colorless, white, gray, but may have any color, depending on impurities Vitreous to greasy luster.	Form, fracture, striation across crystal faces at right angles to long dimension.

Glossary

Abiotic. Said of a compound that is of non-biologic (non-living) origin. (Ch. 3)

Ablation. The loss of mass from a glacier. (Ch. 9)

Absorption (of radiation). The taking up of energy in electromagnetic radiation by the medium through which it is passing. (Ch. 11)

Abyssal plain. A large flat area of the deep seafloor having slopes less than about 1 m/km, and ranging in depth below sea level from 3 to 6 km. (Ch. 5)

Accumulation. The additions of mass of a glaciers. (Ch. 9)

Acid mine drainage. Water from an active or abandoned mine that has been strongly acidified by the decomposition of pyrite (FeS_2) or other sulfide minerals. (Ch. 18)

Acid precipitation. Rainfall that is abnormally acid due to the presence of nitrous and sulfur oxides reacting with water in the atmosphere. (Ch. 19)

Active volcano. A volcano that has erupted within historic times. (Ch. 6)

Adaptation. The process of change in response to an environmental pressure. (Ch. 14)

Adiabatic (lapse rate). The way temperature changes with altitude in rising or falling air. (Ch. 11)

Aerated zone. Region above the water table where pore spaces are not completely filled with water, and where water is held by capillarity. (Ch. 8)

Aerosol. A tiny liquid droplet or tiny solid particle so small it remains suspended in air. (Chs. 3, 11)

Agriculture. Cultivation of the ground and the production of crops and livestock. (Ch. 17)

Air. The invisible, odorless mixture of gases and suspended particles that surrounds the Earth. (Ch. 11)

Air mass. Large volume of air with a fairly homogeneous internal temperature and humidity. (Ch. 12)

Air pressure. The force exerted by the weight of overlying air. (Ch. 11)

Air-pressure gradient. The air pressure drop per unit distance. (Ch. 13)

Albedo. The reflectivity of the surface of a planet. (Ch. 2)

Allopatric speciation. Evolution of a new species in populations that are physically isolated. (Ch. 16)

Alluvial fan. A fan-shaped body of alluvium typically built where a stream leaves a steep mountain valley. (Ch. 8)

Alluvium. Sediment deposited by streams in nonmarine environments. (Ch. 8)

Alpine biomes. Terrestrial biomes in which altitude exerts a major influence. (Ch. 15)

Andesitic magma. One of the three common kinds of magma, containing about 60% SiO_2 by weight. (Ch. 6)

Angiosperm. A plant whose seeds are surrounded by fruit. (Ch. 14)

Antarctic Bottom Water (AABW). The dense, cold water that forms off Antarctica as a result of the freezing of sea ice. (Ch. 10)

Anthropocentric. A human-centered point of view. (Ch. 16)

Anthropogenic greenhouse effect. Human contribution to the greenhouse effect. (Ch. 19)

Anthroposphere. That part of the natural system that has been modified by humans for human purposes, or as a result of human activities. (Ch. 17)

Anticyclone. Air spiraling outward away from a high-pressure center. (Ch. 12)

Aphotic zone. Oceanic zone below the photic zone where little to no light is able to penetrate. (Ch. 15)

Aquaculture. The raising of fish, shellfish, crustaceans, and aquatic plants in captivity. (Ch. 17)

Aqueous. Water-based. (Ch. 3)

Aquiclude. A body of impermeable or distinctly less permeable rock adjacent to an *aquifer*. (Ch. 8)

Aquifer. A body of permeable rock or regolith saturated with water and through which groundwater moves. (Ch. 8)

Arable soil. Soil suited for growing agricultural crops. (Ch. 17)

Archean Eon. The period in Earth's history from 3.8 to 2.5 billion years ago. (Ch. 14)

Asteroid. A sub-planetary object orbiting the Sun. The orbits of most asteroids lie between the orbits of Mars and Jupiter. (Ch. 4)

Asthenosphere. The region of the mantle where rocks become ductile, having little strength, and are easily deformed. It lies at a depth of 100 to 350 km below the surface. (Ch. 3)

Astronomical unit (AU). Unit of measure used by astronomers; it is the average distance from Earth to the Sun; 149,600,000 kilometers. (Ch. 4)

Atmosphere. The mixture of gases, predominantly nitrogen, oxygen, carbon dioxide, and water vapor that surrounds the Earth. (Ch. 11)

Atmospheric blind. The visible-light portion of the spectrum that is absorbed and will not pass through the atmosphere. (Ch. 11)

Atmospheric window. The visible-light portion of the spectrum that is transmitted easily through the atmosphere. (Ch. 11)

Atom. The smallest individual particle that retains all the properties of a given chemical element. (Ch. 3)

Autotrophs. Organisms that can get energy directly from sunlight. (Ch. 14)

Barometer. A device that measures air pressure. (Ch. 11)

Basaltic magma. One of the three common forms of magma, containing approximately 50% SiO_2 by weight. (Ch. 6)

Batholith. The largest kind of pluton. A very large, igneous body of irregular shape that cuts across the layering of the rock it intrudes. (Ch. 6)

Beach. Wave-washed sediment along a coast, extending throughout the surf zone. (Ch. 10)

Bed. The smallest formal unit of a body of sediment or sedimentary rock. (Ch. 9)

Bedding. The layered arrangement of strata in a body of sediment or sedimentary rock. (Ch. 7)

Benthic zone. The ocean-floor environment. (Ch. 15)

Benthos. Bottom-dwelling marine organisms. (Ch. 15)

Bioaccumulation. The build-up of a substance in the tissues of an organism. (Ch. 15)

Bioavailability. A measure of the ease with which a substance can be absorbed and used by organisms. (Ch. 15)

Bioconcentration. The consumption of a substance by an organism faster than it can excrete the substance. (Ch. 15)

Biodiversity. The overall variety of life forms. (Chs. 14, 16)

Biodiversity "hot spots". The clustering of biodiversity in certain highly productive areas. (Ch. 19)

Biogenic sediment. Chemical sediment formed as a result of biochemical reactions in water. (Ch. 7)

Biogeochemical cycle. A natural cycle describing the movements and interactions through the Earth's spheres of the chemicals essential to life. (Ch. 15)

Biological production. The transformation of energy into matter by biological processes. (Ch. 15)

Biomagnification. Accumulation of a substance by organisms in higher trophic levels. Also called food-chain concentration. (Ch. 15)

Biomass. Usually used to mean the amount of living material, both as live and dead material, as in the leaves (live) and stem wood (dead) of trees. (Ch. 15)

Biomass energy. Energy derived from plant life. (Ch. 18)

Biome. A kind of ecosystem. The rain forest is an example of a biome; rain forests occur in many parts of the world but are not all connected with each other. (Ch. 15)

Biopolymer. Polymers that consist of organic compounds and are of biologic origin. (Ch. 3)

Biosphere. The totality of the Earth's organisms and, in addition, organic matter that has not yet been completely decomposed. (Ch. 14)

Biotic. A compound that is of biologic (living) origin. (Ch. 3)

Biotic potential (of a population). The growth of a population under optimal environmental conditions and in the absence of resource limitations. (Ch. 16)

Birth rate. Births in a population per unit time. (Ch. 16)

Black body radiator. A (hypothetical) perfect radiator of light that absorbs all light that strikes it and reflects none; its light output depends only on its temperature. (Ch. 2)

Black smoker. Hot-spring of superheated water on the seafloor. (Ch. 14)

Body waves. Seismic waves that travel outward from an earthquake focus and pass through the Earth. (Ch. 6)

Bond. The electrical forces that draw two atoms together. (Ch. 3)

Boreal forest. The terrestrial biome that lies to the south of the tundra in North America and Eurasia. Also called the *taiga.* (Ch. 15)

Box model. A simple, convenient graphical representation of a system. (Ch. 1)

Braided stream. A channel system consisting of a tangled network of two or more smaller branching and reuniting channels that are separated by islands or bars. (Ch. 8)

Burial metamorphism. Metamorphism caused solely by the burial of sedimentary or pyroclastic rocks. (Ch. 7)

Caldera. A roughly circular, steep-walled volcanic basin several kilometers or more in diameter. (Ch. 6)

Calving. The progressive breaking off of icebergs from a glacier that terminates in deep water. (Ch. 9)

Cambrian radiation. The short time period at the beginning of the Phanerozoic Eon when the pace of evolution increased dramatically. (Ch. 14)

Capture fisheries. Fish caught in the wild. (Ch. 17)

Carbohydrates. Organic compounds composed of carbon, hydrogen and oxygen, of which sugars, starches and cellulose are examples. Carbohydrates are formed by all green plants and constitute a major source of food for animals. (Ch. 3)

Carnivore. Consumers that only eat other animals. (Ch. 16)

Carnivores. Meat-eating heterotrophs. (Ch. 15)

Carrying capacity. The limit on the population that an ecosystem can carry, imposed by the limited resources of that ecosystem. (Ch. 16)

Cave. Natural underground open space, generally with an opening to the surface. (Ch. 8)

Cell. The basic structural unit of all living organisms. (Ch. 14)

Celsius temperature scale. The temperature scale in which the zero reference point is the freezing temperature of water at surface atmospheric pressure and the boiling point of water is assigned a temperature of 100° C. (Ch. 2)

Cementation. The joining together of particles in a loose sediment through the addition of a cementing agent. (Ch. 7)

Channel. The passageway in which a stream flows. (Ch. 8)

Channelization. The engineering modification of a stream channel for the purposes of flood control and control of erosion. (Ch. 8)

Chaparral. Biome of the Mediterranean climate characterized by scrubby evergreen bushes and drought resistant trees. (Ch. 15)

Chemical elements. The fundamental substances into which matter can be separated by ordinary chemical means. (Chs. 3, 7)

Chemical sediment. Sediment formed by precipitation of minerals from solutions in water. (Chs. 3, 7)

Chemical weathering. The decomposition of rocks through chemical reactions such as hydration and oxidation. (Ch. 7)

Chemosynthesis. The synthesis of small organic molecules such as amino acids. (Ch. 14)

Chlorofluorocarbons (CFCs). Synthetic industrial gases that cause the breakdown of ozone in the stratosphere. (Ch. 19)

Cirque. A bowl-shaped hollow on a mountainside, open downstream, bounded upstream by a steep slope (headwall), and excavated mainly by frost wedging and by glacial abrasion and plucking. (Ch. 9)

Cirrus clouds. Fine, wispy or feathery clouds formed high in the troposphere, and composed entirely of ice crystals. (Ch. 11)

Clastic sediment. The loose fragmented debris produced by the mechanical breakdown of older rocks. (Chs. 3, 7)

Clay. The smallest clastic particles in sediment. The term is also used for a family of minerals. (Ch. 7)

Climate. The average weather conditions of a place or area over a period of years. (Chs. 12, 13)

Climate forcing. Causes that result in climate change. (Ch. 13)

Climate proxy record. Records of past natural events that are influenced by, and closely mimic, climate. (Ch. 13)

Closed system. A system in which the boundary allows the exchange of energy, but not matter, with the surroundings. (Ch. 1)

Cloud. Visible aggregations of minute water droplets, tiny ice crystals, or both. (Ch. 12)

Coal. A black, combustible, sedimentary or metamorphic rock consisting chiefly of decomposed plant matter and containing more than 50 percent organic matter. (Chs. 8 and 18)

Coalification. The process by which plant matter is converted into coal. (Ch. 18)

Coevolution. A kind of mutualism in which two species evolve together into their dependency on one another. (Ch. 16)

Colloid. A gel consisting of extremely fine particles dispersed in a continuous medium, usually a liquid. (Ch. 3)

Comet. Small solar system body composed primarily of ice with some dust and rock particles, which orbits the Sun in a highly elliptical orbit. (Ch. 4)

Commensalism. A beneficial relationship in which one partner benefits without affecting the other. (Ch. 16)

Common property resource. Resources that are commonly or communally owned, accessed, or managed. (Ch. 17)

Community. A group of interacting, independent organisms of different species that share the same environment. (Ch. 16)

Compaction. Reduction of the volume of sediment as a result of increased pressure. (Ch. 7)

Competition (biology). The circumstance where individuals vie against one another for scarce resources, including food, water, space, and light. (Ch. 16)

Competitive exclusion principle. The idea that two populations of different species with exactly the same requirements cannot persist indefinitely in the same habitat—one will always win out and the other will become extinct. Which one wins depends on the exact environmental conditions. Referred to as a principle, the idea has some basis in observation and experimentation. (Ch. 16)

Compound (chemical). The combination of one or more kinds of anion with one or more kinds of cation in a specific ratio. (Ch. 3)

Concentration factor. The amount of an element within an organism compared to the concentration of that element in the surrounding environment. (Ch. 15)

Condensation. The formation of a more ordered liquid from a less ordered gas. (Chs. 8, 11)

Conduction. The means by which heat is transmitted through solids without deforming the solid. (Ch. 2)

Conservation. Conserving and protecting species and the natural environment. (Ch. 16)

Consumer (biology). Heterotrophs that occupy higher trophic levels. (Ch. 15)

Contact metamorphism. Temperature-driven metamorphism adjacent to an intrusive igneous body. (Ch. 7)

Continental collision zone. A zone of collision between two converging plates capped by continental crust. (Ch. 5)

Continental crust. The part of the Earth's crust that comprises the continents, which has an average thickness of 45 km. (Ch. 3)

Continental divide. A line that separates streams flowing towards opposite sides of a continent, usually into different oceans. (Ch. 8)

Continental drift. Slow movement of continents across the face of Earth. (Ch. 5)

Continental rise. A region of gently changing slope where the floor of the ocean basin meets the margin of a continent. (Ch. 5)

Continental shelf. A submerged platform of variable width that forms a fringe around a continent. (Ch. 5)

Continental shield. An assemblage of cratons and orogens that has reached isostatic equilibrium. (Ch. 5)

Continental slope. A pronounced slope beyond the seaward margin of the continental shelf. (Ch. 5)

Convection. The process by which hot, less dense materials rise upward, being replaced by cold, dense, downward flowing material to create a convection current. (Ch. 2)

Convergence. The coming together of air masses, caused by the inward spiral flow in a cyclone and leading to an upward flow of air at the center of the low-pressure center. (Ch. 12)

Convergent margin. The zone where plates meet as they move toward each other. See *subduction zone*. (Ch. 5)

Coral bleaching. Corals turning white due to sickness or death, as a result of stressful changes in their aquatic environment. (Ch. 19)

Core. The spherical mass, largely metallic iron, at the center of the Earth. (Ch. 3)

Coriolis force. An effect that causes any body that moves freely with respect to the rotating solid Earth to veer toward the right in the northern hemisphere and toward the left in the southern hemisphere, regardless of the initial direction of the moving body. (Ch. 10)

Correlation (stratigraphic). Determination of the equivalence of age of the succession of strata found in two or more different areas. (Ch. 7)

Craton. A core of ancient rock in the continental crust that has attained tectonic and isostatic stability. (Ch. 5)

Crevasse. A deep, gaping fissure in the upper surface of a glacier. (Ch. 9)

Crude oil. Petroleum as it emerges from the ground. (Ch. 18)

Crust. The outermost and thinnest of the Earth's compositional layers, which consists of rocky matter that is less dense than the rocks of the mantle below. (Ch. 3)

Cryosphere. The part of the Earth's surface that remains perennially frozen. (Ch. 9)

Crystal. A solid compound composed of ordered, three-dimensional arrays of atoms or ions chemically bonded together and displaying crystal form. (Ch. 3)

Crystallization (of magma). The set of processes whereby crystals of individual mineral components nucleate and grow in a cooling magma. (Ch. 7)

Cumulative effects. The collective effects of local and regional environmental impacts eventually reaching global proportions. (Ch. 19)

Cumulus clouds. Puffy, globular, individual clouds that form when hot, humid air rises convectively and reaches a level of condensation. (Ch. 11)

Cyanobacteria. The earliest of the photosynthetic prokaryotes. Sometimes incorrectly called blue-green algae. (Ch. 14)

Cycle. The constant, repeated movement of matter or energy from one reservoir to another. (Ch. 1)

Cyclone. Air spiraling inward around a low-pressure center. (Ch. 12)

Dansgaard-Oeschger events. Warm-cool fluctuations of Earth's climate on an approximately 1500-year cycle following the last glacial event (Ch. 13)

Death rate. The number of deaths in a population relative to the total population, per unit of time. (Ch. 16)

Decomposer (biology). Organisms that break down complex organic compounds so that the chemical elements can be recycled. (Ch. 15)

Deforestation. The process of forest clearing. (Ch. 17)

Deformation. The change in shape or size of a solid body. (Ch. 6)

Degradation (of energy). The transformation of energy into a form that is less useful, or less available for work. (Ch. 2)

Delta. A body of sediment deposited by a stream where it flows into standing water. (Ch. 8)

Denitrification. The conversion of nitrate to molecular nitrogen by the action of bacteria—an important step in the nitrogen cycle. (Ch. 15)

Density. The average mass per unit volume. (Ch. 3)

Denudation. The sum of the weathering, mass-wasting, and erosional processes that result in the progressive lowering of the Earth's surface. (Ch. 7)

Deoxyribonucleic acid (DNA). A biopolymer consisting of two twisted, chainlike molecules held together by organic molecules called bases; the genetic material for all organisms except viruses, it stores the information on how to make proteins. (Ch. 14)

Deposition (of sediment). Accumulation of sediment following transport. (Ch. 7)

Desert. Arid land, whether "deserted" or not, in which annual rainfall is less that 250 mm or in which the evaporation rate exceeds the precipitation rate. (Chs. 12, 15)

Desertification. The invasion of desert into nondesert areas. (Ch. 19)

Dew point. The temperature at which the relative humidity reaches 100 percent and condensation starts. (Ch. 11)

Diagenesis. The various low-temperature and low-pressure changes that happen to a sediment after deposition. (Ch. 7)

Discharge. The quantity of water that passes a given point in a stream channel per unit time. (Ch. 8)

Dissolution. The chemical weathering process whereby minerals and rock material pass directly into solution. (Ch. 8)

Divergence. The separation of air masses in different directions, caused by the outward spiral flow in an anticyclone and leading to an outward flow of air from the center of a high-pressure center. (Ch. 12)

Divergent margin (of a plate). A fracture in the lithosphere where two plates move apart. Also called a *spreading center.* (Ch. 5)

Divide. The line that separates adjacent drainage basins. (Ch. 8)

Dobson unit (DU). A measure of the ozone content of the atmosphere. (Ch. 19)

Domain (of life). The highest taxonomic ranking of organisms. Also called superkingdoms. (Ch. 14)

Dormant volcano. A volcano that has not erupted within recent memory, but still exhibits some signs of volcanic activity. (Ch. 6)

Downwelling. The process by which surface water thickens and sinks. (Ch. 10)

Drainage basin. The total area that contributes water to a stream. (Ch. 8)

Drought. Below average rainfall in a region for an extended period of time. (Ch. 12)

Earth system. The whole Earth as a system of many interacting parts. (Part One)

Earth system science. The science that studies the whole Earth as a system of many interacting parts and focuses on the changes within and between these parts. (Ch. 1)

Ecological footprint. A measure of the resources needed to support a particular person or community rendered in terms of land area. (Ch. 19)

Ecosphere. The zone around star within which an orbiting planet and its moons would be just the right distance to allow for the existence of liquid water and thus, possibly, the support of life. (Ch. 14)

Ecosystem. The life-supporting system in which living organisms interact with each other and with the abiotic component of their environment. (Ch. 15)

Ecosystem diversity. A component of biodiversity; the variety of habitat types available in an ecosystem, and the biologic richness and complexity of those habitats. Also called *habitat diversity.* (Ch. 16)

Ecozone. The assemblage of biomes that co-exist in a particular geographic region. (Ch. 15)

Ekman transport. The average flow of water in a current over the full depth of the Ekman spiral. (Ch. 10)

El Niño/Southern oscillation (ENSO). A periodic climatic variation in which tradewinds slacken and surface waters of the central and eastern Pacific become anomalously warm. (Ch. 12)

Elastic energy. Energy stored in an elastically deformed solid. (Ch. 6)

Elastic rebound hypothesis. The hypothesis that earthquakes result from the sudden release of stored elastic energy by slippage on faults. (Ch. 6)

Electromagnetic radiation. A self-propagating electric and magnetic wave, such as light, radio, ultraviolet, or infrared radiation; all types travel at the same speed and differ in wavelength or frequency, which relates to the energy. (Ch. 2)

Emergence. An increase in the area of land exposed above sea level resulting from uplift of the land and/or fall of sea level. (Ch. 10)

Endangered species. A species in imminent danger of extinction. (Ch. 16)

Energy. The capacity to do work. (Ch. 2)

Energy cycle. The flow of energy from the external and internal sources of the planet, that drives the cycles of the Earth system. (Chs. 1, 2)

Entropy. A measure of disorganization. (Ch. 2)Equilibrium. Said of the state of a system that is in balance. (Ch. 1)

Epicenter. That point on the Earth's surface that lies vertically above the focus of an earthquake. (Ch. 6)

Equilibrium line. A line that marks the level on a glacier where net mass loss equals net gain. (Ch. 9)

Erosion. The complex group of related processes by which rock is broken down physically and chemically and the products are moved. (Ch. 7)

Eruption column. Hot, turbulent mixture of volcanic gasses and fine pyroclasts rising upward above an erupting volcano. (Ch. 6)

Essential nutrient. The 24 chemical elements that are known to be required for life. (Ch. 15)

Estuary. A semienclosed body of coastal water within which seawater is diluted with fresh water. (Ch. 10)

Eucaryotic cell (eucaryotes). A cell that includes a nucleus with a membrane, as well as other membrane-bound organelles. (Ch. 14)

Eutrophication. Bodies of water with a high level of plant nutrients and consequently high levels of algae growth. (Ch. 8)

Evaporation. The process by which a liquid is converted to its vapor. (Chs. 8, 11)

Evaporite deposits. Layers of salts that precipitate as a consequence of evaporation. (Ch. 18)

Evolution. The changes that species undergo through time, eventually leading to the formation of new species. (Ch. 14)

***Ex situ* conservation.** Preservation of species through a captive breeding program at a zoo or aquarium. (Ch. 16)

Experimentation. The act of experimenting or testing a hypothesis. (Ch. 1)

Exponential growth. Geometric growth in which a value increases at a given rate (the *growth rate*) per unit of time. (Ch. 16)

Extinct volcano. A volcano that has not erupted in historic times and that shows no sign of any volcanic activity. (Ch. 6)

Extinction. The permanent disappearance of a species. (Ch. 14)

Facies (metamorphic). The assemblage of minerals formed during metamorphism of a rock of a given composition subjected to a given temperature and pressure. (Ch. 7)

Fault. A fracture in a rock along which movement occurs. (Ch. 6)

Feedback. A system response that occurs when the output of the system also serves as an input and leads to changes in the state of the system. (Ch. 1)

Ferrel cells. In each hemisphere, the cell of air circulation that lies poleward of the Hadley cell. (Ch. 12)

Fission. Controlled radioactive transformation. (Chs. 2, 18)

Fissure eruption. Lava that reaches Earth's surface through an elongate fracture in the crust. (Ch. 6)

Fjord. A deep, glacially carved valley submerged by the sea. Also spelled *fiord*. (Ch. 9)

Flood. When a stream's discharge exceeds the capacity of the channel. (Ch. 8)

Flood plain. A wide valley. (Ch. 8)

Flux. The amount of energy flowing through a given area in a given time. (Ch. 3)

Focus. The point where energy is first released during an earthquake, and from which seismic waves travel outward. (Ch. 6)

Foliation. The planar texture of mineral grains, principally micas, produced by metamorphism. (Ch. 7)

Food chains. The pathways by which energy (as food) is moved from one trophic level to another. (Chs. 15 and 16)

Food web. The map of all interconnections among food chains for an ecosystem. (Ch. 15)

Fossil. The naturally preserved remains or traces of an animal or a plant. (Ch. 14)

Fossil fuel. Remains of plants and animals trapped in sediment that may be used for fuel. (Ch. 18)

Fractionation (of isotopes). The separation and differential concentration of isotopes of slightly different mass. (Ch. 13)

Front. The boundary between air masses of different temperature and humidity, and therefore different density. (Ch. 11)

Fuel cell. An electrochemical device that coverts a fuel such as hydrogen into electricity. (Ch. 18)

Fundamental niche. The set of all environmental conditions in which a species could survive in the absence of crowding or competition. (Ch. 16)

Fusion (of nuclei). The merging of the nuclei of lightweight chemical elements, particularly hydrogen, to form heavier elements such as helium and carbon. (Chs. 2, 18)

Galaxy. A cluster of a billion or more stars, plus gas and dust, that is held together by gravity. (Ch. 4)

Gangue. The nonvaluable minerals of an ore. (Ch. 18)

Gas. State of matter that takes on the shape of the container in which it is contained, filling the container completely (or escaping into space if it is not contained), while its constituent atoms move freely and acquire a uniform distribution within the container. (Ch. 3)

Gene. Regions of DNA coded for specific proteins that perform particular functions. (Chs. 3, 14)

General Circulation Model (GCM). A mathematical model used to simulate present and past climate conditions on the Earth. (Ch. 19)

Generalist. Species that have a broad range of tolerance and that occupy broadly defined niches. (Ch. 16)

Genetic bottleneck. Vulnerability of a species when the genetic diversity is extremely limited; an extreme decline in population can cause a genetic bottleneck, even if the number of individuals rebounds, if the genetic diversity among the survivors is limited. (Ch. 16)

Genetic diversity. The variability available among the DNA of individuals within a population or species. (Ch. 16)

Genetic drift. Changes in the frequency of a gene in a population as a result of chance rather than of mutation, selection, or migration. (Ch. 14)

Geo-engineering. Human intervention to change some portion of the Earth system, typically applied to broad-scale interventions. (Ch. 19)

Geographic Information System (GIS). Computer-based software programs which allow massive amounts of spatially referenced data points to be stored along with their characteristics. (Ch. 1)

Geologic column. A composite diagram combining in chronological order the succession of known strata, fitted together on the basis of their fossils or other evidence of relative or actual age. (Ch. 4)

Geologic time scale. Numerical time scale fitted to the relative ages of the geologic column. (Ch. 7)

Geosphere. The solid Earth. (Ch. 5)

Geostrophic wind. A wind that results from a balance between pressure-gradient flow and the Coriolis deflection. (Ch. 12)

Geothermal energy. Heat energy drawn from the Earth's internal heat. (Ch. 18)

Geothermal gradient. The rate of increase of temperature downward in the Earth. (Ch. 2)

Glacial period. An interval of time when Earth's global ice cover greatly exceeded that of today. (Ch. 13)

Glaciation. The modification of the land surface by the action of glacier ice. (Ch. 10)

Glacier. A permanent body of ice, consisting largely of recrystallized snow, that shows evidence of downslope or outward movement, due to the stress of its own weight. (Ch. 9)

Glacier ice. Snow that gradually becomes denser and denser until it is no longer permeable to air. (Ch. 9)

Glass. Non-crystalline, amorphous solid. (Ch. 3)

Global change. The changes produced in the Earth system as a result of human activities. (Ch. 19)

Global commons. A common property owned by all people on Earth; for example, the atmosphere. (Ch. 19)

Grade (of an ore). The percentage by weight of a mineral or chemical element on an ore being mined. (Ch. 18)

Gradient. A measure of the vertical drop over a given horizontal distance. (Chs. 2, 8)

Grassland. Temperate climate biome characterized by extensive prairies typical of Midwestern United States and Canada. (Ch. 15)

Gravel. The coarsest particles of clastic sediment. (Ch. 7)

Gravity. The mutual physical attraction between any two masses, such as Earth and the Moon. (Ch. 2)

Gravity anomaly. Variations in the pull of gravity after correction for latitude and altitude. (Ch. 5)

Green Revolution. The period following the 1950s when agricultural production increased dramatically. (Ch. 17)

Greenhouse effect. The property of the Earth's atmosphere by which long wavelength heat rays from the Earth's surface are trapped or reflected back by the atmosphere. (Ch. 2)

Gross production (biology). Production before respiration losses are subtracted. (Ch. 15)

Groundwater. All the water contained in the spaces within bedrock and regolith. (Ch. 8)

Growth rate (of a population). The change in number of individuals in a population over a given interval of time. (Ch. 16)

Gymnosperms. Naked-seed plants. (Ch. 14)

Gyre. A large subcircular current system of which each major ocean current is a part. (Ch. 10)

Habitat. Where an individual, population, or species exists or can exist. For example, the habitat of the Joshua tree is the Mojave Desert of North America. (Ch. 16)

Habitat diversity. The variety of habitat types in an ecosystem and the biologic richness of those habitats. Also called *ecosystem diversity*. (Ch. 16)

Habitat fragmentation. Large tracts of natural areas broken up into smaller patches by roads and other disruptions. (Ch. 16)

Hadean Eon. The period in Earth's history from the beginning at 4.56 billion years ago, to 3.8 billion years. (Ch. 14)

Hadley cell. Convection cells on both sides of the equator that dominate the winds in tropical and equatorial regions. (Ch. 12)

Half-life. The time needed for the number of parent atoms of a radioactive isotope to be reduced by one-half. (Ch. 4)

Heat. The energy a body has due to the motions of its atoms. (Ch. 2)

Heinrich events. Massive discharges of icebergs during glacial ages that left distinctive layers of ice-rafted sediment in the North Atlantic. (Ch. 13)

Herbivores. Plant-eating heterotrophs. (Ch. 16)

Heterotrophs. Organisms that are unable to use the energy from sunlight directly and so must get their energy by eating autotrophs or other heterotrophs. (Ch. 14)

High (H). An area of relatively high air pressure, characterized by diverging winds. (Ch. 12)

Hotspot. Point on Earth's surface where lava erupts above a plume. (Ch. 6)

Humidity. The amount of water vapor in the air. (Ch. 11)

Humus. The constituent of soil that is partially decayed organic matter. (Ch. 15)

Hurricane. A tropical cyclonic storm with wind speeds that exceed 119 km/h. (Ch. 12)

Hydrocarbon gas hydrate. Hydrocarbon gases trapped in a tiny cage made of water-ice molecules. Also called *clathrate hydrate*. (Ch. 13)

Hydrocarbons. Organic compounds that contain hydrogen and form carbon-hydrogen bonds. (Chs. 3, 13, 18)

Hydroelectric power. Power captured from the kinetic energy of a flowing stream of water. (Ch. 18)

Hydrogen fuel. Gaseous hydrogen used as a fuel in internal combustion engines. (Ch. 18)

Hydrologic cycle. The movement of water between the various reservoirs of the hydrosphere. (Ch. 8)

Hydrosphere. The totality of the Earth's water, including the oceans, lakes, streams, water underground, and all the snow and ice, including glaciers. (Ch. 8)

Hydrothermal solution. Hot, aqueous, metal-saturated fluid involved in the formation of many minerals deposits. (Ch. 18)

Hypothesis. An unproved explanation for the way things happen. (Ch. 1)

Ice. The solid form of H_2O. (Ch. 9)

Ice cap. A mass of ice that covers mountain highlands, or low-lying lands in high latitudes. (Ch. 9)

Ice sheet. Continent-sized mass of ice that covers nearly all the land surface within its margins. (Ch. 9)

Ice shelf. Floating sheets of ice, hundreds of meters thick, that occupy large embayments along the coast of Antarctica. (Ch. 9)

Igneous rock. Rock formed by the cooling and consolidation of magma. (Chs. 3, 7)

Infiltration. Water that falls as rain, then penetrates into the soil where it becomes part of the groundwater. (Ch. 8)

Inner core. The central, solid portion of the Earth's core. (Ch. 3)

Inorganic. Chemical compounds of non-biologic origin. (Ch. 3)

Insectivore. A carnivore that only eats insects. (Ch. 16)

Interglacial period. A time in the past when both the climate and global ice cover were similar to those of today. (Chs. 9, 13)

Intergovernmental Panel on Climate Change (IPCC). Panel established in 1988 by the United Nations and the World Meteorological Organization to review and report on the scientific evidence concerning climate change. (Ch. 19)

Intertropical convergence zone. A low-pressure zone of convergent air masses caused by warm air rising in the tropics. (Ch. 12)

Intrinsic value. The value of a species, wholly separate from any value it may have, or service it may provide, to humans. (Ch. 16)

Invasive species. An introduced or foreign species that out-competes and displaces native species. (Ch. 16)

Ion. An atom that has excess positive or negative charges caused by electron transfer. (Ch. 3)

IPAT equation. A relationship that expresses the human impact (I) on the environment as a function of population

(P), consumption and waste generation (A), and technology (T). (Ch. 19)

Island biogeography. The study of biodiversity on islands, applicable, in some circumstances, to fragmented terrestrial habitats. (Ch. 16)

Isobar. Places of equal air pressure. (Ch. 12)

Isolated system. A system in which the boundary prevents the system from exchanging either energy or matter with its surroundings. (Ch. 1)

Isostasy. The ideal property of flotational balance among segments of the lithosphere. (Ch. 5)

Isotopes. Atoms of an element having the same atopic number but differing mass numbers. (Ch. 3)

Jet stream. An upper-atmosphere westerly wind associated with a steep gradient in the height of the tropopause. (Ch. 12)

Joint. Sheet-like fractures in rock. (Ch. 7)

Jovian planets. Giant planets in the outer regions of the solar system that are characterized by great masses, low densities, and thick atmospheres consisting primarily of hydrogen and helium. (Ch. 4)

Karst. Topography formed on limestones due to solution by groundwater; characterized by sinkholes and caves. (Ch. 8)

Kelvin temperature scale. The absolute temperature scale in which the foundation is the point where entropy is zero. (Ch. 2)

Keystone species. A species that plays a fundamental role in an ecosystem. (Ch. 16)

Kinetic energy. Energy that is expressed in the movement of matter. (Ch. 2)

Kingdom (of life). One of the major classes of living things. (Ch. 14)

Köppen system of climate classification. A system using the distribution of native vegetation types, based on the premise that vegetation is the best indicator of climate. (Ch. 13)

***K*-strategist.** Organisms that produce few offspring but invest heavily in their care and upbringing. (Ch. 16)

Lake. Inland body of water in a depression on the Earth's surface; the water may be fresh or saline. (Ch. 8)

Landslide. Any perceptible downslope movement of a mass of bedrock or regolith, or a mixture of the two. (Ch. 7)

Lava. Magma that reaches the Earth's surface through a volcanic vent. (Chs. 3, 6)

Law (scientific). A statement that some aspect of nature is always observed to happen in the same way and that no deviations have ever been seen. (Ch. 1)

Life zone. The volume of the Earth system within which all life exists, approximately 10 km. above and below the surface of the geosphere. (Ch. 14)

Limiting factor (biology). A limitation in the supply of a chemical element that prevents the growth of an individual or a species, or that can even cause local extinction. (Ch. 15)

Limnetic zone. Open-water environment of a lake down to the depth where light can penetrate. (Ch. 15)

Linear growth. Constant amount of growth per unit time. (Ch. 16)

Lipid. A chemically diverse group of compounds that do not dissolve in water, including fats, oils, waxes, and steroids. (Ch. 3)

Liquefaction. Transformation of sediment or soil into a liquid-like quicksand due to a sudden disturbance. (Ch. 6)

Liquid. State of matter that has definite volume but its constituent atoms are able to flow freely past one another; the material does not retain its own shape but conforms to the shape of its container. (Ch. 3)

Lithification. The processes by which sediment and soil become rock. (Ch. 7)

Lithology. Description of a rock on the basis of color, mineralogical composition, and grain size. (Ch. 7)

Lithosphere. The outer 100 km of the solid Earth, where rocks are harder and more rigid than those in the plastic asthenosphere. (Ch. 3)

Little Ice Age. The interval of generally cool climate between the middle thirteenth and middle nineteenth centuries. (Ch. 13)

Littoral zone. The warm, near-shore zone of a lake or sea. (Ch. 15)

Load. The material that is moved or carried by a natural transporting agent, such as a stream, the wind, a glacier, or waves, tides, and currents. (Ch. 8)

Loess. Wind-deposited silt, sometimes accompanied by some clay and fine sand. (Ch. 9)

Logistic growth curve. The S-shaped curve describing the growth of a population over time under conditions of constant environmental conditions and constant food supply. (Ch. 16)

Longshore current. A current, within the surf zone, that flows parallel to the coast. (Ch. 10)

Low (L). An area of relatively low air pressure, characterized by converging winds, ascending air, and precipitation. (Ch. 12)

Luminosity. The total amount of energy radiated outward each second by the Sun or any other star. (Ch. 2)

Macronutrients. Elements required in large amounts by living things. These include the big six—carbon, hydrogen, oxygen, nitrogen, phosphorus, and sulfur. (Ch. 15)

Magma. Molten rock, together with any suspended mineral grains and dissolved gases, that forms when temperatures rise and melting occurs in the mantle or crust. (Chs. 3, 6)

Magnetic field (Earth's). The region surrounding Earth in which magnetic forces are exerted on any magnetized body or electric field. (Ch. 5)

Magnetic reversal. The reversal of the polarity of Earth's magnetic field. (Ch. 5)

Magnitude (of an earthquake). The amount of energy released during an earthquake. (Ch. 6)

Main sequence. The principal series of stars in the Hertzsprung-Russell diagram, which includes stars that are converting hydrogen to helium. (Ch. 4)

Managed (domesticated) ecosystem. Environments such as parks, aquaria, and botanical gardens. (Ch. 16)

Mantle. The thick shell of dense, rocky matter that surrounds the core. (Ch. 3)

Mass extinction. Event in which many types of organisms die out over a very short period. (Ch. 14)

Mass-wasting. The movement of regolith downslope by gravity without the aid of a transporting medium. (Ch. 7)

Matter. Substance that has mass and occupies space. (Ch. 3)

Maturation. The processes by which organic matter is transformed into various forms of petroleum. (Ch. 18)

Meander. A looplike bend of a stream channel. (Ch. 8)

Medieval Warm Period. An episode of relatively mild climate during the Middle Ages. (Ch. 13)

Mesosphere (atmospheric science). One of the four thermal layers of the atmosphere, lying above the stratosphere. (Ch. 12)

Mesosphere (geology). The region between the base of the asthenosphere and the core/mantle boundary. (Chs. 3, and 11)

Metabolism. The sum of all the chemical reactions in an organism, by which it grows and maintains itself. (Ch. 14)

Metallic mineral. Material mined specifically for the metal that can be extracted. (Ch. 18)

Metamorphic rock. Rock whose original compounds or textures, or both, have been transformed to new compounds and new textures by reactions in the solid state as a result of high temperature, high pressure, or both. (Chs. 3, 7)

Metamorphism. All changes in mineral assemblage and rock texture, or both, that take place in sedimentary and igneous rocks in the solid state within the Earth's crust as a result of changes in temperature and pressure. (Ch. 7)

Metasomatism. The metamorphic process in which abundant fluids change the composition and texture of a rock. (Ch. 7)

Meteorite. Piece of natural debris that falls to Earth. (Ch. 4)

Micronutrients. Chemical elements required in very small amounts by at least some forms of life. Boron, copper, and molybdenum are examples of micronutrients. (Ch. 15)

Milankovitch cycles. The combined influences of astronomical factors that produce changes in Earth's climate. (Ch. 13)

Mineral. Any naturally formed, crystalline solid with a definite chemical composition and a characteristic crystal structure. (Ch. 3)

Mineral assemblage. The variety and abundance of minerals present in a rock. (Ch. 3)

Mineral deposit. Any volume of rock containing an enrichment of one or more minerals. (Ch. 18)

Minesite decommissioning. The series of steps that must be followed when a mine is closed. (Ch. 18)

Mining. The set of processes by which useful resources are withdrawn from the stock of any non-renewable resource. (Ch. 18)

Model. A representation of something. (Ch. 1)

Modified Mercalli Scale. A scale used to compare earthquakes based on the intensity of damage caused by the quake. (Ch. 6)

Moho. See *Mohorovičić discontinuity*. (Ch. 6)

Molecule. The smallest unit that retains all the properties of a compound. (Ch. 3)

Monsoon. Seasonally reversing wind system. (Ch. 12)

Moon. A natural object in a regular orbit around a planet. (Ch. 4)

Moraine. An accumulation of drift deposited beneath or at the margin of a glacier and having a surface form that is unrelated to the underlying bedrock. (Ch. 9)

Mutation. Stated most simply, a chemical change in a DNA molecule. It means that the DNA carries a different message than it did before, and this change can affect the expressed characteristics when cells or individual organisms reproduce. (Ch. 14)

Mutualism. A relationship that is beneficial to both interacting organisms. (Ch. 16)

Natural gas. Naturally occurring hydrocarbon that is gaseous at ordinary temperature and pressure. (Ch. 18)

Natural selection. A process by which organisms whose biological characteristics better fit them to the environment are better represented by descendants in future generations than those whose characteristics are less fit for the environment. (Ch. 14)

Nebular hypothesis. The proposition that the Sun and planets formed from a huge, swirling cloud of cosmic gas and dust. (Ch. 4)

Negative feedback. The influence of a product on the process that produces it, such that production decreases with the growth of the product. (Ch. 15)

Nekton. The active swimming pelagic organisms. (Ch. 15)

Net production (biology). The biomass left from gross production after it has been used to fuel the processes of life. (Ch. 15)

New renewable (energy sources). Alternative energy sources based on new and developing technologies. (Ch. 18)

Niche. An organism's fundamental role and use of resources in a community. (Ch. 16)

Niche differentiation. Specialization of species in order to fill slightly different niches. (Ch. 16)

Nitrogen fixation. The process by which atmospheric nitrogen is converted to ammonia, nitrate ion, or amino acids. Microorganisms perform most of the conversion but a small amount is also converted by lightning. (Ch. 15)

Nonmetallic mineral. Mineral mined for its chemical or physical properties, not for the metal it contains. (Ch. 18)

Nonrenewable resource. Resource that is fixed in total quantity in Earth's crust. (Ch. 17)

North Atlantic Deep Water (NADW). A deep-ocean mass in the North Atlantic that extends from the intermediate water to the ocean floor; dense and cold, it originates at several sites near the surface of the North Atlantic, flows downward, and spreads southward into the South Atlantic. (Ch. 10)

Nuclear energy. Energy released by the controlled breakdown of a large radioactive isotope into two or more smaller isotopes. (Ch. 18)

Nucleic acid. Giant organic polymers built from molecules called nucleotides, each of which contains a sugar group, a phosphate group, and a nitrogenous base. (Ch. 3)

Numerical age. The time in years when a specific event happened or a specific material formed or was deposited. (Ch. 4)

Oceanic crust. The crust beneath the oceans. (Ch. 3)

Oceanic trench. Deep trench in the ocean floor where oceanic lithosphere sinks into the asthenosphere. (Ch. 5)

Oil. The liquid form of petroleum. (Ch. 18)

Oil shale. Fine-grained sedimentary rock containing much bituminous organic matter. (Ch. 18)

Old-growth forest. A forest that has endured for hundreds or thousands of years. (Ch. 17)

Omnivores. Heterotrophs that eat both meat and plants. (Ch. 16)

Open system. A system in which the boundary allows the exchange of both energy and matter with the surroundings. (Ch. 1)

Ore. An aggregate of minerals from which one or more minerals can be extracted profitably. (Ch. 18)

Organic. Said of compounds consisting of carbon atoms that are joined to other carbon atoms by a covalent bond. (Ch. 3)

Orogens. Elongate regions of the crust that have been intensively folded, faulted, and thickened as a result of continental collisions. (Ch. 5)

Outer core. The outer portion of the Earth's core, which is molten. (Ch. 3)

"Ozone hole". A region centered above the poles in which the ozone content of the stratosphere has been severely reduced. (Ch. 19)

P waves. Seismic body waves transmitted by alternating pulses of compression and expansion. *P* waves pass through solids, liquids, and gases. (Ch. 6)

Paleoclimate. Climate of the ancient past. (Ch. 13)

Paleomagnetism. Remanent magnetism in ancient rock recording the direction of the magnetic poles at some time in the past. (Ch. 5)

Parasitism. Relationship in which one partner benefits while the other is harmed. (Ch. 16)

Parent rock. The initial rock from which a given soil is formed. (Ch. 15)

Partial melt. Melting of a rock into a liquid (magma) and a residual solid. Also called a *fractional melt.* (Ch. 6)

Pause. The boundaries that separate the four principal thermal layers of the atmosphere. (Ch. 11)

Peat. An unconsolidated deposit of plant remains that is the first stage in the conversion of plant matter to coal. (Ch. 18)

Pelagic zone. The open-water environment of the ocean. (Ch. 15)

Percolation. The flow of groundwater, including the vertical flow down from the surface and the lateral flow of water in the saturated zone. (Ch. 8)

Periglacial. A land area beyond the limit of glaciers where low temperature and frost action are important factors in determining landscape characteristics. (Ch. 9)

Permafrost. Sediment, soil, or bedrock that remains continuously at a temperature below 0°C for an extended time. (Ch. 9)

Permeability. A measure of how easily a solid allows a fluid to pass through it. (Ch. 9)

Perpetual (inexhaustible) resource. Resource that is renewed on such a continuous and long-term basis that it can be considered inexhaustible. (Ch. 17)

Petroleum. Gaseous, liquid, and semisolid substances occurring naturally and consisting chiefly of chemical compounds of carbon and hydrogen. (Ch. 18)

Petroleum trap. An impermeable cap rock, such as shale, that prevents the migration and escape of petroleum to the surface. (Ch. 18)

Phanerozoic Eon. The period in Earth's history from 542 million years ago to the present day. (Ch. 14)

Phases. Masses of material that can be separated from one another by a definable boundary. (Ch. 3)

Photic zone. The uppermost zone of the ocean where light can penetrate and photosynthesis can occur. (Ch. 15)

Photochemical smog. Fog caused by the accumulation in the lower troposphere of ozone, nitrous oxides, and carbon-bearing compounds. (Ch. 19)

Photosynthesis. Synthesis of sugars from carbon dioxide and water by living organisms using light as energy. Oxygen is given off as a by-product. (Ch. 14)

Phylogenetic tree. Taxonomy that organizes the forms of life into groupings according to their genetic and evolutionary relationships. (Ch. 14)

Physical (mechanical) weathering. The disintegration (physical breakup) of rocks. (Ch. 7)

Phytoplankton. Floating or drifting (planktonic) organisms which are mainly single-celled plants. (Ch. 10)

Pile. A device in which nuclear fission can be controlled. (Ch. 18)

Placer deposit. Mineral deposit formed by mechanical concentration of mineral particles from weathered debris. (Ch. 18)

Planet. A natural body in orbit around a star that is massive enough to be spherical and to have cleared its orbital path of other objects. (Ch. 4)

Planetary accretion. The process by which bits of condensed solid matter were gathered to form the planets. (Ch. 4)

Planetary differentiation. The process of chemical segregation by which a planet separates into a core of dense matter surrounded by one or more layers of less dense rocky matter. (Ch. 4)

Plantation (forest). Managed forest where trees are planted of eventual harvesting. (Ch. 17)

Plasma. Ionized (electrically charged) gas with unique properties and characteristics. (Ch. 3)

Plate tectonics. The special branch of tectonics that deals with the processes by which the lithosphere is moved laterally over the asthenosphere. (Ch. 5)

Plume. Upwelling mass of hot rock from deep in the mantle. The surface indication of a plume is a *hotspot.* (Ch. 6)

Pluton. Any body of intrusive igneous rock, regardless of shape or size. (Ch. 6)

Plutonic rock. Rock formed by the crystallization of magma underground. (Ch. 3)

Polar cells. In each hemisphere, the cell of air circulation that lies poleward of the *Ferrel cell.* (Ch. 12)

Polar front. The region where equatorward-moving polar easterlies meet poleward-moving westerlies. (Ch. 12)

Polar glacier. A glacier of high altitude or high latitude within which the temperature remains below the pressure melting point. (Ch. 9)

Polarity. The north-south directionality of Earth's magnetic field. (Ch. 5)

Polymer. Large chain or network structures formed by the linking together of small molecules (called *monomers*). (Ch. 3)

Population. A group of individuals of the same species living in the same area or interbreeding and sharing genetic information. (Ch. 16)

Population density. The number of individuals in a population per unit area. (Ch. 16)

Porosity. The proportion (in percent) of the total volume of a given body of bedrock or regolith that consists of pore spaces. (Ch. 8)

Positive feedback. The influence of a product on the process that produces it, such that production increases the growth of the product. (Ch. 15)

Potential energy. The energy stored in a system. (Ch. 2)

Power. The amount of work done per unit time. (Ch. 2)

Precautionary principle. The concept that if the potential consequences of an anticipated event are unacceptably severe, those in authority have a responsibility to take action to avoid or mitigate those consequences. (Ch. 19)

Precipitation (of rain or snow). The process by which condensed water gathers into droplets or particles and falls under the pull of gravity. (Ch. 8, 11)

Predation. Circumstance in which one organism (predator) eats another (prey). (Ch. 16)

Pressure gradient force. The drop in air pressure per unit of distance. (Ch. 12)

Primary (primordial) atmosphere. The original envelopes of hydrogen and helium which surrounded the terrestrial planets early in the history of the solar system. (Ch. 4)

Primary producer (biology). Species such as algae and photosynthetic bacteria that form the first trophic level. (Ch. 15)

Primary production. Carbohydrate production and the consequent building of body mass by autotrophs. (Ch. 15)

Primary waves. See *P waves*. (Ch. 5)

Principle (scientific). See *Law (scientific)*. (Ch. 1)

Principle of Uniformitarianism. The same external and internal processes we recognize in action today have been operating unchanged, though at different rates, throughout most of the Earth's history. (Ch. 4)

Procaryotic cell (procaryotes). Cells without a nucleus; refers to single-celled organisms that have no membrane separating their DNA from the cytoplasm. (Ch. 14)

Profundal zone. Region of a lake below the limnetic zone, where light cannot penetrate. (Ch. 15)

Protein. Molecule formed through the polymerization of an amino acid. (Ch. 3)

Proterozoic Eon. The period in Earth's history from 2.5 billion to 542 million years ago. (Ch. 14)

Pyroclast. A fragment of rock ejected during a volcanic eruption. (Ch. 6)

Pyroclastic flow. A mobile mass of tephra, denser than air, which rushes down the flank of an erupting volcano. (Ch. 6)

Pyroclastic rocks. Rocks formed from pyroclasts. (Ch. 6)

Radiation. Transmission of heat energy through the passage of electromagnetic waves. (Ch. 2)

Radiatively active gas. Gases in the lower part of the atmosphere that absorb outgoing radiation, thus preventing the radiative loss of heat. (Ch. 2)

Radioactive. Isotopes of certain chemical elements that transform spontaneously into another isotope of the same element, or to an isotope of a different element. (Ch. 3)

Radiogenic heat. Heat energy produced by the spontaneous breakdown, or decay, of radioactive elements. (Ch. 2)

Radiometric dating. Determination of the time in years since the formation of a rock or other natural object using the contained radioactive isotopes. (Ch. 4)

Realized niche. Subset of the fundamental niche in which a species persists in the presence of a competing species. (Ch.16)

Recharge. The addition of water to the saturated zone of a groundwater system. (Ch. 8)

Recrystallization. The formation, in the solid state, of new crystalline minerals grains in a rock. (Ch. 7)

Recycling. The reuse of metals, glass and other materials. (Ch. 18)

Reef. A generally ridge-like structure composed chiefly of the calcareous remains of sedentary marine organisms such as corals and algae. (Ch. 10)

Reflection. The bouncing of a wave off the surface between two media. (Ch. 6)

Refraction. The change in velocity when a wave passes from one medium to another; the process by which the path of a beam of light is bent when the beam crosses from one transparent material to another. (Ch. 6)

Regional metamorphism. Metamorphism affecting large volumes of crust and involving both mechanical and chemical changes. (Ch. 7)

Regolith. The irregular blanket of loose, noncemented rock particles that covers the Earth. (Chs. 3, 7)

Relative age. The age of an object, material, or event, as determined by comparison to an older or younger object or event. (Ch. 4)

Relative humidity. The ratio of the vapor pressure in a sample of air to the saturation vapor pressure at the same temperature, expressed as a percentage. (Ch. 4)

Remote sensing. Continuous or repetitive collection of information about a target from a distance. (Ch. 1)

Renewable resource. A resource which is replenished by new growth each season. (Ch. 17)

Reserve. That portion of a resource that has been identified and is economically extractable using current technologies. (Ch. 18)

Reservoir. A storage place; a place in the Earth system where material or energy resides for some period of time. (Ch. 1)

Residence time. The average length of time *a given* material spends in a reservoir. (Ch. 1)

Residual deposit (mineral). Any local concentration of minerals formed as a result of weathering. (Ch. 18)

Residual mineral deposit. Any local concentration of minerals formed as a result of weathering. (Ch. 18)

Resource. Something that is useful or necessary. (Ch. 17)

Resource cycle. The extraction, processing, use, and disposal of raw materials. (Ch. 17)

Resource partitioning. Sharing of resources in order to avoid direct competition. (Ch. 16)

Respiration. The use of oxygen by aerobic heterotrophs to break down carbohydrates, releasing carbon dioxide, water, and energy. (Ch. 14)

Rhyolitic magma. Magma with a silica content of approximately 70% by weight and the overall composition of granite. (Ch. 6)

Ribonucleic acid (RNA). A single-stranded molecule similar to the DNA molecule, but with a slightly different chemical composition; it reads and executes the codes contained in the DNA. (Ch. 14)

Richter magnitude scale. A scale, based on the recorded amplitudes of seismic body waves, for comparing the amounts of energy released by earthquakes. (Ch. 6)

Rift. Depression along the center of a spreading center. (Ch. 5)

RNA. See *ribonucleic acid*. (Ch. 14)

Rock. Any naturally formed, nonliving, firm, and coherent aggregate mass of mineral matter that constitutes part of a planet. (Ch. 3)

Rock cycle. The cyclic movement of rock material, in the course of which rock is created, destroyed, and altered through the operation of internal and external Earth processes. (Ch. 7)

***r*-strategist.** An organism that reproduces rapidly, in large numbers, but invests little energy in the care and upbringing of the offspring. (Ch. 16)

S waves. Seismic body waves transmitted by an alternating series of sideways (shear) movements in a solid. S waves cause a change of shape and cannot be transmitted through liquids and gases. (Ch. 6)

Salinity. The measure of the sea's saltiness; expressed in parts per thousand. (Ch. 10)

Sand. Clastic particles intermediate in size between gravel and silt. (Ch. 7)

Saturated zone. The groundwater zone in which all openings are filled with water. (Ch. 8)

Savanna. Open forest consisting of broad, grassy plains and scattered trees. (Ch. 15)

Scattering (of radiation). A reflection phenomenon that involves the dispersal of radiation in all directions by tiny particles, droplets, and gas molecules. (Ch. 11)

Scientific method. The use of evidence that can be seen and tested by anyone who has the means to do so, consisting often of observation, formation of a hypothesis, testing of that hypothesis and formation of a theory, formation of a law, and continued reexamination. (Ch. 1)

Sea ice. A thin veneer of ice on polar oceans that covers approximately two-thirds of the area of Earth's persistant ice cover. (Ch. 9)

Seafloor spreading (theory of). A theory proposed during the early 1960s in which lateral movement of the oceanic crust away from midocean ridges was postulated. (Ch. 5)

Secondary atmosphere. The envelope of gaseous volatile elements that leaked from the interior of a terrestrial planet via volcanoes and was trapped by the planet's gravity. (Ch. 4)

Secondary production. Production of body mass of heterotrophs that derive their food energy by eating other organisms. (Ch. 15)

Secondary waves. See *S waves*. (Ch. 6)

Sediment. Regolith that has been transported by any of the external processes. (Chs. 3, 7)

Sedimentary rock. Any rock formed by chemical precipitation or by sedimentation and cementation of mineral grains transported to a site of deposition by water, wind, ice, or gravity. (Chs. 3, 7)

Seismic discontinuity. A boundary within Earth where the velocity of a seismic wave travel changes suddenly rather than smoothly. (Ch. 6)

Seismic gap. A gap along a tectonically active plate margin where earthquakes have not occurred for a long time. (Ch. 6)

Seismic moment magnitude. A measure of the energy released by an earthquake, taking account of the fact that energy may be released over a large area of a fault. (Ch. 6)

Seismic waves. Elastic disturbances spreading outward from an earthquake focus. (Ch. 6)

Seismicity. Earthquake activity. (Ch. 6)

Seismogram. Record of seismic waves made by a *seismograph*. (Ch. 6)

Seismograph. A device for continuously detecting and recording seismic waves. (Ch. 6)

Sequestration. Materials that have such long residence times in a reservoir they are isolated from the rest of the Earth system for long periods of time. (Ch. 1)

Shield volcano. A volcano that emits fluid lava and builds up a broad dome-shaped edifice with a surface slope of only a few degrees. (Ch. 7)

Silt. Fine clastic particle intermediate in size between sand and clay. (Ch. 7)

Sink. A reservoir in which the inward flux of matter exceeds the outward flux. The opposite of a source. (Ch. 1)

Sinkhole. A large solution cavity open to the sky. (Ch. 8)

Snow. Precipitation that consists of solid H_2O in crystalline form. (Ch. 9)

Snowline. The lower limit of perennial snow. (Ch. 9)

Soil. The part of the regolith that can support rooted plants. (Ch. 15)

Soil fertility. The ability of a soil to provide nutrients to growing plants. (Ch. 15)

Soil horizon. The sub-horizontal weathered zones formed as a soil develops. (Ch. 15)

Soil moisture. Water that fills or partially fills the pore spaces between minerals grains in a soil. (Ch. 15)

Soil order. Highest classification rank for soils. (Ch. 15)

Soil profile. A vertical section through a soil that displays its component horizons. (Ch. 15)

Soil structure. The aggregation of soil particles into clusters. (Chs. 3, 15)

Soil texture. The relative proportions of the various constituents of soil. (Ch. 15)

Solar energy. Power derived from the electromagnetic energy radiated by the Sun. (Ch. 18)

Solar nebula. A flattened rotating disc of gas and dust surrounding the Sun. (Ch. 4)

Solar system. The group of planets, moons, asteroids, comets, and other natural objects in orbit around the Sun. (Ch. 4)

Solid. State of matter that is firm or compact in substance with a definite volume and density, and that tends to retain its shape even if it is not confined, because its constituent atoms are fixed in position relative to each other. (Ch. 3)

Source. A reservoir in which more of a substance is coming from the reservoir than is flowing into it. The opposite of a sink. (Ch. 1)

Specialist. Species with a narrow range of tolerance that occupy small, narrowly defined niches. (Ch. 16)

Speciation. The emergence of a new species. (Ch. 14)

Species. A population of individuals that can interbreed to produce offspring that are, in turn, interfertile with each other. (Ch. 14)

Species diversity. The number and variety of different species in an area and their relative abundances (indicated by *species evenness* and *species richness*). (Ch. 16)

Species evenness. The distribution, or relative abundance of individuals of different species in a community. (Ch. 16)

Species richness. The number of different species in a given area. (Ch. 16)

Spectrum. A group of electromagnetic rays arranged in order of increasing or decreasing wavelength. (Ch. 2)

Spreading center. A fracture in the lithosphere where two plates move apart. Also called a *divergent plate margin*. (Ch. 5)

Spring. A flow of groundwater emerging naturally at the ground surface. (Ch. 8)

Star. A large spherical mass of ionized gas that radiates heat as a result of thermonuclear reactions in its core. (Ch. 2)

Steady state. A state in which the flux of matter into a reservoir exactly balances the flux of mater out of the reservoir. (Ch. 1)

Stock. A small, irregular body of intrusive igneous rock, smaller than a batholith, that cuts across the layering of the intruded rock. (Ch. 17)

Stratigraphy. The study of strata. (Ch. 7)

Stratosphere. One of the four thermal layers of the atmosphere, lying above the troposphere and reaching a maximum of about 50 km. (Ch. 11)

Stratospheric ozone layer. The region of the stratosphere within which ozone (O_3) absorbs radiation in the short-wavelength region of the electromagnetic radiation spectrum. (Ch. 2)

Stratovolcano(es). Volcanoes that emit both tephra and viscous lava, and that build up steep conical mounds. (Ch. 6)

Stratum (plural = *strata*). A distinct layer of sediment that accumulated at the Earth's surface. (Ch. 7)

Stratus clouds. Uniform dull grey clouds with a cloud base less than two km above sea level. (Ch. 11)

Stream. A body of water that carries detrital particles and dissolved substances and flows down a slope in a definite channel. (Ch. 8)

Streamflow. Overland flow that is concentrated into well-defined conduits. It consists of *storm flow* and *base flow*. (Ch. 8)

Stress. Force per unit area acting on a body. (Ch. 7)

Strike-slip. Motion on a fault parallel to the strike of the fault. (Ch. 5)

Subduction. The sinking of old, cold oceanic lithosphere into the asthenosphere. (Ch. 5)

Subduction zone (also called a *convergent margin*). The linear zone along which a plate of lithosphere sinks down into the asthenosphere. (Chs. 5 and 6)

Submergence. A rise of water level relative to the land so that areas formerly dry are inundated. (Ch. 10)

Supercontinent. Assemblage of cratons into a large continental complex. (Ch. 5)

Surf. Wave activity between the line of breakers and the shore. (Ch. 10)

Surface runoff. Water that drains off the surface of the land after rain. (Ch. 8)

Surface waves. Seismic waves that are guided by the Earth's surface and do not pass through the body of the Earth. (Ch. 6)

Surge. An unusually rapid movement of a glacier marked by dramatic changes in glacier flow and form. (Ch. 9)

Sustainability. The ability to support a population at a particular level of abundance on a long-term basis without incurring serious or permanent damage to the ecosystem. (Ch. 16)

Sustainable development. The cautious utilization of Earth's resources to meet current needs without degrading ecosystems or jeopardizing the future availability of those resources. (Ch. 17)

Sustainable yield. A harvest, removed from the stock of a renewable resource, that is smaller than the amount of stock thatis replaced by natural growth each year. (Ch. 17)

Symbiosis. A close, long-term relationship between individuals of different species. (Ch. 15)

Symbiosis. The close, long-term relationship between individuals of different species. (Ch. 16)

Sympatric speciation. Emergence of a new species in the absence of any geographic barrier. (Ch. 16)

System. Any portion of the universe that can be isolated from the rest of the universe for the purpose of observing and measuring changes. (Ch. 1)

Tar sand. Thick, dense, asphalt-like oil that cements sand grains together. (Ch. 18)

Taxonomic rank. The position of a species in the hierarchy of life. (Ch. 14)

Tectonic cycle. The processes by which Earth's major geologic features are formed. (Ch. 1)

Temperate deciduous forest. Forest of mainly deciduous trees characterized by distinct seasonal changes from summer to winter. (Ch. 15)

Temperate (warm) glacier. A glacier in which the ice is at the pressure-melting point and water and ice coexist in equilibrium. (Ch. 10)

Temperate rain forest. A type of forest, either broadleaf or coniferous, that is characteristic of areas with temperate climate and high precipitation. (Ch. 15)

Temperature. A measure of the average kinetic energy of all the atoms in a body. (Ch. 2)

Tephra. A loose assemblage of pyroclasts. (Ch. 6)

Tephra cone. A cone-shaped pile of tephra deposited around a volcanic vent. (Ch. 6)

Terminus. The outer, lower margin of a glacier. (Ch. 9)

Terrestrial planets. The innermost planets of the solar system (Mercury, Venus, Earth, and Mars), which have high densities and rocky compositions. (Ch. 4)

Texture. The overall appearance that a rock has because of the size, shape, and arrangement of its constituent mineral grains. (Ch. 3)

Theory. A hypothesis that has been examined and found to withstand numerous tests. (Ch. 1)

Thermocline. A zone of ocean water lying beneath the surface zone, characterized by a marked decrease in temperature. (Ch. 10)

Thermodynamics. The set of natural laws that govern the transfer of energy from one body to another. (Ch. 2)

Thermohaline circulation. Global patterns of water circulation propelled by the sinking of dense cold and salty water. (Ch. 11)

Thermosphere. One of the four thermal layers of the atmosphere, reaching out to about 500 km. (Ch. 11)

Threatened species. A species that has had a significant decrease in population or range. (Ch. 16)

Thunderstorm. Updrafts of warm, humid, air (a cell) that release a lot of latent heat very quickly and become unstable. (Ch. 12)

Tidal energy. Power derived from the kinetic energy involved in the movement of water during the rise and fall of tides. (Ch. 18)

Tide. The twice-daily rise and fall of the ocean surface resulting from the gravitational attraction of the Moon and Sun. (Chs. 2, 10)

Till. Unsorted sediment deposited by a glacier. (Ch. 9)

Topsoil. The uppermost horizon of a soil profile. (Ch. 15)

Tornado. Violent windstorm produced by a spiraling column of air that extends downward from a cumulonimbus cloud. (Ch. 12)

Tradewinds. Low level easterly winds that return air in a Hadley cell toward the tropics. (Ch. 12)

Transform fault margin (of a plate). A fracture in the lithosphere along which two plates slide past each other. (Ch. 5)

Transmission (of energy). Movement of energy through a system by the combined effects of radiation, convection, and conduction. (Ch. 11)

Transpiration. Water vapor released from the surface of a leaf. (Ch. 8)

Tributary. A stream that joins a larger stream. (Ch. 9)

Trophic level. In an ecological community, all the organisms that are the same number of food-chain steps from the primary source of energy. For example, in a grassland the green grasses are on the first trophic level, grasshoppers are on the second, birds that feed on grasshoppers are on the third, and so forth. (Ch. 15)

Trophic pyramid. The hierarchy of organisms in which energy is moved from one level to the next. (Ch. 15)

Tropical deciduous forest. Deciduous forest of the equatorial region. (Ch. 15)

Tropical rain forest. Non-deciduous forest of the equatorial region. (Ch. 15)

Troposphere. One of the four thermal layers of the atmosphere, which extends from the surface of the Earth to variable altitudes of 10 to 16 km. (Ch. 11)

Tsunami. See *seismic sea waves*. (Chs. 6, 10)

Tundra. A terrestrial biome of the high latitudes. (Ch. 15)

Unconformity. A substantial break or gap in a stratigraphic sequence that marks the absence of part of the rock record. (Ch. 7)

Upwelling. The process by which subsurface waters flow upward and replace the water moving away. (Ch. 10)

Utilitarian (value of biodiversity). Argument for the value of biodiversity based on the usefulness or potential usefulness of species to humans. (Ch. 16)

Vapor. Gas—see entry for *Gas*. (Ch. 3)

Varve. A pair of sedimentary layers deposited during the seasonal cycle of a single year. (Ch. 13)

Vascular plants. Plants with specialized conductive tissues in organs such as roots, stems, and leaves that are characteristic of all higher plants. (Ch. 14)

Viscosity. The internal property of a substance that offers resistance to flow. (Ch. 6)

Volcano. The vent from which igneous matter, solid rock, debris, and gases are erupted. (Ch. 6)

Volcanic ash. The smallest tephra particles emitted by an erupting volcano. (Ch. 6)

Volcanic gas. Gaseous emissions from an active volcano. (Ch. 6)

Volcanic rock. Rock formed from the volcanic eruption of lava or tephra; often very fertile. (Ch. 3)

Water table. The upper surface of the saturated zone of groundwater. (Ch. 8)

Water-scarce. Country or region where the annual renewable water supplies are below 1000 m^3 per person. (Ch. 17)

Water-stressed. A country or region with annual renewable water supplies between 1000 and 2000 m^3 per person. (Ch. 17)

Wave base. The effective lower limit of wave motion, which is half of the wavelength. (Ch. 10)

Wave cyclone. Mid-altitude cyclones that form between 30° and 60° N and S, as a result of the interaction between warm and cold air along the polar front, (Ch. 12)

Wave energy. Power drawn from the kinetic energy of waves. (Ch. 18)

Wave refraction. The process by which the direction of a series of waves, moving into shallow water at an angle to the shoreline, is changed. (Ch. 10)

Wave refraction. The process by which the direction of a series of waves, moving into shallow water at an angle to the shoreline, is changed. (Ch. 10)

Wavelength. The distance between the crests or troughs of adjacent waves. (Ch. 10)

Weather. The state of the atmosphere at a given time and place. (Ch. 12)

Weathering. The chemical alteration and mechanical breakdown of rock materials during exposure to air, moisture, and organic matter. (Ch. 7)

Wetland. An area that is either permanently or intermittently moist. (Ch. 8)

Wilderness. Area where natural forces are more significant that human intervention and where people do not live. (Ch. 17)

Wind. Horizontal air movement arising from differences in air pressure. (Ch. 12)

Wind energy. Power drawn from the kinetic energy of wind. (Ch. 18)

Windchill. Heat loss from exposed skin as a result of the combined effects of low temperature and wind speed. (Ch. 12)

Windchill factor. The heat loss from exposed skin as a result of the combined effects of low temperature and wind speed. (Ch. 13)

Work. The addition or subtraction to the internal energy of a system. (Ch. 2)

World ocean. All of the world's interconnected bodies of seawater. (Ch. 10)

Younger Dryas event. A rapid cooling of Earth's climate between 11,000 and 10,000 years ago. (Ch. 13)

Zooplankton. Floating or drifting (planktonic) organisms that are tiny animals. (Ch. 10)

Photo Credits

PART 1

Opener: Image Source/Photolibrary.

CHAPTER 1

Opener: NASA. **Figure 1.1a:** Weatherstock/Peter Arnold Images/Photolibrary. **Figure 1.1b:** Antoine Beyeler/iStockphoto. **Figure 1.1c:** Lyle Leduc/Workbook/Getty Images, Inc. **Figure 1.1d:** Jochen Schienker/Photolibrary. **Figure 1.1e:** Andrew Reese/iStockphoto. **Figure 1.2:** European Space Agency/Photo Researchers. **Figure C1.1a:** NASA/Goddard Sace Flight Center. **Figure C1.1b:** RSMAS/NASA. **Figure C1.1c:** NASA/GSFC. **Figure C1.2:** NASA. **Figure 1.10b (left):** Keith Brofsky/Age Fotostock America, Inc. **Figure 1.10b (center):** Daniel Barillot/Masterfile. **Figure 1.10b (right):** Keith Brofsky/Age Fotostock America, Inc. **Figure 1.14a:** Arias Francisco/Sipa/NewsCom. **Figure 1.14b:** Arctic-Images/Getty Images, Inc. **Figure 1.15a:** NASA. **Figure 1.15b:** Remi Benali/Getty Images, Inc. **Figure 1.15c:** J-L. Klein & M-L. Hubert/Photolibrary. **Figure B1.2 (bottom left):** S.C. Porter. **Figure B1.2 (top):** Tom & Susan Bean, Inc. **Figure B1.2 (bottom center):** S.C. Porter. **Figure B1.2 (bottom right):** John S. Shelton/University of Washington Libraries, Special Collections.

CHAPTER 2

Opener: Marko Heuver/iStockphoto. **Figure 2.1a:** Radius Images/Getty Images, Inc. **Figure 2.1b:** Jim Lopes/iStockphoto. **Figure 2.1c:** Tom Pfeiffer/VolcanoDiscovery/Getty Images, Inc. **Figure 2.1d:** iStockphoto. **Figure 2.1e:** Steve Allen/Brand X Pictures/Getty Images, Inc. **Figure 2.3b:** Charles O'Rear/Corbis. **Figure 2.6b:** Francois Gohier/Photo Researchers, Inc. **Figure C2.1a:** Martin Bond/Still Pictures/Photolibrary. **Figure C2.1b:** Jorgen Schytte/Photolibrary.

CHAPTER 3

Opener: Stephen & Donna O'Meara/Photolibrary. **Figure 3.1a:** Stephen & Donna O'Meara/Photo Researchers, Inc. **Figure 3.1b:** Zen Shui/Age Fotostock America, Inc. **Figure 3.3a:** US Department of Energy/Medical Images, Inc./Photo Researchers, Inc. **Figure 3.3b:** Carsten Peter/NG Image Collection. **Figure 3.11:** Breck P. Kent (Augite, Amphibole and Mica), Joel Arem/Photo Researchers, Inc.(Beryl),. **Figure 3.12:** William Sacco. **Figure 3.13a:** Wiilliam Sacco. **Figure 3.13b:** Brian J. Skinner. **Figure 3.13c:** William Sacco. **Figure 3.13d:** William Sacco. **Figure 3.14a:** William Sacco. **Figure 3.14b:** John Buitenkant/Photo Researchers, Inc. **Figure 3.15:** William Sacco. **Figure 3.17:** Courtesy Brian J. Skinner. **Figure 3.18a:** Dirk Wiersma/Photo Researchers, Inc. **Figure 3.18b:** Doug Martin/Photo Researchers, Inc. **Figure 3.18c:** Dirk Wiersma/Photo Researchers, Inc. **Figure 3.18d:** Andrew J. Martinez/Photo Researchers, Inc. **Figure 3.18e:** Stephen J. Krasemann/Photo Researchers, Inc. **Figure 3.19a:** William E. Ferguson. **Figure 3.19b:** Image Science & Analysis Laboratory, NASA Johnson Space Center./NASA/Johnson Space Center. **Figure 3.19c:** Jim Richardson/NG Image Collection.

CHAPTER 4

Opener: ESO/L. Calcada. **Figure 4.1:** NASA, ESA and The Hubble Heritage Team (STScI/AURA). **Figure 4.2:** NASA and European Space Agency. **Figure 4.4:** NASA. **Figure 4.6:** Mary Evans Picture Library/Photo Researchers, Inc. **Figure 4.9:** NASA/Johns Hopkins University Applied Physics Laboratory/Smithsonian Institution/Carnegie Institution of Washington. **Figure 4.7b:** NASA, ESA, A.V. Filippenko (University of California, Berkeley), P. Challis (Harvard-Smithsonian Center for Astrophysics), et al. **Figure 4.10a:** Richard Bizley/Photo Researchers. **Figure 4.10b:** NASA. **Figure 4.10c:** Nasa Jet Propulsion Laboratory. **Figure 4.10d:** Caltech/Nasa Jet Propulsion Laboratory. **Figure 4.10e:** Stephen Dalton/Minden Pictures, Inc. **Figure 4.11:** John S. Shelton. **Figure 4.12a:** University of Arizona/Nasa Jet Propulsion Laboratory. **Figure 4.12b:** NASA. **Figure 4.13:** NASA. **Figure 4.16a:** Public Domain. **Figure 4.16b:** USGS/Nasa Jet Propulsion Laboratory. **Figure 4.17:** NASA. **Figure 4.18:** Dan Schechter/Photo Researchers, Inc. **Figure 4.19:** Jon Sanford/Science Photo Library/Photo Researchers, Inc. **Figure 4.22:** Caltech/Nasa Jet Propulsion Laboratory. **Figure 4.26a:** John S. Shelton/University of Washington Libraries, Special Collections. **Figure 4.26b:** Patrick Frischknecht/Photolibrary. **Figure C4.1:** Nasa Jet Propulsion Laboratory.

PART 2

Opener: Per-Andre Hoffmann/Aurora Photos Inc.

CHAPTER 5

Opener: TAO Images Limited/Photolibrary. **Figure 5.1:** NASA/Johnson Space Center. **Figure 5.2a:** Peter Ryan/NGS/Getty Images, Inc. **Figure 5.2b:** Jeff Schmaltz, MODIS Rapid Response Team/NASA/GFSC. **Figure 5.4b:** National Library of Wales. **Figure 5.1:** Science Source/Photo Researchers, Inc. **Figure 5.16:** National Geophysical Data/Courtesy NOAA. **Figure 5.21:** Dr. Morley Read/Shutterstock. **Figure 5.22:** James P. Blair/NG Image Collection.

CHAPTER 6

Opener: Brynjar Gauti/AP World WIde. **Figure 6.02b:** NewsCom. **Figure 6.04a:** Carsten Peter/NG Image Collection. **Figure 6.07:** DuBin/The New York Times/Redux Pictures. **Figure 6.08a:** Ramon Espinosa/AP/Wide World Photos. **Figure 6.08b:** Reza/Getty Images, Inc. **Figure 6.08c:** David Walters/Miami Herald/MCT/NewsCom.

Figure 6.08d: George Plafker, U.S. Geological Survey. **Figure 6.08e:** Courtesy NOAA/NGDC. **Figure C6.1:** Courtesy DigitalGlobe. **Figure C6.2:** Bay Ismoyo/AFP/Stone/Getty Images. **Figure 6.13:** J.D. Griggs/USGS. **Figure 6.14a:** Krafft Explorer/Photo Researchers, Inc. **Figure 6.14b:** Science VU-ASIS/Visuals Unlimited. **Figure B6.4a:** G. Brad Lewis Photography. **Figure B6.4b:** J.D. Griggs/USGS. **Figure 6.18a:** Paul Chesley/NG Image Collection. **Figure 6.18b:** Courtesy Brian J. Skinner. **Figure 6.19a:** Francois Gohier/Photo Researchers, Inc. **Figure 6.19b:** J.P. Lockwood//USGS. **Figure 6.19c:** Bjarki Reyr/Age Fotostock America, Inc. **Figure 6.20:** Roger Werth/Woodfin Camp. **Figure 6.22:** S.C. Porter. **Figure 6.23a:** Krafft Explorer/Photo Researchers. **Figure 6.23b:** Tom Bean/Alamy. **Figure 6.24:** Jon Arnold/Alamy. **Figure 6.25:** Greg Vaughn/Alamy. **Figure 6.26:** Explorer/Photo Researchers, Inc. **Figure 6.28:** Krafft/Photo Researchers, Inc. **Figure 6.29:** NRSC Ltd./Science Photo Library/Photo Researchers, Inc. **Figure 6.30:** Lyn Topinka/USGS. **Figure 6.31a:** Jim Wark/Peter Arnold Images/Photolibrary. **Figure 6.33a:** OAR/National Undersea Research/Photo Researchers, Inc. **Figure 6.33b:** Philippe Bourseiller/The Image Bank/Getty Images. **Figure 6.33c:** Raymond Gehman/NG Image Collection. **Figure 6.33d:** Pablo Corral Vega/NG Image Collection. **Figure 6.33e:** Michael Falzone/Jon Arnold Travel/Photolibrary.

CHAPTER 7

Opener: Gerhard Zwerger-Schoner/Photolibrary. **Figure 7.2:** David Kilpatrick/Alamy. **Figure 7.3:** Warren Kovach/Alamy. **Figure 7.4:** Courtesy Brian J. Skinner. **Figure 7.5:** Scoutingstock/Alamy. **Figure 7.6a:** Phil Schermeister/NG Image Collection. **Figure 7.6b:** George F. Mobley/NG Image Collection. **Figure 7.6c:** Kenneth W. Fink/Masterfile. **Figure 7.11a:** Bill Hatcher/NG Image Collection. **Figure 7.11b:** Courtesy Stephen C. Porter. **Figure 7.11c:** Dr. Marli Miller/ Visuals Unlimited. **Figure 7.12a:** W.E. Garrett R./NG Image Collection. **Figure 7.12b:** Robert Madden/NG Image Collection. **Figure 7.14:** Russ Bishop/Stock Connection/Aurora Photos Inc. **Figure 7.16a:** Dirk Wiersma/Photo Researchers, Inc. **Figure 7.16b:** Courtesy Richard J. Stewart. **Figure C7.1:** Courtesy Brian J. Skinner. **Figure 7.17a:** Annie Griffiths Belt/ NG Image Collection. **Figure 7.17b:** Sam Abell/NG Image Collection. **Figure 7.17c:** Robert Harding Images//Masterfile. **Figure 7.21a:** William Sacco. **Figure 7.21b:** William Sacco. **Figure 7.25a:** Patricia Tye/Photo Researchers, Inc. **Figure 7.25b:** A.J. Copley/Visuals Unlimited. **Figure 7.26:** Brian Skinner. **Figure 7.27:** William Sacco. **Figure 7.29:** S.C. Porter. **Figure B7.3:** Mannfred Gottschalk/Tom Stack & Associates.

PART 3

Opener: Robert Harding Images/Masterfile.

CHAPTER 8

Opener: Caroline Penn/Photolibrary. **Figure B8.1:** Courtesy C. Clapperton. **Figure B8.3a:** Tomasz Pietryszek/iStockphoto. **Figure B8.3b:** Dennis Drenner/Visuals Unlimited. **Figure 8.004b:** Carr Clifton/Minden Pictures, Inc. **Figure 8.007:** S.C. Porter. **Figure 8.009:** Colin Monteath/Age Fotostock. **Figure 8.012:** Hiroji Kubota/Magnum Photos, Inc. **Figure 8.013b:** S.C. Porter. **Figure 8.013c:** Marli Bryant Miller. **Figure 8.013d:** Courtesy Visible Earth/NASA. **Figure 8.016:** Eric Baccega/ Minden Pictures, Inc. **Figure 8.017:** Gary Randall/Getty Images, Inc. **Figure 8.018a:** MDA Federal, Inc. **Figure 8.018b:** MDA Federal, Inc. **Figure 8.019:** NASA image courtesy Lawrence Ong, EO-l Mission Science Office, NASA GSFC. **Figure 8.025:** Courtesy Brian J. Skinner. **Figure 8.026:** Michael Nichols/Magnum Photos, Inc. **Figure 8.027:** Hiroji Kubota/ Magnum Photos, Inc. **Figure 8.029:** Don Johnston/ Photolibrary. **Figure C8.001a:** USGS. **Figure C8.001b:** Courtesy NASA. **Figure C8.002:** Kelly Cheng Travel Photography/Getty Images, Inc.

CHAPTER 9

Opener: Kevin G Smith/Photolibrary. **Figure 9.03:** S.C. Porter. **Figure 9.05a:** Melissa Farlow/NG Image Collection. **Figure 9.05b:** All Canada Photos/Getty Images, Inc. **Figure 9.05c:** Marli Miller/Visuals Unlimited. **Figure 9.05d:** Gavin Heillier/Getty Images, Inc. **Figure 9.05e:** Science VU/ESIAL/ NASA/JSC/Visuals Unlimited. **Figure 9.06:** Courtesy USGS. **Figure 9.07:** Courtesy NASA. **Figure B9.001:** Sovin Zankl/ Visuals Unlimited. **Figure 9.09a:** S.C. Porter. **Figure 9.09b:** S.C. Porter. **Figure 9.13:** Alsakastock/Masterfile. **Figure 9.15:** Courtesy Austin Post. **Figure 9.16:** Michio Hoshino/Minden Pictures, Inc. **Figure 9.18:** Bryan & Cherry Alexander/Photo Researchers. **Figure 9.20a:** Mark Burnett/Photo Researchers, Inc. **Figure 9.20b:** Grant Dixon/Hedgehog House/Minden Pictures, Inc. **Figure 9.20c:** James P. Blair/NG Image Collection. **Figure 9.21:** E.R. Degginger/Photo Reseachers. **Figure 9.22a:** Raymond Gehman/NG Image Collection. **Figure 9.22b:** Courtesy Gerald Osborn, University of Calgary. **Figure 9.22c:** LOOK Die Bildagentur der Fotografen GmbH/Alamy. **Figure 9.22d:** NG Image Collection. **Figure 9.23:** S.C. Porter. **Figure 9.24:** Steve McCutcheon/Visuals Unlimited. **Figure 9.25b:** Emma Pike/Wikimedia. **Figure 9.26a:** Peter Essick/ Aurora Photos Inc. **Figure 9.26b:** Courtesy Stephen C. Porter. **Figure 9.27:** NASA. **Figure 9.28:** Greg Dimijian/Photo Researchers. **Figure C9.001:** Courtesy NASA. **Figure C9.002:** Stapleton Collection/Corbis.

CHAPTER 10

Opener: Jerry Dodrill/Aurora/Getty Images, Inc. **Figure 10.1:** Courtesy Walter H.F. Smith and David T. Sandwell. **Figure 10.7:** Deep Sea Drilling Project/Scripps Institution of Oceanography. **Figure 10.15:** NASA. **Figure B10.3b:** Matthias Kulka/ Corbis. **Figure 10.18:** Sean Davey/Aurora Photos/Corbis. **Figure 10.19:** John S. Shelton. **Figure 10.24a and b:** Laszlo Podor/Alamy. **Figure 10.25:** S.C. Porter. **Figure 10.26a:** NASA. **Figure 10.26b:** NASA. **Figure 10.27b:** NASA. **Figure 10.28:** M-Sat Ltd/Photo Researchers, Inc. **Figure 10.31:** Alaska Stock LLC/. **Figure C10.1:** Stocktrek Images/Getty Images, Inc.

PART 4

Opener: Shigeru Ueki/Getty Images.

CHAPTER 11

Opener: NASA. **Figure 11.1:** Chris Butler/Photo Researchers. **Figure 11.3a:** NASA/Photo Researchers. **Figure 11.3b:** ESA/ Courtesy NASA. **Figure 11.4:** Bertrand Sordet/ Bios/©Photolibrary. **Figure C11.1b:** Dan Suzio/Photo Researchers. **Figure C11.1c:** Dr. Gerald L. Fisher/Photo Researchers. **Figure B11.3:** Nigel Pavitt/JAI/Corbis. **Figure 11.10:** Pekka Parviainen/Science Photo/Photo Researchers. **Figure 11.12a:** Jaubert Images/Alamy. **Figure 11.12b:** NASA. **Figure 11.22a:** GIPPhotostock/Photo Researchers.

Figure 11.22b: David R. Frazier/Photo Researchers. **Figure 11.22c:** Doug Millar/Photo Researchers. **Figure 11.22d:** image100/SuperStock. **Figure 11.22e:** John Mead/Photo Researchers. **Figure 11.22f:** Photo Researchers.

CHAPTER 12

Opener: Gene Rhoden/Still Pictures/Photolibrary. **Figure 12.1:** Steve Starr/Corbis. **Figure 12.7:** National Oceanic and Atmospheric Administration/DLR–FRG/Starlight. **Figure 12.12a:** NASA. **Figure 12.12b:** Ginny Lloyd Photo/Shutterpoint. **Figure 12.18b:** S.C. Porter. **Figure 12.21:** NASA Earth Observatory. **Figure 12.22a:** Courtesy NASA. **Figure 12.23a:** Gene Blevins/LA Daily News/Corbis. **Figure 12.23b:** John H. Hoffman/Bruce Coleman, Inc. **Figure 12.24:** Gene & Karen Rhoden/Peter Arnold, Inc./Alamy. **Figure 12.26a:** Carsten Peter/NG Image Collection. **Figure 12.26b:** Dr. Cynthia M. Beall & Dr. Melvyn C. Goldstein/NG Image Collection. **Figure 12.26c:** Marc Moritsch/NG Image Collection. **Figure 12.26d:** Annie Griffiths Belt/NG Image Collection. **Figure 12.26e:** David Keaton/Corbis. **Figure 12.27:** Steve McCurry/Magnum Photos, Inc. **Figure 12.28:** Courtesy NOAA George E. Marsh Album. **Figure 12.29b:** Frans Lemmens/SuperStock.

CHAPTER 13

Opener: Antonio Lopez Roman/Agefotostock/Photolibrary. **Figure 13.3:** S.C. Porter. **Figure 13.4:** Royal British Columbia Museum/Reuters/Corbis. **Figure 13.6a:** Peter Essick/Aurora Photos Inc. **Figure 13.6b:** Accent Alaska.com/Alamy. **Figure 13.7a:** Taylor S. Kennedy/NG Image Collection. **Figure 13.7b:** Courtesy Rob Dunbar, Stanford University. **Figure 13.8:** Peter Arnold, Inc./Alamy. **Figure 13.9:** NASA. **Figure 13.12:** S.C. Porter. **Figure 13.15:** Carl Purcell/Photo Researchers. **Figure 13.22a:** Grant Dixon/Lonely Planet Images/Getty. **Figure 13.22c:** Nasa Jet Propulsion Laboratory. **Figure 13.24:** age fotostock/SuperStock, Inc.

PART 5

Opener: Mark Moffet/Minden Pictures, Inc.

CHAPTER 14

Opener: Ronald Toms/Photolibrary. **Figure 14.1:** NASA/Johnson Space Center. **Figure 14.2a:** Dr. Torsten Wittmann/Photo Researchers. **Figure 14.2b:** Dr. John D. Cunningham/Visuals Unlimited/Corbis. **Figure 14.2c:** David Tipling/Getty Images, Inc. **Figure 14.3a:** David Allan Brandt/Stone/Getty Images, Inc. **Figure 14.3b:** Alain Julien/AFP/Getty Images. **Figure 14.4a:** Dr. Jeremy Burgess/Photo Researchers, Inc. **Figure 14.4b:** CNRI/Photo Researchers, Inc. **Figure 14.8:** Steve Munsinger/Photo Researchers. **Figure 14.9:** Courtesy J. William Schopf, Dept. of Earth and Planetary Sciences, UCLA. **Figure C14.1:** AFP/Fisheries And Oceans Canada/UVIC-Verna Tunnicliffe/NewsCom. **Figure 14.11a:** Ralph Lee Hopkins/NG Image Collection. **Figure 14.11b:** Getty Images, Inc. **Figure 14.11c:** Gerald & Buff Corsi/Visuals Unlimited. **Figure 14.11d:** Fritz Polking/Visuals Unlimited. **Figure 14.11e:** Joe & Mary Ann McDonald/Visuals Unlimited. **Figure 14.11f:** Tierbild Okapia/Photo Researchers, Inc. **Figure 14.11g:** Eric Hosking/Photo Researchers, Inc. **Figure 14.13a:** Ted Kinsman/photo Researchers. **Figure 14.13b:** Frans Lanting Studio/Alamy. **Figure 14.14a:** The Natural History Museum/Alamy. **Figure 14.14b:** O. Louis Mazzatenta/National Geographic/Getty Images, Inc. **Figure 14.16b:** Albert Copley/Visuals Unlimited. **Figure 14.17:** Raymond Gehman/NG Image Collection. **Figure 14.18a:** Ken Lucas/Visuals Unlimited. **Figure 14.18b:** D. Hurst/Alamy. **Figure 14.18c:** Marli Miller/Visuals Unlimited. **Figure 14.18d:** Theodore Clutter/Photo Researchers, Inc. **Figure 14.19:** Courtey Peabody Museum of Natural History, Yale University. **Figure 14.20:** Courtesy Smithsonian Institution. **Figure 14.21:** Gerald Hoberman/Photolibrary. **Figure 14.22a:** Francois Gohier/Photo Researchers, Inc. **Figure 14.22b:** Courtesy Carnegie Museum of Natural History. **Figure 14.23a:** Mauricio Anton/Photo Researchers, Inc. **Figure 14.23b:** Lealisa Westerhoff/AFP/Getty Images. **Figure 14.23c:** John Reader/Science Photo Library/Photo Researchers, Inc. **Figure 14.25:** Mark Garlick/Photo Researchers.

CHAPTER 15

Opener: Don Johnston/Photolibrary. **Figure 15.1:** Joseph Sohm/Visions of America/Corbis. **Figure 15.2:** Joe Sohm/Photo Researchers, Inc. **Figure 15.7:** Randy Morse/Tom Stack & Associates, Inc. **Figure C15.1:** Michael S. Yamashita/Corbis. **Figure 15.13c:** Dr. Jeremy Burgess/Photo Researchers, Inc. **Figure 15.15a:** Photo Researchers, Inc. **Figure 15.15b:** Photo Researchers, Inc. **Figure 15.18:** Jim Richardson/NG Image Collection. **Figure 15.020a:** Herve Donnezan/Photo Researchers, Inc. **Figure 15.020b:** Jim Wark/Photolibrary. **Figure 15.23a:** John E Marriott/Alamy. **Figure 15.23b:** Minden Pictures/Masterfile. **Figure 15.23c:** Boyd & Evans/Corbis. **Figure 15.23d:** Robert Harding Images/Masterfile. **Figure 15.23e:** Frans Lanting/Corbis. **Figure 15.23f:** Wildlife GmbH/Alamy. **Figure 15.23g:** James Steinberg/Photo Researchers. **Figure 15.23h:** Grant Faint/The Image Bank/Getty Images, Inc. **Figure 15.25a:** Shaun Lowe/iStockphoto. **Figure 15.25b:** MvH/iStockphoto. **Figure 15.25c:** Frank Krahmer/Masterfile. **Figure 15.27a:** Mark Conlin/Alamy. **Figure 15.27b:** Joe Albias/Gallo Images/Getty Images, Inc. **Figure 15.27c:** Flip Nicklin/Minden Pictures, Inc.

CHAPTER 16

Opener: Heidi & Hans-Juergen Koch/Minden Pictures, Inc. **Figure 16.1:** Accent Alaska.com/Alamy. **Figure 16.2a:** Paul E Tessier/Photodisc/Getty Images. **Figure 16.4a-inset:** Bios/T & S Allofs/Photolibrary. **Figure 16.4b-inset:** Wim van Egmond/Visuals Unlimited. **Figure 16.5 inset:** Joe McDonald/Visuals Unlimited/Getty Images. **Figure 16.6:** Jeff Vanuga/Corbis. **Figure C16.1a:** Malerapaso/iStockphoto. **Figure C16.1b:** Jim Brandenburg/Minden Pictures, Inc. **Figure C16.1c:** Cal Crary/Getty Images. **Figure C16.1d:** Konrad Wothe/Getty Images. **Figure 16.7a:** Oxford Scientific/Photolibrary. **Figure 16.7b:** Anthony Bannister; Gallo Images/Corbis. **Figure 16.7c:** Gerry Bishop/Visuals Unlimited. **Figure 16.8a:** Gerry Bishop/Visuals Unlimited/Getty. **Figure 16.8b:** Jean-Jacques Alcalay/Photolibrary. **Figure 16.9:** David Courtenay/Photolibrary. **Figure 16.13a:** Nture Picture Library/Photolibrary. **Figure 16.13b:** Achim Mittler, Frankfurt am Main/Getty Images. **Figure 16.16:** Mike Hill/Alamy. **Figure 16.18b:** Angelo Cavalli/Corbis. **Figure 16.18c:** David R. Frazier/. **Figure 16.19a:** Jany Sauvent/Visuals Unlimited. **Figure 16.19b:** Theo Allofs/Visuals Unlimited. **Figure 16.19c:** Alain Compost/Photolibrary. **Figure 16.19d:** Karl Weidmann/Photo Researchers, Inc. **Figure B16.2:** AP/Wide World Photos. **Figure 16.22a:** Nigel Pavitt/JAI/Corbis. **Figure 16.22b:** Russell Cheyne/Reuters/Corbis. **Figure 16.22c:** Frans Lanting/Corbis. **Figure 16.26:** Corbis. **Figure 16.27:** Paulo Santos/Reuters/Corbis.

PART 6

Opener: NASA.

CHAPTER 17

Opener: Lite Productions/Photolibrary. **Figure 17.1:** Dung V0 Trung/Corbis. **Figure 17.2:** Wendy Stone/Corbis. **Figure 17.3:** Francois Gohier/Photo Researchers, Inc. **Figure 17.5:** Samsam Panthaky/AFP/Getty Images. **Figure B17.1a:** David Gowans/ Alamy. **Figure B17.1b:** Siephoto/Masterfile. **Figure B17.1c:** Altrendo Images/Getty Images. **Figure B17.1d:** Radius Images/ Corbis. **Figure B17.1e:** Calvin Larsen/Photo Researchers, Inc. **Figure B17.1f:** Crown Copyright/Health & Safety Laboratory/ Photo Researchers, Inc. **Figure 17.9a:** John Sohlden/Visuals Unlimited. **Figure 17.9b:** Ippei Naoi/Getty Images. **Figure 17.11:** NASA. **Figure 17.14:** Bettmann/Corbis. **Figure 17.15:** Philippe Lissac/Godong/Corbis. **Figure C17.1a:** Martin Hanke/Alamy. **Figure C17.1b:** Israel Images/Alamy. **Figure 17.16:** Wiskerke/ Alamy. **Figure 17.17:** USDA/Soil Conservation Service. **Figure 17.18a:** Altrendo Images/Getty Images. **Figure 17.18b:** Mick Rock/Alamy.

CHAPTER 18

Opener: Gillianne Tedder/Photolibrary. **Figure 18.1:** Design Pics/SuperStock. **Figure 18.3a:** Courtesy Brian J. Skinner. **Figure 18.3b:** William Sacco. **Figure 18.3c:** Courtesy Brian J. Skinner. **Figure 18.3d:** Marli Bryant Miller. **Figure 18.3e:** Courtesy Brian J. Skinner. **Figure 18.3f:** Dirk Wiersma/Photo Researchers, Inc. **Figure 18.4b:** Ken Lucas/Visuals Unlimited. **Figure 18.8:** Peter Treiber/Corbis. **Figure 18.9a:** Rob Crandall/ The Image Works. **Figure 18.9b:** Mark Karrass/Alamy. **Figure 18.9c:** Science Faction/SuperStock. **Figure 18.12:** Michael Nichols/NG Image Collection. **Figure 18.14:** Robert Sisson/NG Image Collection. **Figure 18.16:** SuperStock. **Figure 18.18a:** Sarah Leen/NG Image Collection. **Figure 18.18b:** Jonathan P. Blair/NG Image Collection. **Figure 18.02:** Sarah Leen/NG Image Collection. **Figure 18.20:** Jonathan Blair/Corbis. **Figure 18.21:** Marc Moritsch/NG Image Collection. **Figure 18.22:** Greenshoots Communications/Alamy. **Figure C18.1a:** Roger Ressmeyer/Corbis. **Figure C18.1b:** Peter Essick/Aurora Photos Inc. **Figure 18.24a:** Justin Stumberg/AP. **Figure 18.24b:** Julie Dermansky/Photo Researchers, Inc.

CHAPTER 19

Opener: NASA. **Figure 19.1:** Bettman/Corbis. **Figure 19.3:** George Steinmetz/Corbis. **Figure 19.4b:** Frans Lemmens/Alamy. **Figure 19.5:** Eco Images/Universal Images Group/Getty Images, Inc. **Figure 19.7a:** Lloyd Cluff/Corbis. **Figure 19.8:** Wes C. Skiles/ NG Image Collection. **Figure 19.9:** Igor Kostin/Corbis. **Figure 19.10a:** NASA. **Figure 19.11:** Ted Spiegel/Corbis. **Figure 19.14b:** 24BY36/Alamy. **Figure 19.15a:** Reinhard Dirscherl/Visuals Unlimited. **Figure 19.15b:** Reinhard Dirscherl/Visuals Unlimited. **Figure 19.19:** GISS/NASA. **Figure 19.20:** GISS/NASA. **Figure 19.21:** Courtesy IPCC. **Figure 19.22:** Courtesy IPCC. **Figure 19.23:** NASA **Figure 19.25:** NASA.

Figure and Table Credits

The following figures are reprinted or adapted by permission of John Wiley & Sons: B1.1 (p.11), 1.7 (p. 13), 2.7 (p. 40), 2.8 (p. 41), 4.15 (p. 94), B5.1 (p. 116), 5.4A & C (p. 114), 5.8 (p. 119), 5.17 (p. 130), 5.23 (p. 133), 5.24 (p. 134), 6.2a (p. 145), 6.3 (p. 148), 6.10 (p. 159), 6.12 (p. 160), 6.16 (p. 163), 6.17 (p. 167), 6.27 (p. 174), B7.1 (p. 200), B7.2 (p. 201), 7.1 (p. 186), 7.4 (p. 188), 7.9 (p. 191), 7.10 (p. 192), 7.13 (p. 194), 7.19 (p. 202), 7.23 (p.206), 7.24 (p. 206), 7.28 (p. 212), 8.1 (p.224), 8.3 (p.225), 8.5b (p. 229), 8.13a (p. 225), 8.14 (p. 236), 8.20 (p. 240), 8.21 (p. 241), 8.22 (p. 242), 8.23 (p. 243), 8.24 (p.245), 8.25b (p.246), 9.5 (p. 261), 9.10 (p.265), 11.16 (p. 338), 12.15 (p. 361), C13.2B (p. 391), 13.11 (p. 393), 13.22A & B (p. 404) 13.23 (p. 405), 15.4B (p. 454), B16.1 (p.492), 16.5 (p. 491), 16.23 (p. 508), 16.24 (p. 509), 17.4 (p. 521) 17.10 (p. 527), 17.12 (p.529), 18.4a (p. 549), 18.6 (548), 18.7 (p.549), 18.10 (p. 552), 18.18 (p. 559), 18.19 (p. 561), 19.12 (p. 587), 19.16 (p. 591).

Figure 2.2 (p. 27): From the Energy Information Administration (EIA).

Figure 2.6A (p. 33): Based on John Gribbin (1993), Inside the Sun, New Scientist, p. 1; Edward J. Tarbuck and Fredrick J. Lutgens (1994), Earth Science, 7th ed. (New York, Macmillian), Fig. 19.18, p. 656; Michael Zeilik (1991), Astronomy: The Evolving Universe, 6th ed. (Wiley), Fig. 13.31, P. 284; Cesare Emilani (1992), Planet Earth, Cosmology, Geology, and the Evolution of Life and Environment (Cambridge, Cambridge University Press), Fig. 6.3.

Figure 2.13 (p. 39): Adapted with permission from the U.S. Committee from the Global Atmospheric Research Program, *Understanding Climate Change: A Program for Action*, Fig. 3.2. Copyright 1975 by the National Academy Press, Washington, D.C.

Figure 2.14 (p. 41): From Leonardo- Energy.

Figure 7.30 (p. 214): From W. Ludwig and J. L. Probst (1998), River Sediment Discharge to the Oceans: Present Day Controls and Global Budgets, *American Journal of Science*, 298, Fig. 9, p. 287. Reprinted by permission of the *American Journal of Science*.

Figure 7.31 (p. 216): Reprinted by permission of Dr. Pat Behling from W.F. Ruddiman and J.E. Kutzbach (1991), Plateau Uplift and Climate Change, *Scientific American*, 264, pp. 70–71.

Figure 8.2 (p. 225): Used by permission of Go Green, from U.S. Geological Survey.

Figure 8.4A (p. 228): Based on World Water Balance, 1978; L'vovich, 1979; Ponce, 2006.

Figure 8.11 (p. 233): Adapted from P.E. Potter et al. (1988), Teaching and field guide to alluvial processes and sedimentation of the Mississippi River, Fulton County, Kentucky, and lake County Tennessee. *Kentucky Geological Survey,* Fig. 8.

Figure 9.4 (p. 262): Adapted by permission from M.F. Meier and A.S. Post (1962), Recent variations in mass net budgets of glaciers in western North America, in *Symposium of Obergurl: Variations of the Regime of Existing Glaciers*, IAHS pub. 58, pp. 63–77.

Figure C10.2 (p. 313): Used by permission of Steve Earle, Vancouver Island University.

Figure 10.4 (p. 291): Based on plate 3 in the *Times Atlas of the World*, Comprehensive edition (1967), published by Houghton Mifflin; © John Bartholomew & Sons Ltd., Edinburgh.

Figure 10.5 (p. 292): Adapted by permission of Andrew McIntyre from CLIMAP Project Members (1981), Map and Chart Series MC-36, Map 6A, published by the Geological Society of America.

Figure 10.12 (p. 300): Adapted by permission from A.L. Gordon (1990/91), The Role of Thermohaline Circulation in Global Climate Change, *Lamont-Doherty Geological Observatory (Columbia University) Report 1990/91*, Fig. 3, p. 48.

Figure 10.14 (p. 301): Based on information from *Lamont-Doherty Geological Observatory (Columbia University) Report 1990/91*, Fig. 4, p. 50; and from J. Imbrie et al. (1992), On the structure and Origin of Major Glaciation Cycles, 1: Linear responses to Milankovitch forcing. *Paleoceanography* 7, Fig. 1b, p.704. Published by American Geophysical Union, Washington D.C.

Figure B11.1 (p. 330): Based on J.M. Moran and M.D. Morgan (1991), *Meteorology,*, 3rd ed. (New York Macmillan); and R.G. Fleagle and J. Businger (1963), *An Introduction to atmospheric physics* (New York, Academic Press), p. 153.

Figure 11.2 (p.323): From MacKenzie, Fred, T. Our Changing Planet: An Introduction to Earth System Science and Global Environmental Change, 3rd Edition, (c) 2003, p. 205. Reprinted with permission of Pearson education, Inc., Upper Saddle River, NJ.

Figure 12.5 (p. 354): From National Weather Service.

Figure 12.13 (p. 360): Based on information from National Weather Service; and H. Riehl (1962), *Jet Streams of the Atmosphere* (Fort Collins, Colorado State University).

Figure 12.18A (p. 363): From the National Weather Service.

Figure 12.22B (p. 366): Based on Information from Zeeya Merali and Brian Skinner, *Visualizing Earth Science*. Copyright © John Wiley & Sons, 2009.

Figure 12.29A (p. 372): From K. Pye and H. Tsoar (1990), *Aeolia and Sand Dunes* (London, Uwin Hyman), Fig. 2.17. Reprinted with kind permission of Springer Science and Business Media.

C13.1 (p. 390): From J. Jouzel, 2004. EPICA Dome C Ice Cores Deuterium Data. IGBP PAGES, World Data Center for Paleoclimatology, Data Contribution Series # 2004 - 038. NOAA/NGDC Paleoclimatology Program, Boulder CO, USA.

C13.2A (p. 391): From Integrated Ocean Drilling Program-Management International (IODP-MI)

Figure 13.2 (p. 381): From Climate Change 2001: The Scientific Basis. Contribution of Working Group I to the Third Assessment Report of the Intergovernmental Panel on Climate Change, Figure 2–3, Cambridge University Press.

Figure 13.5A (p. 386): Adapted by permission from Zhang De-er (1982), Analysis of Dust Rain in the Historic Times of China, *Kexue Tongbao*, vol. 27, pp. 294–297.

Figure 13.5B&C: Adapted by permission of Taylor & Francis (Routledge) from H.H. Lamb (1977), *Climate-Present, Past, and Future*, vol. 2: *Climate History and the Future* (London, Methuen).

Figure 13.5D: Adapted by permission of Taylor & Francis (Routledge) from H.H. Lamb (1966), *The Changing Climate: Selected Papers* (London, Methuen), app. 3, p. 223.

Figure 13.12 (p.394): Reprinted by permission of Taylor & Francis (Routledge) from H.H. Lamb (1977), *Climate-Present, Past, and Future*, vol. 2: *Climate History and Future* (London, Methuen), Fig. 17.17, p. 462; derived from an unpublished paper by L.M. Libby, R and D. Associates, Santa Monica, CA, 1974.

Figure 13.16 (p. 397): Adapted by permission from Thompson Webb III et al. (1993), Vegetation, Lake Levels and Climate in Eastern North America from the Past 18,000 Years in H.E. Wright, Jr., et al. (eds.), Global Climates Since the Last Glacial Maximum: Copyright (c) 1993 by the Regents of the University of Minnesota.

Figure 13.18 (p. 399): Adapted from Fig 8.3 of *Paleoclimatology*, by Thomas J. Crowley and Gerald R. North. Copyright © 1991 by Oxford University Press, Inc. Used by permission of Oxford University Press, Inc. And adapted by permission of Elsevier Science Publishers and E.J. Barron from E.J. Barron and W.M. Washington (1982), Createaceous Climate: A Comparison of atmospheric simulations and with the geological record. *Paleogeography, Paleoclimatology, Paleoecology*, vol. 40, Fig. 9.

Figure B14.1 (p.420): Adapted by permission from Gil Brum, Larry McKane, and Gerry Karp (1994), *Biology: Explaining Life* 2d ed. (Wiley), Fig. 14.4.

Figure 14.7A (p. 426): Based on G. Brum, L. McKane, and G. Karp (1994), *Biology: Explaining Life*, 2d ed. (Wiley), Fig. 35.3.

Figure 14.7B: Adapted from E. Barghoorn, The Oldest Fossil, 1971. Copyright © *Scientific American*, a division of Nature America, Inc. All rights reserved.

Figure 14.10 (p. 430): From Falkowski, Paul G. et al, "The Rise of Oxygen over the Past 205 Million Years and the Evolution of Large Placental Mammals." *Science 30* Sep. 2005, Vol. 309, no. 5744, pages 2202–2204. Reprinted with permission of AAAS.

Figure 14.15 (p. 436): Composite drawing adapted by permission from Gil Brum, Larry McKane, and Gerry Karp (1994), *Biology: Explaining Life* 2d ed. (Wiley), Figs. 35.9, 35.10, and 35.12.

Figure 14.16A (p.437): Based on a drawing by Virge Kask in Derek E.G. Briggs (1991), Extraordinary fossils, American Scientist, vol. 79, no. 2, p.133.

Figure 15.10 & 15.11 (p. 462): From G.E. Likens, F.H. Bormann, R.S. Pierce, J.S. Eaton, and N.M Johnson (1995), *The Biogeochemistry of a Forested Ecosystem*, 2d ed. With kind permission of Springer Science and Business Media.

Figure 15.12 (p.464): Modified from G. Lambert (1987), *La Recherche*, 18, pp. 782–783, with some data from R. Houghton (1993), *Bulletin of the Ecological Society of America*, 74(4), pp. 355–356.

Figure 15.13A (p. 466): Data from R. Soderlund and T. Rosswall (1982), in O. Hutzinger (ed), *The Handbook of Environmental Chemistry*, vol. 1, pt B. With kind permission of Springer Science and Business Media.

Figure 16.18A (p. 504): Reprinted by permission of Yale University Press from F.G. Stehli (1968), Taxonomic diversity gradients in pole location: The recent model in E.T. Drake (ed), *Evolution and Environment*, Fig 6, p. 170. Copyright © 1968 by Yale University.

Figure 16.13C (p.500): From the UK Red Squirrel Group.

Figure 16.25 (p.510): From UNEP-WCMC (2003) Extent of the World's protected areas and location of World Heritage sites (produced from the World Database on Protected Areas). UNEP-WCMC, Cambridge, UK.

Figure 17.13 (p. 531): World Capture Fisheries Production, Figure 3, from State of World Fisheries and Aquaculture 2008-Food and Agriculture Organization of the United Nations.

Figure 18.13 (p. 554): Adapted from Bookout, J.F., in *Episodes*, vol. 12 (4), 1989, figure 1. Used with permission of the International Union of Geological Sciences.

Figure B19.1 (p.575): Adapted with the permission of Cambridge University Press from B.L. Turner et al. (eds.) (1990), *The Earth as Transformed by Human Action*, Fig. 3.1, p. 43.

Figure 19.7 (p. 581): From Climate Change 1990: The IPPC Scientific Assessment – Report Prepared by Working Group I, Table 5.1. Cambridge University Press.

Figure 19.10B&D (p.585): From the National Weather Service.

Figure 19.13 (p.588): Adapted by permission of the World Resource Institute-Global Forest Watch, 2005.

Figure 19.14 (p.589): From GreenFacts.com

Figure 19.15C (p. 590): From Earth Trends.

Figure 19.18 (593): From Climate Change 2007: Synthesis Report. Contribution of Working Groups I, II, and III to the Fourth Assessment Report of the Intergovernmental Panel on Climate Change, IPCC, Geneva Switzerland.

Figure 19.19 (p.594): From the American Geophysical Union

Figure 19.20 (p. 595): From Climate Change 2001: The Scientific Basis. Contribution of Working Group I to the Third Assessment Report of the Intergovernmental Panel on Climate Change, Figure 1. Cambridge University Press.

Figure 19.21 (p. 596): Climate Change 2001: Synthesis Report. A Contribution of Working Groups I, II, and III to the Third Assessment Report of the Intergovernmental Panel on Climate Change, Figure SPM-2. Cambridge University Press.

Figure 19.24 (p. 600): Modified from Rockstrom et al. (2009), Nature and IStockphoto. Copyright © The International Geosphere-Biosphere Programme.

Index

Page references followed by "f" indicate entries that appear within figures. Page references followed by a "t" indicate entries that appear within tables.

B

D

S